ASTRONOMY AND ASTROPHYSICS ABSTRACTS

A Publication of the Astronomisches Rechen-Institut Heidelberg
Member of the Abstracting Board of the International
Council of Scientific Unions

Volume 17
Literature 1976, Part 1

Edited by S. Böhme U. Esser W. Fricke
U. Güntzel-Lingner I. Heinrich F. Henn D. Krahn
L. D. Schmadel H. Scholl G. Zech

Springer-Verlag Berlin Heidelberg GmbH 1976

Astronomisches Rechen-Institut
Heidelberg

Director: Professor Dr. Walter Fricke

Astronomy and Astrophysics Abstracts
Editors-in-Chief: Ute Esser, Dr. Lutz D. Schmadel

Astronomy and Astrophysics Abstracts
is prepared under the auspices
of the International Astronomical Union

ISBN 978-3-662-12306-5   ISBN 978-3-662-12304-1 (eBook)
DOI 10.1007/978-3-662-12304-1

Library of Congress Catalog Card Number 72-104650.

Frieda Henn
1915–1976

The Astronomisches Rechen-Institut has lost its wellknown senior staff member, the Editor-in-Chief of *Astronomy and Astrophysics Abstracts*. Frieda Henn, who died on 4 June, 1976 from a serious disease, was born on 24 February, 1915 in Weinheim near Heidelberg, where her father was the station-master of the railroad station. She grew up in the station house, where the family lived, with her sister and brother. The three children learned very early that service to the human society, reliability in one's work and responsibility for others belong to the pillars of society. All three children got a university education. Frieda Henn studied mathematics and natural sciences at the University of Heidelberg and completed her studies with the Staatsexamen. She decided to devote her life to astronomy.

In 1947 she began her career as an astronomer at the Astronomisches Rechen-Institut. She took part in the computations of astronomical ephemerides and of data for the Astronomisch-Geodätisches Jahrbuch. Moreover, she was always ready to help her colleagues whenever urgent scientific programs of the Institute required help for a punctual completion. In 1955 Frieda Henn entered the division of the Institute concerned with the production of the *Astronomischer Jahresbericht* thus increasing the scientific staff from two to three astronomers. The work in astronomical bibliography became not only her favored occupation but also the field in which she increasingly developed highest standards of expertness. In 1969, she became Editor-in-Chief of *Astronomy and Astrophysics Abstracts*, a series of semi-annual volumes, which succeeded the *Astronomischer Jahresbericht*. Her knowledge of the current problems in astronomy, a unique knowledge of the literature and of the demands of astronomers for an abstracting service contributed to the establishment of *Astronomy and Astrophysics Abstracts*.

Open-minded for new developments and suggestions she served in the International Astronomical Union, the sponsoring organisation of *Astronomy and Astrophysics Abstracts*, and in the Abstracting Board of the International Council of Scientific Unions. Frieda Henn died when the volume covering the second half of 1975 was just about to appear and after she with her colleagues had completed the Index Volume with the General Index of Authors and Subjects covering the five-year period 1969 to 1973.

The community of astronomers and the Institute have lost one of their most unusual members. Frieda Henn's home was the Institute where she used to be in her office for twelve hours a day and where she was available for advice and help besides her work on the *Abstracts*. She was a completely reliable colleague and staunch friend. The procedures and attitudes that she set for her colleagues will surely live on.

Walter Fricke

Preface

Astronomy and Astrophysics Abstracts, which has appeared in semi-annual volumes since 1969, is devoted to the recording, summarizing and indexing of astronomical publications throughout the world. It is prepared under the auspices of the International Astronomical Union (according to a resolution adopted at the 14th General Assembly in 1970).

Astronomy and Astrophysics Abstracts aims to present a comprehensive documentation of literature in all fields of astronomy and astrophysics. Every effort will be made to ensure that the average time interval between the date of receipt of the original literature and publication of the abstracts will not exceed eight months. This time interval is near to that achieved by monthly abstracting journals, compared to which our system of accumulating abstracts for about six months offers the advantage of greater convenience for the user.

Volume 17 contains literature published in 1976 and received before August 15, 1976; some older literature which was received late and which is not recorded in earlier volumes is also included.

We acknowledge with thanks contributions to this volume by Dr. J. Bouška, who surveyed journals and publications in the Czech language and supplied us with abstracts in English, and by the Commonwealth Scientific and Industrial Research Organization (C.S.I.R.O.), Sydney, for providing titles and abstracts of papers on radio astronomy. We want to acknowledge valuable contributions to this volume by Zentralstelle für Atomkernenergie-Dokumentation, Leopoldshafen, which supported our abstracting service by sending us retrospective literature searches.

It is a pleasure to express our warmest thanks again to Ms. Helga Ballmann, Ms. Monika Betz, Ms. Karola Gudé, Ms. Lore Kiefert, and Ms. Ingrid Wolf, who typed the text of this volume on IBM 72 Composers and compiled the pages from abstract slips in a perfect form for offset reproduction. We are indebted to Ms. Elisabeth Feigenbutz for punching material for the author index and for the subject index which finally were printed with a TN chain on a 1403 IBM high-speed printer.

Heidelberg, September 1976

Siegfried Böhme	Frieda Henn †
Ute Esser	Dietlinde Krahn
Walter Fricke	Lutz D. Schmadel
Ulrich Güntzel-Lingner	Hans Scholl
Inge Heinrich	Gert Zech

Contents

Positional Astronomy, Celestial Mechanics

Space Research

Theoretical Astrophysics

Sun

Earth

Planetary System

Stars

Interstellar Matter, Gaseous Nebulae, Planetary Nebulae

Radio Sources, Quasars, Pulsars, Infrared, X-ray, Gamma-Ray Sources, Cosmic Radiation

Stellar Systems

Introduction

Astronomical bibliographies

Astronomy and Astrophysics Abstracts begins documentation and abstracting from the year 1969. For information on astronomical literature before this date consultation of one of the following bibliographies is suggested:

(1) J. J. de Lalande, Bibliographie Astronomique, Paris 1803 (this work covers the time from 480 B. C. to the year 1803, VIII + 966 pages).

(2) J. C. Houzeau, A. Lancaster, Bibliographie générale de l'astronomie, Volume I (in two parts), Bruxelles 1882, 1887, Volume II, Bruxelles 1889. The complete title of Volume II is "Bibliographie générale de l'astronomie ou catalogue méthodique des ouvrages, des mémoires et des observations astronomiques, publiés depuis l'origine de l'imprimerie jusqu'en 1880". A new edition of these volumes was prepared by D. W. Dewhirst in 1964.

(3) Bibliography of Astronomy, 1881 - 1898. The literature of this period was recorded on standard slips by the Observatoire Royal de Belgique. From the material (some 52,000 items) a microfilm version was produced by University Microfilms Limited, Tylers Green, High Wycombe, Buckinghamshire, England, in 1970.

(4) Astronomischer Jahresbericht, 1899 gegründet von Walter Wislicenus, herausgegeben vom Astronomischen Rechen-Institut in Heidelberg (formerly in Berlin), Verlag W. de Gruyter, Berlin. For the period from 1899 to 1968 sixty-eight volumes were published, each of which, in general, covers the literature of one year.

(5) Bulletin Signalétique — Section 120: Astronomie, Physique Spatiale, Géophysique. Published by Centre de Documentation du Centre National de la Recherche Scientifique, Paris. This publication is a continuation of "Bibliographie Mensuelle de l'Astronomie" founded in 1933 by the Société Astronomique de France. The publication is continued.

(6) Referativnyj Zhurnal. Founded in 1953 and published by Vsesoyuznyj Institut Nauchnoj i Tekhnicheskoj Informatsii, Akademiya Nauk, Moskva. The publication is continued.

Concept of Astronomy and Astrophysics Abstracts

This abstracting service aims to present a comprehensive documentation of the literature in all fields of astronomy and astrophysics. It appears in semi-annual volumes, two of which cover the literature of a calendar year. The half-yearly period of issue is regarded as an optimal period of time for summarizing papers into subject categories and for the presentation of abstracts as quickly as possible after the publication of the original literature. The time limits at which the documentation begins and ends for a volume are not sharply defined, except in the sense that all literature will be covered which was received by the editors within these limits.

Vol. 17 is devoted to the recording, summarizing and indexing of astronomical publications of the year 1976 received from January 1, 1976 to August 15, 1976; it also records a number of papers issued before 1976 but received within the period of time.

The main characteristics of the concept of Astronomy and Astrophysics Abstracts may be summarized briefly.

(1) Titles of papers are given in the language of their authors whenever possible. If they are not in English but supplied with English translations they will be given in English. Abstracts are presented in English, French or German. Titles of papers in Russian are given in English.

(2) Authors' abstracts are used whenever possible. As a rule, popular articles were not abstracted; however their titles are usually given with the notation "Popular article".

(3) As a rule, each paper has been classified into one of 108 numbered subject categories and allocated a serial number within the category. In this way each item is numbered by six figures, the first three of which indicate the number of the category. Three further figures indicate the serial number within the category, which was allocated in the order of the receipt of the abstract. Reference to an abstract in Volume 1 is indicated by "01" before the number of the category; for example, 01.074.028, denotes Volume 1, category 074, abstract 028, Vol. 2 is indicated by "02", etc., Vol. 17 by "17".

A paper may have been classified into more than one category. Then its abstract has been allocated a number in one of the categories involved, and in the other category (or categories) the paper has been indicated by the title and a reference to the abstract number.

Papers whose authors are not named were treated like those with authors' names, with one exception: reports from correspondents of journals whose names were unknown were not numbered.

(4) Border fields of astronomy and astrophysics have been taken into account by presenting titles of papers occasionally without abstracts. The selection of papers for inclusion has been made according to the degree of relevance to astronomical research.

Transliteration of the Russian alphabet

The transliteration of the Russian alphabet in use in Astronomy and Astrophysics Abstracts is presented here.

А	а	a	Р	р	r
Б	б	b	С	с	s
В	в	v	Т	т	t
Г	г	g	У	у	u
Д	д	d	Ф	ф	f
Е	е	e	Х	х	kh
Ё	ё	e	Ц	ц	ts
Ж	ж	zh	Ч	ч	ch
З	з	z	Ш	ш	sh
И	и	i	Щ	щ	shch
Й	й	j	Ъ	ъ	''
К	к	k	Ы	ы	y
Л	л	l	Ь	ь	'
М	м	m	Э	э	eh
Н	н	n	Ю	ю	yu
О	о	o	Я	я	ya
П	п	p			

This transliteration was recommended by the Abstracting Board of the International Council of Scientific Unions in 1969. It is essentially the same as the transliteration proposed by the Academy of Sciences, Moscow, and used by the Referativnyj Zhurnal. It may be noted that the letters can be read and printed by usual data processing machines.

If the names of Russian authors in the literature are transliterated very different from this scheme we present the names in the form in which they are given in the references cited and in addition in round brackets according to our transliteration table.

Sources of information

The majority of sources of information for this volume are given in section **001 Periodicals** and in section **008 Observatories, Institutes**. The term "periodical" has been used in its widest sense for publications in a sequence of undetermined duration, even if the intervals of appearance are not regular. Section 001 records 426 periodicals with their full titles and with abbreviations which are in use in Astronomy and Astrophysics Abstracts. It may be noted that the titles of the periodicals are given in their original languages, and that Russian titles have been transliterated applying the transliteration given above. Section 008 records 107 periodicals; these are publication series of observatories and astronomical institutes. The abbreviations of the titles of the periodicals have been given so that in most cases they permit recognition of the full title without recourse to the key in section 001. The steadily growing number of periodicals makes it necessary to use more extensive abbreviations and to abandon the use of very condensed ones.

Other abstracting journals have been consulted in order to examine the degree of completeness of our service. Occasionally, in particular in Physics Abstracts, Referativnyj Zhurnal, and Bulletin Signalétique abstracts of papers were found which had not come to our attention. In such cases Astronomy and Astrophysics Abstracts cites these paper and gives in general reference to the abstracting service which acted as the source.

Classification into a scheme of subject categories

The subdivision of astronomy and its border fields into subject categories is facilitated by the fact that the astronomical objects appear to be particularly well suited for the formation of categories. Sun, moon, earth, planets, comets, and meteorites, the various kinds of stars, galaxies, radio sources, quasars, and pulsars etc. suggest natural subdivisions. It may be assumed that such subdivisions can be maintained for long periods of time. Experience shows, however, that progress in research may imply changes in the classification scheme, in particular, in fields where the expansion of knowledge is explosive.

A few explanatory remarks may be in order on some of the subject categories. In section 003 books on astronomy and astrophysics and its border fields are listed which came to our notice from January 1976 to August 1976. References to book reviews are given if the review appeared quickly.

For completeness of documentation, personal notes (section 006) and obituaries (section 007) are listed. In section 012 (Proceedings of Colloquia, Congresses, Meetings, and Symposia) the proceedings etc. are listed with titles and editors. The individual papers are classified into their corresponding subject categories, but mostly not included in the subject index.

The main subjects of the symposia are cited in the index under section 012.

Errata to papers communicated by the authors are listed at the end of the corresponding subject categories.

Author index and subject index

The subject category and the serial number forming six figures for each abstract have been used as a means of reference in the author index and the subject index. These references are more precise than page references. They offer considerable advantages in indexing by means of data processing machines, and they are more convenient for the user.

The author index of this volume contains 8283 names. A complete reference comprises six figures, three for the subject category and three for the serial number within the category. In the case of more than one reference to abstracts in one category, the number of the category is given only once and not repeated in the immediately following references. The total number of papers (some do not give names of authors) recorded in this volume is about 6690.

We consider the subject index as an approximation to an optimal index covering all fields of astronomy and astrophysics and their border fields. The assigning of one or more key words to a paper is undoubtedly a difficult task. Some journals have started giving key words together with the titles of papers. These key words are chosen by the authors themselves and are in many cases identical with our designations of subject categories with no additional specification. In fact, in some cases it may be more useful to refer to a subject category as a whole than to an item number, in particular, if the total number of abstracts in a category is very small, and if more specific key words do not provide a proper description of the paper.

While each volume is scheduled to contain an author index and a subject index, the magnetic tapes containing the index information will be used to produce separate index volumes (authors and subjects) at intervals of a few years.

The text of the publication was typed on IBM 72 Composers in the editorial office, and it was given to the printer in a form ready for offset reproduction.

The sorting program for the author and subject indexes is based on the IBM SORT/MERGE Program. This SORT-program sorts blank before hyphen (−) and before letters. Apostrophes are ignored by a special routine.

Examples: a) De Laeter

:

Deacon

:

DeLaeter

b) A Stars

:

Aberration Constant

c) Solar X Rays

:

Solar–Terrestrial Relations

d) Boehm–Vitense

Boehme

The introduction of small and capital letters in the layout caused some difficulties. Special programs had to code the capital letters into small ones. For the layout, a TN chain for a 1403 IBM high-speed printer was used. All the programs were written in PL/I. The computations and printing were carried out on an IBM 360/44.

Abbreviations

AAS	American Astronomical Society	Geogr.	Geography, etc.
AAVSO	American Association of Variable Star Observers	Geophys.	Geophysics, etc.
		Ges.	Gesellschaft
Abh.	Abhandlungen	Glav.	Glavnyj (Main)
Abstr.	Abstract	Gos.	Gosudarstvennyj (State)
Abt.	Abteilung	HRD	Herzsprung-Russell diagram
Acad.	Academy, etc.	Hydrogr.	Hydrography, etc.
Accad.	Accademia	IAF	International Astronautical Federation
Adv.	Advances	IAU	International Astronomical Union
AG	Astronomische Gesellschaft	ICSU	International Council of Scientific Unions
AIAA	American Institute of Aeronautics and Astronautics	IEEE	Institute of Electrical and Electronics Engineers
AJB	Astronomischer Jahresbericht	Industr.	Industry, etc.
Akad.	Akademie	Inform.	Information
An.	Anales, etc.	Inst.	Institute, etc.
Ann.	Annals, etc.	Instn.	Institution
Arch.	Archiv, etc.	Ionosph.	Ionosphere, etc.
Ark.	Arkiv	Issled.	Issledovaniya (Research)
ASA	Astronomical Society of Australia	Ist.	Istituto
Asoc.	Asociación	Izv.	Izvestiya (News)
ASP	Astronomical Society of the Pacific	Jb.	Jahrbuch
Ass.	Association	JO	Journal des Observateurs
ASSA	Astronomical Society of Southern Africa	Journ.	Journal
Astrofis.	Astrofisica, etc.	Kl.	Klasse
Astrofiz.	Astrofizika, etc.	Lab.	Laboratory
Astron.	Astronomy, etc.	Mag.	Magazine
Astronaut.	Astronautics, etc.	Mat.	Matematica, etc.
Astrophys.	Astrophysics, etc.	Math.	Mathematics, etc.
ASV	Astronomical Society of Victoria	Mech.	Mechanics, etc.
ASWA	Astronomical Society of Western Australia	Med.	Mededelingen
Atmosph.	Atmosphere, etc.	Medd.	Meddelande, Meddelser
BA	Bulletin Astronomique	Mekhan.	Mekhanika, etc.
BAA	British Astronomical Association	Mém.	Mémoires
BAN	Bulletin of the Astronomical Institutes of the Netherlands	Mem.	Memoirs, Memorandum, etc.
		Meteorol.	Meteorology, etc.
Ber.	Berichte	MIT	Massachusetts Institute of Technology
BIH	Bureau International de l'Heure (Paris)	Mitt.	Mitteilungen
Bol.	Boletin	MVS Sonneberg	Mitteilungen über Veränderliche Sterne, Sonneberg
Boll.	Bolletino		
Bull.	Bulletin	Nachr.	Nachrichten
Byull.	Byulleten' (Bulletin)	Nat.	Naturwissenschaftlich, etc.
Circ.	Circular	Naut.	Nautics, etc.
Cl.	Classe	NBS	National Bureau of Standards
Coll.	Collection	NRAO	National Radio Astronomy Observatory (Green Bank)
Commun.	Communication		
Comun.	Comunicazioni	NRL	Naval Research Laboratory (Washington)
Contr.	Contributions, etc.	NASA	National Aeronautics and Space Administration
COSPAR	Committee on Space Research		
C.S.I.R.O.	Commonwealth Scientific Industrial Research Organization	Obs.	Observatory, etc.
		OSA	Optical Society of America
Dep.	Department	Oss.	Osservatorio, Osservazioni, etc.
Diss.	Dissertation	Ped.	Pedagogika, etc. (Pedagogics)
Div.	Division	Phil.	Philosophical
Dokl.	Doklady (Reports)	Phys.	Physics, etc.
ESA	European Space Agency	Planet.	Planetary
ESO	European Southern Observatory	Priklad.	Prikladnoj (Applied)
Fis.	Fisica, etc.	Proc.	Proceedings
Fiz.	Fizika, etc.	Progr.	Progress, etc.
Fys.	Fysica, etc.	Pubbl.	Pubblicazioni
Géod.	Géodésie, etc.	Publ.	Publications
Geod.	Geodesy, etc.	Rap.	Raportoj
Geofis.	Geofisica, etc.	RAS	Royal Astronomical Society
Geofiz.	Geofizika, etc.	RAS Canada	Royal Astronomical Society of Canada
Geofys.	Geofysik, etc.	Rech.	Recherches
Geol.	Geology, etc.	Rend.	Rendiconti

Rep.	Report	Techn.	Technics, etc.
Repr.	Reprint	Tekhn.	Tekhnika, etc.
Res.	Research	Teor.	Teoreticheskij
Rev.	Review, etc.	Terr.	Terrestrial, etc.
Ric.	Ricerche	TH	Technische Hochschule
Roy.	Royal, etc.	Theor.	Theoretical
SAAO	South African Astronomical Observatory	Tidssk.	Tidsskrift
SAF	Société Astronomique de France	Trans.	Transactions
SAI	Società Astronomica Italiana	Trudy	Trudy (Publications)
SAO	Smithsonian Astrophysical Observatory	Tsentr.	Tsentral'nyj (Central)
SAS	Société Astronomique de Suisse	Tsirk.	Tsirkulyar (Circular)
Sci.	Science, etc.	TU	Technical University
Sect.	Section	Uch. Zap.	Uchenye Zapiski (Treatise)
Ser.	Series, etc.	Univ.	University, etc.
S. I. R.	Service International Rapide des Latitudes	URSI	Union Radio Scientifique Internationale
Sitzungsber.	Sitzungsberichte	USNO	US Naval Observatory
Soc.	Society	Verh.	Verhandlungen
Soobshch.	Soobshcheniya (Communications)	Veröff.	Veröffentlichungen
Sternw.	Sternwarte	Wet.	Wetenschappen
Stud. Cerc.	Studii şi Cercetari	Wiss.	Wissenschaften, etc.
Supl.	Suplemento	Zeitschr.	Zeitschrift
Suppl.	Supplement	ZfA	Zeitschrift für Astrophysik
SuW	Sterne und Weltraum	Zhurn.	Zhurnal (Journal)

Periodicals, Proceedings, Books, Activities

001 Periodicals

AAS Photo-Bull.
AAS (American Astronomical Society) Photo-Bulletin. Published by the Working Group on Photographic Materials. Produced by Eastman Kodak Co., Rochester, N. Y.

AAVSO Bull.
Bulletin of the American Association of Variable Star Observers, 187 Concord Avenue, Cambridge, Mass., 02138, U.S.A.

Acad. Roy. Belgique, Bull. Cl. Sci.
Académie Royale de Belgique, Bulletin de la Classe des Sciences (Koninklijke Academie van België, Mededelingen van de Klasse der Wetenschappen). 5ᵉ Série, Palais des Académies, Bruxelles.

Acta Astron.
Acta Astronomica. An international quarterly journal. Publisher: Polska Akademia Nauk, Komitet Astronomii (Polish Academy of Sciences, Committee of Astronomy), Warszawa – Wrocław.

Acta Astron. Sinica
Acta Astronomica Sinica. Published by Purple Mountain Observatory, Academia Sinica, Nanking, China.

Acta Astronaut.
Acta Astronautica. Journal of the International Academy of Astronautics. Publisher: Pergamon Press Inc., Elmsford, New York, U.S.A.; Pergamon Press Ltd., Oxford, England.

Acta Cosmologica
Acta Cosmologica. Published by Obserwatorium Astronomiczne Uniwersytetu Jagiellońskiego, Kraków, Poland.

Acta Crystallogr. A
Acta Crystallographica, Section A: Crystal Physics, Diffraction, Theoretical and General Crystallography. Munksgaard International Booksellers and Publishers Ltd., 35 Norre Sogade, DK 1370 Kobenhavn K, Denmark.

Acta Electronica
Acta Electronica. 3 Avenue Descartes, BP 15, 94 450 Limeil-Brevannes, (Val-de-Marne), France.

Acta Geophys. Sinica
Acta Geophysica Sinica. Chinese Academy of Sciences, Department of Geophysical Research. Published by Science Press, Peking, Peoples Republic of China.

Acta Phys. Acad. Sci. Hungaricae
Acta Physica Academiae Scientiarum Hungaricae. Postafiok 24, Budapest 502, Hungary.

Acta Phys. Austriaca
Acta Physica Austriaca. Springer-Verlag, A-1011 Wien, Molkerbastei 5, Postfach 367, Austria.

Acta Phys. Polonica B
Acta Physica Polonica B. ARS Polona-Ruch, Warszawa 1, P.O. Box 154, Poland.

Acta Phys. Sinica
Acta Physica Sinica. Chinese Academy of Sciences, Institute of Physics, Peking, Peoples Republic of China.

Acta Techn. CSAV
Acta Technica Ceskoslovenska Akademie Ved. Academia Publishing House of the Czechoslovak Academy of Sciences, Vodickova 40, 112 29 Praha 1, Czechoslovakia.

Acta Univ. Carolinae Math. Phys.
Acta Universitatis Carolinae, Mathematica et Physica. Administrace: Matematicko-fyzikální fakulta University Karlovy, Praha.

Actas Acad. Nacional Cienc. Lima
Actas de la Academia Nacional de Ciencias Exactas, Fisicas y Naturales de Lima. Lima - Peru.

Adv. Astron. Astrophys.
Advances in Astronomy and Astrophysics. Publisher: Academic Press, New York – London.

Aerotecn. Missili Spazio
L'Aerotecnica Missili e Spazio. Tamburini Editore S.P.A., Via Pascoli 55, 20133 Milano, Italy.

A.F.O.E.V. Bull.
Association Française des Observateurs d'Étoiles Variables. Bulletin. Redaction and publication: M. Émile Schweitzer, 1, rue Beethoven, Strasbourg.

AIAA Journ.
AIAA Journal. A Publication of the American Institute of Aeronautics and Astronautics devoted to Aerospace Research and Development. Published by the American Institute of Aeronautics and Astronautics, New York, N.Y.

AIP Conf. Proc.
AIP Conference Proceedings. American Institute of Physics, 335 East 45th Street, New York, N.Y. 10017, USA.

American Journ. Phys.
American Journal of Physics. Published for the American Association of Physics Teachers by the American Institute of Physics, 335 East 47th Street, New York, N.Y. 10017, USA.

American Mineral.
American Mineralogist. Mineralogical Society of America, 1707 L Street, N.W., Washington, DC 20036, USA.

American Scient.
American Scientist. Society of Sigma Xi, New Haven, Conn.

An. Acad. Brasil. Ci.
Anais da Academia Brasileira de Ciencias. Caixa Postal 229, ZC-00 Rio de Janeiro gb, Brazil.

An. Stiint. 'Al. I. Cuza' Iasi (Ser. Noua) I
Analele Stiintifice ale Universitatu 'Al. I. Cuza' din Iasi (Serie Noua), Sectiunea I Fizica. Calea 23 August, Iasi, Rumania.

Ann. Acad. Sci. Fenn., Ser. A VI
Annales Academiae Scientiarum Fennicae, Series A VI (Physica). Snellmaninkatu 9–11, 00170 Helsinki-17, Finland.

Ann. Françaises Chronométrie Micromécanique
Annales Françaises de Chronométrie et de Micromécanique, publication annuelle de l'Observatoire de Besançon, du Centre Technique de l'Industrie Horlogère et de la Société Française de Chronométrie et de Micromécanique. Rédaction et administration: Observatoire de Besançon. Publiées avec le concours du Centre National de la Recherche Scientifique et des organismes corporatifs.

Ann. Géophys.
Annales de Géophysique. Revue Internationale trimestrielle, publiée par le Centre National de la Recherche Scientifique, Paris.

Ann. Inst. Henri Poincaré A
Annales de l'Institut Henri Poincaré, Section A (Physique Theorique). 11 Rue Pierre-Curie, Paris 5, France.

Ann. Obs. Astron. Météorol. Toulouse
Annales de l'Observatoire Astronomique et Météorologique de Toulouse. Publisher: Gauthier-Villars, Paris.

Ann. Physics
Annals of Physics. Publisher: Academic Press Inc., New York, N.Y.

Ann. Physik
Annalen der Physik. 7. Folge. Publisher: Johann Ambrosius Barth, Leipzig.

Ann. Physique
Annales de Physique. Publisher: Masson et Cie., Paris.

Ann. Sci.
Annals of Science. Taylor & Francis Ltd., 10–14 Macklin Street, London, WC2B 5NF, England.

Ann. Soc. Sci. Bruxelles
Annales de la Société Scientifique de Bruxelles. Série I: Sciences Mathématiques, Astronomiques et Physiques. Published by Institut de Physique, Heverlé-Louvain.

Ann. Télécommun.
Annales des Télécommunications. Centre National d'Études des Télécommunications, 38 rue du Général Leclerc, 92 Issy-les-Moulineaux, France.

Ann. Tokyo Astron. Obs.
Annals of the Tokyo Astronomical Observatory. University of Tokyo, Mitaka, Tokyo, Japan.

Ann. Univ.-Sternw. Wien
Annalen der Universitäts-Sternwarte Wien. In Kommission bei Ferd. Dümmlers Verlag, Bonn.

Annual Rep. Astron. Inst. Greece
Annual Reports of the Astronomical Institutes of Greece. Published by the Greek National Committee for Astronomy. Academy of Athens, Research Center for Astronomy and Applied Mathematics.

Annual Rev. Astron. Astrophys.
Annual Review of Astronomy and Astrophysics. Publisher: Annual Reviews Inc., Palo Alto, California.

Anzeiger. Österreich. Akad. Wiss. Math.-Nat. Kl.
Anzeiger. Österreichische Akademie der Wissenschaften. Mathematisch-Naturwissenschaftliche Klasse. Publisher: Springer-Verlag, Wien.

Applied Optics
Applied Optics. A monthly publication of the Optical Society of America. Published for the Optical Society of America by the American Institute of Physics, New York, N. Y.

Applied Phys.
Applied Physics. Springer-Verlag, Heidelberger Platz 3, 1 Berlin 33, F.R. Germany.

Applied Phys. Letters
Applied Physics Letters. American Institute of Physics, 335 East 45th Street, New York, N.Y. 10017, USA.

Applied Spectroscopy
Applied Spectroscopy. 428 East Preston Street, Baltimore, MD 21202, USA.

Arch. Sci.
Archives des Sciences, éditées par la Société de Physique et d'Histoire Naturelle de Genève. Publisher: Imprimerie Kundig, Genève. Subscription address: Librairie Payot, Genève.

Arch. Ration. Mech. Analysis
Archive for Rational Mechanics and Analysis. Springer-Verlag, Heidelberger Platz 3, D 1000 Berlin 33, F. R. Germany.

Ark. Astron.
Arkiv för Astronomi. Utgivet av Kungliga Svenska Vetenskapsakademien, Stockholm. Printed by Almqvist & Wiksell, Stockholm.

Ark. Geofys.
Arkiv för Geofysik. Kungliga Svenska Vetenskapsakademien, Stockholm. Printed by Almqvist & Wiksell, Stockholm

Artificial Satellites
Artificial Satellites. Publication of Polish Scientific Institutions. Polish Academy of Sciences, National Committee of Geophysics and Geodesy, National Committee for Space Research, Warsaw. Publishing Office: Palac Kultury i Nauki, Warszawa.

Asoc. Argentina Astron. Bol.
Asociación Argentina de Astronomía. Boletin. Editor: Instituto Argentino de Radioastronomía, Provincia de Buenos Aires, Argentina. Printer: Talleres Gráficos "Renovación", La Plata, República Argentina.

Astrofizika
Astrofizika. Izdatel'stvo Akademii Nauk Armyanskoj SSR, Erevan. [An English translation is published in "Astrophysics".]

Astrofiz. Issled. Izv. Spets. Astrofiz. Obs.
Astrofizicheskie Issledovaniya. Izvestiya Spetsial'noj Astrofizicheskoj Observatorii. Akademiya Nauk SSSR. Publishers: Izdatel'stvo "Nauka", Leningradskoe Otdelenie, Leningrad.

Astrometr. Astrofiz.
Astrometriya i Astrofizika. Respublikanskij Mezhvedomstvennyj Sbornik. Akademiya Nauk Ukrainskoj SSR, Glavnaya Astronomicheskaya Observatoriya. Naukova Dumka, Kiev.

Astron. Astrophys.
Astronomy and Astrophysics. A European Journal. Published by Springer-Verlag, Berlin – Heidelberg–New York.

Astron. Astrophys. Suppl. Ser.
Astronomy and Astrophysics. Supplement Series. A European Journal. Published by the Astronomical Institute Lausanne and Geneva Observatory, Switzerland, on behalf of the Board of Directors.

Astron. Herald
Astronomical Herald. Astronomical Society of Japan, Tokyo Astronomical Observatory, Oosawa Mitaka, Tokyo, Japan.

Astron. in der Schule
Astronomie in der Schule. Zeitschrift für die Hand des Astronomielehrers. Herausgegeben vom Verlag Volk und Wissen, Berlin. Redaktion: Sternwarte Bautzen.

Astron. Journ.
The Astronomical Journal. Published for the American Astronomical Society by the American Institute of Physics, New York, N. Y. Editorial Office: Department of Astronomy, Columbia University, New York, N. Y.

Astron. Nachr.
Astronomische Nachrichten. Publisher: Akademie-Verlag, Berlin.

Astron. Rep.
The Astronomical Reports. Polish Amateur Astronomical Society. Polskie Towarzystwo Miłośników Astronomii, Kraków, Poland.

Astron. Tidsskr.
Astronomisk Tidsskrift. Edited by Astronomisk Selskab, København; Norsk Astronomisk Selskap, Oslo; Svenska Astronomiska Sällskapet, Stockholm. Printed by John Griegs Boktrykkeri, Bergen.

Astron. Tsirk.
Astronomicheskij Tsirkulyar, izdavaemyj Byuro Astronomicheskikh Soobshchenij Akademii Nauk SSSR. Tipografiya Astrosoveta AN SSSR, Moskva.

Astron. Vestn.
Astronomicheskij Vestnik. Publishers: Izdatel'stvo "Nauka", Moskva.

Astron. Zhurn. Akad. Nauk SSSR
Astronomicheskij Zhurnal. Akademiya Nauk SSSR. Publishers: Izdatel'stvo "Nauka", Moskva. [An English translation is published in "Soviet Astronomy AJ"].

Astronautik
Astronautik. Organ der Hermann-Oberth-Gesellschaft e.V. Astronautik-Verlag, Druckerei H. Brandt, Delmenhorst, F.R. Germany.

Astrophys. Journ.
The Astrophysical Journal. Published for the American Astronomical Society by the University of Chicago Press, Chicago, Illinois.

Astrophys. Journ., (*Letters*)
The Astrophysical Journal. Letters to the Editors. Published for the American Astronomical Society by the University of Chicago Press, Chicago, Illinois.

Astrophys. Journ., Suppl. Ser.
The Astrophysical Journal. Supplement Series. Published for the American Astronomical Society by the University of Chicago Press, Chicago, Illinois.

Astrophys. Letters
Astrophysical Letters. An International *EXPRESS* Journal. Published monthly by Gordon and Breach Science Publishers Ltd., New York – London – Paris.

Astrophys. Space Sci.
Astrophysics and Space Science. An International Journal of Cosmic Physics. Published by D. Reidel Publishing Company, Dordrecht – Holland.

Astrophysics
Astrophysics. A cover-to-cover translation of Astrofizika (USSR). Consultants Bureau, New York, N.Y.

Atomkernenergie
Atomkernenergie. Verlag Karl Thiemig, Pilgersheimerstrasse 38, 8 München 90, Postfach 900740, F.R. Germany.

Atti Accad. Nazionale Lincei. Mem.
Atti della Accademia Nazionale dei Lincei. Serie Ottava. Memorie. Classe di Scienze fisiche, matematiche e naturali. Sezione I: Matematica, Meccanica, Astronomia, Geodesia e Geofisica. Published by Accademia Nazionale dei Lincei, Roma.

Atti Accad. Nazionale Lincei. Rend.
Atti della Accademia Nazionale dei Lincei. Serie Ottava. Rendiconti. Classe di Scienze fisiche, matematiche e naturali. Published by Accademia Nazionale dei Lincei, Roma.

Atti Fond. G. Ronchi, Contr. Ist. Nazionale Ottica
Atti della Fondazione Giorgio Ronchi e Contributi dell' Istituto Nazionale di Ottica. Largo Enrico Fermi 6, 50125 Firenze, Italy.

Atti Soc. Astron. Italiana
Atti della Società Astronomica Italiana. Publisher: Tipografia Baccini & Chiappi, Firenze (Italy).

Australian Journ. Phys.
Australian Journal of Physics. Published by the Commonwealth Scientific and Industrial Research Organization, East Melbourne, Victoria.

Australian Journ. Phys. Astrophys. Suppl.
Australian Journal of Physics, Astrophysical Supplement. Published by Commonwealth Scientific and Industrial Research Organization, East Melbourne, Victoria.

BAV Rundbrief
BAV Rundbrief. Mitteilungsblatt der Berliner Arbeitsgemeinschaft für Veränderliche Sterne. Editor: BAV Berliner Arbeitsgemeinschaft für Veränderliche Sterne eV., Berlin.

BBSAG Bull.
Bedeckungsveränderlichen Beobachter der Schweizerischen Astronomischen Gesellschaft, [Swiss Astronomical Society's Eclipsing Variable Observers], Bulletin. To be

obtained from R. Diethelm, Winterthur, Switzerland.

Bild der Wiss.
Bild der Wissenschaft. Zeitschrift über die Naturwissenschaften und die Technik in unserer Zeit. Publisher: Deutsche Verlagsanstalt, Stuttgart.

Blick in das Weltall
Astronomische Veranstaltungen und Mitteilungen für Sternfreunde. Archenhold-Sternwarte Berlin-Treptow. Gesamtherstellung: Betriebsschule Rudi Arndi, Bildungsstätte der Druckerei Neues Deutschland.

Bol. Astron. Obs. Madrid
Boletín Astronómico del Observatorio de Madrid. Editor: Instituto Geografico y Catastral. General Ibáñez de Ibero, Madrid.

Bol. Inst. Mat., Astron., Fis. Univ. Nacional Córdoba
Boletin del Instituto de Matematica, Astronomia y Fisica, Universidad Nacional de Córdoba (R. A.).Dirección General de Publicaciones, Córdoba (Argentina).

Bol. Inst. Tonantzintla
Boletin del Instituto de Tonantzintla. Instituto Nacional de Astrofisica, Optica y Electronica, Apartados Postales Nos. 216 y 51, Puebla, Pue, Mexico.

Bol. Liga Latinoamericana Astron.
Boletin de la Liga Latinoamericana de Astronomia. Publicado por la Asociacion Argentina Amigos de la Astronomia, Buenos Aires, Argentina.

Bol. Obs. Ebro
Boletín del Observatorio del Ebro, Tortosa. Printed by Cooperativa Gráfica Dertosense, Tortosa.

Boll. Geod. Sci. Affini
Bolletino di Geodesia e Scienze Affini. Pubblicazione dell'Istituto Geografico Militare, Firenze.

British Astron. Ass. Circ.
British Astronomical Association, Circular. Editorial Office: 97 Hawkswood Drive, Hailsham, Sussex.

British Journ. Philosoph. Sci.
British Journal for the Philosophy of Science. Cambridge University Press, Bentley House, 200 Euston Road, London, NW1 2DB, England.

Bull. American Astron. Soc.
Bulletin of the American Astronomical Society. Published for the American Astronomical Society by the American Institute of Physics Inc., New York, N. Y.

Bull. Astron. Inst. Czechoslovakia
Bulletin of the Astronomical Institutes of Czechoslovakia. Published under the auspices of the Czechoslovak Academy of Sciences by Academia, Praha. Editor: Astronomical Institutes of the Czechoslovak Academy of Sciences, Praha.

Bull. Astron. Soc. India
Bulletin of the Astronomical Society of India. Edited and published by M. S. Vardya, Tata Institute of Fundamental Research, Bombay on behalf of the Astronomical Society of India, Osmania University, Hyderabad.

Bull. Géod.
Bulletin Géodésique, being the Journal of the International Association of Geodesy. Nouvelle Série. Publié par le Bureau Central de l'Association Internationale de Géodésie, Paris.

Bull. Geograph. Survey Inst.
Bulletin of the Geographical Survey Institute. Published by the Geographical Survey Institute, Ministry of Construction, Tokyo, Japan.

Bull. Groupe Recherches Géod. Spatiale
Groupe de Recherches de Géodésie Spatiale. Bulletin. CNES/Toulouse, France.

Bull. Inst. Space Aeronaut. Sci., Univ. Tokyo, B
Bulletin of the Institute of Space and Aeronautical Science, University of Tokyo, B. Tokyo, Japan.

Bull. Obs. Astron. Beograd
Bulletin de l'Observatoire Astronomique de Béograd. Editor: Observatoire Astronomique de Béograd. Printed by Naucna delo, Béograd.

Bull. Sci. Yougoslavie
Bulletin Scientifique. Conseil des Academies des Sciences et des Arts de la RSF de Yougoslavie. Section A: Sciences Naturelles, Techniques et Médicales. Rédaction et Administration: Opatička ul. 18/II, Zagreb, Yougoslavie.

Bull. Signal.
Bulletin Signalétique. Section 120: Astronomie, Physique spatiale, Géophysique. Centre de Documentation du Centre Nationale de la Recherche Scientifique, Paris.

Bull. Signal.
Bulletin Signalétique. Bibliographie des Sciences de la Terre. Section 220, Cahier A: Minéralogie, Géochimie, Géologie extraterrestre. Centre de Documentation du C.N.R.S., Paris; Département Documentation du B.R.G.M., Orléans.

Bull. Soc. Roy. Sci. Liège
Bulletin de la Société Royale des Sciences de Liège. L'Université, 15 Avenue des Tilleurs, Liège, Belgium.

Byull. Abastuman. Astrofiz. Obs.
Abastumanskaya Astrofizicheskaya Observatoriya, Gora Kanobili. Byulleten'. Akademiya Nauk Gruzinskoj SSR. Publishers: Izdatel'stvo "Metsniereba", Tbilisi.

Byull. Inst. Astrofiz.
Byulleten' Instituta Astrofiziki, Akademiya Nauk Tadzhikskoj SSR. Izdatel'stvo Donish, Dushanbe.

Byull. Inst. Teor. Astron.
Byulleten' Instituta Teoreticheskoj Astronomii. Izdatel'stvo Nauka, Leningradskoe Otdelenie, Leningrad.

Canadian Journ. Phys.
Canadian Journal of Physics. Published by the National Research Council of Canada, Ottawa. Printed in Canada by the University of Toronto Press, Toronto, Ont.

Carter Obs. Astron. Bull.
Carter Observatory, Astronomical Bulletin. Carter Observatory, P.O. Box 2909, Wellington 1, New Zealand.

Celestial Mechanics
Celestial Mechanics. An International Journal of Space Dynamics. Publishers: D. Reidel Publishing Company, Dordrecht—Holland.

Center Astrophys. Prepr. Ser.

Center for Astrophysics, Preprint Series. Harvard College Observatory, Smithsonian Astrophysical Observatory. Center for Astrophysics, 60 Garden St., Cambridge, Mass. 02138.

Centre Données Stellaires, Inform. Bull.
Centre de Données Stellaires. Information Bulletin. Compiled at Observatoire de Strasbourg, Strasbourg, France.

Cesk. Casopis Fis. A
Ceskoslovensky Casopis pro Fisiku. Sekce A. Academia Publishing House of the Czechoslovak Academy of Sciences, Vodickova 40, 112 29 Praha 1, Czechoslovakia.

Chem. Phys. Letters
Chemical Physics Letters. North-Holland Publishing Co., P.O. Box 211, Amsterdam-C, Netherlands.

Chinese Journ. Phys. (*Taiwan*)
Chinese Journal of Physics. Physical Society of the Republic of China, Physics Department, National Taiwan University, Taipei, Taiwan, China.

Ciel et Terre
Ciel et Terre. Bulletin de la Société Belge d'Astronomie, de Météorologie et de Physique du Globe. Administration: Avenue Circulaire, 3, Bruxelles. Printed by Imprimerie R. Louis, Bruxelles.

Circ. d'Information
Circulaire d'Information. Union Astronomique Internationale. Commission des Etoiles Doubles. Address: Observatoire de Meudon, Meudon, France.

Circ. Stazione Astron. Internazionale Latitudine, Carloforte–Cagliari.
Circolari della Stazione Astronomica Internazionale di Latitudine, Carloforte–Cagliari. Serie A printed by Tipo-Offset "3 T", Cagliari. Serie B printed by Multi Copy, Milano.

Coelum
Coelum. Periodico bimestrale per la Divulgazione dell' Astronomia. Editor: Osservatorio Astronomico Universitario di Bologna.

Comment. Phys.-Math.
Commentationes Physico-Mathematicae. Societas Scientiarum Fennica, Helsinki–Helsingfors. Printed by Keskuskirjapaino–Centraltryckeriet, Helsinki–Helsingfors.

Comments Astrophys. Space Phys.
Comments on Astrophysics and Space Physics. A Journal of Critical Discussion of the Current Literature. Comments on Modern Physics: Part C. Publishers: Gordon and Breach Science Publishers, Inc., New York – London

Comments Nuclear Particle Phys.
Comments on Nuclear and Particle Physics. Gordon & Breach Science Publishers Ltd., 41 and 42 William IV Street, London, WC2, England.

Commun. Math. Phys.
Communications in Mathematical Physics. Springer-Verlag, Postfach 1780, 6900 Heidelberg 1, F.R. Germany.

Comptes Rendus Acad. Bulg. Sci.
Comptes Rendus de l'Académie bulgare des Sciences. (Doklady Bolgarskoj Akademii Nauk). Sofia, Bulgaria.

Comptes Rendus Acad. Sci. Paris
Comptes Rendus hebdomadaires des Séances de l'Académie des Sciences, publié avec le concours du Centre National de la Recherche Scientifique. Imprimerie: Gauthier-Villars, Paris.

Contr. Atmosph. Phys.
Contributions to Atmospheric Physics – Beiträge zur Physik der Atmosphäre. Publisher: Friedrich Vieweg & Sohn, Braunschweig.

COSPAR Inform. Bull.
COSPAR. Information Bulletin. Address: COSPAR Secretariat, Paris.

Current Sci.
Current Science, Current Science Association, Raman Research Institute, Bangalore 6, India.

Deutsche Geod. Kommission Bayer. Akad. Wiss.
Deutsche Geodätische Kommission bei der Bayerischen Akademie der Wissenschaften. Reihe A: Höhere Geodäsie; Reihe B: Angewandte Geodäsie; Reihe C: Dissertationen; Reihe D: Tafelwerke; Reihe E: Geschichte und Entwicklung der Geodäsie. Published by Verlag der Bayerischen Akademie der Wissenschaften, München.

Dokl. Akad. Nauk
Doklady Akademii Nauk SSSR. Seriya Matematika, Fizika. Publishers: Izdatel'stvo "Nauka", Moskva.

Dunsink Obs. Publ.
Dunsink Observatory Publications. The Observatory of the School of Cosmic Physics, Dublin Institute for Advanced Studies, Dublin.

Earth Extraterr. Sci.
Earth and Extraterrestrial Sciences. Gordon & Breach Science Publishers Ltd., 41 and 42 William IV Street, London, WC2, England.

Earth Planet. Sci. Letters
Earth and Planetary Science Letters. A Letter Journal devoted to the Development in Time of the Earth and Planetary System. Publisher: North-Holland Publishing Company, Amsterdam, Netherlands.

El Universo
El Universo. Organo de la Sociedad Astronomica de Mexico, Mexico, D. F.

Electronic Production Methods Equipment
Electronic Production Methods and Equipment. Kiver-Patterson Publishing, 322 St. John Street, London, EC1, England.

Electronics (*USA*)
Electronics. Published by McGraw-Hill Publishing Company, New York, N.Y., U.S.A.

Electronics Australia
Electronics Australia. 12th floor, 235–242 Jones Street, Broadway, Sydney, N.S.W., Australia.

Electronics Letters
Electronics Letters. Institution of Electrical Engineers, Savoy Place, London, WC2R 0BL, England.

Elektrotechn. Zeitschr. B
Elektrotechnische Zeitschrift. Ausgabe B: Der Elektrotechniker. Verband Deutscher Elektrotechniker

Publication address: VDE-Verlag, Bismarckstrasse 33, 1000 Berlin 12, F. R. Germany.

Endeavour
Endeavour. A review of the progress of science, published in four languages by Imperial Chemical Industries Limited, London.

ESO Bull.
European Southern Observatory, Bulletin. Edited by European Southern Observatory. Office of the Director: Hamburg.

ESO Techn. Rep.
European Southern Observatory, (ESO), Technical Report. Published by the European Southern Observatory Telescope Project Division, CERN, Geneva, Switzerland.

Experim. Techn. Phys.
Experimentelle Technik der Physik, VEB Deutscher Verlag der Wissenschaften, Traubenstrasse 10, 108 Berlin 8, German Democratic Republic.

Feinwerktechn. & Messtechn.
F & M. Feinwerktechnik und Messtechnik. Fusion of "Feinwerktechnik" and "Messtechnik" (formerly Zeitschrift für Instrumentenkunde) beginning with Jahrgang 82, No. 5 (1974). Publishers: Karl Hanser Verlag, München, Germany.

Fiz. Szemle
Fizikai Szemle. Kiadja a Lapkiado Vallalat, Budapest VII, Lenin korut 9–11, Hungary.

Found. Phys.
Foundations of Physics. Plenum Publishing Co., 8 Scrubs Lane, Harlesden, London, NW10 6SE, England.

General Relativ. Gravitation
General Relativity and Gravitation. Published under the auspices of the International Committee on General Relativity and Gravitation GRG. Publishing Office: Plenum Publishing Company Limited, London.

Geochim. Cosmochim. Acta
Geochimica et Cosmochimica Acta. Journal of the Geochemical Society. Publishing House: Pergamon Press, Ltd., Oxford.

Geodezja Kartografia
Geodezja i Kartografia. Komitet Geodezji Polskiej Akademii Nauk. Publisher: Państwowe Wydawnictwo Naukowe, Warszawa.

Geomagn. Aeronom.
Geomagnetizm i Aehronomiya. Akademiya Nauk SSSR. Izdatel'stvo "Nauka", Moskva [An English translation is published in "Geomagnetism and Aeronomy", American Geophysical Union, Washington, D.C.].

Geophys. Journ. Roy. Astron. Soc.
The Geophysical Journal of the Royal Astronomical Society. Published for the Royal Astronomical Society by Blackwell Scientific Publications, Oxford – Edinburgh.

Geophys. Res. Letters
Geophysical Research Letters. Published monthly by the American Geophysical Union, Washington, D.C., U.S.A.

GEOS
GEOS. Department of Energy, Mines and Resources, Ottawa, Canada.

Gerlands Beiträge Geophys.
Gerlands Beiträge zur Geophysik. Publisher: Akademische Verlagsgesellschaft Geest & Portig K.-G., Leipzig.

Giorn. A.A.B.
Giornale dell'A.A.B. Notiziario trimestrale delle attività culturali e scientifiche della Associazione Astrofili Bolognesi, Bologna, Italy.

Giorn. Astron.
Giornale di Astronomia, Pubblicazione della Società Astronomica Italiana. Printed by Tipolitografia Lodigraf S.p.A. Lodi (MI).

Glasnik Mat.
Glasnik Matematicki. Published by the Society of Mathematicians and Physicists of the S. R. of Croatia. Publisher: Drustvo Matematicara i Fizicara S. R. Hrvatske, Zagreb.

GSFC Document
Goddard Space Flight Center, Greenbelt, Maryland. Available from Technical Information Division, Code 250, Goddard Space Flight Center, Greenbelt, Maryland 20771.

Helvetica Phys. Acta
Helvetica Physica Acta. Schweizerische Physikalische Gesellschaft. Publisher: E. Birkhäuser, Basel.

IAU Circ.
International Astronomical Union, Circular. Central Bureau for Astronomical Telegrams, Smithsonian Astrophysical Observatory, Cambridge, Mass.

Icarus
Icarus. International Journal of Solar System Studies. Publisher: Academic Press, New York – London.

ICSU Bull.
ICSU Bulletin. International Council of Scientific Unions. Secretariat: 51, Bd de Montmorency, Paris, France.

IEEE Spectrum
IEEE Spectrum. Published monthly by the Institute of Electrical and Electronics Engineers, Inc., New York, N. Y.

IEEE Trans. Aerospace Electron. Systems
IEEE Transactions on Aerospace and Electronic Systems. Published by the Institute of Electrical and Electronics Engineers, 345 East 47th Street, New York, N.Y. 10017, USA.

IEEE Trans. Antennas Propagation
IEEE Transactions on Antennas and Propagation. Published by the Institute of Electrical and Electronics Engineers, 345 East 47th Street, New York, N.Y. 10017, USA.

IEEE Trans. Electron Devices
IEEE Transactions on Electron Devices. Published by the Institute of Electrical and Electronics Engineers, 345 East 47th Street, New York, N.Y. 10017, USA.

IEEE Trans. Geosci. Electronics
IEEE Transactions on Geoscience Electronics. Published

by the Institute of Electrical and Electronics Engineers, 345 East 47th Street, New York, N.Y. 10017, USA.

IEEE Trans. Instrument. Measurement
IEEE Transactions on Instrumentation and Measurement. Published by the Institute of Electrical and Electronics Engineers, 345 East 47th Street, New York, N.Y. 10017, USA.

IEEE Trans. Microwave Theory Techn.
IEEE Transactions on Microwave Theory and Techniques. Published by the Institute of Electrical and Electronics Engineers, 345 East 47th Street, New York, N.Y. 10017, USA.

Indian Journ. Meteorol. Geophys.
Indian Journal of Meteorology and Geophysics. Indian Meteorological Department. Publication address: Civil Lines, Delhi-6, India.

Indian Journ. Phys.
Indian Journal of Physics. 2 and 3 Lady Willingdon Road, Raja Subodhchandra, Mallick Road, Calcutta 32, India.

Indian Journ. Pure Applied Math.
Indian Journal of Pure and Applied Mathematics. National Institute of Sciences India, Bahadur Shah Zafar Marg, New Delhi 1, India.

Indian Journ. Pure Applied Phys.
Indian Journal of Pure and Applied Physics. Council of Scientific and Industrial Research, New Delhi, India.

Indian Journ. Radio Space Phys.
Indian Journal of Radio & Space Physics. Council of Scientific & Industrial Research. Editorial address: Publications & Information Directorate, Hillside Road, New Delhi-12, India.

Industr. Math.
Industrial Mathematics. Industrial Mathematics Society, P.O. Box 159, Roseville, MI 48066, USA.

Inform. Bull. Southern Hemisphere
Information Bulletin for the Southern Hemisphere. Editorial Office: Observatorio Astronómico, La Plata, Argentina.

Inform. Bull. Variable Stars
Commission 27 of the I.A.U. Information Bulletin on Variable Stars. Konkoly Observatory, Budapest.

Infrared Physics
An International Research Journal. Publisher: Pergamon Press Ltd., Oxford – London – New York.

Inst. Theor. Astrophys., Blindern–Oslo, Rep.
Institute of Theoretical Astrophysics, Blindern–Oslo, Report. Universitetsforlagets trykningssentral, Oslo.

Internat. Elektron. Rundschau
Internationale Elektronische Rundschau. Elektro-Welt-Verlag Dr. Hüthig, D-6900 Heidelberg 1, Postfach 102869, Wilckensstrasse 3–5, F.R. Germany.

Internat. Journ. Electronics
International Journal of Electronics. Taylor and Francis Ltd., 10–14 Macklin Street, London, WC2B 5BF, England.

Internat. Journ. Theor. Phys.
International Journal of Theoretical Physics. Plenum Publishing Co. Ltd., Davis House, 8 Scrubs Lane, London, NW10 6SE, England.

Irish Astron. Journ.
The Irish Astronomical Journal. A Quarterly Publication under the auspices of the Observatories of Armagh and Dunsink. Subscription address: Managing Editor, Irish Astronomical Journal, Armagh Observatory, Northern Ireland.

Izv. Akad. Nauk Armyan. SSR
Izvestiya Akademii Nauk Armyanskoj SSR. Fizika. Publisher: Izdatel'stvo AN Armyanskoj SSR, Erevan.

Izv. Astron. Ehngel'gardt. Obs.
Izvestiya Astronomicheskoj Ehngel'gardtovskoj Observatorii. Izdatel'stvo Kazanskogo Universiteta, Kazan.

Izv. Glav. Astron. Obs. Pulkovo
Izvestiya Glavnoj Astronomicheskoj Observatorii v Pulkove. Akademiya Nauk SSSR. Izdanie Glavnoj astronomicheskoj observatorii v Pulkove, Leningrad.

Izv. Krymskoj Astrofiz. Obs.
Izvestiya Krymskoj Astrofizicheskoj Observatorii. Akademiya Nauk SSR. Publishers: Izdatel'stvo "Nauka", Moskva.

Jenaer Rundschau (Jena Review).
Jenaer Rundschau (Jena Review). Publisher: VEB Verlag Technik, Berlin, German Democratic Republic.

JETP Letters
JETP Letters. A translation of JETP Pis'ma v Redaktsiyu of the Academy of Sciences in the USSR. Published semimonthly by the American Institute of Physics, Lancaster, Pennsylvania.

Journ. American Ass. Variable Star Observers
The Journal of the American Association of Variable Star Observers. Published by The American Association of Variable Star Observers, Cambridge, Mass.

Journ. Astron. Soc. Victoria
The Journal of the Astronomical Society of Victoria. Printed by D. Buscombe Printers, Glen Waverly, Victoria.

Journ. Astron. Soc. Western Australia
The Journal of the Astronomical Society of Western Australia. Edited by the Astronomical Society of Western Australia, Perth, W. A.

Journ. Astronaut. Sci.
Journal of the Astronautical Sciences. American Astronautical Society, 6060 Duke Street, Alexandria, VA 22304, USA.

Journ. Atmosph. Sci.
Journal of the Atmospheric Sciences. American Meteorological Society, 45 Beacon Street, Boston, MA 02108, USA.

Journ. Atmosph. Terr. Phys.
Journal of Atmospheric and Terrestrial Physics. Publishers: Pergamon Press, Oxford – London – New York.

Journ. British Astron. Ass.
Journal of the British Astronomical Association. Subscription address: British Astronomical Association,

Burlington House, Piccadilly, London.

Journ. British Interplanet. Soc.
Journal of the British Interplanetary Society. Printed by Unwin Brothers Ltd., at the Gresham Press, Old Woking, and published for the British Interplanetary Society Ltd., by Space Educational Aids Ltd., London.

Journ. Colloid Interface Sci.
Journal of Colloid and Interface Science. Academic Press Inc., 111 Fifth Avenue, New York, N.Y. 10003, USA.

Journ. Fluid Mechanics
Journal of Fluid Mechanics. Cambridge University Press, Bentley House, 200 Euston Road, London, NW1 2DB, England.

Journ. Geomagn. Geoelectr.
Journal of Geomagnetism and Geoelectricity. Society of Terrestrial Magnetism and Electricity of Japan, Geophysical Institute, Tokyo University, Tokyo 113, Japan.

Journ. Geophys.
Journal of Geophysics / Zeitschrift für Geophysik. Publisher: Springer-Verlag, Berlin–Heidelberg–New York

Journ. Geophys. Res.
Journal of Geophysical Research. An International Scientific Publication. Published three times a month by the American Geophysical Union, Washington, D. C. First section: Space Physics; Second section: Physics and chemistry of the solid earth, planetology, geodesy; Third section: Oceans and atmospheres.

Journ. History Astron.
Journal for the History of Astronomy. Publisher: Science History Publications Ltd., Chalfont St Giles, Buckinghamshire, England. American Representative: Neale Watson Academic Publications, Inc., New York City, U.S.A.

Journ. Indian Inst. Sci.
Journal of the Indian Institute of Science. Bangalore 12, India.

Journ. Instn. Electronics Telecommun. Engineers
Journal of the Institution of Electronics and Telecommunication Engineers. 2 Institutional Area, Lodi Road, New Delhi - 110003, India.

Journ. Math. Phys.
Journal of Mathematical Physics. American Institute of Physics, 335 East 45th Street, New York, N.Y. 10017, USA.

Journ. Mechanical Engineering Lab.
Journal of Mechanical Engineering Laboratory. Agency of Industrial Science and Technology, Igusa Suginami-ku, Tokyo, Japan.

Journ. Navigation
The Journal of Navigation. Published quarterly by The Royal Institute of Navigation at the Royal Geographical Society, London. To be obtained from Scottish Academic Press Ltd., Edinburgh, Scotland.

Journ. Optical Soc. America
Journal of the Optical Society of America. American Institute of Physics, 335 East 45th Street, New York, N.Y. 10017, USA.

Journ. Optics
Journal of Optics. Optical Society of India, Department of Applied Physics, University of Calcutta, 92 Acharya Prafulla Chandra Road, Calcutta-9, India.

Journ. Phys. A
Journal of Physics A, (Mathematical, Nuclear and General). Europhysics Journal. Published by the Institute of Physics and Physical Society, London, England, in association with the American Institute of Physics, New York.

Journ. Phys. B
Journal of Physics B, (Atomic and Molecular Physics). Institute of Physics, 47 Belgrave Square, London, SW1X 8QX, England.

Journ. Physique
Journal de Physique. Publication de la Société Française de Physique, Paris.

Journ. Plasma Phys.
Journal of Plasma Physics. Publishers: Cambridge University Press, London.

Journ. Proc. Roy. Soc. New South Wales
Journal and Proceedings of the Royal Society of New South Wales. Publisher: Science House, Sydney, N.S.W. (Australia).

Journ. Quant. Spectrosc. Radiat. Transfer
Journal of Quantitative Spectroscopy & Radiative Transfer. Publisher: Pergamon Press, Oxford – New York.

Journ. Radio Res. Lab.
Journal of the Radio Research Laboratories. Chief Planning Section, Radio Research Laboratories, Ministry of Posts & Telecommunications, Nukui-Kitamachi, Konganei-shi, Tokyo 184, Japan.

Journ. Roy. Astron. Soc. Canada
The Journal of the Royal Astronomical Society of Canada, devoted to the advancement of astronomy and allied sciences. Printed by the University of Toronto Press, Toronto, Ontario, Canada.

Journ. Soc. Motion Picture Television Engineers
Journal of the Society of Motion Picture and Television Engineers. 862 Searsdale Avenue, Searsdale, New York, N.Y. 10583, USA.

Journ. Spacecraft Rockets
Journal of Spacecraft and Rockets. American Institute of Aeronautics and Astronautics, 1290 Avenue of the Americas, New York, N.Y. 10019, USA.

Journ. Testing Evaluation
Journal of Testing and Evaluation. American Society for Testing and Materials, 1916 Race Street, Philadelphia, PA 19103, USA.

JPL Techn. Rep.
Jet Propulsion Laboratory, California Institute of Technology, Pasadena, California. National Aeronautics and Space Administration. Technical Report.

Kometn. Tsirk. *Kiev*
Kometnyj Tsirkulyar. Gruppa po Issledovaniyu Komet Astrosoveta i Mezhduvedomstvennyj Geofizicheskij Komitet Akademii Nauk SSSR. Kievskij Universitet im. T. G. Shevchenko.

Komety i Meteory
Komety i Meteory. Akademiya Nauk Tadzhikskoj SSR. Astronomicheskij Sovet Akademii Nauk SSSR. Publishers: Izdatel'stvo "Donish", Dushanbe.

Kosmich. Issled.
Kosmicheskie Issledovaniya. Akademiya Nauk SSSR. Publishers: Izdatel'stvo "Nauka", Moskva [An English translation is published as "Cosmic Research", Consultants Bureau, New York, N.Y.].

Kozmos
Kozmos. Popular Astronomical Journal of the Slovak Central Observatory in Hurbanovo. Publisher: Slovenská ústredná hvezdáren v Hurbanove.

L'Astronomie
L'Astronomie et Bulletin de la Société Astronomique de France. Revue mensuelle. Rédaction: Société Astronomique de France, Paris.

L'Universo
L'Universo. Rivista dell'Instituto Geografico Militare. Direzione, Redazione e Amministrazione: Istituto Geografico Militare, Firenze.

Magnitnye Polya Solnech. Pyaten
Magnitnye Polya Solnechnykh Pyaten. (Supplements to Solnechnye Dannye. Byulleten' (*Solar Data*)). Publishers: Izdatel'stvo "Nauka", Leningrad.

Marconi Rev.
Marconi Review. Marconi Co., Marconi House, Chelmsford, Essex, England.

Materiały i Prace
Publications of the Institute of Geophysics, Polish Academy of Sciences. Edited by Państwowe Wydawnictwo Naukowe, Warszawa.

Math. Rev.
Mathematical Reviews. Published by the American Mathematical Society, Providence, R. I.

Mem. Fac. Sci. Kyoto Univ.
Memoirs of the Faculty of Science, Kyoto University. Series of Physics, Astrophysics, Geophysics, and Chemistry. Printed by Yamashiro Printing Publishing Co. Ltd., Kamigyo, Kyoto.

Mem. Roy. Astron. Soc.
Memoirs of the Royal Astronomical Society. Published for the Royal Astronomical Society by Blackwell Scientific Publications, Oxford – Edinburgh.

Mem. Soc. Astron. Italiana
Memorie della Società Astronomica Italiana. Nuova Serie. Pubblicate sotto gli auspici del Consiglio Nazionale dell Ricerche. Publisher: Tipografia Baccini & Chiappi, Firenze.

Mercury
Mercury. The Journal of the Astronomical Society of the Pacific. Published by the Astronomical Society of the Pacific, San Francisco, California.

Meteoritics
Meteoritics. The Journal of the Meteoritical Society. Published quarterly by The Meteoritical Society and Arizona State University Bureau of Publications. Editorial address: Center for Meteorite Studies, The Arizona State University, Tempe, Arizona.

Meteoritika
Akademiya Nauk SSSR. Komitet po Meteoritam. Publishers: Izdatel'stvo "Nauka", Moskva.

Microwave Journ.
Microwave Journal. To be obtained from 610 Washington Street, Dedham Plaza, Dedham, Massachusetts, U.S.A.

Microwaves
Microwaves. Hayden Publishing Co., 50 Essex Street, Rochelle Park, NJ 07662, USA.

Minor Planet. Bull.
The Minor Planet Bulletin. Bulletin of the Minor Planets Section of the Association of Lunar and Planetary Observers. Editorial Office: R. G. Hodgson, Dordt College, Sioux Center, Iowa, U.S.A.

Mitt. Astron. Ges.
Mitteilungen der Astronomischen Gesellschaft, Hamburg. Printed by G. Braun, GmbH, Karlsruhe.

Modern Geol.
Modern Geology. Gordon & Breach Science Publishers Ltd., 41 and 42 William IV Street, London WC2, England.

Monthly Notes Astron. Soc. Southern Africa
Monthly Notes of the Royal Astronomical Society of Southern Africa. Published by the Astronomical Society of Southern Africa, Royal Observatory, Cape Province, South Africa.

Monthly Notes Internat. Polar Motion Service
Monthly Notes of the International Polar Motion Service. Published by the Central Bureau, International Latitude Observatory of Mizusawa, Mizusawa-shi, Iwate-ken, Japan.

Monthly Notices Roy. Astron. Soc.
Monthly Notices of the Royal Astronomical Society. Published for the Royal Astronomical Society by Blackwell Scientific Publications, Oxford – Edinburgh.

Moon
The Moon. An International Journal of Lunar Studies. Publisher: D. Reidel Publishing Company, Dordrecht – Holland.

MVS Sonneberg
Mitteilungen über Veränderliche Sterne. Edited by Sternwarte Sonneberg der Akademie der Wissenschaften der DDR, Zentralinstitut für Astrophysik.

Nablyud. Iskusstv. Nebesn. Tel
Nablyudeniya Iskusstvennykh Nebesnykh Tel. Published by Astronomicheskij Sovet Akademii Nauk SSSR, Moskva.

Nachr. Akad. Wiss. Göttingen
Nachrichten der Akademie der Wissenschaften in Göttingen. II. Mathematisch-Physikalische Klasse. Vandenhoeck & Ruprecht, Göttingen.

Nachr. Karten-, Vermessungswesen
Nachrichten aus dem Karten- und Vermessungswesen. Editor: Institut für Angewandte Geodäsie (Abt. II des Deutschen Geodätischen Forschungsinstituts). Published by Verlag des Instituts für Angewandte Geodäsie, Frankfurt a. M.

Nature
Nature. Editorial and Publishing Offices: Macmillan Journals Limited, 4 Little Essex Street, London; 711 National Press Building, Washington, D. C.

Naturwissenschaften
Die Naturwissenschaften. Publisher: Springer-Verlag, Berlin – Heidelberg – New York.

Nauchn. Informatsii
Nauchnye Informatsii. Astronomicheskij Sovet Akademii Nauk SSSR, Moskva.

Navigation (*France*)
Navigation. Institut Française de Navigation, 3 avenue Octave-Greard, Paris 7, France.

New Scient.
New Scientist. New Science Publications, 128 Long Acre, London, WC2E 9QH, England.

Nouv. Rev. Optique
Nouvelle Revue d'Optique. Masson, 120 Boulevard Saint-Germain, F-75280 Paris Cedex 06, France.

Nuclear Instruments Methods
Nuclear Instruments and Methods. North-Holland Publishing Co., P.O. Box 211, Amsterdam, Netherlands.

Nuclear Phys. A
Nuclear Physics, Volume A. North-Holland Publishing Co., P. O. Box 211, Amsterdam, Netherlands.

Numer. Math.
Numerische Mathematik. Springer-Verlag, Berlin–Heidelberg–New York.

Nuovo Cimento
Il Nuovo Cimento. Rivista Internazionale e Organo della Società Italiana di Fisica, Series A, B. Publisher: Nicola Zanichelli, Editore, Bologna.

Nuovo Cimento Lettere
Lettere al Nuovo Cimento, a Cura della Società Italiana di Física. Editrice Compositori, Bologna.

Nuovo Cimento Rivista
Rivista del Nuovo Cimento. Società Italiana di Física, Via Degli Andalo 2, Bologna 40124, Italy.

Nuovo Cimento Suppl.
Supplemento al Nuovo Cimento. Publisher: Nicola Zanichelli, Editore, Bologna.

Observations Artificial Earth Satellites
Observations of Artificial Satellites of the Earth (Nablyudeniya Iskusstvennykh Sputnikov Zemli). Magyar Tudományos Akadémia Csillagvizsgáló Intézete, Budapest.

Observatory
The Observatory. A Review of Astronomy. Publishers: The Editors of "The Observatory", Royal Greenwich Observatory, Herstmonceux Castle, Hailsham, Sussex, England.

Österreich. Zeitschr. Vermessungswesen
Österreichische Zeitschrift für Vermessungswesen und Photogrammetrie. Editor and Publisher: Österreichischer Verein für Vermessungswesen und Photogrammetrie, Wien, Austria.

Optica Acta
Optica Acta. Taylor and Francis Ltd., 10 - 14 Macklin Street, London, WC2B 5NF, England.

Optical Engineering
Optical Engineering. Society of Photo-Optical Instrumentation Engineers, 337 Tejon Place, Palos Verdes Estates, CA 90274, USA.

Optics Commun.
Optics Communications. North-Holland Publishing Co., P.O. Box 211, Amsterdam, Netherlands.

Optics News
Optics News. Publication of the Optical Society, Washington, D.C.

Optik
Optik. Zeitschrift für das gesamte Gebiet der Licht- und Elektronenoptik. Publishers: Wissenschaftliche Verlagsgesellschaft mbH., Stuttgart, F. R. Germany.

Origins of Life
Origins of Life (Formerly Space Life Sciences). An International Journal. Publisher: D. Reidel Publishing Company, Dordrecht–Holland.

Orion
Orion. Zeitschrift der Schweizerischen Astronomischen Gesellschaft (SAG). Bulletin de la Société Astronomique de Suïsse (SAS). Printed by A. Schudel & Co. AG, Riehen, Suisse.

Peremennye Zvezdy, Byull.
Peremennye Zvezdy, Byulleten', izdavaemyj Astronomicheskim Sovetom Akademii Nauk SSSR. Published by Astronomicheskij Sovet Akademii Nauk SSSR, Moskva.

Peremennye Zvezdy, Prilozhenie
Peremennye Zvezdy, Prilozhenie (The Variable Stars, Supplement). Astronomicheskij Sovet Akademii Nauk SSSR, Moskva.

Oss. Astrofis. Catania, Pubbl.
Osservatorio Astrofisico di Catania, Pubblicazione. Printed by Scuola Salesiana del Libro, Catania.

Phil. Mag.
The Philosophical Magazine. A Journal of Theoretical, Experimental and Applied Physics. Eighth Series. Publisher: Taylor & Francis, Ltd., London.

Phil. Trans. Roy. Soc. London
Philosophical Transactions of the Royal Society of London. Series A, Mathematical and Physical Sciences. Published by the Royal Society, London.

Phys. Abstr.
Physics Abstracts. Science Abstracts, Series A. An INSPEC Publication, published by The Institution of Electrical Engineers, London.

Phys. Ber.
Physikalische Berichte. Herausgegeben von der Deutschen Physikalischen Gesellschaft e.V. und von der Deutschen Akademie der Wissenschaften zu Berlin. Physik-Verlag, Weinheim, F.R. Germany.

Phys. Blätter
Physikalische Blätter. Physik-Verlag, Mosbach/Baden.

Phys. Bull.
Physics Bulletin. Published by the Institute of Physics and the Physical Society, London, England.

Phys. Earth Planet. Interiors
Physics of the Earth and Planetary Interiors. A journal devoted to observational and experimental studies of the Earth and Planetary interiors and their theoretical interpretation by the physical sciences. Publisher: North-Holland Publishing Company, Amsterdam, Netherlands.

Phys. Fluids
The Physics of Fluids. Published by the American Institute of Physics, New York, N.Y.

Phys. Letters
Physics Letters. Volumes A and B. Publisher: North-Holland Publishing Company, Amsterdam.

Phys. Med. Biol.
Physics in Medicine and Biology. Institute of Physics, 47 Belgrave Square, London, SW1X 8QX, England.

Phys. Rev. A
Physical Review A, General Physics. Published for the American Physical Society by the American Institute of Physics, Lancaster, Pa., and New York, N.Y.

Phys. Rev. B
Physical Review B, Solid State. Published for the American Physical Society by the American Institute of Physics, Lancaster, Pa., and New York, N. Y.

Phys. Rev. C
Physical Review C, Nuclear Physics. Published for the American Physical Society by the American Institute of Physics, Lancaster, Pa., and New York, N.Y.

Phys. Rev. D
Physical Review D, Particles and Fields. Published for the American Physical Society by the American Institute of Physics, Lancaster, Pa., and New York, N.Y.

Phys. Rev. Letters
Physical Review Letters. Published weekly by The American Physical Society, New York, N. Y.

Phys. Today
Physics Today. Published by the American Institute of Physics, New York, N.Y.

Physica
Physica. Publishers: North-Holland Publishing Company, Amsterdam, The Netherlands, on request of the Foundation "Physica", Utrecht.

Physica Scripta
Physica Scripta. (Formerly Arkiv för Fysik). Published by the Royal Swedish Academy of Sciences, Stockholm.

Pis'ma v Astron. Zhurn.
Pis'ma v Astronomicheskij Zhurnal. Akademiya Nauk SSSR. Publishers: Izdatel'stvo 'Nauka', Moskva.

Planet. Space Sci.
Planetary and Space Science. Pergamon Press, Oxford – London – New York.

Plasma Physics
Plasma Physics. Publisher: Pergamon Press, Oxford, England.

Pokroky
Pokroky matematiky, fyziky a astronomie. Editor: Jednota čs. matematiků a fyziků. Publisher: Academia, Praha.

Postępy Astron.
Postępy Astronomii. Czasopismo Poświecone Upowszechnianiu Wiedzy Astronomicznej. Polskie Towarzystwo Astronomiczne, Warszawa. Printed in Poland by Pánstwowe Wydawnictwo Naukowe, Lódź.

Pramāņa
Pramāņa. Indian Academy of Sciences, Bangalore 560006, India.

Priroda
Priroda. Publishers: Izdatel'stvo "Nauka", Moskva.

Probl. Kosm. Fiz.
Problemy Kosmichskoj Fiziki. Mezhvedomstvennyj Nauchnoj Sbornik. Izdatel'skoe Obedinenie Vishcha Shkola. Izdatel'stvo pri Kievskom Universitete, Kiev.

Proc. Astron. Soc. Australia
Proceedings of the Astronomical Society of Australia. Published for the Society by Sydney University Press, Sydney.

Proc. Cambridge Phil. Soc.
Proceedings of the Cambridge Philosophical Society (Mathematical and Physical Sciences). Publishers: Cambridge University Press, London.

Proc. IEEE
Proceedings of the IEEE. Published monthly by the Institute of Electrical and Electronics Engineers, Inc., New York, N. Y.

Proc. Indian Acad. Sci. A
Proceedings of the Indian Academy of Sciences, Section A. Bangalore 560006, India.

Proc. Instn. Electr. Engineers
Proceedings of the Institution of Electrical Engineers. Institution of Electrical Engineers, Savoy Place, London, WC2R OBL, England.

Proc. Instn. Radio Electron. Engineers Australia
Proceedings of the Institution of Radio and Electronics Engineers, Australia. Science House, 157 Gloucester Street, Sydney, N.S.W. 2000, Australia.

Proc. Japan Acad.
Proceedings of the Japan Academy. Ueno Park, Tokyo, Japan.

Proc. Koninkl. Nederl. Akad. Wet.
Koninklijke Nederlandse Akademie van Wetenschappen. Proceedings. Series B, Physical Sciences. Publishers: North-Holland Publishing Company, Amsterdam.

Proc. National Acad. Sci. U.S.A.
Proceedings of the National Academy of Sciences of the United States of America. Published monthly by the National Academy of Sciences, Washington, D.C.

Proc. Roy. Soc. London
Proceedings of the Royal Society of London. Series A: Mathematical and Physical Sciences. Published by the Royal Society, London.

Progr. Theor. Phys.
Progress of Theoretical Physics. Published for the Research Institute for Fundamental Physics and the Physical Society of Japan. Publication Office: Progress of Theoretical Physics, Yukawa Hall, Kyoto University, Kyoto, Japan.

Progr. Theor. Phys. Suppl.
Supplement of the Progress of Theoretical Physics. Published for the Research Institute for Fundamental Physics and The Physical Society of Japan. Publication Office: Progress of Theoretical Physics, Yukawa Hall, Kyoto University, Kyoto, Japan.

PTB Mitt.
PTB Mitteilungen. Amts- und Mitteilungsblatt der Physikalisch-Technischen Bundesanstalt, Braunschweig – Berlin.

Publ. Astron. Soc. Japan
Publications of the Astronomical Society of Japan. Published by the Astronomical Society of Japan. Office of the Society: Tokyo Astronomical Observatory, Mitaka, Tokyo. Agent: Maruzen Co. Ltd. (Export Department), Nihonbashi, Tokyo, Japan.

Publ. Astron. Soc. Pacific
Publications of the Astronomical Society of the Pacific. Published in Provo, Utah, by the Astronomical Society of the Pacific, San Francisco, California. Printed by Brigham Young University Press, Provo, Utah.

Publ. Dominion Astrophys. Obs.
Publications of the Dominion Astrophysical Observatory, Victoria, B.C. National Research Council of Canada.

Publ. Eidgen. Sternw. Zürich
Publikationen der Eidgenössischen Sternwarte Zürich. Schulthess Polygraphischer Verlag, Zürich.

Publ. Inst. Roy. Meteorol. Belgique, Ser. A
Publications, Institut Royal Météorologique de Belgique. Serie A. 3 Avenue Circulaire, Uccle-Bruxelles 1180, Belgium.

Publ. Roy. Obs. Edinburgh
Publications of the Royal Observatory, Edinburgh. Published by The Royal Observatory, Edinburgh, Scotland.

Publ. Tartu Astrofiz. Obs.
W. Struve nimelise Tartu Astrofüüsika Observatooriumi, Publikatsioonid. Eesti NSV Teaduste Akadeemia, Tartu.

Publ. Variable Star Section, Roy. Astron. Soc. New Zealand
Publications of the Variable Star Section, Royal Astronomical Society of New Zealand. Director: F. M. Bateson, Greerton, Tauranga, New Zealand.

Quarterly Bull. Solar Activity
International Astronomical Union, Quarterly Bulletin on Solar Activity. Published by the Eidgenössische Sternwarte in Zürich with financial support from UNESCO.

Quarterly Journ. Roy. Astron. Soc.
Quarterly Journal of the Royal Astronomical Society. Published for the Royal Astronomical Society by Blackwell Scientific Publications, Oxford.

Radio Sci.
Radio Science. American Geophysical Union, 2901 Byrdhill Road, Richmond, VA 23228, USA.

Radiotekhn. Ehlektron.
Radiotekhnika i Ehlektronika. Moskva TSP-3, Pr. Karl Marx 18, USSR.

Recherche
Recherche, 4 Place de l'Odéon, Paris 6, France.

Referativ. Zhurn. 51. Astron.
Referativnyj Zhurnal. 51. Astronomiya. Vsesoyuznyj Institut Nachnoj i Tekhnicheskoj Informatsii. Moskva.

Referativ. Zhurn. 52. Geod. i Aehrosemka
Referativnyj Zhurnal. 52. Geodeziya i Aehrosemka. Vsesoyuznyj Institut Nauchnoj i Tekhnicheskoj Informatsii. Moskva.

Referativ. Zhurn. 62. Issled. kosm. prostranstva
Referativnyj Zhurnal. 62. Issledovanie Kosmicheskogo Prostranstva. Vsesoyuznyj Institut Nauchnoj i Tekhnicheskoj Informatsii. Moskva.

Rep. Inst. Phys. Chem. Res.
Reports of the Institute of Physical and Chemical Research. Rikagaku Kenkyushu, Wako-shi, Saitama 351, Japan.

Rep. Ionosph. Space Res. Japan
Report of Ionosphere and Space Research in Japan. Institute of Space and Aeronautical Science, University of Tokyo, Komaba, Meguro-ku, Tokyo 153, Japan.

Rep. Progr. Phys.
Reports on Progress in Physics. Published by The Institute of Physics and the Physical Society, London.

Rev. Astron.
Revista Astronomica. Organo de la Asociación Argentina Amigos de la Astronomia, Buenos Aires.

Rev. Brasil. Fis.
Revista Brasileira de Fisica. Sociedade Brasileira de Fisica, Cx. Postal 20553, Sao Paolo SP, Brazil.

Rev. Geofis.
Revista de Geofisica, Instituto Nacional de Geofisica, Serrano 123, Madrid 2, Spain.

Rev. Geophys. Space Phys.
Reviews of Geophysics and Space Physics (formerly Reviews of Geophysics). Published by the American Geophysical Union, Richmond, Virginia.

Rev. Mexicana Astron. Astrofis.
Revista Mexicana de Astronomia y Astrofisica. Dirección: Instituto de Astronomia, Universidad Nacional Autónoma de México, México, D.F.

Rev. Mexicana Fis.
Revista Mexicana de Fisica. Sociedad Mexicana de Fisica, Apartado Postal No. 20-364, Mexico 20, D.F, Mexico.

Rev. Modern Phys.
Reviews of Modern Physics. Published for the American Physical Society by the American Institute of Physics,

Lancaster, Pa., and New York, N.Y.

Rev. Radio Res. Lab.
Review of the Radio Research Laboratories. Tokyo, Japan.

Rev. Sci. Instruments
Review of Scientific Instruments. American Institute of Physics, 335 East 45th Street, New York, NY 10017, USA.

Rezul'taty Nablyud. Iskusstv. Sputnikov Zemli
Rezul'taty Nablyudenij Iskusstvennykh Sputnikov Zemli. Published by Astronomicheskij Sovet Akademii Nauk SSSR, Ryazanskij Gosudarstvennyj Pedagogicheskij Institut, Ryazan'.

Rezul'taty Nablyud. Sovet. Iskusstv. Sputnikov Zemli
Rezul'taty Nablyudenij Sovetskikh Iskusstvennykh Sputnikov Zemli. Published by Astronomicheskij Sovet Akademii Nauk SSSR, Moskva. Replaced after No. 140 by Rezul'taty Nablyudenij Iskusstvennykh Sputnikov Zemli.

Říše hvězd
Říše hvězd. Czechoslovak popular astronomical journal. Publisher: Orbis, Praha.

Roy Astron. Soc. New Zealand Publ.
Publications of Variable Star Section, Royal Astronomical Society of New Zealand. Publication Office: Greerton, Tauranga, New Zealand.

Rumanian Sci. Abstr.
Rumanian Scientific Abstracts. Natural Sciences. Publishers: The Scientific Documentation Centre of the Academy of the Socialist Republic of Romania, Bucureşti.

Sci. American
Scientific American. Published monthly by Scientific American, Inc., New York, N.Y.

Sci. Dimension
Science Dimension. National Research Council of Canada, Ottawa K1A OR6, Canada.

Sci. Papers Inst. Phys. Chem. Res.
Scientific Papers of the Institute of Physical and Chemical Research. Rikagaku Kenkyusho, Wako-shi, Saitama 351, Japan.

Sci. Progr. Découverte
Science Progrès Découverte (formerly Science Progrès, La Nature). Revue publiée avec la participation du Palais de la Découverte. Published by Dunod, Editeur, Paris. Imprimerie Bayeusaine, Bayeux.

Sci. Rep. Tôhoku Univ.
The Science Reports of the Tôhoku University. First Series (Physics, Chemistry, Astronomy). Published by the Faculty of Science, Tôhoku University, Sendai, Japan.

Science
Science. American Association for the Advancement of Science, Washington, D.C.

Scient. Sinica
Scientia Sinica. Edited by Editorial Committee of Scientia Sinica, Peking. Published by Science Press, Peking, China.

Sdelovaci Techn.
Sdelovaci Technika. Publishers of Technical Literature, Spalena 51, 11302 Praha 1, Czechoslovakia.

SIAM Journ. Applied Math.
SIAM Journal on Applied Mathematics. Society for Industrial and Applied Mathematics, 33 South 17th Street, Philadelphia, PA 19103, USA.

Siemens Rev.
Siemens Review. Siemens-Aktiengesellschaft, Postfach 325, 8520 Erlangen 2, F. R. Germany.

Simon Stevin
Simon Stevin. De Natuur - en Geneeskundige Vennootschap, Rozier 44, B-9000 Gent, Belgium.

Sitzungsber. Akad. Wiss. Berlin
Sitzungsberichte der Akademie der Wissenschaften der DDR. Klasse für Mathematik, Physik und Technik. Publisher: Akademie-Verlag, Berlin.

Sitzungsber. Bayer. Akad. Wiss.
Bayerische Akademie der Wissenschaften. Mathematisch-Naturwissenschaftliche Klasse. Sitzungsberichte. Publisher: Verlag der Bayerischen Akademie der Wissenschaften, München.

Sitzungsber. Heidelberger Akad. Wiss.
Sitzungsberichte der Heidelberger Akademie der Wissenschaften. Mathematisch-Naturwissenschaftliche Klasse. Publisher: Springer-Verlag, Heidelberg.

Sitzungsber. Österreich. Akad. Wiss.
Sitzungsberichte. Österreichische Akademie der Wissenschaften. Mathematisch-Naturwissenschaftliche Klasse. Abteilung II: Mathematik, Astronomie, Meteorologie und Technik. Publisher: Springer-Verlag, Wien.

Sky Telescope
Sky and Telescope. Published by Sky Publishing Corporation, Cambridge, Mass.

Smithsonian Contr. Astrophys.
Smithsonian Contributions to Astrophysics. Smithsonian Institution Astrophysical Observatory, Cambridge, Mass. Printed by Smithsonian Institution Press, City of Washington. For sale by the Superintendent of Documents, U. S. Government Printing Office, Washington, D. C.

Smithsonian Year
Smithsonian Year. Annual Report of the Smithsonian Institution, including the financial report of the Executive Committee of the Boards of Regents. Published by the Smithsonian Institution, Washington, D.C.

Solar Physics
Solar Physics. A Journal for Solar Research and the Study of Solar Terrestrial Physics. Publishers: D. Reidel Publishing Company, Dordrecht—Holland.

Solnechnye Dannye Byull.
Solnechnye Dannye. Byulleten'. *(Solar Data)*. Publishers: Izdatel'stvo "Nauka", Leningradskoe Otdelenie, Leningrad.

Soobshch. Byurakan. Obs.
Soobshcheniya Byurakanskoj Observatorii. Akademiya Nauk Armyanskoj SSR, Erevan.

Soobshch. Gos. Astron. Inst. Shternberg

Soobshcheniya Gosudarstvennogo Astronomicheskogo Instituta im P.K. Shternberga. Publishers: Izdatel'stvo Moskovskogo Universiteta, Moskva.

Soobshch. Spets. Astrofiz. Obs.
Soobshcheniya Spetsial'noj Astrofizicheskoj Observatorii. Izdanie Spetsial'noj Astrofizicheskoj Observatorii AN SSSR.

Southern Stars
Southern Stars. The Journal of the Royal Astronomical Society of New Zealand (Inc.). Address of the Society: P.O. Box 3181, Wellington C1, New Zealand.

Soviet Astron.
Soviet Astronomy. A translation of Astronomicheskij Zhurnal (Astronomical Journal). Published by the American Institute of Physics, New York, N.Y.

Space Sci. Instrum.
Space Science Instrumentation. An International Journal of Scientific Instruments for Aircraft, Balloons, Sounding Rockets, and Spacecraft. Published by D. Reidel Publishing Company, Dordrecht—Holland.

Space Sci. Rev.
Space Science Reviews. Publishers: D. Reidel Publishing Company, Dordrecht – Holland.

Spaceflight
Spaceflight. A Publication of the British Interplanetary Society. Printed by Eyre & Spottiswoode Limited at Grosvenor Press, Portsmouth, and published by the British Interplanetary Society, London.

Spaceworld
Spaceworld. Palmer Publications Inc., Amherst, WI 54406, USA.

Sterne
Die Sterne. Zeitschrift für alle Gebiete der Himmelskunde. Johann Ambrosius Barth, Leipzig.

Sternenbote
Sternenbote. Monatsschrift für Österreichs Amateurastronomen. Publisher: Astronomisches Büro, Hermann Mucke, Wien.

Stockholms Obs. Ann.
Stockholms Observatoriums Annaler. Printed by Almquist & Wiksell, Stockholm.

Stockholms Obs. Rep.
Stockholms Observatorium, Saltsjöbaden, Sweden, Report.

Strolling Astronomer
The Strolling Astronomer. The Journal of The Association of Lunar and Planetary Observers, Publication Office: The Strolling Astronomer, Box 3 AZ, University Park, New Mexico.

Stud. Cerc. Astron.
Studii şi Cercetări de Astronomie. Editura Academiei Republicii Socialiste România. Editorial Office: Observatorul Astronomic, Bucureşti.

Stud. Cerc. Fiz.
Studii si Cercetari de Fizica. Academia Republicii Populare Romine. P.O. Box 134-5, Calca Victoriei 126, Bucuresti, Rumania.

Stud. Geophys. Geod.
Studia geophysica et geodaetica. Published for the Geophysical Institute of the Czechoslovak Academy of Sciences by Academia, Praha.

Stud. Soc. Sci. Torunensis
Studia Societatis Scientiarum Torunensis, Toruń – Polonia. Sectio F (Astronomia).

Stud. Univ. Babeş-Bolyai
Studia Universitatis Babeş-Bolyai. Series Mathematica-Physica. Publishers: Intreprinderea Poligrafica, Cluj.

SuW
Sterne und Weltraum. Astronomische Monatsschrift. Publisher: Verlag Sterne und Weltraum Dr. Vehrenberg, Düsseldorf, F.R. Germany.

Tartu Astron. Obs. Teated
Tartu Astronoomia Observatoorium Teated. Eesti NSV Teaduste Akadeemia W. Struve nim. Tartu Astrofüüsika Observatoorium, Tartu.

Tellus
Tellus, a bi-monthly Journal of Geophysics. Svenska Geofysiska Foreningen. Printed in Sweden by Almqvist & Wiksells Boktryckeri AB, Uppsala.

Tokyo Astron. Bull.
Tokyo Astronomical Observatory, Tokyo Astronomical Bulletin.

Tokyo Astron. Obs. Rep.
University of Tokyo, Tokyo Astronomical Observatory, Report.

Trans. Astron. Obs. Yale Univ.
Transactions of the Astronomical Observatory of Yale University. Published by the Observatory, New Haven.

Trans. IAU
Transactions of the International Astronomical Union. Published and distributed for the IAU (UAI) by D. Reidel Publishing Company, Dordrecht – Holland/Boston – U.S.A.

Trans. Roy. Soc. Canada
Transactions of the Royal Society of Canada. Published by the Royal Society of Canada, National Research Building, Ottawa.

Trudy Astrofiz. Inst. Alma-Ata
Trudy Astrofizicheskogo Instituta, Alma-Ata. Akademiya Nauk Kazakhskoj SSR. Publishers: Izdatel'stvo "Nauka" Kazakhskoj SSR, Alma-Ata.

Trudy Astron. Obs., *Leningrad*
Uchenye Zapiski Gosudarstvennogo Universiteta im. A. A. Zhdanova, Seriya matematicheskikh nauk = Trudy Astronomicheskoj Observatorii. Izdatel'stvo Leningradskogo Universiteta, Leningrad.

Trudy Glav. Astron. Obs. Pulkovo
Trudy Glavnoj Astronomicheskoj Observatorii v Pulkove. Akademiya Nauk SSSR. Izdanie Glavnoj astronomicheskoj observatorii v Pulkove, Leningrad.

Trudy Inst. Teor. Astron.,*Leningrad*
Trudy Instituta Teoreticheskoj Astronomii. Akademiya Nauk SSSR. Publishers: Izdatel'stvo "Nauka", Leningrad.

Trudy Kazan. Gorod. Astron. Obs.
Trudy Kazanskoj Gorodskoj Astronomicheskoj Observatorii. Izdatel'stvo Kazanskogo Universiteta, Kazan.

Trudy Tashkent. Astron. Obs.
Trudy Tashkentskoj Astronomicheskoj Observatorii. Akademiya Nauk Uzbekskoj SSR. Publishers: Izdatel'stvo "FAN" Uzbekskoj SSR, Tashkent.

Tsirk. Astron. Inst. Tashkent
Tsirkulyar Astronomicheskogo Instituta. Akademiya Nauk Uzbekskoj SSR. Izdatel'stvo "FAN" Uzbekskoj SSR, Tashkent.

Tsirk. Astron. Obs. L'vov
Tsirkulyar. Astronomicheskaya Observatoriya. L'vovskij Ordena Lenina Gosudarstvennyj Universitet imeni Ivana Franko. Publisher: Izdatel'stvo L'vovskogo Universiteta, L'vov.

Umschau
Umschau in Wissenschaft und Technik, vereinigt mit Weltraumfahrt – Raketentechnik. Umschau Verlag Breidenstein KG, Frankfurt am Main.

Urania Barcelona
Urania. Revista de Astronomia y Ciencias Afines. Organo de la Sociedad Astronómica de España y América, Barcelona; Unión Nacional de Astronomia y Ciencias Afines, Madrid.

Urania Kraków
Urania. Miesiecznik Polskiego Towarzystwa Miłośników Astronomii, Kraków. Publisher: Krakowska Drukarnia Prasowa, Kraków.

Vasiona
Vasiona. Revue d'Astronomie et d'Astronautique. Bulletin de la Société Astronomique "R. Bosković", Beograd.

Veröff. Astron. Rechen-Inst. Heidelberg
Veröffentlichungen des Astronomischen Rechen-Instituts Heidelberg. Verlag G. Braun, Karlsruhe.

Veröff. Sternw. Sonneberg
Akademie der Wissenschaften der DDR, Zentralinstitut für Astrophysik, Veröffentlichungen der Sternwarte in Sonneberg. Publisher: Akademie-Verlag, Berlin.

Veröff. Zentralinst. Phys. Erde
Akademie der Wissenschaften der DDR, Forschungsbereich Geo- und Kosmowissenschaften. Veröffentlichungen des Zentralinstituts für Physik der Erde, Potsdam.

Vesmír
Vesmír. Přírodovědecky časopis Čs. akadmie věd. Publisher: Academia, Praha.

Vestn. Khar'kov. Univ.
Vestnik Khar'kovskogo Universiteta. Seriya Astronomicheskaya. Publishers: Izdatel'stvo Khar'kovskogo Universiteta, Khar'kov.

Vestn. Kiev. Univ.
Vestnik Kievskogo Universiteta. Seriya Astronomii. Publishers: Izdatel'stvo Kievskogo Universiteta, Kiev.

VJS Naturforsch. Ges. Zürich
Vierteljahresschrift der Naturforschenden Gesellschaft in Zürich. Printer and Publisher: Leeman AG, Zürich.

Wiss. Zeitschr. Friedrich-Schiller Univ. Jena
Wissenschaftliche Zeitschrift der Friedrich-Schiller-Universität. Jena. Mathematisch-Naturwissenschaftliche Reihe. Edited by the Rektor der Friedrich-Schiller-Universität Jena.

Wiss. Zeitschr. Humboldt-Univ. Berlin
Wissenschaftliche Zeitschrift der Humboldt-Universität zu Berlin. Mathematisch-Naturwissenschaftliche Reihe. Edited by the Rektor der Humboldt-Universität, Berlin.

Zeitschr. angew. Math. Mech.
Zeitschrift für angewandte Mathematik und Mechanik. Akademie-Verlag GmbH, 108 Berlin, Leipziger Strasse 3–4, German Democratic Republic.

Zeitschr. Naturforschung
Zeitschrift für Naturforschung. Europhysics Journal. Teil a: Astrophysik, Physik, Physikalische Chemie. Published by Verlag der Zeitschrift für Naturforschung, Tübingen, Germany.

Zeitschr. Physik A
Atoms and Nuclei. Zeitschrift für Physik A. Springer-Verlag, Berlin–Heidelberg–New York.

Zeitschr. Physik B
Condensed Matter. Zeitschrift für Physik B. Springer-Verlag, Berlin–Heidelberg–New York.

Zemlya i Vselennaya
Zemlya i Vselennaya. Astronomiya, Geofizika, Issledovaniya Kosmicheskogo Prostranstva. Nauchno-Populyarnyj Zhurnal Akademii Nauk SSSR. Publishers: Izdatel'stvo "Nauka", Moskva.

Zenit
Populair wetenschappelijk maandblad over sterrenkunde/weerkunde/ruimtevaart/ruimte-onderzoek/aanverwante wetenschappen en technieken. Bureau: Stichting De Koepel, Utrecht.

Zentralbl. Math. Grenzgebiete – Math. Abstr.
Zentralblatt für Mathematik und ihre Grenzgebiete – Mathematics Abstracts. Publisher: Springer-Verlag, Berlin–Heidelberg–New York.

Zvaigžņota Debess
Latvijas PSR Zinātņu Akadēmijas Radioastrofizikas Observatorijas Populārzinatniks Gadalaiku Izdevums. Izdevnieciba "Zinātne", Riga.

002 Bibliographical Publications

002.001 **Képler et Copernic.** J.-C. Pecker.
L'Astronomie, Vol. 90, 207 - 210 (1976). – Livres d'astronomie.

002.002 **Annotations on astrophysical papers published in the journal "Radiofizika", Vol. 18, Nos. 2, 6, 7, 10, 12 for the year 1975.**
Astron. Zhurn. Akad. Nauk SSSR, Vol. 53, 676 (1976).
In Russian. English translation in Soviet Astron., Vol. 20, No. 3.

002.003 **Astronomy and Astrophysics Abstracts. Vol. 14, Literature 1975, Part II.**
S. Böhme, U. Esser, W. Fricke, U. Güntzel-Lingner, I. Heinrich, F. Henn, D. Krahn, L. Schmadel, H. Scholl, G. Zech (Editors).
Published for Astronomisches Rechen-Institut, Heidelberg by Springer-Verlag, Berlin–Heidelberg–New York. 10 + 747 pp. Price DM 86.00; (ca. US $ 35.30) [Subscription price DM 68.80; (ca. US $ 28.30)] (1976).

002.004 **Status of availability of Mariner 10 (1973-085A) TV picture data.**
Data Announcement Bull., National Space Sci. Data Center, NASA, Goddard Space Flight Center, Greenbelt, Maryland, 4 + 24 + 6 pp. (1975).
This Bulletin describes the Mariner 10 TV data now available from the National Space Science Data Center (NSS DC) and explains the procedures for ordering these data.

002.005 **Annotated literature survey of microwave ferrite control components and materials for 1968–1974.**
L. R. Whicker, D. M. Bolle.
IEEE Trans. Microwave Theory Techn., Vol. MTT-23, 908 - 918 (1975).

002.006 **Bibliography.** Z. Kopal, M. Moutsoulas, J. W. Salisbury, F. B. Waranius (Editors).
The Moon, Vol. 15, 183 - 201 (1976).

003 Books (Astronomy and Astrophysics)

003.001 **Stellar physics and evolution.**
Yu. L. Frantsman (Editor).
Astronomicheskij Sovet Akademii Nauk SSSR, Moskva. 135 pp. = Nauchnye Informatsii, vypusk (No.) 31 (1974). In Russian. – The individual contributions are included in their corresponding subject categories – see abstracts 065.049 - 065.051, 117.010, 122.030, 153.013, 153.014, 154.012.

003.002 **Astrometriya i Astrofizika, Vypusk 28.**
Eh. G. Yanovitskij (Editor).
Respublikanskij Mezhvedomstvennyj Sbornik. Akademiya Nauk Ukrainskoj SSR, Glavnaya Astronomicheskaya Observatoriya. Izdatel'stvo "Naukova dumka", Kiev. 123 pp. Price 97 Kop. (1976). In Russian. – The papers included are abstracted in their corresponding subject categories – see abstracts 031.005, 041.011, 041.012, 044.005, 044.006, 046.014, 064.040, 065.056, 071.020, 073.042, 082.035 - 082.037, 103.102.

003.003 **Astrometriya i Astrofizika, Vypusk 29.**
D. P. Duma (Editor).
Respublikanskij Mezhvedomstvennyj Sbornik. Akademiya Nauk Ukrainskoj SSR, Glavnaya Astronomicheskaya Observatoriya. Izdatel'stvo "Naukova dumka", Kiev. 140 pp. Price 1 Rbl. 18 Kop. (1976). In Russian. – The papers included are abstracted in their corresponding subject categories – see abstracts 034.018, 041.013 - 041.015, 045.005, 054.006,

064.041, 071.021, 071.022, 081.012, 099.032, 102.016, 103.101, 103.102, 114.329, 114.330, 122.038.

003.004 **Problems of meteoritics.** V. S. Sobolev (Editor).
AN SSSR. Sib. otd. In-t geol. i geofiz. Vses. astron.-geod. o-vo. Tomsk. un-t. Nauka, Novosibirsk. 148 pp. Price 91 Kop. (1975). In Russian. – See abstracts 105.048 - 105.058.

003.005 **Solar phenomena research.**
AN SSSR. Dal'nevost. nauch. tsentr. Ussur. st. Sluzhby Solntsa. Vladivostok. 180 pp. Price 81 Kop. (1975).
In Russian. – See abstracts 072.041 - 072.046, 072.050, 073.077, 077.032, 077.033.

003.006 **Cosmic rays.** Results of researches on the International Geophysical Projects. No. 15.
S. N. Vernov, L. I. Dorman (Editors).
Publishing House "Nauka", Moscow. 220 pp. Price 1 Rbl. 84 Kop. (1975). In Russian. – Individual papers within the subject scope of Astronomy and Astrophysics Abstracts are included in their corresponding subject categories – see abstracts 009.013, 009.014, 031.251, 032.019, 032.020, 072.047, 074.047, 074.048, 078.026, 078.027, 083.064, 084.021, 084.261, 084.409, 106.039, 143.037 - 143.046.

003.007 **Methods in computational physics. Vol. 14. Radio**

astronomy.
B. Alder, S. Fernbach, M. Rotenberg (Editors).
Academic Press, New York—San Francisco—London. 10 + 239
pp. Price DM 72.30 (1975). — The individual contributions are
included in their corresponding subject categories — see
abstracts 031.252 - 031.254, 141.343.

003.008 **Astrophysics. Part B: Radio telescopes.**
M. L. Meeks (Editor), with a foreword by L. Marton.
Methods of experimental physics: Vol. 12. Academic Press,
New York — San Francisco — London. 22 + 309 pp. Price
DM 94.90 (1976). — The individual contributions are included
in their corresponding subject categories — see abstracts 033.
016 - 033.026, 082.061 - 082.064, 083.065.

003.009 **Investigation of the sun and red stars. 4.**
A. Balklavs (Editor).
Latvijas PSR Zinātņu akādemija, Radioastrofizikas observatorija. Akademiya nauk Latvijskoj SSR, Radioastrofiziches-
kaya observatoriya. Izdatel'stvo "Zinātne", Riga. 88 pp.
Price 27 Kop. (1976). In Russian. — The individual contributions are included in their corresponding subject categories —
see abstracts 034.068, 064.061, 065.080, 113.044, 114.360.

003.010 **Galaxies and the universe.**
A. Sandage, M. Sandage, J. Kristian (Editors),
Index prepared by G. A. Tammann, with a preface by G. P.
Kuiper, B. M. Middlehurst.
Stars and stellar systems, Vol. 9. The University of Chicago
Press, Chicago — London, 22 + 818 pp. Price $ 36.00, £ 18.00
respectively (1975). — The individual contributions are in-
cluded in their corresponding subject categories — see abstracts
141.105 - 141.107, 151.064, 151.065, 158.129 - 158.139,
160.037, 162.069 - 162.071.

003.011 **Tadeáš Hájek z Hájku (Hagecius) 1525—1600.**
J. Bouška (Editor).
Univerzita Karlova, Praha. 40 pp. Price Kčs 11 (1976). — The
individual contributions within the subject scope of Astronomy and Astrophysics Abstracts are included in their corresponding categories — see abstracts 005.008 - 005.010.

003.012 **Annual review of earth and planetary sciences,**
Vol. 4.
F. A. Donath, F. G. Stehli, G. W. Wetherill (Editors).
Annual Reviews Inc., Palo Alto, California. 9 + 484 pp., with
indexes for volumes 1—4 (1976). — The individual papers
within the subject scope of Astronomy and Astrophysics
Abstracts are included in their corresponding categories — see
abstracts 083.075, 091.015, 091.016, 094.590, 097.059,
105.091.

003.013 **Infrared: the new astronomy.** D. A. Allen.
Keith Reid Ltd., Shaldon, Devon. 228 pp. Price
£ 6.50 (1975). ISBN-0-904094-13-8. — Reviews in Nature,
Vol. 261, 81; 1976 (*M. Rowan-Robinson*); Spaceflight, Vol.
18, 187; 1976 (*J. L. Ball*).

003.014 **Introduction to cosmic radiation.**
O. C. Allkofer.
Buchreihe der Atomkernenergie, Vol. 10.
Thiemig, München. 221 pp. (1975).

003.015 **The wanderers in the year of the elder fire dragon**
1976. D. E. Alter, Jr.
The Wanderers, Silver Spring Md. 48 pp. Price $ 4.95 (1975).
Review in Sky Telescope, Vol. 51, 53 (1976).

003.016 **The solar chromosphere and corona: Quiet sun.**
R. G. Athay.
Astrophysics and Space Science Library, Vol. 53. D. Reidel

Publishing Company, Dordrecht, Holland — Boston, U.S.A.
11 + 504 pp. Price Dfl. 150.00, $ 59.00 respectively (1976).
ISBN 90-277-0244-6.

003.017 **Evolution of stars and galaxies.** W. Baade.
The MIT Press, Cambridge, Mass. — London. 13 +
321 pp. Price $ 5.95 (1975). — Review in Sky Telescope, Vol.
51, 125 (1976).

003.018 **Verschiedene einem Doktor der Sorbonne mitge-**
teilte Gedanken über den Kometen, der im Monat
Dezember 1680 erschienen ist. P. Bayle.
Reclams Universalbibliothek Bd. 592, Verlag Philipp Reclam
Jr., Leipzig. 576 pp. Price M. 3.00 (1975). — Review in Blick
in das Weltall, 24. Jahrgang, p. 8 (1976).

003.019 **Thinkers and tinkers.** S. A. Bedini.
Charles Scribner's Sons, New York. 520 pp. Price
$ 17.50 (1975). — Review in Sky Telescope, Vol. 51, 53 (1976).

003.020 **Dynamic light scattering.** With applications to
chemistry, biology, and physics.
B. J. Berne, R. Pecora.
Wiley-Interscience, New York. 8 + 376 pp. Price $ 24.95
(1976). — (From Science, Vol. 192, 883 (1976)).

003.021 **The southern universe.** L. Bickel.
The Macmillan Company of Australia (Pty), Ltd.,
Melbourne and Sydney; Macmillan London Ltd., London.
96 pp. Price £ 4.95 (1975). — (From Nature, Vol. 261, No.
5562, p. ix (1976)).

003.022 **Zeit und Zahl. Studien zur Zeittheorie bei Platon,**
Aristoteles, Leibniz und Kant. G. Böhme.
Philosophische Abhandlungen Bd. 45. Vittorio Klostermann,
Frankfurt a.M., 7 + 281 pp. Price DM 48.00 (1974). — Review
in Deutsche Literaturzeitung, Jahrgang 96, p. 820 - 821; 1975
(*H.-J. Treder*).

003.023 **Geodesy.** 3rd Ed. G. Bomford.
At the Clarendon Press, Oxford, 10 + 731 pp. Price
£ 15.00 net (1975). ISBN 0-19-85-1919-2.

003.024 **The new astronomies.** B. Bova.
Translated from the English edition by B. N.
Panovkin.
Mir, Moskva. 232 pp. Price 53 Kop. (1976). In Russian.

003.025 **Gravitation and relativity.** M. G. Bowler.
Pergamon Press, Oxford, England — New York,
U.S.A. 182 pp. Price $ 14.50, £ 7.25; $ 8.50, £ 4.25 respectively (1976). ISBN 0-08-020567-4, ISBN 0-08-020408-2 f
respectively.

003.026 **Gravitational perturbation theory and synchrotron**
radiation. R. A. Breuer.
Lecture Notes in Physics, Vol. 44. Springer-Verlag, Berlin —
New York. 6 + 196 pp. Price DM 20.00, $ 8.20 respectively
(1975). ISBN-3-540-07530-5. — (From Science, Vol. 192,
1151 (1976); Nature, Vol. 261, No. 5555, p. IX (1976)).

003.027 **Noctilucent clouds.**
V. A. Bronshten (*Bronshtehn*), N. I. Grishin.
Published by IPST Keter. Distributed by John Wiley and Sons
Ltd., Chichester, England. 244 pp. Price £ 15.45, $ 28.55
respectively (1976). — (From Nature, Vol. 261, No. 5556,
p. vii (1976)).

003.028 **Bogen om astronomi.** P. L. Brown.
Politikens Forlag. København. 271 pp. Price DKr.
49.75 (1975). — Review in Astron. Tidssk., Årg. 9, p. 47 (1976).

003.029 **Sky and sextant, practical celestial navigation.**
J. P. Budlong.
Van Nostrand Reinhold Company, New York – London – Canada – Australia. 6 + 151 pp. Price £ 3.75, $ 7.95 respectively (1975). – Reviews in Journ. Navigation, Vol. 29, 109 - 110; 1976 (*D. H. Sadler*); Sky Telescope, Vol. 51, 125 (1976).

003.030 **Lunar soil science: physicomechanical properties of lunar soils.** I. I. Cherkasov, V. V. Shvarev.
Translated from Russian.
Israel Program for Scientific Translations, Jerusalem; John Wiley and Sons, Ltd., Chichester. 4 + 170 pp. Price £ 9.35, $ 18.70 respectively (1975). ISBN 0-7065-1539-0. – (From Nature, Vol. 261, No. 5562, p. ix,(1976)).

003.031 **The dark night sky: a personal adventure in cosmology.** D. D. Clayton.
A Demeter Press Book. Quadrangel/New York Times Book Co., New York. 12 + 206 pp. Price $ 9.95 (1975). – Reviews in Nature, Vol. 260, 203; 1976 (*P. Davies*); Phys. Today, Vol. 29, No. 1, p. 78; No. 5, p. 71 - 72; 1976 (*E. R. Harrison*); Sky Telescope, Vol. 51, 125; 190 - 192; 1976 (*P. Rizzo*).

003.032 **The universe and its structure.** B. E. Clotfelter.
McGraw-Hill Book Company, New York. 8 + 438 pp. Price $ 12.95 (1976). – Review in Sky Telescope, Vol. 51, 421 (1976).

003.033 **Isaac Newton's "Theory of the moon's motion" (1702).** With a bibliographical and historical introduction by I. B. Cohen.
Wm. Dawson and Sons, Ltd., Folkestone, Kent, GB; Neale Watson Academic Publications, New York. 8 + 170 pp. Price £ 12.00, $ 25.00 respectively (1975). ISBN-0-7129-0642-8. Review in Sky Telescope, Vol. 51, 276 (1976).

003.034 **Apollo expeditions to the moon.**
E. M. Cortright (Editor).
National Aeronautics and Space Administration, Washington, D.C. NASA SP-350. 12 + 314 pp. Price $ 8.90 (1975). Review in Sky Telescope, Vol. 51, 196 (1976).

003.035 **Atomic inner-shell processes.**
B. Crasemann (Editor).
Academic Press,Inc., New York – San Francisco – London. Vol. I. 468 pp. Vol. II. 220 pp. Price $ 47.50 (1975). – Review in Space Sci. Rev., Vol. 18, 542; 1976 (*W. van Rensbergen*).

003.036 **Laser speckle and related phenomena.**
J. C. Dainty (Editor).
Springer Verlag, Berlin (1975). – Review in Phys. Abstr., Vol. 79, A037699 (1976).

003.037 **Gravitazione universale e sistema solare attraverso le diverse fasi storiche.**
G. Dalpozzo, G. Fiorito, L. Nuvoli, A. Prat, C. Tessitori.
Libreria Ed. Univ. Levrotto & Bella, Torino. 87 pp. Price L. 550 (1974). – Review in Giorn. Astron., Vol. 1, 241 - 242; 1975 (*M. Rigutti*).

003.038 **Astronomy. A popular history.**
J. Dorschner, C. Friedemann, S. Marx, W. Pfau.
Illustrations by G. Löffler.
Translated from the German edition (Leipzig 1975).
Van Nostrand Reinhold Company, New York. 208 pp. Price $ 22.50 (1975). – Review in Sky Telescope, Vol. 52, 52 - 54; 1976 (*E. E. Both*).

003.039 **Evolution of the earth.** R. H. Dott, Jr., R. L. Batten. Maps and diagrams by R. D. Sale.
McGraw-Hill Book Company, New York. 2nd edition, 8 + 504 pp. Price $ 14.95 (1975).

003.040 **Solar noise storms.** Ø. Elgaroy.
Pergamon Press, Oxford, England – New York, U.S.A. ca 226 pp. Price approx. $ 17.50, £ 8.70 (1976). ISBN 0-08-021039-2.

003.041 **Maximiliana. Die widerrechtliche Ausübung der Astronomie.**
M. Ernst. Compiled by P. Schamoni.
Verlag F. Bruckmann K.G., München. 88 pp. (1974). – Review in Blick in das Weltall, 24. Jahrgang, p. 73 - 74; 1976 (*D. Wattenberg*).

003.042 **Sojus-Apollo 1975.** K.-H. Eyermann.
„akzent-Bändchen", Urania-Verlag Leipzig Jena Berlin, Leipzig. 128 pp. Price M 4.50 (1975). – Review in Astron. in der Schule, 13. Jahrgang, p. 44; 1976 (*E.-H. Schmidt*).

003.043 **The Gresham lectures of John Flamsteed.**
Edited and introduced by E. G. Forbes.
Mansell Information/Publishing Limited, London. 17 + 479 pp. Price £ 18.00, $ 43.50 respectively (1975). ISBN-0-7201-0518-8. – (From Nature, Vol. 259, No. 5542, p. XV (1976)).

003.044 **The amazing universe.** H. Friedman
National Geographic Society, Washington, D.C.
200 pp. Price $ 4.25 (1975). – Reviews in Nature, Vol. 260, 654; 1976 (*J. Gribbin*); Sky Telescope, Vol. 51, 276 (1976).

003.045 **Dictionary of Scientific Biography, Vol. XIII: Hermann Staudinger – Giuseppe Veronese.**
C. C. Gillispie (Editor in chief).
Charles Scribner's Sons, New York. 13 + 623 pp. Price DM 128.00 (1976). ISBN 0-684-12925-6.

003.046 **Theoretical physics and astrophysics. Additional chapters.** V. L. Ginzburg.
Nauka, Moskva. 416 pp. Price 1 Rbl. 79 Kop. (1975). In Russian. – Review in Referativ. Zhurn. 51. Astron., 3.51.35 (1976).

003.047 **Guide de l'astronome amateur.** D. Godillon.
Librairie Maloine S.A., Paris. 574 pp. Price F 135.00 (1975). – Review in Sky Telescope, Vol. 51, 52 (1976).

003.048 **Quasars, pulsars and black holes.** F. Golden.
Charles Scribner's Sons, New York, 16 + 206 pp. Price $ 7.95 (1976). – (From Science, Vol. 192, 922 (1976)).

003.049 **The observation and analysis of stellar photospheres.**
D. F. Gray.
Wiley-Interscience, New York – Chichester, Sussex, England. 16 + 472 pp. Price $ 24.95 (1976). – Reviews in Phys. Abstr., Vol. 79, A033154 (1976); Sky Telescope, Vol. 51, 276 (1976).

003.050 **Search for and discoveries of planets.**
E. A. Grebenikov, Yu. A. Ryabov.
Nauka, Moskva. 215 pp. Price 33 Kop. (1975). In Russian.

003.051 **Galaxy formation: a personal view.** J. Gribbin.
The Macmillan Press, Ltd., London – Basingstoke. 8 + 79 pp. Price hardcover £ 5.95, paper £ 2.95 (1976). (From Nature, Vol. 261, No. 5562, p. ix (1976)).

003.052 **Dynamical processes in solar flares.**
R. Eh. Gusejnov.
Akademiya Nauk Azerbajdzhanskoj SSR, Shemakhinskaya Astrofizicheskaya Observatoriya. Izdatel'stvo "EhLM" AN Azerbajdzhanskoj SSR, Izdatel'stvo AN Armyanskoj SSR,

Baku – Erevan. 200 pp. Price 1 Rbl. 32 Kop. (1975). In Russian. – Review in Referativ. Zhurn. 51. Astron., 6.51.416 (1976).

003.053 **Circles and standing stones.** E. Hadingham.
Walker and Company, New York. 240 pp. Price $ 12.50 (1975). – Review in Journ. Roy. Astron. Soc. Canada, Vol. 70, 91; 1976 (*J.-R. Roy*).

003.054 **Astronomie – ein modernes Hobby.** H.-M. Hahn.
Arena Verlag, Würzburg, 168 pp. Price DM 18.00 (1976). – Review in SuW, 15. Jahrgang, p. 212; 1976 (*J. Krautter*).

003.055 **Atlas of optical transformers.**
G. Harburn, C. A. Taylor, T. R. Welberry.
Cornell University Press, Ithaca. 33 pp. Price $ 15.00 (1975). Review in Strolling Astronomer, Vol. 26, 37; 1976 (*J. R. Smith*).

003.056 **Canvases of a cosmecologist.** C. W. Hetzler.
Vantage Press, Inc., New York. 154 pp. Price $ 4.95 (1975). – Review in Sky Telescope, Vol. 51, 276 (1976).

003.057 **Die Erde aus dem All.** Satellitengeographie unseres Planeten.
H. Heuseler, A. Brucker (Editors).
Deutsche Verlags-Anstalt, Georg Westermann Verlag, Stuttgart – Braunschweig, 160 pp. Price DM 86.00 (1976). ISBN 3-421-02681-5, ISBN 3-14-160333-2.

003.058 **Glossary of astronomy and astrophysics.**
J. Hopkins, with a foreword by S. Chandrasekhar.
The University of Chicago Press, Chicago – London. 7 + 169 pp. Price $ 10.95 (1976). ISBN 0-226-35172-6.

003.059 **The telescope handbook and star atlas.**
N. E. Howard.
Thomas Y. Crowell Company, New York. 12 + 226 pp. Price $ 14.95 (1975). – Review in Sky Telescope, Vol. 51, 51 (1976).

003.060 **Planets, stars, and galaxies.** S. J. Inglis.
John Wiley and Sons Inc., New York. 4th edition, 10 + 336 pp. Price $ 11.95 (1976). – Review in Sky Telescope, Vol. 51, 421 (1976).

003.061 **Des astres, de la vie et des hommes.** R. Jastrow.
Translated from the English edition by C. de Richemont.
Éditions du Seuil, Paris. 256 pp. Price F 11.40 (1975). – (From Science, Vol. 191, 878 (1976)).

003.062 **Die Sonne, Licht und Leben.** J. Jobé (Editor).
Herder Verlag, Freiburg. Price DM 138.00 (1975).
Review in SuW, 15. Jahrgang, p. 140; 1976 (*C. Möllenhoff*).

003.063 **The search for the nebulae.** K. G. Jones.
Alpha Academic (Science History Publications), Halfpenny Furze, Mill Lane, Chalfont St. Giles, Buckinghamshire, England; Neale Watson Academic Publications, New York. 11 + 84 pp. Price £ 3.50, $ 8.50 respectively (1975). Reviews in Journ. British Astron. Ass., Vol. 86, 349; 1976 (*E. A. Beet*); Journ. History Astron., Vol. 7, 67; 1976 (*J. A. Bennett*); Sky Telescope, Vol. 51, 421 (1976).

003.064 **The problem CETI [communication with extraterrestrial intelligence].** S. A. Kaplan (Editor).
Mir, Moskva. 350 pp. Price 2 Rbl. 18 Kop. (1975). In Russian.
Review in Priroda, 1976, No. 4, p. 157.

003.065 **Non-linear waves in dispersive media.**

V. I. Karpman.
Translated from the Russian by F. F. Cap.
Pergamon Press, Oxford. 183 pp. (1975).

003.066 **Advances in plasma physics. Vol. 6.**
P. K. Kaw, W. L. Kruer, C. S. Liu, K. Nishikawa.
John Wiley and Sons Inc., New York. 600 pp. (1976).

003.067 **Advances in image pickup and display. Vol. 1.**
B. Kazan (Editor).
Academic Press, Inc., New York. 308 pp. (1974).

003.068 **The universe unfolding.** I. R. King.
W. H. Freeman and Company, San Francisco.
8 + 504 pp. Price $ 14.95 (1976). ISBN 0-7167-0521-4.
Review in Sky Telescope, Vol. 51, 421 (1976).

003.069 **American astronauts and spacecraft.**
D. C. Knight (Editor).
Franklin Watts, Inc., New York. 208 pp. Price $ 8.87 (1975).
Review in Sky Telescope, Vol. 51, 51 (1976).

003.070 **Über die möglichen Formen des Lebens auf anderen Planeten.** H. W. Koepcke.
Goecke & Evers, Krefeld. 148 pp. (1975). – (From SuW, 15. Jahrgang, p. 69 (1976)).

003.071 **Zpráva o vesmíru.** Z. Kopal.
Translated from the English (Man and his universe) by M. Kopal.
Mladá fronta, Praha. 292 + 36 pp. Price Kčs 24.00 (1976).

003.072 **Pulsating stars.** B. V. Kukarkin.
Translated from the Russian edition (Moscow, 1970) by R. Hardin.
Halsted Press, a Division of John Wiley and Sons, New York – Chichester; IPST Astrophysics Library, Israel Program for Scientific Translations, Jerusalem. 16 + 320 pp. Price $ 37.50, £ 19.10 respectively (1975). ISBN 0-470-51035-8. – Reviews in Sky Telescope, Vol. 51, 276; 343 - 345; 1976 (*L. G. Jacchia*); Strolling Astronomer, Vol. 26, 37; 1976 (*J. R. Smith*).

003.073 **Elements of analytical dynamics.** R. Kurth.
Pergamon Press, Oxford, England – New York, U.S.A.
180 pp. Price $ 15.00, £ 7.50 (1976). ISBN 0-08-019848-1.

003.074 **Blanketed model atmospheres for early-type stars.**
R. L. Kurucz, E. Peytremann, E. H. Avrett.
Smithsonian Astrophysical Observatory, Smithsonian Institution, Washington, D.C. (for sale by US Government Printing Office). 6 + 186 pp. Price $ 7.60 (1974). – (From Nature, Vol. 260, No. 5551, p. VII (1976); Science, Vol. 192, 135 (1976)).

003.075 **Advances in geophysics, Vol. 19.**
H. E. Landsberg, J. Van Mieghem (Editors).
Academic Press, Inc., New York – London. 328 pp. Price $ 29.00, £ 15.95 (1976). ISBN 0-12-018819-8. – (From Nature, Vol. 261, No. 5559, p. VIII (1976)).

003.076 **Problem book in relativity and gravitation.**
A. P. Lightman, W. H. Press, R. H. Price, S. A. Teukolsky.
Princeton University Press, Princeton, New Jersey – London.
14 + 599 pp. Price cloth $ 20.00, £ 11.70, paper $ 7.50, £ 4.40 respectively (1975). – (From Nature, Vol. 259, No. 5544, p. V (1976); Phys. Today, Vol. 29, No. 3, p. 65 (1976)).

003.077 **The decision to go to the moon.** Project Apollo and the national interest. J. M. Logsdon.
University of Chicago Press, Chicago. 14 + 188 pp. Price

$ 3.95 (1976). Reprint of the 1970 edition.

003.078 Man's relation to the universe. B. Lovell.
W. H. Freeman and Company, San Francisco, USA—
Reading, England. 6 + 118 pp. Price £ 3.60, $ 5.95 respectively
(1975). — Reviews in Astron. Nachr., Vol. 297, 166 - 167;
1976 (*K. Fritze*); Astron. Tidssk., Årg. 9, p. 47 (1976);
Nature, Vol. 261, 82; 1976 (*P. Davies*); Strolling Astronomer,
Vol. 25, 257; 1976 (*B. M. Frank*).

003.079 Drehbare Sternkarte. (Äquinoktium 2000.0). Kreis-
förmige Scheibe mit drehbarem Zeiger und Hin-
weisen für die Benutzung. S. Marx, W. Pfau.
Johann Ambrosius Barth, Leipzig. Price M 19.00 (1975). —
Reviews in Blick in das Weltall, 24. Jahrg., p. 10; 1976
(*D. Wattenberg*); Sterne, Vol. 52, 121 - 122; 1976 (*K. Lindner*).

003.080 Detection and spectrometry of faint light.
J. Meaburn.
D. Reidel Publishing Company, Dordrecht, Holland—Boston,
U.S.A. 9 + 266 pp. Price Dfl. 90.00, $ 34.00 (1976). ISBN
90-277-0678-6.

003.081 1976 yearbook of astronomy.
P. Moore (Editor).
Sidgwick and Jackson Ltd., London, W. W. Norton and Co.,
Inc., New York. 216 pp. Price $ 9.95 (1975). — Review in
Sky Telescope, Vol. 51, 125 (1976).

003.082 The sky at night, 5. P. Moore.
The British Broadcasting Corporation Publications,
London, GB. Price £ 4.00 (1975). —Reviews in Journ. British
Astron. Ass., Vol. 86, 344; 1976 (*W. E. Fox*); Sky Telescope,
Vol. 51, 421 (1976).

003.083 The next fifty years in space.
P. Moore, with drawings by A. Farmer.
Taplinger, New York; William Luscombe, London. 144 pp.
Price $ 12.95, £ 4.25 respectively (1976). — Reviews in Journ.
British Astron. Ass., Vol. 86, 346 - 347; 1976 (*H. Miles*); Sky
Telescope, Vol. 51, 421 (1976).

003.084 The universe: its beginning and end. L. Motz.
Charles Scribner's Sons, New York. 14 + 343 pp.
Price $ 14.95 (1975). — Reviews in Sky Telescope, Vol. 51,
196; Vol. 52, 51 - 52; 1976 (*C. A. Federer, Jr.*).

**003.085 Some problems of solar activity influence on the
lower atmosphere.**
Eh. R. Mustel', V. F. Loginov (Editors).
Trudy VNII gidrometeorol. inform.-Mirovoj tsentr dannykh,
vyp. (No.) 23. Moskva, Gidrometeoizdat, 1975. 80 pp. Price
31 Kop. In Russian. — Review in Referativ. Zhurn. 51. Astron.,
3.51.352 (1976).

003.086 Telescope of an amateur astronomer.
M. S. Navashin.
Nauka, Moskva. 3rd revised and enlarged edition. 432 pp.
Price 91 Kop. (1975). In Russian. — Review in Referativ.
Zhurn. 51. Astron., 6.51.97 (1976).

**003.087 Interplanetary encounters: close-range gravitational
interactions.** E. J. Öpik.
Developments in solar system and space science, 2.
Elsevier Scientific Publishing Company, Amsterdam—Oxford—
New York. 7 + 155 pp. Price $ 26.95, Dfl. 67.00 respectively
(1976). — (From Nature, Vol. 261, No. 5562, p. ix (1976)).

003.088 The theory of relativity. R. K. Pathria.
Pergamon Press, Oxford, England—New York, USA.
2. Edition, 327 pp. Price $ 24.00, £ 12.00; $ 15.00, £ 7.50

respectively (1974). ISBN 0-08-018032-9, ISBN 0-08-018995-
4f respectively.

003.089 The satellite spin-off. The achievements of space
flight. G. Paul.
Translated from the German edition by A. Lacy, B. Lacy.
Luce, Washington, D.C. Distributor: McKay, New York. 272
pp. Price $ 10.00 (1975). — (From Science, Vol. 192, 168
(1976)).

003.090 Chemical evolution of the giant planets.
C. Ponnamperuma.
Academic Press, Inc., New York—London. 232 pp. Price
$ 11.50, £ 6.65 respectively (1976). ISBN 0-12-561350-4.
(From Nature, Vol. 261, No. 5559, p. VIII (1976)).

003.091 Graze observer's handbook. H. R. Povenmire.
Vantage Press, Inc., New York. 134 pp. Price
$ 4.95 (1975). — Review in Sky Telescope, Vol. 51, 189 - 190;
1976 (*R. Nolthenius*).

003.092 Astronomie heute und morgen. H. Rohr.
P. Meili und Co., Schaffhausen. 90 pp. Price Fr.
18.00 (1976). — Review in Orion, 33. Jahrgang, p. 223; 1975
(*E. Wiedemann*).

003.093 Space investigation. R. Z. Sagdeev.
Nauka, Moskva. 16 pp. (1975). In Russian.

003.094 Galileo Galilei. E. Schmutzer, W. Schütze.
BSB B. G. Teubner Verlagsgesellschaft, Leipzig.
152 pp. Price M 6.90 (1975). — Review in Astron. in der
Schule, 13. Jahrgang, p. 44; 1976 (*H. Bernhard*).

**003.095 Entwicklungsphasen der Erforschung der leuchten-
den Nachtwolken.** W. Schröder.
Akademie-Verlag, Berlin. 64 pp. Price DM 12.00 (1975).
(From SuW, 15. Jahrgang, p. 69 (1976)).

003.096 Moon morphology. P. H. Schultz.
University of Texas Press, Austin, Texas. Price
$ 35.00 (1976). — Review in Sky Telescope, Vol. 52, 56
(1976).

**003.097 Mathematical cosmology and extragalactic astron-
omy.** I. E. Segal.
Pure and applied mathematics. A series of monographs and
textbooks. Academic Press, New York — San Francisco —
London. 9 + 204 pp. Price DM 63.78 (1976). ISBN 0-12-
635250-X. — Review in Sky Telescope, Vol. 51, 169 (1976).
Contents: (1) General introduction; (2) Mathematical
development; (3) Physical theory; (4) Astronomical applications; (5) Discussion.

003.098 Structure and evolution of galaxies. G. Setti (Editor).
NATO Advanced Study Institutes Series. D. Reidel
Publishing Company, Dordrecht, Holland—Boston, USA. 8 +
334 pp. Price $ 32.00 (1975). — Reviews in Journ. British
Astron. Ass., Vol. 86, 168; 1976 (*S. Mitton*); SuW, 15. Jahr-
gang, 213; 1976 (*H. J. Staude*).

003.099 Relativität Gruppen Teilchen. Spezielle Relativitäts-
theorie als Grundlage der Feld- und Teilchenphysik.
R. U. Sexl, H. K. Urbantke.
Springer-Verlag, Wien — New York. 9 + 301 pp. Price DM
58.00 (1976). ISBN 3-211-81364-0/ ISBN 0-387-81364-0.

003.100 Astronomy. Volume 11. Celestial mechanics.
I. S. Shcherbina-Samojlova (Editor).
Itogi nauki i tekhniki. Seriya Astronomiya, tom 11. Moskva.
156 pp. Price 87 Kop. (1975). In Russian. — Contents: 1.

Qualitative methods of celestial mechanics; 2. Periodic solutions and resonances; 3. Theory of perturbations.

003.101   **Spallation nuclear reactions and their applications.**
   B. S. P. Shen, M. Merker (Editors).
D. Reidel Publishing Company, Dordrecht, Holland—Boston, U.S.A. 7 + 227 pp. Price Dfl. 75.00, $ 27.50 (1976). ISBN 90-277-0746-4.

003.102   **Light scattering in planetary atmospheres.**
   V. V. Sobolev. Translated by W. M. Irvine, with the collaboration of M. Gendel and A. P. Lane.
International Series of monographs in natural philosophy, Vol. 76. Pergamon Press, Oxford—New York—Toronto—Sydney—Braunschweig. 17 + 256 pp. Price $ 25.00, DM 91.00 respectively (1975). ISBN 0-08-017934-7. – Review in Journ. British Astron. Ass., Vol. 86, 254 - 255; 1976 (*S. Mitton*).
   Contents: (1) Basic equations; (2) Semi-infinite atmospheres; (3) Atmospheres of finite optical thickness; (4) Atmospheres overlying a reflecting surface; (5) General theory; (6) General theory (continued); (7) Linear integral equations for the reflection and transmission coefficients; (8) Approximate formulas; (9) The radiation emerging from a planet; (10) Optical properties of planetary atmospheres; (11) Spherical atmospheres.

003.103   **Space Shuttles – die neue Brücke ins All.**
   B. Stanek, L. Pesek.
Hallwag Verlag, Bern—Stuttgart. 47 pp. (1975). – (From SuW, 15. Jahrgang, p. 69 (1976)).

003.104   **Linear and regular celestial mechanics. Perturbed two-body motion. Numerical methods. Canonical theory.**   E. L. Stiefel, G. Scheifele.
Translated from the English edition. Nauka, Moskva. 303 pp. Price 2 Rbl. 5 Kop. (1975). In Russian. – Review in Referativ. Zhurn. 51. Astron., 3.51.74 (1976).

003.105   **Geodäsie.**   W. Torge.
   Walter de Gruyter, Berlin – New York. 268 pp. Price DM 19.80 (1975). ISBN 3-11-004394-7.

003.106   **Radiation processes in astrophysics.**
   W. H. Tucker.
MIT Press, Cambridge Mass.—London. 11 + 311 pp. Price $ 12.50 (1975). ISBN 0-262-20021-X. – Review in Sky Telescope, Vol. 51, 196 (1976).

003.107   **Van Marum's scientific instruments in Teyler's museum. Part II. Descriptive catalogue.**
   G. L'E. Turner.
Nordhoff International Publishing, Leyden. pp. 125 - 401. Price Dfl. 40.00 (1973). – Review in Journ. History Astron., Vol. 7, 70 (1976).

003.108   **Die Kunst Sonnenuhren auf das Papier oder eine Mauer zu zeichnen.**   L. Voch.
Zentralantiquariat der DDR, Leipzig, DDR. 62 pp. Price M 30.00 (1975). – Review in Sky Telescope, Vol. 52, 56 (1976).

003.109   **Mensch und Wetter.**   P. von Eynern.
   Wilhelm Heyne Verlag, München. 220 pp. Price DM 7.80 (1975). – Review in SuW, 15. Jahrgang, p. 32; 1976 (*G. D. Roth*).

003.110   **Dark matters in open clusters.**   A. Wallenquist.
   Almqvist & Wiksell Publ., Stockholm. Price Kr. 94.00 (1975). – (From SuW, 15. Jahrgang, p. 103 (1976)).

003.111   **Erschröckliche und warhafftige Wunderzeichen**

1543 - 1586. Faksimiledruck von Einblattdrucken aus der Sammlung Wikiana in der Zentralbibliothek Zürich. B. Weber (Editor).
(folio volume of plates, 46 leaves).
Commentary volume: **Wunderzeichen und Winkeldrucker 1543 - 1586.**   B. Weber.
Urs Graf Verlag, Dietikon-Zürich. 153 pp. Price Swiss fr. 1070; Swiss fr. 250 for the commentary volume alone. (Distributor for Canada, Japan, and the United States: B. M. Rosenthal, Inc., 251 Post Street, San Francisco, Ca. 94108, $ 402; $ 94). Essay review "Sixteenth-Century Broadsides" in Journ. History Astron., Vol. 7, 145 - 150; 1976 (*O. Gingerich*).

003.112   **Aberrations of the symmetrical optical system.**
   W. T. Welford.
Academic Press, London. 10 + 240 pp. Price £ 7.60 (1974). Review in Phys. Blätter, 32. Jahrgang, 237 - 238; 1976 (*H. Köhler*).

003.113   **Scientific instruments.**   H. Wynter, A. Turner.
   Charles Scribner's Sons, New York. 239 pp. Price $ 27.50 (1975). – Review in Sky Telescope, Vol. 51, 421 (1976).

003.114   **Astronomy for the amateur: Vol. 1, Planetary astronomy; Vol. 2, Stellar astronomy.**
   R. P. Van Zandt.
R. P. Van Zandt, Box 3013, Peoria, Ill., Vol. 1, 195 pp., Vol. 2, 187 pp. Price $ 6.50 each (1975). – Review in Sky Telescope, Vol. 51, 196 (1976).

003.115   **Astronomy: the evolving universe.**   M. Zeilik.
   Harper and Row, Publishers, New York – London. 14 + 530 pp. Price $ 13.95 (1976). ISBN 0-06-047383-5. (From Nature, Vol. 261, No. 5557, p. IX (1976)).

003.116   **Cosmic rays in the stratosphere and the circumterrestrial cosmic space.**
Trudy Fiz. in-ta. AN SSSR, 88. Nauka, Moskva. 191 pp. Price 1 Rbl. 63 Kop. (1976). In Russian. – Review in Referativ. Zhurn. 62. Issled. kosmich. prostranstva, 6.62.187 (1976).

003.117   **Geodynamics today: a review of the earth's dynamic processes.**
The Royal Society, London. 197 pp. Price £ 2.75 (1975). Review in Nature, Vol. 260, 812; 1976 (*P. J. Smith*).

003.118   **Objections to astrology, 1975.**
   Prometheus Books, Buffalo, N.Y. 62 pp. Price $ 2.95 (1975). – Review in Sky Telescope, Vol. 51, 125 (1976).

003.119   **L'oeuvre astronomique de Thémon Juif, maître parisien du XIVe siècle.**
Centre de Recherches d'Histoire et de Philologie de la IVe Section de l'École Practique des Hautes Études, Série V: Hautes Études Médiévales et Modernes, 16.
Henri Hugonnard-Roche Droz, Geneva, and Minard, Paris. 429 pp. (1973). – Review in Journ. History Astron., Vol. 7, 68 - 69; 1976 (*A. G. Molland*).

003.120   **Questions of modelling the ionosphere.**
   Kaliningr. un-t. Kaliningrad. 111 pp. Price 68 Kop. (1975). In Russian. – Review in Referativ. Zhurn. 62. Issled. kosmich. prostranstva, 4.62.267 (1976).

003.121   **The Skylab results.**
   American Astronautical Society. 2 parts. 1146 pp. Price $ 60.00 per set (1975). – Review in Journ. British Interplanet. Soc., Vol. 29, 215 - 216; 1976 (*D. Baker*).

003.122 **The solar system.** Collected articles from the 1975 September issue of The Scientific American.
W. H. Freeman &Co., San Francisco—Reading. 145 pp. Price cloth $ 8.50, £ 5.10, paper $ 4.50, £ 2.25 (1976). — Reviews in Journ. British Astron. Ass., Vol. 86, 346; 1976 (*P. Moore*); Sky Telescope, Vol. 51, 276 (1976); Spaceflight, Vol. 18, 233; 1976 (*A. J. Jeffries*).

003.123 **Weltraumphysik** (Herder Lexikon).

Herder, Freiburg—Basel—Wien. 239 pp. Price DM 19.80 (1975). — Review in Naturwissenschaften, 63. Jahrgang, p. 249; 1976 (*G. Traving*).

003.124 **J. C. Poggendorff: Biographisch-literarisches Handwörterbuch der exakten Naturwissenschaften.**
Band VIIb, Teil 5.
Sächsische Akademie der Wissenschaften zu Leipzig. Akademie-Verlag, Berlin, pp. 3145 - 3304 (1975).

004 History of Astronomy, Chronology

004.001 **Zur Geschichte des Meridiankreises und verwandter Instrumente. Von Römer bis Repsold und Reichenbach. I. II.** L. Brandt.
SuW, 15. Jahrgang, p. 10 - 12, 52 - 55 (1976).

004.002 **The two-headed Janus of French astronomy (J. D. Cassini, 1625 - 1712).** A. A. Gurshtejn.
Zemlya i Vselennaya, 1976, No. 1, p. 60 - 66. In Russian.

004.003 **The importance of the transit of Mercury of 1631.** A. van Helden.
Journ. History Astron., Vol. 7, 1 - 10 (1976).

004.004 **The two Megalithic lunar observatories at Carnac.** A. Thom, A. S. Thom, J. M. Gorrie.
Journ. History Astron., Vol. 7, 11 - 26 (1976).

004.005 **S Andromedae, 1885: an analysis of contemporary reports and a reconstruction.** K. G. Jones.
Journ. History Astron., Vol. 7, 27 - 40 (1976).

004.006 **The double-argument lunar tables of Cyriacus.** G. A. Saliba.
Journ. History Astron., Vol. 7, 41 - 46 (1976).

004.007 **Ritchey, Curtis and the discovery of novae in spiral nebulae.** M. A. Hoskin.
Journ. History Astron., Vol. 7, 47 - 53 (1976).

004.008 **Chinese cosmology.** N. Sivin.
Nature, Vol. 259, 249 (1976).

004.009 **A study of records of the solar and lunar eclipses in scripts on tortoise-shells or ox-bones.** P.-y. Zhang.
Acta Astron. Sinica, Vol. 16, 210 - 224 (1975). In Chinese.

004.010 **The California-Wisconsin axis in American astronomy – I, II.** D. E. Osterbrock.
Sky Telescope, Vol. 51, 9 - 14, 91 - 97 (1976).

004.011 **America's last king and his observatory.** J. Ashbrook.
Sky Telescope, Vol. 51, 163 - 164 (1976).

004.012 **Practical astronomy in Russia before and during the reign of Peter I.** V. L. Chenakal.
Vopr. geogr. petrovsk. vremeni. Leningrad, Gidrometeoizdat, 1975, p. 39 - 59. In Russian. – Abstr. in Referativ. Zhurn. 51. Astron., 3.51.11 (1976).

004.013 **Astronomical-geodetical problems at the Petersburg Academy of Sciences (the early period).** Yu. Kh. Kopelevich.
Vopr. istorii astron. No. 3. Moskva, 1974, p. 43 - 60, 196 - 197. In Russian. – Abstr. in Referativ. Zhurn. 51. Astron., 3.51.12 (1976).

004.014 **I. I. Islen'ev – astronomer-geographer and cartographer of the 18th century.** F. A. Shibanov.
Vestn. Leningr. un-ta, 1975, No. 18, p. 133 - 139. In Russian. Abstr. in Referativ. Zhurn. 51. Astron., 3.51.15 (1976).

004.015 **Die Versammlung der Astronomischen Gesellschaft im August 1921 in Potsdam.** H. Lambrecht.
Sterne, Vol. 52, 47 - 50 (1976).

004.016 **Ergänzende Bemerkungen zum Beitrag "Die Kosmogonie Immanuel Kants".** H. Lambrecht.
Sterne, Vol. 52, 51 - 53 (1976). – See also 11.004.055 and 13.004.014.

004.017 **The Royal Greenwich Observatory 1675 - 1975. Some of its external relations.** W. H. McCrea.
Quarterly Journ. Roy. Astron. Soc., Vol. 17, 4 - 24 (1976). Paper based on a lecture given on 1975 November 6 in the 'Great Centenaries' series organized by the Centre for Continuing Education, University of Sussex.
After a sketch of the background history of the Oberv Observatory, some of its extensional – or fringe – activities during the past 300 years are illustrated. It is then shown that the sort of activities that have been 'fringe' in the past are now in process of becoming central for the foreseeable future.

004.018 **Building Chaucer's astrolabe – II, III.** S. Eisner.
Journ. British Astron. Ass., Vol. 86, 125 - 132, 219 - 227 (1976).

004.019 **On the solar eclipse of 1860.** J. E. Kennedy.
Journ. Roy. Astron. Soc. Canada, Vol. 70, 74 - 76 (1976). – Paper presented at the 1975 General Assembly of the R.A.S.C., Halifax.

004.020 **Regiomontanus, 1436–1476.** A. V. Douglas.
Journ. Roy. Astron. Soc. Canada, Vol. 70, 79 - 80 (1976).

004.021 **Der Urmaßstab Christian Mayers.** H. Schmidt.
SuW, 15. Jahrgang, p. 148 - 150 (1976).

004.022 **Sur l'histoire du manuscrit Copernicien: "De revolutionibus orbium coelestium".** K. Hujer.
L'Astronomie, Vol. 90, 181 - 190 (1976).

004.023 **Astronomy in the ancient Americas.** R. D. Hicks III.
Sky Telescope, Vol. 51, 372 - 377 (1976).

004.024 **De wetten van Kepler en Newton: grondslag der hemelmechanica.** T. Dethier.
Zenit, 3e jaargang, p. 178 - 180 (1976).

004.025 **The astronomers of the Orient.** T. Z. Dworak.
Urania Kraków, Vol. 47, 13 - 17 (1976). In Polish.

004.026 **Fabels en feiten rond de rode planeet.** G. W. E. Beekman.
Zenit, 3e jaargang, p. 194 - 199 (1976).

004.027 **15 Kometenflugblätter des 17. und 18. Jahrhunderts. I.** J. Classen.
Sterne, Vol. 52, 98 - 106 (1976).

004.028 **15 Kometenflugblätter des 17. und 18. Jahrhunderts. II.** J. Classen.
Sterne, Vol. 52, 107 - 114 (1976).

004.029 **Mayan chronology and 'the spectrum of time'.** D. J. Schove.
Nature, Vol. 261, 471 - 473 (1976).
By applying methods used in varve and tree-ring analysis to link the floating Mayan chronology with the anchored chronology of calculated planetary positions a new correlation has been derived. The intervals between certain Mayan dates

are known to be multiples of the synodic periods of several planets. Mayan astronomers are assumed to have observed close conjunctions of two or more planets; this is tested by comparisons with observed or calculated conjunctions. The author concludes that the Mayan dates as conventionally expressed are 27.3 yr too old. The suggested correlation number is 594 250 ± 1 d, the difference between the Julian and the Mayan day numbers.

004.030 Ash-Shirazi's ideas on the nature of the ashen light of the moon and the luminosity of planets.
M. Shermatov.
Trudy XVI Nauch. konf. aspirantov i ml. nauch. sotrudn. Inta istorii estestvozn. i tekhn. Sekts. istorii fiz. In-t istorii estestvozn. i tekhn. AN SSSR. Moskva, 1973, p. 43 - 49. In Russian. — Abstr. in Referativ. Zhurn. 51. Astron., 6. 51. 9 (1976).

004.031 Solar eclipses mentioned in old Serbian annals.
J. L. Simovljević.
Glas Srpske akad. nauka i umetn., Vol. 291, 71 - 80 (1974). In Serbo-Croatian. — Abstr. in Bull. Sci. Yougoslavie, Sect. A, Vol. 21, 18 (1976).

004.032 Astronomija Latvijā 18. un 19. gadsimtā.
I. Daube.
Zvaigžņotā debess, 1975./76. gada ziema, p. 36 - 44.

004.033 Gaismas ātruma 300 gadi. M. Zepe.
Zvaigžņotā debess, 1975./76. gada ziema, p. 44 - 46.

004.034 Vēsturiskas etīdes astronomijā (1, 2).
I. Rabinovičs.
Zvaigžņotā debess, 1976. gada pavasaris, p. 39 - 44.

004.035 Zur Frühentwicklung der Astrophysik. Das internationale Forscherkollektiv 1865 - 1899.
D. B. Herrmann, J. Hamel.
NTM, Schriftenr. Gesch., Naturwiss., Techn., Med., Band 12, 25 - 30 = Mitt. Archenhold-Sternw. Berlin-Treptow No. 113 (1975).

004.036 'On the power of penetrating into space': the telescopes of William Herschel. J. A. Bennett.
Journ. History Astron., Vol. 7, 75 - 108 (1976).

004.037 The recovery of early Greek astronomy from India.
D. Pingree.
Journ. History Astron., Vol. 7, 109 - 123 (1976).

004.038 Early astronomical researches of John Flamsteed.
E. G. Forbes.
Journ. History Astron., Vol. 7, 124 - 138 (1976).

004.039 The position of supernova 1006 and the St Gallen chronicle. C. M. Botley.
Journ. History Astron., Vol. 7, 139 - 140 (1976).

004.040 A possible Pythagorean triangle at Stonehenge.
W. E. Dibble.
Journ. History Astron., Vol. 7, 141 - 142 (1976).

004.041 The Stonehenge Stations. R. J. C. Atkinson.
Journ. History Astron., Vol. 7, 142 - 144 (1976).

004.042 Cosmological theories in ancient China.
Y.-t. Cheng.
Scient. Sinica, Vol. 19, 291 - 309 (1976).

004.043 Il sistema copernicano dopo Galileo e l'ultimo conflitto per la sua affermazione. P. Maffei.

Giorn. Astron., Vol. 1, 5 - 12 (1975).

004.044 15 Kometenflugblätter des 17. und 18. Jahrhunderts. III. J. Classen.
Sterne, Vol. 52, 172 - 184 (1976).

004.045 Die Dynamik der Kreisbewegungen der Himmelskörper und des freien Falls bei Aristoteles, Copernicus, Kepler und Descartes. H.-J. Treder.
Colloquia Copernicana IV, (Ossolineum 1975), p. 105 - 150.

004.046 Die Karfreitagsfinsternis. K. Ferrari d'Occhieppo.
Sternenbote, 19. Jahrgang, p. 74 - 77 (1976).

004.047 J. E. de Villiers's Observatory at Sea Point.
B. Warner, R. F. Hurly.
Monthly Notes Astron. Soc. Southern Africa, Vol. 35, 57 - 62 (1976).

004.048 L'opera astronomica di Niccolò Copernico.
M. Cimino.
Oss. Astron. Roma, Contr. sci., Ser. III, No. 140, 42 pp. (1975).

004.049 Ein wichtiger Fund zur Geschichte der Mathematik, Astronomie und Optik. B. A. Rozenfel'd.
Vopr. Istor. Estestvozn. 1974, p. 123 - 124 (1974). In Russian. Abstr. in Zentralbl. Math. Grenzgebiete — Math. Abstr., Band 307, No. 01003 (1976).

004.050 L'eliocentrismo di Aristarco da Archimede a Copernico. G. Derenzini.
Physis. Vol. 16, 289 - 308 (1975). — Abstr. in Zentralbl. Math. Grenzgebiete — Math. Abstr., Band 308, No. 01004 (1976).

004.051 Clairaut's critique of Newtonian attraction: some insights into his philosophy of science.
P. Chandler.
Ann. Sci., (GB), Vol. 32, 369 - 378 (1975). — Abstr. in Zentralbl. Math. Grenzgebiete — Math. Abstr., Band 313, No. 01004 (1976).

004.052 Dal sistema copernicano alla visione moderna dell'universo. A. Pignedoli.
Atti Accad. Sci. Ist. Bologna, Cl. Sci. Fis., Anno 262, Rend. 13, Ser. 1, No. 1, p. 10 - 37 (1974).

004.053 Nicolas Copernic. J. P. Palewski.
Acad. Polon. Sci., Centre sci. Paris, Conférences 106, 20 pp. (1974). — Abstr. in Zentralbl. Math. Grenzgebiete — Math. Abstr., Band 315, No. 01001 (1976).

004.054 Raja Sawai Jai Singh II: an 18th century mediaeval astronomer. W. A. Blanpied.
American Journ. Phys., Vol. 43, 1025 - 1035 (1975). — Abstr. in Phys. Abstr., Vol. 79, A027691 (1976)

Verschiedene einem Doktor der Sorbonne mitgeteilte Gedanken über den Kometen, der im Monat Dezember 1680 erschienen ist. See Abstr. 003.018.

Thinkers and tinkers. See Abstr. 003.019.

Zeit und Zahl. Studien zur Zeittheorie bei Platon, Aristoteles, Leibniz und Kant. See Abstr. 003.022.

Isaac Newton's "Theory of the moon's motion" (1702). See Abstr. 003.033.

The Gresham lectures of John Flamsteed.

See Abstr. 003.043.

Circles and standing stones.   See Abstr. 003.053.

Galileo Galilei.   See Abstr. 003.094.

005   Biography

005.001   **Edward Charles Howard and an early British contribution to meteoritics.**   D. W. Sears.
Journ. British Astron. Ass., Vol. 86, 133 - 139 (1976).

005.002   **Joseph von Fraunhofer – zu seinem 150. Todestag.**
D. B. Herrmann.
Astron. in der Schule, 13. Jahrgang, p. 33 - 35 (1976).

005.003   **Over het werk van Rudolph Minkowski.**
J. H. Oort.
Zenit, 3e jaargang, p. 173 - 174 (1976).

005.004   **A Victorian amateur astronomer Roger Langdon (1825 - 1894).**   P. Moore.
Journ. British Astron. Ass., Vol. 86, 309 - 311 (1976).

005.005   **J. C. Kapteyn – ein Pionier der Erforschung des Sternsystems. – Zum 125. Geburtstag des Astronomen.**   D. B. Herrmann.
Blick in das Weltall, 24. Jahrg., p. 18 - 21 (1976).

005.006   **Einstein und die Atombombe.**   F. Herneck.
Archenhold-Sternw. Berlin-Treptow, Vorträge und Schriften, No. 51, 31 pp. (1976).

005.007   **Die Copernicus-Biographie von Georg Christoph Lichtenberg.**   D. B. Herrmann.
NTM, Schriftenr. Gesch., Naturwiss., Techn., Med., Band 11, 40 - 45 = Mitt. Archenhold-Sternw. Berlin-Treptow No. 112 (1974).

005.008   **Czech astronomer, botanist and physician Tadeáš Hájek z Hájku (Hagecius).**   J. Bouška.
Tadeáš Hájek z Hájku (Hagecius) 1525–1600 (see 003.011), p. 11 - 12 (1976). In Czech.

005.009   **Tadeáš Hájek z Hájku (Hagecius) as astronomer.**
V. Guth.
Tadeáš Hájek z Hájku (Hagecius) 1525–1600 (see 003.011), p. 29 - 34 (1976). In Czech.

005.010   **Tadeáš Hájek z Hájku (Hagecius) and the bright comet 1577.**   V. Vanýsek.
Tadeáš Hájek z Hájku (Hagecius) 1525–1600 (see 003.011),

p. 35 - 38 (1976). In Czech.

005.011   **Camille Flammarion (1842–1925).**   J. Bouška.
Vesmír, Vol. 55, 158 (1976). In Czech.

005.012   **J, von Fraunhofer (1787–1826).**
Říše hvězd, Vol. 57, 134 (1976). In Czech.

005.013   **Zum Einfluß von F. von Zach auf Alexander von Humboldt.**   K.-R. Biermann.
Sterne, Vol. 52, 166 - 171 (1976).

005.014   **Eric Lindsay and the southern heavens.**
B. J. Bok.
Irish Astron. Journ., Vol. 12, 73 - 76 (1975).

005.015   **Dr. Eric Lindsay in South Africa.**   A. H. Jarrett.
Irish Astron. Journ., Vol. 12, 89 - 92 (1975).

005.016   **E. M. Lindsay, an all-Ireland astronomer.**
D. J. Mullan.
Irish Astron. Journ., Vol. 12, 110 - 111 (1975).

005.017   **Eric Lindsay as I knew him.**   G. G. Cillié.
Irish Astron. Journ., Vol. 12, 112 - 113 (1975).

005.018   **Eric Lindsay – an appreciation.**   D. McNally.
Irish Astron. Journ., Vol. 12, 114 - 115 (1975).

005.019   **Eric Lindsay – educationalist astronomer.**
P. Moore.
Irish Astron. Journ., Vol. 12, 116 - 117 (1975).

005.020   **A noble dreamer. A personal recollection of Dr. Eric Lindsay.**   T. Murtagh.
Irish Astron. Journ., Vol. 12, 118 - 123 (1975).

005.021   **Eric Mervyn Lindsay – an amateur appreciation.**
D. E. Beesley.
Irish Astron. Journ., Vol. 12, 123 - 124 (1975).

005.022   **Eric Mervyn Lindsay at Armagh.**   C. D. Trimble.
Irish Astron. Journ., Vol. 12, 127 - 128 (1975).

005.023 **Eric Mervyn Lindsay.** D. O'Connell.
Irish Astron. Journ., Vol. 12, 129 (1975).

005.024 **Eric Mervyn Lindsay (1907–1974). Some recollections.** P. G. Corvan.
Irish Astron. Journ., Vol. 12, 130 - 137 (1975).

005.025 **Eric Mervyn Lindsay. A personal tribute.**
H. A. Brück.
Irish Astron. Journ., Vol. 12, 146 - 147 (1975).

005.026 **Eric Mervyn Lindsay.** S. Grew.
Irish Astron. Journ., Vol. 12, 148 - 150 (1975).

005.027 **Eric Mervyn Lindsay: list of publications.**
Irish Astron. Journ., Vol. 12, 151 - 153 (1975).

005.028 **Memories of work and play.** E. Öpik.

Irish Astron. Journ., Vol. 12, 154 - 155 (1975).
Concerning E. M. Lindsay.

005.029 **Eric Mervyn Lindsay – some recollections.**
P. M. Millman.
Irish Astron. Journ., Vol. 12, 155 - 157 (1975).

005.030 **Outlines of the history of astronomy at the Petersburg – Leningrad University. V. K. Vishnevskij, S. I. Zelenoj, A. N. Savich.** P. M. Gorshkov.
Trudy Astron. Obs., *Leningrad*, Vol. 32 (= Uchenye Zapiski Leningr. Un-ta, No. 385 = Seriya matem. nauk, vyp. (No.) 52), p. 166 - 184 (1976). In Russian.
The history of life and scientific activity of the professors of Petersburg University V. K. Vishnevskij (1781 – 1855), S. I. Zelenoj (1812 – 1892) and A. N. Savich (1810 – 1883) is described.

006 Personal Notes

M. K. V. Bappu received the Shanti Swarup Bhatnagar Prize of 1970 for Physical Science.
Bull. Astron. Soc. India, Vol. 3, 39 (1975).

G. A. Bazilevskaya received the Lenin prize.
Nature, Vol. 261, 175 (1976).

A. Beer, 75th birthday. W. Hartner.
SuW, 15. Jahrgang, p. 117 (1976).

A. N. Charakhch'yan received the Lenin prize.
Nature, Vol. 261, 175 (1976).

T. N. Charakhch'yan received the Lenin prize.
Nature, Vol. 261, 175 (1976).

Ľ. Pajdušáková, 60th birthday. M. Bélik.
Kozmos, Vol. 7, 80 - 81 (1976). In Slovak.

Ľ. Pajdušáková, 60th birthday.
Říše hvězd, Vol. 57, 115 - 116 (1976). In Czech.

I. I. Shapiro received the Albert A. Michelson medal.
Phys. Today, Vol. 29, No. 1, p. 81 (1976).

L. Spitzer, Jr. received the Maxwell Plasma prize.
Phys. Today, Vol. 29, No. 2, p. 63 - 64 (1976).

Yu. I. Stozhkov received the Lenin prize.
Nature, Vol. 261, 175 (1976).

G. Swarup received the Shanti Swarup Bhatnagar Prize of 1972 for Engineering Sciences.
Bull. Astron. Soc. India, Vol. 3, 39 (1975).

H. C. Urey received the V. M. Goldschmidt Medal.
G. J. Wasserburg.
Geochim. Cosmochim. Acta, Vol. 40, 569 - 570 (1976).

007 Obituaries

A. A. **Aristarkhov**, 1947, July 25 - 1975, March 9.
Astron. Tsirk., No. 895, p. 6 - 8 (1975). In Russian.

P. M. S. **Blackett**, 1897 November 18 - 1974
July 13.   B. Lovell.
Quarterly Journ. Roy. Astron. Soc., Vol. 17, 68 - 79 (1976).

P. **Bok** died 1975 November 19.
Sky Telescope, Vol. 51, 25 (1976).

W. F. **Bushell**, 1885 April 18 - 1974 November 21.
D. W. Dewhirst.
Quarterly Journ. Roy. Astron. Soc., Vol. 17, 189 - 190 (1976).

G. **Chebotarev**, 1913, August 1 - 1975, August 4.
M. Dīriķis.
Zvaigžņotā debess, 1976. gada pavasaris, p. 52 - 53. In Latvian.

P. **Collinder** died 1975 December 6.   P. Nilson.
Astron. Tidssk., Årg. 9, 95 - 96 (1976).

L. **Detre**, 1906 April 18 - 1974 October 15.
B. Szeidl.
Mitt. Astron. Ges., No. 38, p. 7 - 9 (1976).

G. **Florsch** died.   E. Schweitzer.
A.F.O.E.V. Bull., Tome 10, 27 (1976).

B. J. **Harris** died 1974 December 23.
I. Nikoloff.
Inform. Bull. Southern Hemisph., No. 26, p. 31 - 32 (1975).

W. **Heisenberg** died February 1, 1976.
E. Teller.
Nature, Vol. 260, 657 - 658 (1976).

W. K. **Heisenberg**, 1901 - 1976 February 1.
E. P. Wigner.
Phys. Today, Vol. 29, No. 4, p. 86 - 87 (1976).

W. **Heisenberg** died 1976 February 1.
H.-P. Dürr.
Zeitschr. Naturforschung, Band 31a, 507 - 516 (1976).

J. **Hopmann**, 22 December 1890 - 11 October 1975.
K. Ferrari d'Occhieppo.
Sternenbote, 19. Jahrgang, p. 22 - 30 (1976).

J. **Hopmann** died 1975 October 11.
SuW, 15. Jahrgang, p. 21 (1976).

R. M. G. **Inglis**, 1900 - 1975 June 25.   P. Moore.
Journ. British Astron. Ass., Vol. 86, 148 (1976).

N. S. **Kalikhevich**, 1932, February 23 - 1975, March 9.
Astron. Tsirk., No. 895, p. 6 - 8 (1975). In Russian.

H. **Kienle**, 1895 October 22 – 1975 February 15.
J. Wempe.
Astron. Nachr., Vol. 297, 99 - 105 (1976).

H. **Kienle**, 1895 October 22 - 1975 February 15.
O. Heckmann.
Mitt. Astron. Ges., No. 38, p. 9 - 11 (1976).

K.-O. **Kiepenheuer**, 1910 November 10 - 1975
May 23.   W. Mattig.

Mitt. Astron. Ges., No. 38, p. 11 - 13 (1976).

R. V. **Kunitskij**, 1890 - 1975, October 31.
Zemlya i Vselennaya, 1976, No. 1, p. 55. In Russian.

R. **Minkowski**, 1895 May 28 - 1976 January 4.
J. G. Phillips.
Mercury (Journ. Astron. Soc. Pacific), Vol. 5, No. 1, p. 2 (1976).

R. **Minkowski** died 1976 January 4.
I. R. King.
Nature, Vol. 260, 377 (1976).

R. L. **Minkowski** died 1976 January 4.
L. V. Kuhi.
Phys. Today, Vol. 29, No. 3, p. 78, 80 (1976).

R. **Minkowski**, 1895 - 1976 January 4.
Sky Telescope, Vol. 51, 147, 158 (1976).

O. **Nögel**, 1907 - 1976 January 20.   G. D. Roth.
SuW, 15. Jahrgang, p. 98 (1976).

C. P. **Olivier**, 1884 - 1975 August 14.
Strolling Astronomer, Vol. 25, 259 (1976).

S. B. **Pikel'ner**, 1921, February 7 - 1975, November 19.
Astron. Tsirk., No. 896, p. 7 - 9 (1975). In Russian.

S. B. **Pikel'ner**, 1921 February 7 - 1975 November 19.
Astron. Zhurn. Akad. Nauk SSSR, Vol. 53, 233 - 235 (1976).
In Russian. English translation in Soviet Astron., Vol. 20, No. 1.

S. B. **Pikel'ner** died 1975, November 19.
Z. Kopal, L. M. Ozernoy.
Astrophys. Space Sci., Vol. 40, 241 - 242 (1976).

S. **Pikel'ner** died 1975 November 19.
I. S. Shklovskij, S. A. Kaplan.
Phys. Today, Vol. 29, No. 2, p. 65 (1976).

S. B. **Pikel'ner** died 1975 November 19.
Sky Telescope, Vol. 51, 98 (1976).

S. B. **Pikel'ner**, 1921 February 7 - 1975 November 19.
Zemlya i Vselennaya, 1976, No. 2, p. 71 - 72. In Russian.

J. **Proudman** died 1976 June 28.
G. E. R. Deacon.
Quarterly Journ. Roy. Astron. Soc., Vol. 17, 187 - 188 (1976).

R. O. **Redman**, 1905 - 1975 March 6.
C. S. Beals.
Journ. Roy. Astron. Soc. Canada, Vol. 70, 34 - 39 (1976).

R. O. **Redman**, 1905 July 17 - 1975 March 6.
D. E. Blackwell, D. W. Dewhirst.
Quarterly Journ. Roy. Astron. Soc., Vol. 17, 80 - 86 (1976).

E. W. **Salpeter** died 1976 January 6.
P. J. Treanor.
Specola Vaticana, Annual Rep. 1975, p. 3 - 4 (1976).

W. M. Smart died 1975 September 17.
D. H. Sadler.
Journ. Navigation, Vol. 29, 104 - 105 (1976).

J. Verbaandert, 1901 December 26 - 1974 September 4.   P. Melchior.
Ciel et Terre, Vol. 92, 1 - 6 (1976).

M. Waitz, 1903 - 1975.   B. Morando.
L'Astronomie, Vol. 90, 2 (1976).

R. Wildt died 1976 January 9.

Phys. Today, Vol. 29, No. 4, p. 89 (1976).

R. Wildt, 1905 June 25 - 1976 January 9.
Sky Telescope, Vol. 51, 156 (1976).

K. Wurm, 1899 July 21 - 1975 February 16.
V. Vanýsek.
Mitt. Astron. Ges., No. 38, p. 14 - 16 (1976).

V. Zonns, 1905, Nov. 14 - 1975, February 28.
J. Francmanis.
Zvaigžņotā debess, 1976. gada pavasaris, p. 50 - 51. In Latvian.

008 Observatories, Institutes

Reports, communications and publications of observatories and astronomical institutes are recorded in this section; included are numbered series of reprints. Whenever possible, the numbers of the abstracts referring to the publications are given. Observatories and institutes are listed in alphabetical order of their towns. In some cases observatory publications do not give the name of the town; the following list which gives names and towns of some institutions may serve as an aid in such cases.

Aarne Karjalainen Observatory	**Oulu,** Finland
Algonquin Radio Observatory	**Lake Traverse,** Ontario, Canada
Allegheny Observatory	**Pittsburgh,** Pennsylvania, USA
Archenhold-Sternwarte	**Berlin-Treptow,** Germany
Argentine Radioastronomy Institute	**Pereyra Iraola,** Argentina
Arthur J. Dyer Observatory	**Nashville,** Tennessee, USA
Astronomical Latitude Station, Polish Academy of Sciences	**Borowiec,** Poland
Bell Telephone Laboratories	**Holmdel,** New Jersey, USA
Bosscha Observatory	**Lembang,** Indonesia
Boyden Observatory	**Bloemfontein,** South Africa
Bureau International de l'Heure	**Paris,** France
Cajigal Observatory	**Caracas,** Venezuela
California Institute of Technology	**Pasadena,** California, USA
Carter Observatory	**Wellington,** New Zealand
Catalina Station	**Tucson,** Arizona, USA
Cavendish Laboratory	**Cambridge,** England
Cerro Tololo Interamerican Observatory	**La Serena,** Chile
Ceskoslovenská Akademie Ved Astronomický Ustav	**Praha,** Czechoslovakia
Chamberlin Observatory, University of Denver	**Denver,** Colorado, USA
Columbia University, Department of Astronomy	**New York,** New York, USA
Commonwealth Observatory	**Canberra,** Australia
Cornell University, Center for Radiophysics and Space Research	**Ithaca,** New York, USA
Corralitos Observatory	**Las Cruces,** New Mexico, USA
Crawford Hill Laboratory	**Holmdel,** New Jersey, USA
David Dunlap Observatory, University of Toronto	**Richmond Hill,** Ontario, Canada
Dearborn Observatory	**Evanston,** Illinois, USA
Department of Astronomy and Observatory, Univ. California	**Los Angeles,** California, USA
Department of Astronomy University of Texas	**Austin,** Texas, USA
Division Radiophysics, C.S.I.R.O. University Grounds	**Sydney,** Australia
Dominion Astrophysical Observatory	**Victoria,** B.C., Canada
Dominion Observatory	**Ottawa,** Ontario, Canada
Dominion Radio Astrophysical Observatory	**Penticton,** B. C., Canada,
Dudley Observatory	**Albany,** New York, USA
Dunsink Observatory	**Dublin,** Ireland
Dyer Observatory, Vanderbilt University	**Nashville,** Tennessee, USA
Engelhardt Observatory	**Kazan,** USSR
Erwin W. Fick Observatory, Iowa State University	**Ames,** Iowa, USA

European Southern Observatory	**La Silla,** Chile
Felix Aguilar Observatory	**San Juan,** Argentina
Fernbank Observatory	**Atlanta,** Georgia, USA
Five College Observatories	**Amherst,** Massachusetts, USA
Florida State University Radio Observatory	**Tallahassee,** Florida, USA
Flower and Cook Observatories, University of Pennsylvania	**Philadelphia,** Pennsylvania, USA
Fraunhofer Institut	**Freiburg,** F.R. Germany
George R. Wallace Jr. Astrophysical Observatory	**Cambridge,** Massachusetts, USA
Georgetown Observatory	**Washington,** D.C., USA
Glavnaya Astronomicheskaya Observatoriya AN SSSR	**Pulkovo,** USSR
Goddard Space Flight Center	**Greenbelt,** Maryland, USA
Goethe Link Observatory, Indiana University	**Bloomington,** Indiana, USA
"Guido Horn d'Arturo" Observatory	**Bologna,** Italy
Hale Observatories	**Pasadena,** California, USA
Harvard College Observatory	**Cambridge,** Massachusetts, USA
Harvard Radio Astronomy Station	**Cambridge,** Massachusetts, USA
Haute Provence Observatory	**Saint Michel,** France
Haystack Observatory	**Westford,** Massachusetts, USA
Heinrich-Hertz-Institut	**Berlin-Adlershof,** Germany
Herzberg Institute of Astrophysics	**Victoria,** B. C., Canada
High Altitude Observatory, University of Colorado	**Boulder,** Colorado, USA
Hopkins Observatory	**Williamstown,** Massachusetts, USA
Horn d'Arturo Observatory	**Bologna,** Italy
IBM Thomas J. Watson Research Center	**Yorktown Heights,** New York, USA
Institute for Astronomy, University of Hawaii	**Honolulu,** Hawaii, USA
Institute for Theoretical Astronomy (Institut Teoreticheskoj Astronomii)	**Leningrad,** USSR
Institute of Astronomy and Space Science, University of British Columbia	**Vancouver,** B. C., Canada
Institute of Theoretical Astrophysics, Blindern	**Oslo,** Norway
Instituto Argentino de Radioastronomia	**Pereyra Iraola,** Argentina
Instituto Venezolano de Astronomia	**Merida,** Venezuela
Instituto y Observatorio de Marina	**San Fernando (Cádiz),** Spain
Inter-American Observatory	**Cerro Tololo,** La Serena, Chile
International Latitude Observatory	**Mizusawa,** Japan
Instituto de Astronomía y Física del Espacio (IAFE)	**Buenos Aires,** Argentina
Joint Institute for Laboratory Astrophysics (JILA)	**Boulder,** Colorado, USA
Kandilli Observatory	**Istanbul,** Turkey
Kansas University Observatory	**Lawrence,** Kansas, USA

Kapteyn Astronomical Laboratory	**Groningen**, Netherlands
Karl-Schwarzschild-Observatorium	**Tautenburg**, German Democratic Republic
Kenneth Mees Observatory	**Rochester**, New York, USA
Kitt Peak National Observatory	**Tucson**, Arizona, USA
Kwasan Observatory	**Kyoto**, Japan
Lamont-Hussey Observatory	**Bloemfontein**, South Africa
Lawrence Livermore Laboratory, University of California	**Livermore**, California, USA
Leander McCormick Observatory, University of Virginia	**Charlottesville**, Virginia, USA
Lee Observatory	**Beirut**, Lebanon
Leopold-Figl-Observatorium	**Wien**, Austria
Leuschner Observatory	**Berkeley**, California, USA
Lick Observatory	**Santa Cruz**, (Mount Hamilton), California, USA
Lindheimer Astronomical Research Center	**Evanston**, Illinois, USA
Lockheed Palo Alto Research Laboratory	**Palo Alto**, California, USA
Lockheed Solar Observatory	**Saugus**, California, USA
Lohrmann-Observatorium der Technischen Universität Dresden	**Dresden**, German Democratic Republic
Louisiana State University Observatory	**Baton Rouge**, Louisiana, USA
Lowell Observatory	**Flagstaff**, Arizona, USA
Lunar and Planetary Laboratory	**Tucson**, Arizona, USA
Max-Planck-Institut für Astronomie	**Heidelberg**, F. R. Germany
Max-Planck-Institut für Physik und Astrophysik	**München**, F. R. Germany
Max-Planck-Institut für Radioastronomie	**Bonn**, F. R. Germany
McDonald Observatory	**Fort Davis**, Texas, USA
McGraw-Hill Observatory	**Kitt Peak**, Arizona, USA
McMath Hulbert Observatory	**Pontiac**, Michigan, USA
C.E.K. Mees Observatory, University of Rochester	**Rochester**, New York, USA
Michigan State University Observatory	**East Lansing**, Michigan, USA
Molonglo Radio Observatory, University of Sydney	**Sydney**, Australia
Monterey Institute for Research in Astronomy	**Carmel Valley**, California, USA
Mount Cuba Observatory	**Wilmington**, Delaware, USA
Mount John Observatory	**Lake Tekapo**, New Zealand
Mount Palomar Observatory	**Pasadena**, California, USA
Mount Wilson Observatory	**Pasadena**, California, USA
Mullard Radio Astronomy Observatory	**Cambridge**, England
Narrabri Observatory, University of Sydney	**Sydney**, Australia
National Bureau of Standards	**Washington**, D.C., USA
National Observatory, USA	**Kitt Peak**, Arizona, USA
National Research Council of Canada	**Ottawa**, Ontario, Canada
National Radio Astronomy Observatory	**Charlottesville**, Virginia, USA **Green Bank**, West Virginia, USA **Socorro**, New Mexico, USA **Tucson**, Arizona, USA
New Mexico State University Observatory	**Las Cruces**, New Mexico, USA
Nicholas Copernicus Observatory and Planetarium	**Brno**, Czechoslovakia
Nizamiah Observatory	**Hyderabad**, India
Nuffield Radio Astronomy Laboratories, Jodrell Bank University of Manchester	**Manchester**, England
Observatoire Royal de Belgique	**Uccle**, Belgium
Observatories of the University of Western Ontario	**London**, Canada
Observatório Astronômico do Instituto de Física da Universidade Federal do Rio Grande do Sul	**Porto Alegre**, Rio Grande do Sul, Brazil
Observatorio de Cartuja	**Granada**, Spain
Observatorio del Ebro	**Tortosa**, Spain
Observatorio Fabra	**Barcelona**, Spain
Observatorio Nacional	**Rio de Janeiro**, Brazil
Observatorio Nacional de Física Cósmica	**San Miguel**, Argentina
Observatory, University of Michigan	**Ann Arbor**, Michigan, USA
Ohio State University Radio Observatory	**Columbus**, Ohio, USA
Ole Roemer-Observatoriet	**Aarhus**, Denmark
Onsala Space Observatory	**Gothenburg**, Sweden
Owens Valley Radio Observatory	**Big Pine**, California, USA
Perkins Observatory, Ohio State and Wesleyan Universities	**Delaware**, Ohio, USA
Purple Mountain Observatory	**Nanking**, China
Radcliffe Observatory	**Pretoria**, South Africa
Remeis-Sternwarte	**Bamberg**, F. R. Germany
Ritter Astrophysical Research Center of the University of Toledo	**Toledo**, Ohio, USA
Rosemary Hill Observatory	**Gainesville**, Florida, USA
Royal Radar Establishment, Radio Astronomy Division	**Malvern**, England
Sagamore Hill Radio Observatory	**Bedford**, Massachusetts, USA
San Fernando Observatory	**El Segundo**, California, USA
Smithsonian Astrophysical Observatory	**Cambridge**, Massachusetts, USA
Sonnenobservatorium Kanzelhöhe	**Graz**, Austria
South African Astronomical Observatory	**Cape Town**, South Africa
Specola Astronomica Vaticana	**Castel Gandolfo**, Italy
Specola di Padova	**Asiago**, Italy
Sproul Observatory	**Swarthmore**, Pennsylvania, USA
Stellar Data Center	**Strasbourg**, France
Sternberg Observatory	**Moskva**, USSR
Steward Observatory, University of Arizona	**Tucson**, Arizona, USA
W. Struve Tartu Astrophysical Observatory	**Tartu**, USSR
United States Naval Observatory	**Washington**, D. C., USA
University of California	**Berkeley**, California, USA
University of Florida Observatories	**Gainesville**, Florida, USA
University of Florida, Radio Observatory	**Gainesville**, Florida, USA
University of Hawaii	**Honolulu**, Hawaii, USA
University of Illinois Observatory	**Urbana**, Illinois, USA
University of Kansas Observatory	**Lawrence**, Kansas, USA
University of Maryland	**College Park**, Maryland, USA
University of Michigan Observatories	**Ann Arbor**, Michigan, USA
University of Minnesota	**Minneapolis**, Minnesota, USA

University of South Florida
Observatory — **Tampa**, Florida, USA
University of Washington,
Astronomy Department — **Seattle**, Washington, USA
Uttar Pradesh State Observatory — **Naini Tal**, India
Van Vleck Observatory — **Middletown**, Connecticut, USA
Vatican Observatory — **Castel Gandolfo**, Italy
Venezuelan Astronomical
Institute — **Merida**, Venezuela
Wallace Astrophysical
Observatory — **Cambridge**, Massachusetts, USA

Warner and Swasey Observatory — **Cleveland**, Ohio, USA
Washburn Observatory,
University of Wisconsin — **Madison**, Wisconsin, USA
West Melton Observatory — **Christchurch**, New Zealand
Wilhelm-Förster Sternwarte — **Berlin**, Germany
Yale University Observatory — **New Haven**, Connecticut, USA
Yerkes Observatory — **Williams Bay**, Wisconsin, USA
Zentralinstitut für Astrophysik
Sternwarte Babelsberg, (Fach-
bereich Kosmische Physik) — **Potsdam-Babelsberg**, German Democratic Republic

008.001 **Abastumani**

Abastumanskaya Astrofizicheskaya Observatoriya, Gora Kanobili, Byulleten', No. 47 (M. V. Dolidze, 17.114.045; M. V. Dolidze, 17.114.046; M. V. Dolidze, 17.122.112; M. V. Dolidze, 17.114.047).

008.002 **Alger**

Université d'Alger. Annales de l'Observatoire Astronomique d'Alger, Tome 4, Fasc. 2 (17.041.024, 17.031. 261, 17.158.127, 17.158.128, 17.158.316), 3 (17.103.102, 17.103.142).

008.003 **Alma-Ata**

L'Observatoire d'Alma-Ata. J. Heidmann. L'Astronomie, Vol. 90, 214 - 216 (1976).

008.004 **Ames, Iowa**

Erwin W. Fick Observatory, Iowa State University, Ames, Iowa. — Observatory report. W. I. Beavers. Bull. American Astron. Soc., Vol. 8, 80 - 81 (1976).

008.005 **Amherst, Mass.**

Five College Astronomy Department; Amherst College, Amherst, Massachusetts; Hampshire College, Amherst, Massachusetts; Mount Holyoke College, South Hadley, Massachusetts; Smith College, Northampton, Massachusetts; University of Massachusetts, Amherst, Massachusetts. — Observatory reports. T. Arny. Bull. American Astron. Soc., Vol. 8, 81 - 85 (1976).

008.006 **Ann Arbor, Mich.**

University of Michigan, Ann Arbor, Michigan. — Observatory report. W. A. Hiltner. Bull. American Astron. Soc., Vol. 8, 200 - 202 (1976).

008.007 **Arecibo, Puerto Rico**

Arecibo Observatory, Arecibo, Puerto Rico. — Observatory report. H. D. Craft Jr. Bull. American Astron. Soc., Vol. 8, 8 - 11 (1976).

008.008 **Armagh**

Armagh Astronomy Centre. T. Murtagh. Journ. British Astron. Ass., Vol. 86, 312 - 315 (1976).

008.009 **Athens**

Astronomical Institute, National Observatory of Athens. — Annual report 1975. G. Contopoulos. Annual Rep. Astron. Inst. Greece 1975, p. 3 - 6 (1976).

The new Greek 48-inch Cassegrain-coudé telescope. See Abstr. 032.006.

Memoirs of the National Observatory of Athens, Series I. Astronomy, No. 18 (17.079.102).

Department of Astronomy, University of Athens. Annual report 1975. G. Contopoulos. Annual Rep. Astron. Inst. Greece 1975, p. 7 - 11 (1976).

Chair of Astrophysics, University of Athens. Annual report 1975. S. Svolopoulos. Annual Rep. Astron. Inst. Greece 1975, p. 28 (1976).

Department of Astronomy, Technical University of Athens. — Annual report 1975. J. Argyrakos. Annual Rep. Astron. Inst. Greece 1975, p. 16 (1976).

Research Center for Astronomy and Applied Mathematics, Academy of Athens. — Annual report 1975. C. Macris. Annual Rep. Astron. Inst. Greece 1975, p. 17 - 20 (1976).

Research Center for Astronomy and Applied Mathematics, Academy of Athens, Contributions Series I (Astronomy), Nos. 33 (10.085.011), 34 (10.097.086), 35 (10.072.052), 39 (17.085.019), 40 (17.073.079), 41 (17.085. 020), 42 (17.073.080), 44 (17.080.055).

008.010 **Atlanta, Ga.**

Fernbank Observatory, Fernbank Science Center, Atlanta, Georgia. — Observatory report. R. R. Hayward. Bull. American Astron. Soc., Vol. 8, 79 (1976).

008.011 **Baton Rouge, La.**

Louisiana State University Observatory, Baton Rouge, Louisiana. — Observatory report. A. U. Landolt. Bull. American Astron. Soc., Vol. 8, 183 - 185 (1976).

008.012 Bedford, Mass.

Sagamore Hill Radio Observatory, Air Force Cambridge Research Laboratories, Hanscom AFB, Massachusetts. — Observatory report.   D. A. Guidice. Bull. American Astron. Soc., Vol. 8, 249 - 251 (1976).

008.013 Berkeley, Calif.

University of California, Berkeley, Los Angeles, San Diego, and Lick Observatories. I. Berkeley Campus. — Observatory report.   L. V. Kuhi. Bull. American Astron. Soc., Vol. 8, 27 - 36 (1976).

008.014 Berlin

Veröffentlichungen der Wilhelm-Foerster-Sternwarte Berlin, Nos. 40, 41 (17.153.025), 42 (14.123.008).

008.015 Berlin-Adlershof

Heinrich-Hertz-Institut, Solare Beobachtungsergebnisse. Akademie der Wissenschaften der DDR, Zentralinstitut für Solar-Terrestrische Physik (Heinrich-Hertz-Institut), Berlin-Adlershof. HHI Solar Data, Vol. 26, 1975 October - December; Vol. 27, 1976 January - April (17.075.006).

008.016 Berlin-Treptow

Achtzig Jahre Archenhold-Sternwarte. Die Chronik des Riesenfernrohrs von 1896.   See Abstr. 032.025.

Blick in das Weltall, Archenhold-Sternwarte Berlin-Treptow. Astronomische Veranstaltungen und Mitteilungen für Sternfreunde, 24. Jahrgang, Nos. 1 - 6 (1976).

Mitteilungen der Archenhold-Sternwarte Berlin-Treptow Nos. 99 (14.004.082), 112 (17.005.007), 113 (17.004.035), 114 (14.015.027), 126 (17.015.019).

Archenhold-Sternwarte Berlin-Treptow, Sonderdruck No. 22 (17.032.025).

Archenhold-Sternwarte, Berlin-Treptow, Vorträge und Schriften, No. 51 (17.005.006).

008.017 Bloemfontein

Boyden Observatory. Inform. Bull. Southern Hemisph., No. 26, p. 20 - 21 (1975). Current research report.

University of the Orange Free State, Boyden Observatory, Department of Astronomy.   A. H. Jarrett. Monthly Notes Astron. Soc. Southern Africa, Vol. 35. 26 - 27 (1976). — Report for 1975.

008.018 Bloomington, Ind.

Goethe Link Observatory, Indiana University, Bloomington, Indiana. — Observatory report. F. K. Edmondson. Bull. American Astron. Soc., Vol. 8, 169 - 171 (1976).

008.019 Bologna

Otto anni di lavoro all'Osservatorio "Guido Horn d'Arturo".   A. Maitan. Giron. A.A.B., Anno 11, No. 41, p. 7 - 8 (1976).

008.020 Bonn

Die Universitäts-Sternwarte in Bonn (1839–1845). P. Müller. SuW, 15. Jahrgang, p. 81 - 85 (1976).

008.021 Borowiec

Polish Academy of Sciences, Astronomical Latitude Observatory, Borowiec—Poland, Circular Nos. 136 - 137 (17.044.014).

008.022 Boston, Mass.

Boston University, Department of Astronomy and the Judson B. Coit Observatory, Boston, Massachusetts. — Observatory report.   M. D. Papagiannis. Bull. American Astron. Soc., Vol. 8, 20 - 22 (1976).

008.023 Boulder, Colo.

Joint Institute for Laboratory Astrophysics of the National Bureau of Standards and the University of Colorado, Boulder, Colorado. — Observatory report. R. H. Garstang. Bull. American Astron. Soc., Vol. 8, 118 - 128 (1976).

008.024 Brno

Contributions of the Nicholas Copernicus Observatory and Planetarium in Brno, Nos. 19 (17.104.058), 20 (17.121.102, 17.121.103, 17.121.104).

008.025 Buenos Aires

Instituto de Astronomía y Física del Espacio (Institute of Astronomy and Space Physics). Inform. Bull. Southern Hemisph., No. 26, p. 1 (1975). Current research report.

Instituto de Astronomía y Física del Espacio (Institute of Astronomy and Space Physics). Inform. Bull. Southern Hemisph., No. 27, p. 2 (1975). Current research report.

008.026 Byurakan

Byurakan Astrophysical Observatory, Armenia, USSR, Reprints, Nos. 163 (V. A. Ambartsumyan, H. C. Arp, A. A. Hoag, L. V. Mirzoyan, 17.158.099), 164 (K. A. Saakyan, Eh. E. Khachikyan, 17.158.100), 165 (J. Heidmann, A. T. Kalloglyan, 17.158.102), 166 (R. M. Muradyan, 17.158.103), 167 (R. A. Vardanyan, 17.141.628), 168 (B. A. Vorontsov-Vel'yaminov, 17.158.104; K. A. Saakyan, 17.122.063; F. Börngen, V. E. Karachentseva, I. P. Kostyuk, 17.158.105; S. G. Iskudaryan, 17.158.106; A. M. Ehjgenson, N. N. Samus', 17.154.022).

008.027 Cambridge, England

University of Cambridge, Institute of Astronomy. D. Lynden-Bell. Quarterly Journ. Roy. Astron. Soc., Vol. 17, 43 - 57 (1976). Report for the year ending 1975 August 31.

008.028 Cambridge, Mass.

The George R. Wallace Jr. Astrophysical Observatory, Massachusetts Institute of Technology, Cambridge, Massachusetts. – Observatory report. T. B. McCord. Bull. American Astron. Soc., Vol. 8, 267 - 268 (1976).

Smithsonian Institution. Astrophysical Observatory. Research in Space Science. SAO Special Reports, Nos. 370 (17.122.134), 372 (17.081.033), 373 (17.052.031).

Center for Astrophysics, Harvard College Observatory, and Smithsonian Astrophysical Observatory, Cambridge, Massachusetts. – Observatory reports. G. B. Field. Bull. American Astron. Soc., Vol. 8, 91 - 99 (1976).

Center for Astrophysics, Preprint Series, Nos. 443 (17.044.018), 493 (17.042.077).

008.029 Cape Town

University of Cape Town: Department of Astronomy (*B. Warner*), **Department of Applied Mathematics** (*G. F. R. Ellis*). Monthly Notes Astron. Soc. Southern Africa, Vol. 35, 22 - 26 (1976). – Report for 1975.

008.030 Carloforte

Rapporto annuale per l'anno 1974. E. Proverbio. Pubbl. Stazione Astron. Internazionale Latitudine, Carloforte–Cagliari, Nuova Ser., No. 52, 19 pp. (1975).

Circolari della Stazione Astronomica Internazionale di Latitudine, Carloforte–Cagliari, Serie A (3), No. 11 (17.044.015).

Circolari della Stazione Astronomica Internazionale di Latitudine, Carloforte–Cagliari, Serie B (7), No. 9 (17.045.007), B (8), No. 10 (17.045.008).

Pubblicazioni della Stazione Astronomica Internazionale di Latitudine, Carloforte–Cagliari, Nuova Serie, Nos.

35 (17.102.031), 36 (17.044.009), 37 (13.031.256), 40 (14.035.005), 44 (14.035.006), 45 (14.045.021), 46 (13.044.016), 48 (17.045.001), 49 (14.041.011), 51 (17.009.021), 52 (17.008.030).

008.031 Carmel Valley, Calif.

Monterey Institute for Research in Astronomy, Carmel Valley, California. – Observatory report. W. B. Weaver. Bull. American Astron. Soc., Vol. 8, 209 (1976).

008.032 Castel Gandolfo

Specola Vaticana. Annual report 1975: Report of the Astronomical Observatory; Report of the Astrophysical Laboratory. P. J. Treanor, J. Junkes. Printed in Vatican City, 12 pp. (1976).

Specola Vaticana, *Castel Gandolfo,* **Comunicazione,** Nos. 64 (12.126.034), 65 (12.034.105), 66 (13.113.046), 67 (14.114.078).

Vatican Observatory Publications, Specola Vaticana, Città del Vaticano, Vol. 1, Nos. 6 (17.114.051), 7 (17.114.052), 8 (17.114.053).

Ricerche Astronomiche, Specola Vaticana, Città del Vaticano, Vol. 8, No. 28 (17.114.050).

008.033 Catania

Report from the Catania Astrophysical Observatory (1975). List of the papers sent to the publisher during 1975. G. Godoli. Oss. Astrofis. Catania, Pubbl. Nuova Ser. No. 156, 9 pp. (1976).

Osservatorio Astrofisico di Catania, Pubblicazione Nuova Serie Nos. 156 (17.008.033), 157 (17.075.007).

008.034 Charlottesville, Va.

National Radio Astronomy Observatory, Charlottesville, Virginia; Green Bank, West Virginia; Socorro, New Mexico; Tucson, Arizona. – Observatory report covering the period July 1974 – June 1975. D. S. Heeschen, D. E. Hogg. Bull. American Astron. Soc., Vol. 8, 229 - 239 (1976).

008.035 Cleveland, Ohio

Warner and Swasey Observatory, Case Western Reserve University, Cleveland, Ohio. – Observatory report for the period 1 July 1974 – 30 June 1975. W. P. Bidelman. Bull. American Astron. Soc., Vol. 8, 268 - 271 (1976).

Warner and Swasey Observatory, Case Western Reserve University, Reprints, Nos. 267 (13.114.324), 268 (13.114.325), 270 (13.121.063), 271 (13.114.359), 272 (13.114.026), 273 (13.114.022), 274 (13.159.010), 276 (13.141.091), 277 (13.155.028), 278 (13.114.342),

008.036   College Park, Md.

University of Maryland, College Park, Maryland.
Observatory report covering the period 16 August 1974 – 15
August 1975.   E. v. P. Smith.
Bull. American Astron. Soc., Vol. 8, 189 - 195 (1976).

008.037   Columbus, Ohio

Ohio State University Radio Observatory, Columbus,
Ohio. – Observatory report.   J. Kraus.
Bull. American Astron. Soc., Vol. 8, 239 - 240 (1976).

008.038   Copenhagen

Copenhagen University Observatory Reprint Nos.
261 (11.032.021), 262 (14.032.044), 263 (13.032.030), 264
(14.114.013), 265 (14.122.118), 266 (14.113.033), 267
(14.121.042), 268 (14.031.290), 269 (13.121.021), 273
(14.121.079), 274 (14.121.083), 275 (14.121.102), 276
(14.121.082), 277 (14.121.080), 278 (14.121.081), 279
14.121.103), 280 (17.121.003), 281 (17.121.005), 282
(17.121.016), 283 (17.151.020), 284 (17.121.017).

008.039   Córdoba

Observatorio Astronómico (Astronomical Observa-
tory, National University of Córdoba).
Inform. Bull. Southern Hemisph., No. 26, p. 1 - 2 (1975).
Current research report.

008.040   Dresden

Mitteilungen des Lohrmann-Observatoriums der
Technischen Universität Dresden, Nos. 30 (17.031.262),
31 (14.046.052).

Technische Universität Dresden, Lohrmann-Obser-
vatorium, Zirkular, Nos. 72 - 76 (17.045.010).

008.041   Dushanbe

Byulleten' Instituta Astrofiziki, Akademiya Nauk
Tadzhikskoj SSR, Nos. 63 (S. G. Pomagaev, 17.062.031;
N. G. Ptitsyna, 17.151.045; O. P. Vasil'yanovskaya, 17.122.
062; L. N. Rubtsov, B. G. Solovej, 17.104.035; T. K. Kiseleva,
17.113.039), 64 (M. N. Maksumov, 17.151.046; M. N. Maksumov, Yu. N. Mishurov, 17.151.047; M. N. Maksumov, 17.151.
048; M. N. Maksumov, 17.151.049; M. N. Maksumov, 17.151.
050; Eh. Ya.Maldybaeva, N. G. Ptitsyna, 17.151.051), 65
(L. N. Rubtsov, A. E. Epishova, 17.083.048; B. G. Solovej,
A. E. Epishova, 17.083.049; L. N. Rubtsov, B. G. Solovej,
S. P. Rogozhkina, 17.083.050; N. P. Lyakhova, B. G. Solovej,
17.083.051; D. Latipov, L. N. Rubtsov, 17.083.052; D.
Latipov, L. N. Rubtsov, 17.104.036; O. Alimov, D. Latipov,
17.083.053; O. Alimov, D. Latipov, 17.083.054; P. B. Babadzhanov, A. F. Zausaev,17.104.037; U. Shodiev, 17.085.015).

008.042   East Lansing, Mich.

Michigan State University, Department of Astron-
omy and Astrophysics and the Observatory, East Lansing,
Michigan. – Observatory report.   A. P. Linnell.
Bull. American Astron. Soc., Vol. 8, 203 - 205 (1976).

008.043   Edinburgh

Royal Observatory, Edinburgh.   V. C. Reddish.
Quarterly Journ. Roy. Astron. Soc., Vol. 17, 146 - 153 (1976).
Report for the year ending 1975 December 31.

008.044   El Segundo, Calif.

The Aerospace Corporation, El Segundo, California.
– Observatory reports.   G. A. Paulikas.
Bull. American Astron. Soc., Vol. 8, 1 - 5 (1976).
This report reviews the astronomy research carried out at
The Aerospace Corporation during 1975. The report describes
the activities of the San Fernando Observatory, the research in
millimeter-wave radio astronomy, as well as the space astron-
omy research.

008.045   Evanston, Ill.

Lindheimer Astronomical Research Center and
Dearborn Observatory, Evanston, Illinois; Corralitos Observa-
tory, Las Cruces, New Mexico. – Observatory reports.
J. D. R. Bahng.
Bull. American Astron. Soc., Vol. 8, 167 - 169 (1976).

008.046   Flagstaff, Ariz.

Lowell Observatory, Flagstaff, Arizona. – Observatory report.   J. S. Hall.
Bull. American Astron. Soc., Vol. 8, 185 - 189 (1976).

008.047   Gainesville, Fla.

University of Florida Observatories, Department of
Physics and Astronomy, Gainesville, Florida. – Observatory
report.   A. G. Smith.
Bull. American Astron. Soc., Vol. 8, 85 - 89 (1976).

008.048   Gent

Universiteit te Gent, Sterrenkundig Instituut,
Mededelingen, Nos. 38 (17.071.037), 39 (17.141.024).

008.049   Goloseevo

The Main Astronomical Observatory of the
Ukrainian SSR.   A. K. Korol'.
Zemlya i Vselennaya, 1976, No. 1, p. 67 - 72. In Russian.

008.050 Green Bank

National Radio Astronomy Observatory, Charlottesville, Virginia; Green Bank, West Virginia; Socorro, New Mexico; Tucson, Arizona. – Observatory report covering the period July 1974 – June 1975.   D. S. Heeschen, D. E. Hogg.
Bull. American Astron. Soc., Vol. 8, 229 - 239 (1976).

National Radio Astronomy Observatory, *Green Bank,* Reprints, Series A, Nos. 494 (14.033.079), 495 (14.131.026), 496 (14.158.169), 497 (14.141.090), 498 (14.131.515), 499 (14.131.055), 500 (14.121.028), 501 (14.131.052), 502 (14.158.045), 503 (14.131.512), 504 (14.142.060), 505 (14.141.028), 506 (14.131.046), 507 (14.133.021), 508 (14.131.513), 509 (14.141.059), 510 (14.131.086), 511 (14.131.087), 512 (14.132.029), 513 (14.132.030), 514 (14.033.007), 515 (14.141.073), 516 (14.158.127), 517 (14.162.024), 518 (14.022.056), 519 (14.131.109), 520 (14.141.055), 521 (14.131.091), 522 (14.141.074), 523 (14.131.104), 524 (14.141.625), 525 (14.158.146), 526 (14.131.117), 527 (14.131.123), 528 (14.141.143), 529 (14.131.111), 530 (14.158.172), 531 (14.158.190), 532 (14.131.141), 533 (14.131.138), 534 (14.141.128), 535 (14.131.160), 536 (14.131.548), 537 (14.141.129), 538 (14.141.105), 539 (14.131.158), 540 (17.033.030), 541 (17.142.008), 542 (17.141.015), 543 (17.131.020), 544 (17.033.031), 545 (17.141.037), 546 (17.141.014), 547 (17.131.007), 548 (17.158.011), 549 (17.151.010), 550 (17.131.012), 551 (17.160.008), 552 (17.158.018), 553 (17.131.011), 554 (14.141.141), 555 (17.141.021).

National Radio Astronomy Observatory, *Green Bank,* Reprints, Series B, Nos. 461 (14.015.009), 462 (14.158.061), 463 (14.158.065), 464 (13.141.041).

008.051 Greenbelt

Goddard Space Flight Center, Greenbelt, Maryland, GSFC Documents X-660-76-19 (17.143.345), X-660-76-1 (17.142.237), X-660-76-25 (17.106.100), X-660-76-26 (17.106.101), X-660-76-27 (17.143.346), X-660-76-58 (17.073.081), X-661-76-8 (17.142.231), X-661-76-37 (17.142.232), X-661-76-66 (17.142.233), X-661-76-83 (17.160.041), X-900-75-216 (17.081.035), X-921-75-176 (17.081.036), X-921-75-300 (17.081.037), X-921-76-20 (17.081.038).

008.052 Greenwich

Tercentenary of the Royal Greenwich Observatory.
A. V. Douglas.
Journ. Roy. Astron. Soc. Canada, Vol. 70, 31 - 33 (1976).

Tercentenary of the Royal Observatory Greenwich.
D. B. Herrmann.
Sterne, Vol. 52, 92 - 97 (1976).

300 years of the Greenwich Observatory.
L. Zajdler.
Urania Kraków, Vol. 47, 50 - 52 (1976). In Polish.

300 jaar Royal (Greenwich) Observatory.
G. W. E. Beekman.
Zenit, 3e jaargang, p. 74 - 82 (1976).

The Royal Greenwich Observatory 1675 - 1975. Some of its external relations.   See Abstr. 004.017.

008.053 Groningen

Nederlandse Vereniging voor Weer- en Sterrenkunde. **Observations of Variable Stars. Report** (Kapteyn Astronomical Laboratory, Groningen–Netherlands), No. 29 (17.123.032).

008.054 Hamburg

Deutsches Hydrographisches Institut, Hamburg, **Zeit- und Breitendienst,** 1975 April - June (17.044.016).

008.055 Heidelberg

Astronomy and Astrophysics Abstracts, Vol. 14 (17.002.003).

Astronomisches Rechen-Institut in Heidelberg, **Mitteilungen** Serie A, Nos. 96 (17.041.005), 97 (17.041.006), 98 (17.041.009), 99 (17.151.024).

Astronomisches Rechen-Institut in Heidelberg, **Mitteilungen** Serie B, Nos. 56 (14.098.041), 57 (17.044.003).

Max-Planck-Institut für Kernphysik. – Jahresbericht 1975.
[Available from MPI Kernphys., Bibliothek, Postfach 103980, 6900 Heidelberg 1, W-Germany], 201 pp. (1976).

008.056 Holmdel, N.J.

Bell Telephone Laboratories, Crawford Hill Laboratory, Holmdel, New Jersey. – Observatory report.
A. A. Penzias.
Bull. American Astron. Soc., Vol. 8, 19 - 20 (1976).

008.057 Honolulu, Hawaii

University of Hawaii, Institute for Astronomy, Honolulu, Hawaii. – Observatory report.   J. T. Jefferies.
Bull. American Astron. Soc., Vol. 8, 99 - 106 (1976).

Institute for Astronomy, University of Hawaii, Honolulu, Hawaii, **Reprints** Nos. 84 (06.124.001), 85 (06.113.033), 86 (AJB 68.1667), 87 (06.034.025), 88 (07.074.086), 89 (06.073.027), 90 (06.073.028), 91 (06.074.015), 92 (06.074.014), 93 (05.073.073), 94 (05.034.095), 96 (06.074.059), 97 (06.031.037), 99 (07.063.008), 100 (07.063.009), 101 (07.008.069), 102 (06.011.036), 103 (05.093.059), 104 (05.099.063), 105 (07.022.003), 106 (07.074.024), 107 (07.074.025), 108 (07.094.094), 109 (07.114.072), 110 (07.158.111), 111 (07.099.040), 112 (07.071.021), 113 (07.131.137), 114 (08.114.010), 115 (08.100.002), 116 (09.080.044), 117 (08.063.007), 118 (08.063.004), 119 (08.122.024), 120 (08.101.014), 121 (08.131.051), 122 (08.074.012), 123 (08.116.005), 124 (08.116.006), 125 (08.116.015), 126 (08.100.021), 127 (08.113.026), 128 (08.141.066), 129 (08.125.102), 130 (08.101.015), 168 (10.100.020), 169 (10.113.092), 170 (10.116.021), 171 (11.121.001), 172 (11.098.005), 173 (10.099.063), 174 (11.158.006), 175 (11.116.001), 176 (11.114.043), 177 (11.122.023), 178 (10.098.070), 179 (11.155.020), 180 (10.071.080), 181 (10.158.066), 182 (11.093.020), 183 (11.099.006), 184 (11.116.003), 185 (11.117.020), 186 (11.116.008), 188

(06.071.055), 189 (11.114.064), 190 (11.114.043), 191 (11.
080.009), 192 (11.005.002), 193 (11.071.009), 194 (11.091.
004),195 (11.158.103), 196 (11.022.108), 197 (11.082.122),
198 (11.131.211), 199 (11.116.018), 200 (12.091.060), 201
(12.080.061), 202 (11.116.013), 203 (11.119.009), 204 (11.
119.010), 205 (11.119.011), 206 (11.100.208), 207 (11.071.
020), 208 (11.100.011), 209 (11.008.061), 210 (10.155.043),
212 (11.126.008), 213 (11.126.010), 214 (11.132.025), 215
(11.100.013), 216 (11.114.156), 217 (12.158.021), 218 (12.
141.002), 219 (12.098.010), 220 (12.131.004), 221 (11.116.
009), 222 (12.113.030), 223 (12.074.090), 224 (11.080.026),
225 (12.124.103), 226 (12.124.103), 227 (12.114.062), 228
(11.092.034), 229 (12.099.234), 230 (12.072.020), 231 (12.
100.210), 232 (12.073.063), 233 (12.141.620), 234 (12.126.
028), 235 (12.114.107), 236 (12.098.017), 238 (12.082.047),
239 (12.101.010), 240 (12.072.041), 241 (12.103.104), 242
(12.114.147), 243 (13.064.009), 244 (13.142.016), 245(Abstr.
see 11.034.067), 246 (12.114.166), 247 (13.100.205), 248
(12.072.053), 249 (13.142.028), 250 (13.116.004), 251 (13.
008.058), 252 (13.122.089), 253 (13.126.017), 254 (13.114.
355), 255 (13.113.050), 256 (12.114.148), 257 (12.074.090),
258 (13.076.009), 259 (13.098.050), 260 (13.114.350), 261
(14.132.002), 262 (14.122.072), 263 (12.141.002), 264 (13.
141.601), 265 (13.072.063), 268 (13.064.068), 269 (13.126.
008), 270 (13.114.302), 271 (13.113.009), 272 (13.158.035),
273 (13.131.007), 274 (10.071.024), 275 (13.114.030).

008.058 Ioannina

Department of Astronomy, University of Ioannina.
Annual report 1975.   G. Banos.
Annual Rep. Astron. Inst. Greece 1975, p. 25 (1976).

008.059 Iowa City, Iowa

University of Iowa, Iowa City, Iowa. – Observatory
report for the period 1 October 1974 – 30 September 1975.
J. S. Neff.
Bull. American Astron. Soc., Vol. 8, 114 - 118 (1976).

008.060 Ithaca, N.Y.

Cornell University, Center for Radiophysics and
Space Research, Ithaca, New York. – Research report.
P. J. Gierasch.
Bull. American Astron. Soc., Vol. 8, 64 - 69 (1976).

008.061 Kiev

Astrometriya i Astrofizika, *Kiev,* Vyp. (Nos.) 28
(Eh. G. Yanovitskij, 17.003.002), 29 (D. P. Duma, 17.003.
003).

Kievskij ordena Lenina gosudarstvennyj universitet
im. T. G. Shevshenko, Astronomicheskaya Observatoriya.
(Astron. Obs.), Preprint Nos. 7 (P. N. Polupan, 17.073.040),
8 (P. R. Romanchuk, Yu. N. Kudrya, 17.072.022), 9 (V. I.
Efimenko, V. M. Efimenko, V. V. Tel'nyuk-Adamchuk,
17.073.041), 10 (P. R. Romanchuk, 17.072.023), 11 (V. B.
Gumanitskij, V. M. Efimenko, P. R. Romanchuk, A. N. Sergeeva, V. V. Tel' nyuk-Adamchuk, 17.072.024).

008.062 Kitt Peak, Ariz.

Kitt Peak National Observatory, Tucson, Arizona,
Cerro Tololo Inter-American Observatory, La Serena, Chile.

Observatory reports.   L. Goldberg.
Bull. American Astron. Soc., Vol. 8, 129 - 167 (1976).

The new McGraw-Hill Observatory on Kitt Peak.
D. E. Mook.
Sky Telescope, Vol. 51, 23 - 25 (1976).

008.063 Krim

Chronicle.
Izv. Krymskoj Astrofiz. Obs., Vol. 54, 350 - 351 (1976).
In Russian.

Izvestiya Krymskoj Astrofizicheskoj Observatorii,
Akademiya Nauk SSSR, Tom (Vol.) 54 (V. B. Nikonov, 17.
113.033; V. I. Burnashev, V. B. Nikonov, 17.113.034; V. B.
Nikonov, G. A. Terez, 17.114.030; P. P. Petrov, 17.122.044;
P. P. Petrov, 17.122.045; P. F. Chugajnov, 17.122.046; P. F.
Chugajnov, 17.122.047; N. M. Shakovskoj, Yu. S. Efimov, 17.
142.080; N. S. Polosukhina, 17.116.007; T. S. Galkina, 17.
119.012; V. A. Efanov, I. G. Moiseev, N. S. Nesterov, 17.141.
076; V. I. Pronik, 17.158.091; N. B. Grigor'eva, 17.158.092;
V. P. Grinin, 17.064.050; V. A. Kotov, 17.072.035; M. B.
Ogir', L. I. Yurovskaya, 17.080.035; Yu. F. Yurovskij, L. I.
Yurovskaya, O. Alvares, 17.077.025; A. F. Bachurin, A. S.
Dvoryashin, N. N. Eryushev, 17.077.026; Dzh. I. Irgashev, 17.
072.036; A. F. Bachurin, N. N. Eryushev, 17.077.027; V. A.
Slepyan, T. L. Slutskaya, N. N. Stepanyan, 17.072.037; L. S.
Lyubimkov, 17.080.036; V. K. Prokof'ev, 17.031.011; G. M.
Popov, M. B. Popova, 17.031.012; G. M. Popov, 17.031.013;
17.012.009; O. M. Kovrizhnykh, M. I. Kudryavtsev, A. S.
Melioranskij, I. A. Savenko, V. M. Shamolin, L. N. Chupova,
17.076.019; L. S. Bratolyubova-Tsulukidze, M. I. Kudryav-
tsev, A. S. Melioranskij, I. A. Savenko, B. Yu. Yushkov, 17.
142.085; R. A. Sunyaev, 17.142.086; L. M. Ozernoj, 17.151.
067; A. M. Gal'per, V. G. Kirillov-Ugryumov, B. I. Luchkov,
17.142.087; G. E. Kocharov, 17.061.025; P. I. Fomin, 17.143.
026; A. I. Belyaevskij, V. L. Bokov, V. K. Bocharkin, I. F.
Bugakov, Yu. G. Derevitskij, B. A. Dmitriev, G. M. Gorodinskij,
E. M. Kruglov, E. V. Myakinin, G. A. Pyatigorskij, E. I.
Chujkin, 17.143.027; S. A. Volobuev, L. V. Kurnosova, B. I.
Luchkov, L. A. Razorenov, V. I. Ryabenkov, M. I. Fradkin,
17.142.088).

008.064 La Plata

Observatorio Astronómico (Astronomical Observa-
tory, National University of La Plata).
Inform. Bull. Southern Hemisph., No. 26, p. 2 (1975).
Current research report.

Observatorio Astronómico (Astronomical Observa-
tory, National University of La Plata).
Inform. Bull. Southern Hemisph., No. 27, p. 2 - 3 (1975).
Current research report.

008.065 La Serena

Kitt Peak National Observatory, Tucson, Arizona;
Cerro Tololo Inter-American Observatory, La Serena, Chile.
Observatory reports.   L. Goldberg.
Bull. American Astron. Soc., Vol. 8, 129 - 167 (1976).

Observatorio Interamericano (Cerro Tololo Inter-
american Observatory).
Inform. Bull. Southern Hemisph., No. 26, p. 12 - 16 (1975).

Current research report.

Observatorio Interamericano (Cerro Tololo Inter-american Observatory).
Inform. Bull. Southern Hemisph., No. 27, p. 16 - 20 (1975).
Current research report.

Visiting Cerro Tololo.
Sky Telescope, Vol. 51, 231 - 234 (1976).

008.066   La Silla

Observatorio Europeo Austral (European Southern Observatory).
Inform. Bull. Southern Hemisph., No. 26, p. 17 - 18 (1975).
Current research report.

Observatorio Europeo Austral (European Southern Observatory).
Inform. Bull. Southern Hemisph., No. 27, p. 21 - 22 (1975).
Current research report.

The Messenger – El Mensajero.
Edited by European Southern Observatory (ESO), Hamburg, No. 4, 8 pp; No. 5, 12 pp. (1976).

European Southern Observatory, (ESO), Technical Report Nos. 4 (17.032.032), 5 (17.031.409), 7 (17.032.033).

008.067   Las Campanas

Las Campanas Observatory (CARSO), Carnegie Institution of Washington.
Inform. Bull. Southern Hemisph., No. 27, p. 21 (1975).
Current research report.

008.068   Las Cruces

Lindheimer Astronomical Research Center and Dearborn Observatory, Evanston, Illinois; Corralitos Observatory, Las Cruces, New Mexico. – Observatory reports.
J. D. R. Bahng.
Bull. American Astron. Soc., Vol. 8, 167 - 169 (1976).

008.069   Lawrence, Kans.

University of Kansas Observatory, Lawrence, Kansas. – Observatory report.   S. J. Shawl.
Bull. American Astron. Soc., Vol. 8, 128 - 129 (1976).

008.070   Leningrad

Trudy Astronomicheskoj Observatorii, (Transactions of the Astronomical Observatory), *Leningrad*, Vol. 32 (V. V. Ivanov, 17.063.033; V. V. Ivanov, 17.063.034; A. K. Kolesov, 17.063.035; M. K. Babadzhanyants, S. K. Vinokurov, V. A. Hagen-Thorn, E. V. Semenova, 17.141.085; G. V. Khozov, T. N. Khudyakova, S. N. Nikitin, 17.113.038; V. A. Antonov, 17.151.043; A. A. V'yuga, V. S. Kaliberda, I. V. Petrovskaya, 17.151.044; R. A. Lyakh, 17.042.051; M. S. Zverev, 17.041. 016; P. M. Gorshkov, 17.005.025).

008.071   Livermore, Calif.

Lawrence Livermore Laboratory, University of California, Livermore, California. – Observatory report.
C. B. Tarter.
Bull. American Astron. Soc., Vol. 8, 172 - 174 (1976).

008.072   London, England

University College London, Mullard Space Science Laboratory.   R. L. F. Boyd.
Quarterly Journ. Roy. Astron. Soc., Vol. 17, 174 - 186 (1976).
Report for the period 1974 October 1 - 1975 September 30.

008.073   London, Canada

The Observatories of the University of Western Ontario, London, Canada. – Observatory report.
W. Wehlau.
Bull. American Astron. Soc., Vol. 8, 278 - 280 (1976).

University of Western Ontario, London, Ontario.
A. Wehlau.
Journ. Roy. Astron. Soc. Canada, Vol. 70, 144 - 145 (1976).

008.074   Los Alamos, N. Mex.

Los Alamos Scientific Laboratory, University of California, Los Alamos, New Mexico. – Research report.
M. T. Sandford II, A. N. Cox, E. M. Jones.
Bull. American Astron. Soc., Vol. 8, 175 - 183 (1976).

008.075   Los Angeles

University of California, Berkeley, Los Angeles, San Diego, and Lick Observatories. II. Los Angeles Campus. Observatory report.   M. Plavec.
Bull. American Astron. Soc., Vol. 8, 36 - 42 (1976).

University of California, Los Angeles, Astronomical Papers, Vol. 14, Nos. 1 (13.133.012), 2 (13.114.034), 3 (13. 114.021), 4.(11.100.011), 5 (12.112.014), 6 (12.080.034), 7 (12.133.006), 8 (13.159.019), 9 (17.015.021), 10 (13.121. 021), 11 (13.133.011), 12 (13.074.011), 13 (13.132.019), 14 (13.117.001), 15 (14.031.221), 16 (13.133.028), 17 (13. 121.042), 18 (14.131.044), 19 (13.065.025), 20 (14.065. 036), 21 (14.160.031), 22 (17.080.059), 23 (14.158.189), 24 (17.133.011), 25 (14.076.055), 26 (17.071.036), 27 (17. 121.069), 28 (14.031.232), 29 (12.133.025).

008.076   L'vov

Tsirkulyar, Astronomicheskaya Observatoriya, L'vov, No. 49 (I. A. Klimishin, A. F. Novak, 17.064.039; M. B. Girnyak, V. V. Golovatyj, 17.123.004; V. V. Golovatyj, O. S. Yatsyk, 17.131.102; I. V. Shpychka, 17.123.005; I. S. Laba, 17.073.039; T. L. Mandrykina, 17.072.021; E. B. Vovchik, I. I. Vovchik, V. M. Luk'yanets, R. F. Fedoriv, 17.034.017; I. I. Vovchik, D. I. Galych, Yu. V. Fridel', 17.096. 004; M. V. Bratijchuk, Ya. M. Motrunich, 17.054.005).

008.077 **Lyon**

Observatoire de Lyon, Reprint Nos. 1 (03.124.100), 4 (03.112.010), 6 (08.113.055), 6' (06.131.046), 7 (05.155.026), 8 (07.141.015), 9 (07.113.004), 10 (10.155.021), 11 (08.141.039), 12 (07.114.119), 13 (08.133.004), 14 (07.154.009), 15 (07.123.002), 16 (07.124.102), 18 (10.155.021), 19 (10.034.028), 20 (17.034.071), 21 (09.117.007), 22 (10.113.013), 24 (10.123.016), 25 (10.123.017), 26 (10.153.024), 27 (11.153.025), 28 (12.113.028), 29 (11.114.131), 30 (14.158.163), 31 (12.141.606), 32 (11.159.003), 33 (12.133.013), 34 (11.131.504), 35 (12.022.036), 36 (11.124.103), 37 (11.123.040), 38 (12.154.014), 39 (11.122.094), 41 (13.154.017), 42 (14.159.001), 43 (14.122.133), 44 (14.122.181), 45 (13.114.311), 46 (13.122.116), 47 (13.113.021), 48 (14.122.037), 49 (13.158.107), 50 (13.122.024), 52 (13.141.618), 53 (14.158.202), 54 (13.141.619), 55 (13.031.245).

008.078 **Madison, Wis.**

Washburn Observatory, University of Wisconsin, Madison, Wisconsin. – Observatory report.   R. C. Bless. Bull. American Astron. Soc., Vol. 8, 271 - 275 (1976).

008.079 **Madrid**

Boletin Astronómico del Observatorio de Madrid, Instituto Geografico y Catastral, Vol. 9, Nos. 2 (17.141.109, 17.118.019, 17.114.362, 17.098.073, 17.103.102, 17.096.009, 17.042.076), 3 (17.075.008), 4 (17.075.009).

008.080 **Manchester**

Astronomical Contributions from the University of Manchester, Series II, Jodrell Bank Reprints, Nos. 514 (12.141.001), 519 (12.131.033), 520 (13.143.053), 521 (12.141.020), 522 (12.141.049), 523 (12.131.110), 524 (13.131.501), 525 (17.083.076), 526 (13.158.024), 527 (12.141.070), 528 (13.008.079), 529 (12.158.149), 530 (13.141.026), 531 (13.157.008), 532 (13.159.005), 533 (13.131.511), 534 (13.141.349).

008.081 **Merida**

Instituto Venezolano de Astronomía (Venezuelan Astronomical Institute). Inform. Bull. Southern Hemisph., No. 27, p. 23 (1975). Current research report.

008.082 **Middletown, Conn.**

Van Vleck Observatory, Wesleyan University, Middletown, Connecticut. – Observatory report. A. R. Upgren. Bull. American Astron. Soc., Vol. 8, 263 - 264 (1976).

008.083 **Minneapolis, Minn.**

University of Minnesota, Minneapolis, Minnesota.

Observatory report.   E. P. Ney. Bull. American Astron. Soc., Vol. 8, 205 - 208 (1976).

University of Minnesota, Minneapolis, Minnesota, Separate prints (17.112.005, 17.112.006, 17.112.007).

008.084 **Mizusawa**

Monthly Notes of the International Polar Motion Service, 1975 No. 12, 1976 Nos. 1 - 5 (17.045.011).

008.085 **Moskva**

Chronicle. [Dissertations at the Sternberg Astronomical Institute]. Astron. Tsirk., No. 897, p. 7 - 8 (1975). In Russian.

008.086 **Naini Tal**

Uttar Pradesh State Observatory, *Naini Tal,* **Reprints,** Nos. 73 (10.122.128), 74 (10.072.016), 75 (10.122.086), 76 (10.123.015), 77 (10.071.040), 78 (11.115.003), 79 (11.114.065), 80 (12.123.014), 81 (11.071.017), 82 (11.071.016), 83 (11.072.017), 84 (11.122.122), 85 (12.122.051), 86 (12.072.001), 87 (12.073.001), 88 (12.073.002), 89 (12.114.006), 90 (12.121.002), 91 (12.064.016), 93 (13.072.003), 94 (13.121.012), 95 (13.121.013), 96 (13.153.028), 97 (13.122.079), 98 (14.121.001).

008.087 **Nashville, Tenn.**

Dyer Observatory, Vanderbilt University, Nashville, Tennessee. – Observatory report for the period 1 October 1974 – 30 September 1975.   A. M. Heiser. Bull. American Astron. Soc., Vol. 8, 76 - 78 (1976).

008.088 **Neuchâtel**

Rapport d'activité pour l'exercice 1975. J. Bonanomi, G. Fischer. Obs. Cantonal de Neuchâtel, 23 pp. (1976).

008.089 **New Haven, Conn.**

Yale University Observatory, New Haven, Connecticut. – Observatory report.   W. van Altena. Bull. American Astron. Soc., Vol. 8, 280 - 288 (1976).

008.090 **New York. N.Y.**

Columbia University, Department of Astronomy and Department of Physics, New York, New York. – Research report.   L. B. Lucy, N. H. Baker. Bull. American Astron. Soc., Vol. 8, 61 - 64 (1976).

008.091 Oslo

Institute of Theoretical Astrophysics, Blindern—Oslo. **Report**, No. 45 (17.071.039).

Institutt for Teoretisk Astrofysikk, Blindern—Oslo. **Småtrykk**, Nos. 85 (17.066.089), 86 (17.162.078), 87 (13.066.018), 88 (14.162.031), 89 (14.099.071), 90 (17.071.007).

008.092 Ottawa

National Research Council of Canada, Ottawa, Ontario, Canada, NRC Nos. 14270 (12.141.100), 14879 (14.131.511), 15038 (14.022.110).

008.093 Oxford

Universtiy of Oxford, Department of Theoretical Physics (Theoretical Astrophysics Group). Quarterly Journ. Roy. Astron. Soc., Vol. 17, 58 - 60 (1976). Report for period ending December 1974.

008.094 Padova-Asiago

Attrezzature e ricerche all'Osservatorio astrofisico di Padova-Asiago. L. Rosino. Giorn. Astron., Vol. 1, 119 - 133 (1975).

008.095 Palo Alto, Calif.

Lockheed Palo Alto Research Laboratory, Palo Alto, California. − Observatory report. H. M. Johnson. Bull. American Astron. Soc., Vol. 8, 174 - 175 (1976).

008.096 Paris

Bureau International de l'Heure. **Rapport annuel pour 1975.** R. Michard (Editor). Printed by Observatoire de Paris, 5 + A16 + B69 + C17 pp. (1976).

Bureau International de l'Heure, (B.I.H.), **Circular D 110 - D 116** (17.044.020).

008.097 Pasadena, Calif.

Hale Observatories, operated by Carnegie Institution of Washington and California Institute of Technology, Pasadena, California. Annual report of the director, 1974—1975. H. W. Babcock, J. B. Oke. Reprinted from Carnegie Institution, Washington, Year Book, Vol. 74, 307 - 378 (1975).

Jet Propulsion Laboratory, California Institute of Technology, Pasadena, California, **Technical Report 32-1603** (17.091.017).

008.098 Patras

Department of Astronomy, University of Patras. Annual report 1975. B. Barbanis. Ann. Rep. Astron. Inst. Greece 1975, p. 26 - 27 (1976).

008.099 Pereyra Iraola

Instituto Argentino de Radioastronomía (Argentine Radioastronomy Institute). Inform. Bull. Southern Hemisph., No. 27, p. 3 - 4 (1975). Current research report.

008.100 Philadelphia, Penn.

Flower and Cook Observatory, University of Pennsylvania, Philadelphia, Pennsylvania. − Observatory report. B. S. P. Shen. Bull. American Astron. Soc., Vol. 8, 89 - 91 (1976).

008.101 Pittsburgh, Penn.

Allegheny Observatory, University of Pittsburgh, Pittsburgh, Pennsylvania. − Observatory report covering the period 1 July 1974 − 30 June 1975. J. H. Kiewiet de Jonge. Bull. American Astron. Soc., Vol. 8, 5 - 8 (1976).

008.102 Porto Alegre

Observatório Astronômico do Instituto de Física da Universidade Federal do Rio Grande do Sul (Astronomical Observatory, Institute of Physics of the Federal University of Rio Grande do Sul). Inform. Bull. Southern Hemisph., No. 26, p. 8 (1975). Current research report.

Observatório Astronômico do Instituto de Física da Universidade Federal do Rio Grande do Sul (Astronomical Observatory, Institute of Physics of the Federal University of Rio Grande do Sul). Inform. Bull. Southern Hemisph., No. 27, p. 10 (1975). Current research report.

008.103 Potsdam

Mitteilungen des Astrophysikalischen Observatoriums Potsdam, Nos. 158 (04.063.044), 159 (05.063.021), 160 (06.063.003), 161 (06.122.051), 162 (10.063.074), 165 (12.119.015), 166 (12.062.059), 167 (12.062.058), 168 (12.114.043), 169 (12.062.019), 170 (12.065.036), 171 (13.116.002), 172 (12.005.020), 173 (17.063.042), 174 (14.008.081).

008.104 Prague

Académie Tchécoslovaque des Sciences, Institut Astronomique, **Station de l'Heure à Prague**, Série 7, No. 7 (17.044.021).

008.105 Pretoria

National Physical Research Laboratory (C.S.I.R.) Time Standard Section. — Report by J. Hers for the period 1972–1974.
Inform. Bull. Southern Hemisph., No. 26, p. 21 - 22 (1975).

008.106 Princeton, N.J.

Princeton University Observatory, Princeton, New Jersey. — Observatory report.   L. Spitzer, Jr.
Bull. American Astron. Soc., Vol. 8, 241 - 246 (1976).

008.107 Pulkovo

Glavnaya Astronomicheskaya Observatoriya AN SSSR, Katalog Solnechnoj Deyatel'nosti (za 1974 god, R. S. Gnevysheva, 17.075.004).

008.108 Richmond Hill

David Dunlap Observatory, University of Toronto, Richmond Hill, Ontario, Canada. — Observatory report for the period 1 July 1974 — 30 June 1975.   D. A. MacRae.
Bull. American Astron. Soc., Vol. 8, 69 - 76 (1976).

008.109 Rio de Janeiro

Observatorio Nacional (National Observatory).
Inform. Bull. Southern Hemisph., No. 26, p. 8 - 9 (1975).
Current research report.

Observatorio Nacional (National Observatory).
Inform. Bull. Southern Hemisph., No. 27, p. 10 - 12 (1975).
Current research report.

008.110 Rochester, N.Y.

C.E.K. Mees Observatory, University of Rochester, Rochester, New York. — Observatory report.
J. G. Duthie.
Bull. American Astron. Soc., Vol. 8, 195 - 199 (1976).

C. E. Kenneth Mees Observatory, University of Rochester, Rochester, N. Y., Reprint, No. 58 (17.008.110).

008.111 Roma

Osservatorio Astronomico di Roma, Monte Mario – Monte Porzio – Stazione Astrofisica sul Gran Sasso. Contributi scientifici, Serie III, Nos. 124 (08.065.055), 125 (17.009.023), 126 (07.123.014), 127 (09.117.027), 128 (09.115.024), 130 (09.041.015), 131 (10.065.027), 132 (10.122.069), 133 (10.117.039), 134 (11.121.015), 135 (11.117.002), 136 (11.066.122), 137 (17.084.284), 138 (11.114.052), 139 (12.031.058), 140 (17.004.048), 141 (11.034.058), 142 (17.082.085), 143 (17.064.065).

Monthly Bulletin.   Osservatorio Astronomico di Roma, Nos. 209 - 212 (17.075.010).

Photographic Journal of the Sun. Osservatorio Astronomico di Roma, Nos. 99 - 106 (17.075.011).

008.112 Saint Michel, France

The Haute Provence Observatory.
C. Fehrenbach.
Zemlya i Vselennaya, 1976, No. 1, p. 73 - 76. In Russian. Translated from French by P. G. Kulikovskij.

008.113 San Diego, Calif.

University of California, Berkeley, Los Angeles, San Diego, and Lick Observatories. III. San Diego Campus. Observatory report.   E. M. Burbidge.
Bull. American Astron. Soc., Vol. 8, 42 - 46 (1976).

008.114 San Fernando (Cádiz), España

Memoria de las actividades en 1975.
Inst. y Obs. de Marina, San Fernando (Cádiz), España, 15 pp. (1976).

Instituto y Observatorio de Marina, San Fernando (Cádiz), España, Serie C, No. 77 (17.044.023).

008.115 San Juan

Observatorio "Félix Aguilar" ("Félix Aguilar" Observatory, School of Engineering, National University of Cuyo).
Inform. Bull. Southern Hemisph., No. 26, p. 3 - 4 (1975).
Current research report.

Observatorio "Félix Aguilar" ("Félix Aguilar" Observatory, School of Engineering, National University of Cuyo).
Inform. Bull. Southern Hemisph., No. 27, p. 4 - 6 (1975).
Current research report.

008.116 San Miguel

Observatorio Nacional de Física Cósmica (National Observatory of Cosmic Physics).
Inform. Bull. Southern Hemisph., No. 27, p. 6 - 7 (1975).
Current research report.

008.117 Santa Cruz (Mount Hamilton)

University of California, Berkeley, Los Angeles, San Diego, and Lick Observatories. IV. Lick Observatory, D. E. Osterbrock; V. Board of Studies in Astronomy and Astrophysics, W. G. Mathews. — Observatory reports.
Bull. American Astron. Soc., Vol. 8, 46 - 54 (1976).

The University of California, Santa Cruz. Publications of the Lick Observatory, Vol. 22, Parts 4 (17.111.006), 5 (17.111.007).

008.118 São Paulo

Instituto Astronômico e Geofísico da Universidade de São Paulo (Astronomical and Geophysical Institute of São Paulo University).
Inform. Bull. Southern Hemisph., No. 26, p. 10 - 11 (1975).
Current research report.

Instituto Astronômico e Geofísico da Universidade de São Paulo (Astronomical and Geophysical Institute of São Paulo University).
P. Benevides Soares.
Inform. Bull. Southern Hemisph., No. 27, p. 12 - 14 (1975).
Current research report.

Centro de Rádioastronomia e Astrofísica da Universidade Mackenzie (Center of Radioastronomy and Astrophysics, Mackenzie University).
Inform. Bull. Southern Hemisph., No. 26, p. 9 - 10 (1975).
Current research report.

Centro de Rádioastronomia e Astrofísica da Universidade Mackenzie (Center of Radioastronomy and Astrophysics, Mackenzie University).
Inform. Bull. Southern Hemisph., No. 27, p. 15 (1975).
Current research report.

Instituto Tecnológico de Aeronáutica (Aeronautical Technologic Institute).
Inform. Bull. Southern Hemisph., No. 26, p. 11 (1975).
Current research report.

008.119 Seattle, Wash.

University of Washington Astronomy Department, Seattle, Washington. – Observatory report. G. Wallerstein.
Bull. American Astron. Soc., Vol. 8, 275 - 278 (1976).

008.120 Shanghai

Time Service Annual Report 1973 (17.044.022).

008.121 Socorro, N. Mex.

National Radio Astronomy Observatory, Charlottesville, Virginia; Green Bank, West Virginia; Socorro, New Mexico; Tucson, Arizona. – Observatory report covering the period July 1974 – June 1975. D. S. Heeschen, D. E. Hogg.
Bull. American Astron. Soc., Vol. 8, 229 - 239 (1976).

008.122 Sonneberg

Akademie der Wissenschaften der DDR, Zentralinstitut für Astrophysik. Veröffentlichungen der Sternwarte in Sonneberg, Vol. 8, No. 5 (17.123.037).

Zentralinstitut für Astrophysik. Mitteilungen über Veränderliche Sterne, *Sonneberg,* Band 7, No. 4 (017.122. 140), 5 (17.121.116, 17.121.117, 17.122.141, 17.124.105, 17.142.234, 17.123.038, 17.121.118, 17.123.039, 17.123. 040, 17.122.142, 17.123.041, 17.124.102, 17.122.143), 6 (17.121.119, 17.121.120, 17.121.121, 17.142.235, 17.123. 042, 17.123.043, 17.123.044).

008.123 Stanford, Calif.

Stanford Radio Astronomy Institute, Stanford University, Stanford, California. – Observatory report. R. N. Bracewell.
Bull. American Astron. Soc., Vol. 8, 252 - 253 (1976).

008.124 Stockholm

Stockholms Observatorium, Saltsjöbaden, Sweden, Report Nos. 1 (17.114.060), 2 (17.155.060), 3 (17.034.075), 4 (17.071.040), 5 (17.122.139), 6 (17.034.076), 7 (17.114. 061), 8 (17.061.043), 9 (17.131.165), 10 (17.062.081).

008.125 Stony Brook, N.Y.

State University of New York at Stony Brook, Department of Earth and Space Sciences, Stony Brook, New York. – Observatory report. R. F. Knacke.
Bull. American Astron. Soc., Vol. 8, 253 - 257 (1976).

008.126 Sunspot, N. Mex.

Sacramento Peak Observatory, Contribution No. 263 (17.080.056).

008.127 Swarthmore, Penn.

Sproul Observatory, Swarthmore College, Swarthmore, Pennsylvania. – Observatory report for the period 1 July 1974 – 30 June 1975. S. L. Lippincott, W. D. Heintz.
Bull. American Astron. Soc., Vol. 8, 251 - 252 (1976).

008.128 Sydney

Division of Radiophysics, C.S.I.R.O.
Inform. Bull. Southern Hemisph., No. 26, p. 5 - 7 (1975).
Current research report.

Division of Radiophysics, CSIRO, Sydney, Australia, Radiophysics Publication RPP 1763 (11.158.024), 1813 (13.141.040), 1822 (17.077.037), 1825 (14.141.082), 1837 (14.099.003), 1851 (14.141.036), 1853 (14.131.120), 1856 (14.131.516), 1920 (14.158.192).

008.129 Tartu

Eesti NSV Teaduste Akadeemia, W. Struve nim. Tartu Astrofüüsika Observatoorium (Academy of Sciences of the Estonian SSR, W. Struve Tartu Astrophysical Observatory), **Preprints**, Nos. 8 (17.158.153), 9 (17.158.154), 10 (17.158.155), 11 (17.158.156), 12 (17.158.157).

008.130 Tauranga

Variable Star Section, R.A.S.N.Z.
Inform. Bull. Southern Hemisph., No. 26, p. 19 (1975).
Current research report.

008.131 Thessaloniki

Astronomy Department, University of Thessaloniki.
Annual report 1975.   S. Persides.
Annual Rep. Astron. Inst. Greece 1975, p. 12 - 15 (1976).

Department of Geodetic Astronomy, University of Thessaloniki. – Annual report 1975.   L. N. Mavridis.
Annual Rep. Astron. Inst. Greece 1975, p. 21 - 24 (1976).

008.132 Tokyo

Annual report from Tokyo Astronomical Observatory, 1975.
Separate print Tokyo Astron. Obs., 82 pp. (1976). In Japanese.

Annals of the Tokyo Astronomical Observatory,
University of Tokyo, Second Series, Vol. 15, No. 4 (17.131. 166).

Tokyo Astronomical Bulletin, Tokyo Astronomical Observatory, Second Series, Nos. 241 (17.124.102), 242 (17.124.102), 243 (17.131.545).

University of Tokyo, Tokyo Astronomical Observatory, Report (No. 67), Vol. 17, No. 4 (17.032.034 - 17.032. 037; 17.034.077; 17.046.035, 17.046.036; 17.081.040; 17. 17.082.086).

Tokyo Astronomical Observatory, Reprints Nos. 483 (17.151.026), 484 (17.113.022), 485 (17.158.303), 486 (17.158.304), 487 (17.122.028), 488 (17.124.102), 489 (17.117.031), 490 (17.045.013).

Time and Latitude Bulletins, Tokyo Astronomical Observatory, Vol. 49, Nos. 5 - 12 (17.044.025).

Data Report of Hydrographic Observations. Series of Astronomy and Geodesy, Maritime Safety Agency, Tokyo, Japan, No. 10 (17.096.010; 17.046.034).

008.133 Toledo, Ohio

Ritter Astrophysical Research Center of the University of Toledo, Toledo, Ohio. – Observatory report.
A. N. Witt.
Bull. American Astron. Soc., Vol. 8, 246 - 248 (1976).

008.134 Torino

Osservatorio Astronomico di Torino, Pino Torinese.
Time Service, Bulletin Nos. 11 - 12 (17.044.024, 17.044.026, 17.044.027).

Contributi dell'Osservatorio Astronomico di Torino (Pino Torinese), Nos. 86 (17.098.083), 88 (17.098. 002), 89 (17.098.020), 92 (17.102.033).

Pubblicazioni Varie Fuori Serie dell'Osservatorio Astronomico di Torino (Pino Torinese), Nos. 61 (17.047.010), 62 (17.031.267), 63 (17.066.047).

008.135 Toronto

David Dunlap Observatory, University of Toronto.
D. A. MacRae.
Quarterly Journ. Roy. Astron. Soc., Vol. 17, 154 - 173 (1976).
Report for the period 1974 July 1 to 1975 June 30.

008.136 Tortosa

Boletin del Observatorio del Ebro, Vol. 56 (17.075. 012), 57 (17.075.013).

008.137 Tucson, Arizona

National Radio Astronomy Observatory, Charlottesville, Virginia; Green Bank, West Virginia; Socorro, New Mexico; Tucson, Arizona. – Observatory report covering the period July 1974 – June 1975.   D. S. Heeschen, D. E. Hogg.
Bull. American Astron. Soc., Vol. 8, 229 - 239 (1976).

University of Arizona, Department of Planetary Sciences and Lunar and Planetary Laboratory, Tucson, Arizona. – Observatory report covering the period from 1 October 1974 to 30 September 1975.   C. P. Sonett.
Bull. American Astron. Soc., Vol. 8, 11 - 19 (1976).

008.138 Utrecht

Utrechtse Sterrekundige Overdrukken, Sterrewacht "Sonnenborgh", Utrecht, Nos. 296 (12.077.064), 297 (12.071. 017), 298 (12.034.060), 299 (12.077.040), 300 (12.077.050), 301 (17.077.038), 302 (12.034.061), 303 (13.073.003), 304 (13.114.303), 305 (13.072.006), 306 (13.073.025), 307 (13.114.318), 308 (13.131.092), 309 (13.073.034), 310 (13.064.038), 311 (17.073.084), 312 (13.071.044), 313 (17.051.021), 314 (13.031.240), 315 (13.012.009), 316 (12.034.057), 317 (13.034.063), 318 (17.034.080), 319 (13.074.090), 320 (13.073.090), 321 (13.076.055), 322 (13.022.088), 323 (13.022.087), 324 (13.114.057), 325 (14.114.117), 326 (14.064.071), 327 (14.064.072), 328 (14.114.401), 329 (13.077.065), 330 (14.114.017), 331 (14.114.018), 332 (14.076.024), 333 (14.142.014), 334 (14.064.002), 335 (17.034.081), 336 (14.031.281),

337 (14.031.416), 338 (14.034.149), 339 (14.031.418),
340 (14.031.289), 341 (17.079.106), 342 (14.142.045),
343 (14.114.353), 344 (14.079.100), 345 (14.122.167),
346 (14.077.040), 347 (14.063.045), 348 (14.064.048),
349 (14.114.090), 350 (14.114.361).

008.139 Vancouver, B.C.

  Institute of Astronomy and Space Science, University of British Columbia, Vancouver, British Columbia, Canada. – Current research report. G. A. H. Walker. Bull. American Astron. Soc., Vol. 8, 22 - 27 (1976).

008.140 Victoria, B.C.

  The Dominion Astrophysical Observatory, Victoria, B.C. J. B. Hutchings. Journ. Roy. Astron. Soc. Canada, Vol. 70, 90 (1976).

  Contributions from the Dominion Astrophysical Observatory, Victoria, B.C., Nos. 248 (14.114.315), 257 (17.114.308), 260 (14.119.002), 261 (14.153.013), 262 (14.118.020), 263 (14.114.043), 264 (14.142.112), 265 (14.119.013), 266 (14.115.013), 267 (17.142.236), 268 (14.113.029), 269 (14.114.047), 270 (17.114.006), 271 (17.114.012), 274 (14.114.337), 276 (14.034.148), 277 (14.142.115), 278 (14.114.355), 279 (14.013.013), 286 (17.114.309), 290 (17.119.003), 301 (17.116.004),

  Publications of the Dominion Astrophysical Observatory, Victoria, B.C., Vol. 14, Nos. 15 (17.118.024), 16 (17.114.063), 17 (17.112.010).

  The Herzberg Institute of Astrophysics – tuned to the wavelengths of the universe. W. J. Cherwinski. Sci. Dimension, Vol. 7, No. 4, p. 16 - 19 (1975).

  University of Victoria, Department of Physics, Victoria, British Columbia, Canada. – Observatory report. R. M. Pearce. Bull. American Astron. Soc., Vol. 8, 265 - 266 (1976).

008.141 Villanova

  Villanova University, Department of Astronomy, Villanova, Pennsylvania. – Observatory report for the period September 1974 – September 1975. G. P. McCook. Bull. American Astron. Soc., Vol. 8, 266 - 267 (1976).

008.142 Warszawa

  Politechnika Warszawska, Obserwatorium Astronomiczno-Geodezyjne w Józefosławiu, (Warsaw Technical University, Astronomic-Geodetical Observatory at Józefosław), Latitude Circular, Nos. 55 - 57 (17.045.014).

008.143 Washington, D.C.

  U.S. Naval Observatory, Washington, D.C. – Observatory report for the period 1 July 1974 – 30 June 1975.

K. A. Strand. Bull. American Astron. Soc., Vol. 8, 257 - 262 (1976).

  U. S. Naval Observatory, Washington, D.C. Time Service Publications, Series 4, Nos. 466 - 490; Series 7, Nos. 418 - 443; Series 11, No. 223.1 (17.044.033, 17.044.034, 17.044.901).

  National Aeronautics and Space Administration, Washington, D.C. N. Boggess. Bull. American Astron. Soc., Vol. 8, 215 - 228 (1976).

  National Bureau of Standards, Washington, D.C. L. Hagan. Bull. American Astron. Soc., Vol. 8, 210 - 214 (1976).

008.144 Wellington, New Zealand

  Carter Observatory. Inform. Bull. Southern Hemisph., No. 26, p. 19 (1975). Current research report.

  Carter Observatory, Wellington, New Zealand, Astronomical Bulletin, No. 85 (17.047.009).

008.145 Westford, Mass.

  Haystack Observatory, Northeast Radio Observatory Corporation (NEROC), Westford, Massachusetts. – Observatory report covering the period 1 July 1974 – 30 June 1975. P. B. Sebring. Bull. American Astron. Soc., Vol. 8, 107 - 111 (1976).

008.146 Wien

  Annalen der Universitäts-Sternware Wien, Band 31, Nos. 1 (17.098.085), 2 (17.103.146), 3 (17.112.009)

008.147 Williams Bay, Wis.

  University of Chicago, Department of Astronomy and Astrophysics, Chicago, Illinois; The Yerkes Observatory, Williams Bay, Wisconsin. – Observatory reports. E. N. Parker, L. M. Hobbs. Bull. American Astron. Soc., Vol. 8, 54 - 61 (1976).

008.148 Williamstown, Mass.

  Hopkins Observatory, Williamstown, Massachusetts. Observatory report. J. M. Pasachoff. Bull. American Astron. Soc., Vol. 8, 111 - 113 (1976).

008.149 Wrocław

  Wrocław Astronomical Observatory, Reprint Nos. 97 (14.122.029), 98 (14.124.100).

008.150 **Yorktown Heights, N.Y.**

   IBM Thomas J. Watson Research Center, Yorktown Heights, New York.   G. Lasher.
Bull. American Astron. Soc., Vol. 8, 113 (1976).

008.151 **Zelenchukskaya**

   Chronicle.
Astrofiz. Issled., Izv. Spets. Astrofiz. Obs., Vol. 8, 144 - 145 (1976). In Russian.

   Astrofizicheskie Issledovaniya. Izvestiya Spetsial'noj Astrofizicheskoj Observatorii, Vol. 8 (N. F. Vojkhanskaya, 17.122.039; L. I. Snezhko, 17. 114.024; V. S. Lebedev, 17.065.059; I. A. Zenina, O. A. Zenina, V. V. Leushin, 17.114.331; B. P. Artamonov, F. Börngen, A. I. Shapovalova, 17.158.068; A. I. Shapovalova, 17.158.069; G. N. Alekseev, G. M. Beskin, 17.031.222; L. G. Antropova, V. S. Rylov, M. F. Shabanov, A. Ch. Uzdenov, 17.036.002; Eh. B. Gazhur, M. M. Kononov, V. B. Nebelitskij, A. F. Fomenko, 17.032. 009; V. S. Rylov, 17.034.019; N. F. Ryzhkov, 17.031.223;

N. L. Kajdanovskij, 17.033.009; E. L. Chentsov, 17.114.332; R. N. Kumajgorodskaya, N. M. Chunakova, 17.122.040; A. A. Korovyakovskaya, V. S. Rylov, 17.034.020; I. D. Najdenov, G. A.Chuntonov, 17.034.021; A. B. Basaev, G. A. Chuntonov, 17.034.022; 17.008.151).

008.152 **Zürich**

   Tätigkeitsbericht der Eidgenössischen Sternwarte Zürich für das Jahr 1975.   M. Waldmeier.
Zürich, 8 pp. (1976).

   Astronomische Mitteilungen der Eidgenössischen Sternwarte Zürich, Nos. 340 (14.074.120), 343 (17.073.085), 344 (17.072.054).

   Publikationen der Eidgenössischen Sternwarte Zürich, Band 14, No. 5 (17.071.041).

   Quarterly Bulletin on Solar Activity (Zürich), Nos. 189 - 190 (17.075.014).

009 Notes on Observatories, Planetaria, and Exhibitions

009.001 **The Ottawa River Solar Observatory.**
   V. Gaizauskas.
Journ. Roy. Astron. Soc. Canada, Vol. 70, 1 - 22 (1976).
   An observatory has been built in Canada for high resolution cinematography of active regions in the solar photosphere and chromosphere. The installation on the shore of the Ottawa River is the successor to solar facilities maintained at the Dominion Observatory from 1905 to 1970. The building, telescope, and automated control system are described with comments on the factors that influenced their design.

009.002 **An astronomical mural at Flandrau Planetarium.**
   Sky Telescope, Vol. 51, 165 - 168 (1976).

009.003 **A roundup of observatories by amateurs.**
   N. D. Coleman, T. D. Burns, G. M. Collenberg, S. Hartsema, R. Tynemouth.
Sky Telescope, Vol. 51, 220 - 224 (1976).

009.004 **Planetarium für Fachleute.**   A. Kunert.
   Mitt. Astron. Ges., No. 38, p. 55 - 57 (1976).

009.005 **Aberdeen University, Department of Natural Philosophy.**   R. V. Jones.
Quarterly Journ. Roy. Astron. Soc., Vol. 17, 61 - 62 (1976). Report for the year ending 1975 September 30.

009.006 **Kitab International Latitude Station.**
   A. M. Kalmykov.
Zemlya i Vselennaya, 1976, No. 2, p. 83 - 87. In Russian.

009.007 **The planetarium as a research and educational tool.**
   R. Subramanian.
Bull. Astron. Soc. India, Vol. 3, 29 (1975). – Abstract of paper presented at the A.S.I. meeting 1975.

009.008 **20 Jahre Planetarium Chorzow.**   M. Pańków.
   Astron. in der Schule, 13. Jahrgang, p. 41 - 42 (1976).

009.009 **Rhodes University, Department of Physics.**
   E. E. Baart.
Monthly Notes Astron. Soc. Southern Africa, Vol. 35, 27 - 28 (1976). – Report for 1975.

009.010 **University of South Africa, Department of Mathematics and Astronomy.**
Monthly Notes Astron. Soc. Southern Africa, Vol. 35, 28 - 29 (1976). – Report for 1975.

009.011 **University of Witwatersrand, Departments of Applied Mathematics and Physics.**   B. L. Fanaroff.
Monthly Notes Astron. Soc. Southern Africa, Vol. 35, 29 (1976).– Report for 1975.

009.012 **The Dominion Radio Astrophysical Observatory, Penticton, B.C.**   J. A. Galt.
Journ. Roy. Astron. Soc. Canada, Vol. 70, 89 (1976).

009.013 **Experience of cosmic ray station operating in Apatity.**

M. A. Vetrina, I. A. Kuz'min, V. T. Ustinovich.
Cosmic rays, No. 15, (see 003.006), p. 178 - 180 (1975). In Russian.

009.014 **Complete block-schema of the cosmic ray station of the Ionospheric Department of the Kazakh Academy of Sciences.** Yu. A. Egorov.
Cosmic rays, No. 15, (see 003.006), p. 181 - 183 (1975). In Russian.

009.015 **A new cosmic ray observatory at Mawson, Antarctica.** R. M. Jacklyn, A. Vrana, D. J. Cooke.
14th Intern. Cosmic Ray Conf., (see 012.011), Vol. 4, 1497 - 1502 (1975).

009.016 **Zur Didaktik von Planetariumsvorführungen.** H.-U. Keller.
Separate print Sternw. Bochum, Planetarium, 64 pp. (1976).

009.017 **The Anglo-Australian Observatory.**
Inform. Bull. Southern Hemisph., No. 27, p. 8 - 9 (1975).

009.018 **Observatory Hvar.** J. Sýkora.
Kozmos, Vol. 7, 21 - 22 (1976). In Slovak.

009.019 **Haute Provence observatoriet.** P. E. Nissen.
Astron. Tidssk., Årg. 9, 49 - 58 (1976).

009.020 **Pikdimidi Observatorija.** J. Francmanis.
Zvaigžņotā debess, 1976. gada pavasaris, p. 20 - 25.

009.021 **A laser range tracking station for geodynamic satellites.** L. Cugusi, F. Messina, E. Proverbio.
Repr. from Second Workshop on laser tracking instrumentation, Prague 11 - 16 August 1975, 30 pp. = Pubbl. Stazione Astron. Internazionale Latitudine, Carloforte—Cagliari, Nuova Ser., No. 51 (1975).

009.022 **Yale University Observatory at the turn of the century.** D. Hoffleit.
Irish Astron. Journ., Vol. 12, 104 - 108 (1975).

009.023 **Il Museo Astronomico e Copernicano dell'Osservatorio Astronomico di Roma.** M. Cimino.
Oss. Astron. Roma, Contr. sci., Ser. III, No. 125, 35 pp. (1973).

009.024 **"Research at Sacramento Peak Observatory".** R. C. Altrock.
Proc. Southwest Regional Conf., Vol. 1, (see 012.021), 89 - 90 (1976). — Conference abstract.

009.025 **The New Mexico State University Meteor Observatory.** E. F. Tedesco.
Proc. Southwest Regional Conf., Vol. 1, (see 012.021), 97 - 103 (1976).

009.026 **Le Bureau des Longitudes.** B. Morando.
L'Astronomie, Vol. 90, 279 - 294 (1976).

Scientific astronomical centres in Poland.
Postępy Astron., Vol. 24, 69, 131 (1976). In Polish.

On the summit of Colorado's Mount Evans.
Sky Telescope, Vol. 51, 172 (1976).

010 Societies, Associations, Organizations

010.001 **American Association of Variable Star Observers (AAVSO)**

Annual report of the director, October 1974 - September 1975. J. A. Mattei.
Journ. American Ass. Variable Star Observers, Vol. 4, 114 - 120 (1975/76).

Minutes of the general meeting of the AAVSO held at the Concord Academy, Concord, Mass. October 18, 1975. C. B. Ford.
Journ. American Ass. Variable Star Observers, Vol. 4, 105 - 113 (1975/76).

AAVSO Bulletin 39.
(See 123.031).

010.002 **American Astronomical Society (AAS)**

No publication received.

010.003 **Association of Lunar and Planetary Observers (ALPO)**

Current events on Jupiter and Jupiter Section activities.
Strolling Astronomer, Vol. 25, 257 - 258 (1976).

010.004 **Astronomical Society of Australia (ASA)**

No publication received.

010.005 **Astronomical Society of Czechoslovakia**

No publication received.

010.006 **Astronomical Society of the Pacific (ASP)**

No publication received.

010.007 **Astronomical Society of Southern Africa (ASSA)**

Notices.
Monthly Notes Astron. Soc. Southern Africa, Vol. 35, 1 - 2, 19, 47 (1976).

Transvaal Centre's new home.
Monthly Notes Astron. Soc. Southern Africa, Vol. 35, 14 - 15 (1976).

010.008 **Astronomical Society of Victoria (ASV)**

No publication received.

010.009 **Astronomical Society of Western Australia (ASWA)**

No publication received.

010.010 **Astronomische Gesellschaft (AG)**

Wissenschaftliche Tagung der Astronomischen Gesellschaft mit 55. ordentlicher Mitgliederversammlung in Berlin, 15.–19. September 1975. H. Mauder.
Mitt. Astron. Ges., No. 38, p. 17 - 18 (1976). — Bericht über die Tagung.

010.011 **Astronomisk Selskab København**

No publication received.

010.012 **British Astronomical Association (BAA)**

Notices.
Journ. British Astron. Ass., Vol. 86, 101 - 102, 182 - 185, 261 - 263 (1976).

The annual general meeting of the Association held on 1975 October 29. H. G. Miles, G. Stone, H. R. Hatfield.
Journ. British Astron. Ass., Vol. 86, 103 - 105 (1976).

The association's observational astronomy course at Horncastle – September 1976.
British Astron. Ass. Circ., No. 573 (1976).

Meetings of the Association.
Journ. British Astron. Ass., Vol. 86, 106 - 111, 185 - 191, 264- 273 (1976).

New members elected.
Journ. British Astron. Ass., Vol. 86, 169 - 174, 353 - 354 (1976).

Jupiter Section. W. E. Fox.
Journ. British Astron. Ass., Vol. 86, 162 - 163 (1976).

Historical Section. E. A. Beet.
Journ. British Astron. Ass., Vol. 86, 333 - 334 (1976).

Lunar Section. P. Moore.
Journ. British Astron. Ass., Vol. 86, 241 - 244 (1976).

Mercury and Venus Section. J. H. Robinson.
Journ. British Astron. Ass., Vol. 86, 155 - 161 (1976).

Saturn Section. A. W. Heath.
Journ. British Astron. Ass., Vol. 86, 111 (1976).

Solar Section. V. Barocas.
Journ. British Astron. Ass., Vol. 86, 149 - 150, 228 - 237, 317 - 326 (1976).

Variable Star Section. J. E. Isles.
Journ. British Astron. Ass., Vol. 86, 245 - 249, 327 - 332 (1976).

010.013   British Interplanetary Society

The report of the council for the year ended 31 December 1975.   K. W. Gatland.
Spaceflight, Vol. 18, 148 - 149 (1976).

010.014   Committee on Space Research (COSPAR)

No publication received.

010.015   European Space Agency (ESA)

Europe's 'NASA' gets off the ground.
R. Gibson, W. J. Kleen.
IEEE Spectrum, Vol. 13, No. 2, p. 66 - 70 (1976). — The new Paris-based European Space Agency has ten member countries, a $ 500 million annual budget, and a wide range of future programs.

010.016   International Astronautical Federation (IAF)

No publication received.

010.017   International Astronomical Union (IAU)

Report of the working group on numerical data of IAU Commission 5.   G. A. Wilkins.
Centre Données Stellaires, Inform. Bull. No. 10, p. 17 - 27 (1976).
A brief account is given of the relevant activities of each of the Commissions of the Union, and of other international organisations. Then, all data centres and major data projects reported to the IAU Working Group on Numerical Data are listed under the name of the country in which the centre or project is based. Finally, a short list of publications concerned with data of relevance to astronomy is given.

International Astronomical Union, Information Bulletin, Nos. 35, 36. 35 + 43 pp. (1976).
G. Contopoulos.
Printed by D. Reidel, Dordrecht, Holland.
The sixteenth general assembly; Executive committee; Commissions; IAU symposia and colloquia; Other scientific meetings; IAU publications; Other publications; International organizations; Membership; Instructions for preparing camera-ready copies for IAU symposia and 'Highlights of Astronomy'.

010.018   Meteoritical Society

No publication received.

010.019   Nederlandse Vereniging voor Weer- en Sterrenkunde

No publication received.

010.020   Polskie Towarzystwo Astronomiczne (PTA)

No publication received.

010.021   Polskie Towarzystwo Miłośników Astronomii (PTMA)

No publication received.

010.022   Royal Astronomical Society (RAS)

Meetings of the Society.
Observatory, Vol. 96, 33 - 40, 73 - 89 (1976).

Meetings of the Society.
Quarterly Journ. Roy. Astron. Soc., Vol. 17, 87 - 92; 191 - 199 (1976).

Royal Astronomical Society MIST meeting on the rotation of planetary atmospheres.
Quarterly Journ. Roy. Astron. Soc., Vol. 17, 139 - 145 (1976).
1975 November 28.

010.023   Royal Astronomical Society of Canada (RAS Canada)

No publication received.

010.024   Royal Astronomical Society of New Zealand (RAS New Zealand)

53rd annual report of council for the year ended 1975 September 30.   N. J. Rumsey, R. McIntosh.
Southern Stars, Vol. 26, 116 - 120 (1976).

Variable star section. Report for the year ended 1975 September 30.   F. M. Bateson.
Southern Stars, Vol. 26, 120 - 126 (1976).

010.025   Schweizerische Astronomische Gesellschaft (SAG)

No publication received.

010.026   Sociedad Astronómica de México

No publication received.

010.027   Societá Astronomica Italiana (SAI)

Notizie della Società.
Giorn. Astron., Vol. 1, 87 - 90, 163 - 168, 245 - 255 (1975).

Gli astrofili nella Società Astronomica Italiana.
L. Baldinelli.
Giorn. A.A.B., Anno 11, No. 41, p. 18 - 20 (1976).

010.028   Société Astronomique de France (SAF)

  Assemblée Générale Extraordinaire. B. Clouet.
L'Astronomie, Vol. 90, 269 (1976).

  Les séances de la Société.
P. de La Cotardière, B. Clouet, L. Tartois.
L'Astronomie, Vol. 90, 118, 268 - 273 (1976).

  Une réunion commune à toutes les commissions de
la Société Astronomique de France. J.-C. Pecker.
L'Astronomie, Vol. 90, 51 (1976).

  Commission des Cadrans Solaires. R. Sagot.
L'Astronomie, Vol. 90, 53 - 57 (1976).

  Commission des Instruments et de la Photographie.
J. Funel.
L'Astronomie, Vol. 90, 58 - 64 (1976).

  Commission des Relations Extérieures.
P. de La Cotardière.
L'Astronomie, Vol. 90, 52 (1976).

  Commission du Soleil.
L'Astronomie, Vol. 90, 89 - 99 (1976).

  Commission des Surfaces Planétaires.
C. Botton.
L'Astronomie, Vol. 90, 65 - 88 (1976).

010.029   Société Astronomique "R. Boškovic"

  No publication received.

010.030   Société Chronométrique de France

  No publication received.

010.031   Société Belge d'Astronomie, de Météorologie et de
  Physique du Globe

  No publication received.

010.032   Svenska Astronomiska Sällskapet

  No publication received.

010.033   VAGO (Astronomical-Geodetical Society of the
  USSR)

  Congress of VAGO in the capital of Armenia.
L. S. Khrenov, E. P. Levitan.
Zemlya i Vselennaya, 1976, No. 3, p. 64 - 69. In Russian.
Erevan, 1975, October 21 - 24.

010.034   Vereniging voor Sterrenkunde, België

  No publication received.

010.035   Association Française des Observateurs d'Étoiles
  Variables (A.F.O.E.V.)

  Activité de l'A.F.O.E.V. E. Schweitzer.
L'Astronomie, Vol. 90, 106 - 107, 254 - 255 (1976).

  La vie de l'association. E. Schweitzer.
A.F.O.E.V. Bull., Tome 10, 26 - 27 (1976).

  Le programme A.F.O.E.V.
See Abstr. 122.146.

010.036   Abstracts of the papers presented at the second
  meeting of the Astronomical Society of India held
at the Indian Institute of Astrophysics, Kodaikanal, 12 - 14
March 1975.
Bull. Astron. Soc. India, Vol. 3, 27 - 37 (1975).

010.037   Nachrichten der Vereinigung der Sternfreunde e.V.
  SuW, 15. Jahrgang, p. 24 - 26, 28 - 30, 64, 98 - 100,
136 - 139, 172 - 174, 176, 209 - 211 (1976).

010.038   Société Royale d'Astronomie d'Anvers [Koninklijk
  Sterrenkundig Genootschap van Antwerpen].
Cinquante-Sixième rapport 1975. J. Storms.
Imprimerie: «La Prévoyance», Antwerpen. 14 + 21 pp. (1976).
In French and Flemish.

010.039   L'Associazione Astrofili Bolognesi nel tempo.
  A. Betti.
Giorn. A.A.B., Anno 11, No. 41, p. 5 - 6 (1976).

011 Reports on Colloquia, Congresses, Meetings, Symposia, and Expeditions

011.001 **Jupiter: I.A.U. Colloquium No. 30, University of Arizona, Tucson, Arizona, May 18–23, 1975.**
Part I: Atmosphere and clouds. Part II: Fields and particles.
Icarus, Vol. 27, 171 - 179 (1976). – Meeting review by G. E. Hunt (part I) and T. R. McDonough (part II).

011.002 **IAU Colloquium No. 26 "On reference coordinate systems for earth dynamics".Toruń, 1974 August 26 - 31.** W. Dobaczewska, B. Kołaczek, J. Śledziński.
Geodezja Kartografia, Vol. 25, 71 - 77 (1976). In Polish.

011.003 **IAU Colloquium No. 26. – Reference coordinate systems for earth dynamics, Toruń, Poland, 1974 August 26 - 31.** B. Kołaczek, J. Śledziński.
Postępy Astron., Vol. 23, 289 - 292 (1975). In Polish.

011.004 **A magnetic monopole?** Z. Kobyliński.
Postępy Astron., Vol. 23, 293 (1975). In Polish. – Conference 1975 August 15 - 29.

011.005 **Rotation of planetary atmospheres.**
Nature, Vol. 259, 16 (1976). – A meeting of the Royal Astronomical Society, Institute of Physics and the Royal Meteorological Society, held in London on November 28, 1975, provided interdisciplinary discussions of the atmospheric motions at all atmospheric levels.

011.006 **Progress with X rays.**
A. Fabian, J. Pringle.
Nature, Vol. 260, 14 (1976). – Report on a meeting of the AAS on X-ray astronomy, held 1976, January 27 - 29 at the Mass. Inst. of Technology.

011.007 **Galactic gas dynamics.**
Nature, Vol. 260, 195 (1976).
Report on a meeting on 'Galactic gas dynamics', held by the Royal Astronomical Society, London, February 13, 1976.

011.008 **Colloquium on solar research.** S.-z. Zhang.
Acta Astron. Sinica, Vol. 16, 231 (1975). In Chinese. Brief note.

011.009 **Conference on time and latitude service.**
S.-z. Zhang.
Acta Astron. Sinica, Vol. 16, 232 (1975). In Chinese. – Brief note.

011.010 **Summaries of papers presented at Royal Astronomical Society specialist discussion on "The principle and practice of star formation",** held on 1975 May 9, with opening and concluding remarks by D. McNally.
Observatory, Vol. 96, 1 - 9 (1976). – For the individual contributions – see abstracts 065.028 - 065.030, 131.057, 131.058, 132.008, 141.609.

011.011 **Astronomy east and west.** D. Trombino.
Sky Telescope, Vol. 51, 16 - 22 (1976).

011.012 **Chronicle. The 20th Astrometrical Conference of the USSR.**
M. S. Zverev, K. G. Gnevyshiva, E. G. Zhilinskij.
Astron. Zhurn. Akad. Nauk SSSR, Vol. 53, 229 - 232 (1976). In Russian. English translation in Soviet Astron., Vol. 20, No. 1.
Pulkovo, 1975, May 20–22.

011.013 **Report on the first symposium "Use of artificial satellites for geodesy and geodynamics" in Athens, 1973, May 14 - 21.** L. Stange.
Dynamical questions of satellite geodesy, 1973, (see 012.006), p. 234 - 239 (1974). In Russian and German.

011.014 **The first European conference on astronomy.**
V. Trimble.
Quarterly Journ. Roy. Astron. Soc., Vol. 17, 25 - 42 (1976).

011.015 **Symposium 'Observations and interpretation of the secular variations of the earth's magnetic field'.**
G. N. Petrova, A. N. Pushkov.
Geomagn. Aeronom., Vol. 16, 392 - 394 (1976). In Russian.
Potsdam, 1975, March 3 - 7.

011.016 **Report on a symposium on X-ray binaries.**
M. S. Vardya.
Bull. Astron. Soc. India, Vol. 3, 81 - 83 (1975). – The symposium was held at Goddard Space Flight Center, Greenbelt, Maryland, U.S.A. between October 20 - 22, 1975.

011.017 **Report on the colloquium on the physics of Ap stars.**
D. S. Leckrone.
Bull. Astron. Soc. India, Vol. 3, 83 - 85 (1975). – IAU Colloquium No. 32, held at the University of Vienna, September 8 - 11, 1975.

011.018 **International seminar on 'Corpuscular streams of the sun and the radiation belts of the earth and Jupiter'.** G. E. Kocharov.
Vestn. AN SSSR, 1975, No. 11, p. 114 - 116. In Russian. Abstr. in Referativ. Zhurn. 62. Issled. kosmich. prostranstva, 4.62.11 (1976).

011.019 **The structure and origin of comets.**
Observatory, Vol. 96, 41 - 49 (1976). – Meeting, held at the University of York, 1975 March 25 - 26.

011.020 **Astronomische Tagung Berlin 1975.**
M. Grewing.
Phys. Blätter, 32. Jahrgang, 232 - 233 (1976).

011.021 **Solar system origin.** D. W. Hughes.
Nature, Vol. 261, 15 - 16 (1976).
Advanced study institute on the origin of the solar system, held in the Institute of Lunar and Planetary Science of the University of Newcastle-upon-Tyne between March 29 - April 9, 1976.

011.022 **IIIrd European astronomical conference.**
G. N. Salukvadze.
Zemlya i Vselennaya, 1976, No. 3, p. 62 - 63. In Russian.
Tbilisi, 1975, July 1 - 5.

011.023 **Meeting of directors of planetaria.**
K. A. Portsevskij.
Zemlya i Vselennaya, 1976, No. 3, p. 84 - 86. In Russian.
Prag, 1975, August 18 - 23.

011.024 **Scientific session of the Department of General Physics and Astronomy of the USSR Academy of Sciences, April 23, 1975.**
Uspekhi fiz. nauk, Vol. 117, 563 - 576 (1975). In Russian.

011.025 Exploding black holes.   P. C. W. Davies.
Nature, Vol. 261, 280 (1976). — Symposium held in
Oxford on April 20 and 21, 1976.

011.026 Scientific session of the Department of Physical-
Mathematical Sciences of the Armenian SSR.
1975, March 25.
Izv. Akad. Nauk Armyanskoj SSR, Fizika, Vol. 10, 495 - 497
(1975). In Russian.

011.027 IXth All-Union conference on physics and dynamics of comets and asteroids.
Kometn. Tsirk., *Kiev,* No. 195 (1976). In Russian. — Kiev,
1976, October.

011.028 Konferences un sanāksmes.
Zvaigžņotā debess, 1975./76. gada ziema, p. 26 - 35.
X Baltijas zinātnes vēstures konferencē (*J. Stradiņš*), p. 26 -
31; Pirmā apspriede par infrasarkano astronomiju (*J. Kižla*),
p. 31 - 33; XX PSRS astrometrijas konference (*L. Roze,
M. Dīriķis*), p. 33 - 35.

011.029 Pie Ungāru astronomiem.   A. Alksnis.
Zvaigžņotā debess, 1976. gada pavasaris, p. 25 - 31.
Budapest, 1975 Sept.

011.030 Konferences un sanāksmes.
Zvaigžņotā debess, 1976. gada pavasaris, p. 32 - 38.
8. Vissavienības radioastronomijas konference (*A. Balklavs*),
p. 32 - 36; Eiropas astronomu sanāksme Gruzijā (*A. Alksnis*),
p. 36 - 38.

011.031 XX meeting of the Argentina Astronomical Association, I.A.R., Pereyra, December 4 - 6, 1974.
Inform. Bull. Southern Hemisph., No. 26, p. 25 - 27 (1975).

011.032 II scientific meeting of the Asociación Venezolana
de Astronomía, Mérida, Venezuela, January 16 - 18,
1975.
Inform. Bull. Southern Hemisph., No. 26, p. 27 - 28 (1975).

011.033 First regional meeting on extragalactic astronomy,
Córdoba, April 24 - 26, 1975.   J. L. Sérsic.
Inform. Bull. Southern Hemisph., No. 27, p. 33 - 35 (1975).

011.034 Czechoslovak conference on stellar astronomy.
J. Zverko.
Kozmos, Vol. 7, 56 - 57 (1976). In Slovak.

011.035 71st IAU symposium on basic mechanism of solar
activity.   P. Macák.
Říše hvězd, Vol. 57, 8 - 10 (1976). In Czech.

011.036 The third European regional astronomical conference, Tbilisi 1975.   V. Bahýl, D. Chochol.
Kozmos, Vol. 7, 13 - 14 (1976). In Slovak.

011.037 International seminar on new methods in space
geodesy, November 24 - 30, 1975. Leningrad,
U.S.S.R.   V. K. Abalakin.
Bull. Géod., Vol. 50, 193 - 195 (1976).

011.038 S.A.A.O. conference 1976.
Monthly Notes Astron. Soc. Southern Africa,
Vol. 35, 48 - 54 (1976).

011.039 5th International Planetarium Directors
Conference.   M. Pańkow.
Urania Kraków, Vol. 47, 18 - 21 (1976). In Polish.

011.040 Workshop on laser tracking instrumentation,
Prague, 11–15 August 1975.   L. Cugusi.
Mem. Soc. Astron. Italiana, Vol. 46, 505 - 511 (1975).

011.041 Report from the plenary meeting of the Polish
Astronomical Society, Gdańsk, September 17,
1975.   E. Basińska-Grzesik.
Postępy Astron., Vol. 24, 71 - 73 (1976). In Polish.

011.042 Report of the executive council of the Polish
Astronomical Society for the period 1973–1975.
R. Głębocki, T. Jarzębowski.
Postępy Astron., Vol. 24, 73 - 74 (1976). In Polish.

011.043 5th international conference of planetarium
directors.   M. Pańków.
Postępy Astron., Vol. 24, 74 - 75 (1976). In Polish.

011.044 5th Advanced Astronomy School in Erice-Trapani,
Sicily: Astronomy in the ultraviolet.
T. Z. Dworak, G. Sęk.
Postępy Astron., Vol. 24, 76 - 77 (1976). In Polish.

011.045 Progress in Solar System Studies as reported at the
18th COSPAR Planetary Conference in Varna, May
29 - June 6, 1975.   O. Wołczek.
Postępy Astron., Vol. 24, 133 - 140 (1976). In Polish.

011.046 The symposium "Results of coordinated pro-
grammes at upper atmosphere measurements" in
Varna, May 1975.   A. W. Wernik.
Postępy Astron., Vol. 24, 141 - 142 (1976). In Polish.

011.047 18th COSPAR meeting, Varna, June 2 - 7, 1975.
Z. Kłos.
Postępy Astron., Vol. 24, 143 - 144 (1976). In Polish.

011.048 Conference on "Problems of extragalactic research",
Potsdam, October 28 - November 1, 1975.
K. Rudnicki.
Postępy Astron., Vol. 24, 145 (1976). In Polish.

011.049 Earth's particles and fields, 1973. Summer advanced
study institute. Held at the University Sheffield,
England, August 13 - 24, 1973.
B. M. McCormac, J. E. Evans.
Earth Extraterr. Sci., Vol. 2, No. 5, p. 161 - 179 (1975).
Abstr. in Phys. Abstr., Vol. 79, A037284 (1976).

011.050 Solar physics. Plasma physics workshop.
P. J. Baum, J. M. Beckers, C. E. Newman, E. R.
Priest, H. Rosenberg, D. F. Smith, P. A. Sturrock, D. G.
Wentzel.
Separate print Stanford Univ., California, USA, Inst. Plasma
Res., 35 pp. (1974).

011.051 Conferences and meetings: III Congress of the
International Union of Amateur Astronomers.
K. Ziołkowski.
Urania Kraków, Vol. 47, 85 - 87 (1976). In Polish.

012 Proceedings of Colloquia, Congresses, Meetings, and Symposia

012.001 **Solid state astrophysics.** Proceedings of a symposium held at the University College, Cardiff, Wales, 9–12 July 1974.
N. C. Wickramasinghe, D. J. Morgan (Editors), with a preface by F. Hoyle.
D. Reidel Publishing Company, Dordrecht, Holland – Boston, U.S.A. Astrophysics and Space Science Library, Vol. 55 (Proceedings). 12 + 314 pp. Price Dfl. 95.00, US $ 37.00 respectively (1976). – The individual contributions are included in their corresponding subject categories – see abstracts 022.038, 064.026 - 064.028, 065.040 - 065.043, 131.072 - 131.088, 131.524, 131.525, 132.011, 133.013, 161.003.

012.002 **The Scientific Satellite Programme during the International Magnetospheric Study.** Proceedings of the 10th ESLAB Symposium, held at Vienna, Austria, 10 - 13 June 1975.
K. Knott, B. Battrick (editors) with a summary status report on the International Magnetospheric Study by J. G. Roederer.
D. Reidel Publishing Company, Dordrecht, Holland – Boston, U.S.A.
Astrophysics and Space Science Library, Vol. 57, 15 + 463 pp. Price Dfl. 110.00, US $ 39.50 respectively (1976). – For the individual contributions within the subject scope of Astronomy and Astrophysics Abstracts – see 083.022, 084.223 - 084.232, 085.002.

012.003 **Space Research XV.** Proceedings of open meetings of Working Groups on Physical Sciences of the seventeenth plenary meeting of COSPAR, São Paulo, Brazil – June 1974.
M. J. Rycroft (Editor), with a foreword by C. de Jager.
Akademie-Verlag, Berlin. 15 + 737 pp. Price DM 160.00 (1975). – The individual contributions within the subject scope of Astronomy and Astrophysics Abstracts are included in their corresponding categories – see abstracts 041.008, 046.008 - 046.010, 061.012, 061.013, 073.023, 074.019, 076.011 - 076.015, 081.007, 082.020 - 082.023, 083.023 - 083.025, 084.233, 093.010, 094.113 - 094.116, 094.405 - 094.408, 097.020, 099.013, 106.015 - 106.022, 114.322, 131.089, 142.054, 142.055.

012.004 **Interaction of meteor matter with the earth and estimate of meteor matter influx on the earth and on the moon,** Dushanbe, September 24 - 28, 1974.
Astron. sovet AN SSSR, Probl. rabochaya gruppa po issled. meteorn. veshchestva. AN TadzhSSR. In-t astrofiz. Donish, Dushanbe. 168 pp. Price 1 Rbl. (1975). In Russian. – See abstracts 104.012 - 104.023, 104.030 - 104.033, 106.023 - 106.025.

012.005 **Long-time predictions in dynamics.** Proceedings of the NATO Advanced Study Institute held in Cortina d'Ampezzo, Italy, August 3 - 16, 1975.
V. Szebehely, B. D. Tapley (Editors).
NATO Advanced Study Institutes Series, Ser. C, Math. and Phys. Sci., Vol. 26. D. Reidel Publishing Company, Dordrecht, Holland – Boston, U.S.A. 18 + 358 pp. Price Dfl. 85.00, $ 33.00 respectively (1976). ISBN 90-277-0692-1. – The individual contributions are included in their corresponding subject categories – see abstracts 021.004, 021.005, 022.043 - 022.046, 042.011 - 042.036, 044.004, 046.012, 052.009 - 052.012, 084.237, 084.238, 091.008, 094.006, 107.005, 162.027.

012.006 **Publications of the seminar "Dynamical questions of satellite geodesy".** Belogradchik, 1973, June 20 - 25.

L. V. Rykhlova (Editor).
Nablyud. Iskusstv. Sputnikov Zemli, No. 13, 243 pp. Astronomicheskij Sovet Akademii Nauk SSSR, Moskva (1974). In Russian. – The individual contributions are included in their corresponding subject categories – see abstracts 011.013, 045.003, 045.004, 046.013, 046.018, 052.013 - 052.020, 081.008 - 081.011.

012.007 **VII Leningrad international seminar.** Materials of the international seminar. "Corpuscular streams of the sun and the radiation belts of the earth and Jupiter", Leningrad, 1975, May 25 - 28.
Fiz.-tekhn. in-t AN SSSR. NII yader. fiz. Mosk. un-ta. Leningrad. 401 pp. Price 1 Rbl. 50 Kop. (1975). In Russian. – See abstracts 062.019 - 062.021, 073.045, 073.046, 073.048, 073.049, 076.017, 078.012 - 078.019, 080.031, 080.032, 084.013, 084.246, 084.247, 084.406, 084.407, 085.011, 085.012, 099.033 - 099.036, 106.030 - 106.032, 106.034, 143.019.

012.008 **Problems of gravitation.** Collection of reports of the plenary meetings of the Third Soviet Gravitational Conference, Erevan, 11 - 14 October 1972.
G. S. Saakyan (Editor).
Sektsiya gravitatsii Ministerstva vysshego i srednego spetsial'nogo obrazovaniya SSSR, Erevanskij ordena Trudovogo Krasnogo Znameni gosudarstvennyj universitet, Akademiya Nauk Arm. SSR, Erevan. 226 pp. Price 63 Kop. (1975). In Russian. – The individual contributions are included in their corresponding subject categories – see abstracts 061.024, 066.036 - 066.042, 141.075, 155.036, 158.090, 162.042.

012.009 **First All-Union seminar on X-ray and gamma-ray astronomy.** Crimean Astrophysical Observatory, 1974, October 17 - 19.
Izv. Krymskoj Astrofiz. Obs., Vol. 54, 315 - 349 (1976). In Russian. – The individual contributions are included in their corresponding subject categories – see abstracts 061.025, 076.019, 142.085 - 142.088, 143.026, 143.027, 151.067.

012.010 **Proceedings of the Sixth Lunar Science Conference,** Houston, Texas, March 17–21, 1975.
R. B. Merrill (managing editor).
Vol. 1: Mineralogical and petrological studies; Vol. 2: Chemical and isotopic studies; Vol. 3: Physical studies, (with lunar orbital data maps, 12 pp., and errata to Proc. Apollo 11 Lunar Sci. Conf., Proc. 2nd–5th Lunar Sci. Conf.).
Pergamon Press, New York – Oxford – Toronto – Sydney – Braunschweig. Geochim. Cosmochim. Acta, Suppl. 6, 12 + 12 + 25 + 3637 + 26 + 26 + 42 pp. Price $ 130.00 (1975). The individual papers are included in their corresponding subject categories – see abstracts 032.518, 074.049, 094.010, 094.011, 094.132 - 094.179, 094.415 - 094.576, 104.045, 105.060 - 105.062, 106.040.

012.011 **14th International Cosmic Ray Conference.** Conference Papers. Proceedings of the 14th International Cosmic Ray Conference, München, 1975 August 15 - 29.
Vol. 1, 2: OG (*origin*) sessions; Vol. 3: MG (*modulation and geomagnetic effects*) session; Vol. 4: MG session and Pioneer symposium; Vol. 5: SP (*solar particles*) session and Helios symposium; Vol. 6: MN (*muons and neutrinos*) session; Vol. 7: HE (*high energy physics*) session; Vol. 8: EA (*extensive air showers*) session; Vol. 9: T (*techniques*) session; Vol. 10: Table of contents and author index; Vol. 11: Invited lectures and rapporteur papers; Vol. 12: Miscellany: Late papers, corrections and list of attendees. With a preface by K. Pinkau. Edited by Max-Planck-Inst. Extraterr. Phys., München,

Germany.
Printed by Print KG, München. 144 + 4032 pp. (1975). − The individual contributions within the subject scope of Astronomy and Astrophysics Abstracts are included in their corresponding subject categories − see abstracts 009.015, 022.076 - 022.078, 061.033 - 061.035, 062.037 - 062.040, 066.054, 073.069 - 073.075, 074.051 - 074.059, 076.021 - 076.024, 078.028 - 078.100, 080.044 - 080.051, 084.265, 094.586, 099.048 - 099.053, 106.044 - 106.050, 122.110, 122.111, 125.030 - 125.032, 131.152 - 131.154, 141.347, 142.119 - 142.161, 143.047 - 143.334, 155.041 - 155.051, 158.122.

012.012 Growth rhythms and the history of the earth's rotation. Interdisciplinary Winter Conference on Biological Clocks and Changes in the Earth's Rotation: Geophysical and Astronomical Consequences, Newcastle upon Tyne, January 8 - 10, 1974.
G. D. Rosenberg, S. K. Runcorn (Editors).
John Wiley & Sons, London−New York−Sydney−Toronto. 16 + 559 pp. Price £ 22.50 (1975). ISBN 0-471-73616-3. The individual papers within the subject scope of Astronomy and Astrophysics Abstracts are included in their corresponding categories − see abstracts 044.008 - 044.012, 081.021 - 081.024, 084.264, 085.017, 107.011.

012.013 International conference on infrared physics (CIRP). Proceedings of a conference held in Zürich, 11 - 15 August 1975.
Infrared Physics, Vol. 16, (Nos. 1/2), 1 - 329 (1976). − The individual contributions within the subject scope of Astronomy and Astrophysics Abstracts are included in their corresponding categories − see abstracts 022.079, 031.259, 031.260, 034.063 - 034.066, 071.034, 082.072.

012.014 Computing in plasma physics and astrophysics. Proceedings of the 2nd European Conference on Computational Physics. Contributed papers. April 27 - 30, 1976. D. Biskamp (Editor).
Edited by Max-Planck-Institut für Plasmaphysik, Garching, Germany, 124 pp. (1976). − The individual contributions are included in their corresponding subject categories − see abstracts 021.006 - 021.012, 022.080, 062.042 - 062.077, 063.040, 064.060, 066.059, 066.060, 074.061, 106.052, 141. 348, 151.059, 151.060.

012.015 Interplanetary dust and zodiacal light. Proceedings of the International Astronomical Union Colloquium No. 31, Heidelberg, June 10 - 13, 1975.
H. Elsässer, H. Fechtig (Editors), with final remarks by F. L. Whipple.
Lecture Notes in Physics. Vol. 48. Springer-Verlag, Berlin − Heidelberg − New York. 12 + 496 pp. Price DM 39.00, ca. $ 16.00 respectively (1976). ISBN 3-540-07615-8, ISBN 0-387-07615-8 respectively. − The individual contributions are included in their corresponding subject categories − see abstracts 022.082, 022.083, 032.520, 032.521, 053.018, 063.041, 074.063 - 074.066, 080.054, 082.076, 082.077, 094.014, 094.184 - 094.188, 102.024 - 102.030, 103.102, 104.048 - 104.057, 105.082 - 105.085, 106.054 - 106.096, 115.016, 131.159, 131.160.

012.016 COSPAR Symposium on Fast Transients in X- and Gamma-Rays, held at Varna, Bulgaria, 29 - 31 May, 1975.
Astrophys. Space Sci., Vol. 42, (No. 1), 3 - 254 (1976). − The individual contributions are included in their corresponding subject categories − see abstracts 032.522 - 032.524, 083.068, 083.069, 141.350, 142.191 - 142.216.

012.017 Proceedings of the Second European Regional Meeting in Astronomy, Trieste, Italy: 1974, Sep- tember, 2 - 5. Part II.
Mem. Soc. Astron. Italiana, Vol. 45, No. 3/4, p. 489 - 996 (1974). − The individual contributions are included in their corresponding subject categories − see abstracts 013.016 - 013.020, 031.266, 061.040 - 061.042, 064.064, 065.083, 065.084, 066.084 - 066.088, 113.048 - 113.051, 114.057 - 114.059, 114.366, 117.027 - 117.030, 119.019, 121.113 - 121.115, 122.138, 124.008, 124.009, 125.038, 126.020, 131.544, 141.111 - 141.120, 141.353, 142.222 - 142.230, 158.142 - 158.149, 160.038 - 160.040, 162.077.

012.018 On reference coordinate systems for earth dynamics. International Astronomical Union, Colloquium No. 26, held in Toruń, Poland, 26 - 31 August 1974.
B. Kołaczek, G. Weiffenbach (Editors).
Wykonano w Zakładzie Graficznym Politechniki, Warsaw, Poland. 478 pp. (1975). − The individual contributions are included in their corresponding subject categories − see abstracts 031.268, 032.038, 041.031 - 041.043, 044.031, 044.032, 045.015 - 045.020, 046.037 - 046.046, 052.032, 081.041 - 081.043, 094.015.

012.019 Atomic and molecular processes in astrophysics. Swiss Society of Astronomy and Astrophysics. Fifth advanced course.
M. C. E. Huber, H. Nussbaumer (Editors).
Published and sold by Geneva Observatory, CH-1290 Sauverny/ Switzerland, 11 + 308 pp. (1975). − The individual contributions are included in their corresponding subject categories − see abstracts 022.090 - 022.092.

012.020 Explosive nucleosynthesis. Proceedings of a conference held in Austin, Texas, April 2−3, 1973.
D. N. Schramm, W. D. Arnett (Editors).
University of Texas Press, Austin − London. 12 + 301 pp. Price $ 4.95 (1973). − The individual contributions are included in their corresponding subject categories − see abstracts 022.093, 022.094, 061.045 - 061.053, 065.086 - 065.089, 125.039 - 125.044, 155.061.

012.021 Proceedings of the Southwest Regional Conference for Astronomy and Astrophysics (Lubbock, Texas, July 12, 1975). Vol. 1.
P. F. Gott, P. S. Riherd (Editors).
Available from the editors, Dep. Phys., Texas Tech Univ., Lubbock, Texas, USA. 179 pp. (1976). − The individual papers are included in their corresponding subject categories − see abstracts 009.024, 009.025, 014.008, 031.269, 031.270. 032.039, 033.032, 034.079, 082.087, 114.062, 117.033, 131.546, 133.025 - 133.027, 143.349, 158.151, 158.152.

012.022 1975 AAS/AIAA Astrodynamic Conference. American Astron. Soc./American Inst. Aeronautics and Astronautics Conference, Nassau, Bahamas, 28 - 30 July 1975.
American Astron. Soc., Tarzana, Calif., USA. 1694 pp. (1975). Preprints. − See abstr. 041.045, 042.091, 042.092, 051.023, 052.037 - 052.048, 052.050 - 052.053, 052.055, 052.057, 053.020, 053.024, 053.025, 054.013, 097.203, 100.203.

012.023 Proceedings of the seminar on 'Infrared and millimeter range astronomy'.
S. M. Alladin, K. D. Abhyankar (Editors).
Osmania University, Hyderabad, India. (1974). − Review in Bull. Astron. Soc. India, Vol. 3, 80, 85; 1975 (*K. S. Krishna Swamy*).

012.024 Atomic and molecular physics and the interstellar matter. (Proceedings of the XXVI Session of the Les Houches Summer School of Theoretical Physics, Grenoble, France, July - August 1974).

R. Balian, P. Encrenaz, J. Lequeux (Editors).
North-Holland Publishing Company, Amsterdam — Oxford;
American Elsevier Publishing Company, Inc., New York.
22 + 632 pp. Price Dfl 160.00, $ 66.75 respectively (1975).
ISBN-0-7204-0330-8. — Review in Sky Telescope, Vol. 51,
421 (1976).

**012.025 Possible relationships between solar activity and
meteorological phenomena.**
W. R. Bandeen, S. P. Maran (Editors).
Proceedings of a symposium, Greenbelt, Md., Nov. 1973.
National Aeronautics and Space Administration, Washington,
NASA SP-366 (available from the Superintendent of Documents Washington, D.C.) 10 + 264 pp. Price $ 4.00 (1975).
Review in Nature, Vol. 261, 80; 1976 (*J. Gribbin*).

012.026 Role of magnetic fields in physics and astrophysics.
Proceedings of a conference held at Copenhagen in
June, 1974. V. Canuto (Editor).
Annals of the New York Academy of Sciences, Vol. 257.
New York Academy of Sciences, New York. 226 pp. Price
$ 24.00 (1975). — Reviews in Journ. British Astron. Ass.,
Vol. 86, 351 - 352; 1976 (*S. Mitton*); Journ. Roy. Astron. Soc.
Canada, Vol. 70, 47; 1976 (*R. N. Henriksen*); Sky Telescope,
Vol. 51, 126 (1976).

012.027 The dusty universe. Conference held at Cambridge,
Mass., October 1973.
G. B. Field, A. G. W. Cameron (Editors), with a commentary
by F. L. Whipple.
Neale Watson Academic Publications, New York. 10 + 323 pp.
Price $ 15.00 (1975). — Review in Sky Telescope, Vol. 51, 53
(1975).

012.028 The nature of scientific discovery.
O. Gingerich (Editor).
A symposium commemorating the 500th anniversary of the
birth of Nicolaus Copernicus.
Smithsonian International Symposia Series 5. Smithsonian
Institution: Washington, D.C. 616 pp. Price $ 15.00 (1975).
Reviews in Nature, Vol. 260, 81 - 82; 1976 (*J. Watling*);
Science, Vol. 191, 1164 - 1165; 1976 (*B. Lovell*).

012.029 Physics of the hot plasma in the magnetosphere.
Proceedings of a symposium, Kiruna, Sweden, April
1975. B. Hultqvist, L. Stenflo (Editors).
Nobel Symposium No. 30. Plenum Press, New York — London.
10 + 370 pp. Price $ 25.00, $ 30.00 respectively (1975).
Review in Science, Vol. 192, 133 - 134; 1976 (*D. P. Stern*).

012.030 5th European microwave conference. Conference
Hamburg, Germany, 1 - 4 Sept. 1975.
Microwave Exhibitions and Publishers, Sevenoaks, Kent,
England. 11 + 736 pp. (1975). — (From Phys. Abstr., Vol. 79,
A003448 (1976)).

012.031 Magnetospheric particles and fields. Proceedings of
the Summer Advanced Study Institute, Graz,
Austria, August 4 - 15, 1975. B. M. McCormac (Editor).
D. Reidel Publishing Company, Dordrecht, Holland — Boston,
U.S.A. 340 pp. Price Dfl. 110.00, $ 39.50 (1976). ISBN 90-
277-0702-2.

**012.032 Sixth Symposium on Photo-Electronic Image
Devices,** held at Imperial College, London, September 1974. B. Morgan (Editor).
Parts A/B (Advances in Electronics and Electron Physics).
Academic Press, London, 100 pp. (1975). — (From SuW, 15.
Jahrgang, p. 141 (1976)).

012.033 Far infrared astronomy.
M. Rowan-Robinson (Editor).
Proceedings of a Conference held at Cumberland Lodge,
Windsor, U.K., on July 11th - 13th, 1975. Sponsored by the
Royal Astronomical Society. (Supplement to "Vistas in
Astronomy").
Pergamon Press, Oxford — New York. 13 + 335 pp. Price
£ 10.00 (1976). — (From Nature, Vol. 261, No. 5562, p. ix
(1976)).

012.034 General relativity and gravitation. Proceedings of the
Seventh International Conference (GR7) Tel-Aviv
University, June 23 - 28, 1974.
G. Shaviv, J. Rosen (Editors).
Halsted Press, John Wiley and Sons, New York — Toronto;
Israel Universities Press, Jerusalem. 7 + 344 pp. Price £ 17.60,
$ 35.20 respectively (1975). ISBN-0-470-77939-X. — (From
Nature, Vol. 261, No. 5558, p. IX (1976)).

012.035 Primary cosmic radiation. (Proceedings (trudy) of
the P.N. Lebedev Physics Institute, Vol. 64).
D. V. Skobel'tsyn (Editor). Translated from the Russian
edition by J. B. Barbour.
Consultants Bureau (Plenum), New York — London. 6 + 110
pp. Price $ 32.50, $ 39.00 respectively (1975). ISBN-0-306-
10914-X. — (From Nature, Vol. 260, No. 5551, p. VII (1976);
Phys. Today, Vol. 29, No. 2, p. 61 (1976)).

**012.036 17th Yugoslav. symposium and summer school on
the physics of ionised gases.** Conference Rovinj,
Yugoslavia, 16 - 21 September 1974. V. Vujnovic (Editor).
Councils for Sci. Res. Socialists Republics of Yugoslavia.
Zagreb, Yugoslavia. Inst. Phys., Univ. Zagreb. 18 + 1025 pp.
(1975). — (From Phys. Abstr., Vol. 79, A015295 (1976)).

012.037 The Copernican Achievement. Papers from a symposium, Los Angeles, November 1973.
R. S. Westman (Editor).
UCLA Center for Medieval and Renaissance Studies, 7.
University of California Press, Berkeley. 16 + 406 pp. Price
$ 14.50 (1976). — (From Science, Vol. 192, 883 (1976)).

012.038 H II regions and related topics. Proceedings of a
symposium, Mittelberg, Austria, Jan. 1975.
T. L. Wilson, D. Downes (Editors).
Lecture Notes in Physics, Vol. 42.
Springer-Verlag, Berlin — New York — Heidelberg. 12 + 488
pp. Price $ 16.00, DM 39.00 respectively (1975). — Review in
Astron. Tidssk., Årg. 9, p. 48 (1976).

012.039 25th International Astronautical Congress. Conference Amsterdam, Netherlands, 30 September -
5 October 1975.
Acta Astronaut., Vol. 2, No. 7 - 8 (1975). — Review in Phys.
Abstr., Vol. 79, A028048 (1976).

013 Reports on Astronomy in Various Countries and Particular Fields, International Cooperation

013.001 **Om astronomin i Kina.** B. Stenholm.
Astron. Tidssk., Årg. 9, p. 1 - 10 (1976).

013.002 **New highlights in Soviet astronomy.**
Astron. Zhurn. Akad. Nauk SSSR, Vol. 53, 3 - 5
(1976). In Russian. English translation in Soviet Astron., Vol.
20, No. 1.

013.003 **Comet hunters in the land of the rising sun.**
J. Bortle.
Sky Telescope, Vol. 51, 159 - 162 (1976).

013.004 **Whatever happened to British optical astronomy?**
A. Hunter.
Journ. British Astron. Ass., Vol. 86, 192 - 202 (1976).

013.005 **Our interviews.**
Zemlya i Vselennaya, 1976, No. 2, p. 2 - 10. In
Russian. – Interviews with A. A. Mikhajlov, V, A.
Ambartsumyan, Ya. B. Zel'dovich, V. P. Tsesevich, Yu. V.
Batrakov, A. B. Severnyj, M. S. Zverev, E. K. Kharadze, V. V.
Sobolev, and G. A. Vasyuk on the main tasks of space
research, astronomy and geophysics.

013.006 **State, perspectives and problems of development of Soviet astronomy.** G. S. Khromov.
Zemlya i Vselennaya, 1976, No. 2, p. 14 - 20. In Russian.

013.007 **Cooperation of socialist countries in investigation of problems of stellar evolution.** Eh. Ehrgma.
Zemlya i Vselennaya, 1976, No. 2, p. 73 - 74. In Russian.

013.008 **Space and international cooperation.**
E. F. Chugunov.
Zemlya i Vselennaya, 1976, No. 3, p. 28 - 37. In Russian.

013.009 **Meteorbeobachtung in der Tschechoslowakei.**
H. Nováková.
SuW, 15. Jahrgang, 203 (1976).

013.010 **The beginnings of American astronomy.**
T. E. Bell.
Sky Telescope, Vol. 52, 26 - 31 (1976).

013.011 **Astronomy at the Jagellonian University.**
P. Heinzel.
Říše hvězd, Vol. 57, 81 - 84 (1976). In Czech.

013.012 **Actual trends of the development of astronomy.**
O. Melnikov, V. Popov.
Říše hvězd, Vol. 57, 1 - 8 (1976). In Czech.

013.013 **Advances in astronomy in the year 1975.**
J. Grygar.
Říše hvězd, Vol. 57, 41 - 49, 69 - 74, 86 - 92, 111 - 115, 130 -
133 (1976). In Czech.

013.014 **Problemi e prospettive di ricerca nell'astronomia stellare ottica.** P. Giannone.
Giorn. Astron., Vol. 1, 99 - 108 (1975).

013.015 **L'astronomie au Québec.** R. Racine.
Journ. Roy. Astron. Soc. Canada, Vol. 70, 138 -
142 (1976).

013.016 **Large optical telescope projects in European context.**
A. Blaauw.
Mem. Soc. Astron. Italiana, Vol. 45, (see 012.017), 921 - 928
(1974).

013.017 **Cooperation in solar astronomy in Europe.**
C. Zwaan.
Mem. Soc. Astron. Italiana, Vol. 45, (see 012.017), 929 - 933
(1974).

013.018 **Ultraviolet stellar astronomy in Europe.**
L. Houziaux.
Mem. Soc. Astron. Italiana, Vol. 45, (see 012.017), 937 - 943
(1974).

013.019 **Development of infrared astronomy in Europe.**
K. W. Michel.
Mem. Soc. Astron. Italiana, Vol. 45, (see 012.017), 945 - 962
(1974).

013.020 **Radioastronomy.** E. J. Blum.
Mem. Soc. Astron. Italiana, Vol. 45, (see 012.017),
963 - 965 (1974).

**The California-Wisconsin axis in American astron-
omy – I, II.** See Abstr. 004.010.

014 Teaching in Astronomy

014.001 New branch at pedagogical institutes.
V. V. Porfir'ev, O. D. Shebalin.
Zemlya i Vselennaya, 1976, No. 1, p. 81 - 82. In Russian.

014.002 The 1974–1976 Chautauqua program of astronomy workshops for college teachers.
R. R. Robbins.
Proc. Southwest Regional Conf., Vol. 1, (see 012.021), 13 - 21 (1976).

014.003 Astronomie – Schule – Öffentlichkeit.
H. H. Voigt.
Mitt. Astron. Ges., No. 38, p. 19 - 25 (1976). – Lecture on the occasion of the opening of the meeting of the Astron. Ges., Berlin 1975.

014.004 Astronomie als Wahlpflichtfach des Gymnasiums.
R. H. Giese.
Mitt. Astron. Ges., No. 38, p. 128 - 132 (1976). – Short report.

014.005 Problems of astronomy in the course of physics for the 6th and 8th classes. N. V. Lisina.
Puti povysheniya ehffektivn. ucheb. protsessa. Joshkar-Ola, 1975, p. 71 - 74. In Russian. – Abstr. in Referativ. Zhurn. 51. Astron., 4.51.23 (1976).

014.006 Teachers in astronomy are studying.
A. P. Barkov.
Zemlya i Vselennaya, 1976, No. 3, p. 61. In Russian.

014.007 Zur Einführung einiger Begriffe im Astronomieunterricht. W. Deutschmann.
Astron. in der Schule, 13. Jahrgang, p. 64 - 66 (1976).

015 Miscellanea

015.001 Call signals of space civilizations.
P. V. Makovetskij.
Astron. Zhurn. Akad. Nauk SSSR, Vol. 53, 221 - 224 (1976). In Russian. English translation in Soviet Astron., Vol. 20, No. 1.
A new class of call signals for communication with space civilizations is proposed: a strictly monochromatic wave with the frequency of interstellar hydrogen f_H multiplied by a mathematical irrational constant (πf_H, $e f_H$, $\sqrt{2} f_H$...).

015.002 Influence of ancient solar-proton events on the evolution of life.
G. C. Reid, I. S. A. Isaksen, T. E. Holzer, P. J. Crutzen.
Nature, Vol. 259, 177 - 179 (1976).
There is mounting evidence that past extinctions of faunal species have occurred in near coincidence with reversals in polarity of the geomagnetic field. Could the link lie in catastrophic depletions of stratospheric ozone caused by solar-proton irradiation over a reduced geomagnetic field?

015.003 An ear to the universe. R. E. Machol.
IEEE Spectrum, Vol. 13, No. 3, p. 42 - 47 (1976).
Sensitive antennas, pattern recognizing computers, and informed conjecture are being tuned to possible signals from the stars.

015.004 Calling M 13 on 12.6 cm – do you read?
A. T. Lawton.
Spaceflight, Vol. 18, 10 - 12 (1976).

015.005 The coup against entropy. A. M. G. Moore.
Spaceflight, Vol. 18, 126 - 129 (1976).

015.006 New interpretation of a mysterious radioecho.
A. V. Shpilevskij.
Zemlya i Vselennaya, 1976, No. 2, p. 74 - 77. In Russian.

015.007 Model of a contact, but no evidence for a sonde.
L. M. Gindilis.
Zemlya i Vselennaya, 1976, No. 2, p. 78 - 82. In Russian.

015.008 Traveling near the speed of light.
W. J. Kaufmann III.
Mercury (Journ. Astron. Soc. Pacific), Vol. 5, No. 1, p. 4 - 8 (1976).

015.009 Lunar nomenclature: a dissenting note.
D. W. G. Arthur.
Icarus, Vol. 27, 571 - 573 (1976).
This note reviews the nature of the traditional (Mädler) lunar nomenclature and the recent developments based on the use of more than 2000 named provinces. It appears that the new nomenclature is less efficient than the old in many cases and may lead to an impossible publication situation. The unnecessary break with the past is especially criticized.

015.010 On solar system nomenclature. C. Sagan.
Icarus, Vol. 27, 575 - 576 (1976).
Arguments are presented for naming topographic features on other solar system objects after human beings other than astronomers; and to institute a more consistent scheme for Jovian satellite nomenclature.

015.011 Disharmony of the spheres: recent trends in planetary surface nomenclature. R. J. Pike.
Icarus, Vol. 27, 577 - 583 (1976).
Inadvisable departures from tradition in naming newly

mapped features on Mars, Mercury, and the moon have been implemented and proposed since 1970. The Mädler scheme for designating smaller craters on the moon should be retained and extended to the farside. Names of surface features on other bodies might best reflect the traditional connotations of planet and satellite names: for example, most craters on Mars would be named for mythical heroes and military personalities in ancient history, craters on Mercury might commemorate explorers or commercial luminaries, and features on Venus would bear the names of famous women.

015.012 **Whatever became of tungol-craft: some notes on the origin of astronomical words.** D. W. Sida.
Journ. Roy. Astron. Soc. Canada, Vol. 70, 67 - 73 (1976).

015.013 **Glaciations and dense interstellar clouds.** B. Dennison, V. N. Mansfield.
Nature, Vol. 261, 32 - 34 (1976). — Letter.

015.014 **Development of the satellite solar power station.** P. E. Glaser.
Spaceflight, Vol. 18, 198 - 208 (1976).

015.015 **Can cosmic clouds cause climatic catastrophes?** M. C. Begelman, M. J. Rees.
Nature, Vol. 261, 298 - 299 (1976). — Letter.

015.016 **Sunspots and earthquakes.** J. Meeus, with a reply by J. Gribbin.
Phys. Today, Vol. 29, No. 6, p. 11, 13 (1976).

015.017 **George Abell on astrology and astronomy.** R. Reis.
Mercury, (Journ. Astron. Soc. Pacific), Vol. 5, No. 2, p. 8 - 13 (1976). — An interview.

015.018 **Populāri par P. Bola kvaziperiodiskām funkcijām.** I. Rabinovičs.
Zvaigžņotā debess, 1975./76. gada ziema, p. 8 - 10.

015.019 **Eine Methode zur Messung der Bedeutung von Naturwissenschaftlern. (A method for mensuration of scientist's importance).** D. B. Herrmann.
Mitt. Archenhold-Sternw. Berlin-Treptow, (Band 6), No. 126, 20 pp. (1976).

015.020 **Forschung an den Grenzen der Astronomie.** J. L. Greenstein.
Umschau, 76. Jahrgang, p. 429 - 432 (1976).

015.021 **Astrology: its principles and relation and nonrelation to science.** G. O. Abell.
The Science Teacher, Vol. 41, No. 9 = Univ. Calif., Los Angeles, Astron. Papers, No. 9 (1974).

015.022 **The problem of communication with extraterrestrial civilizations.** S. Lubertowicz.
Urania Kraków, Vol. 47, 130 - 142 (1976). In Polish.

015.023 **Venerable Martians.** D. G. Lahoz.
Available from the author: Dep. Phys., Univ. Alberta, Edmonton, Alberta, Canada. 35 pp. (1976).
The author shows the astrophysical content of some mythologies.

015.024 **Are solar neutrinos detected by living things.** M. Ruderfer.
Phys. Letters A, Vol. 54A, 363 - 364 (1975). — Abstr. in Phys. Abstr., Vol. 79, A011362 (1976).

Applied Mathematics, Physics

021 Mathematics, Computing

021.001 Calculation of Fourier transforms by the Backus-Gilbert method. D. W. Oldenburg.
Geophys. Journ. Roy. Astron. Soc., Vol. 44, 413 - 431 (1976).
   The linear inverse theory of Backus & Gilbert has been applied to the problem of calculating the Fourier transform of digitized data with the objective of assessing the effects of missing portions of the data series and of contamination of the signal by 'noise'. The effects of data gaps are easily treated and it is shown that it may sometimes be desirable to interpolate these gaps even though a large variance must be ascribed to the fabricated data. The Backus-Gilbert technique is also applied to the calculation of the reverse Fourier transform, and an application to the downward continuation of potential field data is given.

021.002 Multistep methods of numerical integration using back-corrections. T. Feagin, P. R. Beaudet.
Celestial Mechanics, Vol. 13, 111 - 120 (1976).
   A new class of linear multistep methods is proposed for the solution of the equations of motion of certain dynamical systems encountered in celestial mechanics and astrodynamics. The enhanced numerical stability of these methods allows the meaningful application of high-order algorithms. Consequently, stepsizes larger than those attainable with the classical methods may be adopted and thus greater over-all efficiency may be realized. The application of these methods to the problem of determining the orbit of an artificial satellite is accomplished and the results are compared with those obtained using classical methods.

021.003 Pocket astrodynamics. P. Burton.
Journ. British Interplanet. Soc., Vol. 29, 185 - 194 (1976).

021.004 On the interpretation of least squares collocation. B. D. Tapley.
Long-time predictions in dynamics, (see 012.005), p. 165 - 172 (1976).

021.005 Dynamical systems involving uncertainty. G. Adomian.
Long-time predictions in dynamics, (see 012.005), p. 329 (1976). – Abstract.

021.006 DEQTRAN – a program generator for the solution of 1D systems of partial differential equations. D. Düchs, W. Schneider.
Computing in plasma physics and astrophysics, (see 012.014), E5, 2 pp. (1976).

021.007 General ideas of structured programming. J. Nadrchal.
Computing in plasma physics and astrophysics, (see 012.014), E6, 2 pp. (1976).

021.008 A new implicit method for solving two-dimensional partial differential equations in plasma physics. U. Schwenn.
Computing in plasma physics and astrophysics, (see 012.014), F1, 2 pp. (1976).

021.009 Hymniablock, an efficient eigenvalue solver. R. Gruber.
Computing in plasma physics and astrophysics, (see 012.014), F2, 2 pp. (1976).

021.010 Application of nonlinear optimization to solution of differential and integral equations. T. Takeda.
Computing in plasma physics and astrophysics, (see 012.014), F3, 2 pp. (1976).

021.011 Inverse iteration in MHD equilibrium computations. R. Meyer-Spasche.
Computing in plasma physics and astrophysics, (see 012.014), F4, 2 pp. (1976).

021.012 Numerical and analytical alternative to the Krylov-Bogolyubov method. Application to slightly nonlinear autonomous and nonautonomous systems.
A. Nadeau, J. P. Veyrier, M. R. Feix.
Computing in plasma physics and astrophysics, (see 012.014), F5, 2 pp. (1976).

021.013 Energy and momentum conserving methods of arbitrary order for the numerical integration of equations of motion. II. Motion of a system of particles.
R. A. LaBudde, D. Greenspan.
Numer. Math., Vol. 26, 1 - 16 (1976).

021.014 Development of a computer code for the calculation of stellar evolution, with applications to solar models of low neutrino flux. M. J. Newman.
Thesis Rice Univ., Houston, Texas, USA, 165 pp. (1975).
(Available from: Univ. Microfilms, Order No. 75-22,050).

   Statistische Probleme bei der Ausgleichung direkter, unabhängiger, normalverteilter Beobachtungen mit geschätzten Gewichten.   See Abstr. 031.263.

   About the function $(d^n/dt^n) \varphi (r) = f(dr/dt, d^2r/dt^2, ..., d^n r/dt^n)$.   See Abstr. 031.401.

   General numerical fluid dynamics algorithm for astrophysical applications.   See Abstr. 062.094.

   Rapid computation of the Voigt profile.   See Abstr. 063.031.

   SOLPRO: a computer code to calculate probabilistic energetic solar proton fluxes.   See Abstr. 078.108.

   On the application of maximum entropy spectral analysis to study the nearly diurnal free nutation.   See Abstr. 081.012.

Errata

021.901 Erratum: 'Fourier analysis with unequally-spaced data' [Astrophys. Space Sci., Vol. 36, 137 - 158 (1975)].   T. J. Deeming.
Astrophys. Space Sci., Vol. 42, 257 (1976).

022 Physical Papers Related to Astronomy and Astrophysics

022.001 The free-free continuous absorption coefficient of the negative carbon ion.
T. L. John, R. J. Williams.
Monthly Notices Roy. Astron. Soc., Vol. 174, 253 - 258 (1976).

The free-free continuous absorption coefficient of C^- is calculated on a multi-channel basis using Hartree-Fock target wave functions including the ^3P, ^1D and ^1S atomic states. The authors give a table of the absorption coefficient for a wide range of temperatures and frequencies.

022.002 Investigation of forbidden transitions in argon ions.
N. W. Jalufka.
Astrophys. Journ., Vol. 203, 279 - 283 (1976).

This paper presents the results of an attempt to observe the four forbidden transitions in argon ions that are observed in the solar corona. These lines are listed in a table with the corresponding transition and transition probability.

022.003 Laboratory band strengths of methane and their application to the atmospheres of Jupiter, Saturn, Uranus, Neptune, and Titan.
B. L. Lutz, T. Owen, R. D. Cess.
Astrophys. Journ., Vol. 203, 541 - 551 (1976).

The authors report laboratory studies of the visible spectrum of methane at column densities between 0.4 and 5 km-am and confirm the identification of bands at 4410, 4590, 4860, 5090, 5430, 5760, and 5970 Å as caused by methane. Detailed equivalent-width measurements at 15 different pressure path lengths allow us to determine curves of growth and band strengths for the bands at 4410, 4860, 5430, and 5760 Å. Using their curve-of-growth measurements in the reduction of planetary observations, the authors find methane abundances in the atmospheres of Jupiter and Saturn to be between a factor of 3 and 4 larger than previously accepted values, while the amount on Titan is significantly less.

022.004 Physical conditions in a hydrogen gas heated by suprathermal protons. E. Kimmer.
Astrophys. Journ., Vol. 203, 674 - 679 (1976).

A self-consistent calculation has been made of the physical conditions in a pure hydrogen gas that is heated by suprathermal protons (energies greater than 200 keV). The geometry adopted is that of a plane-parallel slab with suprathermal particles incident normally. Lower and upper limits were set on the degree of ionization and temperature of the gas by calculating conditions as a function of penetration distance for (1) an optically thin gas where all radiation immediately escapes, and (2) an optically thick gas where $L\alpha$ photons are trapped and destroyed on the spot by two-photon emission. For both cases, when the gas is not completely ionized, the calculated electron temperatures are on the order of 10^4 K and the degrees of ionization are usually less than 10 percent.

022.005 Classification of $d^4 - d^3$ 4p Co VI spectra.
H. F. Henrichs, B. C. Fawcett.
Astron. Astrophys., Suppl. Ser., Vol. 23, 139 - 146 (1976).

Precise wavelength measurements and a detailed analysis of the $d^4 - d^3$ 4p transition array of Co VI are presented. One hundred and nineteen energy levels, including those of the d^4 ground configuration, are located with an accuracy of ± 1 cm^{-1}. Theoretical energy level calculations are applied to the data. Energy values for the ground configuration aid the prediction of levels in Ni VII of astrophysical interest.

022.006 Tables of damping constants of spectral lines broadened by H and He.

G. Deridder, W. van Rensbergen.
Astron. Astrophys., Suppl. Ser., Vol. 23, 147 - 165 (1976).

Damping constants γ of lines from atoms and singly charged ions, broadened by H and He atoms in the ground state, are calculated for s–p, p–d and d–f transitions, using a Smirnov-Roueff potential. Tables of parameters α and β enable astrophysicists to calculate $\gamma = N \alpha T^\beta$ for temperatures ranging from 4000 to 10000 K.

022.007 The spectrum of five-times-ionized vanadium.
J. O. Ekberg.
Physica Scripta, Vol. 13, 111 - 116 (1976).

022.008 On the classical theory of equilibrium between matter and radiation. J. J. Monaghan.
Australian Journ. Phys., Vol. 28, 715 - 730 (1975).

022.009 H_2O rotation line emission between 10 and 50 μ due to collisions with O. J. L. Kulander.
Journ. Quant. Spectrosc. Radiat. Transfer, Vol. 16, 21 - 25 (1976).

Line emission from the pure rotation bands of H_2O is calculated for single particle collisions with O atoms. The cascading of the initially excited levels is followed in detail for levels up to $J = 12$. Individual line energies and band energies for wavelengths between 10 and 50 μ are given.

022.010 On the calculation of scattering cross sections from absorption cross sections.
W. D. Barfield, W. F. Huebner.
Journ. Quant. Spectrosc. Radiat. Transfer, Vol. 16, 27 - 34 (1976).

022.011 Optical thickness in $\Sigma - \Sigma$ molecular bands.
E. Young.
Journ. Quant. Spectrosc. Radiat. Transfer, Vol. 16, 35 - 48 (1976).

The single line resonance results of Malinowski and others are extended to include overlapping neighboring lines. This result is combined with the molecular line strengths and energy levels of a permitted $\Sigma - \Sigma$ transition to calculate both relative and absolute optical depths of individual lines, multiplets, and vibrational bands within a $\Sigma - \Sigma$ molecular band system. An average optical depth is also calculated for the entire band system. The results will apply to self absorption in laboratory and auroral emissions within appropriate ranges of temperature, pressure, optical depth, and molecular constants.

022.012 Hartree-Fock wave functions and oscillator strengths for the helium isoelectronic sequence.
D. K. Datta, S. K. Ghoshal, S. Sengupta.
Journ. Quant. Spectrosc. Radiat. Transfer, Vol. 16, 49 - 52 (1976).

022.013 Line strength and halfwidth measurements from far infrared absorption spectra: carbon monoxide.
J. W. Fleming.
Journ. Quant. Spectrosc. Radiat. Transfer, Vol. 16, 63 - 68 (1976).

A method for deriving Lorentzian line shape parameters (line strengths and halfwidths) from absorption spectra taken at two different path lengths is discussed. Far i.r. absorption spectra of carbon monoxide are analysed to yield line strengths and halfwidths for the rotational transitions $J'' = 3$ through $J'' = 9$. Experimental errors are discussed, and the value for the dipole moment derived from the line strengths, 0.108 $\pm$ 0.005 D, is in agreement with the microwave value within experimental uncertainty.

022.014 Criteria for local thermal equilibrium in non-hydrogenic plasmas.   J. D. Hey.
Journ. Quant. Spectrosc. Radiat. Transfer, Vol. 16, 69 - 75 (1976).

Rate equations for collisional processes are formulated and used to obtain the minimum electron density necessary to ensure LTE of a particular atomic species, with particular attention to the existence of low-lying metastable levels. The final expression is compared with an earlier criterion of Griem. The results of various experiments on plasmas containing one or more ionization stages of helium, nitrogen, oxygen, and argon are discussed. The importance of making reliable estimates of collision strengths, for excitation involving metastable levels, is emphasized.

022.015 Equivalent widths of superfluorescent lines with combined Doppler and collision broadening.
P. Pinson, J. Dupre-Maquaire.
Journ. Quant. Spectrosc. Radiat. Transfer, Vol. 16, 101 - 104 (1976).

The equivalent widths for lines with Voigt contour have been evaluated numerically in the case of amplification. The results of these calculations are presented in tabular form.

022.016 A theoretical study of the np^k configurations.
K. Rahimullah, M. S. Z. Chaghtai.
Journ. Quant. Spectrosc. Radiat. Transfer, Vol. 16, 105 - 112 (1976).

022.017 Theoretical oscillator strengths for ultraviolet lines of doubly-ionized elements of astrophysical interest.   E. Biémont.
Journ. Quant. Spectrosc. Radiat. Transfer, Vol. 16, 137 - 142 (1976).

New oscillator strengths of ultraviolet lines ($1464 \leqslant \lambda \leqslant 2734$ Å) in the spectra of Sc(III), Ti(III), V(III), Cr(III), Mn(III), Fe(III), Co(III), Ni(III), Cu(III) and Zn(III) are calculated. The lines studied belong to the $3d^n 4s - 3d^n 4p$ transition array (n = 0 to 9). The radial wavefunctions are obtained by using an adaptation of the Scaled Thomas-Fermi method. A detailed comparison of oscillator strengths computed by this method and by the Hartree-Fock method (in both dipole length and dipole velocity formalisms) is also presented.

022.018 Rotational dependence of Franck-Condon factors for molecules of astrophysical interest.
R. A. Bell, D. Branch, W. L. Upson II.
Journ. Quant. Spectrosc. Radiat. Transfer, Vol. 16, 177 - 184 (1976).

The rotational dependence of Franck-Condon factors has been computed for selected bands of CN, CaH, C_2, NH, TiO, MgH and CH using Morse-Pekeris eigenfunctions. Some bands show a rotational dependence which may be significant for the determination of rotational temperatures and abundances of elements and isotopes, and for opacity distribution function ("giant line") calculations. The predicted rotational dependence for the (0, 1) band of the CH 4300 Å system is supported by an analysis of observed solar intensities.

022.019 Excitation of the $3p\ ^3\Pi_u$ level of H_2 by electron impact; pressure effects.
P. Baltayan, O. Nedelec.
Journ. Quant. Spectrosc. Radiat. Transfer, Vol. 16, 207 - 211 (1976).

The excitation cross-section of the $3p\ ^3\Pi_u$ level of H_2 has been measured for electrons from energy threshold to 100 eV. The authors compare the intensities of the emitted lines to the Balmer lines whose excitation cross-sections from H_2 are known. The effects of dissociation, cascade, polarisation and pressure are discussed. The maximum excitation cross-section is found to be 3.30×10^{-18} cm^2 $\pm$ 50% at 22.5 eV.

022.020 The role of $\Delta K = \pm 3$ transitions in collision-broadened NH_3.   J. Bonamy.
Journ. Quant. Spectrosc. Radiat. Transfer, Vol. 16, 213 - 215 (1976).

The half-widths of ammonia lines collision-broadened by H_2, N_2 and CO_2 have been calculated for microwave and far-i.r. spectra. The R^{-7} dispersion potential, which induced $\Delta K = \pm 3$ transitions, was included in the calculations. A comparison is then made with available experimental data.

022.021 Theoretical oscillator strengths for some transitions in Si(II), Ge(II), Sn(II) and Pb(II) spectra.
J. Migdałek.
Journ. Quant. Spectrosc. Radiat. Transfer, Vol. 16, 265 - 272 (1976).

022.022 Theoretical oscillator strengths for the NI and OI resonance transitions.   D. R. Beck, C. A. Nicolaides.
Journ. Quant. Spectrosc. Radiat. Transfer, Vol. 16, 297 - 300 (1976).

022.023 Determination of the oscillator strengths of Sn(I) in the ultraviolet (2400–4000 Å).
J. Lotrian, J. Cariou, A. Johannin-Gilles.
Journ. Quant. Spectrosc. Radiat. Transfer, Vol. 16, 315 - 319 (1976).

022.024 Radiative-lifetime measurements for sulfur and silicon transitions observed in interstellar absorption spectra.
A. E. Livingston, H. Garnir, Y. Baudinet-Robinet, P. D. Dumont, E. Biémont, N. Grevesse.
Astrophys. Letters, Vol. 17, 23 - 25 (1976).

The authors have used the beam-foil technique to measure radiative lifetimes and derive oscillator strengths for several transitions in S III, Si II and Si III that are observed in interstellar absorption spectra. They report the first experimental f-value for the 1018 Å multiplet in S III (f = 0.036 $\pm$ 0.004), and they confirm a recent revision of the f-value for 1194 Å in Si II.

022.025 Results of infrared reflectivity measurements on astronomically interesting silicates.   J. M. Penman.
Monthly Notices Roy. Astron. Soc., Vol. 175, 149 - 156 (1976).

Middle infrared reflectivities have been measured for five silicate minerals. Kramers–Kronig analysis and Mie theory have been used to calculate absorption cross-sections for small particles of the minerals. The results of these calculations are compared with the astronomical 'silicate' feature as observed against W3/IRS5. It is proposed that small particles of a hydrated silicate are responsible for the astronomical 10-μ feature.

022.026 The optical constants of polyoxymethylene.
D. C. B. Whittet, I. J. Dayawansa, P. M Dickinson, J. P. Marsden, B. Thomas.
Monthly Notices Roy. Astron. Soc., Vol. 175, 197 - 207 (1976).

Polyoxymethylene (POM) has recently been suggested as a possible constituent of interstellar grains. This paper presents laboratory measurements of the optical constants n and k for POM in the wavelength range 0.4–14 μ. The principal absorption bands are centred at 3.48 and 9.09 μ with peak k values of approximately 0.1 and 0.2, respectively. There is a series of closely-spaced bands between 7 and 12 μ.

022.027 Theoretical rates for electron excitation of highly-excited atoms.
C. S. Gee, I. C. Percival, J. G. Lodge, D. Richards.

Monthly Notices Roy. Astron. Soc., Vol. 175, 209 - 215 (1976).

Semi-empirical cross-sections and rates are presented for transitions between highly-excited levels of hydrogen atoms induced by electron impact. The results are valid whenever the initial and final quantum numbers satisfy n, $n' \geqslant 5$.

022.028 **Oscillator strengths and collision strengths in Si X.**
W. Dankwort, E. Trefftz.
Astron. Astrophys., Vol. 47, 365 - 370 (1976).

Using the multiconfiguration Hartree-Fock approximation, energy levels and wave functions are calculated for the low lying configurations in Si^{+9}. Dipole oscillator strengths in intermediate coupling are determined for transitions between several of these terms. Collision strengths for electron collision excitation from the ground state are calculated in the Coulomb Born I approximation.

022.029 **The C_4 molecule.**
W. R. M. Graham, K. I. Dismuke, W. Weltner, Jr.
Astrophys. Journ., Vol. 204, 301 - 310 (1976).

A carbon molecule formed by (1) the photolytic dissociation of C_4H_2 and (2) the evaporation of graphite, and trapped at 4 K in solid neon and argon, has been identified as C_4 on the basis of electron spin resonance and optical spectra. Molecular constants are predicted for C_4 in the gas-phase.

022.030 **Experimental oscillator strengths in the $C_2(A^1\Pi_u - X^1\Sigma_g^+)$ Phillips band system.**
F. Roux, D. Cerny, J. d'Incan.
Astrophys. Journ., Vol. 204, 940 - 943 (1976).

Absolute line radiancies in the 0−0 band of the $^1\Pi−^1\Sigma$ Phillips system of C_2 molecule have been determined using a high-resolution infrared grill spectrometer, a blackbody, and an oxyacetylene flame in thermal equilibrium. From these data the oscillator band strength $f_{00} = (2.5 \pm 0.2) \times 10^{-3}$ is calculated.

022.031 **The microwave spectrum of hydrogen isocyanide.**
R. J. Saykally, P. G. Szanto, T. G. Anderson, R. C. Woods.
Astrophys. Journ., (*Letters*), Vol. 204, L143 - L145 (1976).

The authors have observed the interstellar U90.663 line in laboratory dc glow discharges in mixtures of cyanogen and hydrogen, cyanogen and acetylene, and nitrogen and acetylene, unambiguously confirming Snyder and Buhl's original assignment of this transition to the HNC molecule. The precise measurement of the rest frequency ($90,663.568 \pm 0.004$ MHz) is in excellent agreement with recent astrophysical values.

022.032 **Partially elastic collisions in dense matter.**
L. Nygrén.
Astrophys. Space Sci., Vol. 39, 313 - 319 (1976).

The effects of partially elastic collisions on a dense system of particles moving in Keplerian orbits are studied. As in the case of a low-density system (Hämeen-Anttila, 1975), evolution leads to the formation of separate ringlets. The results reveal an anisotropic structure of matter, which may explain some peculiarities in the photometry of Saturn's rings.

022.033 **Extinction properties of porous spheres.**
H. Abadi, N. C. Wickramasinghe.
Astrophys. Space Sci., Vol. 39, L31 - L32 (1976).

An approximate theory is developed for computing the optics of porous grains. Numerical results are given for extinction efficiencies of spheres comprised of material with refractive index $m = 1.4−0.05i$ with varying degrees of porosity.

022.034 **Detection of the millimeter wave spectrum of hydrogen isocyanide, HNC.**
R. A. Creswell, E. F. Pearson, M. Winnewisser, G. Winnewisser.
Zeitschr. Naturforschung, Vol. 31a, 221 - 224 (1976).

The rotational spectra of three isotopic species of HNC have been observed in the millimeter wave region. In the case of the parent species, $H^{14}N^{12}C$, the authors' measurement of the $J = 1 \leftarrow 0$ transition unequivocally confirms the assignment of the U90.7 interstellar line to HNC. For the parent species the molecular constants obtained are $B_0 = 45332.005$ (40) MHz and $D_0 = 0.1019$ (50) MHz. Structural parameters derived from the ground state rotational constants of this linear molecule are $r(H−N) = 0.987$ (3) Å and $r(N−C) = 1.171$ (4) Å.

022.035 **Determination of the electron density in a helium plasma.** D. Einfeld, G. Sauerbrey.
Zeitschr. Naturforschung, Vol. 31a, 310 - 315 (1976).
In German.

022.036 **Teilchenbeschleunigung in starken elektromagnetischen Feldern: theoretische Energiespektren.**
M. Grewing, E. Schrüfer.
Mitt. Astron. Ges., No. 38, p. 105 - 107 (1976). − Short report.

022.037 **Properties of matter in strong and superstrong magnetic fields.** E. Müller, W. Hillebrandt.
Mitt. Astron. Ges., No. 38, p. 120 - 121 (1976). − Abstract.

022.038 **Optical properties of particulates.**
D. R. Huffman.
Astrophys. Space Sci. Library, Vol. 55, (see 012.001), 191 - 200 (1976).

Optical properties of small particles of olivine (less than 0.1 μ) have been studied in the ultraviolet as an example of an insulating solid. The general trend of optical properties of graphite is surprisingly similar to the behavior required to explain all features of the interstellar extinction and albedo curves from near visible to 1000 Å. Measured extinction of small olivine particles in the infrared agrees with calculations based on newly measured optical constants, but dominant sharp structure in the 10 μ region still presents a bit of a problem in explaining 'silicate' features in astronomical data.

022.039 **On the EUV spectrum of Fe X and Ni XII.**
H. Nussbaumer.
Astron. Astrophys., Vol. 48, 93 - 99 (1976).

Transition probabilities and energy levels for Fe X and Ni XII are calculated in a multiconfiguration approximation. Apparent discrepancies between the observed solar Fe X spectrum and calculated intensity ratios are investigated and partly resolved.

022.040 **Transition probabilities within $2s^2$-$2s2p$-$2p^2$ in the Be I sequence, Be I−Ni XXV.**
H. P. Mühlethaler, H. Nussbaumer.
Astron. Astrophys., Vol. 48, 109 - 114 (1976).

The authors have calculated probabilities for electric and magnetic dipole (including intercombination), and electric and magnetic quadrupole transitions for several members of the Be I sequence in intermediate coupling.

022.041 **Relativistic motor dynamics of systems of mass points and rigid bodies.** V. S. Nuotio-Antar.
Astron. Nachr., Vol. 297, 65 - 69 (1976).

A relativistic motor model for the dynamics of systems of mass points and rigid bodies, partly satisfying the program of Kustaanheimo (1971) for a relativistic motor mechanics, is given. The conservation laws are valid independently of the type of dependence between mass and velocity.

022.042 **The gravitational field of a homogeneous plane layer.**
R. M. Avakyan, J. Horsky.
Astrofizika, Vol. 11, 689 - 697 (1975). In Russian. English translation in Astrophysics, Vol. 11, No. 4.

The gravitational field generated by a static homogeneous

plane layer is studied. The metric tensor inside such a configuration is found as well as the pressure distribution in the layer. The authors' internal solution and Taub's external solution are joined on the boundary. For small central pressure the metric tensor is presented in analytical form and the physical meaning of the constant in the external solution is given.

022.043 **Catastrophes in Lagrangian systems.**
C. DeWitt-Morette.
Long-time predictions in dynamics, (see 012.005), p. 57 - 65 (1976).

022.044 **Statistical behavior in conservative classical systems.**
L. E. Reichl.
Long-time predictions in dynamics, (see 012.005), p. 71 - 98 (1976).

022.045 **Analytical and numerical study of a perturbed pendulum.** C. Simó.
Long-time predictions in dynamics, (see 012.005), p. 347 - 348 (1976). — Abstract.

022.046 **Algebraical aspects of perturbation theories.**
F. W. Spirig.
Long-time predictions in dynamics, (see 012.005), p. 349 - 351 (1976). — Short report.

022.047 **Statistical mechanics of light elements at high pressure. IV. A model free energy for the metallic phase.** H. E. DeWitt, W. B. Hubbard.
Astrophys. Journ., Vol. 205, 295 - 301 (1976).

A Monte Carlo program has been used to evaluate the thermodynamic properties of dense ionized hydrogen and hydrogen helium mixtures for a very wide range of temperature and density. Electron screening effects around each ion are taken into account. Approximately 150 data points were run, giving results as functions of the classical Coulomb interaction parameter Γ, electron screening as a function of r_s, and the hydrogen-helium ratio Y. With this result the energy and pressure of metallic hydrogen and hydrogen and helium mixtures may be computed to within about 2 percent accuracy over the range $0 < \Gamma < 150, 0 < r_s < 1$, and $0 < Y < 0.5$.

022.048 **Thermonuclear reaction rates derived from thick target yields.** N. A. Roughton, M. J. Fritts, R. J. Peterson, C. S. Zaidins, C. J. Hansen.
Astrophys. Journ., Vol. 205, 302 - 307 (1976).

A new technique to extract thermonuclear reaction rates from thick target yield data is described. Reported here are (p, γ) reactions on the nuclei ^{10}B, ^{12}C, $^{28, 29}Si$, ^{40}Ca, $^{46, 47}Ti$, ^{50}Cr, $^{54, 56}Fe$, $^{58, 60, 61}Ni$, $^{64, 67}Zn$, ^{70}Ge, and $^{92, 95, 98, 100}Mo$; the (p, α) reaction on ^{10}B; and (p, n) reactions on ^{68}Zn, ^{74}Ge, ^{76}Ge, and $^{95, 96}Mo$.

022.049 **Remarks concerning the dispersion equation of electromagnetic waves in a magnetized cold plasma.**
K. Rawer, K. Suchy.
Journ. Atmosph. Terr. Phys., Vol. 38, 395 - 398 (1976).

022.050 **A comparison of foreign gas broadening and shift of neon and argon emission lines.** G. H. Copley.
Journ. Quant. Spectrosc. Radiat. Transfer, Vol. 16, 377 - 384 (1976).

Foreign gas broadening and shift coefficients for Ne and Ar emission lines produced in a high pressure (0.5−2.0 atm) He discharge have been measured. The shift coefficients vary significantly for different transitions. This variation results from the differing excited state potentials for the He−Ne* and He−Ar* collision pairs, which must be included in any reasonable quantum-mechanical calculation of line shifts.

022.051 **The role of molecular rotation in inelastic vibrational collisions.** F. A. Hopf, D. Rogovin.
Journ. Quant. Spectrosc. Radiat. Transfer, Vol. 16, 389 - 393 (1976).

The authors examine the effects of molecular rotation on vibrational relaxation processes and show that such collisions are dominated by channels in which there are large changes in the rotational quantum number.

022.052 **Measurement of Ar(I) and Ar(II) transition probabilities in LTE arc plasmas.** H. Nubbemeyer.
Journ. Quant. Spectrosc. Radiat. Transfer, Vol. 16, 395 - 404 (1976).

022.053 **Measurement of hydrogen- and self-broadened half-widths of ammonia at 200 and 300 K.**
J. S. Margolis, S. Sarangi.
Journ. Quant. Spectrosc. Radiat. Transfer, Vol. 16, 405 - 408 (1976).

The hydrogen- and self-broadened half-widths have been measured for the $(\nu_1 + \nu_2)$- and $(\nu_2 + \nu_3)$-bands of ammonia at 300 and 207 K. Measurement of hydrogen-broadened widths has been restricted to $J, K \leqslant 6$, but that of self-broadened widths is done for a few lines outside that range. Assuming a power-law dependence of half-width on temperature given by $\gamma(T) = \gamma(T_0)(T/T_0)^\alpha$, the average value of the index α for the lines measured is found to be 0.57 for hydrogen broadening.

022.054 **Ultraviolet oscillator strengths for carbon, nitrogen and oxygen ions.** J. V. Mallow, P. S. Bagus.
Journ. Quant. Spectrosc. Radiat. Transfer, Vol. 16, 409 - 414 (1976).

Oscillator strengths were calculated for UV transitions in carbon, nitrogen and oxygen ions. Single and multiconfiguration numerical Hartree-Fock wavefunctions were used to represent initial and final states of the various transitions, and both dipole length and dipole velocity transition matrix elements were calculated. A careful choice of a few important configurations leads in most cases to multiconfiguration results which are in good agreement with experiment, and with other theoretical calculations.

022.055 **Scattering of solar Lyman alpha by the (14,0) band of the fourth positive system of CO.** T. T. Kassal.
Journ. Geophys. Res., Vol. 81, 1411 - 1412 (1976).

A calculation of the fluorescent scattering spectrum of the solar Lyman alpha line by the (14,0) band of the CO 4(+) system is presented. It is shown that emissions from the $v' = 14$ level of the $A^1 \Pi$ state dominate the total scattering spectrum at column densities of 10^{18} molecules/cm^2 and at ambient temperatures.

022.056 **Excitation of forbidden lines of Fe III and Fe VI.**
R. H. Garstang, W. D. Robb, S. P. Rountree.
Bull. American Astron. Soc., Vol. 8, 291 (1976). — Abstr. AAS.

022.057 **Limit on the variation of the proper mass of an electron with time.** T. L. Chow.
Astrophys. Letters, Vol. 17, 57 - 59 (1976).

Recently Thompson speculated that the electron mass may evolve with time, while the mass of other atomic particles (e.g. protons) remain constant. The results of the two independent measurements of the spontaneous fission decay constant λ_f of U^{238} set an upper limit on any possible time dependence of the electron mass m_e: $\delta m_e/m_e \leqslant 1.47 \times 10^{-3}$ during the last two billion years.

022.058 **Photochemistry of neutral and ionized magnesium.**
O. Ovezgel'dyev, L. P. Korsunova.
Geomagn. Aeronom., Vol. 16, 188 - 191 (1976). In Russian.
Brief information.

022.059 Calculations of pressure induced widths and shifts in microwave spectra. P. L. Hewitt.
Journ. Quant. Spectrosc. Radiat. Transfer, Vol. 16, 499 - 503 (1976).

Calculations of the width and shift of pressure induced spectral lines in gases at low pressures in the microwave region of the electromagnetic spectrum are presented. These calculations are based on an extension of the Anderson-Tsao-Curnutte theory and are compared with those based on the quantum mechanical model developed by Di Giacomo and Cattani.

022.060 Plasma effects in the spectrum of high Balmer lines. G. Himmel.
Journ. Quant. Spectrosc. Radiat. Transfer, Vol. 16, 529 - 536 (1976).

Measurements of high Balmer lines (n_{upper} = 12 to 19) emitted from a hydrogen discharge of relatively low plasma density show that the spectrum may be well reproduced by a superposition of symmetric Stark profiles, which accounts for the broadening effect generated by ions and quasistatic electrons. The small deviations which were found most probably involve inelastic electron collisions connecting adjacent upper levels. Within experimental accuracy, a shift is not detectable either for odd or for even series members. This finding is inconsistent with the observations of other authors according to which odd Balmer lines are shifted to the red.

022.061 Transition probabilites in Pr(II) and the solar praseodymium abundance. C. S. Lage, W. Whaling.
Journ. Quant. Spectrosc. Radiat. Transfer, Vol. 16, 537 - 542 (1976).

The authors have measured branching ratios for all of the known transitions from five levels in Pr⁺. They use the equivalent widths of Moore et al. and the solar model parameters of Righini and Rigutti to compute the photospheric Pr abundance: $\log (N_{Pr}/N_H) + 12 = 0.66 \pm 0.15$.

022.062 Measurement and significance of the reaction $^{13}C^+ + ^{12}CO \rightleftharpoons ^{12}C^+ + ^{13}CO$ for alteration of the $^{13}C/^{12}C$ ratio in interstellar molecules.
W. D. Watson, V. G. Anicich, W. T. Huntress, Jr.
Astrophys. Journ.,(*Letters*), Vol. 205, L165 - L168 (1976).

Laboratory measurements presented here using the ion cyclotron resonance technique yield a rate constant 2×10^{-10} cm³s⁻¹ at 300 K for the isotope exchange $^{13}C^+ + ^{12}CO \rightarrow ^{12}C^+ + ^{13}CO$. According to the usual ideas about ion-molecule reactions, this rate constant should also be appropriate at temperatures $\lesssim 100$ K. Then the observed $^{13}C/^{12}C$ ratio obtained from radio observation of interstellar molecules may be either larger or smaller than the actual value in the interstellar medium by factors of 2 or so.

022.063 Calculation of the cross section for C IV–H charge exchange: significance for interstellar X-rays/cosmic-ray particles. R. J. Blint, W. D. Watson, R. B. Christensen.
Astrophys. Journ., Vol. 205, 634 - 637 (1976).

The authors present here a calculation for the charge-exchange cross section of atomic hydrogen and C^{+3}. Recent observations have focused on the abundance of C^{+3} as an indicator for the low-energy X-ray and cosmic-ray particle flux in the Galaxy. Inclusion of charge exchange with the rate calculated here invalidates these interpretations.

022.064 The infrared optical constants of sulfuric acid at 250 K. L. W. Pinkley, D. Williams.
Journ. Optical Soc. America, Vol. 66, 122 - 124 (1976).

022.065 Observations of $3p^5 3d-3p^5 4f$ and $3p^6 3d-3p^5 3d 4s$ transitions in iron, cobalt, nickel, and copper.
M. Swartz, S. O. Kastner, L. Goldsmith, W. M. Neupert.
Journ. Optical Soc. America, Vol. 66, 240 - 244 (1976).

022.066 A comparison of self broadening and shift of helium, neon and argon emission lines. G. H. Copley.
Journ. Quant. Spectrosc. Radiat. Transfer, Vol. 16, 553 - 558 (1976).

022.067 Estimates of Stark broadening of nitrogen ion lines. J. D. Hey.
Journ. Quant. Spectrosc. Radiat. Transfer, Vol. 16, 575 - 577 (1976).

Recent measurements of spectral line widths for singly- and doubly-ionized nitrogen atoms, are compared with the corresponding values predicted by the semi-empirical method of Griem. The overall agreement, for representative transitions from 6 multiplets of N(II) and 3 multiplets of N(III), tends to confirm the usefulness and general validity of Griem's approach

022.068 Measurement of ultraviolet argon (II) transition probabilities. K. Behringer, P. Thoma.
Journ. Quant. Spectrosc. Radiat. Transfer, Vol. 16, 605 - 609 (1976).

022.069 Structure of Stark broadened Hβ lines at low electron densities. J. Ramette, H. W. Drawin.
Zeitschr. Naturforschung, Band 31a, 401 - 407 (1976).

The profile of the hydrogen Hβ line emitted by a hydrogen low-temperature afterglow plasma has been measured at electron densities 1.8×10^{15}, 1×10^{15} and 5×10^{14} cm⁻³ by means of a 10-channel analyzer system with a spectral resolution of 0.1 Å.

022.070 The rotational spectrum of monothioformic acid. I. cis- and trans-HC(:O)SH.
W. H. Hocking, G. Winnewisser.
Zeitschr. Naturforschung, Band 31a, 422 - 437 (1976).
Presented in part at the "Fourth Colloquium on High Resolution Molecular Spectroscopy", Tours, France, September 15 - 19, 1975.

022.071 The rotational spectrum of monothioformic acid. II. cis- and trans-HD(:O)SD, DC(:O)SH, HC(:O)³⁴ SH. W. H. Hocking, G. Winnewisser.
Zeitschr. Naturforschung, Band 31a, 438 - 453 (1976).
Presented in part at the "Fourth Colloquium on High Resolution Molecular Spectroscopy", Tours, France, September 15 - 19, 1975.

022.072 A new $^1\Sigma - ^1\Sigma$ system of the ZrO molecule. J. G. Phillips, S. P. Davis.
Astrophys. Journ., Vol. 206, 632 - 639 (1976).

The λ5860 band of ZrO is found to be the 0−0 band of a new $^1\Sigma^+ - X^1\Sigma^+$ transition. Additional bands of the system have R heads at λλ5893, 5926, and 5961. A weak rotational perturbation in the λ5860 band is probably produced by coincidences with rotational states of the ν = 2 vibrational level of the $^1\Sigma$ electronic state.

022.073 Collisionally induced hyperfine-structure transitions of OH. S.-I Chu.
Astrophys. Journ., Vol. 206, 640 - 651 (1976).

The theory of excitation of the internal structures of OH by electron or heavy-particle impact is formulated within the quantum-mechanical close-coupling approximation, incorporating spin-orbit and hyperfine interactions. In the case of electron impact, explicit Born formulae are derived and the selection rules are discussed for the dipole- and the quadrupole-induced transitions. Numerical results are reported for the electron excitation cross sections of the hyperfine-structure transitions among the lower-lying rotational ladders of the $X^2\Pi^{\pm}_{3/2}(\nu = 0)$ and $X^2\Pi^{\pm}_{1/2}(\nu = 0)$ states.

022.074 Fine-structure excitation of carbon by atomic hydro-

gen impact.   A. W. Yau, A. Dalgarno.
Astrophys. Journ., Vol. 206, 652 - 657 (1976).

The cross sections of the fine-structure excitations of carbon by collisions with atomic hydrogen, $H(^2S) + C(^3P_J) \rightarrow H(^2S) + C(^3P_{J'})$ have been calculated using an elastic scattering model that incorporates an approximate correction for the inelasticity. Maxwellian-averaged cross sections are presented graphically and rate coefficients are tabulated for temperatures less than 1000 K. Analytical fits for the rate coefficients near the thresholds are given.

022.075 Laboratory measurements of the (0.0), (1.0) and (2.1) bands of the red system of the $C^{13}N^{14}$ molecule.   G. Hosinsky, B. Lindgren.
Astron. Astrophys., Suppl. Ser., Vol. 25, 1 - 8 (1976).

The (0.0), (1.0) and (2.1) bands of the $A^2\Pi_i - X^2\Sigma^+$ transition of the $C^{13}N^{14}$ molecule have been recorded in the laboratory. Quantum numbers, wavelengths and wavenumbers are given for 633 lines. The accuracy in the measurements is estimated to be 0.02 cm^{-1} for unblended lines.

022.076 Charged particle propagation in strong disordered magnetic fields.   F. C. Jones.
14th Intern. Cosmic Ray Conf., (see 012.011), Vol. 3, 856 - 860 (1975).

022.077 Simulation of the velocity diffusion of charged particles in turbulent magnetic fields.   T. B. Kaiser.
14th Intern. Cosmic Ray Conf., (see 012.011), Vol. 3, 861 - 865 (1976).

022.078 Numerical experiments on energetic particle scattering in pitch angle.
X. Moussas, J. J. Quenby, S. Webb.
14th Intern. Cosmic Ray Conf., (see 012.011), Vol. 3, 866 - 871 (1975).

022.079 A quantitative heterodyne experiment with extrinsic silicon at 10.6 μm.   T. de Graauw, P. Norton.
Infrared Physics, Vol. 16, (see 012.013), 51 - 54 (1976).

022.080 A new Poisson solver for isolated charge distributions and its applications to galactic problems.
R. A. James, J. A. Sellwood.
Computing in plasma physics and astrophysics, (see 012.014), F6, 2 pp. (1976).

022.081 Forbidden helium radio lines.   N. N. Bulatov.
Astron. Zhurn. Akad. Nauk SSSR, Vol. 53, 558 - 559 (1976). In Russian. English translation in Soviet Astron., Vol. 20, No. 3.

The transition probabilities and intensity of forbidden helium radio lines arising from the transitions between the fine-structure levels of the 2^3P state are calculated.

022.082 Microcraters produced by oblique incidence of projectiles.   V. Stähle, K. Nagel, E. Schneider.
IAU Colloquium No. 31, (see 012.015), p. 241 (1976). Abstract.

022.083 Impact light flash studies: temperature, ejecta, vaporization.   G. Eichhorn.
IAU Colloquium No. 31, (see 012.015), p. 243 - 247 (1976).

022.084 Application of an artificial satellite of the earth to determine the velocity of the gravitational interaction within Newtonian gravitational fields.   G. Cristea.
Separate print Comit. Stat Pentru Energia Nucleara, Inst. Fiz. Atomica, Bucharest, Romania, FR-123-1975, 8 pp. (1975).

In the first part of this paper, additional data are given concerning a gravimeter consisting in a pendulum-laser set

proposed in a previous paper of the author. The second part of the paper points out the advantages resulting from the experiment of determining the velocity of the gravitational reaction in an artificial earth satellite.

022.085 On ionization equilibrium of helium.
U. Narain, N. K. Jain, S. Chandra.
Zeitschr. Naturforschung, Band 31 a, 565 - 568 (1976).

The ionization equilibrium of helium is investigated using new ionization and modified dielectronic recombination rate coefficients. The results are discussed in the light of available data.

022.086 Maximum rate for the proton-proton reaction compatible with conventional solar models.
M. J. Newman, W. A. Fowler.
Phys. Rev. Letters, Vol. 36, 895 - 897 (1976).

022.087 Lifetime of the $c^1\Phi$ state of TiO.
J. Feinberg, M. G. Bilal, S. P. Davis, J. G. Phillips.
Astrophys. Letters, Vol. 17, 147 - 149 (1976).

The lifetime of the $v' = 0$ level of the $c^1\Phi$ state of TiO is 17.5 ± 1.0 nsec as measured by the decay of resonant fluorescent radiation of vapor in a furnace.

022.088 Oscillator strengths in the Mg isoelectronic sequence.
G. A. Victor, R. F. Stewart, C. Laughlin.
Astrophys. Journ., Suppl. Ser., Vol. 31, 237 - 247 (1976).

Configuration interaction calculations, within a semiempirical model potential framework, are used to calculate bound-bound oscillator strengths for members of the Mg isoelectronic sequence up to Cl^{+5}.

022.089 The stability of electrons in a unified field theory.
H.-J. Treder.
Experim. Techn. Phys., Vol. 23, 577 - 582 (1975).

022.090 Molecular processes in interstellar clouds.
A. Dalgarno.
Atomic and molecular processes in Astrophys., (see 012.019), p. 1 - 98 (1975).

022.091 Introduction to collision theory and to some astrophysical applications.   F. Masnou-Seeuws.
Atomic and molecular processes in Astrophys., (see 012.019), p. 101 - 184 (1975).

022.092 The contribution of laboratory measurements to the interpretation of astronomical spectra.
R. W. P. McWhirter.
Atomic and molecular processes in Astrophys., (see 012.019), p. 187 - 308 (1975).

022.093 Measurement of the $^{12}C(\alpha, \gamma)^{16}O$ cross section.
P. Dyer.
Explosive nucleosynthesis, (see 012.020), p. 195 - 202 (1973).

022.094 Experiments of relevance to explosive nucleosynthesis. Summary by the editors of the talk given by
W. A. Fowler.
Explosive nucleosynthesis, (see 012.020), p. 297 - 301 (1973).

022.095 A new method of measuring the life-time of excited atomic and molecular states.
J. Heldt, H. Sałajczyk.
Postępy Astron., Vol. 24, 33 - 40 (1976). In Polish.

022.096 Rotational excitation of CN molecules by proton impact.
M. J. Jamieson, P. M. Kalaghan, A. Dalgarno.

Journ. Phys. B, Vol. 8, 2140 - 2148 (1975).

022.097 **R-centroids and Franck-Condon factors of CH$^+$ molecule.**   M. L. P. Rao, A. A. N. Murty, D. V. K. Rao, P. T. Rao.
Phys. Letters A, Vol. 54A, 177 - 178 (1972). – Abstr. in Phys. Abstr., Vol. 79, A001211 (1976).

022.098 **Photoproduction of gravitons in static electromagnetic fields.**   G. Papini, S.-R. Valluri.
Canadian Journ. Phys., Vol. 53, 2306 - 2311 (1975). – Abstr. in Phys. Abstr., Vol. 79, A020304 (1976).

022.099 **The direction of time.**   R. Mirman.
Found. Phys., Vol. 5, 491 - 511 (1975). – Abstr. in Phys. Abstr., Vol. 79, A024626 (1976).

022.100 **New two-stage accelerator for hypervelocity impact simulation (micrometeoroid environment simulation).**   E. B. Igenbergs, D. W. Jex, E. L. Shriver.
AIAA Journ., Vol. 13, 1024 - 1030 (1975). – Abstr. in Phys. Abstr., Vol. 79, A037981 (1976).

022.101 **Recent progress in the classification of the spectra of highly ionized atoms.**   B. C. Fawcett.
Adv. At. Mol. Phys., Vol. 10, 223 - 293 (1974).

Errata

022.901 **Erratum: 'Radiative and dielectronic recombination coefficients for complex ions [Astron. Astrophys., Vol. 25, 137 - 140 (1973)].**
S. M. V. Aldrovandi, D. Péquignot.
Astron. Astrophys., Vol. 47, 321 (1976).

022.902 **Erratum: 'Energy spectrum of hydrogen-like atoms in a strong magnetic field' [Astrophys. Journ., Vol. 190, 741 - 742 (1974)].**
G. L. Surmelian, R. F. O'Connell.
Astrophys. Journ., Vol. 204, 311 (1976).

Astronomical Instruments and Techniques

031 Astronomical Optics, Methods of Observation and Reduction, Data Processing, Automation

Astronomical Optics

031.001 **Binoculars and astronomical observations with them.** A. D. Marlenskij.
Zemlya i Vselennaya, 1976, No. 1, p. 83 - 87. In Russian.

031.002 **Identification algorithms of azimuthal mirror systems.** Yu. L. Bronshtejn.
Trudy TsNII geod., aehrosemki i kartogr., 1975, vyp. (No.) 210, p. 4 - 19. In Russian. − Abstr. in Referativ. Zhurn. 51. Astron., 3.51.55 (1976).

031.003 **Elasticité et miroirs à focale variable.** G. Lemaître.
Comptes Rendus Acad. Sci. Paris, Sér. B, Vol. 282, 87 - 89 (1976).

On recherche les configurations de charges et d'appuis associés à un dioptre de symétric axiale dont le profil d'épaisseur, fonction du rayon, permet d'engendrer des variations de courbure par flexion. Trois configurations simples permettent d'obtenir des méridiennes flexibles dont la flèche est une expression purement parabolique.

031.004 **Artificial star design.** C. R. Burch.
Journ. British Astron. Ass., Vol. 86, 203 - 209 (1976).

031.005 **A study of star image vibrations by the photographic method under day time conditions.**
A. G. Zhestkov, B. I. Kozarenko, L. P. Sedova.
Astrometriya i Astrofizika, Kiev, vyp. (No.) 28, (see 003.002), p. 40 - 41 (1976). In Russian.

031.006 **Dependence of solar image vibration on telescope aperture.**
P. G. Kovadlo, V. I. Ivanov, Sh. P. Darchiya.
Issled. po geomagnetizmu, aehron. i fiz. Solntsa. Vyp. (No.) 37. Moskva, Nauka, 1975, p. 136 - 140. In Russian. − Abstr. in Referativ. Zhurn. 51. Astron., 4.51.92 (1976).

031.007 **Computer generated holograms for testing aspheric lenses.**
T. Takahashi, K. Konno, M. Kawai, M. Isshiki.
Applied Optics, Vol. 15, 546 - 549 (1976).

031.008 **Dynamic Hartmann test: comments.** T. L. Williams.
Applied Optics, Vol. 15, 599 (1976). − Letter.

031.009 **Echelle gratings: their testing and improvement.** G. R. Harrison, E. G. Loewen, R. S. Wiley.
Applied Optics, Vol. 15, 971 - 976 (1976).

031.010 **Optical antenna gain. 3: The effect of secondary element support struts on transmitter gain.**
B. J. Klein, J. J. Degnan.
Applied Optics, Vol. 15, 977 - 979 (1976).

The effect of a secondary element spider support structure on optical antenna transmitter gain is analyzed. An expression describing the influence of the struts on the axial gain, in both the near and far fields, is derived as a function of the number of struts and their width. It is found that, for typical systems, the struts degrade the on-axis gain by less than 0.4 dB, and the first side-lobe level is not increased significantly. Contour plots have also been included to show the symmetry of the far-field distributions for three and four support members.

031.011 **Meridional focal curves of the Seya−Namioka mounting of the concave grating.**
V. K. Prokof'ev.
Izv. Krymskoj Astrofiz. Obs., Vol. 54, 265 - 271 (1976). In Russian.

The meridional focal curves of the Seya−Namioka mounting of the concave grating are discussed. It is possible to approximate large parts of these curves as arcs of a circle with suitable radius if the angle of incidence is constant, and to use the Seya−Namioka monochromator as spectrograph. It is possible also to place some exit slits and transform the Seya−Namioka monochromator to a polychromator. During the rotation of the grating each slit will register different parts of the spectrum.

031.012 **New two-mirror systems.** G. M. Popov, M. B. Popova.
Izv. Krymskoj Astrofiz. Obs., Vol. 54, 272 - 288 (1976). In Russian.

Two-mirror coma-free systems with zero spherical aberration are examined. Special kinds of two-mirror aplanats are investigated, which can be used in X-ray and UV regions. Some new fast-speed aplanatic two-mirror systems are described, which cannot be found using third-order aberration and Gauss-optics. These systems can be used in fast-speed spectrographs, radio telescopes and for exploration the far UV region. A Gregorian-type system of two concave elliptic mirrors is examined; this system can be used for tracking an object by means of tilting the secondary mirror and can be used on balloons.

031.013 **Equivalent concentric systems.** G. M. Popov.
Izv. Krymskoj Astrofiz. Obs., Vol. 54, 289 - 314 (1976). In Russian.

Simple methods of constructing concentric systems (equivalent systems) are discussed; these systems have the same aberrations (for given λ) as a system with known aberrations. Two main kinds of equivalent systems are studied: systems with the same equivalence property for changing distance to the object, and systems which do not have this property. The equivalent menisci are discussed in detail. Simple methods of looking for equivalent systems for special cases are discussed. Some examples of fast-speed equivalent systems are given.

031.014 **Surface-contribution algorithms for analysis and optimization.** S. H. Brewer.
Journ. Optical Soc. America, Vol. 66, 8 - 13 (1976).

A set of algorithms is developed to compute partial derivatives of transverse-ray errors with repect to the optical parameters. These are based on a generalized version of the Aldis equations. Results are given for a test case that demonstrates feasibility of application of these derivatives to lens optimization.

031.015 Geometrical optics and dynamic programming.
A. Noyola-Isgleas.
Journ. Optical Soc. America, Vol. 66, 60 - 62 (1976).

Dynamic programming applied to geometrical optics leads to a nonlinear partial differential equation for the path taken by a light ray in any optical system.

031.016 Error calculation of polarization measurements.
M. A. F. Thiel.
Journ. Optical Soc. America, Vol. 66, 65 - 67 (1976).

A mathematical procedure is presented for the determination of the transmission properties (instrumental operator) of an instrument with respect to the polarization state of any arbitrarily polarized radiation. This method is applied to a complete polarimeter system in order to find the systematic error of the polarization measurements. If the operator of the polarimeter is known, the systematic error can be removed.

031.017 Caustic surface and the Coddington equations.
D. L. Shealy.
Journ. Optical Soc. America, Vol. 66, 76 - 77 (1976).

The caustic surface, which is the locus of the sagittal and tangential focal points for general skew rays, is shown to be given by the Coddington equations for the special case of meridional rays reflected or refracted by a spherical surface.

031.018 Polarization effects on overcoated mirrors.
V. Williams, J. B. Goodell.
Journ. Optical Soc. America, Vol. 66, 79 (1976). — Conference abstract.

031.019 A general theory of the aberrations of diffraction gratings and gratinglike optical instruments.
C. H. F. Velzel.
Journ. Optical Soc. America, Vol. 66, 346 - 353 (1976).

031.020 Applications of adaptive optical techniques to figure control of large telescopes. R. M. Scott.
Journ. Optical Soc. America, Vol. 66, 391 (1976). — Conference abstract.

031.021 "Zerodur" — ein neuer glaskeramischer Werkstoff für die reflektierende Optik. J. Petzoldt.
SuW, 15. Jahrgang, p. 156 - 161 (1976).

031.022 Field error of NA-MK-25 cameras.
I. M. Khaimov.
Komety i Meteory, *Dushanbe*, No. 23, p. 46 - 53 (1974). In Russian.

The photometric error of the field of NA-MK-25 cameras depending on position, velocity, orientation of meteor trails on the plates and on brightness of the object are investigated.

031.023 Proximate ray tracing and optical aberration coefficients. G. W. Hopkins.
Journ. Optical Soc. America, Vol. 66, 405 - 410 (1976).

Algebraic ray-trace equations for axially symmetric optical systems are expanded in terms of system parameters and paraxial variables. The results are the paraxial ray-trace equations and equations for third-, fifth-, and seventh-order deviations from the paraxial ray. These equations, which approximate real rays in an extended region about the axis, are used to trace a selected set of rays to the image space. Devia-

tions of given order are then related to equal-order terms in the aberration expansion. The resulting set of linear equations is solved for aberration coefficients.

031.024 Collimation test by double grating shearing interferometer.
K. Patorski, S. Yokozeki, T. Suzuki.
Applied Optics, Vol. 15, 1234 - 1240 (1976).

The double grating shearing interferometry method for determination of the degree of light collimation is described. High accuracy is obtained by performing the observation of fringes in the area of the size twice as big as the one usually assumed in shearing interferometry experiments. The conditions under which such a detection mode is feasible are derived They represent at the same time a very strong argument proving the highly diffractive (wave optics) character of the classical Ronchi test.

031.025 Cassegrain telescopes: limits of secondary movement in secondary focusing. R. N. Clark.
Applied Optics, Vol. 15, 1266 - 1269 (1976).

The maximum distance to which the secondary mirror can be moved with respect to its primary in a true Cassegrain telescope with limited image deterioration is found to be proportional to the fourth power of the focal ratio of the primary mirror. This limit is independent of all other parameters describing the system when the magnification of the secondary is greater than about 3.

031.026 Investigation of the objective of the Zeiss short-focus astrograph. P. P. Pavlenko.
Vestn. Khar'kov. Univ., No. 129 (Ser. Astron., vyp. (No.) 10), p. 9 - 14 (1975). In Russian.

031.027 Chromatic aberration of the magnification of the Moscow PZT objective.
A. A. Volchkov, G. A. Gutsalo.
Astron. Tsirk., No. 901, p. 7 - 8 (1976). In Russian.

031.028 Cores in star images. T. S. McKechnie.
Journ. Opt. Soc. America, Vol. 66, 635 (1976).
Conference abstract.

031.029 The optical transfer function of decentered optical systems. B. H. Beyeler, H. J. Tiziani.
Optik, Band 44, 317 - 328 (1976). In German.

The influence of defects due to the manufacturing process, under special consideration of centring errors, on the optical transfer function is discussed.

031.030 Production of aspherical surfaces by mechanical means and by replication. E. Heynacher.
Optik, Band 45, 249 - 267 (1976). In German.

031.031 Far infrared measurements of selected optical materials at $1.6°K$.
J. A. Alvarez, R. E. Jennings, A. F. M. Moorwood.
Infrared Physics, Vol. 15, 45 - 49 (1975).

Measurements at $1.6°K$ of the transmittance and refractive index of quartz, polyethylene, polytetrafluoroethylene and polychlorotrifluoroethylene have been made, using a Michelson interferometer operating in the phase modulated mode.

031.032 Schmidt systems for the near-infrared.
A. Greve.
Infrared Physics, Vol. 15, 239 - 241 (1975).

The formulae which define the parameters (focal ratio, field) of a seeing-limited Schmidt system are given. Near-infrared ($5000 Å \leqslant \lambda \leqslant 15,000 Å$) Schmidt systems are discussed for nine different corrector plate materials.

031.033 **Methods of testing astronomical optics on limitations of Hartmann's method.** Eh. A. Vitrichenko.
Astron. Zhurn. Akad. Nauk SSSR, Vol. 53, 660 - 670 (1976).
In Russian. English translation in Soviet Astron., Vol. 20, No. 3.

An analysis of conditions which restrict Hartmann test application is given. Three most important circumstances are considered: (1) diameters of the diaphragm holes are essentially limited from below and from above by the effects of diffraction and elasticity of Hartmann diaphragm, (2) a general number of holes is limited from below by the accuracy adopted, (3) an adequate description of the optical surface tested is limited by the surface smoothness.

031.034 **On the self-weight sag of arch-like structures in the context of light-weight mirror design.**
D. C. Talapatra.
Optica Acta, Vol. 22, 745 - 759 (1975). — Abstr. in Phys. Abstr., Vol. 79, A000085 (1976).

031.035 **Holographic construction of open structure, dispersion transmission gratings.**
J. H. Dijkstra, L. J. Lantwaard.
Optics Commun., Vol. 15, 300 - 305 (1975). — Abstr. in Phys. Abstr., Vol. 79, A000086 (1976).

031.036 **Optical schemes for segmented telescopes.**
G. I. Tsukanova.
Izv. vyssh. ucheb. zavedenij priborostr., Vol. 18, No. 7, p. 116 - 119 (1975). In Russian. — Abstr. in Phys. Abstr., Vol. 79, A003435 (1976).

031.037 **Analysis of the possibility of interferometric laser control of the surface of an optical telescope mirror. I.** V. S. Zuev, E. P. Orlov, V. A. Sautkin.
Kvantovaya Ehlektron., Moskva, Vol. 2, No. 1, p. 78 - 91 (1975). — Abstr. in Phys. Abstr., Vol. 79, A003443 (1976).

031.038 **Analysis of the possibility of interferometric laser control of the surface of an optical telescope mirror. II.** V. S. Zuev, E. P. Orlov, V. A. Sautkin.
Kvantovaya Ehlektron., Moskva, Vol. 2, No. 1, p. 92 - 98 (1975). — Abstr. in Phys. Abstr., Vol. 79, A003444 (1976).

031.039 **Optical fabrication of a large lightweight mirror of unusual shape.** B. L. Barlow.
Optical Engineering, Vol. 14, 514 - 519 (1975). — Abstr. in Phys. Abstr., Vol. 79, A038044 (1976).

031.040 **Asymptote analysis of a large optical system.** J. Bisbee.
Optical Engineering, Vol. 14, 536 - 538 (1975). — Abstr. in Phys. Abstr., Vol. 79, A042460 (1976).

031.041 **Fabrication and test of 1.8-meter-diameter high quality ULE mirror.** R. J. Wollensak, C. A. Rose.
Optical Engineering, Vol. 14, 539 - 543 (1975). — Abstr. in Phys. Abstr., Vol. 79, A042461 (1976).

Aberrations of the symmetrical optical system. See Abstr. 003.112.

A Baranne-type spectrograph with echelle-grating. See Abstr. 034.062.

Methods of Observation and Reduction

031.201 **Feasibility of a search for planets around solar-type stars with a polarimetric radial velocity meter.**
K. Serkowski.
Icarus, Vol. 27, 13 - 24 (1976).

A method of wavelength calibration is proposed which may enable measuring changes in radial velocity of bright solar-type stars to an accuracy of about 5 meters per second. Such accuracy would be sufficient for detecting Jupiter-like planets around these stars. The stellar spectrum is imaged by a slitless echelle spectrograph onto a 100-channel Digicon image tube. Instrumental profiles of Digicon diodes are narrowed down by a Fabry–Perot etalon, making the profiles less dependent on atmospheric seeing. The spectrograph and the etalon act merely as a series of narrow band filters for the individual diodes.

031.202 **Measurement of lunar and planetary magnetic fields by reflection of low energy electrons.**
K. A. Anderson, R. P. Lin, R. E. McGuire, J. E. McCoy.
Space Sci. Instrum., Vol. 1, 439 - 470 (1975).

The authors describe a method to measure the magnitude, direction and scale size of magnetic fields near the surface of the moon and other planetary bodies. The method is especially useful when the planetary surface fields are weak and of small scale size. The method to measure magnetic fields by this technique can be described as follows: (1) The moon is situated in an ambient magnetic field (e.g., the interplanetary magnetic field) which is always populated by electrons in the energy range 15 eV to 15 keV. These electrons move along the field lines in helical trajectories. (2) Electrons approaching the moon's surface are reflected if they encounter magnetic fields there. The fraction of electrons reflected depends on the strength, spatial extent and direction of the lunar magnetic field. The electron reflection method has the following capabilities: (1) The spatial extent of lunar surface magnetic features as small as approximately 1 km can be measured. (2) The sensitivity of the method permits lunar magnetic fields at the moon's surface as low as about 0.01 γ to be measured. (3) The direction of lunar surface magnetic field features can be classified as predominantly pointed upward, downward or horizontal.

031.203 **A demonstration of an independent-station radio interferometry system with 4-cm precision on a 16-km base line.** J. B. Thomas, J. L. Fanselow, P. F. MacDoran, L. J. Skjerve, D. J. Spitzmesser, H. F. Fliegel.
Journ. Geophys. Res., Vol. 81, 995 - 1005 (1976).

031.204 **Solar seeing and the statistical properties of the photospheric solar granulation. I. Noise in Michelson and speckle interferometry.** C. Aime.
Astron. Astrophys., Vol. 47, 5 - 7 (1976).

Signal to noise ratios in speckle and Michelson experiments are compared from measurements of the spatial power spectrum of the solar granulation. The computation is restricted to the case where a calibration of the atmospheric turbulence can be carried out. A discussion concerning the seeing leads to the conclusions that, except for very good seeing, the signal to noise ratio is higher in speckle techniques than in Michelson experiments.

031.205 **The application of a new method for the determination of relative radial velocities with the Fehrenbach-prism.** F. Gieseking.
Astron. Astrophys., Vol. 47, 43 - 47 (1976). In German.

A new method for the determination of relative radial velocities is presented. On the basis of observations with the GPO-telescope of the European Southern Observatory it is shown that this method permits radial velocities to be measured with a mean error of the order of ±8 km/s for spectral types B and A, and ±4.5 km/s for late spectral types. This accuracy is about 2.5 times better than the accuracy obtained with the conventional method and even much better than that one expected for slit spectrographs of comparable dispersion. Finally some remarks on the great efficiency of the GPO are given: In 2 times 2 min exposure time the limiting photographic magnitude of measurable spectra is already 9^m. The absolute limiting magnitude of the GPO is about 12^{m}5. An exposure of 2 times 40 min yields up to 350 measurable spectra.

031.206 **An error analysis of power spectra.** P. Hoyng.
Astron. Astrophys., Vol. 47, 449 - 452 (1976).

An expression is derived for the uncertainty in the Fourier power spectrum due to noise present in the data. The author considers both series with Poisson-distributed noise as well as with normal-distributed noise. The expressions for the errors involve only the power spectrum itself and they can be readily generalized to arbitrarily distributed noise. A working example is discussed.

031.207 **Least-squares frequency analysis of unequally spaced data.** N. R. Lomb.
Astrophys. Space Sci., Vol. 39, 447 - 462 (1976).

The statistical properties of least-squares frequency analysis of unequally spaced data are examined. It is shown that, in the least-squares spectrum of gaussian noise, the reduction in the sum of squares at a particular frequency is a χ_2^2 variable. The reductions at different frequencies are not independent, as there is a correlation between the height of the spectrum at any two frequencies, f_1 and f_2, which is equal to the mean height of the spectrum due to a sinusoidal signal of frequency f_1, at the frequency f_2. These correlations reduce the distortion in the spectrum of a signal affected by noise. Some numerical illustrations of the properties of least-squares frequency spectra are also given.

031.208 **Photometrische Periodensuche bei Ap-Sternen.** H. M. Maitzen.
Mitt. Astron. Ges., No. 38, p. 177 (1976). – Abstract.

031.209 **Grundsätzliche Aspekte der Mehrfarbenphotometrie.** R. Buser, U. Steinlin.
Mitt. Astron. Ges., No. 38, p. 223 - 225 (1976). – Abstract.

031.210 **Verwendung eines Interferenzverlauffilters für relative spektralphotometrische Messungen.** D. Oberlerchner.
Mitt. Astron. Ges., No. 38, p. 234 - 237 (1976).

031.211 **Radialgeschwindigkeitsmessungen mit einer SEC-Vidicon am Cassegrain-Gitterspektrographen des Observatorium Hoher List der Universitätssternwarte Bonn.** M. Hoffmann, J. D. Schumann.
Mitt. Astron. Ges., No. 38, p. 237 (1976). – Abstract.

031.212 **Messung veränderlicher Radialgeschwindigkeiten mit dem Fehrenbach-Prisma.** F. Gieseking.
Mitt. Astron. Ges., No. 38, p. 238 (1976). – Abstract.

031.213 **Absorptionsverlauf in Sternfeldern aus dreifarben-photometrischen Messungen.** H. Hartl, J. Pfleiderer.
Mitt. Astron. Ges., No. 38, p. 247 - 249 (1976). – Short report.

031.214 **Investigation of the fluctuational character of the amplitude of stellar image vibrations.** M. M. Osipenko, I. V. Shvalagin.
Materialy nauch. konf. aspirantov 1 molodykh uchenykh. Uzhgorod. un-t. Sekts. astron. n. Uzhgorod, 1975, p. 13 - 18.

In Ukrainian. − Abstr. in Referativ. Zhurn. 51. Astron., 3.51.69 (1976).

031.215 **On the accuracy of astronomical measurements of objects and individual errors of measurers.**
P. P. Zhila, V. P. Epishev, A. G. Kirichenko, K. A. Kudak, V. A. Fenchak, L. S. Khokhlova.
Materialy nauch. konf. aspirantov i molodykh uchenykh. Uzhgorod. un-t. Sekts. astron. n. Uzhgorod, 1975, p. 2 - 7. In Ukrainian. − Abstr. in Referativ. Zhurn. 51. Astron., 3.51. 123 (1976).

031.216 **On the analysis of blended spectra.** A. Cassatella.
Astron. Astrophys., Vol. 48, 281 - 286 (1976).

A least-square curve-fitting method is presented which allows to resolve strongly overlapping lines in observed spectra. The application to blended stellar spectra for line identification, radial velocity determination and equivalent width measurement is tested and discussed.

031.217 **Aperture synthesis without phase measurements.**
J. E. Baldwin, P. J. Warner.
Monthly Notices Roy. Astron. Soc., Vol. 175, 345 - 353 (1976).

Under certain circumstances, aperture synthesis techniques may be used to obtain an unambiguous map of an area of sky without any measurements of phase. The necessary conditions, the accuracy of the map and the circumstances under which the method may prove useful are discussed. An experimental evaluation of the technique is described.

031.218 **Two dimensional photon counting echelle spectroscopy.** D. Bardas, J. E. McClintock, P. Peterson, G. W. Clark, C. R. Canizares.
Bull. American Astron. Soc., Vol. 8, 290 (1976). − Abstr. AAS.

031.219 **The statistically rigorous non-iterative computation of parallaxes.** H. Eichhorn, J. Russell.
Bull. American Astron. Soc., Vol. 8, 292 (1976). − Abstr. AAS.

031.220 **Algorithm for the restoration of infrared raster scans.** T. Simon.
Astron. Journ., Vol. 81, 135 - 137 (1976).

A simple algebraic procedure is described for the restoration of spatial scans obtained with a double-beam infrared photometer. The restoration method is well suited for use at the telescope, especially in survey work, because it requires little knowledge of the instrumental profile and only modest computer programming. However, unlike most Fourier transform deconvolution techniques, the algorithm leaves the restored spatial distribution of flux uncorrected for beam-smearing effects. The method is applied to a scan at $20\,\mu$ of the molecular line maser OH 26.5 + 0.6; the associated infrared source is unresolved with a 4.5-arcsec-diameter aperture.

031.221 **Absolute calibration of millimeter-wavelength spectral lines.** B. L. Ulich, R. W. Haas.
Astrophys. Journ., Suppl. Ser., Vol. 30, 247 - 258 (1976).

A detailed analysis of the chopper-wheel method of calibrating the intensity of a millimeter-wavelength spectral line is presented. The zenith atmospheric extinction between 3.5 mm and 2.6 mm wavelength was measured, and the intensities of six spectral lines in this range were absolutely calibrated.

031.222 **Observations of rapidly fluctuating objects. II. Mathematical data processing.**
G. N. Alekseev, G. M. Beskin.
Astrofiz. Issled., Izv. Spets. Astrofiz. Obs., Vol. 8, 53 - 63 (1976). In Russian.

A complex of mathematical methods of observational

data for detection and investigation of rapidly fluctuating objects is described. An indirect Fourier analysis using constricting windows is adopted as the principal method. The problems of accuracy and sensitivity of the method are considered.

031.223 **Instrumental methods of radio spectroscopy of the interstellar medium. II. Comparative estimate of methods.** N. F. Ryzhkov.
Astrofiz. Issled., Izv. Spets. Astrofiz. Obs., Vol. 8, 89 - 119 (1976). In Russian.

A comparative estimate of the basic radiometric methods of spectral measurements is performed. Expressions are obtained to estimate false spectral signals caused by the instability of characteristics of low-noise microwave frequency amplifiers and by the level variations of the received continuous spectrum radiation. Quantitative estimates of the relative amount of false signals and of the relative decrease in sensitivity which can be obtained using different measuring techniques are presented.

031.224 **Stigmatic mounting of holographic plane grating.** M. Singh.
Bull. Astron. Soc., India, Vol. 3, 28 (1975). − Abstract of paper presented at the A.S.I. meeting 1975.

031.225 **A new holographic plane grating monochromator.** M. Singh, K. C. A. Raheem.
Bull. Astron. Soc. India, Vol. 3, 28 (1975). − Abstract of paper presented at the A.S.I. meeting 1975.

031.226 **Preliminary observations with the new grating monochromator.** D. B. Jadhav, A. D. Tillu.
Bull. Astron. Soc. India, Vol. 3, 28 (1975). − Abstract of paper presented at the A.S.I. meeting 1975.

031.227 **Application of a Fabry-Perot spectrometer to the measurement of spectral line shifts much smaller than line width.** H. F. Döbele, J. H. Massig.
Applied Optics, Vol. 15, 69 - 72 (1976).

031.228 **Chopping factors for circular and square apertures.** C. S. Guenzer.
Applied Optics, Vol. 15, 80 - 83 (1976).

For the cases of circular and square apertures, calculations are presented for the beam size dependence of the chopping factor that is the factor relating beam intensity to that component of the intensity passing a filter tuned to the chopping frequency. For small beams, a value of 0.45 applies, but as the beam size approaches the chopping blade size, its value decreases and becomes geometry dependent. Complete numerical calculations are given for the case of a circular chopping wheel, but it is shown that analytic results of an approximation are usually sufficient. Harmonics of the chopping factor are also presented.

031.229 **Vacuum ultraviolet absolute radiometry utilizing total blackbody radiation.** G. Marette.
Applied Optics, Vol. 15, 440 - 444 (1976).

A new calibration technique is described which allows a monochromatic source and detectors to be obtained as secondary standards in the vacuum ultraviolet spectral range. The method of operation is based on the comparison, against a thermopile, between an ultraviolet flux from a monochromator and the total flux from a blackbody simulator. The over-all uncertainty does not exceed 20 %.

031.230 **Une heureuse combinaison de filtre et d'émulsion astronomiques.** A. Heck.
Orion, 34. Jahrgang, p. 32 - 34 (1976).

031.231 **Image processing in radio astronomy.** L. E. Somers.

Journ. Optical Soc. America, Vol. 66, 171 (1976). Conference abstract.

031.232 Maximum entropy restorations of Ganymede.
B. R. Frieden, W. Swindell.
Journ. Optical Soc. America, Vol. 66, 172 (1976). – Conference abstract.

031.233 Recent developments at JPL in the application of digital image processing techniques to astronomical images. J. J. Lorre, D. J. Lynn, W. D. Benton.
Journ. Optical Soc. America, Vol. 66, 174 - 175 (1976). Conference abstract.

031.234 Astronomical holography through the turbulent atmosphere. J. B. Breckinridge.
Journ. Optical Soc. America, Vol. 66, 175 (1976). – Conference abstract.

031.235 A posteriori restoration of atmospherically degraded images using multiframe imagery. J. W. Sherman.
Journ. Optical Soc. America, Vol. 66, 175 (1976). – Conference abstract.

031.236 Speckle interferometry on the 2.5 m Isaac Newton Telescope.
R. Bedows, J. C. Dainty, B. L. Morgan, R. J. Scaddan.
Journ. Optical Soc. America, Vol. 66, 181 (1976). – Conference abstract.

031.237 Effects of the atmosphere in stellar speckle interferometry. F. Roddier, C. Roddier.
Journ. Optical Soc. America, Vol. 66, 181 (1976). – Conference abstract.

031.238 Reconstructed images of Alpha Orionis using stellar speckle interferometry.
S. P. Worden, J. W. Harvey, C. R. Lynds.
Journ. Optical Soc. America, Vol. 66, 181 (1976). – Conference abstract.

031.239 Fundamental limitations of stellar speckle interferometry. M. G. Miller.
Journ. Optical Soc. America, Vol. 66, 181 (1976). – Conference abstract.

031.240 Lens–atmosphere MTF measurements.
D. P. Karo, A. M. Schneiderman.
Journ. Optical Soc. America, Vol. 66, 182 (1976). – Conference abstract.

031.241 Image retrieval from astronomical speckle patterns.
K. T. Knox.
Journ. Optical Soc. America, Vol. 66, 182 (1976). – Conference abstract.

031.242 Evaluation of photometric spectral line profiles by digital computer methods. H. Schwiecker.
Journ. Quant. Spectrosc. Radiat. Transfer, Vol. 16, 623 - 625 (1976).
A computer method for the transformation of photographic spectra into the intensity scale has been derived and applied to the radial distribution of spectral line shapes and continuum intensities as obtained from the side-on spectra of a plasma light-source of rotational symmetry by the use of Abel's inversion.

031.243 Nomographs for the preliminary solution of spectroscopic binary radial-velocity curves.
J. P. Oliver, T. R. Flesch.
Publ. Astron. Soc. Pacific, Vol. 88, 148 - 153 = Contr. Rose-

mary Hill Obs., Univ. Florida, *Gainesville*, No. 63 (1976).
Nomographs are presented as an extension of Irwin's method for the solution of radial-velocity curves. Slight modifications of Irwin's original set of parameters have reduced the number of nomographs through an increase in the symmetry of the functions. The final nomographic technique is simple, fast, and reasonably accurate. Results of tests of the method on velocity curves from the literature are presented.

031.244 Principi generali della fotometria fotoelettrica.
S. Ghedini.
Coelum, Vol. 44, 14 - 26 (1976).

031.245 Voor het eerst: steroppervlak in beeld.
G. W. E. Beekman.
Zenit, 3e jaargang, p. 164 - 166 (1976).

031.246 Tekenen achter de telescoop: leuk en leerzaam.
A. van der Jeugt.
Zenit, 3e jaargang, p. 175 - 177 (1976).

031.247 Automation of photographing during astronomical observations. M. B. Kerimbekov, Ch. A. Ehfendiev.
Solnechnye Dannye 1976 Byull., No. 2, p. 98 - 102 (1976). In Russian.
The authors offer a simple system providing high efficiency of photographing during ground observations. Two schemes are given to be used as key system for launching registrating cameras. The system operates in case the atmospheric turbulence interference is reduced.

031.248 Algorithm of the problem of determination of the parameters of the curve of the photostream in observations on a transit instrument.
A. F. Vantsan, A. D. Egorov, Z. A. Rytova.
Vestn. Khar'kov. Univ., No. 129 (Ser. Astron., vyp. (No.) 10), p. 20 - 24 (1975). In Russian.

031.249 A method of blink comparison using standard 35 mm photographic equipment. C. R. Martys.
Journ. British Astron. Ass., Vol. 86, 277 - 279 (1976).

031.250 Lunar map making. R. J. Livesey.
Journ. British Astron. Ass., Vol. 86, 296 - 298 (1976).

031.251 Primary processing of cosmic ray station data in a real time scale by means of the Nairi computer.
Ya. L. Blokh, R. A. Simsar'yan, F. A. Starkov, A. P. Tipikin.
Cosmic rays, No. 15, (see 003.006), p. 203 - 206 (1975). In Russian.

031.252 Radioheliography.
N. R. Labrum, D. J. McLean, J. P. Wild.
Methods comput. phys., Vol. 14, (see 003.007), 1 - 53 (1975).
The sun and its radio image; The principles of radioheliography; The evolution of the radioheliograph; The Culgoora radioheliograph; Culgoora data processing; Solar radio astronomy with the radioheliograph; Future developments in radioheliography.

031.253 Aperture synthesis. W. N. Brouw.
Methods comput. phys., Vol. 14, (see 003.007), 131 - 175 (1975).
Aperture synthesis; Earth rotation aperture synthesis; Data processing.

031.254 Computations in radio-frequency spectroscopy.
J. A. Ball.
Methods comput. phys., Vol. 14, (see 003.007), 177 - 219 (1975).

Introduction to spectroscopy in radio astronomy; Power spectra; Selected problems in calibration and observing techniques; Selected problems in interpretation of spectra.

031.255 **Absolute calibration of the Ultraviolet Sky Survey Telescope in satellite TD 1.**
C. M. Humphries, C. Jamar, D. Malaise, H. Wroe.
Astron. Astrophys., Vol. 49, 389 - 406 (1976).

Absolute determinations of ultraviolet stellar energy distributions with the S 2/68 Ultraviolet Sky Survey Telescope in satellite TD 1 are based on laboratory calibration measurements performed before launch of the spacecraft. The present paper gives the results of these measurements and describes the procedures which were used to obtain them. The calibrations were performed independently by groups in the UK and Belgium using an absolute radiometric detector and a black body source, respectively, as primary absolute standards.

031.256 **L'observation des galaxies avec la machine COSMOS.**
P. R. Williams, N. M. Pratt.
Orion, 34. Jahrgang, p. 58 - 63 (1976).

La machine COSMOS détecte et mesure à grande vitesse les positions et les détails de structure de nombreuses images d'étoiles et de galaxies précédemment enregistrées sur plaques photographiques.

031.257 **Some results on the imaging of incoherent sources through turbulence.** R. L. Fante.
Journ. Opt. Soc. America, Vol. 66, 574 - 580 (1976).

031.258 **Limitations of ground based near and far infrared astronomy.**
G. Dall'Oglio, I. Guidi, B. Melchiorri, F. Melchiorri, V. Natale.
Infrared Physics, Vol. 15, 13 - 17 (1975).

The possibilities of ground based infrared astronomy are discussed. It is shown that telescopes used up to now are not optimized and that the Diffraction Limited Detectivity conditions (DLD) cannot be reached, in the near infrared, by the larger telescopes. The maximum useful diameters for near i.r. and far i.r. telescopes have been evaluated and a new parameter (i.e. visibility) is introduced in order to describe the performances of i.r. ground based telescopes.

031.259 **Upconversion spectrometry for astrophysical applications.**
T. Kostiuk, M. Abbas, K. W. Ogilvie, M. Mumma.
Infrared Physics, Vol. 16, (see 012.013), 55 - 59 (1976).

031.260 **Atmospheric spectroscopy at very high resolution in the middle infrared with a rapid-scan Michelson interferometer.** M. Anderegg, J. E. Beckman, A. F. M.
Moorwood, J. P. Baluteau, E. Bussoletti, N. Coron.
Infrared Physics, Vol. 16, (see 012.013), 329 (1976).
Abstract.

031.261 **Sur la réduction des clichés astronomiques.**
N. G. Rizvanov, L. A. Ourassine (Urasin).
Ann. Obs. Astron. Alger, Vol. 4, Fasc. 2, p. 11 - 16 (1975).

031.262 **Über thermisch bedingte Einflüsse bei Transitbeobachtungen im Meridian.** J. Dittrich.
Veröff. Zentralinst. Physik Erde, Akad. Wiss. DDR, No. 36, 71 pp. = Mitt. Lohrmann-Obs. Techn. Univ. Dresden, No. 30 (1976).

031.263 **Statistische Probleme bei der Ausgleichung direkter, unabhängiger, normalverteilter Beobachtungen mit geschätzten Gewichten.** H. Sellge.
Deutsche Geod. Komm. Bayer. Akad. Wiss., Reihe C, Nr. 213, Diss. Techn. Univ. Berlin, 113 pp. München (1975).

031.264 **Photographic position determination by means of the method of Rinia.** A. T. Son.
Minor Planet. Bull., Vol. 3, 39 - 42 (1976).

031.265 **Photographing minor planets and measuring their positions.** D. S. Lynn.
Minor Planet Bull., Vol. 3, 51 - 52 (1976).

031.266 **Astrophysical parameters of close visual binaries, obtained with an area scanner.** H. Jenkner.
Mem. Soc. Astron. Italiana, Vol. 45, (see 012.017), 815 - 816 (1974).

031.267 **Determinazione degli istanti di transito di una stella oraria nei confronti della scala di tempo nazionale TUC (IEN).** C. Moranzino, V. Pettiti, G. Rovera.
Alta Frequenza, Vol. 45, 177 - 180 = Pubbl. Varie Fuori Ser., Oss. Astron. Torino, (Pino Torinese), No. 62 (1976).

031.268 **The role of VLBI in the establishment of coordinate systems.** I. Zhongolovitch (Zhongolovich).
IAU Colloquium No. 26, (see 012.018), p. 293 - 295 (1975).

031.269 **Color composite photographs of the Magellanic Clouds.** R. J. Dufour, R. A. Goodding.
Proc. Southwest Regional Conf., Vol. 1, (see 012.021), 71 - 77 (1976).

Simulated color photographs of the Large and Small Magellanic Clouds were produced from black and white photographs taken with the Curtis Schmidt telescope at Cerro Tololo using multichannel projection techniques. A brief discussion of the techniques employed, the color photographs obtained, and the potential of future work of this type is given.

031.270 **Simultaneous multicolor planetary photography.** C. F. Capen.
Proc. Southwest Regional Conf., Vol. 1, (see 012.021), 173 - 177 (1976).

031.271 **Far infra-red background astronomy at ground.**
G. Dall'Oglio, S. Fonti, I. Guidi, B. Melchiorri, F. Melchiorri, V. Natale.
Nuovo Cimento B, Ser. 11, Vol. 27B, 27 - 40 (1975). – Abstr. in Phys. Abstr., Vol. 79, A003439 (1976).

031.272 **A coincidence long baseline celestial gamma-ray burst detection system.**
D. J. Fegan, B. McBreen, C. O'Sullivan, V. Ruddy.
Nucl. Instruments Methods, Vol. 129, 613 - 616 (1975).
Abstr. in Phys. Abstr., Vol. 79, A020051 (1976).

031.273 **The treatment of astronomical images degraded by atmospheric fluctuations.** F. Roddier.
Ann. Telecommun., Vol. 30, 304 - 308 (1975). In French.
Abstr. in Phys. Abstr., Vol. 79, A023905 (1976).

031.274 **On measurements with pendulum astrolabes and their evaluation.** K. Deichl.
Zeitschr. Vermessungswesen, Vol. 100, 499 - 509 (1975). In German. – Abstr. in Phys. Abstr., Vol. 79, A037698 (1976).

031.275 **Stellar speckle interferometry.** J. C. Dainty.
Laser speckle and related phenomena, Springer-Verlag, Berlin, p. 255 - 280 (1975). – Abstr. in Phys. Abstr., Vol. 79, A037699 (1976).

031.276 **Tips für die Astropraxis.**
SuW, 15. Jahrgang, p. 58 - 63, 132 - 134, 204 - 206 (1976).

On the use of Tautenburg Schmidt for improvement of astronomical reference system.   See Abstr. 041.038.

Geodetic and astrometric results of very-long baseline interferometric measurements of natural radio sources. See Abstr. 041.039.

On constructing the inertial system of high accuracy by VLBI methods.   See Abstr. 041.041.

The application of radio interferometers in astrometry.   See Abstr. 041.043.

Infrared photometry at the Stockholm Observatory. See Abstr. 061.043.

Theoretical atmospheric transmission in the mid- and far-infrared at four altitudes.   See Abstr. 082.043.

Propagation-medium limitations on phase-compensated atmospheric imaging.   See Abstr. 082.055.

Diffraction-limited atmospheric imaging of extended objects.   See Abstr. 082.056.

Measurements of the atmospheric attenuation of the spectral components of astronomical images. See Abstr. 082.057.

On the fringe visibility in a Michelson stellar interferometer.   See Abstr. 082.067.

Imaging spectrometer study of Jupiter and Saturn. See Abstr. 099.020.

Coupe photométrique des quatre anneaux de Saturne suivant leur grand axe, à partir d'enregistrements bi-dimensionnels de photographies de la planète. Les mesures. See Abstr. 100.007.

Can 'invisible' bodies be observed in the solar system?   See Abstr. 102.006.

Utvärdering av objektivprismespektra av tidiga stjärnor med hjälp av dator.   See Abstr. 114.060.

Digital analysis of speckle photographs: the angular diameter of Arcturus.   See Abstr. 115.004.

High resolution observations of NGC 1275 with a four-element intercontinental radio interferometer. See Abstr. 141.037.

Radiointerferometrie mit transatlantischen Basislinien: Galaxienkerne und Quasare.   See Abstr. 141.047.

Fast 8550 MHz survey technique and digital computer processing.   See Abstr. 141.070.

Timing of the Crab pulsar. II. Method of analysis. See Abstr. 141.315.

Pulsar signal processing.   See Abstr. 141.343.

Data Processing, Automation

031.401   About the function $(d^n/dt^n) \varphi (r) = f (dr/dt, d^2 r/dt^2, ..., d^n r/dt^n)$.   E. Fichera.
Astron. Nachr., Vol. 297, 59 - 60 (1976). — In French.
The modern means of computation require mathematical relations in forms which take care of the data directly furnished to the computer by observational instruments. In this communication the general relation of the n*th* derivative of the function $\varphi(r)$ is given.

031.402   Computer-unterstützte Bestimmung von Linienpositionen in Sternspektren.   H. Jenkner.
Mitt. Astron. Ges., No. 38, p. 238 - 241 (1976). — Short report.

031.403   Computer program for the determination of photographic magnitudes from iris measurements.
G. Auner.
Mitt. Astron. Ges., No. 38, p. 251 - 252 (1976). — Abstract.

031.404   A system for recording fast varying light curves.
J. C. Bhattacharyya, A. Sundareswaran.
Bull. Astron. Soc. India, Vol. 3, 28 (1975). — Abstract of paper presented at the A.S.I. meeting 1975.

031.405   Oversampling of digitized images.   D. Fischel.
Astron. Journ., Vol. 81, 285 - 291 (1976).
Oversampling is defined by the author as sampling with a device whose characteristic width is greater than the interval between samples. This paper shows why oversampling should be avoided and discusses the limitations in the data processing if circumstances dictate that oversampling cannot be circumvented. Principally, oversampling should not be used to provide interpolating data points. Rather, the time spent oversampling should be used to obtain more signal with less relative error and the Sampling Theorem used to provide any desired interpolated values. The concepts are applicable to single element and multielement detectors.

031.406   Determination of periods of variable astronomical objects with the aid of a computer.   M. M. Basko.
Peremennye Zvezdy, Prilozhenie, Vol. 2, 337 - 347 (1976). In Russian.
Description and text of a program for the computer BESM-6 is presented. The program is created on the basis of two independent algorithms and is intended to determine periods of variable astronomical objects.

031.407   Search for and approximate determination of the periods of variable stars by using the computer

"MIR-2".   Yu. A. Fadeev.
Peremennye Zvezdy, Prilozhenie, Vol. 2, 375 - 379 (1976).
In Russian.
    The results of searching for periods of variable stars with the "MIR-2" computer are described. The description of the programme in "ANALITIC" language is given.

031.408   Computer programs for transforming the data of an iris photometer into magnitudes.
U. A. Nurmanova.
Peremennye Zvezdy, Prilozhenie, Vol. 2, 381 - 392 (1976).
In Russian.
    The description of two computer programs to obtain magnitudes from iris photometer data is given. The programs are written in ALGOL-60 computer language.

031.409   Automatic measurement of photographic plates with a photo-diode array.
E. Høg, D. Wiskott, with a preface by S. Laustsen.
ESO Techn. Rep. No. 5, 3 + 46 pp. (1974).

031.410   Observatory aims for the stars with COSMOS.
    Electronics Australia, Vol. 37, No. 5, p. 28 - 29 (1975). — Abstr. in Phys. Abstr., Vol. 79, A003433 (1976).

031.411   Digital reconstruction of point sources imaged by a zone plate camera.
D. R. Dance, B. G. Wilson, R. P. Parker.
Phys. Med. Biol., Vol. 20, 747 - 756 (1975). — Abstr. in Phys. Abstr., Vol. 79, A019802 (1976).

    Radioheliography.   See Abstr. 031.252.

    Aperture synthesis.   See Abstr. 031.253.

    A possible data handling system for a scientific spacecraft.   See Abstr. 032.513.

    On a semi-automatical iris-type photometer. See Abstr. 034.016.

    Electrophotometer—computer system and its application to astronomical observations. See Abstr. 034.018.

    A digital method of spectrosensitogram processing for investigation of astronomical photoemulsions. See Abstr. 036.002.

    A programme to calculate the coordinates of tracking stations by means of the semi-dynamical method. See Abstr. 046.018.

    A computer-based programme for the improvement of orbital elements of artificial satellites. See Abstr. 052.020.

    The Ariel 5 post launch control and data handling system.   See Abstr. 054.008.

    Operational experience with the ANS on-board computer.   See Abstr. 054.009.

    The computer system for Spacelab. See Abstr. 054.010.

    A data processing algorithm for comparing theory with observations in the case of cometary motions. See Abstr. 102.013.

    Complex of programs for the analysis of the light curves of eclipsing binary systems by a direct method. See Abstr. 121.059.

    Determination of the periods of the Blazhko-effect (amplitude variation) of two RR Lyrae variables in globular clusters by computers.   See Abstr. 122.095.

    Fast 8550 MHz survey technique and digital computer processing.   See Abstr. 141.070.

    Pulsar signal processing.   See Abstr. 141.343.

032 Astronomical Instruments, Space Instrumentation

Astronomical Instruments

032.001 **A narrow-band Hα telescope for visual and photo-graphic solar observations.** A. Gabriël.
Journ. British Astron. Ass., Vol. 86, 140 - 143 (1976).

032.002 **Developing production of astronomical instruments by self-reliance. – How China's photoelectric astrolabes were produced.**
The Photoelectric Astrolabe Research and Production Group, Academia Sinica.
Acta Astron. Sinica, Vol. 16, 101 - 103 (1975). In Chinese.

032.003 **Preliminary analysis of observational data obtained with a photoelectric astrolabe type I.**
Photoelectric Astrolabe Section and Data Analysis Section, First Laboratory, Shansi Observatory, Academia Sinica.
Acta Astron. Sinica, Vol. 16, 104 - 114 (1975). In Chinese.
The photoelectric astrolabe type I is a new type instrument for time determination and latitude determination. In this paper, the observational results during the period 1973 October to 1974 December are analysed, and the quality of the instrument is estimated. The result of analysis shows that this instrument is entirely free of the effect of personal equation, and the precision of observation is high. Using the data obtained during this period, the problems of determining the systematic error of the FK4 and improving the constant of aberration are investigated.

032.004 **The photoelectric astrolabe type II.**
The Photoelectric Astrolabe Research and Production Group, Academia Sinica.
Acta Astron. Sinica, Vol. 16, 115 - 122 (1975). In Chinese.
The photoelectric astrolabe type I was made in China in 1971; several improved photoelectric astrolabes, of a second model, were constructed from 1972 to 1974. The principles and a description of the main parts of the new instruments are given. At Shanghai Observatory, observations have been carried out with the new instrument since Sep. 1974.

032.005 **Offset guider for the prime focus of the 3.9-meter Anglo-Australian Telescope.** H. Kobler, P. T. Wallace.
Publ. Astron. Soc. Pacific, Vol. 88, 80 - 85 (1976).
The automatic offset guider at the prime focus of the AAT is described. A description of the guider and its organization in the telescope's computerized drive system is followed by the initial results obtained with it.

032.006 **The new Greek 48-inch Cassegrain-coudé telescope.** G. Contopoulos, C. Banos.
Sky Telescope, Vol. 51, 154 - 155 (1976).

032.007 **Das 75-cm-Metallspiegelteleskop der Wilhelm-Foerster-Sternwarte.** B. Wedel.
Mitt. Astron. Ges., No. 38, p. 124 - 128 (1976). – Short report.

032.008 **Double-astrograph on the international parallel.** A. A. Latypov, Yu. M. Ivanov.
Zemlya i Vselennaya, 1976, No. 2, p. 62 - 63. In Russian.

032.009 **The 600-millimeter television telescope of the Special Astrophysical Observatory of the USSR Academy of Sciences.** Eh. B. Gazhur, M. M. Kononov, V. B. Nebelitskij, A. F. Fomenko.
Astrofiz. Issled., Izv. Spets. Astrofiz. Obs., Vol. 8, 69 - 76 (1976). In Russian.
The 600-millimeter television telescope constructed at the Special Astrophysical Observatory of the USSR Academy of Sciences is described. The penetrability limit is 15^m. The prospects of application of the TV equipment to astrophysical investigations are discussed.

032.010 **An instrument for astrometric observations from the lunar surface.** A. A. Gurshtejn, A. A. Konopikhin.
Astron. Zhurn. Akad. Nauk SSSR, Vol. 53, 441 - 445 (1976). In Russian. English translation in Soviet Astron., Vol. 20, No. 2.
A construction principle for a high-precision astrometrical automatic instrument for carrying out position observations from the lunar surface according to the equal altitudes method is discussed.

032.011 **Kitt Peak 60-cm vacuum telescope.** W. C. Livingston, J. Harvey, A. K. Pierce, D. Schrage, B. Gillespie, J. Simmons, C. Slaughter.
Applied Optics, Vol. 15, 33 - 39 (1976).
Described is a major new tool for solar research. The observation of magnetic and velocity (circulation) field structure on a synoptic basis and with diffraction-limited resolution is the aim. Test data in the form of the system MTF and optical transmission, together with examples of full disk magnetograms and photoheliograms, show present performance capability. Measured MTF indicates a response of 0.2 at 1 sec of arc (whereas diffraction-limited response would be ~0.8).

032.012 **Coatings for the far uv region.** E. Spiller.
Journ. Optical Soc. America, Vol. 66, 161 (1976). Conference abstract.

032.013 **Britten krijgen nieuwe sterrenwacht op het noordelijk halfrond.** F. G. Smith.
Zenit, 3e jaargang, p. 162 - 164 (1976).

032.014 **Portable refractor.** S. Yu. Zhukhovitskij.
Zemlya i Vselennaya, 1976, No. 3, p. 90 - 91. In Russian.

032.015 **On the control of astronomical telescopes.** B. I. Kontorovich, V. N. Medvedev.
Issled. po geomagnetizmu, aehron. i fiz. Solntsa. Vyp. (No.) 37. Moskva, Nauka, 1975, p. 182 - 189. In Rumanian. Abstr. in Referativ. Zhurn. 51. Astron., 5.51.112 (1976).

032.016 **Optical center and inclination of the plate of the Zeiss short-focus astrograph.** P. P. Pavlenko.
Vestn. Khar'kov. Univ., No. 129 (Ser. Astron., vyp. (No.) 10), p. 3 - 9 (1975). In Russian.

032.017 **Essai des télescopes Schmidt-Cassegrain Celestron.** J. Dragesco.
L'Astronomie, Vol. 90, 238 - 245 (1976).

032.018 **Les plus grandes lunettes.** P. Kohler.
L'Astronomie, Vol. 90, 247 - 248 (1976).

032.019 **Gamma telescope with an acoustic spark camera.** V. T. Barsukov, V. K. Bocharkin, I. F. Bugakov, P. A. Volkov, G. A. Vorob'ev, G. M. Gorodinskij, P. I. Gorshkov, L. I. Zhukova, E. M. Kruglov, G. A. Pyatigorskij, A. M. Romanov, I. A. Sokolov, E. I. Chujkin.
Cosmic rays, No. 15, (see 003.006), p. 187 - 189 (1975). In Russian.

032.020 Azimuthal telescopes of large effective area based on liquid scintillation detectors.
V. N. Bakatanov, A. V. Voevodskij, V. A. Kozyarivskij, S. P. Mikheev, V. I. Stepanov, T. I. Tulupova.
Cosmic rays, No. 15, (see 003.006), p. 207 - 210 (1975). In Russian.

032.021 The heat regime in the pavilion of the Moscow meridian circle GOMZ. V. G. Shamaev.
Astron. Tsirk., No. 899, p. 3 - 5 (1976). In Russian.

032.022 On the aerodynamics of astronomical domes.
Yu. A. Panov, P. V. Shcheglov.
Astron. Tsirk., No. 900, p. 1 - 3 (1976). In Russian.

032.023 On the use of acoustic methods for studying temperature and wind fields near astronomical instruments. P. V. Shcheglov.
Astron. Tsirk., No. 900, p. 3 - 5 (1976). In Russian.

032.024 Statistical analysis of the Danjon astrolabe at São Paulo. L. B. F. Clauzet.
Astron. Astrophys., Suppl. Ser., Vol. 25, 55 - 64 (1976).

After the first seven years of observations with the astrolabe the author analysed the zenith distance residuals and mean standard deviations of a large number of transits in order to get a better knowledge of the instrument and the observational station. In this paper he presents this analysis and also the curves $\Delta\alpha_\delta$ and $\Delta\delta_\delta$ of the systematic errors of the catalogue and the individual corrections $\Delta\alpha$ and $\Delta\delta$ for the double transit stars of the programme. The results are compared with those of other observatories.

032.025 Achtzig Jahre Archenhold-Sternwarte. Die Chronik des Riesenfernrohrs von 1896. D. Wattenberg.
Blick in das Weltall, 24. Jahrgang, p. 48 - 62 (1976).

032.026 The MMT (*Multiple Mirror Telescope*) Observatory on Mount Hopkins. N. P. Carleton, T. E. Hoffman.
Sky Telescope, Vol. 52, 14 - 21 (1976).

032.027 Il telescopio da 152 cm dell'Istituto di astronomia di Bologna. A. Braccesi.
Giorn. Astron., Vol. 1, 135 - 145 (1975).

032.028 Un semplice strumento per l'osservazione.
P. Andrenelli.
Giorn. Astron., Vol. 1, 217 - 234 (1975).

032.029 An astronomical eye of the world.
S. R. Brzostkiewicz.
Urania Kraków, Vol. 47, 162 - 168 (1976). In Polish.
Concerning the Soviet 6 m telescope.

032.030 First results of the tests of the photometric twin-reflector. K. Panov, G. Hildebrandt, W. Schöneich, E. Želwanowa (*Shelvanova*).
Astron. Nachr., Band 297, 213 - 214 (1976). In German.

First measurements of the short time variations of Ap stars with the twin telescope show a nonlinear drift caused by a sensibility of the amplifier depending on temperature. Moreover the authors found a dependence of the impulse rate on the position of the star image in the diaphragm. Special test measurements with a simple incandescent lamp show that this telescope gives results with an accuracy of ±0.002 mag after elimination of these effects.

032.031 Comments on instrumentation at the Ondřejov Observatory. J. Grygar.

Astron. Nachr., Band 297, 215 - 216 (1976).
A short description of the instrumentation of the Astronomical Institute of the Czechoslovak Academy of Sciences at Ondřejov is given.

032.032 Supplementary tests of the optics of the ESO 3.6 m telescope by the method of Lytle. R. N. Wilson.
ESO Techn. Rep. No. 4, 21 pp. (1974).

032.033 Acceleration feedback applied to the 3.6 m telescope servosystem. T. Andersen, R. Zurbuchen.
ESO Techn. Rep. No. 7, 3 + 26 pp. (1976).

032.034 On some inprovements of the control system of the PZT. M. Yoshinari.
Tokyo Astron. Obs. Rep., Vol. 17, 440 - 447 (1976). In Japanese.

032.035 Measurements of the dynamic characteristics of a pulse motor as applied to the PZT.
M. Yoshinari, K. Fujiwara.
Tokyo Astron. Obs. Rep., Vol. 17, 464 - 470 (1976). In Japanese.

032.036 The instrumental constants of meridian circle and the personal equations. R. Fukaya.
Toyko Astron. Obs. Rep., Vol. 17, 471 - 482 (1976). In Japanese.

032.037 Tilt of the pillars of meridian circle measured by the TEM tiltmeter. H. Hara.
Tokyo Astron. Obs. Rep., Vol. 17, 483 - 493 (1976). In Japanese.

032.038 The 65 cm PZT of the U.S. Naval Observatory.
W. J. Klepczyński.
IAU Colloquium No. 26, (see 012.018), p. 309 - 313 (1975).

032.039 A 41″ telescope/observatory for North Texas.
T. G. Harrison.
Proc. Southwest Regional Conf., Vol. 1, (see 012.021), 27 - 30 (1976).

032.040 Mechanical deformation studies of an astronomical instrument using an interferometric method.
Nouv. Rev. Optique, Vol. 7, 63 - 68 (1976). In French.
Abstr. in Phys. Abstr., Vol. 79, A033144 (1976).

032.041 Large-aperture ground-based telescope design and fabrication. D. Dodgen.
Optical Engineering, Vol. 14, 520 - 524 (1975). – Abstr. in Phys. Abstr., Vol. 79, A042457 (1976).

032.042 Current status of the MMT (*multiple mirror telescope*) optics. R. G. Shannon, G. M. Sanger.
Optical Engineering, Vol. 14, 544 - 551 (1975). – Abstr. in Phys. Abstr., Vol. 79, A042462 (1976).

Zur Geschichte des Meridiankreises und verwandter Instrumente. Von Römer bis Repsold und Reichenbach. I. II.
See Abstr. 004.001.

Some preliminary results from current observational programs with the new 23-cm zone astrograph.
See Abstr. 041.004.

Untersuchung eines astronomischen Theodolits mit automatischer Fernrohrnachführung durch Schrittmotoren.
See Abstr. 046.031.

Space Instrumentation

032.501 **Large Space Telescope: astronomers go into orbit.**
J. Walsh.
Science, Vol. 191, 544 - 545 (1976).

032.502 **Pulse pile-up in hard X-ray detector systems.**
D. W. Datlowe.
Space Sci. Instrum., Vol. 1, 389 - 406 (1975).

When pulse-height spectra are measured by a nuclear detection system at high counting rates, the probability that two or more pulses will arrive within the resolving time of the system is significant. This phenomenon, pulse pile-up, distorts the pulse-height spectrum and must be considered in the interpretation of spectra taken at high counting rates. A computational technique for the simulation of pile-up is developed here. The technique is used to model the UCSD solar hard X-ray experiment on the OSO-7 satellite.

032.503 **The Skylab ten color photoelectric polarimeter.**
J. L. Weinberg, J. G. Sparrow, R. C. Hahn.
Space Sci. Instrum., Vol. 1, 407 - 418 (1975).

A 10-color photoelectric polarimeter was used during Skylab missions SL-2 and SL-3 to measure sky brightness and polarization associated with zodiacal light, background starlight, and the spacecraft corona. A description is given of the instrument and observing routines together with initial results on the spacecraft corona and on the polarization of the zodiacal light.

032.504 **Design of a servo-mechanism to control the position of a mirror in a space experiment.**
A. Labeque, M. Vite, G. Bourguignon.
Space Sci. Instrum., Vol: 1, 419 - 437 (1975).

An electro-mechanical actuator designed to move the secondary mirror of a Cassegrainian telescope has been constructed for an experiment on board the NASA Orbiting Solar Observatory (OSO-I) for rastering the image of the sun at the focus. The main originality of this device is its internal servo-control which considerably increases its performance. The construction and the computation of the various parts of the actuator are described in detail. Results are presented.

032.505 **Techniques for measuring bulk gas-motions from satellites.** W. B. Hanson, R. A. Heelis.
Space Sci. Instrum., Vol. 1, 493 - 524 (1975).

There is a growing appreciation for the dynamic nature of the upper atmosphere. This paper is concerned principally with discussing the methods for making in-situ measurements of the ambient gas velocities from satellite. Both neutral and ion instrumentation is discussed, but most attention is paid to the ion sensors on Atmosphere-Explorer-C. It seems reasonable to expect that both neutral and ion gas velocities will be measured to an accuracy of the order of 20 m s^{-1} in future satellites.

032.506 **The Viking orbiter cameras' potential for photometric measurement.** T. E. Thorpe.
Icarus, Vol. 27, 229 - 239 (1976).

Although photometry of Mars is not listed as a major mission objective, the Viking Project has provided the orbiter Imaging Team with cameras exhibiting significant improvement in photometric measurement as compared with past Mariners. Sample calibration data are described, together with predicted performance capabilities.

032.507 **Der Flug von Spektro-Stratoskop.**
J. P. Mehltretter.
SuW, 15. Jahrgang, p. 44 - 47 (1976).

032.508 **The mirror X-ray telescope of the Salyut-4 station.**
I. P. Tindo.
Zemlya i Vselennaya, 1976, No. 1, p. 23 - 28. In Russian.

032.509 **Der Flug von Spektro-Stratoskop.**
J. P. Mehltretter.
Mitt. Astron. Ges., No. 38, p. 137 (1976). – Abstract.

032.510 **Orbital infrared telescope-spectrometer piloted by the Salyut 4 station.** M. N. Markov, V. S. Petrov, Yu. M. Ivanov, V. Yu. Modenov, V. K. Surovov.
Fiz. in-t AN SSSR. Preprint No. 148. Moskva, 1975. 22 pp. In Russian. – Abstr. in Referativ. Zhurn. 62. Issled. kosmich. prostranstva, 3.62.119 (1976).

032.511 **The electrostatic spectrometer of Cosmos 426.**
Yu. P. Gordeev, V. V. Mel'nikov, T. I. Pervaya, B. I. Savin, D. P. Sukhoj, B. M. Yakovlev.
Issled. kosmich. luchej. Moskva, Nauka, 1975, p. 254 - 258. In Russian. – Abstr. in Referativ. Zhurn. 62. Issled. kosmich. prostranstva, 4.62.121 (1976).

032.512 **Spectrometer for local sources of cosmic γ-radiation in the energy region of 0.03 - 2.5 MeV.**
N. L. Grigorov, M. I. Kudryavtsev, A. S. Melioranskij, I. A. Savenko.
Issled. kosmich. luchej. Moskva, Nauka, 1975, p. 228 - 230. In Russian. – Abstr. in Referativ. Zhurn. 62. Issled. kosmich. prostranstva, 4.62.122 (1976).

032.513 **A possible data handling system for a scientific spacecraft.** D. C. Thomas.
Journ. British Interplanet. Soc., Vol. 29, 363 - 385 (1976).

A preliminary design is described for a space-borne data handling system based on the use of the SAAB-SCANIA computer. A major design aim was to replace as much conventional on-board hardware as possible by the computer system, with the computer system providing a central service for several user sub-systems simultaneously.

032.514 **A modularized computer system for European space projects.** J. A. Hanje.
Journ. British Interplanet. Soc., Vol. 29, 387 - 401 (1976).

032.515 **Guidance law for the Ariane on-board computer.**
M. D. Waterman.
Journ. British Interplanet. Soc., Vol. 29, 402 - 407 (1976).

Ariane, a European Satellite Launch vehicle, will be controlled by a small on-board digital computer with navigation data from an inertial platform. This paper describes the computer program to be used, in particular the guidance law, and the reasons for choosing that form of guidance. The methods of testing the program are described, and a typical trajectory is given.

032.516 **Secondary electron conduction camera tube for space applications.**
C. Kunze, H. Samuelsson.
Applied Optics, Vol. 15, 661 - 667 (1976).

032.517 **Röntgen-Fernrohr in einer Erdumlaufbahn.**
E. Hintsches.
SuW, 15. Jahrgang, p. 162 - 163 (1976).

032.518 **Preliminary design and performance of an advanced gamma-ray spectrometer for future orbiter missions.**
A. E. Metzger, R. H. Parker, J. R. Arnold, R. C. Reedy, J. I. Trombka.
Proc. Sixth Lunar Sci. Conference, (see 012.010), Vol. 3, 2769 - 2784 (1975).

032.519 **Investigation of the sporadic radio radiation of the sun and of the parameters of the earth's ionosphere aboard the satellite Intercosmos-Copernicus 500. I. Scientific apparatus and method of the experiments.**
V. I. Aksenov, S. Aleśkewicz, H. Wełnowski, B. Wikierski, S. Gorgolewski, G. P. Komrakov, S. Krawczyk, A. P. Modestov, I. V. Popkov, L. Yu. Sokolov, V. A. Timofeev, J. Hanasz.
Kosmich. Issled., Vol. 14, 392 - 399 (1976). In Russian.

032.520 **Helios zodiacal light experiment.**
E. Pitz, C. Leinert, H. Link, N. Salm.
IAU Colloquium No. 31, (see 012.015), p. 19 - 23 (1976).

032.521 **Presentation of zodiacal light instrument aboard the D2B astronomical satellite.**
M. Maucherat, P. Cruvellier.
IAU Colloquium No. 31, (see 012.015), p. 74 - 77 (1976).

032.522 **Localization of gamma-ray bursts with wide field multiple pinhole camera system in near earth orbit.**
P. Gorenstein, H. Helmken, H. Gursky.
Astrophys. Space Sci., Vol. 42, 89 - 97 (1976). – See 012.016.
    A multiple pinhole camera system has been designed and proposed for a small satellite of the SAS type for the detection and localization of gamma-ray bursts. The instrument consists of a three unit array of detectors each of which includes a semi-cylindrical collimator surrounding a two-dimensional position-sensitive detector.

032.523 **The French-Russian 3 satellite γ-burst experiment.**
M. Niel, K. Hurley, G. Vedrenne, I. V. Esthulin (*Ehstulin*), A. S. Melioransky (*Melioranskij*).
Astrophys. Space Sci., Vol. 42, 99 - 102 (1976). – See 012. 016.
    A γ-burst monitoring network, involving an eccentric satellite and two interplanetary spacecraft, is planned for completion in 1977.

032.524 **The ability of COS-B to measure gamma-ray bursts.**
G. Boella, M. Gorisse, J. Paul, B. G. Taylor, R. D. Wills.
Astrophys. Space Sci., Vol. 42, 103 - 110 (1976). – See 012.016.

032.525 **A high sensitivity balloon-borne X-ray telescope system.** M. R. Pelling.
Separate print Washington Univ., St. Louis, Mo., USA, 14 pp. (1974). – Abstr. in Phys. Abstr., Vol. 78, A087376 (1975).

032.526 **The airoscope pointing and stabilization system.**
J. P. Murphy, K. R. Lorell.
Separate print NASA, Moffett Field, Calif., USA, 29 pp. (1974). – Abstr. in Phys. Abstr., Vol. 78, A087378 (1975).

032.527 **Spectrophotometric calibration design for the ultraviolet from 700 to 3500 Å.** M. Saisse.
Nouv. Rev. Optique, Vol. 6, 231 - 240 (1975). In French. Abstr. in Phys. Abstr., Vol. 79, A007088 (1976).

032.528 **Communications in space – a system outline for a space laboratory.** G. D. Priewe.
Internat. Elektron. Rundschau, Vol. 29, 211 - 215 (1975). In German. – Abstr. in Phys. Abstr., Vol. 79, A015546 (1976).

032.529 **Slow-scan television system for a balloon-borne telescope.** N. Niwa.
Journ. Soc. Motion Picture Television Engineering, Vol. 84, 794 - 797 (1975). – Abstr. in Phys. Abstr., Vol. 79, A015550 (1976).

032.530 **The solar backscatter ultraviolet and total ozone mapping spectrometer (SBUV/TOMS) for Nimbus G.**
D. F. Heath, A. J. Krueger, H. A. Roeder, B. D. Henderson.
Optical Engineering, Vol. 14, 323 - 331 (1975). – Abstr. in Phys. Abstr., Vol. 79, A023554 (1976).

032.531 **Spectral calibration in infrared space sensors.**
R. H. Meier, A. B. Dauger.
Optical Engineering, Vol. 14, SR144 - 147 (1975). – Abstr. in Phys. Abstr., Vol. 79, A033145 (1976).

032.532 **Soft X-ray astronomy payload for use with satellites.** K. Kasturirangan, U. R. Rao, A. K. Jain.
Indian Journ. Radio Space Phys., Vol. 4, 144 - 145 (1975). – Abstr. in Phys. Abstr., Vol. 79, A037678 (1976).

032.533 **Errors induced in the measurement and azimuth directions of morphological features imaged on oblique lunar orbiter photographs.** B. S. Siegal.
Modern Geol., Vol. 5, 55 - 64 (1974). – Abstr. in Phys. Abstr., Vol. 79, A042106 (1976).

032.534 **A three-mirror space telescope.** D. Korsch.
Optical Engineering, Vol. 14, 533 - 535 (1975). Abstr. in Phys. Abstr., Vol. 79, A042459 (1976).

032.535 **Design of highly stable optical support structure (for the large space telescope).** H. H. Krim.
Optical Engineering, Vol. 14, 552 - 558 (1975). – Abstr. in Phys. Abstr., Vol. 79, A042463 (1976).

032.536 **Spectral calibration of infrared space sensors. II.**
R. H. Meier, A. B. Danger.
Optical Engineering, Vol. 14, SR182 - SR188 (1975). – Abstr. in Phys. Abstr., Vol. 79, A042464 (1976).

032.537 **The low-energy experiment on EXOSAT.**
J. A. M. Bleeker.
ESRO colloquium X-ray astron. related topics, Noordwijk, Netherlands, 25 - 26 Feb. 1975, p. 89 - 108 (1975).

032.538 **Description of a satellite experiment for isotopic-composition measurement of cosmic nuclei by the slowing-down method.** M. Bouffard, J. Engelmann.
ESRO workshop res. goals cosmic-ray astrophys. in the 1980's, Frascati, Italy, 24 - 25 Oct. 1974, p. 141 - 156 (1975).

032.539 **A proposed ultra-heavy cosmic-ray detector for space-shuttle exposure.** D. Henshaw.
ESRO workshop res. goals cosmic-ray astrophys. in the 1980's, Frascati, Italy, 24 - 25 Oct. 1974, 163 - 166 (1975).

032.540 **The medium energy experiment for EXOSAT.**
J. A. Hoffman.
ESRO colloquium X-ray astron. related topics, Noordwijk, Netherlands, 25 - 26 Feb. 1975, p. 75 - 87 (1975).

032.541 **The SAS-C X-ray observatory.** J. E. McClintock.
ESRO colloquium X-ray astron. related topics, Noordwijk, Netherlands, 25 - 26 Feb. 1975, p. 61 - 67 (1975).

Absolute calibration of the Ultraviolet Sky Survey Telescope in satellite TD 1. See Abstr. 031.255.

Rocket-borne baffled photometer: design and calibration. See Abstr. 034.042.

The stellar and solar tracking system of the Geneva Observatory gondola. See Abstr. 034.083.

A far-ultraviolet photometer for planetary surface analysis. See Abstr. 094.103.

"Filin" investigates X-ray stars. See Abstr. 142.022.

The Leicester X-ray crystal spectrometer on Ariel V and some early results on Cas A, Tycho and Sco X-1. See Abstr. 142.096.

033 Radio Telescopes and Equipment

033.001 The use of the Westerbork Synthesis Radio Telescope for solar observations at 21 cm.
J. D. Bregman, M. Felli.
Astron. Astrophys., Vol. 46, 41 - 48 (1976).
   A description is given of the use of the Westerbork Synthesis Radio Telescope for solar observation. The modification to the standard use of the telescope, the calibration and the programs needed to obtain a two dimensional brightness map are illustrated. The total intensity map and the circular polarization map of the whole sun are presented as a preliminary result. The one dimensional description of the evolution of a solar burst is also included.

033.002 Computer controlled radiospectrograph for recording the active solar radio emission in the range from 100 - 1,000 MHz. H. K. Asper.
Naturwissenschaften, 63. Jahrgang, p. 111 - 118 (1976). In German.
   A newly designed computer-controlled receiver measures and records the active solar radio emission (radiobursts) in the frequency range of 100–1000 MHz, corresponding to a wavelength of 3 m to 30 cm. The data stored on magnetic tape are corrected off line (frequency, intensity, and time) and are processed further in a large, high-speed computer. After choosing the appropriate data display (frequency- and time profiles, isodensity line charts, etc.) astrophysicists can use the data in theoretical investigations.

033.003 Millimeterwellen-Radioteleskop. E. Hintsches.
SuW, 15. Jahrgang, p. 92 - 93 (1976).

033.004 A radiometer of a sub-millimeter and short-wave part of millimeter wave region for radioastronomical observations. N. G. Afonchenkov, A. N. Vystavkin, E. M. Gershenzon, V. F. Zabolotnyj, A. V. Pavlov, V. I. Slysh, G. B. Sholomitskij, V. D. Shtykov.
Astron. Zhurn. Akad. Nauk SSSR, Vol. 53, 178 - 184 (1976). In Russian. English translation in Soviet Astron., Vol. 20, No. 1.
   A construction of a radiometer with a detector from n-InSb, cooled with liquid helium, is described. The fluctuational radiometer sensitivity measured from black-body radiation is 4×10^{-3} °K at constant time of 1 sec.

033.005 Interplanetary scintillation observations with the Cocoa Cross radio telescope. W. M. Cronyn, S. D. Shawhan, F. T. Erskine, A. H. Huneke, D. G. Mitchell.
Journ. Geophys. Res., Vol. 81, 695 - 697 (1976).
   Physical and electrical parameters for the 34.3-MHz

Cocoa Cross radio telescope are given. The telescope is dedicated to the determination of solar wind characteristics in and out of the ecliptic plane through measurement of electron density irregularity structure as determined from IPS (interplanetary scintillation) of natural radio sources. The collecting area (7.2×10^4 m²), angular resolution (0.4° EW by 0.6° NS), and spatial extent (1.2 km EW by 0.8 km NS) make the telescope well suited for measurements of IPS index and frequency scale for hundreds of weak radio sources without serious confusion effets.

033.006 Erfahrungsbericht über die Nutzung des 100-m-Radioteleskops des Max-Planck-Instituts für Radioastronomie. R. Schwartz.
Mitt. Astron. Ges., No. 38, p. 85 - 87 (1976). – Short report.

033.007 Radio telescopes of large resolving power. M. Ryle.
Uspekhi fiz. nauk, Vol. 117, 211 - 226 (1975). In Russian. Abstr. in Referativ. Zhurn. 51. Astron., 3.51.66 (1976).

033.008 On a method of automatic control of radio telescope antennas. P. V. Belyanskij, Yu. M. Maksimov.
Dinamika sistem. Vyp. (No.) 6. Gor'kij, 1975, p. 129 - 135. In Russian. – Abstr. in Referativ. Zhurn. 51. Astron., 3.51.67 (1976).

033.009 Radio source tracking by the RATAN-600 radio telescope with feed on a radial way. N. L. Kajdanovskij.
Astrofiz. Issled., Izv. Spets. Astrofiz. Obs., Vol. 8, 120 - 127 (1976). In Russian.
   A possibility of tracking radio sources by means of altering the primary mirror profile and moving the feed along the radial way is considered. The gain in sensitivity of the RATAN-600 as compared to the regime of the source transit across the antenna pattern is evaluated.

033.010 Some aspects of the decametre wave antenna system at Gauribidanur. C. V. Sastry.
Bull. Astron. Soc. India, Vol. 3, 36 (1975). – Abstract of a paper presented at the A.S.I. meeting 1975.

033.011 On restauration of radio images smoothed by the radio telescope directional pattern on small digital computers. V. A. Putilov.
Issled. po geomagnetizmu, aehron. i fiz. Solntsa. Vyp. (No.) 37. Moskva, Nauka, 1975, p. 214 - 219. In Russian. – Abstr.

in Referativ. Zhurn. 51. Astron., 4.51.86 (1976).

033.012 Optimum arrangement of readings of two-dimensional radio brightness distribution.
A. G. Obukhov.
Issled. po geomagnetizmu, aehron. i fiz. Solntsa. Vyp. (No.) 37. Moskva, Nauka, 1975, p. 220 - 224. In Russian. — Abstr. in Referativ. Zhurn. 51. Astron., 4.51.87 (1976).

033.013 Analog-digital filter of lower frequencies as an integrating device at the output of a radio telescope receiver. L. M. Risover.
Issled. po geomagnetizmu, aehron. i fiz. Solntsa. Vyp. (No.) 37. Moskva, Nauka, 1975, p. 225 - 232. In Russian. — Abstr. in Referativ. Zhurn. 51. Astron., 4.51.88 (1976).

033.014 Relative calibration of multichannel interferometer gains by digital methods. V. G. Miller.
Issled. po geomagnetizmu, aehron. i fiz. Solntsa. Vyp. (No.) 37. Moskva, Nauka, 1975, p. 233 - 235. In Russian. — Abstr. in Referativ. Zhurn. 51. Astron., 4.51.191 (1976).

033.015 Parasitic modulation in interferometers.
B. B. Krissinel'.
Issled. po geomagnetizmu, aehron. i fiz. Solntsa. Vyp. (No.) 37. Moskva, Nauka, 1975, p. 236 - 242. In Russian. — Abstr. in Referativ. Zhurn. 51. Astron., 4.51.192 (1976).

033.016 Essentials of radiometric measurements.
M. L. Meeks.
Astrophysics, Part B, (see 003.008), p. 1 - 6 (1976).
Antenna considerations; Radiometer considerations.

033.017 Types of astronomical antennas. W. J. Welch.
Astrophysics, Part B, (see 003.008), p. 7 - 28 (1976).
Pencil-beam antennas; Aperture synthesis.

033.018 Analysis of paraboloidal-reflector systems.
W. V. T. Rusch.
Astrophysics, Part B, (see 003.008), p. 29 - 63 (1976).
Analysis of the paraboloid with a prime-focus feed; Analysis of multireflector systems.

033.019 Feed systems for paraboloidal reflectors.
J. Ruze.
Astrophysics, Part B, (see 003.008), p. 64 - 81 (1976).
The purpose of this chapter is to acquaint the radio astronomer with various types of feeds for parabolic reflectors.

033.020 Antenna calibration. R. Wielebinski.
Astrophysics, Part B, (see 003.008), p. 82 - 97 (1976).
In this section the criteria for the selection of calibration sources are described. Lists of calibrators, both for positional and for flux scale calibration, are given. A discussion of present-day accuracy of the flux scale is undertaken. The techniques of calibration are described. A description of the aims of a complete pointing theory is discussed, which is an attempt to obtain the ultimate in positional accuracy in a radio telescope.

033.021 Practical problems of antenna arrays.
J. C. James.
Astrophysics, Part B, (see 003.008), p. 98 - 118 (1976).
Practical problems and techniques of array construction are discussed in this chapter. Emphasis is on antenna arrays in the HF and VHF frequency bands, roughly from 3 to 300 MHz.

033.022 Radiometer fundamentals. R. M. Price.
Astrophysics, Part B, (see 003.008), p. 201 - 224 (1976).

Types of signals in radio astronomy; Measurement of radio astronomy signals; The basic receiver system; Practical receiver configurations; Special-purpose receivers; Present trends in receiver systems; Considerations in radiometer system design.

033.023 Parametric amplifiers. J. Edrich.
Astrophysics, Part B, (see 003.008), p. 225 - 245 (1976).
Fundamentals of nonlinear reactances; Fundamentals of parametric amplifiers; Design considerations and practical parametric amplifiers.

033.024 Maser amplifiers. K. S. Yngvesson.
Astrophysics, Part B, (see 003.008), p. 246 - 265 (1976).
Basic properties of maser amplifiers; Systems and operational considerations; Data for specific maser amplifiers and systems; Millimeter wave masers.

·033.025 Multichannel-filter spectrometers. H. Penfield.
Astrophysics, Part B, (see 003.008), p. 266 - 279 (1976).
General description; Filter characteristics; Detectors; Integrators; Special operating features; Output devices; Calibration.

033.026 Autocorrelation spectrometers. B. F. C. Cooper.
Astrophysics, Part B, (see 003.008), p. 280 - 298 (1976).
Sampling and quantizing considerations; Computation of the power spectrum; Standard deviation of spectral estimate; Spectrum denormalization; Prefilters and video converters; Digital correlator logic; A scheme for optional one-bit or two-bit correlation; Extension to cross-correlation spectrometry; Some examples of correlation spectrometers.

033.027 Synchronization of time scales at sites of very long baseline interferometers from observations of maser sources of cosmic radio emission. V. A. Alekseev, Eh. D. Gatehlyuk, B. N. Lipatov, V. N. Nikonov, A. S. Sizov, A. I. Chikin, B. V. Shchekotov.
Izv. vyssh. ucheb. zavedenij. Radiofizika, Vol. 18, 1777 - 1785 (1975). In Russian. — Abstr. in Referativ. Zhurn. 51. Astron., 6.51.110 (1976).

033.028 De nijmeegse radio-interferometer.
I. Atanasijevic, H. Bluyssen, L. v. Dongen, L. ten Horn, L. Maas, J. van Nieuwkoop.
Separate print Res.-Praktikum Natuurk., Fac. Wis. Nat. Nijmegen, 48 pp. (1975).

033.029 Radioastronomia per astrofili e radioamatori.
G. Sinigaglia.
Giorn. Astron.,Vol. 1, 203 - 215 (1975).

033.030 New method of measuring the shape of precise antenna reflectors.
J. M. Payne, J. M. Hollis, J. W. Findlay.
Rev. Sci. Instruments, Vol. 47, 50 - 55 = National Radio Astron. Obs., Green Bank, Repr. Ser. A, No. 540 (1976).

033.031 Switching subreflector for millimeter wave radio astronomy. J. M. Payne.
Rev. Sci. Instruments, Vol. 47, 222 - 223 = National Radio Astron. Obs., Green Bank, Repr. Ser. A, No. 544 (1976).
A method of beam switching a Cassegrain antenna used for radio astronomy by tilting the hyperbolic secondary reflector is described. The system described has been in use for over a year at the 11-m millimeter-wave antenna operated by

the National Radio Astronomy Observatory at Kitt Peak, Arizona.

033.032 **Feeds for spherical radio telescopes.**   T. F. Trost.
Proc. Southwest Regional Conf., Vol. 1, (see 012.021), 91 - 95 (1976).
Improved design of feed antennas for use in the focal region of spherical reflector antennas such as the one at Arecibo has resulted in high aperture efficiencies of about 90 %. The feeds consist of long waveguide structures specially perforated and tapered.

033.033 **The calculation of varactor frequency multipliers by computers.**   A. N. Bruevich.
Radiotekhn. Ehlektron., Vol. 20, 1634 - 1645 (1975). In Russian.

033.034 **An ultrasonic image-forming system for use with multi-element antenna arrays.**   N. E. Holmes.
Proc. Instn. Radio Electron. Engineers Australia, Vol. 36, 321 - 328 (1975).

033.035 **Broadband parametric amplifiers for small earth stations – design survey.**   R. K. Flavin.
Telecom. Aust. Res. Labs. Rep., No. 6964, 69 pp. (1975).

033.036 **A new broadband square law detector.**
M. S. Reid, R. A. Gardner, C. T. Stelzried.
JPL Techn. Rep. 32–1599, 13 pp. (1975).

033.037 **Electrostatic electron-cyclotron waves with an anti-loss-cone distribution function.**
M. Nambu, T. Watanabe.
Geophys. Res. Letters, Vol. 2, 176 - 178 (1975).

033.038 **Antenna pattern corrections to microwave radiometer temperature calculations.**   F. B. Beck.
Radio Sci., Vol. 10, 839 - 845 (1975).

033.039 **Design of coupled combline directional couplers.**
D. J. Gunton.
Electronics Letters, Vol. 11, 607 - 608 (1975).

033.040 **Doppler-independent selective heterodyne radiometry for detection of remote species.**   M. C. Teich.
Rev. Sci. Instruments, Vol. 46, 1313 - 1317 (1975).

033.041 **Novel smoothing algorithm.**   G. K. Wertheim.
Rev. Sci. Instruments, Vol. 46, 1414 - 1415 (1975).

033.042 **"Aether drag" in a transversely moving medium.**
R. V. Jones.
Proc. Roy. Soc. London, Vol. 345, 351 - 364 (1975).

033.043 **Microstrip varactor-tuned millimeter-wave impatt diode oscillators.**
E. J. Denlinger, J. Rosen, E. Mykietyn, E. C. McDermott, Jr.
IEEE Trans. Microwave Theory Techn., Vol. MTT-23, 953 - 958 (1975).

033.044 **Impatt pump sideband noise and its effect on parametric amplifier noise temperature.**
C. A. Tearle, K. R. Heath.
IEEE Trans. Microwave Theory Techn., Vol. MTT-23, 1036 - 1042 (1975).

033.045 **A millimeter wave radiometer for cosmic background radiation measurements.**
R. J. Pedersen, F. L. Vernon, Jr.
IEEE S-MTT Internat. Microwave Symposium, 1974 June 12–14, p. 118 - 119 (1974).

033.046 **Development and testing of a receiver at 230 GHz.**
M. V. Schneider, G. T. Wrixon.
IEEE S-MTT Internat. Microwave Symposium, 1974 June 12–14, p. 120 - 122 (1974).

033.047 **Radiometer microwave receiver for the measurement of atmospheric attenuation at 11 and 14 GHz.**
R. K. Flavin.
Proc. Instn. Radio Electron. Engineers, Australia, Vol. 36, 353 - 358 (1975).

033.048 **The focal-region fields of paraboloidal reflectors of arbitrary f/d ratio.**   D. E. Baker.
Diss. Graduate School Ohio State Univ. 277 pp. (1974).
[Available from Ann. Arbor, Mich., Univ. Microfilms (1975)].

033.049 **InP Schottky-gate field-effect transistors.**
J. S. Barrera, R. J. Archer.
IEEE Trans. Electron. Devices, Vol. ED-22, 1023 - 1030 (1975).

033.050 **The microwave mesfet and its applications.**
R. Horton.
Aust. Post Office Res. Labs Rep., No. 6904, 7 pp. (1974).

033.051 **Graphs simplify load impedance measurements.**
K. M. Keen.
Microwaves, Vol. 14, No. 11, p. 53 - 54 (1975).

033.052 **Antenna measurement techniques at 11 GHz.**
J. W. M. Baars, C. D. Hughes.
Intern. Conf. Satellite Commun. Systems Techn., London, 1975 April 7–10, p. 253 - 260 (1975).

033.053 **Wide-band varactor-tuned coaxial oscillators.**
C. D. Corbey, R. Davies, R. A. Gough.
IEEE Trans. Microwave Theory Techn., Vol. MTT-24, 31 - 39 (1976).

033.054 **Cross-polarization effects of paraboloidal reflector antennas surface errors.**   S. I. Ghobrial.
Intern. Conf. Satellite Commun. Systems Techn., London, 1975 April 7–10, p. 246 - 252 (1975).

033.055 **Analysis of the radiation patterns of reflector antennas.**
J. F. Kauffman, W. F. Croswell, L. J. Jowers.
IEEE Trans. Antennas Propagation, Vol. AP-24, 53 - 65 (1976).

033.056 **Millimeter-wave spiral antenna.**
Microwave Journ., Vol. 18, No. 12, p. 28 (1975).

033.057 **Design technique for microwave-transistor power amplifiers.**   K. L. Kotzebue.
Electronics Letters, Vol. 12, 74 - 75 (1976).

033.058 **A cooled mixer-receiver for the CSIRO millimeter-wave radio telescope.**
M. Balister, R. A. Batchelor.
Instn. Radio Electronic Engineers, Intern. Electronics Convention 1975 August 25 - 29, p. 276 - 278 (1975).

033.059 **A multi-frequency low-noise receiving system for radio astronomy.**   M. Balister.
Instn. Radio Electronic Engineers, Intern. Electronics Convention 1975 August 25 - 29, p. 273 - 275 (1975).

033.060 **A phase stable radio link for radio astronomy.**
A. Deutsch.
Instn. Radio Electronic Engineers, Intern. Electronics Conven-

tion 1975 August 25 - 29, p. 217 - 220 (1975).

033.061 **Transistor amplifier design up to 12 GHz.**
R. Horton.
Instn. Radio Electronic Engineers, Intern. Electronics Convention 1975 August 25 - 29, p. 38 - 40 (1975).

033.062 **Feeds with low cross-polarisation for prime-focus and Cassegrain reflectors.** B. M. Thomas.
Instn. Radio Electronic Engineers, Intern. Electronics Convention 1975 August 25 - 29, p. 208 - 210 (1975).

033.063 **Improvement of the Parkes 64-m radio telescope surface efficiency to provide useful performance at 1.35 cm wavelength.** D. E. Yabsley.
Instn. Radio Electronic Engineers, Intern. Electronics Convention 1975 August 25 - 29, p. 203 - 205 (1975).

033.064 **Comment on choice of grid configuration of array antennas.** P. K. Bondyopadhyay.
Electronics Letters, Vol. 12, 97 - 98 (1976).

033.065 **Effective area of an antenna.** G. Dubost, J. Dupuy.
Electronics Letters, Vol. 12, 98 - 99 (1976).

033.066 **Pattern synthesis for antennas with multiple primary beams by minimax optimisation.**
L. W. Pearson, R. Mittra.
Electronics Letters, Vol. 12, 100 - 102 (1976).

033.067 **Power conservation for reflector antennas with truncated feed patterns.** W. V. T. Rusch, P. Cramer.
Electronics Letters, Vol. 12, 130 - 131 (1976).

033.068 **Influence of a metallic cylinder on the radiation of a microwave antenna.** E. Schweicher, A. Laloux.
Proc. Instn. Electr. Engineers, Vol. 123, 200 - 202 (1976).

033.069 **Design of short E-plane sectoral horns.**
F. Reisdorf, H. Henke.
Electronics Letters, Vol. 12, 164 - 165 (1976).

033.070 **Delay-lock tracking of stochastic signals.**
H. Meyr.
IEEE Trans. Commun., Vol. COM-24, 331 - 339 (1976).

033.071 **A method of sensitivity improvement of radio astronomy receivers.**
J. N. Gupta, G. J. M. Aitken.
Internat. Journ. Electronics, Vol. 39, 513 - 523 (1975).
Abstr. in Phys. Abstr., Vol. 79, A011325 (1976).

033.072 **The calibration of a large paraboloidal antenna using a cryogenic black-body enclosure.** J. Keen.
5th European Microwave Conference, Sevenoaks, Kent, England, p. 668 - 671 (1975). − Abstr. in Phys. Abstr.,

Vol. 79, A015565 (1976).

033.073 **Phase and gain adjustment of the Ooty radio telescope.**
N. V. G. Sarma, M. N. Joshi, S. Ananthakrishnan.
Journ. Instn. Electronics Telecommun. Engineers, Vol. 21, 107 - 109 (1975). − Abstr. in Phys. Abstr., Vol. 79, A023910 (1976).

033.074 **An electrically steerable array of 968 dipoles for the Ooty radio telescope.**
Journ. Instn. Electronics Telecommun. Engineers, Vol. 21, 117 - 122 (1975). − Abstr. in Phys. Abstr., Vol. 79, A023912 (1976).

033.075 **Receiver system of the Ooty radio telescope.**
N. V. G. Sarma, M. N. Joshi, D. S. Bagri, S. Ananthakrishnan.
Journ. Instn. Electronics Telecommun. Engineers, Vol. 21, 110 - 116 (1975). − Abstr. in Phys. Abstr., Vol. 79, A023911 (1976).

033.076 **An ultrasonic image-forming system for use with multi-element antenna arrays.** N. E. Holmes.
Proc. Instn. Radio Electron. Engineers, Australia, Vol. 36, 321 - 328 (1975). − Abstr. in Phys. Abstr., Vol. 79, A037693 (1976).

033.077 **New method of measuring the shape of precise antenna reflectors.**
J. M. Payne, J. M. Hollis, J. W. Findlay.
Rev. Sci. Instruments, Vol. 47, 50 - 55 (1976).

033.078 **Performance degradation due to wind deflection of billboard antennas.** J. R. Mark, L. W. Gill.
Marconi Rev., Vol. 38, 105 - 137 (1975).

033.079 **Crystal oscillator design employing digital integrated circuits as the active element.** M. F. Lane.
Telecom. Aust. Res. Labs Rep., No. 6949, 7 pp. (1975).

Progress report on the VLA.
Phys. Today, Vol. 29, No. 2, p. 18 - 19 (1976). − Concerning the Very Large Array telescope project.

Annotated literature survey of microwave ferrite control components and materials for 1968−1974.
See Abstr. 002.005.

Image processing in radio astronomy.
See Abstr. 031.231.

Radioheliography. See Abstr. 031.252.

Aperture synthesis. See Abstr. 031.253.

Computations in radio-frequency spectroscopy.
See Abstr. 031.254.

034 Astronomical Accessories (Spectrometers, Photometers, etc.)

034.001 A computer controlled spectrum scanner.
        A. R. Walker.
Monthly Notices Roy. Astron. Soc., Vol. 174, 555 - 561
(1976).
        A Nova-minicomputer controlled spectrum scanner is de-
scribed. A small transmission diffraction grating is used in an
f/15 Cassegrain beam to achieve a nominal resolution of 50 Å
over a wavelength range of up to 3000 Å. The performance of
the scanner is evaluated and reduction methods described
briefly.

034.002 Astrophotographie im Infrarot.
        M. Beetz, H. Elsässer, R. Weinberger.
SuW, 15. Jahrgang, p. 4 - 9 (1976).

034.003 Holographische Beugungsgitter.
        G. Schmahl, D. Rudolph.
SuW, 15. Jahrgang, p. 86 - 91 (1976).

034.004 Investigation of photomultipliers for registration of
        weak signals by pulse counting.   A. Eh. Gur'yanov.
Astron. Zhurn. Akad. Nauk SSSR, Vol. 53, 219 - 221 (1976).
In Russian. English translation in Soviet Astron., Vol. 20, No. 1.

034.005 400 mm horizontal solar telescope and multiple
        spectral region spectrographs.
The Solar Telescope Research and Production Group.
Acta Astron. Sinica, Vol. 16, 225 - 228 (1975). In Chinese.
Research note.

034.006 A photometer for the study of H II regions.
        J. Caplan, G. Grec.
Astron. Astrophys., Vol. 47, 59 - 63 (1976).
        The authors have designed and put into operation a pho-
toelectric photometer which is used with interference filters
for the measurement of line and continuum intensities in emission nebulae. An offset system allows accurate pointing and
guiding. A preselected sequence of measurements using six filters is controlled by an electronic system and the data, in the
form of pulse counts, is recorded on punched tape. The filters
are temperature controlled. The methods of measurement of
nebulae and of calibration against standard stars, and the reduction principles, are then presented.

034.007 The design, construction and performance of a large
        Hα and [N II] interference filter for the SRC 48-inch
Schmidt.   K. H. Elliott, J. Meaburn.
Astrophys. Space Sci., Vol. 39, 437 - 445 (1976).
        The design, construction and performance of a 15-in.
square mosaic interference filter is described. This is designed
to improve the sensitivity of the SRC 48-in. Schmidt for the
detection of gaseous nebulae in the light of Hα and [N II] by
approximately seven times. Its bandwidth with Kodak 098-04
emulsion is 105 Å but when combined with 103 aE emulsion
the effective bandwidth is around 80 Å.

034.008 Ein projizierender Blinkkomparator.   J. de Kort.
        Mitt. Astron. Ges., No. 38, p. 222 (1976).
Abstract.

034.009 Ein lichtelektrisches Photometer mit Photodiode.
        J. Dengel.
Mitt. Astron. Ges., No. 38, p. 249 - 251 (1976). – Abstract.

034.010 Initial operation of a scanning Stokes polarimeter.
        L. L. House, T. G. Baur, H. K. Hull.
Solar Physics, Vol. 45, 495 - 500 (1975).

A polarimeter capable of obtaining line profiles in the four
Stokes parameters is described in its initial operation. A brief
description of the instrument and the first sunspot observation
is given.

034.011 Another chopper design for astronomical infrared
        photometry.   J. Koornneef, J. van Overbeeke.
Astron. Astrophys., Vol. 48, 33 - 37 (1976).
        The chopper described was designed for use near the
focal plane. Amplitude and frequency of the modulation function can be varied without any additional electronic adjust-
ments. The basic design characteristics are: moving coil
principle with resonance free coil suspension; frequency
independent velocity feedback with additional position control
of one of the two end positions; computer calculated input
signal stored in a programmable read only memory (PROM).

034.012 Spectrometer for investigation of cosmic X-ray
        radiation.
M. M. Anisimov, V. P. Belyaev, N. L. Grigorov, A. S.
Melioranskij, N. I. Nazarova, V. M. Pankov, I. A. Savenko.
Issled. kosmich. luchej. Moskva, Nauka, 1975, p. 223 - 227.
In Russian. – Abstr. in Referativ. Zhurn. 62. Issled. kosmich.
prostranstva, 3.62.122 (1976).

034.013 Spectrometer for high-energy electrons and its
        calibration on accelerators.   O. B. Ben'kovskij,
N. L. Grigorov, L. F. Kalinkin, A. V. Klyachko, G. V. Lupenko,
Yu. I. Nagornykh, B. K. Podrezan, I. A. Savenko.
Issled. kosmich. luchej. Moskva, Nauka, 1975, p. 231 - 242.
In Russian. – Abstr. in Referativ. Zhurn. 62. Issled. kosmich.
prostranstva, 3.62.123 (1976).

034.014 Instrument for measurement of the spectrum of
        high-energy electrons in cosmic radiation.
N. L. Grigorov, L. F. Kalinkin, G. I. Pugacheva, I. A. Savenko,
N. M. Safronova, V. Ya. Shiryaeva.
Issled. kosmich. luchej. Moskva, Nauka, 1975, p. 243 - 253.
In Russian. – Abstr. in Referativ. Zhurn. 62. Issled. kosmich.
prostranstva, 3.62.124 (1976).

034.015 The effectiveness of recording electrons with
        energies between 1 - 2 MeV in silicon semicon-
ducting detectors.   N. V. Alekseev, P. V. Vakulov,
Yu. A. Gribov, Yu. I. Denisov, B. Ya. Shcherbovskij.
Issled. kosmich. luchej. Moskva, Nauka, 1975, p. 284 - 288.
In Russian. – Abstr. in Referativ. Zhurn. 62. Issled. kosmich.
prostranstva, 3.62.125 (1976).

034.016 On a semi-automatical iris-type photometer.
        R. Schielicke.
Astron. Nachr., Vol. 297, 91 - 98 = Mitt. Univ.-Sternw. Jena
No. 126 (1976). In German.
        A method automatically giving the equality-information
in iris-type photometers is described. It bases on digital techniques. Its resolution is 0.001 mag for brighter stars and 0.004
mag for fainter ones. For a single measurement of a star image
the mean error is ± 0.004 mag and ± 0.010 mag, respectively,
including errors in centering, equalization, and reading-out. The
range of a reduction curve extends over more than 10 magnitudes. The rate of measurements is 200 to 360 stars per hour
depending on the density of stars on the plate and the meas-
urer's practice.

034.017 Electrophotometer for observations of artificial
        earth satellites.   E. B. Vovchik, I. I. Vovchik, V.
M. Luk'yanets, R. F. Fedoriv.

Tsirk. Astron. Obs. L'vov, No. 49, p. 28 - 31 (1974). In Russian.

034.018 **Electrophotometer—computer system and its application to astronomical observations.**
V. N. Ivanov, V. M. Grigorevskij, I. I. Krishchuk, N. S. Zgonyajko, M. P. Petrov, A. V. Dobrovol'skij.
Astrometriya i Astrofizika, *Kiev*, vyp. (No.) 29, (see 003.003), p. 121 - 131 (1976). In Russian.
   A semiautomatic stellar electrophotometer with a photon counter is described. The results of an electrophotometer—computer system test are given.

034.019 **A multichannel spectrometer with a Fabry-Perot interferometer for the 6-meter telescope (BTA).**
V. S. Rylov.
Astrofiz. Issled., Izv. Spets. Astrofiz. Obs., Vol. 8, 77 - 88 (1976). In Russian.
   A multichannel spectrometer based on a pneumatic scanning Fabry-Perot interferometer intended for the resolution 0.1–0.01 Å in the region 3600–6600 Å is described. A comparison is made with other spectral instruments used in astronomy. The penetrating power of the instrument is calculated in terms of visual magnitudes.

034.020 **On the resolving power of image-converter tubes.**
A. A. Korovyakovskaya, V. S. Rylov.
Astrofiz. Issled., Izv. Spets. Astrofiz. Obs., Vol. 8, 135 - 138 (1976). In Russian.
   A calculation of the limiting resolving power of a single chamber image tube for different magnetic fields is given. Electron trajectories are determined. Data of the acceptable photocathode-to-screen angles of inclination and demands on the stability of the electric and magnetic fields are presented.

034.021 **A device for stepwise change of pressure in the chamber of a Fabry-Perot interferometer.**
I. D. Najdenov, G. A. Chuntonov.
Astrofiz. Issled., Izv. Spets. Astrofiz. Obs., Vol. 8, 139 - 140 (1976). In Russian.
   An auxiliary Fabry-Perot interferometer in the pressure chamber of the main interferometer is used for pressure step control and measurement.

034.022 **On the possibility to use organic dyes solution quantum amplifiers for increasing optical detector efficiency in the near infrared region.**
A. B. Basaev, G. A. Chuntonov.
Astrofiz. Issled., Izv. Spets. Astrofiz. Obs., Vol. 8, 141 - 143 (1976). In Russian.
   Broad band light amplification in organic dyes may be used in astronomy for increasing the efficiency of optical detectors in the visible and IR regions.

034.023 **The fabrication techniques of certain wide angle catadioptric systems.** A. P. Jayarajan.
Bull. Astron. Soc. India, Vol. 3, 28 (1975). — Abstract of paper presented at the A.S.I. meeting 1975.

034.024 **The instrumental profile of the 74-inch Radcliffe coudé spectrograph (48-inch camera).**
J. C. Blades, A. L. T. Powell.
Monthly Notes Astron. Soc. Southern Africa, Vol. 35, 30 - 37 (1976).

034.025 **Discrete electromagnetic gear with small angular step.**
V. I. Kruglov, L. V. Granitskij, V. E. Stepanov, L. E. Rishe.
Issled. po geomagnetizmu, aehron. i fiz. Solntsa. Vyp. (No.) 37. Moskva, Nauka, 1975, p. 193 - 195. In Russian. — Abstr. in Referativ. Zhurn. 51. Astron., 4.51.77 (1976).

034.026 **On practical use of the symmetry of light characteristic in an electro-optical shutter.**
V. M. Grigor'ev, N. I. Kobanov.
Issled. po geomagnetizmu, aehron. i fiz. Solntsa. Vyp. (No.) 37. Moskva, Nauka, 1975, p. 141 - 146. In Russian. — Abstr. in Referativ. Zhurn. 51. Astron., 4.51.79 (1976).

034.027 **Electro-optic analyzer of polarization with code-pulse control.**
V. E. Stepanov, V. M. Grigor'ev, N. I. Kobanov, B. F. Osak.
Issled. po geomagnetizmu, aehron. i fiz. Solntsa. Vyp. (No.) 37. Moskva, Nauka, 1975, p. 147 - 152. In Russian. — Abstr. in Referativ. Zhurn. 51. Astron., 4.51.80 (1976).

034.028 **Electronic measuring complex of a solar magnetograph with code-pulse modulation.**
B. F. Osak, S. A. Druzhinin.
Issled. po geomagnetizmu, aehron. i fiz. Solntsa. Vyp. (No.) 37. Moskva, Nauka, 1975, p. 153 - 158. In Russian. — Abstr. in Referativ. Zhurn. 51. Astron., 4.51.81 (1976).

034.029 **Automatic compensation of brightness fluctuations in photoelectric recording of solar magnetic fields.**
N. I. Kobanov, N. V. Klochek.
Issled. po geomagnetizmu, aehron. i fiz. Solntsa. Vyp. (No.) 37. Moskva, Nauka, 1975, p. 159 - 168. In Russian. — Abstr. in Referativ. Zhurn. 51. Astron., 4.51.82 (1976).

034.030 **Photoelectric tracking system of an extra-eclipse coronograph.**
L. V. Granitskij, Yu. A. Kuznetsov, P. G. Papushev.
Issled. po geomagnetizmu, aehron. i fiz. Solntsa. Vyp. (No.) 37. Moskva, Nauka, 1975, p. 190 - 192. In Russian. — Abstr. in Referativ. Zhurn. 51. Astron., 4.51.83 (1976).

034.031 **Recorder of solar image quality.**
L. V. Granitskij, Yu. M. Palachev.
Issled. po geomagnetizmu, aehron. i fiz. Solntsa. Vyp. (No.) 37. Moskva, Nauka, 1975, p. 209 - 213. In Russian. — Abstr. in Referativ. Zhurn. 51. Astron., 4.51.84 (1976).

034.032 **Photoelectric recorder of solar image vibration.**
P. G. Kovadlo, V. I. Ivanov, Sh. P. Darchiya.
Issled. po geomagnetizmu, aehron. i fiz. Solntsa. Vyp. (No.) 37. Moskva, Nauka, 1975, p. 196 - 202. In Russian. — Abstr. in Referativ. Zhurn. 51. Astron., 4.51.93 (1976).

034.033 **Photoelectric recorder for vibration of star images.**
P. G. Kovadlo, V. I. Ivanov, Sh. P. Darchiya.
Issled. po geomagnetizmu, aehron. i fiz. Solntsa. Vyp. (No.) 37. Moskva, Nauka, 1975, p. 203 - 208. In Russian. — Abstr. in Referativ. Zhurn. 51. Astron., 4.51.94 (1976).

034.034 **Physical mapping technique.** A. N. Korol'.
Soobshch. AN GruzSSR, Vol. 79, 577 - 579 (1975). In Russian. — Abstr. in Referativ. Zhurn. 51. Astron., 4.51. 195 (1976).

034.035 **A device for automatic sensitivity regulation of the photomultiplier of a multi-channel photometer.**
B. F. Osak, N. F. Zolotuev, S. A. Druzhinin.
Issled. po geomagnetizmu, aehron. i fiz. Solntsa. Vyp. (No.) 37. Moskva, Nauka, 1975, p. 169 - 173. In Russian. — Abstr. in Referativ. Zhurn. 51. Astron., 4.51.199 (1976).

034.036 **The optimum use of a photomultiplier for recording weak light signals.**
N. N. Lebedev, V. I. Levkovskij.
Issled. po geomagnetizmu, aehron. i fiz. Solntsa. Vyp. (No.) 37. Moskva, Nauka, 1975, p. 174 - 181. In Russian. — Abstr. in Referativ. Zhurn. 51. Astron., 4.51.200 (1976).

034.037 **Control of the transparence of a multi-layer dielectric system in the process of vacuum evaporation.**
S. V. Aleksandrovich, N. I. Kobanov, B. F. Osak, R. I. Parfenova, V. I. Skomorovskij.
Issled. po geomagnetizmu, aehron. i fiz. Solntsa. Vyp. (No.) 37. Moskva, Nauka, 1975, p. 243 - 246. In Russian. — Abstr. in Referativ. Zhurn. 51. Astron., 4.51.201 (1976).

034.038 **Optics for viewing a spectrograph slit or the unobstructed field.** R. V. Willstrop.
Observatory, Vol. 96, 64 - 66 (1976).

034.039 **Solar magnetograph employing integrated diode arrays.**
W. C. Livingston, J. Harvey, C. Slaughter, D. Trumbo.
Applied Optics, Vol. 15, 40 - 52 (1976).
A solar magnetograph employing as detectors a pair of self-scanning 512-element integrated diode arrays is described. Coupled to a 1.5-m telescope, photospheric flux as small as $5(10^{16})$ maxwells is detected, corresponding in intensity to $\Delta I/I = 3(10^{-4})$ at $\lambda 0.8688 \, \mu m$.

034.040 **Heat trap: an optimized far infrared field optics system.**
D. A. Harper, R. H. Hildebrand, R. Stiening, R. Winston.
Applied Optics, Vol. 15, 53 - 60 (1976).
An ir field optics system has been designed that achieves the maximum flux concentration allowed by the Abbé sine inequality and provides efficient coupling to bolometer-type detectors.

034.041 **Sensitivity limits of an infrared heterodyne spectrometer for astrophysical applications.**
M. M. Abbas, M. J. Mumma, T. Kostiuk, D. Buhl.
Applied Optics, Vol. 15, 427 - 436 (1976).
The sensitivity of an ideal heterodyne spectrometer approaches the quantum detection limit provided the local oscillator power is sufficiently large and the shot noise dominates all other sources of noise. For astronomical observations a number of factors (Δ_i) tend to degrade the sensitivity. It is shown that the minimum achievable degradation $[\pi_t(\Delta_t)]$ in the sensitivity of a practical astronomical heterodyne spectrometer is $\sim$30. Estimates of SNR's with which ir line emission from astronomical sources of interest may be detected are given.

034.042 **Rocket-borne baffled photometer: design and calibration.** G. Marette, J.-C. Gérard.
Applied Optics, Vol. 15, 437 - 439 (1976).
The design of an interference filter twin photometer suitable for the observation by rocket of the daytime magnetospheric cleft aurora at 2761 Å and 5200 Å is presented. The two-stage baffling system test and the absolute calibration procedure are described with some detail.

034.043 **Photometer calibration error using extended standard sources.** M. R. Torr, P. B. Hays, B. C. Kennedy, D. G. Torr.
Applied Optics, Vol. 15, 600 - 602 (1976). — Letter.

034.044 **Linear polarimeter with rapid modulation, achromatic in the $0.3 - 1.1 \, \mu m$ range.**
J. E. Frecker, K. Serkowski.
Applied Optics, Vol. 15, 605 - 606 (1976). — Letter.

034.045 **Spectroscopic photoelectric imaging Fabry-Perot interferometer: its development and preliminary observational results.**
W. H. Smith, J. Born, W. D. Cochran, J. Gelfand.
Applied Optics, Vol. 15, 717 - 724 (1976).

034.046 **Infrared detector Dewars: increased LN_2 hold time and vacuum jacket life spans.**
D. E. Jennings, W. J. Boyd, W. E. Blass.
Applied Optics, Vol. 15, 836 - 837 (1976). — Letter.

034.047 **Passive infrared sensors: limitations on performance.**
J. A. Jamieson.
Applied Optics, Vol. 15, 891 - 909 (1976).

034.048 **Infrared upconversion for astronomical applications.**
M. M. Abbas, T. Kostiuk, K. W. Ogilvie.
Applied Optics, Vol. 15, 961 - 970 (1976).
The performance of an upconversion system is examined for observation of astronomical sources in the low to middle ir spectral range. Theoretical values for the performance parameters of an upconversion system for astronomical observations are evaluated in terms of the conversion efficiencies, spectral resolution, field of view minimum detectable source brightness, and source flux. Experimental results of blackbody measurements and molecular absorption spectrum measurements using a lithium niobate upconverter with an argon-ion laser as the pump are presented. Estimates are given of the expected optimum sensitivity of an upconversion device that may be built with presently available components.

034.049 **A study of magnetically focused electronic optics in the case of imperfect fields.** J. P. Picat.
Astron. Astrophys., Vol. 49, 21 - 28 (1976).
The aberrations in magnetically focused electron optics have been studied. The computation of aberrations using the analytical formulas is much shorter and in good agreement with numerical integration of the equations of motion. Experimental results obtained by using a strong magnetic field of 5600 G are presented.

034.050 **Broad-band dielectric mirror coatings for Fabry-Perot spectroscopy.** J. Trauger.
Journ. Optical Soc. America, Vol. 66, 160 (1976). — Conference abstract.

034.051 **Ultra-precision photometry of antireflection coatings.**
D. D. Preonas.
Journ. Optical Soc. America, Vol. 66, 165 (1976). — Conference abstract.

034.052 **A sky-compensating filter photometer.**
J. K. Davidson, J. S. Neff, D. C. Enemark.
Publ. Astron. Soc. Pacific, Vol. 88, 209 - 212 (1976).
A filter photometer, incorporating a number of improvements over previous photometers, has been constructed for the new 32-cm telescope of the University of Iowa Observatory. These improvements include a substantial reduction in weight. Sky compensation is accomplished with an oscillating aperture in a mirror so that continuous guiding is possible. The optical and mechanical, as well as the electronic components of the photometer, are described.

034.053 **A "blink" comparator for variable stars.**
H. Rumball-Petre.
Journ. American Ass. Variable Star Observers, Vol. 4, 55 - 57 (1975/76).
A device for displacing telescope images is described and its use in comparing stellar brightnesses is outlined.

034.054 **On the question of the possibility of application of thermoresistors for measuring temperature gradients in astronomical observations.** V. I. Kiyaev.
Vestn. Leningr. un-ta, 1975, No. 19, p. 137 - 143. In Russian. Abstr. in Referativ. Zhurn. 51. Astron., 5.51.251 (1976).

034.055 **Ultraviolet polarimeter to record resonance-line polarization in the solar spectrum around 130—150**

nm.   J. O. Stenflo, H. Biverot, L. Stenmark.
Applied Optics, Vol. 15, 1188 - 1198 (1976).

A Swedish-built uv spectropolarimeter to be launched on a Soviet satellite in the Intercosmos series is described. The scientific objective is to record linear polarization across monochromatic solar images formed in resonance lines in the 130 – 150-nm wavelength region. This polarization arises by coherent scattering in the chromosphere-corona transition region of the sun. The instrument uses two parallel optical channels with highly different polarizing properties. The polarization analysis is done by reflection at gold-coated mirrors. The uv calibrations of the two flight models are described.

034.056   Channel electron multipliers: detection efficiencies with opaque MgF$_2$ photocathodes at XUV wavelengths.   L. B. Lapson, J. G. Timothy.
Applied Optics, Vol. 15, 1218 - 1221 (1976).

The detection efficiencies of channel electron multipliers with opaque MgF$_2$ photocathodes have been measured at wavelengths between 44 Å and 900 Å. Efficiencies a factor of 2 greater than those of uncoated channel electron multipliers were obtained over the wavelength range from 50 Å to 350 Å. The absolute detection efficiencies were greater than 10% in the range from 67 Å to 990 Å for photocathodes illuminated at an angle of incidence of 45°, with additional increases in sensitivity being obtained at short wavelengths using higher angles of incidence. Following an initial aging period, the photocathodes showed no degradation of response during storage under vacuum or in air.

034.057   A miniature spectrograph – its construction and application.   C. J. Watkis.
Journ. British Astron. Ass., Vol. 86, 280 - 284 (1976).

034.058   On possible instrumental errors in photoelectric spectrophotometry.   G. A. Terez.
Astron. Tsirk., No. 892, p. 4 - 6 (1975). In Russian.

034.059   Method of measurement of scattered light in spectrophotometers.   G. A. Terez.
Astron. Tsirk., No. 894, p. 5 - 7 (1975). In Russian.

034.060   A new two-mirror focal plane chopper for infrared astronomy.
P. R. Jorden, J. F. Long, A. D. MacGregor, M. J. Selby.
Astron. Astrophys., Vol. 49, 421 - 424 (1976).

A new focal plane chopper is described which uses two mirrors oscillating in phase to give an action equivalent to a single mirror chopper undergoing parallel motion. A sinusoidal image displacement is achieved which can be maintained at high frequencies using small power dissipation; the resultant detector microphonics are many times lower than those of a conventional chopper for the same image displacement.

034.061   A transmission filter for the far infared.
J. J. Wijnbergen.
Astron. Astrophys., Vol. 49, 467 - 469 (1976).

A transmission filter is described to select a spectral band at 30 μm for astronomical photometry. The peak transmission is 50%. There is practically no change in the transmission profile for temperatures down to 4 K.

034.062   A Baranne-type spectrograph with echelle-grating.   W. W. Weiss.
Optik, Band 44, 203 - 216 (1976). In German.

Many astrophysical problems can only be solved by using high dispersion spectrographs. Until recently, coudé-spectrographs have been the most common used type of spectrographs for this purpose. But with modern grating ruling techniques it is possible to rule larger echelle gratings and this brings up new possibilities for the spectrograph design.

034.063   Infrared detectors.   T. S. Moss.
Infrared Physics, Vol. 16, (see 012.013), 29 - 36 (1976). – Review paper.

034.064   Sensitivity of an astronomical infrared heterodyne spectrometer.
T. Kostiuk, M. J. Mumma. M. M. Abbas, D. Buhl.
Infrared Physics, Vol. 16, (see 012.013), 61 - 64 (1976).

034.065   Materials suitable for making filters beyond 300 μm.   B. Melchiorri, V. Natale, B. Fiscella, P. Lombardini.
Infrared Physics, Vol. 16, (see 012.013), 253 - 255 (1976).

Materials suitable for making filters in the far infrared between 300 and 2000 μm are investigated by means of a Michelson interferometer and monochromatic radiation obtained by a klystron with a GaAs distorcer.

034.066   A helium cooled Fabry-Perot interferometer for infrared astronomical spectroscopy.
M. J. Selby, P. R. Jorden, A. D. MacGregor.
Infrared Physics, Vol. 16 (see 012.013), 317 - 323 (1976).

The authors have built a cooled, scanning Fabry-Perot interferometer for astronomical spectroscopy in the ten micron atmospheric window. The reasons for choosing this approach are discussed, and details are given of the construction and performance of the prototype instrument.

034.067   A new infrared modulator.
B. de Batz, S. Bensammar, M. Chiron.
Infrared Physics, Vol. 16, 335 - 338 (1976).

The authors describe here a new infrared modulator designed for astronomical work. Several modes of modulation as well as a T.V. type scanning are possible. This system is the heart of a prototype two channel i.r. photometer in use in their laboratory.

034.068   Two-channel electrophotometer.
A. Avotiņš, G. Spulgis.
Investigation of the sun and red stars. 4, (see 003.009), p. 75 - 83 (1976). In Russian.

034.069   On sensitivity of a stellar interferometer with non-coherent accumulation of a signal.
D. V. Korol'kov, O. I. Krat.
Astron. Zhurn. Akad. Nauk SSSR, Vol. 53, 655 - 659 (1976). In Russian. English translation in Soviet Astron., Vol. 20, No. 3.

A comparison of a photoelectric accumulating stellar interferometer with a photographic accumulating stellar interferometer (speckle interferometer) is made. It is shown that the sensitivities of a photoelectric accumulating stellar interferometer and of a speckle interferometer are equal. A conclusion that the sensitivity of a speckle interferometer is less than that of a Michelson stellar interferometer is made.

034.070   The three channel microphotometer.
S. Krawczyk, J. Smoliński, Z. Turło.
Acta Astron., Vol. 26, 183 - 188 (1976).

A three channel microphotometer simultaneously scanning stellar and comparison spectra has been built. The paper presents pertinent details of this instrument constructed as an extension of the single channel Zeiss Schnell microphotometer.

034.071   Un appareil automatique d'acquisition de données en astronomie.   J.-H. Bigay, D. Dubet.
Obs. Lyon, Repr. No. 20 (1973).

034.072   An integrating, centroid timing, receiver for satellite ranging.   A. B. Sharma.
Rep. Finnish Geod. Inst., Helsinki, No. 75:10, 2 + 25 pp.

(1975).

034.073 **A high power Q-switched ruby laser for satellite laser ranging.** M. V. Paunonen.
Rep. Finnish Geod. Inst., Helsinki, No. 75:11, 2 + 10 pp. (1975).

034.074 **Ein Interferometer für geodätische Basismessungen nach dem Väisälä-Prinzip.**
K. Kühne, J. Rauhut.
Veröff. Zentralinst. Phys. Erde, No. 26, 83 pp. (1975).

034.075 **Improved sensitivity for cooled silicon pin photodiode.** G. Olofsson.
Stockholms Obs. Rep. No. 3, 34 pp. (1974).

034.076 **A rapid scanning prism monochromator for the wavelength region 0.4–0.95 μ.** L. Nordh.
Stockholms Obs. Rep. No. 6, 2 + 44 pp. (1974).

034.077 **Multi-channel spectrophotometer.**
M. Shimizu, S. Nishimura.
Tokyo Astron. Obs. Rep., Vol. 17, 506 - 528 (1976). In Japanese.

034.078 **Uno spettrofotometro infrarosso di tipo Fabry-Pérot ad alta risoluzione (1–0.15 Å) per il telescopio da 182 cm di Cima Ekar.** C. B. Cosmovici,
E. D'Anna, F. Strafella, C. Barbieri, A. Bianchini, G. Canton, K.-W. Michel, T. Nishimura.
Mem. Soc. Astron. Italiana, Vol. 46, 417 - 441 (1975).
A new kind of high resolution Fabry-Pérot tilting-filter photometer was constructed for the 182 cm Asiago telescope. It was successfully used for observations on comet Kohoutek and on interstellar nebulae. In this paper the efficiency and the different kinds of application are described.

034.079 **Color filter techniques for bright comets.**
C. F. Capen.
Proc. Southwest Regional Conf., Vol. 1, (see 012.021), 177 - 179 (1976).

034.080 **A single-sideband optical balanced mixed.**
T. De Graauw, H. Nieuwenhuijzen, H. Van de Stadt, R. Goebel.
Optics Commun., Vol. 14, 276 - 279 = Utrechtse Sterrekundige Overdrukken No. 318 (1975).

034.081 **Holographic construction of open structure, dispersion transmission gratings.**
J. H. Dijkstra, L. J. Lantwaard.
Optics Commun., Vol. 15, 300 - 305 = Utrechtse Sterrekundige Overdrukken No. 335 (1975).

034.082 **New methods of measuring radial velocities: redshifts of clusters of galaxies and search for planets around stars.** K. Serkowski.
Postępy Astron., Vol. 24, 3 - 13 (1976). In Polish.
Two new instruments for measuring radial velocities, designed by the author, are described.

034.083 **The stellar and solar tracking system of the Geneva Observatory gondola.** D. Huguenin.
Separate print Geneva Obs., Switzerland, 9 pp. (1974).
Abstr. in Phys. Abstr., Vol. 78, A087375 (1975).

034.084 **Method of analyzing a star angle data sensor with a tuning-fork light-beam modulator.**
E. E. Gopp, N. V. Zotikov.
Opt.-mekh. promyshlennost', Vol. 41, 18 - 19 (1974). – Abstr. in Phys. Abstr., Vol. 79, A007090 (1976).

034.085 **A scanning Fabry-Perot interferometer and its application to space research. IV. Tandem piezoelectric scanning Fabry-Perot interference spectrometer.**
T. Kohno, N. Yajima.
Journ. Mechanical Engineering Lab., Vol. 29, No. 5, p. 171 - 181 (1975). In Japanese. – Abstr. in Phys. Abstr., Vol. 79, A011655 (1976).

034.086 **A sampling unit for X-Y recorder.** M. Pracka.
Sdelovaci Techn., Vol. 23, 381 - 384 (1975). In Czech. – Abstr. in Phys. Abstr., Vol. 79, A011671 (1976).

034.087 **A scanning Fabry-Perot interferometer and its application to space research. V. Ultra-violet high resolution interference spectrometer.**
T. Kohno, N. Yajima.
Journ. Mechanical Engineering Lab., Vol. 29, No. 5, p. 182 - 188 (1975). In Japanese. – Abstr. in Phys. Abstr., Vol. 79, A011656 (1976).

034.088 **High-performance multilayer interference filters for the region 12 - 50 μm.**
C. S. Evans, R. Hunneman, J. S. Seeley.
Journ. Phys. D, Vol. 9, 309 - 320 (1976). – Abstr. in Phys. Abstr., Vol. 79, A024310 (1976).

034.089 **Optical thickness changes in freshly deposited layers of lead telluride.** C. S. Evans,
R. Hunneman, J. S. Seeley.
Journ. Phys. D, Vol. 9, 321 - 328 (1976). – Abstr. in Phys. Abstr., Vol. 79, A024311 (1976).

Detection and spectrometry of faint light.
See Abstr. 003.080.

Meridional focal curves of the Seya–Namioka mounting of the concave grating. See Abstr. 031.011.

L'observation des galaxies avec la machine COSMOS.
See Abstr. 031.256.

Preliminary design and performance of an advanced gamma-ray spectrometer for future orbiter missions.
See Abstr. 032.518.

Introduction à la photométrie astronomique.
See Abstr. 061.054.

Initial results from the EUV spectroheliometer on ATM. See Abstr. 076.029.

Vidicon spectral imaging: color enhancement and digital maps. See Abstr. 094.152.

A correlation method for measurement of variable magnetic fields. See Abstr. 106.109.

Veränderlichenbeobachtung mit einem lichtelektrischen Photometer. See Abstr. 122.132.

035   Clocks and Frequency Standards

035.001   **De analemmatische zonnewijzer met verplaatsbare gnomon.**   M. J. Hagen.
Zenit, 3e jaargang, p. 9 - 12 (1976).

035.002   **Digital coding clock and calendar.**   M. Xu, D.-l. Lai, X.-r. Ning, L.-f. Zhang, X.-b. Wang, X.-h. Liu.
Acta Geophys. Sinica, Vol. 19, 147 - 150 (1976). In Chinese.

The time accuracy of the clock is $\leqslant 3 \times 10^{-9}$/day. It is built of MOS IC and the display consists of a group of miniature fluorescent display tubes. The clock and calendar data are also given in series code format as well as the normal output of hour, minute and second time marks.

035.003   **A polar analemmic noon mark.**   L. M. Dougherty.
Journ. British Astron. Ass., Vol. 86, 299 - 302 (1976).

035.004   **A digital clock for sideral time.**   F. Reid, K. Honeycutt.
Sky Telescope, Vol. 52, 59 - 63 (1976).

035.005   **A device for automatic time marking.**   G. Mardirosyan.
Izv. Geofiz. Inst., (*Bulgaria*), Vol. 20, 105 - 112 (1974). In Russian. — Abstr. in Phys. Abstr., Vol. 79, A041944 (1976).

Errata

035.901   Erratum: 'Un cadran solaire oublié'[Orion, 33. Jahrgang, p. 179 - 182 (1975)].   L. Janin.
Orion, 34. Jahrgang, p. 74 (1976).

036   Photographic Auxiliaries

036.001   **La photographie des planètes.**   P. Campiche.
Orion, 34. Jahrgang, p. 15 - 18 (1976).

036.002   **A digital method of spectrosensitogram processing for investigation of astronomical photoemulsions.**
L. G. Antropova, V. S. Rylov, M. F. Shabanov, A. Ch. Uzdenov.
Astrofiz. Issled., Izv. Spets. Astrofiz. Obs., Vol. 8, 64 - 68 (1976). In Russian.

A digital method of spectrosensitogram processing used to investigate the informative properties of photoemulsions is described. The method is notable for its high accuracy and allows to obtain readily the photographic noise and the signal-to-noise ratios depending on the illumination, the wavelength, and the photometric area. The first results of investigation of Kodak 103aO, 103aF and A-500, A-600, A-700 photoemulsions are presented.

036.003   **Hypersensitization by baking of certain Kodak photographic plates for use in high-resolution Raman spectroscopy.**   K. S. Jammu, H. L. Welsh.
Applied Optics, Vol. 15, 423 - 426 (1976).

Hypersensitization by baking of various Kodak photographic emulsions is studied from the point of view of high-resolution Raman spectroscopy. In the 4400–4900-Å region baking can increase the sensitivity of IIa-O plates to ~4 times that of 103a-O plates for long exposures. In the Raman region for Ar⁺ laser excitation, 4700–5800 Å, baked IIa-D plates show good sensitivity and low reciprocity failure. Baked IIIa-J plates have better resolving power than the IIa-D type, but the reciprocity failure is greater, and the useful region does not extend beyond 5300 Å.

036.004   **L'emulsione fotografica e il meccanismo di formazione dell'immagine latente.**
F. Bònoli, F. Fusi-Pecci.
Coelum, Vol. 44, 49 - 57 (1976).

036.005   **Photographie und Photometrie des Planetensystems.**   W. Gruschel.
Orion, 34. Jahrgang, p. 70 - 74 (1976).

036.006   **L'emulsione fotografica e il meccanismo di formazione dell'immagine latente.**
F. Bònoli, F. Fusi-Pecci.
Coelum, Vol. 44, 118 - 126 (1976).

Simultaneous multicolor planetary photography.
See Abstr. 031.270.

Positional Astronomy. Celestial Mechanics

041 Positional Astronomy, Astrometry, Star Catalogues and Atlases

041.001 **Saturn observations with Besançon astrolabe (winter 1972–1973).**   J. Colin, P. Grudler, E. Oblak.
Astron. Astrophys., Suppl. Ser., Vol. 24, 139 - 141 (1976). In French.
.Results of Saturn observations with Danjon astrolabe during winter 1972–1973 at Besançon Observatory are given. A table contains residuals in zenith distance R, calculated with group corrections, for each transit (east or west), corrections φ_1 and φ_2 for defective illumination, Julian dates (UT) and the observer's name.

041.002 **Abschlußbericht zum AGK 3-Unternehmen.**
W. Dieckvoss.
Mitt. Astron. Ges., No. 38, p. 177 - 178 (1976). – Abstract.

041.003 **Positions of 54 southern circumpolar FK4-stars.**
E. Høg.
Mitt. Astron. Ges., No. 38, p. 178 - 181 (1976). – Short report.

041.004 **Some preliminary results from current observational programs with the new 23-cm zone astrograph.**
C. de Vegt.
Mitt. Astron. Ges., No. 38, p. 181 - 185 (1976). – Short report.

041.005 **Erweiterung des astronomischen Fundamentalkatalogs durch Radioquellen.**   H. G. Walter.
Mitt. Astron. Ges., No. 38, p. 185 - 188 (1976). – Short report.

041.006 **Herleitung und Erprobung eines erweiterten Verfahrens zur Bestimmung systematischer Differenzen zwischen verschiedenen fundamentalen Beobachtungen.**
H. Schwan.
Mitt. Astron. Ges., No. 38, p. 188 - 189 (1976). – Abstract.

041.007 **Interactions between positional and dynamical astronomy.**   C. A. Williams.
1975 AAS/AIAA Astrodyn. Conf., (see 012.022), AAS75-076, 15 pp. (1975). – Abstr. in Phys. Abstr., Vol. 79, A023582 (1976).

041.008 **Geodetic and astrometric results of very long baseline interferometric measurements of natural radio sources.**   ⸫. M. Moran.
Space Research XV, (see 012.003), p. 33 - 47 (1975).
The technique of very long baseline interferometry, initially developed in 1967 to study the angular structure of compact radio sources of the order of a thousandth of a second of arc, has been used to measure to great accuracy geometric quantities of geodetic and geophysical importance. The baseline vector between various radio telescopes has been measured to an accuracy of about 1 m. Variations in UT 1 have been measured to an accuracy of 1 msec, polar motion to 500 cm, and the coordinates of radio sources to about 0.1 arcsec. Improvement of about 1 order of magnitude is expected in the next 5 years.

041.009 **Stellare Kinematik aus Eigenbewegungen in bezug auf Galaxien.**   B. du Mont.
Mitt. Astron. Ges., No. 38, p. 252 - 253 (1976). – Abstract.

041.010 **Effect of the selection of catalogues on the density** functions derived from them.   G. G. Borzov.
Issled. ehkstremal'no molodykh zvezdn. kompleksov. Tashkent, Fan, 1975, p. 127 - 138. In Russian. – Abstr. in Referativ. Zhurn. 51. Astron., 3.51.595 (1976).

041.011 **A study of the accuracy of determination of the zero points and periodic errors of a fundamental catalogue from observations of planets.**
D. P. Duma, N. F. Minyajlo.
Astrometriya i Astrofizika, *Kiev*, vyp. (No.) 28, (see 003.002), p. 3 - 10 (1976). In Russian.
A conclusion is made on the dependence of the accuracy of determination of zero points and periodic errors of star catalogues on the mean distance earth–planet, the distribution of observations in the orbit and on other factors. It is noted that very accurate observations of minor planets cannot compete with less accurate observations of major planets when estimating the systematic errors of a fundamental catalogue.

041.012 **On the influence of possible refraction anomalies on the absolute system of declinations.**   A. S. Kharin.
Astrometriya i Astrofizika, *Kiev*, vyp. (No.) 28, (see 003.002), p. 11 - 13 (1976). In Russian.
Solving artificial examples, the author shows that the classical method of obtaining latitude corrections and the refraction constant from observations of stars in two culminations for certain cases of refraction anomalies may give false results, even if systematic instrumental errors are fully excluded.

041.013 **Determination of FK4 equinox and equator point corrections from the Washington meridian observations of major planets from 1949 to 1971.**
Yu. I. Safronov.
Astrometriya i Astrofizika, *Kiev*, vyp. (No.) 29, (see 003.003), p. 9 - 14 (1976). In Russian.
The results are given of the processing of 4927 Washington meridian observations of Mercury, Venus and Mars from 1949 to 1971. The corrections to the FK4 equinox and equator point are determined for the epoch 1960.0. The conclusion is confirmed that the accuracy of the corrections to the catalogue zero points depends on the mean geocentric distance of the planet.

041.014 **Determination of the FK4 zero points from the Washington meridian observations of the minor planets Ceres, Pallas, Juno and Vesta from 1949 to 1971.**
D. P. Duma, L. N. Kizyun.
Astrometriya i Astrofizika, *Kiev*, vyp. (No.) 29, (see 003.003), p. 15 - 17 (1976). In Russian.
The paper deals with the determination of corrections to the equinox and equator of the FK4 obtained from the Washington observations of Ceres, Pallas Juno and Vesta from 1949 to 1971. It is noted that the accuracy of the Fundamental Catalogue zero points is increased when using Duncombe's tables and continuous series of observations of minor planets.

041.015 **On spherical coordinate systems.**
E. P. Fedorov.
Astrometriya i Astrofizika, *Kiev*, vyp. (No.) 29, (see 003.003),

p. 18 - 23 (1976). In Russian.

It is proposed to designate different celestial coordinate systems used in astronomy by symbols codifying information on the direction of the coordinate axes. This enables transformation equations to be written in a very compact form.

041.016 **The principles of constructing an astronomical coordinate system in classical astrometry.**
M. S. Zverev.
Trudy Astron. Obs., *Leningrad*, Vol. 32 (= Uchenye Zapiski Leningr. Un-ta, No. 385 = Seriya matem. nauk, vyp. (No.) 52), p. 131 - 165 (1976). In Russian.

The classical method of constructing a system of celestial coordinates based on the earth's rotation is considered; the method is realized by meridian observations of stars and the solar system bodies. The fundamental star catalogues and the methods of absolute determination of star coordinates are criticized. More extensive application of observations with various meridian instruments and the data of the Time and Latitude Services are shown to be expedient for constructing a system of coordinates. A development of radio astrometry (especially of radiointerferometry) is very perspective for fundamental astrometry.

041.017 **New possibilities for solving problems of astrometry, geodynamics and geodesy by methods of radio interferometry with superlong baseline.**
V. S. Troitskij, V. A. Alekseev, V. N. Nikonov.
Uspekhi fiz. nauk, Vol. 117, 363 - 368 (1975). In Russian.
Abstr. in Referativ. Zhurn. 51. Astron., 5.51.172; 52. Geodeziya i Aehrosemka, 5.52.136 (1976).

041.018 **The effect of the personal equation on the results of coordinate determination by azimuthal methods.**
A. P. Gerasimov.
Geod. i kartografiya, 1975, No. 10, p. 18 - 23. In Russian.
Abstr. in Referativ. Zhurn. 51. Astron., 5.51.182; 52. Geodeziya i Aehrosemka, 5.52.135 (1976).

041.019 **On the influence of the "evening error" on the determination of right ascensions of the FKSZ.**
K. N. Kuz'menko, V. Kh. Pluzhnikov.
Vestn. Khar'kov. Univ., No. 129 (Ser. Astron., vyp. (No.) 10), p. 14 - 17 (1975). In Russian.

041.020 **Investigation of the influence of the mean system of an instrument on the system of the differential catalogue of right ascensions of 1355 bright stars in the zone +30° to +90°.**
N. G. Zuev, V. M. Kirpatovskij, R. M. Shut'eva.
Vestn. Khar'kov. Univ., No. 129 (Ser. Astron., vyp. (No.) 10), p. 17 - 20 (1975). In Russian.

041.021 **Perspectives spatiales pour l'astrométrie.**
P. Lacroute.
L'Astronomie, Vol. 90, 223 - 236 (1976).

041.022 **Catalogue of right ascensions of 89 circumpolar stars in the FK4 system.** I. N. Nabokov.
Astron. Tsirk., No. 894, p. 3 - 4 (1975). In Russian.

041.023 **Astrolabe stars catalogues of San Fernando.**
M. Sánchez.
Astron. Astrophys., Suppl. Ser., Vol. 25, 9 - 23 (1976). In French.

This paper shows the corrections to the adopted coordinates of the stars observed with the astrolabe at San Fernando, Spain. The first San Fernando Astrolabe Catalogue (CASF1) contains 144 stars of FK4 and 46 stars of FK4 Supp. The second one (CASF2) contains 160 stars of FK4 and 66 stars of FK4 Supp. The deduction of group corrections and magni-

tude-colour equations is considered in some detail.

041.024 **Résultats des observations faites à Alger avec l'astrolabe impersonnel A. Danjon OPL 8. Temps et latitude 1974.**
A. Ghezloun, M. Benhocine, J. Pham-Van, M. Haroun.
Ann. Obs. Astron. Alger, Vol. 4, Fasc. 2, p. 1 - 9 (1975).

041.025 **The astrometric reference frame.** C. A. Murray.
Observatory, Vol. 96, 90 - 97 (1976).

041.026 **Observations of the position of Barnard's star.**
R. Fangor.
Urania Kraków, Vol. 47, 54 - 56 (1976). In Polish.

041.027 **Sternkoordinatenkorrektionen für den FK4 aus Beobachtungsmaterial am Astrolab Danjon.**
J. Höpfner.
Veröff. Zentralinst. Phys. Erde, No. 32, 55 pp. (1975).

041.028 **Sternkatalog für das Potsdamer PZT.**
M. Meinig.
Veröff. Zentralinst. Phys. Erde, No. 37, 22 pp. (1976).

041.029 **Errors in the Henry Draper Catalogue.**
D. Hoffleit.
Centre Données Stellaires, Inform. Bull. No. 10, p. 2 - 13 (1976).

The list includes the errors that had already been reported in Volumes 98 - 100 of the Harvard Annals (HD and HDE). A few were found in the course of work on Yale Zone Catalogues.

041.030 **Star catalogs and files available at the Stellar Data Center.**
Centre Données Stellaires, Inform. Bull. No. 10, p. 35 - 36 (1976).

041.031 **Practical realization of the reference coordinate systems.**
IAU Colloquium No. 26, (see 012.018), p. 27 - 38 (1975).

041.032 **Relations between different reference systems.**
IAU Colloquium No. 26, (see 012.018), p. 39 - 46 (1975).

041.033 **Nature of the requirements for reference coordinate systems.** C. A. Lundquist.
IAU Colloquium No. 26, (see 012.018), p. 51 - 62 (1975).
Review paper.

041.034 **Reference coordinate system requirements for geophysics.** P. L. Bender.
IAU Colloquium No. 26, (see 012.018), p. 85 - 92 (1975).

041.035 **Some problems related to the definition of reference systems.** J. Kovalevsky.
IAU Colloquium No. 26, (see 012.018), p. 123 - 132 (1975).
Review paper.

041.036 **Definition of the celestial reference coordinate system in fundamental catalogues.** W. Fricke.
IAU Colloquium No. 26, (see 012.018), p. 201 - 222 (1975).

Discussed are (1) the methods which, in practice, provide the equator, equinox and precession, (2) the prospects of progress in the direction of higher accuracy, (3) the realization of a homogeneous system on the sky, and (4) the extension of the system to a reference frame represented by extragalactic objects. The basic observational material available for the project of an improvement of the Fourth Fundamental Catalogue (FK4) is discussed, and the plan of establishing FK5 outlined.

041.037 Celestial reference systems derivable from solar system dynamics.
R. L. Duncombe, P. K. Seidelmann, T. C. Van Flandern.
IAU Colloquium No. 26, (see 012.018), p. 223 - 233 (1975).

041.038 On the use of Tautenburg Schmidt for improvement of astronomical reference system. H. U. Sandig.
IAU Colloquium No. 26, (see 012.018), p. 241 - 246 (1975).

041.039 Geodetic and astrometric results of very-long-base-line interferometric measurements of natural radio sources. J. M. Moran.
IAU Colloquium No. 26, (see 012.018), p. 269 - 292 (1975).

041.040 The determination of the equator-point for observing star declinations.
A. A. Mikhailov (*Mikhajlov*).
IAU Colloquium No. 26, (see 012.018), p. 315 - 317 (1975).

041.041 On constructing the inertial system of high accuracy by VLBI methods. C. A. Krasinsky.
IAU Colloquium No. 26, (see 012.018), p. 381 - 393 (1975).

041.042 Note on selection of coordinate systems for geodynamics. S. Henriksen.
IAU Colloquium No. 26, (see 012.018), p. 395 - 407 (1975).

041.043 The application of radio interferometers in astrometry. S. Gorgolewski.
IAU Colloquium No. 26, (see 012.018), p. 453 - 456 (1975).

041.044 Identification of reference stars in the old positional observations. J. Bem.
Postępy Astron., Vol. 24, 65 - 68 (1976). In Polish.

The role of VLBI in the establishment of coordinate systems. See Abstr. 031.268.

Preliminary analysis of observational data obtained with a photoelectric astrolabe type I.
See Abstr. 032.003.

Statistical analysis of the Danjon astrolabe at São Paulo. See Abstr. 032.024.

Systematic differences between global or local solutions of polar motion derived from astronomical and spatial measurements. See Abstr. 045.018.

Requirements for coordinate systems for earth dynamics. See Abstr. 046.037.

Conceptual definitions of reference coordinate systems for earth dynamics. See Abstr. 046.038.

Reference coordinate system requirements for geodesy and ocean dynamics. See Abstr. 046.039.

Magnitudes and spectra of important dynamical phenomena. See Abstr. 081.041.

Coordinate systems in lunar ranging.
See Abstr. 094.015.

Further astrometric observations with the 5-km radio telescope. See Abstr. 141.008.

Bestimmungen der galaktischen Rotation und der Präzession aus fundamentalen Eigenbewegungen und aus Eigenbewegungen relativ zu Galaxien. See Abstr. 155.053.

042 Celestial Mechanics, Figure of Celestial Bodies

042.001 The use of the concept of the force centre for a generalized three-body problem.   M. Zelený.
Bull. Astron. Inst. Czechoslovakia, Vol. 27, 61 - 63 (1976).

This paper deals, on the one hand, with deriving the relations which are conditioned by the existence of the force centre in systems of three non-colinear particles with arbitrary masses and interacting by means of general central forces of interaction; and on the other, with employing these relations to study systems in which it is true that the force centre is identical with the centre of mass of the system at any time. In this case three integrals of motion have been found.

042.002 A statistical theory of resonance motion in the sun—Jupiter system.   W. E. Wiesel.
Celestial Mechanics, Vol. 13, 3 - 37 (1976).

The author considers the application of the statistical method of phase mixing to the approximate Poincaré solution to resonant motion. The theory of the phase mixing of an initial *ad hoc* distribution of particles is then developed for this dynamical system, and the absence of significant evolution of the system far from resonance is verified. A selection of results is given for the $2:1$, $3:1$, and $5:2$ resonances, which show in general a peak on the low side of exact resonance and a gap on the high side. Comparison with large numbers of numerically integrated orbits gives good agreement with the model, at least for small eccentricities. However, the model is unable to exhibit the clean gaps shown by the real asteroid belt. Hence, a purely statistical model of the Kirkwood gaps is ruled out.

042.003 Perturbation formulations for satellite attitude dynamics.   L. G. Kraige, J. L. Junkins.
Celestial Mechanics, Vol. 13, 39 - 64 (1976).

Two analytical developments for the arbitrarily torqued motion of an asymmetric rigid body, both of which utilize a new torque-free solution as the reference motion, are presented. The first is an Encke-type perturbation formulation in which differential equations for the angular velocity and orientation departures from Poinsot motion are derived. The second technique is a variation of parameters scheme in which an analogue of Herrick's two-body perturbative differentiation technique is employed.

042.004 Stabilized Kepler motion connected with analytic step adaptation.   J. Baumgarte.
Celestial Mechanics, Vol. 13, 105 - 109 (1976).

In the publication Baumgarte and Stiefel (1974, see Abstr. 012.042.007) a stabilization of the Keplerian motion was offered by making use of a manipulation of the Hamiltonian. By this stabilization technique the given Hamiltonian $H(p_i, q_i)$ is replaced by a new Hamiltonian H^*, which leads to Lyapunov-stable differential equations of motion. Whereas, in the quoted publication, the physical time t was used as the independent variable the author now developes a generalization which allows to combine the stabilization with the introduction of a new independent variable s. Such a fictitious time s is popular for achieving an analytic step-size adaptation. Perturbations of Kepler motion are discussed.

042.005 On the stability of regular motions of a solid body in a Newtonian gravitational field.   Eh. A. Vagner.
Kosmich. Issled., Vol. 14, 133 - 134 (1976). In Russian.
Brief information.

042.006 Estimates of osculating orbit elements for the two-body problem with variable mass.   L. G. Glikman.
Astron. Zhurn. Akad. Nauk SSSR, Vol. 53, 185 - 190 (1976).

In Russian. English translation in Soviet Astron., Vol. 20, No. 1.

The two-body problem with variable mass for any law of mass variation is considered. Slowness of mass variation is not supposed. Two-sided estimates of osculating orbit elements were obtained.

042.007 Circular and rectilinear motions of an axisymmetric body under the effect of attraction of two spherical bodies.   L. S. Troitskaya.
Astron. Zhurn. Akad. Nauk SSSR, Vol. 53, 191 - 197 (1976).
In Russian. English translation in Soviet Astron., Vol. 20, No. 1.

Certain exact solutions of the problem are found. These solutions describe circular motions of the centres of mass of the spherical bodies and rectilinear motions of the centre of mass of an axisymmetric body together with the regular precession of its axis. It is also proved that the axisymmetric body leaves to infinity for a finite time interval.

042.008 Numerische Versuche zum Einfangproblem.   R. Dvorak.
Mitt. Astron. Ges., No. 38, p. 190 - 192 (1976). — Abstract.

042.009 Trigonometric series of near-parabolic motion.   P. Petroškevičius.
Geod. darbai. Vilniaus inz. stat. inst., Trudy po geod. Vil'nyus. inzh. stroit. in-t, Vol. 7, 106 - 109 (1974). In Russian. — Abstr. in Referativ. Zhurn. 51. Astron., 3.51.76 (1976).

042.010 On the secular and periodic part of precession in the Euler—Poinsot problem.   M. L. Pivovarov.
Opredelenie dvizheniya kosmich. apparatov. Moskva, Nauka, 1975, p. 107 - 110. In Russian. — Abstr. in Referativ. Zhurn. 51. Astron., 3.51.82; 62. Issled. kosmich. prostranstva, 3.62. 246 (1976).

042.011 From the theory of numbers via gyroscopes and Lie algebras to linear celestial mechanics.   E. Stiefel.
Long-time predictions in dynamics, (see 012.005), p. 3 - 15 (1976).

042.012 Lectures on linearizing transformations of dynamical systems.   V. Szebehely.
Long-time predictions in dynamics, (see 012.005), p. 17 - 42 (1976).

(1) One-degree of freedom dynamical systems. The direct problem; (2) One-degree of freedom dynamical systems. The inverse problem; (3) N-degrees of freedom dynamical systems. The direct problem; (4) N-degrees of freedom dynamical systems. The inverse problem.

042.013 Strongly perturbed dynamical systems.   G. Contopoulos.
Long-time predictions in dynamics, (see 012.005), p. 43 - 56 (1976).

042.014 Global stability of area-preserving mappings.   J. H. Bartlett.
Long-time predictions in dynamics, (see 012.005), p. 99 - 110 (1976).

042.015 On the method of averaging.   U. Kirchgraber.
Long-time predictions in dynamics, (see 012.005), p. 111 - 117 (1976).

In this paper the author presents several new aspects of the method of averages: first he describes some formal properties of the method, second he applies it in order to reprove Hopf's bifurcation theorem (and obtain a direction of bifurcation formula which is similar to that of Hsü and Kazarinoff), thirdly he offers a theorem concerning error bounds. Since the details will appear elsewhere only the results are described.

042.016 On the theory of averaging. F. Verhulst.
Long-time predictions in dynamics, (see 012.005), p. 119 - 140 (1976).
The purpose of this paper is to present a number of recently obtained results in averaging, both in theory and in applications, and to show at the same time what kind of pitfalls there are in applying the theory to problems in dynamics.

042.017 Qualitative investigation of almost separable Hamiltonian system of two degrees of freedom.
N. Sigrist.
Long-time predictions in dynamics, (see 012.005), p. 141 - 150 (1976).

042.018 Stabilization, manipulation and analytic step adaption. J. Baumgarte.
Long-time predictions in dynamics, (see 012.005), p. 153 - 163 (1976).
In order to reduce the error propagation in the numerical calculation of orbits two methods of stabilization of the differential equations of the perturbed Keplerian motion are offered: (1) a nonconservative method by asymptotical stabilization of the energy relation, (2) a conservative method by manipulation of the Hamiltonian. In both methods the stabilization is combined with the introduction of a new independent variable in order to achieve an analytic step size adaption.

042.019 A note on stabilization in three-body regularization.
S. J. Aarseth.
Long-time predictions in dynamics, (see 012.005), p. 173 - 177 (1976).
The time transformation $t' = R_1 R_2$ in the three-body regularization method of Aarseth and Zare is modified to include the potential energy, thereby enabling the physical time to be obtained explicitly. A new recommendation for stabilizing the equations of motion further is considered briefly. It is concluded that a third type of time transformation, $t' = R_1 R_2/(R_1 + R_2)^{1/2}$, appears to offer the best prospects in practical calculations unless the time itself is required very accurately.

042.020 Qualitative methods and results in celestial mechanics. C. Marchal.
Long-time predictions in dynamics, (see 012.005), p. 181 - 208 (1976).

042.021 On the characteristic exponents of the general three-body problem. R. Broucke.
Long-time predictions in dynamics, (see 012.005), p. 209 - 222 (1976).

042.022 Families of periodic orbits in the general n-body problem. J. D. Hadjidemetriou.
Long-time predictions in dynamics, (see 012.005), p. 223 - 240 (1976).
The author shows that families of periodic orbits of the general n-body problem ($n \geq 3$) exist in a rotating frame of reference and he develops a systematic way to obtain them.

042.023 Triple collision. J. Waldvogel.
Long-time predictions in dynamics, (see 012.005), p. 241 - 258 (1976).

042.024 An invariant measure for the planar restricted three body problem. F. Nahon.
Long-time predictions in dynamics, (see 012.005), p. 259 - 266 (1976).

In the study of hyperbolic comets, von Zeipel has introduced an integral invariant which is related but not equivalent to Khinchin's invariant integral of statistical mechanics. The author gives a simple expression of this invariant in the planar case and studies some of its properties.

042.025 Formal expressions for the motion of n planets in the plane, with the secular variations included, and an extension to Poisson's theorem. P. J. Message.
Long-time predictions in dynamics, (see 012.005), p. 279 - 293 (1976).

042.026 The principle of least interaction action.
M. W. Ovenden.
Long-time predictions in dynamics, (see 012.005), p. 295 - 305 (1976).

042.027 Periodic orbits around the equilibrium points of a central gravitational potential perturbed by the J_{22} term. P. A. Baxa.
Long-time predictions in dynamics, (see 012.005), p. 330 (1976). — Abstract.

042.028 A new transformation for the perturbed problem of two bodies. T. Feagin.
Long-time predictions in dynamics, (see 012.005), p. 336 (1976). — Abstract.

042.029 Three-dimensional branchings of plane periodic solutions. V. V. Markellos, P. G. Kazantzis.
Long-time predictions in dynamics, (see 012.005), p. 341 (1976).— Abstract.

042.030 On stabilization using the energy integral.
P. E. Nacozy.
Long-time predictions in dynamics, (see 012.005), p. 342 - 343 (1976). — Abstract.

042.031 Resonance effects in satellite orbits.
J. Osorio.
Long-time predictions in dynamics, (see 012.005), p. 344 (1976). — Abstract.

042.032 On the resonance 1 - 1 for Hamiltonian systems near an equilibrium point. J. Roels.
Long-time predictions in dynamics, (see 012.005), p. 345 (1976).— Abstract.

042.033 Periodic orbits near an equilibrium point: the Liapunov center theorem. D. S. Schmidt.
Long-time predictions in dynamics, (see 012.005), p. 346 (1976). — Abstract.

042.034 Stabilization and time elements for Kepler's problem. A. Stokes.
Long-time predictions in dynamics, (see 012.005), p. 352 - 353 (1976). — Abstract.

042.035 A note on the Hénon-Heiles problem.
F. Verhulst.
Long-time predictions in dynamics, (see 012.005), p. 354 - 355 (1976). — Abstract.

042.036 The effects of integrals on the totality of solutions of dynamical systems. K. Zare.
Long-time predictions in dynamics, (see 012.005), p. 358 (1976). — Abstract.

042.037 Existence of the second integral in the restricted problem. R. Sergysels.
Astron. Astrophys., Vol. 48, 275 - 280 (1976). In French.

The author obtains the canonical formulation of the differential equations of the trajectories of the same Jacobi constant in the planar restricted problem. The one degree of freedom Hamiltonian is periodic in the independent variable (the polar angle). By applying to the perturbed Hamiltonian the canonical averaging method the author gets another formal development of the second integral (Contopoulos, 1965), in powers of the inverse of the square root of the distance between the two primaries.

042.038 **New treatment of the critical argument in resonance problems.** W. H. Jefferys.
Astron. Journ., Vol. 81, 132 - 134 (1976).

Traditional methods for isolating the critical argument in Hamiltonian mechanics involve the introduction of a special set of action variables which are both complicated in form and arbitrary in nature. The new variables are in general obscure in their interpretation. This paper introduces a new canonical system corresponding to the original one, but with one more degree of freedom, in which all the variables of the original Hamiltonian can be treated on the same footing. It appears to have significant advantages for machine calculations.

042.039 **Families of periodic planetary-type orbits in the three-body problem and their stability.**
J. D. Hadjidemetriou.
Astrophys. Space Sci., Vol. 40, 201 - 224 (1976).

Several families of the planar general three-body problem for fixed values of the three masses are found, in a rotating frame of reference, where the mass of two of the bodies is small compared to the mass of the third body. The above families represent planetary systems with the body with the large mass representing the sun and the two small bodies representing two planets or comets. One section of a family is shown to represent the Jupiter family of comets and also a model for the sun-Jupiter-Saturn system is found. The Jupiter family of comets, for small masses of the two small bodies, and the sun-Jupiter-Saturn system are proved to be stable.

042.040 **On the expansion of the perturbing function of a planet.** Yu. V. Plakhov.
Izv. vyssh. ucheb. zavedenij. Geod. i aehrofotosemka, 1975, No. 1, p. 31 - 34. In Russian. – Abstr. in Referativ. Zhurn. 51. Astron., 4.51.127 (1976).

042.041 **On the continuation of periodic orbits from the restricted to the general three-body problem.**
G. Bozis, J. D. Hadjidemetriou.
Celestial Mechanics, Vol. 13, 127 - 136 (1976).

Some properties of the characteristic surface of a family of symmetric periodic orbits of the general three-body problem, corresponding to a fixed value of the ratio of the masses of two of the bodies, are studied in view of recent theoretical and numerical results. Periodic orbits of the planar circular restricted problem with period equal to an integer multiple of 2π are of special interest for the structure of a characteristic surface.

042.042 **Stationary solutions and their characteristic exponents in the restricted three-body problem when the more massive primary is an oblate spheroid.**
R. K. Sharma, P. V. Subba Rao.
Celestial Mechanics, Vol. 13, 137 - 149 (1976).

This paper deals with the numerical investigations of the locations of the five equilibrium points by taking into consideration the effect of oblateness of the more massive primary for some systems of astronomical interest. This note is further concerned with the periodic solutions of the linearized equations of motion around the five equilibrium points. Interesting differences in the trends of the angular frequencies of these motions have been noticed.

042.043 **Mouvement d'un satellite à rotation lente dans le champ magnétique terrestre.** I. Stellmacher.
Celestial Mechanics, Vol. 13, 177 - 201 (1976).

The movement, in the earth's magnetic field, of a magnetized satellite around its center of gravity is investigated. The magnetic moment is of the order of 10 Amp m², and its direction is that of the principal axis of smallest inertia $0z$ of the satellite. The initial spin is low. The movement around the center of mass can be described if the angle between the angular moment L and the axis $0z$ is small.

042.044 **Libration effects for retrograde satellites in the restricted three-body problem. I: Circular plane Hill's case.** D. Benest.
Celestial Mechanics, Vol. 13, 203 - 215 (1976).

Numerical explorations of the restricted problem have shown that for stable large non-periodic retrograde satellite orbits, the motion can be decomposed into a fast 'reference motion' and a slow libration around B_2. The author studies here this libration in the circular plane Hill's case, for which the 'reference motion' is elliptic. He finds two integrals of motion: the first is the semi-major axis of the ellipse; the second is essentially Jacobi's integral, translated into the new coordinates.

042.045 **Perturbations des systèmes différentiels et résultats de géométrie différentielle.** M. Rapaport.
Celestial Mechanics, Vol. 13, 217 - 227 (1976).

The author uses classical definitions and results of differential geometry in studying properties of transformations depending on a small parameter, acting on differential systems. Hori's and Deprit's algorithms can be defined for these systems. A lemma is given to show that these algorithms are equivalent.

042.046 **A theory of libration.** B. Garfinkel.
Celestial Mechanics, Vol. 13, 229 - 246 (1976).

It is shown here that many problems of libration in celestial mechanics can be reduced to a perturbation of an intermediary defined by the Hamiltonian $F = B(y) + 2\mu^2 A(y)f(x)$. This generalization of the Ideal Resonance Problem, with a periodic function $f(x)$ replacing $\sin^2 x$, is solved here to $O(\mu^2)$ by an algorithm that is essentially the same as the one used in the original formulation. Libration may be simple or multiple, depending on the nature of the function $f(x)$ and on the initial conditions. Double libration is illustrated here by the horseshoe-shaped orbits enclosing two libration centers.

042.047 **Stabilization by modification of the Lagrangian.** J. W. Baumgarte.
Celestial Mechanics, Vol. 13, 247 - 251 (1976). – Paper presented at the Flight Mechanics/Estimation Theory Symposium, Goddard Space Flight Center, Greenbelt, Md., April 15 - 16, 1975.

In order to reduce the error growth during a numerical integration, a method of stabilization of the differential equations of the Keplerian motion is offered. It is characterized by the use of the eccentric anomaly as independent variable in such a way that the time transformation is given by a generalized Lagrange formalism. The control terms in the equations of motion obtained by this modified Lagrangian give immediately a completely Lyapunov-stable set of differential equations. The equation of time integration is modified by a control term which leads to an integral which defined the time element for the perturbed Keplerian motion.

042.048 **Ideal frames for perturbed Keplerian motions.** A. Deprit.
Celestial Mechanics, Vol. 13, 253 - 263 (1976).

Cartan's exterior calculus is used to refer a perturbed

Keplerian motion to an ideal frame by means of either the Eulerian parameters or the Eulerian angles, in which case the equations are given a Hamiltonian form. The results are compared with the corresponding systems in the orbital and nodal frames.

042.049 On the stability of the earth's motion.
P. I. Sajdov, R. I. Vcherashnij, Yu. P. Sajdov.
Izv. Leningr. ehlektrotekhn. in-ta, 1975, vyp. (No.) 175, p. 52 - 57. In Russian. — Abstr. in Referativ. Zhurn. 51. Astron., 4.51.111 (1976).

042.050 A note on the regularization of the restricted three-body problem.   E. Leimanis.
Astron. Nachr., Vol. 297, 141 - 143 (1976).

The requirement that near a singular point of the equations of motion the power series expansions of the old variables in terms of the new ones start with second order terms leads to the transformation $z = \sin^2 {}^1\!/_2 \omega$ related to that of Thiele-Burrau. Using this new transformation, a derivation of the regularized equations of motion is given. The original as well as the regularized equations of motion are of interest.

042.051 Differential improvement of the orbital elements in the undisturbed problem of two fixed centres.
R. A. Lyakh.
Trudy Astron. Obs., *Leningrad*, Vol. 32 (= Uchenye Zapiski Leningr. Un-ta, No. 385 = Seriya matem. nauk, vyp. (No.) 52), p. 119 - 130 (1976). In Russian.

Analytical expressions for the coefficients of equations of condition are derived which can be used for the improvement of orbital elements in the generalized problem of two fixed centers.

042.052 A special capture by Neptune.   R. Dvorak.
Astron. Astrophys., Vol. 49, 293 - 298 (1976).

In calculations with artificial comets by E. I. Kazimirchak-Polonskaya there happens to be the very interesting case of a capture by Neptune. In trying to confirm her results the author found a very good agreement in his numerical integration concerning the stable orbit of the captured comet. In computations back the expected escape did not occur. Theoretically a capture is possible when we regard it as an elliptic restricted three body problem, but the possibility for finding the initial conditions and for realizing them is very small.

042.053 A family of periodic solutions of the planar three-body problem, and their stability.   M. Hénon.
Celestial Mechanics, Vol. 13, 267 - 285 (1976).

The author describes a one-parameter family of periodic orbits in the planar problem of three bodies with equal masses. This family begins with Schubart's (1956) rectilinear orbit and ends in retrograde revolution, i. e. a hierarchy of two binaries rotating in opposite directions. The first-order stability of the orbits in the plane is also computed. Orbits of the retrograde revolution type are stable; more unexpectedly, orbits of the 'interplay' type at the other end of the family are also stable. This indicates the possible existence of triple stars with a motion entirely different from the usual hierarchical arrangement.

042.054 The development of the Poincaré-similar elements with eccentric anomaly as the independent variable.
V. R. Bond.
Celestial Mechanics, Vol. 13, 287 - 311 (1976).

A new set of element differential equations for the perturbed two-body motion is derived. The elements are canonical and are similar to the classical canonical Poincaré elements, which have time as the independent variable. The phase space is extended by introducing the total energy and time as canonically conjugated variables. The new independent variable is,

to within an additive constant, the eccentric anomaly. These elements are compared to the Kustaanheimo-Stiefel element differential equations, which also have the eccentric anomaly as the independent variable.

042.055 On the stability of circular 'asteroid' orbits in an N-planetary system.
M. A. Vashkovjak (*Vashkov'yak*).
Celestial Mechanics, Vol. 13, 313 - 324 (1976).

The restricted problem of the motion of a point of negligible mass ('asteroid') in an N-planetary system is considered. It is assumed that all the planets move about the central body ('sun') along circular orbits in the same plane and the mean motions of the asteroid and the planets are incommensurable. The asteroid orbit evolution is described as a first approximation by secular equations with the perturbing function averaged by the mean longitudes of the asteroid and the planets. For small values of the asteroid orbit eccentricity an expression for the secular part of the perturbing function has been obtained.

042.056 Lagrangian and near-Lagrangian solutions of the translatory-rotary motion of the three-body problem.   V. V. Vidyakin.
Celestial Mechanics, Vol. 13, 325 - 361 (1976).   In Russian.

In the present paper the following problem is solved: whether there exist exact special solutions in the general problem of the translatory-rotary motion of three rigid bodies, which have three mutually perpendicular symmetry planes, analogous to the Lagrangian solutions in the problem of three point bodies. It is proved that such solutions exist if the bodies possess a definite structure and orientation. In particular, the conditions for the existence of Lagrangian solutions of the problem of three rigid bodies have been found.

042.057 Application of the Jacobi integral for the investigation of satellite librations around the centre of mass in a gravitational field.
N. F. Martinova (*Martynova*).
Celestial Mechanics, Vol. 13, 363 - 381 (1976).

This paper deals with a three-dimensional rotationally and dynamically symmetrical satellite. The centre of mass of the satellite moves in a circular orbit. The existence of two first integrals of motion enables one to transform the system of differential equations to a special form facilitating the choice of the zero-approximation solution. The first approximation solution is constructed for the case of spatial libration of the satellite axis of dynamical symmetry about the position of stable equilibrium.

042.058 The bifurcation from the Maclaurin to the Jacobi sequence as a second-order phase transition.
G. Bertin, L. A. Radicati.
Astrophys. Journ., Vol. 206, 815 - 821 (1976).

It is proven that under suitable assumptions the change of symmetry associated with the bifurcation from the axisymmetric Maclaurin to the triaxial Jacobi ellipsoids can be described as a second-order phase transition in the entropy-volume plane. The proof is given by showing explicitly that the value of the eccentricity and other physical quantities at the bifurcation point calculated according to Landau's theory agree with those calculated in the ordinary way. These values are independent of the equation of state.

042.059 Simplified method of deriving equations of motion of a material point in a central field of force (Kepler's equation).   L. P. Orfanitskaya.
Sb. nauch.-metod. statej po teor. mekh. M-vo vyssh. i sredn. spets. obrazovaniya SSSR, 1975, vyp. (No.) 5, p. 119 - 122. In Russian. — Abstr. in Referativ. Zhurn. 51. Astron., 6. 51. 117 (1976).

042.060 Translatory-rotational motion of two dynamically symmetric bodies under their mutual attraction.
E. V. Bibik.
Probl. mekh. upravlyaemogo dvizheniya. Vyp. (No.) 7. Perm',
1975, p. 14 - 16. In Russian. − Abstr. in Referativ. Zhurn. 51.
Astron., 6.51.124; 62. Issled. kosmich. prostranstva, 6.62.283
(1976).

042.061 Periodic satellite motions in the plane of an elliptical orbit. V. A. Zlatoustov.
Uspekhi mat. nauk, Vol. 30, 200 - 201 (1975). In Russian.
Abstr. in Referativ. Zhurn. 51. Astron., 6.51.126 (1976).

042.062 Some problems of the theory of Hamiltonian systems and their application to celestial mechanics.
A. P. Markeev.
Uspekhi mat. nauk, Vol. 30, 203 - 204 (1975). In Russian.
Abstr. in Referativ. Zhurn. 51. Astron., 6.51.127 (1976).

042.063 On the problem of motion of a rigid body around a fixed point. M. G. Markevich.
Sb. nauch.-metod. statej po teor. mekh. M-vo vyssh. i sredn.
spets. obrazovaniya SSSR, 1975, vyp. (No.) 5, p. 68 - 72. In
Russian. − Abstr. in Referativ. Zhurn. 51. Astron., 6.51.128
(1976).

042.064 On the inverse problem of dynamics of a rigid body with one fixed point in a Newtonian field of forces.
I. A. Galiullin.
Sb. nauch.-metod. statej po teor. mekh. M-vo vyssh. i sredn.
spets. obrazovaniya SSSR, 1975, vyp. (No.) 5, p. 49 - 53.
In Russian. − Abstr. in Referativ. Zhurn. 51. Astron., 6. 51.
129 (1976).

042.065 Construction of periodic solutions in the problem of motion of a rigid body with one fixed point in a central Newtonian field of forces. S. Bularkhiev, F. I. Kiselev.
Sb. nauch.-metod. statej po teor. mekh. M-vo vyssh. i sredn.
spets. obrazovaniya SSSR, 1975, vyp. (No.) 5, p. 143 - 147. In
Russian. − Abstr. in Referativ. Zhurn. 51. Astron., 6.51.130
(1976).

042.066 Analysis of some schemes of determination of the estimates of the parameters of motion of celestial bodies. A. M. Chernitsov.
Materialy nauch.-prakt. konf. 'Molodye uchenye i spetsialisty
Tomsk. obl. v 9-oj pyatiletke. (Sekts. mekh.-mat)'. Tomsk,
1975, p. 105 - 107. In Russian. − Abstr. in Referativ. Zhurn.
51. Astron., 6. 51. 137 (1976).

042.067 Stability vs inclination of motions in E^3.
A. Halioulias, G. A. Katsiaris, V. V. Markellos.
Astrophys. Space Sci., Vol. 41, 417 - 422 (1976).
Five families of three-dimensional doubly symmetric
motions are computed after establishing their existence by
means of a grid-search technique. It is confirmed that within
the same family orbits of lower inclination with respect to the
plane of motion of the primaries are stable while the critical
inclination at which instability occurs varies between families.
The maximum inclination at which stable motions of the type
presented here were found is about 52°.

042.068 Approximate solution of the Euler-Lambert problem for nearly circular orbits.
A. I. Averbukh, B. V. Girshovich.
Kosmich. Issled., Vol. 14, 454 - 457 (1976). In Russian.
Brief information.

042.069 On the solution of the Clairaut-Laplace-Lyapunov problem.
V. P. Trubitsyn, P. P. Vasil'ev, A. B. Efimov.

Astron. Zhurn. Akad. Nauk SSSR, Vol. 53, 626 - 633 (1976).
In Russian. English translation in Soviet Astron., Vol. 20,
No. 3.
A successive solution of the Clairaut-Laplace-Lyapunov
problem is reported. A proof of validity of the Laplace expansion for the derivation of the equations of the figures of
planets and stars is given. An equation for the P_{12} harmonic
and values of the first seven computed non-zero even coefficients of the figure and field of planets for linear density
distribution are given.

042.070 The semi-restricted problem of three bodies.
V. V. Radzievskij.
Astron. Zhurn. Akad. Nauk SSSR, Vol. 53, 634 - 638 (1976).
In Russian. English translation in Soviet Astron., Vol. 20,
No. 3.
The problem of three bodies m_1, m_2 and m_3 when the
centre of masses m_2 and m_3 moves on an elliptic orbit about
m_1 is considered. The angular momentum and energy integrals
are obtained. The accuracy of determination of the velocity of
the moon with the help of the energy integral of this problem
is much higher than with the help of Jacobi's integral.

042.071 On the problem of three fixed centres.
G. T. Arazov.
Astron. Zhurn. Akad. Nauk SSSR, Vol. 53, 639 - 646 (1976).
In Russian. English translation in Soviet Astron., Vol. 20,
No. 3.
The exact solution of three variants of the plane problem
of three fixed centres has been obtained.

042.072 Sur une généralisation des équations différentielles de forme canonique. J. Meffroy.
Comptes Rendus Acad. Sci. Paris, Sér. A, Vol. 282, 801 - 804
(1976).
Les équations différentielles canoniques de la dynamique
classique résultent d'un lagrangien ne contenant pas les dérivées
des paramètres de position d'ordre supérieur au premier. On
étend la méthode au cas où le lagrangien contient les dérivées
des paramètres de position jusqu'à l'ordre k, l'entier naturel k
étant quelconque.

042.073 Sur les équations invariantes du problème des trois corps alignés. F. Nahon.
Comptes Rendus Acad. Sci. Paris, Sér. A, Vol. 282, 927 - 929
(1976).
Conformément à une idée de E. Cartan, l'auteur cherche
les équations du problème des trois corps alignés invariantes
par une homothétie dont le rapport est une fonction arbitraire
des positions et des vitesses. Il généralise ainsi les résultats
déjà obtenus pour le cas d'énergie nulle et trouve une indication d'une liaison entre les orbites de collision triple et cer-
taines orbites d'énergie nulle.

042.074 Sur une généralisation de la théorie de Hamilton-Jacobi. J. Meffroy.
Comptes Rendus Acad. Sci. Paris, Sér. A, Vol. 282, 1243 -
1246 (1976).
Généralisant la théorie de Hamilton-Jacobi, on étand la
notion de champ d'extrémales et la notion de surface trans-
versale au cas où le lagrangien contient les dérivées des paramètres de position jusqu'à l'ordre k, l'entier naturel k étand
quelconque.

042.075 Perturbation theory for strongly perturbed dynamical systems. II. Canonical coordinates.
W. H. Jefferys.
Astron. Journ., Vol. 81, 485 - 488 (1976).
A method is given for replacing the quasicanonical co-
ordinates in Jefferys' (1968) theory for strongly perturbed
dynamical systems with canonical coordinates. A method of

successive approximations is employed. The method has been programmed for the formula manipulation language Trigman, and its application to a two-degree-of-freedom system is discussed.

042.076 Relación entre las funciones determinantes de los métodos de perturbaciones de Von Zeipel y Hori. J. F. Lahulla.
Bol. Astron. Obs. Madrid, Vol. 9, No. 2, p. 87 - 96 (1976).

042.077 A note on expansions of functions of velocity in the two-body problem. H. Kinoshita.
Center Astrophys. Prepr. Ser., No. 493, 9 pp. (1976).

042.078 Aspects topologiques en mécanique céleste. M. C. Simó.
Exposé de la table ronde de Mécanique Céleste de l'Institut Henri Poincaré, Paris, 17 pp. (1975).

042.079 General vector solution of the differential equation describing the motion of a heavy particle near the surface of the earth. F. Verheest, O. Leroy.
Simon Stevin (Belgium), Vol. 46, 153 - 157 (1973). – Abstr. in Zentralbl. Math. Grenzgebiete – Math. Abstr., Band 305, No. 70006 (1976).

042.080 Equations of motion of nonholonomic dynamical systems in Poincaré-Četaev variables. Q. K. Ghori, M. Hussain.
Zeitschr. angew. Math. Mech., Vol. 54, 311 - 318 (1974). Abstr. in Zentralbl. Math. Grenzgebiete – Math. Abstr., Band 305, No. 70016 (1976).

042.081 Regularization in celestial mechanics. V. Szebehely.
Comput. Methods nonlin. Mech., (Conference Austin 1974), p. 257 - 263 (1976). – Abstr. in Zentralbl. Math. Grenzgebiete – Math. Abstr., Band 307, No. 70006 (1976).

042.082 Zur Konstruktion partikulärer Lösungen des restringierten ebenen kreisförmigen Dreikörperproblems. A. G. Iljuhin, A. S. Korneeva.
Mat. Fiz., respubl. mezhvedomstv. Sbornik, Vol. 14, 58 - 62 (1973). In Russian. – Abstr. in Zentralbl. Math. Grenzgebiete – Math. Abstr., Band 308, No. 70004 (1976).

042.083 Geschlossene Trajektorien des Zusammenstosses im räumlichen kreisförmigen restringierten Dreikörperproblem. I. V. Kurcheeva.
Vestnik Leningrad. Univ. 1973, No. 13 (Mat. Mekh. Astron. 3), p. 108 - 113 (1973). In Russian. – Abstr. in Zentralbl. Math. Grenzgebiete – Math. Abstr., Band 309, No. 70012 (1976).

042.084 Die Zerlegungen der Störungen der Bahnelemente eines Satelliten, die durch die Kompression des Planeten hervorgerufen werden, bis zur neunten Potenz der Exzentrizität. D. Z. Koenov.
Izv. Akad. Nauk TadzhSSR, Otd. fiz.-mat. geol.-khim. Nauk 1971, No. 1(39), p. 16 - 23 (1971). In Russian. – Abstr. in Zentralbl. Math. Grenzgebiete – Math. Abstr., Band 309, No. 70019 (1976).

042.085 A recent comparison of numerical methods for solving ordinary differential equations in celestial mechanics. D. G. Bettis, K. Sepehrnoori.
Comput. Methods nonlin. Mech., (Conference Austin 1974), p. 207 - 212 (1974).

042.086 Geometrical singularities in perturbation theory. J. Henrard.
Comput. Methods nonlin. Mech.,(Conference Austin1974),

p. 213 - 220 (1974). – Abstr. in Zentralbl. Math. Grenzgebiete – Math. Abstr., Band 312, No. 70022 (1976).

042.087 The magic of elliptical orbits. M. Berrondo, J. Flores, O. Novaro.
Rev. Mexicana Fis., Vol. 23 Suppl., E13 - E26 (1974). In Spanish. – Abstr. in Phys. Abstr., Vol. 79, A000011 (1976).

042.088 The laws of classical mechanics concerning the conservation of energy and impulse for variable masses. I. Nemedi.
Fiz. Szemle, Vol. 25, 251 - 257 (1975). In Hungarian. –Abstr. in Phys. Abstr., Vol. 79, A000380 (1976).

042.089 Concerning the dependence of secular and periodic terms. H. W. Milnes.
Industr. Math., Vol. 25, Pt. 1, p. 29 - 36 (1975). – Abstr. in Phys. Abstr., Vol. 79, A011971 (1976).

042.090 On a class of escape orbits without collisions in the general three-body problem. M. Breitenecker.
Acta Phys. Austriaca, Vol. 43, 255 - 265 (1975). – Abstr. in Phys. Abstr., Vol. 79, A016192 (1976).

042.091 A uniformly valid solution for motion about the interior libration point of the perturbed elliptic-restricted problem. D. L. Richardson, N. D. Cary.
1975 AAS/AIAA Astrodyn. Conf., (see 012.022), AAS75-021, 28 pp. (1975). – Abstr. in Phys. Abstr., Vol. 79, A023590 (1976).

042.092 Introduction to the restricted Jupiter orbiter problem. T. A. Heppenheimer.
1975 AAS/AIAA Astrodyn. Conf., (see 012.022), AAS75-037, 42 pp. (1975). – Abstr. in Phys. Abstr., Vol. 79, A023593 (1976).

042.093 On the motion of a particle in orbit around a growing mass. R. Greenberg.
Journ. Astronaut. Sci., Vol. 23, 75 - 77 (1975). – Abstr. in Phys. Abstr., Vol. 79, A023609 (1976).

Astronomy. Volume 11. Celestial mechanics.
See Abstr. 003.100.

Linear and regular celestial mechanics. Perturbed two-body motion. Numerical methods. Canonical theory. See Abstr. 003.104.

Die Dynamik der Kreisbewegungen der Himmelskörper und des freien Falls bei Aristoteles, Copernicus, Kepler und Descartes. See Abstr. 004.045.

Multistep methods of numerical integration using back-corrections. See Abstr. 021.002.

Relativistic motor dynamics of systems of mass points and rigid bodies. See Abstr. 022.041.

Statistical behavior in conservative classical systems. See Abstr. 022.044.

Determination of elements of sub-synchronous or super-synchronous orbits and their application to operational satellites. See Abstr. 052.036.

Sur la répartition de la masse équivalente à l'énergie potentielle et ses conséquences. See Abstr. 066.022.

Effects of low-frequency gravitational waves upon

orbital elements (planets). See Abstr. 091.023.

Relativistic theory of the moon's motion.
See Abstr. 094.008.

An orbital improvement of Vesta.
See Abstr. 098.012.

On the motion of Hecuba-type asteroids.

See Abstr. 098.049.

The strange case of the missing apocentric librators in the 3 : 2 resonance. See Abstr. 098.071.

The theory of Enceladus and Dione — an application of computerized algebra in dynamical astronomy.
See Abstr. 100.203.

043 Astronomical Constants

043.001 **Wie veränderlich ist die Gravitationskonstante?**
H. G. Walter.
SuW, 15. Jahrgang, p. 41 - 43 (1976).

043.002 **Secular variation of the obliquity of the ecliptic.**
R. L. Duncombe, T. C. Van Flandern.
Astron. Journ., Vol. 81, 281 - 284 (1976).
  This paper summarizes the observational determinations of the secular variation of the obliquity of the ecliptic. The sun and planet determinations show a discordance between observation and theory amounting to −0.3 arcsec/century, but determinations from lunar occultations agree with the theoretical variation. It is shown that incomplete reduction of sun, Mercury, and Venus observations to the fundamental star

reference system may be the cause of the apparent observed correction to the secular rate of the obliquity.

Stellare Kinematik aus Eigenbewegungen in bezug auf Galaxien. See Abstr. 041.009.

Changing gravitational constant and white dwarfs.
See Abstr. 126.013.

Bestimmungen der galaktischen Rotation und der Präzession aus fundamentalen Eigenbewegungen und aus Eigenbewegungen relativ zu Galaxien. See Abstr. 155.053.

044 Time, Rotation of the Earth

044.001 **Le temps physique: qu'est-ce que c'est?**
A. Peton.
L'Astronomie, Vol. 90, 129 - 143 (1976).

044.002 **Determinations of the k-Love number and the factor Λ which affect the observations of Universal Time.**
D. Djurovic.
Astron. Astrophys., Vol. 47, 325 - 332 (1976). In French.
A mean value TU 1-TUC is computed for each day during a five-year observation period (1967–1971) at 49 observatories. A tidal terms research in the rotation of the earth was carried out starting from TU 1-TUC and the residuals obtained directly from TU 0-TUC, after the elimination of the secular and main harmonic terms. A certain number of tidal terms, among which are O_1, M_2, M_f and M_m were identified. Two values were obtained for Love's number k:0.301 and 0.343. The amplitude factor $\Lambda = 1 + k + l$ was also determined: from O_1 component $\Lambda = 0.807$ and from M_2 component $\Lambda = 0.675$.

044.003 **Effect of semi-annual solar nutation error in longitude variation.** K. Yokoyama.
Astron. Astrophys., Vol. 47, 333 - 340 (1976).
An analysis of the global time observations supports Wako's (1970) interpretation that Kimura's (1902, 1940) annual z-term can be explained by the theoretical semi-annual solar nutation term based on the hypothesis of a liquid core of the earth. A new term which describes the effect of the error due to this nutation term in the observed longitude variation is added to the equation of observation. A stable annual variation found in the newly added term has revealed a good correlation with the annual z-term.

044.004 **Theory of the rotation of the rigid earth.**
H. Kinoshita.
Long-time predictions in dynamics, (see 012.005), p. 339 (1976). – Abstract.

044.005 **Estimate of the irregularity in the earth's rotation caused by magnetic disturbances of the sun.**
Yu. A. Bilde.
Astrometriya i Astrofizika, Kiev, vyp. (No.) 28, (see 003.002), p. 14 - 21 (1976). In Russian.
An estimate of the irregularity is given for the earth's rotation caused by magnetic disturbances of the sun in circumterrestrial space. The case of both periodic and nonperiodic disturbances is considered. The model used has the following assumptions: 1) the earth is considered to be a perfectly solid spherical top with constant moment of inertia; 2) the geomagnetic field is assumed to be a magnetic dipole with centered magnetic moment rigidly connected with the earth; 3) the magnetic field evolution and the influence of magnetic disturbances are not taken into consideration.

044.006 **Seasonal irregularity of the earth's rotation for 1956–1973.** A. A. Korsun', N. S. Sidorenkov.
Astrometriya i Astrofizika, Kiev, vyp. (No.) 28, (see 003.002), p. 22 - 29 (1976). In Russian.
The mean monthly parameters are given for annual, semiannual and quarterly waves in seasonal variations of relative changes of the earth's angular velocity for 17 years. From general analysis of the main components of seasonal irregularity of the earth's rotation a formula is suggested which takes into account the amplitude modulations of annual and semiannual waves.

044.007 **Zwanzig Jahre Ephemeridenzeit.**
H.-J. Felber.

Sterne, Vol. 52, 83 - 91 (1976).

044.008 **Palaeontological and astronomical observations on the rotational history of the earth and moon.**
S. K. Runcorn.
Growth rhythms and the history of the earth's rotation, (see 012.012), p. 285 - 291 (1975).

044.009 **Astronomical evidence of change in the rate of the earth's rotation and continental motion.**
E. Proverbio, A. Poma.
Growth rhythms and the history of the earth's rotation, (see 012.012), p. 385 - 395 (1975).

044.010 **The detection of recent changes in the earth's rotation.** N. P. J. O'Hora.
Growth rhythms and the history of the earth's rotation, (see 012.012), p. 427 - 444 (1975).

044.011 **Changes in the earth's rotation from astronomical observations.** L. V. Morrison.
Growth rhythms and the history of the earth's rotation, (see 012.012), p. 445 - 457 (1975).

044.012 **The accelerations of the earth and moon from early astronomical observations.**
P. M. Muller, F. R. Stephenson.
Growth rhythms and the history of the earth's rotation, (see 012.012), p. 459 - 534 (1975).

044.013 **The velocity variations in the rotation of the earth.** D. Djurović.
Glas Srpske akad. nauka i umetn., Vol. 291, 55 - 69 (1974). In Serbo-Croatian. – Abstr. in Bull. Sci. Yougoslavie, Sect. A, Vol. 21, 18 (1976).

044.014 **Time and latitude service.**
Polish Acad. Sci., Astron. Latitude Obs., Borowiec, Poland, Circ. Nos. 136 - 137, 9 + 9 pp. (1975/76). – 1975 October - 1976 March.

044.015 **Time service for 1974.** V. Quesada.
Circ. Stazione Astron. Internazionale Latitudine, Carloforte–Cagliari, Ser. A (3), No. 11, 38 pp. (1975).

044.016 **Zeit- und Breitendienst.**
Edited by Deutsches Hydrographisches Institut, Hamburg, 8 pp. (1975). – 1975 April - June.

044.017 **Modern time service.** I. Domiński.
Urania Kraków, Vol. 47, 34 - 47 (1976). In Polish.

044.018 **Theory of the rotation of the rigid earth.**
H. Kinoshita.
Center Astrophys. Prepr. Ser., No. 443, 85 pp. (1975).

044.019 **Astronomische Zeit- und Breitenbestimmungen, Empfangszeiten von Zeitsignalen, Präzisionszeitvergleiche.**
Akad. Wiss. DDR, Zentralinst. Phys. Erde, Abt. Geod. Astron., Jahrgang 1974, Nos. 5 - 6, 16 + 16 pp. (1976). – 1974 September - December.

044.020 **Universal time and coordinates of the pole; Emission time of time signals; Universal time (coordinated); Independent local atomic time scales AT(i); Information.**

Bureau International de l'Heure, (B.I.H.), Paris, Circ. D 110 - D 116, with notices to users of Circular D (1976). — 1976 January - July.

044.021 Détermination astronomique de l'heure et heures demi-définitives de réception des signaux horaires.
L. Webrová, V. Ptáček.
Acad. Tchécoslov. Sci., Inst. Astron., Station de l'Heure, Prague, Sér. 7, No. 7, 15 pp. (1976). — 1975 January - February.

044.022 Time Service Annual Report 1973.
Zi-Ka-Wei Section, Shanghai Obs., Acad. Sinica, Shanghai, China, 88 pp.

044.023 Rotacion de la tierra año 1974. Resultados obtenidos en San Fernando con el Astrolabio Impersonal Danjon OPL No. 37. I.—Tiempo y latitud. II.—Observaciones de Saturno. L. Rohat-Jullien, A. Orte, M. Sánchez, I. Vitini, J. B. Fernández, F. Parra.
Published by Inst. y Obs. de Marina, San Fernando (Cádiz), Ser. C. No. 77, 27 pp. (1976).
  This bulletin contains definitive results of time and latitude observations made at San Fernando with the Danjon astrolabe during 1974. BIH corrections and astronomical coordinates of the astrolabe are shown. Results of Saturn observations during the winter 1973—74 are given.

044.024 Time service. C. Moranzino (Editor).
Oss. Astron. Torino (Pino Torinese), Bull. Nos. 11 - 12 (1975). — Results of the time determinations 1975 May - December.

044.025 International Time and Latitude Service at the Tokyo Astronomical Observatory during 1975.
K. Osawa.
Tokyo Astron. Obs., Time and Latitude Bull., Vol. 49, Nos. 5 - 12, p. 31 - 84 (1975). — Results of the time determinations 1975 May - December.

044.026 A comparison of transit measurements of clock stars at instrumental meridian.
C. Moranzino, G. Chiumiento.
Oss. Astron. Torino (Pino Torinese), Time Service Bull. No. 11, p. 6 - 13 (1975).

044.027 Measuring set-up for time determinations at the Turin Observatory.
C. Moranzino, V. Pettiti, G. Rovera.
Oss. Astron. Torino (Pino Torinese), Time Service Bull. No. 12, p. 7 - 11 (1975).

044.028 Les termes de marée dans le temps universel.
D. Djurovic.
Obs. Roy. Belgique, Bull. Observations, Vol. 4, Fasc. 3, p. 1 - 16 (1975).

044.029 Coordonnées du pôle et TU1-TUC pour l'intervalle 1967—1974. D. Djurovic.
Obs. Roy. Belgique, Bull. Observations, Vol. 4, Fasc. 3, p. 17 - 64 (1975).

044.030 An interpretation of the multiple—peak spectra of the polar wobble of the earth. N. Sekiguchi.
Publ. Astron. Soc. Japan, Vol. 28, 277 - 291 (1976).
  It is shown that the increase rate of the revolution angle of the polar wobble has a tendency to take one of discrete values, among which 0.845 and 0.82 yr^{-1} are predominant. The variations of the increase rate of the revolution angle can occur either due to change of the intrinsic Chandlerian frequency inherent in the earth or through any nature of ex-

ternal excitation mechanism, but if the excitation process has a stationary and random character, it should be possible to separate the effects from each other, and the results favor the former possibility.

044.031 Classical methods for the determination of universal time and polar motion.
N. P. J. O'Hora, B. D. Yallop.
IAU Colloquium No. 26, (see 012.018), p. 303 - 308 (1975).

044.032 On some problems of BIH and IPMS services.
P. Melchior.
IAU Colloquium No. 26, (see 012.018), p. 409 - 413 (1975).

044.033 Daily phase values and time differences.
U.S. Naval Obs., Washington, D.C., Time Service Publ., Ser. 4, Nos. 466 - 490 (1976). — 1976 January 8 - June 24.

044.034 Preliminary times and coordinates of the pole.
U.S. Naval Obs., Washington, D.C., Time Service Publ., Ser. 7, Nos. 418 - 443 (1976). — 1976 January 1 - June 24.

044.035 Comparison of various fundamental methods of defining time, with special reference to astronomical applications. A. Orte.
Navigation (*France*), Vol. 24, 72 - 84 (1976). — Abstr. in Phys. Abstr., Vol. 79, A028228 (1976).

  Aperture synthesis. See Abstr. 031.253.

  The 65 cm PZT of the U.S. Naval Observatory. See Abstr. 032.038.

  Equal hour angle method for rapid and precise simultaneous determination of astronomical time, longitude, latitude and azimuth. See Abstr. 045.002.

  The use of Gauss' frequency curve for smoothing observed values of the geographic latitude and time corrections. See Abstr. 045.006.

  Variations of the angles between the plumb lines of the astronomical instruments. See Abstr. 045.015.

  Comparison of the Uccle—Brussels coordinates in the BIH, NWL and EUR 50 systems. See Abstr. 045.019.

  Tides and the rotation of the earth. See Abstr. 081.021.

  The earth's interior and the earth's rotation. See Abstr. 081.022.

  Gravity and the earth's rotation. See Abstr. 081.023.

  On a tentative correlation between changes in the geomagnetic polarity bias and reversal frequency and the earth's rotation through Phanerozoic time. See Abstr. 084.264.

Errata

044.901 Erratum: Polar coordinates, corrections for seasonal and polar (longitude) variations.
U.S. Naval Obs., Washington, D.C., Time Service Publ., Ser. 11, No. 223.1 (1976).

045 Latitude Determination, Polar Motion

045.001 The secular motion of the pole from BIH results.
A. Poma, E. Proverbio.
Astron. Astrophys., Vol. 47, 105 - 111 (1976).

The secular motion of the earth's pole emphasized by latitude observations carried out in the International Latitude Service (ILS) is briefly discussed. The authors have successively calculated the secular motion of the mean pole starting from the values of the co-ordinates x and y of the new system of the Bureau Internationale de l'Heure established in 1968 (1968 BIH system). This system is based on time and latitude observations independent of the ILS-IPMS system. The only difference between the two paths can be attributed to differences in the origins of the Conventional International Origin and the 1968 BIH system. Explanations of the observed secular polar motion are given in terms of large lithospheric plate motions combined with a slow movement of the axis of inertia of the earth.

045.002 Equal hour angle method for rapid and precise simultaneous determination of astronomical time, longitude, latitude and azimuth. Y.-h. Yu.
Scient. Sinica, Vol. 19, 45 - 64 (1976).

The equal hour angle method presented in this paper consists in noting horizontal direction when three stars cross successively the same hour circle within ten minutes or more, thus achieving easily a simultaneous determination of precise longitude, latitude and azimuth. It is unnecessary to know rough values of these elements, nor is there the need to prepare a star observing list. If only latitude and azimuth are determined, a stop watch will be sufficient; no chronometer or radio receiver is needed. Theoretical study and experimental results have proved the advantages of this method.

045.003 Use of observations of artificial satellites for the solution of problems of geodynamics.
L. V. Rykhlova.
Dynamical questions of satellite geodesy, 1973, (see 012.006), p. 194 - 218 (1974). In Russian.

045.004 Betrachtungen zur Ableitung von Polbewegungen aus Satellitenbeobachtungen. L. Stange.
Dynamical questions of satellite geodesy, 1973, (see 012.006), p. 219 - 233 (1974). In Russian and German.

045.005 Nearly diurnal free nutation from latitude observations at Greenwich from 1911 to 1935.
Ya. S. Yatskiv.
Astrometriya i Astrofizika, *Kiev,* vyp. (No.) 29, (see 003.003), p. 24 - 32 (1976). In Russian.

Spectral and regression analyses of the latitude observations at Greenwich from 1911 to 1935 were carried out for studying the nearly diurnal free nutation. Latitude variations with periods of 196 and 206 mean days were found in the frequency region under consideration. These variations are shown to be of diurnal type.

045.006 The use of Gauss' frequency curve for smoothing observed values of the geographic latitude and time corrections. J. Kabeláč.
Bull. Astron. Inst. Czechoslovakia, Vol. 27, 143 - 152 (1976).

045.007 Results of latitude observations for the year 1973 with VZT. E. Proverbio, S. Uras.
Circ. Stazione Astron. Internazionale Latitudine, Carloforte-Cagliari, Ser. B (7), No. 9, 11 + 209 + 13 pp. (1974).

045.008 Statistical analysis of latitude observations in the

1973. E. Proverbio, S. Uras.
Circ. Stazione Astron. Internazionale Latitudine, Carloforte-Cagliari, Ser. B (8), No. 10, 2 + 45 pp. (1974).

045.009 Observations of the fortnightly nutation terms and the 'dynamical variation of latitude' with photographic zenith tubes. D. D. McCarthy.
Astron. Journ., Vol. 81, 482 - 484 (1976).

Observations made with three photographic zenith tubes show fortnightly nutation terms in good agreement with values determined from earth tide observations. The radius of the circular 'dynamical variation of latitude' occurring with the argument $(a - 2L)$ is found to be less than the rigid earth values.

045.010 Breitenbestimmungen.
Techn. Univ. Dresden, Lohrmann-Obs., Zirk. Nos. 72 - 76 (1974/75). − 1974 November - 1975 August.

045.011 Monthly Notes of the International Polar Motion Service.
IPMS Monthly Notes, International Latitude Obs. Mizusawa (Japan). 1975 No. 12, p. 109 - 117; 1976 Nos. 1 - 5, p. 1 - 46 (1976). − Announces the values of latitudes observed at the collaborating statins during 1975 December − 1976 May.

045.012 Phasenbeziehungen als Indikatoren für die Erregung der Polbewegung. H. Jochmann.
Vermessungstechn., 23. Jahrg., p. 99 - 101 = Mitt. Zentralinst. Phys. Erde, *Potsdam*, No. 414 (1975).

045.013 On the latitude variations of the interval between 1830 and 1860. N. Sekiguchi.
Journ. Geod. Soc. Japan, Vol. 21, 131 - 141 = Tokyo Astron. Obs. Repr. No. 490 (1975).

045.014 Results of the determination of latitude in Józefosław by observations of the Horrebow-Talcott pairs. L. Pieczyński.
Warsaw Tech. Univ., Astron. - Geod. Obs. Józefosław. Latitude Circ. Nos. 55 - 57 (1975/76). − 1975 July - 1976 March.

045.015 Variations of the angles between the plumb lines of the astronomical instruments. N. T. Mironov.
IAU Colloquium No. 26, (see 012.018), p. 79 - 84 (1975).

045.016 Pole coordinates by the IMPS stations.
S. Yumi, K. Yokoyama, H. Ishii.
IAU Colloquium No. 26, (see 012.018), p. 115 - 122 (1975).

045.017 Latitude star system of the ILS and the conventional International Origin. K. Sato, Y. Wako.
IAU Colloquium No. 26, (see 012.018), p. 319 (1975). Abstract.

045.018 Systematic differences between global or local solutions of polar motion derived from astronomical and spatial measurements. N. Capitaine, M. Feissel.
IAU Colloquium No. 26, (see 012.018), p. 415 - 432 (1975).

045.019 Comparison of the Uccle–Brussels coordinates in the BIH, NWL and EUR 50 systems.
P. Pâquet, R. Dejaiffe, W. De Rop, R. Verbeiren.
IAU Colloquium No. 26, (see 012.018), p. 457 - 459 (1975).

045.020 Results of latitude observations with the floating

zenith telescope at Mizusawa: comparisons between the results obtained from the astronomical and Doppler satellite observations.   S. Takagi.
IAU Colloquium No. 26, (see 012.018), p. 461 - 468 (1975).

045.021   Analysis of periodic deviations of the zenith point of a number of observatories from observational data λ, φ.   J. Moczko.
Materiały i Prace, Vol. 94, 119 - 134 (1975).

Effect of semi-annual solar nutation error in longitude variation.   See Abstr. 044.003.

Time and latitude service.   See Abstr. 044.014.

Zeit- und Breitendienst.   See Abstr. 044.016.

Astronomische Zeit- und Breitenbestimmungen, Empfangszeiten von Zeitsignalen, Präzisionszeitvergleiche. See Abstr. 044.019.

Rotacion de la tierra año 1974.
See Abstr. 044.023.

Coordonnées du pôle et TU1-TUC pour l'intervalle 1967–1974.   See Abstr. 044.029.

Classical methods for the determination of universal time and polar motion.   See Abstr. 044.031.

On some problems of BIH and IPMS services.
See Abstr. 044.032.

Gravimetric and astrogeodetic reference systems in space and time.   See Abstr. 046.040.

046   Astronomical Geodesy, Satellite Geodesy, Navigation

046.001   Estimation of the significance of control-tests of the results of geodetic measurements.
M. K. Szacherska, W. Markowski.
Geodezja Kartografia, Vol. 24, 269 - 274 (1975). In Polish.
   The authors point out the necessity of an estimation of the significance of the results of control-tests and propose a method based on the statistical analysis of the results of measurements. They present an example of the analysis of control-tests which have been effectuated in connexion with measurements of astronomical azimuths.

046.002   Examen des méthodes principales de l'identification des astres d'appui à l'aide d'ordinateurs.
E. Woźny.
Geodezja Kartografia, Vol. 25, 35 - 49 (1976). In Polish.
   L'auteur présente brièvement la théorie des méthodes principales de l'identification des astres d'appui sur les prises de vue des satellites artificiels de la terre et examine leur utilité à l'application générale. Il propose à la base de cet examen une modification de la méthode analytique de Kapilewicz qui assure une efficacité augmentée de l'identification.

046.003   Contributions to the National Geodetic Satellite Program by Goddard Space Flight Center.
D. E. Smith, F. J. Lerch, J. G. Marsh, C. A. Wagner, R. Kolenkiewicz, M. A. Khan.
Journ. Geophys. Res., Vol. 81, 1006 - 1026 (1976).

046.004   Nautical astronomy and the Mercator principle.
C. H. Cotter.
Journ. Navigation, Vol. 29, 14 - 20 (1976).

046.005   The course of navigation.   D. J. Lindsay.
Journ. Navigation, Vol. 29, 49 - 56 (1976).

046.006   Multiple sun sights with one reduction.
J. P. Budlong.
Journ. Navigation, Vol. 29, 95 - 96 (1976).

046.007   Spherical hyperbolae and ellipses.
H. C. Freiesleben.
Journ. Navigation, Vol. 29, 194 - 199 (1976).

046.008   A tabular method for star-sight reduction.
V. Nastro, A. Russo, with comments by D. H. Sadler.
Journ. Navigation, Vol. 29, 205 - 207 (1976).

046.009   The determination of the coordinates of station 1147 at Ondřejov.
G. Karský, J. Kostelecký, V. Skoupý, I. Synek.
Space Research XV, (see 012.003), p. 21 - 24 (1975).

046.010   Analysis of satellite altimetry data.   W. Benning.
Space Research XV, (see 012.003), p. 59 - 63 (1975).

046.011   New method of tabular solution of the problem of two heights.   V. F. D'yakonov.
Sudovozhdenie. Vyp. (No.) 17. Moskva, 1975, p. 58 - 70. In Russian. – Abstr. in Referativ. Zhurn. 51. Astron., 3.51.120 (1976).

046.012   Geodynamic investigations using very long term observation of near-earth satellites.
B. C. Douglas, C. C. Goad.
Long-time predictions in dynamics, (see 012.005), p. 335 (1976). – Abstract.

046.013   Dynamical methods of satellite geodesy.
S. K. Tatevyan.

Dynamical questions of satellite geodesy, 1973, (see 012.006), p. 184 - 189 (1974). In Russian.

046.014 **On seasonal variations of the angles between plumb lines of astronomical instruments.** N. T. Mironov. Astrometriya i Astrofizika, *Kiev*, vyp. (No.) 28, (see 003.002), p. 30 - 39 (1976). In Russian.

The seasonal variations of 66 angles between plumb lines of 12 astronomical instruments are calculated for the period 1968 II–1971 XII. Their amplitudes and phases are obtained. The annual term in these variations is found for all the cases, even for neighbouring observatories and for instruments mounted at one observatory. This indicates that the seasonal variations are due to instrumental errors and systematic errors of catalogues, but not to periodic motion of continental blocks.

046.015 **Analysis of the accuracy of "Sight reduction tables for marine navigation" HO-229 for altitude and azimuth computation.** G. K. Prikhod'ko, V. V. Tomson. Sudovozhdenie. Vyp. (No.) 17. Moskva, 1975, p. 70 - 80. In Russian. – Abstr. in Referativ. Zhurn. 51. Astron., 4.51. 143 (1976).

046.016 **Semi-dynamical Doppler satellite positioning.** J. Kouba, D. E. Wells. Bull. Géod., Vol. 50, 27 - 42 (1976).

In semi-dynamical satellite geodesy, the satellite orbit is determined beforehand by dynamical methods, and tracking station coordinates are determined geometrically from satellite observations and the accepted satellite coordinates. This paper analyzes the use of the U.S. Navy Navigational Satellite Doppler (formerly Transit) system in the semi-dynamical mode, presenting some test network numerical results.

046.017 **Error model for geodetic positions derived from Doppler satellite observations.** R. J. Anderle. Bull. Géod., Vol. 50, 43 - 77 (1976).

Doppler observations of Navy Navigation Satellites have been used to strengthen and extend many terrestrial geodetic networks. The main sources of errors in positions determined from these observations are random error of observations, random and systematic errors in satellite positions due to uncertainties in the gravity field, and biases in the coordinate system in which the satellite ephemeris is given.

046.018 **A programme to calculate the coordinates of tracking stations by means of the semi-dynamical method.** Yu. V. Batrakov, T. K. Nikol'skaya. Dynamical questions of satellite geodesy, 1973, (see 012.006), p. 190 - 192 (1974). In Russian.

046.019 **Intrinsic geodesy. Part I. Elements of tensor calculus. Part II. Geodetic applications.** A. Marussi. Boll. Geod. Sci. Affini, Anno 35, p. 1 - 35 (1976).

046.020 **The experimental Doppler campaign 1974 in Italy.** L. Ciraolo, L. Mezzani. Boll. Geod. Sci. Affini, Anno 35, p. 37 - 48 (1976).

046.021 **Das Universal Theo 002; sein Einsatz bei zwei genauen Verfahren der Geodätischen Astronomie.** R. Sigl, A. Bauch. Allgem. Vermessungs-Nachr., Heft 4/1976, p. 113 - 128 = Mitt. Inst. Astron. Physik Geod. Techn. Hochschule München No. 134.

046.022 **Coordinate determination in a navigational system for vectorial aeromagnetic measurements.** M. A. Sergeev, V. I. Yushchenko.

Izv. vyssh. ucheb. zavedenij. Priborostroenie, Vol. 18, No. 11, p. 71 - 75 (1975). In Russian. – Abstr. in Referativ. Zhurn. 51. Astron., 5.51.183 (1976).

046.023 **On a comparison of methods of direct determination of the geodetic azimuth of direction from astronomical observations.** A. N. Kuznetsov. Izv. vyssh. ucheb. zavedenij. Geod. i aehrofotosemka, 1975, No. 2, p. 61 - 65. In Russian. – Abstr. in Referativ. Zhurn. 52. Geodeziya i Aehrosemka, 5.52.134 (1976).

046.024 **Results of the "ISAGEX" campaign.** J. Kovalevsky. Space Research XV, (see 012.003), p. 3 - 15 (1975).

During the International Satellite Geodetic Experiment (ISAGEX) campaign, organized in 1971 with the sponsorship of COSPAR, a great number of observations of positions of artificial satellites was gathered by about 50 stations. These data are described and analyzed and their value in geodetic and geodynamic research is estimated. Examples of their use are given.

046.025 **News in planetary geodesy.** K. Macháčková, J. Klokočník. Říše hvězd, Vol. 57, 124 - 127 (1976). In Czech.

046.026 **Positionsbestimmung zur See und in der Luft.** G. Gerstbach. Sternenbote, 19. Jahrgang, p. 102 - 110, 112 - 113 (1976).

046.027 **Bestimmung von Zeitkorrekturen für die Rotations-verschlüsse ballistischer Messkammern.** W. Böhler. Beiträge zur Satellitengeodäsie. Veröff. Bayer. Komm. Internat. Erdmessung. Astron.-geod. Arbeiten, No. 34, p. 3 - 12, München (1975).

046.028 **Ein Verfahren zur Reduktion von BC4-Aufnahmen.** K. Kaniuth. Beiträge zur Satellitengeodäsie. Veröff. Bayer. Komm. Internat. Erdmessung. Astron.-geod. Arbeiten, No. 34, p. 13 - 27, München (1975).

046.029 **Numerische Untersuchungen über den Einfluß von Korrelationen in der Satellitentriangulation.** M. Näbauer, H. Seifers. Beiträge zur Satellitengeodäsie. Veröff. Bayer. Komm. Internat. Erdmessung. Astron.-geod. Arbeiten, No. 34, p. 28 - 43, München (1975).

046.030 **Untersuchungen zum Nachführverhalten von siderisch nachgeführten Satellitenmeßkammern.** F. M. Scherer. Deutsche Geod. Komm. Bayer. Akad. Wiss., Reihe C, Nr. 214. Diss. Univ. Bonn, 131 pp. München (1975).

046.031 **Untersuchung eines astronomischen Theodolits mit automatischer Fernrohrnachführung durch Schritt-motoren.** R. Joeckel. Deutsche Geod. Komm. Bayer. Akad. Wiss., Reihe C, Nr. 215. Diss. Univ. Stuttgart, 86 pp. München (1975).

046.032 **Der Beitrag der Photogrammetrie zum heutigen Stand der Geodäsie.** H. H. Schmid. Eidgen. Techn. Hochschule Zürich, Inst. Geod. Photogrammetrie, Mitt. No. 18, 23 pp. (1975).

046.033 **Die Nachweisgrenze von Laserechos bei Satelliten-entfernungsmessungen.** H. Fischer. Vermessungstechn., 22. Jahrg., p. 244 - 247 = Mitt. Zentralinst. Phys. Erde, *Potsdam*, No. 385 (1974).

046.034 Geodetic positions of astronomical stations of the Hydrographic Department.
Data Rep. Hydrographic Observations, Ser. Astron. Geod., *Tokyo*, No. 10, p. 42 - 48 (1976).

046.035 Determination of longitude and latitude at the Okayama Astrophysical Observatory.
S. Fujii, A. Takechi, M. Yoshinari, Y. Niimi.
Tokyo Astron. Obs. Rep., Vol. 17, 430 - 439 (1976). In Japanese.

046.036 On the secular variation of longitude as obtained from recent time observations made with PZT during 12 years from 1962 to 1973. S. Fujii, S. Iijima.
Tokyo Astron. Obs. Rep., Vol. 17, 448 - 463 (1976). In Japanese.

046.037 Requirements for coordinate systems for earth dynamics.
IAU Colloquium No. 26, (see 012.018), p. 17 - 20 (1975).

046.038 Conceptual definitions of reference coordinate systems for earth dynamics.
IAU Colloquium No. 26, (see 012.018), p. 21 - 26 (1975).

046.039 Reference coordinate system requirements for geodesy and ocean dynamics. R. S. Mather.
IAU Colloquium No. 26, (see 012.018), p. 93 - 114 (1975).

046.040 Gravimetric and astrogeodetic reference systems in space and time. H. Moritz.
IAU Colloquium No. 26, (see 012.018), p. 161 - 179 (1975).

046.041 Coordinates used in range or range-rate systems and their extension to a dynamic earth.
R. R. Newton.
IAU Colloquium No. 26, (see 012.018), p. 181 - 200 (1975).

046.042 Some problems related to geodetic datum definitions for terrestrial measurements. E. Groten.
IAU Colloquium No. 26, (see 012.018), p. 247 - 259 (1975).

046.043 General principles for the realization of reference systems for earth dynamics. G. Veis.
IAU Colloquium No. 26, (see 012.018), p. 261 - 267 (1975).
Review paper.

046.044 Prospects for rapid realization of a quasi-earth-fixed coordinate system. P. L. Bender.
IAU Colloquium No. 26, (see 012.018), p. 297 - 302 (1975).

046.045 Practical realization of a reference system for earth dynamics by satellite methods.
R. J. Anderle, M. C. Tanenbaum.
IAU Colloquium No. 26, (see 012.018), p. 341 - 380 (1975).

046.046 On the interrelation of classical and space geodetic systems of position reference. L. Aardoom.
IAU Colloquium No. 26, (see 012.018), p. 445 - 452 (1975).

046.047 Sul problema della localizzazione geografica di satelliti. G. D'Elia.
Aerotecn. Missili Spazio, Vol. 53, 323 - 335 (1974). — Abstr. in Zentralbl. Math. Grenzgebiete — Math. Abstr., Band 315, No. 70026 (1976).

046.048 Alignment of geodetic and satellite coordinate systems to the average terrestrial system.
D. E. Wells, P. Vanicek.
Bull. Géod., No. 117, p. 241 - 257 (1975). — Abstr. in Phys. Abstr., Vol. 79, A032162 (1976).

046.049 A catalog of station coordinates for GEOS-C orbit determination.
J. G. Marsh, B. C. Douglas, D. M. Walls.
Bull. Géod., No. 117, p. 259 - 266 (1975). — Abstr. in Phys. Abstr., Vol. 79, A032648 (1976).

A laser range tracking station for geodynamic satellites. See Abstr. 009.021.

A demonstration of an independent-station radio interferometry system with 4-cm precision on a 16-km base line. See Abstr. 031.203.

Geodetic and astrometric results of very long baseline interferometric measurements of natural radio sources. See Abstr. 041.008.

New possibilities for solving problems of astrometry, geodynamics and geodesy by methods of radio interferometry with superlong baseline. See Abstr. 041.017.

Nature of the requirements for reference coordinate systems. See Abstr. 041.033.

Reference coordinate system requirements for geophysics. See Abstr. 041.034.

Note on selection of coordinate systems for geodynamics. See Abstr. 041.042.

Systematic differences between global or local solutions of polar motion derived from astronomical and spatial measurements. See Abstr. 045.018.

Long-term orbit determination and prediction for geodynamic investigations. See Abstr. 052.009.

Orientation of the earth by numerical integration. See Abstr. 081.032.

Review of the classical methods for the determination of geodetic datums. See Abstr. 081.043.

047 Ephemerides, Almanacs, Calendars

047.001 **Events of 1976 in the graphic time table.**
Sky Telescope, Vol. 51, 35 - 37 (1976).

047.002 **Astronomical calendar 1976.** G. Ottewell.
Furman University Physics Dept., Greenville. 56 pp.
Price $ 4.95 (1975).

047.003 **Effect of apsidal motion on solar transits and decay**
of lunar months. S. D. Sharma.
Bull. Astron. Soc. India, Vol. 3, 34 (1975). – Abstract of a
paper presented at the A.S.I. meeting 1975.

047.004 **The Air Almanac 1976, September - December.**
Her Majesty's Stationery Office, London; United
States Naval Observatory, Washington, p. 489 - 734, A84 +
F4 pp. Price £ 3.00 (1976).

047.005 **Connaissance des temps ou des mouvements cé-**
lestes pour l'an 1977 à l'usage des astronomes et des
navigateurs.
Publiée par le Bureau des Longitudes under the supervision of
B. Morando.
Gauthier-Villars Editeur, Paris. 42 + 497 ± A 145 pp. (1976).
ISBN 2-04-00-3947-9.

047.006 **Éphémérides Nautiques pour l'an 1977.**
Ouvrage publié par le Bureau des Longitudes
spécialement à l'usage des marins.
Gauthier-Villars Éditeur, Paris. 478 pp. (1976).

047.007 **Japanese Ephemeris 1977.**
Compiled under the supervision of A. M. Sinzi, by
T. Uniwa, T. Mori, A. Senda, Y. Harada, K. Inoue, K. Naga-
mori, Y. Suzuki, T. Jojo.
Astronomical Division, Hydrographic Department, Tokyo,
Japan. Pub. No. 684, 6 + 451 pp. (1975).

047.008 **Grafikoni izlaza i zalaza Sunca i Mjeseca 1976.**
Edited by Hidrografski Institut Jugoslavenske Ratne
Mornarice, Split, HI–N–32, 27 pp. (1975).

047.009 **Astronomical handbook for 1976.**
A. C. Gilmore, P. M. Kilmartin, B. M. Lewis.
Carter Obs., Astron. Bull. No. 85, 40 pp. (1976).

047.010 **Annuario 1976 (bisestile).**
Edited by Osservatorio Astronomico di Torino.
Pubbl. Varie Fuori Ser., Oss. Astron. Torino, (Pino Torinese),
No. 61, 91 pp. (1975).
Cronologia; Coordinate dell'osservatorio; Calendario ed
effemeridi del sole e della luna; I pianeti nel 1976; Eclissi ed
occultazioni; Attività dell'osservatorio (*M. G. Fracastoro*);
Riflessioni sul limite e sulle superfici di Roche nelle applica-
zioni astronomiche ed astrofisiche (*V. Banfi*); L'eclisse anulare
di sole del 29 Aprile 1976 (*R. Pannunzio*); Ricordo della
Scuola astronomica di Padova (*M. Navarria Viaro*); Insolazione
a Pino Torinese (*A. Di Battista, M. G. Fracastoro*).

047.011 **Nautisches Jahrbuch 1977.**
Edited by Seehydrographischer Dienst der Deutschen Demokratischen Republik, Rostock, 27th year. 29 +
365 pp. (1976).

The wanderers in the year of the elder fire dragon
1976.   See Abstr. 003.015.

The telescope handbook and star atlas.
See Abstr. 003.059.

JPL Development Ephemeris No. 96.
See Abstr. 091.017.

Space Research

051 Extraterrestrial Research, Spaceflight Related to Astronomy and Astrophysics

051.001 **Skylab experiment results.** R. Edgar.
Spaceflight, Vol. 18, 59 - 67 (1976).

051.002 **Project Daedalus: astronomical data on nearby stellar systems.** H. R. Mattinson.
Journ. British Interplanet. Soc., Vol. 29, 76 - 93 (1976).

The design of any interstellar mission depends upon accurate information on astronomical data pertinent to the mission objectives and direction. Because of the large differences in published astronomical data this report has been compiled to give a standard set of data to be used in mission planning for the British Interplanetary Society's Project Daedalus study. The report also contains a brief description of astronomical terminology and a discussion of the nearby stellar systems.

051.003 **Project Daedalus: the ranking of nearby stellar systems for exploration.** A. R. Martin.
Journ. British Interplanet. Soc., Vol. 29, 94 - 100 (1976).

A simple analysis is presented which assigns weights to various factors relevant to near-stellar exploration, and which allows the importance of missions to the nearby stars to be assessed relative to each other.

051.004 **Project Daedalus: the mission profile.**
A. Bond, A. R. Martin.
Journ. British Interplanet. Soc., Vol. 29, 101 - 112 (1976).

This paper describes the evolution and final outcome of the mission profile for the Project Daedalus Starship, designed to carry out a one-way undecelerated flyby mission to Barnard's star.

051.005 **An inside look at NASA planetology.**
S. E. Dwornik.
Sky Telescope, Vol. 51, 4 - 8 (1976)

051.006 **Solar-electric rocket propulsion.**
Sky Telescope, Vol. 51, 88 - 90 (1976).

051.007 **Space highlights from around the world.**
Sky Telescope, Vol. 51, 169 - 171 (1976).

051.008 **Helios A nach dem ersten Perihel.** C. Leinert.
Mitt. Astron. Ges., No. 38, p. 59 - 76 (1976). – Invited paper.

051.009 **Über die Zuverlässigkeit der HELIOS-Experimente.**
E. Seiler.
PTB Mitt., 86. Jahrgang, p. 31 - 35 (1976).

A short description of the actual behaviour of the experiments on board of the solar probe HELIOS A is given. Necessary actions to achieve high reliability for experiments designed for space research are described.

051.010 **Entwicklungsetappen der sowjetischen Raumfahrt.**
H. Hoffmann.
Astronomie in der Schule, 13. Jahrgang, p. 3 - 6 (1976).

051.011 **Telescopes above the clouds.** A. A. Chernov.
Zemlya i Vselennaya, 1976, No. 2, p. 52 - 62. In Russian.

051.012 **On the investigation of the sun with automatic sondes.** V. P. Kaznevskij, Yu. M. Fatkin, V. M. Zavadskij. `
Trudy devyatykh chtenij, posvyashch. razrabotke nauch. naslediya i razvitiyu idej K. Eh. Tsiolkovskogo, Kaluga, 1974. Moskva, 1975, p. 55 - 70. In Russian. – Abstr. in Referativ. Zhurn. 62. Issled. kosmich. prostranstva, 4.62.72 (1976).

051.013 **Space astrophysical observatory 'Orion-2'.**
G. A. Gurzadyan, A. L. Jarakyan, M. N. Krmoyan, A. L. Kashin, G. M. Loretsyan, J. B. Ohanesyan (*Dzh. B. Oganesyan*).
Astrophys. Space Sci., Vol. 40, 393 - 446 (1976).

Ultraviolet spectrograms of a large number of faint stars up to 13^m were obtained in the wavelengths 2000–5000 Å by means of the space observatory 'Orion-2' installed in the spaceship 'Soyuz-13' with two spacemen on board. The paper deals with a description of the operation modes of this observatory, the designs and basic schemes of the scientific and auxiliary device and the method of combining the work of the flight engineer and the automation system of the observatory itself.

051.014 **Investigations of radio wave propagation in the solar system.** M. A. Kolosov, O. I. Yakovlev.
Rasprostr. radiovoln. Moskva, Nauka, 1975, p. 291 - 311. In Russian. – Abstr. in Referativ. Zhurn. 51. Astron., 5.51.69 (1976).

051.015 **Space probes versus comets.** D. W. Hughes.
Nature, Vol. 261, 453 - 454 (1976).

051.016 **Current trends in astronautics.** M. Grün.
Vesmír, Vol. 55, 137 - 143 (1976). In Czech.

051.017 **Advance in astronautics.** I. Hudec, R. Hudec.
Kozmos, Vol. 7, 2 - 4 (1976). In Czech.

051.018 **Astronautics in the year 1975.**
M. Grün, P. Koubský.
Říše hvězd, Vol. 57, 105 - 110 (1976). In Czech.

051.019 **The Vikings are coming.** C. R. Spitzer.
IEEE Spectrum, Vol. 13, No. 6, p. 48 - 53 (1976).

051.020 **Raumfähre, Planeten-Sonden und Anwendungs-Satelliten.** H. W. Köhler.
Umschau, 76. Jahrgang, p. 433 - 438 (1976).

051.021 **Sterrekundig onderzoek in de ruimte.**
C. de Jager.
Repr. from Jaarboek 1974 Koninkl. Nederlandse Akad. Wet., 10 pp. = Utrechtse Sterrekundige Overdrukken No. 313 (1975).

051.022 **The French neutron and gamma ray detector Signe I, launched aboard the Soviet satellite Prognoz II.**
F, Albernhe, M. Cassignol, F. Cotin, C. Doulade, R. Talon, G. Vedrenne.
Nuclear Instruments Methods, Vol. 129, 295 - 301 (1975).
Abstr. in Phys. Abstr., Vol. 79, A019706 (1976).

051.023 **Shuttle-launched multi-comet mission 1985.**
R. W. Farquhar, D. P. Muhonen, F. I. Mann, W. H. Wooden, D. K. Yeomans.
1975 AAS/AIAA Astrodyn. Conf., (see 012.022), AAS75-085, 23 pp. (1975). − Abstr. in Phys. Abstr., Vol. 79, A023599 (1976).

051.024 **Planetary exploration: earth's new horizon.**
H. M. Schurmeier.
Journ. Spacecraft Rockets, Vol. 12, 385 - 405 (1975). − Abstr. in Phys. Abstr., Vol. 79, A033142 (1976).

051.025 **Magellan: a balloon-borne collection technique for large cosmic dust particles.** R. Wlochowicz, C. L. Hemenway, D. S. Hallgren, C. D. Tackett.

Canadian Journ. Phys., Vol. 54, 317 - 321 (1976). − Abstr. in Phys. Abstr., Vol. 79, A042441 (1976).

051.026 **Research in space physics at the University of Iowa.**
J. A. Van Allen.
Separate print Iowa Univ., Iowa City, USA, Dept. Phys. Astron., 35 pp. (1974).

051.027 **Space report.**
Journ. British Interplanet. Soc., Vol. 29, 211 - 214, 286 - 287 (1976).

051.028 **Space report.**
Spaceflight, Vol. 18, 19 - 22, 40, 53 - 55, 92 - 98, 136 - 138, 176 - 178, 196, 214 - 222 (1976).

The satellite spin-off. The achievements of space flight. See Abstr. 003.089.

Sonnenforschung mit Stratosphärenballons.
See Abstr. 071.025.

052 Astrodynamics and Navigation of Space Vehicles

052.001 **An exact and a new first-order solution for the relative trajectories of a probe ejected from a space station.** T. F. Berreen, J. D. C. Crisp.
Celestial Mechanics, Vol. 13, 75 - 88 (1976).
The motion of an unpowered probe, ejected into an elliptic orbit in the orbital plane of a space station in circular orbit, is investigated from a reference system attached to the space station. A new first-order solution is derived from an exact solution: no restriction is placed on the relative displacement and the solution osculates with the exact solution at points corresponding to perigee and apogee of the probe's orbit.

052.002 **Short-period and long-period perturbations of a spherical satellite due to direct solar radiation.**
K. Aksnes.
Celestial Mechanics, Vol. 13, 89 - 104 (1976).
On the basis of expressions derived by Kozai, and new ones developed here, a detailed, semianalytic algorithm is presented for calculating radiation-pressure perturbations in the Keplerian elements. The algorithm is tested by means of numerical integration of the equations of motion and through comparisons with observations of the balloon satellite 1963 30D during a 200-day interval.

052.003 **Some problems of optimization of the process of**

trajectory measurements. N. N. Kozlov.
Kosmich. Issled., Vol. 14, 5 - 16 (1976). In Russian.

052.004 **Investigation of the problem of estimating the parameters of motion of space objects in the presence of a certain kind of errors of trajectory measurements.**
P. E. Ehl'yasberg, R. R. Nazirov.
Kosmich. Issled., Vol. 14, 17 - 22 (1976). In Russian.

052.005 **On the solution of the generalized problem of two fixed centres by Weierstrassian functions.**
P. Andrle.
Bull. Astron. Inst. Czechoslovakia, Vol. 27, 118 - 125 (1976).
Similarly to Aksenov et al. (1963) the generalized problem of two fixed centres is studied. Weierstrassian elliptic functions (in a generalized sense) are used instead of Jacobian functions. All results, which could be applied in the theory of artificial earth satellites, are expressed uniformly (by one general equation) for each coordinate.

052.006 **Statistical estimate of experimental data on deceleration of artificial earth satellites.**
E. I. Bushuev, A. I. Vasil'eva, E. Ya. Egortsev, A. A. Krasovskij, V. Ya. Mashtak.
Opredelenie dvizheniya kosmich. apparatov. Moskva, Nauka, 1975, p. 182 - 198. In Russian. − Abstr. in Referativ. Zhurn.

62. Issled. kosmich. prostranstva, 3.62.197 (1976).

052.007 **A class of periodic translatory-rotational motions of a satellite.** E. B. Bibik.
Dnepropetr. un-t. Dnepropetrovsk, 1975. 21 pp. In Russian. Abstr. in Referativ. Zhurn. 62. Issled. kosmich. prostranstva, 3.62.247 (1976).

052.008 **Some stationary motions of a satellite with gyroscope.** A. Kh. Akhmetshin.
Trudy Kazan. aviats. in-ta, 1975, vyp. (No.) 184, p. 9 - 14. In Russian. − Abstr. in Referativ. Zhurn. 62. Issled. kosmich. prostranstva, 3.62.248 (1976).

052.009 **Long-term orbit determination and prediction for geodynamic investigations.** B. E. Douglas.
Long-time predictions in dynamics, (see 012.005), p. 321 - 326 (1976).

052.010 **Time regularization and stabilization of an Adams-Moulton-Cowell algorithm.** N. Borderies.
Long-time predictions in dynamics, (see 012.005), p. 331 - 332 (1976).− Abstract.

052.011 **On the theory of an artificial satellite.** H. Claes.
Long-time predictions in dynamics, (see 012.005), p. 333 (1976). − Abstract.

052.012 **Kozai's theory of an artificial satellite.** J. Henrard.
Long-time predictions in dynamics, (see 012.005), p. 337 (1976). − Abstract.

052.013 **Secular and long-period perturbations in the motion of artificial satellites.**
E. P. Aksenov, I. P. Prokhorova.
Dynamical questions of satellite geodesy, 1973, (see 012.006), p. 18 - 41 (1974). In Russian.

052.014 **Short-period perturbations due to zonal harmonics in the motion of artificial satellites.** N. A. Sorokin.
Dynamical questions of satellite geodesy, 1973, (see 012.006), p. 42 - 50 (1974). In Russian.

052.015 **The effect of non-gravitational forces on the motion of artificial satellites.** L. Sehnal.
Dynamical questions of satellite geodesy, 1973, (see 012.006), p. 91 - 109 (1974). In Russian.

052.016 **Secular and short-period perturbations in the elements of an intermediate orbit.** B. N. Noskov.
Dynamical questions of satellite geodesy, 1973, (see 012.006), p. 110 - 123 (1974). In Russian.

052.017 **The effect of radiation pressure on the motion of artificial satellites.** S. N. Vashkov'yak.
Dynamical questions of satellite geodesy, 1973, (see 012.006), p. 124 - 143 (1974). In Russian.

052.018 **Calculation of air-drag perturbations in orbital elements of artificial satellites on computers.**
A. M. Fominov.
Dynamical questions of satellite geodesy, 1973, (see 012.006), p. 144 - 164 (1974). In Russian.

052.019 **Method of improving orbital elements of artificial satellites applied in different socialist countries.**
W. Dobaczewska, J. Krinsky.
Dynamical questions of satellite geodesy, 1973, (see 012.006), p. 166 - 177 (1974). In Russian.

052.020 **A computer-based programme for the improvement of orbital elements of artificial satellites.**
J. Krinsky.
Dynamical questions of satellite geodesy, 1973, (see 012.006), p. 178 - 182 (1974). In Russian.

052.021 **Étude analytique de l'influence du frottement atmosphérique sur le mouvement des satellites artificiels.**
R. Fogliani.
Bull. Groupe Recherches Géod. Spatiale, No. 15, p. 47 - 57 (1976).
The problem of the motion of an artificial satellite undergoing atmospheric drag is solved analytically in this paper, using the gravitational model of two fixed centers. The equations of motion are written and, after some approximations, a method can be applied for the generally adopted form of density models of the higher atmosphere.

052.022 **On the accuracy of the semi-analytical method of calculation of the motion of a space vehicle in the vicinity of a lunar libration point.**
M. L. Lidov, M. A. Vashkov'yak, A. P. Markeev.
In-t prikl. mat. AN SSSR. Preprint No. 85, Moskva, 1975. 48 pp. Price 18 Kop. In Russian. − Abstr. in Referativ. Zhurn. 62. Issled. kosmich. prostranstva, 4.62.301 (1976).

052.023 **Variational orbit of equatorial and nearly equatorial artificial earth satellites.** N. B. Batueva.
Probl. mekh. upravlyaemogo dvizheniya. Vyp. (No.) 7. Perm', 1975, p. 8 - 13. In Russian. − Abstr. in Referativ. Zhurn. 62. Issled. kosmich. prostranstva, 4.62.310 (1976).

052.024 **On a method of deceleration of motion of a space vehicle around the centre of mass.**
N. N. Povoraznikov.
Kosmich. Issled., Vol. 14, 175 - 183 (1976). In Russian.

052.025 **Analytical estimates of the accuracy of determining the orbital parameters of artificial earth satellites from measurements of the relative motion satellite−probe.**
V. I. Kuznetsov, L. F. Porfir'ev.
Kosmich Issled., Vol. 14, 184 - 190 (1976). In Russian.

052.026 **On a method of numerical integration of differential equations.** M. Yu. Belyaev, V. P. Semenko.
Kosmich. Issled., Vol. 14, 300 - 301 (1976). In Russian. Brief information.

052.027 **On the stability of a rotating space station containing a movable element.** G. R. Salimov.
Izv. AN SSSR. Mekh. tverd. tela, 1975, No. 5, p. 52 - 56. In Russian. − Abstr. in Referativ. Zhurn. 62. Issled. kosmich. prostranstva, 5.62.333 (1976).

052.028 **Optimal damping of the nutating motion of spin-stabilized satellites.**
V. A. Sarychev, V. V. Sazonov.
Celestial Mechanics, Vol. 13, 383 - 405 (1976). In Russian.
This article investigates the dynamics of a system for damping the nutating motion of a spin-stabilized satellite. The equations of motion of the satellite-damper system are derived omitting consideration of the influence of external torques. The conditions of stability of the stationary spinning are obtained and the optimal parameters of the satellite and the damper ensuring a maximal rate of damping the nutation motion are determined.

052.029 **The influence of the earth's shadow on the motion of a space vehicle with a solar sail.**
L. K: Grinevitskaya, E. N. Polyakhova.
Vestn. Leningr. un-ta, 1976, No. 1, p. 133 - 141. In Russian.

Abstr. in Referativ. Zhurn. 62. Issled. kosmich. prostranstva, 6.62.282 (1976).

052.030　The influence of correlation of measurement errors on the optimum program of trajectory measurements.　V. B. Britkov.
Kosmich. Issled., Vol. 14, 330 - 335 (1976). In Russian.

052.031　On the tesseral-harmonics resonance in artificial-satellite theory. Part II.　B. A. Romanowicz.
Smithsonian Astrophys. Obs., *Cambridge, Mass.*, Special Rep. 373, 5 + 23 + A4 + B3 (1976).
　　The author introduces some modifications to the perturbations on an artificial satellite, which now result in better agreement with numerical integration.

052.032　Hybrid systems for use in the dynamics of artificial satellites.　Y. Kozai.
IAU Colloquium No. 26, (see 012.018), p. 235 - 240 (1975).

052.033　Die Gleichungen der relativen orbitalen Näherungs-und Rendezvousbewegung unter Berücksichtigung der Asphärizität des Planeten.
A. Marinescu, S. Staicu, F. Atanasiu.
Bul. Inst. politehn. Bucureşti, Vol. 35, Nr. 4, p. 69 - 86 (1973). In Rumanian.

052.034　Thermally induced nutational body motion of a spinning spacecraft with flexible appendages.
K. Tsuchiya.
AIAA Journ., Vol. 13, 448 - 453 (1975). – Abstr. in Zentralbl. Math. Grenzgebiete – Math. Abstr., Band 313, No. 70033 (1976).

052.035　Navigation and guidance in interstellar space.　D. G. Hoag, W. Wrigley.
Acta Astronaut., Vol. 2, 513 - 533 (1975). – Abstr. in Phys. Abstr., Vol. 79, A007082 (1976).

052.036　Determination of elements of sub-synchronous or super-synchronous orbits and their application to operational satellites.　K. Takahashi.
Journ. Radio Res. Lab., Vol. 22, 45 - 57 (1975). – Abstr. in Phys. Abstr., Vol. 79, A015147 (1976).

052.037　The motion of a satellite in resonance with the longitude-dependent harmonics.　S. S. Dallas.
1975 AAS/AIAA Astrodyn. Conf., (see 012.022), AAS75-022, 49 pp. (1975). – Abstr. in Phys. Abstr., Vol. 79, A023591 (1976).

052.038　Lunar and solar perturbations on the orbit of a geo-synchronous satellite.　O. F. Graf, Jr.
1975 AAS/AIAA Astrodyn. Conf., (see 012.022), AAS75-023, 28 pp. (1975). – Abstr. in Phys. Abstr., Vol. 79, A023592 (1976).

052.039　Jupiter orbiter lifetime. The hazard of Galilean satellite collision.　A. L. Friedlander.
1975 AAS/AIAA Astrodyn. Conf., (see 012.022), AAS75-038, 25 pp. (1975). – Abstr. in Phys. Abstr., Vol. 79, A023594 (1976).

052.040　Station-keeping requirements.　J. B. Eades, Jr.
　　1975 AAS/AIAA Astrodyn. Conf., (see 012.022), AAS75-039, 36 pp. (1975). – Abstr. in Phys. Abstr., Vol. 79, A023595 (1976).

052.041　Satellite disintegration dynamics.
　　R. R. Dasenbrock, B. Kaufman, W. B. Heard.
1975 AAS/AIAA Astrodyn. Conf., (see 012.022), AAS75-040,

11 pp. (1975). – Abstr. in Phys. Abstr., Vol. 79, A023596 (1976).

052.042　Relativity mission with two counter-orbiting polar satellites.　R. A. Van Patten, C. W. F. Everitt.
1975 AAS/AIAA Astrodyn. Conf., (see 012.022), AAS75-041, 34 pp. (1975). – Abstr. in Phys. Abstr., Vol. 79, A023597 (1976).

052.043　Collision avoidance for two counter-orbiting polar satellites.　D. Schaechter, J. V. Breakwell.
1975 AAS/AIAA Astrodyn. Conf., (see 012.022), AAS75-042, 15 pp. (1975). – Abstr. in Phys. Abstr., Vol. 79, A023598 (1976). – See Abstr. 052.061.

052.044　Multi-asteroid flyby trajectories using Venus–earth gravity assists.　D. F. Bender, A. L. Friedlander.
1975 AAS/AIAA Astrodyn. Conf., (see 012.022), AAS75-086, 34 pp. (1975). – Abstr. in Phys. Abstr., Vol. 79, A023600 (1976).

052.045　New flight techniques for outer planet missions.　G. R. Hollenbeck.
1975 AAS/AIAA Astrodyn. Conf., (see 012.022), AAS75-087, 25 pp. (1975). – Abstr. in Phys. Abstr., Vol. 79, A023601 (1976).

052.046　The relativistic dynamics of a sub-light speed interstellar ramjet probe.　W. B. Roberts.
1975 AAS/AIAA Astrodyn. Conf., (see 012.022), AAS75-020, 34 pp. (1975). – Abstr. in Phys. Abstr., Vol. 79, A023915 (1976).

052.047　Preliminary Mariner Jupiter-Saturn 1977 launch/encounter strategy.　R. A. Wallace.
1975 AAS/AIAA Astrodyn. Conf., (see 012.022), AAS75-071, 29 pp. (1975). – Abstr. in Phys. Abstr., Vol. 79, A023917 (1976).

052.048　The powered swingby and its application to Jupiter orbit mission design.　C. Uphoff.
1975 AAS/AIAA Astrodyn. Conf., (see 012.022), AAS75-084, 45 pp. (1975). – Abstr. in Phys. Abstr., Vol. 79, A023919 (1976).

052.049　The application of general perturbation theories to the artificial satellite problem.
K. T. Alfriend, C. E. Velez.
Acta Astronaut., Vol. 2, 577 - 591 (1975). – Abstr. in Phys. Abstr., Vol. 79, A027693 (1976).

052.050　Optimal rendezvous in the neighborhood of a circular orbit.　J. B. Jones.
1975 AAS/AIAA Astrodyn. Conf., (see 012.022), AAS75-061, 51 pp. (1975). – Abstr. in Phys. Abstr., Vol. 79, A027694 (1976).

052.051　A method for handling coast arcs in low-thrust interplanetary trajectory optimization.
P. M. O'Neill, W. T. Fowler.
1975 AAS/AIAA Astrodyn. Conf., (see 012.022), AAS75-062, 21 pp. (1975). – Abstr. in Phys. Abstr., Vol. 79, A028065 (1976).

052.052　The navigation of a satellite re-encounter in a Jupiter orbiter mission.
G. N. Sherman, J. B. Jones.
1975 AAS/AIAA Astrodyn. Conf., (see 012.022), AAS75-072, 23 pp. (1975). – Abstr. in Phys. Abstr., Vol. 79, A023918 (1976).

052.053 **Model for solar torque effects on DSCS II.**
T. E. Suttles, R. E. Beverly.
1975 AAS/AIAA Astrodyn. Conf., (see 012.022), AAS75-095, 24 pp. (1975). − Abstr. in Phys. Abstr., Vol. 79, A023920 (1976).

052.054 **Optimum three-dimensional atmospheric entry.**
N. X. Vinh, A. Busemann, R. D. Culp.
Acta Astronaut., Vol. 2, 593 - 611 (1975). − Abstr. in Phys. Abstr., Vol. 79, A028049 (1976).

052.055 **A linear feedback guidance law for an aeromaneuvering orbit-to-orbit shuttle.** J. J. Rehder.
1975 AAS/AIAA Astrodyn. Conf., (see 012.022), AAS75-065, 31 pp. (1975). − Abstr. in Phys. Abstr., Vol. 79, A028066 (1976).

052.056 **Set of orbits having variable radius vectors of start.**
K. Mison.
Acta Techn. CSAV, Vol. 20, 634 - 644 (1975). − Abstr. in Phys. Abstr., Vol. 79, A032646 (1976).

052.057 **A study of orbit determination accuracies for future earth observatory missions.**
W. C. Bryant, Jr., C. C. Goad.
1975 AAS/AIAA Astrodyn. Conf., (see 012.022), AAS75-049, 12 pp. (1975). − Abstr. in Phys. Abstr., Vol. 79, A032661 (1976).

052.058 **Fine pointing control for a large spinning spacecraft in earth orbit.** H. Salzwedel, J. V. Breakwell.
Journ. Spacecraft Rockets, Vol. 12, 414 - 419 (1975). − Abstr. in Phys. Abstr., Vol. 79, A033143 (1976).

052.059 **Determination of elements of sub-synchronous or super-synchronous orbits and their application to operational satellites.** K. Takahashi.
Rev. Radio Res. Lab., Vol. 21, 19 - 26 (1975). − Abstr. in Phys. Abstr., Vol. 79, A037408 (1976).

052.060 **Perturbations of satellite orbits due to $\gamma(x^3 + y^3)$.**
K. M. Khanna, P. K. Phillips.
Indian Journ. Radio Space Phys., Vol. 4, 187 - 189 (1975). Abstr. in Phys. Abstr., Vol. 79, A042000 (1976).

052.061 **Collision avoidance for two counter-orbiting polar satellites.** D. B. Schaechter, J. V. Breakwell, R. A. Van Patten, C. W. F. Everitt.
66th annual conf. internat. district heating assoc., Skytop, Pennsylvania, USA, 23 June 1975, 17 pp. (1975).

The satellite spin-off. The achievements of space flight. See Abstr. 003.089.

Multistep methods of numerical integration using back-corrections. See Abstr. 021.002.

Pocket astrodynamics. See Abstr. 021.003.

Perturbation formulations for satellite attitude dynamics. See Abstr. 042.003.

Mouvement d'un satellite à rotation lente dans le champ magnétique terrestre. See Abstr. 042.043.

Application of the Jacobi integral for the investigation of satellite librations around the centre of mass in a gravitational field. See Abstr. 042.057.

Approximate solution of the Euler-Lambert problem for nearly circular orbits. See Abstr. 042.068.

Introduction to the restricted Jupiter orbiter problem. See Abstr. 042.092.

Investigation of the geopotential and of gravitational perturbations in the motion of an artificial satellite. See Abstr. 081.008.

Investigation of the density of the upper atmosphere and aerodynamics of satellites from orbital evolution data. See Abstr. 082.026.

053 Lunar and Planetary Probes and Satellites

053.001 **The Viking landing sites.**   R. Edgar.
Spaceflight, Vol. 18, 30 - 31 (1976).

053.002 **Why explore?**
G. A. Soffen, R. S. Young, T. Owen.
Spaceflight, Vol. 18, 78 - 80 (1976).
    Mars awaits two robot visitors now winging their way around the sun — the Vikings. They seek answers to basic questions regarding the origins of the solar system, of earth — and of life.

053.003 **Viking: the orbiter.**   D. Baker.
Spaceflight, Vol. 18, 84 - 86 (1976).

053.004 **Viking: Orbiter science equipment.**   D. Baker.
Spaceflight, Vol. 18, 124 - 125 (1976).

053.005 **Orbits of Soviet deep space probes — 1, 2.**
P. S. Clark.
Spaceflight, Vol. 18, 139 - 141, 188 - 189 (1976).

053.006 **In-flight utilization of the Mariner 10 spacecraft computer.**   A. J. Hooke.
Journ. British Interplanet. Soc., Vol. 29, 273 - 285 (1976).

053.007 **Helios: Die ersten bursts auf der Sonne.**
E. Hintsches.
Umschau, 76. Jahrgang, p. 249 - 251 (1976).
    The space probes Helios A and B have reported the first bursts from the sun. Some most outstanding results of the experiments installed are just dropping on the earth.

053.008 **Venus na de landingen van Venera 9 en 10.**
P. Smolders.
Zenit, 3e jaargang, p. 83 - 85 (1976).

053.009 **The voyages of Viking — 3. Viking: the lander.**
D. Baker.
Spaceflight, Vol. 18, 158 - 161 (1976).

053.010 **Projekt Viking.**   E. Moser.
Orion, 34. Jahrgang, p. 27 - 31 (1976).

053.011 **Viking: lander science equipment - 1.**
D. Baker.
Spaceflight, Vol. 18, 211 - 213 (1976).

053.012 **Surveyor on the moon. 6.**   D. Baker.
Spaceflight, Vol. 18, 228 (1976).

053.013 **Pioneer 11 op juiste koers naar Saturnus.**
G. W. E. Beekman.
Zenit, 3e jaargang, p. 174 (1976).

053.014 **Vikings started to Mars.**   S. A. Nikitin.
Priroda, 1976, No. 6, p. 140 - 143. In Russian.

053.015 **Vikings are flying to Mars.**   D. Yu. Gol'dovskij.
Zemlya i Vselennaya, 1976, No. 3, p. 16 - 20. In Russian.

053.016 **Doppellandung auf dem Mars.**   H. W. Köhler.
SuW, 15. Jahrgang, 184 - 186 (1976).

053.017 **Navigation between the planets.**   W. G. Melbourne.
Sci. American, Vol. 234, No. 6, p. 58 - 64, 68 - 74 (1976).

053.018 **Mariner mission to Encke 1980.**
C. M. Yeates, K. T. Nock, R. L. Newburn.
IAU Colloquium No. 31, (see 012.015), p. 346 - 355 (1976).

053.019 **Viking and Mars: a prelude.**   J. K. Beatty.
Sky Telescope, Vol. 52, 4 - 9 (1976).

053.020 **Mission analysis and operations of the US-German solar-probe Helios.**   H. J. Panitz.
1975 AAS/AIAA Astrodyn. Conf., (see 012.022), AAS75-101, 37 pp. (1975). — Abstr. in Phys. Abstr., Vol. 79, A023921 (1976).

053.021 **Effect of the Jovian oblateness on Pioneer 10/11 radio occultations.**
W. B. Hubbard, D. M. Hunten, A. Kliore.
Geophys. Res. Letters, Vol. 2, 265 - 269 (1975). — Abstr. in Phys. Abstr., Vol. 79, A027796 (1976).

053.022 **Mariner 10 mission to Venus and Mercury.**
W. E. Giberson, N. W. Cunningham.
Acta Astronaut., Vol. 2, 715 - 743 (1975). — Abstr. in Phys. Abstr., Vol. 79, A028052 (1976).

053.023 **Viking.**
Spaceworld, Vol. L-8-140, 4 - 19 (1975). — Abstr. in Phys. Abstr., Vol. 79, A028063 (1976).

053.024 **Trajectory, atmosphere, and wind reconstruction from Viking entry measurements.**   F. W. Hopper.
1975 AAS/AIAA Astrodyn. Conf., (see 012.022), AAS75-068, 39 pp. (1975). — Abstr. in Phys. Abstr., Vol. 79, A028067 (1976).

053.025 **Autonomous rendezvous and docking in Mars orbit.**
F. Vandenberg.
1975 AAS/AIAA Astrodyn. Conf., (see 012.022), AAS75-070, 12 pp. (1975). — Abstr. in Phys. Abstr., Vol. 79, A023916 (1976).

053.026 **Orbit determination for Mariner 9 using radio and optical data.**   G. H. Born, S. N. Mohan.
Journ. Spacecraft Rockets, Vol. 12, 439 - 441 (1975). — Abstr. in Phys. Abstr., Vol. 79, A032658 (1976).

053.027 **Space shuttle landing system.**
Spaceworld, Vol. L-9-141, 9 - 15 (1975). — Abstr. in Phys. Abstr., Vol. 79, A033150 (1976).

053.028 **Pioneer spacecraft reliability and performance.**
W. J. Dixon.
Acta Astronaut., Vol. 2, 801 - 817 (1975). — Abstr. in Phys. Abstr., Vol. 79, A037671 (1976).

053.029 **Mariner Venus/Mercury 1973 solar radiation force and torques.**   R. M. Georgevic.
Separate print Jet Propulsion Lab., Pasadena, California, USA, 162 pp. (1974).

053.030 **Directional discontinuities from Mariner 5 and solar wind structure.**   J. M. Turner.
Separate print National Aeronaut. Space Administr., Greenbelt, Maryland, USA, Goddard Space Flight Center, 36 pp. (1974).

Status of availability of Mariner 10 (1973-085A) TV picture data.   See Abstr. 002.004.

Apollo expeditions to the moon.
See Abstr. 003.034.

Helios A nach dem ersten Perihel.
See Abstr. 051.008.

Preliminary Mariner Jupiter-Saturn 1977 launch/ encounter strategy.   See Abstr. 052.047.

The powered swingby and its application to Jupiter orbit mission design.   See Abstr. 052.048.

The challenge of Viking: is there life on Mars?
See Abstr. 097.048.

The journey to Jupiter.
See Abstr. 099.005.

054   Artificial Earth Satellites

054.001   Analysis of the orbit of 1970-114F in its last 20 days.   D. G. King-Hele.
Planet. Space Sci., Vol. 24, 1 - 16 (1976).

054.002   On the spin motion of the rocket 1970−97 B.
G. Chiş, V. Mioc.
Astron. Nachr., Vol. 297, 61 - 62 (1976).
   The spin period of Cosmos 378 rocket was studied during all its lifetime. An exponential increase of the spin period values is pointed out.

054.003   ANS on-board software: functional integration as a key to unattended spacecraft operation.
V. H. C. M. Evers.
Journ. British Interplanet. Soc., Vol. 29, 255 - 272 (1976).

054.004   Identification of a navigation satellite system within the Cosmos programme.   G. E. Perry, C. D. Wood.
Journ. British Interplanet. Soc., Vol. 29, 307 - 316 (1976).
   Analysis of orbital plane spacing shows that Cosmos satellites in near-circular orbits of 105 minute period provide the global coverage needed for a navigation system.

054.005   On the possibilities of photographic photometry of artificial earth satellites.
M. V. Bratijchuk, Ya. M. Motrunich.
Tsirk. Astron. Obs. L'vov, No. 49, p. 34 - 36 (1974). In Russian.

054.006   Results of an electrophotometry of some artificial earth satellites.
M. V. Bratijchuk, Ya. M. Motrunich, T. I. Laslo.
Astrometriya i Astrofizika, *Kiev*, vyp. (No.) 29, (see 003.003), p. 107 - 120 (1976). In Russian.
   The paper deals with the results of electrophotometry of 24 satellites carried out from 20. 4. 1971 to 27. 3. 1974. The code numbers of the satellites, the time during which observa-

tions were made, the period of brightness variation are given in tables.

054.007   Aryabhata—eight months of life in orbit.
U. R. Rao, K. Kasturirangan.
Bull. Astron. Soc. India, Vol. 3, 75 - 78 (1975).
   'Aryabhata', India's first artificial earth satellite has completed eight months of satisfactory performance in orbit, thus fulfilling most of the planned mission goals.

054.008   The Ariel 5 post launch control and data handling system.   A. J. Rogers.
Journ. British Interplanet. Soc., Vol. 29, 408 - 415 (1976).

054.009   Operational experience with the ANS on-board computer.   V. H. C. M. Evers.
Journ. British Interplanet. Soc., Vol. 29, 417 - 427 (1976).

054.010   The computer system for Spacelab.
G. R. Bolton.
Journ. British Interplanet. Soc., Vol. 29, 429 - 438 (1976).

054.011   Prognoz 4.   S. A. Nikitin.
Priroda, 1976, No. 6, p. 144 - 145. In Russian.

054.012   Breakup of Pageos and other satellites.
R. D. Eberst.
Journ. British Astron. Ass., Vol. 86, 274 - 276 (1976).

054.013   The mission of the German aeronomy satellite Aeros B goals and achievements.   H. Wanke.
1975 AAS/AIAA Astrodyn. Conf., (see 012.022), AAS75-097, 21 pp. (1975). − Abstr. in Phys. Abstr., Vol. 79, A023559 (1976).

054.014   Explorer 2 − X-ray astronomy satellite.
Spaceworld, Vol. L-9-141, 4 - 8 (1975). − Abstr. in

Phys. Abstr., Vol. 79, A033149 (1976).

054.015 **Satellite digest.**
Compiled by R. D. Christy.
Spaceflight, Vol. 18, 32 - 33, 72, 110 - 111, 152, 156, 192, 229 - 230 (1976).

Guidance law for the Ariane on-board computer.
See Abstr. 032.515.

Analyses of the solid earth and ocean tidal perturbations on the orbits of the Geos 1 and Geos 2 satellites.
See Abstr. 081.030.

Photometry of eclipses of artificial earth satellites observed in situ. (Part I).   See Abstr. 082.004.

055   Observations of Earth Satellites, Lunar and Planetary Probes

055.001 **Geocentric coordinates of artificial earth satellites calculated from visual observations. Vyp. (No.) 2.**
Satellite 1963-43-1, June - August 1969.
V. V. Kondrashin, I. E. Kupriyanova, I. A. Kuznetsov, V. A. Smirnov.
Nablyud. Iskusstv. Nebesn. Tel, No. 67, 97 pp. (1975). In Russian.

055.002 **Geocentric coordinates of artificial earth satellites calculated from visual observations. Vyp. (No.) 4.**

Satellite 65-11-4, February - July 1967.
V. V. Kondrashin, I. E. Kupriyanova, I. A. Kuznetsov, V. A. Smirnov.
Nablyud. Iskusstv. Nebesn. Tel., No. 68, 66 pp. (1975). In Russian.

055.003 **Representation and interpolation of predicted satellite ephemerides.**   J. D. Vedder.
Journ. Spacecraft Rockets, Vol. 12, 499 - 501 (1975).
Abstr. in Phys. Abstr., Vol. 79, A042001 (1976).

Theoretical Astrophysics

061 General Theoretical Problems of Astrophysics, Gravitational Instability, Neutrino Astronomy, Infrared, X-Ray, Gamma-Ray Astronomy, Abundances and Origin of Elements

061.001 Possibility of synthesis of proton-rich nuclei in highly evolved stars. II. M. Arnould.
Astron. Astrophys., Vol. 46, 117 - 125 (1976).

This paper reports on exploratory calculations performed in order to examine the possibility of production of p-elements in conditions which might be typical of certain hydrostatic oxygen burning phases. Some results derived on grounds of the assumption that temperature, density and neutron concentration are constant in time seem to present rather interesting features, at least if some enhancement of the seed r and/or s-nuclei is assumed with respect to the solar abundances. The problem of the transport of the p-nuclei from the zones of production into the interstellar medium at the right time and without any further significant nuclear processing constitutes one of the major difficulties of the model described in this paper.

061.002 Cosmic radiation and fundamental problems in physics. W. Heisenberg.
Naturwissenschaften, 63. Jahrgang, 63 - 67 (1976). — Lecture given at the 14th International Cosmic Ray Conference, August 18, 1975, Munich. — Review article.

061.003 Enrichment of heavy elements in the Galaxy: a simple formula. J. Dorschner, J. Gürtler.
Astron. Nachr., Vol. 297, 25 - 27 = Mitt. Univ. Sternw. Jena No. 125 (1976).

A simple algebraic formula describing the enrichment of heavy elements in the interstellar gas is derived. Already very simple assumptions lead to the known result that the amount of heavy elements depends critically on the shape of the luminosity function.

061.004 On the limit polarization of a wave passing a weakly anisotropic layer with random irregularities.
L. A. Apresyan.
Astron. Zhurn. Akad. Nauk SSSR, Vol. 53, 53 - 62 (1976). In Russian. English translation in Soviet Astron., Vol. 20, No. 1.

The problem of limit polarization of a wave passing a statistically uniform anisotropic thick layer with random irregularities is considered. The dependence of polarization on the parameters of the medium is investigated.

061.005 The integral polarization of hydrogen spectral lines in a strong magnetic field.
G. G. Pavlov, Yu. A. Shibanov.
Astron. Zhurn. Akad. Nauk SSSR, Vol. 53, 63 - 73 (1976). In Russian. English translation in Soviet Astron., Vol. 20, No. 1.

The circular and linear integral polarizations of hydrogen lines in a strong magnetic field are calculated. The integral circular polarization is caused by different populations of Zeeman sublevels and by different oscillator forces for σ_V and σ_R components of Zeeman features. The integral linear polarization is mainly produced by different quadratic Zeeman shifts of π and σ components. The effects under consideration might be responsible for the polarization observed from magnetic

white dwarfs.

061.006 Importance of isotopic composition of iron in cosmic rays. S. E. Woosley.
Astrophys. Space Sci., Vol. 39, 103 - 121 (1976).

Measurement of the isotopic composition of iron in primary cosmic rays will yield valuable information about the site (or sites) of its production. Enhancements of ^{54}Fe or ^{58}Fe relative to ^{56}Fe would be indicative of acceleration of material from regions typically more neutron-rich than those responsible for solar abundances. Slight enhancements of ^{54}Fe could be explained by taking as a cosmic-ray source supernovae of a different typical mass or higher initial metallicity than those responsible for solar nucleosynthesis. Dominance of ^{54}Fe or ^{58}Fe would be indicative of very neutron-rich matter such as can only occur in the deep interior of a highly evolved star and would be strong evidence for the acceleration of relatively unmixed material from deep inside a supernova.

061.007 Local gamma ray events as tests of the antimatter theory of gamma ray bursts.
S. Sofia, R. E. Wilson.
Astrophys. Space Sci., Vol. 39, L7 - L11 (1976).

Nearby examples of the antimatter 'chunks' postulated by Sofia and Van Horn to explain the cosmic gamma ray bursts may produce detectable gamma ray events when struck by solar system meteoroids. These events would have a much shorter time scale and higher energy spectrum than the bursts already observed. In order to have a reasonably high event rate, the local meteoroid population must extend to a distance from the sun of the order of 0.1 pc, but the required distance could become much lower if the instrumental threshold is improved. The authors also examine the expected gamma ray flux for interaction of the antimatter bodies with the solar wind, and find it far below present instrumental capabilities.

061.008 Production of heavy elements in nature.
V. M. Chechotkin, M. Kowalski.
Nature, Vol. 259, 643 - 644 (1976).

The authors describe here a possible mechanism for the production of heavy elements (with $A > 60$): these elements, they suggest, may be formed in the matter ejected from the envelope of a neutron star during the starquake process.

061.009 Low-energy elastic neutrino-nucleon and nuclear scattering and its relevance for supernovae.
J. Bernabéu.
Astron. Astrophys., Vol. 47, 375 - 379 (1976).

Low-energy ($E_\nu \lesssim 20$ MeV) elastic neutrino-nuclear scattering, from the neutron to the Fe-Ni region, is analysed by using both the conventional extension to hadrons of the Weinberg model and the recent vector model. The calculation is relevant to supernova theory. The author notes that the vector model gives more dramatic effects in the supernova "strategy", both lowering the cross-section in the inner core and increasing it in the outer region.

061.010 Scattering functions for neutrino transport.
W. R. Yueh, J. R. Buchler.
Astrophys. Space Sci., Vol. 39, 429 - 435 (1976).

The authors derive an expression for the scattering functions for electron-neutrino and electron-anti-neutrino Compton scattering in a form suitable for a numerical solution of the neutrino transfer equations. An analytical expression is given for the case of large electron degeneracy. The modification due to possible neutral currents is discussed.

061.011 Explosive nucleosynthesis of ^{7}Li.
K. J. Fricke, H. Nørgaard.
Mitt. Astron. Ges., No. 38, p. 176 - 177 (1976). – Abstract.

061.012 Gamma ray astronomy (0.1–1000 MeV).
K. Pinkau.
Space Research XV, (see 012.003), p. 681 - 697 (1975).

The results of gamma ray astronomy between 0.1 and 1000 MeV are reviewed. After touching very briefly on properties of instruments used in gamma ray astronomy, the presentation concerns itself with the following topics: (1) gamma ray pulses and their origin; (2) diffuse gamma radiation; (3) gamma rays from the galactic plane; (4) emission from localized regions.

061.013 High energy gamma ray astronomy.
C. E. Fichtel.
Space Research XV, (see 012.003), p. 699 - 714 (1975).
Review paper.

061.014 Ionising flux of cosmic background radiation.
J. Silk, R. A. Sunyaev.
Nature, Vol. 260, 508 - 509 (1976).

An indirect means of estimating the diffuse metagalactic flux of ionising radiation was proposed by Sunyaev, who studied the interaction of this radiation with neutral hydrogen at the peripheries of galaxies. More locally, the intensity of this radiation within ~1 kpc of the sun can be probed by studying the ionisation of heavy elements in the interstellar medium. Here the authors re-examine Sunyaev's proposal in the context of more detailed studies of the interaction of metagalactic ionising radiation with galactic gas. Their principal conclusion is that further than ~30 kpc from the centre of the Galaxy, self-shielding by H II is possible only if the flux of ionising radiation is below a certain critical value.

061.015 Light elements and the isotropy of the universe.
J. Barrow.
Monthly Notices Roy. Astron. Soc., Vol. 175, 359 - 370 (1976).

Nucleosynthesis in homogeneous Bianchi-type universes admitting Robertson-Walker solutions is investigated. The requirement that helium and deuterium are synthesized cosmologically in the abundance range observed is shown to constrain the ratio of the shear to the expansion rate of the universe today more severely than does the measured iostropy of the microwave background radiation. The local shear induced by large scale density perturbations is also considered together with some qualitative discussion of the physical effects of inhomogeneity.

061.016 Origin of the diffuse γ-ray background.
A. W. Strong, A. W. Wolfendale, D. M. Worrall.
Monthly Notices Roy. Astron. Soc., Vol. 175, 23P - 27P (1976).

An attempt is made to assess the contributions to the diffuse γ-ray background from normal galaxies, radio galaxies and Seyfert galaxies. Normal galaxies produce about 4 per cent of the background above 100 MeV. Radio galaxies can produce most and perhaps all of the background in the 1 - 10 MeV range provided their evolution as γ-ray sources is similar to that for their radio properties. Seyfert galaxies will also be important if inverse Compton models are applicable to them, particularly at energies above about 100 MeV.

061.017 The low temperature photonuclear nucleosynthesis of the anomalously low abundance isotopes ^{50}V, ^{138}La and ^{180}Ta. T. G. Harrison.
Astrophys. Letters, Vol. 17, 61 - 68 (1976).

The rare proton rich odd–odd nuclei ^{50}V, ^{138}La and ^{180}Ta have traditionally been attributed to the p-process. In this paper the author shows where the 14 and 17 MeV photons from the p-p II hydrogen cycle member ^{7}Li(p, γ)^{8}Be can cause photoneutron emission from the seed nuclei ^{51}V, ^{139}La, and ^{181}Ta, resulting in the synthesis of the anomalously low abundance isotopes ^{50}V, ^{138}La and ^{180}Ta. The astrophysical significance of the low temperature synthesis of these isotopes is discussed and the author also examines the effects of nuclear decay schemes, mass dilution and high temperature photonuclear erosion of ^{138}La and ^{180}Ta as possible mechanisms in reducing the potentially large overabundances to better match the observed solar system values, commenting upon the resultant simultaneous abundance alteration of ^{50}V.

061.018 Neutral currents and neutrino emission of stars.
S. S. Gershtejn, V. N. Folomeshkin, M. Yu. Khlopov, R. A. Ehramzhyan.
Zhurn. ehksperim. i teor. fiz., Vol. 69, 1121 - 1126 (1975).
In Russian. – Abstr. in Referativ. Zhurn. 51. Astron., 4.51. 168 (1976).

061.019 Laboratory microwave spectrum and rest frequencies of the N$_2$H$^+$ ion. R. J. Saykally, T. A. Dixon, T. G. Anderson, P. G. Szanto, R. C. Woods.
Astrophys. Journ., Letters, Vol. 205, L101 - L103 (1976).

The $J = 1 \leftarrow 0$ multiplet of N$_2$H$^+$ has been detected in laboratory glow discharges in mixtures of hydrogen and nitrogen. This constitutes the first observation of this species in the laboratory by any spectroscopic method. A pattern of three lines, due to the quadrupole hyperfine splitting of the outer nitrogen, was observed, and the frequencies of these three lines were measured, and are in very good agreement with previous astronomical measurements.

061.020 Dielectronic recombination to form helium-like ions.
A. Burgess, A. S. Tworkowski.
Astrophys. Journ., Letters, Vol. 205, L105 - L107 (1976).

The discrepancy between the results of Shore for helium-like ions, and the general formula for dielectronic recombination given by Burgess, is shown to be due to errors in the former. Further results on these ions are given.

061.021 Cosmic-ray spallative origin of the rare odd-odd nuclei, consistent with light-element production.
K. L. Hainebach, D. N. Schramm, J. B. Blake.
Astrophys. Journ., Vol. 205, 920 - 930 (1976).

Reeves, Fowler, and Hoyle have suggested that the light nuclei could be synthesized by mutual spallation of the cosmic rays and the interstellar medium. The authors now perform an integral over all cosmic-ray energies, utilize the newly available more accurate cross sections for peripheral reactions, and consider the effect of large fluxes of low-energy (5–25 MeV) particles in the region where (p, n) cross sections are very high, to show that one might be able in some specific models to produce not only additional ^{7}Li and ^{11}B, which are not fitted well in standard models, but also some additional ^{138}La.

061.022 Interstellar abundances of light elements revisited.
J. Audouze, J. Lequeux.
Astron. Astrophys., Vol. 49, 133 - 136 (1976).

As many other elements lithium and boron have been

claimed to be depleted by rather important factors (> 30) in the interstellar gas relative to the solar system (Morton, 1975). These depletion factors are reexamined in this note and found to be not as large and possibly non-existent. Some consequences on the origin of the light elements (Li, Be, B) are presented.

061.023 Neutrino processes in dense matter.
W. R. Yueh, J. R. Buchler.
Astrophys. Space Sci., Vol. 41, 221 - 251 (1976).

Expressions for the source and collision terms of the neutrino transport equation relevant to the neutrino transport supernova model are derived in the framework of the theory of neutral currents. In particular the authors study the capture and emission of neutrini (and anti-neutrini) by free nucleons, the inelastic scattering by free nucleons, the coherent scattering by nuclei, as well as the corresponding muonic processes.

061.024 Modern state of the theory of superdense celestial bodies. G. S. Saakyan.
Problems of gravitation. Third Soviet Gravitational Conference, Erevan, 1972, (see 012.008), p. 208 - 224 (1975). In Russian.

061.025 Gamma astronomy and nuclear astrophysics.
G. E. Kocharov.
Izv. Krymskoj Astrofiz. Obs., Vol. 54, 335 - 339 (1976). In Russian. – Abstract of a paper presented at the seminar on "X-ray and gamma-ray astronomy" at the Crimean Astrophys. Obs., 1974, (see 012.009).

061.026 Stellar neutrino pair emission from de-excitation of nuclear states via weak neutral currents.
J. P. Crawford, C. J. Hansen, K. T. Mahanthappa.
Astrophys. Journ., Vol. 206, 208 - 212 (1976).

The rates of neutrino-antineutrino emission from direct de-excitation of excited nuclei at high temperatures are calculated in a theory of the weak interaction containing neutral currents. The results, as applied to a sample astrophysical environment, indicate that these neutrino loss rates are small compared to more usual processes.

061.027 On the cascade process in strong magnetic and electric fields under astrophysical conditions.
Ya. I. Al'ber, Z. N. Krotova, V. Ya. Ehjdman.
Astrofizika, Vol. 11, 283 - 292 (1975). In Russian.
English translation in Astrophysics, Vol. 11, No. 2.

A self-supporting process of electron-positron cascade is considered in connection with the pulsar phenomenon. The electric and magnetic fields in an electron-positron plasma around a neutron star have been considered. The radio emission from such a system has been calculated.

061.028 Fundamental limitations of X-ray spectra as diagnostics of plasma temperature structure.
I. J. D. Craig, J. C. Brown.
Astron. Astrophys., Vol. 49, 239 - 250 (1976).

The problem and physical importance of deriving the temperature distribution in hot optically thin plasmas from their X-ray spectra are discussed and it is contended that these spectra are extremely limited in their usefulness in this respect, however high the spectral resolution.

061.029 Possible role of strongly interacting fields in astrophysics and relativistic cosmology. C.-G. Källman.
Astron. Astrophys., Vol. 49, 387 - 388 (1976).

A renormalizable scalar field theory with strong self-coupling and interacting with nucleons is considered. At a sufficiently high density, around twice nuclear matter density, nucleons become massless. Saturation occurs at a slightly higher density around 10^{15} g cm^{-3} and matter is at the saturation point in a bound state. The possibility of a third family of degenerate stars, "clumps" of matter in cosmic rays and possible violation of the energy condition for collapse to a singularity is discussed.

061.030 s-process studies: branching and the time scale.
R. A. Ward, M. J. Newman, D. D. Clayton.
Astrophys. Journ., Suppl. Ser., Vol. 31, 33 - 59 (1976).

The theory of s-process heavy-element formation is reformulated to allow for competition between beta decay and neutron capture at various nuclei along the path. Solutions to the resulting branching network equations are presented (under the assumption of constant temperature and neutron flux) that do not require steady flow for the neutron current.

061.031 Production of galactic ^{7}Li by slow mass loss.
J. M. Scalo.
Astrophys. Journ., Vol. 206, 795 - 799 (1976).

The mass loss rate in super-lithium stars required to account for the galactic abundance of ^{7}Li is calculated. In the first approach the required rate depends only on observational quantities, and the main uncertainties are the fraction of cool giants which are carbon stars and the Li abundances in the super-lithium stars. In the second approach these uncertain quantities are replaced by the integral of the Li abundance over time, taken from models of nucleosynthesis in hot-bottom convective envelopes, and the lifetime of the M giant phase of evolution, taken from considerations of the evolution of degenerate carbon-oxygen cores and mass loss rates.

061.032 The long-lived radioisotopes as monitors of stellar, galactic, and cosmological phenomena.
H. Reeves, O. Johns.
Astrophys. Journ., Vol. 206, 958 - 962 (1976).

Observational data on long-lived radioactivities, on stellar metal abundances, and on galactic deuterium-to-hydrogen ratio can be related and interpreted in terms of a simplified model of galactic nucleosynthesis. The data are consistent with an approximate time constancy of the astration rate and the r-process yields during most of the life of the Galaxy but does not preclude the occurrence of a large—but brief — increase of the yields in the early days of the galactic life.

061.033 Observation of radioactive background in the OSO-7 gamma ray monitor.
P. P. Dunphy, D. J. Forrest, E. L. Chupp, C. S. Dyer.
14th Intern. Cosmic Ray Conf., (see 012.011), Vol. 9, 3116 - 3121 (1975).

061.034 Interpretation of the radioactive background observed in the OSO-7 gamma-ray monitor.
C. S. Dyer, P. Dunphy, D. J. Forrest, E. L. Chupp.
14th Intern. Cosmic Ray Conf.; (see 012.011), Vol. 9, 3122 - 3127 (1975).

061.035 Nuclear experimentation relevant to cosmic ray composition. W. A. Fowler.
14th Intern. Cosmic Ray Conf., (see 012.011), Vol. 11, 3550 - 3565 (1975).

061.036 X-ray astronomy. J. L. Culhane.
Vistas Astron., Vol. 19, 1 - 67 (1975).

The author examines the production of X-radiation and its interaction with matter and describes the rather specialized instrumentation that is required to carry out the observations. He then discusses the study of solar X-rays, the X-ray sources in our own Galaxy and in the universe as a whole and finally the diffuse X-ray flux whose origin remains somewhat obscure.

061.037 Clarification on the law of angular momentum.
J. N. Tokis, V. S. Geroyannis.

Astrophys. Space Sci., Vol. 41, 475 - 480 (1976).

In most courses on continuum mechanics the law of angular momentum is studied about a fixed point (usually about the origin of the frame). In this article the authors clarify this law about any point in three-dimensional Euclidean space and discuss the law, about any point, for an observer in a rotating frame.

061.038 On the determination of the magnetic field under astrophysical conditions from the polarization of atomic fluorescence. A. A. Ruzmajkin.
Astron. Zhurn. Akad. Nauk SSSR, Vol. 53, 550 - 557 (1976). In Russian. English translation in Soviet Astron., Vol. 20, No. 3.

The polarization of the atomic fluorescence is sensitive to the influence of weak magnetic fields. If the existing light has a broad spectrum, the interference of the σ-components of the quasi-degenerate level gives a linearly polarized line with rotating plane of polarization. The rotation angle and the degree of the polarization are determined by the value of the magnetic field. This effect can be used in astrophysics for recording of weak magnetic fields. Estimates of the magnitude of the effect for various astronomical objects are given.

061.039 Röntgen-Astronomie — ein Überblick. I. T. Schmidt-Kaler.
Sterne, Vol. 52, 129 - 154 (1976).

061.040 X-ray astronomy — a status report. K. A. Pounds.
Mem. Soc. Astron. Italiana, Vol. 45, (see 012.017), 495 - 522 (1974).

061.041 On the nucleosynthesis of lithium, beryllium, and boron. R. Canal.
Mem. Soc. Astron. Italiana, Vol. 45, (see 012.017), 773 - 780 (1974).

061.042 Galactic abundances of ^{18}O, ^{21,22}Ne, ^{25,26}Mg and s-elements by depletion of N during helium burning phases. R. Gallino, A. Masani.
Mem. Soc. Astron. Italiana, Vol. 45, (see 012.017), 781 (1974) Abstract.

061.043 Infrared photometry at the Stockholm Observatory. L. Nordh, G. Olofsson.
Stockholms Obs. Rep. No. 8, 82 pp. (1974).

Some considerations involved in infrared measurements and technology; Technical description of the final photometer; "Quasi-simultaneous" infrared filter photometry; The influence of the thermal background; The observations; The astronomical interpretation of the observations.

061.044 Infrared pumping for SiO masers. S. Deguchi, T. Iguchi.
Publ. Astron. Soc. Japan, Vol. 28, 307 - 315 (1976).

A pumping mechanism for SiO masers by the infrared continuum radiation is proposed, where the population inversion is due to the anisotropic trapping of radiation in the expanding gas with a large velocity gradient. The numerical calculations show that the masers of the transitions $J = 1-0$ and $2-1$ in the first and the second vibrationally excited states of SiO occur in the envelope of infrared stars. The observed radial velocities of the maser emissions and the acceleration mechanism of the gas are discussed.

061.045 A critical discussion of the abundances of nuclei. Summary by the editors of the talk given by A. G. W. Cameron.
Explosive nucleosynthesis, (see 012.020), p. 3 - 21 (1973).

061.046 On the origin and evolution of the light elements.

B. M. Tinsley.
Explosive nucleosynthesis, (see 012.020), p. 22 - 33 (1973).

061.047 Hot CNO process. J. Audouze.
Explosive nucleosynthesis, (see 012.020), p. 47 - 59 (1973).

061.048 Explosive carbon burning. W. M. Howard.
Explosive nucleosynthesis, (see 012.020), p. 60 - 69 (1973).

061.049 The dynamic r-process. D. N. Schramm.
Explosive nucleosynthesis, (see 012.020), p. 84 - 101 (1973).

061.050 p-process nucleosynthesis. J. W. Truran.
Explosive nucleosynthesis, (see 012.020), p. 102 - 111 (1973).

061.051 The s-process in stars. R. K. Ulrich.
Explosive nucleosynthesis, (see 012.020), p. 139 - 167 (1973).

061.052 Confirming explosive nucleosynthesis with gamma-ray telescopes. D. D. Clayton.
Explosive nucleosynthesis, (see 012.020), p. 264 - 282 (1973).

061.053 Problems with nuclear reaction rates. G. Michaud.
Explosive nucleosynthesis, (see 012.020), p. 284 - 296 (1973).

061.054 Introduction à la photométrie astronomique. G. Florsch.
A.F.O.E.V. Bull.,Tome 9, 130 - 161; Tome 10, 1 - 16 (1975/ 76).

(1) Prolégomènes: grandeurs caractérisant le rayonnement; (2) Emission du rayonnement; (3) Transmission du rayonnement; (4) Réception du rayonnement.

061.055 Ultraviolet astrophysics. T. Jarzębowski.
Postępy Astron., Vol. 24, 85 - 108 (1976). In Polish.

This article presents a review of astrophysical observations in the ultraviolet: (1) the transparency of the earth atmosphere, (2) absorption and extinction in interstellar space, (3) the instruments on boards of the satellites OAO-2, OAO-3, TD-1A and ANS, (4) observational results on molecular hydrogen and on ultraviolet energy distribution of stars.

061.056 Hydrogen burning of ^{17}O in the CNO cycle. C. Rolfs, W. S. Rodney.
Nuclear Phys. A, Vol. A250, 295 - 308 (1975). — Abstr. in Phys. Abstr., Vol. 79, A000972 (1976).

061.057 Massive stars: a source of cosmic rays. J. Audouze, C. Cesarsky.
Recherche, Vol. 6, 762 - 763 (1975). In French. — Abstr. in Phys. Abstr., Vol. 79, A015138 (1976).

061.058 Spherically symmetric neutrino radiating stars. I. Dmaiao Soares, M. Novello.
Phys. Letters A, Vol. 55A, 5 - 6 (1975). — Abstr. in Phys. Abstr., Vol. 79, A015268 (1976).

061.059 A new redshift mechanism with possible applications to astrophysical problems such as quasars. C. F. Gallo.
Internat. Journ. Theor. Phys., Vol. 13, 417 - 418 (1975). Abstr. in Phys. Abstr., Vol. 79, A015491 (1976).

061.060 The deformation of the Fermi surface and magnetic susceptibility of neutron matter.

P. Haensel, J. Dabrowski.
Nuclear Phys. A, Vol. A254, 211 - 220 (1975). — Abstr. in
Phys. Abstr., Vol. 79, A016695 (1976).

061.061 Realistic quark models and astrophysics.
L. B. Okun, Ya. B. Zel'dovich.
Comments Nuclear Particle Phys., Vol. 6, No. 3, p. 69 - 73
(1976). — Abstr. in Phys. Abstr., Vol. 79, A020627 (1976).

**061.062 Nuclear structure: weak and electromagnetic inter-
actions in nuclei.**   H. Frauenfelder.
AIP Conf. Proc., No. 26, p. 448 - 453 (1975). — Abstr. in
Phys. Abstr., Vol. 79, A020732 (1976).

061.063 Abnormal states of nuclear matter.   L. Castillejo.
AIP Conf. Proc., No. 26, p. 686 - 699 (1975).
Abstr. in Phys. Abstr., Vol. 79, A020788 (1976).

**061.064 The $^{16}O(p, \gamma)^{17}F$ direct capture cross section with
an extrapolation to astrophysical energies.**
H. C. Chow, G. M. Griffiths, T. H. Hall.
Canadian Journ. Phys., Vol. 53, 1672 - 1686 (1975). — Abstr.
in Phys. Abstr., Vol. 79, A020836 (1976).

**061.065 Additional results concerning nuclei in the terminal
parts of the r-process path.**
R. Boleu, W. Stepien-Rudzka.
Nuclear Phys. A, Vol. A255, 204 - 220 (1975). — Abstr. in
Phys. Abstr., Vol. 79, A023745 (1976).

**061.066 Tables of theoretical ultraviolet fluxes and colours,
including line opacities.**
Bull. Soc. Roy. Sci. Liège, Vol. 44, 119 - 156 (1975). — Abstr.
in Phys. Abstr., Vol. 79, A023751 (1976).

**061.067 Light-fragment production in p-nucleus interactions
at 600 MeV. Astrophysical application.**
J. P. Alard, A. Baldit, R. Brun, J. P. Costilhes, J. Dhermain,
J. Fargeix, L. Fraysse, J. Pellet, G. Roche, J. C. Tamain,
A. Cordaillat, A. Pasinetti.
Nuovo Cimento A, Ser. 11, Vol. 30A, 320 - 344 (1975).
Abstr. in Phys. Abstr., Vol. 79, A025098 (1976).

**061.068 A study of the $^{23}Na + p$ reactions to determine ^{23}Mg
consumption in stars.**   F. M. Mann, D. W. Kneff,
Z. E. Switkowski, S. E. Woosley.
Nuclear Phys. A, Vol. A256, 163 - 172 (1976). — Abstr. in
Phys. Abstr., Vol. 79, A025107 (1976).

**061.069 The temperature dependences of some types of
gaseous ionic reactions of astrochemical interest.**
M. Meot-Ner, F. H. Field.
Origins of Life, Vol. 6, 377 - 393 (1975). — Abstr. in Phys.
Abstr., Vol. 79, A028029 (1976).

**061.070 Subsidiary conditions and variational calculations of
nuclear and neutron matters.**   O. Benhar,
C. Ciofi degli Atti, A. Kallio, L. Lantto, P. Toropainen.
Phys. Letters B, Vol. 60B, 129 - 133 (1976). — Abstr. in Phys.
Abstr., Vol. 79, A029264 (1976).

**061.071 Renormalisation of the πNN-vertex in a nuclear
medium.**   E. Oset, W. Weise.
Phys. Letters B, Vol. 60B, 141 - 145 (1976). — Abstr. in Phys.
Abstr., Vol. 79, A029265 (1976).

**061.072 Neutrino production in stellar matter by photons in
a renormalizable scalar-boson-exchange model of**
weak interactions.   S. N. Biswas, S. R. Chaudhuri,
J. K. S. Taank, J. A. Campbell.
Phys. Rev. D, Vol. 12, 2523 - 2525 (1975). — Abstr. in Phys.

Abstr., Vol. 79, A032674 (1976).

061.073 Solidification of neutron matter.
V. Canuto, J. Lodenquai.
Phys. Rev. C, Vol. 12, 2033 - 2037 (1975). — Abstr. in Phys.
Abstr., Vol. 79, A034387 (1976).

**061.074 The capture of charged particles by transverse
electromagnetic waves.**
V. Ya. Davydovskii (*Davydovskij*), V. G. Sapogin, A. S. Ukolov.
Izv. vyssh. ucheb. zavedenij fiz., No. 12, p. 110 - 115 (1975).
In Russian. — Abstr. in Phys. Abstr., Vol. 79, A037440 (1976).

061.075 Infra-red astronomy: a fresh look at the stars.
A. Tucker.
Electronics Australia, Vol. 37, No. 10, p. 26 (1976).
Abstr. in Phys. Abstr., Vol. 79, A037674 (1976).

061.076 Vibrational theory of pion condensation.
O. Nalcioglu, A. Goswami.
Nuovo Cimento Lettere, Ser. 2, Vol. 15, 359 - 363 (1976).
Abstr. in Phys. Abstr., Vol. 79, A038979 (1976).

**061.077 The mechanisms of amino acids synthesis by high
temperature shock-waves.**
I. Barak, A. Bar-Nun.
Origins of Life, Vol. 6, 483 - 506 (1975). — Abstr. in Phys.
Abstr., Vol. 79, A041386 (1976).

**061.078 Study on the photochemical reaction of HCN and
its polymer products relating to primary chemical
evolution.**
H. Mizutani, H. Mikuni, M. Takahasi, H. Noda.
Origins of Life, Vol. 6, 513 - 525 (1975). — Abstr. in Phys.
Abstr., Vol. 79, A041412 (1976).

**061.079 Explosive nucleosynthesis in zones rich in hydrogen
and helium.**   J. Toussaint.
Thèse Paris-11 Univ., 91 - Orsay, France, 115 pp. (1975).
In French.

061.080 Introduction to nucleosynthesis.   J. Guasp.
Separate print Junta de Energia Nuclear, Madrid,
Spain, 92 pp. (1975). In Spanish.

**061.081 Cooling blackbody: a mechanism for cosmic gamma-
ray bursts.**   R. Ramaty.
Symposium significant accomplishments sci. technol., Greenbelt, Maryland, USA, 18 Dec. 1973, p. 44 - 47 (1975).

Spallation nuclear reactions and their applications.
See Abstr. 003.101.

Role of magnetic fields in physics and astrophysics.
Proceedings of a conference held at Copenhagen in June, 1974.
See Abstr. 012.026.

Classification of $d^4 - d^3 4p$ Co VI spectra.
See Abstr. 022.005.

Tables of damping constants of spectral lines broadened by H and He.   See Abstr. 022.006.

Introduction to collision theory and to some astrophysical applications.   See Abstr. 022.091.

Experiments of relevance to explosive nucleosynthesis. See Abstr. 022.094.

Neutrino pair emission from finite-temperature neutron superfluid and the cooling of young neutron stars.

See Abstr. 065.062.

Neutrino flow and gravitational collapse.
See Abstr. 066.153.

Spectral distribution and origin of the X- and gamma-ray diffuse background. See Abstr. 066.157.

Type I supernovae and galactic production of iron.
See Abstr. 125.018.

Synthesis of the light elements in supernovae.
See Abstr. 125.028.

Nucleosynthesis during explosive oxygen and silicon burning. See Abstr. 125.039.

The abundance of deuterium relative to hydrogen in interstellar space. See Abstr. 131.008.

Chemical composition of H II regions in the Small Magellanic Cloud and the pregalactic helium abundance.
See Abstr. 131.505.

Gamma-ray emission and nucleosynthesis of lithium by young pulsars. See Abstr. 141.340.

Features in the brightness distribution and spectra of the soft X-ray background. See Abstr. 142.060.

Photonuclear interactions of ultrahigh energy cosmic rays and their astrophysical consequences.

See Abstr. 143.022.

Chemical evolution of the Galaxy: coefficients from stellar evolution and alternative solutions to the problem of few metal-poor stars. See Abstr. 155.061.

Expected rate of transient events from stellar deaths in other galaxies. See Abstr. 158.082.

Frühphasen des Kosmos und die Entstehung der chemischen Elemente. See Abstr. 162.023.

Big-bang nucleosynthesis with nonzero lepton numbers. See Abstr. 162.044.

Leptonic numbers and the neutron to proton ratio in the hot Big Bang model. See Abstr. 162.058.

Errata

061.901 **Errata: 'Neutrino opacities at high temperatures and densities'** [Astrophys. Journ., Vol. 201, 467 - 488 (1975)]. D. L. Tubbs, D. N. Schramm.
Astrophys. Journ., Vol. 205, 308 (1976).

061.902 **Errata: 'Does astronomy need 'new physics'? '** [Quarterly Journ. Roy. Astron. Soc., Vol. 16, 265 - 281 (1975)]. V. L. Ginzburg.
Quarterly Journ. Roy. Astron. Soc., Vol. 17, 209 (1976).

062 Hydrodynamics, Magnetohydrodynamics, Plasma

062.001 The recombination and level populations of ions–I. Hydrogen and hydrogenic ions.
A. Burgess, H. P. Summers.
Monthly Notices Roy. Astron. Soc., Vol. 174, 345 - 391 (1976).

A description is given of the atomic processes occurring in an unbounded plasma containing ions, electrons and protons and permeated by external radiation fields. A method of numerical solution of the statistical balance equations for the collisional radiative coefficients and the level populations (including the very highly-excited levels) is given. Results are presented for hydrogen and hydrogenic ions in both graphical and numerical form. Tabulations of level populations are given appropriate to H I regions and to fairly dense H II regions. Simple expressions for bound-bound and bound-free Gaunt factors and for impact parameter cross-sections are given as Appendices.

062.002 Radiation from cosmic blast waves.
D. Bollea, A. Cavaliere.
Astron. Astrophys., Vol. 46, 219 - 228 (1976).

A study is made of the dynamics of blast waves, of their internal structure and of the Bremsstrahlung radiation from the shocked plasma in conditions when the cooling time is longer than the expansion time. Models appropriate for the interpretation of the X-ray sources associated with supernova remnants and with the explosive activity of the radiogalaxies in clusters of galaxies are computed and discussed; the inverse Compton emission from continuously injected relativistic electrons is included.

062.003 On the temperature modulation of the thermal convection instability in hydromagnetics.
K. M. Srivastava.
Astron. Astrophys., Vol. 46, 361 - 368 (1976).

The thermal convection instability of a plane fluid layer heated from below, when the temperature gradient has both a steady and a time periodic component, in the presence of a uniform magnetic field parallel to the gravity and normal to the boundaries is investigated.

062.004 Parametric excitation of standing electromagnetic waves. E. Leer.
Physica Scripta, Vol. 13, 47 - 50 (1976).

062.005 Nonresonant wave-coupling and wave-particle interactions. D. Anderson.
Physica Scripta, Vol. 13, 117 - 121 (1976).

062.006 Deficiencies of the asymptotic solutions commonly found in the quasilinear relaxation theory.
R. J.-M. Grognard.
Australian Journ. Phys., Vol. 28, 731 - 753 (1975).

062.007 Continuous emission of argon, krypton and xenon plasmas. D. Meiners, C. O. Weiss.
Journ. Quant. Spectrosc. Radiat. Transfer, Vol. 16, 273 - 280 (1976).

062.008 Parametric instabilities in relativistic plasma.
N. L. Tsintsadze, E. G. Tsikarishvili.
Astrophys. Space Sci., Vol. 39, 181 - 190, 191 - 199 (1976). In Russian and English.

Parametric instabilities in the relativistic plasma are considered. It is shown that in the electron relativistic plasma ($T_e \geqslant m_{0e} c^2$) the electron mass oscillation in the external electrical field leads to the instability of Langmuir and low frequency aperiodic oscillations as well. In the case of the hot electron ion plasma with relativistic electron temperature the low frequency aperiodic and periodic oscillations are studied. The wave increments for all considered cases are obtained.

062.009 The interaction of an obliquely incident plane electromagnetic wave with an anisotropic moving conducting half-space. P. K. Mukherjee, S. P. Talwar.
Astrophys. Space Sci., Vol. 39, 213 - 234 (1976).

The interaction of an obliquely incident plane electromagnetic wave with an anisotropic moving conducting medium described by tensor constitutive parameters ($\vec{\vec{\epsilon}}, \vec{\vec{\mu}}$, and $\vec{\vec{\sigma}}$) is studied. Starting from the Maxwell-Minkowski equations the wave solutions in the laboratory frame, the modified law of refraction and the reflecton and the transmission coefficients are obtained both for incident E- and H-waves, corresponding to the two specific orientations of the plane of incidence relative to the direction of motion of the medium. Numerical results for the reflection and the transmission coefficients are presented for a range of the parameters characterizing the anisotropy and the velocity of the moving medium.

062.010 Radiative cooling of a low-density plasma.
J. C. Raymond, D. P. Cox, B. W. Smith.
Astrophys. Journ., Vol. 204, 290 - 292 (1976).

The authors have calculated the radiative cooling coefficient for a low-density, optically thin gas of cosmic abundances in the range $10^4–10^8$ K incorporating significant elements through nickel and many recently improved rate calculations.

062.011 A nonlinear theory of cosmic-ray pitch-angle diffusion in homogeneous magnetostatic turbulence.
M. L. Goldstein.
Astrophys. Journ., Vol. 204, 900 - 919 (1976).

A plasma strong turbulence, weak coupling, theory is applied to the problem of cosmic-ray pitch-angle scattering in magnetostatic turbulence. The theory used contains no ad hoc approximations. A detailed calculation is presented for a model of "slab" turbulence with an exponential correlation function. The results agree well with numerical simulations. The rigidity dependence of the pitch-angle scattering coefficient differs from that found by previous researchers. The differences result from an inadequate treatment of particle trajectories near 90° pitch angle in earlier work.

062.012 A comment on the damping of magnetohydrodynamic waves. Bibhas R. De.
Astrophys. Space Sci., Vol. 39, L43 - L46 (1976).

It is shown that a revision is necessary in the commonly used expression for the damping scale-length of a magnetohydrodynamic wave in a plasma when the plasma exhibits anisotropic conductivity. The consequence of this revision may be quite drastic in some astrophysical situations.

062.013 Über die Bedeutung des Nichols–Tolman-Effektes für das Zustandekommen kosmischer Magnetfelder.
F. Krause, K.-H. Rädler, G. Rüdiger.
Gerlands Beiträge Geophys., Vol. 85, 26 - 34 (1976). – Paper presented at the International Symposium on "Theoretical Research Methods of Geophysics, Geology and Astrophysics", Eisenach, G.D.R., January 13–18, 1975.

The Nichols–Tolman effect is discussed in connection with the question as to the origin of cosmic magnetic fields. An estimation of the contribution of this effect to the magnetic field of the earth is given by using a relation for the magnetic moment which has been deduced from Ohm's law. In conformity with former results, this contribution turns out to be negligibly small. Furthermore, it is shown that an explanation of the origin of the earth's magnetic field given by Schmutzer cannot be correct, especially for reasons connected with Cowling's theorem.

062.014 A type of motion in magnetohydrodynamics.
V. I. Zhukov.
Solnechnye Dannye 1975 Byull., No. 12, p. 55 - 58 (1976). In Russian.

In some cases the structure of the magnetic field is shown to be determined by the density distribution, not by the velocity field.

062.015 Mouvements sphériques des fluides visqueux im-compressibles.
A. Avez, Y. Bamberger.
Comptes Rendus Acad. Sci. Paris, Sér. A, Vol. 282, 341 - 344 (1976).

On étudie un modèle météorologique: l'atmosphère est un fluide visqueux incompressible dont la vitesse est horizontale et proportionnelle à la distance au centre de la terre.

062.016 Element separation effects in the boundary region of a plasma surrounded by neutral gas.
B. Lehnert.
Astrophys. Space Sci., Vol. 40, 225 - 229 (1976).

A one-dimensional model is being considered where a fully ionized plasma is separated from a neutral gas by a homogeneous magnetic field directed along the plasma boundary. The plasma and the neutral gas consist of two different types of ions and neutral particles. It is found that the ratio between the ion densities in the fully ionized region will in general differ from the density ratio of the two types of neutrals being present in the gas region.

062.017 Ion-ion collisions and element separation effects in the boundary region of a plasma surrounded by neutral gas. B. Bonnevier.
Astrophys. Space Sci., Vol. 40, 231 - 240 (1976).

In the boundary region between a plasma and neutral gas, the temperature is low (some thousand degrees). Then the Coulomb cross-section is large compared with other collision cross-sections. The influence of ion-ion collisions in the boundary region is discussed. In particular a simple formula is derived for the separation between different kinds of constituents.

062.018 Stability of a degenerate electron plasma in a strong magnetic field. S. Ames.
Astrophys. Letters, Vol. 17, 77 - 79 (1976).

A degenerate electron plasma in a strong magnetic field is shown to be unstable at temperatures less than approximately 5×10^8 °K for densities less than 10^6 gm/cm^3, if the Canuto-Chiu equation of state is used.

062.019 The process of magnetic field line reconnection and its role in cosmic plasma. S. I. Syrovatskij.
VIIth Leningrad seminar. "Corpuscular streams of the sun and the radiation belts of the earth and Jupiter", 1975, (see 012. 007), p. 63 - 80 (1975). In Russian. – Abstr. in Referativ. Zhurn. 51. Astron., 4.51.345; 62. Issled. kosmich. prostranstva, 4.62.201 (1976).

062.020 Experimental investigations of the behaviour of plasma in the vicinity of magnetic zero lines.
A. G. Frants.
VIIth Leningrad seminar. "Corpuscular streams of the sun and the radiation belts of the earth and Jupiter", 1975, (see 012. 007), p. 81 - 92 (1975). In Russian. – Abstr. in Referativ. Zhurn. 62. Issled. kosmich. prostranstva, 4.62.202 (1976).

062.021 Instationary convection and precipitation of electron particles. S. A. Rumyantsev, V. S. Smirnov.
VIIth Leningrad seminar. "Corpuscular streams of the sun and the radiation belts of the earth and Jupiter", 1975, (see 012. 007), p. 237 - 238 (1975). In Russian. – Abstr. in Referativ. Zhurn. 62. Issled. kosmich. prostranstva, 4.62.262 (1976).

062.022 The solution of the radio transfer equation in a plasma with constant pressure and conductive flux.
G. Noci.
Astron. Astrophys., Vol. 48, 359 - 365 (1976).

The thermal radio emission of a layer of plasma in conductive equilibrium with constant pressure is studied for propagation perpendicular to the layer. The temperature dependence of the optical depth is given and the effect of the deviation of the refractive index from unity discussed. Finally the transfer equation is solved, the dependence of the brightness temperature on frequency, pressure, conductive flux and highest temperature is discussed, and some approximate formulae are given.

062.023 On the stationary solution of the equation system of $\vec{v} = \vec{h}/\sqrt{4\pi\rho}$ type in magnetohydrodynamics.
V. I. Zhukov.
Solnechnye Dannye 1976 Byull., No. 1, p. 91 - 92 (1976). In Russian.

062.024 Instabilités rayonnées dans un plasma par un faisceau d'électrons. R. Pellat, J. Lavergnat, A. Saint-Marc.
Comptes Rendus Acad Sci. Paris, Sér. B, Vol. 282, 201 - 203 (1976).

On montre qu'en choisissant de bonnes conditions aux limites, les instabilités d'un système plasma-faisceau aux dimensions finies peuvent être rayonnées à l'extérieur de ce système.

062.025 Depolarization of electromagnetic waves in a turbulent cosmic plasma.
V. G. Gavrilenko, N. S. Stepanov.
Astron. Zhurn. Akad. Nauk SSSR, Vol. 53, 291 - 294 (1976). In Russian. English translation in Soviet Astron., Vol. 20, No. 2.

The problem of radio wave depolarization effect in a cosmic plasma caused by vortex medium motion is discussed. The general expression connecting the dispersion of the polarization angle with the space-time spectrum of the velocity field is obtained. Estimates are presented for some models of this spectrum.

062.026 The effect of adiabatic focusing upon charged-particle propagation in random magnetic fields.
J. A. Earl.
Astrophys. Journ., Vol. 205, 900 - 919 (1976).

The charged particles considered in this paper are scattered by random fields while they propagate along the diverging lines of force of a spatially inhomogeneous guiding field. Their longitudinal transport is described in terms of the eigenfunctions of a Sturm-Liouville operator which incorporates the effect of adiabatic focusing along with that of scattering. The relaxation times and characteristic velocities, which appear in this matrix formulation of the transport problem, are graphed and tabulated.

062.027 Einige Wirkungen zweidimensionaler Turbulenz in der Hydrodynamik für inkompressible Materie.
G. Rüdiger.
Gerlands Beiträge Geophys., Band 85, 157 - 165 (1976).

062.028 Approximations for gyrosynchrotron emissivity in a weak, isotropic plasma. G. L. Tarnstrom.
Astron. Astrophys., Vol. 49, 31 - 38 (1976).

The Wild and Hill approximations for gyrosynchrotron emissivity per electron in vacuo and in a weak, isotropic plasma allows one to derive an approximation for the frequency of maximum emissivity as a function of aspect angle, particle speed, particle pitch angle, and local plasma and cyclotron frequencies. Given sufficient suppression by the Razin effect, the emissivity may be separated into frequency-dependent and mode-dependent parts, providing a simple expression for narrowband, Razin-suppressed spectra.

062.029 **The hydrodynamics of accretion discs. II: Turbulent models.** J. M. Stewart.
Astron. Astrophys., Vol. 49, 39 - 51 (1976).

Turbulent models are discussed here and two new points are made. The crucial problem for the horizontal structure of the disc, the prescription of the Reynolds stress gradient is resolved by a direct calculation from first principles. A preliminary attempt is also made to describe the vertical structure, leading to a sandwich model. The predictions of this theory are shown to be consistent with the fine scale structure of Cyg X-1.

062.030 **Guided wave propagation in a moving magnetized plasma bounded by parallel perfectly conducting planes.** P. K. Mukherjee.
Astrophys. Space Sci., Vol. 41, 97 - 104 (1976).

Guided wave propagation in a moving magnetoplasma, bounded by parallel perfectly conducting planes, has been studied. The problem is formulated for the general case of a three-dimensional (rectangular) waveguide magnetized along a transverse direction. It is found that the modes carried by the waveguide are in general coupled modes of mixed E and H type.

062.031 **Small perturbations in a viscous heat-conducting hot expanding medium.** S. G. Pomagaev.
Byull. Inst. Astrofiz., *Dushanbe*, No. 63, p. 3 - 14 (1974). In Russian.

The characteristic equation for small perturbations in a viscous heat-conducting hot expanding medium in the presence of a magnetic field is derived and studied. The criterion for the stability of the perturbations is determined, and the regions of wave numbers where small perturbations are stable are found.

062.032 **On the non-stationary Rayleigh problem for a strongly magnetized plasma.**
L. A. Abramov, L. S. Al'perovich.
Astrofizika, Vol. 11, 293 - 303 (1975). In Russian.
English translation in Astrophysics, Vol. 11, No. 2.

The problem concerning oscillatory onset of convection in a horizontal plasma layer immersed in constant vertical gravitational and magnetic fields has been solved. When thermal conductivity anisotropy is significant, viscosity is isotropic (only the electrons are magnetized) and also the Hartman number is large, an instability can occur if the "magnetic Prandtl number" is large in comparison with unity. The formulae deduced are a natural generalization of Chandrasekhar's criteria and may by used in the theory of hydromagnetic wave generation.

062.033 **Electromagnetic waves in a relativistic plasma with a strong magnetic field.**
E. V. Suvorov, Yu. V. Chugunov.
Astrofizika, Vol. 11, 305 - 318 (1975). In Russian.
English translation in Astrophysics, Vol. 11, No. 2.

The properties of electromagnetic modes are investigated for a relativistic plasma with one-dimensional distribution function which establishes radiative losses in a strong magnetic field.

062.034 **Compressible magnetic field reconnection: a slow wave model.** C.-K. Yang, B. U. Ö. Sonnerup.
Astrophys. Journ., Vol. 206, 570 - 582 (1976).

The incompressible infinite-conductivity reconnection model, consisting of wedges of uniform flow and magnetization separated by large-amplitude Alfvén waves, is generalized to compressible flow. The leading waves become slow-mode isentropic expansion fans centered not at the reconnection point, but at corners located upstream in the flow field. The trailing waves are slow shocks emanating from the reconnection point. By numerical integration across the expansion fans and by use of the slow-shock jump conditions, the properties of the two plasma jets emitted from the reconnection region are obtained as a function of plasma properties and flow speed in the inflow. The possibility of fast transverse shocks in the exit jets is discussed.

062.035 **Radiation from a hot, thin plasma from 1 to 250 Å.** T. Kato.
Astrophys. Journ., Suppl. Ser., Vol. 30, 397 - 449 (1976).

A calculation of emission spectrum of a hot, low-density plasma in the region 1–250 Å is presented. The mechanisms considered are electron collision-induced line emission, bremsstrahlung, and radiative recombination; and the temperature range studied is 10^5– 10^7 K. The author has included 795 lines. The elemental abundances of the ions of He, C, N, O, Ne, Mg, Si, S, Ca, Fe, and Ni were taken to be as in the solar corona. The line emission of Fe ions produces a maximum in the curve of an emission power between 1 and 250 Å versus temperature around 10^6 K. The emission rate around 10^6 K is larger than the results calculated by Cox and Tucker and Tucker and Koren.

062.036 **Nonhomologous contraction and equilibria of self-gravitating, magnetic interstellar clouds embedded in an intercloud medium: star formation. I. Formulation of the problem and method of solution.** T. C. Mouschovias.
Astrophys. Journ., Vol. 206, 753 - 767 (1976).

In this paper the author formulates the equilibrium problem for isothermal clouds with a frozen-in magnetic field, and he describes a method for its solution. Self-gravity and the pressure of the hot and tenuous intercloud medium bind a cloud against the disruptive effects of its internal pressure and magnetic stresses. The surface of a cloud is a free boundary determined by the requirement that there exist pressure balance across it. The equilibrium problem is characterized by two free parameters and a free function, which describes the amount of mass in each of the flux tubes of the system. In principle, this function can be obtained from high-resolution observations of dense clouds. If one assumes that an interstellar cloud contracted nonhomologously from some initial uniform state while the magnetic field remained frozen in the matter, then the free function reduces to a single parameter.

062.037 **On diffusive-convective solutions of the Fokker-Planck equation.** J. Kunstmann, W. Alpers.
14th Intern. Cosmic Ray Conf., (see 012.011), Vol. 5, 1728 - 1732 (1975).

062.038 **The effect of adiabatic focusing upon charged particle propagation in random magnetic fields.**
J. A. Earl.
14th Intern. Cosmic Ray Conf., (see 012.011), Vol. 5, 1733 - 1738 (1975).

062.039 **Acceleration of charged particles in oblique MHD shocks.** G. Chen, T. P. Armstrong.
14th Intern. Cosmic Ray Conf., (see 012.011), Vol. 5, 1814 - 1819 (1975).

062.040 **Preliminary results of the Helios plasma experiment.**

H. Rosenbauer, B. Meyer, H. Miggenrieder, M. Montgomery, R. Schwenn.
14th Intern. Cosmic Ray Conf., (see 012.011), Vol. 5, 1857 (1975). — Abstract.

062.041 **On radiation production in a sheared medium.**
I. Lerche.
Astrophys. Space Sci., Vol. 41, 387 - 406 (1976).
The author gives an expression for the radiation produced by a uniformly charged particle when it traverses normally a semi-infinite boundary between two media, both of the same constant refractive index n, measured in the appropriate rest frames of the media. The media are taken to slip relative to each other with constant velocity parallel to the boundary. The author computes the differential power output and shows that (a) the emitted radiation has a flat spectrum up to a frequency such that n can no longer be considered constant; (b) the angular dependence of the emitted radiation is peaked at an angle to the direction of motion of the particle; (c) there is a back-scattered component to the radiation. In view of the complexity of the analytic formulae for the differential power output he gives some numerical examples for $n > 1$ and $n < 1$ to illustrate the different angular dependences of the power output in both cases.

062.042 **The hydrodynamics of the helium flash.**
A. J. Wickett.
Computing in plasma physics and astrophysics, (see 012.014), A1, 2 pp. (1976).

062.043 **A magnetostatic particle code and its application to studies of anomalous current penetration of a plasma.** A. T. Lin, P. L. Pritchett, J. M. Dawson.
Computing in plasma physics and astrophysics, (see 012.014), B1, 2 pp. (1976).

062.044 **Theory and simulation on the anomalous transport due to drift wave instabilities.**
Y. Y. Kuo, W. W. Lee, H. Okuda, M. True.
Computing in plasma physics and astrophysics, (see 012.014), B2, 2 pp. (1976).

062.045 **Numerical simulation of dissipative trapped-electron instability.**
Y. Matsuda, H. Okuda.
Computing in plasma physics and astrophysics, (see 012.014), B3, 2 pp. (1976).

062.046 **Numerical simulation of neutral beam injection into a tokamak.**
J. Sampson, Y. Matsuda, H. Okuda.
Computing in plasma physics and astrophysics, (see 012.014), B4, 2 pp. (1976).

062.047 **High beta equilibria in beam heated tokamak plasmas.** R. A. Dory, Y.-K. M. Peng.
Computing in plasma physics and astrophysics, (see 012.014), C1, 2 pp. (1976).

062.048 **In search of stable 3 D MHD-equilibria.**
R. Chodura, A. Schlüter.
Computing in plasma physics and astrophysics, (see 012.014), C2, 2 pp. (1976).

062.049 **A two-dimensional MHD stability code with a finite hybrid expansion in both directions.**
D. Berger, R. Gruber, F. Troyon.
Computing in plasma physics and astrophysics, (see 012.014), C3, 2 pp. (1976).

062.050 **Axisymmetric stability of numerically computed tokamak equilibria.** K. Lackner.
Computing in plasma physics and astrophysics, (see 012.014), C4, 2 pp. (1976).

062.051 **MHD instabilities of 2D diffuse high-beta equilibria as an initial-boundary-value problem.**
F. Herrnegger, W. Schneider.
Computing in plasma physics and astrophysics, (see 012.014), C5, 2 pp. (1976).

062.052 **Analytical and numerical study of MHD instabilities in rotating plasmas.** G. O. Spies.
Computing in plasma physics and astrophysics, (see 012.014), C6, 2 pp. (1976).

062.053 **Three-dimensional, nonlinear magnetohydrodynamic computations of the postimplosion dynamics**
of the Los Alamos Scyllac experiment.
D. C. Barnes, J. U. Brackbill.
Computing in plasma physics and astrophysics, (see 012.014), D1, 2 pp. (1976).

062.054 **Numerical studies of resistive instabilities.**
D. Biskamp, H. Welter.
Computing in plasma physics and astrophysics, (see 012.014), D2, 2 pp. (1976).

062.055 **Numerical solutions of the two-dimensional, non-stationary, nonlinear, compressible, resistive MHD**
equations for astrophysical and laboratory applications.
G. Knorr.
Computing in plasma physics and astrophysics, (see 012.014), D3, 2 pp. (1976).

062.056 **Numerical study of anisotropic magnetohydrodynamic turbulence.** S. J. Schwartz.
Computing in plasma physics and astrophysics, (see 012.014), D5, 2 pp. (1976).

062.057 **MUSCL, a new approach to numerical gas dynamics.**
B. van Leer.
Computing in plasma physics and astrophysics, (see 012.014), E1, 2 pp. (1976).

062.058 **Stability of vorticity distributions using the water-bag model.** M. L. Noyer.
Computing in plasma physics and astrophysics, (see 012.014), E2, 2 pp. (1976).

062.059 **Exact magnetic energy conservation for a non-uniform mesh.** J. P. Christiansen, K. V. Roberts.
Computing in plasma physics and astrophysics, (see 012.014), E3, 2 pp. (1976).

062.060 **The influence of aliasing errors on nonlinear wave propagation.** H. Schamel, K. Elsässer.
Computing in plasma physics and astrophysics, (see 012.014), E4, 2 pp. (1976).

062.061 **Calculation of neutral beam heating of tokamaks.**
D. E. Post, P. H. Rutherford, H. P. Furth, R. R. Smith.
Computing in plasma physics and astrophysics, (see 012.014), G1, 1 p. (1976).

062.062 **Effects of impurity transport on neutral injection in tokamaks.** M. H. Hughes.
Computing in plasma physics and astrophysics, (see 012.014), G2, 2 pp. (1976).

062.063 **Turbulent transport in reversed field pinches.**

J. P. Christiansen, K. V. Roberts.
Computing in plasma physics and astrophysics, (see 012.014), G3, 2 pp. (1976).

062.064 Diffusive transition from belt pinch to doublet.
H. Grad, P. N. Hu, D. C. Stevens, E. Turkel.
Computing in plasma physics and astrophysics, (see 012.014), G4, 2 pp. (1976).

062.065 A general approach to solving the rate equations for ionisation-recombination processes in a plasma.
J. Magill, E. W. Laing, N. Tahir.
Computing in plasma physics and astrophysics, (see 012.014), G5, 2 pp (1976).

062.066 A simple routine for reaction products transport in laser-produced plasma.   A. D. Krumbein,
M. Rosenblum, H. Zmora, Y. Kelson.
Computing in plasma physics and astrophysics, (see 012.014), G6, 2 pp. (1976).

062.067 Numerical models of astrophysical convection.
D. R. Moore.
Computing in plasma physics and astrophyscis, (see 012.014), H1, 2 pp. (1976).

062.068 Low Prandtl number thermal convection.
R. S. Peckover, C. S. Garland.
Computing in plasma physics and astrophysics, (see 012.014), H2, 2 pp. (1976).

062.069 The application of fast Fourier transforms to the primitive equations of Boussinesq convection.
A. K. Parrott.
Computing in plasma physics and astrophysics, (see 012.014), H3, 2 pp. (1976).

062.070 Two-dimensional modeling of tokamaks.
H. H. Klein, R. N. Byrne, N. A. Krall.
Computing in plasma physics and astrophysics, (see 012.014), P1, 2 pp. (1976).

062.071 Influence of neutral injection on impurity transport in TEXTOR.   A. Nicolai.
Computing in plasma physics and astrophysics, (see 012.014), P2, 2 pp. (1976).

062.072 Finite element approach to the quasi-linear evolution of the gentle bump instability in a 2-D plasma.
K. Appert, T. M. Tran, J. Vaclavik.
Computing in plasma physics and astrophysics, (see 012.014), P3, 2 pp. (1976).

062.073 The resistive tearing mode by a two-dimensional finite difference method.   J. Birn.
Computing in plasma physics and astrophysics, (see 012.014), P4, 2 pp. (1976).

062.074 Sheath near a conducting half-plane parallel to a collisionless hypersonic plasma.
P. Chenevier, J. M. Dolique.
Computing in plasma physics and astrophysics, (see 012.014), P6, 2 pp. (1976).

062.075 Numerical simulation of resistive instabilities of charged particle beams.   E. Messerschmid.
Computing in plasma physics and astrophysics, (see 012.014), P8, 2 pp. (1976).

062.076 Collisional damping of Bernstein echoes.
E. Fijalkow, J. R. Burgan, M. R. Feix.

Computing in plasma physics and astrophysics, (see 012.014), P9, 2 pp. (1976).

062.077 Numerical studies of particle confinement in high beta equilibrium mirror plasmas.
D. V. Anderson, B. M. Johnston, J. Breazeal.
Computing in plasma physics and astrophysics, (see 012.014), P10, 2 pp. (1976).

062.078 Synchrotron radiation from magnetic field gaps.
A. A. Rumyantsev, A. A. Zotov.
Astron. Zhurn. Akad. Nauk SSSR, Vol. 53, 560 - 567 (1976). In Russian. English translation in Soviet Astron., Vol. 20, No. 3.
Synchrotron emission from magnetohydrodynamic shock wave fronts propagating in a cosmic plasma is considered. The radiation spectra and angular distribution are calculated. The radiation features related to the magnetic field gap are analyzed in detail. The results obtained are applied to the radiation of electrons which are accelerated on shock wave fronts of strong compression propagating in supernova shells and in the solar chromosphere.

062.079 Populations of excited levels in rarefied plasma with recombination mechanisms.
I. L. Bejgman, E. D. Mikhal'chi.
Astron. Zhurn. Akad. Nauk SSSR, Vol. 53, 568 - 571 (1976). In Russian. English translation in Soviet Astron., Vol. 20, No. 3.
Populations of the excited levels are calculated due to the photorecombination mechanism and to the mechanism associated to dielectronic recombination.

062.080 The equations of radiative hydrodynamics.
J. L. Anderson.
General Relativ. Gravitation, Vol. 7, 53 - 67 (1976).
The hydrodynamic equations of motion for an electron gas in interaction with a radiation field are derived from the kinetic theory description of this system. Variational expressions for the transport coefficients appearing in these equations are obtained. The expressions for the transport coefficients are compared to the previously obtained expressions for these quantities and the differences are noted.

062.081 Laboratory simulation of erosion by space plasma.
L. Kristoferson, K. Fredga.
Stockholms Obs. Rep. No. 10, 2 + 31 pp. (1976).

062.082 Theory of homogeneous dynamos in a rotating liquid sphere.   C. L. Pekeris, Y. Accad.
Proc. National Acad. Sci., USA, Vol. 72, 1496 - 1500 (1975). Abstr. in Zentralbl. Math. Grenzgebiete — Math. Abstr., Band 309, No. 76078 (1976).

062.083 Nonlinear interaction of magnetic field and convection.   F. H. Busse.
Journ. Fluid Mechanics, Vol. 71, 193 - 206 (1975). – Abstr. in Phys. Abstr., Vol. 79, A001351 (1976).

062.084 A wave model of near wakes.   V. C. Liu.
Geophys. Res. Letters, Vol. 2, 485 - 488 (1975). Abstr. in Phys. Abstr., Vol. 79, A025814 (1976).

062.085 Thermodynamics and phase separation of dense fully ionized hydrogen-helium fluid mixtures.
D. J. Stevenson.
Phys. Rev. B, Vol. 12, 3999 - 4007 (1975). – Abstr. in Phys. Abstr., Vol. 79, A027809 (1976). – See 062.092.

062.086 Radiation processes from hot plasmas in a strong magnetic field.   M. El-Khishen, A. El-Gowhari,

A. El-Lakani.
Atomkernenergie, Vol. 27, No. 1, p. 49 - 52 (1976). — Abstr. in Phys. Abstr., Vol. 79, A035018 (1976).

062.087   **Two-fluid analysis of radiation-matter coupling via Compton scattering.**   S.-H. Hsieh.
Thesis Columbia Univ., New York, USA, 142 pp. (1975). (Available from Univ. Microfilms, Ann Arbor, Mich., USA. Order No. 75-18388). — Abstr. in Phys. Abstr., Vol. 79, A037556 (1976).

062.088   **Quasi-linear spectrum of current driven ion cyclotron waves.**   R. W. Harvey.
Phys. Fluids, Vol. 18, 1790 - 1799 (1975). — Abstr. in Phys. Abstr., Vol. 79, A039842 (1976).

062.089   **A theory of strong and weak scintillations with applications to astrophysics.**   L.-C. Lee.
Thesis California Inst. Technol., Pasadena, USA, 297 pp. (1975). (Available from Univ. Microfilms, Ann Arbor, Mich., USA, Order No. 75-20000). — Abstr. in Phys. Abstr., Vol. 79, A042155 (1976).

062.090   **An association among planets, sunspots and solar plasma physics.**   G. L. Goodwin.
AIP National Congress, Adelaide, 1974, p. 13 (1974). —Abstr.

062.091   **Magnetohydrodynamics near a black hole.**
J. R. Wilson.
Meeting recent progr. fundamentals general relativ., Miramare, Trieste, Italy, 7 July 1975, 33 pp. (1975).

062.092   **Thermodynamics and phase separation of dense fully-ionized hydrogen-helium fluid mixtures.**
D. J. Stevenson.
Separate print Cornell Univ., Ithaca, New York, USA, Lab. Atomic Solid State Phys., 34 pp. (1975). — See 062.085.

062.093   **Hydromagnetic turbulence in the direct interaction approximation.**   S. Nagarajan.
Proefschrift Techn. Hogeschool, Eindhoven, Netherlands, 127 pp. (1975).

062.094   **General numerical fluid dynamics algorithm for astrophysical applications.**
H. M. Ruppel, L. D. Cloutman.
Separate print Los Alamos Scientific Lab., New Mexico, USA, 15 pp. (1975).

062.095   **The hydrodynamics of accretion discs. I. Stability. II. Turbulent models.**   J. M. Stewart.
Separate print Max-Planck-Inst. Phys. Astrophys., München, F. R. Germany (1975). — See 14.062.015, 17.062.029.

062.096   **On the average effect of a highly turbulent gravitohydrodynamic field in the hadron era of the universe. 2. Consideration of dissipative processes.**
H. Nariai.
Separate print Hiroshima Univ., Takehara, Japan, Res. Inst. Theoret. Phys., 26 pp. (1974).

A new implicit method for solving two-dimensional partial differential equations in plasma physics.
See Abstr. 021.008.

Inverse iteration in MHD equilibrium computations.
See Abstr. 021.011.

Criteria for local thermal equilibrium in non-hydrogenic plasmas.   See Abstr. 022.014.

Determination of the electron density in a helium plasma.   See Abstr. 022.035.

Remarks concerning the dispersion equation of electromagnetic waves in a magnetized cold plasma.
See Abstr. 022.049.

On the cascade process in strong magnetic and electric fields under astrophysical conditions.
See Abstr. 061.027.

Fundamental limitations of X-ray spectra as diagnostics of plasma temperature structure.
See Abstr. 061.028.

Radiative shock dynamics. I. The Lyman continuum.
See Abstr. 064.047.

Relativistic magnetohydrodynamical effects of plasma accreting into a black hole.   See Abstr. 066.137.

Studies on hard X-ray emission from solar flares and on cyclotron radiation from a cold magnetoplasma.
See Abstr. 073.084.

Collective wave-particle interactions in solar type IV radio sources.   See Abstr. 077.038.

A local diffusion process associated with the sweeping of energetic particles by Io.   See Abstr. 099.001.

Interaction of interplanetary magnetic fields with cometary plasma.   See Abstr. 106.052.

Self-consistent equilibria in the pulsar magnetosphere.   See Abstr. 141.302.

The pulsar equation including the inertial term: its first integrals and its Alfvénic singularity.   See Abstr. 141.325.

The existence of an ultrarelativistic plasma beyond the Alfvén cylinder of a pulsar.   See Abstr. 141.344.

Hydromagnetic waves and cosmic ray diffusion theory.   See Abstr. 143.177.

Errata

062.901   **Erratum: 'Magnetohydrodynamic accretion and the instability of smooth trans-Alfvénic flow'** [Monthly Notices Roy. Astron. Soc., Vol. 171, 537 - 549 (1975)].
D. J. Williams.
Monthly Notices Roy. Astron. Soc., Vol. 174, 725 (1976).

062.902   **Erratum: "On an equilibrium magnetohydrodynamical configuration with axial symmetry. I. "**
[Solnechnye Dannye 1975 Byull., No. 5, p. 70 - 76 (1975)].
V. A. Osherovich.
Solnechnye Dannye 1976 Byull., No. 2, p. 106 (1976). In Russian.

063 Radiative Transfer, Scattering

063.001 Solution of the comoving-frame equation of transfer in spherically symmetric flows. II. Picket-fence models.
D. Mihalas, P. B. Kunasz, D. G. Hummer.
Astrophys. Journ., Vol. 203, 647 - 659 (1976).
To examine the effect of the radial flow of atmospheric material on the temperature distribution in a stellar atmosphere, a picket-fence model with Gaussian lines is formulated and solved numerically in the comoving frame of the gas, which is assumed to move with a prescribed velocity law. Extensive results have been obtained for both static and dynamical models, with planar and moderately extended spherical geometries. The results suggest that the deposition of energy arising from the intrusion of line opacity into the continuum, caused by velocity gradients, could influence the dynamics of the flow.

063.002 Formation of spectral lines with partial frequency redistribution. J. N. Heasley, F. Kneer.
Astrophys. Journ., Vol. 203, 660 - 666 (1976).
The authors present a method for treating the effects of partial frequency redistribution in non-LTE transfer problems. This formulation is appropriate to spectral lines arising from sharp lower levels. Numerical examples of the solar $L\alpha$ and Ca II K line profiles are presented and compared with the profiles obtained by Milkey and his collaborators.

063.003 Resonance-line polarization. I. A non-LTE theory for the transport of polarized radiation in spectral lines in the case of zero magnetic field. J. O. Stenflo.
Astron. Astrophys., Vol. 46, 61 - 68 (1976).
A non-LTE theory for resonance-line polarization due to coherent scattering is developed for the magnetic field-free case. The equation of radiative transfer for the Stokes vector is written in a form analogous to the formalism normally used for the well-explored non-LTE case of isotropic scattering and complete redistribution. This makes it possible to apply the powerful numerical methods developed in the past to solve non-LTE problems and extend them to calculate the transport of polarized light for multi-level atoms. The explicit formulas for the special cases of a spherically symmetric atmosphere and for a two-level atom are also given.

063.004 Resonance-line polarization. II. Calculations of linear polarization in solar UV emission lines.
J. O. Stenflo, L. Stenholm.
Astron. Astrophys., Vol. 46, 69 - 79 (1976).
Resonance-line polarization in a spherically symmetric, plane-parallel atmosphere has been explored by making numerical calculations of the center-to-limb variation of the intensity and polarization line profiles for the two solar UV emission lines Si IV 139.38 nm and C III 97.70 nm. The polarization turns out to depend on the model atmosphere only through τ_d, the optical thickness of the line-forming layer. When τ_d is small, the polarization is constant throughout the core of the line but decreases in the far wings due to the increased contribution from the unpolarized continuum. For large values of τ_d the polarization is reduced in the opaque line core, and maxima develop in the wings where the line-forming layer becomes transparent. It is indicated how observations of resonance-line polarization can be used to determine the structure of the solar atmosphere, including its small-scale geometry.

063.005 Compton reflected spectra of X-ray illuminated stellar atmospheres. J. Felsteiner, R. Opher.
Astron. Astrophys., Vol. 46, 189 - 195 (1976).
The reflected X-ray spectra from a planar stellar atmosphere are calculated, taking into account Compton scatter-

ing, atomic coherent scattering, and photo-electric absorption. The polarization of the reflected spectra, as well as the heating of the stellar atmosphere as a function of depth are evaluated. Some qualitative implications of the results to X-ray binaries are discussed.

063.006 Line formation in turbulent media: comparison between continuous and discontinuous velocity fields. H.-P. Gail, E. Sedlmayr, G. Traving.
Astron. Astrophys., Vol. 46, 441 - 445 (1976).
This is a numerical study of spectral lines formed in a turbulent medium with correlated velocities. In a comparison of two kinematic models which are identical in their one-point velocity distribution functions and in their two-point velocity correlations it turned out that the lines which are formed in a medium with discontinuous fields (generated by a Kubo-Anderson process) are always closer to the macro-turbulent limit than lines which are formed in velocity fields which vary continuously along the ray (generated by an Uhlenbeck-Ornstein process). This systematic difference can be understood as the consequence of the influence of the two point correlations of higher order velocity moments.

063.007 CO line formation in turbulent interstellar clouds where a small velocity gradient is present. R. Lucas.
Astron. Astrophys., Vol. 46, 473 - 475 (1976).
The coupled equations of radiative transfer and statistical equilibrium are solved for CO millimeter-wavelength lines in spherical clouds, in the intermediate case where both microturbulence and a small velocity gradient are present; predicted CO profiles are presented. The implications for the interpretation of observations are discussed.

063.008 Radiative transfer and rotation effects on thermal-convective instability.
R. C. Sharma, K. Prakash.
Indian Journ. Phys., Vol. 49, 557 - 560 (1975). – Abstr. in Phys. Abstr., Vol. 79, A042161 (1976).

063.009 Equation of transfer in an inhomogeneous magnetic field. M. Šidlichovský.
Bull. Astron. Inst. Czechoslovakia, Vol. 27, 1 - 6 (1976).
The equation of radiation transfer in lines arising in multipole transitions in an inhomogeneous magnetic field is given. Polarized radiation is described in terms of the Stokes parameters. The parameter j appears in the equation of transfer for 2^j-pole transitions as an index of associated Legendre polynomials. In the special case $j = 1$ the general equation changes into the well-known equation for dipole radiation.

063.010 Non-isotropic redistribution effects on spectral line formation. I. M. Vardavas.
Journ. Quant. Spectrosc. Radiat. Transfer, Vol. 16, 1 - 13 (1976).
The combined effects of frequency and angular redistribution are examined in slab and semi-infinite static isothermal atmospheres for the cases of pure Doppler damping and Doppler with natural damping using both the dipole and isotropic phase function.

063.011 Analytical approach to radiation transport for non-thermal plasma slabs. S. Suckewer, A. Kuszell.
Journ. Quant. Spectrosc. Radiat. Transfer, Vol. 16, 53 - 62 (1976).
Approximate formulae for the radiation escape factor and for the line source function of a plasma slab are presented. The anisotropic transport equation for a two-level atomic

model is solved by using a modified Schuster-Schwarzschild method. The slowly varying first and second "moments" of the line profiles are introduced. In the present approach, consistent boundary conditions are obtained. The formulae give satisfactory agreement with the numerical results of other authors. A generalization to the case of multi-level atoms is proposed.

063.012 Migration of excitation in transfer of spectral-line radiation. D. G. Hummer, P. B. Kunasz.
Journ. Quant. Spectrosc. Radiat. Transfer, Vol. 16, 77 - 96 (1976).

A simple mathematical model is developed for the transfer of energy through a gas by the combined effect of radiative transfer and migration of excitation. The theory developed here is valid when the distance traveled by an atom while excited is much larger than the typical distance at which two atoms can exchange excitation (roughly 10^{-6} cm). Extensive numerical results are obtained and discussed in terms of characteristic lengths for the various transfer processes.

063.013 Non-gray radiative transfer in plane-parallel media with reflecting boundaries.
Y. Yener, M. N. Özişik, C. E. Siewert.
Journ. Quant. Spectrosc. Radiat. Transfer, Vol. 16, 165 - 175 (1976).

A generalized equation of radiative transfer in the two-group picket-fence model is analyzed for a plane parallel, emitting, absorbing and isotropically scattering medium containing uniform heat sources and having boundary surfaces which are diffuse emitters and diffuse reflectors and are maintained at uniform but arbitrary temperatures. The solution of the general problem is expressed by the superposition of simpler problems which are solved by the application of the normal-mode-expansion technique. Highly accurate numerical results are presented for the temperature distribution and the radiative heat flux in the medium.

063.014 An approximation for solving the non-LTE line transfer in a spatially correlated random velocity field. C. Magnan.
Journ. Quant. Spectrosc. Radiat. Transfer, Vol. 16, 281 - 296 (1976).

The author solves the non-LTE line transfer problem in the presence of random, non-thermal velocities. The correlation length of the turbulence is taken into account and thus he fills the gap between the microturbulence (vanishing correlation length) and the macroturbulence (infinite correlation length). Illustrative numerical examples are given for two-level atoms in finite slabs with two directions of propagation. The results are easily described as being a continuous sequence from the macroturbulent case up to the microturbulent case when the number of turbulent eddies increases. The transitions from one case to the other involve the consideration of three scales: the eddy size, the thermalization length and the mean free path of photon at the center of the line.

063.015 A recurrence relation for transmission and absorption functions of infrared radiating gases.
D. A. Nelson.
Journ. Quant. Spectrosc. Radiat. Transfer, Vol. 16, 321 - 323 (1976).

A new recurrence relation for transmission and absorption functions of i.r. radiating gases analogous to the familiar recurrence relations for the exponential integral functions is derived. Applications of this to the determination of transmission and absorption functions for a band model including line structure effects illustrate the usefulness of the result.

063.016 A perturbation solution of the radiative transfer equation in a differentially moving atmosphere.

L. E. Cram, P. B. Lopert.
Journ. Quant. Spectrosc. Radiat. Transfer, Vol. 16, 347 - 355 (1976).

The authors present a new technique for solving the radiative transfer equation in a differentially moving atmosphere. The method is based on a perturbation of the solution of the transfer problem in a static atmosphere. The perturbation technique may be applied with any method for solving the static atmosphere problem and leads to significant reductions in computer time and storage requirements. The method is flexible and may be used to solve problems involving depth dependence in any of the parameters of the transfer equation.

063.017 An analytic solution for the line formation in a magnetic field. I. M. Šidlichovský.
Bull. Astron. Inst. Czechoslovakia, Vol. 27, 71 - 74 (1976).

An analytic solution for LTE line formation in a homogeneous magnetic field is presented without assuming Milne-Eddington model of atmosphere. Faraday rotation is taken into account.

063.018 Non-LTE transfer − II. Two-level atom with stochastic velocity field. H. Frisch, U. Frisch.
Monthly Notices Roy. Astron. Soc., Vol. 175, 157 - 175 (1976).

Following previous work on LTE stochastic transfer (Auvergne et al.; Frisch), transfer with incoherent scattering is considered for two-level atoms in the presence of turbulent velocity fields with finite eddy-size. It is shown that finite eddy-size effects can change the effective source function and the emergent profile by a factor 2 or more.

063.019 Radiative hydrodynamics of chromospheric transients. F. Kneer, Y. Nakagawa.
Astron. Astrophys., Vol. 47, 65 - 76 (1976).

A self-consistent method for the numerical modeling of transient phenomena in stellar atmospheres is developed. The non-linear equations of hydrodynamics are solved together with the equations of radiative transfer and non-LTE ionization equilibrium. As the first example, the authors calculate the effects of a thermal pulse introduced at the bottom of a model solar chromosphere. The temperature perturbation produces a pressure pulse which runs upwards through the atmosphere and generates complex variations of temperature, velocity, and radiation. The most unexpected results are: (1) For an appreciable time duration after the passage of the pulse, large parts of the chromosphere remain cooler than the initial state. (2) The passage of the pulse with its high temperature variations is hardly seen at the band-head of the Lyman continuum.

063.020 Probabilistic radiative transfer: an integral-equation approach. R. G. Athay.
Astrophys. Journ., Vol. 204, 160 - 167 (1976).

The kinetic equilibrium equations of radiative transfer and steady-state populations of energy levels are formulated in terms of the probabilities for photon creation, destruction, and transport. This new formulation is used to derive an approximate equation, a special case of which has been used elsewhere. The probabilistic formulation leads to new approaches for obtaining rapid, approximate solutions, and, in addition, is amenable to solution by lambda iteration.

063.021 Quasi-asymptotic solutions of the radiative transfer problem in an optically finite shell. I. Conservative scattering. M. A. Mnatsakanyan.
Astrofizika, Vol. 11, 659 - 678 (1975). In Russian. English translation in Astrophysics, Vol. 11, No. 4.

Quasi-asymptotic solutions of the radiative transfer problem in an optically finite shell in the case of monochromatic isotropic conservative scattering are obtained. These solutions are in fact asymptotic ones, but they are more correct than Sobolev's well-known asymptotic solutions and can be practically applied to a shell of arbitrary thickness. If the thickness is very large, they turn to Sobolev's asymptotics.

**063.022 An alternative formulation of the complete lineariza-
tion method for the solution of non-LTE transfer
problems.**   L. H. Auer, J. N. Heasley.
Astrophys. Journ., Vol. 205, 165 - 171 (1976).

The complete linearization method is reformulated so that the timing is linearly proportional to the number of frequencies. The resulting system of equations may be solved by a simple block iterative method. Numerical examples are given which show that although the new formulation requires the same amount of work per iteration as the "equivalent two-level atom" approach, it converges in cases where the latter fails.

**063.023 An initial-value method for internal radiation field
in generalized Chandrasekhar's planetary problem.**
S. Ueno, A. P. Wang.
Bull. American Astron. Soc., Vol. 8, 291 - 292 (1976).
Abstr. AAS.

**063.024 Extension of the doubling method to inhomogene-
ous sources.**   W. J. Wiscombe.
Journ. Quant. Spectrosc. Radiat. Transfer, Vol. 16, 477 - 489 (1976).

This paper gives a general theory for applying the doubling method to spatially inhomogeneous radiation sources whose angular and spatial variations separate. In particular, in-homogeneous sources of thermal radiation may be efficiently treated by the methods herein, as well as direct and specular-ly-reflected beams of radiation which do not lie along a quadrature direction for the intensity.

**063.025 Infrared radiative transfer in planetary atmo-
spheres – I. Effects of computational and spectroscopic economies on thermal heating/cooling rates.**
G. E. Hunt, S. R. Mattingly.
Journ. Quant. Spectrosc. Radiat. Transfer, Vol. 16, 505 - 520 (1976).

The availability of accurate spectroscopic data and improved knowledge of continuum absorption characteristics has required a further detailed study of i.r. radiative transfer. A sophisticated radiative transfer scheme based upon the Mayer-Goody random band model is developed and the de-rived transmissivities for H_2O, CO_2 and O_3 are calibrated against laboratory measurements. This calibrated scheme is used to assess the effects on calculated heating/cooling rate profiles of introducing computational and spectroscopic economies.

**063.026 Resonance-line transfer with partial redistribution.
VIII. Solution in the comoving frame for moving
atmospheres.**
D. Mihalas, R. A. Shine, P. B. Kunasz, D. G. Hummer.
Astrophys. Journ., Vol. 205, 492 - 498 (1976).

An analysis of the effects of partial frequency redistribu-tion in the scattering process for lines formed in moving atmo-spheres has been performed using a flexible and general meth-od which allows solution of the transfer equation in the comoving frame of the gas. As a specific example, the authors consider the same chromospheric and atomic model, with the same velocity field, that was studied by Cannon and Vardavas. They find that the large changes in the profiles obtained by those authors, between the cases of complete and partial re-distribution are spurious effects of angle-averaging in the ob-server's frame instead of the comoving frame.

**063.027 A tensor formulation of the equation of transfer
for spherically symmetric flows.**   B. M. Haisch.

Astrophys. Journ., Vol. 205, 520 - 526 (1976).

A tensor formulation of the equation of radiative transfer is derived in a seven-dimensional Riemann space such that the resulting equation constitutes a divergence in any coordinate system. The divergence theorem may be applied to yield a numerical differencing scheme which is expected to be stable and conserve luminosity. It is shown that the equation of transfer derived by this method in a Lagrangian coordinate system may be reduced to that given by Castor, although it is, of course, desirable to leave the equation in divergence form.

**063.028 Multiple scattered radiation emerging from Rayleigh
and continental haze layers. 1: Radiance, polariza-
tion, and neutral points.**   G. W. Kattawar, G. N. Plass, S. J. Hitzfelder.
Applied Optics, Vol. 15, 632 - 647 (1976).

**063.029 Multiple scattered radiation emerging from Rayleigh
and continental haze layers. 2: Ellipticity and direc-
tion of polarization.**   G. N. Plass, G. W. Kattawar, S. J. Hitz-felder.
Applied Optics, Vol. 15, 1003 - 1011 (1976).

**063.030 Numerical solution of the radiative transfer equation
in spherically symmetric dust shells.**
C. M. Leung.
Journ. Quant. Spectrosc. Radiat. Transfer., Vol. 16, 559 - 574 (1976).

An efficient numerical method is presented for solving problems of radiative transfer in dust shells with spherical sym-metry. The method can be applied to a wide variety of radia-tive transfer problems in planetary atmospheres, circumstellar dust envelopes and extended stellar atmospheres. As an illus-tration, the method is used to determine the grain temperature distribution and far i.r. emission from interstellar dust clouds associated with compact H II regions.

063.031 Rapid computation of the Voigt profile.
S. R. Drayson.
Journ. Quant. Spectrosc. Radiat. Transfer, Vol. 16, 611 - 614 (1976). – Note.

063.032 Notes on probabilistic radiative transfer.
G. D. Finn.
Journ. Quant. Spectrosc. Radiat. Transfer, Vol. 16, 615 - 618 (1976).

Some further applications are made of Frischs' theorem on the surface values of functions satisfying linear integral equations of transfer. The meaning of the surface values is discussed in terms of random walks.

**063.033 Radiative transfer in a multilayer optically thick
atmosphere. I.**   V. V. Ivanov.
Trudy Astron. Obs., *Leningrad*, Vol. 32 (= Uchenye Zapiski Leningr. Un-ta, No. 385 = Seriya matem. nauk, vyp. (No.) 52), p. 3 - 23 (1976). In Russian.

Asymptotic forms are studied of the reflection and trans-mission functions of a plane-parallel atmosphere composed of several optically thick layers with different optical properties. The paper begins with a summary of the results pertaining to homogeneous atmospheres. A study is given of a medium composed of two adjacent half-spaces with different optical properties. It is assumed that the radiation field in the medium is due to a source at infinite depth. The angular dependence of the intensity at the boundary of the half-spaces and the asymp-totic regime far from the boundary are studied in detail.

**063.034 Radiative transfer in a multilayer optically thick
atmosphere. II.**   V. V. Ivanov.
Trudy Astron. Obs., *Leningrad*, Vol. 32 (= Uchenye Zapiski Leningr. Un-ta, No. 385 = Seriya matem. nauk, vyp. (No.)

52), p. 23 - 39 (1976). In Russian.

Asymptotically exact expressions are given for the reflection and transmission functions of an atmosphere composed of several optically thick layers. Each of the layers has its own optical properties, i.e. probability of photon survival and phase function.

063.035 The reflection of radiation from a semi-infinite two-layer atmosphere.   A. K. Kolesov.
Trudy Astron. Obs., *Leningrad*, Vol. 32 (= Uchenye Zapiski Leningr. Un-ta, No. 385 = Seriya matem. nauk, vyp. (No.) 52), p. 39 - 51 (1976). In Russian.

Isotropic scattering of radiation in an atmosphere consisting of a semi-infinite plane-parallel layer and an upper layer of finite optical thickness is investigated. The values of the albedo for single scattering in the layers are different. The emergent intensity and the intensities of radiation at the boundary of the layers are expressed in terms of six auxiliary functions of one angular variable.

063.036 Solution of the comoving-frame equation of transfer in spherically symmetric flows. III. Effect of aberration and advection terms.
D. Mihalas, P. B. Kunasz, D. G. Hummer.
Astrophys. Journ., Vol. 206, 515 - 524 (1976).

The authors investigate the importance of the advection and aberration terms, which are of order V/c, in the comoving-frame transfer equation in spherical geometry. Characteristic trajectories are found which reduce the spatial derivatives to a perfect differential, and a generalization of the numerical procedure developed in the earlier papers of this series that permits the integration of the transfer equation on these characteristics is presented.

063.037 Scattering by spheres with nonisotropic refractive indices.   R. T. Wang, J. M. Greenberg.
Applied Optics, Vol. 15, 1212 - 1217 (1976).

Some results of experimental studies on microwave scattering by artificially constructed axially symmetric spheres with anisotropic refractive indices are presented. An approximation using Mie theory for spheres with appropriate orientation dependent indices of refraction is shown to provide a good explanation of the observed dependence of the complex forward-scattering amplitudes on target orientation.

063.038 Transfer of X-rays through a spherically symmetric gas cloud.
S. Hatchett, J. Buff, R. McCray.
Astrophys. Journ., Vol. 206, 847 - 860 (1976).

The authors present approximate solutions of the transfer of radiation through spherically symmetric gas clouds surrounding a point source of X-rays. The temperature and ionization structure are sensitive to the source spectrum and the solutions are not unique if soft X-rays are deficient. The emergent spectrum is rich in optical, ultraviolet, and X-ray emission lines. The radiation force due to photoelectric absorption of X-rays may exceed the force due to Compton scattering by a factor of order 10 for the radiation fields and densities likely to be encountered in galactic binary X-ray sources.

063.039 Über den Strahlungstransport in zirkumstellaren Staubhüllen und seinen Einfluß auf Temperaturverteilung und Spektrum.   D. Hermann.
Diss. Nat. Gesamtfakultät, Ruprecht-Karl-Univ., Heidelberg, 106 pp. (1976).

063.040 Perturbation solutions of the equation of radiative transfer in a stellar atmosphere.   L. E. Cram.
Computing in plasma physics and astrophysics, (see 012.014), H4, 2 pp. (1976).

063.041 Scattering functions of dielectric and absorbing irregular particles.   R. Zerull.
IAU Colloquium No. 31, (see 012.015), p. 130 - 134 (1976).

063.042 Lösung des stationären Strahlungstransportproblems für Energiestreuung mit Hilfe von Matrizenfunktionen.   H. Melcher, E. Gerth.
Kernenergie, 16. Jahrgang., p. 47 - 52 = Mitt. Astrophys. Obs. Potsdam No. 173 (1973).

063.043 The Eddington approximation generalized for radiative transfer in spherically symmetric systems. I. Basic method.   W. Unno, M. Kondo.
Publ. Astron. Soc. Japan, Vol. 28, 347 - 354 (1976).

A modified Eddington approximation proper to spherically symmetric systems is formulated by use of the concept of radiation streams in two solid angle ranges approximating the radiation field. Application to the gray and conservative cases shows that simplicity and considerable accuracy are achieved with the method.

063.044 Radiative transfer and collisional effects on thermal-convective instability of a composite medium.
R. C. Sharma, K. Prakash.
Progr. Theor. Phys., Vol. 54, 409 - 414 (1975). — Abstr. in Phys. Abstr., Vol. 79, A012191 (1976).

063.045 Analytically solvable problems in radiative transfer. IV.   C. von Trigt.
Phys. Rev. A, Vol. 13, 726 - 733 (1976). — Abstr. in Phys. Abstr., Vol. 79, A042011 (1976).

Dynamic light scattering. With applications to chemistry, biology, and physics.   See Abstr. 003.020.

Light scattering in planetary atmospheres. See Abstr. 003.102.

Radiation processes in astrophysics. See Abstr. 003.106.

Extinction properties of porous spheres. See Abstr. 022.033.

The solution of the radio transfer equation in a plasma with constant pressure and conductive flux. See Abstr. 062.022.

Linear polarization from rotating extended atmospheres of stars.   See Abstr. 064.006.

Radiative transport in circumstellar dust shells. See Abstr. 064.018.

Radiative shock dynamics. I. The Lyman continuum. See Abstr. 064.047.

Radiative transfer in spherical circumstellar dust envelopes.   See Abstr. 064.066.

Light scattering properties of naked singularities. See Abstr. 066.125.

A hybrid mode of diffuse and specular reflector for computation of the emergent radiation by the adding method. See Abstr. 082.039.

Absorption and emission by atmospheric gases. See Abstr. 082.062.

Extinction by condensed water.

See Abstr. 082.063.

The optics of spherically stratified graphite grains.
See Abstr. 131.028.

Radiative transfer in circumstellar dust.
See Abstr. 131.180.

Exact evolution of photons in an anisotropic cosmology with scattering. See Abstr. 162.032.

064   Stellar Atmospheres, Stellar Envelopes, Mass Loss

064.001   An attempt to determine the circumstellar reddening law.   D. A. Allen.
Monthly Notices Roy. Astron. Soc., Vol. 174, 29P - 33P (1976).

The reddening law for circumstellar dust can in principle be determined if it is assumed that the energy radiated at infrared wavelengths by the dust exactly balances that absorbed from the star. The method is attempted for a number of early-type stars, and no evidence is found that the ratio of total to selective absorption, A_V/E_{B-V}, differs from that of interstellar reddening. An apparent reduction of A_V/E_{V-I} relative to the interstellar case is probably explained by free-free emission from the circumstellar gas.

064.002   Accelerated gas outflow in early-type emission-line stars.   K. A. Marsh.
Astrophys. Journ., Vol. 203, 552 - 555 (1976).

The viscous force of dust grains driven by radiation pressure is likely to provide an effective accelerating mechanism for the gaseous envelopes of early-type emission-line stars such as V1016 Cygni, provided the dust grains are uniformly mixed with the gas. The form of the observed radio spectrum of V1016 Cygni is consistent with such an accelerated expansion.

064.003   Evidence for mass loss at moderate to high velocity in Be stars.   T. P. Snow, Jr., J. M. Marlborough.
Astrophys. Journ., (*Letters*), Vol. 203, L87 - L90 (1976).

Ultraviolet spectra of intermediate resolution have been obtained with Copernicus of 12 objects classified as Be or shell stars, and 19 additional early B dwarfs. Direct evidence for mass loss is seen for the first time in dwarf stars as late as B1.5; the only objects later than B0.5 which show this effect are Be or shell stars. The mass loss rate for one of the most active Be stars, 59 Cyg, is crudely estimated to be $10^{-10} - 10^{-9} \, M_\odot \, \mathrm{yr}^{-1}$. The data are suggestive that the extended atmospheres associated with Be star phenomena may be formed by mass ejection.

064.004   Applications of Fourier analysis to broadening of stellar line profiles. II. A saturated line in 32 Aquarii and in Sirius.   M. A. Smith.
Astrophys. Journ., Vol. 203, 603 - 609 (1976).

An application of Gray's Fourier analysis technique demonstrates that the shape of the Fe I λ 4476 profile in 32 Aqr is consistent with a Gaussian microturbulent velocity field in the atmosphere of the Am star. Its magnitude lies in the range 3 ± 1 km s^{-1}. The rotational velocity of 32 Aqr is 10 ± 1 km s^{-1}. The amount of Gaussian macroturbulence appears to be negligible (<2 km s^{-1}). An extremely well observed profile of a saturated line in Sirius is used to derive a rotational velocity of $V \sin i = 17 \pm 1$ km s^{-1}. The roundedness of this observed profile suggests, in addition to the rotation, the presence of a small amount of turbulence, probably macroturbulence.

064.005   Rotating and selfgravitating winds.   R. D. Weidelt.
Astron. Astrophys., Vol. 46, 213 - 218 (1976).

The author considers a rotating star with an isothermal and stationary expanding corona whose gravitational potential cannot be neglected. As in ordinary wind theories there exists a critical sonic point. The position of this critical point, the initial velocity of the gas and the mass loss of the critical solution are calculated. As long as the density of the corona is 'small' the mass loss is found to increase with growing κ and η while for 'high' density a decrease occurs.

064.006   Linear polarization from rotating extended atmospheres of stars.   A. Peraiah.
Astron. Astrophys., Vol. 46, 237 - 241 (1976).

Linear polarization from the extended atmospheres of rotating stars has been calculated. Two types of atmospheres have been considered: (1) the atmospheres of early type supergiants with electron scattering (Thomson scattering) and (2) the atmospheres of late type supergiants with scattering due to H_2 molecules (Rayleigh scattering). In the latter type of atmosphere, the wavelength dependence of linear polarization has been calculated and it appears to be in accordance with the observational result that the polarization increases with decreasing wavelength. It is also shown that, in the case of an electron scattering atmosphere, polarization always increases with the size of the atmosphere, while this need not be so in the case of an atmosphere with molecular scattering.

064.007   On total mass loss in stars with $M \lesssim 8 \, M_\odot$.   F. Fusi-Pecci, A. Renzini.
Astron. Astrophys., Vol. 46, 447 - 454 (1976).

In two previous papers the authors have obtained an expression for the mass loss rate from red giants and supergiants, assuming that the physical process responsible for the mass loss is similar to that leading to the solar wind. In this paper it is shown that this mass loss expression coincides, within observational uncertainty, with that experimentally determined by Reimers (1975). The two mass loss expressions coupled with the available evolutionary rates, are used to predict the total mass loss suffered by stars (with $M \lesssim 8 \, M_\odot$) evolving along the first and second red giant branch (single shell and double shell models respectively). The problem of mass loss by radiation pressure on dust grains and by an overall envelope instability are also discussed.

064.008   A grid of model atmospheres for metal-deficient giant stars. II.   R. A. Bell, K. Eriksson, B. Gustafsson, Å. Nordlund.
Astron. Astrophys., Suppl. Ser., Vol. 23, 37 - 95 (1976).

The details of a grid of flux-constant model atmospheres, described in a previous paper, are given. The grid covers the general range $3750 \, \mathrm{K} \leq T_{eff} \leq 6000 \, \mathrm{K}$, $0.75 \leq \log g \leq 3.0$, $-3.0 \leq [\mathrm{A/H}] \leq 0.0$. Line blanketing due to metal lines and molecular lines is considered in great detail. Convection is treated in the mixing-length approximation. A solar model, consistent with the rest of the models, is also presented.

064.009   Model atmospheres for cool hydrogen-rich white dwarfs.   R. Wehrse.
Astron. Astrophys., Suppl. Ser., Vol. 24, 95 - 110 (1976).

Model atmospheres for cool ($7000 \leq T_{eff} \leq 12000 \, \mathrm{K}$) hydrogen-rich white dwarfs (spectral type DA) are presented for two gravities ($\log g = 8.0$ and $\log g = 8.5$). For most models the helium and CNO abundances are assumed to be solar, while the abundances of the heavy elements are reduced by a factor of 100.

064.010   Study of turbulence in the atmosphere of Procyon and Arcturus by the Goldberg-Unno method.   J. Sikorski.
Acta Astron., Vol. 26, 1 - 14 (1976).

The method as proposed by Goldberg has been applied to the stars α Boo and α CMi. The Doppler width and the turbulent velocity have been determined as a function of depth in the atmosphere of the star. An increase of the ξ parameter with height in both of those stars has been obtained.

064.011   Stellar chromospheres.   R. Głębocki.
Postępy Astron., Vol. 23, 229 - 242 (1975). In Polish.

Principal properties of stellar chromospheres for late-type stars are presented. A brief discussion is related mainly to the problems of: observational indicators of stellar chromospheres, correlations between stellar chromospheres and other atmospheric parameters, evolutionary effects and possibilities of construction of chromospheric models.

064.012 Propagation of an optically thin isothermal perturbation in an atmosphere traversed by a radiation field.
G. Berthomieu, J. Provost, A. Rocca.
Astron. Astrophys., Vol. 47, 413 - 416 (1976).

The propagation of optically thin isothermal perturbations in an isothermal slab of an atmosphere traversed by a radiation flux is considered. Such perturbations are amplified during their propagation up to a finite limit which is a function of the effective gravity, and which becomes infinite only in the rather unrealistic case where this effective gravity is zero. Although this result depends on the assumption made for the absorption coefficient, it shows that a non local analysis of the amplification is needed in order to be able to apply it to the chromospheric heating of hot stars, a mechanism suggested by Hearn (1972).

064.013 A statistical method for treating molecular line opacities.
C. Sneden, H. R. Johnson, B. M. Krupp.
Astrophys. Journ., Vol. 204, 281 - 289 = Publ. Goethe Link Obs., Indiana Univ., *Bloomington*, No. 173 (1976).

A method for treating atomic and molecular line opacities in cool stellar atmospheres by a statistical opacity sampling is investigated. The authors investigate the number of frequencies needed to allow an accurate integration of the energy flux over a given spectral interval as a function of the depth and including opacity for both CN and C_2. They extend this method to the calculation of a model atmosphere of a star and study the effect of the number and placement of frequency points. The authors finally apply the method to treating molecular lines of CO, C_2, and CN in a cool carbon star.

064.014 Forbidden and permitted emission lines of singly ionized iron as a diagnostic in the investigation of stellar emission-line spectra.   R. Viotti.
Astrophys. Journ., Vol. 204, 293 - 300 (1976).

The author presents a diagnostic approach to the problem of the origin of the [Fe II] and Fe II emission lines in the extended atmospheres of hot stars. The rates of excitation and de-excitation of Fe$^+$ levels by electron impact, spontaneous decay, and photoexcitation are studied. Radiative recombination of doubly ionized iron is also considered in an approximate form. The two relevant parameters which describe the physical conditions in the envelopes of hot stars are $N_e T_e^{-1/2}$ and the dilution factor W. The author then introduces a two-parameter diagram $[W, N_e T_e^{-1/2}]$, and discusses the position in this diagram of Be stars, novae, composite-spectrum variables, and other peculiar stars.

064.015 Mass loss in red giants and supergiants.
F. Sanner.
Astrophys. Journ., (*Letters*), Vol. 204, L41 - L45 (1976).

The circumstellar envelopes surrounding 13 late-type giants and supergiants have been studied using a homogeneous collection of high-resolution, photoelectric scans of strong optical resonance lines. Various properties of the envelopes, including the mass loss rate, dilution factor, hydrogen density, and degree of ionization, have been determined quantitatively.

064.016 The geometry of VY Canis Majoris derived from SiO maser lines.   D. Van Blerkom, L. Auer.
Astrophys. Journ., Vol. 204, 775 - 780 = Contr. Five Coll. Obs., Univ. Mass., *Amherst*, No. 213 (1976).

The authors find by means of a radiation transfer calcula-

tion that the SiO maser lines observed in the spectrum of VY CMa are formed in a rotating equatorial disk seen nearly edge-on. Other geometries are considered and eliminated. It is suggested that the SiO lines of NML Cyg also show evidence that they are formed in a disk.

064.017 The chemical compositions of two subgiant CH stars.
C. Sneden, H. E. Bond.
Astrophys. Journ., Vol. 204, 810 - 817 = Lick Obs. Bull., No. 717 = Contr. Louisiana State Univ. Obs., *Baton Rouge*, No. 110 (1976).

The authors present model-atmosphere analyses of the atmospheres of two "subgiant CH stars," HD 176021 and HD 204613. These stars possess small, but definite, iron-peak-element deficiencies; substantial overabundances of the s-process elements; and large, nearly solar, surface gravities. Through spectrum-synthesis techniques, the authors derive overabundances of carbon but normal abundances of nitrogen. The derived log g values lead to absolute magnitudes near M_v = +4. HD 176021 and HD 204613 are interpreted as highly evolved population II stars that have recently returned to the vicinity of the main sequence as a consequence of extensive internal mixing.

064.018 Radiative transport in circumstellar dust shells.
R. E. Taam, R. D. Schwartz.
Astrophys. Journ., Vol. 204, 842 - 852 = Lick Obs. Bull., No. 713 (1976).

A calculation of a nongray, extended spherical circumstellar dust shell is presented. The study has been limited to an exploration of the following parameters: (1) angular dependence of the incident radiation field, (2) density distribution, (3) opacities, (4) relative extension of the dust shell, and (5) distance of the inner radius of the shell from the stellar surface. Anisotropic scattering has been included in an approximate manner. The results are applied to VY CMa.

064.019 Theoretische Chromosphären später Sterne.
F. Schmitz, P. Ulmschneider.
Mitt. Astron. Ges., No. 38, p. 174 - 175 (1976). – Abstract.

064.020 Ausgedehnte statische Sternatmosphären im HR-Diagramm.   J. Schmid-Burgk, M. Scholz.
Mitt. Astron. Ges., No. 38, p. 175 (1976). – Abstract.

064.021 Konvektive Durchmischung in der Hülle von Weißen Zwergen.   D. Koester.
Mitt. Astron. Ges., No. 38, p. 175 - 176 (1976). – Abstract.

064.022 Non-LTE model atmospheres of subluminous O-stars.
R. P. Kudritzki.
Mitt. Astron. Ges., No. 38, p. 200 - 201 (1976). – Abstract.

064.023 H II-Regionen rasch rotierender O-Sterne.
R. Kippenhahn, E. Krügel.
Mitt. Astron. Ges., No. 38, p. 201 (1976). – Abstract.

064.024 Bildung von Fraunhoferlinien in turbulenten Atmosphären bei Abweichungen vom lokalen thermodynamischen Gleichgewicht.
H. P. Gail, E. Sedlmayr, G. Traving.
Mitt. Astron. Ges., No. 38, p. 201 (1976). – Abstract.

064.025 Entmischung in Ap-Stern-Atmosphären durch selektive Diffusionen.   H.-P. Gail, E. Sedlmayr.
Mitt. Astron. Ges., No. 38, p. 201 (1976). – Abstract.

064.026 Nucleation and growth of dust grains.
E. E. Salpeter.
Astrophys. Space Sci. Library, Vol. 55, (see 012.001), 67 (1976). – Abstr. of a paper published in Astrophys. Journ.,

Vol. 193, 579 - 584 (1974) — see 012.064.034.

064.027 Formation and flow of dust grains in cool stellar atmospheres. E. E. Salpeter.
Astrophys. Space Sci. Library, Vol. 55, (see 012.001), 69 (1976). — Abstr. of a paper published in Astrophys. Journ., Vol. 193, 585 - 592 (1974) — see 012.064.035.

064.028 Polarization properties of silicate-like grains in circumstellar envelopes of late-type stars due to temperature variations. J. Svatoš, M. Šolc, V. Vanýsek.
Astrophys. Space Sci. Library, Vol. 55, (see 012.001), 201 - 206 (1976).

The influence of temperature changes in circumstellar silicate-like envelopes upon the polarization effects is investigated. It is shown that under the assumption that $\Delta T_g > 50°$ and conductivity of silicate grains is indirectly proportional to T_g this mechanism can be responsible for the observed dependence of intensity vs polarization in some late-type stars, e.g. V CVn. The same effects can be produced by dirty ices and graphite grains. It is suggested that irradiation by electrons and/or protons can affect the circumstellar envelopes in a similar way, especially those of early-type stars, and irradiation by neutrons can exert an influence on the envelopes of supernovae.

064.029 Mass loss from Mira variables by the action of radiation pressure on molecules. W. J. Maciel.
Astron. Astrophys., Vol. 48, 27 - 31 (1976).

The mass loss rate from late-type giants is still poorly known. The author develops a simple model for the stellar envelope and compute dM/dt owing to the action of the radiation pressure on molecular bands of CO, H_2O and OH. He estimates the efficiency of the mechanism and concludes that it is more important in mass ejection than was thought before.

064.030 Distribution of dust around η Carinae. C. D. Andriesse.
Astron. Astrophys., Vol. 48, 137 - 139 (1976).

A simple model is presented for the wavelength dependent extent of the infrared envelope of η Car. The mass density of dust is postulated to vary with a power of the distance to the star between the inner and outer edges at 5×10^{14} m and 4×10^{15} m. At the inner edge the mass density is found to be 1.2×10^{-20} kg m^{-3} and at the outer edge higher 2.8×10^{-19} kg m^{-3}.

064.031 Model of the dust envelope of an early-type star in a young cluster. V. I. Kardopolov.
Issled. ehkstremal'no molodykh zvezdn. kompleksov. Tashkent, Fan, 1975, p. 83 - 95. In Russian. — Abstr. in Referativ. Zhurn. 51. Astron., 3.51.399 (1976).

064.032 Model atmospheres for normal stars (a survey from 1965 to 1973). L. S. Lyubimkov.
Astrofizika, Vol. 11, 703 - 739 (1975). In Russian. English translation in Astrophysics, Vol. 11, No. 4.

Basic data are given on model atmospheres for O-M type stars calculated in 1965–1973. The influence of blanketing effect, convection, departures from LTE and abundance anomalies on the atmospheric structure and emergent radiation is discussed. The most accurate models for each spectral type are pointed out. Comparison with observations is carried out.

064.033 OH—IR stars. I. Physical properties of circumstellar envelopes. P. Goldreich, N. Scoville.
Astrophys. Journ., Vol. 205, 144 - 154 = Contr. Div. Geol. Planet. Sci., California Inst. Technol., No. 2623 (1976).

A theoretical model of the circumstellar envelope which surrounds a OH—IR star is developed. The circumstellar gas is ejected by radiation pressure which acts on dust grains that condense in the atmosphere of the central star. The dust grains transfer momentum to the gas by collisions with the gas molecules. These collisions are the dominant source of heat input to the circumstellar gas. The OH molecule abundance in the circumstellar envelope is controlled by chemical exchange reactions and by the dissociation of H_2O molecules. In the outer region of the circumstellar envelope, OH molecules are produced from the photodissociation of H_2O molecules by the interstellar ultraviolet radiation and from the dissociation of H_2O molecules by collisions with dust grains.

064.034 Properties of the chromosphere-corona transition region in Capella. B. M. Haisch, J. L. Linsky.
Astrophys. Journ., (*Letters*), Vol. 205, L39 - L42 (1976).

Analysis of recent ultraviolet observations of the Capella binary system (α Aur) indicates a dense, geometrically narrow chromosphere-corona transition region in the Capella system primary (G5 III) similar in many respects to a solar active region. An examination of the coronal energy balance, together with the coronal base pressure derived from the line fluxes, predicts a corona with a mean temperature of 1.2×10^6 K and a large stellar wind consistent with observations.

064.035 The fundamental bands of CO as chromospheric indicators in late-type giant stars.
J. N. Heasley, R. W. Milkey.
Astrophys. Journ., (*Letters*), Vol. 205, L43 - L45 (1976).

The authors present synthetic spectra for the vibrational-rotational fundamental transitions ($\Delta V = 1$) in the ground electronic state of CO for the upper-photosphere–lower-chromosphere atmospheric model of Arcturus derived by Ayres. They find that the CO spectrum is formed in LTE and the strongest molecular lines exhibit emission cores reflecting the chromospheric temperature rise. The CO fundamental bands offer an excellent observational probe for the presence of stellar chromospheres and a consistency check for chromospheric models derived from traditional chromospheric indicators.

064.036 The theoretical near ultraviolet spectrum of B type stars. M. Burger, K. A. van der Hucht.
Astron. Astrophys., Vol. 48, 173 - 185 (1976).

Theoretical spectra are computed for a grid of stellar model atmospheres with 10000 K $\lesssim T_{eff} \lesssim 30000$ K and $2.5 \lesssim \log g \lesssim 4$ in three wavelength regions: 2060–2160 Å, 2495–2595 Å, 2770–2870 Å. The results show a good agreement with observations. Lines of singly ionized elements are insensitive to gravity around 16000 K, lines of twice ionized elements around 25000 K. The microturbulent velocity is certainly much lower than 4 km/s in main sequence B stars cooler than 20000 K.

064.037 Synthetic spectra of red giants. I. Representative band head profiles of diatomic molecules.
J. M. Scalo, J. E. Ross.
Astron. Astrophys., Vol. 48, 219 - 234 (1976).

Synthetic band head profiles of the $^2\Sigma - ^2\Sigma$ systems of AlO, YO and CN, the $^4\Sigma - ^4\Sigma$ system of VO, and the $^2\Delta - ^2\Pi$ system of SiH are calculated using model atmospheres with varying (O–C)/H and temperature. The results indicate that most of the observed properties of the molecular spectra of M, MS, S, SC and N stars can be accounted for by a simple increase in the carbon abundance from its solar system value towards unity and above.

064.038 Predicted line strengths of the CH infrared bands in stellar spectra. B. M. Krupp.
Bull. American Astron. Soc., Vol. 8, 292 (1976). — Abstr. AAS.

064.039 Note on the theory of shock waves in stellar envelopes. I. A. Klimishin, A. F. Novak.

Tsirk. astron. Obs. L'vov, No. 49, p. 3 - 5 (1974). In Russian.

064.040 Theoretical values of the limb darkening coefficients for dwarf stars in the B, V system.
A. A. Rubashevskij.
Astrometriya i Astrofizika, *Kiev*, vyp. (No.) 28, (see 003.002), p. 89 - 92 (1976). In Russian.

Theoretical values of limb darkening coefficients for classical stars in the B, V system were calculated by means of reaction curves for the B, V system and the tables of monochromatic coefficients of limb darkening. It is shown that the values of monochromatic darkening coefficients for the wavelengths 4400 Å and 5500 Å respectively, can be used when interpreting light curves of eclipsing binaries in the B, V system.

064.041 Evaluation of the line blanketing effect on theoretical limb darkening coefficients for late dwarf stars.
A. A. Rubashevskij.
Astrometriya i Astrofizika, *Kiev,* vyp. (No.) 29, (see 003.003), p. 71 - 75 (1976). In Russian.

Taking into account the line blanketing effect at wavelength $\lambda = 3646$ Å is shown to reduce the values of theoretical limb darkening coefficients by 0.01–0.02, 0.02, 0.04, 0.04 for four stellar atmosphere models with effective temperatures 7200, 6300, 5700 and 5600°K respectively. Within the range of $3648 \leqslant \lambda \leqslant 6500$ Å the decrease in the limb darkening coefficients is appreciably smaller and practically independent of the wavelength. It is equal to 0.005, 0.008, 0.015 and 0.016 for the same sequence of temperatures. The limb darkening coefficients in the B and V system decrease by the same amount.

064.042 A study of M dwarfs. II. A grid of model atmospheres. J. R. Mould.
Astron. Astrophys., Vol. 48, 443 - 459 (1976).

A simple opacity distribution function treatment of line blanketing by atoms and molecules is described based on data from statistical spectroscopy and the smeared line model for diatomic molecules. Reasonable agreement between models and observations is apparent in the red and infrared regions of M dwarfs. The emergent fluxes of a grid of models are dominated by oxide bands (TiO and H_2O) which contain abundance information available even at low spectral resolution. The relative flux calibration of red and infrared colors is examined, and a calibration adopted which is probably accurate to within a few percent. Broadband colors are calculated from the models, using the calibration.

064.043 Stellar accretion disks. B. Warner.
Observatory, Vol. 96, 49 - 53 (1976).

064.044 The application of the diffusion hypothesis to extreme overabundance factors in Ap stars.
C. R. Cowley, C. A. Day.
Astrophys. Journ., Vol. 205, 440 - 445 (1976).

The authors investigate the possibility of producing large overabundances ($\sim 10^5$) in Ap stars by radiation-driven diffusion. The emphasis is on the time scale for the particles to diffuse from the depths necessary to produce the overabundances. A number of the relevant parameters can be estimated only very crudely, but if a reasonable choice, favorable to the diffusion, is made, the mechanism can create large overabundances in a time much shorter than the lifetime of the star. The element mercury is specifically discussed.

064.045 Mass loss from dwarf M stars through stellar flaring.
G. D. Coleman, S. P. Worden.
Astrophys. Journ., Vol. 205, 475 - 481 (1976).

It is shown that mass loss from dwarf M stars arising from flaring and stellar winds may lead to a significant mass

and energy input into the interstellar medium. It is further demonstrated that the stellar flares on these stars may be producing through nuclear reactions a large fraction of the observed interstellar deuterium. The same mass loss would also give rise to substantial galactic winds in galaxies with small ambient interstellar gas components. These results show that stellar flaring can account for many effects previously ascribed to supernovae.

064.046 Mass loss by cool carbon stars. L. B. Lucy.
Astrophys. Journ., Vol. 205, 482 - 491 (1976).

Extended model atmospheres, both static and expanding, are calculated for cool carbon stars. Expanding atmospheres are constructed, the basic assumptions being (1) that the outflow is stationary; (2) that sudden formation of small grains occurs when the supersaturation ratio S reaches a critical value, S_c; and (3) that there is momentum-coupling between the grains and the gas. With these assumptions, grain formation occurs at the sonic point, and the resulting increase in the gradient of radiation pressure then drives the gas-grain mixture from the star.

064.047 Radiative shock dynamics. I. The Lyman continuum.
R. I. Klein, R. F. Stein, W. Kalkofen.
Astrophys. Journ., Vol. 205, 499 - 519 (1976).

The authors investigate the shock produced by a constant-velocity piston moving into an atmosphere in radiative, hydrostatic, and statistical equilibrium. Self-consistent numerical solutions are obtained to the equations of hydrodynamics, radiative transfer, and level population. Only Lyman continuum radiation is considered. The results are interpreted in terms of the relaxation lengths for the collisional and radiative processes, and by comparing the "radiative" case both with one in which only collisional transitions occur and with the adiabatic case.

064.048 Model atmospheres and effective temperatures of K- and M-type giants. J. van Paradijs.
Astron. Astrophys., Vol. 49, 53 - 56 (1976).

Using empirical line blocking coefficients and a set of simple model atmospheres (closely resembling scaled solar models) a relationship between $R-I$ and effective temperature is obtained for K 0–M 3 III stars. This relationship agrees well with empirical data.

064.049 On the evolution of ionized gas around hot stars.
J. Manfroid.
Astrophys. Space Sci., Vol. 41, 39 - 56 (1976).

The initial evolution of a uniform H II region around a star of 36 solar masses is described. The general equations of the problem are solved numerically by a finite difference method. Forbidden line emission by ions of C, N, O, and Ne is taken into account.

064.050 The transfer of resonance radiation in extended envelopes with differential rotation and expansion. Diffuse approximation. V. P. Grinin.
Izv. Krymskoj Astrofiz. Obs., Vol. 54, 176 - 183 (1976). In Russian.

The diffuse type equation for the determination of the density of excited atoms in extended envelopes with differential rotation and expansion is obtained. Two kinds of motion are considered: 1) radial-symmetric expansion, 2) axisymmetric motion – differential rotation and expansion. The region of application is discussed.

064.051 Neutral helium emission in Wolf-Rayet envelopes.
W. R. Oegerle, D. Van Blerkom.
Astrophys. Journ., Vol. 206, 150 - 155 (1976).

A non-LTE analysis of neutral helium emission is performed for Wolf-Rayet stars. The statistical equilibrium equations are derived and simplified by using the escape probability method. Using a thirty-six-bound-level model helium atom, the relative intensities of allowed transitions are calculated for various helium abundances at T_e= 10,000 and 20,000 K and are compared with the observations of MR 119.

064.052 Treatment of atomic and molecular line blanketing by opacity sampling.
H. R. Johnson, B. M. Krupp.
Astrophys. Journ., Vol. 206, 201 - 207 = Publ. Goethe Link Obs., Indiana Univ., *Bloomington*, No. 177 (1976).

A sampling technique for treating the radiative opacity of large numbers of atomic and molecular lines in cool stellar atmospheres is subjected to several tests. In this opacity sampling (OS) technique, the global opacity is sampled at only a selected set of frequencies, and at each of these frequencies the total monochromatic opacity is obtained by summing the contribution of every relevant atomic and molecular line. The effects of atomic and molecular lines are separately studied. A test model computed using the OS method agrees very well with a model having identical atmospheric parameters, but computed with the giant line (opacity distribution function) method.

064.053 Convective instability in a compressible atmosphere. II.
D. O. Gough, D. R. Moore, E. A. Spiegel, N. O. Weiss.
Astrophys. Journ., Vol. 206, 536 - 542 (1976).

The onset of steady convection in a polytropic atmosphere with constant viscosity is studied numerically.

064.054 Effect of the absorbers upon the thermal structure of an LTE atmosphere (II). C, Si, Mg, Fe and Al.
S. Dumont, N. Heidmann.
Astron. Astrophys., Vol. 49, 271 - 275 (1976).

The introduction of so-called "metallic" absorbers C I, Si I, Mg I, Fe I and Al I in LTE model atmospheres having only H and He as absorbers, produces different, opposite, effects according to the effective temperature of the models. From a detailed study of the 15000 K-model, the authors explain a rise in T_e.

064.055 A study of unstable acoustic waves in a convective zone.
P. Graff.
Astron. Astrophys., Vol. 49, 299 - 305 (1976).

Several mechanisms may occur and give rise to overstable oscillations in a convective zone. One of these instabilities, previously described by Cowling and by Spiegel, is due to the joint effect of the superadiabatic temperature gradient and of the dissipation of acoustic waves. A semi-analytical investigation of some properties of this instability is presented.

064.056 Mass loss in early stages of stellar evolution.
D. S. P. Dearborn, M. Kozlowski, D. Schramm.
Nature, Vol. 261, 210 - 211 (1976). – Letter.

064.057 Not so wealthy stars?   M. G. Edmunds.
Nature, Vol. 261, 365 (1976).

064.058 Model atmospheres synthetic spectra: H/C/N/O abundance ratio in carbon stars.
M. Querci, F. Querci.
Astron. Astrophys., Vol. 49, 443 - 462 (1976).

The authors examine the influence of the effective temperature, gravity and abundance ratio on the spectrum of a given molecule: its behaviour is indicative of the information they could obtain from each selected spectral range, when they compare the synthetic and stellar spectra. The various synthetic spectra are matched to the spectra of the carbon stars UU Aur and Y CVn. From the best fit between the computed and the stellar spectra, they have attempted to

determine which nucleosynthetic process can be expected to produce the light elements C, N and O in these cool stars. Model atmosphere emergent fluxes against the published observed photometric ones are compared.

064.059 Mass and angular momentum effluxes of stellar winds.   T. Yeh.
Astrophys. Journ., Vol. 206, 768 - 776 (1976).

The mass and angular momentum effluxes of a stellar wind are determined by the mass and radius of the star and the mass density, temperature, rotation, and magnetic field of the stellar corona. A parametric study, under the assumption of a polytrope relationship, indicates that the mass efflux will be large if the star has small mass and large radius and the stellar corona has high mass density and temperature. The angular momentum efflux increases in more than linear proportion to the stellar rotation and magnetic field.

064.060 Computing nonlinear periodic motions of stellar envelopes and their stability.   K. von Sengbusch.
Computing in plasma physics and astrophysics, (see 012.014), A2, 2 pp. (1976).

064.061 Structure of stellar atmospheres of late-type stars.
J. I. Straume.
Investigation of the sun and red stars. 4, (see 003.009), p. 31 - 59 (1976). In Russian.

Model atmospheres have been constructed for effective temperatures of 2500, 3000, 3500, 4000, 4500 K for solar abundances and for surface gravities log g = 0.0, 1.0, 2.0, 3.0, 4.0 and 5.0 based on LTE and plane-parallel horizontally homogeneous structure. Results are given in tables.

064.062 Effects of back warming in cocoon stars.
J. R. Donnison, I. P. Williams.
Nature, Vol. 261, 674 - 675 (1976).

There now exists general agreement among astronomers that dust shells surrounding young stars are a relatively frequent occurrence, and attempts have been made to calculate some of the properties of such shells. The authors discuss the evolution of the central star only, the dust being regarded merely as a means of redirecting radiation back on to the surface of this star. The effects of back warming by a dust shell will depend very much on the thickness of the shell, with a thin shell having very little effect so that for any shell where some light from the parent central star is observed there will be no significant effect. A thick shell, however, will mask the main characteristics of the star and substantially slow down its evolution.

064.063 Empirical model atmospheres for explanation of the light variations of magnetic stars.   W. Schöneich, A. A. Krivosheina, V. L. Khokhlova, I. A. Aslanov.
Astron. Nachr., Band 297, 207 - 211 (1976). In German.

Empirical model atmospheres on the basis of Mihalas models were investigated. If the temperature of the outer layers in a spot is higher than in the surrounding normal atmosphere and lower than in the deeper layers, many characteristics of the observed amplitude-wavelength relation of magnetic stars can be explained.

064.064 Mass-loss from ultraviolet observations of ζ Puppis.
F. Macchetto, A. Natta, A. Preite-Martinez, N. Panagia.
Mem. Soc. Astron. Italiana, Vol. 45, (see 012.017), 747 - 756 (1974).

064.065 Sul moto di materia gassosa emessa da stelle o da ammassi stellari.   M. Cimino.
Rend. Mat., Vol. 8, (Ser. VI), 311 - 335 = Oss. Astron. Roma, Contr. sci.,Ser. III, No. 143 (1975).

064.066 Radiative transfer in spherical circumstellar dust envelopes. J. P. Apruzese.
Thesis Wisconsin Univ., Madison, USA, 117 pp. (1974).
(Available from: Univ. Microfilms, Order No. 74-28,790).

064.067 Coronas with bremsstrahlung cooling.
N. M. Hoffman III.
Thesis Wisconsin Univ., Madison USA, 92 pp. (1974). (Available from: Univ. Microfilms, Order No. 75-9977).

064.068 Effects of titanium oxide upon the atmospheres of late-type stars. J. G. Collins.
Thesis Indiana Univ., Bloomington, USA, 159 pp. (1975).
(Available from: Univ. Microfilms, Order No. 75-23,423).

064.069 Hydrodynamic models of a cepheid atmosphere.
A. H. Karp.
Thesis Maryland Univ., College Park, USA, 195 pp. (1974).
(Available from: Univ. Microfilms, Order No. 75-17,885).
See 14.065.009, 14.122.082, 14.064.036.

064.070 Comparison of Lyman alpha and He λ 10830 line structures and variations in early-type star atmospheres. Final technical report, January–October 1974.
D. D. Meisel.
Separate print State Univ. of New York, Genesco, USA, Dept. Phys. Astron., 27 pp. (1974).

The observation and analysis of stellar photospheres.
See Abstr. 003.049.

Blanketed model atmospheres for early-type stars.
See Abstr. 003.074.

Production of galactic ⁷Li by slow mass loss.
See Abstr. 061.031.

Infrared pumping for SiO masers.
See Abstr. 061.044.

Solution of the comoving-frame equation of transfer in spherically symmetric flows. II. Picket-fence models.
See Abstr. 063.001.

Formation of spectral lines with partial frequency redistribution. See Abstr. 063.002.

Compton reflected spectra of X-ray illuminated stellar atmospheres. See Abstr. 063.005.

Line formation in turbulent media: comparison between continuous and discontinuous velocity fields.
See Abstr. 063.006.

Radiative hydrodynamics of chromospheric transients. See Abstr. 063.019.

Resonance-line transfer with partial redistribution. VIII. Solution in the comoving frame for moving atmospheres.
See Abstr. 063.026.

A tensor formulation of the equation of transfer for spherically symmetric flows. See Abstr. 063.027.

Numerical solution of the radiative transfer equation in spherically symmetric dust shells. See Abstr. 063.030.

Über den Strahlungstransport in zirkumstellaren Staubhüllen und seinen Einfluß auf Temperaturverteilung und Spektrum. See Abstr. 063.039.

Perturbation solutions of the equation of radiative transfer in a stellar atmosphere. See Abstr. 063.040.

Excitation of pulsations in the CNO ionization zone of luminous stars. See Abstr. 065.025.

Steady mass loss and the minimum stellar mass for carbon ignition. See Abstr. 065.044.

Possible properties of pre-outburst FU Orionis stars.
See Abstr. 065.066.

Cooling times, luminosity functions and progenitor masses of degenerate dwarfs. See Abstr. 065.076.

Models of the upper photospheres of the sun and Arcturus based on molecular spectra.
See Abstr. 071.045.

Linear simulations of Boussinesq convection in a deep rotating spherical shell (sun). See Abstr. 080.062.

U Cephei: a mass-transfer event. II. Observations.
See Abstr. 113.043.

Stellar winds from hot supergiants.
See Abstr. 114.006.

Line blocking in the visual and near infrared spectra of G-, K- and M-type giants. See Abstr. 114.009.

The spectra of peculiar Be stars with infrared excesses. See Abstr. 114.014.

Absolute fluxes of K chromospheric emission on the H–R diagram. See Abstr. 114.018.

The nature of the objects of Joy: a study of the T Tauri phenomenon. See Abstr. 114.023.

The complex outer shell of Eta Carinae.
See Abstr. 114.314.

Spectroscopic survey of the far-ultraviolet (1160–1700 Å) emissions of Capella. See Abstr. 114.326.

The chemical composition of Gamma Pegasi.
See Abstr. 114.352.

Shock-wave interpretation of emission lines in long-period variable stars. I. The velocity of the shock.
See Abstr. 122.034.

A study of the velocity pattern of maser emission from infrared stars. See Abstr. 141.626.

Infrared sources in molecular clouds.
See Abstr. 141.629.

Stellar winds and accretion in massive X-ray binaries.
See Abstr. 142.107.

Mass loss in globular-cluster red giants.
See Abstr. 154.003.

064.901 Errata: "Transfer of line radiation in differentially expanding atmospheres. IV. The two-level atom in plane-parallel geometry solved by the Feautrier method"
[Astrophys. Journ., Vol. 193, 651 - 676 (1974)].
P. D. Noerdlinger, G. B. Rybicki.
Astrophys. Journ., Vol. 203, 769 (1976).

065   Star Formation, Stellar Structure and Evolution, Neutron Stars

065.001   On isolated stars in non-radial oscillation.
J. Christensen-Dalsgaard.
Monthly Notices Roy. Astron. Soc., Vol. 174, 87 - 89 (1976).
Isolated stars in non-radial oscillation do not have their centres of mass displaced.

065.002   Force-free stellar magnetospheres.
F. Milsom, G. A. E. Wright.
Monthly Notices Roy. Astron. Soc., Vol. 174, 307 - 317 (1976).
A series of force-free axially-symmetric magnetospheric models is constructed using a power law to represent the relation between the toroidal magnetic field and the poloidal magnetic 'stream function'. It is found that when the toroidal field exceeds a certain critical intensity (depending on the power index) no continuous, physically plausible models exist. Reasons are given why the existence of a toroidal field in the magnetosphere may help to prevent the accretion of interstellar matter and thus assist the formation of 'peculiar' spectral lines.

065.003   Pulsation of high luminosity helium stars.
P. R. Wood.
Monthly Notices Roy. Astron. Soc., Vol. 174, 531 - 539 (1976).
A study has been made of the radial pulsation of high luminosity helium stars in the region of the HR diagram occupied by the R Coronae Borealis and other hydrogen-deficient carbon stars. As well as instability of the fundamental mode when $\log T_{eff} < 3.88$, overtone instability is found in the region $3.93 < \log T_{eff} < 4.12$, and possibly when $\log T_{eff} > 4.2$ if the luminosity to mass ratio is high enough. Non-adiabatic effects are very large and there is no one-to-one correspondence between linear adiabatic and non-adiabatic modes. Non-linear calculations show the fundamental pulsation to be very violent but the full amplitude overtone pulsation is well behaved.

065.004   The cepheid loop as a threshold effect.
R. A. Durand, J. G. Eoll, B. M. Schlesinger.
Monthly Notices Roy. Astron. Soc., Vol. 174, 671 - 677 = Publ. Goethe Link Obs., Indiana Univ., *Bloomington,* No. 175 (1976).
Pairs of evolutionary tracks have been calculated for stars of 3 $^1/_4$, 5 and 7 solar masses; one track of each pair uses the Reeves rate for the $N^{14}(\alpha, \gamma) F^{18}$ reaction, in the other track this reaction is ignored. At 5 and 7 solar masses, the models with deeper mixing had extended cepheid loops; the less deeply-mixed models showed none. No loop appeared at 3 $^1/_4 M_\odot$. This predicted absence is consistent with observations.

065.005 $^{12}C/^{13}C$ ratios in stars ascending the giant branch the first time.
D. S. P. Dearborn, P. P. Eggleton, D. N. Schramm.
Astrophys. Journ., Vol. 203, 455 - 462 (1976).
Stellar evolution calculations were carried out for 1 $M_\odot$ and 2 $M_\odot$ red giants in order to understand the observed $^{12}C/^{13}C$ ratios. In addition to the standard mixing at the base of the giant branch, which can give $^{12}C/^{13}C$ ratios in the region of 25 to 30, the authors consider: variation in initial composition; meridional mixing; mass loss; zero-age shell. Of these, only the mass loss mechanism and the hydrogen shell instability produce values below ~ 20.

065.006   Carbon stars in the Large Magellanic Cloud.
D. R. Crabtree, H. B. Richer, B. E. Westerlund.
Astrophys. Journ., (*Letters*), Vol. 203, L81 - L85 (1976).

The properties of seven carbon stars which are very likely members of the Large Magellanic Cloud are discussed. Three exhibit enhanced molecular bands of ^{13}C while one bears a spectroscopic resemblance to the peculiar galactic carbon star WZ Cas. The range in bolometric absolute magnitude for the seven stars is found to be from -5.7 to -6.6. By placing the stars in a theoretical H-R diagram, it is concluded that (1) all seven are double-shell burning stars, (2) four of the seven have just recently begun helium shell flashes and (3) the mass range for the objects observed is from 3 to 9 $M_\odot$.

065.007   Fluorine power and helium-shell flashes.
J. M. Scalo, K. H. Despain.
Astrophys. Journ., Vol. 203, 667 - 673 (1976).
During high-temperature helium burning in convective regions the ^{18}F created from $^{14}N(\alpha, \gamma)^{18}F$ may be destroyed by the $^{18}F(\alpha, p)^{21}Ne$ reaction rather than $^{18}F(\beta^+ \nu)^{18}O$. The conditions for such burning are met in advanced helium-shell flashes. The resulting nucleosynthesis is examined using a simple one-zone model.

065.008   Synchronization in binaries and age.   H. Levato.
Astrophys. Journ., Vol. 203, 680 - 688 (1976).
Analyzing the published data on field binary systems, the author found that the periods below which the orbital and rotational motions are synchronized are systematically longer for evolved main-sequence stars than for ones near the zero-age main sequence. This is interpreted as showing that part of the tendency for tidal interaction to achieve synchronization occurs during the main-sequence lifetime of stars; the remainder occurred during contraction to the zero-age main sequence.

065.009   Coherent scattering of neutrinos from a shock wave at high densities.   R. Opher.
Astron. Astrophys., Vol. 46, 253 - 255 (1976).
Neutrino radiation pressure on a shock wave in stellar collapse is evaluated. It is shown that at high densities ($\rho \sim 10^{13}$ gm/cm³), due to coherent scattering, the neutrino radiation pressure is very large. The implications of this large pressure are discussed.

065.010   Non-radial secular stability of the hydrogen burning shell in stars of 1.1 $M_\odot$ and 8 $M_\odot$.
M. Gabriel, A. Noels.
Astron. Astrophys., Vol. 46, 313 - 316 (1976).
The secular stability of hydrogen shell burning against non-radial perturbations for stars of 1.1 $M_\odot$ and 8 $M_\odot$ is studied. All models proved to be very stable. The results are discussed from both a mathematical and an intuitive point of view.

065.011   Laboratory simulation of thermal convection in rotating planets and stars.
F. H. Busse, C. R. Carrigan.
Science, Vol. 191, 81 - 83 (1976).
Because of dynamical constraints in a rotating system, the component of gravity perpendicular to the axis of rotation is the dominant driving force of convection in liquid planetary cores and in stars. Except for the sign, the centrifugal force closely resembles the perpendicular component of gravity. Convection processes in stars and planets can therefore be modeled in laboratory experiments by using the centrifugal force with a reversed temperature gradient.

065.012   Evolution of red-giant stars.   A. V. Sweigart.
Phys. Today, Vol. 29, No. 1, p. 25 - 27, 29, 31 - 32 (1976).

Thermal instabilities in helium-burning shells are linked to the origin of certain elements heavier than iron—in one star the surface abundances of these elements increased by 25 times in the last ten years.

065.013 Some peculiarities of population II stars formation.
A. G. Doroshkevich, I. G. Kolesnik.
Astron. Zhurn. Akad. Nauk SSSR, Vol. 53, 10 - 19 (1976). In Russian. English translation in Soviet Astron., Vol. 20, No. 1.

Some features of the dynamical collapse of population II protostars containing little or no heavy elements are considered It is shown that Ly-α radiation pressure sets an upper limit of the order of 10 $M_\odot$ on the mass of such stars. For population I, stars containing normal heavy elements with much larger masses may be possible.

065.014 Electron and proton regions in the magnetosphere of a rotating charged neutron star with strong magnetic field. Yu. A. Rylov.
Astron. Zhurn. Akad. Nauk SSSR, Vol. 53, 44 - 52 (1976). In Russian. English translation in Soviet Astron., Vol. 20, No. 1.

Within the framework of the neutron star magnetosphere model suggested by Goldreich and Julian the action of particle ejection from the star surface and that of electric charge of the star upon the shape of magnetosphere regions filled by charged particles is considered. The shape of the equatorial belt and the cap are calculated.

065.015 Evidence on neutron star structure from pulsars and related objects. M. J. Rees.
Astrophys. Space Sci., Vol. 39, 3 - 7 (1976). − Paper presented at the Symposium on Solid State Astrophysics, held at the University College, Cardiff, Wales, between 9 - 12 July 1974. − See 065.041.

065.016 The melting of neutron stars' crystalline cores and gamma-ray bursts. Ju. M. Bruk (*Yu. M. Bruk*), K. I. Kugel.'
Astrophys. Space Sci., Vol. 39, 235 - 241, 243 - 249 (1976). In Russian and English.

Possible phase transitions in neutron star matter, particularly the melting of neutron stars' crystalline cores, are discussed. Such processes may explain the observed luminosity of pulsars. They are used also as a basis for an explanation of the origin of low energy gamma-ray bursts which have been intensively studied for the last few years. The authors discuss the structure of gamma-ray bursts and the possibility of obtaining from observational data some information on thermal evolution of neutron stars and dynamic processes in the pulsar crust.

065.017 Evolution of helium stars: a self-consistent determination of the boundary of a helium burning convective core. G. J. Savonije, R. J. Takens.
Astron. Astrophys., Vol. 47, 231 - 241 (1976).

A generalization of the Henyey-scheme is given that introduces the mass of the convective core and the density at the outer edge of the convective core boundary as unknowns which have to be solved for simultaneously with the other unknowns. Using this scheme, the evolution of helium stars (X=0, Z= 0.03) was followed up to (non-degenerate) carbon ignition for a number of stellar masses: 2.5 $M_\odot$, 3 $M_\odot$, 4 $M_\odot$ and 8 $M_\odot$. As compared with some earlier investigations, the calculations show a rather large increase in mass of the convective cores during core helium burning. Evolutionary calculations for a 2 $M_\odot$ helium star show that the critical mass for which a helium star ignites carbon non-degenerately lies near 2 $M_\odot$ (corresponding to a C/O core of 0.99 $M_\odot$).

065.018 Stellar evolution V: Evolutionary models of population I stars with or without overshooting from con-
vective cores. A. Maeder.
Astron. Astrophys., Vol. 47, 389 - 400 (1976).

Three sets of evolutionary models from the main-sequence to the red-giant branch are computed; one has the usual treatment of convective cores and the other sets take into account non-local effects of convection in stellar cores with 2 values of the mixing-length ratio. The composition adopted is X = 0.70 and Z = 0.03; the models presented cover the range of 0.85 $M_\odot$ to 3 $M_\odot$, i.e. the range of A- to G-type stars. The effects of changes of the mixing-length ratio in the core on the extension of the zone of overshooting, the stellar structure, the evolutionary tracks, the lifetimes, the isochrones, the gap parameters and the age and mass determinations are investigated.

065.019 A method for the calculation of axially symmetric dynamical systems.
K. J. Fricke, C. Möllenhoff, W. Tscharnuter.
Astron. Astrophys., Vol. 47, 407 - 412 (1976).

A two dimensional computer code is described which is suitable for hydrodynamic selfgravitating systems with axial symmetry. While the radial coordinate is treated by a difference scheme the angular dependence of every function is expanded into Legendre polynomials. The expansion coefficients are computed using a conventional Henyey scheme. The results of test computations for the collapse of rotating isothermal protostellar clouds with two different values of angular momentum are presented.

065.020 The evolutionary status of population II cepheids.
R. A. Gingold.
Astrophys. Journ., Vol. 204, 116 - 130 (1976).

Several grids of post−horizontal-branch evolutionary models have been constructed assuming a range in either total mass or initial core mass on the horizontal branch. The less luminous population II cepheids (BL Herculis stars) are identified with stars of small envelope mass undergoing " blueward noses" from the asymptotic giant branch. The more luminous cepheids (W Virginis stars) are identified with stars undergoing blueward loops from the second giant branch. The models predict relatively too few variables spread over too large a range in luminosity in the brighter group. The assumption of a range in total mass on the zero-age horizontal branch produces a slightly better fit to the observed luminosity distribution of population II cepheids than does the assumption of a range in initial core mass.

065.021 Slow mass transfer in semidetached binaries.
J. P. Pratt, P. A. Strittmatter.
Astrophys. Journ., (*Letters*), Vol. 204, L29 - L33 (1976).

It is shown that, if the usual assumption of synchronous rotation (infinitely effective tidal coupling) in semidetached binary systems is replaced by that of detailed conservation of angular momentum with negligible tidal torques, the system is predicted to evolve in a qualitatively different manner after the onset of mass loss. It is suggested that the actual evolution of the system is determined by the time scale of tidal interaction and that the mass transfer rate probably lies between the classical values and those found here. Possible implications of the present results for X-ray sources such as HZ Her are discussed briefly.

065.022 Stellar evolution at high mass with semiconvective mixing according to the Schwarzschild criterion.
R. Stothers, C.-w. Chin.
Astrophys. Journ., Vol. 204, 472 - 480 (1976).

New evolutionary sequences of models for stars of 5−60 $M_\odot$ with four different initial chemical compositions have been constructed with the use of the Schwarzschild criterion for convection. This criterion leads to semiconvection outside the convective core (1) just after the zero-age main-sequence stage for masses greater than $\sim 12\,M_\odot$ and (2) just before the stage of central hydrogen exhaustion for masses greater than $\sim 6\,M_\odot$. The tracks for the more massive stars are found to be relatively insensitive to the uncertainties in the nuclear reaction rates. An observational test utilizing the surface hydrogen abundance of stellar remnants in binary systems is at present inconclusive in deciding whether the Schwarzschild criterion or the Ledoux criterion is the correct criterion for convection to use. It is strongly suggested that neither criterion with any reasonable adopted initial chemical composition can at present lead to model predictions that satisfy all the observational requirements.

065.023 **Effects of convective overshoot on lithium depletion in main-sequence stars.** J. M. Straus, J. B. Blake, D. N. Schramm.
Astrophys. Journ., Vol. 204, 481 - 487 (1976).
A current problem in stellar evolution is to understand the lithium depletion in low-mass main-sequence stars. Standard stellar models do not produce temperatures in the outer convective zone high enough to allow lithium burning to occur. Convective overshoot could extend the mixing region deep enough to allow lithium burning. A gradual dependence of the degree of convective overshoot with stellar mass can be obtained if one uses a dynamical approach for describing the convective process. Thus both the mass and age dependence of lithium depletion might be understood. Specific examples of the Hyades and Pleiades clusters as well as the sun are discussed.

065.024 **A binary hypothesis for the subdwarf B stars.** J. G. Mengel, J. Norris, P. G. Gross.
Astrophys. Journ., Vol. 204, 488 - 492 (1976).
Stellar evolution computations pertinent to the hypothesis that subdwarf B stars are formed in close binary systems are presented. Under the assumption that total mass and orbital angular momentum are conserved, the authors find that for systems with initial masses ($0.80\,M_\odot$, $0.78\,M_\odot$) and composition $X = 0.73$, $Z = 0.001$, there exists a range of initial separations for which core helium-burning (horizontal-branch) stars of mass $\sim 0.5\,M_\odot$ are formed. The authors identify these with the subdwarf B stars. For smaller separations, white dwarfs are formed.

065.025 **Excitation of pulsations in the CNO ionization zone of luminous stars.** R. Stothers.
Astrophys. Journ., Vol. 204, 853 - 868 (1976).
A new source of excitation for stellar pulsations has been discovered by employing Carson's new radiative opacities in models of chemically homogeneous stars. These opacities have a "bump" arising from the ultimate ionization of the CNO elements at moderate temperatures and very low densities. Such high opacities induce in the envelopes of the more massive stars a strong local convection zone, a high central condensation (which renders the pulsational ϵ-mechanism ineffective), and pulsational instability via the κ-mechanism above a certain stellar mass.

065.026 **Magnetohydrodynamic phenomena in collapsing stellar cores.**
D. L. Meier, R. I. Epstein, W. D. Arnett, D. N. Schramm.
Astrophys. Journ., Vol. 204, 869 - 878 (1976).
The collapse and explosion of a rotating, magnetized, iron-nickel stellar core is described. The core collapse and the amplification of the seed magnetic field are characterized by several dimensionless parameters whose values are inferred from numerical computations. Parametrization of the problem in this way enables the authors to estimate (1) the necessary conditions for MHD explosions or instabilities, (2) the mass and energy released in such events, and (3) the effects of the ejected material on the overlying stellar matter. The implica-

tions of these MHD phenomena for the dynamics of supernovae and for the chemical composition of the interstellar gas and the cosmic rays are discussed.

065.027 **Variational analysis of rotating neutron stars.**
M. A. Abramowicz, R. V. Wagoner.
Astrophys. Journ., Vol. 204, 896 - 899 (1976).
The authors generalize the variational method of Nauenberg and Chapline to obtain explicit but approximate formulae for the total mass, baryon number, and angular momentum of slowly rotating neutron stars. The trial configurations employed have uniform density and angular velocity (as measured locally by zero-angular-momentum observers), but an arbitrary zero-entropy equation of state $\rho = \rho(n)$. It is found that rotation can increase the pressure in a body which is sufficiently relativistic.

065.028 **The theoretical basis of star formation.**
P. Bodenheimer.
Observatory, Vol. 96, 1 - 3 (1976).

065.029 **Recent numerical studies of collapse.** D. C. Black.
Observatory, Vol. 96, 3 - 4 (1976).

065.030 **Optical studies of young stellar objects evolving to the main sequence.** G. F. Gahm.
Observatory, Vol. 96, 7 - 8 (1976).

065.031 **Stellar oscillations and magnetic field perturbations of the boundary conditions.**
M. Goossens, P. Smeyers, J. Denis.
Astrophys. Space Sci., Vol. 39, 257 - 272 (1976).
The effect of a weak magnetic field on the adiabatic radial and non-radial oscillations of a stellar configuration is studied by means of a perturbation method. Special attention is devoted to the perturbation of the oscillation frequencies resulting from the change of the boundary conditions caused by the magnetic field. This change is related to the fact that the introduction of a magnetic field removes the singularity at the surface of the equilibrium configuration. The perturbation method is applied to Ferraro's model and the influence of a magnetic field on the frequencies of the different types of oscillation modes is discussed.

065.032 **The URCA process in convective cores.**
S. Tsuruta, A. G. W. Cameron.
Astrophys. Space Sci., Vol. 39, 397 - 400 (1976).
A Monte Carlo simulation has been carried out for URCA processes in a degenerate convective stellar core, such as would be obtained following degenerate carbon ignition. The authors find energy loss rates substantially less than obtained by Paczynski in his vibrational approximation for this process. They conclude that the loss rates should be computed by direct integration over the convective core assuming a uniform ratio of the abundances of the URCA pair.

065.033 **Stability of g^+ modes in a $30\,M_\odot$ star.**
R. Scuflaire, A. Noels, M. Gabriel, A. Boury.
Astrophys. Space Sci., Vol. 39, 463 - 475 (1976).
The evolution of a population I $30\,M_\odot$ star has been computed during the main sequence phases without taking semiconvection into account. These models have a temperature gradient larger than the adiabatic value in the inhomogeneous region. The models have been tested for stability towards g^+ modes of non-radial oscillations to see whether Kato's mechanism leads to an instability. Whereas the models are stable during the early main sequence phases, they become unstable for low enough central hydrogen abundance.

065.034 **Klassische und moderne Formulierungen des Vogt-Russell-Theorems.** D. Lauterborn.

Mitt. Astron. Ges., No. 38, p. 155 - 162 (1976). – Short report.

065.035 **Spectral appearance of a 1 $M_\odot$ protostar.**
C. Bertout.
Mitt. Astron. Ges., No. 38, p. 162 - 165 (1976). – Short report.

065.036 **Über die physikalischen Ursachen der Existenz von Mehrfachlösungen bei statischen Sternmodellen mit Heliumkernen.**   D. Lauterborn.
Mitt. Astron. Ges., No. 38, p. 166 - 173 (1976). – Short report. In English.

065.037 **Die Mannigfaltigkeit von Sternmodellen.**
H. Kähler.
Mitt. Astron. Ges., No. 38, p. 173 (1976). – Abstract.

065.038 **The dynamic evolution of a massive protostar envelope.**   H. W. Yorke, E. Krügel.
Mitt. Astron. Ges., No. 38, p. 222 - 223 (1976). – Abstract.

065.039 **Equilibrium problems in a rotating convection zone.**
Yu. V. Vandakurov.
Solar Physics, Vol. 45, 501 - 520 (1975).

This paper is a continuation of the author's work on stellar convection (Vandakurov, 1975). The approximate equations for convective perturbations are somewhat corrected and generalized to include both nonlinear terms and possible variations in molecular weight. A crude estimate of the nonlinear terms is given by means of an expansion of the solution in powers of the perturbation amplitude. An expansion in powers of the angular velocity is also performed. It is shown that an azimuth-averaged azimuthal force is created by the unstable perturbations. A simple formula for the above azimuthal force is derived in the case of a latitude-dependent angular velocity and a small viscosity of the medium.

065.040 **Solid state physics and cooling of neutron stars.**
S. Tsuruta.
Astrophys. Space Sci. Library, Vol. 55, (see 012.001), 263 - 272 (1976).

The author shows the possible effect of the 'magnetic' condensation on cooling of neutron stars. Its observational significance (especially for younger pulsars such as the Crab pulsar) is emphasized. Other effects of solid state physics on cooling are also discussed.

065.041 **Evidence on neutron star structure from pulsars and related objects.**   M. J. Rees.
Astrophys. Space Sci. Library, Vol. 55, (see 012.001), 273 - 277 (1976).

Properties of pulsars and binary X-ray sources which seem particularly relevant to theories of neutron star structure and strong magnetic fields are reviewed and discussed.

065.042 **Neutron star cores.**   R. G. Palmer.
Astrophys. Space Sci. Library, Vol. 55, (see 012.001), 279 - 292 (1976).

Current theories, and the astrophysical implications, of the nature of high density neutron star matter are reviewed. Suggestions are made for a compromise between the alternatives of neutron crystallization and pion condensation.

065.043 **Structure of neutron star cores.**
V. Canuto, B. Datta, J. Lodenquai.
Astrophys. Space Sci. Library, Vol. 55, (see 012.001), 293 - 299 (1976).

After reviewing the outer and central regions of a neutron star, the authors discuss the central region and the possibility that the core has a solid structure.

065.044 **Steady mass loss and the minimum stellar mass for carbon ignition.**   J. G. Mengel.
Astron. Astrophys., Vol. 48, 83 - 86 (1976).

An assessment is made of the importance of steady mass loss for the evolution of stars of intermediate mass during the double-shell phase. The basis is an empirical formula due to Reimers for the mass-loss rate from red giants. It is combined with a linear core mass-luminosity law and the location of the evolutionary tracks in the H−R diagram to obtain a relationship between the total mass and luminosity of stars evolving towards carbon ignition. This relationship yields estimates of the minimum initial mass M_{WD} required for ignition to occur. The results suggest that steady mass loss determines the fate of stars of intermediate mass. The possibility exists that degenerate carbon ignition is precluded in single stars.

065.045 **Approach to thermal equilibrium in a contracting pure hydrogen cloud.**   Y. Nakada, T. Yoneyama.
Publ. Astron. Soc. Japan, Vol. 28, 61 - 68 (1976).

The formation of a radiation field in a contracting pure hydrogen cloud is followed until thermal equilibrium is achieved between gas and radiation. It is shown that the specific entropy in the cloud is much lower than that required for the main-sequence state.

065.046 **Mixing of carbon and s-process elements during thermal pulses of helium-burning shell.**
M. Y. Fujimoto, K. Nomoto, D. Sugimoto.
Publ. Astron. Soc. Japan, Vol. 28, 89 - 103 (1976).

Five thermal pulses have been computed for a star of intermediate mass, which has an electron-degenerate carbon-oxygen core of mass around $1.26\,M_\odot$. The authors have found that a convective envelope penetrates very deeply into a helium zone in a declining phase of each thermal pulse. Because of this penetration, the material, which has been processed by the unstable helium shell-burning and associated nuclear reactions, is mixed into the envelope. The depth of penetration depends only slightly upon a parameter of the mixing-length theory and upon the stellar mass.

065.047 **Vibrational stability against nonradial oscillations of a 10 solar mass star during the secondary contraction and the shell hydrogen-ignition stages.**   Y. Osaki.
Publ. Astron. Soc. Japan, Vol. 28, 105 - 115 (1976).

Vibrational stability against nonradial quadrupole ($l = 2$) oscillations is examined for massive stars with a 10 solar mass during the secondary contraction and the shell hydrogen-ignition stages of evolution. No vibrational instability is found. A local stability analysis indicates that vibrational instability of nonradial g-modes seems unlikely provided that the temperature gradient in the chemically inhomogeneous zone is subadiabatic.

065.048 **Zwei Endstadien der Sternentwicklung. Weiße Zwerge und Schwarze Löcher (Teil I).**
R. Wehrse, J. Schmid-Burgk.
SuW, 15. Jahrgang, p. 114 - 117 (1976).

065.049 **The evolution of magnetic stars.**
A. V. Tutukov, G. V. Ruben.
Stellar physics and evolution, (see 003.001), p. 5 - 16 (1974). In Russian.

If it is assumed that the magnetic field has a "fossil" nature and was not destroyed in the course of the collapse of the star, then it is reasonable to assume the following relation between the strength of the magnetic field and the density of stellar matter: $H = A\rho^{2/3}$. The value of A was determined from an examination of the physical conditions of collapse. In the first approximation $A \approx 10^7 (M/M_\odot)^{4/3}$ gauss/gm$^{2/3}$. Homogeneous models of $4\,M_\odot$ and $32\,M_\odot$ stars were obtained for different values of A. Sequences of evolutionary models in the stage of hydrogen burning were computed for two values of A for the 32 $M_\odot$ star and for one value of A for the 4 $M_\odot$ star. It appears that qualitatively the influence of the magnetic field on the structure of a homogeneous model and on the evolution of stars is similar to that of rotation. Results of calculations are presented in tables and figures.

065.050 An application of the "method of partition" for shock wave propagation in stellar interiors.
B. M. Shustov, A. L. Vinogradov.
Stellar physics and evolution, (see 003.001), p. 17 - 26 (1974). In Russian.

For the problem of shock wave propagation in stellar interiors the "method of partition" can be applied successfully. All the equations are divided into two groups: the dynamical group and another consisting of the equations for energy, transfer, ionisation, etc. After linearisation each group is reduced to one difference equation and is solved by an iterative method. Several advantages of the above mentioned method are discussed. For this purpose an astrophysical pattern — the propagation of strong shock waves in a polytrope was calculated. The star (4 $M_\odot$, 2 $R_\odot$) was considered as a polytropic configuration with n = 3 and a mean molecular weight μ = 1.634. The results show that the "method of partition" is very useful for solving hydrodynamical problems in astrophysics.

065.051 Sequences of equilibrium models for helium-burning stars in the blue supergiant region. II.
Yu. L. Frantsman, E. I. Popova, J. Ziólkowski.
Stellar physics and evolution, (see 003.001), p. 27 - 39 (1974). In Russian.

Sequences of equilibrium stellar models in the He-burning stage are presented. The initial masses are 30 $M_\odot$ and 15 $M_\odot$. The chemical composition of the envelope of the initial model is $X_0 = 0.70$, $Z_0 = 0.03$. Two sequences with helium content in the center $Y_c = 0.97$ and 0.49 were computed for 15 $M_\odot$ and three sequences with helium content in the center $Y_c = 0.97$, 0.60 and 0.35 for 30 $M_\odot$.

065.052 The thermal effects of H_2 molecules in rotating and collapsing spheroidal gas clouds.
J. B. Hutchins.
Astrophys. Journ., Vol. 205, 103 - 121 (1976).

The thermal effect of H_2 molecules has been studied for a range of uniform rotating spheroidal gas clouds undergoing pressure-free homologous collapse, with the goal of finding the Jeans mass, which governs star formation during the collapse of clouds to form galaxies. It is found that, for prolate clouds which rotate sufficiently slowly about their long axes, the Jeans mass can become as small as 0.014 $M_\odot$ in the cases considered, suggesting an early period of formation of objects in the mass range of ordinary stars. The association of low rotation with efficient early star formation is suggestive of the supposed early history of elliptical galaxies.

065.053 Boundary and initial conditions in protostar calculations. M. J. Disney.
Monthly Notices Roy. Astron. Soc., Vol. 175, 323 - 333 (1976).

On first fragmentation protostars probably form part of a larger protocluster cloud already in a state of dynamic collapse. In that case it is argued that the protostar boundary is initially collapsing at supersonic speed relative to the core. This prevents information from the boundary reaching the core and calls into question models like Larson's, which start homogeneously but become centrally condensed due to the propagation of a rarefaction wave from the boundary.

065.054 Magnetic braking of collapsing interstellar clouds.
R. C. Fleck, J. H. Hunter.
Monthly Notices Roy. Astron. Soc., Vol. 175, 335 - 343

(1976).

The purpose of this investigation is to show that recourse to anisotropic compression along a magnetic field is not a necessary condition for star formation within large collapsing interstellar gas clouds. The authors examine angular momentum transfer from magnetically braked, cool interstellar gas clouds of 10^2, 10^3 and 10^4 $M_\odot$. Initially, the braking constrains the clouds to co-rotate with the galactic background until just prior to the epoch of magnetic field uncoupling, when the braking mechanism becomes inefficient and the clouds contract conserving angular momentum thereafter. The results are shown to be consistent with observations of stellar rotational velocities, and, also, with the angular momentum of the protosun.

065.055 The limiting luminosity of accreting neutron stars with magentic fields. M. M. Basko, R. A. Sunyaev.
Monthly Notices Roy. Astron. Soc., Vol. 175, 395 - 417 (1976).

Accretion on to a magnetized neutron star for high accretion rates, when one can no longer ignore the back-reaction of emergent light on the infalling material, is discussed in detail. The luminosity L^* is evaluated beyond which one should allow for the dynamic effect of emergent light on the infalling material. The limiting X-ray luminosity L^{**} of accreting magnetized neutron stars is shown to depend crucially on the geometry of the accretion channel. In the framework of the adopted model, regular pulsations of hard and soft X-rays are discussed.

065.056 Chemical transformations during the collapse of protostars. Qualitative analysis. I. G. Kolesnik.
Astrometriya i Astrofizika, Kiev, vyp. (No.) 28, (see 003.002), p. 79 - 88 (1976). In Russian.

The significance of chemical reactions proceeding during a protostar collapse on abundance of elements contributing to cooling processes was studied. Variations of molecular hydrogen abundance both in a gas medium and in the case of dust grain impurities, the ratio between ionized and neutral carbon, absorption in dust particles and growth of dust grain radius were considered. The influence of these processes on cooling rates is discussed.

065.057 Cloud collapse and star formation. M. Jura.
Astron. Journ., Vol. 81, 178 - 181 (1976).

Using a model for cloud heating, the author calculates minimum masses for interstellar clouds to collapse as a function of external pressure, and he finds a strong dependence of this critical mass M_{cr} upon the external pressure. In the model, M_{cr} is approximately independent of metal abundance, but metal-enhanced star formation may still occur if clouds are preferentially formed out of metal-rich material.

065.058 Evolution of massive close binaries and formation of neutron stars and black holes.
A. G. Massevitch (Masevich), A. V. Tutukov, L. R. Yungelson (Yungel'son).
Astrophys. Space Sci., Vol. 40, 115 - 133 (1976).

Main results of computations of evolution for massive close binaries (10 $M_\odot$ + 9.4 $M_\odot$, 16 $M_\odot$ + 15 $M_\odot$, 32 $M_\odot$ + 30 $M_\odot$, 64 $M_\odot$ + 60 $M_\odot$) up to oxygen exhaustion in the core are described. The results obtained allow to outline the following evolutionary chain: two detached main-sequence stars — mass exchange — Wolf-Rayet star or blue supergiant plus main-sequence star — explosion of the initially more massive star appearing as a supernova event — collapsed or neutron star plus main-sequence star, that may be observed as a 'runaway star' — mass exchange leading to X-rays emission — collapsed or neutron star plus WR-star or blue supergiant — second explosion of supernova that preferentially disrupts the system and gives birth to two single high spatial velocity pulsars.

065.059 Spot analysis of rigidly rotating Ap stars.
V. S. Lebedev.
Astrofiz. Issled., Izv. Spets. Astrofiz. Obs., Vol. 8, 20 - 24 (1976). In Russian.

A method of search for the distribution of chemical elements over the surface of rotating stars is suggested within the framework of a spot oblique rotator model. The object is to find the expansion coefficients of the observed curves of the equivalent width and radial velocity variation of a spectral line in series in certain eigenfunctions. Tables of the values for these functions are calculated for various orientations of a star, diverse positions and sizes of spots. The distribution of Eu in the magnetic star 21 Per is found as having the form of two spots with diameters $130°$ and $150°$ each.

065.060 On convection in late-type stars. M. S. Vardya.
Bull. Astron. Soc. India, Vol. 3, 29 (1975).
Abstract of paper presented at the A.S.I. meeting 1975.

065.061 Nucleosynthesis and star formation of the Galaxy and Magellanic Clouds. G. L. Olson, J. H. Peña.
Astrophys. Journ., Vol. 205, 527 - 534 (1976).

A simple model of galactic nucleosynthesis is presented and extended for an examination of the Magellanic Clouds. The authors make the following conclusions: (a) The average stellar birthrate in the Magellanic Clouds has been nearly constant. (b) Unless stars evolve differently in the Clouds, the initial mass function (IMF) for star formation in the Clouds must be significantly different from the Galaxy. (c) To completely explain the colors of the Clouds when using a power-law IMF, it is necessary to postulate that the stellar birthrate is not a monotonically decreasing function of time but is interrupted by periods with higher than average star formation.

065.062 Neutrino pair emission from finite-temperature neutron superfluid and the cooling of young neutron stars. E. Flowers, M. Ruderman, P. Sutherland.
Astrophys. Journ., Vol. 205, 541 - 544 (1976).

The neutrons inside neutron stars are almost certainly superfluid below a critical temperature $T_c \sim 10^{10}$ K. Below T_c, pairs of excited neutron quasiparticles may recombine, resulting, if weak neutral currents exist, in the emission of neutrino-antineutrino pairs. The authors calculate the emissivity associated with this process and compare it with other neutrino emissivities. For neutron star interior temperatures in the range $10^9 - 10^{10}$ K the recombination emissivity can dominate all others.

065.063 Axisymmetric stars with both poloidal and toroidal magnetic fields. J. J. Monaghan.
Astrophys. Space Sci., Vol. 40, 385 - 391 (1976).

Chandrasekhar and Prendergast have established a result which has been assumed to imply that axisymmetric stars with an internal toroidal magnetic field should have zero external poloidal field. By considering mildly singular functions, the range of solutions is increased, and models can then be constructed which have toroidal and poloidal fields in the interior and a non-zero, external, poloidal field. Both the magnetic field and its associated current are continuous everywhere.

065.064 The mass-luminosity relationship for cepheids in the Small Magellanic Cloud. B. C. Cogan.
Astron. Astrophys., Vol. 49, 17 - 20 (1976).

Results of evolutionary calculations by several authors are combined to show that the slope of the mass-luminosity relationship increases significantly above $10 M_\odot$. This is in good agreement with the relationship derived from SMC cepheids, using pulsation masses. Thus there appears to be no "mass discrepancy" for these stars when a color-temperature relationship derived from model atmospheres (Bell and Parsons, 1972) with $Z = 0.005$ is used.

065.065 Vibrational stability towards non-radial oscillations during central hydrogen burning.
A. Noels, A. Boury, M. Gabriel, R. Scuflaire.
Astron. Astrophys., Vol. 49, 103 - 106 (1976).

Vibrational stability towards low order g^+ modes ($l = 1$) have been tested in the case of three evolutionary sequences corresponding to 0.5, 0.6 and 1.1 $M_\odot$. The instability detected in the solar case and associated with an evolution with recurrent mixing and thermal imbalance phases, is also present in 0.5 and 0.6 $M_\odot$. On the other hand, all models of 1.1 $M_\odot$ are stable. The presence of a convective core in stars more massive than 1 $M_\odot$ fixes to that value the upper boundary of the range of mass concerned by this instability.

065.066 Possible properties of pre-outburst FU Orionis stars.
G. Welin.
Astron. Astrophys., Vol. 49, 145 - 148 (1976).

Under the assumption that the rapid brightenings of V 1057 Cygni and FU Orionis by more than five magnitudes can be largely explained in terms of the dissipation of a circumstellar dust envelope, observable properties of stars in an evolutionary stage immediately preceding such outbursts are discussed, and criteria for finding them are proposed. These criteria are low-amplitude variability, advanced T Tauri type spectra, strong infrared excesses, and, possibly, OH, CO and H_2O radio emission.

065.067 Angular momentum transport by tidal acoustic wave.
T. Sakurai.
Astrophys. Space Sci., Vol. 41, 15 - 25 (1976).

The author gives an analytic expression of the braking torque on a Jacobian ellipsoid rotating steadily in an environmental gas, based on the assumption that the ellipsoid rotates around its shortest principal axis with an angular momentum slightly larger than that at the bifurcation point of the Maclaurin spheroid. The results are applied to the pre- and post-main sequence phases of a rotating star, and relating astrophysical problems are discussed.

065.068 Transport properties of dense matter.
E. Flowers, N. Itoh.
Astrophys. Journ., Vol. 206, 218 - 242 (1976).

Transport properties of stellar matter are important ingredients in understanding stellar evolution and stellar structure. Using theoretical techniques that have proven useful in solid state physics, Fermi liquid theory, and the theory of liquid metals, the authors have calculated the electron contribution to the electrical conductivity, the thermal conductivity, and the viscosity of neutron star matter in the absence of magnetic fields for densities less than 2×10^{14} g cm^{-3}. The results are also applicable to high-density white dwarf matter. Variational solutions of the transport equations are used except where exact solutions exist. Transport by electrons of any energy both in the presence of a lattice and without the lattice present are considered. Relativistic electron-electron scattering is considered, as is electron-impurity and electron-neutron scattering.

065.069 The role of turbulent pressure in mixing length convection. R. F. Stellingwerf.
Astrophys. Journ., Vol. 206, 543 - 547 (1976).

The mathematical behavior of the equations of mixing length convection including turbulent pressure is discussed using a simplified physical picture. As expected, the modification of the static stellar structure is small except in the case of appreciable convective velocities. These correction terms can cause severe numerical difficulties, however, greatly complicating a straightforward analysis. The nature of this problem is discussed and a remedy is proposed.

065.070 **Comments on the diffusion model of turbulent mixing.** L. D. Cloutman, J. G. Eoll.
Astrophys. Journ., Vol. 206, 548 - 554 (1976).

The authors justify the diffusion model in the framework of a statistical theory of turbulence and show that the difference between the two diffusion equations that have been proposed is related to an ambiguity in the definition of a Lagrangian hydrodynamical calculation by finite differences in the presence of turbulence. They present difference equations for solving the diffusion equation which theory and experience indicate are free of numerical difficulties.

065.071 **Evolution and secular stability of stellar models crossing the Hertzsprung gap for $M = 4, 6$ and $8 M_\odot$.**
D. Lauterborn, R. A. Siquig.
Astron. Astrophys., Vol. 49, 285 - 292 (1976).

A real unstable eigenvalue is detected in the secular eigenvalue spectrum of stars with mass $M = 4, 6$ and $8 M_\odot$ evolving through the Hertzsprung gap. It is argued that this instability is related to the existence of multiple solutions for stars in thermal imbalance in a way similar to the well-known multiplicity of solutions in the case of static stellar models.

065.072 **Tightening hyperon systems and stellar evolution.** V. A. Filimonov.
Izv. Tomsk. politekhn. in-ta, Vol. 278, 7 - 10 (1975). In Russian. − Abstr. in Referativ. Zhurn. 51. Astron., 5.51.540 (1976).

065.073 **On the possibility of nuclear explosion in a contracting protostar.** V. S. Shevchenko.
Peremennye Zvezdy, Vol. 20, 179 - 195 (1975). In Russian.

The high abundance of deuterium and lithium isotopes, beryllium and boron in protostellar clouds is supposed to change essentially the evolution of a protostar and under some conditions to cause a nuclear explosion with disastrous results.

065.074 **Zwei Endstadien der Sternentwicklung: II. Schwarze Löcher.** J. Schmid-Burgk, R. Wehrse.
SuW, 15. Jahrgang, 189 - 194 (1976).

065.075 **^{7}Li production in bouncing supermassive stars.** H. Nørgaard, K. J. Fricke.
Astron. Astrophys., Vol. 49, 337 - 342 (1976).

Nucleosynthesis in detailed models for bouncing supermassive stars is investigated. The authors consider a non-rotating $5.2 \times 10^5 M_\odot$ and a rotating $3 \times 10^6 M_\odot$ star and follow the time evolution of the abundances throughout the quasistatic contraction phase as well as through the implosion-explosion. The numerical network integrations show that explosions of such objects cause predominantly the enrichment of ^{7}Li. The implications of the results concerning the amount of galactic matter which might be processed in such objects is discussed.

065.076 **Cooling times, luminosity functions and progenitor masses of degenerate dwarfs.** M. A. Sweeney.
Astron. Astrophys., Vol. 49, 375 - 385 (1976).

In the present work, cooling models of degenerate dwarfs have been obtained for a wide range of masses and luminosities. The method of determining the cooling times is described. Cooling characteristics of the models are discussed and the theoretically derived luminosity functions are compared with Weidemann's (1967) observationally derived ones. Estimates of the progenitor masses of degenerate dwarfs in clusters and of those of known mass in binary systems are obtained, using recent scanner and four-color observations.

065.077 **Constraints on nucleosynthesis imposed by extremely metal-poor stars.** R. C. Peterson.
Astrophys. Journ., Vol. 206, 800 - 808 (1976).

The author derives accurate relative abundances of elements in very metal-poor stars of high space motion from new high-resolution spectra. The abundance results for the elements through the iron peak support explosive nucleosynthesis with a constant ratio of the products of carbon-, oxygen-, and silicon-burning at all [Fe/H], and with low values of the neutron excess at [Fe/H] = −1.5, consistent with helium burning but not hydrostatic carbon burning as the preceding stage of stellar evolution.

065.078 **Neutral currents and neutrino emission from cool neutron stars.** V. P. Frolov.
Kratkie soobshch. po fiz., 1975, No. 9, p. 14 - 17. In Russian. Abstr. in Referativ. Zhurn. 51. Astron., 6. 51. 502 (1976).

065.079 **Possible effects of internal rotation in low-mass stars.** J. G. Mengel, P. G. Gross.
Astrophys. Space Sci., Vol. 41, 407 - 415 (1976).

The influence of internal rotation on the evolution of a $0.85 M_\odot$ star is investigated by the construction of model sequences. The calculations assume solid-body rotation on the zero-age main sequence, followed by conservation of angular momentum in shells. The 4 cases considered have the initial angular velocities $0, 2 \times 10^{-4}, 6 \times 10^{-4}$, and 8×10^{-4}/sec. All cases but the last are followed to helium ignition.

065.080 **On the evolutionary state of blue and red supergiants.** Y. Frantsman (*J. Francmanis*).
Investigation of the sun and red stars. 4, (see 003.009), p. 60 - 74 (1976). In Russian.

065.081 **Supernova explosions, the new leptons, and right-handed neutrinos.** K. O. Mikaelian.
Phys. Rev. Letters, Vol. 36, 1089 - 1092 (1976).

New leptons and right-handed neutrinos, created by weak neutral currents in stellar interiors, will do much to produce a supernova explosion. The author studies their production and absorption mechanisms, and notes that angular momentum carried by neutrinos might affect the rotation of the subsequent neutron star.

065.082 **Radiative opacity tables for 40 stellar mixtures.** A. N. Cox, J. E. Tabor.
Astrophys. Journ., Suppl. Ser., Vol. 31, 271 - 312 (1976).

Using improved methods, radiative opacities for 40 mixtures of elements are given for use in calculations of stellar structure, stellar evolution, and stellar pulsation.

065.083 **A neutron star as the main product of supernova explosions.** O. H. Guseinov (*O. Kh. Gusejnov*), F. K. Kasumov.
Mem. Soc. Astron. Italiana, Vol. 45, (see 012.017), 723 - 726 (1974).

065.084 **Non explosive collapse of white dwarfs.** R. Canal, E. Schatzman.
Mem. Soc. Astron. Italiana, Vol. 45, (see 012.017), 763 - 768 (1974).

The possibility, for the neutron stars present in binary X-ray sources, of being formed through a non explosive process is examined. It is shown that accretion by a white dwarf close to its critical mass can initiate a collapse with a relatively long time scale. So, neutronization would prevent the explosive nuclear burning. Accretion by white dwarfs in binary systems might then explain dwarf novae, ordinary novae, type I supernovae and non explosive collapses (leading to X-ray sources), the sequence being with a decreasing probability of occurrence.

065.085 **Overstability of gravity modes in massive stars with the semiconvective zone.** H. Shibahashi, Y. Osaki.

Publ. Astron. Soc. Japan, Vol. 28, 199 - 214 (1976).

The vibrational stability of nonradial modes in massive stars with the semiconvective zone is studied. It is shown that nonradial oscillations with a large l of spherical harmonics $Y_l{}^m(\theta, \varphi)$ in massive evolved stars can be distinguished very clearly into two types, one type being a gravity mode trapped in the zone of varying molecular weight (i.e., the μ-gradient zone) and the other being a mode trapped in the envelope. It is argued that the overstability of nonradial modes found in this paper is not directly responsible for the Beta Cephei pulsations.

065.086 **Nucleosynthesis in red giants.**
R. L. Smith, I-J. Sackmann, K. H. Despain.
Explosive nucleosynthesis, (see 012.020), p. 168 - 175 (1973).

065.087 **Degenerate carbon burning.**
R. G. Couch, W. D. Arnett.
Explosive nucleosynthesis, (see 012.020), p. 179 - 185 (1973).

065.088 **The influence of screening effects on carbon ignition.** H. C. Graboske, Jr.
Explosive nucleosynthesis, (see 012.020), p. 186 - 194 (1973).

065.089 **Some quantitative calculations of final stages of stellar evolution.** W. D. Arnett.
Explosive nucleosynthesis, (see 012.020), p. 236 - 247 (1973).

065.090 **A variational description of the stability and equilibrium of cold configurations.** M. A. Abramowicz.
Postępy Astron., Vol. 24, 41 - 49 (1976). In Polish.

065.091 **The Chandrasekhar limit for a stable configuration and the critical ratio M/R.** P. Bucci.
Acta Phys. Austriaca, Vol. 42, 271 (1975). – Abstr. in Phys. Abstr., Vol. 79, A003309 (1976).

065.092 **On the role of ^{14}N in helium burning phase in stellar evolution.** K. Kaminisi, K. Arai, K. Yoshinaga.
Progr. Theor. Phys., Vol. 53, 1855 - 1856 (1975). – Abstr. in Phys. Abstr., Vol. 79, A003318 (1976).

065.093 **About critical radius and critical density of neutron stars.** P. Bucci.
Acta Phys. Austriaca, Vol. 42, 269 (1975). – Abstr. in Phys. Abstr., Vol. 79, A003322 (1976).

065.094 **Neutron star models with condensed π^- phase.** D. Ray.
Indian Journ. Phys., Vol. 49, 333 - 337 (1975). – Abstr. in Phys. Abstr., Vol. 79, A011193 (1976).

065.095 **Neutrino pair bremsstrahlung by nucleons in neutron-star matter.**
E. G. Flowers, P. G. Sutherland, J. R. Bond.
Phys. Rev. D, Vol. 12, 315 - 318 (1975). – Abstr. in Phys. Abstr., Vol. 79, A012475 (1976).

065.096 **The neutron polarizability and dense neutron matter.** S. Ragusa.
Nuovo Cimento Lettere, Ser. 2, Vol. 16, 338 - 340 (1975). Abstr. in Phys. Abstr., Vol. 79, A012633 (1976).

065.097 **MHD instabilities in stellar interiors with toroidal magnetic field.** V. Cadez.
17th Yugoslav. symposium and summer school on the physics of ionised gases, p. 1011 - 1025 (1975). – Abstr. in Phys. Abstr., Vol. 79, A015295 (1976).

065.098 **Electric field in neutron stars.**
B. Bertotti, S. Boffi, G. Giusti.

Nuovo Cimento Lettere, Ser. 2, Vol. 14, 365 - 368 (1975). Abstr. in Phys. Abstr., Vol. 79, A023793 (1976).

065.099 **Relativistic formulation of the neutron starquake theory of pulsar glitches.**
B. Carter, H. Quintana.
Ann. Physics, Vol. 95, 74 - 89 (1975). – Abstr. in Phys. Abstr., Vol. 79, A027907 (1976).

065.100 **Dielectric properties of the magnetic surface of neutron stars.** J. Ventura.
Thesis City Univ., New York, USA, 180 pp. (1975). (Available from: Univ. Microfilms, Ann Arbor, Mich., USA. Order No. 75-13643). – Abstr. in Phys. Abstr., Vol. 79, A027971 (1976).

065.101 **Stellar stability in Newtonian theory of gravitation with shadow effect.** C. C. Chiang, V. H. Hamity.
Nuovo Cimento B, Ser. 11, Vol. 30B, 280 - 290 (1975). Abstr. in Phys. Abstr., Vol. 79, A032858 (1976).

065.102 **Electron capture in highly evolved stars.**
Y. Egawa, K. Yokoi, M. Yamada.
Progr. Theor. Phys., Vol. 54, 1339 - 1355 (1975). – Abstr. in Phys. Abstr., Vol. 79, A032859 (1976).

065.103 **A new theory on the generation of the magnetic field of celestial bodies. III. Aspects of application.**
E. Schmutzer.
Experim. Techn. Phys., Vol. 23, 445 - 453 (1975). – Abstr. in Phys. Abstr., Vol. 79, A032871 (1976).

065.104 **Magnetized Fermi gas in the outer shell of neutron stars.** K. Y. Fu.
Chinese Journ. Phys. (*Taiwan*), Vol. 13, 183 - 187 (1975). Abstr. in Phys. Abstr., Vol. 79, A032930 (1976).

065.105 **Instability of neutron star matter for pion condensation.**
A. Suzuki, Y. Takahashi, Y. Futami.
Progr. Theor. Phys., Vol. 54, 1429 - 1439 (1975). – Abstr. in Phys. Abstr., Vol. 79, A032951 (1976).

065.106 **Higgs meson emission from a star and a constraint on its mass.** K. Sato, H. Sato.
Progr. Theor. Phys., Vol. 54, 1564 - 1565 (1975). – Abstr. in Phys. Abstr., Vol. 79, A034225 (1976).

065.107 **Spontaneous symmetry breaking in the interior of a neutron star.** M. M. Janson.
Nuovo Cimento Lettere, Ser. 2, Vol. 15, 231 - 234 (1976). Abstr. in Phys. Abstr., Vol. 79, A037588 (1976).

065.108 **Neutron stars.** G. Baym, C. Pethick.
Annual Rev. Nuclear Sci., Vol. 25, 27 - 77 (1975). – Abstr. in Phys. Abstr., Vol. 79, A037603 (1976).

065.109 **Magnetic properties of neutron-star matter.**
N. C. Chao, J. W. Clark.
Rev. Brasil. Fis., Vol. 5, 169 - 187 (1975). – Abstr. in Phys. Abstr., Vol. 79, A042271 (1976).

065.110 **Neutrino astrophysics.** S. A. Bludman.
7th Texas symposium relativistic astrophys., Dallas, Texas, USA, 16 Dec. 1974, 36 pp. (1975).

065.111 **Early neutron-star matter.** E. R. Hilf.
Mesonic effects in nuclear structure, Symposium Bonn, F. R. Germany, 29 Jan. 1974, p. 57 - 79 (1975). In German.

065.112 **Carbon detonations in rapidly rotating stellar cores.**

J. H. Mahafey.
Thesis Colorado Univ., Boulder, USA, 126 pp. (1974). (Available from Univ. Microfilms, Order No. 75-13,447). – See 14.065.113.

065.113 **Structure and stability of relativistic, differentially rotating stars.** F. H. Seguin.
Thesis California Inst. Tech., Pasadena, USA, 105 pp. (1975). (Available from Univ. Microfilms, Order No. 75–12, 518).

065.114 **Electric field in neutron stars.**
B. Bertotti, S. Boffi, C. Giusti.
Proc. internat. conf. nuclear self-consistent fields, Trieste, Italy, 24 - 28 Feb 1975, p. 402 (1975).

065.115 **Evolution of helium rich stars with hydrogen burning.** M. Roeser.
Separate print Max-Planck-Inst. Phys. Astrophys., München, F. R. Germany, 14 pp. (1975). (Available from ZAED). – See 14.065.173.

065.116 **Numerical study of rotating relativistic stars.**
J. R. Wilson.
Meeting phys. astrophys. neutron stars black holes, Varenna, Italy, 14 July 1975, 54 pp. (1975).

065.117 **Many-particle theory of nuclear systems with application to neutron star matter. Research report,**
1 December 1973 - 30 November 1974.
D. A. Chakkalakal, C. H. Yang.
Separate print Southern Univ., Baton Rouge, Louisiana, USA, Dept. Phys., 25 pp. (1974).

065.118 **Advanced evolution of a 15 solar mass star.**
A. S. Endal.
Thesis Florida Univ., Gainesville, USA, 256 pp. (1974). (Available from: Univ. Microfilms, Order No. 75-16,378).

065.119 **Non-adiabatic radial oscillations and pulsational stability of hot degenerate dwarfs.** E. M. Sion.
Thesis Pennsylvania Univ., Philadelphia, USA, 110 pp. (1975). (Available from: Univ. Microfilms, Order No. 75-24, 129).

Development of a computer code for the calculation of stellar evolution, with applications to solar models of low neutrino flux. See Abstr. 021.014.

Possibility of synthesis of proton-rich nuclei in highly evolved stars. II. See Abstr. 061.001.

Production of heavy elements in nature.
See Abstr. 061.008.

The s-process in stars. See Abstr. 061.051.

Neutrino production in stellar matter by photons in a renormalizable scalar-boson-exchange model of weak interactions. See Abstr. 061.072.

Introduction to nucleosynthesis.
See Abstr. 061.080.

Cooling blackbody: a mechanism for cosmic gamma-ray bursts. See Abstr. 061.081.

The hydrodynamics of the helium flash.
See Abstr. 062.042.

Numerical models of astrophysical convection.
See Abstr. 062.067.

General numerical fluid dynamics algorithm for astrophysical applications. See Abstr. 062.094.

Mass loss in early stages of stellar evolution.
See Abstr. 064.056.

On dynamical stability of supermassive objects.
See Abstr. 066.001.

Interaction of neutron stars with black holes.
See Abstr. 066.004.

On the structure and stability of rapidly rotating fluid bodies in general relativity. I. The numerical method for computing structure and its application to uniformly rotating homogeneous bodies. See Abstr. 066.015.

A higher stability limit for neutron stars.
See Abstr. 066.023.

Application of continuous variation formulas and discrete invariance principles to black holes and neutron star models. See Abstr. 066.120.

On the evolution of a 1 $M_\odot$ star with a periodically mixed core. See Abstr. 080.013.

The nature of the objects of Joy: a study of the T Tauri phenomenon. See Abstr. 114.023.

P Cygni profiles in Zeta Ophiuchi and Zeta Puppis.
See Abstr. 114.303.

Luminosity functions and the evolution of low-mass population I giants. See Abstr. 115.015.

Rotating stellar models according to the quasi-dynamic method. See Abstr. 116.010.

Multiplicity among solar-type stars.
See Abstr. 117.014.

Evolution of carbon-oxygen dwarfs in binary systems. See Abstr. 117.026.

Evolutionary problems of cepheids and other giants investigated with new radiative opacities. See Abstr. 122.024.

Evidence favoring nonevolutionary cepheid masses.
See Abstr. 122.033.

Supernova remnants. See Abstr. 125.032.

Presupernova evolution. See Abstr. 125.040.

How to make metal-poor stars, redden OB associations and grow mantles on grains. See Abstr. 131.086.

Evolution of rotating interstellar clouds. II. The collapse of protostars of 1, 2, and 5 $M_\odot$.
See Abstr. 131.124.

H_2O maser emission associated with T Tauri and other regions of star formation. See Abstr. 131.147.

Fragmentation of magnetic interstellar clouds by ambipolar diffusion. I. See Abstr. 131.168.

Emission-line spectra of individual condensations of Herbig-Haro objects. See Abstr. 132.004.

The Orion nebula and other regions of star formation
See Abstr. 132.008.

Stellar evolution and planetary nebula ejection.
See Abstr. 133.028.

A new pulsar atmospheric model. I. Aligned magnetic and rotational axes.   See Abstr. 141.345.

Infrared sources and star formation.
See Abstr. 141.609.

CRL 2688: a post–carbon-star object and probable planetary nebula progenitor.   See Abstr. 141.616.

A neutron star crustquake origin for γ-ray bursts.
See Abstr. 142.199.

X-rays from neutron stars heated by accretion.
See Abstr. 142.230.

A study of Be stars in clusters.
See Abstr. 153.010.

Evolved stars in open clusters.
See Abstr. 153.023.

Stervorming in het merkwaardige radiostelsel
NGC 5128.   See Abstr. 158.049.

Star formation and the structure of the Large Magellanic Cloud.   See Abstr. 159.004.

Errata

065.901 Errata: "The triple-alpha rate, screening factors, and the helium flash" [Astrophys. Journ., Vol. 199, 443 - 447 (1975)].   T. D. Tarbell, R. T. Rood.
Astrophys. Journ., Vol. 203, 770 (1976).

065.902 Erratum: 'Corequakes of neutron stars and soft γ-ray bursts' [Astron. Astrophys., Vol. 44, 21 - 24 (1975)].   A. I. Tsygan.
Astron. Astrophys., Vol. 49, 159 (1976).

066 Relativistic Astrophysics (without Cosmology), Background Radiation, Gravitation Theory

066.001 On dynamical stability of supermassive objects.
S. Yabushita.
Monthly Notices Roy. Astron. Soc., Vol. 174, 637 - 647 (1976).

Calculation is made, within the framework of general relativity, of the central redshift of a linear series of configurations with an isothermal core and an envelope. Two kinds of envelope are discussed; adiabatically stable ones ($dp = d\rho$) and a thin mass shell. The stability analysis shows that as z_c is increased, the massive sphere becomes dynamically unstable. The critical value is $z_c = 1.48$ or 4.43 (according as core equation of state $p = \rho/3$ or $p = \rho$, with a thin mass shell envelope) and $z_c = 2.82$ ($p = \rho/3$ in core, $dp = d\rho$ in envelope). The present investigation has an important implication regarding the interpretation of QSO redshifts.

066.002 The electromagnetic background: limitations on models of unseen matter. D. Eichler, A. Solinger.
Astrophys. Journ., Vol. 203, 1 - 5 (1976).

The existence of bound systems implies the release of energy associated with their formation. Estimating the fraction of this energy which appears in observable diffuse form places restrictions on various theories involving high binding energy densities of seen or unseen matter. The potency of pulsars and white dwarfs as cosmic-ray sources is severely limited by these considerations. The authors also find: (1) The amount of unseen matter in the form of burned-out stars or neutron stars must be well below that needed to bind the universe. (2) Black holes, if not primordial, are restricted to an extremely narrow range of parameters if they are supposed to bind the universe, and are thus unlikely candidates.

066.003 On the detectability of gravitational waves from W Ursae Majoris binary stars. W. L. Burke.
Astrophys. Journ., Vol. 203, 694 - 696 = Lick Obs. Bull., No. 707 (1976).

The possibility of using the collective gravitational radiation from all of the W Ursae Majoris binary stars in the Galaxy is considered, and such radiation is shown to be undetectable because of interference from local gravity gradients.

066.004 Interaction of neutron stars with black holes.
R. M. Misra.
Astrophys. Journ., Vol. 203, 704 - 713 (1976).

An object like a neutron star, whose crust and possibly whose core have a lattice structure, and which is in close orbit around a black hole, gets deformed and/or fractures/breaks up due to tidal interaction with the black hole. The author discusses the elastic deformations of such orbiting components. It is found that the mode of fracture of the orbiting body in stable circular orbits around a Schwarzschild black hole is similar qualitatively to that for a Kerr black hole. In the case of vanishing modulus of rigidity the earlier results for fluid bodies are recovered.

066.005 Spherical winds and accretion in general relativity.
G. R. Blumenthal, W. G. Mathews.
Astrophys. Journ., Vol. 203, 714 - 719 = Lick Obs. Bull., No. 706 (1976).

Analytic general solutions are obtained for time-independent, adiabatic, spherically symmetric wind and accretion flows in Schwarzschild coordinates. Various equations of state applicable for both relativistic and nonrelativistic temperatures are considered. The general solution is shown to strongly resemble the nonrelativistic stellar wind solution. Finally, shock transitions in special relativity are briefly considered, and it is shown how they can be fitted onto the full solution to accommodate a broader class of boundary conditions.

066.006 Remarks on Khalatnikov/Lifshitz type theorems.
H.-H. von Borzeszkowski, U. Kasper.
Astron. Nachr., Vol. 297, 29 - 31 (1976).

It is shown that the proof of the $g = 0$ theorem established by Khalatnikov and Lifshitz in general relativity has to be supplemented by additional considerations. The same investigations are made for Treder's tetrad theory of gravitation.

066.007 Is gravity getting weaker? T. C. Van Flandern.
Sci. American, Vol. 234, No. 2, p. 44 - 52 (1976).

Several theories of gravitation predict that the force of gravity diminishes as the universe expands. Preliminary results of timing eclipses of stars by the moon suggest that it may well be the case.

066.008 Black and white holes.
Ya. B. Zel'dovich, I. D. Novikov, A. A. Starobinskij.
Priroda, 1976, No. 1, p. 34 - 42. In Russian.

066.009 Stability of a polytropic plate in general relativity.
J. Horský, E. V. Chubaryan, V. V. Papoyan.
Bull. Astron. Inst. Czechoslovakia, Vol. 27, 115 - 118 (1976).

The problem of stability with respect to small perturbations of a thick plane plate, formed by a polytropic substance, is studied. It is proved that these configurations are more stable with respect to small perturbations as compared to the spherically symmetric case.

066.010 Entropy production by black holes. W. Kundt.
Nature, Vol. 259, 30 - 31 (1976). – Letter.

066.011 Long conductors as antennae for gravitational radiation. R. J. Adler.
Nature, Vol. 259, 296 - 297 (1976).

To date, the antennae used in the search for gravitational radiation have been mechanical in nature, that is they involve the detection of small displacements or strains in mechanical systems. Here the author demonstrates that a conceptually simple system of two long conducting wires of different metallic composition constitutes, in principle, a non-mechanical antenna for gravitational radiation, and that if the conductors are sufficiently long such a system may be of practical use. The principle of operation of the system is that gravitational radiation produces a greater force on the ions in a metal than on the electrons, which, to preserve charge neutrality, necessitates a compensating electric field which is, in principle, detectable.

066.012 Obese 'neutron' stars.
K. Brecher, G. Caporaso.
Nature, Vol. 259, 377 - 378 (1976).

The authors show here that non-rotating, spherically symmetric, general relativistic 'neutron' stars of mass $\geq 3\ M_\odot$ are consistent with the known laws of physics. In particular, for the 'bag' model of hadrons, they find $M_{max} \cong 3\ M_\odot$. The maximum mass for polytropic equations of state is found to be $\cong 5\ M_\odot$. This is particularly interesting since it has been repeatedly claimed that the X-ray source Cygnus X-1 is a binary star system containing a black hole accreting matter from its relatively normal companion star HDE226868.

066.013 Annihilation of matter and antimatter and the cosmic X-ray background.
P. Carlqvist, B. Laurent.
Nature, Vol. 260, 225 - 226 (1976).

In cosmological models where matter and antimatter occur symmetrically in the universe, their annihilation is likely to take place also at the present time. At the annihilation of a proton and antiproton, ~1.6 negative electrons and an equal number of positrons, each with an energy of ~10−100 MeV are produced, together with γ rays and neutrinos. The initial electron energies can be expected to decrease gradually because of their interaction with cosmic magnetic fields, plasmas, and electromagnetic radiation. As a result of these interactions synchrotron radiation, bremsstrahlung, and inverse Compton radiation are all emitted. When optical photons are scattered against the relativistic electrons, X rays are produced by the inverse Compton effect. This mechanism may be an important source of cosmic X rays. The authors purpose is to study how efficient this X-ray mechanism can be in intergalactic space (where most of the annihilation electrons are expected to be found) and to see whether it can offer an explanation to the cosmic X-ray background observed.

066.014 Spheroidal and toroidal configurations as sources of the Kerr metric? I. A kinematical approach.
F. de Felice, L. Nobili, M. Calvani.
Astron. Astrophys., Vol. 47, 309 - 313 (1976).

The existence of extended Kerr metric sources of perfect fluid is taken as work-hypothesis to investigate the structure of the boundaries which derive from the Boyer's surface condition (Boyer, 1965). The authors find closed spheroidal configurations which hide an internal cavity as well as toroidal configurations; however both the boundary of the internal cavity and the tori touch the ring singularity. They judge this feature non physical and conclude that the Boyer's condition is not sufficient to completely define a well behaved physical source.

066.015 On the structure and stability of rapidly rotating fluid bodies in general relativity. I. The numerical method for computing structure and its application to uniformly rotating homogeneous bodies.
E. M. Butterworth, J. R. Ipser.
Astrophys. Journ., Vol. 204, 200 - 223 (1976).

A new numerical method for computing the structure of rapidly rotating fluid bodies in general relativity is presented. The method is a Henyey-type relaxation method of the kind previously used by Stoeckly in Newtonian theory. The method is used to construct sequences of uniformly rotating homogeneous bodies, the relativistic analogs of the classical Maclaurin spheroids. The results reveal that, in contrast to the Newtonian sequence, most, and probably all, of the relativistic sequences terminate at nonzero ratios of proper polar radius to proper equatorial radius where centrifugal and gravitational accelerations balance at the equator. Other relativistic effects, including those associated with the formation of regions within which observers must rotate relative to infinity, are discussed. The computational results are applied to uniformly rotating neutron stars.

066.016 Active mass in relativistic gravity: theoretical interpretation of the Kreuzer experiment. C. M. Will.
Astrophys. Journ., Vol. 204, 224 - 234 (1976).

A 1966 experiment performed by Kreuzer set an upper limit of 5 parts in 10^5 on the difference in the ratio of active to passive mass between fluorine and bromine. (Active mass of a body is the mass that generates gravity, while passive mass is the mass that responds to gravity). The author introduces a parametrized post-Newtonian formalism for the spacetime metric of a system of charged point particles using parameters whose values vary from gravitation theory to gravitation theory. He shows that any theory of gravity that possesses post-Newtonian integral conservation laws for total momentum automatically agrees with the Kreuzer experiment.

066.017 Gravitational-wave bursts from the nuclei of distant galaxies and quasars: proposal for detection using Doppler tracking of interplanetary spacecraft.
K. S. Thorne, V. B. Braginsky (Braginskij).
Astrophys. Journ., (Letters), Vol. 204, L1 - L6 (1976).

It is likely that supermassive black holes ($M \approx 10^6$ to $10^{10} M_\odot$) exist in the nuclei of many quasars and galaxies. The collapse which forms these holes and subsequent collisions between them should produce strong, broad-band bursts of gravitational waves: for a source of mass M at the Hubble distance of ~10^{10} light-years, the dimensionless amplitude would be $h \sim 2 \times 10^{-17} \times (M/10^6 M_\odot)$, and the duration of the burst would be $\tau \sim (90 \text{ s}) \times (M/10^6 M_\odot)$. Such bursts might arrive at earth as often as 50 times per year – or as rarely as once each 300 years. The detection of such bursts may be possible within the next few years using dual-frequency Doppler tracking of interplanetary spacecraft.

066.018 An observation of the diffuse soft X-ray/extreme-ultraviolet background.
W. Cash, R. Malina, R. Stern.
Astrophys. Journ., (Letters), Vol. 204, L7 - L11 (1976).

Observations of diffuse radiation in the 114−150 Å and 44−120 Å bands have been made with rocket-borne proportional counters. The small collecting area, large field of view (55° FWHM) counters observed areas of sky near $l = 150°$, $b = −15°$ and $l = 189°$, $b = 3°$. The authors observed a flux of 0.125 photons cm^{-2} s^{-1} sr^{-1} eV^{-1} at 180 eV, consistent with previous measurements, and a flux of 3.2 photons cm^{-2} s^{-1} sr^{-1} eV^{-1} at 100 eV, indicating a steeply rising spectrum. The experimental data have been used to place constraints on simple models for galactic origin of the 50−150Å background. An upper limit on the space density of white dwarfs at $T \gtrsim 10^5$ K is obtained.

066.019 On the structure and stability of rapidly rotating fluid bodies in general relativity. II. The structure of uniformly rotating pseudopolytropes. E. M. Butterworth.
Astrophys. Journ., Vol. 204, 561 - 572 (1976).

A method is described for obtaining numerical solutions to the exact Einstein field equations that represent uniformly rotating perfect fluid bodies which are stationary and obey equations of state of the form (pressure) ∝ (energy density)$^{1+1/n}$. Sequences parametrized by the rate of rotation are generated for polytropic indices n between 0.5 and 3 and for varying strengths of relativity. All are found to terminate at surface velocities which are approximately 10 percent or more of the velocity of light.

066.020 Gravitational radiation from a particle falling into a Kerr black hole. I. G. Dymnikova.
Astrophys. Space Sci., Vol. 39, L25 - L29 (1976).

The gravitational radiation from a highly relativistic test particle spiralling inward toward a Kerr black hole along a conical surface is estimated. The spectra of several lowest multipoles depending on the polar angle of the falling particle at infinity are obtained for the frequency band $\omega \lesssim c^3 GM^{-1}$.

066.021 Neuer Anlauf bei der Suche nach Schwerkraftwellen.
R. Breuer.
SuW, 15. Jahrgang, p. 130 - 131 (1976).

066.022 Sur la répartition de la masse équivalente à l'énergie potentielle et ses conséquences.
R. Lucas.
Comptes Rendus Acad. Sci. Paris, Sér. B, Vol. 282, 43 - 45

(1976).

Le problème de la localisation de la masse attribuable à l'énergie potentielle d'interaction entre deux corps de masses M_1 et M_2 est étudié. La répartition proposée par de Broglie est confirmée. Des applications de ces résultats sont faites à la loi de déplacement vers le rouge des raies spectrales sous l'action d'un potentiel de gravitation ainsi qu'au problème de la rotation du périhélie d'une planète gravitant autour du soleil.

066.023 A higher stability limit for neutron stars.
R. L. Bowers, A. M. Gleeson, R. D. Pedigo.
Astrophys. Journ., Vol. 205, 261 - 267 (1976).

A fully relativistic equation of state for hyperons based on strong interactions fitted to nuclear matter has been used to construct neutron stars. The model yields a maximum stable mass of $2.39\,M_\odot$ and a maximum moment of inertia of 3.13×10^{45} g cm^2. The significance of these results and their relation to observations are discussed. The sensitivity of the results, particularly the upper mass limit, due to variations of the input parameters is considered.

066.024 Experimental examination of the gravitational inverse square law. D. R. Long.
Nature, Vol. 260, 417 - 418 (1976).

The author suggests that the gravitational force law should be modified at laboratory dimensions by $G(R) = G_0[1+(0.002)\ln R]$. It should be remarked that while these results indicate an inverse square law failure at small distance, it is also clear that the experiment was carried out in the earth's gravitational field and that we may, in fact, have a failure of the superposition principle.

066.025 The fall of the shell of dust on to a rotating black hole. J. Bičák, Z. Stuchlík.
Monthly Notices Roy. Astron. Soc., Vol. 175, 381 - 393 (1976).

The motion of particles falling radially from rest at infinity with zero total angular momentum on to a rotating (Kerr) black hole is studied. The motion is examined analytically for large distances as well as near the horizon and also numerically in the case of an extreme Kerr black hole. The approach of the particles towards the horizon in terms of the arrival times of these photons to a distant observer, the redshift of the radiation and its intensity show dependence exponentially on the observer's proper time as in the non-rotating case, however the characteristic e-folding times become infinite as the hole's angular momentum approaches the extreme value. In the case of an extreme Kerr black hole these exponential laws go over into power laws.

066.026 The gravitational perturbation of the cosmic background radiation by density concentrations.
C. C. Dyer.
Monthly Notices Roy. Astron. Soc., Vol. 175, 429 - 447 (1976).

The gravitational effect of density concentrations in the universe on the temperature distribution of the cosmic blackbody background radiation is considered, using the Swiss cheese model universe, and supposing each hole to contain an expanding, homogeneous dust sphere at its centre. The temperature profile across such a hole differs in an essential way from that obtained earlier by Rees & Sciama. The evolution of this effect with the expansion of the universe is considered for 'relatively increasing' density contrasts emerging from the same initial singular state as the rest of the universe.

066.027 On the second integrals of geodesics in the Kerr field.
O. P. Krivenko, K. A. Pyragas, I. T. Zhuk.
Astrophys. Space Sci., Vol. 40, 39 - 61 (1976).

Qualitative analysis is given of the second integrals of the equations of geodesics of the Kerr metric expressed in the Boyer-Lindquist coordinates.

066.028 The gravitational searchlight effect.
S. M. Chitre, J. V. Narlikar, R. C. Kapoor.
Bull. Astron. Soc. India, Vol. 3, 33 (1975). – Abstract of paper presented at the A.S.I. meeting 1975.

066.029 Motion of spinning particles in relativity.
E. A. Lord, R. C. Kapoor.
Bull. Astron. Soc. India, Vol. 3, 33 (1975). – Abstract of paper presented at the A.S.I. meeting 1975.

066.030 Self-gravitating astronomical objects filled with magnetofluid. T. H. Date.
Bull. Astron. Soc. India, Vol. 3, 34 (1975). – Abstract of paper presented at the A.S.I. meeting 1975.

066.031 The center of mass in the post-Newtonian approximation of general relativity.
G. Contopoulos, N. Spyrou.
Astrophys. Journ., Vol. 205, 592 - 598 (1976).

The uniform motion of the center of mass of a bounded perfect fluid mass up to the first post-Newtonian approximation of general relativity is derived, by proving that the total linear momentum of the fluid can be put in the form of the total time-derivative of a certain function. The same formula applies in the case that the fluid mass is composed of N perfect fluid bodies of finite dimensions. The authors define a conserved mass and a center of mass for each body in the post-Newtonian approximation, such that the equations of the motion of the center of mass of the system reduce to those of a system of N point masses. The mass so defined is the total mass-energy of each body.

066.032 Métrique 2-post-newtonienne pour un système de N-corps fluide parfait et sphériques. Calcul de g_{0i} et g_{ij}. E. Verdaguer, L. Mas.
Comptes Rendus Acad. Sci. Paris, Sér. A, Vol. 282, 551 - 554 (1976).

En partant des équations données par Chandrasekhar, nous calculons la métrique d'un système limité à une région finie de l'espace de N-corps sphériques, composés de fluide parfait, de petites dimensions et qui ont des vitesses petites par rapport à c. On donne aussi une forme de la métrique pour un fluide parfait en coordonnées harmoniques, mais elle a des termes divergents pour grandes distances.

066.033 Sur la positivité de la masse.
Y. Choquet-Bruhat, J. E. Marsden.
Comptes Rendus Acad. Sci. Paris, Sér. A, Vol. 282, 609 - 612 (1976).

Nous démontrons qu'il existe un voisinage (au sens d'espaces fonctionnels convenables) de l'espace-temps de Minkowski tel que tout espace-temps de ce voisinage, solution des équations d'Einstein du vide, a une masse positive. Nous indiquons quelques résultats et difficultés pour la résolution du problème global.

066.034 The Einstein effect in astrometry.
A. A. Mikhajlov.
Astron. Zhurn. Akad. Nauk SSSR, Vol. 53, 436 - 440 (1976). In Russian. English translation in Soviet Astron., Vol. 20, No. 2.

The apparent displacement of a star by relativistic bending of light rays in the sun's gravitational field is derived by a geometrical method. The corresponding correction to the stars's apparent place is given.

066.035 A test of post-Newtonian conservation laws in the binary system PSR 1913+16. C. M. Will.

Astrophys. Journ., Vol. 205, 861 - 867 (1976).

Observations that set upper limits on secular changes in the pulsar period and orbital period in the binary system PSR 1913+16 may provide a test of post-Newtonian conservation laws. According to some metric theories of gravitation, the center of mass of a binary system may be accelerated in the direction of the periastron of the orbit because of a violation of post-Newtonian momentum conservation. In the binary system PSR 1913+16, this effect could produce secular changes in both pulsar and orbital periods (changing overall Doppler shift) as large as two parts in 10^6 per year.

066.036 **Search for gravitational radiation from cosmic sources.**
V. B. Braginskij, A. B. Manukin, E. I. Popov, V. N. Rudenko.
Problems of gravitation. Third Soviet Gravitational Conference, Erevan, 1972, (see 012.008), p. 18 - 31 (1975). In Russian.

066.037 **Modern state of the theory of gravitational collapse and astrophysical discoveries.** I. D. Novikov.
Problems of gravitation. Third Soviet Gravitational Conference, Erevan, 1972, (see 012.008), p. 37 - 41 (1975). In Russian.

066.038 **Physics in rotating reference systems.**
E. Schmutzer.
Problems of gravitation. Third Soviet Gravitational Conference, Erevan, 1972, (see 012.008), p. 71 - 89 (1975). In Russian.

066.039 **Equivalence of canonical and covariant approaches to quantization of an asymptotically plane field of Einstein gravitation.** L. D. Faddeev.
Problems of gravitation. Third Soviet Gravitational Conference, Erevan, 1972, (see 012.008), p. 90 - 103 (1975). In Russian.

066.040 **Matrix generalization of the Cartan form in general relativity.** N. V. Mitskevich.
Problems of gravitation. Third Soviet Gravitational Conference, Erevan, 1972, (see 012.008), p. 104 - 112 (1975). In Russian.

066.041 **Ultrarelativistic particles and the tetrad theory of some optical effects in a Schwarzschild field.**
A. E. Levashev.
Problems of gravitation. Third Soviet Gravitational Conference, Erevan, 1972, (see 012.008), p. 113 - 135 (1975). In Russian.

066.042 **Global properties of matter in collapsed state.**
M. A. Markov.
Problems of gravitation. Third Soviet Gravitational Conference, Erevan, 1972, (see 012.008), p. 165 - 207 (1975). In Russian.

066.043 **Gamma rays from primordial black holes.**
D. N. Page, S. W. Hawking.
Astrophys. Journ., Vol. 206, 1 - 7 (1976).

This paper examines the possibilities of detecting hard γ-rays produced by the quantum-mechanical decay of small black holes created by inhomogeneities in the early universe. The best prospect for detecting a primordial black hole seems to be to look for the burst of hard γ-rays that would be expected in the final stages of the evaporation of the black hole.

066.044 **The effect of radiation pressure on accretion disks around black holes.**
L. Maraschi, C. Reina, A. Treves.
Astrophys. Journ., Vol. 206, 295 - 300 (1976).

Stationary disk accretion onto a black hole is studied for high accretion rates $\dot{M} \gtrsim \dot{M}_c = 2r_0 L_E/GM$ (L_E is the Eddington luminosity) for which the dynamic effect of radiation pressure is important. The rotation of the disk is not assumed to be Keplerian but is considered as an unknown in the Newtonian dynamic equation. It is found that stationary solutions without mass-outflow exist for $\dot{M} > \dot{M}_c$. The radiated luminosity,

however, is always of the order of the Eddington luminosity. For increasing accretion rates, the kinetic energy swallowed by the hole and the size of the radiating region increase.

066.045 **A theory of the instability of disk accretion on to black holes and the variability of binary X-ray sources, galactic nuclei and quasars.**
N. I. Shakura, R. A. Sunyaev.
Monthly Notices Roy. Astron. Soc., Vol. 175, 613 - 632 (1976).

A linear stability theory is constructed for stationary disk accretion on to black holes. The inner region of the disk, where radiation pressure dominates, is unstable against small perturbations. Growing short wave perturbations take the form of travelling concentric waves. For long wavelengths there are two branches of growing standing waves. The growth rate of one branch decreases rapidly with increasing wavelength. The above instabilities can explain the observed variability of radiation from binary X-ray sources, galactic nuclei and quasars, assuming these objects really do contain accreting black holes.

066.046 **Astronomical tests for the presence of an ether.**
T. W. Cole.
Monthly Notices Roy. Astron. Soc., Vol. 175, 93P - 96P (1976).

The classical Michelson-Morley experiment and all subsequent ether-drift experiments have been sensitive to only second-order effects. It is possible to perform several astronomical tests which are sensitive to the first-order effects and these are described. Existing observations are used in such tests to confirm that no ether effect is observed.

066.047 **Fotografia di un campo stellare durante un'eclisse totale di sole.** M. G. Fracastoro.
Coelum, Vol. 44, 69 - 75 (1976). − Concerning the light deviation in the gravitation field of the sun.

066.048 **Trägheitsfreie Dynamik und Hamiltonsche Bewegungsgleichung.** H.-J. Treder.
Astron. Nachr., Vol. 297, 113 - 116 (1976).

We formulate the canonical equations of motion for particles with an (post-Newtonian) interaction potential and the Hamiltonian form of our Machian dynamics without inertia.

066.049 **Theoretical frameworks for testing relativistic gravity. V. Post-Newtonian limit of Rosen's theory.**
D. L. Lee, C. M. Caves, W.-T. Ni, C. M. Will.
Astrophys. Journ., Vol. 206, 555 - 558 (1976).

The post-Newtonian limit of Rosen's theory of gravity is evaluated and is shown to be identical to that of general relativity, except for the PPN parameter α_2. Both the value of α_2 and the value of the Newtonian gravitational constant depend on the present cosmological structure of the universe.

066.050 **On the latitudinal and radial motion in the field of a rotating black hole.** J. Bičák, Z. Stuchlík.
Bull. Astron. Inst. Czechoslovakia, Vol. 27, 129 - 133 (1976).

De Felice's and Calvani's analysis of the general features of the latitudinal motion and the features of the radial motion of photons in the Kerr metric is complemented. Special attention is then devoted to photons moving along the geodesics of two null principal congruences.

066.051 **The stability of a superdense plane plate in general relativity.** J. Horský, E. V. Chubaryan.
Bull. Astron. Inst. Czechoslovakia, Vol. 27, 133 - 135 (1976).

066.052 **Electromagnetic radiation, relativity, and anomalous red-shifts.** E. W. Silvertooth.
Applied Optics, Vol. 15, 1100 - 1102 (1976).

066.053 Gravitational analogue of magnetic force.
A. R. Khan, R. F. O'Connell.
Nature, Vol. 261, 480 - 481 (1976). — Letter.

066.054 Particle acceleration near an accreting black hole.
S. Ames.
14th Intern. Cosmic Ray Conf., (see 012.011), Vol. 2, 439 - 442 (1975).

066.055 On the problem of superdense matter.
D. D. Ivanenko, N. I. Maksyukov.
Vestn. Mosk. un-ta. Fiz., astron., Vol. 16, 563 - 568 (1975). In Russian. — Abstr. in Referativ. Zhurn. 51. Astron., 6. 51. 817 (1976).

066.056 Group theory of the massless spin 2 field and gravitation. U. Niederer.
General Relativ. Gravitation, Vol. 6, 433 - 437 (1975).

066.057 Response of Doppler spacecraft tracking to gravitational radiation.
F. B. Estabrook, H. D. Wahlquist.
General Relativ. Gravitation, Vol. 6, 439 - 447 (1975).
A calculation is made of the effect of gravity waves on the observed Doppler shift of a sinusoidal electromagnetic signal transmitted to, and transponded from, a distant spacecraft.

066.058 Blueshift and spectral features of the gravitational searchlight.
S. M. Chitre, J. V. Narlikar, R. C. Kapoor.
General Relativ. Gravitation, Vol. 6, 477 - 487 (1975).
It is shown that a photon emitted in the forward direction by a charged particle moving in an equatorial circular orbit centred on a highly collapsed mass M, the radius being slightly in excess of one and a half times the Schwarzschild radius, is strongly blueshifted when it arrives at a distant receiver. A ring shaped emitting region composed of such orbiting particles has a power law spectrum of the form $d\nu/\nu$ as seen by a distant stationary observer.

066.059 The gravitational collapse of a matter-antimatter symmetric gas sphere. A. H. Nelson.
Computing in plasma physics and astrophysics, (see 012.014), A3, 2 pp. (1976).

066.060 Relativistic gravitational collapse towards the black hole state. J. C. Miller.
Computing in plasma physics and astrophysics, (see 012.014), A4, 2 pp. (1976).

066.061 Calcul du coefficient g_{00} de la métrique 2-post-newtonienne pour un système de N-corps fluide parfait et sphériques. E. Verdaguer, L. Mas.
Comptes Rendus Acad. Sci. Paris, Sér. A, Vol. 282, 1059 - 1062 (1976).

066.062 Internal gravity waves deduced from the HF Doppler data during the April 19, 1958, solar eclipse.
T. Ichinose, T. Ogawa.
Journ. Geophys. Res., Vol. 81, 2401 - 2404 (1976).
Periodic variations observed from the Doppler data of F layer reflection during the solar eclipse day and control days are analyzed by using a numerical Fourier transform. The gravity waves of the 22-min period may be generated by the solar eclipse, although it is hard to confirm the existence of the gravity waves because the period is difficult to distinguish from the periods of the medium-scale TID's in the F region.

066.063 Nonlinear gravitons and curved twistor theory.
R. Penrose.
General Relativ. Gravitation, Vol. 7, 31 - 52 (1976).

066.064 General relativity and astrophysics.
E. L. Schucking.
General Relativ. Gravitation, Vol. 7, 113 - 126 (1976).

066.065 The nonlinear graviton. R. Penrose.
General Relativ. Gravitation, Vol. 7, 171 - 176 (1976).

066.066 On the topology of spacelike hypersurfaces, singularities, and black holes. D. Gannon.
General Relativ. Gravitation, Vol. 7, 219 - 232 (1976).
Sufficient topological conditions and some reasonable geometric properties near infinity are given on a partial Cauchy surface to study the occurrence of singularities and the topological structure of smooth event horizons. Conditions are given on electrovac space-times to show the horizon is a sphere without assuming the solution is stationary.

066.067 Astrophysical applications of the gravitational searchlight. S. M. Chitre, J. V. Narlikar.
General Relativ. Gravitation, Vol. 7, 233 - 238 (1976).
The astrophysical consequences of the blue-shifted radiation emitted in the forward direction by a source moving in an equatorial orbit with radius slightly in excess of 1.5 times the Schwarzschild radius of a highly collapsed central object are examined with special reference to quasistellar objects.

066.068 An improved radiation metric. P. D. Noerdlinger.
General Relativ. Gravitation, Vol. 7, 239 - 249 (1976).

066.069 On Yang's spherically symmetric, static gravitational field. S. Chen, T.-h. Ho, H.-y. Kuo, C.-l. Chou.
Scient. Sinica, Vol. 19, 199 - 206 (1976).

066.070 Integral formalism of gauge fields and general relativity. M. Carmeli.
Phys. Rev. Letters, Vol. 36, 59 - 61 (1976).

066.071 New test of the equivalence principle from lunar laser ranging. J. G. Williams, R. H. Dicke, P. L. Bender, C. O. Alley, W. E. Carter, D. G. Currie, D. H. Eckhardt, J. E. Faller, W. M. Kaula, J. D. Mulholland, H. H. Plotkin, S. K. Poultney, P. J. Shelus, E. C. Silverberg, W. S. Sinclair, M. A. Slade, D. T. Wilkinson.
Phys. Rev. Letters, Vol. 36, 551 - 554 (1976).
An analysis of six years of lunar-laser-ranging data gives a zero amplitude for the Nordtvedt term in the earth-moon distance yielding the Nordvedt parameter $\eta = 0.00 \pm 0.03$. Thus, earth's gravitational self-energy contributes equally, $\pm 3\%$, to its inertial mass and passive gravitational mass. At the 70% confidence level this result is only consistent with the Brans-Dicke theory for $\omega > 29$. The authors obtain $|\beta - 1| \lesssim 0.02$ to 0.05 for five-parameter parametrized post-Newtonian theories of gravitation with energy-momentum conservation, or $-|\beta - 1| \lesssim 0.01$ if only β and γ are considered.

066.072 Verification of the principle of equivalence for massive bodies.
I. I. Shapiro, C. C. Counselman III, R. W. King.
Phys. Rev. Letters, Vol. 36, 555 - 558, with a correction, p. 1068 (1976).
Analysis of 1389 measurements, accumulated between 1970 and 1974, of echo delays of laser signals transmitted from earth and reflected from cube corners on the moon shows gravitational binding energy to contribute equally to earth's inertial and passive gravitational masses to within the estimated uncertainty of 1.5%. The corresponding restriction on the Eddington-Robertson parameters is $4\beta - \gamma - 3 =$

−0.001 ± 0.015. Combination with other results, as if independent, yields $\beta = 1.003 \pm 0.005$ and $\gamma = 1.008 \pm 0.008$, in accord with general relativity.

066.073 Possible experiment with two counter-orbiting drag-free satellites to obtain a new test of Einstein's general theory of relativity and improved measurements in geodesy. R. A. Van Patten, C. W. F. Everitt.
Phys. Rev. Letters, Vol. 36, 629 - 632 (1976).

In 1918, Lense and Thirring calculated that a moon orbiting a rotating planet would experience a nodal dragging effect due to general relativity. The authors describe an experiment to measure this effect to 1% with two counter-orbiting drag-free satellites in polar earth orbit. In addition to tracking data from existing ground stations, satellite-to-satellite Doppler ranging data are taken near the poles. New geophysical information is inherent in the polar data.

066.074 Limit on the secular change of the gravitational constant based on studies of solar evolution.
C. W. Chin, R. Stothers.
Phys. Rev. Letters, Vol. 36, 833 - 835 (1976).

The past and present observable properties of the sun have been theoretically calculated on the assumption that the gravitational constant G increases or decreases with time. Consideration of Davis's experimental upper limit on the present solar neutrino flux and of terrestrial paleontological data sets a limit on the absolute rate of change of G, namely, $|\dot{G}_0/G_0| < 1 \times 10^{-10}$/yr, if other standard constants do not also change with time.

066.075 Black holes in thermal equilibrium.
G. W. Gibbons, M. J. Perry.
Phys. Rev. Letters, Vol. 36, 985 - 987 (1976).

It is argued that a black hole can remain in thermal equilibrium with a heat bath even in the presence of particle interactions. This is achieved by proving the identity of the Hartle-Hawking Feynman propagator and a certain thermal Green's function.

066.076 Particle creation by gravitational fields.
N. M. J. Woodhouse.
Phys. Rev. Letters, Vol. 36, 999 - 1001 (1976).

066.077 Measurements of the solar gravitational deflection of radio waves in agreement with general relativity.
E. B. Fomalont, R. A. Sramek.
Phys. Rev. Letters, Vol. 36, 1475 - 1478 (1976).

In two experiments in 1974 and 1975, utilizing a radio interferometer of 35-km baseline, the relative positions of three radio sources were monitored over a period of a month when the sun was within 10 deg of the sources. The mean gravitational deflection is 1.007 ± 0.009 (standard error) times the value predicted by general relativity. These results exclude the Brans-Dicke theory of gravitation with a scalar coupling constant $\omega < 23$ at the 99% confidence level.

066.078 Gravitational deflection of light: solar eclipse of 30 June 1973 I. Description of procedures and final results. R. A. Brune, Jr., C. L. Cobb, B. S. DeWitt, C. DeWitt-Morette, D. S. Evans, J. E. Floyd, B. F. Jones, R. V. Lazenby, M. Marin, R. A. Matzner, A. H. Mikesell, M. R. Mikesell, R. I. Mitchell, M. P. Ryan, H. J. Smith, A. Sy, C. D. Thompson.
Astron. Journ., Vol. 81, 452 - 454 (1976).

A condensed account is given of the equipment and procedures of the Texas expeditions to Chinguetti Oasis, Mauritania, to observe the Einstein shift at the eclipse of 30 June 1973. Some of the instrumental problems brought to light in the accompanying paper describing the reductions are identified. The final value is $L = (0.95 \pm 0.11) L_E$, where the error is 1σ

and $L_E = 1''.75$ is Einstein's value.

066.079 Gravitational deflection of light: solar eclipse of 30 June 1973 II. Plate reductions. B. F. Jones.
Astron. Journ., Vol. 81, 455 - 463 (1976).

The eclipse plates obtained on 30 June 1973 and the comparison plates obtained in November 1973 at Chinguetti, Mauritania, were measured. A description of the reduction procedure is given and values are determined for probable errors. The final value obtained for the light deflection, extrapolated to the solar limb, is $L = (0.95 \pm 0.11 \text{ std. dev.}) L_E$, where $L_E = 1.75$ arcsec is the value predicted by general relativity theory.

066.080 Nonstatic spherically symmetric solutions for a perfect fluid in general relativity.
N. Chakravarty, S. B. D. Choudhury, A. Banerjee.
Australian Journ. Phys., Vol. 29, 113 - 117 (1976).

A general method is described by which exact solutions of Einstein's field equations are obtained for a nonstatic spherically symmetric distribution of a perfect fluid. In addition to the previously known solutions which are systematically derived, a new set of exact solutions is found, and the dynamical behaviour of the corresponding models is briefly discussed.

066.081 Eine neue Theorie über die Gravitation aus kybernetischer Sicht. J. C. Metz.
Selbstverlag J. C. Metz, Aachen, 31 pp. Price DM 10.00 (1975).

066.082 Active and passive generalizations of the Lorentz-Poincaré transformations and Einstein's principles of light velocity and of special relativity. Part II.
H.-J. Treder.
Experim. Techn. Physik, Vol. 23, 211 - 221 (1975).
In German.

066.083 Die Einsteinsche Verschiebung der Spektrallinien und die Äquivalenz von Trägheit und Schwere beim Licht. H.-J. Treder.
Wiss. Fortschr., Vol. 26, 210 - 213 (1976).

066.084 Gravitational imaging by elliptical galaxies.
M. Jaroszynski, B. Paczynski.
Mem. Soc. Astron. Italiana, Vol. 45, (see 012.017), 673 - 680 (1974).

066.085 The spectrum of the background radiation between 3 mm and 800 μm wavelength. J. E. Beckman, P. E. Clegg, J. S. Huizinga, E. I. Robson, D. G. Vickers.
Mem. Soc. Astron. Italiana, Vol. 45, (see 012.017), 681 - 686 (1974).

066.086 Classical gravity: a status report. R. Penrose.
Mem. Soc. Astron. Italiana, Vol. 45, (see 012.017), 969 - 970 (1974). − Abstract.

066.087 Quantum gravity: a status report. C. J. Isham.
Mem. Soc. Astron. Italiana, Vol. 45, (see 012.017), 971 - 972 (1974). − Abstract.

066.088 Black hole explosions. G. W. Gibbons.
Mem. Soc. Astron. Italiana, Vol. 45, (see 012.017), 973 (1974). − Abstract.

066.089 Sorte hull − en vei ut av universet?
Y. Hartvigsen.
Naturen, Årg. 99, 221 - 226 = Inst. Teor. Astrofys., Blindern−Oslo, Småtr. No. 85 (1975).

066.090 **Zur Gravitationsstrahlung.** H.-J. Treder.
Sitzungsber. Akad. Wiss. Berlin, Jahrg. 1975, No. 1N, p. 31 - 33 (1975).

066.091 **Black holes.** R. C. Kapoor.
Bull. Astron. Soc. India, Vol. 4, 4 - 12 (1976).

066.092 **Le temps physique: qu'est-ce que c'est? II — Le voyage dans le temps.** A. Peton.
L'Astronomie, Vol. 90, 298 - 318 (1976).

066.093 **Gravitation in the light cone gauge.**
J. Scherk, J. H. Schwarz.
General Relativ. Gravitation, Vol. 6, 537 - 550 (1975).

066.094 **Special relativistic mechanics and electrodynamics with applications to synchrotron radiation.**
J. H. Romig.
Thesis Colorado Univ., Boulder, USA, 225 pp. (1975). (Available from Univ. Microfilms, Order No. 75-23, 639).

066.095 **Analysis and development of a very sensitive low temperature gravitational radiation detector.**
H. J. Paik.
Thesis Stanford Univ., Calif., 333 pp. (1975). (Available from Univ. Microfilms, Ann Arbor, Mich., USA. Order No. 75-13573). — Abstr. in Phys. Abstr. Vol. 79, A000518 (1976).

066.096 **Black-hole physics.** R. U. Sexl.
Acta Phys. Austriaca, Vol. 42, 303 - 347 (1975).
Abstr. in Phys. Abstr., Vol. 79, A003323 (1976).

066.097 **Supercritical fields and bald black holes.**
J. M. Irvine.
Journ. Phys. A, Vol. 8, L117 - L120 (1975). — Abstr. in Phys. Abstr., Vol. 79, A003338 (1976).

066.098 **Signal to noise ratio in a satellite gravitational wave experiment.** V. B. Braginsky (*Braginskij*).
Acta Astronaut., Vol. 2, 535 - 537 (1975). — Abstr. in Phys. Abstr., Vol. 79, A004136 (1976).

066.099 **On experimental verification of gravitational and inertial mass equivalency.**
N. I. Kolosnitsin (*Kolosnitsyn*), V. M. Myheev, A. V. Osipova, K. P. Stanyukovich.
Acta Astronaut., Vol. 2, 539 - 542 (1975). — Abstr. in Phys. Abstr., Vol. 79, A004137 (1976).

066.100 **A priori error estimation for the Sorel mission.**
E. A. Roth.
Acta Astronaut., Vol. 2, 543 - 555 (1975). — Abstr. in Phys. Abstr., Vol. 79, A004138 (1976).

066.101 **On the nonexistence of extreme black holes.**
B. H. Voorhees.
Nuovo Cimento B, Ser. 11, Vol. 29B, 63 - 70 (1975). — Abstr. in Phys. Abstr., Vol. 79, A006997 (1976).

066.102 **Ghost neutrinos in the Einstein-Cartan theory of gravitation.** P. S. Letelier.
Phys. Letters A, Vol. 54A, 351 - 352 (1975). — Abstr. in Phys. Abstr., Vol. 79, A007905 (1976).

066.103 **Vacuum polarization and the spontaneous loss of charge by black holes.** G. W. Gibbons.
Commun. Math. Phys., Vol. 44, 245 - 264 (1975). — Abstr. in Phys. Abstr., Vol. 79, A011076 (1976).

066.104 **Gravitational radiation generated by the gravitational scattering of two stars.**
C. Mache, E. Frehland.
Internat. Journ. Theor. Phys., Vol. 13, 410 - 416 (1975).
Abstr. in Phys. Abstr., Vol. 79, A012125 (1976).

066.105 **On particle creation by black holes** R. M. Wald.
Commun. Math. Phys., Vol. 45, 9 - 34 (1975).
Abstr. in Phys. Abstr., Vol. 79, A015350 (1976).

066.106 **Optical appearance of distant objects to observers near and inside a Schwarzschild black hole.**
C. T. Cunningham.
Phys. Rev. D, Vol. 12, 323 - 328 (1975). — Abstr. in Phys. Abstr., Vol. 79, A015355 (1976).

066.107 **Pair creation and the Gowdy model.**
C. P. Winlove, D. J. Raine.
Ann. Physics, Vol. 93, 116 - 124 (1975). — Abstr. in Phys. Abstr., Vol. 79, A016272 (1976).

066.108 **Primordial Higgs mesons and cosmic background radiation.** K. Sato, H. Sato.
Progr. Theor. Phys., Vol. 54, 912 - 913 (1975). — Abstr. in Phys. Abstr., Vol. 79, A019331 (1976).

066.109 **On a transformation of Teukolsky's equation and the electromagnetic perturbations of the Kerr black hole.** S. Chandrasekhar.
Proc. Roy. Soc. London, Ser. A, Vol. 348, 39 - 55 (1976).
Abstr. in Phys. Abstr., Vol. 79, A019523 (1976).

066.110 **Structure and stability of an accretion disk around a black hole.** N. Shibazaki, R. Hoshi.
Progr. Theor. Phys., Vol. 54, 706 - 718 (1975). — Abstr. in Phys. Abstr., Vol. 79, A019524 (1976).

066.111 **Photoproduction of gravitational radiation in static electromagnetic fields. Radiative corrections.**
G. Papini, S.-R. Valluri.
Canadian Journ. Phys., Vol. 53, 2315 - 2320 (1975). — Abstr. in Phys. Abstr., Vol. 79, A020306 (1976).

066.112 **Probability distribution of particles created by a black hole.** L. Parker.
Phys. Rev. D, Vol. 12, 1519 - 1525 (1975). — Abstr. in Phys. Abstr., Vol. 79, A023813 (1976).

066.113 **Gauge-invariant perturbations of Reissner-Nordstrom black holes.** V. Moncrief.
Phys. Rev. D, Vol. 12, 1526 - 1537 (1975). — Abstr. in Phys. Abstr., Vol. 79, A023814 (1976).

066.114 **Electromagnetic field of a current loop around a Kerr black hole.** D. M. Chitre, C. V. Vishveshwara.
Phys. Rev. D, Vol. 12, 1538 - 1543 (1975). — Abstr. in Phys. Abstr., Vol. 79, A023815 (1976).

066.115 **Pair creation in expanding universes.**
D. J. Raine, C. P. Winlove.
Phys. Rev. D, Vol. 12, 946 - 951 (1975). — Abstr. in Phys. Abstr., Vol. 79, A024628 (1976).

066.116 **Electromagnetic radiation damping of charges in external gravitational fields (weak field, slow motion approximation). (Planet orbiting black hole).** E. Rudolph.
Ann. Inst. Henri Poincaré A, Vol. 23, 113 - 136 (1975).
Abstr. in Phys. Abstr., Vol. 79, A024638 (1976).

066.117 **Scattering and absorption of electromagnetic waves by a Schwarzschild black hole.** R. Fabbri.
Phys. Rev. D, Vol. 12, 933 - 942 (1975). — Abstr. in Phys. Abstr., Vol. 79, A024640 (1976).

066.118 **Kinks and extensions.** G. McCollum.
Thesis Yeshiva Univ., New York, USA, 48 pp. (1975)
(Available from: Univ. Microfilms, Ann Arbor, Mich., USA,
Order No. 75-20585). – Abstr. in Phys. Abstr., Vol. 79,
A027708 (1976).

066.119 **Absorption and defocusing of electromagnetic radiation by a Schwarzschild black hole.**
W. Haxton, R. Ruffini.
Ann. Physics, Vol. 95, 1 - 9 (1975). – Abstr. in Phys. Abstr.
Vol. 79, A027905 (1976).

066.120 **Application of continuous variation formulas and discrete invariance principles to black holes and neutron star models.** B. Carter.
Ann. Physics, Vol. 95, 53 - 73 (1975). – Abstr. in Phys. Abstr.
Vol. 79, A027906 (1976).

066.121 **Geodesic motion in the Tomimatsu-Sato space-times.**
M. Calvani, R. Catenacci.
Nuovo Cimento B, Ser. 11, Vol. 31B, 41 - 52 (1976). – Abstr.
in Phys. Abstr., Vol. 79, A028747 (1976).

066.122 **Solutions for gravity coupled to massless gauge fields (black holes).** P. B. Yasskin.
Phys. Rev. D, Vol. 12, 2212 - 2217 (1975). – Abstr. in Phys.
Abstr., Vol. 79, A028751 (1976).

066.123 **Gravitational theory of Edward Milne – achievements and failures.** N. Ionescu-Pallas.
Stud. Cerc. Fiz., Vol. 27, 999 - 1019 (1975). In Rumanian.
Abstr. in Phys. Abstr., Vol. 79, A028761 (1976).

066.124 **Impossibility of collapse under the law of gravitation $R_{44} = 0$.** J. P. Kobus.
Found. Phys., Vol. 5, 649 - 654 (1975). – Abstr. in Phys.
Abstr., Vol. 79, A032669 (1976).

066.125 **Light scattering properties of naked singularities.**
V. D. Sandberg.
Phys. Rev. D, Vol. 12, 2226 - 2229 (1975). – Abstr. in Phys.
Abstr., Vol. 79, A032673 (1976).

066.126 **Interaction of two black holes in the slow-motion limit.** P. D. D'Eath.
Phys. Rev. D, Vol. 12, 2183 - 2199 (1975). – Abstr. in Phys.
Abstr., Vol. 79, A032958 (1976).

066.127 **Stationary axisymmetric electromagnetic fields around a rotating black hole.** J. A. Petterson.
Phys. Rev. D, Vol. 12, 2218 - 2225 (1975). – Abstr. in Phys.
Abstr., Vol. 79, A032959 (1976).

066.128 **Einstein's theory of relativity and Mach's principle.**
H. Okamura, T. Ohta, T. Kimura, K. Hiida.
Progr. Theor. Phys., Vol. 54, 1872 - 1878 (1975). – Abstr. in
Phys. Abstr., Vol. 79, A033857 (1976).

066.129 **Radiation from a moving mirror in two dimensional space-time: conformal anomaly.**
S. A. Fulling, P. C. W. Davies.
Proc. Roy. Soc. London, Ser. A, Vol. 348, 393 - 414 (1976).
Abstr. in Phys. Abstr., Vol. 79, A034162 (1976).

066.130 **Quasispherical gravitational collapse.**
P. Szekeres.
Phys. Rev. D, Vol. 12, 2941 - 2948 (1975). – Abstr. in Phys.
Abstr., Vol. 79, A037413 (1976).

066.131 **Relativistic superdense matter in cold systems: theory.** R. L. Bowers, A. M. Gleeson, R. D.
Pedigo.
Phys. Rev. D, Vol. 12, 3043 - 3055 (1975). – Abstr. in Phys.
Abstr., Vol. 79, A037415 (1976).

066.132 **Relativistic superdense matter in cold systems: applications.** R. L. Bowers, A. M. Gleeson, R.
D. Pedigo.
Phys. Rev. D, Vol. 12, 3056 - 3068 (1975). – Abstr. in Phys.
Abstr., Vol. 79, A037416 (1976).

066.133 **Gravitational collapse of marginally bound spheroids: initial conditions.** D. M. Eardley.
Phys. Rev. D, Vol. 12, 3072 - 3076 (1975). – Abstr. in Phys.
Abstr., Vol. 79, A037417 (1976).

066.134 **On the Hamiltonian treatment of a quantum scalar field in a Bianchi I universe. II. The explicit construction of the Hamiltonian.** E. Pessa.
Nuovo Cimento Lettere, Ser. 2, Vol. 15, 295 - 299 (1976).
Abstr. in Phys. Abstr., Vol. 79, A037424 (1976).

066.135 **Schwarzschild black hole in an asymptotically uniform magnetic field.** R. S. Hanni, R. Ruffini.
Nuovo Cimento Lettere, Ser. 2, Vol. 15, 189 - 194 (1976).
Abstr. in Phys. Abstr., Vol. 79, A037587 (1976).

066.136 **Hadron physics and primordial black holes.**
G. F. Chapline.
Phys. Rev. D, Vol. 12, 2949 - 2954 (1975). – Abstr. in Phys.
Abstr., Vol. 79, A037592 (1976).

066.137 **Relativistic magnetohydrodynamical effects of plasma accreting into a black hole.**
R. Ruffini, J. R. Wilson,
Phys. Rev. D, Vol. 12, 2959 - 2962 (1975). – Abstr. in Phys.
Abstr., Vol. 79, A037593 (1976).

066.138 **Quantisation of a scalar field in the Kerr spacetime.**
L. H. Ford.
Phys. Rev. D, Vol. 12, 2963 - 2977 (1975). – Abstr. in Phys.
Abstr., Vol. 79, A037594 (1976).

066.139 **Black holes and magnetic fields.**
A. R. King, J. P. Lasota, W. Kundt.
Phys. Rev. D, Vol. 12, 3037 - 3042 (1975). – Abstr. in
Phys. Abstr., Vol. 79, A037596 (1976).

066.140 **A solution of coupled Einstein-SO(3) gauge field equations.** M. Y. Wang.
Phys. Rev. D, Vol. 12, 3069 - 3071 (1975). – Abstr. in
Phys. Abstr., Vol. 79, A037597 (1976).

066.141 **Statistical black-hole thermodynamics.**
J. D. Bekenstein.
Phys. Rev. D, Vol. 12, 3077 - 3085 (1975). – Abstr. in
Phys. Abstr., Vol. 79, A037598 (1976).

066.142 **Gauge fields arising from spacetime symmetries and gravitational theories. II.**
A. G. Agnese, P. Calvini.
Phys. Rev. D, Vol. 12, 3804 - 3809 (1975). – Abstr. in
Phys. Abstr., Vol. 79, A038465 (1976).

066.143 **The stability of composite models (of fluid spheres in general relativity).**
S. E. Brown, J. C. Hargreaves.
Journ. Phys. A, Vol. 9, 343 - 356 (1976). – Abstr. in
Phys. Abstr., Vol. 79, A038483 (1976).

066.144 **Photoproduction of gravitational radiation by some astrophysical objects.**

G. Papini, S.-R. Valluri.
Canadian Journ. Phys., Vol. 54, 76 - 79 (1976). — Abstr.
in Phys. Abstr., Vol. 79, A038493 (1976).

066.145 A theory of gravity.   P. Rastall.
Canadian Journ. Phys., Vol. 54, 66 - 75 (1976).
Abstr. in Phys. Abstr., Vol. 79, A038497 (1976).

066.146 On quantum resonances in stationary geometries.
T. Damour, N. Deruelle, R. Ruffini.
Nuovo Cimento Lettere, Ser. 2, Vol. 15, 257 - 262 (1976).
Abstr. in Phys. Abstr., Vol. 79, A038727 (1976).

066.147 Collective features of gravitational light focusing.
G. Mayer.
Optics Commun., Vol. 16, 317 - 319 (1976). — Abstr. in Phys.
Abstr., Vol. 79, A042009 (1976).

066.148 The solution of Maxwell's equations in Kerr geome-
try.   S. Chandrasekhar.
Proc. Roy. Soc. London, Ser. A, Vol. 349, 1 - 8 (1976).
Abstr. in Phys. Abstr., Vol. 79, A042010 (1976).

066.149 Thermodynamic equilibrium and heavy particles
near a black hole.   Ya. B. Zel'dovich.
Phys. Letters A, Vol. 56A, 3 - 4 (1976). — Abstr. in Phys.
Abstr., Vol. 79, A042250 (1976).

066.150 General geodesic motion in the extended Kerr
manifold.   R. H. St John.
Thesis New Mexico Univ., Albuquerque, USA, 90 pp. (1974).
(Available from:Univ. Microfilms, Order No. 75–6977).

066.151 Relativity experiment on Helios: a status report.
J. D. Anderson, W. G. Melbourne, D. L. Cain, E. K.
Lau, S. K. Wong, W. Kundt.
66th annual conf. Internat. district heating assoc., Skytop,
Pennsylvania, USA, 23 June 1975, 37 pp. (1975).

066.152 Magneto-acoustic-gravity waves.
N. S. Hartunian.
Thesis Brandeis Univ., Waltham, Massachusetts, USA, 177 pp.
(1975). (Available from: Univ. Microfilms, Order No. 75-24,
806).

066.153 Neutrino flow and gravitational collapse.
J. R. Wilson.
Meeting phys. astrophys. neutron stars black holes, Varenna,
Italy, 14 July 1975, 36 pp. (1975).

066.154 Cosmic primeval turbulence.   K. L. Chan.
Thesis Princeton Univ., New Jersey, USA, 93 pp.
(1975). (Available from: Univ. Microfilms, Order No. 75-
23,187).

066.155 Gravitational optics: a study of null geodesics and
gravity waves in the geometry produced by a star.
J. F. Cyranski.   •
Thesis Wisconsin Univ. Milwaukee, USA, 167 pp. (1974).
(Available from: Univ. Microfilms, Order No. 75-17,804).

066.156 Black holes — a way out of the universe.
Y. Hartvigsen.
Naturen, Vol. 99, No. 5, p. 221 - 226 (1975). In Norwegian.

066.157 Spectral distribution and origin of the X- and gamma
ray diffuse background.
R. Rocchia, R. Ducros.
Congr. French Physical Soc., Dijon, France, 30 June 1975,
18 pp. (1975). In French.

066.158 Annihilation of matter and antimatter and the
cosmic X-ray background.
P. Carlqvist, B. Laurent.
Separate print Tek. Hoegskolan, Stockholm, Sweden, 9 pp.
(1975). — See 17.066.013.

Quasars, pulsars and black holes.
See Abstr. 003.048.

Wie veränderlich ist die Gravitationskonstante?
See Abstr. 043.001.

Le temps physique: qu'est-ce que c'est?
See Abstr. 044.001.

Collision avoidance for two counter-orbiting polar
satellites.   See Abstr. 052.061.

Modern state of the theory of superdense celestial
bodies.   See Abstr. 061.024.

Magnetohydrodynamics near a black hole.
See Abstr. 062.091.

Evolution of massive close binaries and formation
of neutron stars and black holes.   See Abstr. 065.058.

Zwei Endstadien der Sternentwicklung: II. Schwarze
Löcher.   See Abstr. 065.074.

Some quantitative calculations of final stages of
stellar evolution.   See Abstr. 065.089.

Relativistic formulation of the neutron starquake
theory of pulsar glitches.   See Abstr. 065.099.

Structure and stability of relativistic, differentially
rotating stars.   See Abstr. 065.113.

Numerical study of rotating relativistic stars.
See Abstr. 065.116.

The solar self-gravity-induced electrostatic redshift.
See Abstr. 080.063.

The Tungus event as a small black hole: geophysical
considerations.   See Abstr. 105.028.

Evolution of massive close binaries and the origin
of relativistic objects.   See Abstr. 117.029.

Possible observational evidence for a black hole in
Beta Lyrae system.   See Abstr. 121.114.

Gravitational radiation from supernova explosions.
See Abstr. 125.050.

Analytic supernova models.
See Abstr. 125.051.

On the maximum gravitational redshift of white
dwarfs.   See Abstr. 126.003.

Gravitational collapse of a cold white dwarf.
See Abstr. 126.011.

3C 268.4 — evidence for the presence of a gravita-
tionally-lensed secondary image.   See Abstr. 141.003.

Ejection speed in the slingshot theory of radio
sources. II. General relativistic approximation.

See Abstr. 141.019.

Incompatibility of the continuous steady-state models of pulsar magnetospheres with relativistic magneto-hydrodynamics.   See Abstr. 141.341.

Pulsar slow-down and the temporal change of G. See Abstr. 141.354.

Anti-correlated hard and soft X-ray intensity variations of the black-hole candidates Cyg X-1 and A0620−00. See Abstr. 142.032.

Origin of the black hole in Cyg X-1. See Abstr. 142.035.

A two-temperature accretion disk model for Cygnus X-1: structure and spectrum.   See Abstr. 142.039.

A measurement of fluctuations in the X-ray background by Uhuru.   See Abstr. 142.040.

Black holes in X-ray binaries: marginal existence and rotation reversals of accretion disks.   See Abstr. 142.044.

Scattering model for X-ray bursts: massive black holes in globular clusters.   See Abstr. 142.082.

Gamma-ray bursts from black hole accretion disks. See Abstr. 142.156.

Röntgenquellen, Neutronensterne und schwarze

Löcher.   See Abstr. 142.217.

Cygnus X-1 temporal microstructure: evidence for a black hole.   See Abstr. 142.256.

On a spherical star system with a collapsed core. See Abstr. 151.036.

On the possibility of the existence of black holes in the centres of galaxies.   See Abstr. 158.126.

Clusters of galaxies as gravitational lenses? See Abstr. 160.022.

Multiple image probabilities for a spheroidal gravitational lens.   See Abstr. 162.036.

Thermal radiation produced by the expansion of the universe.   See Abstr. 162.040.

Vorticity perturbations and isotropy of the cosmic microwave background.   See Abstr. 162.052.

Big bang cosmology and the cosmic black-body radiation.   See Abstr. 162.076.

On the initial singularity in the scalar-tensor anisotropic cosmology.   See Abstr. 162.084.

Can the effect of distant matter on physical observables be observed.   See Abstr. 162.085.

Sun

071 Solar Photosphere, Spectrum

071.001 The solar spectrum: wavelengths and identifications from 160 to 770 Angstroms.
W. E. Behring, L. Cohen, U. Feldman, G. A. Doschek.
Astrophys. Journ., Vol. 203, 521 - 527 (1976).

The full-sun solar spectrum from 160 to 770 Å was photographed under quiet solar conditions by a rocket-borne spectrograph flown in 1973 September. The spectral resolution is 0.06 Å or better. The authors present a composite list of spectral lines, including wavelengths, identifications, and approximate intensities.

071.002 Physical conditions in granulation.
R. C. Altrock, S. Musman.
Astrophys. Journ., Vol. 203, 533 - 540 (1976).

The authors analyze high-resolution spectroheliograms made in the Ti I and Fe I 5016 Å lines with the diode array of the tower telescope at Sacramento Peak Observatory. A $10 \times 10''$ region was scanned with $1''$ resolution. The authors find that $|\Delta T|$ increases going downward in the low photosphere. In the middle photosphere $|\Delta T| \leqslant 70$ K. The convective flux at $\tau_{5000} = 3$ is small compared with the total flux. Also, the temperature fluctuations are much larger than those that would be produced by the observed convective velocities alone. Thus the observable low photosphere appears to be near radiative equilibrium.

071.003 Solar limb brightening in submillimeter wavelengths.
C. Lindsey, H. S. Hudson.
Astrophys. Journ., Vol. 203, 753 - 759 (1976).

Differential two-beam scans of the sun in submillimeter wavelengths ($350 \mu - 1$ millimeter) indicate limb brightening approaching 1 percent at $\mu = 0.6$. The observed degree of limb brightening is considerably less than that consistent with spherically symmetric model atmospheres based on continuum brightness–temperature measurements. The suppression of limb brightening suggests the existence of irregular granular structure with both horizontal and vertical characteristic sizes of order 1500 km. High-resolution images in the wings of the K-line show granular structure of about this horizontal scale.

071.004 Spectroscopic evidence for a higher rotation rate of magnetized plasma at the solar photosphere.
P. Foukal.
Astrophys. Journ., (*Letters*), Vol. 203, L145 - L148 (1976).

Simultaneous observations of the photospheric rotation velocity and magnetic field are made with $5''$ spatial resolution. The results show a significant increase in plasma rotation velocity with absolute magnetic intensity. The difference in rotation velocity between the highest and lowest magnetic fields, in areas of quiet photosphere, is 41 ± 6 m s^{-1} in 1975 and 46 ± 11 m s^{-1} in 1974. These results indicate that the magnetic network rotates at least 2 percent faster than the general photosphere, and the flux tubes accelerate the local plasma of the network boundaries.

071.005 Table of solar diatomic molecular lines. II. Spectral range: 6600–7100 Å.
R. Boyer, P. Sotirovski, J. W. Harvey.
Astron. Astrophys., Suppl. Ser., Vol. 24, 111 - 138 (1976).

The present publication of results between 6600 and 7100 Å represents the second part of a work of analysis of molecular lines observed in the spectrum of a sunspot (wavelength range: 6100–8300 Å). It is an extension of results already published concerning the area: 6100–6600 Å.

071.006 Structure of the solar chromosphere. II. The underlying photosphere and temperature-minimum region.
J. E. Vernazza, E. H. Avrett, R. Loeser.
Astrophys. Journ., Suppl. Series, Vol. 30, 1 - 60 (1976).

The authors present a non-LTE empirical model of the quiet solar photosphere and the temperature-minimum region. The continuous spectrum computed from this model is in good overall agreement with available disk-center observations throughout the wavelength range $0.125-500 \mu$.

071.007 Noen resultater av soleksperimentene med ATM, Skylab, O. Engvold.
Astron. Tidsskr., Årg. 9, p. 11 - 20 (1976).

071.008 A propos du spectre solaire. J.-C. Pecker.
L'Astronomie, Vol. 90, 33 - 36 (1976).

071.009 Determining differences of excitation temperatures of a granule - intergranular region.
B. M. Korduba, M. E. Musievskaya, P. A. Olijnyk, V. M. Sobolev.
Solnechnye Dannye 1975 Byull., No. 10, p. 79 - 84 (1976). In Russian.

Results of theoretical determinations of equivalent widths for 8 Fraunhofer lines (Fe I, Cr I and Ca I) are presented. The calculations have been made for a given set of differences of excitation temperatures. Theoretical equivalent widths have been compared with observed ones. It is shown that the difference between the observed euquivalent widths in granules and intergranular space can be explained by excitation temperature differences of the order of $200°-300°$ for Fe I, $110°-120°$ for Cr I and $145°$ for Ca I.

071.010 The asymmetry of oxygen infrared triplet lines and determination of the normalized O^{17} and O^{18} isotope abundance in the solar photosphere. R. I. Kostyk.
Astron. Zhurn. Akad. Nauk SSSR, Vol. 53, 125 - 129 (1976). In Russian. English translation in Soviet Astron., Vol. 20, No. 1.

The normalized solar oxygen isotope abundance determined from the asymmetry of the $\lambda 7774.177$ O I line is equal to $O^{17}/O^{16} = 0.00028$, $O^{18}/O^{16} = 0.0063$.

071.011 The center-to-limb variation of optically thin UV-lines and at centimetric wavelengths and the influence of spicules. L. R. Elzner.
Astron. Astrophys., Vol. 47, 9 - 18 (1976).

It is known that the emission lines below 912 Å are partly absorbed in cool spicules. The author develops a method to calculate the expectation value of the intensity for a given line of sight. The measured limb brightening of many UV-lines can be reproduced by choosing a spicular diameter of 1100...1300 km and a mean number density of unexcited hydrogen of $\geqq 5 \times 10^9$ cm^{-3}. The closest approach to the measured center-to-limb behaviour is achieved with a spicular temperature of 6000...7500 K.

071.012 Photospheric magnetic flux concentrations and the granular velocity field. F.-L. Deubner.
Astron. Astrophys., Vol. 47, 475 - 477 = Mitt. Fraunhofer Inst., *Freiburg*, No. 141 (1976).

The discovery that magnetic flux in the solar photosphere outside of sunspots occurs only in the form of highly concentrated flux elements of subgranular size with high maximum fields of the order of 1.5 kGauss has prompted a continuing discussion of how these fragile structures are being produced and protected against rapid diffusion and dissipation. In this context an adequate quantitative picture of the photospheric (granular) velocity field is of crucial importance. The author reviews some recent observations which indicate that the granular velocities are locally much higher, as up to ~6 km s^{-1}, than the values commonly assumed.

071.013 Welcher Stern gleicht der Sonne? J. Hardorp.
Mitt. Astron. Ges., No. 38, p. 143 - 148 (1976).
Short report.

071.014 5-Minuten-Oszillationen solarer Äquivalentbreiten. H. Holweger, L. Testerman.
Mitt. Astron. Ges., No. 38, p. 204 (1976). − Abstract.

071.015 Infrared observations of supergranule temperature structure. S. P. Worden.
Solar Physics, Vol. 45, 521 - 532 (1975).

One and two-dimensional observations were made at 1.64 μ, the deepest observable level in the solar atmosphere; at 1.72 μ, representing the chromosphere; and at 1.17 μ, representing the visible photosphere. Structures distributed on a supergranular size scale (30000 km) are apparently present at all levels. These structures in the deep photospheric level (1.64 μ) seem to be a 50K–500K temperature decrease over surrounding photosphere confined to the magnetic field elements with horizontal scales less than 4000 km at supergranular boundaries rather than a general temperature structure over the entire supergranule cell appropriate to convective energy transport.

071.016 Synthesis of several solar hydrogen lines. A. Zelenka.
Astron. Astrophys., Vol. 48, 75 - 82 (1976).

A five level plus continuum model hydrogen atom and the HSRA (Harvard Smithsonian Reference Atmosphere) are used to synthesize the solar Hα, Hβ, Hγ, Pα, Pβ and Bα lines under assumption of kinetic equilibrium (non-LTE). For the Hγ line core and Pα in non-LTE, as well as for Pβ in LTE, the calculated profiles lie within the error limits of the observations. The cores of Hα and Hβ show the largest discrepancies: they are too narrow and too deep. The Hβ, Hγ and Bα wings are slightly too bright. These discrepancies are qualitatively discussed.

071.017 Photoelectric spectrophotometry of the absorption line profiles in the centre and on the limb of the solar disk. G. O. Melikov.
Solnechnye Dannye 1975 Byull., No. 12, p. 82 - 88 (1976). In Russian.

Six absorption line profiles have been observed in the centre and on the limb of the solar disk by the photoelectric method. The results are presented. The parameters of the profiles under consideration (r_0, h, W_λ) have been measured. In tables their true and observed values are given. The line profiles are shown to have significant centre-to-limb variations.

071.018 Automatic construction of maps of the conventional relief from photographic observations of the sun. M. V. Kushnir, A. V. Andrejko.
Solnechnye Dannye 1975 Byull., No. 12, p. 88 - 93 (1976). In Russian.

The reduction of the sun's photographic observations made with a rapid-acting microphotometer with digital registration gives much data. With the purpose of presenting the data in a convenient form for further analysis a method of automatized construction of conventional relief maps is proposed. The brightness maps and the brightness Laplacian compiled on the basis of the granulation photograph obtained during the 3d flight of the Soviet automatic solar stratospheric observatory are presented.

071.019 The solar EUV flux between 230 and 1220 Å on November 9, 1971. J. E. Higgins.
Journ. Geophys. Res., Vol. 81, 1301 - 1305 (1976).

071.020 The general photospheric velocity field. Eh. A. Gurtovenko.
Astrometriya i Astrofizika, *Kiev*, vyp. (No.) 28, (see 003.002), p. 93 - 99 (1976). In Russian.

Some problems of the study of the field of photospheric motions and the notion of general photospheric velocity field are considered. The values of radial and tangential components of the general photospheric velocity field in photospheric depth range of $-3.0 \leqslant \lg \tau_s \leqslant +0.5$ are determined.

071.021 On the asymmetry of selected Fraunhofer lines. III. R. I. Kostyk, T. V. Orlova, I. M. Gerbil'skaya.
Astrometriya i Astrofizika, *Kiev*, vyp. (No.) 29, (see 003.003), p. 89 - 93 (1976). In Russian.

19 Fraunhofer lines were observed and, after smoothing and correction for the instrumental profile, were investigated for asymmetry. The asymmetry is shown to be "positive" for all lines, except for oxygen. No correlation between the asymmetry and lower excitation potential was revealed.

071.022 Convective overshoot in the photosphere of the sun described by the Harvard-Smithsonian Reference Atmosphere (HSRA). M. G. Gerbil'skij.
Astrometriya i Astrofizika, *Kiev*, vyp. (No.) 29, (see 003.003), p. 94 - 98 (1976). In Russian.

The balance of radiative and mechanical energy fluxes is considered. It is shown that the diffusion approximation for the HSRA model is valid for $\lg \tau_{5000} \geqslant 0.5$. The velocity field relevant to the average mechanical energy flux is calculated.

071.023 Centre-limb observations of the intensity fluctuations in the Balmer lines in the solar atmosphere. K. R. Sivaraman, P. P. Venkitachalam.
Bull. Astron. Soc, India, Vol. 3, 27 (1975). − Abstract of paper presented at the A.S.I. meeting 1975.

071.024 Fine structure of photospheric magnetic field and Ca$^+$ K$_3$ network. J. C. Bhattacharyya, A. K. Saxena, J. Singh.
Bull. Astron. Soc. India, Vol. 3, 28 (1975). − Abstract of paper presented at the A.S.I. meeting 1975.

071.025 Sonnenforschung mit Stratosphärenballons. A. Wittmann.
Umschau, 76. Jahrgang, 293 - 295 (1976).

Recent observations of the sun with balloon-borne telescopes are summarized. Properties of the solar granulation as obtained from these flights are outlined. The recent flight of the 'Spektrostratoskop', which took place on 17 May 1975, is described. During that flight, a gondola equipped with a telescope of 32 cm aperture and a spectrograph was carried to a height of 27 km, where during 6 hours of observation a large number of both white light pictures and spectrograms of the granulation were obtained.

071.026 The Mg II h and k lines. I. Absolute center and limb measurements of the solar profiles. J. L. Kohl, W. H. Parkinson.
Astrophys. Journ., Vol. 205, 599 - 611 (1976).

In this paper the authors present new observations of the intensity profiles of Mg II *h* and *k* for quiet regions at the center of the solar disk and near the limb.

071.027 The Mg II *h* and *k* lines. II. Comparison with synthesized profiles and Ca II K.
T. R. Ayres, J. L. Linsky.
Astrophys. Journ., Vol. 205, 874 - 894 (1976).

The authors compare the Mg II *h* and *k* resonance line data of Kohl and Parkinson and profiles of the Ca II K line with synthetic spectra computed using a partial redistribution formalism and several single-component solar upper photosphere and lower chromosphere models. They find that the HSRA and Vernazza et al. models predict systematically lower intensities in the *h*, *k*, and K inner wings than are observed, but that models with a somewhat larger minimum temperature ($T_{min} \approx 4450$ K) can reproduce the measured inner wing intensities and limb darkening of these resonance lines.

071.028 Doppler wavelength shifts of transition zone lines measured in Skylab solar spectra.
G. A. Doschek, U. Feldman, J. D. Bohlin.
Astrophys. Journ.,(*Letters*), Vol. 205, L177 - L180 (1976).

Wavelengths of lines of the transition-zone ions Si IV, C IV, O IV, N V, and O V are observed to be redshifted relative to the wavelengths of chromospheric lines in XUV spectra obtained from the NRL normal incidence spectrograph on Skylab. The spectra cover the wavelength range from 1200 to 1565 Å. Only some of the spectra show redshifted transition zone lines. The observed shifts are between 0.03 and 0.08 Å, implying velocities of 15 km s^{-1} or less. The shifts imply that descending plasma in the solar atmosphere produces more emission than ascending plasma at temperatures between $\sim 7 \times 10^4$ K and 2×10^5 K.

071.029 Spectrum synthesis and the solar abundance of cobalt. D. X. Kerola, L. H. Aller.
Publ. Astron. Soc. Pacific, Vol. 88, 122 - 127 (1976).

The method of spectrum synthesis is applied to a number of Co I lines for which high-resolution spectral data and center-limb comparisons are available. The authors find $\log A$ (Co) = 4.97 ± 0.15.

071.030 On contrasts of facular granules. I. V. Yudina.
Solnechnye Dannye 1976 Byull., No. 2, p. 88 - 92 (1976). In Russian.

Contrasts for 52 facular granules in relation to the neighbouring and undisturbed photospheric background were obtained as a result of the photometric reduction of the direct photograph of the solar photosphere taken by the solar stratospheric observatory during its fourth flight on 20 June 1973. It is detected that the pores located near the solar limb are fringed with facular matter only on the limb side. It follows from this fact that facular granules are likely to be semi-transparent "clouds" elevated above the photosphere.

071.031 An analytic solution for the line formation in a magnetic field. II. M. Šidlichovský.
Bull. Astron. Inst. Czechoslovakia, Vol. 27, 155 - 159 (1976).

The $\eta(\tau)$ function calculated by a computer for the HSRA model of the photosphere and for the Stellmacher and Wiehr sunspot model was substituted by a hyperbola with sufficient accuracy. Calculations were performed for the line Fe I 525.3479 nm. The Faraday effect is taken into account.

071.032 Observations of penumbral waves in the photosphere. S. Musman, A. H. Nye, J. H. Thomas.
Astrophys. Journ., (*Letters*), Vol. 206, L175 - L178 (1976).

Simultaneous observations have been made of velocities in the chromosphere (in Hα) and in the photosphere (in the nonmagnetic Fe I line λ5576) of three sunspots. The results reveal waves propagating horizontally outward across the penumbra in the photosphere with about the same period as the running penumbral waves in Hα (250–290 s). The photospheric waves are more intermittent and have higher horizontal phase velocity (by a factor of 2 or more) than the chromospheric penumbral waves.

071.033 Calculation of depths of formation of the central parts of weak Fraunhofer lines by a computer.
N. A. Drake, Yu. A. Solonskij.
Vestn. Leningr. un-ta, 1975, No. 19, 132 - 136. In Russian. Abstr. in Referativ. Zhurn. 51. Astron., 6. 51. 369 (1976).

071.034 High resolution infrared solar observations by balloon. R. Zander.
Infrared Physics, Vol. 16, (see 012.013), 125 - 127 (1976).

071.035 On the difference of the curves of growth for the even-odd and odd-even Fe I transitions.
D. M. Kuli-Zade, K. I. Gusejnov, S. M. Veliev.
Astron. Zhurn. Akad. Nauk SSSR, Vol. 53, 577 - 588 (1976). In Russian. English translation in Soviet Astron., Vol. 20, No. 3.

The paper deals with the excitation temperature of the solar photosphere according to the lines of 1) even-odd transitions, 2) odd-even transitions and 3) both types of Fe I transitions for eight spectral ranges within the wavelength range from λ 3000 Å to λ 9000 Å.

071.036 The solar abundance of platinum.
H. L. Burger, L. H. Aller.
Proc. National Acad. Sci. USA, Vol. 72, 4193 - 4195 = Univ. Calif., Los Angeles, Astron. Papers, No. 26 (1975).

071.037 Analysis of the solar spectrum: results for the range λ 6132.0–λ 6150.0.
H. Steyaert, N. Baeck, P. Dingens.
Accad. Roy. Belgique, Bull. Cl. Sci. 5e Sér., Vol. 61, 881 - 887 = Univ. Gent, Sterrenkundig Inst., Med. No. 38 (1975).

071.038 A solar abundance of titanium, chromium and nickel deduced from a study of weak infrared lines.
E. Biémont.
Astrophys. Letters, Vol. 17, 127 - 130 (1976).

New values of the solar photospheric abundances of titanium, chromium and nickel have been derived from a study of weak infrared transitions of Ti I, Cr I and Ni I for which theoretical oscillator strengths have recently been obtained by Kurucz and Peytremann (1975). The results, $A_{Ti} = 4.82 \pm 0.12$, $A_{Cr} = 5.69 \pm 0.10$ and $A_{Ni} = 6.21 \pm 0.19$ are compared with coronal and meteoritic values.

071.039 The isotopic abundance ratio $^{12}C/^{13}C$ in the solar atmosphere determined from CH-lines in the photospheric spectrum. Ø. Iversen.
Inst. Theor. Astrophys. Blindern–Oslo, Rep. No. 45, 37 pp. (1976).

071.040 A search for spallation reactions in the solar photosphere by a study of the Li I-line at λ 6707 Å.
L. Hultqvist.
Stockholms Obs. Rep. No. 4, 2 + 46 pp. (1974/75).

Details of the performance of the experiment and of the reduction of the spectrograms are described. No evidence was found of spallation reactions. A rough upper limit to the number of protons per cm^2 impinging in the uppermost layers of the photosphere is given.

071.041 Heliographic maps of the photosphere for the year 1975. M. Waldmeier.

Publ. Eidgen. Sternw. Zürich, Band 14, (No. 5), 125 - 153 (1976).

The present publication gives heliographic maps of the photosphere and evolution tables of sunspot-groups for the year 1975. Maps and tables are based on daily drawings of spots and faculae using a projected solar image with a diameter of 25 cm.

071.042 **Solar supergranulation.** S. P. Worden.
Thesis Univ. Arizona, Tucson, USA, 168 pp. (1975). (Available from: Univ. Microfilms, Ann Arbor, Mich., USA. Order No. 75-19576). – Abstr. in Phys. Abstr., Vol. 79, A027780 (1976).

071.043 **Identification of the solar spectra of multicharged iron ions on the basis of laboratory measurements.**
Eh. Ya. Kononov, K. N. Koshelev, L. I. Podobedova, S. V. Chekalin, S. S. Churilov.
Journ. Phys. B, Vol. 9, 565 - 572 (1976). – Abstr. in Phys. Abstr., Vol. 79, A034624 (1976).

071.044 **Solar photospheric velocity field and a new analysis of the five minute oscillations.** D. K. Lynch.
Thesis Texas Univ., Austin, USA, 213 pp. (1975). (Available from: Univ. Microfilms, Order No. 75-24, 906).

071.045 **Models of the upper photospheres of the sun and Arcturus based on molecular spectra.**
G. H. Mount.
Thesis Colorado Univ., Boulder, USA, 228 pp. (1975). (Available from: Univ. Microfilms, Order No. 75-23, 626).

Rotational dependence of Franck-Condon factors for molecules of astrophysical interest.
See Abstr. 022.018.

Transition probabilities within $2s^2$-$2s2p$-$2p^2$ in the Be I sequence, Be I–Ni XXV. See Abstr. 022.040.

Transition probabilities in Pr(II) and the solar praseodymium abundance. See Abstr. 022.061.

Observations of $3p^53d-3p^54f$ and $3p^63d-3p^53d4s$ transitions in iron, cobalt, nickel, and copper.
See Abstr. 022.065.

Solar seeing and the statistical properties of the photospheric solar granulation. I. Noise in Michelson and speckle interferometry. See Abstr. 031.204.

Formation of spectral lines with partial frequency redistribution. See Abstr. 063.002.

Ein empirisches Modell für die tiefen Umbra-Schichten. See Abstr. 072.013.

Emission heights and centre-to-limb variation of some chromospheric lines. See Abstr. 073.024.

High-temperature flare lines in the solar spectrum 171 Å–630 Å. See Abstr. 073.036.

Filtering of acoustic waves in the solar atmosphere. See Abstr. 080.002.

Hydraulic concentration of magnetic fields in the solar photosphere. III. Fields of one or two kilogauss. See Abstr. 080.015.

Fourier spectroscopy in planetary research. See Abstr. 091.020.

Errata

071.901 **Erratum: 'Broadening of some solar Na I lines by atomic hydrogen'** [Astron. Astrophys., Vol. 38, 41 - 44 (1975)]. E. Roueff.
Astron. Astrophys., Vol. 46, 149 (1976).

072 Sunspots, Faculae, Solar Activity Cycles

072.001 What determines sunspot maximum?
    G. M. Brown.
Monthly Notices Roy. Astron. Soc., Vol. 174, 185 - 189 (1976).

An examination of secularly-smoothed sunspot number data shows that the magnitude and time of occurrence of a solar maximum are determined one half-cycle earlier by conditions prevailing at the preceding solar minimum epoch. There is evidence, too, for a progressive increase in response of the magnitude of the maximum to that of the minimum over the last 21 solar cycles.

072.002 The flux of Alfvén waves in sunspots.
    J. M. Beckers.
Astrophys. Journ., Vol. 203, 739 - 752 (1976).

Under the assumption that the horizontal velocities observed in the sunspot photosphere are caused by Alfvén waves, the author derives the flux of these waves in the photosphere to be equal to 20–50 percent of the missing sunspot flux or 2–3 times the spot's flux of electromagnetic radiation. Since the downward-moving Alfvén wave flux is likely to be similar to the upward-moving flux, this observation is consistent with the hypothesis that spots are cooled by their strong emission of Alfvén waves.

072.003 A model of the solar cycle driven by the dynamo action of the global convection in the solar convection zone.   H. Yoshimura.
Astrophys. Journ. Suppl. Ser., No. 294, Vol. 29, 467 - 494 (1975).

The dynamo equation which represents the longitudinally averaged magnetohydrodynamical action of the global convection influenced by the rotation in the solar convection zone is solved numerically to simulate the solar cycle as an initial boundary-value problem. Oscillatory solutions are obtained to simulate the solar cycle with the period of the right order of magnitude and with the patterns of evolution of the latitudinal distribution of the toroidal component of the magnetic field similar to the observed Butterfly Diagram of sunspots. The evolution of the latitudinal distribution of the radial component of the magnetic field shows patterns similar to the Butterfly Diagram, but having two branches of different polarity in each hemisphere. The importance of the poleward migrating branch of the Butterfly Diagram, barely seen for sunspots but clearly seen for polar prominences and for the observed general magnetic field, is emphasized in relation to the relative importance of the role of the latitudinal and radial shears of the differential rotation.

072.004 Does the sun change its rhythms?
    L. I. Miroshnichenko.
Priroda, 1976, No. 2, p. 135 - 137. In Russian.

072.005 Forecast for solar cycle 21.   O. B. Vasil'ev, Yu.
    I. Vitinskij, K. A. Kandaurova.
Solnechnye Dannye 1975 Byull., No. 10, p. 55 - 59 (1976). In Russian.

Forecasting for annual Wolf numbers in 1976 - 1990 is made using the spectral-frequency method. For cycle 21 the epoch of minimum is expected to be 1976.5 ± 0.5 and that of maximum in the first half of 1980. The height of cycle 21 has been estimated.

072.006 The brightness distribution in a facular area.
    I. Ferro, M. Cid.
Solnechnye Dannye 1975 Byull., No. 10, p. 65 - 67 (1976). In Russian.

A study of the brightness distribution in a facular area passing the solar disc has been made. It has been found that bright elements have a tendency to distribute mainly along the meridian. East-west asymmetry in their location with respect to the sunspot has been observed.

072.007 Ohmic dissipation of sunspots.   A. A. Solov'ev.
    Astron. Zhurn. Akad. Nauk SSSR, Vol. 53, 140 - 147 (1976). In Russian. English translation in Soviet Astron., Vol. 20, No. 1.

An important role of ohmic dissipation in the process of sunspots decay is shown.

072.008 General theory of sunspot statistics on a rotating sun.   M. Kopecký, J. Suda, E. Marková.
Bull. Astron. Inst. Czechoslovakia, Vol. 27, 65 - 71 (1976).

General statistical relations, valid for sunspot groups on a rotating sun, are derived under the assumption that both fundamental parameters (the number of sunspots originated per unit time over unit heliographic longitude and their frequency distribution with respect to lifetime) depend on time as well as on heliographic longitude.

072.009 A note on recurrences in the magnetic field distribution during the present cycle of solar activity.
V. Bumba.
Bull. Astron. Inst. Czechoslovakia, Vol. 27, 74 - 76 (1976).

Some results of a preliminary investigation of Mt. Wilson magnetic synoptic charts obtained during the present activity cycle are presented: in this cycle of activity practically no changes in the behaviour of the magnetic field distribution, when compared with the preceding cycle, are observed. Even the roles of the positive and negative polarities seem to remain unexchanged, although their functions of leading and following polarity are exchanged. The importance of the two main recurrences of successions in the large-scale magnetic field distribution with the 27 day and 28–29 day synodic periods of rotation, is emphasized once more. The shifts or "oscillations" of activity in heliographic latitude are mentioned.

072.010 Overstability and cooling in sunspots.
    B. Roberts.
Astrophys. Journ., Vol. 204, 268 - 280 (1976).

The role played by overstable Alfvén modes in magnetic structures such as sunspots is considered in detail for a column of magnetic field. It is demonstrated explicitly that overstable Alfvén waves cool the interior of the magnetic column. It is suggested that these waves account for the cooling in sunspot umbrae, and therefore, in concurrence with Parker, the author concludes that a sunspot is a region of enhanced heat transport. He also suggests that cooling by overstable Alfvén waves may explain the existence of the intense small magnetic flux tubes that constitute the general solar magnetic field.

072.011 Solar magneto-atmospheric waves. II. A model for running penumbral waves.
A. H. Nye, J. H. Thomas.
Astrophys. Journ., Vol. 204, 582 - 588 (1976).

A simple two-layer model of a sunspot penumbra is used to study the mode of running penumbral waves. Exact solutions of the linearized wave equation, not limited to the smallwavelength approximation, are employed in each layer. The lowest 'plus' eigenmode of magneto-atmospheric waves in the model penumbra is in good agreement with observations of running penumbral waves. The results indicate that running penumbral waves should be observable in a photospheric spectral line.

072.012 **Bemerkungen zur jahreszeitlichen Verteilung der Sonnenfleckenmaxima.** K. H. Schroeter.
Mitt. Astron. Ges., No. 38, p. 140 - 143 (1976). – Short report.

072.013 **Ein empirisches Modell für die tiefen Umbra-Schichten.** E. Wiehr, G. Stellmacher, H. Schleicher.
Mitt. Astron. Ges., No. 38, p. 217 (1976). – Abstract.

072.014 **Makroskopische Geschwindigkeiten in der unmittelbaren Umgebung von Sonnenflecken.**
W. Mattig, A. Nesis.
Mitt. Astron. Ges., No. 38, p. 217 - 218 (1976). – Abstract.

072.015 **Spectral analysis of solar and geomagnetic activity indices.** V. F. Loginov.
Trudy VNII gidrometeorol. inform. – Mirovoj tsentr dannykh, 1975, vyp. (No.) 23, p. 63 - 68. In Russian. – Abstr. in Referativ. Zhurn. 51. Astron., 3.51.353 (1976).

072.016 **Properties of motion of some long-lived sunspot groups.** G. P. Shchegoleva.
Solnechnye Dannye 1975 Byull., No. 11, p. 63 - 70 (1976). In Russian.
Proper motions of long-lived sunspot groups consisting of a stable spot and sometimes appearing neighbouring small spots and pores have been studied. Some stable spots are characterized by agreement (duration of about three solar rotations) with the rigid rotation law at certain stages of their life. The situation changes at the moment of group distruction.

072.017 **The split of the π-component of magnetoactive lines and the double-component model of a sunspot.**
A. V. Baranov.
Solnechnye Dannye 1975 Byull., No. 11, p. 86 - 88 (1976). In Russian.
The split of the π-component of the magnetoactive lines can be explained in the frame of sunspot model consisting of regions of strong and weak magnetic field with opposite polarity. It is shown that, if there are convective and wave motions in the weak field region, supposition of low turbulent velocities being present at the weak field region of opposite polarity cannot be made.

072.018 **Fotograferen van zonnevlekken met een 17-cm refractor.**
W. Brückner, with some remarks by C. Zwaan.
Zenit, 3e jaargang, p. 13 - 17 (1976).

072.019 **On short-time variations of the magnetic field of local features in an active region of the sun.**
V. L. Lentsman, L. D. Parfinenko.
Solnechnye Dannye 1975 Byull., No. 12, p. 93 - 94 (1976). In Russian.
Variations of the magnetic field intensity of separate features in an active region with a period of 270 sec and amplitude of 400 - 600 G were found with the Pulkovo TV spectroheliograph.

072.020 **Solar cycles of Sevang.** O. A. Azernikova.
Solnechnye Dannye 1975 Byull., No. 12, p. 96 - 99 (1976). In Russian.
The results of calculations of power spectra and cross-correlation functions of the Sevang parameters and Wolf numbers are presented. An 11-year cycle of these characteristics due to solar activity is found.

072.021 **Distribution of sunspot groups from maximum areas.** T. L. Mandrykina.
Tsirk. Astron. Obs. L'vov, No. 49, p. 24 - 27 (1974). In Russian.

072.022 **On the question of the north-south asymmetry of solar activity.** P. R. Romanchuk, Yu. N. Kudrya.
Astron. Obs., *Kiev*, Preprint No. 8, 11 pp. (1975). In Russian.
Results of testing a possible explanation of the north-south asymmetry by planetary action are presented. A hypothesis is proposed of the explanation of this asymmetry by a drift of the solar nucleus caused by the motion of the major planets relative to its magnetized shell, which is quasi-rigidly bound to the interplanetary plasma.

072.023 **The nature of solar cycles.** P. R. Romanchuk.
Astron. Obs., *Kiev,* Preprint No. 10, 15 pp. (1975). In Russian.
A critical analysis of the existing hypotheses of planetary action on the sun has been carried out. It is shown that planetary tidal and dynamical forces regulate the ascent of convective elements (active centres) on the solar surface, causing resonance phenomena of solar activity.

072.024 **On the question of forecasting the duration of existence of sunspot groups.** V. B. Gumanitskij, V. M. Efimenko, P. R. Romanchuk, A. N. Sergeeva, V. V. Tel'nyuk-Adamchuk.
Astron. Obs., *Kiev,* Preprint No. 11, 12 pp. (1975). In Russian.
On the basis of data about spot groups in 1964 - 1970 the application of seven parameters for the prediction of the duration of sunspot groups existence is investigated. The programs for the forecasting of group appearances on the next revolution are described. The forecast accuracy ranges about 70 per cent.

072.025 **Video-magnetograph observations of moving magnetic features around sunspots.**
A. G. Michalitsanos, A. Bhatnagar.
Bull. Astron. Soc. India, Vol. 3, 27 (1975). – Abstract of paper presented at the A.S.I. meeting 1975.

072.026 **Mean intensity of the H and K Ca II lines emission in a sunspot umbra.**
S. A. Ehfendieva, I. P. Turova.
Issled. po geomagnetizmu, aehron. i fiz. Solntsa. Vyp. (No.) 37. Moskva, Nauka, 1975, p. 82 - 86. In Russian. – Abstr. in Referativ. Zhurn. 51. Astron., 4.51.338 (1976).

072.027 **The influence of the magnetic field on the determination of Doppler width and source functions of H and K Ca II lines in a sunspot umbra.** S. A. Ehfendieva.
Issled. po geomagnetizmu, aehron. i fiz. Solntsa. Vyp. (No.) 37. Moskva, Nauka, 1975, p. 87 - 92. In Russian. – Abstr. in Referativ. Zhurn. 51. Astron., 4.51.339 (1976).

072.028 **Spatial structures of flare ribbons in bipolar sunspot groups of classes E, F, G.**
V. V. Kasinskij, N. N. Lyakhov.
Issled. po geomagnetizmu, aehron. i fiz. Solntsa. Vyp. (No.) 37. Moskva, Nauka, 1975, p. 37 - 45. In Russian. – Abstr. in Referativ. Zhurn. 51. Astron., 4.51.379 (1976).

072.029 **Solar activity forecasting.** P. R. Romanchuk.
Fiz.-mat. i biol. probl. dejstviya ehlektromagnitn. polej i ionizatsii vozdukha. Tom (Vol.) 1. Moskva, Nauka, 1975, p. 73 - 83. In Russian. – Abstr. in Referativ. Zhurn. 51. Astron., 4.51.383 (1976).

072.030 **Forecast of active region evolution by constructing logical recognizing graphs.**
B. G. Dolgoarshinnykh, G. V. Kuklin.
Issled. po geomagnetizmu, aehron. i fiz. Solntsa. Vyp. (No.) 37. Moskva, Nauka, 1975, p. 24 - 31. In Russian. – Abstr. in Referativ. Zhurn. 51. Astron., 4.51.384 (1976).

072.031 **Timing of solar cycles by rigid internal rotations.**
C. L. Wolff.
Astrophys. Journ., Vol. 205, 612 - 621 (1976).

The so-called 11 year cycle of solar activity is really more complex. Periods as long as 178 years and as short as 3.1 years are predicted by a theory based on beats between rigidly rotating, inertially oscillating g-modes inside the sun. The theory receives further support when tested against an independent class of observations — the large scale magnetic sector structure. As a by-product of these successful fits to observation, the mean rotation of the entire solar mass becomes known.

072.032 **On the possibility of analytical description of active region structure.** N. N. Bulatov.
Astron. Zhurn. Akad. Nauk SSSR, Vol. 53, 377 - 379 (1976).
In Russian. English translation in Soviet Astron., Vol. 20, No. 2.

It is shown that Chambe's empirical approximation of differential emission measure is following the assumption of hydrostatic density distribution and linear temperature growth in the coronal condensation. An analytical approximation of the dependence of electron temperature on height in a solar active region is suggested. The free-free radio emission spectrum calculated for the suggested model corresponds to local sources observations at mm-wavelengths.

072.033 **The latitudinal zones, motion of matter and cycle activity on the sun.** N. I. Kozhevnikov.
Astron. Zhurn. Akad. Nauk SSSR, Vol. 53, 389 - 397 (1976).
In Russian. English translation in Soviet Astron., Vol. 20, No. 2.

The distribution of the velocity of motion of spot groups and other parameters with regard to heliographic latitude is considered. It is found that these parameters and the distribution of velocities are correlated with those for spot groups of mean lifetime, and that the velocities of the motion of spot groups change with the phase of solar cycle activity.

072.034 **A magnetohydrodynamic dynamo model of the solar cycle.** T. S. Ivanova, A. A. Ruzmajkin.
Astron. Zhurn. Akad. Nauk SSSR, Vol. 53, 398 - 410 (1976).
In Russian. English translation in Soviet Astron., Vol. 20, No. 2.

A hydromagnetic dynamo model of the cycle process in the sun's convective zone, in which the differential rotation and the gyrotropic turbulent-convective motions play the main role, is investigated.

072.035 **The rotation of matter in a sunspot.** V. A. Kotov.
Izv. Krymskoj Astrofiz. Obs., Vol. 54, 184 - 200 (1976). In Russian.

With the use of the Crimean magnetograph the line-of-sight velocities in a sunspot were measured in the four spectral lines Fe I λ4808, Fe I λ5250, Ca I λ6103 and Hα. The rotational motion was found to be cyclonic (left-handed) at the photospheric layers and anticyclonic in the chromosphere. This agrees with what should be expected for a spot in the southern hemisphere. The eddy structure of the currents and that of the magnetic field lines is discussed.

072.036 **Some peculiarities of the velocity field in active regions.** Dzh. I. Irgashev.
Izv. Krymskoj Astrofiz. Obs., Vol. 54, 233 - 240 (1976). In Russian.

The velocity field on the photospheric level in well developed active regions located near the central meridian has been studied. Assuming equal densities of the downward and upward gas flows, it is shown that disbalance of the matter flows is connected with disbalance between magnetic fluxes. This connection depends on the type of the sunspot group. There is predominantly ascending gas in the places of active regions with weak ($H_\parallel < 20$ Gs) magnetic fields and prevalence of descending gas in the places where the magnetic field strength exceeds 20 gauss.

072.037 **Application of the pattern recognition methods to the problem of forecasting sunspot group development.** V. A. Slepyan, T. L. Slutskaya, N. N. Stepanyan.
Izv. Krymskoj Astrofiz. Obs., Vol. 54, 244 - 247 (1976).
In Russian.

A flare activity forecast of the just emerged sunspot group is given using observations of the active region before and on the day of appearance of the sunspots. Two methods are used for the solution of the problem: the potential functions method and the method of calculation of the values. Three years observations (1967 - 1969) are used as initial data for deriving and checking of the forecasting function. The flare activity forecast is based on 10 initial parameters. The potential functions method gives the right forecast for 82% of the sunspot groups, and the method of calculation of the values only for 67%.

072.038 **The Maunder minimum.**
Sky Telescope, Vol. 51, 394 - 395 (1976).

072.039 **Beziehungen zwischen der Sonnenaktivität und dem Massenzentrum des Sonnensystems.**
T. Landscheidt.
Nachr. Olbers-Ges. Bremen, No. 100, p. 2 - 19 (1976).

072.040 **The intensity of the magnetic field of sunspots during the generation of a penumbra.**
Š. Knoška.
Bull. Astron. Inst. Czechoslovakia, Vol. 27, 159 - 164 (1976).

The maximum values of the intensity of the magnetic field of sunspots prior to and after the generation of a penumbra are studied. The results obtained from observations from various periods of the solar activity cycles (1917–1924 and 1962–1970) made using different methods of observations (visual and photographic) indicate that the intensity of the sunspot magnetic field does not vary continuously in the course of their development, but displays discrete values.

072.041 **Morphological peculiarities of photospheric faculae.** A. P. Kramynin.
Solar phenomena research, Vladivostok, 1975, (see 003.005), p. 111 - 120. In Russian. — Abstr. in Referativ. Zhurn. 51. Astron., 5.51.425 (1976).

072.042 **Umbra brightness of a large sunspot.**
Eh. P. Surkov, L. D. Surkova.
Solar phenomena research, Vladivostok, 1975, (see 003.005), p. 124 - 136. In Russian. — Abstr. in Referativ. Zhurn. 51. Astron., 5.51.426 (1976).

072.043 **Spectrophotometry of an evolving spot.**
Eh. P. Surkov, L. D. Surkova.
Solar phenomena research, Vladivostok, 1975, (see 003.005), p. 137 - 155. In Russian. — Abstr. in Referativ. Zhurn. 51. Astron., 5.51.427 (1976).

072.044 **On the influence of temperature inhomogeneities of the profile of magneto-active lines.**
A. V. Baranov.
Solar phenomena research, Vladivostok, 1975, (see 003.005), p. 156 - 160. In Russian. — Abstr. in Referativ. Zhurn. 51. Astron., 5.51.428 (1976).

072.045 **Comparison of methods for measurement of sunspot magnetic fields.**
A. V. Baranov, V. A. Golubev.
Solar phenomena research, Vladivostok, 1975, (see 003.005), p. 161 - 164. In Russian. — Abstr. in Referativ. Zhurn. 51. Astron., 5.51.429 (1976).

072.046 **Character of the solar limb darkening of sunspot**

umbrae derived from observations of the center and symmetrical wings of the Hα line.
G. I. Kornienko, Eh. P. Surkov.
Solar phenomena research, Vladivostok, 1975, (see 003.005), p. 165 - 179. In Russian. – Abstr. in Referativ. Zhurn. 51. Astron., 5.51.430 (1976).

072.047 On the relation of the development of active regions on the sun to the accompanying phenomena in cosmic ray intensity.   N. P. Milovidova, V. P. Nefed'ev.
Cosmic rays, No. 15, (see 003.006), p. 82 - 85 (1975). In Russian.

072.048 On the time behaviour of oscillations in sunspot umbrae.   E. H. Schröter, D. Soltau.
Astron. Astrophys., Vol. 49, 463 - 465 (1976).
    The authors studied the time behaviour of oscillations in a sunspot umbra from temporarily high resolved Doppler compensator velocity records. A power spectrum analysis yields modes with different periods.

072.049 The Maunder minimum.   J. A. Eddy.
Science, Vol. 192, 1189 - 1202 (1976).

072.050 A review of rapid fluctuations of magnetic fields of sunspots by observations at Pulkovo, Potsdam, and Sverdlovsk.   V. F. Chistyakov.
Solar phenomena research, Vladivostok, 1975, (see 003.005), p. 69 - 91. In Russian. – Abstr. in Referativ. Zhurn. 51. Astron., 6. 51. 450 (1976).

072.051 Structural and spectral studies of sunspots. Final report, 1 Sept. 1967 - 31 Aug. 1974.
A. A. Wyller.
Separate print Bartol Res. Found., Swarthmore, Pennsylvania, USA, 76 pp. (1974).

072.052 Beobachtung der Sonnenflecken-Relativzahlen.
G. Lustig.
Sternenbote, 19. Jahrgang, p. 126 - 132 (1976).

072.053 Weitere Hinweise auf Zusammenhänge zwischen den Sonnenflecken und den Planeten mit ihren Satelliten
R. Liese.
Separate print Techn. Univ. Hannover, 20 pp. (1976).
    The influence on the sunspot activity caused by the planets Mercury, Venus, Earth, Jupiter, Saturn and Uranus as well as by the earth-moon is shown.

072.054 Solar cycle variation of solar wind characteristics.
K. W. Ogilvie.
Symposium significant accomplishments sci. technol., Greenbelt, Maryland, USA, 18 Dec. 1973, p. 97 - 99 (1975).

072.055 Nombres relatifs de Wolf pour 1975.
M. Waldmeier.
L'Astronomie, Vol. 90, 267 (1976).

072.056 On the variation of planetary orbits and the solar activity period.   A. Dauvillier.
Bull. Soc. Roy. Sci. Liège, Vol. 44, 161 - 164 (1975). In French. – Abstr. in Phys. Abstr., Vol. 79, A027785 (1976).

072.057 Notes on solar activity, geomagnetics and the ionosphere.   J. O. Cardús.
Rev. Geofis., Vol. 33, 139 - 151 (1974). In Spanish. – Abstr. in Phys. Abstr., Vol. 79, A032730 (1976).

Initial operation of a scanning Stokes polarimeter. See Abstr. 034.010.

An association among planets, sunspots and solar plasma physics.   See Abstr. 062.090.

Table of solar diatomic molecular lines. II. Spectral range: 6600–7100 Å.   See Abstr. 071.005.

An analytic solution for the line formation in a magnetic field. II.   See Abstr. 071.031.

Observations of penumbral waves in the photosphere.   See Abstr. 071.032.

Heliographic maps of the photosphere for the year 1975.   See Abstr. 071.041.

Changes in X-ray brightness of a solar active region. See Abstr. 073.008.

On some anomalies in the development of solar phenomena at the end of cycle No. 20.   See Abstr. 073.026.

On the relationship between chromospheric flares, configurations of sunspot groups and their magnetic structures. See Abstr. 073.032.

On the relationship between the frequency of flare occurrences in active regions and the value of their magnetic moment.   See Abstr. 073.033.

High-speed fluxes of solar wind plasma and active regions on the sun.   See Abstr. 074.039.

Magnetic fields of sunspots.   See Abstr. 075.017.

On restoring the true radio image of a local source. See Abstr. 077.005.

The spectrum of the continuum of noise storms and its relationship with active regions.   See Abstr. 077.029.

On the relation between background fluctuations of noise storms and strength of sunspot magnetic fields. See Abstr. 077.033.

Solar radio continuum storms and a breathing magnetic field model. Final report.   See Abstr. 077.049.

Long period magnetic activity – 2 to 100 years. See Abstr. 080.006.

The rapid oscillations of the solar rotation. See Abstr. 080.007.

Reliability of sunspots as tracers of solar surface rotation.   See Abstr. 080.011.

Solar magnetic fields and convection. IV: Magnetic flux ropes and their fibres.   See Abstr. 080.029.

Structure of the magnetic field of active regions and noise storms.   See Abstr. 080.035.

Long term changes in the solar diurnal anisotropy. See Abstr. 080.044.

Solar EUV flux variation during a solar cycle as derived from ionospheric modeling considerations. See Abstr. 083.044.

Effects of solar activity on annual means of geomagnetic components.   See Abstr. 084.277.

Decametric radio emission of Jupiter and solar activity.   See Abstr. 099.033.

Meteor radar rates and the solar cycle.
See Abstr. 104.055.

Variations of the spectrum of fluctuations of the interplanetary magnetic field and of diffusion coefficients for cosmic radiation in the solar activity cycle.
See Abstr. 106.033.

Comparative analysis of cosmophysical and geophysical phenomena in the 19th and 20th solar activity cycles.
See Abstr. 106.039.

Modulation of galactic cosmic rays and solar activity distribution over heliocoordinates.
See Abstr. 143.040.

Expected spatial distribution of cosmic rays in interplanetary space including the real heliolatitude distribution of solar activity.   See Abstr. 143.190.

Expected 11-year and annual variations in cosmic rays of various rigidities in terms of the anisotropic diffusion model on the basis of the data on heliolatitude distribution of solar activity.   See Abstr. 143.191.

Solar effects associated with the cosmic ray modulation spectrum.   See Abstr. 143.197.

The time and rigidity dependence of the 11 year modulation of cosmic rays.   See Abstr. 143.198.

The variation in the latitude dependence of cosmic rays at 307 g/cm² during solar cycle No 20.
See Abstr. 143.199.

The solar cycle modulation of the galactic cosmic rays and the solar flare activity.   See Abstr. 143.200.

Cosmic ray modulation and green coronal line activity.   See Abstr. 143.201.

Modulation of low energy electrons and protons near solar maximum.   See Abstr. 143.207.

Long term modulation of cosmic-ray intensity and solar activity cycle.   See Abstr. 143.210.

Rate of production of cosmogenic isotopes in the past and solar activity.   See Abstr. 143.213.

The eleven-year cosmic ray cycle according to stratosphere measurements.   See Abstr. 143.214.

The short term modulation of cosmic radiation.
See Abstr. 143.217.

The power spectra of cosmic ray intensity and the solar activity indices in the period range of 50 - 1000 days.
See Abstr. 143.229.

N–S asymmetry and variations of cosmic ray isotropic flux and solar activity.   See Abstr. 143.239.

Analysis of hysterisis in cosmic rays relative to variations in the heliolatitude index of solar activity for various epochs of solar activity cycle.   See Abstr. 143.266.

27-day recurrences of enhanced daily variations in the cosmic ray intensity during 1973—1975.
See Abstr. 143.273.

Solar effects associated with the cosmic ray modulation spectrum.   See Abstr. 143.330.

Some characteristics of solar tridiurnal variation of cosmic rays.   See Abstr. 143.331.

On the effect of directional medium scale interplanetary variations on the diffusion of galactic cosmic rays and their solar cycle variation.   See Abstr. 143.344.

Dynamic model for the time evolution of the modulated cosmic-ray spectrum.   See Abstr. 143.365.

Errata

072.901   Erratum: 'The deep layers of sunspot umbrae'
[Astron. Astrophys., Vol. 45, 69 - 76 (1975)].
G. Stellmacher, E. Wiehr.
Astron. Astrophys., Vol. 47, 479 (1976).

073 Solar Chromosphere, Flares, Prominences

073.001 Extreme-ultraviolet transients observed at the solar pole. G. L. Withbroe, D. T. Jaffe, P. V. Foukal, M. C. E. Huber, R. W. Noyes, E. M. Reeves, E. J. Schmahl, J. G. Timothy, J. E. Vernazza.
Astrophys. Journ., Vol. 203, 528 - 532 (1976).

Extreme-ultraviolet observations of two polar transient features called macrospicules are described. These features appear to be caused by jets of chromospheric material that shoot upward to a height of 35,000 km above the limb and then fall back into the chromosphere, reaching terminal velocities of about 140 km s^{-1}. On the basis of a model developed from the EUV measurements, the authors find that the energy required to produce each event is about 3×10^{26} ergs, about 2 orders of magnitude more than that required to produce an ordinary spicule. This indicates that macrospicules may be an important factor in the energy balance of the chromosphere and corona.

073.002 Solar gamma-ray lines as probes of accelerated particle directionalities in flares.
R. Ramaty, C. J. Crannell.
Astrophys. Journ., Vol. 203, 766 - 768 (1976).

Anisotropies of charged particles accelerated in solar flares can be studied by observing Doppler shifts of selected γ-ray lines. The authors have calculated the spectral shape of the 6.1 MeV line of ^{16}O. If the accelerated particles are isotropic, the line remains centered at $E_0 = 6129.4$ keV, and its width (FWHM) is about 100 keV. However, for particle anisotropies that may be produced in solar flares, the line is shifted to lower energies by about 30 to 40 keV.

073.003 Statistical properties of chromospheric surges. I. Yu. M. Slonim, K. F. Kuleshova.
Solnechnye Dannye 1975 Byull., No. 10, p. 68 - 75 (1976). In Russian.

305 surges on the solar disc and near the limb are studied. A connection between their kinetic energy and the 11-year cycle phase is found. Statistical relations between the surge properties are analysed. Certain evidence has been obtained that surge rise is caused by forces of electromagnetic character. As to surge accumulation in the chromosphere it is controlled mainly by gravitation.

073.004 On X-rays of solar flares in July-December, 1972. N. N. Evstaf'ev, N. V. Illarionova, A. S. Melioranskij, I. N. Rozantsev, I. A. Savenko.
Solnechnye Dannye 1975 Byull., No. 10, p. 87 - 93 (1976). In Russian.

The correlations between X-ray bursts, Hα flares, radio bursts and sudden ionospheric disturbances are presented.

073.005 On some peculiarities of motions in surges. M. N. Stoyanova.
Solnechnye Dannye 1975 Byull., No. 10, p. 93 - 101 (1976). In Russian.

Hydrogen lines of the Balmer series have been studied in the spectra of surges. It is found that their profiles often deviate from a Gaussian and the Hα line doesn't correspond to the optical thickness determined from the upper terms of the Balmer series. Reduction for depth variations of the source function is shown to influence significantly the line profiles. There is evidence of the existence of proper radiation in surges.

073.006 Measurement of the absorbed radiation portion of the solar flare from August 4, 1972 in the open space.
O. I. Savun, A. I. Sladkova.
Kosmich. Issled., Vol. 14, 135 - 140 (1976). In Russian.

Brief information.

073.007 On the pitch-angle distribution of electrons generating a hard X-ray emission in flaring loops.
V. P. Maksimov.
Astron. Zhurn. Akad. Nauk SSSR, Vol. 53, 215 - 217 (1976). In Russian. English translation in Soviet Astron., Vol. 20, No. 1.

The problem of a non-linear interaction of Langmuir oscillation with magnetic field fluctuations in flaring loops is considered. This interaction can lead to a considerable increase of fluctuations amplitude by anisotropic pitch-angle distribution of electrons in the beam. The interaction is absent in the case of isotropic pitch-angle distribution.

073.008 Changes in X-ray brightness of a solar active region. W. M. Glencross, D. H. Brabban.
Monthly Notices Roy. Astron. Soc., Vol. 175, 33 - 42 (1976).

The soft X-ray flux in the waveband 0.3–0.9 nm has been monitored during most of the solar disk passage of McMath region 12094. These data show how the emission changed during quiet periods as well as during flaring. Throughout the first four days of observations the mean flux showed a gradual decay even though the magnetic region was still growing. At the end of this phase the region remained extremely inactive for almost half a day and then brightened by more than an order of magnitude within an hour. This enhancement lasted nearly one day and marked the onset of the break-up of the region.

073.009 Observation and analysis of low-energy solar particle propagation from discrete flare events.
C. R. Countee, L. J. Lanzerotti.
Journ. Geophys. Res., Vol. 81, 441 - 449 (1976).

The proton, alpha particle, and electron fluxes measured following two west limb solar flare events are analyzed by using numerical solutions to the complete particle transport equation, including considerations of particle convection and energy loss. The values of the obtained mean free paths λ_r suggest that interplanetary propagation conditions were different during these two events. During the November 4, 1968, event the derived values of λ_r were ~0.2–0.3 AU, while during the November 1969 event the derived values of λ_r were ~0.1 AU for particles with energies as low as a few MeV. For these events the rise and decay of the alpha and proton flux profiles can be reasonably reproduced with a distant (10 AU) free escape boundary.

073.010 Production of flare-produced H^2 and He^3. P. L. Rothwell.
Journ. Geophys. Res., Vol. 81, 709 - 714 (1976).

Recent measurements indicate the enriched production of He^3 but not of H^2 in some solar flares. This leads to an apparent paradox if the reaction $H + He^4 \rightarrow H^2 + He^3$ is the dominant mode of production. A detailed Monte Carlo calculation for this reaction is performed to determine the kinematical regions where He^3 dominates. It is concluded that a thick-target model based on primary protons mirroring just below the solar surface is the most likely source of the He^3 enriched events.

073.011 Die Bewegungen des solaren Kalzium-Netzwerkes. H. Wöhl.
Umschau, 76. Jahrgang, p. 222 - 223 (1976).

The differential rotation and motions of the so-called Ca$^+$-mottles, which are bright points in the solar network, is visible in the light of the lines of ionized calcium in the near ultraviolet; these are being investigated since 1974 at the Institute for Solar Research in Locarno, Switzerland, by E. H.

Schröter and by the author. The methods used for measuring as well as the first results are given here.

073.012 **The lithium-like $2s\ ^2S-2p\ ^2P$ transition in solar flares.** K. G. Widing, J. D. Purcell.
Astrophys. Journ., (Letters), Vol. 204, L151 - L153 (1976).

Accurate wavelengths of the lithium-like doublet of Fe XXIV at 192 and 255 Å derived from measurements in a sample of solar flares observed with the NRL spectroheliograph on Skylab are presented. The exceptional strength of the $^2S-^2P$ doublets in solar flares has led to new identifications including Cr XXII, Mn XXIII, and Ni XXVI.

073.013 **Untersuchung einiger starker Fraunhoferlinien in der ruhigen Chromosphäre.** H. Schleicher.
Mitt. Astron. Ges., No. 38, p. 139 - 140 (1976). — Abstract.

073.014 **Chromosphärischer Temperaturanstieg in der Sonne aufgrund akustischer Heizung.**
H. U. Bohn, P. Ulmschneider.
Mitt. Astron. Ges., No. 38, p. 202 - 203 (1976). — Abstract.

073.015 **Wechselwirkung zwischen Strahlungstransport und Gasdynamik in der Sonnenchromosphäre.**
F. Kneer, Y. Nakagawa.
Mitt. Astron. Ges., No. 38, p. 203 (1976). — Abstract.

073.016 **Neue Beobachtungen zur Deutung der Mikroturbulenz.** F. L. Deubner.
Mitt. Astron. Ges., No. 38, p. 203 - 204 (1976). — Abstract.

073.017 **Filamentbeobachtungen im mm- und cm-Wellenlängenbereich.** M. Butz, W. Hirth, E. Fürst.
Mitt. Astron. Ges., No. 38, p. 211 - 216 (1976). — Short report.

073.018 **The calculation of force-free fields from discrete flux distributions.** N. R. Sheeley, Jr., J. W. Harvey.
Solar Physics, Vol. 45, 275 - 290 (1975).

This paper presents particularly simple mathematical formulas for the calculation of force-free fields of constant α from the distribution of discrete sources on a flat surface. The advantage of these formulas lies in their physical simplicity and the fact that they can be easily used in practice to calculate the fields. The disadvantage is that they are limited to fields of 'sufficiently small α'. These formulas may be useful in the study of chromospheric magnetic fields by the comparison of high-resolution Hα photographs and photospheric magnetograms.

073.019 **The prominence radiation theory.**
N. A. Yakovkin, M. Yu. Zeldina (Zel'dina).
Solar Physics, Vol. 45, 319 - 338 (1975).

The present work is a review of papers related to the theory of prominence radiation. Special attention is paid to stationary equations and the theory of radiation diffusion in the lines and continua of hydrogen, helium and metals.

073.020 **Observations of helium and hydrogen emission in quiescent prominences.**
R. M. E. Illing, D. A. Landman, D. L. Mickey.
Solar Physics, Vol. 45, 339 - 349 (1975).

Observations of a number of helium triplet (λλ 10830, 4713, 4471, 3889, 4026) and hydrogen (Hγ, δ, η, θ) emission line intensities in six quiescent prominences are presented. The results are compared with previous observations and theory. In particular, the intensity of the λ 10830 emission relative to the other triplets is found to differ strongly from the predictions of the recent detailed calculations of Heasley et al. (1974) for model quiescent prominences.

073.021 **Studies of the prominence-corona transition zone**

from rocket ultraviolet spectra of the March 1970 eclipse. C. Y. Yang, R. W. Nicholls, F. J. Morgan.
Solar Physics, Vol. 45, 351 - 362 (1975).

Slitless VUV spectra of the eclipsed sun were obtained from a rocket experiment for the first time during the 1970 eclipse. The spatially resolved spectra of a quiescent prominence in the wavelength range 900 Å—2200 Å consist of emission lines from ions formed in the temperature range $3.5 \times 10^4\,K—3.2 \times 10^5\,K$. The spectral intensities have been interpreted in terms of physical parameters which indicate a transition zone of shell-like layers, the inner the cooler and thinner, the outer the hotter and more extended. The transition zone is about 3 km thick for a model thread of 2000 km in diameter.

073.022 **Expansion of chromospheric matter in the gradual phase of solar flares.** K. Ohki.
Solar Physics, Vol. 45, 435 - 452 (1975).

Interferometric radio observations at 17 GHz together with soft X-ray observations are presented here to show that during the growth phase of soft X-ray flares, a large mass increase occurs simultaneously with the creation of an X-ray hot region in the corona. Since even small flares produce coronal hot regions radiating thermal soft X-rays and microwaves, the formation of the hot region may be a basic process in most flares. The new 17 GHz data permit an evaluation of the matter content of a pre-flare active region, by means of the observation of the slowly varying component. Further investigation of the conduction process from the corona to the chromosphere enables us to determine the fine structure of the corona-chromosphere transition region, which emits radiations at various wavelengths, of which the Hα and EUV are the best studied. The author discusses here the temperature distribution in the flaring transition region, including the expanding gas heated by conduction from a coronal energy source.

073.023 **Investigation of solar flare X-ray spectra in the hydrogen-like Fe ion region.**
Yu. I. Grineva, V. I. Karev, V. V. Korneev, V. V. Krutov, S. L. Mandel'stam (Mandel'shtam), U. I. Safronova, A. M. Urnov, L. A. Vainstein (Vajnshtejn), I. A. Zhitnik.
Space Research XV, (see 012.003), p. 637 - 640 (1975).

Results of preliminary studies of solar X-ray flare spectra in the region of the Fe XXVI resonance line are given. The spectra were obtained on board the Intercosmos 4 satellite. The composite spectrum obtained from ten scans reveals some peaks. Three peaks could correspond to the $K\beta$ (Fe), $L\alpha$ (Fe XXVI) and its satellite lines; this identification, however, leads to significant difficulties. A short discussion of the problem is given.

073.024 **Emission heights and centre-to-limb variation of some chromospheric lines.** A. Wittmann.
Astron. Astrophys., Vol. 48, 121 - 127 (1976).

A modified photoelectric drift scan technique has been used to measure emission heights as well as the centre-to-limb variation of some of the strongest lines in the quiet chromosphere (Ca^+-K, $Mg-b_1$, $Na-D$, H_a). The true centre-to-limb variations as derived from the measurements are compared with previous results and with predictions from very recent NLTE calculations. With few exceptions, the agreement is very satisfactory.

073.025 **On the continuous radiation and inhomogeneity in the low chromosphere.** K. Nakayama.
Publ. Astron. Soc. Japan, Vol. 28, 141 - 153 (1976).

The inconsistency between the high intensity of the Balmer continuum just outside the limb, observed at total solar eclipses, and the low brightness temperatures of the far infrared ($\lambda \gtrsim 50\,\mu m$) at the disk center and of the extreme ultraviolet ($\lambda \lesssim 1682$ Å) continuum is interpreted in terms of an inhomogeneous chromospheric model. On the condition that the inhomogeneous model must reproduce the IR and UV observations, the following points become clear: (1) the volume fraction of the boundary region is about 10 % in the low chromosphere and seems to increase in the upper layer, (2) the boundary region is hotter at least 700 K than the inner cell region, and (3) the predicted intensity excess of the boundary in white light is consistent with observations.

073.026 On some anomalies in the development of solar phenomena at the end of cycle No. 20.
V. F. Chistyakov.
Solnechnye Dannye 1975 Byull., No. 11, p. 71 - 77 (1976). In Russian.

Anomalies in the frequency of flare appearance and sunspot motions in the second half of 1972 are described. It is supposed that these anomalies are connected with instability in the interior of the sun.

073.027 Statistical properties of chromospheric surges. II.
Yu. M. Slonim, K. F. Kuleshova.
Solnechnye Dannye 1975 Byull., No. 11, p. 82 - 85 (1976). In Russian.

A close statistical relationship between surges and impulse flares has been found. The connection between their physical and dynamic characteristics has been discussed. It is shown that the surge is to be considered an integral part of the impulse flare process.

073.028 On the spiral motion in a solar prominence.
V. S. Pyatnitskij.
Solnechnye Dannye 1975 Byull., No. 11, p. 88 - 92 (1976). In Russian.

From the analysis of spectrograms and filtergrams of a solar prominence the conclusion is made that its motion is spiral. With increase of the spiral parameter of the lines the configuration of the prominence may become instable. This may lead to a magnetic rope split at the top of the prominence and to reconnection of magnetic force lines.

073.029 The D_1 and D_2 Na I lines in the spectra of prominences.
K. I. Selyakov, K. S. Tavastsherna.
Solnechnye Dannye 1975 Byull., No. 11, p. 92 - 96 (1976). In Russian.

The ratios of the intensities of the D_1 and D_2 Na I lines in the spectra of prominences are determined. A flat layer and a sphere have been taken as geometric models of the prominence. Using the results obtained, it is possible to estimate the optical thickness of the prominence in the D_1 and D_2 Na I lines from observed ratios of the intensities of these lines.

073.030 Determination of some parameters of a magnetic rope of a solar prominence on the basis of radial velocity distribution.
A. A. Solov'ev.
Solnechnye Dannye 1975 Byull., No. 12, p. 51 - 55 (1976). In Russian.

It is shown that using a detailed distribution of radial velocity over a solar prominence with spiral structure one can calculate some parameters of the magnetic rope: the pitch of the spiral, the angle between the line of sight and axis of the spiral and the total velocity vector.

073.031 Absolute spectrophotometry of the chromosphere in the region $\lambda\lambda$ 4588 - 4789 Å from photographs of the total solar eclipse on June 30, 1973.
V. M. Sobolev, G. F. Vyal'shin.
Solnechnye Dannye 1975 Byull., No. 12, p. 59 - 70 (1976). In Russian.

Total intensities and half-widths of 232 chromospheric lines in the region $\lambda\lambda$ 4588–4789 Å are given. Due to high dispersion of the spectrograph most of the line profiles

(He λ4713.07 and He$^+$ λ4685.95 particularly) were obtained. A significant difference between total intensities of some of the lines for various chromospheric places was found, while for most lines that difference did not exceed the range of errors.

073.032 On the relationship between chromospheric flares, configurations of sunspot groups and their magnetic structures.
Lyu Van Lyong.
Solnechnye Dannye 1975 Byull., No. 12, p. 71 - 77 (1976). In Russian.

It is shown that solar flares cause geomagnetic storms independently of the magnetic configuration of the flare region. A new morphological classification of groups is proposed. By this classification one can know in general the sunspot magnetic configuration and the neutral line positions in it.

073.033 On the relationship between the frequency of flare occurrences in active regions and the value of their magnetic moment.
V. P. Mikhajlutsa.
Solnechnye Dannye 1975 Byull., No. 12, p. 78 - 81 (1976). In Russian.

The value of the horizontal component of the vector of the magnetic moment for active regions lies in the range from $\sim 10^{29}$ erg/gauss (for pores) to 10^{31}–10^{32} erg/gauss (for developed groups). The frequency of flare occurrences in the active regions correlates with the horizontal component value.

073.034 Asymmetry of the field of radial velocities in the solar chromosphere.
V. L. Lentsman, L. D. Parfinenko.
Solnechnye Dannye 1975 Byull., No. 12, p. 94 - 95 (1976). In Russian.

From the results of observations with the Pulkovo TV spectroheliograph the conclusion is made that the structure of the field of radial velocities is different for the maximum velocities of the rising and lowering elements.

073.035 Structure and spectrum of quiescent prominences: energy balance and hydrogen spectrum.
J. N. Heasley, D. Mihalas.
Astrophys. Journ., Vol. 205, 273 - 285 (1976).

The authors present theoretical models of quiescent prominences which satisfy the constraints of radiative, magnetohydrostatic, and statistical equilibrium. They obtain reasonable models only if they assume that the exciting ultraviolet radiation field can penetrate diffusely into the slab, or that there is a source of nonradiative energy input. The computed temperatures in the models are in good agreement with observational estimates. The models reproduce most observed features of these objects quite well.

073.036 High-temperature flare lines in the solar spectrum 171 Å–630 Å.
G. D. Sandlin, G. E. Brueckner, V. E. Scherrer, R. Tousey.
Astrophys. Journ., (Letters), Vol. 205, L47 - L50 (1976).

XUV lines characteristic of $3-20 \times 10^6$ K plasma are observed in 1973 August 9 and June 15 ATM flare spectra. New solar identifications and wavelengths in the Li I, Be I, and B I sequences, including Ni XXVI, are proposed. Unidentified lines are classified with respect to their estimated temperatures.

073.037 Positron annihilation in solar flares.
C. J. Crannell, R. Ramaty, C. Werntz.
Bull. American Astron. Soc., Vol. 8, 291 (1976). – Abstr. AAS.

073.038 Upper limits to the quiet-time solar neutron flux from 10 to 100 MeV.
S. Moon, R. S. White, G. M. Simnett.
Bull. American Astron. Soc., Vol. 8, 291 (1976). – Abstr. AAS.

073.039 Brightness distribution on the solar disk in the

chromospheric net in the K_{232} Ca II line.
I. S. Laba.
Tsirk. Astron. Obs. L'vov, No. 49, p. 18 - 23 (1974). In Russian.

073.040 **On the excitation of Ca II in a flare.**
P. N. Polupan.
Astron. Obs., *Kiev*, Preprint No. 7, 9 pp. (1975). In Russian.
The collission excitation mechanism of radiation in the H and K Ca II lines with corpuscular streams into a powerful flare is very effective.

073.041 **The short-term forecast of chromospheric flares at the Kiev University Observatory.** V. I. Efimenko, V. M. Efimenko, V. V. Tel'nyuk-Adamchuk.
Astron. Obs., *Kiev*, Preprint No. 9, 16 pp. (1975). In Russian.
A method for the forecast of chromospheric flares in a sunspot group for the following 1 -2 days is developed and applied. For that purpose parameters are used which characterize the sunspot group development for some preceeding days. The forecast is true for 80 percent of flares for the 1974 year data.

073.042 **Active prominence on April 1, 1970. II. Calcium emission conditions.** A. S. Rakhubovskij.
Astrometriya i Astrofizika, *Kiev*, vyp. (No.) 28, (see 003.002), p. 100 - 108 (1976). In Russian.
The results of a spectrophotometric treatment of an active prominence are presented. The emission lines of ionized calcium H and K $\lambda 3706$, $\lambda 3737$ and the line $\lambda 4227$ of neutral calcium as well as the lines of ionized and neutral metals were observed. The number of calcium atoms along the line of sight is obtained. The ionization degree of Ca II is determined.

073.043 **Time evolution studies of the fine structure of the solar chromosphere.**
K. R. Sivaraman, P. P. Venkitachalam.
Bull. Astron. Soc. India, Vol. 3, 27 (1975). − Abstract of a paper presented at the A.S.I. meeting 1975.

073.044 **Report on the new solar observatory and H-alpha observations of the sun.**
A. Bhatnagar, G. M. Ballabh.
Bull. Astron. Soc. India, Vol. 3, 27 (1975). − Abstract of a paper presented at the A.S.I. meeting 1975.

073.045 **Propagation of solar flare particles from discrete events.** L. J. Lanzerotti.
VIIth Leningrad seminar. "Corpuscular streams of the sun and the radiation belts of the earth and Jupiter", 1975, (see 012. 007), p. 161 - 178 (1975). − Abstr. in Referativ. Zhurn. 51. Astron., 4.51.329; 62. Issled. kosmich. prostranstva, 4.62.208 (1976).

073.046 **Correlated observations of shock wave propagation from the solar flare of August 4, 1972 according to** radio burst of type II and solar wind measurements.
V. P. Grigor'eva, G. N. Zastenker, V. V. Temnyj.
VIIth Leningrad seminar. "Corpuscular streams of the sun and the radiation belts of the earth and Jupiter", 1975, (see 012. 007), p. 215 - 216 (1975). In Russian. − Abstr. in Referativ. Zhurn. 51. Astron., 4.51.334; 62. Issled. kosmich. prostranstva, 4.62.186 (1976).

073.047 **Brightness variations of chromospheric mottles in the neighbourhood of an active region from filter-**grams in the $H\alpha + 0.5$ Å line.
V. I. Polyakov, V. E. Merkulenko, V. I. Skomorovskij.
Issled. po geomagnetizmu, aehron. i fiz. Solntsa. Vyp. (No.) 37. Moskva, Nauka, 1975, p. 55 - 58. In Russian. − Abstr. in Referativ. Zhurn. 51. Astron., 4.51.337 (1976).

073.048 **X-ray and ultraviolet emission of solar flares.**
B. V. Somov.
VIIth Leningrad seminar. "Corpuscular streams of the sun and the radiation belts of the earth and Jupiter", 1975, (see 012. 007), p. 103 - 130 (1975). In Russian. − Abstr. in Referativ. Zhurn. 51. Astron., 4.51.347 (1976).

073.049 **Mechanisms of ^{3}He generation in solar flares.**
J. D. Anglin, J. A. Simpson.
VIIth Leningrad seminar. "Corpuscular streams of the sun and the radiation belts of the earth and Jupiter", 1975, (see 012. 007), p. 33 - 62 (1975). − Abstr. in Referativ. Zhurn. 51. Astron., 4.51.348; 62. Issled. kosmich. prostranstva, 4.62.200 (1976).

073.050 **S-structures of some flares and possible topology of an electric current in plane geometry of a bipolar** magnetic field. V. V. Kasinskij.
Issled. po geomagnetizmu, aehron. i fiz. Solntsa. Vyp. (No.) 37. Moskva, Nauka, 1975, p. 46 - 54. In Russian. − Abstr. in Referativ. Zhurn. 51. Astron., 4.51.361 (1976).

073.051 **The influence of instrumental polarization on mag-netographic measurements of magnetic fields in** prominences. V. S. Bashkirtsev.
Issled. po geomagnetizmu, aehron. i fiz. Solntsa. Vyp. (No.) 37. Moskva, Nauka, 1975, p. 72 - 81. In Russian. − Abstr. in Referativ. Zhurn. 51. Astron., 4.51.365 (1976).

073.052 **Variations of orientation of the magnetic field struc-ture and prominence brightness at the earlier stage** of its activation. G. Ya. Smol'kov.
Issled. po geomagnetizmu, aehron. i fiz. Solntsa. Vyp. (No.) 37. Moskva, Nauka, 1975, p. 59 - 64. In Russian. − Abstr. in Referativ. Zhurn. 51. Astron., 4.51.366 (1976).

073.053 **On the relation between the direction of the line separating polarities and the probability of forma-**tion of large solar flares. V. G. Banin.
Issled. po geomagnetizmu, aehron. i fiz. Solntsa. Vyp. (No.) 37. Moskva, Nauka, 1975, p. 32 - 36. In Russian. − Abstr. in Referativ. Zhurn. 51. Astron., 4.51.378 (1976).

073.054 **Estimate of the original parameters significance for forecasts of flocculus appearance from behind the** limb. T. L. Slutskaya, N. N. Stepanyan.
Solnechnye Dannye 1976 Byull., No. 1, p. 70 - 75 (1976). In Russian.
The problem concerning forecasts of flocculus appear-ance from behind the limb was solved by a method consisting in finding the optimal algorithm from some class of algorithms for estimation calculations and by Sebestian's method. The forecast proved to be true for 92% (the first method) and 88% (the second method). Both the methods permit to deter-mine the significance of the initial parameters for the forecast.

073.055 **On the forecast of appearance of flocculi born on the solar disk from behind the limb.** V. A. Burov.
Solnechnye Dannye 1976 Byull., No. 1, p. 75 - 79 (1976). In Russian.
The pattern recognition method (Benghardt's algorithm) is used for solving problems of forecasting the appearance of flocculi born on the solar disk from behind the limb. The func-tion "R" is introduced to characterize the quality of the solv-ing rule. In the first problem 15 parameters characterizing the flocculus during its passing the disk are used; here R = 0.16 (97% of cases were identified, of which 86% proved to be true). In the second problem 14 parameters were taken into con-sideration; here R = 0.19, which can be considered also satis-factory. The "hierarchy" of the parameters informativeness is as follows: flare activity, character of the sunspot group

development in the flocculus, flocculus area and presence of satellites.

073.056 Absolute spectrophotometry of the chromosphere in the region λλ 3824 - 3991 Å from photographs of the total solar eclipse on June 30, 1973.
V. M. Sobolev, G. F. Vyal'shin.
Solnechnye Dannye 1976 Byull., No. 1, p. 79 - 84 (1976).
In Russian.

The results of the spectrophotometric reduction of an eclipse spectrogram are presented. In a table total intensities and halfwidths are given in the region λλ 3824 - 3991 Å. Because of the high dispersion of the spectrograph many line profiles were obtained, that of He λ 3964.72 Å particularly. It is shown that the total intensities of some lines differ significantly for various tracings, however, this difference is negligible for most lines.

073.057 Research of the prominence of August 23, 1958.
I. Yu. Izotova, Yu. I. Izotov.
Solnechnye Dannye 1976 Byull., No. 1, p. 84 - 90 (1976).
In Russian.

The electron temperature, electron density, effective length, and number of hydrogen atoms in the ground quantum state along the line of sight were determined from the equations of statistical equilibrium for 10 levels and the continuum of the hydrogen atom. The abundances of Ca, Mg, Sr, Cr, Ti, Fe, Sc relative to hydrogen were calculated. The chemical composition of the prominence is shown to be similar to that of the photosphere and chromosphere.

073.058 The chromospheric structure before the proton flare on September 7, 1973 according to the films taken with a birefringent filter in the Hα and KCa⁺ lines.
V. I. Makarov, M. M. Molodenskij.
Solnechnye Dannye 1976 Byull., No. 1, p. 93 - 98 (1976).
In Russian.

The results of an analysis of the chromospheric structure before the 3b flare taken with a two-camera device are presented. The observations cover a period of 8^h before the flare and 2^h40^m after the flare beginning in the Hα and KCa⁺ lines. On the basis of the data on the photospheric magnetic field the conclusion is drawn that there is likely at least one zero-point in the flare region.

073.059 The distribution of neutral sodium density and the non-thermal motions velocity in the lower chromosphere from eclipse observations of the D_1 - D_2 resonance doublet.
G. K. Ajmanova, R. A. Gulyaev.
Astron. Zhurn. Akad. Nauk SSSR, Vol. 53, 353 - 360 (1976).
In Russian. English translation in Soviet Astron., Vol. 20, No. 2.

The Na I D lines intensity distribution in the undisturbed chromosphere is studied on the basis of slitless flash spectrograms taken during the total solar eclipse of July 10, 1972. The height distribution of the Na I atoms density is obtained with allowance for self-absorption. An analysis of the D lines intensities revealed that the mean velocity of non-thermal motions confines within 4 and 5 km/s at heights of 1300−1700 km, in full accordance with Unno's results.

073.060 The hydrogen radiation in gas dynamical models of solar flares.
N. D. Kostyuk.
Astron. Zhurn. Akad. Nauk SSSR, Vol. 53, 361 - 369 (1976).
In Russian. English translation in Soviet Astron., Vol. 20, No. 2.

The hydrogen radiation in Ly α, Ly β, Hα was obtained on the basis of simultaneous solutions of the equations of radiative transfer and statistical equilibrium for a model hydrogen atom including three levels and all corresponding continua. The values of density, temperature, ionization and velocity used in the calculations are the results of the solution of gas dynamical equations for the flare region heated by non-thermal

electrons. The values of parameters describing the physical conditions in the region of Hα excitation agree with observations.

073.061 The triggering and subsequent development of a solar flare.
J. A. Vorpahl.
Astrophys. Journ., Vol. 205, 868 - 873 (1976).

High temporal and spatial resolution solar X-ray pictures of a flare at 1827 UT on 1973 September 5 were taken. Observations suggest that the flare occurred in an entire arcade of loops rather than in any single loop. Sequential brightening of different X-ray features indicates that some excitation moved perpendicular to the magnetic field of the arcade at velocities of 180−280 km s⁻¹. The most intense X-ray features were located in places where the magnetic field composing the arcade had a small radius of curvature with horizontal field gradients higher than the surrounding region and where the axis of the arcade changed direction.

073.062 Spectroscopy from laser-produced plasmas at flare temperatures.
U. Feldman.
Astrophys. Space Sci., Vol. 41, 155 - 181 (1976). – Review paper given at the IAU Colloquium No. 27 on Ultraviolet and X-Ray Spectroscopy of Astrophysical and Laboratory Plasmas, held at Harvard University, 9−11 September, 1974.

A discussion of laboratory spectra similar in temperatures and ion abundances to solar flare spectra is given. The current state of knowledge regarding line identifications in laboratory spectra is reviewed, and some of the results are used to identify lines in the high temperature solar flare spectrum in the 100 Å region and in the 1000 Å region. In addition, the physical dimensions and temperatures of the hot regions in the plasmas produced by the low inductance spark were recently measured, and a short summary of the results is given.

073.063 Methods of investigation and reduction of radio emission of solar filaments.
G. P. Apushkinskij, A. N. Tsyganov, N. A. Topchilo.
Solnechnye Dannye 1976 Byull., No. 2, p. 56 - 62 (1976).
In Russian.

A method of determination of brightness temperatures in filaments from observations at millimetre wavelengths is described. Data on filaments observed with the PT-22 radio telescope are given. A method of restoration of the physical characteristics of filaments from the radio spectrum is suggested.

073.064 Absolute spectrophotometry of the chromospheric lines Hε, H8, H9 and He 3888.65 Å from photographs of the total solar eclipse on June 30, 1973.
V. M. Sobolev, G. F. Vyal'shin.
Solnechnye Dannye 1976 Byull., No. 2, p. 74 - 80 (1976).
In Russian.

The results of spectrophotometric reduction of the lines Hε, H8, H9 and He 3888.65 are given. Total intensities and half-widths of the profiles for various altitudes were obtained. Intensities of the lines relative to the intensity of H8 were calculated. A variation of this ratio for the line He 3888.65 with the altitude was obtained. Gradients for all the lines were determined.

073.065 On the magnetic field in surges on the solar disc.
Kh. I. Abdusamatov.
Solnechnye Dannye 1976 Byull., No. 2, p. 81 - 83 (1976).
In Russian.

The intensities of the magnetic field in some investigated knots of dark surges on the solar disc may reach 1000 ± 200 oersted. The possibility of existence of magnetic fields with intensities to 1000 oersted in some surges and prominences, and to 5000 oersted in separate small areas of some sunspots is discussed. Their existence may be explained as a result of vigorous twisting of the lines of force of the magnetic field and compression of a region of magnetized solar plasma caused by non-stationary processes or phenomena.

073.066 On the rope structure of the magnetic field in the quiet chromosphere. I. Condition for the equilibrium of straight twisting of a magnetic rope. V. E. Merkulenko. Solnechnye Dannye 1976 Byull., No. 2, p. 84 - 87 (1976). In Russian.

The equilibrium of a straight axisymmetric twisting magnetic rope is examined. It is shown that the rope can be in equilibrium, external forces being absent, only when the gradient of the toroidal component flux becomes zero.

073.067 On peculiarities of motion in surges. M. N. Stoyanova, I. Ferro. Solnechnye Dannye 1976 Byull., No. 2, p. 92 - 97 (1976). In Russian.

The distribution of radial velocities in an absorption surge and active prominence is analysed. The conclusion is made that a surge is likely to appear at a point of magnetic field variation and the picture of motion in the surge is complex.

073.068 Further observations of helium and hydrogen emission in quiescent prominences. D. A. Landman, R. M. E. Illing. Astron. Astrophys., Vol. 49, 277 - 283 (1976).

The authors present additional hydrogen and helium emission line intensity observations throughout six quiescent prominences. The data are analyzed with reference to recent theoretical and observational research. Several discrepancies are found to arise in this analysis, and specific suggestions for their resolution are advanced.

073.069 Acceleration of energetic particles in solar flares. K. Sakurai. 14th Intern. Cosmic Ray Conf., (see 012.011), Vol. 5, 1552 - 1556 (1975).

073.070 Source energy spectrum of protons accelerated in a high density medium. J. Pérez-Peraza, J. G. Trejo. 14th Intern. Cosmic Ray Conf., (see 012.011), Vol. 5, 1557 - 1562 (1975).

073.071 Observations of ≤ 1 MeV/nuc protons and ions during the September 1974 series of flares. F. M. Ipavich, G. Gloeckler, C. Y. Fan, D. Hovestadt. 14th Intern. Cosmic Ray Conf., (see 012.011), Vol. 5, 1568 - 1573 (1975).

073.072 Time dependences of the 0.51 and 2.2 MeV lines in solar flares. R. Ramaty, H. T. Wang. 14th Intern. Cosmic Ray Conf., (see 012.011), Vol. 5, 1635 - 1637 (1975).

073.073 Positron annihilation in solar flares. C. J. Crannell, R. Ramaty, C. Werntz. 14th Intern. Cosmic Ray Conf., (see 012.011), Vol. 5, 1656 - 1661 (1975).

073.074 Gamma-ray and radio evidence for the acceleration of high-energy particles in solar flares. T. Bai, R. Ramaty. 14th Intern. Cosmic Ray Conf., (see 012.011), Vol. 5, 1662 - 1667 (1975).

073.075 A search for solar flare positrons. R. A. Mewaldt, E. C. Stone, R. E. Vogt. 14th Intern. Cosmic Ray Conf., (see 012.011), Vol. 5, 1668 - 1673 (1975).

073.076 On some possibilities of forecast of solar geoeffective flares. S. I. Arbuzov, M. M. Kobrin, A. I. Korshunov, V. V. Pokhomov, N. A. Prokof'eva, V. M. Fridman. Fiz.-mat. i biol. probl. dejstviya ehlektromagnitn. polej i ionizatsii vozdukha. Tom 1. Moskva, Nauka, 1975, p. 83. In Russian. – Abstr. in Referativ. Zhurn. 51. Astron., 6. 51. 446 (1976).

073.077 On peculiarities of rapid brightness fluctuations of solar flocculi. M. I. Fisenko, V. F. Chistyakov. Solar phenomena research, Vladivostok, 1975, (see 003.005), p. 37 - 68. In Russian. – Abstr. in Referativ. Zhurn. 51. Astron., 6. 51. 447 (1976).

073.078 The radio-frequency radiation transfer in a solar filament. G. P. Apushkinskij, N. A. Topchilo. Astron. Zhurn. Akad. Nauk SSSR, Vol. 53, 572 - 576 (1976). In Russian. English translation in Soviet Astron., Vol. 20, No. 3.

The distribution of the electron temperature in a solar filament is calculated using radio data.

073.079 On the difference between the intensity of the chromospheric background on the polar and equatorial radii. C. J. Macris. Praktika Acad. Athens, Vol. 50, 214 - 227 = Res. Center Astron. Appl. Math., Acad. Athens, Contr. Ser. I (Astron.), No. 40 (1975).

073.080 Comparison of the dimensions of the flocculi at periods of maximum and minimum solar activity. H. C. Dara, C. J. Macris, T. G. Zachariadis. Praktika Acad. Athens, Vol. 50, 391 - 398 = Res. Center Astron. Appl. Math., Acad. Athens, Contr. Ser. I (Astron.), No. 42 (1975).

073.081 Gamma-ray and microwave evidence for two phases of acceleration in solar flares. T. Bai, R. Ramaty. GSFC Document X-660-76-58, Prepr., 29 pp. (1976).

Relativistic electrons in large solar flares produce gamma-ray continuum by bremsstrahlung and microwave emission by gyrosynchrotron radiation. Using observations of the 1972, August 4 flare, the authors evaluate in detail the electron spectrum and the physical properties (density, magnetic field, size, and temperature) of the common emitting region of these radiation. They also obtain information on energetic protons in this flare by using gamma-ray lines.

073.082 Der Einfluß interplanetarer Stoßwellen auf energetische solare Flareteilchen. G. Morfill, M. Scholer. MPI Kernphys., Heidelberg, Jahresbericht 1975, p. 172 - 173.

073.083 Magnetohydrodynamic interpretation of the motion of prominences. T. Sakurai. Publ. Astron. Soc. Japan, Vol. 28, 177 - 198 (1976).

The author studies three types of prominence eruptions, which he calls the arch type, the loop type, and the gigantic arch type, respectively, from their shapes. If we regard the prominence as a magnetic flux tube, the onset of its ascending motion can be interpreted as the motion due to the screw-mode instability, which is the most unstable mode of instabilities of a magneto-fluid column (pinch). The result shows that the characteristic motion of the arch-, loop-, and gigantic arch-type eruptions may be reproduced by perturbing a model sequence with decreasing pitch angles of the unperturbed helical magnetic field lines.

073.084 Studies on hard X-ray emission from solar flares and on cyclotron radiation from a cold magneto-

plasma.   P. Hoyng.
Utrechtse Sterrekundige Overdrukken No. 311, 160 pp. =
Proefschr. Wis. Nat., Rijksuniv. Utrecht, (1975).

073.085   The ascending prominence of September 4, 1959.
           M. Waldmeier.
Astron. Mitt. Eidgen. Sternw. Zürich No. 343, 9 pp. (1976).

073.086   Solar ^{3}He: information from nuclear reactions in
           flares.   R. Ramaty, B. Kozlovsky.
Separate print National Aeronaut. Space Administr., Greenbelt,
Maryland, USA, Goddard Space Flight Center, 32 pp. (1974).
See 14.073.087.

073.087   Solar flares and magnetic structure of the chromo-
           sphere.   S. Pikel'ner.
Veda. Tech. SSSR, Vol. 3, No. 3, p. 21 - 26 (1975). In Czech.

073.088   Analysis of OGO-5 and OSO-7 X-ray data. Final
           report.   R. L. Moore.
Separate print California Inst. Tech., Pasadena, USA, 8 pp.
(1975).

073.089   Origin and implications of gamma rays from solar
           flares.   R. Ramaty.
Separate print National Aeronaut. Space Administr., Greenbelt,
Maryland, USA, Goddard Space Flight Center, 11 pp. (1975).

073.090   Relationship between solar flares and solar sector
           boundaries.   P. H. Dittmer.
Separate print Stanford Univ., California, USA, Inst. Plasma
Res., 11 pp. (1974). – See 13.073.023.

073.091   Spectral analysis of solar spicules.   K. R. Krall.
           Thesis Wyoming Univ., Laramie, USA, 154 pp.
(1975). (Available from: Univ. Microfilms, Order No. 75-
22,332).

073.092   Radio- and X-emission (and their polarization) at the
           time of early phase of solar flare connected with
γ-line emissions.   L. Krivsky.
7th Czech. seminar plasma phys. technol. Liblice, Czechoslovakia, 24 April 1974, p. 25 - 30 (1974).

The solar chromosphere and corona: Quiet sun.
See Abstr. 003.016.

Dynamical processes in solar flares.
See Abstr. 003.052.

Radiative hydrodynamics of chromospheric transients.   See Abstr. 063.019.

Solar limb brightening in submillimeter wavelengths.
See Abstr. 071.003.

Noen resultater av soleksperimentene med ATM,
Skylab.   See Abstr. 071.007.

The center-to-limb variation of optically thin UV-
lines and at centimetric wavelengths and the influence of
spicules.   See Abstr. 071.011.

Infrared observations of supergranule temperature
structure.   See Abstr. 071.015.

Doppler wavelength shifts of transition zone lines
measured in Skylab solar spectra.   See Abstr. 071.028.

Observations of penumbral waves in the photosphere.   See Abstr. 071.032.

Diffusion in the solar chromosphere-corona transition region.   See Abstr. 074.008.

The sources of material comprising a mass ejection
coronal transient.   See Abstr. 074.017.

Disturbances of the solar corona and of the interplanetary medium accompanying the activation and disappearance of a prominence.   See Abstr. 074.033.

A model of the solar transition region and corona
for the minimum of activity.   See Abstr. 074.035.

Prominences in 1974.   See Abstr. 075.003.

The solar XUV spectrum of He II.
See Abstr. 076.001.

Spatial structure and temporal development of a
solar X-ray flare observed from Skylab on June 15, 1973.
See Abstr. 076.009.

Contribution of electron-electron bremsstrahlung to
solar hard X-radiation during flares.   See Abstr. 076.010.

The solar subflare X-ray spectrum.
See Abstr. 076.014.

Solar hard X-ray bursts and accelerated electrons in
solar flares.   See Abstr. 076.017.

On model representations of a source of flare X rays.
See Abstr. 076.018.

Investigation of hard X-ray radiation of solar flares
aboard the artificial earth satellites of the Prognoz series.
See Abstr. 076.019.

Time history and model calculations of the 2.2 MeV
gamma ray line from the flares of August, 1972.
See Abstr. 076.021.

A hard X ray observation of the 4 July 1974 solar
flare.   See Abstr. 076.023.

X-ray bursts from solar flares behind the limb.
See Abstr. 076.025.

Slow X-ray bursts and chromospheric flares with
filament disruption.   See Abstr. 076.026.

Solar X-ray studies.   See Abstr. 076.027.

OGO-V radio burst analysis.
See Abstr. 077.047.

On the anisotropy of relativistic solar protons from
the flare of September 1, 1971.   See Abstr. 078.008.

On the nuclear reactions during solar flares.
See Abstr. 078.060.

Late-time distribution of solar-flare cosmic rays –
its dependence on the functional form of the diffusion coefficient.   See Abstr. 078.070.

A model for the propagation of solar-flare cosmic
rays.   See Abstr. 078.071.

Propagation characteristics of solar flare particles.
See Abstr. 078.076.

On cosmic ray flares.   See Abstr. 078.078.

Low energy particle composition.
See Abstr. 078.092.

Variation of solar proton energy spectra and size distribution with heliolongitude.   See Abstr. 078.105.

Observations of hydrogen and helium isotopes in solar cosmic rays.   See Abstr. 078.106.

Solar brightness temperature distribution at 350 and 450 microns.   See Abstr. 080.001.

Unusual sodium nightglow behaviour and possible association with solar flares.   See Abstr. 082.003.

The electric potential of the ionosphere as controlled by the solar magnetic sector structure.
See Abstr. 083.032.

Geomagnetic crochets associated with proton flares.
See Abstr. 084.285.

Geomagnetic effects in the dark hemisphere associated with solar flares.   See Abstr. 084.287.

Night time geomagnetic effects of solar flares.
See Abstr. 084.292.

Geomagnetic solar flare effect in the dark hemisphere.   See Abstr. 084.293.

Sudden f_{min} enhancements and sudden cosmic noise absorptions associated with solar X-ray flares.
See Abstr. 085.023.

Studies of solar flares and impact craters in partially protected crystals.   See Abstr. 094.562.

Solar-wind sputter erosion of microcrater populations on the lunar surface.   See Abstr. 094.563.

Track studies bearing on solar-system regoliths.
See Abstr. 094.567.

Solar-wind and solar-flare maturation of the lunar regolith.   See Abstr. 094.568.

Compositional studies of solar-iron group and heavier nuclei of kinetic energy below 10 MeV/nucleon.
See Abstr. 094.571.

Long-term differential energy spectrum for solar-flare iron-group particles.   See Abstr. 094.572.

A shock surface geometry: the February 15–16, 1967, event.   See Abstr. 106.006.

Shock waves from the sun.   See Abstr. 106.113.

Cosmic ray modulation by corotating solar streams in presence of flare plasma clouds.   See Abstr. 143.218.

Modulation and diffusion theory of cosmic rays.
See Abstr. 143.312.

074 Solar Corona, Solar Wind

074.001 A two-region model of the solar wind including azimuthal velocity. M. H. Acuna, Y. C. Whang.
Astrophys. Journ., Vol. 203, 720 - 738 (1976).

The two-region model of the solar wind divides the interplanetary space into two regions: it assumes that the solar wind is one-fluid in an inner region $r < 0.4$ AU and two-fluid in an outer region $r > 0.4$ AU. The second and third moment equations of the Vlasov equation together with other conservation equations are used to describe the solar-wind flow in the two-fluid region. The predicted azimuthal velocity at 1 AU is less than 2 km s^{-1}. The numerical results confirm that when the azimuthal velocity is included in the analysis, the amount of magnetic-field energy converted into kinetic energy in the solar wind is only a small fraction of the total expansion energy flux and has little effect upon the final radial expansion velocity.

074.002 The relative abundance of neon and magnesium in the solar corona. H. R. Rugge, A. B. C. Walker, Jr.
Astrophys. Journ., (Letters), Vol. 203, L139 - L143 (1976).

A relationship between the calculated emission functions for Ne X and Mg XI is discussed and used for determining the relative solar coronal abundance of neon and magnesium. Moderate resolution Bragg crystal spectrometer results from the OV1−10 satellite are used to determine a coronal neon to magnesium relative abundance of 1.47 ± 0.38. The application of this technique to a recent higher resolution rocket observation gives an abundance ratio of ∼0.93 ± 0.15.

074.003 Stochastic scattering and other contributions to the sun's wake in the local hydrogen cloud.
M. K. Wallis.
Celestial Mechanics, Vol. 13, 65 - 74 (1976).

Gas streaming through the solar system experiences both destructive and scattering processes, the latter primarily in collisional interactions with the solar wind protons. The scattering interactions can be important in filling the downstream wake. They may effectively increase the velocity dispersion and also cause discrete orbit changes. The downstream intensity moment is here evaluated analytically for particles suffering a single, discrete collision, and compared with the moment from a thermal velocity dispersion.

074.004 Synchrotron emission from the active K-corona.
V. P. Vasil'ev, V. I. Kucherov.
Solnechnye Dannye 1975 Byull., No. 10, p. 85 - 87 (1976). In Russian.

An attempt is made to interpret the inhomogeneities of light, characteristic for coronal condensation, by means of interaction of high-energy protons with the cores of a magnetic trap target.

074.005 A semi-empirical model of the solar wind.
V. A. Kovalenko, N. P. Korzhov.
Astron. Zhurn. Akad. Nauk SSSR, Vol. 53, 148 - 153 (1976). In Russian. English translation in Soviet Astron., Vol. 20, No. 1.

Within the framework of hydrodynamic approximation, using the experimental data of density distribution at various heliocentric distances, a semi-empirical solar wind model has been developed, in which the basic parameters of the solar wind agree with the data of measurements for all regions of the interplanetary space.

074.006 Polarization of the solar corona on June 30, 1973.
S. A. Jacoub, M. Yu. Nevskij, V. V. Leushin, L. N. Lebedeva.
Astron. Zhurn. Akad. Nauk SSSR, Vol. 53, 154 - 161 (1976).

In Russian. English translation in Soviet Astron., Vol. 20, No. 1.

The distribution of polarization at the distance of 2.5 $R_\odot$ from the edge of the solar disk is found from a photometric study of polarimetric photographs of the corona obtained on June 30, 1973. The scattered light is excluded from the intensity of the solar corona. The degree of polarization for polar regions averaged over position angles has a maximum of 41 % at 0.3 $R_\odot$, for equatorial regions a maximum of 45 % at 0.6 $R_\odot$.

074.007 Quasi-exospheric heat flux of solar-wind electrons.
A. Eviatar, M. Schulz.
Astrophys. Space Sci., Vol. 39, 65 - 73 (1976).

Density, bulk-velocity, and heat-flow moments are calculated for truncated Maxwellian distributions representing the cool and hot populations of solar-wind electrons, as realized at the base of a hypothetical exosphere. The electrostatic potential is thus calculated by requiring charge quasi-neutrality and the absence of electrical current. Plasma-kinetic coupling of the cool-electron and proton bulk velocities leads to an increase in the electrostatic potential and a decrease in the heat-flow moment. If the velocities differ by the Alfvén speed along the magnetic field, for example, the potential rises to 72.6 V and the heat flux falls to 2.72×10^{-2} erg cm^{-2} s^{-1}. In each case the heat flux is carried mainly by the quasi-exospheric hot electrons.

074.008 Diffusion in the solar chromosphere-corona transition region. A. S. Tworkowski.
Astrophys. Letters, Vol. 17, 27 - 30 (1976).

The author examines effects of radial diffusion on local ionization equilibrium in the solar chromosphere-corona transition region. Concentration, thermal, pressure and force diffusion included in the system. The author gives a transition region temperature-density model, which accounts for diffusion. He finds that ions are pushed by diffusion to higher temperatures, although after $10^{6\,\circ}$K some return to their local ionization equilibrium temperature positions. The element-to-proton abundance density ratio varies with height.

074.009 A possible large-scale meridional structure in the helioequatorial solar wind.
B. M. Vladimirsky (Vladimirskij), L. S. Levitsky (Levitskij).
Nature, Vol. 260, 27 - 28 (1976).

Over the past few years observational data for a large-scale meridional structure of the solar wind have been accumulated. Space-probe measurements have revealed heliolatitudinal effects at 1 AU, but little is known about the wind flow pattern at small radial distances, just outside the corona. Here the authors report the existence of a solar equatorial belt, from which few protons are emitted, and discuss its causes.

074.010 October 1972 solar event: the third dimension in solar particle propagation.
V. Domingo, D. E. Page, K.-P. Wenzel.
Journ. Geophys. Res., Vol. 81, 43 - 50 (1976).

From late on October 29 until November 3, 1972, the authors' experiment on the European Space Research Organization satellite Heos 2 recorded the arrival of an enhanced interplanetary particle intensity. A dramatic 'slot' in count rate and other sudden anisotropy and flux changes (measured in and perpendicular to the ecliptic plane) were found to coincide with changes in the θ (north-south) ecliptic direction of the interplanetary magnetic field. It appears that when the solar wind blows three-dimensional snakelike tubes in interplanetary space, MeV particles obediently follow the field line bundles within such tubes and experience considerable difficulty in crossing from one tube to a neighboring tube which encloses a different regime.

074.011 Microscale 'Alfvén waves' in the solar wind at 1 AU.
L. F. Burlaga, J. M. Turner.
Journ. Geophys. Res., Vol. 81, 73 - 77 (1976).

Analysis of Imp 1 (Explorer 43) interplanetary plasma and magnetic field fluctuations on a scale of 1 hour for the period March 18 to April 7, 1971, reveals that linearly and circularly polarized Alfvén waves were rarely present. Fluctuations having most of the characteristics of large amplitude ($\delta B_x/\bar{B} \approx 0.4$) 'Alfvén waves' were observed $\approx 40\%$ of the time and moved away from the sun nearly along $\mathbf{B}$. These were not pure transverse Alfvén waves, however, because they were accompanied by nonzero fluctuations in the magnetic field intensity ($\delta B/\bar{B} \approx 0.06$). No simple relation between 'Alfvén waves' and streams was found.

074.012 The role of Coulomb collisions in limiting differential flow and temperature differences in the solar wind. M. Neugebauer.
Journ. Geophys. Res., Vol. 81, 78 - 82 (1976).

Analysis of data obtained by Ogo 5 confirms the Imp 6 observations of the inverse dependence of the helium to hydrogen temperature ratio T_{He}/T_H on τ_e/τ_c, the ratio of the solar wind expansion time to the Coulomb collision equipartition time.

074.013 On the causes of spectral enhancements in solar wind power spectra. T. Unti, C. T. Russell.
Journ. Geophys. Res., Vol. 81, 469 - 482 (1976).

Enhancements in power spectra of solar wind ion flux in the frequency neighborhood of 0.5 Hz had been noted by Unti et al. (1973). It was speculated that these were due to convected small-scale density irregularities. In this paper, the authors examine 54 flux spectra, calculated from Ogo 5 data. It is seen that the few prominent spectral peaks which occur were not generated by density irregularities but were due to several different causes, including convected discontinuities and propagating transverse waves.

074.014 On the heliographic latitude dependence of the solar wind velocity. L. Diodato, G. Moreno.
Astrophys. Space Sci., Vol. 39, 409 - 414 (1976).

Plasma data from Pioneers 6−7 and from a variety of satellites operating near the earth are used to investigate the heliographic latitude dependence of the solar wind bulk speed near the sunspot maximum. No evidence is found for a latitude effect: the latitudinal gradient, if any, turns out to be $\lesssim 2$ km (sec degree)$^{-1}$, to be compared with the gradient of ~ 10 km (sec degree)$^{-1}$ observed in periods of low or moderate solar activity.

074.015 Erste Messungen von "Coronal Holes" bei cm-Wellen.
W. Hirth, E. Fürst.
Mitt. Astron. Ges., No. 38, p. 208 - 211 (1976). − Short report.

074.016 H and K (Ca II) emissions as observed in coronal spectrum in the July 20, 1963 solar eclipse.
F. Cavallini, A. Righini.
Solar Physics, Vol. 45, 291 - 299 (1975).

From a detailed analysis of a coronal spectrum taken from a DC-8 jet airplane during the eclipse of 20 July, 1963 a rough model of a coronal cold region ($T \approx 10^5$ K) has been obtained. The model explains the presence of the abnormal H and K (Ca II) emissions and the large amount of F corona present in the spectrum.

074.017 The sources of material comprising a mass ejection coronal transient.
E. Hildner, J. T. Gosling, R. T. Hansen, J. D. Bohlin.

Solar Physics, Vol. 45, 363 - 376 (1975).

A disturbance on 26 and 27 August 1973, during which a slowly ascending prominence and a more rapid accompanying coronal transient were simultaneously observed is studied. Prominence images obtained in Hα 6563 Å and in He II 304 Å are nearly identical. The mass ejection transient observed in white light (3700−7000 Å) appeared to be a loop about 1 $R_\odot$ higher than the top of the ascending prominence; it accelerated away from the prominence below it. The observations imply: (1) the bulk of the ejected material did not originate in the ascending prominence; (2) therefore, most of the material must have come from the low corona above the prominence, (and was at coronal temperatures during its outward passage); and (3) the total event − ascending prominence accompanied by coronal mass ejection − was far larger, more energetic, and longer lasting than would be inferred from the prominence observations alone.

074.018 Coronal changes associated with a disappearing filament. N. R. Sheeley, Jr., J. D. Bohlin, G. E. Brueckner, J. D. Purcell, V. E. Scherrer, R. Tousey, J. B. Smith, Jr., D. M. Speich, E. Tandberg-Hanssen, R. M. Wilson, A. C. de Loach, R. B. Hoover, J. P. McGuire.
Solar Physics, Vol. 45, 377 - 392 (1975).

The authors describe the disappearing filament event of January 18, 1974 using XUV spectroheliograms, X-ray filterheliograms, Hα spectroheliograms, and photospheric magnetograms. Subsequent study has produced the unexpected bonus that this event is apparently not isolated, but instead is characteristic of a wide class of long-duration X-ray and microwave events. Now it seems that the Skylab/ATM observations of this and other long-duration events may help to clarify our understanding of flares as well as coronal transients, as will be discussed within the text.

074.019 The solar corona as observed during the 30 June 1973 solar eclipse on board Concorde 001.
P. Léna, D. Hall, A. Soufflot, Y. Viala.
Space Research XV, (see 012.003), p. 579 - 583 (1975).

Observations of the F-corona of the sun have been made in the 10 μm region. Bright features previously observed at shorter wavelengths, notably emission at 4 solar radii are evident on the 10 μm scans, strongly indicating that the radiation is due to thermal emission by dust. The specific intensity in the 4 $R_\odot$ feature is 6.9×10^{-6} W cm^{-2} ster^{-1} μ^{-1} higher than the intensity 22 arc minutes above the ecliptic. Spectra were taken at 3 $R_\odot$ near the ecliptic.

074.020 Contribution of the non linear Landau damping of Alfvén waves to the heating of the solar wind protons.
C. Lacombe.
Astron. Astrophys., Vol. 48, 11 - 17 (1976).

The author gives an estimate of the proton heating by non linear Landau damping of Alfvén waves in a plasma where the Alfvén velocity is much larger than the sound velocity. This non linear heating operates only when backward and forward (with respect to the direction of the general magnetic field B_0) Alfvén waves are both intense. This process can account for the solar wind data at 1 AU.

074.021 Solar wind interaction with the tail of comet Kohoutek. A. I. Ershkovich.
Planet. Space Sci., Vol. 24, 287 - 291 (1976).

Helical waves of large amplitude observed recently in the tail of comet Kohoutek are interpreted as stable waves arising due to non-linear evolution of Kelvin−Helmholtz instability. The dispersion equation for waves of a finite amplitude shows that the phase velocity of these waves should approximately coincide with the velocity of the plasma outflow in the tail rather than with the Alfvén velocity. This fact is shown to be in agreement with observations.

074.022 Solar-wind tritium limit and the mixing rate of the solar atmosphere.   E. L. Fireman.
Astrophys. Journ., Vol. 205, 268 - 272 (1976).

Tritium has been measured in Surveyor 3 samples, some of which were adjacent to those in which solar-wind-implanted ^{4}He had previously been measured. Little of the ^{3}H can be attributed to solar-wind implantation. The upper limit for the ^{3}H/^{4}He ratio in the solar wind is 4×10^{-10} and corresponds to a ^{3}H/^{1}H limit of 2×10^{-11}. This limit imposes a requirement on the mixing rate in the solar atmosphere if the ^{3}H production rate in solar-surface nuclear reactions is greater than 160 cm^{-2} s^{-1}.

074.023 Proton temperature anisotropy instabilities in the solar wind.   S. P. Gary, M. D. Montgomery, W. C. Feldman, D. W. Forslund.
Journ. Geophys. Res., Vol. 81, 1241 - 1246 (1976).

The linear dispersion properties of proton temperature anisotropy instabilities in a homogeneous infinite Vlasov plasma are studied by using a configuration appropriate to the solar wind at 1 AU. The proton distribution is taken to consist of two components, a cooler $T_\perp > T_\parallel$ 'core' and a hotter $T_\parallel > T_\perp$ 'halo'. For the parameters considered the $\mathbf{k} \parallel \mathbf{B}_0$ fire hose and ion cyclotron instabilities are the most important modes. Resonant proton effects enhance both instabilities, and the presence of the cooler component can substantially reduce the threshold anisotropy of the halo-driven fire hose.

074.024 Solar wind structure at large heliocentric distances: an interpretation of Pioneer 10 observations.
A. J. Hundhausen, J. T. Gosling.
Journ. Geophys. Res., Vol. 81, 1436 - 1440 (1976).

Examination of hourly values of the solar wind speed observed by the Pioneer 10 spacecraft beyond a heliocentric distance of 4 AU reveals (1) a prevalent 'sawtoothlike' speed-time profile, most speed fluctuations displaying a rapid rise and a much slower decline, and (2) the nearly universal appearance of abrupt (on the 1-hour time resolution of these data) changes in the speed on the rising portions of the speed fluctuations. The Pioneer 10 observations give the first confirmation of the general concept of solar wind stream evolution employed in these models, i.e., that solar wind speed inhomogeneities appear to steepen to form shock waves and that the 'wave amplitudes' decay slowly as the shock waves propagate outward from the sun.

074.025 Nonstationary flow resulting from field line reconnection following a major coronal transient.
R. Kopp, G. W. Pneuman.
Bull. American Astron. Soc., Vol. 8, 293 (1976). – Abstr. AAS.

074.026 Viscous interaction between the solar wind and cometary plasmas.
H. Pérez de Tejada, M. Dryer.
Bull. American Astron. Soc., Vol. 8, 295 (1976). – Abstr. AAS.

074.027 Latitudinal structure of the solar wind and interplanetary magnetic field.
M. Dobrowolny, G. Moreno.
Space Sci. Rev., Vol. 18, 685 - 748 (1976).

A review is given of both observational and theoretical results concerning the latitudinal structure of the solar wind and interplanetary magnetic field. Observations are reported on the solar wind plasma and magnetic fields. Results of theoretical work, both on three-dimensional modelling of the solar wind and on gas-magnetic field interactions in the solar corona are summarized. Finally, an attempt is made to compare available observations and theories.

074.028 Comment on the paper 'Solution of one-fluid model equations with short range retarding magnetic forces

for the quiet solar wind' by Cuperman and Harten.
A. Barnes.
Astrophys. Space Sci., Vol. 40, 35 - 38 (1976). – Concerning 11.074.021.

074.029 Further discussion of the electron coronal densities.   S. Cuperman, A. Harten.
Astrophys. Space Sci., Vol. 40, 111 - 113 (1976).

074.030 On electron acceleration on the boundary of the solar wind with the interstellar medium.
M. F. Bakhareva.
Geomagn. Aeronom., Vol. 16, 7 - 14 (1976). In Russian.

074.031 On a connection of variations of solar wind velocity with the magnetic field in the magnetosphere's tail.
S. A. Mart'yanov.
Geomagn. Aeronom., Vol. 16, 193 - 194 (1976). In Russian. Brief information.

074.032 On the influence of large-scale inhomogeneities upon the low-frequency wave excitation in the solar wind.   V. V. Lebedev.
Issled. po geomagnetizmu, aehron. i fiz. Solntsa. Vyp. (No.) 37. Moskva, Nauka, 1975, p. 106 - 108. In Russian. – Abstr. in Referativ. Zhurn. 51. Astron., 4.51.317; 62. Issled. kosmich. prostranstva, 4.62.183 (1976).

074.033 Disturbances of the solar corona and of the interplanetary medium accompanying the activation and disappearance of a prominence.   G. Ya. Smol'kov.
Issled. po geomagnetizmu, aehron. i fiz. Solntsa. Vyp. (No.) 37. Moskva, Nauka, 1975, p. 65 - 71. In Russian. – Abstr. in Referativ. Zhurn. 51. Astron., 4.51.370 (1976).

074.034 On a mechanism of fast dissipation of a magnetic field in collision-free plasma.
S. I. Vajnshtejn, V. M. Tomozov.
Issled. po geomagnetizmu, aehron. i fiz. Solntsa. Vyp. (No.) 37. Moskva, Nauka, 1975, p. 100 - 105. In Russian. – Abstr. in Referativ. Zhurn. 62. Issled. kosmich. prostranstva, 4.62. 181 (1976).

074.035 A model of the solar transition region and corona for the minimum of activity.
F. Chiuderi Drago, G. Noci.
Astron. Astrophys., Vol. 48, 367 - 371 (1976).

The hypotheses of constant pressure and conductive flux in the transition region, and of constant temperature and hydrostatic equilibrium in the corona are made to allow the use of simple formulae to derive the brightness temperature, T_b. Contributions to T_b from transition region and corona are compared. Three of the very few available observations in the relevant frequency range are used to deduce a model of these two regions for solar minimum, which turns out to be quite similar to UV and X models of coronal holes.

074.036 On the nonexistence of plane-polarized large amplitude Alfvén waves.   A. Barnes.
Journ. Geophys. Res., Vol. 81, 281 - 282 (1976).

It is shown that purely Alfvénic plane-polarized large amplitude disturbances cannot exist, even in nonplanar fluctuations.

074.037 Long-lived coronal structures and recurrent geomagnetic patterns in 1974.
R. T. Hansen, S. F. Hansen, C. Sawyer.
Planet. Space Sci., Vol. 24, 381 - 388 (1976).

Daily measurements of the intensity distribution of the sun's white-light corona over the height range $1.1 - 2.7\,R_\odot$ show that the global structure became quite stable in late 1973 and throughout 1974, as flares, ascending prominences and other transient activity became less frequent with the decline of the solar activity cycle. A highly persistent pattern of geomagnetic activity prevailed for much of this time. Bright coronal structures in the ecliptic plane were associated with geomagnetically quiet conditions, and faint coronal regions ("holes") with geomagnetic disturbance, after a delay of about three days.

074.038 The spectrum of Fe XVII in the solar corona.
R. J. Hutcheon, J. P. Pye, K. D. Evans.
Monthly Notices Roy. Astron. Soc., Vol. 175, 489 - 499 (1976).

The wavelengths and intensities of $n = 3 \rightarrow 2$ and $n = 4 \rightarrow 2$ transitions in Fe XVII are reduced from measurements of the X-ray spectra of three coronal active regions. New wavelength measurements are given with an uncertainty of ± 0.003 Å. A feature near 11.0 Å in the measured spectra is tentatively identified with a blend of two previously unknown inner shell transitions.

074.039 High-speed fluxes of solar wind plasma and active regions on the sun.
M. A. Zel'dovich, B. M. Kuzhevskij.
Kosmich. Issled., Vol. 14, 248 - 256 (1976). In Russian.

074.040 Characteristics of the proton and α-components of the solar wind after the passage of interplanetary shock waves from observations aboard the satellite Prognoz on May 15 and 30, 1972.
A. A. Zertsalov, O. L. Vajsberg, V. V. Temnyj.
Kosmich Issled., Vol. 14, 257 - 264 (1976). In Russian.

074.041 Coronal holes observed by OSO-7 and interplanetary magnetic sector structure. W. J. Wagner.
Astrophys. Journ., Vol. 206, 583 - 588 (1976).

Evidence is presented which suggests that from 1972 May through 1973 October the interplanetary magnetic field structure at earth orbit was correlated with the distribution of high-latitude $(40°–80°)$ coronal holes as observed in $\lambda 284$ of Fe XV from OSO-7. The sense of this relation is consistent with a picture of the holes being magnetically open to the interplanetary medium with "toward" field polarity near the south polar cap and "away" at the north.

074.042 Depletion of solar wind plasma near a planetary boundary. B. J. Zwan, R. A. Wolf.
Journ. Geophys. Res., Vol. 81, 1636 - 1648 (1976).

A mathematical model is presented that describes the squeezing of solar wind plasma out along interplanetary magnetic field lines in the region between the bow shock and the effective planetary boundary (in the case of the earth, the magnetopause). Scaling of the model calculations to Venus and Mars suggests layer thicknesses about $^1/_{10}$ and $^1/_{15}$ those of the earth, respectively, neglecting diffusion and ionospheric effects.

074.043 Collisionless electron heat conduction in the solar wind. J. V. Hollweg.
Journ. Geophys. Res., Vol. 81, 1649 - 1658 (1976).

The point of view that heat-conduction-driven plasma instabilities may not be capable of directly modifying the electron heat conduction flux in the solar wind is explored. The author presents a brief summary of other suggested descriptions of the electron heat conduction flux; and he discusses the physical basis of the collisionless electron heat conduction flux. He presents the model calculations and summarizes and discusses his conclusions.

074.044 Alfvénic acceleration of solar wind helium. 2. Model calculations. S. C. Chang, J. V. Hollweg.

Journ. Geophys. Res., Vol. 81, 1659 - 1663 (1976).

A previous paper (Hollweg, 1974) discussed a new physical mechanism for accelerating α particles in the solar wind, via their interaction with Alfvén waves. This paper presents numerical calculations of a simple three-fluid solar wind model which incorporates the new physical process.

074.045 The transition at a period $T \sim 1$ day in solar wind spectra. N. D'Angelo.
Journ. Geophys. Res., Vol. 81, 1779 - 1781 (1976).

The sharp change in behavior at periods above and below approximately 1 day in the solar wind spectra, reported by Goldstein and Siscoe, is discussed in terms of a damping-growth transition for ion acoustic waves in a density gradient recently observed in laboratory experiments.

074.046 Behavior of low-energy protons and alpha particles during a disturbed time period. F. M. Ipavich, G. Gloeckler, C. Y. Fan, D. Hovestadt.
Journ. Geophys. Res., Vol. 81, 1794 - 1798 (1976).
Brief report.

074.047 Some aspects of solar wind interaction with the interstellar medium.
R. B. Salimzibarov, Yu. G. Shafer.
Cosmic rays, No. 15, (see 003.006), p. 72 - 74 (1975). In Russian.

074.048 Secondary interactions of shock waves and their effect on the development of a geomagnetic storm with sudden commencement. S. A. Grib.
Cosmic rays, No. 15, (see 003.006), p. 103 - 108 (1975). In Russian.

074.049 Solar-wind tritium limit and nuclear processes in the solar atmosphere.
E. L. Fireman, J. D'Amico, J. DeFelice.
Proc. Sixth Lunar Sci. Conference, (see 012.010), Vol. 2, 1811 - 1821 (1975).

074.050 Distributions of tangential discontinuity in the corotating solar wind structure.
O. Saka, T.-i. Kitamura.
Planet. Space Sci., Vol. 24, 621 - 628 (1976).

Distributions of the tangential discontinuity (TD) in the solar wind sector structure are investigated on the basis of the magnetic field data and the ion plasma parameters from the Explorer 33 satellite. It is found that the TD is formed by the thin layered field-aligned currents (the current sheets), and that the TD is predominantly built up in the leading edge of the solar wind where the compression of the plasma and the magnetic field takes place.

074.051 Solar wind structure in the vicinity of the equatorial plane. A. I. Kuzmin, G. F. Krymsky (*Krymskij*), P. A. Krivoshapkin, V. P. Mamrukova, G. V. Skripin.
14th Intern. Cosmic Ray Conf., (see 012.011), Vol. 3, 1025 - 1028 (1975).

074.052 Measurements of large scale turbulence in the solar system. D. S. Intriligator.
14th Intern. Cosmic Ray Conf., (see 012.011), Vol. 3, 1029 - 1032 (1975).

074.053 The solar cycle variation in the solar wind and the modulation of cosmic rays. D. S. Intriligator.
14th Intern. Cosmic Ray Conf., (see 012.011), Vol. 3, 1033 - 1035 (1975).

074.054 Solar wind observations associated with the August 1972 solar events. D. S. Intriligator.

14th Intern. Cosmic Ray Conf., (see 012.011), Vol. 3, 1094 (1975). – Abstract.

074.055 **Relation of large-scale coronal X-ray structure and cosmic rays: 4. Amplitude of the diurnal variation in neutron monitors on interplanetary field lines orbiting above coronal holes.** E. C. Roelof, R. E. Gold, A. S. Krieger, J. T. Nolte, D. Venkatesan.
14th Intern. Cosmic Ray Conf., (see 012.011), Vol. 4, 1138 - 1143 (1975).

074.056 **The solar wind between 0.7 AU and 5.0 AU.** D. S. Intriligator.
14th Intern. Cosmic Ray Conf., (see 012.011), Vol. 4, 1534 - 1539 (1975).

074.057 **Direct observations of the solar wind interaction with Jupiter.** D. S. Intriligator.
14th Intern. Cosmic Ray Conf., (see 012.011), Vol. 4, 1540 - 1545 (1975).

074.058 **Relation of large-scale coronal X-ray structure and cosmic rays: II. Coronal control of interplanetary injection of 300 keV protons.**
E. C. Roelof, R. E. Gold, S. M. Krimigis, A. S. Krieger, J. T. Nolte, P. S. McIntosh, A. J. Lazarus, J. D. Sullivan.
14th Intern. Cosmic Ray Conf., (see 012.011), Vol. 5, 1704 - 1709 (1975).

074.059 **Recent perspectives in solar physics: elemental composition, coronal structure and magnetic fields, solar activity.** G. Newkirk, Jr.
14th Intern. Cosmic Ray Conf., (see 012.011), Vol. 11, 3594 - 3635 (1975).

074.060 **On the model of the solar wind – interstellar medium interaction with two shock waves.**
V. B. Baranov, K. V. Krasnobaev, M. S. Ruderman.
Astrophys. Space Sci., Vol. 41, 481 - 490, 491 - 501 (1976).
In English and Russian.
The model of the solar wind interaction with the interstellar medium suggested by Baranov et al. (1970) is developed. In this model the presence of two shock waves is assumed, through which the solar wind and interstellar gas pass, the latter moving relative to the sun at supersonic speed (20 km s⁻¹). The distribution of the gas parameters in the region between shocks is calculated which, in particular, allows to estimate the possibility of its experimental detection, observing radio-scintillation on interstellar irregularities. The possible influence on the model of neutral atoms of galactic hydrogen penetrating into the interplanetary medium is estimated.

074.061 **Expansion of coronal magnetic fields in the solar wind.** W. J. Weber.
Computing in plasma physics and astrophysics, (see 012.014), D6, 2 pp. (1976).

074.062 **Some results of measurements of the α-component of the solar wind aboard the satellite Prognoz.**
A. A. Zertsalov, Zh. M. Boske, K. Dyuston, A. N. Omel'chenko, O. L. Vajsberg.
Kosmich. Issled., Vol. 14, 463 - 466 (1976). In Russian.
Brief information.

074.063 **Summary of observations of the solar corona/inner zodiacal light from Apollo 15, 16, and 17.**
C. L. Ross.
IAU Colloquium No. 31, (see 012.015), p. 64 (1976).

074.064 **A temporal study of the radiance of the F-corona**

close to the sun. R. H. Munro.
IAU Colloquium No. 31, (see 012.015), p. 65 (1976).

074.065 **Measurements of the F-corona from daily OSO-7 observations.** R. A. Howard, M. J. Koomen.
IAU Colloquium No. 31, (see 012.015), p. 66 (1976).

074.066 **The thermal emission of the dust corona during the eclipse of June 30, 1973.**
P. Léna, Y. Viala, D. Hall, A. Soufflot.
IAU Colloquium No. 31, (see 012.015), p. 67 (1976).
Abstract.

074.067 **Bow-shock-associated proton heating in the upstream solar wind.**
R.-D. Auer, H. Grünwaldt, H. Rosenbauer.
Journ. Geophys. Res., Vol. 81, 2030 - 2040 (1976).
In the following discussion the authors deal especially with the proton component of the solar wind upstream of the earth's bow shock. Starting from theoretical arguments, they present experimental evidence for the proton heating by hydromagnetic waves in the bow-shock-disturbed region. All the data employed were measured on the Heos 2 satellite. The particle measurements were performed by Max-Planck-Institut (MPI) plasma experiment S 210, and the magnetic field data have kindly been made available by P. C. Hedgecock of the Imperial College, London.

074.068 **Solar wind stream evolution at large heliocentric distances: experimental demonstration and the test of a model.** J. T. Gosling, A. J. Hundhausen.
Journ. Geophys. Res., Vol. 81, 2111 - 2122 (1976).
A detailed comparison is made of the speed-time profiles of six solar wind streams observed between June 1 and October 1, 1973, by instruments aboard IMP 7 at 1.0 AU and Pioneer 10 at 4.5 AU. This period includes a short interval when Pioneer 10 was coaligned with IMP 7 and the sun. The comparison provides several vivid illustrations of the phenomena of stream steepening in the solar wind with the attendant formation of forward-reverse shock pairs and the gradual decay of stream amplitudes with increasing heliocentric distance.

074.069 **Evidence of a large-scale gradient in the solar wind velocity.** E. J. Rhodes, Jr., E. J. Smith.
Journ. Geophys. Res., Vol. 81, 2123 - 2134 (1976).
Measurements of the solar wind velocity have been compared at two widely separated locations using plasma data obtained in mid-1967 with Mariner 5, en route to Venus, and with the near-earth satellites Explorer 33, 34, and 35. A previous study of the propagation of interplanetary sector boundaries between Mariner and earth had implied the existence of a large-scale gradient in the velocity which the authors interpreted as a latitude gradient of approximately 13 km/s per degree of latitude.

074.070 **Stability of solar wind double ion streams.** G. S. Lakhina, B. Buti.
Journ. Geophys. Res., Vol. 81, 2135 - 2139 (1976).

074.071 **Nonlinear interaction of discontinuities in the solar wind and the origin of slow shocks.**
F. M. Neubauer.
Journ. Geophys. Res., Vol. 81, 2248 - 2256 (1976).

074.072 **Corrections to and comments on the paper 'The enhancement of solar wind fluctuations at the proton thermal gyroradius'.** M. Neugebauer.
Journ. Geophys. Res., Vol. 81, 2447 - 2448 (1976). – See 13.074.039.

074.073 **Helium and hydrogen velocity differences in the**

solar wind.   J. R. Asbridge, S. J. Bame, W. C. Feldman, M. D. Montgomery.
Journ. Geophys. Res., Vol. 81, 2719 - 2727 (1976).

Scalar and vector velocity differences between helium and hydrogen ions in the solar wind measured with the Los Alamos plasma analyzers on Imp 6 and Imp 7 are presented and interpreted.

074.074   Electromagnetic instabilities driven by unequal
          proton beams in the solar wind.
M. D. Montgomery, S. P. Gary, W. C. Feldman, D. W. Forslund.
Journ. Geophys. Res., Vol. 81, 2743 - 2749 (1976).

The linear dispersion relation for electromagnetic waves is investigated for homogeneous plasmas normally found in the solar wind near 1 AU, including a secondary proton beam of varying strength and drift speed parallel to the ambient magnetic field. Three instabilities are found when the drift speed of the beam approaches the Alfvén speed C_A: (1) field-aligned magnetosonic, (2) oblique magnetosonic, and (3) oblique Alfvén modes.

074.075   Suprathermal protons in the interplanetary solar
          wind.   C. C. Goodrich, A. J. Lazarus.
Journ. Geophys. Res., Vol. 81, 2750 - 2754 (1976).

074.076   There are holes even in the solar corona.
          J. Mergentaler.
Urania Kraków, Vol. 47, 98 - 101 (1976). In Polish.

074.077   Theoretical studies of the large-scale behavior of the
          solar wind.   A. Barnes.
Adv. Electronics Electron. Phys., Vol. 36, 1 - 57 (1974).

074.078   Synoptic maps of solar coronal hole boundaries
          derived from He II 304 Å spectroheliograms from
the manned Skylab missions.
J. D. Bohlin, D. M. Rubenstein.
World Data Center A, Solar-Terrestrial Phys., Rep. UAG-51. 21 pp. (1975).

074.079   On the continuum theory of the one-fluid solar
          wind for small Prandtl number.   R. S. Johnson.
Proc. Roy. Soc. London, Ser. A, Vol. 348, 129 - 142 (1976).
Abstr. in Phys. Abstr., Vol. 79, A019359 (1976).

074.080   Evidence for local ion heating in solar wind high
          speed streams.   S. J. Bame, J. R. Asbridge,
W. C. Feldman, S. P. Gary, M. D. Montgomery.
Geophys. Res. Letters, Vol. 2, 373 - 375 (1975). − Abstr.
in Phys. Abstr., Vol. 79, A027727 (1976).

074.081   On the continuum theory of the two-fluid solar wind
          for small mass ratio.   R. S. Johnson.
Proc. Roy. Soc. London, Ser. A, Vol. 348, 511 - 523 (1976).
Abstr. in Phys. Abstr., Vol. 79, A042049 (1976).

074.082   On the effects of viscosity in one-fluid solar wind
          theory.   D. Summers.
Proc. Roy. Soc. London, Ser. A, Vol. 349, 53 - 62 (1976).
Abstr. in Phys. Abstr., Vol. 79, A042050 (1976).

074.083   Turbulent spectra of the transverse Alfvén waves in
          the corotating solar wind structure.
O. Saka, T.-I. Kitamura.
Rep. Ionosph. Space Res. Japan, Vol. 29, 127 - 132 (1975).
Abstr. in Phys. Abstr., Vol. 79, A042153 (1976).

074.084   Model of coronal holes.
          W. M. Adams, P. A. Sturrock.
Separate print Stanford Univ., California, USA, Inst. Plasma Res., 20 pp. (1974). − See 14.074.060.

074.085   Coordinated rocket and satellite observations of the
          solar corona. Final report.
W. T. Zaumen.
Separate print Lockheed Missiles Space Co., Palo Alto, California, USA, 11 pp. (1974).

074.086   Coronal structures and particle acceleration studies
          from radioelectric and optical observations.
F. Axisa.
Thèse Paris-7 Univ., France (1974). In French.

074.087   Waves in the solar wind and their Doppler shift in
          the frequency range of the Helios-search-coil-experi-
ment E4.   K. U. Denskat.
Separate print Bundesministerium Forschung Technol., Bonn, F. R. Germany, 138 pp. (1975). (Available from ZAED). In German.

074.088   The interaction of a plane interplanetary discontinui-
          ty with the magnetosphere.   B. J. Rigby.
Thesis Queensland Univ., St. Lucia, Australia (1972).

074.089   Microscale Alfvén waves in the solar wind at 1 AU.
          L. F. Burlaga, J. M. Turner.
Separate print National Aeronaut. Space Administr., Greenbelt, Maryland, USA, Goddard Space Flight Center, 20 pp. (1974).

074.090   Termination of the solar wind in the hot, partially
          ionized interstellar medium.   C. K. Lombard.
Thesis Stanford Univ., California, USA, 154 pp. (1975). (Available from: Univ. Microfilms, Order No. 75-13,553).

The solar chromosphere and corona: Quiet sun. See Abstr. 003.016.

Influence of ancient solar-proton events on the evolution of life.   See Abstr. 015.002.

Investigation of forbidden transitions in argon ions. See Abstr. 022.002.

Recent progress in the classification of the spectra of highly ionized atoms.   See Abstr. 022.101.

Directional discontinuities from Mariner 5 and solar wind structure.   See Abstr. 053.030.

The solar spectrum: wavelengths and identifications from 160 to 770 Angstroms.   See Abstr. 071.001.

Solar cycle variation of solar wind characteristics. See Abstr. 072.054.

Extreme-ultraviolet transients observed at the solar pole.   See Abstr. 073.001.

Studies of the prominence-corona transition zone from rocket ultraviolet spectra of the March 1970 eclipse. See Abstr. 073.021.

Expansion of chromospheric matter in the gradual phase of solar flares.   See Abstr. 073.022.

The solar XUV spectrum of He II. See Abstr. 076.001.

ATM observations, X-ray results. See Abstr. 076.002.

The analysis of XUV emission lines.

See Abstr. 076.007.

The interpretation of simultaneous soft X-ray spectroscopic and imaging observations of an active region. See Abstr. 076.008.

Time variability of type II solar bursts and $l \rightarrow t$ conversion processes.   See Abstr. 077.001.

Relation of large-scale coronal X-ray structure and cosmic rays: 5. Solar wind and coronal influence on a Forbush decrease lasting one solar rotation.   See Abstr. 078.037.

Observation using interplanetary scintillations at 34.3 MHz of the effect of a solar wind disturbance on a solar energetic particle event.   See Abstr. 078.066.

Relation of large-scale coronal X-ray structure and cosmic rays: III. Low-intensity solar particle events with enhanced ~3 MeV helium and medium fluxes associated with solar wind streams.   See Abstr. 078.067.

Coronal and interplanetary propagation of solar cosmic rays.   See Abstr. 078.093.

Unusual emission of iron nuclei from the sun. See Abstr. 078.107.

Solar magneto-atmospheric waves. I. An exact solution for a horizontal magnetic field.   See Abstr. 080.016.

Nucleosynthesis of ^{3}He in the sun and the variation of ^{3}He/^{4}He in solar wind.   See Abstr. 080.047.

A model of the solar atmosphere and wind. See Abstr. 080.053.

Polar cap ionospheric electric field modulation by the solar wind sector structure.   See Abstr. 083.071.

Helium isotopes in an aurora. See Abstr. 084.003.

Flapping motions of the tail plasma sheet induced by the interplanetary magnetic field variations. See Abstr. 084.202.

Geomagnetic field variations. See Abstr. 084.204.

Evidence of magnetic field line merging in the solar wind.   See Abstr. 084.211.

A statistical study of the upstream wave boundary outside the earth's bow shock.   See Abstr. 084.212.

Gradients of solar protons in the high-latitude magnetotail and the magnetospheric electric field. See Abstr. 084.219.

Magnetopause position and the reconnection problem.   See Abstr. 084.225.

Propagation characteristics of electromagnetic waves in the magnetosphere.   See Abstr. 084.230.

On the generation of low-frequency waves in the solar wind in the front of the bow shock. See Abstr. 084.235.

Effects on the plasmasphere of irregular electric

fields.   See Abstr. 084.262.

Correlation of bow shock plasma wave turbulence with solar wind parameters.   See Abstr. 084.280.

Anomalous electrical conductivity for the field-line merging current.   See Abstr. 084.294.

Preliminary report of results from the plasma science experiment on Mariner 10.   See Abstr. 093.010.

The Rosiwal Principle and the regolithic distributions of solar-wind elements.   See Abstr. 094.140.

Studies of the formation of acid-hydrolyzable carbon species and finely divided iron by simulated solar-wind implantation.   See Abstr. 094.142.

Implantation of carbon and nitrogen ions into lunar fines: trapping efficiencies and saturation concentrations. See Abstr. 094.143.

A Monte Carlo model for the exposure history of lunar dust grains in the ancient solar wind. See Abstr. 094.181.

Solar-wind induction and lunar conductivity. See Abstr. 094.197.

Solar-wind and solar-flare maturation of the lunar regolith.   See Abstr. 094.568.

Interplanetary stream magnetism: kinematic effects. See Abstr. 106.001.

The interplanetary acceleration of energetic nucleons.   See Abstr. 106.002.

Solar wind electric field modulation in the interplanetary sector structure.   See Abstr. 106.009.

Sun and comets as sources in an external flow. See Abstr. 106.036.

The structure of plasma density irregularities in the interplanetary medium. II. Observations.   See Abstr. 106.111.

On the flux and the energy spectrum of interstellar ions in the solar system.   See Abstr. 131.100.

Solar influence on galactic cosmic ray anisotropy measurements.   See Abstr. 143.011.

A dynamic model for the time evolution of the modulated cosmic ray spectrum.   See Abstr. 143.012.

Propagation and acceleration of cosmic ray electrons in interplanetary space in presence of inhomogeneous structure of solar wind.   See Abstr. 143.176.

On the theory of non-linear cosmic ray modulation by solar wind.   See Abstr. 143.195.

Relation of large-scale coronal X-ray structure and cosmic rays: I. Sources of solar wind streams as defined by X-ray emission and Hα absorption features. See Abstr. 143.283.

Compton-Getting effect for low energy particles. See Abstr. 143.370.

Errata

074.901 Corrigendum: 'Interferometric investigation of the line of sight velocities in λ 5303 during the eclipse of 22 September, 1968' [Solar Physics, Vol. 45, 157 - 168 (1975). See 14.074.121].   A. B. Delone, E. A. Makarova. Solar Physics, Vol. 45, 550 (1975).

075   Solar Patrol

075.001   **L'activité solaire, cartes synoptiques de la chromosphère et des taches.**   M.-J. Martres.
L'Astronomie, Vol. 90, 45 - 46 (1976).

075.002   **Compte rendu d'activité de juin 1974 à juin 1975.**   J. P. Desgrées, R. Boyer, M.-J. Martres, J. R. Frémy, J. Funel.
L'Astronomie, Vol. 90, 89 - 99 (1976).

075.003   **Prominences in 1974.**   M. K. V. Bappu.
Quarterly Journ. Roy. Astron. Soc., Vol. 17, 63 (1976).

075.004   **Catalogue of solar activity for the year 1974.**   R. S. Gnevysheva.
Glav. Astron. Obs., *Pulkovo,* 104 pp. Price 87 Kop. (1976). In Russian.

075.005   **The sun 1965 to 1975.**   B. E. Stonehouse.
Southern Stars, Vol. 26, 111 - 113 (1976).

075.006   **Solare Beobachtungsergebnisse (Solar Data).**   E. A. Lauter, C.-U. Wagner, A. Böhme, F. Fürstenberg, D. Scholz, S. Böhm.
Zentralinst. für Solar-Terrestrische Physik (Heinrich-Hertz-Inst.) Akad. Wiss. DDR, HHI Solar Data, Vol. 26, 75 - 101 (1975); Vol. 27, A-F, 2 - 35 (1976). — 1975 October - 1976 April. — Solar radio emission.

075.007   **Solar observations made at Catania Astrophysical Observatory during 1975.**   G. Godoli.
Oss. Astrofis. Catania, Pubbl. Nuova Ser. No. 157, 77 pp. (1976).
    Sunspots; Hα and K faculae; Hα flares; Hα quiescent prominences; K quiescent prominences; Hα active prominences on disc and at limb; Hα disc and limb patrol hours.

075.008   **Actividad solar en 1974. I. Números relativos de Wolf. II. Estadística de manchas y superficie de las mismas.**   M. Lopez Arroyo.
Bol. Astron. Obs. Madrid, Vol. 9, No. 3, 47 pp. (1976).

075.009   **Actividad solar en 1975. I. Números relativos de Wolf. II. Estadística de manchas y superficie de las mismas.**   M. Lopez Arroyo.
Bol. Astron. Obs. Madrid, Vol. 9, No. 4, 40 pp. (1976).

075.010   **Solar phenomena.**   M. Cimino, M. Torelli.
Oss. Astron. Roma, Monthly Bull. Nos. 209 - 212, 13 pp. (1975). — 1975 September - December: Daily total areas of sunspot-groups; Heliographic position, classification and area of sunspot-groups; Hours of K-line cinematographic patrol; Hours of Hα cinematographic patrol; Explanation.

075.011   **Daily Hα chromosphere pictures, daily K$_{232}$ chromosphere pictures, daily white light photosphere pictures.**   M. Cimino (Editor).
Photographic Journ. of the Sun, Oss. Astron. Roma, Nos. 99 - 106 (1975). — 1975 April 14 - November 18. — Solar rotation 1627 - 1634.

075.012   **Heliofísica 1968.**
Bol. Obs. Ebro, Vol. 56, 63 pp. (1971).

075.013   **Heliofísica 1969.**
Bol. Obs. Ebro, Vol. 57, 68 pp. (1971).

075.014   **Sunspots (sunspot relative-numbers and sunspot-areas); Synoptic charts of solar magnetic fields** (Mount Wilson Observatory); **Eruptions chromosphériques brillantes; Intensité de la couronne solaire; Solar radio emission.**
M. Waldmeier, R. Howard, G. Olivieri, M. Bernot, H. Tanaka.
Quarterly Bull. Solar Activity, Nos. 189 - 190, pp. 1 - 72 (1976). — 1975 January - June.

075.015   **Datos relativos a la actividad solar y geomagnética en 1973.**
Urania Barcelona, Año 60, No. 283, p. 20 (1975).

075.016   **Daily maps of the sun and geophysical graphs.**
Solnechnye Dannye 1975 Byull., No. 10, p. 1 - 52; No. 11, p. 1 - 62; No. 12, p. 1 - 50; 1976, No. 1, p. 1 - 60; No. 2. p. 1 - 43. In Russian.

075.017   **Magnetic fields of sunspots.**
Prilozhenie k Byulletenyu "Solnechnye Dannye", 1975, Nos. 10 - 12; 1976, No. 1.

075.018   **Definitive sunspot-numbers for 1975.**   M. Waldmeier.
Journ. Geophys. Res., Vol. 81, 2451 (1976).

075.019 Sunspot relative numbers for 1975. M. Waldmeier.
Astron. Mitt. Eidgen. Sternw. Zürich No. 344, 10 pp.
(1976).

075.020 Definitive Sonnenflecken-Zahlen für 1975.
M. Waldmeier.
Sterne, Vol. 52, 165 (1976).

075.021 L'activité solaire. M.-J. Martres.
L'Astronomie, Vol. 90, 109 - 110, 170 - 171, 217,
274 - 275 (1976). — Rotations 1631 – 1636.

**075.022 Solar and solar system activity. Radio Astronomy
Section (BAA).** J. R. Smith, R. J. J. Langton.
Journ. British Astron. Ass., Vol. 86, 151 - 154, 237 - 240
(1976).

075.023 Geomagnetic and solar data.
J. V. Lincoln (Editor).
Journ. Geophys. Res., Vol. 81, 296, 720, 1460, 1812, 2449 -
2451, 2939 (1976).

075.024 Sunspot numbers.
Sky Telescope, Vol. 51, 69, 137, 212, 292, 364,
437 (1976).

Beobachtung der Sonnenflecken-Relativzahlen.
See Abstr. 072.052.

Nombres relatifs de Wolf pour 1975.
See Abstr. 072.055.

076 Solar UV, X Rays, Gamma Radiation

076.001 The solar XUV spectrum of He II.
J. L. Linsky, D. L. Glackin, R. D. Chapman, W. M.
Neupert, R. J. Thomas.
Astrophys. Journ., Vol. 203, 509 - 520 (1976).

OSO-7 observations of the first five Lyman lines and the
Lyman continuum of He II are given for the quiet sun, a
coronal hole, prominences, filaments, and the 1972 August 7
flare. These data are calibrated and given in specific intensity
units together with color and brightness temperatures for the
He II continuum. The authors find that He II is overionized in
all features except the flare, and that the continuum is formed
at temperatures near 14,000 K. To account for the intensity
of $L\alpha$, either an implausible 100 km plateau at temperatures
near 80,000 K is needed or, more likely, the incorporation of
diffusion-enhanced collisional excitation into the models.

076.002 ATM observations, X-ray results. G. S. Vaiana,
A. S. Krieger, A. F. Timothy, M. Zombeck.
Astrophys. Space Sci., Vol. 39, 75 - 101 (1976).

Preliminary results of the solar X-ray observations from
Skylab are reviewed, indicating a highly structured nature for
the corona, with closed magnetic loop structures over a wide
range of size scales. A description of the S-054 experiments is
provided, and values are given for parameters, including size,
density, and temperature, describing a variety of typical coronal features. The structures and evolutions of active regions,
coronal holes and bright points are discussed.

076.003 Characteristic figures of solar X-ray plages.
J. Kleczek, B. Růžičková-Topolová.
Bull. Astron. Inst. Czechoslovakia, Vol. 27, 77 - 83 (1976).

Properties of X-ray plages are deduced from 9.1 –10.5 Å
spectroheliograms. Several characteristic figures are introduced, such as the area A_X, the total flux F_X, the maximum
brightness f_{max} and the mean gradient of intensity, grad I.
The characteristic figures for the X-ray plages are then related
to the underlying spot groups and to the corresponding radio
plages on λ 9.1 cm. In general, the radio and X-ray characteristics are well correlated.

076.004 Solar X-ray bursts in relation to the microwave

bursts and SIDs (*sudden ionospheric disturbances*).
R. K. Mitra, S. K. Sarkar, M. K. Das Gupta.
Indian Journ. Radio Space Phys., Vol. 4, 24 - 28 (1975).
Abstr. in Phys. Abstr., Vol. 79, A003254 (1976).

076.005 Full-disk solar fluxes between 1230 and 1940 Å.
L. Heroux, R. A. Swirbalus.
Journ. Geophys. Res., Vol. 81, 436 - 440 (1976).

The intensity of solar radiation in 10-Å intervals in the
wavelength region 1230–1940 Å is presented. The present data
are used to compute the O_2 photodissociation coefficient as a
function of the O_2 column density.

**076.006 Spikulen und die Mitte-Rand-Variation solarer UV-
Linien.** L. Elzner.
Mitt. Astron. Ges., No. 38, p. 216 (1976). — Abstract.

076.007 The analysis of XUV emission lines.
G. L. Withbroe.
Solar Physics, Vol. 45, 301 - 317 (1975).

A technique for analyzing measurements of XUV spectral line intensities is described. Application of the technique
to OSO-4 and OSO-6 spectra indicates that the mean coronal
temperature is 2.1×10^6 K in typical active solar regions and
that the mean coronal temperature in typical quiet regions
ranges from 1.5×10^6 to 2.1×10^6 K.

**076.008 The interpretation of simultaneous soft X-ray spec-
troscopic and imaging observations of an active region.**
J. M. Davis, M. Gerassimenko, A. S. Krieger, G. S. Vaiana.
Solar Physics, Vol. 45, 393 - 410 (1975).

Simultaneous soft X-ray spectroscopic and broad band
imaging observations of an active region have been analyzed
together to determine the parameters which describe the coronal plasma. From the spectroscopic data, models of tempera-
ture-emission measure-elemental abundance have been constructed which provide acceptable statistical fits. Contour maps
of electron temperature and density for the active region have
been constructed from the imaging data. The implications of

the analysis to the determination of coronal abundances and to future satellite experiments are discussed.

076.009 Spatial structure and temporal development of a solar X-ray flare observed from Skylab on June 15, 1973.
R. Pallavicini, G. S. Vaiana, S. W. Kahler, A. S. Krieger.
Solar Physics, Vol. 45, 411 - 433 (1975).

A solar flare on June 15, 1973 has been observed with high spatial and temporal resolution by the S-054 grazing-incidence X-ray telescope on Skylab. Both morphological and quantitative analyses are presented. The implications of these observations for mechanisms of solar flares are discussed. In particular, the flux profiles of different regions of the flare give strong evidence for continued heating during the decay phase, and a multiplicity of flare volumes appears to be present, in all cases consisting of loops of varying lengths.

076.010 Contribution of electron-electron bremsstrahlung to solar hard X-radiation during flares. E. Haug.
Solar Physics, Vol. 45, 453 - 458 (1975).

The bremsstrahlung produced in collisions of energetic electrons with thermal electrons and protons of a hydrogen plasma is calculated. The importance of electron-electron bremsstrahlung to the X-radiation in solar flares is discussed.

076.011 Solar EUV flux models consistent with ionospheric ion composition observations.
P. Chakrabarty, A. P. Mitra.
Space Research XV, (see 012.003), p. 335 - 343 (1975).

076.012 Absolute EUV photon fluxes of aeronomic interest. G. Schmidtke, K. Rawer, W. Fischer, C. Rebstock.
Space Research XV, (see 012.003), p. 345 - 349 (1975).

During the mission of the satellite AEROS-A (18 December 1972–22 August 1973) an EUV spectrometer monitored the solar photon fluxes in the wavelength range from 106 to 16 nm. As a first result a table of absolute photon flux values, representative for 2 March 1973, 1430–1450 UT, is presented.

076.013 Investigation of solar X-rays from the lunar surface, carried out on Lunokhod 2.
G. E. Kocharov, S. V. Victorov (*Viktorov*), V. I. Chesnokov.
Space Research XV, (see 012.003), p. 633 - 636 (1975).

A special detector for measurements of solar X-rays in the energy range 5 to 30 keV has been installed in an X-ray fluorescence spectrometer RIFMA-M aboard Lunokhod 2. The design and characteristics of this instrument are described. Measurements of the quiet sun and the characteristics of a weak X-ray burst (which may or may not be of solar origin) on 17 January 1973 are presented. The duration of the burst was about 10 min, the rise time 1 min, and the total X-ray flux 10^{-7} erg cm^{-2} s^{-1}.

076.014 The solar subflare X-ray spectrum.
I. L. Beigman (*Bejgman*), Yu. I. Grineva, V. V. Korneev, V. V. Krutov, S. L. Mandel'stam (*Mandel'shtam*), L. A. Vainstein (*Vajnshtejn*), B. N. Vasilyev (*Vasil'ev*), I. A. Zhitnik.
Space Research XV, (see 012.003), p. 641 - 650 (1975).

The subflare spectrum in the range 7.8–19.4 Å was obtained with KAP crystal spectrometers aboard the Vertical 2 rocket. The absolute fluxes for the lines are given. The temperature of 5×10^6 K and emission measure of 6×10^{48} cm^{-3} were derived from the Mg XI–XII, Ne IX–X and Na X–XI lines. Analyses of the Lyman decrements for Ne X and O VIII series and the relative intensities of the Fe XVII 2p-nd lines were made. The abundances of Ne, Na, Fe and O were estimated, and are discussed.

076.015 Structure of the sun's polar cap at wavelengths 240–

600 Å.
J. D. Bohlin, N. R. Sheeley, R. Tousey.
Space Research XV, (see 012.003), p. 651 - 656 (1975).

An NRL slitless spectrograph obtained monochromatic images of the sun in the extreme ultraviolet (XUV) 170–630 Å. The solar poles exhibited a complex variety of phenomena: (1) He II 304 Å showed a depressed intensity and lack of supergranular network that is characteristic of so-called coronal holes; (2) 'super spicules' of He II 304 Å exist throughout the polar cap; (3) the limb brightening of Ne VII 465 Å is wider and higher over the polar cap; and (4) Mg IX 368 Å and other coronal lines exhibit polar plumes such as observed at an eclipse.

076.016 Ultraviolet observations of C III transitions in the sun. A. K. Dupree, P. Foukal, C. Jordan.
Bull. American Astron. Soc., Vol. 8, 292 (1976). — Abstr. AAS.

076.017 Solar hard X-ray bursts and accelerated electrons in solar flares. L. D. De Feiter.
VIIth Leningrad seminar. "Corpuscular streams of the sun and the radiation belts of the earth and Jupiter", 1975, (see 012. 007), p. 93 - 102 (1975). — Abstr. in Referativ. Zhurn. 51. Astron., 4.51.346; 62. Issled. kosmich. prostranstva, 4.62.203 (1976).

076.018 On model representations of a source of flare X rays. A. A. Korchak.
Astron. Zhurn. Akad. Nauk SSSR, Vol. 53, 370 - 376 (1976). In Russian. English translation in Soviet Astron., Vol. 20, No. 2.

An analysis of thick-target and thin-target models has been carried out. It is shown that the total number of electrons, their flux, energy and other values, determined by means of the models, can differ from the real ones by more than one order of value because of the impossibility to determine the electron spectrum boundary. The dependence between the values relating to the source and the emitting region has been examined.

076.019 Investigation of hard X-ray radiation of solar flares aboard the artificial earth satellites of the Prognoz series. O. M. Kovrizhnykh, M. I. Kudryavtsev, A. S. Melioranskij, I. A. Savenko, V. M. Shamolin, L. N. Chupova.
Izv. Krymskoj Astrofiz. Obs., Vol. 54, 316 - 319 (1976). In Russian. — Abstract of a paper presented at the seminar on "X-ray and gamma-ray astronomy" at the Crimean Astrophys. Obs., 1974, (see 012.009).

076.020 The solar extreme ultraviolet between 30 and 205 Å on November 9, 1971, compared with previous measurements in this spectral region. J. E. Manson.
Journ. Geophys. Res., Vol. 81, 1629 - 1635 (1976).

The measurement of the full-disk solar EUV flux summed over broad spectral bands in this region is shown to be critically dependent on the determination of the level of scattered radiation and other background. The EUV fluxes of some strong lines are reported, as are the total fluxes summed over 10-Å intervals, based on an Aerobee rocket experiment on November 9, 1971.

076.021 Time history and model calculations of the 2.2 MeV gamma ray line from the flares of August, 1972.
G. Kanbach, C. Reppin, D. J. Forrest, E. L. Chupp.
14th Intern. Cosmic Ray Conf., (see 012.011), Vol. 5, 1644 - 1649 (1975).

076.022 A search for high energy gamma rays from a quiet sun. C. Y. Kim.
14th Intern. Cosmic Ray Conf., (see 012.011), Vol. 5, 1650 - 1654 (1975).

076.023 A hard X ray observation of the 4 July 1974 solar
flare.   K. Hurley.
14th Intern. Cosmic Ray Conf., (see 012.011), Vol. 5, 1674
(1975). − Abstract.

076.024 The X-ray radiation of the solar flare in the loop
magnetic field.
Yu. E. Charikov, Yu. N. Starbunov.
14th Intern. Cosmic Ray Conf., (see 012.011), Vol. 5, 1675 -
1680 (1975).

076.025 X-ray bursts from solar flares behind the limb.
J.-R. Roy, D. W. Datlowe.
Separate print California Inst. Tech., Pasadena, USA, 54 pp.
(1975). − See 13.076.002.

076.026 Slow X-ray bursts and chromospheric flares with
filament disruption.   J.-R. Roy, F. Tang.
Separate print California Inst. Tech., Pasadena, USA, 31 pp.
(1975).

076.027 Solar X-ray studies. Final report.   J. A. Vorpahl.
Separate print Aerospace Corp., Los Angeles,
California, USA, 3 pp. (1975).

076.028 Solar cycle variation of EUV radiation.
R. D. Chapman.
Symposium significant accomplishments sci. technol., Green-
belt, Maryland, USA, 18 Dec. 1973, p. 94 - 96 (1975).

076.029 Initial results from the EUV spectroheliometer on
ATM.   E. M. Reeves, J. G. Timothy, P. V. Foukal,
M. C. E. Huber, R. W. Noyes, E. J. Schmahl, J. E. Vernazza,
G. L. Withbroe.
Conf. sci. experiments Skylab, Huntsville, Alabama, USA, 30
Oct. 1974. AIAA Paper 74-1258, 61 pp. (1974).

On the EUV spectrum of Fe X and Ni XII.
See Abstr. 022.039.

Fundamental limitations of X-ray spectra as diagnostics of plasma temperature structure.
See Abstr. 061.028.

X-ray astronomy.   See Abstr. 061.036.

Resonance-line polarization. II. Calculations of linear
polarization in solar UV emission lines.   See Abstr. 063.004.

The solar spectrum: wavelengths and identifications
from 160 to 770 Angstroms.   See Abstr. 071.001.

The solar EUV flux between 230 and 1220 Å on
November 9, 1971.   See Abstr. 071.019.

Extreme-ultraviolet transients observed at the solar
pole.   See Abstr. 073.001.

Solar gamma-ray lines as probes of accelerated particle directionalities in flares.   See Abstr. 073.002.

On X-rays of solar flares in July-December, 1972.
See Abstr. 073.004.

Changes in X-ray brightness of a solar active region.

See Abstr. 073.008.

Studies of the prominence-corona transition zone
from rocket ultraviolet spectra of the March 1970 eclipse.
See Abstr. 073.021.

Expansion of chromospheric matter in the gradual
phase of solar flares.   See Abstr. 073.022.

Investigation of solar flare X-ray spectra in the hydrogen-like Fe ion region.   See Abstr. 073.023.

X-ray and ultraviolet emission of solar flares.
See Abstr. 073.048.

Time dependences of the 0.51 and 2.2 MeV lines
in solar flares.   See Abstr. 073.072.

Gamma-ray and radio evidence for the acceleration
of high-energy particles in solar flares.   See Abstr. 073.074.

Gamma-ray and microwave evidence for two
phases of acceleration in solar flares.   See Abstr. 073.081.

Studies on hard X-ray emission from solar flares
and on cyclotron radiation from a cold magnetoplasma.
See Abstr. 073.084.

Analysis of OGO-5 and OSO-7 X-ray data. Final
report.   See Abstr. 073.088.

Origin and implications of gamma rays from solar
flares.   See Abstr. 073.089.

Radio- and X-emission (and their polarization) at the
time of early phase of solar flare connected with γ-line emissions.   See Abstr. 073.092.

The relative abundance of neon and magnesium in
the solar corona.   See Abstr. 074.002.

Coronal changes associated with a disappearing filament.   See Abstr. 074.018.

The spectrum of Fe XVII in the solar corona.
See Abstr. 074.038.

Relation of large-scale coronal X-ray structure and
cosmic rays: II. Coronal control of interplanetary injection of
300 keV protons.   See Abstr. 074.058.

Calculation of solar neutron and gamma ray fluxes
during the flares and the quiet periods.
See Abstr. 078.062.

Nuclear astrophysics of the sun.
See Abstr. 080.050.

Relation between X-ray and ultraviolet radiation of
solar flares in the ionization of the ionospheric E-region.
See Abstr. 083.041.

Solar EUV flux variation during a solar cycle as
derived from ionospheric modeling considerations.
See Abstr. 083.044.

077 Solar Radio Radiation

077.001 Time variability of type II solar bursts and $l \rightarrow t$ conversion processes. Y. Leblanc, A. Lecacheux.
Astron. Astrophys., Vol. 46, 257 - 260 (1976).

The authors have shown by means of high resolution observations of type II bursts in the range of 30 to 76 MHz, that the low frequency spectrum of the harmonic component is much wider (3 times) than that of the fundamental emission, whatever the frequency of observation. The origin of these differences is discussed in terms of the propagation of the radio waves in the corona and of the conversion of plasma waves into electromagnetic waves.

077.002 The velocities of type II solar radio bursts. A. Tlamicha, M. Karlický.
Bull. Astron. Inst. Czechoslovakia, Vol. 27, 6 - 10 (1976).

This paper is a summary of type II radio bursts identified at Ondřejov between January 1973 and December 1974 in the frequency range of the dynamic spectrum 70–810 MHz. The velocities of shock waves in the individual cases of the type II bursts are given using the fourfold Newkirk model. Some questions associated with type II radio bursts and with the propagation of the shock wave into interplanetary space and into the region of the earth are also discussed.

077.003 Swept frequency interferometer data analysis for the type IV event of 1972, August 12. A. C. Riddle.
Colorado Univ., Dept. Astro-Geophys., Radio Astron. Obs., Solar System Radio Observations, No. SN-2, 5 pp. (1973).

077.004 Variations of the basic component of solar radio emission in the years 1957–1967. S. Zięba, R. Guła.
Acta Astron., Vol. 26, 55 - 67 (1976).

The basic component of the solar radio emission (B-component) has been investigated using the data obtained at different observing stations at 200, 536, 600, 810, 1000, 2000, 2800, 9400 MHz. Using a statistical method the basic component was estimated for successive periods containing four solar rotations. The calculated values of the basic component show irregular and more or less similar variations independent of frequency. Almost all of these values achieve their minima in the middle of 1964, although some mutual displacments can be observed. The quiet sun spectrum obtained from these smallest values is also discussed.

077.005 On restoring the true radio image of a local source. L. M. Risover.
Solnechnye Dannye 1975 Byull., No. 10, p. 60 - 64 (1976). In Russian.

Results of restoring the true radio image of the radio emission S-component of a local source are presented. The restoration was made by means of mathematical reduction of the radio telescope response, the relationship between the radio brightness distribution and the configuration of the corresponding sunspot group being taken into account. The conclusion is drawn that a restoration like that is expedient for obtaining information on the local source radio brightness distribution (within the sunspot group) from the radio telescope response.

077.006 On the interpretation of solar microwave bursts spectra. V. A. Kovalev, O. S. Korolev.
Astron. Zhurn. Akad. Nauk SSSR, Vol. 53, 130 - 139 (1976). In Russian. English translation in Soviet Astron., Vol. 20, No. 1.

The influence of physical conditions in solar microwave bursts radio sources on the frequency spectrum is considered.

Using the approximation for gyromagnetic emission for isotropic electrons with power law and kinetic energies 20–200 keV the spectra of sources with different magnetic field distributions are computed.

077.007 Type III solar radioburst profiles and the associated electron energy spectra. C. C. Harvey.
Astron. Astrophys., Vol. 47, 31 - 41 (1976).

An attempt is made to explain the observed frequency-time profiles of type III solar radiobursts in terms of a rapid plasma wave decay rate combined with the exciter model recently proposed by Harvey (1975). The model is compared with radio observations by making simple assumptions about the dependence of the radio intensity upon the plasma wave energy. A comparison is made with simultaneous radio and electron observations by further assuming a simple power-law velocity distribution for the electrons at their point of ejection from the sun.

077.008 On the theory of type I solar radio bursts. I. Beam plasma instabilities in a turbulent magnetized plasma. A. Mangeney, P. Veltri.
Astron. Astrophys., Vol. 47, 165 - 180 (1976).

The stability of a beam of mildly relativistic electrons spiralling along the magnetic field in a cold, collisionless spiralling along the magnetic field in a cold, collisionless and turbulent plasma is studied. It is found that the most unstable mode is the whistler mode. As to the high frequency modes, it is shown that the X-mode has a larger growth rate than the O-mode for a given ratio of the background plasma frequency to the electron gyrofrequency. It is shown that non linear coupling of linearly unstable whistler and high frequency X-mode waves with low frequency M.H.D. waves provides an efficient stabilization mechanism for these two modes for low beam densities. The quasilinear relaxation of the remaining O-mode wave instability is studied, and it is suggested that the corresponding characteristics of the emission in the O-mode could explain the polarization, narrow bandwidth and high degree of directivity of type I bursts.

077.009 On the theory of type I solar radio bursts. II. A model for the source. A. Mangeney, P. Veltri.
Astron. Astrophys., Vol. 47, 181 - 192 (1976).

The authors construct a model for type I solar radiobursts. It is shown that, if at some altitude in the corona (1) coronal electrons are accelerated perpendicularly to the magnetic field, (2) $A = \omega_{pe}^2 / \omega_{He}^2$ decreases in the corona just above the acceleration region, one may explain the formation of electron beams with non zero pitch angle and small momentum dispersion. These electron beams become unstable at some height above the acceleration region and produce a burst of radiation, the characteristics of which agree very well with the observed characteristics of type I bursts.

077.010 Synchrotron or plasma process emission in narrow-band type IV $_{dm}$ bursts? A. O. Benz, G. L. Tarnstrom.
Astrophys. Journ., Vol. 204, 597 - 603 (1976).

Extremely narrow-banded type IV $_{dm}$ events have been selected to test the various proposed emission mechanisms. Synchrotron emission in quasi-vacuum is excluded by the observed steep low-frequency spectral slope. Free-free absorption of incoherent synchrotron emission is negligible compared with the Razin effect suppression. A low-frequency cutoff due to cyclotron-resonance absorption is unlikely. Plasma processes seem to be the necessary explanation for the bursts under consideration.

077.011 Solar millimetric bright sources on the quiet sun.
R. Bocchia, F. Poumeyrol.
Astrophys. Journ., (*Letters*), Vol. 204, L107 - L110 (1976).

The radio emission of quiet regions has been investigated with high angular resolution at 35.1 GHz. The observations have shown the existence of bright and evolutionary sources all over the sun's disk. By comparing the observations with a simple model the authors draw the following conclusions: (1) These sources can be described as circular structures, the radius of which increases or decreases with a maximum diameter of 3'. (2) The expansion, or regression, velocities are in the same range: $10-20$ km s^{-1}. (3) The mean brightness temperature T_b is about 3×10^4 K.

077.012 Die Struktur aktiver Gebiete in der Sonnenatmosphäre, abgeleitet aus kombinierten Interferometer- und Einzelantennen-Messungen. E. Fürst, W. Hirth.
Mitt. Astron. Ges., No. 38, p. 204 - 208 (1976). — Short report.

077.013 Solar type III burst profiles at decametre-wave frequencies. I: Observations.
C. H. Barrow, A. Achong.
Solar Physics, Vol. 45, 459 - 465 (1975).

Solar type III bursts have been observed at fixed frequencies of 18, 22, 26 and 36 MHz during the period 1970–1974. 103 profiles have been analyzed in the manner introduced by Aubier and Boischot (1972) whose results are generally confirmed. Average values for the exciter function duration were found for each frequency to be respectively 16,6, 14.7, 10.2 and 8.1 s, and for the decay time constant 3.6, 3.1, 2.3 and 1.8 s. Average peak-to-peak frequency drifts were found to be -1.9, -2.1, and -3.5 MHz s^{-1} for centre frequencies of 20, 22, and 31 MHz, respectively. The decay time-constant appears to bear approximately the same linear relationship to the exciter function duration at each of the four frequencies studied.

077.014 Solar type III burst profiles at decametre-wave frequencies. II: Exciter. A. Achong, C. H. Barrow.
Solar Physics, Vol. 45, 467 - 476 (1975).

The observations (see 077.013) are used to calculate the form of the exciter function. The analysis shows that over the frequency range 18–36 MHz, (1) the exciter function possesses negative skewness, (2) the shapes of burst profiles and exciter profiles are approximately constant and, (3) burst peak time varies linearly with height in the corona. It is suggested that the passage of the exciting electron stream through field-dominated and flow-dominated coronal regions has different effects on the profiles.

077.015 Electron plasma oscillations associated with type III radio emissions and solar electrons.
D. A. Gurnett, L. A. Frank.
Solar Physics, Vol. 45, 477 - 493 (1975).

An extensive study of the IMP-6 and IMP-8 plasma and radio wave data has been performed to try to find electron plasma oscillations associated with type III radio noise bursts and low-energy solar electrons. This study shows that electron plasma oscillations are seldom observed in association with solar electron events and type III radio bursts at 1.0 AU. For the one case in which electron plasma oscillations are definitely produced by the electrons ejected by the solar flare the electric field strength is relatively small, only about $100\,\mu$V m^{-1}. Quantitative calculations of the rate of conversion of the plasma oscillation energy to electromagnetic radiation are presented for plasma oscillations excited by both solar electrons and electrons from the bow shock.

077.016 Spectral peculiarities of a local source of radio emission connected with the August 1972 sunspot group.
Sh. B. Akhmedov, V. M. Bogod, V. N. Borovik, G. B. Gel'frejkh, V. N. Ikhsanova, A. N. Korzhavin, N. G. Peterova.
Solnechnye Dannye 1975 Byull., No. 11, p. 97 - 104 (1976). In Russian.

Some data on radio emission observations taken with the Large Pulkovo Radio Telescope are presented. The observations were made at six wavelengths in the 2.3 - 9.0 cm range of a local source connected with the August 1972 sunspot group. This local source is characterised by high values of the radio flux (30–60 S.U.). The dynamics of changes in the flux and circular polarized component is considered. Peculiarities of radio emission for this local source might be due to an abnormally strong heating and lowering of the corona over sunspots.

077.017 High spectral- and time-resolution observations of decametric solar radio bursts.
H. S. Sawant, S. K. Alurkar, R. V. Bhonsle.
Bull. Astron. Soc. India, Vol. 3, 35 (1975). — Abstract of a paper presented at the A.S.I. meeting 1975.

077.018 Role of mode coupling in polarization of decametric type III solar bursts.
S. K. Mattoo, R. V. Bhonsle.
Bull. Astron. Soc. India, Vol. 3, 35 (1975). — Abstract of a paper presented at the A.S.I. meeting 1975.

077.019 On the origin of pulsations of type IV solar radio emission. Plasma cylinder oscillations (I).
V. V. Zajtsev, A. V. Stepanov.
Issled. po geomagnetizmu, aehron. i fiz. Solntsa. Vyp. (No.) 37. Moskva, Nauka, 1975, p. 3 - 10. In Russian. — Abstr. in Referativ. Zhurn. 51. Astron., 4.51.373 (1976).

077.020 On the origin of pulsations of type IV solar radio emission. Pulsating regime of beam and cone instabilities (II). V. V. Zajtsev, A. V. Stepanov.
Issled. po geomagnetizmu, aehron. i fiz. Solntsa. Vyp. (No.) 37. Moskva, Nauka, 1975, p. 11 - 18. In Russian. — Abstr. in Referativ. Zhurn. 51. Astron., 4.51.374 (1976).

077.021 Fluctuations of polarized solar radio emission at 3.2 cm wavelength and flare activity of sunspot groups. G. B. Gel'frejkh, V. G. Ledenev, V. P. Nefed'ev.
Issled. po geomagnetizmu, aehron. i fiz. Solntsa. Vyp. (No.) 37. Moskva, Nauka, 1975, p. 19 - 23. In Russian. — Abstr. in Referativ. Zhurn. 51. Astron., 4.51.375 (1976).

077.022 On the development of the proton region in August 1971.
M. N. Belovskij, Yu. B. Vedeneev.
Solnechnye Dannye 1976 Byull., No. 1, p. 61 - 65 (1976). In Russian.

The radio events in August 1971 recorded at NIRFI and other world-wide stations are compared with the optic events in the active region, the changes of polarity of the interplanetary magnetic field near the earth, the increase of the solar proton flux near the earth and the principal geomagnetic storms. A model of the development of the active region and the accompanying events is proposed.

077.023 On very fast frequency drifting solar radio bursts with short decay time.
N. N. Gerasimova, I. M. Gordon.
Astron. Zhurn. Akad. Nauk SSSR, Vol. 53, 380 - 388 (1976). In Russian. English translation in Soviet Astron., Vol. 20, No. 2.

An analysis of the radio echo probing of solar data reveals the existence of regions of rapid acceleration and intense heating of the solar wind. Investigations were made to find out what peculiarities are acquired by bursts of type III excited in these regions. Numerical solution of the equations governing the evolution of the Langmuir waves spectrum excited by the electron beam enables to demonstrate that along with a very fast frequency drift these bursts are distinguished by very short time of decay determined by Landau damping.

077.024 Propagation of interplanetary shock waves by observations of type II solar radio bursts on IMP-6.
I. M. Chertok, V. V. Fomichev.
Planet. Space Sci., Vol. 24, 459 - 464 (1976).

A new interpretation of the low frequency type II solar radio bursts of 30 June 1971, and 7−8 August 1972 observed with IMP-6 satellite is suggested. The analysis is carried out for two models of the electron density distribution in the interplanetary medium taking into account that $N \sim 3.5$ cm^{-3} at a distance of 1 a.u. The radio data indicate essential deceleration of the shock waves during propagation from the sun up to 1 a.u. The characteristics of the shock waves obtained from the type II bursts agree with the results of the in situ observations.

077.025 Synchronous observations of solar bursts on Crimea and Cuba.
Yu. F. Yurovskij, L. I. Yurovskaya, O. Alvares.
Izv. Krymskoj Astrofiz. Obs., Vol. 54, 220 - 226 (1976). In Russian.

It is found that there is a difference between fine structure details of solar bursts on two records of the noise storm intensity received simultaneously in two remote points with the same equipment at the frequency of 220 MHz. The observational results could be explained by the high directivity ($\sim 10'$) of the emission of the sources of the fine structure of type I bursts.

077.026 On the solar events observed during August 3 - 6, 1972 at 1.9, 2.5 and 3.5 cm wavelengths.
A. F. Bachurin, A. S. Dvoryashin, N. N. Eryushev.
Izv. Krymskoj Astrofiz. Obs., Vol. 54, 227 - 232 (1976). In Russian.

The data on the solar radio emission obtained at 1.9, 2.5 and 3.5 cm wavelengths are discussed in connection with the complex spot group which appeared in the beginning of August, 1972.

077.027 On the brightness temperature of the quiet sun at centimeter wavelengths.
A. F. Bachurin, N. N. Eryushev.
Izv. Krymskoj Astrofiz. Obs., Vol. 54, 241 - 243 (1976). In Russian.

Measurements made at 1.9, 2.5 and 3.5 cm wavelengths and published data for other wavelengths were used for the determination of the brightness temperature of the quiet sun. It was found that the brightness temperature increases with wavelength according to $\lambda^{3/2}$ in the centimeter spectral range.

077.028 Quasi-periodic components of the fluctuations of the solar radio emission at $\lambda = 3$ cm as reflections of pre-flare states for active regions in July - August 1972.
A. I. Korshunov, N. A. Prokof'eva.
Solnechnye Dannye 1976 Byull., No. 2, p. 52 - 56 (1976). In Russian.

Results of an investigation of the variation of intensity and inclination of the spectrum of the solar radio emission at $\lambda = 3$ cm during July 28 - August 11, 1972 are given. It is shown that the variation of the periods of the fluctuations is due to the pre-flare situation in an active region. A connection of the observed fluctuations with the oscillations in coronal condensations is likely to exist.

077.029 The spectrum of the continuum of noise storms and its relationship with active regions.   G. F. Eliseev.
Solnechnye Dannye 1976 Byull., No. 2, p. 62 - 70 (1976). In Russian.

Power spectra of 17 sources of noise storms backgrounds at the frequency of 208 MHz are calculated. The relationship between periodic components, total area of the sunspot groups, area of the leading sunspot, burst activity and nonstationary processes occurring in the active regions are examined.

077.030 The time and polarization profile of type III solar radio bursts at meter wavelengths.   P. Santin.
Astron. Astrophys., Vol. 49, 193 - 195 (1976).

An analysis of high resolution time profiles of type III bursts has been made with particular care to polarization fine structures. All the bursts may be grouped according to some main features like the time delay between the onset of the two circularly polarized components or the polarization fine structures.

077.031 On the interpretation of the spectra of solar microwave bursts.   V. A. Kovalev, O. S. Korolev.
Issled. po probl. solnech.-zemn. fiz. Moskva, Nauka, 1975, p. 80 - 92. In Russian. − Abstr. in Referativ. Zhurn. 51. Astron., 5.51.483 (1976).

077.032 On the fluctuating structure of solar radio bursts in the m-wave region.   G. F. Eliseev.
Solar phenomena research, Vladivostok, 1975, (see 003.005), p. 92 - 99. In Russian. − Abstr. in Referativ. Zhurn. 51. Astron., 5.51.484 (1976).

077.033 On the relation between background fluctuations of noise storms and strength of sunspot magnetic fields.   G. F. Eliseev, L. F. Lazareva, V. F. Chistyakov.
Solar phenomena research, Vladivostok, 1975, (see 003.005), p. 100 - 110. In Russian. − Abstr. in Referativ. Zhurn. 51. Astron., 5.51.487 (1976).

077.034 On the prediction of the shift of the gravity centre of the S-component of solar radio emission.
V. A. Krylov, A. Kurbanov.
Dokl. AN TadzhSSR, Vol. 18, No. 8, p. 10 - 14 (1975). In Russian. − Abstr. in Referativ. Zhurn. 51. Astron., 5.51.491 (1976).

077.035 Solar activity on August 12, 1975.
B. J. LaBonte.
Nature, Vol. 261, 525 (1976).

077.036 The solar microwave "negative burst" associated with the dark fan of 21 May 1967.   C. Sawyer.
Journ. Roy. Astron. Soc. Canada, Vol. 70, 127 - 134 (1976).

A striking fan of bright and dark surges in Hα and a rare microwave decrease both followed a flare with an associated microwave great burst. Although the optical and radio "absorption" events must surely have been related, the surges occurred earlier than the flux decrease. In this selected event the microwave decrease was a strong effect that apparently accompanied an optical feature that was only weakly visible, and was seen only in the wing of Hα. An explanation of the microwave decrease could be the disappearance or weakening of a radio source, associated with a restructuring of the magnetic fields in the regions.

077.037 Radio investigation of the sun. − Matthew Flinders lecture 1974.   J. P. Wild.
Records Australian Acad. Sci., Vol. 3, (No.1), 93 - 115 = Div. Radiophys., CSIRO, Sydney, Radiophys. Publ. RPP 1822 (1974).

077.038 Collective wave-particle interactions in solar type IV radio sources.   J. Kuijpers.
Utrechtse Sterrekundige OverdrukkenNo. 301, 72 pp. = Proefschr. Wis. Nat., Rijksuniv. Utrecht, (1975).

077.039 **The 127 MHz routine solar observations at the Toruń Observatory.** K. M. Borkowski.
Postępy Astron., Vol. 24, 15 - 31 (1976). In Polish.
In this article methods of solar observations with a simple interferometer are described in detail. A new proposed definition of the variability index as well as an adopted classification of the radio events are presented and discussed. The paper contains valuable information for all those, who use in any form, or wish to use the Toruń 127 MHz results.

077.040 **The variability index for solar radio emission.** K. M. Borkowski.
Postępy Astron., Vol. 24, 115 - 123 (1976). In Polish.

077.041 **Decameter storm and type IV radiation.** T. E. Gergely.
Diss. Fac. Graduate School Univ. Maryland. 161 pp. (1974).

077.042 **An analysis of type III burst positions.** T. B. H. Kuiper.
Diss. Fac. Graduate School Univ. Maryland. 147 pp. (1973).

077.043 **Observations of type IIIb solar radio bursts.** J. J. Riihimaa.
Ann Acad. Sci. Fenn., Ser. A VI,No. 417, 12 pp. (1975).

077.044 **Observation by high temporal and spectral resolution of sporadic emissions from the sun and Jupiter.**
A. Boischot, C. Rosolen.
Ann. Telecommun. (France), Vol. 30, 247 - 250 (1975). In French. – Abstr. in Phys. Abstr., Vol. 79, A028045 (1976).

077.045 **Polarization characteristics of a group of spectral type III solar radio bursts at 25 MHz recorded on 14 July 1969.** S. K. Mattoo, R. V. Bhonsle, S. K. Alurkar.
Indian Journ. Radio Space Phys., Vol. 4, 116 - 119 (1975). Abstr. in Phys. Abstr., Vol. 79, A037441 (1976).

077.046 **Observations of type IIIb solar radio bursts.** J. J. Riihimaa.
Ann. Acad. Sci. Fenn., Ser. A VI, No. 417, p. 1 - 12 (1975). Abstr. in Phys. Abstr., Vol. 79, A042029 (1976).

077.047 **OGO-V radio burst analysis.** F. T. Haddock.
Separate print Michigan Univ., Ann Arbor, USA, 3 pp. (1975).

077.048 **Solar radio continuum storms.**
Separate print Washington Data Processing, Inc., Riverdale, Maryland, USA, 57 pp. (1974). (Available from NTIS).

077.049 **Solar radio continuum storms and a breathing magnetic field model. Final report.**
Separate print Washington Data Processing, Inc., Riverdale, Maryland, USA, 51 pp. (1975). (Available from NTIS).

Solar noise storms. See Abstr. 003.040.

Radioheliography. See Abstr. 031.252.

Investigation of the sporadic radio radiation of the sun and of the parameters of the earth's ionosphere aboard the satellite Intercosmos-Copernicus 500. I. Scientific apparatus and method of the experiments.
See Abstr. 032.519.

The use of the Westerbork Synthesis Radio Telescope for solar observations at 21 cm. See Abstr. 033.001.

Computer controlled radiospectrograph for recording the active solar radio emission in the range from 100–1,000 MHz. See Abstr. 033.002.

The capture of charged particles by transverse electromagnetic waves. See Abstr. 061.074.

Expansion of chromospheric matter in the gradual phase of solar flares. See Abstr. 073.022.

Methods of investigation and reduction of radio emission of solar filaments. See Abstr. 073.063.

Gamma-ray and radio evidence for the acceleration of high-energy particles in solar flares. See Abstr. 073.074.

Radio- and X-emission (and their polarization) at the time of early phase of solar flare connected with γ-line emissions. See Abstr. 073.092.

Coronal changes associated with a disappearing filament. See Abstr. 074.018.

A model of the solar transition region and corona for the minimum of activity. See Abstr. 074.035.

Coronal structures and particle acceleration studies from radioelectric and optical observations.
See Abstr. 074.086.

Solare Beobachtungsergebnisse (Solar Data).
See Abstr. 075.006.

Solar cycle variation of EUV radiation.
See Abstr. 076.028.

The corotation of the magnetic structure associated with Forbush decreases. See Abstr. 078.032.

Propagation characteristics of fast solar electrons in the inner solar system inferred from solar type III radio bursts.
See Abstr. 078.080.

Radio tracking of particles from the sun.
See Abstr. 078.109.

Radio observations of the solar eclipse on December 24, 1973. See Abstr. 079.104.

Propagation characteristics of electromagnetic waves in the magnetosphere. See Abstr. 084.230.

Radiometric studies of sky at centimetre wavelength.
See Abstr. 141.068.

078 Solar Cosmic Radiation

078.001 Abundances of solar cosmic ray protons and alpha particles for 1969–1972. M. A. Stroscio, L. Katz, G. K. Yates, B. Sellers, F. A. Hanser.
Journ. Geophys. Res., Vol. 81, 283 - 286 (1976).

The flare-averaged alpha particle to proton ratio reported here varies as $0.032\hat{\mathcal{E}}^{-0.3}$ in the particle energy range from 2.7 to 8.0 MeV/nucleon. Following the precedent of Lanzerotti and Robbins the alpha particle to proton ratios are determined for (1) equal energy/nucleon, (2) equal rigidity, and (3) equal energy/charge. Finally, a comparison of the yearly average of the total integrated proton and alpha particle fluxes is made for the maximum and descending periods of solar cycle 20.

078.002 Observations of >100-keV protons in the earth's magnetosheath. H. I. West, Jr., R. M. Buck.
Journ. Geophys. Res., Vol. 81, 569 - 584 (1976).

078.003 Anomaly in the quiet-time helium spectrum at 1 MeV per nucleon. G. Gloeckler, D. Hovestadt, B. Klecker, O. Vollmer, C. Y. Fan.
Astrophys. Journ., Vol. 204, 920 - 926 (1976).

An unusual spectral feature and anomalously large abundance of helium between 0.6 and ~2 MeV per nucleon observed during the most quiet time periods in 1974–1975 indicate the presence of low-energy helium of an unknown origin. The energy spectrum of helium below 0.6 MeV per nucleon is consistent with the proton spectrum below 1.5 MeV which has the form $E^{-1.8}$, and the proton-to-alpha ratio is about 30. These $\lesssim 1$ MeV particles are found to be emitted continuously by the sun even during its most inactive periods.

078.004 Forbush decreases in cosmic rays during solar activity increase. S. I. Ayubasheva.
Yaderno-geofiz. issledovaniya, Sverdlovsk, 1975, p. 108 - 120. In Russian. – Abstr. in Referativ. Zhurn. 51. Astron., 3.51. 351 (1976).

078.005 On the acceleration of multiple charged nuclei on the sun. L. V. Kurnosova, V. I. Logachev, L. A. Razorenov, M. I. Fradkin.
Kratkie soobshch. po fiz., 1975, No. 7, p. 18 - 22. In Russian. Abstr. in Referativ. Zhurn. 62. Issled. kosmich. prostranstva, 3.62.174 (1976).

078.006 Anisotropy of solar cosmic rays during the flare of April 29, 1973 from observations of Cosmos 555. G. A. Bazilevskaya, L. V. Kurnosova, V. I. Logachev, L. A. Razorenov, Yu. I. Stozhkov, M. I. Fradkin.
Kratkie soobshch. po fiz., 1975, No. 7, p. 23 - 27. In Russian. Abstr. in Referativ. Zhurn. 62. Issled. kosmich. prostranstva, 3.62.175 (1976).

078.007 The effect of the north-south asymmetry of solar cosmic rays and dynamics of a plasma layer and of the day-time polar cusp.
T. A. Ivanova, Eh. N. Sosnovets, L. V. Tverskaya.
Geomagn. Aeronom., Vol. 16, 159 - 163 (1976). In Russian.

078.008 On the anisotropy of relativistic solar protons from the flare of September 1, 1971.
S. N. Kuznetsov, Yu. I. Logachev, I. V. Petrova, I. A. Savenko.
Geomagn. Aeronom., Vol. 16, 181 - 182 (1976). In Russian. Brief information.

078.009 Stationary modulation of solar cosmic rays in the interplanetary space.
S. N. Vernov, V. V. Smirnova, B. A. Tverskoj.
Geomagn. Aeronom., Vol. 16, 215 - 220 (1976). In Russian.

078.010 Rocket measurements of abundances of solar cosmic ray nuclei of He to Si in January 25, 1971 solar event. S. Biswas, N. Durgaprasad, J. Nevatia.
Bull. Astron. Soc. India, Vol. 3, 37 (1975). – Abstract of a paper presented at the A.S.I. meeting 1975.

078.011 Energetic solar particles. S. Biswas.
Bull. Astron. Soc. India, Vol. 3, 68 - 74 (1975). Invited paper presented at the seminar on Solar Physics, Udaipur Solar Observatory, Vedhshala, Ahmedabad on 20th September, 1975.

078.012 Observation of solar cosmic rays during the flare of April 29, 1973.
G. A. Bazilevskaya, L. V. Kurnosova, V. I. Logachev, L. A. Razorenov, Yu. I. Stozhkov, M. I. Fradkin.
VIIth Leningrad seminar. "Corpuscular streams of the sun and the radiation belts of the earth and Jupiter", 1975, (see 012.007), p. 153 - 154 (1975). In Russian and English. – Abstr. in Referativ. Zhurn. 51. Astron., 4.51.326; 62. Issled. kosmich. prostranstva, 4.62.205 (1976).

078.013 Anisotropy of solar protons and inhomogeneities of the interplanetary medium.
G. P. Lyubimov, N. N. Kontor, N. V. Pereslegina, P. P. Ignat'ev.
VIIth Leningrad seminar. "Corpuscular streams of the sun and the radiation belts of the earth and Jupiter", 1975, (see 012.007), p. 155 - 156. In Russian and English. – Abstr. in Referativ. Zhurn. 51. Astron., 4.51.327; 62. Issled. kosmich. prostranstva, 4.62.206 (1976).

078.014 Quasistationary plasma streams during solar activity decay.
N. V. Pereslegina, G. P. Lyubimov, V. I. Tulupov.
VIIth Leningrad seminar. "Corpuscular streams of the sun and the radiation belts of the earth and Jupiter", 1975, (see 012.007), p. 185 - 192 (1975). In Russian. – Abstr. in Referativ. Zhurn. 51. Astron., 4.51.331; 62. Issled. kosmich. prostranstva, 4.62.185 (1976).

078.015 Studies of strong fluxes of solar plasma by means of cosmic rays. L. I. Dorman.
VIIth Leningrad seminar. "Corpuscular streams of the sun and the radiation belts of the earth and Jupiter", 1975, (see 012.007), p. 193 - 211 (1975). In Russian. – Abstr. in Referativ. Zhurn. 51. Astron., 4.51.332; 62. Issled. kosmich. prostranstva, 4.62.209 (1976).

078.016 Connection of energetic particles during a storm with flare-generated interplanetary shock waves.
S. Pinter.
VIIth Leningrad seminar. "Corpuscular streams of the sun and the radiation belts of the earth and Jupiter", 1975, (see 012.007), p. 212 - 214 (1975). In Russian and English. – Abstr. in Referativ. Zhurn. 51. Astron., 4.51.333; 62. Issled. kosmich. prostranstva, 4.62.210 (1976).

078.017 Corpuscular streams of the sun and radiation belts of the earth and Jupiter. G. E. Kocharov.
VII Leningrad seminar. "Corpuscular streams of the sun and the radiation belts of the earth and Jupiter", 1975, (see 012.007), p. 10 - 19 (1975). In Russian. – Abstr. in Referativ. Zhurn. 62. Issled. kosmich.prostranstva, 4.62.199 (1976).

078.018 On the spatial distribution of solar cosmic rays in the earth's magnetosphere.
N. A. Mikirova, N. A. Mikryukova, M. N. Nazarova, N. K. Pereyaslova, I. E. Petrenko.
VIIth Leningrad seminar. "Corpuscular streams of the sun and the radiation belts of the earth and Jupiter", 1975, (see 012.007), p. 157 - 158 (1975). In Russian. – Abstr. in Referativ. Zhurn. 62. Issled. kosmich. prostranstva, 4.62.207 (1976).

078.019 Anomalies of the 11-year cycle of cosmic radiation.
S. N. Vernov, A. N. Charakhch'yan, G. A. Bazilevskaya, Yu.I. Stozhkov, T. N. Charakhch'yan.
VIIth Leningrad seminar. "Corpuscular streams of the sun and the radiation belts of the earth and Jupiter", 1975, (see 012.007), p. 389 - 398 (1975). In Russian. – Abstr. in Referativ. Zhurn. 62. Issled. kosmich. prostranstva, 4.62.212 (1976).

078.020 North-south asymmetry and anisotropy of solar cosmic radiation during the flare on April 18, 1972.
T. A. Ivanova, S. N. Kuznetsov, Yu. I. Logachev, Eh. N. Sosnovets.
Kosmich. Issled., Vol. 14, 235 - 238 (1976). In Russian.

078.021 On the anomalous anisotropy of the flux of solar cosmic radiation in September 1973.
S. N. Vernov, N. N. Kontor, G. P. Lyubimov, P. P. Ignat'ev, E. V. Gorchakov, B. A. Tverskoj, E. A. Chuchkov, T. E. Shvidkovskaya.
Kosmich. Issled., Vol. 14, 239 - 247 (1976). In Russian.

078.022 Some regularities of the connection of solar cosmic rays in auroral regions with sector structure of the interplanetary magnetic field. N. K. Pereyaslova,
M. N. Nazarova, S. M. Mansurov, L. G. Mansurova.
Geomagn. Aeronom., Vol. 16, 407 - 412 (1976). In Russian.

078.023 Some aspects of particle acceleration on the sun.
L. I. Miroshnichenko.
Issled. po probl. solnech.-zemn. fiz. Moskva, Nauka, 1975, p. 93 - 102. In Russian. – Abstr. in Referativ. Zhurn. 51. Astron., 5.51.417 (1976).

078.024 On the spectrum of particles accelerated through the interaction with fluctuational electromagnetic fields under powerful radio emission propagation in the solar atmosphere. L. V. Rajchenko.
Apparatur. i metod. razrabotki v geofiz. Kiev, Nauk. dumka, 1975, p. 92 - 95. In Russian. – Abstr. in Referativ. Zhurn. 51. Astron., 5.51.485 (1976).

078.025 Isotopic composition of the corpuscular streams of the sun and of the earth's atmosphere.
G. E. Kocharov.
Kosmogen. radioizotopy. Vyp. (No.) 3. Vil'nyus, 1975, p. 5 - 16. In Russian. – Abstr. in Referativ. Zhurn. 62. Issled. kosmich. prostranstva, 5.62.236 (1976).

078.026 Solar cosmic ray diffusion into the polar magnetosphere. M. Yu. Medvedev, V. S. Smirnov.
Cosmic rays, No. 15, (see 003.006), p. 19 - 23 (1975). In Russian.

078.027 Anisotropy and spectrum of solar cosmic rays in the interplanetary space.
L. I. Miroshnichenko, M. O. Sorokin.
Cosmic rays, No. 15, (see 003.006), p. 118 - 122 (1975). In Russian.

078.028 Generation of continuous fluxes of protons, electrons and nuclei with $Z \geqslant 2$ during the different periods of solar activity.

V. A. Kobzev, E. V. Kolomeets, V. P. Shabansky (*Shabanskij*).
14th Intern. Cosmic Ray Conf., (see 012.011), Vol. 2, 764 - 767 (1975).

078.029 The energy spectra of protons and alpha particles above 300 keV/nucleon during quiet times.
G. Gloeckler, D. Hovestadt, B. Klecker, O. Vollmer, C. Y. Fan.
14th Intern. Cosmic Ray Conf., (see 012.011), Vol. 2, 768 - 773 (1975).

078.030 The quiet time spectra of low energy hydrogen and helium nuclei.
R. A. Mewaldt, E. C. Stone, R. E. Vogt.
14th Intern. Cosmic Ray Conf., (see 012.011), Vol. 2, 774 - 779 (1975).

078.031 Studies of solar cosmic ray propagation on the basis of kinetic equation.
L. I. Dorman, M. O. Sorokin, M. E. Katz (*Kats*).
14th Intern. Cosmic Ray Conf., (see 012.011), Vol. 3, 931 - 935 (1975).

078.032 The corotation of the magnetic structure associated with Forbush decreases.
N. Iucci, M. Parisi, M. Storini, G. Villoresi, N. L. Zangrilli.
14th Intern. Cosmic Ray Conf., (see 012.011), Vol. 3, 1064 - 1068 (1975).

078.033 Studies of two-step Forbush-decreases.
A. V. Belov, L. I. Dorman, E. A. Eroshenko, K. G. Ivanov, O. I. Inozemtseva, G. I. Kulanina, N. V. Mikerina.
14th Intern. Cosmic Ray Conf., (see 012.011), Vol. 3, 1069 (1975). – Abstract.

078.034 Isotropic and anisotropic components of cosmic ray increase effect prior to Forbush-decrease.
L. I. Dorman, N. S. Kaminer, Yu. I. Okulov, T. S. Khadakhanova.
14th Intern. Cosmic Ray Conf., (see 012.011), Vol. 3, 1071 - 1075 (1975).

078.035 Asymmetric model of the Forbush-effect recovery phase. A. V. Belov, L. I. Dorman, B. A. Shakhov.
14th Intern. Cosmic Ray Conf., (see 012.011), Vol. 3, 1076 - 1081 (1975).

078.036 The Forbush decrease of November 17, 1966.
H. S. Ahluwalia.
14th Intern. Cosmic Ray Conf., (see 012.011), Vol. 3, 1086 (1975). – Abstract.

078.037 Relation of large-scale coronal X-ray structure and cosmic rays: 5. Solar wind and coronal influence on a Forbush decrease lasting one solar rotation.
R. E. Gold, E. C. Roelof, J. T. Nolte, A. S. Krieger.
14th Intern. Cosmic Ray Conf., (see 012.011), Vol. 3, 1095 - 1100 (1975).

078.038 A simple model for the Forbush decrease in the galactic cosmic ray flux. C. J. Bland.
14th Intern. Cosmic Ray Conf., (see 012.011), Vol. 3, 1101 (1975). – Abstract.

078.039 The possible penetration of solar electrons after solar event on September 24, 1974, on Apatity ground based and stratospheric data. L. P. Borovkov,
M. I. Beloglazov, V. M. Driatsky (*Driatskij*), V. D. Khorkov, L. L. Lazutin, O. I. Shumilov, E. V. Vashenuk.
14th Intern. Cosmic Ray Conf., (see 012.011), Vol. 4, 1273 - 1277 (1975).

078.040 Seasonal and daily variation of directions of approach to ground and balloon stations.
A. Orozco, R. Gall.
14th Intern. Cosmic Ray Conf., (see 012.011), Vol. 4, 1278 - 1283 (1975).

078.041 Cutoff rigidity variations of European mid-latitude stations during the September 1974 Forbush
decrease. E. Flückiger, H. Debrunner, M. Arens, O. Binder.
14th Intern. Cosmic Ray Conf., (see 012.011), Vol. 4, 1331 - 1335 (1975).

078.042 An application of statistical adjustment of data to the energetic solar cosmic ray increase of August 7,
1972. H. Komori.
14th Intern. Cosmic Ray Conf., (see 012.011), Vol. 4, 1336 - 1340 (1975).

078.043 Acceleration of H, He, and heavy ions observed in the magnetosheath, magnetotail, and near-by interplanetary space. C. Y. Fan, G. Gloeckler, D. Hovestadt.
14th Intern. Cosmic Ray Conf., (see 012.011), Vol. 4, 1362 - 1367 (1975).

078.044 Features of the effect of the low-energy solar cosmic rays and precipitating particles on the ionosphere.
L. I. Dorman, I. V. Dorman, T. M. Krupitskaya.
14th Intern. Cosmic Ray Conf., (see 012.011), Vol. 4, 1435 (1975). – Abstract.

078.045 Possibility of studying the solar cosmic rays and the geomagnetic threshold variations at high latitudes
basis of ionospheric effects. L. I. Dorman, T. M. Krupits-
kaya.
14th Intern. Cosmic Ray Conf., (see 012.011), Vol. 4, 1436 - 1440 (1975).

078.046 The time and spatial behavior of solar flare proton anisotropies observed in deep space on Pioneers 10
and 11. J. McCarthy, J. J. O'Gallagher.
14th Intern. Cosmic Ray Conf., (see 012.011), Vol. 4, 1526 - 1531 (1975).

078.047 Solar cosmic ray events at large radial distances from the sun. R. Zwickl, W. R. Webber, F. B.
McDonald, B. J. Teegarden, J. H. Trainor.
14th Intern. Cosmic Ray Conf., (see 012.011), Vol. 4, 1532 (1975). – Abstract.

078.048 Relative abundance of proton to helium nuclei in solar cosmic ray events. M. A. I. Van Hollebeke.
14th Intern. Cosmic Ray Conf., (see 012.011), Vol. 5, 1563 - 1567 (1975).

078.049 Composition of solar energetic particles.
T. T. von Rosenvinge, F. B. McDonald, V. K.
Balasubrahmanyan.
14th Intern. Cosmic Ray Conf., (see 012.011), Vol. 5, 1574 - 1575 (1975).

078.050 Abundances, charge states, and energy spectra of helium and heavy ions during solar particle events.
G. Gloeckler, R. Sciambi, C. Y. Fan, D. Hovestadt.
14th Intern. Cosmic Ray Conf., (see 012.011), Vol. 5, 1576 - 1581 (1975).

078.051 Measurements of He/O and He/Ne ratios in the low energy solar cosmic rays in the January 24, 1971
event. N. Durgaprasad, J. Nevatia, S. Biswas.
14th Intern. Cosmic Ray Conf., (see 012.011), Vol. 5, 1582 - 1585 (1975).

078.052 The abundances of solar O, Ne, Mg and Si nuclei in the January 24, 1971 event and their implications.
J. Nevatia, S. Biswas.
14th Intern. Cosmic Ray Conf., (see 012.011), Vol. 5, 1586 - 1590 (1975).

078.053 On the overabundance of heavy nuclei in the generation process of solar cosmic rays.
J. Pérez-Peraza, J. Galindo Trejo.
14th Intern. Cosmic Ray Conf., (see 012.011), Vol. 5, 1591 (1975). – Abstract.

078.054 Variations in the charge composition of the July 2 - 12, 1974, solar particle event.
T. P. Armstrong, S. M. Krimigis.
14th Intern. Cosmic Ray Conf., (see 012.011), Vol. 5, 1592 - 1596 (1975).

078.055 Solar and galactic cosmic ray abundances – a comparison and some comments. W. R. Webber.
14th Intern. Cosmic Ray Conf., (see 012.011), Vol. 5, 1597 - 1602 (1975).

078.056 The acceleration of heavy nuclei on the sun.
N. A. Dobrotin, L. V. Kurnosova, V. I. Logachev,
L. A. Razorenov, M. I. Fradkin.
14th Intern. Cosmic Ray Conf., (see 012.011), Vol. 5, 1603 - 1606 (1975).

078.057 Enhanced solar cosmic-ray flux in the early solar system. K. Gopalan, M. N. Rao.
14th Intern. Cosmic Ray Conf., (see 012.011), Vol. 5, 1607 - 1612 (1975).

078.058 Heavy particle emission of unusual composition from the sun. D. Hovestadt, B. Klecker, O.
Vollmer, G. Gloeckler, C. Y. Fan.
14th Intern. Cosmic Ray Conf., (see 012.011), Vol. 5, 1613 - 1618 (1975).

078.059 ^{3}He abundance in solar energetic particle events.
F. B. McDonald, T. T. von Rosenvinge, A. T.
Serlemitsos, V. K. Balasubrahmanyan.
14th Intern. Cosmic Ray Conf., (see 012.011), Vol. 5, 1619 (1975). – Abstract.

078.060 On the nuclear reactions during solar flares.
I. A. Ibragimov, G. E. Kocharov.
14th Intern. Cosmic Ray Conf., (see 012.011), Vol. 5, 1620 - 1623 (1975).

078.061 Observations of hydrogen and helium isotopes in solar cosmic rays.
G. J. Hurford, E. C. Stone, R. E. Vogt.
14th Intern. Cosmic Ray Conf., (see 012.011), Vol. 5, 1624 - 1629 (1975).

078.062 Calculation of solar neutron and gamma ray fluxes during the flares and the quiet periods.
A. B. Baisakalova (Bajsakalova), O. A. Bogdanova, E. V.
Kolomeets.
14th Intern. Cosmic Ray Conf., (see 012.011), Vol. 5, 1638 - 1643 (1975).

078.063 Low energy neutrons from sun. H. S. Ahluwalia.
14th Intern. Cosmic Ray Conf., (see 012.011),
Vol. 5, 1655 (1975). – Abstract.

078.064 The exponential decay of solar flare particles. I. Eastern hemisphere events.
G. Wibberenz, R. Reinhard.

14th Intern. Cosmic Ray Conf., (see 012.011), Vol. 5, 1681 - 1686 (1975).

078.065 The exponential decay of solar flare particles. II. Western hemisphere events. R. Reinhard.
14th Intern. Cosmic Ray Conf., (see 012.011), Vol. 5, 1687 - 1691 (1975).

078.066 Observation using interplanetary scintillations at 34.3 MHz of the effect of a solar wind disturbance on a solar energetic particle event.
E. C. Roelof, S. M. Krimigis, W. M. Cronyn, S. D. Shawhan, P. S. McIntosh.
14th Intern. Cosmic Ray Conf., (see 012.011), Vol. 5, 1692 - 1697 (1975).

078.067 Relation of large-scale coronal X-ray structure and cosmic rays: III. Low-intensity solar particle events with enhanced ~3 MeV helium and medium fluxes associated with solar wind streams. R. E. Gold, S. M. Krimigis, E. C. Roelof, A. S. Krieger, J. T. Nolte.
14th Intern. Cosmic Ray Conf., (see 012.011), Vol. 5, 1710 - 1715 (1975).

078.068 Scatter-free collimated convection and cosmic-ray transport at 1 AU. E. C. Roelof.
14th Intern. Cosmic Ray Conf., (see 012.011), Vol. 5, 1716 - 1721 (1975).

078.069 Mathematical formulation of scatter-free propagation of solar cosmic rays.
J. T. Nolte, E. C. Roelof.
14th Intern. Cosmic Ray Conf., (see 012.011), Vol. 5, 1722 - 1727 (1975).

078.070 Late-time distribution of solar-flare cosmic rays — its dependence on the functional form of the diffusion coefficient. C. K. Ng.
14th Intern. Cosmic Ray Conf., (see 012.011), Vol. 5, 1739 - 1743 (1975).

078.071 A model for the propagation of solar-flare cosmic rays. C. K. Ng, L. J. Gleeson.
14th Intern. Cosmic Ray Conf., (see 012.011), Vol. 5, 1744 - 1748 (1975).

078.072 On the influence of injection profiles and of interplanetary propagation on the time-intensity- and time-anisotropy-profiles of solar cosmic-rays at 1 AU.
B.-M. Schulze, A. K. Richter, G. Wibberenz.
14th Intern. Cosmic Ray Conf., (see 012.011), Vol. 5, 1749 - 1753 (1975).

078.073 Solar proton pitch angle distribution for the January 24, 1969 event. A. Balogh, J. J. Quenby.
14th Intern. Cosmic Ray Conf., (see 012.011), Vol. 5, 1754 - 1759 (1975).

078.074 Energy changes and solar cosmic-rays.
G. M. Webb, L. J. Gleeson.
14th Intern. Cosmic Ray Conf., (see 012.011), Vol. 5, 1760 (1975). — Abstract.

078.075 On the propagation of low energy solar protons in the interplanetary medium.
E. Amata, V. Domingo, D. E. Page, K.-P. Wenzel.
14th Intern. Cosmic Ray Conf., (see 012.011), Vol. 5, 1761 - 1766 (1975).

078.076 Propagation characteristics of solar flare particles.
L. S. Ma Sung, M. A. I. Van Hollebeke, F. B.

McDonald.
14th Intern. Cosmic Ray Conf., (see 012.011), Vol. 5, 1767 - 1772 (1975).

078.077 The interplanetary acceleration of energetic nucleons. F. B. McDonald, B. J. Teegarden, J. H. Trainor, T. T. von Rosenvinge, W. R. Webber.
14th Intern. Cosmic Ray Conf., (see 012.011), Vol. 5, 1773 (1975). — Abstract.

078.078 On cosmic ray flares.
N. P. Chirkov, A. T. Filippov.
14th Intern. Cosmic Ray Conf., (see 012.011), Vol. 5, 1774 - 1778 (1975).

078.079 Transport path of solar protons with energies $E_p \sim 1-5$ MeV.
E. V. Gorchakov, G. A. Timofeev, T. I. Morozova.
14th Intern. Cosmic Ray Conf., (see 012.011), Vol. 5, 1779 - 1783 (1975).

078.080 Propagation characteristics of fast solar electrons in the inner solar system inferred from solar type III radio bursts. R. P. Lin, K. A. Anderson, H. Alvarez.
14th Intern. Cosmic Ray Conf., (see 012.011), Vol. 5, 1784 (1975). — Abstract.

078.081 Observation of low energy electrons and protons from solar active regions.
R. E. McGuire, R. P. Lin, K. A. Anderson.
14th Intern. Cosmic Ray Conf., (see 012.011), Vol. 5, 1785 - 1790 (1975).

078.082 Anisotropic arrival of relativistic solar flare particles on the earth.
M. Kodama, M. Wada, K. Murakami.
14th Intern. Cosmic Ray Conf., (see 012.011), Vol. 5, 1791 - 1796 (1975).

078.083 Some data on solar flare protons on the 29th April 1973. N. A. Dobrotin, L. V. Kurnosova, V. I. Logachev, L. A. Razorenov, M. I. Fradkin.
14th Intern. Cosmic Ray Conf., (see 012.011), Vol. 5, 1797 - 1800 (1975).

078.084 Anisotropy measurements of ~50 keV solar protons.
R. E. Gold, C. O. Bostrom, E. C. Roelof, D. J. Williams.
14th Intern. Cosmic Ray Conf., (see 012.011), Vol. 5, 1801 - 1806 (1975).

078.085 The effect of interplanetary discontinuities on the angular distributions of energetic particles.
K. C. Hsieh, Y. C. Lin, J. D. Sullivan, I. Lerche.
14th Intern. Cosmic Ray Conf., (see 012.011), Vol. 5, 1807 (1975). — Abstract.

078.086 Flare-generated interplanetary shock wave deceleration effects in low-energy storm particle events.
S. Pinter.
14th Intern. Cosmic Ray Conf., (see 012.011), Vol. 5, 1808 - 1812 (1975).

078.087 Energetic storm particles and their relation to tangential discontinuities.
R. A. R. Palmeira, R. A. Medrano, I. J. Kantor.
14th Intern. Cosmic Ray Conf., (see 012.011), Vol. 5, 1813 (1975). — Abstract.

078.088 Effect of interplanetary shock waves on solar cosmic rays. M. Scholer, G. Morfill.

14th Intern. Cosmic Ray Conf., (see 012.011), Vol. 5, 1820 - 1822 (1975).

078.089 Upper cutoff in the spectrum of high energy solar particles during cycles 19—20.
Dj. Heristchi, J. Pérez-Peraza, G. Trottet.
14th Intern. Cosmic Ray Conf., (see 012.011), Vol. 5, 1841 - 1846 (1975).

078.090 Angular distribution of solar protons in various distances from the sun.
E. Keppler, E. Nielsen, K. Richter, B. Wilken, D. Williams.
14th Intern. Cosmic Ray Conf., (see 012.011), Vol. 5, 1855 (1975). — Abstract.

078.091 Propagation of solar particles from solar events detected by Helios A.
M. A. I. Van Hollebeke, J. A. Trainor, K. G. McCracken.
14th Intern. Cosmic Ray Conf., (see 012.011), Vol. 5, 1856 (1975). — Abstract.

078.092 Low energy particle composition.
G. Gloeckler.
14th Intern. Cosmic Ray Conf., (see 012.011), Vol. 11, 3784 - 3804 (1975).

078.093 Coronal and interplanetary propagation of solar cosmic rays.   R. A. R. Palmeira.
14th Intern. Cosmic Ray Conf., (see 012.011), Vol. 11, 3805 - 3819 (1975).

078.094 The Forbush decrease of November 17, 1966.
H. S. Ahluwalia.
14th Intern. Cosmic Ray Conf., (see 012.011), Vol. 12, 4201 - 4206 (1975).

078.095 Solar cosmic ray events at large radial distances from the sun.   R. Zwickl, W. R. Webber, F. B. McDonald, B. Teegarden, J. Trainor.
14th Intern. Cosmic Ray Conf., (see 012.011), Vol. 12, 4239 - 4244 (1975).

078.096 Shape of the solar flare proton spectrum in 10—100 MeV region in the past.
N. Bhandari, S. K. Bhattacharya.
14th Intern. Cosmic Ray Conf., (see 012.011), Vol. 12, 4245 - 4246 (1975).

078.097 Low energy neutrons from sun.
H. S. Ahluwalia.
14th Intern. Cosmic Ray Conf., (see 012.011), Vol. 12, 4251 - 4256 (1975).

078.098 Two solar cosmic ray events measured on Helios-1.
G. Green, G. Wibberenz, R. Müller-Mellin, M. Witte, H. Hempe, H. Kunow.
14th Intern. Cosmic Ray Conf., (see 012.011), Vol. 12, 4257 - 4262 (1975).

078.099 Energy spectra of protons and alpha particles on Helios 1 at 0.31 ... 0.4 AU and 0.95 ... 0.98 AU.
M. Witte, H. Hempe, H. Kunow, G. Wibberenz, G. Green, R. Müller-Mellin, B. Iwers.
14th Intern. Cosmic Ray Conf., (see 012.011), Vol. 12, 4263 - 4267 (1975).

078.100 Variations of cosmic ray intensities during the first part of the Helios-1 mission.
H. Kunow, G. Wibberenz, G. Green, R. Müller-Mellin, M. Witte, H. Hempe, H. G. Hasler.
14th Intern. Cosmic Ray Conf., (see 012.011), Vol. 12,

4268 - 4272 (1975).

078.101 On the scattering mechanism of nonrelativistic solar electrons in the interplanetary medium.
V. G. Kurt, Yu. I. Logachev, N. F. Pisarenko.
Kosmich. Issled., Vol. 14, 378 - 382 (1976). In Russian.

078.102 ATS-6 solar cosmic ray and trapped particle experiment.   A. J. Masley, P. R. Satterblom, K. A. Pfitzer.
IEEE Trans. Aerospace Electron. Systems, Vol. AES-11, 1118 - 1124 (1975). — Abstr. in Phys. Abstr., Vol. 79, A037341 (1976).

078.103 Solar cosmic-ray events in the first half of the solar cycle 20.   D. Blanariu.
An. Stiint. 'Al. I. Cuza' Iasi (Ser. Noua) I B, Vol. 21, 77 - 82 (1975). — Abstr. in Phys. Abstr., Vol. 79, A041981 (1976).

078.104 High energy cosmic ray intensity variations associated with the unusual Forbush decrease of August 1972.   M. S. Radha, U. R. Rao, A. G. Ananth.
Indian Journ. Radio Space Phys., Vol. 4, 206 - 209 (1975). Abstr. in Phys. Abstr., Vol. 79, A041983 (1976).

078.105 Variation of solar proton energy spectra and size distribution with heliolongitude.
M. A. I. Van Hollebeke, L. S. Ma Sung, F. B. McDonald.
Separate print National Aeronaut. Space Administr., Greenbelt, Maryland, USA, Goddard Space Flight Center, 72 pp. (1974). See 13.078.006.

078.106 Observations of hydrogen and helium isotopes in solar cosmic rays.   G. J. Hurford.
Thesis California Inst. Tech., Pasadena, USA, 134 pp. (1975). (Available from: Univ. Microfilms, Order No. 75-12,515).

078.107 Unusual emission of iron nuclei from the sun.
G. Gloeckler, D. Hovestadt, O. Vollmer, C. Y. Fan.
Separate print Maryland Univ., College Park, USA, Dept. Phys. Astron., 14 pp. (1975). — See 14.078.004.

078.108 SOLPRO: a computer code to calculate probabilistic energetic solar proton fluxes.
E. G. Stassinopoulos.
Separate print National Aeronaut. Space Administr., Greenbelt, Maryland, USA, Goddard Space Flight Center, 27 pp. (1975).

078.109 Radio tracking of particles from the sun.
J. Fainberg.
Symposium significant accomplishments sci. technol., Greenbelt, Maryland, USA, 18 Dec. 1973, p. 104 - 107 (1975).

The capture of charged particles by transverse electromagnetic waves.   See Abstr. 061.074.

The effect of adiabatic focusing upon charged particle propagation in random magnetic fields.
See Abstr. 062.038.

Observation and analysis of low-energy solar particle propagation from discrete flare events.   See Abstr. 073.009.

Propagation of solar flare particles from discrete events.   See Abstr. 073.045.

Acceleration of energetic particles in solar flares.
See Abstr. 073.069.

Observations of $\lesssim$ 1 MeV/nuc protons and ions

during the September 1974 series of flares.
See Abstr. 073.071.

A search for solar flare positrons.
See Abstr. 073.075.

Gamma-ray and microwave evidence for two
phases of acceleration in solar flares.   See Abstr. 073.081.

Relation of large-scale coronal X-ray structure and
cosmic rays: II. Coronal control of interplanetary injection of
300 keV protons.   See Abstr. 074.058.

Recent perspectives in solar physics: elemental composition, coronal structure and magnetic fields, solar activity.
See Abstr. 074.059.

On the development of the proton region in
August 1971.   See Abstr. 077.022.

Nuclear astrophysics of the sun.
See Abstr. 080.050.

Solar proton and electron precipitation effects
detected by ionosondes.   See Abstr. 083.073.

Gradients of solar protons in the high-latitude
magnetotail and the magnetospheric electric field.
See Abstr. 084.219.

Multisatellite observations of solar protons penetrating the magnetopause.   See Abstr. 084.229.

Particle track record in lunar silicates: long-term
behavior of solar and galactic VH nuclei and lunar surface
dynamics.   See Abstr. 094.202.

The quiet-time low energy nucleon spectrum in the
vicinity of earth.   See Abstr. 099.050.

On the behaviour of the pitch angle diffusion coefficients near pitch angles of 90°.   See Abstr. 106.044.

Enhanced interplanetary magnetic fields as the
cause of Forbush decreases.   See Abstr. 106.046.

The intensity variations of solar and galactic cosmic
rays with azimuthal angle in the polar region.
See Abstr. 143.008.

Anomalous composition and energy spectra of
cosmic rays below 20 MeV/nucleon.   See Abstr. 143.167.

Synthesis of the results of cosmic ray intensity variation during the period of unusual Forbush decrease of
August 1972.   See Abstr. 143.221.

Interplanetary quiet time differential spectra of
protons and alpha particles below 30 MeV/nucleon from
Helios A.   See Abstr. 143.288.

Modulation and diffusion theory of cosmic rays.
See Abstr. 143.312.

Short term modulations and anisotropies.
See Abstr. 143.313.

Hysteresis of primary cosmic rays associated with
Forbush decreases.   See Abstr. 143.343.

Primary cosmic radiation.   See Abstr. 143.361.

Average flux of cosmic-ray nuclei at a distance of
1 AU from the sun.   See Abstr. 143.376.

Current cosmic-ray programmes in Europe.
See Abstr. 143.377.

079   Solar Eclipses

079.001   Observations of Baily's beads from near the northern limit of the total solar eclipse of June 20, 1974.
D. Herald.
The Moon, Vol. 15, 91 - 107 (1976).

Baily beads were observed from near the northern limit of totality during the eclipse of June 20, 1974; no attempt of identification was made at the time of observation. The observations were analysed to identify the features of the lunar profile which gave rise to the beads observed. They were then analysed to derive corrections to the relative positions of the sun and moon, and to investigate certain features of the lunar profile.

079.002   Total solar eclipses in Canada: 1963–2024 AD.
V. Gaizauskas, L. W. Avery.
Journ. Roy. Astron. Soc. Canada, Vol. 70, 135 - 137 (1976).

A map is presented of the six total eclipses of the sun with paths across Canada in the period 1963–2024 AD. Attention is drawn to the eclipse that will pass northwards through central Canada on 26 February 1979.

A study of records of the solar and lunar eclipses in scripts on tortoise-shells or ox-bones.
See Abstr. 004.009.

Solar eclipses mentioned in old Serbian annals.
See Abstr. 004.031.

Solar eclipse effects on HF and VLF propagation.
See Abstr. 083.033.

Periodicities of eclipses.   See Abstr. 095.005.

079.100   Solar eclipse 1974 June 20

Observations of Baily's beads from near the northern limit of the total solar eclipse of June 20, 1974.
See Abstr. 079.001.

079.101   Solar eclipse 1968 September 22

Radio observation of the solar eclipse of September 22, 1968 at the wavelength of 21 cm.
Solar Eclipse Observation Group, Peking Observatory, Academia Sinica.
Acta Astron. Sinica, Vol. 16, 189 - 199 (1975). In Chinese.

The association of radio sources with optical active regions is examined. Source flux densities, one-dimensional sizes, heights and brightness temperatures for these regions are given. The measure of emission from the sources above plages $N^2 L$ is calculated to be 1.8×10^{28} electron^2/cm^5.

Radio observation at the wavelength of 3 cm during the solar eclipse of September 22, 1968.
Eclipse Expedition Group of Peking Observatory, Academia Sinica; Eclipse Expedition Group of Nanking University.
Acta Astron. Sinica, Vol. 16, 200 - 209 (1975). In Chinese.

The following parameters have been obtained: angular size (along the direction of the motion of the moon's center on the sun's disk), flux density, average brightness temperature and height of the S-component sources associated with sunspots, plages and coronal condensations during the eclipse.

079.102   Solar eclipse 1976 April 29

The annular eclipse of the sun on April 29, 1976.
D. P. Duma.
Astron. Tsirk., No. 899, p. 5 - 7 (1976). In Russian.

Solar eclipse of 29 April 1976.   M. Dujnič.
Kozmos, Vol. 7, 23 - 24 (1976). In Slovak.

The annular solar eclipse of April 29, 1976.
D. P. Elias.
Mem. National Obs. Athens, Ser. I, Astron., No. 18, 14 pp. (1975).

Solar eclipse of 29 April 1976.   M. Dujnič.
Říše hvězd, Vol. 57, 50 - 52 (1976). In Slovak.

The solar eclipse of April 29th.
Sky Telescope, Vol. 51, 248 - 249 (1976).

The April solar eclipse – annular and partial.
Sky Telescope, Vol. 52, 69 - 74 (1976).

Observations: Solar eclipse on April 29, 1976.
A. Udalski, M. Zawilski.
Urania Kraków, Vol. 47, 119 - 122 (1976). In Polish.

079.103   Solar eclipse 1963 July 20

H and K (Ca II) emissions as observed in coronal spectrum in the July 20, 1963 solar eclipse.
See Abstr. 074.016.

079.104   Solar eclipse 1973 December 24

Observation of the partial solar eclipse on December 24, 1973 at the wavelength $\lambda = 3.04$ cm.
M. M. Kobrin, E. I. Lebedev, S. D. Snegirev, B. V. Timofeev.
Solnechnye Dannye 1975 Byull., No. 11, p. 77 - 81 (1976). In Russian.

Observations of the solar eclipse on 24 December 1973 were made at the wavelength $\lambda = 3.04$ cm in Santiago-de-Cuba. Data on fluxes and dimensions of the local sources in the southern hemisphere of the sun are presented. The radio radius of the sun is determined to be $(1.03 \pm 0.005) R_{\odot \, optic}$.

Radio observations of the solar eclipse on December 24, 1973.   V. M. Bogod, R. I. Enikeev, V. G. Nagnibeda.
Solnechnye Dannye 1976 Byull., No. 1, p. 66 - 70 (1976). In Russian.

Results of observations of the partial solar eclipse (Santiago de Cuba) at 4.0 cm wavelength are presented. Two elements coinciding in size and position with the leading and following parts of the bi-polar group were detected in a local source over the group. A strong source was found over a small but actively developing group; the source turned out to be appreciably displaced toward the limb in regard to the sunspots. Radio emission fluxes, sizes and brightness temperatures of the radio sources are estimated.

079.105   Solar eclipse 1976 October 23

October's total solar eclipse in Australia.
J. B. Trainor.
Sky Telescope, Vol. 51, 300 - 306 (1976).

Notes on the October total eclipse in Australia.
J. B. Trainor.
Sky Telescope, Vol. 51, 327 - 328 (1976).

079.106   Solar eclipse 1973 June 30

Eclipsspectra van de uiterste zonsrand.
J. Houtgast, O. Namba.
Repr. from Symposium fotonica, Eindhoven, 1975 April 2 - 4,
9 pp. = Utrechtse Sterrekundige Overdrukken No. 341 (1975).

Gravitational deflection of light: solar eclipse of
30 June 1973 I. Description of procedures and final results.
See Abstr. 066.078.

Gravitational deflection of light: solar eclipse of
30 June 1973 II. Plate reductions.   See Abstr. 066.079.

Absolute spectrophotometry of the chromosphere
in the region $\lambda\lambda$ 3824 - 3991 Å from photographs of the total
solar eclipse on June 30, 1973.   See Abstr. 073.056.

Absolute spectrophotometry of the chromospheric
lines Hϵ, H8, H9 and He 3888.65 Å from photographs of the
total solar eclipse on June 30, 1973.   See Abstr. 073.064.

079.107   Solar eclipse 1975 May 11

Saules aptumsums 1975. gada 11. maijā.
A, Maslovskis.
Zvaigžņotā debess, 1976. gada pavasaris, p. 47 - 49. – Solar
eclipse, 1975, May 11.

079.108   Solar eclipse 1958 April 19

Internal gravity waves deduced from the HF Doppler data during the April 19, 1958, solar eclipse.
See Abstr. 066.062.

080   Solar Atmosphere, Figure, Internal Constitution, Neutrinos, Magnetic Fields, Rotation, Miscellanea

080.001   Solar brightness temperature distribution at 350 and 450 microns.   G. Righini, M. Simon.
Astrophys. Journ., (Letters), Vol. 203, L95 - L97 (1976).

Solar scans, passband limited to effective wavelengths 350 and 450 μ, have been obtained with angular resolution of about 1'. The observations show little evidence of the limb brightening that would be expected from spherically symmetric homogeneous atmospheric models and the observed increase in brightness temperature in the regions above the solar atmosphere temperature minimum. The observations suggest that the complex structure that is observed at higher levels in the solar atmosphere extends down to the level of the temperature minimum.

080.002   Filtering of acoustic waves in the solar atmosphere.   J. Provost.
Astron. Astrophys., Vol. 46, 159 - 161 (1976).

The evolution with height of the power spectrum of the response of the solar atmosphere to an excitation in a wide range of frequencies is discussed. It is found that the layers under the photosphere enhance a range of periods around 300 s, while the photospheric and low chromospheric layers enhance 150–200 s periods. The linear filtering appears to be a process relevant for the understanding of major characteristics of the solar atmospheric dynamics.

080.003   Solar neutrinos: a scientific puzzle.   J. N. Bahcall, R. Davis, Jr.
Science, Vol. 191, 264 - 267 (1976).

080.004   On magnetohydrodynamic instability of a force-free field.   M. M. Molodenskij.
Solnechnye Dannye 1975 Byull., No. 10, p. 53 - 54 (1976). In Russian.

Force-free magnetic fields are shown to be unstable as referred to small-scale disturbances at $\nabla \alpha \neq 0$. These disturbances lead to divergence and displacement of lines of force in direction of the current density decrease.

080.005   A method of analytical representation of random fields for investigation of the sun's surface.
V. A. Magerramov.
Solnechnye Dannye 1975 Byull., No. 10, p. 75 - 78 (1976). In Russian.

A method of applying a polynomial of degree n for an analysis of random fields is described. Formulas are given for representation of a square area with n knots as a polynomial of degree 3. The method described can be used in the study of brightness fluctuations on the solar surface.

080.006   Long period magnetic activity – 2 to 100 years.   R. G. Currie.
Astrophys. Space Sci., Vol. 39, 251 - 254 (1976). – Research note.

080.007   The rapid oscillations of the solar rotation.   V. F. Chistyakov.
Bull. Astron. Inst. Czechoslovakia, Vol. 27, 84 - 91 (1976).

Photoheliograms from the years 1955–1963 and 1965–1973 were used for this investigation. The motion of the sunspots was studied. The following facts have been established: (1) the mean law of the solar rotation during 11 years described by the Faye formula $\xi(\varphi) = a - b \sin^2 \varphi$ is equal for the north and south hemispheres; (2) the form of the law of the solar rotation changes from year to year; (3) coefficients

a and b are proportional to each other; (4) the variation of the b coefficients for the northern and southern hemisphere is antiparallel; (5) the rapid oscillations of the solar rotation are accompanied by global meridional motions in the direction of that hemisphere where the rotation velocity is smaller; (6) these peculiarities of the rotation show that torsional oscillations exist on the sun.

080.008   Observations of solar pulsations.
A. B. Severny (Severnyj), V. A. Kotov, T. T. Tsap.
Nature, Vol. 259, 87 - 89 (1976).

The authors have modified their solar magnetograph to measure velocities at the solar surface, rather than magnetic fields. Using this apparatus, they have observed fluctuations of period 2 h 40 min, which are remarkably stable. The interpretation of this phenomenon seems to cause much theoretical difficulty.

080.009   Observation of free oscillations of the sun.
J. R. Brookes, G. R. Isaak, H. B. van der Raay.
Nature, Vol. 259, 92 - 95 (1976).

The Fraunhofer absorption lines for potassium and sodium on the sun are compared with the corresponding lines in the laboratory using a resonant optical scattering method. The observed shifts between the sun and laboratory lines may be interpreted in terms of the gravitational redshift, motion of the laboratory relative to the sun and oscillatory terms which may be related to oscillations of the sun.

080.010   Solar constant during a glaciation.
D. W. Parkin.
Nature, Vol. 260, 28 - 31 (1976). – Letter.

080.011   Reliability of sunspots as tracers of solar surface rotation.   J. M. Beckers.
Nature, Vol. 260, 227 - 229 (1976).

New methods of measuring solar rotation have produced significantly different rotation speeds to those derived from sunspots. Spectroscopic measurements of Doppler displacements give a so-called plasma rotation velocity of 13.76° d⁻¹ (sidereal) at the solar equator as against 14.38° d⁻¹ (sidereal) from sunspot proper motion velocity. Other techniques of measuring solar rotation lead to differences in the differential rotation from a strong variation of the solar rotation with latitude for sunspots and photospheric plasma to hardly any variation at all for coronal holes. The author reports here the result of plasma rotation velocity measurements inside sunspot umbrae and their relation to the rotation velocities derived from spot proper motions. He finds that plasmas inside and outside the spots rotate at similar rates, and conclude that sunspots make poor tracers.

080.012   Solar variability: is the sun an inconstant star?
A. L. Hammond.
Science, Vol. 191, 1159 - 1160 (1976). – Research news.

080.013   On the evolution of a 1 $M_\odot$ star with a periodically mixed core.
M. Gabriel, A. Noels, R. Scuflaire, A. Boury.
Astron. Astrophys., Vol. 47, 137 - 141 (1976).

To solve the neutrino problem, Dilke and Gough have suggested that the vibrational instability of g^+ modes of non radial oscillation may be the cause of recurrent mixing in the sun. Supposing this to be correct, the evolution of the sun is completely different from the standard one. Unmixed solar models are stable when older than 3×10^9 years. It is therefore necessary to check whether in the modified evolution, instabilities still exist at the solar age. They do, provided that the mass fraction of the mixed core is large enough. However, the neutrino flux at its minimum during a thermal pulse occurring at the solar age remains too high. Constraints imposed by ice age records are also discussed.

080.014 Differential rotation and the solar dynamo.
M. Stix.
Astron. Astrophys., Vol. 47, 243 - 254 (1976).

A number of numerical models for the generation of mean magnetic fields is examined and the fields are compared to the mean field of the sun. In particular, $\alpha\omega$-dynamos, which are based on differential rotation ("ω") and cyclonic turbulence ("α-effect"), are studied in the case of cylindrical surfaces of isorotation. Such dynamos have an oscillatory antisymmetric (dipolar) field as the most easily excited mode. A search for oscillatory $\omega \times j$-dynamos, where the α-effect is replaced by a different mean electric field, perpendicular to the rotation vector, ω, and the mean current density, j, is also made.

080.015 Hydraulic concentration of magnetic fields in the solar photosphere. III. Fields of one or two kilogauss.
E. N. Parker.
Astrophys. Journ., Vol. 204, 259 - 267 (1976).

Detailed analysis of weak and strong lines now suggests that the magnetic fields in the isolated intense flux tubes in the supergranule boundaries in the solar photosphere may be as large as 2000 gauss. This paper is a concise systematic review of hydrodynamic effects that might compress a magnetic field to great intensity. The author concludes that only cooling of the gas within the field can produce the high field densities inferred from observation. He also shows that inhibition of convection appears not to possess the necessary qualitative cooling features, with the conclusion that overstability, generating transverse hydromagnetic waves – essentially Alfvén waves – is the only way to account for the cooling and field intensification.

080.016 Solar magneto-atmospheric waves. I. An exact solution for a horizontal magnetic field.
A. H. Nye, J. H. Thomas.
Astrophys. Journ., Vol. 204, 573 - 581 (1976).

The linearized theory of magneto-atmospheric waves (involving the combined restoring forces due to buoyancy, compressibility, and magnetic field) is developed for the case of a horizontal magnetic field. An exact analytical solution to the propagation equation is obtained for the case of an isothermal atmosphere permeated by a uniform horizontal magnetic field, without making the usual short-wavelength assumption.
This solution is applied to an idealized model of the low-corona-chromosphere transition region for comparison with observations of flare-induced coronal waves. The results show that disturbances may propagate horizontally in the low corona in a wave guide formed by the sudden density increase into the chromosphere below and by the rapidly increasing Alfvén velocity with height in the corona.

080.017 On the constancy along cylinders of the angular velocity in the solar convection zone.
B. R. Durney.
Astrophys. Journ., Vol. 204, 589 - 596 (1976).

If, in the absence of rotation, the sun's convection zone is adiabatic and if in the radial and latitudinal equations of motion the main balance of forces is between pressure gradients, Coriolis forces, and buoyancy forces, then the perturbations in the convective flux and the pole-equator differences in flux (ΔF) are very large in the lower half of the convection zone, unless the angular velocity is constant along cylinders. The meridional velocities associated with this rotation law are not small, however, and could generate a significant ΔF.

080.018 Solar oscillations as a guide to solar structure.
I. Iben, Jr.
Astrophys. Journ., (Letters), Vol. 204, L147 - L150 (1976).

The theoretical spectrum of frequencies for the radial oscillations of a solar model is shown to depend on the treatment of convection near the model surface. For example, the period of the fundamental radial mode can vary from 48 minutes to 66 minutes, depending on the treatment. If the longest-period solar oscillation found by Hill, Stebbins, and Brown corresponds to the fundamental radial mode, then our understanding of the structure of the outer layers of the sun has been considerably enhanced.

080.019 High-resolution multichannel drift scans of the sun.
A. Wittmann.
Mitt. Astron. Ges., No. 38, p. 137-139 (1976). – Short report.

080.020 Differentielle Rotation und meridionale Bewegungen des solaren Ca$^+$-Netzwerkes.
H. Wöhl, E. H. Schröter.
Mitt. Astron. Ges., No. 38, p. 208 (1976). – Abstract.

080.021 Variation of the solar atmospheric rotation over the 11 year cycle.   M. El-Raey, R. Amer.
Solar Physics, Vol. 45, 533 - 542 (1975).

The synodic rotation period and power spectra of solar microwave sources are investigated using accurate data in the interval 1956 to 1970. The variation of the approximate 27 day period is obtained over a complete solar cycle and is thought to be a result of the latitude change over the solar cycle of the origins of the radio emissions. High resolution power spectra have also been obtained and revealed the existence of a double peaked line near 160 day period. This line is attributed to changes in either the earth's heliographic latitudes or the earth's inclination to the earth-sun line.

080.022 The chemical composition of the sun.
J. E. Ross, L. H. Aller.
Science, Vol. 191, 1223 - 1229 (1976).

080.023 Die Abplattung der Sonne.   E. Bettwieser.
SuW, 15. Jahrgang, p. 125 - 129 (1976).

080.024 Solare Neutrinos.   P. Gerber.
Orion, 34. Jahrgang, p. 4 - 7 (1976).

080.025 Nonlinear convective motion in shallow convective envelopes.   R. G. Deupree.
Astrophys. Journ., Vol. 205, 286 - 294 (1976).

Numerical solutions of the conservation equation of mass, momentum, and energy in two spatial dimensions and time have been carried out for two shallow convective envelopes. Cells of a preferred width form and have growth rates larger than cells of other widths. Matter is found to flow from the top of the convection zone to the bottom, even when the depth of the convection zone is several pressure scale heights.

080.026 Polymerisation and the solar neutrino problem.
B. Banerjee, S. M. Chitre, P. P. Divakaran, K. S. V. Santhanam, with a reply by K. C. Jacobs.
Nature, Vol. 260, 557 (1976).

080.027 A test of solar atmospheric structure.
L. W. Ramsey.
Bull. American Astron. Soc., Vol. 8, 293 (1976). – Abstr. AAS.

080.028 Towards a heliological inverse problem.
J. Christensen-Dalsgaard, D. O. Gough.
Nature, Vol. 259, 89 - 92 (1976).

Theoretical periods of normal modes of vibration of the

sun are compared with the observed periods of oscillation of the solar surface. It is inferred from the comparison that it may soon be possible to use solar oscillations to measure aspects of the internal structure of the sun.

080.029   Solar magnetic fields and convection. IV: Magnetic flux ropes and their fibres.   J. H. Piddington.
Astrophys. Space Sci., Vol. 40, 73 - 90 (1976).

The flux-rope model of solar magnetic fields is developed further by the use of a variety of observational results. (1) It is confirmed that magnetic fields emerging to form active regions are already in the form of helically twisted flux ropes. (2) A flux rope is not a homogeneous structure but is made up of hundreds or thousands of flux fibres. These are individually twisted and isolated from one another by non-magnetic plasma. They have fields of ≈ 2000 G at the surface, ≈ 4000 G at depths of $\lesssim 700$ km where they are fully compressed and also more tightly packed together. (3) Convection occurs only between the fibres. (4) The dynamics of a sunspot are discussed and a model developed which explains the observed radial decrease of magnetic field strength, umbral flashes and the umbral boundary, the penumbral plasma structure and the Wilson depression. (5) A new, large-scale (up to $\lesssim 10^5$ km) convective motion is described, originating below the penumbra and carrying away the sunspot energy deficit.

080.030   Five-minute oscillatory velocity field in the solar atmosphere.   S. M. Chitre, M. H. Gokhale.
Bull. Astron. Soc. India, Vol. 3, 27 (1975). – Abstract of paper presented at the A.S.I. meeting 1975.

080.031   Nuclear reactions in the interior and on the surface of the sun.   G. E. Kocharov.
VIIth Leningrad seminar. "Corpuscular streams of the sun and the radiation belts of the earth and Jupiter", 1975, (see 012.007), p. 23 - 24 (1975). In Russian and English. – Abstr. in Referativ. Zhurn. 51. Astron., 4.51.300 (1976).

080.032   Can the missing solar neutrinos be explained by a new interpretation of β-decay?   E. Bagge.
VIIth Leningrad seminar. "Corpuscular streams of the sun and the radiation belts of the earth and Jupiter", 1975, (see 012.007), p. 25 - 32 (1975). – Abstr. in Referativ. Zhurn. 51. Astron., 4.51.301 (1976).

080.033   Solar irradiance: total and spectral and its possible variations.   M. P. Thekaekara.
Applied Optics, Vol. 15, 915 - 920 (1976).

080.034   Solar magnetic fields and convection. V: Further interactions.   J. H. Piddington.
Astrophys. Space Sci., Vol. 41, 79 - 95 (1976).

The flux-rope-fibre model of solar magnetic fields is developed further to cover post-spot evolution of the fields, faculae, and the influence of magnetic fields on some convective motions.

080.035   Structure of the magnetic field of active regions and noise storms.   M. B. Ogir', L. I. Yurovskaya.
Izv. Krymskoj Astrofiz. Obs., Vol. 54, 201 - 219 (1976). In Russian.

It is discovered that noise storm radiation arises above active regions only if penetration of the field of one polarity into the field of opposite polarity takes place. The area of the penetrating field must exceed 150×10^6 km². The magnetic flux in the region of field deformations must be greater than 1×10^{21} Mx. The stronger a deformed field, the stronger a noise storm. The arising of noise storm radiation is connected with the stage of growth or maximum development of spot groups. Enhanced activity of surges and dark filaments is observed in the chromosphere above the region of penetration.

080.036   The depth of formation of some lines used for solar magnetic field recording.   L. S. Lyubimkov.
Izv. Krymskoj Astrofiz. Obs., Vol. 54, 248 - 264 (1976). In Russian.

For the atmosphere of the quiet sun effective depths of formation and the contribution function are calculated in different parts of profiles of the following lines: $\lambda 4554$ Ba II, $\lambda 4808$ Fe I, $\lambda 5250$ Fe I, $\lambda 5302$ Fe I, $\lambda 6103$ Ca I and $\lambda 6302$ Fe I. The results obtained are compared with earlier calculations by Buslavskij and Dubov. On the basis of this comparison it is shown that errors in used oscillator strengths, abundances and microturbulent velocities may shift the region of formation of central parts of the lines by several hundreds of kilometres.

080.037   On the theory of five-minute oscillations.   V. I. Zhukov.
Solnechnye Dannye 1976 Byull., No. 2, p. 44 - 50 (1976). In Russian.

The solution of the magnetohydrodynamic equation system found previously (1975) is applied to the problem of wave motions in a system consisting of three flat layers. The fluid in two upper layers is assumed to be non-viscous and infinitely conductive. It is shown that periodic motions can exist in such a system.

080.038   On the variation of the solar rotation with the height of the photosphere.   A. G. Gasanalizade.
Solnechnye Dannye 1976 Byull., No. 2, p. 50 - 52 (1976). In Russian.

A five-percent angular velocity variation has been obtained on the basis of the author's photoelectric observations and the revision of Aslanov's catalogue using the least-squares method. This result coincides with Dicke's values obtained by him from theoretical considerations.

080.039   The scale factor of the lower layers of the solar convective zone in the magnetic field structure of active regions.   V. P. Mikhajlutsa.
Solnechnye Dannye 1976 Byull., No. 2, p. 70 - 73 (1976). In Russian.

It is shown that the configuration of the magnetic field of active regions in non-current approximation has a scale factor equal to $\sim 10^{10}$ cm. The suggestion is made that the scale factor is generated by the magnetic field of the active regions in the lower layers of the solar convective zone.

080.040   Old and new riddles of the sun.   L. I. Miroshnichenko, E. I. Prutenskaya.
Priroda, 1976, No. 6, p. 147. In Russian.

080.041   Thermogravitational instability and 5-min oscillations in the solar atmosphere.   Yu. D. Zhugzhda.
Issled. po probl. solnech.-zemn. fiz. Moskva, Nauka, 1975, p. 71 - 79. In Russian. – Abstr. in Referativ. Zhurn. 51. Astron., 5.51.393 (1976).

080.042   La estructura fina de la rotación diferencial solar.   F.-L. Deubner, M. Vázquez.
Urania Barcelona, Año 60, No. 283, p. 39 - 48 (1975).

Spectroscopic measurements of the solar differential rotation have been made using the lines FeI, 5364.88 and CI 5380.32. A very pronounced minimum of the rotation is found at a mean latitude of 6° in both hemispheres.

080.043   Die Sonne schwingt.   M. Stix.
SuW, 15. Jahrgang, 187 - 188 (1976).

080.044   Long term changes in the solar diurnal anisotropy.   S. P. Duggal, M. A. Pomerantz.
14th Intern. Cosmic Ray Conf., (see 012.011), Vol. 4, 1209 -

1214 (1975).

080.045 **Origin of the twenty year wave in the diurnal variation.** E. H. Levy.
14th Intern. Cosmic Ray Conf., (see 012.011), Vol. 4, 1215 - 1220 (1975).

080.046 **Long-term changes in the solar diurnal variation.** T. Thambyahpillai.
14th Intern. Cosmic Ray Conf., (see 012.011), Vol. 4, 1221 - 1224 (1975).

080.047 **Nucleosynthesis of ^{3}He in the sun and the variation of ^{3}He/^{4}He in solar wind.**
S. A. Stephens, V. K. Balasubrahmanyan.
14th Intern. Cosmic Ray Conf., (see 012.011), Vol. 5, 1630 - 1634 (1975).

080.048 **Statistical significance of the 'counts' observed in the solar neutrino experiment.**
A. M. Aurela, W. S. Pallister, A. W. Wolfendale.
14th Intern. Cosmic Ray Conf., (see 012.011), Vol. 6, 2125 - 2128 (1975).

080.049 **The solar neutrino discrepancy.** B. Kuchowicz.
14th Intern. Cosmic Ray Conf., (see 012.011), Vol. 6, 2129 - 2133 (1975).

080.050 **Nuclear astrophysics of the sun.** G. E. Kocharov.
14th Intern. Cosmic Ray Conf., (see 012.011), Vol. 11, 3521 - 3549 (1975).

080.051 **Neutrino fluxes produced by high energy solar flare particles.** E. V. Kolomeets, V. L. Shmonin.
14th Intern. Cosmic Ray Conf., (see 012.011), Vol. 12, 4247 - 4250 (1975).

080.052 **Neutrinos and the interior structure of the sun.**
Yu. S. Kopysov.
In-t yader. issled. AN SSSR. P-0019. Moskva, 1975, 32 pp. In Russian. – Abstr. in Referativ. Zhurn. 51. Astron., 6. 51. 364 (1976).

080.053 **A model of the solar atmosphere and wind.**
J. H. Piddington.
Astrophys. Space Sci., Vol. 41, 371 - 385 (1976).

Using a combination of solar and interplanetary measurements, a topological model is developed of the overall magnetic and plasma structures. The author combines an earlier attempt to develop a magnetically structured atmosphere (Piddington, 1972, 1974) with the idea of Durney and Pneuman that the interplanetary medium should be largely determined by surface fields and, conversely, that the interplanetary medium should provide information about atmospheric fields and plasma.

080.054 **Submicron particles from the sun.**
C. L. Hemenway.
IAU Colloquium No. 31, (see 012.015), p. 251 - 269 (1976). Invited paper.

080.055 **Power spectrum analysis of the zonal mean annual excess precipitation total over the southern hemisphere.** B. P. Tritakis.
Praktika Acad. Athens, Vol. 50, 478 - 485 = Res. Center Astron. Appl. Math., Acad. Athens, Contr. Ser. I (Astron.), No. 44 (1975).

080.056 **Motions in the solar atmosphere: Spatially resolved motions in the solar atmosphere. Observational evidence for unresolved motions in the solar atmosphere.**

J. M. Beckers, R. C. Canfield.
Sacramento Peak Obs., Project 7649, Air Force Cambridge Res. Lab., Hanscom, Mass., 85 pp. = Sacramento Peak Obs., Contr. No. 263 (1975).

080.057 **L'energia solare e le sue applicazioni.**
C. Calzolari, A. Dumas.
Mem. Soc. Astron. Italiana, Vol. 46, 371 - 395 (1975).

080.058 **Solar pulsations.** A. G. Michalitsanos.
Bull. Astron. Soc. India, Vol. 4, 13 - 14 (1976).

080.059 **Faraday rotation observations during the 1970 Pioneer 9 solar occultation.** A. R. Cannon,
C. T. Stelzried, J. E. Ohlson.
JPL Techn. Rep. 32 - 1526, p. 87 - 93 (1973).

Significant steady-state Faraday rotation was detected between 4 and 12 solar radii at frequency 2.3 GHz. Two transient rotation events were recorded near 6 solar radii.

080.060 **Use of a sextant for the measurement of the temperature of the sun.** V. Mitra.
Journ. Optics, Vol. 3, 28 - 29 (1974). – Abstr. in Phys. Abstr., Vol. 79, A019358 (1976).

080.061 **The solar radiation.** F. Desvignes.
Acta Electronica, Vol. 18, 275 - 294 (1975). In French. – Abstr. in Phys. Abstr., Vol. 79, A023621 (1976).

080.062 **Linear simulations of Boussinesq convection in a deep rotating spherical shell (sun).** P. A. Gilman.
Journ. Atmosph. Sci., Vol. 32, 1331 - 1352 (1975). – Abstr. in Phys. Abstr., Vol. 79, A023629 (1976).

080.063 **The solar self-gravity-induced electrostatic redshift.** A. L. Smith-Hanni.
Helvetica Phys. Acta, Vol. 48, 548 - 551 (1975). – Abstr. in Phys. Abstr., Vol. 79, A027728 (1976).

080.064 **Neutrino magnetic moment, plasmon Čerenkov radiation, and the solar-neutrino problem.**
M. Radomski.
Phys. Rev. D, Vol. 12, 2208 - 2211 (1975). – Abstr. in Phys. Abstr., Vol. 79, A032632 (1976).

080.065 **Objective approach to a value of the solar constant made by indirect determination of the Stefan-Boltzmann constant.** D. Crommelynck.
Publ. Inst. Roy. Meteorol. Belgique, Ser. A, No. 91, p. 95 - 106 (1975). In French. – Abstr. in Phys. Abstr., Vol. 79, A041792 (1976).

080.066 **Solar magneto-atmospheric waves with a horizontal magnetic field.** A. H. Nye.
Thesis Rochester Univ., New York, USA, 118 pp. (1975). (Available from: Univ. Microfilms, Order No. 75-22,766).

Are solar neutrinos detected by living things.
See Abstr. 015.024.

Development of a computer code for the calculation of stellar evolution, with applications to solar models of low neutrino flux. See Abstr. 021.014.

Maximum rate for the proton-proton reaction compatible with conventional solar models.
See Abstr. 022.086.

Nonlinear interaction of magnetic field and convection. See Abstr. 062.083.

A grid of model atmospheres for metal-deficient giant stars. II.   See Abstr. 064.008.

A study of unstable acoustic waves in a convective zone.   See Abstr. 064.055.

Effects of convective overshoot on lithium depletion in main-sequence stars.   See Abstr. 065.023.

Equilibrium problems in a rotating convection zone. See Abstr. 065.039.

Limit on the secular change of the gravitational constant based on studies of solar evolution. See Abstr. 066.074.

Measurements of the solar gravitational deflection of radio waves in agreement with general relativity. See Abstr. 066.077.

Relativity experiment on Helios: a status report. See Abstr. 066.151.

Solar limb brightening in submillimeter wavelengths. See Abstr. 071.003.

Spectroscopic evidence for a higher rotation rate of magnetized plasma at the solar photosphere. See Abstr. 071.004.

Solar photospheric velocity field and a new analysis of the five minute oscillations.   See Abstr. 071.044.

A model of the solar cycle driven by the dynamo action of the global convection in the solar convection zone. See Abstr. 072.003.

A note on recurrences in the magnetic field distribution during the present cycle of solar acitivity. See Abstr. 072.009.

Overstability and cooling in sunspots. See Abstr. 072.010.

Solar magneto-atmospheric waves. II. A model for running penumbral waves.   See Abstr. 072.011.

Timing of solar cycles by rigid internal rotations. See Abstr. 072.031.

Solar-wind tritium limit and the mixing rate of the solar atmosphere.   See Abstr. 074.022.

Recent perspectives in solar physics: elemental composition, coronal structure and magnetic fields, solar activity. See Abstr. 074.059.

The X-ray radiation of the solar flare in the loop magnetic field.   See Abstr. 076.024.

Relations between the solar and interplanetary magnetic field distributions.   See Abstr. 106.037.

Cosmic-ray diurnal anisotropy and the sun's polar magnetic field.   See Abstr. 143.242.

A preliminary idealized network of neutron monitors for the study of solar modulation.   See Abstr. 143.255.

Theory of the solar magnetic cycle wave in the diurnal variation of energetic cosmic rays: physical basis of the anisotropy.   See Abstr. 143.340.

Solar modulation of galactic cosmic rays. 4. Latitude dependent modulation.   See Abstr. 143.346.

Solar daily variation of cosmic ray intensity. See Abstr. 143.349.

Annual modulation of solar diurnal variation of the cosmic ray nucleonic component in three-dimensional space. See Abstr. 143.351.

Earth

081 Figure, Composition, and Gravity of the Earth

081.001 On the definition of gravity anomalies in three-dimensional, potential-invariant representations of the actual gravity field of the earth. F. Bocchio.
Geophys. Journ. Roy. Astron. Soc., Vol. 44, 289 - 291 (1976).

Starting from the metric properties of the normal ellipsoidal field a unequivocal definition of gravity anomalies in connection with three-dimensional, potential-invariant representations of the actual gravity field of the earth on the normal ellipsoidal field can be given. Previous results concerning particular correspondence laws between the ellipsoidal normal co-ordinates (ϕ, λ) of the original points and their image under potential-invariant transformations follow easily from the general results given here.

081.002 The fine resolution of tidal harmonics. M. Amin.
Geophys. Journ. Roy. Astron. Soc., Vol. 44, 293 - 310 (1976).

The tides of deep seas are mainly due to the astronomical forces and these tides can be expressed mathematically by relating the laws of planetary motion with hydrodynamics. Such tides are simple and can be explained in terms of a small number of harmonic components which arise directly from astronomical origins or else indirectly through low order interaction between such terms.

081.003 Satellite altimetry scaling of the geopotential model. M. Burša.
Bull. Astron. Inst. Czechoslovakia, Vol. 27, 57 - 60 (1976).

The scale factor for lengths of the geopotential model is derived theoretically from satellite altimetry. The principal advantage of the proposed solution is that the problem of regularization need not be considered, because observations are carried out above the oceans and seas where there are no external masses relative to the geoid.

081.004 Westwanderung and the lunar tidal couple: modulation of convection by bulge stress.
R. C. Bostrom.
The Moon, Vol. 15, 109 - 117 (1976).

081.005 On the study of continental drift. G. P. Pil'nik.
Astron. Zhurn. Akad. Nauk SSSR, Vol. 53, 198 - 205 (1976). In Russian. English translation in Soviet Astron., Vol. 20, No. 1.

All observations of the International Time Service for 1966–1973 were used for a study of continental drift. It is shown that the accuracy of the best modern astronomical observations is unsufficient for such investigations.

081.006 This planet earth. H. Miles.
Journ. British Astron. Ass., Vol. 86, 112 - 124 (1976). — Presidential address during the annual general meeting of the British Astron. Ass.

081.007 13th-order harmonics in the geopotential from an analysis of four resonant satellites.
G. Balmino, C. Reigber.
Space Research XV, (see 012.003), p. 53 - 57 (1975).

081.008 Investigation of the geopotential and of gravitational perturbations in the motion of an artificial satellite.
K. Hristov.
Dynamical questions of satellite geodesy, 1973, (see 012.006), p. 6 - 17 (1974). In Russian.

081.009 Remarks on the expansion of the geopotential in series of sampling functions. A. Drożyner.
Dynamical questions of satellite geodesy, 1973, (see 012.006), p. 51 - 59 (1974). In Russian.

081.010 Investigation of the earth's gravitational field by combination of satellite and gravimetric data.
P. Biro.
Dynamical questions of satellite geodesy, 1973, (see 012.006), p. 60 - 68 (1974). In Russian.

081.011 On the combination of satellite and gravity observations for determination of the earth's gravitational field. V. G. Shkodrov, Eh. I. Yagudina.
Dynamical questions of satellite geodesy, 1973, (see 012.006), p. 69 - 89 (1974). In Russian.

081.012 On the application of maximum entropy spectral analysis to study the nearly diurnal free nutation.
A. I. Emets, Ya. S. Yatskiv.
Astrometriya i Astrofizika, *Kiev*, vyp. (No.) 29, (see 003.003), p. 3- 8 (1976). In Russian.

The maximum entropy spectral analysis is considered. This method is kown to give better resolution than traditional ones. It may be used for detecting nearby frequencies, for example, in the frequency region of the nearly diurnal free nutation.

081.013 An earth model based on free oscillations and body waves. D. L. Anderson, R. S. Hart.
Journ. Geophys. Res., Vol. 81, 1461 - 1475 (1976).

081.014 The lunar relief of the Tungusic syneclise.
E. S. Kutejnikov, N. S. Kutejnikova.
Priroda, 1976, No. 5, p. 122 - 129. In Russian.

081.015 Application of the solution of the inverse gravitational problem to determination of the figure of the earth. D. Zidarov.
Blg. geofiz. spisanie, Vol. 1, 78 - 95 (1975). In Bulgarian. — Abstr. in Referativ. Zhurn. 52. Geodeziya i Aehrosemka, 4.52.77 (1976).

081.016 On Stokes' formula transformed by means of anomalous vertical gradient of gravity.
J. Georgieva.
Vissh. geod., 1975, No. 1, p. 79 - 85. In Bulgarian. — Abstr. in Referativ. Zhurn. 52. Geodeziya i Aehrosemka, 4.52.80 (1976).

081.017 De drift der continenten. J. Veldkamp.
Zenit, 3e jaargang, p. 150 - 158 (1976).

081.018 Geocentricity and parameters of recent geopotential models. M. Burša.

Bull. Astron. Inst. Czechoslovakia, Vol. 27, 135 - 137 (1976).
Paper presented at the Sixteenth General Assembly of the IAG, Grenoble, Aug. 22, 1975.

081.019 **On the nearly axially-symmetrical model of the hydromagnetic dynamo of the earth.**
S. I. Braginsky (*Braginskij*).
Phys. Earth Planet. Interiors, Vol. 11, 191 - 199 (1976).

081.020 **Rotational deformation of the earth.**
L. Mansinha, P.-Y. Shen.
Phys. Earth Planet. Interiors, Vol. 11, 200 - 206 (1976).
The effects of rotation on two elastic real earth models and the equivalent fluid earth models have been investigated. The present angular velocity has resulted in the increase of the surface radius by about 0.7 km for the real earth and 1.02 km for a liquid earth. The corresponding ellipticities are 0.1052×10^{-2} and 0.3329×10^{-2}, respectively. A surprising result is that the solid inner core of a real-earth model becomes slightly prolate under uniform rotation while the rest of the earth is oblate.

081.021 **Tides and the rotation of the earth.** R. G. Hipkin.
Growth rhythms and the history of the earth's rotation, (see 012.012), p. 319 - 336 (1975).

081.022 **The earth's interior and the earth's rotation.**
J. A. Jacobs, K. D. Aldridge.
Growth rhythms and the history of the earth's rotation, (see 012.012), p. 337 - 352 (1975).

081.023 **Gravity and the earth's rotation.** P. S. Wesson.
Growth rhythms and the history of the earth's rotation, (see 012.012), p. 353 - 375 (1975).

081.024 **Geological processes and the earth's rotation in the past.** D. H. Tarling.
Growth rhythms and the history of the earth's rotation, (see 012.012), p. 397 - 412 (1975).

081.025 **On the regularization of Molodenskij's boundary value problem.** Yu. M. Nejman.
Izv. vyssh. ucheb. zavedenij. Geod. i aehrofotosemka, 1975, No. 3, p. 57 - 64. In Russian. — Abstr. in Referativ. Zhurn. 52. Geodeziya i Aehrosemka, 6.52.92 (1976).

081.026 **Comparison of gravitational fields obtained from satellite data and ground measurements.**
A. B. Bondarenko.
Izv. vyssh. ucheb. zavedenij. Geod. i aehrofotosemka, 1975, No. 3, p. 75 - 78. In Russian. — Abstr. in Referativ. Zhurn. 52. Geodeziya i Aehrosemka, 6.52.97 (1976).

081.027 **The influence of the terrestrial tides on Stokes' constants of the earth.** S. V. Gromov.
Vestn. Leningr. un-ta, 1975, No. 19, p. 126 - 131. In Russian. Abstr. in Referativ. Zhurn. 52. Geodeziya i Aehrosemka, 6.52.98 (1976).

081.028 **Even and odd degree 13th-order harmonics from analysis of stable near-resonant satellite orbits.**
C. Reigber, G. Balmino.
Bull. Groupe Recherches Géod. Spatiale, No. 15, p. 1 - 46 (1976).
Variations of the mean longitude $\lambda = M + \omega + \Omega$ in thirteen 2—3 weeks arcs of the 6 satellites Geos 2, BE-C, D1D, Anna Ib, Oscar 07 and Thor Able Star have been analysed to derive 35 condition equations for even and odd degree harmonics of order 13 in the geopotential.

081.029 **Translational inner core oscillations of a rotating,**

slightly elliptical earth. M. L. Smith.
Journ. Geophys. Res., Vol. 81, 3055 - 3065 (1976).

081.030 **Analyses of the solid earth and ocean tidal perturbations on the orbits of the Geos 1 and Geos 2 satellites.** T. L. Felsentreger, J. G. Marsh, R. W. Agreen.
Journ. Geophys. Res., Vol. 81, 2557 - 2563 (1976).

081.031 **The shape of the earth.** D. King-Hele.
Science, Vol. 192, 1293 - 1300 (1976).

081.032 **Orientation of the earth by numerical integration.**
F. A. Fajemirokun, F. D. Hotter, I. I. Mueller.
Bull. Géod., Vol. 50, 109 - 129 (1976). — Paper presented at the International Symposium on Computational Methods in Geometrical Geodesy, Oxford, September 2 - 8, 1973.

081.033 **Astrogeodetic geoid of Japan.** Y. Ganeko.
Smithsonian Astrophys. Obs., *Cambridge Mass.*, Special Rep. 372, 5 + 34 pp. (1976).

081.034 **Determination of the density of a simple layer as method for the investigation of the outer gravitational field of the earth and celestial bodies.** M. I. Yurkina.
Gerlands Beiträge Geophys., Band 85, 208 - 220 (1976). In Russian.
In the present paper equations relating the density of a simple layer with the following quantities are derived: gravity, astronomical longitude and latitude, height of the quasi-geoid, potential, temporal variations of the orbital parameters of artificial satellites, and mass of the celestial body.

081.035 **Sea surface determination from space. The GSFC geoid.** F. O. Vonbun, J. McGoogan, J. Marsh, F. Lerch.
GSFC Document X-900-75-216, Prepr., 4 + 11 pp. (1975).

081.036 **Tests and comparisons of satellite derived geoids with Skylab altimeter data.** J. G. Marsh, B. C. Douglas, S. Vincent, D. M. Walls.
GSFC Document X-921-75-176, Prepr., 6 + 19 pp. (1975).

081.037 **Power spectra of geoid undulations.** R. D. Brown.
GSFC Document X-921-75-300, Prepr., 5 + 17 pp. (1975).

081.038 **Improvement in the geopotential derived from satellite and surface data (GEM 7 and 8).**
C. A. Wagner, F. J. Lerch, J. E. Brownd, J. A. Richardson.
GSFC Document X-921-76-20, Prepr., 3 + 11 pp. (1976).

081.039 **Marées terrestres.** P. Melchior (Editor).
Bull. d'Informations, (Obs. Roy. Belgique, Bruxelles), No. 73, p. 4186 - 4257 (1976).

081.040 **The horizontal pendulum observations in the polar tube observation room at the Dodaira Station.**
J. Matsumoto, F. Miyamoto.
Tokyo Astron. Obs. Rep., Vol. 17, 425 - 429 (1976). In Japanese.

081.041 **Magnitudes and spectra of important dynamical phenomena.** E. P. Fedorov.
IAU Colloquium No. 26, (see 012.018), p. 63 - 77 (1975).

081.042 **Terrestrial coordinate systems solidly connected with the earth.** M. Bursa.
IAU Colloquium No. 26, (see 012.018), p. 133 - 159 (1975).

081.043 **Review of the classical methods for the determination of geodetic datums.** I. I. Mueller.

IAU Colloquium No. 26, (see 012.018), p. 321 - 339 (1975).

081.044 **Physikalische Erforschung des Erdkörpers mittels künstlicher Erdsatelliten.** H. Kautzleben.
Sitzungsber. Akad. Wiss. Berlin, Jahrg. 1975, No. 1N, p. 55 - 56 (1975).

Geodynamics today: a review of the earth's dynamic processes. See Abstr. 003.117.

Astronomical evidence of change in the rate of the earth's rotation and continental motion.
See Abstr. 044.009.

Use of observations of artificial satellites for the solution of problems of geodynamics. See Abstr. 045.003.

Nearly diurnal free nutation from latitude observations at Greenwich from 1911 to 1935.
See Abstr. 045.005.

Analysis of satellite altimetry data.
See Abstr. 046.010.

Gravimetric and astrogeodetic reference systems in space and time. See Abstr. 046.040.

Some problems related to geodetic datum definitions for terrestrial measurements. See Abstr. 046.042.

Evolution of the moon: the 1974 model.
See Abstr. 094.017.

082 The Earth's Atmosphere Including Refraction, Scintillation, Extinction, Airglow, Site Testing

082.001 Seeing measurements in Greece, Spain, South West Africa, and Chile.
K. Birkle, H. Elsässer, T. Neckel, G. Schnur, B. Schwarze.
Astron. Astrophys., Vol. 46, 397 - 406 (1976).

The methods and results of seeing measurements are presented, which were performed in the course of site testing compaigns for the two observatories of the Max-Planck-Institut für Astronomie to be built in the Mediterranean area and southern hemisphere, respectively. Two independent methods were used to measure image motion: (1) photographic star trails and (2) photoelectric knife-edge seeing monitors. It was found that in South West Africa the seeing conditions are of the same high quality as in Chile.

082.002 Neutral composition changes during a period of increasing magnetic activity.
G. W. Prölss, K. H. Fricke.
Planet. Space Sci., Vol. 24, 61 - 67 (1976).

082.003 Unusual sodium nightglow behaviour and possible association with solar flares. M. C. Isherwood.
Planet. Space Sci., Vol. 24, 99 - 101 (1976).

During many nights in October and November 1970 unusual enhancements in sodium nightglow intensity were observed near Belfast. These are discussed in relation to other atmospheric parameters. It is suggested that instabilities in the atmosphere triggered by a series of solar flares may be the cause, particularly as during this period temperatures in the stratosphere were found to be unstable.

082.004 Photometry of eclipses of artificial earth satellites observed in situ. (Part I). L. Neužil, I. Zacharov.
Bull. Astron. Inst. Czechoslovakia, Vol. 27, 23 - 34 (1976).

The results of observing the eclipses of satellites on board Interkosmos 4 are given. The observed increase of the penumbra cannot be explained using a current model of a high absorbing layer.

082.005 Lunar tide in the upper atmosphere.
A. Palumbo.
Journ. Atmosph. Terr. Phys., Vol. 38, 103 - 106 (1976).
Paper presented at the XVI IUGG general assembly, Grenoble, 1975.

082.006 An explanation of the longitudinal variation of the O^1D (630nm) tropical nightglow intensity.
G. Thuillier, J. W. King, A. J. Slater.
Journ. Atmosph. Terr. Phys., Vol. 38, 155 - 158 (1976).

The intensity of the nightglow 630 nm emission varies with longitude along the tropical arcs at constant local times. This longitudinal fine structure is explained in terms of thermospheric winds which produce different effects at longitudes at which the magnetic declination is different.

082.007 Atmospheric absorption in the range 12 cm^{-1} to 32 cm^{-1} measured in a horizontal path.
G. G. Gimmestad, H. A. Gebbie.
Journ. Atmosph. Terr. Phys., Vol. 38, 325 - 328 (1976).

The temperature dependence of anomalous atmospheric absorption of submillimetre waves is much greater than is expected from the accepted binding energy of water dimers in equilibrium. Non-equilibrium phenomena must be invoked to explain the results.

082.008 Guarantee estimate of the accuracy of constructing a nonspherical model of the upper atmosphere's density from observations of deceleration of artificial earth satellites. P. M. Vingardt.
Kosmich. Issled., Vol. 14, 29 - 35 (1976). In Russian.

082.009 On mid-latitude daily variations of the parameters of the upper atmosphere.
B. A. Mirtov, A. G. Starkova.
Kosmich. Issled., Vol. 14, 48 - 52 (1976). In Russian.

082.010 Distribution of atomic hydrogen in the earth's upper atmosphere taking into account the daily density variations. Yu. M. Gektin, S. D. Chuvakhin.
Kosmich. Issled., Vol. 14, 80 - 84 (1976). In Russian.

082.011 Variations of the corpuscular radiation stream in the mesosphere of mean latitudes during the day.
V. F. Tulinov, L. V. Shibaeva, S. G. Yakovlev.
Kosmich. Issled., Vol. 14, 147 (1976). In Russian. – Brief information.

082.012 Eddy diffusion in the upper atmosphere by global deposition of meteoroids. V. Mitra.
Astrophys. Space Sci., Vol. 39, 133 - 150 (1976).

A theory for the production of eddy diffusion in the upper atmosphere by the global deposition of meteoroids is presented. It is based on the assumption that meteoroids falling on the earth carry, on the average, a greater amount of orbital angular momentum per unit mass than that corresponding to the earth's orbit. This excess of orbital angular momentum of the meteoroids is deposited in some or the other form during their interaction with the earth's atmosphere. It is shown that the other population of meteoroids which is metallic in nature deposits the excess orbital angular momentum below 100 km altitude and produces eddies.

082.013 Far infrared emission spectrum of the stratosphere from balloon altitudes.
T. A. Clark, D. J. W. Kendall.
Nature, Vol. 260, 31 - 32 (1976). – Letter.

082.014 Latitudinal changes of composition in the disturbed thermosphere from Esro 4 measurements.
L. G. Jacchia, J. W. Slowey, U. von Zahn.
Journ. Geophys. Res., Vol. 81, 36 - 42 (1976).

082.015 Photochemically induced departures of [O] and [O₂] from diffusive equilibrium distributions.
E. S. Oran, D. F. Strobel.
Journ. Geophys. Res., Vol. 81, 257 - 259 (1976).

Photochemically induced departures of [O] and [$\overset{\circ}{O}_2$] from diffusive equilibrium distributions are found to depend principally on atmospheric composition, the solar flux between 1250–1750 Å, and the EUV solar flux which produces O_2 dissociation indirectly through odd nitrogen chemistry. Above 200 km the [O_2] concentrations are ~20–35% less than diffusive equilibrium concentrations given by a Jacchia model atmosphere, while the O concentrations are enhanced by ~10%.

082.016 Twilight observation of the forbidden $O^+(^2P-^4S)$ transition at 2470 Å.
P. D. Feldman, P. Z. Takacs.
Journ. Geophys. Res., Vol. 81, 260 - 262 (1976).

082.017 Measurement of the atmospheric agitation at Sheshan and Qingdao by allowing the star image to trail.
Second Laboratory, Shanghai Observatory, Academia Sinica.
Acta Astron. Sinica, Vol. 16, 180 - 188 (1975). In Chinese.

This article describes the measurement of atmospheric agitation by taking photographs of a star trail at Sheshan section of Shanghai Observatory and at Qingdao Observatory.

082.018 Apollo 16 far ultraviolet imagery of the polar auroras, tropical airglow belts, and general airglow.
G. R. Carruthers, T. Page.
Journ. Geophys. Res., Vol. 81, 483 - 496 (1976).

Far ultraviolet imagery of the earth in the 1050- to 1600-Å and 1250- to 1600-Å wavelength ranges was obtained from the lunar surface during the Apollo 16 mission on April 21, 1972. The images have an angular resolution of about 2 arc min (230-km linear resolution) and have been quantitatively analyzed to obtain absolute intensities and spatial distributions of the polar auroras (both wavelength ranges) and of the day and night airglow and tropical airglow belts (1250- to 1600-Å wavelength range). A general night airglow, at least in the northern hemisphere, is indicated.

082.019 Spectrum measurements of star atmospheric scintillation. L. Paternò.
Astron. Astrophys., Vol. 47, 437 - 441 (1976).

Measurements of the atmospheric stellar scintillation time-spectrum were carried out by means of a photoelectric device at two different heights a.s.l., for different zenithal distances and telescope apertures. The instrument used is briefly described. The results given can be considered as representative of mean atmospheric conditions during that period. It seems that lower atmospheric layers contribute substantially to scintillation.

082.020 Optical characteristics of the mesopause and the lower thermosphere on the nightside of the earth.
A. A. Buznikov, K. Ya. Kondratyev (Kondrat'ev), A. I. Lazarev, O. I. Smokty (Smoktij).
Space Research XV, (see 012.003), p. 77 - 80 (1975).

The paper describes and discusses observations made by cosmonauts of a layer near the mesopause which can cause an apparent change of colour and brilliance of stars and planets. They may be explained by an enhanced concentration of aerosol particles near the mesopause at the time.

082.021 High resolution spectra of the stratosphere between 30 and 200 cm^{-1}. J. P. Baluteau, E. Bussoletti.
Space Research XV, (see 012.003), p. 131 - 137 (1975).

082.022 First results of 6300 Å nightglow measurements aboard a rocket launched from Natal, Brazil.
Y. Sahai, A. Drescher, H. Lauche, N. R. Teixeira.
Space Research XV, (see 012.003), p. 251 - 255 (1975).

082.023 Preliminary results of observations of atmospheric ultraviolet twilight emissions by the TD1-A satellite.
A. Monfils, J. C. Gérard.
Space Research XV, (see 012.003), p. 257 - 262 (1975).

082.024 Investigation of the astronomical refraction.
I. I. Motrunich.
Materialy nauch. konf. aspirantov i molodykh uchenykh. Uzhgorod. un-t. Sekts. astron. n. Uzhgorod, 1975, p. 8 - 12. In Ukrainian. – Abstr. in Referativ. Zhurn. 51. Astron., 3.51.117 (1976).

082.025 Estimate of the accuracy of determination of the coefficients of a model of the upper atmosphere's density from deceleration of artificial earth satellites.
P. M. Vingardt, P. E. Ehl'yasberg.
Opredelenie dvizheniya kosmich. apparatov. Moskva, Nauka, 1975, p. 16 - 24. In Russian. – Abstr. in Referativ. Zhurn. 62. Issled. kosmich. prostranstva, 3.62.195 (1976).

082.026 Investigation of the density of the upper atmosphere and aerodynamics of satellites from orbital evolution data. E. I. Bushuev, A. I. Vasil'eva, V. F. Kameko, V. M. Kovtunenko, A. A. Krasovskij, V. Ya. Mashtak, V. A. Shabokhin.
Opredelenie dvizheniya kosmich. apparatov. Moskva, Nauka, 1975, p. 168 - 182. In Russian. – Abstr. in Referativ. Zhurn. 62. Issled. kosmich. prostranstva, 3.62.196 (1976).

082.027 Solar absorption in a stratosphere perturbed by NO_x injection. F. M. Luther.
Science, Vol. 192, 49 - 51 (1976).

The changes in the solar absorption by nitrogen dioxide and ozone induced by the injection of NO_x (oxides of nitrogen) in the stratosphere are complementary, even though the nitrogen dioxide absorption is only a small fraction of the ozone absorption for an unperturbed stratosphere. The factors causing this effect are described, and an analysis is made of the perturbed solar radiation budget.

082.028 Lyman alpha radiation in the night-time mesopause and lower thermosphere – I. Transport of radiation in the absorber-scatterer mixture. G. G. O'Connor.
Journ. Atmosph. Terr. Phys., Vol. 38, 377 - 382 (1976).

082.029 Lyman alpha radiation in the night-time mesopause and lower thermosphere – II. Measurements of atomic hydrogen and molecular oxygen densities in the night-time atmosphere. G. G. O'Connor.
Journ. Atmosph. Terr. Phys., Vol. 38, 383 - 388 (1976).

Measurements of Lyman alpha night airglow made with rocket borne ionisation chambers have been used to derive the atomic hydrogen and molecular oxygen density profiles between 70 and 120 km altitude. The measured profiles are compared with photochemical model and standard atmosphere density profiles respectively.

082.030 Photometric measurements of night and twilight NI (5200 Å) emissions observed from Dar es Salaam.
E. C. Njau, E. H. Carman.
Journ. Atmosph. Terr. Phys., Vol. 38, 439 - 441 (1976).

Zenith measurements of 5200 Å airglow observed from Dar es Salaam during the June–October 1974 period are presented. An analysis of the twilight data reveals that the 5200 Å line has an average photon intensity along the zenith of about $9 \times 10^7 \, cm^{-2} \, sec^{-1}$ in the middle of twilight. It is further shown that the measured 5200 Å night emission rates exhibit no correlation with geomagnetic activity.

082.031 On meteor-generated infrasound. D. O. ReVelle.
Journ. Geophys. Res., Vol. 81, 1217 - 1230 (1976).

An analysis of the generation and propagation characteristics of infrasonic pressure waves excited during meteor entry into the earth's atmosphere is presented.

082.032 Twilight helium 10,830-Å calculations and observations. B. A. Tinsley, A. B. Christensen.
Journ. Geophys. Res., Vol. 81, 1253 - 1263 (1976).

082.033 Fe (3860 Å) emission in the twilight.
A. L. Broadfoot, A. E. Johanson.
Journ. Geophys. Res., Vol. 81, 1331 - 1334 (1976).

Resonance scattered emission of 3860 Å has been detected in the twilight spectrum and is attributed to the Fe atom. Observations of other metallic atoms, Mg, Ca, and their ions, are reviewed with the point of view that the observed presence of one meteoric element in a region of the atmosphere undoubtedley indicates the presence of the other elements as well. It is suggested that ground-based observations of Ca and Ca⁺ could be used to monitor the distribution of elements of meteoric origin through the atmosphere.

082.034 **Triangulation measurements of the hydroxyl emission altitude.**   B. P. Potapov.
Astron. Tsirk., No. 856, p. 5 - 7 (1975). In Russian.

082.035 **Results of an analysis of astronomical refraction anomalies.**   N. A. Vasilenko.
Astrometriya i Astrofizika, *Kiev*, vyp. (No.) 28, (see 003.002), p. 42 - 51 (1976). In Russian.

The results of a qualitative and statistical analysis are presented for anomalies of astronomical refraction. The data were obtained from observations of bright celestial bodies at zenith distances of 80–90°. Approximate estimates of the mean seasonal atmospheric layers inclinations are obtained directly from differences of the measured refraction values.

082.036 **Calculation of the astronomical refraction from aerological data.**   I. G. Kolchinskij.
Astrometriya i Astrofizika, *Kiev*, vyp. (No.) 28, (see 003.002), p. 52 - 65 (1976). In Russian.

Astronomical refraction values obtained by numerical integration for the standard atmosphere were compared with those calculated for the same conditions according to the Pulkovo tables. It is concluded that the standard atmosphere should be used for compiling tables of mean refraction. The method is suggested for calculating astronomical refraction at zenith distances close to 90° by summarizing refraction for several layers.

082.037 **A study of the refractive index of the air from radio-sonde observations.**   V. I. Sergienko.
Astrometriya i Astrofizika, *Kiev*, vyp. (No.) 28, (see 003.002), p. 66 - 78 (1976). In Russian.

The values of the refraction index n calculated by six modern methods were compared. The dependences of Δn on azimuth and their correlation with ΔTU obtained from observations with two astrolabes are studied.

082.038 **Analysis of data on surface and tropospheric water vapour.**   T. O. Aro.
Journ. Atmosph. Terr. Phys., Vol. 38, 565 - 571 (1976).

082.039 **A hybrid mode of diffuse and specular reflector for computation of the emergent radiation by the adding method.**   T. Takashima, C. I. Taggart, E. G. Morrissey.
Astrophys. Space Sci., Vol. 40, 157 - 165 (1976).

A method of computing the diffuse reflection and transmission radiation from an inhomogeneous, plane-parallel planetary atmosphere bounded by the hybrid surface of a diffuse and specular reflector is discussed by using the 'addition' method. It is shown that the method is suitable for numerical computation even if the surface reflects light according to the hybrid mode of the diffuse and specular law.

082.040 **Solar barometer.**   A. S. Gurvich, V. N. Kubasov, A. A. Leonov, A. I. Simonov, T. N. Kharitonova.
Zemlya i Vselennaya, 1976, No. 2, p. 28 - 30. In Russian.

082.041 **Measurement of the parameters of internal gravitational waves in the meteor zone.**
N. M. Gavrilov, I. A. Delov.
Geomagn. Aeronom., Vol. 16, 293 - 297 (1976). In Russian.

082.042 **On a possible mechanism of origin of semi-annual variations of density in the upper atmosphere of the earth.**   L. A. Antonova, V. V. Katyushina.

Geomagn. Aeronom., Vol. 16, 311 - 315 (1976). In Russian.

082.043 **Theoretical atmospheric transmission in the mid- and far-infrared at four altitudes.**
W. A. Traub, M. T. Stier.
Applied Optics, Vol. 15, 364 - 377 (1976).

The ir transmission of the terrestrial atmosphere is calculated at four altitudes of interest: Mauna Kea at 4.2 km (2–1000 μm), aircraft at 14 km (5–1000 μm), and balloon at 28 km and 41 km (10–1000 μm). One salient result for the spectral region around 100 μm is that the absorption (and emissivity) of the atmosphere drops by a factor of 10 for each increase in altitude of 15 km throughout the aircraft and balloon range; thus balloon-borne astronomical photometry and spectroscopy should both enjoy a considerable advantage over aircraft observations in the 30–300-μm region.

082.044 **Global atomic hydrogen density derived from OGO-6 Lyman α measurements.**
G. E. Thomas, D. E. Anderson, Jr.
Planet. SpaceSci., Vol. 24, 303 - 312 (1976).

In this paper the authors present an analysis of an extensive set of Lyman α (Lα) data measured by the u.v. photometer experiment on the OGO-6 spacecraft. This experiment measured the ultraviolet airglow in the local zenith throughout a near-polar orbit at altitudes from 400 to 1100 km.

082.045 **Global atomic oxygen density derived from OGO-6 1304 Å airglow measurements.**
D. J. Strickland, G. E. Thomas.
Planet. Space Sci., Vol. 24, 313 - 326 (1976).

The ultraviolet photometer experiment on the OGO-6 spacecraft measured the hydrogen Lyman-α (1216 Å) and the atomic oxygen zenith emissions at 1302.2 Å, 1304.9 Å and 1306.0 Å in the earth's ultraviolet airglow. In this paper, the authors report an analysis of 1304 Å data measured during the period 15 September - 25 October 1969, which includes both quiet and disturbed conditions. The quantity deduced from the data is the column density of atomic oxygen above the satellite altitude.

082.046 **Determination of atomic oxygen density from rocket borne measurement of hydroxyl airglow.**
R. E. Good.
Planet. Space Sci., Vol. 24, 389 - 395 (1976).

A rocket experiment was conducted which measured the infrared bands of the excited hydroxyl radical in the night airglow. The OH emission was found in a layer centered at 87 km having a half-width of 6 km and a total emission of 1.1 MR. The atomic oxygen altitude profile, ranging from 1.3×10^{10} atoms/cm³ at 83 km to 3×10^{11} atoms/cm³ at 90 km is determined from the hydroxyl airglow measurements.

082.047 **Two-source spherical-wave structure functions in atmospheric turbulence.**   R. L. Fante.
Journ. Optical Soc. America, Vol. 66, 74 (1976). – Letter.

082.048 **Measurement of the amplitude of phase excursions in the earth's atmosphere.**   J. B. Breckinridge.
Journ. Optical Soc. America, Vol. 66, 143 - 144 (1976).

The maximum of the probability density distribution of phase fluctuations in the earth's atmosphere, as a function of point separation, was measured visually in white light for starlight observed from the ground with a 1.5 m astronomical telescope. A wave-front folding interferometer was used.

082.049 **Zernike polynomials and atmospheric turbulence.**   R. J. Noll.
Journ. Optical Soc. America, Vol. 66, 207 - 211 (1976).

This paper discusses some general properties of Zernike polynomials, such as their Fourier transforms, integral representations, and derivatives. A Zernike representation of the Kolmogoroff spectrum of turbulence is given that provides a complete analytical description of the number of independent corrections required in a wavefront compensation system.

082.050   **City sky-glow monitoring at Kitt Peak. II.**
A. A. Hoag.
Publ. Astron. Soc. Pacific, Vol. 88, 207 - 208 (1976).

Continuation of the monitoring from Kitt Peak has shown that the rate of increase in artificial sky glow from Tucson shortward of $\lambda 440$ nm has changed markedly since a regulatory ordinance was adopted.

082.051   **Wavelike irregularities in the mid-latitude 6300 Å airglow.**   M. K. Andrews.
Planet. Space Sci., Vol. 24, 521 - 527 (1976).

082.052   **Observations of the solar and lunar limbs.**
A. N. Demidova.
Solnechnye Dannye 1976 Byull., No. 2, p. 102 - 105 (1976). In Russian.

Invisible thin elevated layers with sharp gradients of the refractive index are usually present daily and nightly in the troposphere. The distribution of their location heights was obtained on the basis of 370 observations of the solar limb and 89 observations of the lunar limb at Pulkovo.

082.053   **Analysis of the vertical component of wind velocity in the meteor zone.**
K. A. Karimov, V. Ya. Ogurtsov.
Komety i Meteory, *Dushanbe*, No. 23, p. 37 - 45 (1974). In Russian.

The work deals with methods and results of measuring the drift of meteoric trails and daily variations of vertical wind components for January and September – October 1971. The results are compared with theoretical calculations based on the thermal tide theory.

082.054   **Sulphate-light scattering ratio as an index of the role of sulphur in tropospheric optics.**
A. P. Waggoner, A. J. Vanderpol, R. J. Charlson, S. Larsen, L. Granat, C. Trägårdh.
Nature, Vol. 261, 120 - 122 (1976). – Letter.

082.055   **Propagation-medium limitations on phase-compensated atmospheric imaging.**   J. H. Shapiro.
Journ. Optical Soc. America, Vol. 66, 460 - 469 (1976).

There has long been interest in reducing or circumventing the limitations imposed by atmospheric turbulence on optical imaging systems. Recent studies have shown that irradiance or speckle interferometry, or real-time atmospheric compensation, may be used to regain diffraction-limited performance. In this paper, the relationship between real-time phase compensation and the optimum channel-matched filter compensator is developed, with emphasis on the fundamental limits imposed by the propagation medium.

082.056   **Diffraction-limited atmospheric imaging of extended objects.**   J. H. Shapiro.
Journ. Optical Soc. America, Vol. 66, 469 - 477 (1976).

082.057   **Measurements of the atmospheric attenuation of the spectral components of astronomical images.**
C. Roddier.
Journ. Optical Soc. America, Vol. 66, 478 - 482 (1976).

Measurements are presented of the long and short exposure visibility of fringes produced by stellar wave fronts, distorted by atmospheric turbulence. A variable shear rotation interferometer is used. Parameters relevant to image degradation are deduced.

082.058   **Apollo 16 far ultraviolet spectra of the terrestrial airglow.**   G. R. Carruthers, T. Page.
Journ. Geophys. Res., Vol. 81, 1683 - 1694 (1976).

Far ultraviolet spectra of the terrestrial airglow were obtained from the lunar surface during the Apollo 16 mission. The spectra cover the wavelength range 490–1600 Å with a resolution of about 40 Å and the wavelength range 1050– 1600 Å with a resolution of 30 Å. Features recorded spectrographically for the first time include He 584 Å, O^+ 834 Å, H Lyman β 1026 Å, and N^+ 1086 Å. Further observations are needed with higher spatial and spectral resolution, particularly in the wavelength range below 1150 Å.

082.059   **On the occurrence of widely observed noctilucent clouds.**   N. D'Angelo, E. Ungstrup.
Journ. Geophys. Res., Vol. 81, 1777 - 1778 (1976).

From available data an anticorrelation is found between high-latitude ionospheric electric fields and the occurrence of 'widely observed noctilucent clouds.' The effect is attributed to heating at the mesopause level arising from dissipation of ionospheric currents.

082.060   **On a possible relation between airglow $\lambda 6300$ Å [OI] enhancement on August 30, 1975 and V1500 Cyg.**   L. M. Fishkova.
Astron. Tsirk., No. 893, p. 1 - 2 (1975). In Russian.

082.061   **Structure of the neutral atmosphere.**
R. K. Crane.
Astrophysics, Part B, (see 003.008), p. 136 - 141 (1976).

082.062   **Absorption and emission by atmospheric gases.**
J. W. Waters.
Astrophysics, Part B, (see 003.008), p. 142 - 176 (1976).

The author first formulates the radiative transfer expressions appropriate for calculating absorption and emission by atmospheric gases. Then the general expressions for the spectral-line absorption coefficient are given, followed by specific expressions for calculating absorption by water vapor and oxygen. Problems associated with the calculation of absorption by these two molecules are also discussed. The absorption by microwave lines of ozone and other minor constituents are considered briefly. Finally, the results of calculations of atmospheric absorption and emission are given, as well as measured values.

082.063   **Extinction by condensed water.**   R. K. Crane.
Astrophysics, Part B, (see 003.008), p. 177 - 185 (1976).

082.064   **Refraction effects in the neutral atmosphere.**
R. K. Crane.
Astrophysics, Part B, (see 003.008), p. 186 - 200 (1976).

In this chapter the author considers the index of refraction at radio frequencies, the bending of rays, and pathlength effects important in interferometry.

082.065   **An interpretation of the rotational temperature of the airglow hydroxyl emissions.**
K. Suzuki, T. Tohmatsu.
Planet. Space Sci., Vol. 24, 665 - 671 (1976).

The rotational temperature of the airglow hydroxyl emissions arising from various schemes of vibrational transitions was obtained by using spectroscopic data from six observational sources. The rotational temperature was found to depend systematically on the quantum number (v') of the upper vibrational level from which the relevant band originates. This v'-dependence of the rotational temperature is in favor of the hypothesis that there are two routes of excitation of the hydroxyl airglow: $O_3 + H = OH(v \lesssim 9) + O_2$, and $HO_2 + O = OH(v \lesssim 6) + O_2$. The present result implies also that the relaxation time of rotation of OH in the upper mesosphere is as long as 0.1 sec.

082.066 Calculations of the equilibrium photoelectron flux in the thermosphere.
G. A. Victor, K. Kirby-Docken, A. Dalgarno.
Planet. Space Sci., Vol. 24, 679 - 681 (1976).

082.067 On the fringe visibility in a Michelson stellar interferometer. C. Roddier, F. Roddier.
Journ. Opt. Soc. America, Vol. 66, 580 - 584 (1976).
Because of atmospheric turbulence, the fringes produced by a Michelson stellar interferometer appear as a random modulation of the two superimposed stellar images. The contribution of the related high spatial frequency peaks in the image Wiener spectrum has been computed as a function of the diameter of the apertures and of the seeing conditions. This contribution appears to be independent of the seeing condition for large apertures.

082.068 Atmospheric transmittance in the far infrared at Testa Grigia.
P. P. Lombardini, F. Melchiorri, G. Salio, L. Dall'Agnola.
Infrared Physics, Vol. 15, 73 - 78 (1975).
Atmospheric transmittance in the band of $300-2000\,\mu m$ was measured in the far i.r. Solar Observatory of Testa Grigia, 3480 m above standard. These measurements are compared with the meteorological data observed at Plateau Rosa in the last 7 yr.

082.069 Synthetic transmission spectra in the far-infrared for the low stratosphere.
B. Rebours, P. Rabache.
Infrared Physics, Vol. 15, 189 - 199 (1975).

082.070 Bands between $9-12\,\mu m$ with total absorption produced by high voltage discharge. T.-C. Li.
Infrared Physics, Vol. 15, 201 - 204 (1975).

082.071 Far infrared atmospheric transmission measurements in North-Norway.
G. Dall'Oglio, I. Pippi, F. Klokkervoll, S. Sivertsen.
Infrared Physics, Vol. 15, 341 (1975). – Letter.

082.072 Infrared spectral radiance of the sky.
W. Tam, R. Corriveau.
Infrared Physics, Vol. 16, (see 012.013), 129 - 134 (1976).

082.073 Synthetic spectra of the atmospheric radiance in the far infrared. P. Rabache.
Infrared Physics, Vol. 16, 339 - 344 (1976).
In this article, the author presents the synthetic spectra of the energy radiated by the fundamental constituents of the atmosphere: H_2O, O_2 and O_3. The computation in the spectral range covers from 10 to 110 cm^{-1}. It is applied to the lower stratosphere, to altitudes between 11.5 and 14.5 km. The computation parameters have been chosen so as to enable a direct comparison to be made between his computed spectra and those measured by other authors.

082.074 Flächenphotometrische Untersuchungen der Ultraviolett- und Infrarot-Nachthimmelshelligkeit.
A. Frey.
Diss. Nat. Gesamtfakultät, Ruprecht-Karl-Univ., Heidelberg, 81 pp. (1976).

082.075 Determination of the height of the optical horizon of the earth from spaceship Soyuz 9.
A. G. Nikolaev, V. I. Sevast'yanov, A. V. Sandomirskij, V. M. Stol'berg.
Kosmich. Issled., Vol. 14, 467 - 469 (1976). In Russian.

Brief information.

082.076 Scattering in the earth's atmosphere: calculations for Milky Way and zodiacal light as extended sources. H. J. Staude.
IAU Colloquium No. 31, (see 012.015), p. 106 (1976).
Abstract of 13.082.052.

082.077 Scattering layer of cosmic dust in the upper atmosphere. F. Link.
IAU Colloquium No. 31, (see 012.015), p. 107 - 111 (1976).

082.078 Sur les éclipses du satellite IK-3. F. Link.
Comptes Rendus Acad. Sci. Paris, Sér. B, Vol. 282, 415 - 416 (1976).
Les mesures photométriques des éclipses du satellite soviétique IK-3 montrent, d'une part, la distribution géographique de l'ozone, et, d'autre part, l'absorption par les aérosols météoriques liés aux essaims des Orionides et des Géminides.

082.079 Mobilité apparente des ions $H^+(H_2O)_4$ et $H^+(H_2O)_5$.
M. Cabane, P. Krien, G. Madelaine, J. Bricard.
Comptes Rendus Acad. Sci. Paris, Sér. B, Vol. 282, 507 - 509 (1976).

082.080 Influenza dell'atmosfera sull'osservazione dei corpi celesti. A. Braccesi, V. Zitelli.
Giorn. Astron., Vol. 1, 39 - 60 (1975).

082.081 Light pollution in southern Ontario. R. L. Berry.
Journ. Roy. Astron. Soc. Canada, Vol. 70, 97 - 115 (1976).
From visual and photoelectric data, a sky brightness – population function for cities and a sky brightness – distance function for average atmospheric conditions are obtained. These functions are used in a simple mathematical model to calculate levels of light pollution over Ontario.

082.082 A simple computer model for the growth of light pollution. R. Pike.
Journ. Roy. Astron. Soc. Canada, Vol. 70, 116 - 126 (1976).
A simple model of the propagation and scattering of light in the atmosphere has been used to calculate the light pollution contributions from each city in southern Ontario and its vicinity.

082.083 Analyse der Gas- und Ionenzusammensetzung der Thermosphäre und deren räumliche und zeitliche Variationen. B. Anweiler, W. Joos, D. Krankowsky, L. R. Lake, P. Lämmerzahl, R. Nord, G. Rühle, H. Schneider, P. Somssich, E. Zettwitz.
MPI Kernphys., Heidelberg, Jahresbericht 1975, p. 160 - 165

082.084 Untersuchungen der Atmosphäre im Höhenbereich 24 km bis 118 km. F. Arnold, H. Böhringer, E. Hettmannsperger, D. Krankowsky, J. Stanislawska, E. Zettwitz.
MPI Kernphys., Heidelberg, Jahresbericht 1975, p. 165 - 168.

082.085 Relazione meteorologica sulle zone dell'Appennino Campano-Lucano e della Sicilia centro-settentrionale. G. B. Baratta, A. Cassatella, R. Viotti.
Oss. Astron. Roma, Contr. sci., Ser. III, No. 142, 5 pp. (1975).

082.086 Sky brightness at Dodaira.
A. Takechi, T. Yamaguchi.
Tokyo Astron. Obs. Rep., Vol. 17, 494 - 505 (1976). In Japanese.

082.087 Image intensifier tube observations of the OH air-

glow during the Space Shuttle simulation.
A. W. Peterson, L. M. Kieffaber.
Proc. Southwest Regional Conf., Vol. 1, (see 012.021), 105 - 114 (1976).

The authors' OH airglow studies have evolved from a lead sulphide cell scanning photometer to photographic techniques to, currently, image tube photography. During the Space Shuttle simulation photography (35-mm and 16-mm) and filter wheel photometry of submicron OH airglow were performed aboard the NASA CV990 Airborne Laboratory. Examples of ground-based and airborne photographs are presented for comparison. Analysis of the aircraft data is described.

082.088 Cooling rate of an electron gas by polar molecules (in planetary and cometary atmospheres).
O. Ashihara.
Rep. Ionosph. Space Res. Japan, Vol. 29, 65 - 68 (1975).
Abstr. in Phys. Abstr., Vol. 79, A032792 (1976).

Noctilucent clouds.   See Abstr. 003.027.

H_2O rotation line emission between 10 and 50 μ due to collisions with O.   See Abstr. 022.009.

Astronomical holography through the turbulent atmosphere.   See Abstr. 031.234.

A posteriori restoration of atmospherically degraded images using multiframe imagery.   See Abstr. 031.235.

Effects of the atmosphere in stellar speckle interferometry.   See Abstr. 031.237.

Lens–atmosphere MTF measurements.
See Abstr. 031.240.

Some results on the imaging of incoherent sources through turbulence.   See Abstr. 031.257.

Far infra-red background astronomy at ground.
See Abstr. 031.271.

Techniques for measuring bulk gas-motions from satellites.   See Abstr. 032.505.

Statistical estimate of experimental data on deceleration of artificial earth satellites.   See Abstr. 052.006.

Analysis of the orbit of 1970-114F in its last 20 days.   See Abstr. 054.001.

Mouvements sphériques des fluides visqueux incompressibles.   See Abstr. 062.015.

Full-disk solar fluxes between 1230 and 1940 Å.
See Abstr. 076.005.

Isotopic composition of the corpuscular streams of the sun and of the earth's atmosphere.   See Abstr. 078.025.

Submicron particles from the sun.
See Abstr. 080.054.

Metastable ^2D atomic nitrogen in the mid-latitude nocturnal ionosphere.   See Abstr. 083.020.

On the intensities of 6300 Å, 5577 Å and 5200 Å emissions from ion composition measurements in the F2 region.
See Abstr. 083.024.

Semi-annual variations in the ionosphere and upper atmosphere at heights of 120 - 500 km.
See Abstr. 083.057.

A search for short-term meteorological effects of solar variability in an atmospheric circulation model.
See Abstr. 085.009.

Martian volatiles: their degassing history and geochemical fate.   See Abstr. 097.035.

Numerical simulation of natural photometric systems.   See Abstr. 113.004.

Errata

082.901 Correction: 'The N I (5200 Å) dayglow' [Journ. Geophys. Res., Vol. 80, 2300 - 2304 (1975)].
D. W. Rusch, A. I. Stewart, P. B. Hays, J. H. Hoffman.
Journ. Geophys. Res., Vol. 81, 295 (1976).

083 Ionosphere

083.001 **A snapshot of the polar ionosphere.**
J. H. Whitteker, L. H. Brace, E. J. Maier, J. R. Burrows, W. H. Dodson, J. D. Winningham.
Planet. Space Sci., Vol. 24, 25 - 32 (1976).

083.002 **Boltzmann–Fokker–Planck model for the electron distribution function in the earth's ionosphere.**
J. R. Jasperse.
Planet. Space Sci., Vol. 24, 33 - 40 (1976).

Using Boltzmann–Fokker–Planck methods and the diffusion approximation, the author derives coupled nonlinear equations for the first two angular moments of the electron distribution function in the earth's ionosphere. The theory includes a phenomenological treatment of photoionization of the neutral species by an externally produced photon flux; electron-ion recombination; electron-neutral particle attachment; elastic, excitation, deexcitation, and ionizing electron-neutral particle collisions; and elastic electron-electron and electron-ion collisions.

083.003 **Ionospheric radio wave absorption in the auroral zone and the interplanetary magnetic field sector structure.**
S. M. Mansurov, L. G. Mansurova, Z. Ts. Rapoport.
Planet. Space Sci., Vol. 24, 55 - 59 (1976).

083.004 **Modeling the midlatitude F-region ionospheric storm using east-west drift and a meridional wind.**
D. N. Anderson.
Planet. Space Sci., Vol. 24, 69 - 77 (1976).

083.005 **Ionospheric composition in SAR-arcs.**
W. J. Raitt, R. W. Schunk, P. M. Banks.
Planet. Space Sci., Vol. 24, 105 - 114 (1976).

083.006 **Some questions of the night-time ionospheric E region formation in middle latitudes.**
Yu. K. Chasovitin, V. P. Nesterov.
Planet. Space Sci., Vol. 24, 139 - 145 (1976).

An attempt is made to analyse known experimental data on electron density and ion composition of the nighttime ionosphere; the main ideas on the night -time E region ionization source are considered; the role of dynamic processes in the irregular structure formation of the night-time ionospheric E region is discussed.

083.007 **Experimental study of ionospheric electron density gradients and their effect on VLF propagation.**
L. Cairó, J. C. Cerisier.
Journ. Atmosph. Terr. Phys., Vol. 38, 27 - 36 (1976).

Ionospheric electron density gradients in the mid-latitude ionosphere have been studied experimentally through their effect on the wave-normal direction of upgoing artificial VLF waves. It is shown that the direction of propagation of the waves is closely related to certain broad features of the ionospheric structure, namely its variations with longitude or local time which act upon the E–W component of the wave normal and its latitudinal variations which influence the N–S component. These data shed some light on the choice of the initial wave-normal directions for use in magnetospheric ray-tracing studies.

083.008 **Increases of equatorial total electron content (TEC) during magnetic storms.**
D. Yeboah-Amankwah.
Journ. Atmosph. Terr. Phys., Vol. 38, 45 - 50 (1976).

This paper is a report on the analysis of equatorial electron content, TEC, during magnetic storms. Storms between 1969 and 1972 have been examined as part of an ongoing study of TEC morphology during magnetically disturbed days. The published magnetic K_p indices and TEC data from the Legon observatory have been employed. The general picture arising from the analysis is that the total electron content of the ionosphere is significantly enhanced during magnetic storms.

083.009 *F-region temperatures from measurements of plasma scale height.*
L. L. Cogger, K. K. Vij, H. C. Carlson.
Journ. Atmosph. Terr. Phys., Vol. 38, 93 - 96 (1976).

From incoherent scatter data collected at the Arecibo Observatory from 1965–1968 a comparison was made between measured values of nighttime F-region electron and ion temperature and the plasma temperature inferred from the topside scale height of the measured electron concentration profile. It was found that these temperatures agreed to within a few per cent, implying that the exospheric temperature can be obtained with reasonable accuracy by either method.

083.010 **The macro-scale structure of equatorial spread-F irregularities.** J. Röttger.
Journ. Atmosph. Terr. Phys., Vol. 38, 97 - 101 (1976).
Short paper.

083.011 **Rocket measurements of electron concentration in the lower ionosphere at two European locations.**
P. H. G. Dickinson, J. E. Hall, F. D. G. Bennett.
Journ. Atmosph. Terr. Phys., Vol. 38, 163 - 173 (1976).

Measurements of electron concentrations in the lower ionosphere are reported for 19 rocket flights. Variations of the electron concentration with season, latitude, degree of ionospheric absorption in winter, degree of auroral activity at night, and with solar zenith angle are examined.

083.012 **Average electron content gradients and nighttime electron fluxes in the mid-latitude ionosphere.**
A. Ebel, G. Schmidt, A. Tauriainen.
Journ. Atmosph. Terr. Phys., Vol. 38, 207 - 215 (1976).

083.013 **A comparison of the relative locations of the mid-latitude electron density trough and the scintillation boundary.** Y. K. Tulunay, O. Demir, A. Tauriainen.
Journ. Atmosph. Terr. Phys., Vol. 38, 217 - 218 (1976).

The mid-latitude electron density trough position and the scintillation boundary have been compared for magnetically quiet periods by using the data returned by Ariel 3 and Explorer 22 satellites. The scintillation boundary is found southward of the trough during daytime, but at night the positions are reversed.

083.014 **Multiple propagation paths between satellites situated in the ionosphere below the F-layer peak.**
E. Woyk (Chvojková).
Journ. Atmosph. Terr. Phys., Vol. 38, 329 - 331 (1976).

Theoretical results which show that up to six different propagation paths are possible at the same frequency between a pair of satellites at an equal height below the peak of a biparabolic concentric model ionosphere are presented. The path changes that arise for different satellite separations are discussed.

083.015 **Rocket measurements of the concentration of positive ions below 90 km.** L. N. Smirnykh.
Kosmich. Issled., Vol. 14, 151 - 152 (1976). In Russian.

Brief information.

083.016 Experimental measurements of the ion concentration in the night D-region of the ionosphere.
L. N. Smirnykh.
Kosmich. Issled., Vol. 14, 153 - 155 (1976). In Russian.
Brief information.

083.017 Simultaneous small and large scale irregularities in the ionospheric F region.
L. A. Hajkowicz, K. L. Jones, W. L. Nowland.
Nature, Vol. 259, 35 - 36 (1976). — Letter.

083.018 Mid-latitude ionospheric scintillation fading of microwave signals.
W. E. Brown III, G. G. Haroules, W. I. Thompson III.
Nature, Vol. 259, 294 - 296 (1976).

083.019 E region ion drifts and winds from incoherent scatter measurements at Arecibo.
R. M. Harper, R. H. Wand, C. J. Zamlutti, D. T. Farley.
Journ. Geophys. Res., Vol. 81, 25 - 35 (1976).
Ion velocity and temperature measurements at Arecibo for two winter days were analyzed to determine the large-scale dynamical structure of the E region. The main results are as follows: (1) Large oscillations in the ion drifts, neutral winds, and temperatures were present. The oscillations possessed a characteristic downward phase progression and dominant periods that were near semidiurnal. (2) Wind and temperature amplitudes were 40—100 m/s and $30°—60°$K, respectively.

083.020 Metastable ^2D atomic nitrogen in the mid-latitude nocturnal ionosphere. M. R. Torr, R. G. Burnside, P. B. Hays, A. I. Stewart, D. G. Torr, J. C. G. Walker.
Journ. Geophys. Res., Vol. 81, 531 - 537 (1976).

083.021 On the production mechanism of electric currents and fields in the ionosphere.
A. D. Richmond, S. Matsushita, J. D. Tarpley.
Journ. Geophys. Res., Vol. 81, 547 - 555 (1976).

083.022 Magnetospheric DC electric fields; present knowledge and outstanding problems to be solved during the IMS. F. S. Mozer.
Astrophys. Space Sci. Library, Vol. 57, (see 012.002), 101 - 131 (1976).
Present knowledge of the relationships between perpendicular and parallel electric fields in the ionosphere and magnetosphere and of their configurations, sources, and variations is summarised.

083.023 Mass spectrometer measurements of the F2 region ion composition from the satellite Cosmos 274.
Yu. A. Romanovsky (*Romanovskij*), V. V. Katyushina, V. G. Istomin.
Space Research XV, (see 012.003), p. 351 - 356 (1975).

083.024 On the intensities of 6300 Å, 5577 Å and 5200 Å emissions from ion composition measurements in the F2 region. M. N. Vlasov, Yu. A. Romanovsky (*Romanovskij*).
Space Research XV, (see 012.003), p. 357 - 361 (1975).

083.025 Measurements of ionospheric parameters at 100—170 km during a total solar eclipse.
A. D. Danilov, U. F. Ivanov, G. S. Ivanov-Kholodny (*Ivanov-Kholodnyj*), T. V. Kazatchevskaya (*Kazachevskaya*), V. K. Semenov, V. V. Selantiev (*Selant'ev*), Yu. K. Chasovitin, V. G. Khryukin.
Space Research XV, (see 012.003), p. 393 - 397 (1975).

083.026 Investigation of ionospheric irregularities by probing technique. G. Gdalevich.
Artificial Satellites, Vol. 10, 11 - 16 (1975).

083.027 Application of the scintillation theory to ionospheric irregularities studies.
A. W. Wernik, C. H. Liu.
Artificial Satellites, Vol. 10, 37 - 59 (1975).
The paper is a review of the present status of the scintillation theory. The limitations of the single scattering theory and some other assumptions of the scintillation theory are reviewed and discussed.

083.028 Calculated daily variations of O^+ and H^+ at mid-latitudes — I. Protonospheric replenishment and F-region behaviour at sunspot minimum.
J. A. Murphy, G. J. Bailey, R. J. Moffett.
Journ. Atmosph. Terr. Phys., Vol. 38, 351 - 364 (1976).

083.029 Satellite measurements of ion composition and temperatures in the topside ionosphere during medium solar activity.
M. K. Goel, B. C. N. Rao, S. Chandra, E. J. Maier.
Journ. Atmosph. Terr. Phys., Vol. 38, 389 - 394 (1976).

083.030 The partially aligned gradient instability in the ionosphere. J. D. Whitehead.
Journ. Geophys. Res., Vol. 81, 1361 - 1368 (1976).
Ionization irregularities which are almost but not exactly aligned along the magnetic field may grow much more rapidly than purely aligned irregularities in the presence of a background gradient of ionization and current flow. The stability of these irregularities is considered, taking into account the finite thickness of the gradient and shear in the ion motion. It is shown in particular that the normal E region of the ionosphere is usually stable, whereas sporadic E is often unstable, and this may limit the compression of metallic ions to form such layers.

083.031 On ionospheric investigations by coherent radio-waves emitted from artificial earth satellites.
Ja. (*Ya.*) L. Al'pert.
Space Sci. Rev., Vol. 18, 551 - 602 (1976).
Results of radio-investigations of the ionosphere with the help of coherent radiowaves emitted by beacons placed on artificial earth satellites are given. Data are given which illustrate results of investigations of local ionospheric characteristics. Such data may help to solve some problems in the present stage of the near earth plasma study. A new possibility of radio-investigation of the near earth plasma with the help of a chain of satellites connected together is pointed out.

083.032 The electric potential of the ionosphere as controlled by the solar magnetic sector structure.
R. Reiter.
Naturwissenschaften, 63. Jahrgang, p. 192 - 193 (1976).

083.033 Solar eclipse effects on HF and VLF propagation.
D. D. Meisel, B. Duke, R. C. Aguglia, N. R. Goldblatt.
Journ. Atmosph. Terr. Phys., Vol. 38, 495 - 502 (1976).

083.034 Ionization enhancement in the middle latitude D-region due to precipitating high energy electrons.
D. S. Wratt.
Journ. Atmosph. Terr. Phys., Vol. 38, 511 - 516 (1976).
For several days following a large solar-terrestrial disturbance in August 1972 increases in energetic electron flux at 1400 km, $L = 2.6$ were observed. These observations, together with calculations of ionization production rates in the D-region based on the observed pitch angle distribution at the

satellite are direct evidence for the importance of precipitating energetic electrons as an ionization source in the lower *D*-region at middle latitudes following magnetic storms.

083.035 Longitudinal change of ionization of the F-region during low and moderate solar activity periods.
M. Berkeliev, Ya. Akyev, A. G. Grigor'yan, E. K. Dubrovskaya.
Izv. AN TurkmSSR. Ser. fiz.-tekhn. khim. Izv. geol. 1., 1975, No. 5, p. 38 - 42. In Russian. – Abstr. in Referativ. Zhurn. 51. Astron., 6. 51. 475 (1976).

083.036 Rocket measurements of mid-latitude ionospheric currents during a magnetic storm. K. Burrows.
Journ. Atmosph. Terr. Phys., Vol. 38, 159 - 162 (1976).

The ionospheric current, flowing during a weak magnetic storm, has been observed during a rocket flight from Woomera, South Australia. It is found to be similar to the normal *Sq* current, both in height and intensity. No significant ionospheric contribution to the magnetic field perturbations at ground level, associated with the storm, can be identified.

083.037 Influence of the upper boundary conditions on modulated ionospheric parameters.
A. A. Namgaladze, K. S. Latyshev.
Geomagn. Aeronom., Vol. 16, 43 - 49 (1976). In Russian.

083.038 The relative contribution of ultraviolet and X-ray radiation to ionization of the F-region.
G. S. Ivanov-Kholodnyj, A. A. Nusinov.
Geomagn. Aeronom., Vol. 16, 76 - 79 (1976). In Russian.

083.039 On the connection of E_s in the near-polar region with the parameters of the interplanetary magnetic field. A. S. Besprozvannaya, A. V. Shirochkov.
Geomagn. Aeronom., Vol. 16, 84 - 87 (1976). In Russian.

083.040 Currents in the high-latitude ionosphere caused by variations of the asymmetrical ring current.
V. V. Denisenko, V. G. Pivovarov.
Geomagn. Aeronom., Vol. 16, 183 - 184 (1976). In Russian. Brief information.

083.041 Relation between X-ray and ultraviolet radiation of solar flares in the ionization of the ionospheric E-region. G. S. Ivanov-Kholodnyj, L. N. Leshchenko, I. N. Odintsova.
Geomagn. Aeronom., Vol. 16, 246 - 250 (1976). In Russian.

083.042 On disturbances of the electron concentration during local heating of the ionospheric F-layer.
S. M. Savel'ev, V. B. Ivanov.
Geomagn. Aeronom., Vol. 16, 356 - 357 (1976). In Russian. Brief information.

083.043 Kinetics of electrons in low-temperature peculiar plasma (ionosphere).
A. V. Gurevich, G. M. Milikh, I. S. Shlyuger.
Zhurn. ehksperim. i teor. fiz., Vol. 69, 1640 - 1653 (1975). In Russian. – Abstr. in Referativ. Zhurn. 62. Issled. kosmich. prostranstva, 4.62.284 (1976).

083.044 Solar EUV flux variation during a solar cycle as derived from ionospheric modeling considerations.
R. G. Roble.
Journ. Geophys. Res., Vol. 81, 265 - 269 (1976).

Calculations in modeling the general ionospheric structure observed during solar minimum and solar maximum suggest that the Hinteregger (1970) solar EUV flux values are adequate for modeling the mid-latitude ionospheric structure during solar minimum. However, these flux values should be doubled for modeling solar maximum conditions. The calcula-

tions also suggest that the atomic oxygen density at 120 km was approximately a factor of 2 lower during solar cycle minimum than during solar cycle maximum.

083.045 On the dependence of the latitudinal variability of the ionospheric radio wave absorption on the solar zenith angle. M. D. Fligel', G. V. Givishvili, Z. Ts. Rapoport.
Gerlands Beiträge Geophys., Band 85, 89 - 92 (1976).

The latitudinal variability of the dependence of the ionospheric radio wave absorption *L* on the solar zenith angle χ is considered, using *L* data obtained by the pulse-echo method (A1). It is shown that the exponent *n* in the formula $L = L_0 \cos{}^n \chi$ decreases monotonically with the latitude – from the dip equator towards the pole. This decrease may be connected with the daily variation of the corpuscular radiation intensity.

083.046 The penetration of soft electrons into the ionosphere.
G. P. Mantas, J. C. G. Walker.
Planet. Space Sci., Vol. 24, 409 - 423 (1976).

083.047 General and local peculiarities of the wind regimes in the lower ionosphere at mean latitudes and near the equator. P. B. Babadzhanov, L. N. Rubtsov, B. G. Solovej.
Komety i Meteory, *Dushanbe,* No. 23, p. 18 - 36 (1974). In Russian.

083.048 Preliminary N(h)-profiles from vertical sounding data of the Station Dushanbe.
L. N. Rubtsov, A. E. Epishova.
Byull. Inst. Astrofiz., *Dushanbe,* No. 65, p. 3 - 7 (1975). In Russian.

083.049 Day-to-day wind-regime variations in the lower ionosphere near the equator.
B. G. Solovej, A. E. Epishova.
Byull. Inst. Astrofiz., *Dushanbe,* No. 65, p. 8 - 9 (1975). In Russian.

083.050 Parameters of the anisotropy of small-scale irregularities in the lower ionosphere of the equatorial region. L. N. Rubtsov, B. G. Solovej, S. P. Rogozhkina.
Byull. Inst. Astrofiz., *Dushanbe,* No. 65, p. 10 - 13 (1975). In Russian.

083.051 The probability of the sporadic E_s-layer appearance and solar activity. N. P. Lyakhova, B. G. Solovej.
Byull. Inst. Astrofiz., *Dushanbe,* No. 65, p. 14 - 19 (1975). In Russian.

The anomalously high solar activity level in 1972 and related ionospheric phenomena are noted. Experimental probabilities PE_s of E_s appearance for 1965, 1971 and 1972 are presented. A negative correlation between PE_s and solar activity is found.

083.052 Results of an experimental investigation of a connection of the E_s-layer with the annual abundance of faint meteors. D. Latipov, L. N. Rubtsov.
Byull. Inst. Astrofiz., *Dushanbe,* No. 65, p. 20 - 25 (1975). In Russian.

A linear connection between the annual means of hourly rates of faint meteors and the annual means of sporadic E layer appearance probabilities is established using ionospheric data of stations Dushanbe, Ashkhabad and Tashkent.

083.053 Characteristics of the sporadic E-layer based on ionospheric data of the Station Dushanbe.
O. Alimov, D. Latipov.
Byull. Inst. Astrofiz., *Dushanbe,* No. 65, p. 30 - 35 (1975). In Russian.

083.054 Variation of the maximum ionization height of the F2-region and corresponding temperature over Dushanbe. O. Alimov, D. Latipov.
Byull. Inst. Astrofiz., *Dushanbe*, No. 65, p. 36 - 39 (1975). In Russian.

083.055 A dynamical model of ionospherically-protonospheric interaction.
M. A. Nikitin, L. P. Zakharov, R. V. Gostrem.
Geomagn. Aeronom., Vol. 16, 423 - 430 (1976). In Russian.

083.056 Transversal conductivity of the earth's ionosphere and magnetosphere. T. N. Soboleva.
Geomagn. Aeronom., Vol. 16, 431 - 436 (1976). In Russian.

083.057 Semi-annual variations in the ionosphere and upper atmosphere at heights of 120 - 500 km.
A. G. Kolesnik.
Geomagn. Aeronom., Vol. 16, 437 - 443 (1976). In Russian.

083.058 Temperature and thickness of the F-region from data of ground vertical sounding and non-coherent scattering. N. I. Potapova, B. S. Shapiro.
Geomagn. Aeronom., Vol. 16, 448 - 453 (1976). In Russian.

083.059 Ionospheric substorms in the subauroral ionosphere. N. P. Ben'kova, G. V. Bukin, N. K. Osipov, N. I. Samorokin.
Geomagn. Aeronom., Vol. 16, 467 - 471 (1976). In Russian.

083.060 Calculations of the ion composition of the mid-latitude ionosphere at day time in the height interval of 100-250 km for solar activity minimum.
A. A. Namgaladze, K. S. Latyshev, L. P. Zakharov.
Vopr. modelir. ionosfery. Kaliningrad, 1975, p. 13 - 18. In Russian. – Abstr. in Referativ. Zhurn. 51. Astron., 5.51.510 (1976).

083.061 Determination of the daily variation of the ionospheric parameters for low solar activity.
A. A. Namgaladze, K. S. Latyshev, L. P. Zakharov.
Vopr. modelir. ionosfery. Kaliningrad, 1975, p. 26 - 35. In Russian. – Abstr. in Referativ. Zhurn. 51. Astron., 5.51.511 (1976).

083.062 Investigations of inhomogeneities of the ionosphere from recording of Doppler and Faraday effects and scintillations of radio signals of artificial earth satellites at mean and high latitudes. V. A. Misyura, N. P. Svetlichnyj, L. V. Bezrodnaya, N. D. Gerasimova, N. D. Zholondkovskij, G. N. Zinchenko, L. B. Volkov, O. F. Tyrnov, A. R. Yagovkin, D. N. Sergeev.
Ionosfer. issledovaniya. No. 23. Moskva, Nauka, 1975, p. 38 - 43. In Russian. – Abstr. in Referativ. Zhurn. 62. Issled. kosmich. prostranstva, 5.62.298 (1976).

083.063 Fluctuations of the electron concentration of the ionosphere from observations of signals of the artificial earth satellite Intercosmos 2.
V. M. Migunov, V. I. Novozhilov, G. K. Solodovnikov.
Ionosfer. issledovaniya. No. 23. Moskva, Nauka, 1975, p. 44 - 46. In Russian. – Abstr. in Referativ. Zhurn. 62. Issled. kosmich. prostranstva, 5.62.299 (1976).

083.064 On the flux of the albedo of X-ray bremsstrahlung according to measurements in the stratosphere.
A. M. Novikov, Yu. G. Shafer.
Cosmic rays, No. 15, (see 003.006), p. 34 - 35 (1975). In Russian.

083.065 The ionosphere. T. Hagfors.

Astrophysics, Part B, (see 003.008), p. 119 - 135 (1976).
It is the purpose of the present chapter to provide the radio astronomer with a brief review of the properties of the ionospheric layers, their origin, their effect on the propagation of electromagnetic waves, and the methods available for observing their properties so that corrective measures can be taken.

083.066 Mutual neutralization rates of ionospherically important ions. D. Smith, N. G. Adams, M. J. Church.
Planet. Space Sci., Vol. 24, 697 - 703 (1976).

083.067 Electric fields and conductivity of the auroral ionosphere in magnetically quiet periods.
I. A. Zhulin, I. M. Kopaev, N. K. Osipov, P. M. Soprunyuk.
Kosmich. Issled., Vol. 14, 372 - 377 (1976). In Russian.

083.068 Ionospheric effects of transient celestial X-ray and gamma-ray events. K. Kasturirangan, U. R. Rao, D. P. Sharma, R. G. Rastogi, S. C. Chakravarty.
Astrophys. Space Sci., Vol. 42, 57 - 62 (1976). – See 012.016.
The paper presents the results on investigations of the ionospheric effects arising out of transient celestial events at X-ray and gamma-ray energies such as that from X-ray novae and cosmic gamma-ray bursts. Theoretical computations are carried out for estimating electron density enhancements using the available data on intensities and energy spectra for these events. Further, the observational results are explained in terms of these theoretical calculations.

083.069 Ionospheric techniques for the detection of transient X- and γ-ray bursts.
E. O'Mongain, G. A. Baird.
Astrophys. Space Sci., Vol. 42, 63 - 67 (1976). – See 012.016.
Calculations are presented of the amount of excess ionization produced in the lower ionosphere by various transient X- and γ-ray bursts under different assumptions about the incident spectrum and the ion recombination rates in the ionosphere. An experiment has been started to measure the power spectrum of the phase and amplitude of the night-time fluctuations of a CW signal in order to determine if it is possible to improve the sensitivity of the ionospheric technique by using the transient nature of the bursts. Preliminary results from this experiment are presented.

083.070 Spectra of ionospheric scintillation. R. K. Crane.
Journ. Geophys. Res., Vol. 81, 2041 - 2050 (1976).

083.071 Polar cap ionospheric electric field modulation by the solar wind sector structure.
N. D'Angelo, M. Møhl Madsen, I. B. Iversen.
Journ. Geophys. Res., Vol. 81, 2417 - 2418 (1976).
Brief report.

083.072 Summer day-time scintillation and sporadic-*E*.
A. Das Gupta, L. Kersley.
Journ. Atmosph. Terr. Phys., Vol. 38, 615 - 618 (1976).

083.073 Solar proton and electron precipitation effects detected by ionosondes. W. R. Piggott, E. Hurst.
Journ. Atmosph. Terr. Phys., Vol. 38, 619 - 622 (1976).

083.074 The day-to-day variability in ionospheric electric fields and currents.
V. W. J. H. Kirchhoff, L. A. Carpenter.
Journ. Geophys. Res., Vol. 81, 2737 - 2742 (1976).

083.075 The topside ionosphere: a region of dynamic transition.

P. M. Banks, R. W. Schunk, W. J. Raitt.
Annual Rev. Earth Planet. Sci., Vol. 4, (see 003.012), 381 - 440 (1976).

083.076 **Observations of radio frequency noise from Ariel 4.**
D. Walsh, A. P. Hayes, V. A. W. Harrison.
Proc. Roy. Soc. London, Ser. A, Vol. 343, 227 - 240 = Astron.
Contr. Univ. Manchester, Ser. II, Jodrell Bank Repr. No. 525 (1975).

Radio frequency observations of noise in the upper iono-sphere were made with a 12 m electric dipole and a narrow-band receiver which either swept from 0.25 to 4.0 MHz every 16 s or operated at a fixed frequency of 2 MHz.

083.077 **Ionospheric effects due to interplanetary neutral helium.** H. J. Fahr, P. W. Blum.
Ann. Géophys., Vol. 31, 271 - 278 (1975). – Abstr. in Phys.
Abstr., Vol. 79, A033074 (1976).

083.078 **Equatorial spread-F configurations and magnetic activity.** J. H. Sastri, B. S. Murphy, K. Sasidharan.
Curr. Sci., Vol. 44, 733 - 734 (1975). – Abstr. in Phys. Abstr.,
Vol. 79, A037265 (1976).

083.079 **Equatorial electrojet and interplanetary plasma parameters.** R. P. Kane.
Indian Journ. Radio Space Phys., Vol. 4, 132 - 137 (1975).
Abstr. in Phys. Abstr., Vol. 79, A037286 (1976).

083.080 **Ionospheric effects of transient celestial X-ray and gamma ray events.** K. Kasturirangan, U. R. Rao,
D. P. Sharma, R. G. Rastogi, S. C. Chakravarty.
Indian Journ. Radio Space Phys., Vol. 4, 203 - 205 (1975).
Abstr. in Phys. Abstr., Vol. 79, A041882 (1976).

083.081 **On the characteristics of spread-F configurations at Kodaikanal.**
J. H. Sastri, B. S. Murthy, K. Sasidharan.
Ann. Géophys., Vol. 31, 409 - 414 (1975). – Abstr. in Phys.
Abstr., Vol. 79, A041891 (1976).

Parametric excitation of standing electromagnetic waves. See Abstr. 062.004.

Instabilités rayonnées dans un plasma par un faisceau d'électrons. See Abstr. 062.024.

Solar EUV flux models consistent with ionospheric ion composition observations. See Abstr. 076.011.

Features of the effect of the low-energy solar cosmic rays and precipitating particles on the ionosphere.
See Abstr. 078.044.

Possiblity of studying the solar cosmic rays and the geomagnetic threshold variations at high latitudes basis of ionospheric effects. See Abstr. 078.045.

Response of electrons in ionosphere and plasma-sphere to magnetic storms. See Abstr. 084.207.

A model current system for the magnetospheric sub-storm. See Abstr. 084.234.

On a case of appearance of magnetic substorms in the mean-latitude night ionosphere. See Abstr. 084.247.

Ionospheric response to gamma-ray bursts of cosmic origin. See Abstr. 142.241.

084   Aurorae, Geomagnetic Field, Radiation Belts

Aurorae

084.001   The power spectrum and spatial structure of complex auroral radio-absorption events.
J. K. Hargreaves, M. G. Berry.
Planet. Space Sci., Vol. 24, 17 - 24 (1976).

084.002   Coordinated ATS 5 electron flux and simultaneous auroral observations.
S. B. Mende, E. G. Shelley.
Journ. Geophys. Res., Vol. 81, 97 - 110 (1976).

084.003   Helium isotopes in an aurora.
F. Bühler, W. I. Axford, H. J. A. Chivers, K. Marti.
Journ. Geophys. Res., Vol. 81, 111 - 115 (1976).
The authors report the detection of ^{3}He and confirm measurements of the ^{4}He flux reported in a previous paper. The ^{4}He: ^{3}He ratio in the first (brighter) aurora was 2950 ± 250. This ratio is only slightly higher than the average solar wind ratio of 2350 and since the atmospheric ratio is grossly different (250 times larger), establishes the solar wind as the principal source of auroral helium.

084.004   Measurements of field-aligned currents in a multiple auroral arc system.   J. Sesiano, P. A. Cloutier.
Journ. Geophys. Res., Vol. 81, 116 - 122 (1976).

084.005   Nonlinear production of suprathermal tails in auroral electrons.
D. L. Matthews, M. Pongratz, K. Papadopoulos.
Journ. Geophys. Res., Vol. 81, 123 - 129 (1976).

084.006   Characteristics of auroral proton precipitation observed from sounding rockets.
J. R. Miller, B. A. Whalen.
Journ. Geophys. Res., Vol. 81, 147 - 154 (1976).
A summary of auroral proton observations from 12 sounding rockets is presented. Energy spectra were peaked in the 5- to 10-keV energy intervals with typical intensities near 10^5 $cm^{-2} s^{-1} sr^{-1} keV^{-1}$. Proton pitch angle distributions in the loss cone were isotropic at all energies less than 20 keV. At higher energies, distributions were either isotropic or anisotropic peaked near 90° pitch angle. No obvious electron-proton correlations were apparent in any of the data. Observations of interactions of the primary proton beam with the neutral atmosphere are presented.

084.007   Quenching of the N_2 Vegard-Kaplan system in aurora.   A. Vallance Jones, R. L. Gattinger.
Journ. Geophys. Res., Vol. 81, 497 - 500 (1976).

084.008   Substorm energy.   S.-I. Akasofu, Y. Kamide.
Planet. Space Sci., Vol. 24, 223 - 227 (1976).

084.009   The charge spectrum of positive ions in a hydrogen aurora.   J. Lynch, D. Pulliam, R. Leach, F. Scherb.
Journ. Geophys. Res., Vol. 81, 1264 - 1268 (1976).

084.010   Gilbert White and the aurora.   J. B. Tyldesley.
Journ. British Astron. Ass., Vol. 86, 214 - 218 (1976).

084.011   Bremsstrahlung effects in auroral electron precipitation event absorption.
B. Sellers, F. A. Hanser, R. P. Vancour.
Journ. Atmosph. Terr. Phys., Vol. 38, 463 - 474 (1976).

084.012   Heating of the auroral ionosphere during an aurora.
V. I. Degtyarev, V. A. Kurilov, B. A. Ferberg.
Geomagn. Aeronom., Vol. 16, 185 - 188 (1976). In Russian. Brief information.

084.013   Convection and electron precipitation in the auroral zone and polar cusp during magnetospheric substorms.   P. Tanskanen.
VIIth Leningrad seminar. "Corpuscular streams of the sun and the radiation belts of the earth and Jupiter", 1975, (see 012. 007), p. 219 - 236 (1975). – Abstr. in Referativ. Zhurn. 62. Issled. kosmich. prostranstva, 4.62.258 (1976).

084.014   Auroral emission at 1084 Å.
G. A. Victor, P. McKenna, A. Dalgarno.
Planet. Space Sci., Vol. 24, 405 - 407 (1976).
Calculations are presented of the auroral emission of a line at 1084 Å of ionized atomic nitrogen that arises from electron impact induced simultaneous ionization and dissociation of molecular nitrogen. Estimates are also presented of the intensities of the argon lines at 1048 and 1067 Å.

084.015   Magnetic pulsation Pi2 and substorm onset.
T. Sakurai, T. Saito.
Planet. Space Sci., Vol. 24, 573 - 575 (1976).

084.016   On proton sources of the daily auroral zone.
N. V. Dzhordzhio, R. A. Kovrazhkin.
Kosmich. Issled., Vol. 14, 230 - 234 (1976). In Russian.

084.017   Emissions of the upper atmosphere during the magnetic storm of March 16 - 17, 1974.
Yu. L. Truttse, V. D. Belyavskaya, A. S. Elokhov, A. A. Kozlova, V. A. Goncharenko.
Geomagn. Aeronom., Vol. 16, 492 - 496 (1976). In Russian.

084.018   Energy spectra of auroral particles and magnetospheric convection.   N. V. Isaev, N. K. Osipov.
Geomagn. Aeronom., Vol. 16, 562 - 564 (1976). In Russian. Brief information.

084.019   Correlated observations of several auroral substorms on February 17, 1971.
E. W. Hones, Jr., S.-I. Akasofu, J. H. Wolcott, S. J. Bame, D. H. Fairfield, C.-I. Meng.
Journ. Geophys. Res., Vol. 81, 1725 - 1736 (1976).
The purpose of this study is to correlate in detail auroral activity with the corresponding disturbances in the magnetotail. The auroral data were recorded by optical instruments aboard an airplane flying over the Arctic Ocean along the Alaska meridian and by the Alaska meridian chain of all-sky cameras.

084.020   Recurrent auroral patterns.   T. Oguti.
Journ. Geophys. Res., Vol. 81, 1782 - 1786 (1976).
Brief report.

084.021   Dynamics of penetration of auroral particles into the midnight sector.
L. L. Lazutin, A. O. Mel'nikov.
Cosmic rays, No. 15, (see 003.006), p. 36 - 39 (1975). In Russian.

084.022   Quasi-periodic poleward propagation of on—off switching aurora and associated geomagnetic pulsations in the dawn.   T. Oguti, T. Watanabe.
Journ. Atmosph. Terr. Phys., Vol. 38, 543 - 551 (1976).

On the basis of TV records of auroras and induction magnetograms obtained in the southern auroral zone, geomagnetic pulsations with periods of a few seconds to several tens of second which often occur in the dawn sector during a post-breakup phase were found to be coherently related to auroral pulsations (or on—off switching) and typically to a quasi-periodic poleward propagation of the on—off switching aurora.

084.023   Auroral surges in southern hemisphere.
F. R. Bond, M. R. Stracey, D. S. Retallack.
Planet. Space Sci., Vol. 24, 611 - 612 (1976). — Research note.

084.024   Auroral electron spectra in the atmosphere.
J. G. Luhmann.
Journ. Atmosph. Terr. Phys., Vol. 38, 605 - 610 (1976).

084.025   Auroral emissions in relation to low energy electron flux in the nightside auroral oval.
G. Gustafsson, A. Egeland.
Journ. Atmosph. Terr. Phys., Vol. 38, 647 - 653 (1976).

084.026   The spatial relationship of auroral electrojets and visible aurora in the evening sector.
D. D. Wallis, C. D. Anger, G. Rostoker.
Journ. Geophys. Res., Vol. 81, 2857 - 2869 (1976).

Optical thickness in $\Sigma - \Sigma$ molecular bands.
See Abstr. 022.011.

The effect of the north-south asymmetry of solar cosmic rays and dynamics of a plasma layer and of the daytime polar cusp.   See Abstr. 078.007.

Apollo 16 far ultraviolet imagery of the polar auroras, tropical airglow belts, and general airglow.
See Abstr. 082.018.

Ionospheric composition in SAR-arcs.
See Abstr. 083.005.

Simultaneous observations of the proton ring current and stable auroral red arcs.   See Abstr. 084.402.

Beobachtung von Meteoren und Nordlichtern mit Hilfe der Amateurfunktechnik.   See Abstr. 104.005.

Geomagnetic Field

084.201 **International geomagnetic reference field 1975.**
IAGA Division I Study Group.
Geophys. Journ. Roy. Astron. Soc., Vol. 44, 733 - 734 (1976).
Short communication.

084.202 **Flapping motions of the tail plasma sheet induced by the interplanetary magnetic field variations.**
T. Toichi, T. Miyazaki.
Planet. Space Sci., Vol. 24, 147 - 159 (1976).

Flapping motions of the magnetotail with an amplitude of several earth radii are studied by analysing the observations made in the near ($x = -25 \sim -30\,R_E$) and the distant ($x \simeq -60\,R_E$) tail regions. It is found that the flapping motions result from fluctuations in the interplanetary magnetic field, especially Alfvénic fluctuations, when the magnitude of the interplanetary magnetic field is larger than $\sim 10\,\gamma$, and they propagate behind the earth with the solar wind flow. The characteristics of the flapping reveal that the geomagnetic tail is a good resonator for the hydromagnetic disturbances in the solar wind.

084.203 **An alkali vapour magnetometer using integrated circuits.** J. M. Stanley, F. C. Ludbey, R. Green.
Space Sci. Instrum., Vol. 1, 471 - 492 (1975).

An optically pumped alkali vapour magnetometer has been constructed which utilizes caesium vapour as the resonant element in the self-oscillator mode of operation. This instrument is capable of monitoring anomalies and fluctuations in the magnetic field with a sensitivity of 0.01 nT over the range of the earth's field (24–75 μT).

084.204 **Geomagnetic field variations.** R. P. Kane.
Space Sci. Rev., Vol. 18, 413 - 540 (1976).

In this review, an attempt has been made to describe the large variety of geomagnetic variations, both regular and irregular. After a brief description of the earth and its environment, different types of quiet-day variations are described and present ideas regarding their possible mechanisms are discussed. In general, periodicities exceeding several tens of years can be attributed to changes in the interior of the earth while periodicities of 22 years or less seem to be related to phenomena connected with the sun, through the interaction of solar wind with the earth's magnetosphere. The morphology of irregular storm-time variations and its relationship with interplanetary plasma parameters is discussed with particular reference to the orientations of interplanetary magnetic field. Various storm-time phenomena occurring in the polar, auroral, mid-latitude and equatorial regions and their interconnections are described. Theoretical models offering explanations of many of these phenomena are discussed, the unsolved problems are outlined, and the direction of the present effort in solving these is indicated.

084.205 **Some observations of electrons with energies > 30 keV made during magnetospheric substorms.**
P. A. Smith, G. R. Thomas.
Journ. Atmosph. Terr. Phys., Vol. 38, 251 - 260 (1976).

084.206 **The time dependence of the geomagnetic activity index A_p on the lunar phases.**
E. M. Apostolov, V. Letfus.
Bull. Astron. Inst. Czechoslovakia, Vol. 27, 110 - 111 (1976).

The effect of the moon on geomagnetic activity, represented by the diurnal values of the A_p-index, was investigated by a new method using cross-correlation analysis for the period 1960–1973. Average percentage deviations from the reference level of annual medians at a phase shift of +5.5 days, derived from the maximum of the cross-correlation function of the lunar phases and the A_p-index, were determined for the four principal phases of the moon. For the low activity period the maximum increase of the A_p-index amounts to +7.6% after full moon and the decrease to −7.3% after new moon.

084.207 **Response of electrons in ionosphere and plasmasphere to magnetic storms.** H. Soicher.
Nature, Vol. 259, 33 - 35 (1976). — Letter.

084.208 **Present trends in the earth's magnetic field.**
J. M. Harwood, S. R. C. Malin.
Nature, Vol. 259, 469 - 471 (1976). — Letter.

084.209 **Lunar effect in the quiet-time D_{st} index.**
T. Kamei, H. Maeda.
Nature, Vol. 259, 644 - 645 (1976). — Letter.

084.210 **Evolution of periodicities exhibited by fluctuations in the intensity of chorus.** S. K. Adjepong.
Nature, Vol. 259, 645 - 648 (1976). — Letter.

084.211 **Evidence of magnetic field line merging in the solar wind.**
B. Bavassano, M. Dobrowolny, F. Mariani.
Journ. Geophys. Res., Vol. 81, 1 - 6 (1976).

An analysis is presented of a few of the sector boundary crossings measured by the Pioneer 8 magnetometer. By using a variance matrix technique, evidence is obtained of some amount of magnetic line reconnection through the sector boundaries. The thicknesses of such structures are around $0.1-2 \times 10^4$ km and are therefore larger than typical ion Larmor radii, in contrast with the magnetic structures characteristic of the geomagnetic tail. A theoretical discussion is given of the possible physical causes of the observed reconnection process. In particular, the possible role of tearing instabilities is analyzed with reference to the results of the observations.

084.212 **A statistical study of the upstream wave boundary outside the earth's bow shock.**
L. Diodato, E. W. Greenstadt, G. Moreno, V. Formisano.
Journ. Geophys. Res., Vol. 81, 199 - 204 (1976).

The forward boundary of the upstream wave region ahead of the earth's bow shock is located statistically by using combined field and plasma data from a single spacecraft, Heos 1. Interpreting the wave as arising from reflected protons traveling upstream with effective velocity $V_{\parallel} = pV$ along the interplanetary field, where V is the solar wind bulk speed, the authors find that the overall average boundary corresponds to a velocity $V_{\parallel} \approx 2V$ for protons reflected from the daylight sector of the bow shock.

084.213 **Neutral sheet observations at 1000 R_E.**
U. Villante.
Journ. Geophys. Res., Vol. 81, 212 - 215 (1976).

A detailed analysis of magnetic field observations internal to the field reversal regions suggests the presence, at $\sim 1000\,R_E$ (Pioneer 7), of a well-defined neutral sheet still preserving most of its near-earth characteristics. The normal direction to the rotation plane of the magnetic field vector is consistent with an aberrated tail model. Observed large values of the reconnection component, occasionally southward, at the reversal point might be interpreted in terms of closed loops of opposite field lines. Order of magnitude estimates of the sheet dimensions confirm a sheet thickness of $0.1-1\,R_E$.

084.214 **Proton flow measurements in the magnetotail plasma sheet made with Imp 6.**

E. W. Hones, Jr., S. J. Bame, J. R. Asbridge.
Journ. Geophys. Res., Vol. 81, 227 - 234 (1976).

Patterns of proton flow in the magnetotail plasma sheet are described. They are derived from observed anisotropies of the flux of protons. These observations, made in the geocentric distance range $r \approx 25-32\,R_E$, are entirely consistent with those made at $r \approx 18\,R_E$ with Vela satellites and reported earlier.

084.215 Repeated sharp flux dropouts observed at 6.6 R_E during a geomagnetic storm.
S.-Y. Su, T. A. Fritz, A. Konradi.
Journ. Geophys. Res., Vol. 81, 245 - 252 (1976).

084.216 The position and shape of the neutral sheet at 30-R_E distance.   S. B. Bowling, C. T. Russell.
Journ. Geophys. Res., Vol. 81, 270 - 272 (1976).

084.217 Comment on 'Second-order statistical structure of geomagnetic field reversals' by P. S. Naidu.
T. J. Ulrych, R. W. Clayton.
Journ. Geophys. Res., Vol. 81, 1033, with a reply by P. S. Naidu, p. 1034 (1976).

084.218 The cusp-magnetosheath interface.
A. M. Hansen, A. Bahnsen, N. D'Angelo.
Journ. Geophys. Res., Vol. 81, 556 - 561 (1976).

Three Heos 2 passes through the distant cusp-magnetosheath transition in summer 1973 are described. By combining data from three different experiments on board Heos 2 it has been possible to identify a distinct transition from a 'stagnant' plasma (the distant cusp) to a region of 'reduced flow' magnetosheath. Closed magnetosphere models are in better agreement with the observations than open ('reconnection') models.

084.219 Gradients of solar protons in the high-latitude magnetotail and the magnetospheric electric field.
I. D. Palmer, P. R. Higbie, E. W. Hones, Jr.
Journ. Geophys. Res., Vol. 81, 562 - 568 (1976).

084.220 Dependence of the latitude of the cleft on the interplanetary magnetic field and substorm activity.
Y. Kamide, J. L. Burch, J. D. Winningham, S.-I. Akasofu.
Journ. Geophys. Res., Vol. 81, 698 - 704 (1976).

The latitudinal motion of the cleft (the polar cusp) associated with the southward interplanetary magnetic field (IMF) and substorm activity is examined. The cleft location is identified on the basis of the location of midday auroras and of electron precipitation (Ogo 4 and Isis 1 satellites). It is found that the IMF and substorm activity control independently the latitude of the cleft and that they can shift the cleft location by $3°-4°$ under average conditions.

084.221 Spectral analysis of geomagnetic reversal time scales.
J. D. Phillips, A. Cox.
Geophys. Journ. Roy. Astron. Soc., Vol. 45, 19 - 33 (1976).

084.222 Magnetic fields of the magnetosheath.
D. H. Fairfield.
Rev. Geophys. Space Phys., Vol. 14, 117 - 134 (1976).
Review paper.

084.223 The IMS (International Magnetospheric Study) satellite programme: scientific objectives.
C. T. Russell.
Astrophys. Space Sci. Library, Vol. 57, (see 012.002), 9 - 42 (1976).

084.224 Magnetospheric convection induced by interplanetary magnetic-field variations.   K. Maezawa.
Astrophys. Space Sci. Library, Vol. 57, (see 012.002), 133 - 166 (1976).

084.225 Magnetopause position and the reconnection problem.   V. Formisano.
Astrophys. Space Sci. Library, Vol. 57, (see 012.002), 167 - 168 (1976).

084.226 New theoretical aspects of the shape of the magnetospheric boundary.
F. F. Cap, M. P. Leubner, F. P. Stössel.
Astrophys. Space Sci. Library, Vol. 57, (see 012.002), 169 - 186 (1976).

084.227 Low-energy heavy-ion observations on the IMP 7 satellite.   C. Y. Fan, D. Hovestadt, G. Gloeckler.
Astrophys. Space Sci. Library, Vol. 57, (see 012.002), 187 - 197 (1976).

H, He, and CNO ion pulses have been detected by an electrostatic deflection spectrometer on the IMP 7 satellite in the magnetotail, the magnetosheath, and in the region upstream from the earth's bow shock. The spatial distribution of the pulses indicates that the ions are accelerated in the upstream region and in the near-earth region of the neutral sheet.

084.228 Low-energy-proton regime in the geomagnetic tail at lunar distance.   H. K. Hills, D. A. Hardy, J. W. Freeman.
Astrophys. Space Sci. Library, Vol. 57, (see 012.002), 199 - 207 (1976).

084.229 Multisatellite observations of solar protons penetrating the magnetospause.   V. Domingo, D. E. Page, K.-P. Wenzel.
Astrophys. Space Sci. Library, Vol. 57 (see 012.002), 225 - 235 (1976).

Simultaneous observations of low-energy solar protons (1 - 40 MeV) by HEOS-2 in the northern high-latitude magnetotail and by other spacecraft in interplanetary space during the solar proton event of 8 June 1972 are presented. The results indicate that solar protons enter that magnetotail lobe which is not well connected to the interplanetary magnetic field – with respect to solar-particle entry – preferentially close to the earth ($|X|< 10\,R_E$) rather than farther downtail.

084.230 Propagation characteristics of electromagnetic waves in the magnetosphere.   D. Jones, R. J. L. Grard.
Astrophys. Space Sci. Library, Vol. 57, (see 012.002) 293 - 302 (1976).

The propagation mechanism of electromagnetic waves in the earth's magnetosphere and magnetotail is investigated using a ray-tracing technique. The present study makes use of macroscopic models for the magnetic-field topology and electron-density distribution. The existence of the bow shock, magnetopause and plasmapause is taken into account; the distortion of the earth's magnetic field by the solar wind is also included in the models. Results are presented in the form of ray paths. This study deals with the penetration of type-III solar bursts into the magnetosphere, the trapping of terrestrial nonthermal radiation, and the propagation of auroral kilometric radiation.

084.231 VLF electrostatic waves in the magnetospheres of the earth and Jupiter.
M. Ashour-Abdalla, C. F. Kennel.
Astrophys. Space Sci. Library, Vol. 57, (see 012.002), 303 - 325 (1976).

084.232 Summary of the La Jolla conference on quantitative magnetospheric models.   W. P. Olson.
Astrophys. Space Sci. Library, Vol. 57, (see 012.002), 367 - 380 (1976).

084.233 Review of selected scientific results of HEOS 2.
D. E. Page, K.-P. Wenzel.
Space Research XV, (see 012.003), p. 433 - 454 (1975).

084.234 A model current system for the magnetospheric sub-
storm. Y. Kamide, F. Yasuhara, S.-I. Akasofu.
Planet. Space Sci., Vol. 24, 215 - 222 (1976).
The authors propose a model three-dimensional current system for the magnetospheric substorm, which can account for the new findings of the field-aligned and ionospheric currents obtained during the last few years by using new techniques.

084.235 On the generation of low-frequency waves in the
solar wind in the front of the bow shock.
M. S. Kovner, V. V. Lebedev, T. A. Plyasova-Bakounina
(*Plyasova-Bakunina*), V. A. Troitskaya.
Planet. Space Sci., Vol. 24, 261 - 267 (1976).
The generation of low-frequency waves in the solar wind by the flux of protons accelerated in the magnetosheath is considered. It is shown that pulsations are produced in two partly overlapping frequency ranges. The growth rate of waves is maximal when the angle θ between the direction of the interplanetary magnetic field and the front of the bow shock is not equal $\pi/2$. The dependence of the increment of perturbation on the solar wind velocity is analysed. A satisfactory agreement between theory and experimental results on the connection of Pc3–4 properties and parameters of the solar wind is obtained.

084.236 Internationaal onderzoek van de magnetosfeer.
E. Aerts.
Zenit, 3e jaargang, p. 6 - 8 (1976).

084.237 Periodic oscillations of the first generation in the
Störmer problem. C. L. Goudas, A. A. Haliou-
lias, V. V. Markellos, G. Macris.
Long-time predictions in dynamics, (see 012.005), p. 267 - 276 (1976).

084.238 Periodic oscillations of the second generation in the
Störmer problem.
V. V. Markellos, S. Klimopoulos, C. L. Goudas.
Long-time predictions in dynamics, (see 012.005), p. 340 (1976). – Abstract.

084.239 An analysis of the spectra of geomagnetic variations
having periods from 5 min to 4 hours.
W. H. Campbell.
Journ. Geophys. Res., Vol. 81, 1369 - 1390 (1976).

084.240 Evidence for a bow shock structure at $\sim 400\, R_E$:
Pioneer 7. U. Villante.
Journ. Geophys. Res., Vol. 81, 1441 - 1446 (1976).
Analysis of simultaneous magnetic field and plasma data by Pioneer 7 reveals the existence of a well-defined bow shock structure at geocentric distances of 350–440 R_E. Bow shock physical parameters are computed by an iterative least squares technique applied to combined plasma and magnetic field data. Observed features suggest that at those distances from the earth the bow wave still preserves its near-earth characteristics.

084.241 Magnetic field variations in the polar region during
magnetically quiet periods and interplanetary
magnetic fields. Ya. I. Feldstein (*Fel'dshtejn*).
Space Sci. Rev., Vol. 18, 777 - 861 (1976). – Review paper.

084.242 Propagation of geomagnetic disturbances in the
waveguide earth-ionosphere. Yu. N. Savchenko.
Geomagn. Aeronom., Vol. 16, 138 - 142 (1976). In Russian.

084.243 Analysis of the space-time structure of the main

geomagnetic field by the method of expansion in
natural orthogonal components. A. N. Pushkov, Eh. B.
Fajnberg, T. A. Chernova, M. V. Fiskina.
Geomagn. Aeronom., Vol. 16, 337 - 343 (1976). In Russian.

084.244 The position of neutral points in a two-dimensional
model of the magnetosphere. P. Oberts.
Geomagn. Aeronom., Vol. 16, 375 - 377 (1976). In Russian.
Brief information.

084.245 Interplanetary magnetic field and long-period
geomagnetic pulsations at mean latitudes.
Ya. M. Gogatishvili.
Geomagn. Aeronom., Vol. 16, 382 - 384 (1976). In Russian.
Brief information.

084.246 Non-adiabatic effects of motion of particles in a
static dipole field and in fields variable with regard
to time. V. D. Il'in, S. N. Kuznetsov.
VIIth Leningrad seminar. "Corpuscular streams of the sun and the radiation belts of the earth and Jupiter", 1975, (see 012. 007), p. 269 - 278 (1975). In Russian. – Abstr. in Referativ. Zhurn. 62. Issled. kosmich. prostranstva, 4.62.256 (1976).

084.247 On a case of appearance of magnetic substorms in
the mean-latitude night ionosphere.
Ts. N. Gogosheva, K. B. Serafimov, M. M. Gogoshev.
VIIth Leningrad seminar. "Corpuscular streams of the sun and the radiation belts of the earth and Jupiter", 1975, (see 012. 007), p. 249 - 263 (1975). In Russian. – Abstr. in Referativ. Zhurn. 62. Issled. kosmich. prostranstva, 4.62.260 (1976).

084.248 On the longitudinal extent of the polar cusp.
S. A. Zaitzeva (*Zajtseva*), M. I. Pudovkin.
Planet. Space Sci., Vol. 24, 518 - 519 (1976).
Variations of the longitudinal extent of the polar cusp are studied in relation to the orientation of the interplanetary magnetic field. In cases when the vertical component of the solar wind magnetic field is positive, the polar cusp is shown to be restricted to a relatively localized region at 12 ± 2 LMT.

084.249 Cyclotron side-band emissions from ring-current
electrons. K. Maeda.
Planet. Space Sci., Vol. 24, 341 - 347 (1976).

084.250 Sudden geomagnetic field variations in the night-
side cusp-region during magnetospheric substorms
accompanied by electron precipitation in the auroral zone.
J. W. Münch, G. Kremser.
Planet. Space Sci., Vol. 24, 365 - 373 (1976).

084.251 Correlation of the intensity of fast electrons in the
earth's magnetosphere and in the interplanetary
space. I. I. Senchuro, P. I. Shavrin.
Kosmich. Issled., Vol. 14, 314 - 316 (1976). In Russian.
Brief information.

084.252 Three-parametric two-dimensional model of the
magnetosphere. P. Oberts.
Geomagn. Aeronom., Vol. 16, 497 - 503 (1976). In Russian.

084.253 On the theory of drift-mirror instability of a mag-
netospheric plasma.
O. A. Pokhotelov, V. A. Pilipenko.
Geomagn. Aeronom., Vol. 16, 504 - 510 (1976). In Russian.

084.254 On a possibility of investigation of discontinuities in
the magnetosphere and interplanetary space by the
method of non-coherent reply.
I. P. Stakhanov, I. V. Kovalevskij.
Geomagn. Aeronom., Vol. 16, 511 - 517 (1976). In Russian.

084.255 **Geomagnetic disturbances caused by shock waves of large meteoritic bodies. II.**   Yu. N. Savchenko.
Geomagn. Aeronom., Vol. 16, 518 - 525 (1976). In Russian.

084.256 **Space-time filtration of geomagnetic fields.**
M. S. Zhdanov, Eh. B. Fajnberg.
Geomagn. Aeronom., Vol. 16, 535 - 541 (1976). In Russian.

084.257 **Archeomagnetic definitions of the geomagnetic field for the territory of Mongolia.**
G. Ayuushzhav, S. P. Burlatskaya, I. E. Nachasova, G. Dodon, I. Balbar.
Geomagn. Aeronom., Vol. 16, 542 - 548 (1976). In Russian.

084.258 **Determinación de la influencia de la Luna en la declinación en Moca (Fernando Poo).**
J. O. Cardús.
Urania Barcelona, Año 60, No. 283, p. 51 - 70 (1975).

084.259 **Pitch angle and radial diffusion of MeV protons in the outer magnetosphere.**
M. Scholer, G. Morfill.
Journ. Geophys. Res., Vol. 81, 1737 - 1743 (1976).

084.260 **Thickness of magnetic structures associated with the earth's bow shock.**   D. L. Morse, E. W. Greenstadt.
Journ. Geophys. Res., Vol. 81, 1791 - 1793 (1976).
Brief report.

084.261 **Height distribution of the geomagnetic field.**
V. F. Nikitin, V. A. Ehl'tekov.
Cosmic rays, No. 15, (see 003.006), p. 13 - 14 (1975). In Russian.

084.262 **Effects on the plasmasphere of irregular electric fields.**   J. M. Grebowsky, A. J. Chen.
Planet. Space Sci., Vol. 24, 689 - 696 (1976).
A conservative convection electric field model developed by Volland (1973) to describe the solar wind induced plasma flow within the inner magnetosphere is modified to include a noisy spatial component.

084.263 **A study of alternative schemes for extrapolation of secular variation at observatories.**
L. R. Alldredge.
Phys. Earth Planet. Interiors, Vol. 11, P18 - P25 (1976).
Letter.

084.264 **On a tentative correlation between changes in the geomagnetic polarity bias and reversal frequency and the earth's rotation through Phanerozoic time.**
K. M. Creer.
Growth rhythms and the history of the earth's rotation, (see 012.012), p. 293 - 317 (1975).

084.265 **The geomagnetic field in the magnetotail.**
D. B. Beard.
14th Intern. Cosmic Ray Conf., (see 012.011), Vol. 12, 4213 - 4214 (1975).

084.266 **On the possibility of using electroreactive accelerators for investigation of the earth's magnetosphere.**
I. M. Podgornyj, A. A. Porotnikov.
Kosmich. Issled., Vol. 14, 461 - 463 (1976). In Russian.
Brief information.

084.267 **Magnetospheric electric fields and their variation with geomagnetic activity.**   M. G. Kivelson.
Rev. Geophys. Space Phys., Vol. 14, 189 - 197 (1976).

084.268 **Representation of magnetic fields in space.**
D. P. Stern.
Rev. Geophys. Space Phys., Vol. 14, 199 - 214 (1976).

084.269 **Day side reconnection between a dipolar geomagnetic field and a uniform interplanetary field.**
T. Yeh.
Journ. Geophys. Res., Vol. 81, 2140 - 2144 (1976).
Field line reconnection on the day side magnetopause is assumed to take place along the separator of field line connectivity in the magnetic topology resulting from the interpermeation of a dipolar geomagnetic field and a uniform interplanetary field. The induced voltage is calculated to show its dependence on the magnitude and direction of the incident magnetic field.

084.270 **Magnetospheric conditions at the time of enhanced wave-particle interactions near the plasmapause.**
J. C. Foster, T. J. Rosenberg, L. J. Lanzerotti.
Journ. Geophys. Res., Vol. 81, 2175 - 2182 (1976).

084.271 **Conduction of thermal energy in the neighborhood of the earth's bow shock.**   R. G. Hohlfeld.
Journ. Geophys. Res., Vol. 81, 2257 - 2260 (1976).

084.272 **Magnetospheric convection induced by the positive and negative Z components of the interplanetary magnetic field: quantitative analysis using polar cap magnetic records.**   K. Maezawa.
Journ. Geophys. Res., Vol. 81, 2289 - 2303 (1976).

084.273 **Fluxes of ≥ 50-keV protons and ≥ 30-keV electrons at $\sim 35\,R_E$. 1. Velocity anisotropies and plasma flow in the magnetotail.**
E. C. Roelof, E. P. Keath, C. O. Bostrom, D. J. Williams.
Journ. Geophys. Res., Vol. 81, 2304 - 2314 (1976).

084.274 **Fluxes of ≥ 50-keV protons and ≥ 30-keV electrons at $\sim 35\,R_E$. 2. Morphology and flow patterns in the magnetotail.**
E. P. Keath, E. C. Roelof, C. O. Bostrom, D. J. Williams.
Journ. Geophys. Res., Vol. 81, 2315 - 2326 (1976).

084.275 **Spectral study of magnetospheric oscillations.**
K. S. Raja Rao.
Journ. Atmosph. Terr. Phys., Vol. 38, 661 - 664 (1976).

084.276 **On the long-period variations of the earth's magnetic field from 2 months to 20 years.**
V. Courtillot, J.-L. Le Mouël.
Journ. Geophys. Res., Vol. 81, 2941 - 2950 (1976).

084.277 **Effects of solar activity on annual means of geomagnetic components.**   L. R. Alldredge.
Journ. Geophys. Res., Vol. 81, 2990 - 2996 (1976).

084.278 **A model of the geomagnetic field for 1975.**
N. W. Peddie, E. B. Fabiano.
Journ. Geophys. Res., Vol. 81, 2539 - 2542 (1976).
A new model of the near-surface geomagnetic main field and its secular variation has been derived. The main field part, a degree 12 spherical harmonic series of 168 internal source terms, was derived from 1248 representative values having a nearly even distribution over the earth.

084.279 **Influence of the interplanetary magnetic field on the occurrence and thickness of the plasma mantle.**
N. Sckopke, G. Paschmann, H. Rosenbauer, D. H. Fairfield.
Journ. Geophys. Res., Vol. 81, 2687 - 2691 (1976).
The response of the plasma mantle to the orientation of

the interplanetary magnetic field (IMF) has been studied by correlating Heos 2 plasma and Imp 6 magnetic field data. The mantle is nearly always present when the IMF has a southward component and often also when the field has a weak northward component. In addition, the mantle appears increasingly thicker with greater southward components. On the other hand, the mantle is thin or missing (from the region where it is normally found) when the average IMF has a strong northward component.

084.280 **Correlation of bow shock plasma wave turbulence with solar wind parameters.**
P. Rodriguez, D. A. Gurnett.
Journ. Geophys. Res., Vol. 81, 2871 - 2882 (1976).

084.281 **Diffusion of photoelectrons along a field line inside the plasmasphere.** G. Lejeune, F. Wormser.
Journ. Geophys. Res., Vol. 81, 2900 - 2916 (1976).

084.282 **Magnetic results 1971, Eskdalemuir, Hartland and Lerwick Observatories.**
Natural Environment Research Council, Inst. Geol. Sci., Geomagnetic Bull. No. 5, [London, Her Majesty's Stationery Office], 6 + 97 pp. Price £ 4.50 (1975).

084.283 **Annual mean values of geomagnetic elements since 1941.** M. P. Fisher.
Natural Environment Research Council, Inst. Geol. Sci., Geomagnetic Bull. No. 6, [London, Her Majesty's Stationery Office], 170 pp. Price £ 7.00 (1976).

084.284 **Selective fading and the propagation of radio waves in the magnetoplasma (ionosphere).** M. Cutolo.
Radio Sci. (*USA*), Vol. 8, 1093 - 1109 = Oss. Astron. Roma, Contr. sci., Ser. III, No. 137 (1973).

084.285 **Geomagnetic crochets associated with proton flares.**
J. H. Sastri, B. S. Murthy, D. Karunakaran.
Indian Journ. Radio Space Phys., Vol. 4, 89 - 92 (1975).
Abstr. in Phys. Abstr., Vol. 79, A003033 (1976).

084.286 **Electromagnetic earth. Theory of the magnetic field of our planet.** E. L. Caesar.
Elektrotechn. Zeitschrift B, Vol. 27, 631 - 636 (1975).
In German. − Abstr. in Phys. Abstr., Vol. 79, A010941 (1976).

084.287 **Geomagnetic effects in the dark hemisphere associated with solar flares.**
J. H. Sastri, B. S. Murthy.
Journ. Geomagn. Geoelectr., Vol. 27, 67 - 73 (1975). − Abstr. in Phys. Abstr., Vol. 79, A010947 (1976).

084.288 **Relationship between solar magnetic sector structure and geomagnetic *Dst*.** R. P. Kane.
Proc. Indian Acad. Sci. A, Vol. 81, 233 - 244 (1975). − Abstr. in Phys. Abstr., Vol. 79, A027437 (1976).

084.289 **The magnetospheric contribution to the quiet-time low energy nucleon spectrum in the vicinity of earth.** S. M. Krimigis, J. W. Kohl, T. P. Armstrong.
Geophys. Res. Letters, Vol. 2, 457 - 460 (1975). − Abstr. in Phys. Abstr., Vol. 79, A027682 (1976).

084.290 **Using the moon to probe the geomagnetic tail lobe plasma.** G. Schubert, C. P. Sonett, B. F. Smith, K. Schwartz, D. S. Colburn.
Geophys. Res. Letters, Vol. 2, 277 - 280 (1975). − Abstr. in Phys. Abstr., Vol. 79, A027815 (1976).

084.291 **Waves and wave-particle interactions in the magnetosphere: a review.** R. Gendrin.

Space Sci. Rev., Vol. 18, 145 - 200 (1975). − Paper presented at the Esro symposium on European sounding rocket and scientific balloon activity at high latitudes with emphasis on the international magnetospheric study, Örenäs Slott, Sweden, 1974.

084.292 **Night time geomagnetic effects of solar flares.**
J. H. Sastri.
Ann. Géophys., Vol. 31, 389 - 393 (1975). − Abstr. in Phys. Abstr., Vol. 79, A041545 (1976).

084.293 **Geomagnetic solar flare effect in the dark hemisphere.** J. H. Sastri.
Indian Journ. Radio Space Phys., Vol. 4, 225 - 227 (1975). Abstr. in Phys. Abstr., Vol. 79, A041883 (1976).

084.294 **Anomalous electrical conductivity for the field-line merging current.** T. Tamao.
Rep. Ionosph. Space Res. Japan, Vol. 29, 140 - 142 (1975). Abstr. in Phys. Abstr., Vol. 79, A041902 (1976).

084.295 **Earth as a radio source: the nonthermal continuum.**
D. A. Gurnett.
Progress report Iowa Univ., Iowa City, USA, Dept. Phys. Astron., 54 pp. (1974). − See 14.084.209.

084.296 **Analysis of proton and electron spectrometer data from OGO-5 spacecraft. Final report, 1 July 1973 - 31 July 1974.** M. A. Pomerantz.
Separate print Bartol Res. Found. Swarthmore, Pennsylvania, USA, 95 pp. (1975).

084.297 **Observation of low-energy protons in the geomagnetic tail at lunar distances.** D. A. Hardy.
Thesis Rice Univ., Houston, Texas, USA, 127 pp. (1974).

Physics of the hot plasma in the magnetosphere.
Proceedings of a symposium, Kiruna, Sweden, April 1975. See Abstr. 012.029.

Magnetospheric particles and fields. Proceedings of the Summer Advanced Study Institute, Graz, Austria, August 4 - 15, 1975. See Abstr. 012.031.

Influence of ancient solar-proton events on the evolution of life. See Abstr. 015.002.

Instationary convection and precipitation of electron particles. See Abstr. 062.021.

Spectral analysis of solar and geomagnetic activity indices. See Abstr. 072.015.

On a connection of variations of solar wind velocity with the magnetic field in the magnetosphere's tail. See Abstr. 074.031.

On a mechanism of fast dissipation of a magnetic field in collision-free plasma. See Abstr. 074.034.

Long-lived coronal structures and recurrent geomagnetic patterns in 1974. See Abstr. 074.037.

Secondary interactions of shock waves and their effect on the development of a geomagnetic storm with sudden commencement. See Abstr. 074.048.

The interaction of a plane interplanetary discontinuity with the magnetosphere. See Abstr. 074.088.

Observations of >100-keV protons in the earth's

magnetosheath.   See Abstr. 078.002.

On the spatial distribution of solar cosmic rays in the earth's magnetosphere.   See Abstr. 078.018.

Solar cosmic ray diffusion into the polar magnetosphere.   See Abstr. 078.026.

ATS-6 solar cosmic ray and trapped particle experiment.   See Abstr. 078.102.

Neutral composition changes during a period of increasing magnetic activity.   See Abstr. 082.002.

Latitudinal changes of composition in the disturbed thermosphere from Esro 4 measurements.
See Abstr. 082.014.

Modeling the midlatitude F-region ionospheric storm using east-west drift and a meridional wind.
See Abstr. 083.004.

Increases of equatorial total electron content (TEC) during magnetic storms.   See Abstr. 083.008.

Rocket measurements of mid-latitude ionospheric currents during a magnetic storm.   See Abstr. 083.010.

Magnetospheric DC electric fields; present knowledge and outstanding problems to be solved during the IMS.
See Abstr. 083.022.

Ionization enhancement in the middle latitude D-region due to precipitating high energy electrons.
See Abstr. 083.034.

Transversal conductivity of the earth's ionosphere and magnetosphere.   See Abstr. 083.056.

Substorm energy.   See Abstr. 084.008.

Energy specta of auroral particles and magnetospheric convection.   See Abstr. 084.018.

Correlated observations of several auroral substorms on February 17, 1971.   See Abstr. 084.019.

Quasi-periodic poleward propagation of on–off switching aurora and associated geomagnetic pulsations in the dawn.   See Abstr. 084.022.

Comparison of magnetospheres and radio emissions of Jupiter with earth.   See Abstr. 099.053.

Is Jupiter's magnetosphere like a pulsar's or earth's? See Abstr. 099.064.

Characteristics of interplanetary plasma near the earth observed during the solar events of August 1972.
See Abstr. 106.015.

Comparative analysis of cosmophysical and geophysical phenomena in the 19th and 20th solar activity cycles.
See Abstr. 106.039.

Investigation of shock waves and tangential discontinuities of the cosmic plasma by the radio interference technique.   See Abstr. 106.042.

The intensity variations of solar and galactic cosmic rays with azimuthal angle in the polar region.
See Abstr. 143.008.

Detection and study of intensity variations of cosmic rays of interplanetary and magnetospheric origin from data of the world network of observatories.
See Abstr. 143.037.

Expected variations in the effective geomagnetic cutoff rigidity of cosmic rays at the stations of the worldwide network for the 11-year, annual, and 27-day variations; solar anisotropy; Forbush-decreases; increase effects prior to magnetic storms; and solar flares.   See Abstr. 143.204.

A comparison of vertical cosmic-ray cutoff rigidities as calculated with different geomagnetic field models.
See Abstr. 143.254.

A five by fifteen degree world grid of calculated cosmic-ray vertical cutoff rigidities for 1965 and 1975.
See Abstr. 143.257.

Experimental measurements of charged particle cutoff latitudes.   See Abstr. 143.262.

Cosmic ray variations due to geomagnetic field variations.   See Abstr. 143.263.

Geomagnetic effects of cosmic rays.
See Abstr. 143.314.

Isotopic analysis of high-energy cosmic-ray nuclei in the geomagnetic field.   See Abstr. 143.383.

Radiation Belts

084.401 **Radial diffusion of inner-zone protons: observations and variational analysis.**
D. R. Croley, Jr., M. Schulz, J. B. Blake.
Journ. Geophys. Res., Vol. 81, 585 - 594 (1976).

084.402 **Simultaneous observations of the proton ring current and stable auroral red arcs.**
D. J. Williams, G. Hernandez, L. R. Lyons.
Journ. Geophys. Res., Vol. 81, 608 - 616 (1976).

084.403 **Disturbing the radiation belts.** K. G. Budden.
Nature, Vol. 260, 392 - 393 (1976).

084.404 **Man-made e.l.f./v.l.f. emissions and the radiation belts.** K. Bullough, A. R. L. Tatnall, M. Denby.
Nature, Vol. 260, 401 - 403 (1976).
Morphological studies of e.l.f./v.l.f. emissions on the Ariel III and IV satellites indicate that man-made electromagnetic emissions, namely power-line harmonics generated in the industrialised regions of North America and, also, v.l.f. transmissions at 17.8 kHz (NAA) and 16.0 kHz (GBR) in the longitude sector which encompasses the South Atlantic Anomaly, are responsible, at least in part, for the formation of the electron slot ($2 < L < 3$) between the inner and outer radiation belts in the magnetosphere.

084.405 **Source of terrestrial non-thermal radiation.**
D. Jones.
Nature, Vol. 260, 686 - 689 (1976).
Observations from satellites have revealed the existence of continuum radiation from just beyond the plasmapause. The author proposes here that this is generated by Cerenkov radiation coupling to the O mode of radiation, which can propagate in low density plasmas. The observations are interpreted on the basis of this theory.

084.406 **Dynamics of energetic electrons of the radiation belts of the earth during magnetic storms.**
P. V. Vakulov, L. M. Kovrygina, O. V. Mineev, L. V. Tverskaya.
VIIth Leningrad seminar. "Corpuscular streams of the sun and the radiation belts of the earth and Jupiter", 1975, (see 012. 007), p. 267 - 269 (1975). In Russian. − Abstr. in Referativ. Zhurn. 62. Issled. kosmich. prostranstva, 4.62.257 (1976).

084.407 **The proton component of the radiation belts of the** earth according to measurements aboard Molniya.
A. S. Kovtyukh, M. I. Panasyuk, Eh. N. Sosnovets.
VIIth Leningrad seminar. "Corpuscular streams of the sun and the radiation belts of the earth and Jupiter", 1975, (see 012. 007), p. 265 - 266 (1975). In Russian. − Abstr. in Referativ. Zhurn. 62. Issled. kosmich. prostranstva, 4.62.259 (1976).

084.408 **Investigation of captured radiation aboard the artificial earth satellite Cosmos 426. III. Dynamics of the outer radiation belt during the magnetic storm of December 17, 1971.** S. N. Vernov, V. A. Vorob'ev, A. V. Zakharov, S. N. Kuznetsov, Yu. I. Logachev, I. A. Savenko, V. G. Stolpovskij.
Kosmich. Issled., Vol. 14, 305 - 313 (1976). In Russian.
Brief information.

084.409 **Detection of the periods of particle oscillations in the radiation belts from Cosmos 137 data.**
R. M. Golynskaya, B. N. Belen'kaya, E. P. Skorokhod, O. I. Savun.
Cosmic rays, No. 15, (see 003.006), p. 15 - 18 (1975). In Russian.

084.410 **Energetic electrons in the inner belt in 1968.**
H. I. West, Jr., R. M. Buck.
Planet. Space Sci., Vol. 24, 643 - 655 (1976).
Pitch-angle data were obtained by the Lawrence Livermore Laboratory's scanning, magnetic electron spectrometer on OGO 5 during its traversals of the inner belt in 1968. Data from the five lowest-energy channels 79−822 keV, were analyzed. The inner-belt electron injection following two storm periods was observed.

084.411 **Insolubility of trapped particle motion in a magnetic dipole field.** A. J. Dragt, J. M. Finn.
Journ. Geophys. Res., Vol. 81, 2327 - 2340 (1976).

084.412 **^{3}He/^{4}He ratios in the lower radiation belt as measured by trapped particles in a recovered satellite.**
R. L. Warasila, O. A. Schaeffer.
Geophys. Res. Letters, Vol. 2, 480 - 482 (1975). − Abstr. in Phys. Abstr., Vol. 79, A027622 (1976).

Corpuscular streams of the sun and radiation belts of the earth and Jupiter. See Abstr. 078.017.

Earth as a radio source: the nonthermal continuum. See Abstr. 084.295.

085 Solar-Terrestrial Relations

085.001 **Physical argument and hypothesis for sun–weather relationships.** K. D. Cole.
Nature, Vol. 260, 229 - 230 (1976). – Letter.

085.002 **Solar phenomena, weather and climate.**
J. W. King.
Astrophys. Space Sci. Library, Vol. 57, (see 012.002), 209 - 222 (1976).
Various ways in which the lower atmosphere is influenced by solar phenomena ranging from short-lived events such as flares to the 11-year and 22-year sunspot cycles are described. Some important consequences of these 'sun-weather relationships' are discussed.

085.003 **On a method of separation of ozone and aerosol contributions in solar radiation attenuation in the 300–350 nm spectral region.** N. N. Ostrogskij, V. M. Rat'kov.
Trudy Mosk. fiz.-tekhn. in-ta. Ser. Obshch. i molekul. fiz. No. 6. Moskva, 1975, p. 84 - 87. In Russian. – Abstr. in Referativ. Zhurn. 51. Astron., 3.51.70 (1976).

085.004 **Long-period oscillations of geoactive solar longitudes.**
V. F. Loginov, B. G. Sherstyukov, A. M. Vysotskij.
Trudy VNII gidrometeorol. inform. – Mirovoj tsentr dannykh, 1975, vyp (No.) 23, p. 69 - 74. In Russian. – Abstr. in Referativ. Zhurn. 51. Astron. 3.51.354 (1976).

085.005 **Objectives and outlooks on further researches of solar-atmospherical relations.** V. F. Loginov.
Trudy VNII gidrometeorol. inform. – Mirovoj tsentr dannykh, 1975, vyp. (No.) 23, p. 3 - 5. In Russian. – Abstr. in Referativ. Zhurn. 51. Astron., 3.51.360 (1976).

085.006 **Pattern of solar activity influence on the earth's atmosphere.** V. F. Loginov.
Trudy VNII gidrometeorol. inform. – Mirovoj tsentr dannykh, 1975, vyp. (No.) 23, p. 6 - 29. In Russian. – Abstr. in Referativ. Zhurn. 51. Astron., 3.51.361 (1976).

085.007 **Oscillations of thermodynamical conditions in the stratosphere in relation with solar activity.**
G. I. Sukhomazova, V. F. Loginov.
Trudy VNII gidrometeorol. inform. – Mirovoj tsentr dannykh, 1975, vyp. (No.) 23, p. 30 - 42. In Russian. – Abstr. in Referativ. Zhurn. 51. Astron., 3.51.362 (1976).

085.008 **Sectorial structure of the interplanetary magnetic field and atmospherical circulation.**
V. F. Loginov, A. M. Vysotskij, B. G. Sherstyukov.
Trudy VNII gidrometeorol. inform. – Mirovoj tsentr dannykh, 1975, vyp. (No.) 23, p. 43 - 49. In Russian. – Abstr. in Referativ. Zhurn. 51. Astron., 3.51.363 (1976).

085.009 **A search for short-term meteorological effects of solar variability in an atmospheric circulation model.**
R. C. J. Somerville, W. J. Quirk, J. E. Hansen, A. A. Lacis, P. H. Stone.
Journ. Geophys. Res., Vol. 81, 1572 - 1576 (1976).
A set of numerical experiments is carried out to test the short-range sensitivity of the Giss (Goddard Institute for Space Studies) global atmospheric general circulation model to changes in solar constant and ozone amount.

085.010 **Expert estimate of similarity between the mean monthly maps of anomalies of precipitations and the AT-500 maps in years of similar solar activity.**
A. G. Sytin.
Solnechnye Dannye 1975 Byull., No. 12, p. 100 - 105 (1976). In Russian.
It was established by a group of specialists with the method of expert estimates that the selection of analogs from the table of epoch superposition in relation to the maximum of solar activity, suggested by Kupetskij, proved to be useful in 80% of cases for making long-term weather forecasts (for a year or more); it was found that similarity was more typical of central regions of Eurasia than of the Atlantic and Pacific Oceans.

085.011 **Determination of cosmic ray variation in the past based on the content of isotopes in the geosphere.**
V. A. Dergachev, G. E. Kocharov.
VIIth Leningrad seminar. "Corpuscular streams of the sun and the radiation belts of the earth and Jupiter", 1975, (see 012.007), p. 347 - 362 (1975). In Russian. – Abstr. in Referativ. Zhurn. 51. Astron., 4.51.388 (1976).

085.012 **Generation of isotopes by solar flare particles in the earth's atmosphere.** P. Povinec.
VIIth Leningrad seminar. "Corpuscular streams of the sun and the radiation belts of the earth and Jupiter", 1975, (see 012.007), p. 363 - 375 (1975). – Abstr. in Referativ. Zhurn. 51. Astron., 4.51.389 (1976).

085.013 **Solar-biological relations and the effect of solar half-rotation.**
L. A. Vitel's, S. A. Karazhaeva, B. A. Ryvkin.
Trudy Glav. geofiz. observ., 1975, vyp. (No.) 330, p. 82 - 91. In Russian. – Abstr. in Referativ. Zhurn. 51. Astron., 4.51.395 (1976).

085.014 **Geomagnetic disturbances caused by solar corpuscular emission and variations in the level of natural radioactivity of the atmosphere.** A. Eh. Shem'i-zade.
Solnechnye Dannye 1976 Byull., No. 1, p. 99 - 102 (1976). In Russian.
Causes of daily non-periodical variations of the level of natural radioactivity of the earth's atmosphere are considered. It is shown that some disturbances of the radioactivity level coincide in time with magnetic storms. The day of maximum radioactivity mostly coincided with that of active period of the magnetic storm. The variations of the level of atmospheric radioactivity with a period of 26 ± 3 days are likely to be due to the recurrent solar corpuscular streams.

085.015 **Correlation between parameters of the upper atmosphere and solar activity.** U. Shodiev.
Byull. Inst. Astrofiz., *Dushanbe,* No. 65, p. 51 - 54 (1975). In Russian.

085.016 **Solar structure and terrestrial weather.**
J. M. Wilcox.
Science, Vol. 192, 745 - 748 (1976).

085.017 **Climate, the earth's rotation and solar variations.**
J. Gribbin.
Growth rhythms and the history of the earth's rotation, (see 012.012), p. 413 - 425 (1975).

085.018 **Some aspects of the occurrence of a quasi-two-year cycle in solar activity indices and in meteorological data.** N. I. Yakovleva.
Trudy Glav. geofiz. observ., 1975, vyp. (No.) 355, p. 94 - 103. In Russian. – Abstr. in Referativ. Zhurn. 51. Astron., 6. 51. 478 (1976).

085.019 **Remarks on high latitude air temperature ranges and solar activity.** C. S. Zerefos, J. Xanthakis.
Praktika Acad. Athens, Vol. 50, 118 - 129 = Res. Center Astron. Appl. Math., Acad. Athens, Contr. Ser. I (Astron.), No. 39 (1975).

085.020 **Probable periodical variations in the frequency of the Etesian winds.** J. Xanthakis.
Repr. from special vol. in memoriam "Demetrios Eginitis", p. 305 - 317 = Res. Center Astron. Appl. Math., Acad. Athens, Contr. Ser. I (Astron.), No. 41 (1975).

085.021 **La théorie astronomique des changements du climat au cours des dernières 5000 années.** Đ. Teleki.
Extrait du Starinar, Revue de l'Institut Archéologique, Nouvelle sér., Vol. 24 - 25, Beograd, 1973–1974. Sep. print. 11 pp. In Serbo-Croatian.

085.022 **Entwicklungstendenzen der solar-terrestrischen Physik.** E. A. Lauter.
Sitzungsber. Akad. Wiss. Berlin, Jahrg. 1975, No. 1N, p. 35 - 54 (1975).

085.023 **Sudden f_{min} enhancements and sudden cosmic noise absorptions associated with solar X-ray flares.** T. Sato.
Journ. Geomagn. Geoelectr., Vol. 27, 95 - 112 (1975). – Abstr. in Phys. Abstr., Vol. 79, A037257 (1976).

085.024 **The relationship of Ca-plage areas and other solar activity indices with ionospheric characteristics in the E and F layers.** P. P. Kuriyan, L. M. Punetha.
Indian Journ. Meteorol. Geophys., Vol. 25, 305 - 309 (1974). Abstr. in Phys. Abstr., Vol. 79, A037268 (1976).

085.025 **Lunar and solar atmospheric tides in surface winds and rainfall.** S. J. Reddy.
Indian Journ. Meteorol. Geophys., Vol. 25, 499 - 502 (1974). Abstr. in Phys. Abstr., Vol. 79, A041712 (1976).

Some problems of solar activity influence on the lower atmosphere. See Abstr. 003.085.

Possible relationships between solar activity and meteorological phenomena. See Abstr. 012.025.

Beziehungen zwischen der Sonnenaktivität und dem Massenzentrum des Sonnensystems.
See Abstr. 072.039.

Notes on solar activity, geomagnetics and the ionosphere. See Abstr. 072.057.

On some possibilities of forecast of solar geoeffective flares. See Abstr. 073.076.

Longitudinal change of ionization of the F-region during low and moderate solar activity periods.
See Abstr. 083.035.

Planetary System

091 Physics of the Planetary System (Planetary Atmospheres, Figure, Interior, Magnetic Fields, Rotation, etc.)

091.001 **Optimum choice of navigation measurements for determination of the position on a planet's surface.**
A. Yu. Kogan, V. N. Khejfets.
Kosmich. Issled., Vol. 14, 36 - 47 (1976). In Russian.

091.002 **L'évolution des atmosphères des planètes.**
G. Moreels.
L'Astronomie, Vol. 90, 145 - 167 (1976).

091.003 **New vistas in planetary radio astronomy.**
J. K. Alexander, Jr.
Sky Telescope, Vol. 51, 148 - 153 (1976).

091.004 **Solution of the inverse problem of radio emission of the atmosphere of planets taking into account the horizontal gradients.** V. M. Ivanov, D. S. Lukin.
Trudy Mosk. fiz.-tekhn. in-ta. Ser. Obshch. i molekul. fiz., No. 6, Moskva, 1975, p. 174 - 180. In Russian. — Abstr. in Referativ. Zhurn. 62. Issled. kosmich. prostranstva, 3.62.204 (1976).

091.005 **On the accuracy of the solution of the inverse problem of radio emission of the atmosphere.**
V. M. Ivanov, D. S. Lukin.
Trudy Mosk. fiz.-tekhn. in-ta. Ser. Obshch. i molekul. fiz., No. 6, Moskva, 1975, p. 181 - 186. In Russian. — Abstr. in Referativ. Zhurn. 62. Issled. kosmich. prostranstva, 3.62.205 (1976).

091.006 **Scintillations during occultations by planets. I. An approximate theory.** A. T. Young.
Icarus, Vol. 27, 335 - 357 (1976).
This paper gives an approximately correct theoretical treatment that is a substantial improvement over published theories, and shows how a more accurate theory could be constructed. Some methods for a more accurate determination of atmospheric structure are proposed.

091.007 **Ray propagation in oblate atmospheres.**
W. B. Hubbard.
Icarus, Vol. 27, 387 - 389 (1976). — Presented at the Jupiter Colloquium (IAU Colloquium 30), Tucson, Arizona, May 18 - 23, 1975.
The author evaluates the departures from Bouguer's law for the case of an oblate atmosphere. He shows that, to lowest order, the plane of refraction is defined by the normal to the atmosphere at closest approach. In next order, however, the ray path is "warped" by the oblateness, which changes slightly the plane of refraction.

091.008 **Rotational motions of the terrestrial planets.**
R. O. Vicente.
Long-time predictions in dynamics, (see 012.005), p. 307 - 320 (1976).

091.009 **Radiographic inspection of a planetary atmosphere: solution of the inverse problem with allowance for horizontal density gradients.** V. M. Ivanov, D. S. Lukin.

Trudy Mosk. fiz.-tekhn. in-ta. Ser. Obshch. i molekul. fiz., No. 6. Moskva, 1975, p. 174 - 180. In Russian. — Abstr. in Referativ. Zhurn. 51. Astron., 4.51.213 (1976).

091.010 **On the accuracy of the solution of the inverse problem of radiographic inspection of an atmosphere.**
V. M. Ivanov, D. S. Lukin.
Trudy Mosk. fiz.-tekhn. in-ta. Ser. Obshch. i molekul. fiz., No. 6. Moskva, 1975, p. 181 - 186. In Russian. — Abstr. in Referativ. Zhurn. 51. Astron., 4.51.214 (1976).

091.011 **On the radiation conditions in the circumplanetary space.** Yu. V. Aleksandrov, V. P. Kulichkin.
Vestn. Khar'kov. Univ., No. 129 (Ser. Astron., vyp. (No.) 10), p. 76 - 81 (1975). In Russian.

091.012 **Spectral line profiles for a planetary corona.**
J. W. Chamberlain.
Journ. Geophys. Res., Vol. 81, 1774 - 1776 (1976).
The Lyman and Balmer emissions of a planetary corona depend on the exospheric temperature, the integrated column density of solar-illuminated hydrogen, and the region of phase space occupied by particles. Measurements of the intensity · alone are incapable of defining the exosphere unambiguously. Line profiles, with high spectral resolution, can show whether a nonthermal component of the escaping hydrogen is present and can indicate at what altitude orbits of hydrogen atoms are depleted.

091.013 **Radiation field in an optically thick planetary atmosphere overlaying a reflecting surface.**
V. V. Ivanov.
Astron. Zhurn. Akad. Nauk SSSR, Vol. 53, 589 - 595 (1976). In Russian. English translation in Soviet Astron., Vol. 20, No. 3.
The radiation field is studied in an externally illuminated optically thick planetary atmosphere overlaying the surface which reflects radiation according to an arbitrary law. Asymptotic expressions are found for the intensity of radiation a) diffusely reflected from the atmosphere, b) illuminating the surface of the planet and c) propagating inside the atmosphere far from the boundaries.

091.014 **Diffraction calculation of occultation light curves in the presence of an isothermal atmosphere.**
R. G. French, P. J. Gierasch.
Astron. Journ., Vol. 81, 445 - 451 (1976).
From diffraction theory, the authors calculate light curves for stellar occultations by a planetary body with an isothermal atmosphere. The character of the resulting curves is determined by the scale height H, the Fresnel zone size l, the surface atmospheric refractivity, and the planetary radius. The authors present an exact general solution and two approximations which are valid when $H \gg l$. Finally, they assess the importance of accounting for diffraction effects of the limb when deducing atmospheric parameters from occultation light curves.

091.015 Generation of planetary magnetic fields.
E. H. Levy.
Annual Rev. Earth Planet. Sci., Vol. 4, (see 003.012), 159 - 185 (1976).

091.016 Hydrogen loss from the terrestrial planets.
D. M. Hunten, T. M. Donahue.
Annual Rev. Earth Planet. Sci., Vol. 4, (see 003.012), 265 - 292 (1976).

091.017 JPL Development Ephemeris No. 96.
E. M. Standish, Jr., M. S. W. Keesey, X X Newhall.
JPL Techn. Rep. 32-1603, 7 + 35 pp. (1976).

The fourth issue of JPL planetary ephemerides, designated JPL Development Ephemeris No. 96 (DE96) is described. The improvements in DE96 include new types of data, better processing of the data, and refined equations of motion. The dynamic evolution of the system given by the sun and the nine planets was obtained by numerical integration.

091.018 Planetary magnetism.   D. J. Stevenson.
AIP Conf. Proc., No. 24, p. 781 - 784 (1974).
Abstr. in Phys. Abstr., Vol. 79, A015202 (1976).

091.019 High energy nuclear reactions in our planetary system.   A. L. Turkevich.
AIP Conf. Proc., No. 26, p. 351 - 363 (1975). – Abstr. in Phys. Abstr., Vol. 79, A023578 (1976).

091.020 Fourier spectroscopy in planetary research.
R. A. Hanel, V. G. Kunde.
Space Sci. Rev., Vol. 18, 201 - 256 (1975).

The application of Fourier transform spectroscopy (FTS) to planetary research is reviewed. The survey includes FTS observations of the sun, all the planets except Uranus and Pluto, the Galilean satellites and Saturn's rings. Instrumentation and scientific results are considered. The prospects and limitations of FTS for planetary research in the forthcoming years are discussed.

091.021 General grid systems of planets.
G. N. Katterfel'd, G. V. Charushin.
Modern Geol., Vol. 4, 253 - 287 (1973). – Abstr. in Phys. Abstr., Vol. 79, A037471 (1976).

091.022 Absorption coefficient of HCl in the region 1400 to 2200 Å (planetary atmospheres).   E. C. Y. Inn.
Journ. Atmosph. Sci., Vol. 32, 2375 - 2377 (1975). – Abstr. in Phys. Abstr., Vol. 79, A042070 (1976).

091.023 Effects of low-frequency gravitational waves upon orbital elements (planets).   F. Salmistraro.
Nuovo Cimento Lettere, Ser. 2, Vol. 15, 343 - 346 (1976). Abstr. in Phys. Abstr., Vol. 79, A042072 (1976).

091.024 The metallic phase of planets and meteorites.
A. P. Vinogradov.
Journ. Phys. Educ., Vol. 2, No. 4, p. 10 - 14 (1975).

Light scattering in planetary atmospheres.
See Abstr. 003.102.

The solar system. Collected articles from the 1975 September issue of The Scientific American.
See Abstr. 003.122.

Statistical mechanics of light elements at high pressure. IV. A model free energy for the metallic phase.
See Abstr. 022.047.

Measurement of hydrogen- and self-broadened half-widths of ammonia at 200 and 300 K.
See Abstr. 022.053.

Scattering of solar Lyman alpha by the (14,0) band of the fourth positive system of CO.   See Abstr. 022.055.

Measurement of lunar and planetary magnetic fields by reflection of low energy electrons.   See Abstr. 031.202.

On the solution of the Clairaut-Laplace-Lyapunov problem.   See Abstr. 042.069.

An inside look at NASA planetology.
See Abstr. 051.005.

An association among planets, sunspots and solar plasma physics.   See Abstr. 062.090.

An initial-value method for internal radiation field in generalized Chandrasekhar's planetary problem.
See Abstr. 063.023.

Extension of the doubling method to inhomogeneous sources.   See Abstr. 063.024.

Infrared radiative transfer in planetary atmospheres – I. Effects of computational and spectroscopic economies on thermal heating/cooling rates.
See Abstr. 063.025.

Multiple scattered radiation emerging from Rayleigh and continental haze layers. 1: Radiance, polarization, and neutral points.   See Abstr. 063.028.

Multiple scattered radiation emerging from Rayleigh and continental haze layers. 2: Ellipticity and direction of polarization.   See Abstr. 063.029.

Numerical solution of the radiative transfer equation in spherically symmetric dust shells.   See Abstr. 063.030.

Laboratory simulation of thermal convection in rotating planets and stars.   See Abstr. 065.011.

On the variation of planetary orbits and the solar activity period.   See Abstr. 072.056.

A hybrid mode of diffuse and specular reflector for computation of the emergent radiation by the adding method.
See Abstr. 082.039.

Cooling rate of an electron gas by polar molecules (in planetary and cometary atmospheres).
See Abstr. 082.088.

Late heavy bombardment of the moon and terrestrial planets.   See Abstr. 094.137.

Cratering and cosmogenic nuclides.
See Abstr. 094.199.

Polarimetric properties of the lunar surface and its interpretation: Part 7 – Other solar system objects.
See Abstr. 094.540.

The meteorology of Jupiter.
See Abstr. 099.004.

Spikes of light during the stellar occultation by planets.   See Abstr. 099.025.

Metallic phase of meteorites and planets.
See Abstr. 105.026.

Presolar grains: isotopic clues to solar system origin.

See Abstr. 107.010.

On the evolution of terrestrial planets.
See Abstr. 107.014.

092 Mercury

092.001 Preliminary geomorphological observations of the photographs of Mercury transmitted by Mariner 10.
P. Leonardi.
Atti. Accad. Nazionale Lincei, Ser. Ottava, Rend. Cl. Sci. fis., mat., nat., Vol. 57, 204 - 212 (1974/75).

092.002 Mercury – brother of the moon.
L. V. Ksanfomaliti.
Zemlya i Vselennaya, 1976, No. 1, p. 40 - 49. In Russian.

092.003 Does Mercury have a molten core? P. E. Fricker, R. T. Reynolds, A. L. Summers, P. M. Cassen.
Nature, Vol. 260, 293 - 294 (1976).
The Mariner 10 mission has discovered a magnetic field associated with the planet Mercury. The preferred explanation of the source of this field based on these observations, is an internal dynamo. A necessary condition for such a dynamo is the existence of an electrically conducting liquid region within the planet. In this letter, the authors investigate whether or not Mercury could contain a molten metallic core.

092.004 Lunar features of Mercury. G. A. Burba.
Priroda, 1976, No. 3, p. 77 - 89. In Russian.

092.005 The significance of the planet Mercury.
W. K. Hartmann.
Sky Telescope, Vol. 51, 307 - 311 (1976).

092.006 Photographic observations of the transit of Mercury on 10 November 1973.
C. Cristescu, A. D. Fiala.
Astron. Astrophys., Vol. 49, 29 - 30 (1976).
The transit of 1973 was photographed at regular intervals to obtain relative positions for astrometric use in studying the motion of Mercury. The reduced positions are presented.

092.007 Results of an improved analysis of observations of the transit of Mercury across the solar disk (November 10, 1973). N. G. Rizvanov, L. A. Urasin.
Astron. Tsirk., No. 892, p. 2 - 4 (1975). In Russian.

092.008 Crustal remanence and the magnetic moment of Mercury. A. Stephenson.
Earth Planet. Sci. Letters, Vol. 28, 454 - 458 (1976).
The magnetic dipole moment of Mercury can be explained on the basis of thermoremanent magnetization acquired by an outer shell in an ancient Mercurian field produced by an internal dipole source such as a core dynamo which is now inactive. Such a shell will give rise to a dipole moment provided that there are differences of permeability between the shell and the interior, or the shell and free space. By assuming that the magnetic properties of the surface rocks of Mercury are similar to those of the moon it is shown that ancient fields of the order of 1 gauss and free iron concentrations of the order of a few percent are sufficient to produce the present dipole moment.

Magnetism of Mercury.
Priroda, 1976, No. 2, p. 139. In Russian.

The importance of the transit of Mercury of 1631.
See Abstr. 004.003.

Mariner 10 mission to Venus and Mercury.
See Abstr. 053.022.

Preliminary report of results from the plasma science experiment on Mariner 10. See Abstr. 093.010.

Cometary impact and the magnetization of the
moon. See Abstr. 094.101.

Untersuchung von Einschlagskrater-Populationen.
See Abstr. 094.596.

093 Venus

093.001 Venus: microwave detection of carbon monoxide.
R. K. Kakar, J. W. Waters, W. J. Wilson.
Science, Vol. 191, 379 - 380 (1976).

The 115-gigahertz microwave line of carbon monoxide has been detected in the spectrum of Venus. The measurement proves that the carbon monoxide mixing ratio increases above an altitude of 85 kilometers in the Venus stratosphere and provides quantitative information on carbon monoxide in the altitude region from 80 to 110 kilometers. This altitude region is well above that which has been previously sensed.

093.002 Infrared imaging of Venus: 8–14 micrometers.
D. J. Diner, J. A. Westphal, F. P. Schloerb.
Icarus, Vol. 27, 191 - 195 = Contr. Div. Geol. Planet. Sci., Calif. Inst. Techn., *Pasadena*, No. 2670 (1976).

High spatial resolution 8–14 μm images of Venus were obtained on March 6, 1974. The planet was at a phase angle of 112° with the morning terminator in view. The images confirm the existence of a previously mapped flux anomaly near the south pole and show evidence of other infrared features, presumably transient in nature. Flux differences of 2–7% were measured, corresponding to brightness temperature variations of roughly 1–3°K. The images also confirm the difference between polar and equatorial limb darkening. Finally, the authors present evidence for a night-day asymmetry in the flux, with the brightness temperature greater by about 2°K on the sunlit side.

093.003 Photometry of Venus below 2200 Å from the Orbiting Astronomical Observatory-2.
E. Shaya, J. Caldwell.
Icarus, Vol. 27, 255 - 264 (1976).

Spectrophotometry of Venus from 2170 to about 1950 Å has been obtained by OAO-2 at 10 Å resolution. The new data confirm and extend previous indications that the geometric albedo decreases continuously below 2500 Å. Secular changes in either the amount or distribution, or both, of absorbing constituents in the upper atmosphere are strongly suggested. A narrow absorption feature is found near 2145 Å.

093.004 Die Venus wird enthüllt. Neue Pläne der amerikanischen Weltraumforschung.   J. von Puttkamer.
Umschau, 76. Jahrgang, p. 69 - 74 (1976).

093.005 How the Venus surface looks like.   S. A. Nikitin.
Priroda, 1976, No. 2, p. 6 - 9. In Russian.

093.006 Thermal radiation streams in the lower Venus atmosphere.   V. P. Shari.
Kosmich. Issled., Vol. 14, 97 - 110 (1976). In Russian.

093.007 Characteristics of the Venus surface measured with a radio altimeter in the landing area of an apparatus of the automatic interplanetary station Venera 8.
M. V. Bashmashnikov, V. T. Guslyakov, V. M. Ezhkov, V. V. Kerzhanovich, M. L. Natalovich, E. E. Tsejtlin.
Kosmich. Issled., Vol. 14, 111 - 114 (1976).   In Russian.

093.008 The Venus oceans problem.   A. Dauvillier.
Journ. British Astron. Ass., Vol. 86, 147 - 148 (1976).

093.009 Report on the elongation of Venus, 1975 June.
J. H. Robinson.
Journ. British Astron. Ass., Vol. 86, 155 - 161 (1976).

093.010 Preliminary report of results from the plasma science experiment on Mariner 10.
H. S. Bridge, A. J. Lazarus, K. W. Ogilvie, J. D. Scudder, R. E. Hartle, J. R. Asbridge, S. J. Bame, W. C. Feldman, G. L. Siscoe, C. M. Yeates.
Space Research XV, (see 012.003), p. 501 - 519 (1975).

Results obtained from the plasma science experiment during the Mariner 10 encounters with Venus and Mercury are described. Results at Venus confirm and extend earlier results by Mariner 5, Venera 4 and Venera 6. They show that the solar wind interaction with the planet most probably involves a bow shock rather than an extended exosphere, but that the "shock" is not a thin boundary at the point where it was traversed by Mariner 10. Near Mercury the data show a fully developed bow shock and "magnetosheath" and provide unambiguous evidence for a strong interaction between Mercury and the solar wind.

093.011 On the possibility of formation of water drops and ice crystals in the atmosphere of Venus.
L. Krystanov, S. Todorova, L. Yuskeselieva.
Blg. geofiz. spisanie, Vol. 1, No. 1, p. 5 - 12 (1975). In Russian. – Abstr. in Referativ. Zhurn. 51. Astron., 3.51.185; 62. Issled. kosmich. prostranstva, 3.62.159 (1976).

093.012 Erste Bilder von der Venusoberfläche.   W. Lüthi.
Orion, 34. Jahrgang, p. 8 - 9 (1976).

093.013 The Mariner 5 ultraviolet photometer experiment: analysis of hydrogen Lyman alpha data.
D. E. Anderson, Jr.
Journ. Geophys. Res., Vol. 81, 1213 - 1216 (1976).

Lyman α measurements of the exosphere of Venus made by the ultraviolet photometer on Mariner 5 on October 19, 1967, are analyzed. Radiative transfer models for a spherical isothermal hydrogen atmosphere with carbon dioxide present as a pure absorber are used to determine the exospheric temperature and density at the bright limb and on the dark disc. It is found that (1) the bright limb data have two components with exospheric temperatures of 275° ± 50°K and 1020° ± 100°K and densities $2 \pm 1 \times 10^5$ cm^{-3} and 1.3×10^3 cm^{-3}, respectively, (2) the dark disc data are best fit by a two-component density model with exospheric temperatures of 150° ± 50°K and 1500° ± 200°K and densities $2 \pm 1 \times 10^5$ cm^{-3} and 10^3 cm^{-3}, respectively, and (3) the dark limb exhibits only a hot component because of the very low temperature of the cold component.

093.014 Venus mesosphere and thermosphere temperature structure. I. Global mean radiative and conductive equilibrium.   R. E. Dickinson.
Icarus, Vol. 27, 479 - 493 (1976).

The author considers the basis for theoretical calculation of global mean temperatures of the upper atmosphere of Venus. Three significant uncertainties in determining thermospheric temperatures are revealed: the efficiency of solar EUV in degrading to heat, the enhancement of vibrational excitation of CO_2 by collisions with O, and the possibility of eddy mixing.

093.015 Variations of reflection characteristics of the equatorial region of Venus.   V. K. Golovkov,
B. I. Kuznetsov, G. M. Petrov, A. F. Khasyanov.
Astron. Zhurn. Akad. Nauk SSSR, Vol. 53, 411 - 417 (1976). In Russian. English translation in Soviet Astron., Vol. 20, No. 2.

Methods and results of radar investigations on the reflection characteristics of the equatorial region of Venus at 39 cm within 290°–340° longitude are presented.

093.016 Erosion and the rocks of Venus.   C. Sagan.

Nature, Vol. 261, 31 (1976). — Letter

093.017 Airborne ultraviolet studies of Venus.
G. G. Sivjee, G. J. Romick.
Nature, Vol. 261, 31 - 32 (1976).
Because of the high atmospheric extinction in the ultraviolet at ground level, an airborne platform offers distinct advantage for planetary optical measurements in the wavelength region below 3,500 Å. The joint NASA–ESO Space Shuttle simulation programme (ASSESS) provided an opportunity for conducting such measurements aboard NASA's Convair 990 jet aircraft, at an altitude of ~40,000 feet. The presented summary of these measurements is a supplement to the published report on the ASSESS experiments.

093.018 Venus: what is known on it today.
M. Ya. Marov.
Zemlya i Vselennaya, 1976, No. 3, p. 3 - 15. In Russian.

093.019 On some characteristics of the cloud layer of Venus from results of measurements with an ionisation densitometer aboard the automatic interplanetary station Venera 4. V. V. Mikhnevich, A. I. Livshits, B. G. Gel'man.
Kosmich. Issled., Vol. 14, 272 - 277 (1976). In Russian.

093.020 Venus: radar maps show evidence of tectonic activity. W. D. Metz.
Science, Vol. 192, 454 - 455 (1976).

093.021 On indirect sounding of the atmosphere of Venus.
M. A. Gruzdeva, Yu. M. Timofeev.
Probl. fiz. atmosf. Leningrad, Leningr. un-t, 1975, p. 9 - 16. In Russian. – Abstr. in Referativ. Zhurn. 51. Astron., 5.51. 268 (1976).

093.022 Method and equipment for determining ammonia in the Venus atmosphere.
Yu. A. Surkov, B. M. Andrejchikov, O. M. Kalinkina.
Zhurn. analit. khimii, Vol. 30, 2422 - 2426 (1975). In Russian. – Abstr. in Referativ. Zhurn. 51. Astron., 5.51.269; 62. Issled. kosmich. prostranstva, 5.62.106 (1976).

093.023 Spectral variation of the contrast of dark formations on Venus. O. M. Starodubtseva.
Vestn. Khar'kov. Univ., No. 129 (Ser. Astron., vyp. (No.) 10), p. 37 - 45 (1975). In Russian.

093.024 Venus could have a large magnetic moment.
P. J. Smith.
Nature, Vol. 261, 543 (1976).

093.025 Calculation of the absorption of infrared radiation in the lower Venus atmosphere.
A. P. Gal'tsev, V. M. Osipov, V. P. Shari.
Kosmich. Issled., Vol. 14, 417 - 427 (1976). In Russian.

093.026 Proposal for observations of the contour of Venus.
R. Baum.
Strolling Astronomer, Vol. 26, 16 - 18 (1976).

093.027 Viscous boundary layer for the Venusian ionopause.
H. Pérez de Tejada, M. Dryer.
Journ. Geophys. Res., Vol. 81, 2023 - 2029 (1976).
A one-fluid model of viscous interaction between the shocked solar wind flow in the Venusian ionosheath and the ionospheric plasma is formulated through a conventional MHD viscous boundary layer theory. The geometry considered applies to the flank regions of the ionosheath, exterior to the ionospheric cavity. Adoption of the proper boundary conditions for this problem leads to velocity profiles which support the interpretation of the Venusian ionopause in terms of a thick mixing region which is gradually forced to taper toward the axis of the ionospheric cavity, as can be inferred from the Mariner 5 and Venera 4 data.

093.028 Spatial variations of the strength of CO_2 absorption and the rotational temperature on Venus.
K. Iwasaki.
Publ. Astron. Soc. Japan, Vol. 28, 215 - 227 (1976).
The CO_2 8689-Å band in the spectrum of Venus was observed with the echelle spectrograph at the coudé focus of the 188-cm reflector at the Okayama Astrophysical Observatory in 1973 and 1974 when the phase angle of Venus was near 90°. The dispersion of the spectrum was 2.5 Å mm^{-1}. The CO_2 absorptions decrease at high latitudes along the terminator and have a maximum on the intensity equator. These spatial variations would be described theoretically with a homogeneous, isotropically scattering atmosphere. The spatial variations of the slope of the curve of growth and the rotational temperature were also analyzed.

093.029 He 584 Å airglow emission from Venus: Mariner 10 observations. S. Kumar, A. L. Broadfoot.
Geophys. Res. Letters, Vol. 2, 357 - 360 (1975). — Abstr. in Phys. Abstr., Vol. 79, A027798 (1976).

093.030 Ionosphere of Venus. D. M. Butler.
Thesis Rice Univ., Houston, Texas, USA, 263 pp. (1975). (Available from: Univ. Microfilms, Order No. 75-21,999).

The infrared optical constants of sulfuric acid at 250 K. See Abstr. 022.064.

Venus na de landingen van Venera 9 en 10.
See Abstr. 053.008.

Mariner 10 mission to Venus and Mercury.
See Abstr. 053.022.

Estimate of the density of surface layer materials of the moon, Mars and Venus. See Abstr. 094.127.

Sources of heat in the Martian and Venus thermospheres owing to absorption of ultraviolet solar radiation. See Abstr. 097.023.

The ionospheres of Mars and Venus.
See Abstr. 097.059.

094 Moon: Dynamics, Global Properties, Local Properties

Moon, Dynamics

094.001 **Laser observations of the moon: normal points for 1973.**
J. D. Mulholland, P. J. Shelus, E. C. Silverberg.
Astron. Journ., Vol. 80, 1087 - 1093 (1975).
McDonald Observatory lunar laser ranging observations for 1973 are presented in the form of compressed normal points and amendments for the 1969−1972 data set are given. Observations of the reflector mounted on the Soviet roving vehicle Lunakhod 2 have also been included.

094.002 **Théorie analytique programmée de la libration physique de la lune.** A. Migus.
The Moon, Vol. 15, 165 - 181 (1976).
The rotation of the moon about its center of mass, taking into account the orbital motion, is treated analytically. A Hamiltonian theory is developed in terms of the Andoyer variables. The periodic parts of departures from three resonances, equivalent to Cassini's laws, are found to be the canonical variables of the problem. The potential is expressed as a function of these new coordinates and the whole Hamiltonian is developed to the second degree in these small variables. One system of equations gives the real center of libration which is found to be near the center defined by Cassini's laws. A second system solved by iterations, gives the libration as analytical series in the constants of the moon's potential, and trigonometric series in Delaunay arguments.

094.003 **Tidal effect in lunar rotation theory.**
V. B. Gurevich.
Astron. Zhurn. Akad. Nauk SSSR, Vol. 53, 170 - 177 (1976). In Russian. English translation in Soviet Astron., Vol. 20, No. 1.
The inertia tensors of tide in an elastic compressible moon and of the moon as a whole relative to the lunar figure axes are calculated. When the earth's selenographic latitude and longitude are not equal to zero, there always exist products of inertia relative to the figure axes, that means deviations of the inertia ellipsoid relative to the figure ellipsoid; it is shown that the amplitude of these deviations is defined by the Love numbers combination $h - 1/6 f = H$ and is commensurable with the physical libration amplitude. The conclusion is drawn that in lunar rotation theory the Liouville equations must be used; the use of Euler's dynamical equations is incorrect.

094.004 **Laser measurements of the earth-moon distance and related investigations of lunar orbit and lunar physical libration.** B. Kołaczek.
Postępy Astron., Vol. 23, 271 - 279 (1975). In Polish.
New laser measurements of the earth-moon distance are discussed together with the determination of lunar libration.

094.005 **Détermination des librations libres de la lune, de l'analyse des mesures de distances par laser.**
O. Calame.
Comptes Rendus Acad. Sci. Paris, Sér. B, Vol. 282, 133 - 135 (1976).
Le problème de l'existence d'oscillations «libres», dans le mouvement de rotation de la lune autour de son centre de masse, a fait l'objet, jusqu'à présent, de nombreuses controverses. Cinq années de mesures de distances lunaires par laser ont désormais rendu possible la mise en évidence de telles librations, avec une détermination quantitative de leurs amplitudes et de leurs phases, pour les trois modes d'oscillation, de période respective 2,9 ans, 27,3 jours et 75 ans, le premier agissant en longitude et les deux autres en latitude.

094.006 **Cassini's laws.** H. Kinoshita.
Long-time predictions in dynamics, (see 012.005), p. 338 (1976). − Abstract.

094.007 **On the origin of the moon.**
Eh. M. Drobyshevskij.
Astron. Tsirk., No. 854, p. 5 - 6 (1975). In Russian.

094.008 **Relativistic theory of the moon's motion.**
A. M. Finkelstein (*Finkel'shtejn*), V. Ja. Kreinovich.
Celestial Mechanics, Vol. 13, 151 - 176 (1976).
Relativistic corrections for the elements and coordinates of the moon have been obtained in the framework of the PPN-formalism. The influence of the coordinate conditions on the observational effects was studied.

094.009 **Excitation and relaxation of the wobble, precession, and libration of the moon.** S. J. Peale.
Journ. Geophys. Res., Vol. 81, 1813 - 1827 (1976).
The rate of impact excitation of each of the free motions of the moon above a given amplitude is compared with the rate of damping from tidal and rotational distortion and from a possible core-mantle interaction.

094.010 **A physical and chemical model of early lunar history.** N. J. Hubbard, J. W. Minear.
Proc. Sixth Lunar Sci. Conference, (see 012.010), Vol. 1, 1057 - 1085 (1975).

094.011 **Genetic relations between the moon and meteorites.**
R. N. Clayton, T. K. Mayeda.
Proc. Sixth Lunar Sci. Conference, (see 012.010), Vol. 2, 1761 - 1769 (1975).

094.012 **The moon: not so different from earth after all.**
A. L. Hammond.
Science, Vol. 192, 875 (1976). − Research news.

094.013 **On the calculation of the moon's moments of inertia.**
G. A. Meshcheryakov, P. M. Zazulyak, V. V. Kirichuk.
Astron. Zhurn. Akad. Nauk SSSR, Vol. 53, 620 - 625 (1976). In Russian. English translation in Soviet Astron., Vol. 20, No. 3.
An adjustment by the method of least squares of the moon's principal moments of inertia and the second-order harmonics of its gravitational potential, dynamical flattenings and parameters of the orbital motion (inequalities in the motions of perigee and node) is carried out. An estimate of the solution is given. The solution is carried out in some variants.

094.014 **A search for forward scattering of sunlight from the lunar libration clouds.** C. L. Ross.
IAU Colloquium No. 31, (see 012.015), p. 73 (1976). Abstract, see 14.094.005.

094.015 **Coordinate systems in lunar ranging.**
J. D. Mulholland.
IAU Colloquium No. 26, (see 012.018), p. 433 - 443 (1975).

094.016 **A determination of the lunar moment of inertia.**
J. P. Gapcynski, W. T. Blackshear, R. H. Tolson, H. R. Compton.
Geophys. Res. Letters, Vol. 2, 353 - 356 (1975). − Abstr. in Phys. Abstr., Vol. 79, A023683 (1976).

094.017 **Evolution of the moon: the 1974 model.**

H. H. Schmitt.
Soviet-American conf. cosmochem. moon planets, Moscow, USSR, 4 June 1974, 33 pp. (1974). – See 12.094.014.

Theory of the rotation of the rigid earth.
See Abstr. 044.004.

Palaeontological and astronomical observations on the rotational history of the earth and moon.
See Abstr. 044.008.

The accelerations of the earth and moon from early astronomical observations. See Abstr. 044.012.

New test of the equivalence principle from lunar laser ranging. See Abstr. 066.071.

Verification of the principle of equivalence for massive bodies. See Abstr. 066.072.

The color characteristics of the earth–moon libration clouds. See Abstr. 106.063.

On the visibility of the libration clouds.
See Abstr. 106.071.

Moon, Global Properties

094.101 Cometary impact and the magnetization of the moon.   T. Gold, S. Soter.
Planet. Space Sci., Vol. 24, 45 - 54 (1976).

Collisions of comets with planetary bodies are capable of impressing patterns of magnetization onto them that match those observed for the moon and possibly for Mercury. The ambient solar wind magnetic field is briefly but strongly enhanced as the large partially ionized cometary atmosphere is compressed against the planetary surface. Just at the time of peak field enhancement, the solid part of the comet collides with the surface and the compressed fields are permanently imprinted by shock magnetization.

094.102 Microcraters formed in hot glass by hypervelocity projectiles.   J. F. Vedder.
The Moon, Vol. 15, 31 - 49 (1976).

Microcraters were formed in heated soda-lime glass by the normal incidence of spheres of plastic or fused silica with diameters between 0.8 and 4.5 μm and velocities between 2.5 and 10 km s^{-1}. The morphology of the craters in targets at temperatures up to 800°C is little different from those formed in unheated glass. The results in conjunction with other evidence suggest that most lunar craters of micrometer size with a smooth central pit, splashed lip, and a spallation zone are the result of primary impacts.

094.103 A far-ultraviolet photometer for planetary surface analysis.
R. C. Henry, W. G. Fastie, R. L. Lucke, B. W. Hapke.
The Moon, Vol. 15, 51 - 65 (1976).

The measurement of local variations in the far-ultraviolet albedo is explored as a means of detecting changes in the refractive index of rocks and dust on the surface of atmosphereless planets and satellites. Far-ultraviolet spectrophotometric measurements of the lunar surface which were obtained on the Apollo 17 orbital mission are presented to demonstrate that significant albedo variations occur in the spectral range 120 to 170 nm. These data also confirm the hypothesis that the albedo variations represent refractive index differences in the surface materials. A three-band photometer is described which, when put in orbit around a solar system object, is capable of providing refractive index maps with a sensitivity of 1 part in the second decimal place and with kilometer resolution. Comparative surface composition and surface history analyses based on such maps are discussed.

094.104 The residual permanent magnetic dipole moment of the moon.   A. Stephenson.
The Moon, Vol. 15, 67 - 81 (1976).

The residual dipole moment of the outer spherical shell of the moon, magnetized in the field of an internal dipole is calculated for the case when the permeability of the shell differs from unity. It is shown that, using an average value of surface magnetization from returned lunar crystalline rock samples and a global figure for the lunar permeability of 1.012, that a residual moment of the order of 10^{15} to 10^{16} Am^2 is expected. At present the magnetic data and the thermal state of the moon are not known with sufficient accuracy to distinguish between a crust magnetized in an internal dipole field of constant polarity and a crust magnetized in the dipole field of a self-reversing core dynamo.

094.105 Random processes as a cause of the lunar asymmetry.
M. Kobrick.
The Moon, Vol. 15, 83 - 89 (1976).

The offset of the center of mass of the moon from its center of figure together with moment of inertia differences are explainable by a lunar crust of randomly varying thickness. The necessity of postulating a method of preferential material transport into a particular lunar hemisphere to explain the lunar asymmetry is eliminated.

094.106 Local lunar topography from the Apollo 17 ALSE radar imagery and altimetry.
C. Elachi, M. Kobrick, L. Roth, M. Tiernan, W. E. Brown, Jr.
The Moon, Vol. 15, 119 - 131 (1976).

The Apollo 17 ALSE (Apollo Lunar Sounder Experiment) VHF radar provided imagery and continuous profiling data around the moon during two revolutions. The imagery data are used to derive depth and diameter measurements of small craters (diameter <30 km). The profiling data are used to study the topography of a few large craters: the bulged floors in Hevelius, Neper, and Aitken; central peaks in Neper and Buisson; and the depressed floor of Maraldi. The same data provided accurate (better than 25 m) profiles of Mare Crisium and Mare Serenitatis.

094.107 Lunar global figure from mare surface elevations.
W. L. Sjogren, W. R. Wollenhaupt.
The Moon, Vol. 15, 143 - 154 (1976).

Laser altimetry data from the Apollo 15, 16, and 17 missions show that the ringed maria surfaces lie on one particular reference surface and that the center of gravity is definitely displaced from the optical center. If these extensive surfaces are assumed to be near hydrostatic surfaces, then there must have existed a time in lunar history when lunar tides and/or internal processes were much different than they are today.

094.108 Lunar cartography with the Apollo 17 ALSE radar imagery.
M. Tiernan, L. Roth, T. W. Thompson, C. Elachi, W. E. Brown, Jr.
The Moon, Vol. 15, 155 - 163 (1976).

Lunar position differences between thirteen lunar craters in Mare Serenitatis were computed from VHF radar-imagery obtained by the Lunar Sounder instrument flown on the Apollo 17 Command Module. The radar-derived position differences agree with those obtained by conventional photogrammetric reductions of Apollo metric photography.

094.109 Lunar glass globules — where do they come from?
M. D. Nusinov, Yu. B. Chernyak.
Zemlya i Vselennaya, 1976, No. 1, p. 50 - 54. In Russian.

094.110 On the three-dimensional structure of the plasma track of the moon.   A. S. Lipatov.
Kosmich. Issled., Vol. 14, 115 - 119 (1976). In Russian.

094.111 Self reversal of thermoremanent magnetisation in basalts and global lunar magnetism.
A. Stephenson.
Nature, Vol. 259, 101 - 102 (1976). — Letter.

094.112 Early lunar magnetism.
S. K. Banerjee, J. P. Mellema.
Nature, Vol. 260, 230 - 231 (1976).

The authors have applied a new method for palaeointensity (ancient magnetic field) to three subsamples of a single, 1-m homogeneous clast (72215) from a recrystallised boulder of lunar breccia. They provide the first credible evidence that the strength of the ambient magnetic field at the Taurus-Littrow region of the moon must have been about 0.4 oersted. 4.0×10^9 yr BP.

094.113 Reference points for selenodetic control.
M. Moutsoulas.
Space Research XV, (see 012.003), p. 65 - 70 (1975).

094.114 Improved coordinates for Lunokhod 2 based on laser observations from McDonald Observatory.
E. S. Barker, O. Calame, J. D. Mulholland, P. J. Shelus.
Space Research XV, (see 012.003), p. 71 - 74 (1975).

094.115 Results of radar experiments performed aboard the Luna 19 and 20 automatic stations. N. N.
Kroupenio (*Krupenio*), A. G. Balo, E. G. Ruzskii (*Ruzskij*), V. A. Ladyghin (*Ladygin*), V. V. Cherkasov, V. S. Fomin.
Space Research XV, (see 012.003), p. 617 - 620 (1975).

The Luna 19 automatic station, the lunar artificial satellite, and the Luna 20 automatic station, which landed on the lunar surface, performed measurements on the reflection characteristics of radiowaves in the 3 cm range. Luna 19 showed that in two regions near the Rümker crater the dielectric constants ϵ are 2.35 ± 0.65 and 3.2 ± 0.2. The rms angles of surface inclination over a base of about 30 cm in these regions σ_a are $10° \pm 1°$ and $8.5° \pm 1°$ respectively. At the Luna 20 landing site (Mare Foecunditatis) $\epsilon = 1.7 \pm 0.2$ and $\sigma_a = 9.5° \pm 1.5°$.

094.116 The lunar magnetic field.
C. T. Russell, P. J. Coleman, Jr., G. Schubert.
Space Research XV, (see 012.003), p. 621 - 628 (1975).

Apollo 15 and 16 subsatellite magnetometer observations in the geomagnetic tail lobes have been used to map the fine scale lunar magnetic field. These data confirm the direct observations of strong localized fields on the lunar farside, and show the existence of many weaker nearside magnetic features. Combining Apollo 15 and 16 measurements of the permanent dipole moment, the authors obtain a southward directed moment inclined at an angle of $34°$ to the lunar rotational pole, with a strength of 11×10^{18} Г cm³.

094.117 Evolution of major mineral compositions and trace element abundances during fractional crystallization of a model lunar composition. M. J. Drake.
Geochim. Cosmochim. Acta, Vol. 40, 401 - 411 (1976).

The evolution of major mineral compositions and trace element abundances during fractional crystallization of a model lunar magma ocean have been calculated. A lunar bulk composition consistent with petrological constraints has been selected. Major mineral compositions have been calculated using published studies of olivine–melt, plagioclase–melt, and pyroxene–olivine equilibria. Trace element abundances have been calculated using experimentally-determined partition coefficients where possible.

094.118 Statistical analysis of the Lunokhod 1 path of motion.
Yu. S. Tyuflin, Yu. S. Timofeev, B. V. Nepoklonov.
Izv. vyssh. ucheb. zavedenij. Geod. i aehrofotosemka, 1975, No. 1, p. 63 - 65. In Russian. − Abstr. in Referativ. Zhurn. 51. Astron., 3.51.219; 62. Issled. kosmich. prostranstva, 3.62. 153 (1976).

094.119 Analysis of information soundness on small-scale general lunar maps.
K. B. Shingareva, V. P. Shashkina.
Izv. vyssh. ucheb. zavedenij. Geod. i aehrofotosemka, 1975, No. 1, p. 107 - 112. In Russian. − Abstr. in Referativ. Zhurn. 51. Astron., 3.51.220; 62. Issled. kosmich. prostranstva, 3.62. 155 (1976).

094.120 Perspectives of lunar mapping on the scale 1 : 5 000 000. G. N. Romankevich.
Izv. vyssh. ucheb. zavedenij. Geod. i aehrofotosemka, 1975, No. 1, p. 113 - 117. In Russian. − Abstr. in Referativ. Zhurn. 51. Astron., 3.51.221; 62. Issled. kosmich. prostranstva, 3.62. 156 (1976).

094.121 Slump phenomena in regolith on slopes of the lunar relief. R. O. Kuz'min.
Izv. AN SSSR. Ser. geol., 1975, No. 10, p. 65 - 72. In Russian. Abstr. in Referativ. Zhurn. 51. Astron., 3.51.222; 62. Issled. kosmich. prostranstva, 3.62.146 (1976).

094.122 Preliminary results of radio sounding of the circumlunar space based on Luna 22 data.
V. A. Vinogradov, M. B. Vasil'ev, A. S. Vyshlov, M. A. Kolosov, N. A. Savich, V. A. Samovol, L. N. Samoznaev, A. I. Sidorenko, D. Ya. Shtern.
XI Vses. konf. po rasprostr. radiovoln. Chast' (Part) 1. Tezisy dokl. Kazan', Kazan. un-t, 1975, p. 177 - 179. In Russian. Abstr. in Referativ. Zhurn. 51. Astron., 3.51.226 (1976).

094.123 Structure of the far side of the moon from observations of the Zond automatic interplanetary station.
Yu. N. Lipskij.
Vestn. AN SSSR, 1975, No. 9, p. 128 - 131. In Russian. Abstr. in Referativ. Zhurn. 62. Issled. kosmich. prostranstva, 3.62.139 (1976).

094.124 Luna incognita in 1976. J. E. Westfall.
Strolling Astronomer, Vol. 25, 227 - 231 (1976).

094.125 Het tekenen van maankraters: geen kunst maar oefening. H. Nieuwenhuis.
Zenit, 3e jaargang, p. 93 - 96 (1976).

094.126 The face of the moon. I. Asimov.
Mercury (Journ. Astron. Soc. Pacific), Vol. 5, No. 1, p. 14 - 18 (1976).

094.127 Estimate of the density of surface layer materials of the moon, Mars and Venus. N. N. Krupenio.
In-t kosmich. issled. AN SSSR. Pr-212. Moskva, 1975, 48 pp. In Russian. − Abstr. in Referativ. Zhurn. 51. Astron., 4.51. 220; 62. Issled. kosmich. prostranstva, 4.62.155 (1976).

094.128 Structure of the far side of the moon from observations of space probe "Zond". Yu. N. Lipskij.
Vestn. AN SSSR, 1975, No. 9, p. 128 - 131. In Russian. Abstr. in Referativ. Zhurn. 51. Astron., 4.51.256 (1976).

094.129 The moon proves to be more complicated.
I. N. Galkin.
Priroda, 1976, No. 4, p. 107 - 109. In Russian.

094.130 On the form of craters formed by high-speed impact.
L. V. Leont'ev.
Kosmich. Issled., Vol. 14, 278 - 286 (1976). In Russian.

094.131 Detailed photometry of the lunar surface from photographs of the automatic interplanetary station Zond 8.
V. A. Ezerskaya, V. I. Ezerskij, N. P. Lavrova, V. A. Psarev.
Vestn. Khar'kov. Univ., No. 129 (Ser. Astron., vyp. (No.) 10), p. 24 - 37 (1975). In Russian.

094.132 Mare volcanism and lunar crustal structure.
S. C. Solomon.
Proc. Sixth Lunar Sci. Conference, (see 012.010), Vol. 1, 1021 - 1042 (1975).

094.133 Differentiation of a very thick magma body and implications for the source regions of mare basalts.
D. Walker, J. Longhi, J. F. Hays.
Proc. Sixth Lunar Sci. Conference, (see 012.010), Vol. 1, 1103 - 1120 (1975).

094.134 Evolution of the lunar highland crust.

S. R. Taylor, A. E. Bence.
Proc. Sixth Lunar Sci. Conference, (see 012.010), Vol. 1, 1121 - 1141 (1975).

094.135 Evolution of the moon between 4.6 and 3.3 AE.
L. S. Hollister.
Proc. Sixth Lunar Sci. Conference, (see 012.010), Vol. 1, 1159 - 1178 (1975).

094.136 A unified trace-element model for the evolution of the lunar crust and mantle.
H. Palme, H. Wänke.
Proc. Sixth Lunar Sci. Conference, (see 012.010), Vol. 1, 1179 - 1202 (1975).

094.137 Late heavy bombardment of the moon and terrestrial planets. G. W. Wetherill.
Proc. Sixth Lunar Sci. Conference, (see 012.010), Vol. 2, 1539 - 1561 (1975).

094.138 The state of meteoritic material on the moon.
H. A. Zook.
Proc. Sixth Lunar Sci. Conference, (see 012.010), Vol. 2, 1653 - 1672 (1975).

094.139 Evidence of gaseous radon-222 between fines grains within lunar regolith.
G. Lambert, J. C. Le Roulley, P. Bristeau.
Proc. Sixth Lunar Sci. Conference, (see 012.010), Vol. 2, 1803 - 1809 (1975).

094.140 The Rosiwal Principle and the regolithic distributions of solar-wind elements. D. R. Criswell.
Proc. Sixth Lunar Sci. Conference, (see 012.010), Vol. 2, 1967 - 1987 (1975).

094.141 Evidence for meteoritic sulfur in the lunar regolith.
J. F. Kerridge, I. R. Kaplan, C. Petrowski.
Proc. Sixth Lunar Sci. Conference, (see 012.010), Vol. 2, 2151 - 2162 (1975).

094.142 Studies of the formation of acid-hydrolyzable carbon species and finely divided iron by simulated solar-wind implantation. P. R. Davis, G. Eglinton, C. T. Pillinger, S. K. Erents, G. M. McCracken, H. Nazari, A. Stephenson, R. M. Housley.
Proc. Sixth Lunar Sci. Conference, (see 012.010), Vol. 2, 2163 - 2178 (1975).

094.143 Implantation of carbon and nitrogen ions into lunar fines: trapping efficiencies and saturation concentrations. S. Chang, K. Lennon.
Proc. Sixth Lunar Sci. Conference, (see 012.010), Vol. 2, 2179 - 2188 (1975).

094.144 Monte Carlo simulation of turnover processes in the lunar regolith. J. R. Arnold.
Proc. Sixth Lunar Sci. Conference, (see 012.010), Vol. 2, 2375 - 2395 (1975).

A Monte Carlo model for the gardening of the lunar surface by meteoritic impact is described, and some representative results are given.

094.145 The simulated depth history of dust grains in the lunar regolith. J. P. Duraud, Y. Langevin, M. Maurette, G. Comstock, A. L. Burlingame.
Proc. Sixth Lunar Sci. Conference, (see 012.010), Vol. 2, 2397 - 2415 (1975).

094.146 The geologic evaluation and regional synthesis of metric and panoramic photographs.

D. H. Scott, J. M. Diaz, J. A. Watkins.
Proc. Sixth Lunar Sci. Conference, (see 012.010), Vol. 3, 2531 - 2540 (1975).

094.147 Lunar crater chains of non-impact origin.
D. Eppler, G. Heiken.
Proc. Sixth Lunar Sci. Conference, (see 012.010), Vol. 3, 2571 - 2583 (1975).

094.148 Relative ages of flow units in Mare Imbrium and Sinus Iridum. J. M. Boyce, A. L. Dial, Jr.
Proc. Sixth Lunar Sci. Conference, (see 012.010), Vol. 3, 2585 - 2595 (1975).

094.149 Cratering in the earth–moon system: consequences for age determination by crater counting.
G. Neukum, B. König, H. Fechtig, D. Storzer.
Proc. Sixth Lunar Sci. Conference, (see 012.010), Vol. 3, 2597 - 2620 (1975).

094.150 · Mare crater size-frequency distributions: implications for relative surface ages and regolith development. R. A. Young.
Proc. Sixth Lunar Sci. Conference, (see 012.010), Vol. 3, 2645 - 2662 (1975).

094.151 Metal → metal charge transfer transitions: interpretation of visible-region spectra of the moon and lunar materials. B. M. Loeffler, R. G. Burns, J. A. Tossell.
Proc. Sixth Lunar Sci. Conference, (see 012.010), Vol. 3, 2663 - 2676 (1975).

094.152 Vidicon spectral imaging: color enhancement and digital maps.
T. V. Johnson, D. L. Matson, R. J. Phillips, R. S. Saunders.
Proc. Sixth Lunar Sci. Conference, (see 012.010), Vol. 3, 2677 - 2688 (1975).

094.153 Infrared orbital mapping of lunar features.
W. W. Mendell, F. J. Low.
Proc. Sixth Lunar Sci. Conference, (see 012.010), Vol. 3, 2711 - 2719 (1975).

094.154 Photogeological, geophysical, and geochemical data on the east side of the moon.
F. El-Baz, D. E. Wilhelms.
Proc. Sixth Lunar Sci. Conference, (see 012.010), Vol. 3, 2721 - 2738 (1975).

094.155 Correlation of Al/Si X-ray fluorescence data with other remote sensing data from the Taurus-Littrow area. C. G. Andre, M. E. Hallam, J. R. Weidner, M. H. Podwysocki, J. A. Philpotts, P. E. Clark, I. Adler.
Proc. Sixth Lunar Sci. Conference, (see 012.010), Vol. 3, 2739 - 2748 (1975).

094.156 Farside lunar gravity from a mass point model.
M. Ananda.
Proc. Sixth Lunar Sci. Conference, (see 012.010), Vol. 3, 2785 - 2796 (1975).

094.157 Negative gravity anomalies on the moon.
C. Bowin.
Proc. Sixth Lunar Sci. Conference, (see 012.010), Vol. 3, 2797 - 2804 (1975).

094.158 Volume of material ejected from major lunar basins and implications for the depth of excavation of lunar samples. J. W. Head, M. Settle, R. S. Stein.
Proc. Sixth Lunar Sci. Conference, (see 012.010), Vol. 3, 2805 - 2829 (1975).

094.159 Shock effects from a large impact on the moon.
J. D. O'Keefe, T. J. Ahrens.
Proc. Sixth Lunar Sci. Conference, (see 012.010), Vol. 3,
2831 - 2844 = Div. Geol. Planet. Sci., California Inst. Techn.,
Pasadena, California, Contr. No. 2616 (1975).

094.160 Seismically induced modification of lunar surface
features. P. H. Schultz, D. E. Gault.
Proc. Sixth Lunar Sci. Conference, (see 012.010), Vol. 3,
2845 - 2862 (1975).

094.161 Energy, frequency, and distance of moonquakes at
the Apollo 17 site. M. R. Cooper, R. L. Kovach.
Proc. Sixth Lunar Sci. Conference, (see 012.010), Vol. 3,
2863 - 2879 (1975).

094.162 Extremal inversion of lunar travel time data.
N. Burkhard, D. D. Jackson.
Proc. Sixth Lunar Sci. Conference, (see 012.010), Vol. 3,
2881 - 2885 (1975).

094.163 Natural lunar seismic events and the structure of the
moon. A. M. Dainty, N. R. Goins, M. N. Toksöz.
Proc. Sixth Lunar Sci. Conference, (see 012.010), Vol. 3,
2887 - 2897 (1975).

094.164 Feldspar electrical conductivity and the lunar interior. A. J. Piwinskii, A. G. Duba.
Proc. Sixth Lunar Sci. Conference, (see 012.010), Vol. 3,
2899 - 2907 (1975).

094.165 Lunar electrical conductivity and magnetic permeability. P. Dyal, C. W. Parkin, W. D. Daily.
Proc. Sixth Lunar Sci. Conference, (see 012.010), Vol. 3,
2909 - 2926 (1975).

094.166 Lunar electromagnetic scattering. 4. Transfer functions in the long-wavelength limit.
K. Schwartz, G. Schubert.
Proc. Sixth Lunar Sci. Conference, (see 012.010), Vol. 3,
2927 - 2941 (1975).

094.167 Solid-state convection and the mechanics of the
moon. S. K. Runcorn.
Proc. Sixth Lunar Sci. Conference, (see 012.010), Vol. 3,
2943 - 2953 (1975).

094.168 The fine-scale lunar magnetic field.
C. T. Russell, P. J. Coleman, Jr., B. K. Fleming,
L. Hilburn, G. Ioannidis, B. R. Lichtenstein, G. Schubert.
Proc. Sixth Lunar Sci. Conference, (see 012.010), Vol. 3,
2955 - 2969 (1975).

094.169 Mapping of lunar surface remanent magnetic fields
by electron scattering. R. P. Lin, R. E.
McGuire, H. C. Howe, K. A. Anderson, J. E. McCoy.
Proc. Sixth Lunar Sci. Conference, (see 012.010), Vol. 3,
2971 - 2973 (1975).

094.170 Magnetic anomalies near Van de Graaff Crater
D. W. Strangway, J. C. Rylaarsdam, A. P. Annan.
Proc. Sixth Lunar Sci. Conference, (see 012.010), Vol. 3,
2975 - 2984 (1975).

094.171 The effect of local magnetic fields on the lunar
photoelectron layer while the moon is in the plasma
sheet. W. J. Burke, P. H. Reiff, D. L. Reasoner.
Proc. Sixth Lunar Sci. Conference, (see 012.010), Vol. 3,
2985 - 2997 (1975).

094.172 On the apparent diamagnetism of the lunar en-

vironment in the geomagnetic tail lobes.
B. E. Goldstein, C. T. Russell.
Proc. Sixth Lunar Sci. Conference, (see 012.010), Vol. 3,
2999 - 3012 (1975).

094.173 The lunar terminator ionosphere.
J. Benson, J. W. Freeman, H. K. Hills.
Proc. Sixth Lunar Sci. Conference, (see 012.010), Vol. 3,
3013 - 3021 (1975).

094.174 Lunar nightside electron fluxes. D. L. Reasoner.
Proc. Sixth Lunar Sci. Conference, (see 012.010),
Vol. 3, 3023 - 3032 (1975).

094.175 Radon concentrations at the Apollo landing sites.
R. L. Brodzinski, J. C. Langford.
Proc. Sixth Lunar Sci. Conference, (see 012.010), Vol. 3,
3033 - 3037 (1975).

094.176 Implications of atmospheric ^{40}Ar escape on the interior structure of the moon.
R. R. Hodges, Jr., J. H. Hoffman.
Proc. Sixth Lunar Sci. Conference, (see 012.010), Vol. 3,
3039 - 3047 (1975).

094.177 On changes in the intensity of the ancient lunar magnetic field.
A. Stephenson, S. K. Runcorn, D. W. Collinson.
Proc. Sixth Lunar Sci. Conference, (see 012.010), Vol. 3,
3049 - 3062 (1975).

094.178 Lunar paleointensity determination using anhys-
teretic remanence (ARM): a critique.
D. J. Dunlop, M. E. Bailey, M. F. Westcott-Lewis.
Proc. Sixth Lunar Sci. Conference, Vol. 3,
3063 - 3069 (1975).

094.179 Some correlation of rock exposure ages and regolith
dynamics. F. Hörz, R. V. Gibbons, D. E. Gault,
J. B. Hartung, D. E. Brownlee.
Proc. Sixth Lunar Sci. Conference, (see 012.010), Vol. 3,
3495 - 3508 (1975).

094.180 Meteoroid storms detected on the moon.
F. K. Duennebier, Y. Nakamura, G. V. Latham,
H. J. Dorman.
Science, Vol. 192, 1000 - 1002 (1976).
Seismometers on the moon have detected several brief
periods of enhanced meteoroid-impact activity, believed to
represent encounters of the moon with "clouds" of objects
in the kilogram range. The latest and most active encounter,
in June 1975, is interpreted as a meteoroid cloud of diameter
0.1 astronomical unit and total mass 10^{13} to 10^{14} grams.

094.181 A Monte Carlo model for the exposure history of
lunar dust grains in the ancient solar wind.
J. Borg, G. M. Comstock, Y. Langevin, M. Maurette,
B. Jouffrey, C. Jouret.
Earth Planet. Sci. Letters, Vol. 29, 161 - 174 (1976).

094.182 Selenonymics and UNO: Scientific-technical and
international-legal aspects of the nomenclature
of the topographic features on the lunar surface.
A. M. Kamkov.
Teoriya i praktika toponim. issled., Moskva, 1975, p. 27 -
34. In Russian. − Abstr. in Referativ. Zhurn. 51. Astron.,
6.51.2 (1976).

094.183 Numerical research for possibilities of lunar probing
from measurements of a vertical electric dipole field.
L. L. Van'yan, V. I. Dmitriev, Eh. A. Fedorova.

Ehlektromagnit. zondir. Zemli i Luny. Moskva, Mosk. un-t.
1975, p. 23 - 35. In Russian. – Abstr. in Referativ. Zhurn. 51.
Astron., 6. 51. 307 (1976).

094.184 Lunar microcraters and interplanetary dust fluxes.
J. B. Hartung.
IAU Colloquium No. 31, (see 012.015), p. 209 - 226 (1976).
Invited paper.

**094.185 The size frequency distribution and rate of produc-
tion of microcraters.** D. A. Morrison, E. Zinner.
IAU Colloquium No. 31, (see 012.015), p. 227 - 231 (1976).

**094.186 Lunar soil movement registered by the Apollo 17
cosmic dust experiment.**
O. E. Berg, H. Wolf, J. W. Rhee.
IAU Colloquium No. 31, (see 012.015), p. 233 - 237 (1976).

094.187 Electrostatic disruption of lunar dust particles.
J. W. Rhee.
IAU Colloquium No. 31, (see 012.015), p. 238 - 240 (1976).

**094.188 Measurements of impact ejecta parameters in
crater simulation experiments.** E. Schneider.
IAU Colloquium No. 31, (see 012.015), p. 242 (1976).
Abstract of 14.094.114.

094.189 Lunar volcanism in space and time. J. W. Head III.
Rev. Geophys. Space Phys., Vol. 14, 265 - 300
(1976).
Lunar volcanic deposits are dominated by areally extensive mare units occurring in regionally low areas predominantly
on the lunar near side. Data obtained from lunar orbit and
earth-based observations have been used to extend the detailed
characterizations derived from Apollo and Luna sample return
missions to other parts of the moon. An early Ti-rich mare
phase (Apollo 11 and Apollo 17 type basalts) flooded large
areas of the eastern portion of the lunar near side in the early
Imbrian Period (about 3.5–3.8 b.y. ago). An intermediate age,
less Ti-rich phase (Apollo 12 and Apollo 15 type basalts)
flooded widespread areas of the moon predominantly in the
middle to late Imbrian Period (about 3.0–3.5 b.y. ago). Finally,
a second Ti-rich phase (unsampled by Apollo and Luna)
flooded portions of Mare Imbrium and the western maria in
the early Eratosthenian Period (about 2.5–3.0 b.y. ago).

**094.190 Sources of lunar limb compressions: topographic
correlations.** L. J. Srnka.
Journ. Geophys. Res., Vol. 81, 2015 - 2022 (1976).
Apollo 15 and 16 subsatellite observations of lunar limb
compressions in the solar wind have been studied statistically,
by correlating occurrence rates for the compressions with
seven topographic parameters derived from Apollo laser altimetry, with the use of a $5° \times 5°$ selenographic grid. Maximum,
rms, and average sunset-facing slopes are significantly correlated with occurrence rates in the source regions, but no correlation with elevation differences is found.

**094.191 On the mechanisms of lunar regolith glass particle
formation.**
Yu. B. Chernyak, M. D. Nussinov (*Nusinov*).
Nature, Vol. 261, 664 - 666 (1976).
The discovery of many particles of glass in the lunar
regolith has resulted in various hypotheses for their origin.
Based on previously unpublished work, the authors here give
an analysis of the principal mechanisms of their formation and
evolution.

**094.192 Volatilisation of elements from melts of lunar
material.** M. D. Nussinov (*Nusinov*), Yu. B.
Chernyak.

Nature, Vol. 261, 666 - 669 (1976).
The authors discuss the theory of the volatilisation of
components of molten rocks and develop a model for this
process, which enables one to draw some conclusions about
such phenomena on the moon.

**094.193 Constraints on the nature of the ancient lunar
magnetic field.** J. N. Goswami.
Nature, Vol. 261, 675 - 677 (1976).– Letter.

094.194 Extreme Zeitmarken in der Mondchronologie.
E. K. Jessberger, T. Kirsten, T. Staudacher.
MPI Kernphys., Heidelberg, Jahresbericht 1975, p. 145 - 146.

**094.195 Physical processes in intense electric fields at the
surface of the moon.** A. A. Vorob'ev.
Ehlektron. Obrab. Material USSR, No. 2,ₙp. 32 - 37 (1975). In
Russian. – Abstr. in Phys. Abstr., Vol. 79, A003282 (1976).

094.196 The interior of the moon. J. Classen.
Veröff. Sternw. Pulsnitz, Germany, No. 10, p. 1 -
23 (1975). – Abstr. in Phys. Abstr., Vol. 79, A003283 (1976).

094.197 Solar-wind induction and lunar conductivity.
C. P. Sonett.
Phys. Earth Planet. Interior, Vol. 10, 313 - 322 (1975). –
Abstr. in Phys. Abstr., Vol. 79, A006944 (1976).

094.198 On the interpretation of lunar magnetism.
S. K. Runcorn.
Phys. Earth Planet Interiors, Vol. 10, 327 - 335 (1975).
Abstr. in Phys. Abstr., Vol. 79, A011111 (1976).

094.199 Cratering and cosmogenic nuclides.
M. L. Blake, G. J. Wasserburg.
Geophys. Res. Letters, Vol. 2, 477 - 479 (1975). – Abstr. in
Phys. Abstr., Vol. 79, A027801 (1976).

**094.200 Variation in the number of meteoroid impacts on
the moon with lunar phase.**
A. M. Dainty, S. Stein, M. N. Toksøz.
Geophys. Res. Letters, Vol. 2, 273 - 276 (1975). – Abstr. in
Phys. Abstr., Vol. 79, A027814 (1976).

**094.201 Energy spectra of decimeter radio waves reflected
by the lunar surface from data of the 'Luna-19'
satellite.** V. I. Kaevitser, S. S. Matyugov, A. G. Pavel'yev.
(*Pavel'ev*), G. M. Petrov, V. I. Rogal'skiy, O. I. Yakovlev.
Radiotekhn. Ehlektron., No. 5, p. 936 - 945 (1974). – Abstr.
in Phys. Abstr., Vol. 79, A032809 (1976).

**094.202 Particle track record in lunar silicates: long-term
behavior of solar and galactic VH nuclei and lunar
surface dynamics.** D. E. Yuhas.
Thesis Washington Univ., Seattle, USA, 282 pp. (1974).
(Available from:Univ. Microfilms, Order No. 75–6627).

**094.203 Clues in the rare gas isotopes to early solar system
history.** J. H. Reynolds.
Soviet-American conf. cosmochem. moon planets, Moscow,
USSR, 4 June 1974, 29 pp. (1974). –See 12.094.214.

**094.204 Lunar elemental analysis obtained from the Apollo
gamma-ray and X-ray remote sensing experiment.**
J. I. Trombka, J. R. Arnold, I. Adler, A. E. Metzger,
R. C. Reedy.
Soviet-American conf. cosmochem. moon planets, Moscow,
USSR, 4 June 1974, 75 pp. (1974). – See 12.094.192.

**094.205 Modification of particle fluxes at the lunar surface
by electric and magnetic fields.** P. H. Reiff.

Thesis Rice Univ., Houston, Texas, USA, 111 pp. (1975). (Available from: Univ. Microfilms, Order No. 75-22,055).

094.206 Lunar neutron source function. J. J. Kornblum. Thesis State Univ. New York, Stony Brook, USA, 192 pp. (1974). (Available from: Univ. Microfilms, Order No. 75-16, 205).

094.207 Accumulation and circulation of gaseous radon between lunar fines. G. Lambert, P. Bristeau, J. C. Le Roulley. Discussion meeting moon, London, UK, 9 June 1975, 14 pp. (1975).

Moon morphology. See Abstr. 003.096.

Lunar nomenclature: a dissenting note. See Abstr. 015.009.

Measurement of lunar and planetary magnetic fields by reflection of low energy electrons. See Abstr. 031.202.

Errors induced in the measurement and azimuth directions of morphological features imaged on oblique lunar orbiter photographs. See Abstr. 032.533.

Solar-wind tritium limit and nuclear processes in the solar atmosphere. See Abstr. 074.049.

Observations of Baily's beads from near the northern limit of the total solar eclipse of June 20, 1974. See Abstr. 079.001.

The lunar relief of the Tungusic syneclise. See Abstr. 081.014.

Using the moon to probe the geomagnetic tail lobe plasma. See Abstr. 084.290.

High energy nuclear reactions in our planetary system. See Abstr. 091.019.

The moon: not so different from earth after all. See Abstr. 094.012.

Chemical, mineralogical and textural systematics of non-mare melt rocks: implications for lunar impact and volcanic processes. See Abstr. 094.433.

Apollo 16 feldspathic melt rocks: clues to the magmatic history of the lunar crust. See Abstr. 094.435.

Petrogenesis of mare basalts: implications for

chemical, mineralogical, and thermal models for the moon. See Abstr. 094.470.

The lunar crust — chemically defined rock groups and their potassium–uranium fractionation. See Abstr. 094.473.

The history of lunar bombardment inferred from ^{40}Ar–^{39}Ar dating of highland rocks. See Abstr. 094.493.

Cl and P_2O_5 systematics: clues to early lunar magmas. See Abstr. 094.499.

Geomorphology of crater and basin deposits — emplacement of the Fra Mauro Formation. See Abstr. 094.535.

Polarimetric properties of the lunar surface and its interpretation: Part 7 — Other solar system objects. See Abstr. 094.540.

Magnetic stratigraphy of the Apollo 15 deep drill core. See Abstr. 094.541.

On the high-temperature stability of magnetite: implications for lunar soil magnetism. See Abstr. 094.542.

Effects of meteorite impact on magnetic properties of Apollo lunar materials. See Abstr. 094.544.

Magnetic effects of shock and their implications for magnetism of lunar samples. See Abstr. 094.545.

Ferromagnetic resonance as a method of studying the micrometeorite bombardment history of the lunar surface. See Abstr. 094.547.

Solar-wind sputter erosion of microcrater populations on the lunar surface. See Abstr. 094.563.

Flux of hyperbolic meteoroids. See Abstr. 104.048.

Activated release of alkalis during the vesiculation of molten basalts under high vacuum: implications for lunar volcanism. See Abstr. 105.038.

In-situ records of interplanetary dust particles — methods and results. See Abstr. 106.073.

Lunar ejecta in heliocentric space. See Abstr. 106.091.

In situ measurements of dust. See Abstr. 106.096.

Moon, Local Properties

094.401 **Study of magnetic field, rock magnetization and lunar electrical conductivity in the Bay Le Monnier.**
Sh. Sh. Dolginov, Ye. G. Yeroshenko (*E. G. Eroshenko*),
V. A. Sharova, T. A. Vnuchkova, L. L. Vanyan, B. A.
Okulessky (*Okulesskij*), A. T. Bazilevsky (*Bazilevskij*).
The Moon, Vol. 15, 3 - 14 (1976).

The data of three-times repeated magnetic survey of the section of Lunokhod-2 route 1.5 km long are analyzed. The results of magnetic survey near the tectonic break of Straight Rille and near the south rim of crater Le Monnier were used for estimation of rock magnetization in situ. It is shown that mare basalts in south-east region of crater Le Monnier have oblique magnetization (at the angle $\sim 30°$ to horizon). The magnitude of magnetization is $\sim 5 \times 10^{-5}$ G cm^3 g^{-1}. The south-east slope of the crater Le Monnier is magnetized roughly vertically, the upper limit of magnetization of the rocks of the rim is $\sim 1 \times 10^{-5}$ G cm^3 g^{-1}.

094.402 **On the spectral reflectance and maturation darkening of lunar soils.**
C. L. Marquardt, D. L. Griscom.
The Moon, Vol. 15, 15 - 30 (1976).

Optical absorption and diffuse reflectance spectra were obtained for simulated lunar glasses of four different compositions, both in their as-quenched (reduced) states and following mild subsolidus oxidation. The transmission spectra, when normalized by the FeO content of the glasses, differed from one another only in the relative intensity of an unresolved band in the UV. It is possible to understand the spectral characteristics both of oxidation darkening of synthetic glass powders and of maturation darkening of lunar soils in terms of (1) the growth of charge transfer band(s) and (2) the development of opaque surface phases. It is shown that mechanism (1) is of primary importance in lunar highland materials and that mechanism (2) dominates in mare materials.

094.403 **A correlation study based on Al/Si X-ray fluorescence data from southwestern part of Mare Sereni-**
tatis. C. G. Andre, M. E. Hallam, J. R. Weidner, J. A.
Philpotts, M. H. Podwysocki, I. Adler.
The Moon, Vol. 15, 133 - 142 (1976).

The Tacquet formation in southwestern Mare Serenitatis has unusually low visible albedos for the Al/Si intensity ratios measured by the Apollo 15 X-ray fluorescence experiment. This is a contradiction of the demonstrated good correlation between Al/Si and visible albedo data. To understand why this situation exists, a correlation study has been undertaken. This study is based on Al/Si intensity ratios and includes such other remote sensing data as visible and near IR spectral measurements. The discrepancy between visible albedos and Al/Si intensities in the vicinity of the Tacquet formation might be due primarily to the addition of highland-type ejecta from the crater, Menelaus, onto a low albedo mare unit.

094.404 **On the influence of the properties of lunar soil on the character of the sampling process** (according to materials of the automatic lunar station Luna 20).
Yu. N. Strelov, V. V. Aksler, D. D. Dryuchenko, M. D. Nusinov, M. I. Smorodinov, V. V. Shvarev.
Kosmich. Issled., Vol. 14, 120 - 125 (1976). In Russian.

094.405 **Chemical composition variations of the lunar surface in the contact zone "mare-highland".**
G. E. Kocharov, S. V. Victorov (*Viktorov*), V. P. Kovalev,
G. A. Matveev, V. I. Chesnokov.
Space Research XV, (see 012.003), p. 587 - 592 (1975).

The chemical composition of the lunar surface in the "mare-highland" boundary zone has been investigated by means of an X-ray fluorescence spectrometer on board Lunokhod 2. Considerable variations in the abundances of some chemical elements in the upper layer of regolith have been discovered. The iron content in the bottom part of crater Le Monnier is about 6 wt %; it falls to 4.0 ± 0.4 wt % in the highland region, while on the slopes of a ~ 20 km tectonic break, where the underlying rocks seem to be bared, it is about 8 wt %. The connection between the chemical composition and geological and morphological peculiarities of the various regions investigated is discussed from the viewpoint of mechanisms responsible for the transport of material on the lunar surface.

094.406 **Abundance of some elements in the regolith of the maria and continental regions of the moon.**
Yu. A. Surkov, G. M. Kolesov, F. F. Kirnozov.
Space Research XV, (see 012.003), p. 593 - 599 (1975).

The determination of the amount of some 30 elements in samples of lunar soil returned by the automatic stations Luna 16 and 20 and Apollo 11, 12, 14, 15, 16, 17 has been carried out by the neutron-activation method, and average values established. This corresponds to the average chemical composition of regolith from conformable regions. It is shown that the content of most of the trace elements in regolith from maria regions is 2–3 times higher than regolith from continental regions, and reflects the composition of the main petrogenic provinces of the moon.

094.407 **Track investigations of lunar soil returned by Luna 20.** L. L. Kashkarov, A. K. Lavrukhina, L. I.
Genaeva, M. K. Antoshin, G. V. Spivak.
Space Research XV, (see 012.003), p. 601 - 606 (1975).

Investigations of the tracks from nuclei ($23 \lesssim Z \lesssim 28$) of solar and galactic cosmic rays in feldspar, olivine and pyroxene crystals have been carried out for samples from different levels of the Luna 20 soil column. The results indicate that even at the full depth (~ 20 cm) particles are found which have traces of the action of solar cosmic rays with $E \gtrsim 10$ MeV nucleon^{-1}. Their energy spectrum may be described by a power law with a power varying from -1.5 to -2. The effective times of irradiation in the layer of regolith deeper than ~ 2 cm for the groups of crystals studied are equal to $(0.2-20) \times 10^6$ years and $(20-200) \times 10^6$ years respectively.

094.408 **Luna 16 and 20 investigations of the physical and mechanical properties of lunar soil.**
A. K. Leonovich, V. V. Gromov, P. S. Semyonov, V. N. Penetrigov, V. V. Shvaryov.
Space Research XV, (see 012.003), p. 607 - 616 (1975).

094.409 **Low pressure radon diffusion: a laboratory study and its implications for lunar venting.**
L. J. Friesen, J. A. S. Adams.
Geochim. Cosmochim. Acta, Vol. 40, 375 - 380 (1976).

The purpose of this paper is to report results of a laboratory study of radon diffusion through soil columns under moderate vacuum conditions, and to point out implications these results have for radon migration on the moon and for lunar gas venting.

094.410 **Investigation of the physical properties of soil returned by Luna 20 and its terrestrial analogues.**
V. V. Rzhevskij, V. V. Shvarev, A. A. Silin, A. R. Golovkin,
N. T. Kruglov, R. G. Petrochenkov, E. A. Dukhovskoj.
Kosmich. Issled., Vol. 14, 287 - 292 (1976). In Russian.

094.411 **Numerical investigation of the possibilities of lunar sounding according to measurements of the field of**
a vertical electrical dipole.
L. L. Van'yan, V. I. Dmitriev, Eh. A. Fedorova.
Ehlektromagnit. zondir. Zemli i Luny. Moskva, Mosk. un-t,

1975, p. 23 - 35. In Russian. — Abstr. in Referativ. Zhurn. 62. Issled. kosmich. prostranstva, 5.62.170 (1976).

094.412 On the colorimetric representation of lunar spectro-photometric measurements.   N. N. Evsyukov.
Vestn. Khar'kov. Univ., No. 129 (Ser. Astron., vyp. (No.) 10), p. 61 - 70 (1975). In Russian.

094.413 On the relation between hot spots on the lunar surface and photometrical inhomogeneities.
V. I. Ezerskij, V. A. Plakhotnichenko.
Vestn. Khar'kov. Univ., No. 129 (Ser. Astron., vyp. (No.) 10), p. 81 - 84 (1975). In Russian.

094.414 Craters of the Orientale region.   P. C. R. Morgan.
Journ. British Astron. Ass., Vol. 86, 288 - 295 (1976).

094.415 Petrology and mineralogy of Apollo 17 mare basalts.   G. M. Brown, A. Peckett, C. H. Emeleus, R. Phillips, R. H. Pinsent.
Proc. Sixth Lunar Sci. Conference, (see 012.010), Vol. 1, 1 - 13 (1975).

094.416 Petrology of the Apollo 16 mare component: Mare Nectaris.   J. W. Delano.
Proc. Sixth Lunar Sci. Conference, (see 012.010), Vol. 1, 15 - 47 (1975).

094.417 Comparative mineralogy and petrology of Apollo 17 mare basalts: samples 70215, 71055, 74255, and 75055.   R. F. Dymek, A. L. Albee, A. A. Chodos.
Proc. Sixth Lunar Sci. Conference, (see 012.010), Vol. 1, 49 - 77 (1975).

094.418 Geology, petrology, and crystallization of Apollo 15 quartz-normative basalts.
G. E. Lofgren, C. H. Donaldson, T. M. Usselman.
Proc. Sixth Lunar Sci. Conference, (see 012.010), Vol. 1, 79 - 99 (1975).

094.419 An estimate of the yield strength of the Imbrium flows.   H. J. Moore, G. G. Schaber.
Proc. Sixth Lunar Sci. Conference, (see 012.010), Vol. 1, 101 - 118 (1975).

094.420 Silica activity in lunar lavas.
W. P. Nash, J. D. Haselton.
Proc. Sixth Lunar Sci. Conference, (see 012.010), Vol. 1, 119 - 130 (1975).

094.421 On high-alumina mare basalts.   W. I. Ridley.
Proc. Sixth Lunar Sci. Conference, (see 012.010), Vol. 1, 131 - 145 (1975).

094.422 Anomalous low-K silicate melt inclusions in ilmenite from Apollo 17 basalts.
E. Roedder, P. W. Weiblen.
Proc. Sixth Lunar Sci. Conference, (see 012.010), Vol. 1, 147 - 164 (1975).

094.423 Pyroxene–phyric basalt 15075: petrography and petrogenesis.   L. A. Taylor, K. C. Misra.
Proc. Sixth Lunar Sci. Conference, (see 012.010), Vol. 1, 165 - 179 (1975).

094.424 Absolute cooling rates of lunar rocks: theory and application.
L. A. Taylor, D. R. Uhlmann, R. W. Hopper, K. C. Misra.
Proc. Sixth Lunar Sci. Conference, (see 012.010), Vol. 1, 181 - 191 (1975).

094.425 Mineralogy, petrology, and chemistry of mare basalts from Apollo 17 rake samples.
R. D. Warner, K. Keil, M. Prinz, J. C. Laul, A. V. Murali, R. A. Schmitt.
Proc. Sixth Lunar Sci. Conference, (see 012.010), Vol. 1, 193 - 220 (1975).

094.426 Petrologic evidence for a plutonic igneous origin of anorthositic norite clasts in 67955 and 77017.
L. D. Ashwal.
Proc. Sixth Lunar Sci. Conference, (see 012.010), Vol. 1, 221 - 230 (1975).

094.427 The problem of the origin of symplectites in olivine-bearing lunar rocks.
P. M. Bell, H. K. Mao, E. Roedder, P. W. Weiblen.
Proc. Sixth Lunar Sci. Conference, (see 012.010), Vol. 1, 231 - 248 (1975).

094.428 Magma genesis by in situ melting within the lunar crust.   M. L. Crawford.
Proc. Sixth Lunar Sci. Conference, (see 012.010), Vol. 1, 249 - 261 (1975).

094.429 Petrology of anorthosites from the Descartes region of the moon: Apollo 16.
J. R. Dixon, J. J. Papike.
Proc. Sixth Lunar Sci. Conference, (see 012.010), Vol. 1, 263 - 291 (1975).

094.430 Lunar anorthosite paradox: an alternative explanation.   M. J. Drake.
Proc. Sixth Lunar Sci. Conference, (see 012.010), Vol. 1, 293 - 299 (1975).

094.431 Comparative petrology of lunar cumulate rocks of possible primary origin: dunite 72415, troctolite 76535, norite 78235, and anorthosite 62237.
R. F. Dymek, A. L. Albee, A. A. Chodos.
Proc. Sixth Lunar Sci. Conference, (see 012.010), Vol. 1, 301 - 341 = Div. Geol. Planet. Sci., California Inst. Techn., Pasadena, California, Contr. No. 2617 (1975).

094.432 The provenance of metal in anorthositic rocks.
R. H. Hewins, J. I. Goldstein.
Proc. Sixth Lunar Sci. Conference, (see 012.010), Vol. 1, 343 - 362 (1975).

094.433 Chemical, mineralogical and textural systematics of non-mare melt rocks: implications for lunar impact and volcanic processes.   A. J. Irving.
Proc. Sixth Lunar Sci. Conference, (see 012.010), Vol. 1, 363 - 394 (1975).

094.434 Petrology of noritic cumulates and a partial melting model for the genesis of Fra Mauro basalts.
I. S. McCallum, E. A. Mathez.
Proc. Sixth Lunar Sci. Conference, (see 012.010), Vol. 1, 395 - 414 (1975).

094.435 Apollo 16 feldspathic melt rocks: clues to the magmatic history of the lunar crust.
B. N. Powell, M. A. Dungan, P. W. Weiblen.
Proc. Sixth Lunar Sci. Conference, (see 012.010), Vol. 1, 415 - 433 (1975).

094.436 Lunar granites with unique ternary feldspars.
G. Ryder, D. B. Stoeser, U. B. Marvin, J. F. Bower.
Proc. Sixth Lunar Sci. Conference, (see 012.010), Vol. 1, 435 - 449 (1975).

094.437 **Minor elements in lunar olivine as a petrologic indicator.** I. M. Steele, J. V. Smith.
Proc. Sixth Lunar Sci. Conference, (see 012.010), Vol. 1, 451 - 467 (1975).

094.438 **Petrography and petrology of basaltic achondrite polymict breccias (howardites).** T. E. Bunch.
Proc. Sixth Lunar Sci. Conference, (see 012.010), Vol. 1, 469 - 492 (1975).

094.439 **The petrogenesis of 77115 and its xenocrysts: description and preliminary interpretation.**
E. C. T. Chao, J. A. Minkin, C. L. Thompson, J. S. Huebner.
Proc. Sixth Lunar Sci. Conference, (see 012.010), Vol. 1, 493 - 515 (1975).

094.440 **Size and modal analyses of fines and ultrafines from some Apollo 17 samples.**
G. M. Greene, D. T. King, Jr., G. S. Banholzer, Jr., E. A. King.
Proc. Sixth Lunar Sci. Conference, (see 012.010), Vol. 1, 517 - 527 (1975).

094.441 **Significance of exsolved pyroxenes from lunar breccia 77215.**
J. S. Huebner, M. Ross, N. Hickling.
Proc. Sixth Lunar Sci. Conference, (see 012.010), Vol. 1, 529 - 546 (1975).

094.442 **Consortium studies of matrix of light gray breccia 73215.**
O. B. James, A. Brecher, D. P. Blanchard, J. W. Jacobs, J. C. Brannon, R. L. Korotev, L. A. Haskin, H. Higuchi, J. W. Morgan, E. Anders, L. T. Silver, K. Marti, D. Braddy, I. D. Hutcheon, T. Kirsten, J. F. Kerridge, I. R. Kaplan, C. T. Pillinger, L. R. Gardiner.
Proc. Sixth Lunar Sci. Conference, (see 012.010), Vol. 1, 547 - 577 (1975).

094.443 **Viscous flow, crystallization behavior, and thermal histories of lunar breccias 70019 and 79155.**
L. Klein, P. I. K. Onorato, D. R. Uhlmann, R. W. Hopper.
Proc. Sixth Lunar Sci. Conference, (see 012.010), Vol. 1, 579 - 593 (1975).

094.444 **Mineralogy and petrology of <1-mm fines from Apollo 16 core sections 60002 and 60004.**
H. O. A. Meyer, R. H. McCallister, H.-M. Tsai.
Proc. Sixth Lunar Sci. Conference, (see 012.010), Vol. 1, 595 - 614 (1975).

094.445 **Characteristics of metal particles in Apollo 16 rocks.** K. C. Misra, L. A. Taylor.
Proc. Sixth Lunar Sci. Conference, (see 012.010), Vol. 1, 615 - 639 (1975).

094.446 **Thermal regimes in impact melts and the petrology of the Apollo 17 Station 6 boulder.**
C. H. Simonds.
Proc. Sixth Lunar Sci. Conference, (see 012.010), Vol. 1, 641 - 672 (1975).

094.447 **Rock 61016: multiphase shock and crystallization history of a polymict troctolitic–anorthositic breccia.** D. Stöffler, S. Schulien, R. Ostertag.
Proc. Sixth Lunar Sci. Conference, (see 012.010), Vol. 1, 673 - 692 (1975).

094.448 **The formation of lunar breccias: sintering and crystallization kinetics.**
D. R. Uhlmann, L. Klein, P. I. K. Onorato, R. W. Hopper.
Proc. Sixth Lunar Sci. Conference, (see 012.010), Vol. 1, 693 - 705 (1975).

094.449 **Origin of the Station 7 boulder: a note.**
S. R. Winzer, D. F. Nava, S. Schuhmann, R. K. L. Lum, J. A. Philpotts.
Proc. Sixth Lunar Sci. Conference, (see 012.010), Vol. 1, 707 - 710 (1975).

094.450 **Surface morphology and chemistry of rusty particle 60002,108.** J. L. Carter.
Proc. Sixth Lunar Sci. Conference, (see 012.010), Vol. 1, 711 - 718 (1975).

094.451 **Morphology and composition of chalcopyrite, chromite, Cu, Ni–Fe, pentlandite, and troilite in vugs of 76015 and 76215.** J. L. Carter, U. S. Clanton, R. Fuhrman, R. B. Laughon, D. S. McKay, T. M. Usselman.
Proc. Sixth Lunar Sci. Conference, (see 012.010), Vol. 1, 719 - 728 (1975).

094.452 **Subsolidus reduction of lunar opaque oxides: textures, assemblages, geochemistry, and evidence for a late-stage endogenic gaseous mixture.**
A. El Goresy, P. Ramdohr.
Proc. Sixth Lunar Sci. Conference, (see 012.010), Vol. 1, 729 - 745 (1975).

094.453 **A model for the evolution of opaques in mare lavas.** J. D. Haselton, W. P. Nash.
Proc. Sixth Lunar Sci. Conference, (see 012.010), Vol. 1, 747 - 755 (1975).

094.454 **Superparamagnetic clusters of Fe^{2+} spins in lunar olivine: dissolution by high-temperature annealing.**
G. P. Huffman, G. R. Dunmyre.
Proc. Sixth Lunar Sci. Conference, (see 012.010), Vol. 1, 757 - 772 (1975).

094.455 **X-ray diffraction and electron microscope studies of clinopyroxenes from lunar basalts 75035 and 75075.**
H. Jagodzinski, M. Korekawa, W. F. Müller, L. Schröpfer.
Proc. Sixth Lunar Sci. Conference, (see 012.010), Vol. 1, 773 - 778 (1975).

094.456 **North Ray Crater breccias: an electron petrographic study.**
G. L. Nord, Jr., J. M. Christie, A. H. Heuer, J. S. Lally.
Proc. Sixth Lunar Sci. Conference, (see 012.010), Vol. 1, 779 - 797 (1975).

094.457 **Shock-induced subsolidus reduction-decomposition of orthopyroxene and shock-induced melting in norite 78235.** C. B. Sclar, J. F. Bauer.
Proc. Sixth Lunar Sci. Conference, (see 012.010), Vol. 1, 799 - 820 (1975).

094.458 **Intracrystalline cation order in a lunar crustal troctolite.** J. R. Smyth.
Proc. Sixth Lunar Sci. Conference, (see 012.010), Vol. 1, 821 - 832 (1975).

094.459 **Shock-induced deformation features in terrestrial peridot and lunar dunite.**
L. W. Snee, T. J. Ahrens.
Proc. Sixth Lunar Sci. Conference, (see 012.010), Vol. 1, 833 - 842 = Div. Geol. Planet. Sci., California Inst. Techn., Pasadena, California, Contr. No. 2574 (1975).

094.460 **Experimental modeling of the cooling history of Apollo 12 olivine basalts.**
C. H. Donaldson, T. M. Usselman, R. J. Williams, G. E. Lofgren.

Proc. Sixth Lunar Sci. Conference, (see 012.010), Vol. 1, 843 - 869 (1975).

094.461 Experimental petrology of Apollo 17 mare basalts. D. H. Green, A. E. Ringwood, W. O. Hibberson, N. G. Ware.
Proc. Sixth Lunar Sci. Conference, (see 012.010), Vol. 1, 871 - 893 (1975).

094.462 Residual products of fractional crystallization of lunar magmas: an experimental study. P. C. Hess, M. J. Rutherford, R. N. Guillemette, F. J. Ryerson, H. A. Tuchfeld.
Proc. Sixth Lunar Sci. Conference, (see 012.010), Vol. 1, 895 - 909 (1975).

094.463 The effects of Al^{3+}, Cr^{3+}, and Ti^{3+} on the stability of armalcolite. S. E. Kesson, D. H. Lindsley.
Proc. Sixth Lunar Sci. Conference, (see 012.010), Vol. 1, 911 - 920 (1975).

094.464 Mare basalts: melting experiments and petrogenetic interpretations. S. E. Kesson.
Proc. Sixth Lunar Sci. Conference, (see 012.010), Vol. 1, 921 - 944 (1975).

094.465 Equilibrium relations among iron–titanium oxides in silicate melts: the system $CaMgSi_2O_6$-"FeO"–TiO_2 in equilibrium with metallic iron. B. R. Lipin, A. Muan.
Proc. Sixth Lunar Sci. Conference, (see 012.010), Vol. 1, 945 - 958 (1975).

094.466 The system anorthite–forsterite–fayalite–silica to 2 kbar with lunar petrologic applications. R. B. Merrill, R. J. Williams.
Proc. Sixth Lunar Sci. Conference, (see 012.010), Vol. 1, 959 - 971 (1975).

094.467 Liquid–solid equilibria involving spinel, ilmenite, and ferropseudobrookite in the system "FeO"–Al_2O_3-TiO_2 in contact with metallic iron. W. A. Schreifels, A. Muan.
Proc. Sixth Lunar Sci. Conference, (see 012.010), Vol. 1, 973 - 985 (1975).

094.468 Relative cooling rates of mare basalts at the Apollo 12 and 15 sites as estimated from pyroxene exsolution data. H. Takeda, M. Miyamoto, T. Ishii, G. E. Lofgren.
Proc. Sixth Lunar Sci. Conference, (see 012.010), Vol. 1, 987 - 996 (1975).

094.469 Experimentally reproduced textures and mineral chemistries of high-titanium mare basalts. T. M. Usselman, G. E. Lofgren, C. H. Donaldson, R. J. Williams.
Proc. Sixth Lunar Sci. Conference, (see 012.010), Vol. 1, 997 - 1020 (1975).

094.470 Petrogenesis of mare basalts: implications for chemical, mineralogical, and thermal models for the moon. M. J. O'Hara, D. J. Humphries, S. Waterston.
Proc. Sixth Lunar Sci. Conference, (see 012.010), Vol. 1, 1043 - 1055 (1975).

094.471 Lunar petrogenesis in a well-stirred magma ocean. J. A. Wood.
Proc. Sixth Lunar Sci. Conference, (see 012.010), Vol. 1, 1087 - 1102 (1975).

094.472 The partitioning of Mg, Fe, Sr, Ce, Sm, Eu, and Yb in lunar igneous systems and a possible origin of KREEP by equilibrium partial melting. D. F. Weill, G. A. McKay.
Proc. Sixth Lunar Sci. Conference, (see 012.010), Vol. 1, 1143 - 1158 (1975).

094.473 The lunar crust — chemically defined rock groups and their potassium–uranium fractionation. J. F. Lovering, D. A. Wark.
Proc. Sixth Lunar Sci. Conference, (see 012.010), Vol. 2, 1203 - 1217 (1975).

094.474 Origin of 78235, a lunar norite cumulate. S. R. Winzer, D. F. Nava, R. K. L. Lum, S. Schuhmann, P. Schuhmann, J. A. Philpotts.
Proc. Sixth Lunar Sci. Conference, (see 012.010), Vol. 2, 1219 - 1229 (1975).

094.475 Dunite 72417: a chemical study and interpretation. J. C. Laul, R. A. Schmitt.
Proc. Sixth Lunar Sci. Conference, (see 012.010), Vol. 2, 1231 - 1254 (1975).

094.476 On the origin of high-Ti mare basalts. C.-y. Shih, L. A. Haskin, H. Wiesmann, B. M. Bansal, J. C. Brannon.
Proc. Sixth Lunar Sci. Conference, (see 012.010), Vol. 2, 1255 - 1285 (1975).

094.477 Sulfur abundances and distributions in mare basalts and their source magmas. E. K. Gibson, Jr., S. Chang, K. Lennon, G. W. Moore, G. W. Pearce.
Proc. Sixth Lunar Sci. Conference, (see 012.010), Vol. 2, 1287 - 1301 (1975).

094.478 Lithophile trace and major elements in Apollo 16 and 17 lunar samples. O. Müller.
Proc. Sixth Lunar Sci. Conference, (see 012.010), Vol. 2, 1303 - 1311 (1975).

094.479 New data on the chemistry of lunar samples: primary matter in the lunar highlands and the bulk composition of the moon. H. Wänke, H. Palme, H. Baddenhausen, G. Dreibus, E. Jagoutz, H. Kruse, C. Palme, B. Spettel, F. Teschke, R. Thacker.
Proc. Sixth Lunar Sci. Conference, (see 012.010), Vol. 2, 1313 - 1340 (1975).

094.480 The role of vaporization processes in lunar rock formation. A. V. Ivanov, K. P. Florensky (*Florenskij*).
Proc. Sixth Lunar Sci. Conference, (see 012.010), Vol. 2, 1341 - 1350 (1975).

094.481 Chemical studies of the lunar regolith with emphasis on zirconium and hafnium. W. D. Ehmann, L. L. Chyi, A. N. Garg, B. R. Hawke, M.-S. Ma, M. D. Miller, W. D. James, Jr., R. A. Pacer.
Proc. Sixth Lunar Sci. Conference, (see 012.010), Vol. 2, 1351 - 1361 (1975).

094.482 Chemical composition of rocks and soils returned by the Apollo 15, 16, and 17 missions. H. J. Rose, Jr., P. A. Baedecker, S. Berman, R. P. Christian, E. J. Dwornik, R. B. Finkelman, M. M. Schnepfe.
Proc. Sixth Lunar Sci. Conference, (see 012.010), Vol. 2, 1363 - 1373 (1975).

094.483 A model for the lunar anorthositic gabbro.

E. Schonfeld.
Proc. Sixth Lunar Sci. Conference, (see 012.010), Vol. 2, 1375 - 1384 (1975).

094.484 Trace-element chemistry and reducing capacity of size fractions from the Apollo 16 regolith.
R. B. Finkelman, P. A. Baedecker, R. P. Christian, S. Berman, M. M. Schnepfe, H. J. Rose, Jr.
Proc. Sixth Lunar Sci. Conference, (see 012.010), Vol. 2, 1385 - 1398 (1975).

094.485 Primordial radionuclide variations in the Apollo 15 and 17 deep core samples and in Apollo 17 igneous rocks and breccias.
J. S. Fruchter, L. A. Rancitelli, R. W. Perkins.
Proc. Sixth Lunar Sci. Conference, (see 012.010), Vol. 2, 1399 - 1406 (1975).

094.486 Primordial and cosmogenic radionuclides in Descartes and Taurus-Littrow materials: extension of studies by nondestructive γ-ray spectrometry.
J. S. Eldridge, G. D. O'Kelley, K. J. Northcutt.
Proc. Sixth Lunar Sci. Conference, (see 012.010), Vol. 2, 1407 - 1418 (1975).

094.487 Sm–Nd age and history of Apollo 17 basalt 75075: evidence for early differentiation of the lunar exterior.
G. W. Lugmair, N. B. Scheinin, K. Marti.
Proc. Sixth Lunar Sci. Conference, (see 012.010), Vol. 2, 1419 - 1429 (1975).

094.488 U–Th–Pb systematics of anorthositic gabbros 78155 and 77017 – implications for early lunar evolution.
P. D. Nunes, M. Tatsumoto, D. M. Unruh.
Proc. Sixth Lunar Sci. Conference, (see 012.010), Vol. 2, 1431 - 1444 (1975).

094.489 Rb–Sr ages and initial $^{87}Sr/^{86}Sr$ for Apollo 17 basalts and KREEP basalt 15386.
L. E. Nyquist, B. M. Bansal, H. Wiesmann.
Proc. Sixth Lunar Sci. Conference, (see 012.010), Vol. 2, 1445 - 1465 (1975).

094.490 Rb–Sr study of a lunar dunite and evidence for early lunar differentiates.
D. A. Papanastassiou, G. J. Wasserburg.
Proc. Sixth Lunar Sci. Conference, (see 012.010), Vol. 2, 1467 - 1489 (1975).

094.491 Pb loss from Apollo 17 glassy samples and Apollo 16 revisited.   P. D. Nunes.
Proc. Sixth Lunar Sci. Conference, (see 012.010), Vol. 2, 1491 - 1499 (1975).

094.492 More on Rb–Sr in lunar breccia 14321.
R. K. Mark, C. Lee-Hu, G. W. Wetherill.
Proc. Sixth Lunar Sci. Conference, (see 012.010), Vol. 2, 1501 - 1507 (1975).

094.493 The history of lunar bombardment inferred from $^{40}Ar-^{39}Ar$ dating of highland rocks.
G. Turner, P. H. Cadogan.
Proc. Sixth Lunar Sci. Conference, (see 012.010), Vol. 2, 1509 - 1538 (1975).

094.494 $^{39}Ar-^{40}Ar$ dating of lunar rocks: effects of grain size and neutron irradiation.
P. Horn, E. K. Jessberger, T. Kirsten, H. Richter.
Proc. Sixth Lunar Sci. Conference, (see 012.010), Vol. 2, 1563 - 1591 (1975).

094.495 $^{40}Ar-^{39}Ar$ dating of Apollo 16 and 17 rocks.
D. Phinney, S. B. Kahl, J. H. Reynolds.
Proc. Sixth Lunar Sci. Conference, (see 012.010), Vol. 2, 1593 - 1608 (1975).

094.496 Ancient meteoritic component in Apollo 17 boulders.   H. Higuchi, J. W. Morgan.
Proc. Sixth Lunar Sci. Conference, (see 012.010), Vol. 2, 1625 - 1651 (1975).

094.497 The source of sublimates on the Apollo 15 green and Apollo 17 orange glass samples.
C. Meyer, Jr., D. S. McKay, D. H. Anderson, P. Butler, Jr.
Proc. Sixth Lunar Sci. Conference, (see 012.010), Vol. 2, 1673 - 1699 (1975).

094.498 Volatiles on the surface of Apollo 15 green glass and trace-element distributions among Apollo 15 soils.
C.-L. Chou, W. V. Boynton, L. L. Sundberg, J. T. Wasson.
Proc. Sixth Lunar Sci. Conference, (see 012.010), Vol. 2, 1701 - 1727 (1975).

094.499 Cl and P_2O_5 systematics: clues to early lunar magmas.   S. Jovanovic, G. W. Reed, Jr.
Proc. Sixth Lunar Sci. Conference, (see 012.010), Vol. 2, 1737 - 1751 (1975).

094.500 Soil breccia relationships and vapor deposits on the moon.   S. Jovanovic, G. W. Reed, Jr.
Proc. Sixth Lunar Sci. Conference, (see 012.010), Vol. 2, 1753 - 1759 (1975).

094.501 Investigation of the carbon, hydrogen, oxygen, and silicon isotope and concentration relationships on the grain surfaces of a variety of lunar soils and in some Apollo 15 and 16 core samples.
S. Epstein, H. P. Taylor, Jr.
Proc. Sixth Lunar Sci. Conference, (see 012.010), Vol. 2, 1771 - 1798 = Div. Geol. Planet. Sci., California Inst. Techn., Pasadena, California, Contr. No. 2622 (1975).

094.502 Oxygen isotope fractionation in Apollo 17 rocks.
T. K. Mayeda, J. Shearer, R. N. Clayton.
Proc. Sixth Lunar Sci. Conference, (see 012.010), Vol. 2, 1799 - 1802 (1975).

094.503 $^{22}Na-^{26}Al$ studies of lunar regolith.
Y. Yokoyama, J.-L. Reyss, F. Guichard.
Proc. Sixth Lunar Sci. Conference, (see 012.010), Vol. 2, 1823 - 1843 (1975).

094.504 The isotopic composition of lithium, potassium, and rubidium in some Apollo 11, 12, 14, 15, and 16 samples.   E. L. Garner, L. A. Machlan, I. L. Barnes.
Proc. Sixth Lunar Sci. Conference, (see 012.010), Vol. 2, 1845 - 1855 (1975).

094.505 Krypton and xenon in Apollo 14 samples: fission and neutron capture effects in gas-rich samples.
R. Drozd, C. Hohenberg, C. Morgan.
Proc. Sixth Lunar Sci. Conference, (see 012.010), Vol. 2, 1857 - 1877 (1975).

094.506 The saturated activities of ^{22}Na, ^{54}Mn, and ^{56}Co and the depth of sampling of soils.
J. E. Keith, R. S. Clark, L. J. Bennett.
Proc. Sixth Lunar Sci. Conference, (see 012.010), Vol. 2, 1879 - 1890 (1975).

094.507 Cosmogenic isotope production in Apollo deep-core

samples.   L. A. Rancitelli, J. S. Fruchter, W. D. Felix, R. W. Perkins, N. A. Wogman.
Proc. Sixth Lunar Sci. Conference, (see 012.010), Vol. 2, 1891 - 1899 (1975).

094.508 Effects of exposure conditions on cosmic-ray records in lunar rocks.
S. K. Bhattacharya, N. Bhandari.
Proc. Sixth Lunar Sci. Conference, (see 012.010), Vol. 2, 1901 - 1912 (1975).

094.509 The surface radioactivity of lunar rocks: implications to solar activity in the past.
N. Bhandari, S. K. Bhattacharya, J. T. Padia.
Proc. Sixth Lunar Sci. Conference, (see 012.010), Vol. 2, 1913 - 1925 (1975).

094.510 600-MeV proton bombardment of an artificial lunar soil core: some implications for the situation on the lunar surface.   W. A. Kaiser, G. Damm, U. Herpers, W. Herr,
H. Kulus, R. Michel, K. P. Rösner, K. Thiel, H. Weigel.
Proc. Sixth Lunar Sci. Conference, (see 012.010), Vol. 2, 1927 - 1951 (1975).

094.511 Trapped xenon in lunar anorthositic breccia 60015.
D. A. Leich, S. Niemeyer.
Proc. Sixth Lunar Sci. Conference, (see 012.010), Vol. 2, 1953 - 1965 (1975).

094.512 Solar-wind-trapped and cosmic-ray-produced noble gases in Luna 20 soil.   O. Eugster, P. Eberhardt,
J. Geiss, N. Grögler, M. Jungck, M. Mörgeli.
Proc. Sixth Lunar Sci. Conference, (see 012.010), Vol. 2, 1989 - 2007 (1975).

094.513 Rare gases in Apollo 17 soils with emphasis on analysis of size and mineral fractions of soil 74241.
W. Hübner, T. Kirsten, J. Kiko.
Proc. Sixth Lunar Sci. Conference, (see 012.010), Vol. 2, 2009 - 2026 (1975).

094.514 Rare gases and Ca, Sr, and Ba in Apollo 17 drill-core fines.   R. O. Pepin, J. C. Dragon, N. L.
Johnson, A. Bates, M. R. Coscio, Jr., V. R. Murthy.
Proc. Sixth Lunar Sci. Conference, (see 012.010), Vol. 2, 2027 - 2055 (1975).

094.515 Noble gas studies on grain size separates of Apollo 15 and 16 deep drill cores.
D. D. Bogard, W. C. Hirsch.
Proc. Sixth Lunar Sci. Conference, (see 012.010), Vol. 2, 2057 - 2083 (1975).

094.516 Rare gases in etched 10084 ilmenite: a search for trapped solar-flare rare gases.
D. A. Leich, S. Niemeyer, R. S. Rajan, B. Srinivasan.
Proc. Sixth Lunar Sci. Conference, (see 012.010), Vol. 2, 2085 - 2096 (1975).

094.517 On the origin of helium, neon, and argon isotopes in sieved mineral separates from an Apollo 15 soil.
U. Frick, H. Baur, H. Ducati, H. Funk, D. Phinney, P. Signer.
Proc. Sixth Lunar Sci. Conference, (see 012.010), Vol. 2, 2097 - 2129 (1975).

094.518 Nitrogen abundances and isotopic compositions in lunar samples.   R. H. Becker, R. N. Clayton.
Proc. Sixth Lunar Sci. Conference, (see 012.010), Vol. 2, 2131 - 2149 (1975).

094.519 Fluorine surface films on lunar samples: evidence
for both lunar and terrestrial origins.
R. H. Goldberg, D. S. Burnett, T. A. Tombrello.
Proc. Sixth Lunar Sci. Conference, (see 012.010), Vol. 2, 2189 - 2200 (1975).

094.520 Inert gases in fines at three levels of the trench at Van Serg Crater.   J. L. Jordan, D. Heymann.
Proc. Sixth Lunar Sci. Conference, (see 012.010), Vol. 2, 2201 - 2218 (1975).

094.521 Exposure histories of Bench Crater rocks.
D. S. Burnett, R. J. Drozd, C. J. Morgan, F. A. Podosek.
Proc. Sixth Lunar Sci. Conference, (see 012.010), Vol. 2, 2219 - 2240 (1975).

094.522 Mixing and transport of lunar surface materials: evidence obtained by the determination of lithophile,
siderophile, and volatile elements.   W. V. Boynton, P. A. Baedecker, C.-L. Chou, K. L. Robinson, J. T. Wasson.
Proc. Sixth Lunar Sci. Conference, (see 012.010), Vol. 2, 2241 - 2259 (1975).

094.523 A comparison of noble gases in lunar fines and soil breccias: implications for the origin of soil breccias.
H. Hintenberger, L. Schultz, H. W. Weber.
Proc. Sixth Lunar Sci. Conference, (see 012.010), Vol. 2, 2261 - 2270 (1975).

094.524 Agglutinates: role in element and isotope chemistry and inferences regarding volatile-rich rock 66095 and glass 74220.
R. O. Allen, Jr., S. Jovanovic, G. W. Reed, Jr.
Proc. Sixth Lunar Sci. Conference, (see 012.010), Vol. 2, 2271 - 2279 (1975).

094.525 Agglutinates as indicators of lunar soil maturity: the rare gas evidence at Apollo 16.
M. P. Charette, J. B. Adams.
Proc. Sixth Lunar Sci. Conference, (see 012.010), Vol. 2, 2281 - 2289 (1975).

094.526 Chemistry of agglutinate fractions in lunar soils.
J. M. Rhodes, J. B. Adams, D. P. Blanchard, M. P. Charette, K. V. Rodgers, J. W. Jacobs, J. C. Brannon, L. A. Haskin.
Proc. Sixth Lunar Sci. Conference, (see 012.010), Vol. 2, 2291 - 2307 (1975).

094.527 Interpretation of the compositional variability of Apollo 15 soils.   A. R. Duncan, M. K. Sher, Y. C.
Abraham, A. J. Erlank, J. P. Willis, L. H. Ahrens.
Proc. Sixth Lunar Sci. Conference, (see 012.010), Vol. 2, 2309 - 2320 (1975).

094.528 A geochemical and petrographic study of 1—2-mm fines from Apollo 17.
D. P. Blanchard, R. L. Korotev, J. C. Brannon, J. W. Jacobs, L. A. Haskin, A. M. Reid, C. H. Donaldson, R. W. Brown.
Proc. Sixth Lunar Sci. Conference, (see 012.010), Vol. 2, 2321 - 2341 (1975).

094.529 High-voltage electron microscope observations of finely divided iron droplets in glassy agglutinates.
C. T. Pillinger, P. R. Davis, L. R. Gardiner, H. Naziri, P. E. Champness.
Proc. Sixth Lunar Sci. Conference, (see 012.010), Vol. 2, 2343 - 2352 (1975).

094.530 Evolution of carbon isotopes, agglutinates, and the lunar regolith.

D. J. DesMarais, A. Basu, J. M. Hayes, W. G. Meinschein.
Proc. Sixth Lunar Sci. Conference, (see 012.010), Vol. 2, 2353 - 2373 (1975).

094.531 **Crater studies in the Apollo 17 region.**
B. K. Lucchitta, A. G. Sanchez.
Proc. Sixth Lunar Sci. Conference, (see 012.010), Vol. 3, 2427 - 2441 (1975).

094.532 **Origin of the Taurus-Littrow massifs.**
V. S. Reed, E. W. Wolfe.
Proc. Sixth Lunar Sci. Conference, (see 012.010), Vol. 3, 2443 - 2461 (1975).

094.533 **Geology of the Taurus-Littrow valley floor.**
E. W. Wolfe, B. K. Lucchitta, V. S. Reed, G. E. Ulrich, A. G. Sanchez.
Proc. Sixth Lunar Sci. Conference, (see 012.010), Vol. 3, 2463 - 2482 (1975).

094.534 **Geology of the Apollo 14 region (Fra Mauro): stratigraphic history and sample provenance.**
J. W. Head, B. R. Hawke.
Proc. Sixth Lunar Sci. Conference, (see 012.010), Vol. 3, 2483 - 2501 (1975).

094.535 **Geomorphology of crater and basin deposits – emplacement of the Fra Mauro Formation.**
R. H. Morrison, V. R. Oberbeck.
Proc. Sixth Lunar Sci. Conference, (see 012.010), Vol. 3, 2503 - 2530 (1975).

094.536 **Photogeology and basin configuration of Mare Smythii.**
H. E. Stewart, J. D. Waskom, R. A. DeHon.
Proc. Sixth Lunar Sci. Conference, (see 012.010), Vol. 3, 2541 - 2551 (1975).

094.537 **Mare Spumans and Mare Undarum: mare thickness and basin floor.** R. A. DeHon.
Proc. Sixth Lunar Sci. Conference, (see 012.010), Vol. 3, 2553 - 2561 (1975).

094.538 **Volcanic and tectonic evolution of crater Goclenius, western Mare Fecunditatis.**
W. B. Bryan, P. A. Jezek, M.-L. Adams.
Proc. Sixth Lunar Sci. Conference, (see 012.010), Vol. 3, 2563 - 2569 (1975).

094.539 **Geochemical and geological units of Mare Humorum: definition using remote sensing and lunar sample information.**
C. Pieters, J. W. Head, T. B. McCord, J. B. Adams, S. Zisk.
Proc. Sixth Lunar Sci. Conference, (see 012.010), Vol. 3, 2689 - 2710 (1975).

094.540 **Polarimetric properties of the lunar surface and its interpretation: Part 7 – Other solar system objects.**
A. Dollfus, J. E. Geake.
Proc. Sixth Lunar Sci. Conference, (see 012.010), Vol. 3, 2749 - 2768 (1975).

094.541 **Magnetic stratigraphy of the Apollo 15 deep drill core.** W. A. Gose, G. W. Pearce, J. F. Lindsay.
Proc. Sixth Lunar Sci. Conference, (see 012.010), Vol. 3, 3071 - 3080 (1975).

094.542 **On the high-temperature stability of magnetite: implications for lunar soil magnetism.**
D. L. Griscom.

Proc. Sixth Lunar Sci. Conference, (see 012.010), Vol. 3, 3081 - 3089 (1975).

094.543 **The effects of heating and subsolidus reduction on lunar materials: an analysis by magnetic methods, optical, Mössbauer, and X-ray diffraction spectroscopy.**
A. Brecher, W. H. Menke, J. B. Adams, M. J. Gaffey.
Proc. Sixth Lunar Sci. Conference, (see 012.010), Vol. 3, 3091 - 3109 (1975).

094.544 **Effects of meteorite impact on magnetic properties of Apollo lunar materials.** T. Nagata, R. M. Fisher, F. C. Schwerer, M. D. Fuller, J. R. Dunn.
Proc. Sixth Lunar Sci. Conference, (see 012.010), Vol. 3, 3111 - 3122 (1975).

094.545 **Magnetic effects of shock and their implications for magnetism of lunar samples.**
S. M. Cisowski, M. D. Fuller, Y. M. Wu, M. F. Rose, P. J. Wasilewski.
Proc. Sixth Lunar Sci. Conference, (see 012.010), Vol. 3, 3123 - 3141 (1975).

094.546 **Petrographic and ferromagnetic resonance studies of experimentally shocked regolith analogs.**
R. V. Gibbons, R. V. Morris, F. Hörz, T. D. Thompson.
Proc. Sixth Lunar Sci. Conference, (see 012.010), Vol. 3, 3143 - 3171 (1975).

094.547 **Ferromagnetic resonance as a method of studying the micrometeorite bombardment history of the lunar surface.** R. M. Housley, E. H. Cirlin, I. B. Goldberg, H. Crowe, R. A. Weeks, R. Perhac.
Proc. Sixth Lunar Sci. Conference, (see 012.010), Vol. 3, 3173 - 3186 (1975).

094.548 **Lunar and terrestrial sample photoconductivity.**
R. Alvarez.
Proc. Sixth Lunar Sci. Conference, (see 012.010), Vol. 3, 3187 - 3197 (1975).

094.549 **Specific gravities of lunar materials using helium pycnometry.** D. A. Cadenhead, J. R. Stetter.
Proc. Sixth Lunar Sci. Conference, (see 012.010), Vol. 3, 3199 - 3206 (1975).

094.550 **Thermal diffusivity of lunar rock sample 12002,85.**
K.-i. Horai, J. L. Winkler, Jr.
Proc. Sixth Lunar Sci. Conference, (see 012.010), Vol. 3, 3207 - 3215 (1975).

094.551 **Internal friction quality factor $Q \geqslant 3100$ achieved in lunar rock 70215,85.**
B. R. Tittmann, J. M. Curnow, R. M. Housley.
Proc. Sixth Lunar Sci. Conference, (see 012.010), Vol. 3, 3217 - 3226 (1975).

094.552 **Characteristics of microcracks in lunar samples.**
G. Simmons, R. Siegfried, D. Richter.
Proc. Sixth Lunar Sci. Conference, (see 012.010), Vol. 3, 3227 - 3254 (1975).

094.553 **Correlation of elastic moduli systematics with texture in lunar materials.**
N. Warren, R. Trice.
Proc. Sixth Lunar Sci. Conference, (see 012.010), Vol. 3, 3255 - 3268 = Inst. Geophys. Planet. Phys., Univ. California, Los Angeles, California, Contr. No. 1475 (1975).

094.554 **ESCA studies of lunar surface chemistry.**

R. M. Housley, R. W. Grant.
Proc. Sixth Lunar Sci. Conference, (see 012.010), Vol. 3,
3269 - 3275 (1975).

094.555 **ESCA studies on solar-wind reduction mechanisms.**
L. Yin, T. Tsang, I. Adler.
Proc. Sixth Lunar Sci. Conference, (see 012.010), Vol. 3,
3277 - 3284 (1975).

094.556 **Auger analysis of the lunar soil: study of processes
which change the surface chemistry and albedo.**
T. Gold, E. Bilson, R. L. Baron.
Proc. Sixth Lunar Sci. Conference, (see 012.010), Vol. 3,
3285 - 3303 (1975).

094.557 **Blocking of the water–lunar fines reaction by air and
water concentration effects.**
R. B. Gammage, H. F. Holmes.
Proc. Sixth Lunar Sci. Conference, (see 012.010), Vol. 3,
3305 - 3316 (1975).

094.558 **Water vapor weathering of Taurus-Littrow orange
soil: a pore-structure analysis.**
D. A. Cadenhead, R. S. Mikhail.
Proc. Sixth Lunar Sci. Conference, (see 012.010), Vol. 3,
3317 - 3331 (1975).

094.559 **Effects of water on electrical properties of lunar
fines.**
G. R. Olhoeft, D. W. Strangway, G. W. Pearce.
Proc. Sixth Lunar Sci. Conference, (see 012.010), Vol. 3,
3333 - 3342 (1975).

094.560 **Interaction of gases with lunar materials: revised
results for Apollo 11.**
H. F. Holmes, R. B. Gammage.
Proc. Sixth Lunar Sci. Conference, (see 012.010), Vol. 3,
3343 - 3350 (1975).

094.561 **Microcrater investigations on lunar rock 12002.**
J. B. Hartung, F. Hodges, F. Hörz, D. Storzer,
Proc. Sixth Lunar Sci. Conference, (see 012.010), Vol. 3,
3351 - 3371 (1975).

094.562 **Studies of solar flares and impact craters in partially
protected crystals.**   D. A. Morrison, E. Zinner.
Proc. Sixth Lunar Sci. Conference, (see 012.010), Vol. 3,
3373 - 3390 (1975).

094.563 **Solar-wind sputter erosion of microcrater populations on the lunar surface.**
J. A. M. McDonnell, W. C. Carey.
Proc. Sixth Lunar Sci. Conference, (see 012.010), Vol. 3,
3391 - 3402 (1975).

094.564 **Microcraters observed on 15015 breccia and micrometeoroid flux.**   J. C. Mandeville.
Proc. Sixth Lunar Sci. Conference, (see 012.010), Vol. 3,
3403 - 3408 (1975).

094.565 **Dependencies of microcrater formation on impact
parameters.**   K. Nagel, G. Neukum, G. Eichhorn,
H. Fechtig, O. Müller, E. Schneider.
Proc. Sixth Lunar Sci. Conference, (see 012.010), Vol. 3,
3417 - 3432 (1975).

094.566 **Surface exposure history of individual crystals in
the lunar regolith.**
G. Poupeau, R. M. Walker, E. Zinner, D. A. Morrison.
Proc. Sixth Lunar Sci. Conference, (see 012.010), Vol. 3,
3433 - 3448 (1975).

094.567 **Track studies bearing on solar-system regoliths.**
P. B. Price, I. D. Hutcheon, D. Braddy, D. Macdougall.
Proc. Sixth Lunar Sci. Conference, (see 012.010), Vol. 3,
3449 - 3469 (1975).

094.568 **Solar-wind and solar-flare maturation of the lunar
regolith.**   J. P. Bibring, J. Borg, A. L. Burlingame,
Y. Langevin, M. Maurette, B. Vassent.
Proc. Sixth Lunar Sci. Conference, (see 012.010), Vol. 3,
3471 - 3493 (1975).

094.569 **Lunar regolith and gas-rich meteorites:characteriza-
tion based on particle tracks and grain-size distributions.**   S. K. Bhattacharya, J. N. Goswami, D. Lal, P. P.
Patel, M. N. Rao.
Proc. Sixth Lunar Sci. Conference, (see 012.010), Vol. 3,
3509 - 3526 (1975).

094.570 **The fission track record of Apennine Front KREEP
basalts.**
E. L. Haines, I. D. Hutcheon, J. R. Weiss.
Proc. Sixth Lunar Sci. Conference, (see 012.010), Vol. 3,
3527 - 3540 (1975).

094.571 **Compositional studies of solar-iron group and
heavier nuclei of kinetic energy below 10 MeV/nu-
cleon.**   J. N. Goswami, D. Lal.
Proc. Sixth Lunar Sci. Conference, (see 012.010), Vol. 3,
3541 - 3555 (1975).

094.572 **Long-term differential energy spectrum for solar-
flare iron-group particles.**
G. E. Blanford, R. M. Fruland, D. A. Morrison.
Proc. Sixth Lunar Sci. Conference, (see 012.010), Vol. 3,
3557 - 3576 (1975).

094.573 **The feasibility of ion identification on cosmic-ray
tracks in lunar feldspars.**
W. Krätschmer, W. Gentner.
Proc. Sixth Lunar Sci. Conference, (see 012.010), Vol. 3,
3577 - 3585 (1975).

094.574 **Crystal chemistry of Pu and U and concordant fission track ages of lunar zircons and whitlockites.**
D. Braddy, I. D. Hutcheon, P. B. Price.
Proc. Sixth Lunar Sci. Conference, (see 012.010), Vol. 3,
3587 - 3600 (1975).

094.575 **Ion-probe studies of artificially implanted ions in
lunar samples.**   E. Zinner, R. M. Walker.
Proc. Sixth Lunar Sci. Conference, (see 012.010), Vol. 3,
3601 - 3617 (1975).

094.576 **Annealing and etching studies of fossil and fresh
tracks in lunar and analogous crystals.**
R. K. Bull, S. A. Durrani.
Proc. Sixth Lunar Sci. Conference, (see 012.010), Vol. 3,
3619 - 3637 (1975).

094.577 **Element distribution in size fractions of Apollo-16
soils: evidence for element mobility during regolith
processes.**   W. V. Boynton, C.-L. Chou, R. W. Bild,
P. A. Baedecker, J. T. Wasson.
Earth Planet. Sci. Letters, Vol. 29, 21 - 33 (1976).

094.578 **Chemistry and structure of lunar and synthetic
armalcolite.**
B. A. Wechsler, C. T. Prewitt, J. J. Papike.
Earth Planet. Sci. Letters, Vol. 29, 91 - 103 (1976).

094.579 Textural remanence: a new model of lunar rock magnetism. A. Brecher.
Earth Planet. Sci. Letters, Vol. 29, 131 - 145 (1976).
In reexamining the accumulated magnetic data on lunar rocks, several common patterns of magnetic behavior are recognized. Their joint occurrence strongly suggests a new model of lunar rock magnetism, which appeals only to partial preferred textural alignment of the spontaneous moments of magnetic grains, without requiring the existence of ancient lunar magnetic fields. The model is supported by a wide variety of direct and indirect evidence and its predictions (e.g. regarding anisotropic susceptibility and remanence acquisition) can be experimentally tested.

094.580 Lunar sample 15405: remnant of a KREEP basalt-granite differentiated pluton. G. Ryder.
Earth Planet. Sci. Letters, Vol. 29, 225 - 268 (1976).

094.581 Heterogeneity in titaniferous lunar basalts. D. Walker, J. Longhi, J. F. Hays.
Earth Planet. Sci. Letters, Vol. 30, 27 - 36 (1976).

094.582 Mare basalts petrogenesis in a dynamic moon. S. E. Kesson, A. E. Ringwood.
Earth Planet. Sci. Letters, Vol. 30, 155 - 163 (1976).

094.583 Elastic properties of anorthite and the nature of the lunar crust. R. C. Liebermann, A. E. Ringwood.
Earth Planet. Sci. Letters, Vol. 31, 69 - 74 (1976).

094.584 A new type of chondritic meteorite found in lunar soil. H. Y. McSween, Jr.
Earth Planet. Sci. Letters, Vol. 31, 193 - 199 (1976).
A fragment found in soil from the Apollo 12 site (12037, from the rim of Bench Crater) appears to be a unique type of chondrite, petrologically and chemically distinct from other chondrites and lunar rocks. The bulk composition of this sample has high Mg/Si and low Fe/Si relative to other chondrites, and P and S are strongly enriched.

094.585 Lateral displacement in the Humorum region. H. D. Tjia.
Phys. Earth Planet. Interiors, Vol. 11, 207 - 215 (1976).

094.586 Fossil tracks in lunar samples: informations on the abundances of VH nuclei in the ancient cosmic radiation. W. Krätschmer, W. Gentner.
14th Intern. Cosmic Ray Conf., (see 012.011), Vol. 1, 296 - 299 (1975).

094.587 Comparative characteristics of the distribution of sizes of lunar soil particles at the landing sites of the automatic lunar stations Luna 16 and Luna 20.
Yu. I. Stakheev, A. V. Ivanov, E. K. Vul'fson.
Kosmich. Issled., Vol. 14, 428 - 434 (1976). In Russian.

094.588 Radar experiments aboard the automatic lunar stations.
N. N. Krupenio, E. G. Ruzskij, V. V. Cherkasov.
Kosmich. Issled., Vol. 14, 460 - 461 (1976). In Russian.
Brief information.

094.589 Diffusion artifacts in dating by stepwise thermal release of rare gases. J. C. Huneke.
Earth Planet. Sci. Letters, Vol. 28, 407 - 417 (1976).

094.590 Petrology of lunar rocks and implication to lunar evolution. W. I. Ridley.
Annual Rev. Earth Planet. Sci., Vol. 4, (see 003.012), 15 - 48 (1976).

094.591 Methodische Untersuchungen zur ^{40}Ar-^{39}Ar-Datierung von Mondgesteinen.
P. Horn, E. K. Jessberger, T. Kirsten.
MPI Kernphys., Heidelberg, Jahresbericht 1975, p. 146 - 147.

094.592 Messung von Tiefenprofilen des Edelgases ^{4}He mit hoher Auflösung.
H. W. Müller, J. Kiko, T. Kirsten.
MPI Kernphys., Heidelberg, Jahresbericht 1975, p. 147 - 148.

094.593 Diffusionsmessungen an simulierten Mondproben. J. Deubner, T. Kirsten.
MPI Kernphys., Heidelberg, Jahresbericht 1975, p. 148 - 150.

094.594 Stickstoffgehalte in lunaren magmatischen Gesteinen. O. Müller, E. Grallath, G. Tölg.
MPI Kernphys., Heidelberg, Jahresbericht 1975, p. 150.

094.595 Bilanz der Reduzierungsprozesse auf dem Mond. A: El Goresy, P. Ramdohr.
MPI Kernphys., Heidelberg, Jahresbericht 1975, p. 150 - 151.

094.596 Untersuchung von Einschlagskrater-Populationen. G. Neukum, B. König, H. Fechtig.
MPI Kernphys., Heidelberg, Jahresbericht 1975, p. 157 - 159.

094.597 Peculiarities of the structure of the lunar highlands iron. R. I. Mints, T. M. Petukhova, V. I. Grokhovskij, E. V. Krivopishina, V. P. Shaldybin.
Meteoritika, vyp. (No.) 35, p. 91 - 96 (1976). In Russian.

094.598 Powder studies of lunar soils. J. I. Langford.
Acta Crystallogr. A, Vol. A31, S201 (1975).
Abstr. in Phys. Abstr., Vol. 79, A015244 (1976).

094.599 Mössbauer and magnetic measurements of iron phase distributions in Apollo lunar samples.
G. P. Huffman, F. C. Schwerer.
AIP Conf. Proc. No. 24, p. 760 (1974). − Abstr. in Phys. Abstr., Vol. 79, A015245 (1976).

094.600 The petrochemistry of the lunar samples. On the differentiation trend of the lunar rocks.
M. Shima, S. Yabuki, A. Okada.
Sci. Papers Inst. Phys. Chem. Res., Vol. 69, No. 2, p. 46 - 49 (1975). − Abstr. in Phys. Abstr., Vol. 79, A023693 (1976).

094.601 Rocks 60618 and 65785: evidence for admixture of KREEP in lunar impact melts. K. Keil, R. D. Warner, M. Prinz, E. Dowty.
Geophys. Res. Letters, Vol. 2, 369 - 372 (1975). − Abstr. in Phys. Abstr., Vol. 79, A027816 (1976).

094.602 FMR (*ferromagnetic resonance*) thermomagnetic studies up to 900°C of lunar soils and potential magnetic analogues. R. V. Morris, R. V. Gibbons, F. Horz.
Geophys. Res. Letters, Vol. 2, 461 - 464 (1975). − Abstr. in Phys. Abstr., Vol. 79, A027817 (1976).

094.603 Mineralogy of lunar norite 78235; second lunar occurrence of $P2_1ca$ pyroxene from Apollo 17 soils.
I. M. Steele.
American Mineral., Vol. 60, 1086 - 1091 (1975). − Abstr. in Phys. Abstr., Vol. 79, A032801 (1976).

094.604 On the stratified rocks of the lunar Apennines photographed by the Apollo 15 astronauts.
P. Leonardi.
Modern Geol., Vol. 4, 245 - 252 (1973). − Abstr. in Phys. Abstr., Vol. 79, A037495 (1976).

094.605 **Metallography of a lunar iron fragment taken from the moon by the Soviet probe Luna-20.**
R. I. Mints, T. M. Petukhova, V. I. Grokhovskii (*Grokhovskij*), V. P. Shaldybin.
Metalloved. i term. obrab. metallov, Vol. 17, No. 1, p. 2 - 5 (1975). − Abstr. in Phys. Abstr., Vol. 79, A037496 (1976).

094.606 **Comments on the shape of pores induced in lunar fines by adsorbed water.**
H. F. Holmes, R. B. Gammage.
Journ. Colloid Interface Sci., Vol. 54, No. 1, p. 151 - 153 (1976). − Abstr. in Phys. Abstr., Vol. 79, A042085 (1976).

094.607 **Phase chemistry of Apollo 14 soil sample 14259.**
F. K. Aitken, D. H. Anderson, M. N. Bass, R. W. Brown, P. Butler, Jr., G. Heiken, P. Jakes, A. M. Reid, W. I. Ridley, H. Takeda, J. Warner, R. J. Williams.
Modern Geol., Vol. 5, 1 - 13 (1974). − Abstr. in Phys. Abstr., Vol. 79, A042104 (1976).

094.608 **Krypton and xenon in lunar and terrestrial samples.**
R. J. Drozd.
Thesis Washington Univ., Seattle, USA, 366 pp. (1974). (Available from: Univ. Microfilms, Order No. 75-6590).

094.609 **Radioactive halos and ion microprobes measurement of Pb isotope ratios.** Final report.
R. V. Gentry.
Separate print Oak Ridge National Lab., Tennessee, USA, 2 pp. (1974).

Lunar soil science: physicomechanical properties of lunar soils. See Abstr. 003.030.

Microcraters produced by oblique incidence of projectiles. See Abstr. 022.082.

Enhanced solar cosmic-ray flux in the early solar system. See Abstr. 078.057.

Shape of the solar flare proton spectrum in 10–100 MeV region in the past. See Abstr. 078.096.

Evolution of the moon between 4.6 and 3.3 AE. See Abstr. 094.135.

Relative ages of flow units in Mare Imbrium and Sinus Iridum. See Abstr. 094.148.

Correlation of Al/Si X-ray fluorescence data with other remote sensing data from the Taurus-Littrow area. See Abstr. 094.155.

Energy, frequency, and distance of moonquakes at the Apollo 17 site. See Abstr. 094.161.

Lithification of gas-rich meteorites. See Abstr. 105.071.

Micrometeoroids and the hyper-velocity impact. See Abstr. 105.125.

Density, chemistry, and size distribution of interplanetary dust. See Abstr. 106.040.

Density and composition of interplanetary dust particles. See Abstr. 106.043.

095 Lunar Eclipses

095.001 **Photométrie visuelle des éclipses totales de lune et mesure du cône d'ombre terrestre.** G. Florsch.
L'Astronomie, Vol. 90, 3 - 24 (1976).

095.002 **Observation of the total lunar eclipse of 18–19 November 1975.** J. Bouška, V. Kováč, L. Kováč, L. Kulčár, V. Wagner, M. Dujnič, L. Hurta.
Říše hvězd, Vol. 57, 53 - 55 (1976). In Czech.

095.003 **November's lunar eclipse: an analysis.** J. Ashbrook.
Sky Telescope, Vol. 51, 76 - 78, 90 (1976).

095.004 **Totaal verduisterde maan bleef opvallend helder.**
Zenit, 3e jaargang, p. 45 - 48 (1976). − Concerning lunar eclipse 1975 November 18.

095.005 **Periodicities of eclipses.** G. P. Können, J. Meeus.
Journ. Roy. Astron. Soc. Canada, Vol. 70, 81 - 83 (1976).

095.006 **Observation of the lunar eclipse of 18–19 November 1975.** L. Hric.
Kozmos, Vol. 7, 61 - 62 (1976). In Slovak.

095.007 **Observation of the lunar eclipse of 18–19 November 1975.** T. Dujničová, P. Rapavý.
Kozmos, Vol. 7, 26 - 27 (1976). In Slovak.

A transatlantic eclipse of the moon.
Sky Telescope, Vol. 51, 62 - 65 (1976). − Concerning 1975 November 18 - 19.

The eclipse in color.
Sky Telescope, Vol. 51, 79 - 82 (1976). − Concerning 1975 November 18 - 19.

A study of records of the solar and lunar eclipses in scripts on tortoise-shells or ox-bones.
See Abstr. 004.009.

096 Lunar Occultations

096.001 Occultation highlights for the year 1976.
D. W. Dunham.
Sky Telescope, Vol. 51, 66 - 67 (1976).

096.002 Two grazing occultations on 9th November 1975.
A. G. F. Morrisby.
Monthly Notes Astron. Soc. Southern Africa, Vol. 35, 9 - 13 (1976).

096.003 Results of photoelectric lunar occultation observations obtained at the Hamburg Observatory during 1969–1973. C. de Vegt, U. K. Gehlich.
Astron. Astrophys., Vol. 48, 245 - 252 (1976).
Observations of lunar occultations have been obtained at the Hamburg Observatory with the 60 cm f : 15 refractor. A total number of 195 observations, including four occultations of the Pleiades stars, has been recorded, 49 of which are reappearance events. For 11 stars a duplicity is indicated and 6 angular diameter determinations could be obtained.

096.004 Observations of lunar occultations at the Lvov Astronomical Observatory in 1971 - 1972.
I. I. Vovchik, D. I. Galych, Yu. V. Fridel'.
Tsirk. Astron. Obs. L'vov, No. 49, p. 32 - 33 (1974). In Russian.

096.005 Lunar occultation of Jupiter's Galilean satellites.
D. W. Dunham, F. Fekel, K. Aksnes.
IAU Circ., No. 2950 (1976).

096.006 Lunar occultation of Jupiter's Galilean satellites.
D. W. Dunham.
IAU Circ., No. 2961 (1976).

096.007 Lunar occultation of Jupiter's Galilean satellites.
D. W. Dunham.
IAU Circ., No. 2967 (1976).

096.008 Ocultación rasante de Spica el 11 de mayo de 1976.
J. Meeus.
Urania Barcelona, Año 60, No. 283, p. 49 - 50 (1976).

096.009 Ocultaciones de estrellas por la luna observadas en Madrid durante el año 1973. M. de Pascual.
Bol. Astron. Obs. Madrid, Vol. 9, No. 2, p. 85 (1976).

096.010 Occultation observations in 1974. T. Mori, Y. Ganeko, Y. Harada, M. Sasaki, M. Yamaguti.
Data Rep. Hydrographic Observations, Ser. Astron. Geod., *Tokyo*, No. 10, p. 1 - 41 (1976).

096.011 Lunar occultation of point X-ray sources.
C. Reppin.
ESRO colloquium X-ray astron. related topics, Noordwijk, Netherlands, 25 - 26 Feb. 1975, p. 119 - 127 (1975).

Asteroidal occultations.
See Abstr. 098.077.

Angular diameters of stars from lunar occultations.
See Abstr. 115.005.

Angular diameter of 31 Leonis from a lunar occultation. See Abstr. 115.010.

Radial-velocity measurements of the lunar-occultation binary HR 2013. See Abstr. 117.016.

Observation of the Crab nebula occultation by the moon on September 10, 1974. See Abstr. 134.005.

On the use of lunar occultations of radio sources for investigation of their angular structure. III.
See Abstr. 141.030.

Occultation of three Markarian galaxies at 327 MHz by the moon. See Abstr. 158.052.

097 Mars, Mars Satellites

Mars

097.001 The volcanoes of Mars. M. H. Carr.
Sci. American, Vol. 234, No. 1, p. 32 - 43 (1976).
Martian volcanoes are apparently formed over the entire span of the planet's history. Their remarkable size may be related to the fact that the crust of Mars is not divided into moving plates.

097.002 Distribution of small channels on the Martian surface. D. Pieri.
Icarus, Vol. 27, 25 - 50 (1976).
The distribution of small channels on Mars has been mapped from Mariner 9 images, at the 1 : 5 000 000 scale, by the author. The small channels referred to here are small valleys ranging in width from the resolution limit of the Mariner 9 wide-angle images (~ 1 km) to about 10 km. The greatest density of small channels occurs in dark cratered terrain. The generally degraded appearance of small channels in the high-resolution images (~ 100 m) imply a major episode of small-channel formation early in Martian geologic history.

097.003 Complex refractive index of Martian dust: Mariner 9 ultraviolet observations.
K. Pang, J. M. Ajello, C. W. Hord, W. G. Egan.
Icarus, Vol. 27, 55 - 67 (1976).
Mariner 9 ultraviolet spectra of the 1971 Mars dust storm were studied to determine the cloud particle size distribution and complex index of refraction. Preliminary results indicate that the effective particle radius is 1 μm with an effective variance (a measure of distribution width) $\gtrsim 0.2$. The real component of the index of refraction is $\gtrsim 1.8$ at both 268 and 305 nm. For the imaginary index, a value of 0.02 was found at 268 nm and 0.01 at 305 nm. Comparison of the Mars ultraviolet refractive indices with laboratory measurements indicates that none of the terrestrial analog samples of limonite, basalt, andesite, or montmorillonite have the required ultraviolet properties.

097.004 The evolution of the surface temperature of Mars.
A. Henderson-Sellers, A. J. Meadows.
Planet. Space Sci., Vol. 24, 41 - 44 (1976).
Changes in the average surface temperature of Mars are studied as a function of the time that has elapsed since the origin of the planet. Time variations in the factors influencing the surface temperature are investigated: and approximate methods for computing the effect of such variations are discussed. Three possible degassing sequences are postulated, and their likely effects for the presence of liquid water on the Martian surface are assessed.

097.005 Diurnal behaviour of water on Mars.
F. M. Flasar, R. M. Goody.
Planet. Space Sci., Vol. 24, 161 - 181 (1976).
The authors have developed a numerical model of the diurnal transport of water across the Martian surface. The atmospheric boundary layer is modelled in terms of local radiative-convective processes. The radiative effects of ice fogs near the surface are included in the model. The diffusion of water in the ground is treated for the cases of adsorption and condensation.

097.006 The determination of the rheological properties and effusion rate of an Olympus Mons lava. G. Hulme.
Icarus, Vol. 27, 207 - 213 (1976).
A new technique for the interpretation of lava flow morphology was applied to a lava flow on Olympus Mons. The yield stress of the flowing lava was determined subject to uncertainties in the estimates of the slope of Olympus Mons. The lava is most probably more silicic than the basaltic lavas of the Hawaiian shield volcanoes and its effusion rate appears to have been greater than those of typical Hawaiian flows.

097.007 Martian glaciation and the flow of solid CO_2.
B. R. Clark, R. P. Mullin.
Icarus, Vol. 27, 215 - 228 (1976).
The flow law determined experimentally for solid CO_2 establishes that a hypothesis of glacial flow of CO_2 at the Martian poles is not physically unrealistic. A plausible glacial model for the Martian polar caps can be constructed and is helpful in explaining the unique character of the polar regions. CO_2-rich layers deposited near the pole would have flowed outward laterally to relieve high internal shear stresses. The topography of the polar caps, the uniform layering of the layered deposits, and the general extent of the polar 'sediments' could all be explained using this model. Flow of CO_2 rather than water ice greatly reduces the problems with Martian glaciation.

097.008 Variable features on Mars. VI. An unusual crater streak in Mesogaea. J. Veverka, C. Sagan, R. Greeley.
Icarus, Vol. 27, 241 - 253 (1976).
An unusual, prominent dark streak located in Mesogaea (near 8°N, 191°W) is described. Its appearance is unlike that of most dark streaks on Mars, many of which have ragged outlines, are variable on short time-scales, and are presumed to be erosional. The Mesogaea streak has a tapered, smooth outline, and no changes within it were observed. Two possible origins for the dark material are suggested: (1) deflation from a recently exposed, relatively unconsolidated subsurface deposit, and (2) production of ash by a volcanic vent.

097.009 The geology and geophysics of Mars.
R. S. Saunders.
Endeavour, No. 124, Vol. 35, 15 - 20 (1976).
The exploration of Mars by orbiting spacecraft — as a preliminary to a landing — has yielded a wealth of new knowledge about the configuration and structure of the planet. Ideas that prevailed up to 1965, ascribing to Mars many earth-like features, have had to be radically revised. This article reviews new knowledge gained predominantly from the Mariner series of space flights.

097.010 The air mass on the beam in a spherical model of the Martian atmosphere.
V. D. Davydov, E. B. Shesterikova.
Kosmich. Issled., Vol. 14, 61 - 64 (1976). In Russian.

097.011 Radioastronomical measurements aboard the automatic interplanetary station Mars 5.
A. E. Basharinov, Yu. N. Vetukhnovskaya, V. N. Galaktionov, S. T. Egorov, M. A. Kolosov, N. N. Krupenio, A. D. Kuz'min, V. A. Ladygin, L. I. Malafeev, E. I. Omel'chenko, V. S. Troitskij, N. Ya. Shapirovskaya, A. M. Shutko.
Kosmich. Issled., Vol. 14, 73 - 79 (1976). In Russian.

097.012 On the structure of the Martian soil from optical and infrared observations. V. I. Moroz.
Kosmich. Issled., Vol. 14, 85 - 96 (1976). In Russian.

097.013 Influence of the regular horizontal gradients in the Martian ionosphere on phase radio measurements.

A. V. Plotnikov.
Kosmich. Issled., Vol. 14, 140 - 143 (1976). In Russian.
Brief information.

097.014 Optical properties of the Martian atmosphere during the 1971 dust storm.
Yu. V. Aleksandrov, D. F. Lupishko.
Astron. Zhurn. Akad. Nauk SSSR, Vol. 53, 162 - 169 (1976).
In Russian. English translation in Soviet Astron., Vol. 20, No. 1.

Under the assumption of the silicate nature of the 1971 dust storm particles on Mars two possibilities are considered: 1) the mean radius of particles $r \sim 1 \, \mu$m; 2) $r \sim 10 \, \mu$m. It is shown that only the case 2) does not contradict the results of photometrical observations.

097.015 Mars — the historical perspective. S. J. Hynes.
Spaceflight, Vol. 18, 81 - 83 (1976).

097.016 The voyages of Viking — 2. Is there life on Mars?
R. S. Young, H. P. Klein.
Spaceflight, Vol. 18, 118 - 123 (1976).

This second instalment of the series on 'The voyages of Viking', published in conjunction with Martin Marietta, considers prospects for finding life on Mars and obtaining a better understanding of the planet's weather and climate.

097.017 Saltation and Martian sandstorms.
W. J. Maegley.
Rev. Geophys. Space Phys., Vol. 14, 135 - 142 (1976).

A brief topical review of current knowledge available for predicting the characteristics of sandstorms is presented. Particular emphasis is placed on extrapolations to Mars where saltation of loose surface material appears to be a necessary condition for initiation of observed dust storms. Extrapolation of current knowledge to Mars is probably sufficient only for estimating purposes and landing craft design.

097.018 Mars to occult Epsilon Geminorum.
Sky Telescope, Vol. 51, 162 (1976).

097.019 Zum Argongehalt der Marsatmosphäre. H. Zimmer.
Mitt. Astron. Ges., No. 38, p. 132 (1976). — Abstract.

097.020 Problems of Martian paleoclimatology.
W. K. Hartmann.
Space Research XV, (see 012.003), p. 629 - 630 (1975).

097.021 Olympus Mons, continued: 1976. J. E. Westfall.
Strolling Astronomer, Vol. 25, 217 (1976).

097.022 Looking at Mars — inside and out.
H. H. Kieffer, T. A. Mutch, R. W. Shorthill,
R. B. Hargraves, D. L. Anderson
Spaceflight, Vol. 18, 162 - 167 (1976).

097.023 Sources of heat in the Martian and Venus thermospheres owing to absorption of ultraviolet solar radiation. M. N. Izakov, O. P. Krasitskij.
Geomagn. Aeronom., Vol. 16, 209 - 214 (1976). In Russian.

097.024 Variable features on Mars. VII. Dark filamentary markings on Mars. J. Veverka.
Icarus, Vol. 27, 495 - 502 (1976).

Some Mariner 9 B-frames show networks of criss-crossing rectilinear albedo markings typically 10 km long by 100 m wide. This paper discusses the location, variability and possible nature of these dark filamentary markings. It is unlikely that the markings are linear dunes.

097.025 Latitudinal variation of wind erosion of crater ejecta deposits on Mars.
R. E. Arvidson, M. Coradini, A. Carusi, A. Coradini, M. Fulchignoni, C. Federico, R. Funiciello, M. Salomone.
Icarus, Vol. 27, 503 - 516 (1976).

Wind erosion seems to be the dominant process eroding crater ejecta deposits and surrounding materials on Mars. In the equatorial zone, ejecta deposits are eroded back by scarp recession, where scarp heights appear to be approximately equivalent to ejecta thicknesses. An empirical model developed for wind erosion of ejecta deposits in nonmantled areas suggests that removal of ejecta materials on the average is exceedingly slow ($\sim 10^{-5}$ m/yr for 10 m high scarp). On the other hand, rapid deflation of aeolian debris around crater ejecta is implied. Results suggest high differential aeolian erosion rates that are a function of both grain sizes and large-scale surface roughness.

097.026 Martian atmospheric lee waves. J. A. Pirraglia.
Icarus, Vol. 27, 517 - 530 (1976).

The Mariner 9 television pictures of Mars showed areas of extensive mountain lee wave phenomenon in the northern mid-latitudes during winter. The cloud patterns resulting from the waves generated by the flow across a mountain or crater are dependent upon the velocity profile of the air stream and the vertical stability of the atmosphere. Using the stability as inferred by the temperature structure obtained from the infrared spectrometer data, a two layer velocity model of the air stream is used in calculations based on the theory of mountain lee waves. The parameters that yield a pattern similar to that in a picture with a well-defined wave configuration are a lower 11 km deep air stream of 40 m/sec and an upper air stream of 85 m/sec.

097.027 Albedo boundaries on Mars in 1972: results from Mariner 9. R. M. Batson, J. L. Inge.
Icarus, Vol. 27, 531 - 536 (1976).

A map of "albedo" boundaries (light and dark markings) on Mars was prepared from Mariner 9 images. After special digital processing, these pictures provide detailed locations of albedo boundaries, which is significant in interpreting recent eolian activity. Derivation of absolute albedo values from the spacecraft data was not attempted. The map correlates well with telescopic observations of Mars after the 1971 dust storm.

097.028 Surface oxidation: a major sink for water on Mars.
R. L. Huguenin.
Science, Vol. 192, 138 - 139 (1976).

Surface oxidation irreversibly removes both oxygen and hydrogen from the martian atmosphere at a rate of 10^8 to 10^{11} per square centimeter per second. This rate corresponds to a net loss of 10^{25} to 10^{28} per square centimeter (10^2 to 10^5 grams per square centimeter) of H_2O, if it is assumed that the loss rate is uniform over geologic time. Heretofore, exospheric escape was considered to be the principal irreversible sink for H_2O, but the loss rate was estimated to be only 10^8 per square centimeter per second. It is possible that surface oxidation may have had a minor effect on the supply of H_2O in the regolith and polar caps.

097.029 The voyages of Viking — 4. The chemistry of Mars.
P. Toulmin III, K. Biemann, T. Owen.
Spaceflight, Vol. 18, 209 - 211 (1976).

097.030 Calculations of the thermal conditions of the Martian atmosphere. V. I. Vorob'ev,
A. P. Gal'tsev, S. D. Gutshabash, V. M. Osipov.
Kosmich. Issled., Vol. 14, 265 - 271 (1976). In Russian.

097.031 Argon in the Martian atmosphere: evidence from the Mars 6 descent module.
V. G. Istomin, K. V. Grechnev.
Icarus, Vol. 28, 155 - 158 (1976).

In the descent of the automatic interplanetary station

Mars 6, after parachute deployment, an analysis was carried out by means of a mass spectrometer, the analyzer of which was pumped out by a getter-ion pump. The dynamics of the pump current at the end of descent operations imply that an inert gas is a basic component of the Martian atmosphere, along with CO_2. Laboratory post-experiment calibrations of other getter-ion pumps, performed with various mixtures of CO_2 and argon, result in a probable value of the argon content in the Martian atmosphere of 35 ± 10% by volume.

097.032 Argon in the Martian atmosphere: do the results of Mars 6 agree with the optical and radio occultation measurements? V. I. Moroz.
Icarus, Vol. 28, 159 - 163 (1976).

Mars 6 discovered an inert gas (probably argon) in the Martian atmosphere. An analysis is carried out for the available spectroscopic observations, radio occultation results, and other data with the aim of determining the maximum argon content with which they are consistent. Possible seasonal variations of pressure are taken into account.

097.033 A new estimate of volatile outgassing on Mars. J. S. Levine.
Icarus, Vol. 28, 165 - 169 (1976).

The presence of 28% argon on Mars as calculated by Levine and Riegler and indirectly inferred from Soviet Mars-6 lander data has important implications for the outgassing history of H_2O, CO_2, and N_2 on Mars. Even if the terrestrial volatile outgassing ratio is only approximately valid for Mars, then large quantities of H_2O, CO_2, and N_2 may have outgassed over the history of Mars.

097.034 Volatile inventories on Mars. T. Owen.
Icarus, Vol. 28, 171 - 177 (1976).

Predictions for the total inventory of outgassed volatiles on Mars can be developed by studying volatiles in meteorites, terrestrial rocks, and the atmospheres of Venus, the moon, and the earth. Two models are presented following the basic assumption that the devolatilization of Mars has been analogous to that of the earth. The recent discovery of a high abundance of argon in the Martian atmosphere appears to indicate that Mars has outgassed as completely as the earth, but present uncertainties and lacunae in the essential data set permit several other interpretations.

097.035 Martian volatiles: their degassing history and geochemical fate. F. P. Fanale.
Icarus, Vol. 28, 179 - 202 (1976).

Contents: Bulk volatile and semivolatile contents of Mars and earth; A simple earth-analogous Mars degassing model; Physical storage of volatiles in the Martian regolith; Chemically bound volatiles in the regolith; Exospheric escape as a massive volatile sink; Effect of the time history of degassing on the "surface" volatile inventory; Earth degassing history: implications for Mars; Mars degassing models; Future testing of Mars volatile history models.

097.036 Mars: chemical weathering as a massive volatile sink. R. L. Huguenin.
Icarus, Vol. 28, 203 - 212 (1976).

In this manuscript the amounts of volatiles incorporated in the regolith by photostimulated oxidation weathering are estimated.

097.037 Martian albedo feature variations with season: data of 1971 and 1973. C. F. Capen, Jr.
Icarus, Vol. 28, 213 - 230 (1976).

This paper is a qualitative and quantitative investigation of recent seasonal and secular albedo feature variations on Mars using the improved red-filter photography of the International Planetary Patrol Program obtained during the 1971 and 1973 apparitions.

097.038 The regulation of hydrogen and oxygen escape from Mars. S. C. Liu, T. M. Donahue.
Icarus, Vol. 28, 231 - 246 (1976).

Martian photochemistry that recombines CO and O to CO_2 through a water catalytic cycle enables the nonthermal escape of oxygen to govern the thermal escape of hydrogen. The result is that water escapes from Mars, not hydrogen alone, and there is no pronounced tendency for H or O to accumulate as a result of escape.

097.039 Martian atmospheric water vapor observations: 1972−74 apparition. E. S. Barker.
Icarus, Vol. 28, 247 - 268 (1976).

This paper presents the results that were obtained during the 1972−74 apparition. It completes the apparition coverage and discusses these apparition data and earlier data sets obtained as a whole to describe the various aspects of the behavior of Martian water vapor.

097.040 The vertical distribution of water vapor in the atmosphere of Mars. S. L. Hess.
Icarus, Vol. 28, 269 - 278 (1976).

Calculations are performed of the vertical distribution of water vapor and condensate in an adiabatic atmosphere on Mars taking into account turbulent diffusion and terminal velocity. The distributions are found to be substantially different when terminal velocity is included.

097.041 Liquid water on Mars. C. B. Farmer.
Icarus, Vol. 28, 279 - 289 (1976).

The factors which affect fusion and evaporation of ice under a variety of Martian surface conditions are examined. Current knowledge of the seasonal and diurnal behavior of the atmospheric vapor is summarized and discussed as it relates to the availability of surface ice at temperate latitudes.

097.042 The prospects for life on Mars: a pre-Viking assessment. C. Sagan, J. Lederberg.
Icarus, Vol. 28, 291 - 300 (1976).

Mariner 9 has provided a refutation or reinterpretation of several historical claims for Martian biology, and has permitted an important further characterization of the environmental constraints on possible Martian organisms. Four classes of conceivable Martian organisms are identified, depending on the environmental temperature and water activity.

097.043 Surface slope probabilities from the spectra of weak radar echoes: application to Mars. B. Lipa, G. L. Tyler.
Icarus, Vol. 28, 301 - 306 (1976).

Slope probability densities were derived from the power spectra of radar echoes from Mars using integral inversion. The authors describe a method of stabilizing the inversion, which was necessary for echoes with signal-to-noise power spectral densities on the order of unity, and for those with broad spectral distributions. The resulting slope probabilities usually consisted of a component due to quasi-specular reflection which decreased rapidly with tilt, plus a broad, slowly decreasing, "diffuse" component due to scattering from (1) surface scales small compared with a radar wavelength, or (2) larger features with high slopes. Root mean square tilts have been determined separately for the two cases.

097.044 Spectrum of thermal radiation of Mars from data of the automatic interplanetary station Mariner 9 and peculiarities of the vertical temperature profile. A. M. Bunakova, K. Ya. Kondrat'ev.
Probl. fiz. atmosf. Leningrad, Leningr. un-t, No. 195, p. 3 - 9. In Russian. − Abstr. in Referativ. Zhurn. 62. Issled. kosmich.

prostranstva, 5.62.180 (1976).

097.045 Results of photoelectric observations of Mars during the opposition in 1973.
L. A. Akimov, D. F. Lupishko, T. A. Lupishko.
Vestn. Khar'kov. Univ., No. 129 (Ser. Astron., vyp. (No.) 10), p. 45 - 53 (1975). In Russian.

097.046 Conditions for performing astronomical observations from the surface of Mars.
Yu. V. Aleksandrov, V. M. Litvinov.
Vestn. Khar'kov. Univ., No. 129 (Ser. Astron., vyp. (No.) 10), p. 70 - 76 (1975). In Russian.

097.047 The British Astronomical Association and the controversy over canals on Mars.
N. S. Hetherington.
Journ. British Astron. Ass., Vol. 86, 303 - 308 (1976).

097.048 The challenge of Viking: is there life on Mars?
R. I. Palsson.
Mercury, (Journ. Astron. Soc. Pacific), Vol. 5, No. 2, p. 14 - 18 (1976).

097.049 Panorama survey of Mars. A. S. Selivanov, V. M. Govorov, V. P. Chemodanov, M. K. Naraeva, M. V. Gitlits.
Tekhnika kino i televideniya, 1976, No. 1, p. 35 - 37. In Russian. − Abstr. in Referativ. Zhurn. 51. Astron., 6. 51. 269 (1976).

097.050 Determination of water vapour content in the Martian atmosphere from the spectrum of emitted thermal radiation. Yu. M. Timofeev, S. P. Obraztsov.
Probl. fiz. atmosf. Leningrad, Leningr. un-t, 1975, p. 87 - 96. In Russian. − Abstr. in Referativ. Zhurn. 51. Astron., 6. 51. 270; 62. Issled. kosmich. prostranstva, 6. 62. 164 (1976).

097.051 Dielectric properties of the material of the upper layer of Mars from the results of ground-based measurements of its radio emission.
A. G. Kislyakov, V. D. Krotikov, O. B. Shchuko.
Izv. vyssh. ucheb. zavedenij. Radiofizika, Vol. 18, 1770 - 1776 (1975). In Russian. − Abstr. in Referativ. Zhurn. 51. Astron., 6. 51. 272 (1976).

097.052 Investigation of the solar plasma near Mars and on the path earth−Mars by means of traps of charged particles aboard Soviet space vehicles in 1971−1973. IV.
Comparison of the results of synchronous plasma and magnetic measurements aboard Mars 2. T. K. Breus, M. I. Verigin.
Kosmich. Issled., Vol. 14, 400 - 405 (1976). In Russian.

097.053 Clouds on Mars: some conclusions from observations aboard Mars 3. V. I. Moroz.
Kosmich. Issled., Vol. 14, 406 - 416 (1976). In Russian.

097.054 Structure and dynamics of the equatorial thermosphere of Mars. M. N. Izakov, S. K. Morozov.
Kosmich. Issled., Vol. 14, 476 - 478 (1976). In Russian. Brief information.

097.055 Gibt es Leben auf dem Mars? J. von Puttkamer.
Umschau, 76. Jahrgang, p. 368, 370, 372 (1976).

097.056 A unique airborne observation. J. L. Elliot,

E. Dunham, C. Church.
Sky Telescope, Vol. 52, 23 - 25 (1976).

097.057 Mars 1971 apparition −Martian polar hoods − ALPO report II. C. F. Capen, R. B. Rhoads.
Strolling Astronomer, Vol. 26, 1 - 8 (1976).

097.058 On a temporary darkening of the south polar cap of Mars. T. Osawa.
Strolling Astronomer, Vol. 26, 12, 14 - 16 (1976).

097.059 The ionospheres of Mars and Venus.
J. C. McConnell.
Annual Rev. Earth Planet. Sci., Vol. 4, (see 003.012), 319 - 346 (1976).

097.060 A surface effect in the photodissociation of CO_2 and its significance for the theory of the Martian atmosphere. P. Papacosta, S. J. B. Corrigan.
Chem. Phys. Letters, Vol. 36, 674 - 676 (1975). − Abstr. in Phys. Abstr., Vol. 79, A015031 (1976).

097.061 Numerical modelling of the diurnal winds near the Martian polar caps. S. D. Burk.
Thesis Univ. Arizona, Tucson, USA, 148 pp. (1975). (Available from: Univ. Microfilms, Ann Arbor, Mich., USA. Order No. 75-19607). − Abstr. in Phys. Abstr., Vol. 79, A027811 (1976).

097.062 Martian sandstorms and their effects on the 1975 Viking lander system. W. J. Maegley, D. P. Diederich.
Journ. Testing Evaluation, Vol. 3, 380 - 388 (1975). − Abstr. in Phys. Abstr., Vol. 79, A037470 (1976).

097.063 To Mars to look for life... S. R. Brzostkiewicz.
Urania Kraków, Vol. 47, 74 - 79 (1976). In Polish.

Mars occults a bright star.
Sky Telescope, Vol. 51, 434 - 436 (1976).

Navigation between the planets.
See Abstr. 053.017.

Viking and Mars: a prelude. See Abstr. 053.019.

Viking. See Abstr. 053.023.

Trajectory, atmosphere, and wind reconstruction from Viking entry measurements. See Abstr. 053.024.

Solution of the inverse problem of radio emission of the atmosphere of planets taking into account the horizontal gradients. See Abstr. 091.004.

Radiographic inspection of a planetary atmosphere: solution of the inverse problem with allowance for horizontal density gradients. See Abstr. 091.009.

Estimate of the density of surface layer materials of the moon, Mars and Venus. See Abstr. 094.127.

Untersuchung von Einschlagskrater-Populationen. See Abstr. 094.596.

Secular acceleration of Phobos and Q of Mars. See Abstr. 097.201.

Mars, Satellites

097.201 **Secular acceleration of Phobos and Q of Mars.**
J. C. Smith, G. H. Born.
Icarus, Vol. 27, 51 - 53 (1976).

Reductions of Mariner 9 TV data of Phobos and Deimos tend to corroborate the existence of a secular acceleration of Phobos commensurate with two recently reported values based on a reprocessing of earth-based data. These values of secular acceleration have been used together with Mariner 9 data on the physical size of Phobos and earth-based photoelectric observations which infer a carbonaceous composition for Phobos to place bounds of $50 < Q < 150$ on the tidal dissipation function of Mars. The corresponding bounds on the tidal lag angle are $0.19° < \Phi < 0.57°$.

097.202 **Occultation of ϵ Geminorum by Phobos on 1976 April 8.**   D. W. Dunham.
IAU Circ., No. 2935 (1976).

097.203 **Accuracy of estimating the masses of Phobos and Deimos from multiple Viking orbiter encounters.**
R. H. Tolson, M. L. Mason.
1975 AAS/AIAA Astrodyn. Conf., (see 012.022), AAS75-089, 22 pp. (1975). — Abstr. in Phys. Abstr., Vol. 79, A023679 (1976).

098 Minor Planets

098.001 Positions of minor planets.   G. Soulié, R. Dumont,
Dupouy, Teulet, Broqua, Dulou.
Astron. Astrophys., Suppl. Ser., Vol. 23, 115 - 124 (1976).
In French.

Bordeaux Observatory is sharing in photographic observations of asteroids, the purpose being to point out systematic errors in equatorial coordinates of stars. The asteroids belong to the "Selected Minor Planets" list. The equatorial coordinates of the observed minor planets are published. The negatives were obtained on Kodak II A O plates, with the 13" photographic refractor. On each plate are made three exposures, separated by one interval of five minutes, and timed according to the brightness of the observed minor planet. Each plate is measured twice on an Asco-Zeiss.

**098.002 A photometric study of the minor planets 192
Nausikaa and 79 Eurynome.**
F. Scaltriti, V. Zappalà.
Astron. Astrophys., Suppl. Ser., Vol. 23, 167 - 179 (1976).

A series of photoelectric observations of the asteroids 192 Nausikaa and 79 Eurynome was made at the Astronomical Observatory of Torino, during the 1974 opposition. The synodic period, the absolute magnitude $V(1,0)$, the phase coefficient, the albedo, the mean radius, and the mass were derived. An attempt was made in order to obtain the pole of 192 Nausikaa. 79 Eurynome shows an evident opposition effect.

098.003 Rotation and photometric characteristics of Pallas.
A. Schroll, H. F. Haupt, H. M. Maitzen.
Icarus, Vol. 27, 147 - 156 (1976).

Pallas was observed in 1968, 1969, and 1971. From these observations the synodic period could be derived to be $P_{syn} = 7^h 48^m 27^s$. The magnitude–phase relation has also been computed. An opposition effect of $0^m 32$ was noticed. Preliminary values for the coordinates of the pole ($\lambda_0 = 228°$, $\beta_0 = +43°$) and for the sidereal period ($P_{sid} = 7^h 48^m 38^s 0$) are given.

098.004 Photographic photometry of the asteroid 291 Alice.
C.-I. Lagerkvist.
Icarus, Vol. 27, 157 - 160 (1976).

During its 1974 opposition the asteroid 291 Alice was observed photographically with the Schmidt telescope at the Kvistaberg Observatory. This article describes the reduction method of the iris readings. The synodic period of 291 Alice is found to be $4^h 18^m 9$ and the amplitude of the light curves $0^m 25$.

**098.005 Minor planets and related objects. XIX. Shape and
pole orientation of (39) Laetitia.**   R. E. Sather.
Astron. Journ., Vol. 81, 67 - 73 (1976).

UBV photometry and lightcurves of Laetitia have been obtained between June 1968 and December 1974. The magnitudes at lightcurve maximum are consistent with a triaxial ellipsoidal figure of axial ratios 15 : 9 : 5 and a linear phase law with coefficient 0.026 ± 0.002 mag/deg. The lightcurves indicate a surface roughness on the order of 10 km. The pole determined by a new method of reducing the scatter in the magnitude phase relation lies near $\lambda_0 = 121° \pm 10°$ and $\beta_0 = +37° \pm 10°$. The absolute magnitude of primary maximum is then $V_0(1,0,90) = 6.56 \pm 0.03$ at 90° aspect and $V_0(1,0,0) = 6.01 \pm 0.03$ when viewed pole-on. Colors at zero phase are $B-V = +0.85 \pm 0.02$ and $U-B = +0.51 \pm 0.02$ with no substantial evidence of variation over the surface.

**098.006 Radii and albedos of 84 asteroids from visual and
infrared photometry.**   O. L. Hansen.
Astron. Journ., Vol. 81, 74 - 84 (1976).

A sample of 84 asteroids has been studied using nearly simultaneous infrared and visual photometry. From the requirement of energy balance, radii and visual geometric albedos p_v have been derived. Statistical analysis based on the results has led to the following conclusions: (1) Asteroids form at least two classes with distinctly different p_v. Respectively, the dark and light classes have typical p_v values of about 0.02 and 0.08. (2) Samples selected according to visual brightness alone will favor light asteroids in the ratio of at least 3:2. (3) The frequency of dark asteroids in the outer belt ($a > 3$ AU), is at least five, and possibly ten, times as high as the frequency of light asteroids. As a consequence, counting statistics and combined-mass estimates may be in error.

098.007 Betulia et Geographos.   J. Meeus.
L'Astronomie, Vol. 90, 37 - 39 (1976).

098.008 A photometric study of Eros 433.
D.-h. Chen, X.-y. Yang, Z.-x. Wu.
Acta Astron. Sinica, Vol. 16, 131 - 137 (1975). In Chinese.

Three photoelectric observations of Eros made in Jan. - Feb. 1975 are presented. A new photometric theory of Eros based upon a three-axis ellipsoid model at full phase obeying Lommel-Seeliger's law is presented. Comparing the theoretical and observational light curves, the authors obtain the ratio of the axes of the ellipsoid model as a:b:c = 1:0.334:0.148, if the pole of Veseley is used.

098.009 Radiometric diameters for an additional 22 asteroids.
D. Morrison, C. R. Chapman.
Astrophys. Journ., Vol. 204, 934 - 939 (1976).

New 10 μ radiometry is presented for 24 asteroids, 22 of which have not previously had their diameters and albedos determined. Half are dark objects ($0.025 < p_v < 0.06$), apparently of the C (or carbonaceous) type, which appears to be by far the largest class of minor planets. One of these, 65 Cybele, is one of the 10 largest asteroids. Unusually high albedos ($p_v \geqslant 0.25$) are found for asteroids 64, 349, and 863. The high albedo for 349 Dembowska reinforces the conclusion from spectrophotometry that is is a metamorphosed ordinary chondrite. 64 Angelina may be an aubrite. The fainter objects in the sample are the smallest main-belt asteroids to have measured diameters and albedos.

098.010 The coming flyby of asteroid Betulia.
Sky Telescope, Vol. 51, 287 - 290 (1976).

**098.011 654 Zelinda – ein Kleinplanet mit ungewöhnlich
langer Rotationsperiode.**   H. J. Schober.
Mitt. Astron. Ges., No. 38, p. 189 - 190 (1976). – Abstract.

098.012 An orbital improvement of Vesta.
R. Dvorak, C. Edelman.
Mitt. Astron. Ges., No. 38, p. 192 - 195 (1976). – Short report.

098.013 Apollo asteroid Helin.   D. W. Hughes.
Nature, Vol. 260, 665 (1976).

**098.014 Radiometric diameter and albedo of the remarkable
asteroid 1976AA.**
D. Morrison, J. C. Gradie, G. H. Rieke.
Nature, Vol. 260, 691 (1976).

The minor planet 1976AA (Fast-Moving Object Helin), discovered on January 7, 1976, is the first asteroid found with an orbital period of < 1 yr (orbital elements $a = 0.97$, $e = 0.18$; $i = 19°$). This letter reports a determination by infrared radiometry of the diameter and albedo of 1976AA. A straightforward solution for diameter and visual geometric albedo yields $D = 900 \pm 200$ m and $p_v = 0.20 \pm 0.07$.

098.015 Thermal emission spectra of 24 asteroids and the Galilean satellites. O. L. Hansen.
Icarus, Vol. 27, 463 - 471 (1976).

Thermal emission spectra between 8 and 23 μm have been obtained for the Galilean satellites and 24 bright asteroids. No significant emission or absorption feature has been found in any of these spectra.

098.016 On the dynamical topology of the Kirkwood gaps. C. Froeschlé, H. Scholl.
Astron. Astrophys., Vol. 48, 389 - 393 (1976).

Some particular orbits of the 2/1 Kirkwood gap are studied using the surface of section method. Direct and averaged numerical methods are compared and a rough picture of the Kirkwood gap topology is given.

098.017 Radar observations at 3.5 and 12.6 cm wavelength of asteroid 433 Eros.
R. F. Jurgens, R. M. Goldstein.
Icarus, Vol. 28, 1 - 15 (1976).

A study of the asteroid 433 Eros using 3.5 and 12.6 cm radar waves indicates that the surface is very much rougher than any planetary or lunar surface observed by this method. A model based on a rough rotating triaxial ellipsoid having radii in the rotation equator of 18.6 and 7.9 km agrees well with the data.

098.018 70-cm radar observations of 433 Eros.
D. B. Campbell, G. H. Pettengill, I. I. Shapiro.
Icarus, Vol. 28, 17 - 20 (1976).

Radar observations of 433 Eros were made at the Arecibo Observatory using a wavelength of 70 cm during the close approach of Eros to earth in mid-January, 1975. A peak radar cross section of 39 ± 15 km^2 was observed. The surface of Eros appears to be relatively rough at the scale of a wavelength as compared to the surfaces of the terrestrial planets and the moon.

098.019 UBV light curves of asteroid 433 Eros.
E. F. Tedesco.
Icarus, Vol. 28, 21 - 28 (1976).

UBV photometry and light curves of Eros were obtained on nine dates between October 1974 and March 1975. The absolute V magnitude at photometric maximum extrapolated to a solar phase angle of zero is 10.75 while the linear phase coefficient is 0.026 mag deg^{-1}. The mean colors at solar phase angles greater than 30° are $B-V = 0.92$ and $U-B = 0.52$ mag. No significant color variations over the surface were detected.

098.020 Photometric light curves and pole determination of 433 Eros. F. Scaltriti, V. Zappalà.
Icarus, Vol. 28, 29 - 35 (1976).

Fourteen photometric light curves of 433 Eros were made at the Astronomical Observatory of Torino during the 1974–75 close passage. The absolute magnitude of the primary maximum (10^m78), the phase coefficient (0.023 mag/degree), the synodic and sidereal period of rotation (0^d21956 and 0^d21959, respectively) and the ecliptic coordinates of the pole ($\lambda = 17°, \beta = 10°$) were deduced.

098.021 Photoelectric light curves of asteroid 433 Eros.
V. Pop, D. Chis.
Icarus, Vol. 28, 37 - 38 (1976).

Photoelectric observations of the asteroid 433 Eros were carried out with the 50-cm reflector of the Astronomical Observatory of the University of Cluj-Napoca on January 21 and February 4 and 5, 1975.

098.022 Photoelectric light curves of asteroid 433 Eros.
C. Cristescu.
Icarus, Vol. 28, 39 - 42 (1976).

During January and February, 1975, four light curves of asteroid 433 Eros were obtained at the Bucharest Observatory with a 50-cm reflector, in integrated light. A sizeable variation of the amplitude (from 1.3 to 0.9 magnitudes) was observed.

098.023 Five-color photoelectric photometry of asteroid 433 Eros. E. Miner, J. Young.
Icarus, Vol. 28, 43 - 51 (1976).

Five-color photoelectric light curves of asteroid 433 Eros were obtained on 9 nights during the 1974/75 apparition. Although color differences due to changing solar phase angle were detected, color differences during a single rotation of Eros are less than 1 %. Amplitudes of up to 1^m44 were measured. The absolute visual magnitude at primary maximum, corrected to zero phase and to one AU from earth and sun, is about $V_0(1,0) = 10^m8$.

098.024 UBV photometry of asteroid 433 Eros.
R. L. Millis, E. Bowell, D. T. Thompson.
Icarus, Vol. 28, 53 - 67 (1976).

UBV observations of asteroid 433 Eros were conducted on 17 nights during the winter of 1974/75. The peak-to-peak amplitude of the light curve varied from about 0.3 mag to nearly 1.4 mag. The absolute V mag at maximum light, extrapolated to zero phase, is 10.85. Phase coefficients of 0.0233 mag/degree, 0.0009 mag/degree, and 0.0004 mag/degree were derived for V, B–V, and U–B, respectively. The photometric behavior of Eros can be modeled by a cylinder with rounded ends having an axial ratio of about 2.3 : 1. The asteroid is rotating about a short axis with the north pole at $\lambda_0 = 15°$ and $\beta_0 = 9°$.

098.025 Light curves and the axis of rotation of 433 Eros.
J. L. Dunlap.
Icarus, Vol. 28, 69 - 78 (1976).

Ten light curves and UBV photometry of 433 Eros were obtained between August 1972 and May 1975. The absolute magnitude of the light curve maximum is 10.75 and the phase coefficient is 0.025 mag/deg. The pole of the axis of rotation is directed toward $\lambda_0 = 16°, \beta_0 = 12°$, ecliptic longitude and latitude, respectively, and the rotation is direct with a sidereal period of $0^d219599$ or $5^h16^m13^s4 \pm 0^s2$. The dimensions derived from the polarimetric albedo and the light curve amplitudes are 12 km $\times$ 12 km $\times$ 31 km for a smooth cylinder with hemispherical ends.

098.026 Photometry of 433 Eros from 0.65 to 2.2 μm.
G. J. Veeder, D. L. Matson, J. T. Bergstralh, T. V. Johnson.
Icarus, Vol. 28, 79 - 85 (1976).

Light curves of 433 Eros are reported for 11 bandpasses ranging from 0.65 to 2.2 μm in wavelength. The relative spectral reflectance, $R(\lambda)$, was not seen to vary during the observations. Eros has $R(1.6 \mu m) = 1.5 \pm 0.1$ and $R(2.2 \mu m) = 1.7 \pm 0.1$, where $R(\lambda)$ is the spectral reflectance scaled to unity at $\lambda = 0.56 \mu$m. This spectral reflectance is suggestive of a mixture of silicates and material with high infrared reflectance, perhaps a metallic phase such as meteoritic "iron".

098.027 Spectrophotometry and UBVRI photometry of Eros.
W. Z. Wisniewski.
Icarus, Vol. 28, 87 - 90 (1976).

Spectrophotometric and broad-band photoelectric observations of Eros are reported. No existing meteorite spectrum matches the asteroid data directly. Assemblages of iron or stony-iron with ordinary-chondrite material generate the best match for the reflectivity curve of Eros. Variability was found in the 0.6 micron band and V–R and V–I colors.

098.028 *J, H, K* photometry of 433 Eros and other asteroids.
C. R. Chapman, D. Morrison.
Icarus, Vol. 28, 91 - 94 (1976).

The authors report photometry for nine asteroids at wavelengths of 1.25, 1.65, and 2.22 μm. Three *C*-type objects (88, 129, and 511) seem slightly redder in $H-K$ and $J-H$ color indices than four *S*-type objects (5, 6, 7, and 116) and an *E*-type object (64). Eros has an unusually red $J-H$ color index; its infrared spectral reflectance is consistent with an appreciable quantity of metallic iron plus some pyroxene and olivine on its surface.

098.029 The infrared spectrum of asteroid 433 Eros.
H. P. Larson, U. Fink, R. R. Treffers, T. N. Gautier III.
Icarus, Vol. 28, 95 - 103 (1976).

The mineralogical composition of asteroid Eros has been determined from its infrared spectrum (0.9–2.7 μm; 28 cm^{-1} resolution). Major minerals include metallic Ni–Fe and pyroxene; no spectroscopic evidence for olivine or plagioclase feldspar was found. The IR spectrum of Eros is most consistent with a stony-iron composition.

098.030 Spectrophotometry (0.33 to 1.07 μm) of 433 Eros and compositional implications.
C. Pieters, M. J. Gaffey, C. R. Chapman, T. B. McCord.
Icarus, Vol. 28, 105 - 115 (1976).

The spectral reflectance (0.33–1.07 μm) for the asteroid 433 Eros was determined as a function of rotational phase during January 28–30, and February 15, 1975. Interpretation of absorption features suggests Eros is composed of an undifferentiated assemblage of moderate to high temperature minerals (iron, pyroxene, and olivine, but no carbon).

098.031 Polarization of the reflected light of asteroid 433 Eros. B. Zellner, J. Gradie.
Icarus, Vol. 28, 117 - 123 (1976).

Linear polarizations measured for asteroid 433 Eros at various wavelengths and at solar phase angles ranging from 9° to 53° are presented. The polarization results are entirely typical of main-belt S asteroids, and indicate a dusty surface with geometric albedo 0.20. The derived effective diameter at photometric maximum is 21 km.

098.032 The diameter and thermal inertia of 433 Eros.
D. Morrison.
Icarus, Vol. 28, 125 - 132 (1976).

Radiometry of Eros at 10 and 20 μm demonstrates that the thermal conductivity of the upper centimeter of the surface is approximately as low as that of the moon, suggesting that the asteroid has a regolith of highly porous rocky material. When combined with photoelectric photometry, these infrared measurements yield an effective diameter of Eros at maximum light of $D_0 = 22 \pm 2$ km and a geometric albedo of $p_v = 0.18 \pm 0.03$.

098.033 The occultation of κ Geminorum by Eros.
B. O'Leary, B. G. Marsden, R. Dragon, E. Hauser, M. McGrath, P. Backus, H. Robkoff.
Icarus, Vol. 28, 133 - 146 (1976).

The predictions and observations of the occultation of κ Gem by Eros on January 24, 1975, are described. A circular cross section of diameter up to 23 km gives a good fit to most of the available data. Consideration of a crucial, unconfirmed negative observation indicates that an elliptical solution 17 by 7 km (with the long axis in the direction of motion of the "shadow" of Eros) may be preferable. It is suggested either that the circle should be warped into an ellipse of dimensions 21 by 13 km, or that the profile is a kind of dumbbell.

098.034 Upper limit to the 2 cm brightness temperature of asteroid 433 Eros.
I. I. K. Pauliny-Toth, A. Witzel, J. R. Dickel.
Icarus, Vol. 28, 147 (1976).

The brightness temperature of 433 Eros was less than 460°K during the close opposition in January 1975.

098.035 Physical properties of asteroid 433 Eros.
B. Zellner.
Icarus, Vol. 28, 149 - 153 (1976).

Newly available photometric, polarimetric, spectroscopic, thermal-radiometric, radar, and occultation results are synthesized in order to derive a coherent model for Eros. The geometric albedo is 0.19 ± 0.01 at the visual wavelength, and the overall dimensions are approximately 13 X 15 X 36 km. The rotation is about the short axis, in the direct sense, with a sidereal period of $5^h 16^m 13\!.\!4$. The pole of rotation lies within a few degrees of ecliptic coordinates $\lambda = 16°$ and $\beta = +11°$. Eros is uniformly coated with a particulate surface layer several millimeters thick. It has an iron-bearing silicate composition, similar to that of a minority of main-belt asteroids, and probably identifiable with *H*-type ordinary chondrites.

098.036 Minor planets and related objects. XX. Polarimetric evidence for the albedos and compositions of 94 asteroids. B. Zellner, J. Gradie.
Astron. Journ., Vol. 81, 262 - 280 (1976).

New observations of asteroids bring the total to 94 for which useful polarimetry has been obtained, and to 52 for which polarimetric albedos and hence diameters can be computed. All asteroids so far observed, ranging over a more than a factor of 50 in diameter, show the polarimetric signatures of microscopically rough or particulate surface textures. Compositional types identifiable in the sample include 48 asteroids belonging to the broad *S* class, with iron-bearing silicate surfaces resembling ordinary chondrites or stony-iron meteorites; 34 objects of the *C* type, probably corresponding to carbonaceous chondrites; at least three and possibly five *M* asteroids with surfaces rich in free metal; two with high albedos (class *E*) attributable to pure enstatite; and five of other types.

098.037 The nature of asteroids. V. A. Bronshtehn.
Priroda, 1976, No. 5, p. 88 - 95. In Russian.

098.038 Asteroids as meteorite parent-bodies: the astronomical perspective. C. R. Chapman.
Geochim. Cosmochim. Acta, Vol. 40, 701 - 719 (1976).

A scenario for the origin and evolution of meteorite parent-bodies is presented which includes: (1) interruption of planet-formation by processes due to Jupiter; (2) substantial asteroidal collisions during the first 0.5 b.y.; and (3) formation of most meteorite types within the differentiated bodies.

098.039 Asteroid photometry and the amateur.
A. Porter.
Journ. American Ass. Variable Star Observers, Vol. 4, 92 - 94 (1975/76).

Amateurs are invited to participate in a program of visual asteroid photometry. Results of studies of 18 Melpomene, 233 Asterope, and 270 Anahita indicate the potential of such research.

098.040 Observations of Eros at the Gissar Astronomical Observatory of the Institute of Astrophysics of the Academy of Sciences of the Tadzhik SSR.
N. N. Kiselev, N. V. Narizhnaya, I. I. Gavrilova.
Kometn. Tsirk., *Kiev*, No. 189 (1975). In Russian.

098.041 Topocentric coordinates of Eros from observations in Pulkovo. N. Bronnikova.
Kometn. Tsirk., *Kiev*, No. 190 (1975). In Russian.

098.042 **Observations of Eros at the Gissar Astronomical Observatory of the Institute of Astrophysics of the Academy of Sciences of the Tadzhik SSR.**
N. N. Kiselev, N. V. Narizhnaya, I. I. Gavrilova.
Kometn. Tsirk., *Kiev*, No. 191 (1976). In Russian.

098.043 **Positions of Eros from observations at the Charkov Observatory of the Charkov State University.**
P. P. Pavlenko, L. S. Pavlenko.
Kometn. Tsirk., *Kiev*, No. 192 (1976). In Russian.

098.044 **La planète d'Olbers et les astéroïdes.**
A. Dauvillier.
Ciel et Terre, Vol. 92, 105 - 116 (1976).

098.045 **Fast-moving object Helin, 1976 AA.**
Kometn. Tsirk., *Kiev*, No. 193 (1976). In Russian.

098.046 **Fast-moving object Kowal, 1975 YA.**
Kometn. Tsirk., *Kiev*, No. 193 (1976). In Russian.

098.047 **Observations of Ceres in Uzhgorod in 1975.**
A. G. Kirichenko, K. A. Kudak.
Kometn. Tsirk., *Kiev*, No. 197 (1976). In Russian.

098.048 **Ephemeris of Hidalgo (944) for physical observations** (B. G. Marsden).
Kometn. Tsirk., *Kiev*, No. 197 (1976). In Russian.

098.049 **On the motion of Hecuba-type asteroids.**
I. A. Gerasimov.
Vestn. Mosk. un-ta. Fiz., astron., 1975, No. 6, p. 749 - 752. In Russian. – Abstr. in Referativ. Zhurn. 51. Astron., 6. 51. 132 (1976).

098.050 **Fast-moving object Kowal.**
R. E. McCrosky, C. Y. Shao, C. Kowal.
IAU Circ., No. 2895 (1976).

098.051 **1975 YA (fast-moving object Kowal).**
C. Kowal, H. L. Giclas, M. L. Kantz, B. G. Marsden.
IAU Circ., No. 2897 (1976).

098.052 **1950 LA.**   E. Roemer, D. Daniels, G. Schwartz, R. E. McCrosky, C. Y. Shao.
IAU Circ., No. 2898 (1976).

098.053 **Fast-moving object Helin.**
E. Helin, S. J. Bus, C. Pryor.
IAU Circ., No. 2899 (1976).

098.054 **1976 AA (fast-moving object Helin).**
E. Helin, S. J. Bus, T. Urata, B. G. Marsden.
IAU Circ., No. 2901 (1976).

098.055 **1975 YA.**
C. Kowal, R. E. McCrosky, G. Schwartz, C. Y. Shao.
IAU Circ., No. 2902 (1976).

098.056 **1975 YA.**   C. Kowal, R. E. McCrosky, G. Schwartz, C. Y. Shao, B. G. Marsden.
IAU Circ., No. 2903 (1976).

098.057 **1976 AA.**   H. L. Giclas, M. L. Kantz.
IAU Circ., No. 2903 (1976).

098.058 **1976 AA.**   M. S. Burkhead, J. Sanders, J. H. Bulger, R. E. McCrosky, C. Y. Shao, T. Urata, J. Young, S. J. Bus, B. G. Marsden.
IAU Circ., No. 2905 (1976).

098.059 **1976 AA.**   T. Seki, H. L. Giclas, M. L. Kantz, B. Milet, J. Young, S. J. Bus, K. Tomita, D. Morrison, J. C. Gradie.
IAU Circ., No. 2909 (1976).

098.060 **1975 YA.**   G. Schwartz, J. H. Bulger, C. Y. Shao, E. Roemer, D. Daniels, B. G. Marsden.
IAU Circ., No. 2913 (1976).

098.061 **1976 AA.**   T. Urata, H. L. Giclas, M. L. Kantz, E. Helin, S. J. Bus, B. G. Marsden.
IAU Circ., No. 2915 (1976).

098.062 **(944) Hidalgo.**   B. G. Marsden.
IAU Circ., No. 2939 (1976).

098.063 **1950 LA.**   E. Roemer, H.-E. Schuster, R. M. West.
IAU Circ., No. 2940 (1976).

098.064 **1963 UA.**   B. G. Marsden.
IAU Circ., No. 2941 (1976).

098.065 **1975 YA.**   E. Roemer, D. Daniels, W. Smith, C. A. Heller, C. D. Vesely, B. G. Marsden.
IAU Circ., No. 2944 (1976).

098.066 **1976 AA.**   C. Y. Shao, R. E. McCrosky, B. G. Marsden.
IAU Circ., No. 2950 (1976).

098.067 **1973 NA.**   A. C. Gilmore, P. M. Kilmartin.
IAU Circ., No. 2960 (1976).

098.068 **1975 YA.**
E. Roemer, C. D. Vesely, C. A. Heller, B. G. Marsden.
IAU Circ., No. 2966 (1976).

098.069 **1976 AA.**   E. Roemer, R. A. McCallister, W. Smith, C. A. Heller, C. D. Vesely, B. G. Marsden.
IAU Circ., No. 2966 (1976).

098.070 **Minor Planet Circulars, (MPC), Nos. 3909 - 4000** (1976).
Edited by Cincinnati Observatory under the supervision of P. Herget.
    A repository of nearly all new data for numbered and unnumbered minor planets: Observations, elements and ephemerides, identifications, newly assigned numbers and names, occultations.

098.071 **The strange case of the missing apocentric librators in the 3 : 2 resonance.**   W.-H. Ip.
Astrophys. Space Sci., Vol. 42, L1 - L3 (1976).
    From a comparison of the 2 : 1 and 3 : 2 resonances (in the asteroidal belt) two possible explanations to the absence of 3 : 2 apocentric librators are suggested. The first one is that such 3 : 2 resonant motion is dynamically unstable. The second interpretation requires the absence of near-circular orbits originally at 4 AU. The latter view, if correct, is inconsistent with cosmogonic models which predict the original orbits of the asteroids to be nearly circular.

098.072 **The A.L.P.O. minor planets section with notes on the 1976 apparition of 1580 Betulia.**
D. N. Wallentine.
Strolling Astronomer, Vol. 26, 18 - 33 (1976).

098.073 **Posiciones exactas de pequeños planetas obtenidas en el observatorio astronómico de Madrid (1973).**
M. de Pascual.
Bol. Astron. Obs. Madrid, Vol. 9, No. 2, p. 44 - 83 (1976).

C98.074 **Distribution of the discoveries of minor planets within the year.** M.-A. Combes.
Minor Planet. Bull., Vol. 3, 35 - 36 (1976).

098.075 **Mean values of the orbital elements for the first nineteen hundred numbered minor planets.**
M.-A. Combes.
Minor Planet. Bull., Vol. 3, 36 - 37 (1976).

098.076 **Evolution of the magnitudes for the first 19 hundreds of minor planets.** M.-A. Combes.
Minor Planet. Bull., Vol. 3, 37 (1976).

098.077 **Asteroidal occultations.** D. Wallentine.
Minor Planet. Bull., Vol. 3, 41 - 44 (1976).

098.078 **Light curves and other physical observations of 233 Asterope in 1975.**
A. Porter, D. Wallentine, E. F. Tedesco, F. Pilcher.
Minor Planet Bull., Vol. 3, 47 - 48 (1976).

098.079 **Observations of planet 99 Dike.** F. Pilcher.
Minor Planet Bull., Vol. 3, 51 (1976).

098.080 **Photographic position determinations for 4 Vesta, 6 Hebe, 7 Iris and 433 Eros.** A. T. Son.
Minor Planet Bull., Vol. 3, 52 (1976).

098.081 **Where have the minor planets been discovered?**
M.-A. Combes.
Minor Planet Bull., Vol. 3, 53 - 54 (1976).

098.082 **The 1976 apparition of 1580 Betulia.**
D. Wallentine.
Minor Planet Bull., Vol. 3, 54 - 55 (1976).
The 1976 apparition of 1580 Betulia is described, with emphasis on occultation and photometry opportunities.

098.083 **A study of the large oscillation stability of Trojan asteroids around libration points.** V. Banfi.
Rend. Accad. Nazionale Lincei, Cl. Sci. fis. mat. nat., Ser. 8, Vol. 58, 220 - 229 = Contr. Oss. Astron. Torino, (Pino Tori-

nese), No. 86 (1975).

098.084 **Photoelectric observations of the minor planet 704 Interamnia during its 1969 opposition.**
P. Tempesti.
Mem. Soc. Astron. Italiana, Vol. 46, 397 - 405 (1975).
Photoelectric observations performed in 22 nights from August to November 1969 allow to ascertain the existence of brightness fluctuations having the period $8^h 42^m 8$ and the amplitude $0^m 05$ in V; the colour index B-V is $+0^m 64$. From the reduced magnitudes $\overline{V}(1, \alpha)$, determined in the range $22°6 \geq \alpha \geq 9°2$, the phase coefficient 0.044 magn/degree and the standard magnitude $\overline{V}(1, 0) = 6^m 31$ have been derived; assuming some already known statistical relations, the value 0.029 for the Bond albedo has been derived.

098.085 **Die Bestimmung der Albedo und des Durchmessers von Asteroiden durch einen Infrarotindex.**
T. Widorn.
Ann. Univ.-Sternw. Wien, Band 31, (No. 1), 1 - 20 (1974).

098.086 **Minor planet rotations reported in 1975.**
A. Porter, D. Wallentine.
Minor Planet Bull., Vol. 3, 48 - 50 (1976).

098.087 **Photometry opportunities: second quarter, 1976.**
Minor Planet Bull., Vol. 3, 50 - 51 (1976).

Photographing minor planets and measuring their positions. See Abstr. 031.265.

A statistical theory of resonance motion in the sun-Jupiter system. See Abstr. 042.002.

Observations of Ganymede, Callisto, Ceres, Uranus, and Neptune at 3.33 mm wavelength. See Abstr. 099.203.

Observations of comets, minor planets, Pluto, and satellites. See Abstr. 103.002.

Sources of interplanetary dust: asteroids.
See Abstr. 106.079.

099 Jupiter, Jupiter Satellites

Jupiter

099.001 A local diffusion process associated with the sweeping of energetic particles by Io.
J. D. Huba, C. S. Wu.
Astrophys. Journ., Vol. 203, 268 - 278 (1976).

A local, turbulent diffusion process which may rapidly refill the partial cavity of energetic particles created by Io's passage through a flux tube is discussed. A model is developed of Io's sweeping action on energetic particles in which a wake is formed behind the flux tube threading Io. It is shown to be possible that this wake contains an excess of positive charge, thereby creating an electric field. The particle drift produced by this field, coupled with the density gradient of energetic protons, can generate an instability. It is this instability that is responsible for the turbulent diffusion process. The sweeping of energetic particles is also discussed in relation to particle energy and pitch angle distribution.

099.002 On Jovian temperature profiles obtained by inverting thermal spectra.   L. Wallace, G. R. Smith.
Astrophys. Journ., Vol. 203, 760 - 763 (1976).

Defects in the inversion of observed thermal infrared spectra of Jupiter to obtain temperature structures are reviewed. An ambiguity is noted when the 7.8 μ band of CH_4 is used and alternative temperature structures are obtained.

099.003 The free oscillations of Jupiter and Saturn.
S. V. Vorontsov, V. N. Zharkov, V. M. Lubimov.
Icarus, Vol. 27, 109 - 118 (1976).

The periods and the eigenfunctions have been calculated for the fundamental modes and for the overtones of spheroidal oscillations of Jupiter and Saturn. Along with ordinary oscillations, the core oscillations appear; these eigenfunctions are localized in the planetary core and have an exponential attenuation to the surface. The rotational splittings of the free oscillations have been calculated. Because of rapid planetary rotation, these splittings were found to be of the order of differences between undisturbed frequencies. The idea is proposed that the rotational splitting of the spectrum of Saturn may present in the future the unique possibility of determining the bodily rotation period of the planet.

099.004 The meteorology of Jupiter.   A. P. Ingersoll.
Sci. American, Vol. 234, No. 3, p. 46 - 56 (1976).

The visible features of the giant planet reflect the circulation of its atmosphere. A model reproducing those features should apply to other planetary atmospheres, including the earth's.

099.005 The journey to Jupiter.
C. F. Hall, H. Mark, J. H. Wolfe.
Endeavour, No. 124, Vol. 35, 9 - 14 (1976).

The two Pioneer spacecraft missions described have now greatly advanced our knowledge of Jupiter, its satellites and interplanetary space.

099.006 Jupiter en 1974.   C. Botton, M. Alecsescu, G. Farroni, J. M. Gomez, R. Reginaldo.
L'Astronomie, Vol. 90, 65 - 83 (1976).

099.007 L'apport des Pioneer 10 et 11 à la compréhension des phénomènes observés à la surface de Jupiter.
C. Botton.
L'Astronomie, Vol. 90, 84 - 88 (1976).

099.008 On the possibility of nuclear energy sources in Jupiter.   A. A. Suchkov.
Astron. Zhurn. Akad. Nauk SSSR, Vol. 53, 217 - 218 (1976).
In Russian. English translation in Soviet Astron., Vol. 20, No. 1.

The mass and radius of Jupiter are such as to make possible deuterium burning in its interior. The energy reserves in deuterium are more than sufficient to keep up the presently observed luminosity of Jupiter during its life time.

099.009 Jovian electron bursts: correlation with the interplanetary field direction and hydromagnetic waves.
E. J. Smith, B. T. Tsurutani, D. L. Chenette, T. F. Conlon, J. A. Simpson.
Journ. Geophys. Res., Vol. 81, 65 - 72 (1976).

The bursts of relativistic electrons detected on Pioneer 10 upstream from Jupiter and within $400r_J$ of the planet have been found to be correlated with the interplanetary magnetic field.

099.010 Observations of plasmas in the Jovian magnetosphere.
L. A. Frank, K. L. Ackerson, J. H. Wolfe, J. D. Mihalov.
Journ. Geophys. Res., Vol. 81, 457 - 468 (1976).

Large intensities of low-energy protons were observed deep within the Jovian magnetosphere with the plasma instrumentation on Pioneer 10. The energy range of the electrostatic analyzer was 108 eV to 4.80 keV during encounter. Inside the flux tubes of the Galilean moon Io is a 'plasmasphere' of protons with relatively high densities, 50–100 cm^{-3}, extending toward the planet to at least 2.8 R_J. The characteristic thermal energies of these protons are about 100 eV. The flux tubes of Io are positioned on a severe decrease of these densities with increasing Jovicentric radial distances — a plasmapause. At greater distances, beyond the plasmapause, is found a great torus of plasma encircling Jupiter with densities in the range 10–15 cm^{-3} and thermal energies about 400 eV. The moon Europa is embedded in this torus, or 'ring current'. The mechanisms for the formation of the Jovian plasmasphere, ring current, and plasma disc must differ substantially from those dominantly participating in the terrestrial magnetosphere.

099.011 Energetic electrons in the Jovian magnetosphere.
D. N. Baker, J. A. Van Allen.
Journ. Geophys. Res., Vol. 81, 617 - 632 (1976).

Detailed analysis of Pioneer 10 data shows that the observed counting rates within the magnetosphere of Jupiter were caused primarily by electrons, $E_e > 0.06$ MeV. This identification holds for the magnetodisc region ($r \gtrsim 20 R_J$) as well as for the central magnetosphere ($r \lesssim 20 R_J$). A model electron differential energy spectrum of the form $dJ/dE = KE^{-15}[1 + (E/H)]^{-n}$ is found to fit the observed intensities throughout most of the encounter trajectory, where J is the omnidirectional electron intensity; E is the kinetic energy; and K, H, and n are fitting parameters dependent on radial distance from the planet and magnetic latitude. It is suggested that the observed spectral shape may be interpreted as resulting from losses of high-energy electrons by pitch angle scattering.

099.012 A comparison of the great red spot with temporary spots on Jupiter.   G. C. Browne, A. J. Meadows.
Observatory, Vol. 96, 16 - 18 (1976).

099.013 Electrons and protons in Jupiter's radiation belts.
A. Mogro-Campero, R. W. Fillius, C. E. McIlwain.
Space Research XV, (see 012.003), p. 521 - 528 (1975).

099.014 Diurnal variation of the Jovian ionosphere.

A. Tan, L. A. Capone.
Publ. Astron. Soc. Japan, Vol. 28, 155 - 161 (1976).

The time-dependent structure of the Jovian ionosphere is examined. Diurnal variation of appreciable magnitude is revealed in the lower ionosphere. The upper ionosphere remains more or less intact at nighttime as in the case of the earth's ionosphere. There is considerable difference in the height-integrated electrical conductivities on the day and night sides.

099.015 Jupiter in 1973—74: rotation periods.
P. W. Budine.
Strolling Astronomer, Vol. 25, 221 - 227 (1976).

099.016 Further notes on Jupiter in 1973—74: measured latitudes, a satellite egress, and referenced illustrations. P. W. Budine.
Strolling Astronomer, Vol. 25, 231 - 232 (1976).

099.017 Stellar occultation spikes as probes of atmospheric structure and composition.
J. L. Elliot, J. Veverka.
Icarus, Vol. 27, 359 - 386 (1976). — Presented at the Jupiter Colloquium (IAU Colloquium 30), Tucson, Arizona, May 18 - 23, 1975.

The characteristics of spikes observed in the occultation light curves of β Scorpii by Jupiter are reviewed and discussed. Using a model in which the refractivity (density) gradients in the Jovian atmosphere are parallel to the local gravitational field, the spikes are shown to yield information about (i) the $[He]/[H_2]$ ratio in the atmosphere, (ii) the fine scale density structure of the atmosphere and (iii) high-resolution images of the occulted stars. The spikes also serve as indicators for ray crossing.

099.018 Pressure-induced absorption by H_2 in the atmospheres of Jupiter and Saturn.
T. Z. Martin, D. P. Cruikshank, C. B. Pilcher, W. M. Sinton.
Icarus, Vol. 27, 391 - 406 (1976).

The S(1) line of the pressure-induced fundamental band of H_2 was identified and measured in the spectra of Saturn and Jupiter. The authors compare the observed line shape to the predictions of both a reflecting-layer model (RLM) and a homogeneous scattering model (HSM). The RLM provides a good fit to the Saturn line profile for temperatures near 150 K. The Jupiter line profile is fit by both the RLM and HSM, but for widely differing temperatures, neither of which seems probable. The precise fitting of the observed S(1) line profile to computed models depends critically on the determination of the true continuum level; difficulties encountered in finding the continuum, especially for Jupiter, are discussed.

099.019 Limb-darkening scans of Jupiter.
C. B. Pilcher, T. D. Kunkle.
Icarus, Vol. 27, 407 - 415 (1976).

An area scanning photometer has been used to obtain photometrically calibrated limb-darkening scans of Jupiter at four wavelengths: 6190, 6300, 7250, and 8200 Å. The first and third of these correspond to methane absorptions near the 4—0 and second and fourth to continuum regions near the 4—0 and 3—0 H_2 quadrupole bands, respectively. Single-scattering albedos have been calculated for several areas on the planet at all four wavelengths assuming a semi-infinite, homogeneous, isotropically scattering atmosphere.

099.020 Imaging spectrometer study of Jupiter and Saturn.
R. B. Wattson, S. A. Rappaport, E. E. Frederick.
Icarus, Vol. 27, 417 - 423 (1976).

Observations with a new near infrared imaging spectrometer with ~15 Å resolution are presented. Twelve spectral images of Saturn in the vicinity of the 8900 Å CH_4 absorption complex were obtained and their interpretation discussed.

Spectral images of Jupiter were also obtained and several of these at widely separated wavelengths were subjected to a Minnaert analysis.

099.021 Recent observations of the decimetric radio emission from Jupiter. D. Stannard, R. G. Conway.
Icarus, Vol. 27, 447 - 452 (1976).

The position angle of linear polarization and the degree of circular polarization of the decimetric emission show no marked variations with time. High resolution observations of the radiation belts show that the asymmetry in the emission found in synthesis maps originates from an extended region. The presence of this feature accounts well for the observed variation in the position of the centroid of the emission with rotation.

099.022 CMA propagation diagrams for the Jovian magnetosphere. B. Melander, H. Liemohn.
Icarus, Vol. 27, 453 - 456 (1976).

Local electromagnetic and hydromagnetic noise in the Jovian magnetosphere is expected to be intense due to the variety of wave-particle interactions and plasma instabilities that may be present. In order to qualitatively assess the nature of the radio noise, configuration space analogues of the well-known Clemmow—Mullaly—Allis (CMA) propagation diagrams have been prepared, based on recent models of the magnetic field and plasma density. These diagrams identify the loci of electron and ion resonances and cutoffs where absorption and reflection of wave energy occur, and specify the propagation modes and frequency bands that are anticipated in various regions.

099.023 Redefinition of system III longitude.
A. C. Riddle, J. W. Warwick.
Icarus, Vol. 27, 457 - 459 (1976). — Presented at the Jupiter Colloquium (IAU Colloquium 30), Tucson, Arizona, May 18 - 23, 1975.

There is a current need for a redefinition of the Jovian system III longitude measure. The authors report on a proposed new definition which has been widely circulated among users and has met with general acceptance. Some errors in current calculations of system III [1957.0] are noted so that these errors can be avoided in future calculations.

099.024 Detection of ionized sulfur in the Jovian magnetosphere.
I. Kupo, Y. (*Yu.*) Mekler, A. Eviatar.
Astrophys. Journ., (*Letters*), Vol. 205, L51 - L53 (1976).

Detection of a pair of emission lines at 6717 and 6730 Å in the near vicinity of Jupiter is reported. It is proposed that the lines be identified with the forbidden doublet of S II. The astrophysical implications of this identification are discussed.

099.025 Spikes of light during the stellar occultation by planets. F. Link.
Astron. Astrophys., Vol. 48, 263 - 268 (1976). In French.

During stellar occultations by Neptune and Jupiter several spikes of light were observed on the photometrical curves. The differential refraction in the planetary atmosphere with some local perturbations due to the discret stratifications is generally considered responsible for these features. The author presents the photometrical theory of these phenomena especially of very intense spikes based on his theory of lunar eclipses.

099.026 New theory of the Great Red Spot from solitary waves in the Jovian atmosphere.
T. Maxworthy, L. G. Redekopp.
Nature, Vol. 260, 509 - 511 (1976). — Letter.

099.027 The absorption of trapped particles by the inner satellites of Jupiter and the radial diffusion coeffi-

cient of particle transport.　A. Mogro-Campero, W. Fillius.
Journ. Geophys. Res., Vol. 81, 1289 - 1295 (1976).

　　The process of trapped particle absorption by the inner Jovian satellites is considered in detail by taking into account both the particle and the satellite motions in a magnetic dipole field which is displaced from the center of the planet and tilted with respect to the planetary rotation axis in the manner found by magnetic field measurements on Pioneer 10. It is assumed that particle motion is controlled exclusively by the planetary field and that a particle is removed from the trapped particle population when its trajectory intersects the physical boundary of a satellite. Measurements performed in the Jovian magnetosphere on Pioneer 10 are then used to obtain estimates of the diffusion coefficient at the orbits of Io $(L \simeq 6)$ and Europa $(L \simeq 9.5)$. We find that the diffusion coefficient is a function of energy and L for electrons in the energy range ~ 0.7–14 MeV.

099.028　**Angular distributions of electrons of energy E_e > 0.06 MeV in the Jovian magnetosphere.**
D. D. Sentman, J. A. Van Allen.
Journ. Geophys. Res., Vol. 81, 1350 - 1360 (1976).

　　Results of an angular distribution analysis of electron intensity data are presented. It is found that the central core of the magnetosphere for radial distances less than $12\,R_J$ (Jovian radii) is dominated by pitch angle distributions strongly peaked at $\alpha = 90°$, while the region of radial distances 12–$25\,R_J$ shows bidirectional and approximately equal maxima at $\alpha = 0°$ and $180°$.

099.029　**Interpretation of the methane absorption bands in the spectrum of Jupiter.**　A. S. Anikonov.
Astron. Tsirk., No. 854, p. 3 - 4 (1975). In Russian.

099.030　**The atmosphere of Jupiter.**　A. P. Ingersoll.
　　Space Sci. Rev., Vol. 18, 603 - 639 = Contr. Div.
Geol. Planet. Sci., California Inst. Technol., *Pasadena*, No. 2652 (1976).

　　Current information on the neutral atmosphere of Jupiter is reviewed, with approximately equal emphasis on composition and thermal structure on the one hand, and markings and dynamics on the other. Studies based on Pioneer 10 and 11 data are used to refine the atmospheric model. Data on the interior are reviewed for the information they provide on the deep atmosphere. The markings and dynamics are discussed with emphasis on qualitative relationships and analogies with phenomena in the earth's atmosphere.

099.031　**Sodium in the Jovian magnetosphere.**
　　Yu. Mekler, A. Eviatar, F. V. Coroniti.
Astrophys. Space Sci., Vol. 40, 63 - 72 = Publ. Inst. Geophys. Planet. Phys., Univ. California, Los Angeles, No. 1410 (1976).

　　Observations of sodium D-line emission from Io and the magnetosphere of Jupiter are reported. A disk-shaped cloud of sodium is found to exist in the Jovian magnetosphere with an inner edge at about $4\,R_{Jupiter}$ and an outer edge at about $10\,R_{Jupiter}$. The gravitational scale height above the equatorial plane is a few Jovian radii. The data are interpreted in terms of a sputtering model, in which the sodium required to maintain the cloud is sputtered off the surface of Io by trapped energetic radiation-belt protons.

099.032　**Estimate of the methane content in the atmospheres of giant planets.**　L. A. Bugaenko.
Astrometriya i Astrofizika, *Kiev*, vyp. (No.) 29, (see 003.003), p. 99 - 106 (1976). In Russian.

　　The observed monochromatic brightness coefficients of Jupiter, Saturn and also the geometrical albedo of Uranus were used to determine the spectral values of the singly scattering albedo in the homogeneous scattering model for line formation. The methane content in the upper layers of the atmosphere of

these planets is estimated.

099.033　**Decametric radio emission of Jupiter and solar activity.**　B. M. Vladimirskij, L. S. Levitskij.
VIIth Leningrad seminar. "Corpuscular streams of the sun and the radiation belts of the earth and Jupiter", 1975, (see 012. 007), p. 337 - 346 (1975). In Russian. – Abstr. in Referativ. Zhurn. 51. Astron., 4.51.381 (1976).

099.034　**The radiation belt of Jupiter according to radio observations from the earth.**　A. A. Korchak.
VIIth Leningrad seminar. "Corpuscular streams of the sun and the radiation belts of the earth and Jupiter", 1975, (see 012. 007), p. 279 - 298 (1975). In Russian. – Abstr. in Referativ. Zhurn. 62. Issled. kosmich. prostranstva, 4.62.170 (1976).

099.035　**Summary of initial results from the GSFC fluxgate magnetometer on Pioneer 11.**
M. H. Acuna, N. F. Ness.
VIIth Leningrad seminar. "Corpuscular streams of the sun and the radiation belts of the earth and Jupiter", 1975, (see 012.007), p. 299 - 325 (1975). – Abstr. in Referativ. Zhurn. 62. Issled. kosmich. prostranstva, 4.62.171 (1976).

099.036　**An analytical model illustrating the effects of rotation on a magnetosphere containing low-energy plasma.**　L. J. Gleeson, W. I. Axford.
VIIth Leningrad seminar. "Corpuscular streams of the sun and the radiation belts of the earth and Jupiter", 1975, (see 012. 007), p. 326 - 336 (1975). – Abstr. in Referativ. Zhurn. 62. Issled. kosmich. prostranstva, 4.62.172 (1976).

099.037　**Jovian sodium plasma.**
　　A. Eviatar, Y. (*Yu.*) Mekler, F. V. Coroniti.
Astrophys. Journ., Vol. 205, 622 - 633 = Publ. Inst. Geophys. Planet. Phys., Univ. California, Los Angeles, No. 1474 (1976).

　　The nature of the sodium plasma created by ionization of the sodium emitted into the environment of Jupiter by Io is discussed. The authors show that the ions form a three-component plasma: cold, thermal, and energetic. Observational evidence for the effect of the plasma on the neutral sodium cloud is presented. They predict that the energetic component will generate ion cyclotron turbulence near the sodium gyrofrequency.

099.038　**Differential rotation of the Jovian atmosphere. I. A criterion of convective instability and geostrophic motion regime.**　R. S. Iroshnikov.
Astron. Zhurn. Akad. Nauk SSSR, Vol. 53, 418 - 428 (1976). In Russian. English translation in Soviet Astron., Vol. 20, No. 2.

　　The aim of this paper is the construction of a theoretical model for the rotation of an envelope with turbulent mixing in it, caused by heat transfer from inside to outside. The averaged motion is supposed to be stationary and axisymmetrical. A criterion of convective instability in a fastly rotating baroclinic medium is deduced. The limiting geostrophic regime of a "pure" rotation is investigated.

099.039　**Jovian magnetotail stretches past Saturn.**
　　J. Gribbin.
Nature, Vol. 261, 12 (1976).

099.040　**Spectral study of the polarization of the Jovian decametric radiobursts.**　A. Lecacheux.
Astron. Astrophys., Vol. 49, 197 - 204 (1976).

　　The author studies the polarization of the Jovian decametric radiobursts by using the Faraday fringes present on four high resolution dynamic spectra recorded at Nançay, over the range 27–35 MHz. He shows that the shape of the polarization ellipse changes from day to day, within a given storm and also depends on the frequency. The author concludes that the previous determinations of the polarization angle at the source must be taken with caution.

099.041 Analysis of some properties of Jupiter's decameter radiation. Z. Pokorný.
Bull. Astron. Inst. Czechoslovakia, Vol. 27, 137 - 143 (1976).

Some of the properties of Jupiter's decameter radiation are analysed with a view to the differences between the source and non-source emissions. The non-source emission, which predominates at frequencies $\lesssim 25$ MHz, is strongly affected by solar activity events. Examples are given of the stimulation of the non-source decameter emission by recurrent high-speed solar wind streams and by solar cosmic radiation, generated by proton flares.

099.042 Outward diffusion of energetic particles from the Jovian radiation belt. A. Nishida.
Journ. Geophys. Res., Vol. 81, 1771 - 1773 (1976).

It is suggested that energetic particles populating the Jovian outer magnetosphere and neighboring interplanetary space are flowing out from the Jovian radiation belt by the outward diffusion process that violates both second and third adiabatic invariants. This cross-field diffusion process operates preferentially at low altitudes, and it does not cause serious degradation of particle energy. A model is constructed which combines this process with ordinary inward diffusion and pitch angle scattering, and it is found to compare favorably with Pioneer 10 observations in the Jovian magnetosphere.

099.043 Some comments on the whistler mode instability at Jupiter. F. L. Scarf, N. L. Sanders.
Journ. Geophys. Res., Vol. 81, 1787 - 1790 (1976).

The possibility that $E_e > 21$ MeV electrons at Jupiter interact resonantly with obliquely propagating whistler mode waves is considered. The Pioneer 10 and 11 data points of Van Allen et al. (1975) are used to deduce the equatorial pitch angle distributions.

099.044 Identifizierung des im 4. vorchristlichen Jahrhundert von Jupiter bedeckten Sterns in den Zwillingen.
K. Locher.
Orion, 34. Jahrgang, p. 55 - 58 (1976).

099.045 Non-Io-related radio emissions and the modulation of relativistic electrons in the Jovian magnetosphere.
K. Sakurai.
Planet. Space Sci., Vol. 24, 657 - 659 (1976).

The modulations of the non-Io-related radio emissions in hectometric and decametric wave frequencies are examined, and compared with the observed variation of the MeV electron fluxes in the morning sector of the Jovian magnetosphere. It is suggested that these radio emissions are controlled by the behaviour of these electrons in this sector.

099.046 A theorectical model of the bow shock and the magnetosphere of Jupiter. K. Sakurai.
Planet. Space Sci., Vol. 24, 661 - 664 (1976).

Using the data obtained from the Pioneer 10 and 11 observations, a theoretical model is proposed for the bow shock and the magnetosphere of Jupiter. This indicates that the distance of the magnetopause from Jupiter on the sunlit side is $(50-55) \times r_J$ and that the ratio of the stand-off distance is about equal to or slightly larger than unity. Hence the Mach number of the solar wind seems to be less than 1.5 at Jupiter's orbit. This result necessarily leads to a blunt body model of the Jovian magnetosphere, the tail region of which is not as extended as observed in the earth's case.

099.047 A model of Jupiter's sulfur nebula.
R. A. Brown.
Astrophys. Journ., (*Letters*), Vol. 206, L179 - L183 (1976).

The observations by Kupo et al. of possibly transient S II forbidden line emission from the Jovian plasmasphere are interpreted using a uniform model with electron collisions as the excitation mechanism. This nebula seems to have been characterized by $\log_{10} T_e = 4.4 \pm 0.6$ K and $\log_{10} [n_e] = 3.5 \pm 0.3$ cm^{-3} at the time of the observations. Forbidden-line emission from S II is a sensitive remote probe of the plasma conditions in the Jovian environment.

099.048 Observations of quiet-time interplanetary electron enhancements of Jovian origin.
S. M. Krimigis, E. T. Sarris, T. P. Armstrong.
14th Intern. Cosmic Ray Conf., (see 012.011), Vol. 2, 752 - 757 (1975).

099.049 Observations of low energy interplanetary electrons. R. A. Mewaldt, E. C. Stone, R. E. Vogt.
14th Intern. Cosmic Ray Conf., (see 012.011), Vol. 2, 758 - 763 (1975).

099.050 The quiet-time low energy nucleon spectrum in the vicinity of earth.
S. M. Krimigis, J. W. Kohl, T. P. Armstrong.
14th Intern. Cosmic Ray Conf., (see 012.011), Vol. 2, 780 - 785 (1975).

099.051 Cosmic rays and Jupiter. D. B. Swinson.
14th Intern. Cosmic Ray Conf., (see 012.011), Vol. 4, 1486 - 1491 (1975).

099.052 Simultaneous observations of MeV electrons of Jovian origin at 1 and 3 AU. F. B. McDonald, B. J. Teegarden, J. H. Trainor, T. T. von Rosenvinge, W. R. Webber.
14th Intern. Cosmic Ray Conf., (see 012.011), Vol. 4, 1533 (1975). — Abstract.

099.053 Comparison of magnetospheres and radio emissions of Jupiter with earth. L. M. Libby, W. F. Libby.
14th Intern. Cosmic Ray Conf., (see 012.011), Vol. 4, 1546 - 1551 (1975).

099.054 Investigation of the two-phase stratiform clouds in the Jovian atmosphere. M. V. Bujkov, K. Yu. Ibragimov, A. M. Pirnach, L. P. Sorokina.
Astron. Zhurn. Akad. Nauk SSSR, Vol. 53, 596 - 602 (1976). In Russian. English translation in Soviet Astron., Vol. 20, No. 3.

Seven models of the NH$_3$ clouds in the Jovian atmosphere are calculated for different values of the parameters of the atmosphere. The relation between the parameters of the atmosphere and the characteristics of the microstructure of the clouds is found. The optical depths of the clouds are calculated.

099.055 Differential rotation of the Jovian atmosphere. II. Equations of the impulse variation in the case of fast rotation of the convective envelope. R. S. Iroshnikov.
Astron. Zhurn. Akad. Nauk SSSR, Vol. 53, 603 - 611 (1976). In Russian. English translation in Soviet Astron., Vol. 20, No. 3.

099.056 Two Jupiter photographs and a suggested comparison.
Strolling Astronomer, Vol. 26, 18 - 19 (1976).

099.057 Plasma in the Jovian magnetosphere.
C. K. Goertz.
Journ. Geophys. Res., Vol. 81, 2007 - 2014 (1976).

The plasma in Jupiter's ionosphere is collisionless above a certain level. In the outer magnetosphere, where the rotational force dominates the gravitational force, the collisionless plasma has a beamlike distribution and gives rise to a two-stream instability. This leads to trapping of plasma in the centrifugally dominated region of the magnetosphere. Plasma is lost through recombination. The equilibrium concentration of trapped particles is calculated by assuming a balance between trapping by wave-particle interaction and loss by recombination.

099.058 Observations of Jovian electrons at 1 AU.
R. A. Mewaldt, E. C. Stone, R. E. Vogt.
Journ. Geophys. Res., Vol. 81, 2397 - 2400 (1976).
Electrons of Jovian origin are responsible for the 'quiet time increases' in the >3-MeV electron intensity observed at 1 AU. Using data from the California Institute of Technology electron/isotope spectrometers on IMP 7 and 8, the authors have studied the temporal behavior of quiet time electrons at 1 AU over the period October 1972 through May 1975. It is suggested that the interconnection of the interplanetary field with an extended Jovian magnetotail of ~2 AU in length could result in the observed longitudinal distribution of Jovian electrons at 1 AU.

099.059 The main magnetic field of Jupiter.
M. H. Acuna, N. F. Ness.
Journ. Geophys. Res., Vol. 81, 2917 - 2922 (1976).
The main magnetic field of Jupiter has been measured by the Goddard Space Flight Center flux gate magnetometer on Pioneer 11, and analysis of the data yields a more detailed model than that obtained from Pioneer 10 results. In a spherical harmonic octupole representation the dipole term (with opposite polarity to earth's) has a magnitude of 4.28 G R_J^3 at a tilt angle of 9.6° and a system III longitude of 232°. The quadrupole and octupole moments are significant, 24% and 21% of the dipole, respectively.

099.060 An analytical model of the Jovian magnetosphere.
F. D. Barish, R. A. Smith.
Geophys. Res. Letters, Vol. 2, 269 - 272 (1975). – Abstr. in Phys. Abstr., Vol. 79, A027797 (1976).

099.061 Recirculation of energetic particles in Jupiter's magnetosphere. D. D. Sentman, J. A. Van Allen, C. K. Goertz.
Geophys. Res. Letters, Vol. 2, 465 - 468 (1975). – Abstr. in Phys. Abstr., Vol. 79, A027799 (1976).

099.062 Discussion of the paper by S. S. Prasad and A. Tan 'The Jovian ionosphere'.
D. F. Strobel, S. Prasad.
Geophys. Res. Letters, Vol. 2, 521 - 524 (1975). – Abstr. in Phys. Abstr., Vol. 79, A027802 (1976).

099.063 1. Spatially resolved absolute spectral reflectivity of Jupiter: 3390–8400 Ångstrøms. 2. The Jovian thermal structure from Pioneer 10 infrared radiometer data. 3. Observations and analysis of 8–14 micron thermal emission of Jupiter: a model of thermal structure and cloud properties. G. S. Orton.
Thesis California Inst. Technol., Pasadena, USA, 187 pp. (1975). (Available from: Univ. Microfilms, Ann Arbor, Mich., USA. Order No. 75-20007). – Abstr. in Phys. Abstr., Vol. 79, A027812 (1976).

099.064 Is Jupiter's magnetosphere like a pulsar's or earth's?
C. F. Kennel, F. V. Coroniti.
Separate print California Univ., Los Angeles, USA, Plasma Physics Group, 67 pp. (1974). – See 14.099.009, 14.099.030.

099.065 Jovian electron spectrum and synchrotron radiation at 375 cm. T. J. Birmingham.
Separate print National Aeronaut. Space Administr., Greenbelt,

Maryland, USA, Goddard Space Flight Center, 15 pp. (1975).

099.066 Observations of plasmas in the Jovian magnetosphere.
L. A. Frank, E. L. Ackerson, J. H. Wolfe, J. D. Mihalov.
Separate print Iowa Univ., Iowa City, USA, Dept. Phys. Astron., 43 pp. (1975).

099.067 Diffusion models for Jupiter's radiation belt.
S. A. Jacques, L. Davis, Jr.
Separate print California Inst. Tech., Pasadena, USA, Dept. Phys., 34 pp. (1972).

099.068 Energetic particle fluxes and spectra in the Jovian magnetosphere. D. N. Baker.
Thesis Iowa Univ., Iowa City, USA, 219 pp. (1974). (Available from: Univ. Microfilms, Order No. 75-13,722).

099.069 Jupiter radiation belt models (July 1974).
N. Divine.
Separate print Jet Propulsion Lab., Pasadena, California, USA, 15 pp. (1974).

099.070 Jupiter's radiation belts and their effects on spacecraft. R. H. Parker, E. L. Divita, G. Gigas.
Separate print Jet Propulsion Lab., Pasadena, California, USA, 42 pp. (1974).

Gewaltiger Jupiter-Magnetfeldschweif entdeckt.
Umschau, 76. Jahrgang, p. 365 (1976).

Jupiter: I.A.U. Colloquium No. 30, University of Arizona, Tucson, Arizona, May 18–23, 1975. Part I: Atmosphere and clouds. Part II: Fields and particles.
See Abstr. 011.001.

Laboratory band strengths of methane and their application to the atmospheres of Jupiter, Saturn, Uranus, Neptune, and Titan. See Abstr. 022.003.

Effect of the Jovian oblateness on Pioneer 10/11 radio occultations. See Abstr. 053.021.

Thermodynamics and phase separation of dense fully ionized hydrogen-helium fluid mixtures. See Abstr. 062.085.

Thermodynamics and phase separation of dense fully-ionized hydrogen-helium fluid mixtures. See Abstr. 062.092.

Special relativistic mechanics and electrodynamics with applications to synchrotron radiation. See Abstr. 066.094.

Direct observations of the solar wind interaction with Jupiter. See Abstr. 074.057.

Observation by high temporal and spectral resolution of sporadic emissions from the sun and Jupiter. See Abstr. 077.044.

Corpuscular streams of the sun and radiation belts of the earth and Jupiter. See Abstr. 078.017.

VLF electrostatic waves in the magnetospheres of the earth and Jupiter. See Abstr. 084.231.

Non-adiabatic effects of motion of particles in a static dipole field and in fields variable with regard to time. See Abstr. 084.246.

Mapping of the sodium emission associated with Io and Jupiter.   See Abstr. 099.208.

Interplanetary dust particles near Jupiter. See Abstr. 106.003.

The interplanetary and near Jovian dust environment: some experimental results.   See Abstr. 106.018.

The cosmic dust environment at earth, Jupiter and interplanetary space: results from Langley experiments on MTS, Pioneer 10, and Pioneer 11.   See Abstr. 106.077.

Photographic studies of quasi-stellar objects and other active radio sources.   See Abstr. 141.128.

On the quiet-time increases of low energy cosmic ray electrons.   See Abstr. 143.166.

Jupiter, Satellites

099.201 On the inversion of mutual occultation light curves.
R. T. Brinkmann.
Icarus, Vol. 27, 69 - 89 (1976).

The extent to which mutual occultation light curves, of the type that occur among the Galilean satellites of Jupiter, can be inverted to produce albedo maps of the surfaces has been investigated. Model calculations have been done utilizing the earth's own moon as the object of study. It is found that the inversion process can be successfully accomplished even in the presence of amounts of random noise typical of routine high-quality photometry of the Galilean satellites. Unless the limb-darkening distribution is pathological or the geometric elements adopted are seriously in error it should be possible to produce fairly high resolution maps of at least some of the Galilean satellites.

099.202 Galilean satellites: observations of mutual occultations and eclipses in 1973.
L. H. Wasserman, J. L. Elliot, J. Veverka, W. Liller.
Icarus, Vol. 27, 91 - 107 (1976).

During the fall of 1973 the authors observed seven Galilean satellite mutual events: two occultations and two eclipses of Europa and three eclipses of Io. The observations were carried out simultaneously at three wavelengths (0.35, 0.50, and 0.91 μm) with a time resolution of 0.1 sec. Their principal aim was to obtain color information about albedo distributions and limb darkening on the satellites. The observations do not yield any conclusive information about Io's limb darkening. For Europa, however, the best data indicate that this satellite is limb-darkened at both 0.50 and 0.91 μm. A variety of computer-generated model light curves are presented.

099.203 Observations of Ganymede, Callisto, Ceres, Uranus, and Neptune at 3.33 mm wavelength.
B. L. Ulich, E. K. Conklin.
Icarus, Vol. 27, 183 - 189 (1976).

The authors have measured the 3.33 mm wavelength disk brightness temperatures of Ganymede (136 ± 21°K), Callisto (95 ± 17°K), Ceres (137 ± 25°K), Uranus (125 ± 9°K), and Neptune (126 ± 9°K). Their observations of Ganymede are consistent with the radiation from a blackbody in solar equilibrium, whereas Callisto's microwave spectrum indicates a surface similar to that of the moon. The disk temperature for Ceres agrees with that expected from a rapidly rotating blackbody. The millimeter temperatures of Uranus and Neptune greatly exceed solar equilibrium values, implying atmospheres with large temperature gradients.

099.204 Photoelectric observations of mutual eclipses and occultations of the Galilean satellites in 1973.
R. M. Williamon.
Publ. Astron. Soc. Pacific, Vol. 88, 73 - 76 (1976).

Four mutual occultations and three mutual eclipses of the Galilean satellites were observed from Fernbank Science Center Observatory between July and December 1973. The observed midtime and duration of each event as well as the magnitude loss in yellow light is presented. Cause for small discrepancies between the observations and predictions are discussed.

099.205 All the outer satellites of Jupiter. C. T. Kowal.
Sky Telescope, Vol. 51, 242 - 243 (1976).

099.206 Restored pictures of Ganymede, moon of Jupiter.
B. R. Frieden, W. Swindell.
Science, Vol. 191, 1237 - 1241 (1976).

Digital restoration of two space pictures of Ganymede has revealed some interesting surface features.

099.207 Die Jupitermonde 1971 bis 1974.
P. Ahnert.
Sterne, Vol. 52, 39 - 46 (1976).

099.208 Mapping of the sodium emission associated with Io and Jupiter.
P. A. Wehinger, S. Wyckoff, A. Frohlich.
Icarus, Vol. 27, 425 - 428 (1976).

The sodium D-lines are observed in emission in a disklike distribution surrounding Io and extending outward in the orbital plane of the Galilean satellites to at least 23 R_J from Jupiter. A scale length for the sodium emission cloud in the orbital plane and the thickness of the sodium disk perpendicular to the orbital plane are determined. Weak D-line emission is also detected over the poles of Jupiter.

099.209 A search for emission features in Io's extended cloud.
L. Trafton.
Icarus, Vol. 27, 429 - 437 (1976). – Presented at the Jupiter Colloquium (IAU Colloquium 30), Tucson, Arizona, May 18 - 23, 1975.

The author presents spectra in the range from 3100 Å to 8700 Å of a portion of Io's extended cloud where Io's scattered continuum is weak but where the sodium emission is still strong. Aluminum and calcium are found to be underabundant relative to sodium. Upper limits are set to some other cosmically abundant elements. In addition, he detected the 10830 Å feature over various parts of the cloud but found it to have an intensity comparable to that observed elsewhere in the night sky.

099.210 **New upper limits for atmospheric constituents on Io.**
U. Fink, H. P. Larson, T. N. Gautier III.
Icarus, Vol. 27, 439 - 446 (1976).

A spectrum of the satellite of Jupiter, Io, from 0.86 to 2.7 µm is presented. No absorptions due to any atmospheric constituents on Io could be found on the spectrum. Upper limits for NH_3, CH_4, N_2O, and H_2S were determined. Laboratory spectra of ammonia frosts as a function of temperature were compared with the spectrum of Io and showed this frost not to be present at the surface of Io. A search for possible resonance lines of carbon, silicon, and sulfur as well as the 1.08 µm line of helium proved negative and upper emission limits were established for these lines.

099.211 **Discussion of the photographic observations of the Galilean satellites in the period 1930–1970.**
S. Ferraz-Mello, L. R. de Paula.
Astron. Journ., Vol. 81, 127 - 131 (1976).

All available series of photographic observations of the four great satellites of Jupiter made in the period 1930–1970 are discussed. It is shown that the time scale of Sampson's tables departs significantly from the ephemeris time. The formula $ST-ET = 0.006$ $(t-1900.0)$ min is a good representation for this difference.

099.212 **The spatial distribution of long lived gas clouds emitted by satellites in the outer solar system.**
T.-M. Fang, W. H. Smyth, M. B. McElroy.
Planet. Space Sci., Vol. 24, 577 - 588 (1976).

Simple models are presented for the spatial distribution of gases emitted by satellites in the outer solar system with emphasis on Io and Titan. The models, valid for long lived species in regions of space outside the gravitational zone of the parent satellite are applied to observed hydrogen and sodium clouds orbiting Jupiter and to an expected hydrogen cloud around Saturn.

099.213 **The Galilean satellites of Jupiter.**
D. P. Cruikshank, D. Morrison.
Sci. American, Vol. 234, No. 5, p. 108 - 116 (1976).

First observed by Galileo in 1610, these four largest of Jupiter's moons have begun to be explored by passing spacecraft. What is being revealed is unlike anything seen so far in the solar system.

099.214 **Jupiter XIII.** P. J. Shelus, J. D. Mulholland.
IAU Circ., No. 2897 (1976).

099.215 **Probable new satellite of Jupiter.**
IAU Circ., No. 2899 (1976).

099.216 **De små Jupitermåner: hvorfor studere dem?**
H. Nielsen.
Astron. Tidssk., Årg. 9, 59 - 63 (1976).

099.217 **Masses of the Galilean satellites of Jupiter.**
S. Ferraz-Mello.
Science, Vol. 192, 1127 - 1128 (1976).

Numerical data derived from the observation of the four great satellites of Jupiter are compared with the values obtained through Sampson's theory by using the new JPL (Jet Propulsion Laboratory) system of masses. It is not possible to fit the coefficient of the free oscillation in the longitude of Ganymede, whose argument is $l_3 - \bar{\omega}_4$ (the mean longitude of Ganymede referred to the proper apse of Callisto), and the mass of Callisto derived from the path of Pioneer 10.

099.218 **Mutual phenomena of the Galilean satellites in 1973. III. Final results from 91 light curves.**

K. Aksnes, F. A. Franklin.
Astron. Journ., Vol. 81, 464 - 481 (1976).

This paper presents an analysis of photoelectric light curves of 91 mutual eclipses and occultations of the Galilean satellites that occurred from June to December 1973. From the deepest curves, the authors deduce the following radii and standard errors: 1533 ± 27 (J2), 2608 ± 32 (J3), and 2445 ± 75 km (J4). They show that the substantial differences between the depths and midtimes observed and those predicted by Sampson's theory can be used in a clear and definite way to revise the orbital constants of that theory. Their analysis, which reduces all the data provided by these mutual events in one body, diminishes the latitude and longitude residuals of the unrevised theory by as much as a factor of about 10.

099.219 **Io's atmosphere and ionosphere: new limits on surface pressure from plasma models.**
T. V. Johnson, D. L. Matson, R. W. Carlson.
Geophys. Res. Letters, Vol. 3, 293 - 296 (1976).

The authors have studied charge particle impact as a mechanism for the production of Io's ionosphere. Pioneer 10 thermal plasma measurements and magnetospheric plasma models imply electron fluxes of $\sim 10^{10}$ cm^{-2} sec^{-1}. The fluxes and the temperature (~ 100 eV) of this plasma suggest that electron impact ionization is the dominant process in forming the ionosphere of Io. It is found that the surface number density of the neutral species required to match the observed electron density profiles is $\sim 10^9$ cm^{-3} or less. This value is two orders of magnitude lower than previous estimates.

099.220 **Analysis of mutual phenomena of Galilean satellites in 1973.** T. Nakamura.
Publ. Astron. Soc. Japan, Vol. 28, 239 - 257 (1976).

The observed data for mutual phenomena of the Galilean satellites in 1973 are analyzed based on Sampson's (1910, 1921) theory. The time corrections are derived for J1 and J2 satellites from the measured midtimes of the phenomena. From the durations of the J1OJ2 events the radius of J2 satellite is found to be 1533 ± 26 km (p.e.).

099.221 **Electron impact ionization of Io's sodium emission cloud.**
R. W. Carlson, D. L. Matson, T. V. Johnson.
Geophys. Res. Letters, Vol. 2, 469 - 472 (1975). – Abstr. in Phys. Abstr., Vol. 79, A027800 (1976).

Maximum entropy restorations of Ganymede.
See Abstr. 031.232.

Introduction to the restricted Jupiter orbiter problem. See Abstr. 042.092.

Jupiter orbiter lifetime. The hazard of Galilean satellite collision. See Abstr. 052.039.

Lunar occultation of Jupiter's Galilean satellites.
See Abstr. 096.005.

Lunar occultation of Jupiter's Galilean satellites.
See Abstr. 096.006.

Lunar occultation of Jupiter's Galilean satellites.
See Abstr. 096.007.

Thermal emission spectra of 24 asteroids and the Galilean satellites. See Abstr. 098.015.

A local diffusion process associated with the sweeping of energetic particles by Io. See Abstr. 099.001.

The absorption of trapped particles by the inner

satellites of Jupiter and the radial diffusion coefficient of particle transport.   See Abstr. 099.027.

Sodium in the Jovian magnetosphere.
See Abstr. 099.031.

Jovian sodium plasma.   See Abstr. 099.037.

Observations of plasmas in the Jovian magnetosphere.
See Abstr. 099.066.

100   Saturn, Saturn Satellites

Saturn

100.001 **Azimuthal brightness variations of Saturn's rings.**
K. Lumme, W. M. Irvine.
Astrophys. Journ., (*Letters*), Vol. 204, L55 - L57 = Contr.
Five College Obs., Univ. Mass., *Amherst*, No. 214 (1976).
    Saturn's ring A exhibits a variation in brightness with azimuth (with orbital phase of the ring particles). Maximum brightness occurs at about $\theta = 45°$ and $\theta = 225°$ and minimum brightness at about $135°$ and $315°$, where the eastern ansa is taken as $\theta = 0°$. No such effect appears to exist for ring B.

100.002 **Saturn central meridian ephemeris: 1976.**
J. E. Westfall.
Strolling Astronomer, Vol. 25, 217 - 220 (1976).

100.003 **Saturnian drift rates and periods.**   J. E. Westfall.
Strolling Astronomer, Vol. 25, 220 - 221 (1976).

100.004 **The 1966—67 apparition of Saturn.**
J. L. Benton, Jr.
Strolling Astronomer, Vol. 25, 232 - 250 (1976).

100.005 **Azimuthal brightness variations in Saturn's rings.**
H. J. Reitsema, R. F. Beebe, B. A. Smith.
Astron. Journ., Vol. 81, 209 - 215 (1976).
    Azimuthal variations in brightness in the rings of Saturn have been studied on recent photographs. A ring-brightness model has been developed which is used in conjunction with a two-dimensional numerical smearing process to investigate the effects of seeing and instrumental smearing on the data. The results of this investigation show that the two ansae of the rings are of equal brightness. Furthermore, the observed surface brightness of ring B, decreasing from the ansae towards the minor axis of the ring ellipse, can be fully accounted for as a result of smearing processes on an initial intensity distribution which is independent of azimuthal angle. In contrast, ring A shows a roughly sinusoidal variation in brightness with minima in the quadrants following geocentric conjunctions and maxima in the others. The smearing is approximately removed from ring A to yield the intrinsic surface brightness of the ring.

100.006 **Infrared thermal models for Saturn's ring.**
M. J. Price.
Icarus, Vol. 27, 537 - 544 (1976).
    Infrared (10 and 20 μm) thermal emission data for Saturn's rings are discussed in terms of simple isothermal radiative transfer models of finite optical thickness. Recent brightness temperature measurements, corresponding to essentially maximum ring tilt, indicate that optical single scattering albedos less than 0.75 are required to provide sufficient heating of the ring material. Reconciliation with analyses of the optical scattering properties of the ring requires the backscattering efficiency to be even higher than for a macroscopic sphere.

100.007 **Coupe photométrique des quatre anneaux de Saturne suivant leur grand axe, à partir d'enregistrements bi-dimensionnels de photographies de la planète. Les mesures.**   P. Guérin.
Comptes Rendus Acad. Sci. Paris, Sér. B, Vol. 282, 317 - 320 (1976).
    On décrit une méthode de dépouillement à deux dimensions des photographies de Saturne, consistant en un balayage par lignes, suivi d'un redressement de la courbure des anneaux, qui fournit une coupe photométrique de ceux-ci, suivant leur grand-axe, tenant compte des mesures faites de part et d'autre.

100.008 **East-west asymmetry of Saturn's rings.**
K. A. Hämeen-Anttila, H. Itävuo.
Astrophys. Space Sci., Vol. 41, 57 - 61 (1976).
    Measurements made in 1974—1975 at the Aarne Karjalainen Observatory do not show any east-west asymmetry of Saturn's rings. Combining the photometric data available for 1913—1975, one finds that the asymmetry varies in an irregular manner without correlating with Saturn's orbital period.

100.009 **The ammonia profile in the atmosphere of Saturn from inversion of its microwave emission spectrum.**
G. Ohring, A. Lacser.
Astrophys. Journ., Vol. 206, 622 - 626 (1976).
    The ammonia profile in the atmosphere of Saturn is inferred directly from inversion of the planet's microwave emission spectrum between 1 and 20 cm wavelength. The inferred profile is characterized by an ammonia mixing ratio of about 1×10^{-4} and the presence of a saturated layer with a base temperature and pressure of about 154 K and 4 atm, respectively.

100.010 **Increasing the spatial resolution on spectrograms of Saturn by methods of numerical linear filtration. I.**
A. M. Gretskij, V. N. Dudinov.
Vestn. Khar'kov. Univ., No. 129 (Ser. Astron., vyp. (No.) 10), p. 53 - 61 (1975). In Russian.

100.011 **On the influence of Titan on the atmosphere of**

Saturn. M. F. Khodyachikh.
Vestn. Khar'kov. Univ., No. 129 (Ser. Astron., vyp. (No.) 10), p. 84 - 86 (1975). In Russian.

100.012 **Brightness variations of Saturn's rings.**
D. W. Hughes.
Nature, Vol. 261, 191 (1976).

100.013 **Coupe photométrique des quatre anneaux de Saturne suivant leur grand-axe, à partir d'enregistrements bi-dimensionnels de photographies de la planète. Les résultats.** P. Guérin.
Comptes Rendus Acad. Sci. Paris, Sér. B, Vol. 282, 365 - 367 (1976).

La méthode de dépouillement utilisée diminue grandement le «bruit» dû la granulation photographique, et améliore la résolution. La précision des coupes photomètriques s'en trouve accrue. L'anneau D est clairement mis en évidence, et sa réalité physique est confirmée.

100.014 **Some systematic observations of Saturn during its 1974—75 apparition.**
E. Sassone-Corsi, P. Sassone-Corsi.
Strolling Astronomer, Vol. 26, 8 - 12 (1976).

100.015 **The rings of Saturn.** J. B. Pollack.
Space Sci. Rev., Vol. 18, 3 - 93 (1975).
This paper reviews observations of the rings of Saturn at visual, infrared, and radio wavelengths. Critical assessments are made of attempts to derive the physical characteristics of the rings from these measurements. A discussion is also given of the origin and evolution of the rings.

100.016 **The Saturnian rings.** H. Alfvén.
Separate print Kungl. Tekn. Hoegsk., Stockholm, Sweden, Inst. Plasmafys., 14 pp. (1975).

Laboratory band strengths of methane and their application to the atmospheres of Jupiter, Saturn, Uranus, Neptune, and Titan. See Abstr. 022.003.

Partially elastic collisions in dense matter.
See Abstr. 022.032.

Saturn observations with Besançon astrolabe (winter 1972—1973). See Abstr. 041.001.

The free oscillations of Jupiter and Saturn.
See Abstr. 099.003.

Pressure-induced absorption by H_2 in the atmospheres of Jupiter and Saturn. See Abstr. 099.018.

Imaging spectrometer study of Jupiter and Saturn.
See Abstr. 099.020.

Estimate of the methane content in the atmospheres of giant planets. See Abstr. 099.032.

Numerical simulation of a system of colliding bodies in a gravitational field. Astrophysical applications.
See Abstr. 151.059.

Saturn, Satellites

100.201 **The satellites of Saturn.**
Strolling Astronomer, Vol. 25, 251 - 252 (1976).

100.202 **Occultation of SAO 80046 by Iapetus on 1976 June 16.** G. E. Taylor.
IAU Circ., No. 2948 (1976).

100.203 **The theory of Enceladus and Dione — an application of computerized algebra in dynamical astronomy.**
W. H. Jefferys, L. M. Ries.
1975 AAS/AIAA Astrodyn. Conf., (see 012.022), AAS75-074, 12 pp. (1975). – Abstr. in Phys. Abstr., Vol. 79, A027695 (1976).

The spatial distribution of long lived gas clouds emitted by satellites in the outer solar system.
See Abstr. 099.212.

On the influence of Titan on the atmosphere of Saturn. See Abstr. 100.011.

Observations of comets, minor planets, Pluto, and satellites. See Abstr. 103.002.

101 Uranus, Neptune, Pluto, Transplutonian Planets

101.001 A measurement of the relative reflectance of Pluto at 0.86 micron. W. A. Lane, J. S. Neff, J. D. Fix.
Publ. Astron. Soc. Pacific, Vol. 88, 77 - 79 (1976).

The authors have examined Pluto's near infrared reflectance by means of photoelectric photometry using a filter whose effective wavelength is 0.86 micron. The ratio of Pluto's reflectance at 0.86 micron to that at 0.55 micron is 1.33 ± 0.04 Pluto's reflectance in the spectral region resembles the reflectance of the asteroid 5 Astraea as well as some stony-iron meteorites and iron meteorites having low nickel content.

101.002 A new constraint on the Uranian satellites' masses. R. Greenberg.
Bull. American Astron. Soc., Vol. 8, 292 (1976). – Abstr. AAS.

101.003 Methane rich models of Uranus. M. Podolak.
Icarus, Vol. 27, 473 - 477 (1976).

A series of models of Uranus is computed assuming that Uranus consists of a core of rocky material surrounded by a convecting envelope rich in H_2O, NH_3, and CH_4. It is found that good fits are obtained to the observed parameters when the $CH_4 : H_2$ ratio is of the order of 0.1. It is suggested that the rotational period of Uranus is roughly 18 hours.

101.004 Possibility of detecting magnetospheric radio bursts from Uranus and Neptune.
C. F. Kennel, J. E. Maggs.
Nature, Vol. 261, 299 - 301 (1976). – Letter.

101.005 Neptune. R. R. Joyce.
IAU Circ., No. 2949 (1976).

101.006 Neptune II (Nereid). K. Aksnes.
IAU Circ., No. 2949 (1976).

101.007 On the farthest planets of the solar system. T. Z. Dworak.
Urania Kraków, Vol. 47, 168 - 171 (1976). In Polish.

The coldest planet: methane ice found on Pluto.
Science, Vol. 192, 362 (1976). – Research note.

Laboratory band strengths of methane and their application to the atmospheres of Jupiter, Saturn, Uranus, Neptune, and Titan. See Abstr. 022.003.

A special capture by Neptune. See Abstr. 042.052.

Stellar occultation spikes as probes of atmospheric structure and composition. See Abstr. 099.017.

Spikes of light during the stellar occultation by planets. See Abstr. 099.025.

Estimate of the methane content in the atmospheres of giant planets. See Abstr. 099.032.

Observations of Ganymede, Callisto, Ceres, Uranus, and Neptune at 3.33 mm wavelength. See Abstr. 099.203.

Observations of comets, minor planets, Pluto, and satellites. See Abstr. 103.002.

102 Comets (Origin, Structure, Atmospheres, Dynamics)

102.001 On the detection of a cometary mass distribution.
A. P. Boss, S. J. Peale.
Icarus, Vol. 27, 119 - 121 (1976).

Precision tracking of deep space probes is less sensitive by three orders of magnitude for detecting an unseen cometary mass distribution at the fringes of the solar system than are the secular perturbations of long-period comets.

102.002 A probability of encounter with interstellar comets and the likelihood of their existence.
Z. Sekanina.
Icarus, Vol. 27, 123 - 133 (1976).

A theory of the probability of encounter of the sun with an interstellar comet at a distance comparable to the earth-sun distance is formulated, and a general expression is derived establishing the relationship among the influx rate of interstellar comets, the perihelion distance, the space density of the comets, the Maxwellian distribution of comet velocities in the interstellar cloud, and the cloud's systematic velocity relative to the sun. The theoretical distribution of semimajor axes of interstellar comets is derived to show that a strong hyperbolic excess must be present in the orbits of a majority of interstellar comets regardless of the dynamical characteristics of the comet cloud, except when the cloud is moving along with the sun and the distribution of individual velocities has a very low dispersion.

102.003 Orbital evolution of the dust streams released from comets. L'. Kresák.
Bull. Astron. Inst. Czechoslovakia, Vol. 27, 35 - 46 (1976).

The orbital evolution of cometary dust grains with characteristic sizes of 1 to 100 μ is discussed. It is shown that the nongravitational perturbations produced by solar radiation strongly depend on the progressive abrasion and fragmentation of individual particles. Since each particle starts in spiralling from an orbit which is determined by the initial blowing-off by direct radiation pressure, the Poynting-Robertson lifetimes have a definite lower limit depending on the orbital elements of the parent body. Even for short-period comets, this limit is relatively high compared with the survival against destruction and perturbational dispersion. Some quantitative estimates of the dynamical evolution are made for the dust streams released from comet Encke, and these are interfaced with the conditions in comets of longer revolution periods. There is apparently no way of maintaining a compact dust stream over a number of revolutions, or displacing it inside the orbit of its parent comet.

102.004 Collision of comets with meteoroids.
O. T. Matsuura.
Icarus, Vol. 27, 323 - 329 (1976).

Statistical analysis of the quantity of dust in the cometary atmosphere in relation to the direction of motion of the comet about the sun suggests an excess of dust for the retrograde comets. This excess is analyzed in the light of Harwit's theory of the cloud of 'boulders' and of Öpik's impact theory. A comparison is also made between these excesses and other cometary phenomena such as splittings and outbursts.

102.005 Stellar perturbations of orbits of long-period comets and their significance for cometary capture.
H. Rickman.
Bull. Astron. Inst. Czechoslovakia, Vol. 27, 92 - 105 (1976).

Approximate expressions for the heliocentric impulse gained by a long-period comet from stellar passages at different distances are derived. The frequency of stellar passages and its dependence on the passage distance is investigated.

The relative importances of different passage distances over very long time intervals are discussed.

102.006 Can 'invisible' bodies be observed in the solar system? M. E. Bailey.
Nature, Vol. 259, 290 - 291 (1976).

Many theories of the origin of comets predict that there are $\sim 10^{11}$ comets in the solar system; most of them, unfortunately, at such a distance that they can never be observed. Some theories go even further and predict that, in addition to these comets in the Oort cloud, the remains of a primeval comet belt may still exist at a distance of ~ 50 AU. The absence of observations of strongly hyperbolic orbits places a limit on the total mass of interstellar comets, but there remains the possibility that a great many comets may be permanently bound to their 'parent' star, and will thus always remain unobserved. It is clear that techniques based on the gravitational effect of hypothetical comets on observed ones are not yet sufficiently well developed to limit speculation, and it is the purpose of the author to suggest a way in which this might be done.

102.007 Les comètes. H. Debehogne.
Ciel et Terre, Vol. 92, 21 - 39 (1976) — with three annexes.

102.008 On the structure of cometary dust tails.
H. Kimura, C.-p. Liu.
Acta Astron. Sinica, Vol. 16, 138 - 166 (1975). In Chinese.

The structure of cometary dust tails is studied in the frame of mechanical theory with special regard to a three-dimensional treatment of the problem. The authors develop a new method for numerical analysis of tail brightness. The basic idea of this method is to combine exact treatment of the motion of a large number of sample particles and counting-technique to estimate the surface brightness integral, taking account of the dust emission characteristics of comets which may be expressed by three source functions, namely, the emission rate $\dot{N}_d(t)$, the modified size-distribution $f(\gamma; t)$, and the velocity distribution where $\Psi(v; \gamma, t)$ ($\gamma = 1 - \mu$). Distribution of tail brightness thus obtained gives essentially the exact solution for the assigned source functions, in the sense that it is not affected by any auxiliary approximations.

102.009 Molecular ions in comet tails.
S. Wyckoff, P. A. Wehinger.
Astrophys. Journ., Vol. 204, 604 - 615 (1976).

Band intensities of the molecular ions CH^+, CO^+, N_2^+, and H_2O^+ have been determined on an absolute scale from tail spectra of comet Kohoutek (1973f) and comet Bradfield (1974b). It is shown that resonance fluorescence is the dominant excitation mechanism for observed comet tail ions at $r \approx 1$ AU. Band system luminosities and molecular ion abundances within a projected nuclear distance $\rho < 10^4$ km have been determined for CH^+, CO^+, N_2^+, and H_2O^+ in comet Kohoutek, and for H_2O^+ in comet Bradfield. The observed H_2O^+ column densities were found to be roughly the same in comet Kohoutek and comet Bradfield at equal heliocentric distances. Finally, the relative abundances of the observed ions and of the presumed parent neutral species are briefly discussed.

102.010 The neutral atmospheres of comets.
D. A. Mendis, W.-H. Ip.
Astrophys. Space Sci., Vol. 39, 335 - 385, with a correction, Vol. 42, 255 (1976).

Review paper. – Contents: (1) Introduction; (2) Observations; (3) Excitation mechanisms; (4) Parent molecules; (5)

Gas phase chemistry; (6) Dynamic models of the neutral atmosphere; (7) The chemical composition of comets; (8) Conclusions.

102.011 Kurzperiodische Kometen als Quellen für interplanetaren Staub. S. Röser.
Mitt. Astron. Ges., No. 38, p. 149 (1976). − Abstract.

102.012 Thermalization of cometary hydrogen.
H. U. Keller.
Mitt. Astron. Ges., No. 38, p. 150 - 152 (1976). − Short report.

102.013 A data processing algorithm for comparing theory with observations in the case of cometary motions.
K. Ziołkowski.
Postępy Astron., Vol. 24, 125 - 130 (1976). In Polish.

An algorithm for processing the ephemeris data is presented. This algorithm was developed in the course of work on the Catalogue of One-Apparition Comets.

102.014 The interpretations of ultraviolet observations of comets. H. U. Keller.
Space Sci. Rev., Vol. 18, 641 - 684 (1976). − Review paper.

102.015 The structure of cometary atmospheres. I: Temperature distribution. M. Shimizu.
Astrophys. Space Sci., Vol. 40, 149 - 155 (1976).

The temperature distributions in cometary atmospheres at various heliocentric distances for comets of Bennett and Encke types have been calculated by taking into account heating due to the absorption of solar ultraviolet radiation, cooling by H_2O far infrared emission, and various dynamical processes (expansion, advection, and thermal conduction). The agreement of the results with the observations is in general satisfactory.

102.016 The dependence "oblateness − size" of isophotes of the coma of comets. V. L. Afanas'ev.
Astrometriya i Astrofizika, *Kiev*, vyp. (No.) 29, (see 003.003), p. 76 - 80 (1976). In Russian.

In the framework of a model of a cool collision-free coma of a comet the dependence oblateness − size of the isophotes of the coma is obtained which connects the parameters of a neutral gas flow into the coma in dependence on the asymmetry of the isophotes. These dependences are constructed using the isophotes of the comets 1902b, 1903c, 1911c, and 1942g and compared with the theoretical ones. The dependence of the velocity of the neutral gas flow into the coma on the heliocentric distance is determined.

102.017 The structure of cometary atmospheres. II. Ion distribution. M. Shimizu.
Astrophys. Space Sci., Vol. 40, 243 - 251 (1976).

The distributions of various kinds of molecular ions in the atmospheres of new and old comets made up from dirty ice of the second kind (H_2O ice and hydrate clathrates of CO and N_2) have been computed at various heliocentric distances, by taking into account photoionization, ion-molecular reactions, electron-ion recombinations, and some transport effects. The results have been compared with observations and other computations. It is argued that dirty ice of the second kind model will impose a restriction on the theory of the origin of the solar system.

102.018 On the possibility of H_2-near-infrared detection in comets with large gas production.
K. W. Michel, T. Nishimura, C. B. Cosmovici.
Astrophys. Space Sci., Vol. 40, 253 - 280 (1976).

Comets with large gas production offer a unique chance to observe a H_2-flux of about 10^5 photon $cm^{-2} s^{-1} sr^{-1}$

(1 Rayleigh) at wavelengths 8497.4 Å, 8560.2 Å and 8747.9 Å − i.e., where photon counting methods are still applicable. In the following it will be shown that population of the vibrational levels, giving rise to these quadrupole overtone transitions, is dominated by photodissociation of methane, and that the emission even of quadrupole lines is not attenuated by collisional quenching.

102.019 Production, in a stationary state, of solid material in a coma of cometary nuclei under the action of solar radiation. O. T. Matsuura.
Astrophys. Space Sci., Vol. 41, 195 - 211 (1976).

A stationary state of production of solid material in the coma of a comet is sought by assuming a production and dynamics of solid grains liberated during the vaporization of cometary nuclei under the action of solar radiation.

102.020 Distribution of neutral particles in cometary atmospheres taking into account the optical depth in the UV region. G. G. Novikov, N. Tujchiev.
Komety i Meteory, *Dushanbe*, No. 23, p. 3 - 7 (1974). In Russian.

A formula for density of particles taking into account the optical depth of the cometary atmospheres in the ultraviolet is obtained. Asymptotic formulae for the density of the cometary particles are given.

102.021 Statistics of cometary orbits. V. P. Tomanov.
Komety i Meteory, *Dushanbe*, No. 23, p. 8 - 11 (1974). In Russian.

New statistical regularities are presented in the system of almost parabolic comets: the increasing concentration of the perihelia towards the solar apex, the dependence of the perihelion distance and average absolute magnitude of comets on the angular distance of the perihelion from the solar apex.

102.022 Determination of the temperature of a cometary nucleus by brightness observations.
M. Z. Markovich.
Komety i Meteory, *Dushanbe*, No. 23, p. 12 - 17 (1974). In Russian.

A dependence has been established between the generalized photometric parameter $n(r)$ and the average temperature of a cometary nucleus for two most simple versions of the ice model. The structure of the surface layer of the nucleus, the temperature and the sublimation heat of ice and other characteristics have been determined by visual observational data on comets 1955 IV and 1955 V.

102.023 Solar apex in relation to the protocometary cloud. V. P. Tomanov.
Astron. Zhurn. Akad. Nauk SSSR, Vol. 53, 647 - 654 (1976). In Russian. English translation in Soviet Astron., Vol. 20, No. 3.

Coordinates of the apex of the peculiar solar motion in relation to the complex of perihelions of cometary orbits have been obtained on the basis of the theories of gravitational focusing and interstellar origin of Lyttleton comets. New arguments in favour of the interstellar origin of comets have been advanced. Conclusions on the evolution of cometary orbits that are in agreement with the results of the diffusion theory have been made.

102.024 Dust in comets and interplanetary matter. V. Vanýsek.
IAU Colloquium No. 31, (see 012.015), p. 299 - 313 (1976). Invited paper.

102.025 The production rate of dust by comets.

A. H. Delsemme.
IAU Colloquium No. 31, (see 012.015), p. 314 - 318 (1976).

102.026 **Can short period comets maintain the zodiacal cloud?** S. Röser.
IAU Colloquium No. 31, (see 012.015), p. 319 - 322 (1976).

102.027 **Optical properties of cometary dust.**
S. Hayakawa, T. Matsumoto, T. Ono.
IAU Colloquium No. 31, (see 012.015), p. 323 - 327 (1976).

102.028 **The dust coma of comets.**
K. W. Michel, T. Nishimura.
IAU Colloquium No. 31, (see 012.015), p. 328 - 333 (1976).

102.029 **Predicted favorable visibility conditions for anomalous tails of comets.** Z. Sekanina.
IAU Colloquium No. 31, (see 012.015), p. 339 - 342 (1976).

102.030 **Can comets be the only source of interplanetary dust?** A. H. Delsemme.
IAU Colloquium No. 31, (see 012.015), p. 481 - 484 (1976).

102.031 **Le comete.** E. Proverbio.
Repr. from Giorn. Fis., Vol. 16, No. 1, 22 pp. = Pubbl. Stazione Astron. Internazionale Latitudine, Carloforte–Cagliari, Nuova Ser., No. 35 (1975).

102.032 **Mass accretion and variations of elements of comets' orbits.** S. L. Piotrowski, G. Sitarski.
Acta Astron., Vol. 26, 77 - 81 (1976).
The effects of mass accretion on elements of a comet approaching the sun along a parabolic or slightly hyperbolic orbit are studied by numerical methods. As compared to results obtained by linearization of analytical formulae the present note confirms for the case of a collision with a very small mass the conclusions about an increase of the chance of discovery for cosmical comets with retrograde motion as compared with comets moving along direct orbits. For increasing accreted mass the orbits of initially retrograde orbits show a tendency to become direct. The statistics of inclinations of elongated cometary orbits supports the results of computations.

102.033 **Sulla struttura e la dinamica delle comete di tipo I.**
A. Santarelli.
Repr. from Mem. Accad. Sci. Torino, 29 pp. = Contr. Oss. Astron. Torino, (Pino Torinese), No. 92 (1976).

102.034 **On the optical properties of cometary dust.**
T. Ono.
Publ. Astron. Soc. Japan, Vol. 28, 229 - 238 (1976).
Referring to photometric observations of comet 1973 XII Kohoutek and comet 1974 III Bradfield over a wide wavelength range and to polarimetric measurements of comet Kohoutek in the optical and the near-infrared range, a grain model in the comas is constructed. The facts that the spectrum of scattered light is similar to the solar spectrum and that the spectrum of thermal emission is nearly Planckian favor micron-size grains.

102.035 **Distribution of the aphelia of long-period comets.**
I. Hasegawa.
Publ. Astron. Soc. Japan, Vol. 28, 259 - 276 (1976).
By examining the distribution of aphelia of all the comets observed since 1800, the author finds that the aphelia of long-period comets show a remarkable concentration near the antapex of the solar motion. Some possible interpretations of this phenomenon are discussed in connection with the origin of long-period comets.

102.036 **The problems of data processing for the comparison of theory with observations in the case of cometary motions.** M. Bielicki.
Postępy Astron., Vol. 24, 59 - 64 (1976). In Polish.

The structure and origin of comets.
See Abstr. 011.019.

R-centroids and Franck-Condon factors of CH⁺ molecule. See Abstr. 022.097.

A special capture by Neptune. See Abstr. 042.052.

Space probes versus comets.
See Abstr. 051.015.

Shuttle-launched multi-comet mission 1985.
See Abstr. 051.023.

Viscous interaction between the solar wind and cometary plasmas. See Abstr. 074.026.

Cometary impact and the magnetization of the moon. See Abstr. 094.101.

The cometary contribution to cosmic dust.
See Abstr. 106.020.

Sun and comets as sources in an external flow.
See Abstr. 106.036.

Zum Einfluss des solaren Strahlungsdrucks auf die Bewegung kleiner Teilchen im Sonnensystem.
See Abstr. 106.051.

Interaction of interplanetary magnetic fields with cometary plasma. See Abstr. 106.052.

On the structure of hyperbolic interplanetary dust streams. See Abstr. 106.084.

103   Comets: Listed Objects

103.001   **Conditions d'observation en Belgique des comètes périodiques en 1976.**
H. Debehogne, A. J. Sauval.
Ciel et Terre, Vol. 92, 7 - 20 (1976).

103.002   **Observations of comets, minor planets, Pluto, and satellites.**
G. Van Biesbroeck, C. D. Vesely, K. Aksnes, B. G. Marsden.
Astron. Journ., Vol. 81, 122 - 124 (1976).
Further astrometric observations made of solar system objects by the late G. Van Biesbroeck are presented. In addition to observations of comets, minor planets, and Pluto, the present paper includes observations of Phoebe, Nereid, and the satellites of Uranus.

103.003   **Search for comets Perrine, Mrkos and Westphal.**
Kometn. Tsirk., *Kiev*, No. 190 (1975). In Russian.

103.004   **Definitive designations of comets of 1974 (B. G. Marsden, IAU Circ. No. 2898).**
Kometn. Tsirk., *Kiev,* No. 193 (1976). In Russian.

103.005   **Is the accretion of masses of comets possible?**
S. K. Vsekhsvyatskij.
Kometn. Tsirk., *Kiev,* No. 194 (1976). In Russian.

103.006   **Observations of comets at the Kleť Observatory.**
Kometn. Tsirk., *Kiev,* No. 188 (1975). In Russian.
Outburst of comet Schwassmann-Wachmann 1; Comet Suzuki-Saigusa-Mori, 1975k; Comet Mori-Sato-Fujikawa, 1975j (*A. Mrkos*).

103.007   **Observations of comets.**   C. Y. Shao, G. Schwartz, R. E. McCrosky, J. H. Bulger.
IAU Circ., No. 2898 (1976). – Concerning P/Schwassmann-Wachmann 1, P/Encke, Kojima (1973 II), van den Bergh (1974g), P/Smirnova-Chernykh (1975e).

103.008   **Roman numeral designations of comets in 1974.**
IAU Circ., No. 2898 (1976).

103.009   **Observations of comets.**
G. Schwartz, R. E. McCrosky, C. Y. Shao, J. H. Bulger.
IAU Circ., No. 2967 (1976). – Concerning P/Schwassmann-Wachmann 1, Bradfield (1975d), P/Smirnova-Chernykh (1975e), P/Arend (1975m), P/Wolf (1975f).

103.010   **Comets in the year 1975.**   J. Bouška.
Vesmír, Vol. 55, 124 (1976). In Czech.

103.011   **Recoveries of periodic comets.**
E. Roemer, C. A. Heller.
British Astron. Ass. Circ., No. 573 (1976).
Concerning: P/d'Arrest 1976e, P/Pons-Winnecke 1976f, P/Johnson 1976h, P/Faye 1976i.

. 15 Kometenflugblätter des 17. und 18. Jahrhunderts
I.   See Abstr. 004.027.

15 Kometenflugblätter des 17. und 18. Jahrhunderts.
II.   See Abstr. 004.028.

Color filter techniques for bright comets.
See Abstr. 034.079.

103.100   **Comet 1969 IX Tago-Sato-Kosaka**

Monochromatic observations of comet Tago-Sato-Kosaka 1969g (1969 IX).
J. Rahe, C. W. McCracken, K. L. Hallam, B. D. Donn.
Astron. Astrophys., Suppl. Ser., Vol. 23, 1 - 12 (1976).
Isophotes have been determined from 10 narrow-band filter photographs of comet Tago-Sato-Kosaka 1969g (1969 IX) taken between February 11 and 14, 1970. The five interference filters used were centered on the CN λ3883Å, C_2 λ4737Å, C_2 λ5165Å, C_3 λ4050Å sequences, and the continuum at λ5300Å. Gradients of intensity in various directions from the nucleus have been derived from the isophotes.

103.101   **Comet 1970 II Bennett**

Certain physical characteristics of comets Bennett 1969i and Mrkos 1957d based on their absolute spectrophotometry.   E. B. Kostyakova.
Astrometriya i Astrofizika, *Kiev,* vyp. (No.) 29, (see 003.003), p. 81 - 86 (1976). In Russian.
The absolute energy flux of these comets was measured in the band λ 3883 CN and some other nearby bands. The mass of radiating gas was estimated for the inner parts of the comets.

Monochromatic and white-light observations of comet Bennett 1969i (1970 II).
J. Rahe, C. W. McCracken, B. D. Donn.
Astron. Astrophys., Suppl. Ser., Vol. 23, 13 - 35 (1976).
Isophotes have been determined from 32 photographs of comet Bennett 1969i (1970 II) taken during the period March 28 to April 18, 1970. The six interference filters used were centered on the CN λ3883Å, C_2 λ4737Å, C_2 λ5165Å, CO^+ λ4277Å sequences, on the sodium-D-lines at λ5893Å, and on the continuum at λ5300Å. Intensity gradients have been derived from these isophotes.

103.102   **Comet 1973 XII Kohoutek**

Les observations photométriques des comètes: La comète Kohoutek 1973f.   D. Andrienko.
Ann. Obs. Astron. Alger, Vol. 4, Fasc. 3, p. 1 - 15 (1975).

Absolute spectrophotometry of comet Kohoutek 1973f. I. Instruments and methods of measurements.
V. P. Tarashchuk, G. A. Terez, Eh. I. Terez.
Astrometriya i Astrofizika, *Kiev,* vyp. (No.) 28, (see 003.002), p. 109 - 116 (1976). In Russian.
The observations were carried out on January 14, 19 and 30, 1974 at the Crimean Astrophysical Observatory. The method of observations, reductions and the accuracy of results are described. The curves of spectral flux density $E(\lambda)$ produced by the comet beyond the earth's atmosphere are given as functions of λ.

Spectrophotometric study of comet Kohoutek 1973f.
Yu. V. Sizonenko.
Astrometriya i Astrofizika, *Kiev,* vyp. (No.) 29, (see 003.003), p. 87 - 88 (1976). In Russian.
The values of luminosity for 9 emission bands as well as the temperature of radiation scattered by dust are given.

On the ionization and excitation of H_2O^+ in comet Kohoutek (1973f).   S. Wyckoff, P. A. Wehinger.
Astrophys. Journ., Vol. 204, 616 - 625 (1976).

Intensities have been measured for the H_2O^+ $\tilde{A}^2A_1 \rightarrow$ $\tilde{X}^2B_1$ bands in tail spectra of comet Kohoutek within a region of approximately 10^4 km of the nucleus. Accurate H_2O photo-decomposition rates have been calculated and H_2O ionization rates for solar-wind interactions (electron impact and charge exchange) and gas-phase reactions estimated. The dominant excitation mechanism was shown to be resonance fluorescence. A dominant ionization process could not be determined. Column densities for H_2O^+ ions in the tail at projected nuclear distances, $\rho \approx 10^4$ km, were found to be $\sim 10^{10\pm1}$ to $\sim 10^{11\pm1}$ cm^{-2} for heliocentric distances in the range 0.5 to 0.9 AU; while an upper limit for the H_2O^+ column density on the sunward side of the coma, $\rho \gtrsim 5 \times 10^3$ km, was estimated to be $\lesssim 10^{9\pm1}$ cm^{-2}.

Posiciones exactas del cometa Kohoutek (1973f) obtenidas en el observatorio astronómico de Madrid.
M. de Pascual.
Bol. Astron. Obs. Madrid, Vol. 9, No. 2, p. 84 (1976).

Dust emission from comet Kohoutek (1973f) at large distances from the sun.
E. Grün, J. Kissel, H.-J. Hoffmann.
IAU Colloquium No. 31, (see 012.015), p. 334 - 338 (1976).

Study of the anti-tail of comet Kohoutek from an observation on 17 January 1974.
P. L. Lamy, S. Koutchmy.
IAU Colloquium No. 31, (see 012.015), p. 343 - 344 (1976). Abstract.

On the nature of the anti-tail of comet Kohoutek (1973f). II. Comparison of the working model with ground-based photographic observations.
Z. Sekanina, F. D. Miller.
Icarus, Vol. 27, 135 - 146 (1976).

Photographic observations of the anti-tail of comet Kohoutek (1973f), obtained at the Cerro Tololo Inter-American Observatory, are photometrically reduced and the results compared with a recently formulated working model of the anti-tail. The most important result reached so far is a quantitative confirmation of the previously suggested hypothesis, arguing that dust particles in the anti-tail suffered a significant loss in radius due to evaporation near the perihelion passage. The authors find that only particles initially larger than 100−150 μm in diameter (at an assumed density of 1 g cm^{-3}) survived.

Absolute spectrophotometry of comet Kohoutek 1973f. I. Column densities of cyanogen.
J. S. Neff, D. Ketelsen, G. D. Schmidt, J. B. Tatum.
Icarus, Vol. 27, 545 - 551 (1976).

Absolute spectrophotometry of the coma of comet Kohoutek 1973f is discussed for the nights of January 24 and 26, 1974. Specific intensities are measured for spectral features and a continuum band in the wavelength region $\lambda\lambda$ 3460−6062 Å. The (0, 0) band of the $\Delta v = 0$ sequence of the violet system of the cyanogen molecule is analyzed and column densities of 1.7×10^{15} m^{-2} and 3.4×10^{14} m^{-2} are found for January 24 and 26, 1974, respectively.

Photoelectric observations of comet Kohoutek 1973f.   J. Svoreň, J. Tremko.
Kozmos, Vol. 7, 54 - 55 (1976). In Slovak.

Comparison of Lyman alpha and He λ 10830 line structures and variations in early-type star atmospheres. Final technical report, January - October 1974.
See Abstr. 064.070.

Solar wind interaction with the tail of comet Kohoutek.   See Abstr. 074.021.

103.103   Comet 1975n West

La comète West (1975n).   J. Meeus.
L'Astronomie, Vol. 90, 40 - 41 (1976).

Comet West (1975n).   K. S. Krishna Swamy.
Bull. Astron. Soc. India, Vol. 4, 13 (1976).

Comet West (1975n).
R. R. D. Austin, P. M. Kilmartin, A. C. Gilmore.
IAU Circ., No. 2894, with a correction, No. 2905 (1976).

Comet West (1975n).
IAU Circ., Nos. 2902, 2914, 2916 (1976).

Comet West (1975n).
J. Hers, A. C. Gilmore, P. M. Kilmartin.
IAU Circ., No. 2905, with a correction, No. 2912 (1976).

Comet West (1975n).
A. C. Gilmore, P. M. Kilmartin, D. Herald, B. G. Marsden, E. P. Ney, K. M. Merrill, E. Everhart.
IAU Circ., No. 2910, with a correction, No. 2928 (1976).

Comet West (1975n).   A. Gomez.
IAU Circ., No. 2912 (1976).

Comet West (1975n).
IAU Circ., No. 2916 (1976).

Comet West (1975n).   E. P. Ney, J. Stoddart.
IAU Circ., Nos. 2917, 2919, with a correction, No. 2924 (1976).

Comet West (1975n).   G. Babu, D. A. Ketelsen, J. S. Neff, J. Young, Z. Sekanina, D. Elmore, S. Koutchmy.
IAU Circ., No. 2924 (1976).

Comet West (1975n).   J. Bortle.
IAU Circ., No. 2927 (1976).

Comet West (1975n).   D. Herald, A. C. Gilmore, P. M. Kilmartin, D. J. Gans, M. P. Candy, H. L. Giclas, M. L. Kantz, B. Milet, J. C. Webber, L. E. Snyder, R. M. Crutcher, G. W. Swenson, C. Barth, G. Lawrence, W. Weller, S. Jeffers, L. Danylewych, J. P. Swings, R. E. McCrosky, G. Schwartz, E. M. Leibowitz, M. Rosenkrantz, A. Levite, T. L. Rokoske, J. Young, J. A. Farrell, J. Bortle, E. P. Ney, K. M. Merrill, J. S. Neff, D. A. Ketelsen, V. V. Smith.
IAU Circ., No. 2928 (1976).

Comet West (1975n).   J. Young, R. Newburn, P. Maley, J. Bortle, Z. Sekanina.
IAU Circ., No. 2930 (1976).

Comet West (1975n).   H. L. Giclas, M. L. Kantz, K. Tomita, B. G. Marsden, J. Bortle, Z. Sekanina, S. Wyckoff, P. A. Wehinger.
IAU Circ., No. 2931 (1976).

Comet West (1975n).   J. M. Codina, J. M. Mundet, N. Torras, T. Seki, B. Milet, U. Surawski, U. Hopp, H. L. Giclas, M. L. Kantz, J. Bortle, S. D. Sinvhal, G. Babu, P. Bowers.
IAU Circ., No. 2937 (1976).

Comet West (1975n).   M. Bielicki, R. R. de
Freitas Mourão, J. M. Codina, B. Milet, K. Tomita, H. L.
Giclas, M. L. Kantz, Z. Sekanina, E. Gérard, I. Kazès, R.
Lauque.
IAU Circ., No. 2943 (1976).

Comet West (1975n).   C. Torres, H. Debehogne,
S. Vaghi, G. De Sanctis, J. M. Codina, N. Torras, M. Bielicki,
D. Bielicki, H. L. Giclas, M. L. Kantz, K. Tomita, J. W.
Christy, J. Bortle.
IAU Circ., No. 2948, with a correction, No. 2956 (1976).

Comet West (1975n).   D. Eaton, A. Peaceman,
H. K. Raudsaar, T. Seki, H. L. Giclas, M. L. Kantz.
IAU Circ., No. 2954 (1976).

Comet West (1975n).
M. P. Candy, P. Jekabsons, K. Tomita, Z. Sekanina.
IAU Circ., No. 2956 (1976).

Comet West (1975n).   D. Gans, M. P. Candy,
C. Jekabsons, R. R. de Freitas Mourão, M. R. Dykes, K. A.
Haddow, R. L. Waterfield, A. C. Gilmore, P. M. Kilmartin.
IAU Circ., No. 2964 (1976).

New comet West, 1975n.   B. G. Marsden.
Kometn. Tsirk., *Kiev*, No. 187 (1975). In Russian.

Comet West, 1975n.
Kometn. Tsirk., *Kiev*, No. 188 (1975). In Russian.

Comet West, 1975n.
Kometn. Tsirk., *Kiev*, No. 189 (1975). In Russian.

Comet West, 1975n.
Kometn. Tsirk., *Kiev*, No. 191 (1976). In Russian.

Comet West, 1975n.
Kometn. Tsirk., *Kiev*, No. 194 (1976). In Russian.

Comet West, 1975n.
Kometn. Tsirk., *Kiev*, No. 195 (1976). In Russian. – Narrow-
band photometry of the head of comet West, 1975n (*D. I.
Gorodetskij*); Spectrum of comet West obtained in Lesnikakh
(*K. I. Churyumov, L. V. Yurevich*); Observations of comet
West at the Ussurijsk Solar Station (*V. A. Golubev, V. D.
D'yakonova*).

Comet West, 1975n.
Kometn. Tsirk., *Kiev*, No. 196 (1976). In Russian. – Brightness and length of the tail from observations at the Ussurijsk
Solar Station (*V. A. Golubev, V. D. D'yakonova*); Observations in Barvukha-2, Vitebsk region (*Yu. V. Selenok*); Estimates in Balyasnoe, Poltava region (*V. Kiva*); Brightness estimates
in Kiev on a light background (*S. K. Vsekhsvyatskij*); Observations of the comet in Stormville (*J. Bortle*).

Comet West, 1975n.
Kometn Tsirk., *Kiev*, No. 197 (1976). In Russian. – Observa-
tions at the Ussurijsk Solar Station (*V. A. Golubev, V. D.
D'yakonova*); Observations in Pulkovo (*N. M. Bronnikova*);
Polarization observations at the Gissar Observatory (*N. N.
Kiselev, G. P. Chernova*); Report of B. G. Marsden (IAU Circ.
No. 2943).

Comet West 1975n.   J. C. Bennett.
Monthly Notes Astron. Soc. Southern Africa, Vol. 35, 54 - 55
(1976).

Komet West (1975 n).
Orion, 34. Jahrgang, p. 64 - 69 (1976).

Observation of comet West 1975n.   M. Dujnič.
Říše hvězd, Vol. 57, 94 (1976). In Slovak.

Observation of comet West 1975n.
J. Květoň.
Říše hvězd, Vol. 57, 139 (1976). In Czech.

The stage is set for comet West.
Sky Telescope, Vol. 51, 173 (1976).

Comet West in the morning sky.
Sky Telescope, Vol. 51, 219, 240 - 241 (1976).

Comet West's fine performance.
Sky Telescope, Vol. 51, 312 - 321 (1976).

Disintegration phenomena in comet West.
Z. Sekanina.
Sky Telescope, Vol. 51, 386 - 393 (1976).

Drawings and photographs of comet West (1975n).
Strolling Astronomer, Vol. 26, 33 - 38 (1976).

Komet West.   H. Vehrenberg.
SuW, 15. Jahrgang, p. 112 - 113 (1976).

Neues vom Kometen West.
SuW, 15. Jahrgang, p. 171 (1976).

**Spectaculaire komeet aan de ochtendhemel. De
eerste waarnemingen van komeet West (1975n).**
Zenit, 3e jaargang, p. 159 - 161 (1976).

103.104   Comet 1965 VIII Ikeya-Seki

**Polarization reversal in the tail of comet Ikeya-Seki
(1965 VIII).**   J. L. Weinberg, D. E. Beeson.
Astron. Astrophys., Vol. 48, 151 - 153 (1976).
    Multicolor observations of brightness and polarization in
the continuum throughout the tail of comet 1965 VIII reveal
the presence of positive and negative polarization (polarization
reversal) characteristic of slightly absorbing particles. The
phase angle of the neutral point is found to vary with color and
with time, the latter indicating short-term changes in the
properties of the particles after perihelion.

103.105   Comet 1976c Schuster

Comet Schuster (1976c).   H.-E. Schuster.
IAU Circ., No. 2923 (1976).

Comet Schuster (1976c).   E. Roemer.
IAU Circ., No. 2924 (1976).

Comet Schuster (1976c).   H.-E. Schuster,
A. C. Gilmore, P. M. Kilmartin, E. Roemer, B. G. Marsden.
IAU Circ., No. 2926 (1976).

Comet Schuster (1976c).
B. M. Blanco, H.-E. Schuster, G. Pizarro, O. Pizarro, R. M.
West, E. Roemer, L. M. Vaughn, B. G. Marsden.
IAU Circ., No. 2941 (1976).

Comet Schuster (1976c).
A. C. Gilmore, P. M. Kilmartin, B. G. Marsden.
IAU Circ., No. 2958 (1976).

New comet Schuster, 1976c.
Kometn. Tsirk., *Kiev,* No. 194 (1976). In Russian.

Comet Schuster, 1976c.
Kometn. Tsirk., *Kiev,* No. 196 (1976). In Russian.

Eine weitere Kometenentdeckung bei ESO.
H.-E. Schuster.
SuW, 15. Jahrgang, p. 113 (1976). − Concerning 1976c.

103.106 Comet 1975h Kobayashi-Berger-Milon

Comet Kobayashi-Berger-Milon (1975h).
N. A. Mis'kin, V. D. Motrich, N. M. Shiper, K. Suzuki, T.
Urata, T. Seki, J. Hers.
IAU Circ., No. 2894 (1976).

Comet Kobayashi-Berger-Milon (1975h).
A. C. Gilmore, R. R. D. Austin, P. M. Kilmartin.
IAU Circ., No. 2912 (1976).

Comet Kobayashi-Berger-Milon (1975h).
J. A. Bruwer, P. L. Fischer, P. Jackson, T. J. Kreidl, W. Ferreri,
C. Torres, S. Barros.
IAU Circ., No. 2950 (1976).

Comet Kobayashi-Berger-Milon, 1975h.
Kometn. Tsirk., *Kiev,* No. 187 (1975). In Russian. − Observations of the comet in Tartu.

Photographic observations of comet 1975h in Odessa.
G. R. Kastel', N. A. Mis'kin, V. D. Motrich, N. M. Shiper.
Kometn. Tsirk., *Kiev,* No. 188 (1975). In Russian.

Comet Kobayashi−Berger−Milon, 1975h.
Kometn. Tsirk., *Kiev,* No. 189 (1975). In Russian. − Observations at the Skalnaté Pleso Observatory (*M. Antal*); Observations of the spectrum in Abastumani (*I. R. Bejtrishvili*); Estimates of the integral brightness in Stormville.

Observations of comet Kobayashi−Berger−Milon 1975h at the Saratov University.
L. N. Berdnikov, A. V. Prokhorov.
Kometn. Tsirk., *Kiev,* No. 190 (1975). In Russian.

Comet Kobayashi−Berger−Milon, 1975h.
Kometn. Tsirk., *Kiev,* No. 192 (1976). In Russian.

Observations of comets in Pulkovo.
Kometn. Tsirk., *Kiev,* No. 195 (1976). In Russian. − Comet Kobayashi-Berger-Milon, 1975h; Comet Suzuki-Saigusa-Mori, 1975k (*N. M. Bronnikova*).

Comet Kobayashi-Berger-Milon, 1975h.
Kometn. Tsirk., *Kiev,* No. 196 (1976). In Russian. − Observations at the Engelhardt Observatory (*G. V. Zhukov, M. I. Kibardina, L. A. Urasin*); Observations of the brightness and form of comet 1975h (*J. Květoň*); Observations of the comet in the Novosibirsk region (*L. L. Sikoruk*).

Comet Kobayashi-Berger-Milon 1975h.
J. Bouška.
Říše hvězd, Vol. 57, 25 - 30 (1976). In Czech.

Observation of comet 1975h.
V. Novotný, J. Mach.
Říše hvězd, Vol. 57, 57 - 60 (1976). In Czech.

Komet Kobayashi-Berger-Milon (1975 h).

F. Börngen.
Sterne, Vol. 52, 162 - 164 (1976).

Observations of comet 1975 h Kobayashi-Berger-Milon. J. Speil.
Urania Kraków, Vol. 47, 25 - 26 (1976). In Polish.

Comet Kobayashi-Berger-Milon and its discoverers.
K. I. Churyumov.
Zemlya i Vselennaya, 1976, No. 3, p. 42 - 49. In Russian.

Komeet Kobayashi-Berger-Milon 1975 h.
H. Feijth.
Zenit, 3e jaargang, p. 125 - 129 (1976).

103.107 Comet 1975k Suzuki-Saigusa-Mori

Comet Suzuki-Saigusa-Mori (1975k).
K. Suzuki, J. Hers.
IAU Circ., No. 2903 (1976).

Comet Suzuki-Saigusa-Mori (1975k).
W. Ferreri, P. M. Kilmartin, A. C. Gilmore, R. R. D. Austin,
C. Torres.
IAU Circ., No. 2913 (1976).

Comet Suzuki-Saigusa-Mori (1975k).
Jekabsons, D. Gans, M. P. Candy, S. Barros, C. Torres.
IAU Circ., No. 2958 (1976).

Comet Suzuki-Saigusa-Mori, 1975k.
Kometn. Tsirk., *Kiev,* No. 187 (1975). In Russian.

Comet Suzuki−Saigusa−Mori, 1975k.
Kometn. Tsirk., *Kiev,* No. 192 (1976). In Russian.

Observations of comets in Pulkovo.
See Abstr. 103.106.

Brightness of comets Mori−Sato−Fujikawa and Suzuki−Saigusa−Mori. See Abstr. 103.108.

103.108 Comet 1975j Mori-Sato-Fujikawa

Comet Mori-Sato-Fujikawa (1975j).
D. Herald, K. Suzuki, T. Kurosaki, J. Hers.
IAU Circ., No. 2894 (1976).

Comet Mori-Sato-Fujikawa (1975j).
R. R. D. Austin, A. C. Gilmore, P. M. Kilmartin, D. Herald,
J. Hers.
IAU Circ., No. 2908 (1976).

Comet Mori-Sato-Fujikawa (1975j).
W. Ferreri, C. Torres, H. Debehogne, R. R. de Freitas Mourão,
J. Hers.
IAU Circ., No. 2922 (1976).

Comet Mori-Sato-Fujikawa (1975j).
K. Suzuki, T. Urata, D. Herald.
IAU Circ., No. 2940 (1976).

Comet Mori-Sato-Fujikawa (1975j).
S. Barros, A. C. Gilmore, P. M. Kilmartin, B. G. Marsden.
IAU Circ., No. 2955 (1976).

Comet Mori-Sato-Fujikawa, 1975j.

Kometn. Tsirk., *Kiev,* No. 187 (1975). In Russian.

Brightness of comets Mori—Sato—Fujikawa and Suzuki—Saigusa—Mori.   J. E. Bortle.
Kometn. Tsirk., *Kiev,* No. 189 (1975). In Russian.

Comet Mori—Sato—Fujikawa, 1975j.
Kometn. Tsirk., *Kiev,* No. 192 (1976). In Russian.

103.109   Comet 1975*l* Harrington-Abell

Periodic comet Harrington-Abell (1975*l*).
G. Schwartz, C. Y. Shao, E. Roemer, M. A. Daniel.
IAU Circ., No. 2894 (1976).

Periodic comet Harrington-Abell (1975*l*).
C. Y. Shao, J. H. Bulger, G. Schwartz, R. E. McCrosky.
IAU Circ., No. 2965 (1976).

Rediscovery of comet Harrington-Abell, 1975*l*.
Kometn Tsirk., *Kiev,* No. 187 (1975). In Russian.

103.110   Comet 1975e Smirnova-Chernykh

Periodic comet Smirnova-Chernykh (1975e).
T. Seki, C. Y. Shao, B. G. Marsden.
IAU Circ., No. 2918 (1976).

Periodic comet Smirnova-Chernykh (1975e).
C. Y. Shao, N. S. Chernykh.
IAU Circ., No. 2945 (1976).

Periodic comet Smirnova-Chernykh, 1975e.
Kometn. Tsirk., *Kiev,* No. 187 (1975). In Russian.

Comet Smirnova-Chernykh, 1975e.
N. S. Chernykh.
Kometn. Tsirk., *Kiev,* No. 195 (1976). In Russian.

103.111   Comet 1975i Churyumov-Gerasimenko

Comet Churyumov-Gerasimenko, 1975i.
Kometn. Tsirk., *Kiev,* No. 187 (1975). In Russian. — Observations of comet 1975i at the Byurakan Observatory (*K. I. Churyumov*).

103.112   Comet 1975f Wolf 1

Periodic comet Wolf (1975f).
IAU Circ., No. 2952 (1976).

Periodic comet Wolf, 1975f.
Kometn. Tsirk., *Kiev,* No. 187 (1975). In Russian.

Comet Wolf-Kamieński.
E. I. Kazimirchak-Polonskaya.
Kometn. Tsirk., *Kiev,* No. 193 (1976). In Russian.

103.113   Comet 1975m Arend

Periodic comet Arend (1975m).   T. Seki.

IAU Circ., No. 2926 (1976).

Rediscovery of the short-periodic comet Arend, 1975m.
Kometn. Tsirk., *Kiev,* No. 188 (1975). In Russian.

103.114   Comet 1975p Bradfield

Comet Bradfield (1975p).
A. C. Gilmore, P. M. Kilmartin.
IAU Circ., No. 2895 (1976).

Comet Bradfield (1975p).
IAU Circ., No. 2899 (1976).

Comet Bradfield (1975p).
J. Hers, T. Seki, J. M. Codina, N. Torras, T. Urata, B. Milet, R. L. Waterfield.
IAU Circ., No. 2904, with a correction, No. 2908 (1976).

Comet Bradfield (1975p).   D. Herald, A. C. Gilmore, P. M. Kilmartin, J. M. Mundet, J. M. Codina, R. H. S. South, R. L. Waterfield, T. Seki, A. Mrkos, B. Milet, T. Urata.
IAU Circ., No. 2908 (1976).

Comet Bradfield (1975p).   C. Torres, J. M. Codina, N. Torras, R. H. S. South, R. L. Waterfield.
IAU Circ., No. 2916 (1976).

Comet Bradfield (1975p).
C. Jekabsons, P. Jekabsons, D. Gans, M. P. Candy.
IAU Circ., No. 2960 (1976).

New comet Bradfield, 1975p.
Kometn. Tsirk., *Kiev,* No. 188 (1975). In Russian.

Ephemeris of comet Bradfield, 1975p.
Kometn. Tsirk., *Kiev,* No. 188 (1975). In Russian.

Comet Bradfield, 1975p.
Kometn. Tsirk., *Kiev,* No. 189 (1975). In Russian.

Comet Bradfield, 1975p.
Kometn. Tsirk., *Kiev,* No. 190 (1975). In Russian.

Comet Bradfield, 1975p.
Kometn. Tsirk., *Kiev,* No. 191 (1976). In Russian.

Comet Bradfield, 1975p.
Kometn. Tsirk., *Kiev,* No. 192 (1976). In Russian. — Observations at the Ussurijsk Solar Station (*V. Golubev*).

Comet Bradfield, 1975p.
Kometn. Tsirk., *Kiev,* No. 193 (1976). In Russian. — Observations at the Ussurijsk Solar Station (*V. A. Golubev, V. D. D'yakonova*); Ephemeris given by B. G. Marsden.

103.115   Comet 1975o Gehrels 3

Periodic comet Gehrels 3 (1975o).
E. Roemer, D. Daniels, W. Smith, C. D. Vesely, C. Y. Shao, G. Schwartz, J. H. Bulger, B. G. Marsden.
IAU Circ., No. 2962 (1976).

New comet Gehrels, 1975o.
Kometn. Tsirk., *Kiev,* No. 188 (1975). In Russian.

Elements and ephemeris of comet Gehrels, 1975o.
Kometn. Tsirk., *Kiev*, No. 188 (1975). In Russian.

Short-periodic comet Gehrels 3, 1975o.
Kometn. Tsirk., *Kiev*, No. 192 (1976). In Russian.

103.116   Comet 1892 VI Brooks = Comet 1911 V Brooks

Orbits of comets 1892 VI and 1911 V.
G. Van Biesbroeck, C. D. Vesely, B. G. Marsden.
Astron. Journ., Vol. 81, 125 - 126 (1976).
   Computations of definitive orbits for Brooks' comets 1892 VI and 1911 V, begun by the late G. Van Biesbroeck, have been completed. The "original" values of the reciprocal semimajor axes are found to be -0.000027 ± 0.000011 and $+0.006285 \pm 0.000004$ AU^{-1}, respectively, although the original hyperbolic nature of the orbit of the former comet is not regarded as proven.

103.117   Comet 1957 V Mrkos

Certain physical characteristics of comets Bennett 1969i and Mrkos 1957d based on their absolute spectrophotometry.   See Abstr. 103.101.

103.118   Comet 1959 I Burnham-Slaughter

Definitive Bahnbestimmung des Kometen 1959 I (Burnham-Slaughter).   K. F. Ölsböck.
Ann. Univ.-Sternw. Wien, Band 31, (No. 2), 21 - 44 (1974).

103.119   Comet 1975q Sato

New comet Sato 1975q.
M. R. Dykes, R. L. Waterfield, M. P. Candy.
British Astron. Ass. Circ., No. 571 (1976).

Comet Sato (1975q).
K. Suzuki, T. Urata, M. P. Candy, P. Jekabsons, C. Jekabsons.
IAU Circ., No. 2896 (1976).

Comet Sato (1975q).
N. Kojima, T. Seki, A. C. Gilmore, P. M. Kilmartin.
IAU Circ., No. 2906 (1976).

Comet Sato (1975q).   K. Suzuki, M. Takeishi,
T. Urata, T. Kurosaki, D. Herald.
IAU Circ., No. 2935 (1976).

Comet Sato (1975q).
M. P. Candy, P. Jekabsons, C. Jekabsons.
IAU Circ., No. 2958 (1976).

New comet Sato, 1975q.
Kometn. Tsirk., *Kiev*, No. 189 (1975). In Russian.

New comet Sato, 1975q.
Kometn. Tsirk., *Kiev*, No. 191 (1976). In Russian.

103.120   Comet 1974 II Schwassmann-Wachmann 1

Periodic comet Schwassmann-Wachmann 1.
A. Mrkos.
IAU Circ., No. 2907 (1976).

Periodic comet Schwassmann-Wachmann 1.
P. Herget.
IAU Circ., No. 2962 (1976).

Ephemeris of comet Schwassmann–Wachmann 1.
Kometn. Tsirk., *Kiev*, No. 190 (1975). In Russian.

Observations of comet Schwassmann-Wachmann 1.
A. Mrkos.
Kometn. Tsirk., *Kiev*, No. 193 (1976). In Russian.

103.121   Comet 1913 VI Westphal

Ephemeris of comet Westphal.
Kometn. Tsirk., *Kiev*, No. 190 (1975). In Russian.

103.122   Comet 1976f Pons-Winnecke

Periodic comet Pons-Winnecke.
E. A. Reznikov.
IAU Circ., No. 2903 (1976).

Periodic comet Pons-Winnecke (1976f).
E. Roemer, C. A. Heller.
IAU Circ., No. 2934 (1976).

Ephemeris of comet Pons-Winnecke.
Kometn. Tsirk., *Kiev*, No. 191 (1976). In Russian.

Rediscovery of comet Pons-Winnecke, 1976f.
Kometn. Tsirk., *Kiev*, No. 196 (1976). In Russian.

Non-gravitational effects in the motion of comet Pons-Winnecke.   E. A. Reznikov.
Kometn. Tsirk., *Kiev*, No. 196 (1976). In Russian.

103.123   Comet 1976b Kopff

Periodic comet Kopff (1976b).
E. Roemer, C. A. Heller.
IAU Circ., No. 2919 (1976).

Elements and ephemeris of short-periodic comet Kopff.   Yu. Chernetenko.
Kometn. Tsirk., *Kiev*, No. 192 (1976). In Russian.

Rediscovery of comet Kopff, 1976b.
Kometn. Tsirk., *Kiev*, No. 195 (1976). In Russian.

103.124   Comet 1975d Bradfield

Comet Bradfield (1975d).
T. Seki, E. Roemer, D. Daniels.
IAU Circ., No. 2908 (1976).

Comet Bradfield (1975d).
IAU Circ., No. 2915 (1976).

Comet Bradfield (1975d).　　C. Y. Shao.
IAU Circ., No. 2938 (1976).

103.125　Comet 1976a Bradfield

New comet Bradfield 1976a.
W. A. Bradfield, B. G. Marsden.
British Astron. Ass. Circ., No. 572, with a correction, 573 (1976).

Comet Bradfield (1976a).
W. A. Bradfield, T. B. Tregaskis, Curnick.
IAU Circ., No. 2914 (1976).

Comet Bradfield (1976a).　　D. Herald, J. Hers.
IAU Circ., No. 2916 (1976).

Comet Bradfield (1976a).　　M. P. Candy.
IAU Circ., No. 2917 (1976).

Comet Bradfield (1976a).　　D. Herald.
IAU Circ., No. 2919 (1976).

Comet Bradfield (1976a).
C. Jekabsons, P. Jekabsons, M. P. Candy, H. L. Giclas, M. L. Kantz, M. Zemelman, B. M. Blanco, B. G. Marsden.
IAU Circ., No. 2921 (1976).

Comet Bradfield (1976a).　　T. Seki.
IAU Circ., No. 2925 (1976).

Comet Bradfield (1976a).　　R. R. D. Austin,
A. C. Gilmore, P. M. Kilmartin, M. P. Candy, P. Jekabsons, D. Harwood, V. Guth.
IAU Circ., No. 2933 (1976).

Comet Bradfield (1976a).
T. Seki, B. Milet, B. G. Marsden, E. Roemer, L. M. Vaughn.
IAU Circ., No. 2942 (1976).

Comet Bradfield (1976a).
R. R. D. Austin, A. C. Gilmore, P. M. Kilmartin, D. Harwood, C. Jekabsons, T. Urata, M. Akiyama.
IAU Circ., No. 2956 (1976).

Comet Bradfield (1976a).
E. Roemer, L. M. Vaughn, C. D. Vesely.
IAU Circ., No. 2960 (1976).

New comet Bradfield, 1976a.
Kometn. Tsirk., *Kiev,* No. 194 (1976). In Russian.

Comet Bradfield, 1976a.
Kometn. Tsirk., *Kiev,* No. 195 (1976). In Russian.

Comet Bradfield, 1976a.
Kometn. Tsirk., *Kiev,* No. 197 (1976). In Russian.

103.126　Comet 1976h Johnson

Periodic comet Johnson.　　S. W. Milbourn, G. Lea.
IAU Circ., No. 2911 (1976).

Periodic comet Johnson (1976h).

E. Roemer, C. A. Heller.
IAU Circ., No. 2947 (1976).

103.127　Comet 1976d Bradfield

Comet Bradfield (1976d).
W. A. Bradfield, M. P. Candy.
IAU Circ., No. 2923 (1976).

Comet Bradfield (1976d).　　M. P. Candy.
IAU Circ., No. 2924 (1976).

Comet Bradfield (1976d).
M. P. Candy, C. Jekabsons, B. G. Marsden.
IAU Circ., Nos. 2926, 2931 (1976).

Comet Bradfield (1976d).
M. P. Candy, P. Jekabsons, A. C. Gilmore, P. M. Kilmartin.
IAU Circ., No. 2956 (1976).

New comet Bradfield (1976d).
Kometn. Tsirk., *Kiev,* No. 194 (1976). In Russian.

Comet Bradfield, 1976d.
Kometn. Tsirk., *Kiev,* No. 196 (1976). In Russian.

103.128　Comet 1916 II Neujmin 2

Finding ephemeris of comet Neujmin 2, 1916 V
(1916 II ?).
Kometn. Tsirk., *Kiev,* No. 197 (1976). In Russian.

103.129　Comet 1976i Faye

Periodic comet Faye.
IAU Circ., No. 2896 (1976).

Periodic comet Faye (1976i).
E. Roemer, C. A. Heller.
IAU Circ., No. 2947 (1976).

103.130　Comet 1976e d'Arrest

La comète périodique d'Arrest et son apparition en 1976.　　J. Meeus.
L'Astronomie, Vol. 90, 210 - 213 (1976).

Periodic comet d'Arrest.
B. G. Marsden, J. Bortle.
IAU Circ., No. 2900 (1976).

Periodic comet d'Arrest (1976e).
E. Roemer, C. A. Heller.
IAU Circ., No. 2934 (1976).

Periodic comet d'Arrest (1976e).
E. Roemer, C. A. Heller, L. M. Vaughn, C. D. Vesely, G. Schwartz, C. Y. Shao, B. G. Marsden.
IAU Circ., No. 2964 (1976).

Elements and ephemeris of comet d'Arrest.
Kometn. Tsirk., *Kiev,* No. 193 (1976). In Russian.

Rediscovery of the short-period comet d'Arrest, 1976e.
Kometn. Tsirk., *Kiev,* No. 196 (1976). In Russian.

Comet d'Arrest approaches the earth.
B. G. Marsden, J. E. Bortle.
Sky Telescope, Vol. 52, 10 - 13 (1976).

Gunstige wederverschijning van periodieke komeet d'Arrest.   J. Meeus.
Zenit, 3e jaargang, p. 201 - 202 (1976).

103.131   Comet 1974c Lovas

Comet Lovas (1974c).
C. Torres, S. Barros, M. Wischnjewsky, H. Wroblewski, A. C. Gilmore, P. M. Kilmartin, B. G. Marsden.
IAU Circ., No. 2920 (1976).

Comet Lovas (1974c).   J. A. Bruwer.
IAU Circ., No. 2951 (1976).

Comet Lovas (1974c).   D. Gans, F. Jekabsons, C. Jekabsons, M. P. Candy, D. Harwood.
IAU Circ., No. 2963 (1976).

103.132   Comet 1974 XVI Honda-Mrkos-Pajdušáková

Periodic comet Honda-Mrkos-Pajdušáková (1974 XVI).   C. Torres.
IAU Circ., No. 2920 (1976).

103.133   Comet 1974 XIII Schwassmann-Wachmann 2

Periodic comet Schwassmann-Wachmann 2 (1974 XIII).   C. Torres.
IAU Circ., No. 2923 (1976).

103.134   Comet 1974 XIV Longmore

Periodic comet Longmore (1974 XIV).
V. M. Blanco, B. M. Blanco.
IAU Circ., No. 2938 (1976).

103.135   Comet 1975b West-Kohoutek-Ikemura

Periodic comet West-Kohoutek-Ikemura (1975b).
R. M. West, M. Dieckvoss.
IAU Circ., No. 2944 (1976).

103.136   Comet 1965 VI Klemola

Periodic comet Klemola (1965 VI).
IAU Circ., No. 2945 (1976).

103.137   Comet 1976g Harlan

Comet Harlan (1976g).
A. R. Klemola, C. T. Kowal, E. A. Harlan.
IAU Circ., No. 2947 (1976).

Comet Harlan (1976g).
C. T. Kowal, E. A. Harlan, A. R. Klemola, B. G. Marsden.
IAU Circ., No. 2951 (1976).

Comet Harlan (1976g).
L. Giclas, M. L. Kantz, T. Seki.
IAU Circ., No. 2958 (1976).

Comet Harlan (1976g).
H. L. Giclas, M. L. Kantz, B. G. Marsden, E. Roemer.
IAU Circ., No. 2960 (1976).

103.138   Comet 1974 XII van den Bergh

Comet van den Bergh (1974 XII).
G. Schwartz, C. Y. Shao, R. E. McCrosky.
IAU Circ., No. 2952 (1976).

103.139   Comet 1975c Kohoutek

Periodic comet Kohoutek (1975c).
E. Roemer, J. Stocke, L. M. Vaughn, C. D. Vesely.
IAU Circ., No. 2963 (1976).

103.140   Comet 1969 II Gunn

Periodic comet Gunn.
A. C. Gilmore, P. M. Kilmartin.
IAU Circ., No. 2967 (1976).

103.141   Comet 1974 V Encke

Mariner mission to Encke 1980.
See Abstr. 053.018.

Sources of interplanetary dust.
See Abstr. 106.085.

103.142   Comet 1974 III Bradfield

Les observations photométriques des comètes: La comète Bradfield 1974b.   D. Andrienko.
Ann. Obs. Astron. Alger, Vol. 4, Fasc. 3, p. 16 - 28 (1975).

103.143   Comet 1910 II Halley

Halleys komet.   Ø. Hauge.
Astron. Tidssk., Årg. 9, 64 - 69 (1976).

103.144 **Comet 1963 III Alcock**

Spectral analysis of the outburst of the comet 1963 III (Alcock 1963b).   S. Grudzińska.
Acta Astron., Vol. 26, 117 - 145 (1976).

From the objective prism spectra of comet 1963 III (Alcock 1963b) the gas (CN, C_3 and C_2) and dust particles masses, densities and distributions were determined. It follows that: (1) the ratio of the total energies of the bands and continuous spectra is constant for the observations before and after the outburst; (2) the intensity distribution of the cometary continuous spectrum before and after the outburst is the same.

103.145 **Comet 1968 I Ikeya-Seki**

Positions of comet Ikeya-Seki 1967n.
H. Brancewicz, J. M. Kreiner, M. Kurpińska, B. Lipska, M. Winiarski.
Acta Astron., Vol. 26, 189 - 191 (1976).

107 positions of comet Ikeya-Seki 1967n obtained in Cracow during the period February 1 - May 4, 1968 are given.

104 Meteors, Meteor Streams

104.001 The role of the oxides of meteoric species as a source of meteor train luminosity.
W. J. Baggaley.
Monthly Notices Roy. Astron. Soc., Vol. 174, 617 - 620 (1976).

The energetics of the processes leading to the formation of excited monoxides of meteoric elements are examined. It is shown that formation of the excited molecules AlO, FeO, CaO and MgO may occur in an expanding meteor column and that the emission of their associated band systems provides a possible contributor to long-lived meteor train luminosity.

104.002 Is fragmentation the answer to the difference in the dynamic and photometric masses of fireballs?
V. Padevět.
Bull. Astron. Inst. Czechoslovakia, Vol. 27, 11 - 18 (1976).

The author attempted to find an ablation mechanism in large meteors which would be independent of the differences in the densities and strength of meteoric bodies and which would also be capable of explaining the order-of-magnitude difference observed in the determination of fireballs masses by the photometric and dynamic methods. The fragmentation of the meteoric body and the motion of the fragments in the perturbed medium behind the body would be capable of explaining the experimentally found difference in masses only at large altitudes in the atmosphere.

104.003 Statistical model of meteor streams. IV. A study of radio streams from the synoptic year.
Z. Sekanina.
Icarus, Vol. 27, 265 - 321 (1976).

With the use of the most powerful stream-search technique in existence, the author has detected 275 streams in the synoptic-year (December 2, 1968 — December 14, 1969) sample of 19 698 radio meteors, most of which were not reported before. Some of the new streams have most uncommon orbits. A new, rich stream with a revolution period of about 30 yr has been discovered. Streams of low inclination are often detected at both nodes. A computer technique developed, for determining the two parameters of the D-distribution of meteor orbits in a stream, has been applied to the 275 streams. A number of known comet-meteor associations have been confirmed, and a few new possible associations established. Plots on a height-velocity diagram of both the individual radio meteors and the radio streams fail to exhibit the discrete-level structure known to exist for photographic meteors.

104.004 Radio echoes from meteor trains at a radio frequency of 1.98 MHz. N. Brown.
Journ. Atmosph. Terr. Phys., Vol. 38, 83 - 87 (1976).

Results are given of the distribution in height of radio echoes from meteor trains at a frequency of 1.98 MHz. The variations with height of their decay time constants and their durations are also shown. A comparison is made between wind measurements made with these echoes and with a meteor wind system operating on a frequency of 27 MHz at a nearby site.

104.005 Beobachtung von Meteoren und Nordlichtern mit Hilfe der Amateurfunktechnik.
H. U. Finkenzeller.
SuW, 15. Jahrgang, p. 22 - 23 (1976).

104.006 Möglichkeiten praktischer Meteorastronomie. Auswertebeispiel Juli/August-Radianten höherer Deklinationen. H. J. Becker.
SuW, 15. Jahrgang, p. 76 - 80 (1976).

104.007 A note on meteor and micrometeoroid orbits determined from rough velocity data.
L. Kresák, M. Kresáková.
Bull. Astron. Inst. Czechoslovakia, Vol. 27, 106 - 109 (1976).

Some effects of the errors in velocity measurements on the determination of meteor orbits are discussed. Acceptable limits of measuring accuracy for discriminating between different types of the parent bodies are examined. A correct interpretation of the origin of the interplanetary dust sets high requirements on the accuracy of the velocity data, which have not yet been satisfied by spacecraft observations.

104.008 Variation of meteor echo characteristics with the direction of the antenna beam. A. Hajduk.
Bull. Astron. Inst. Czechoslovakia, Vol. 27, 112 - 115 (1976).

The results of an experiment recording meteor echoes with the antenna beam alternatively directed to the shower radiant and 90° from the radiant at the same elevation are presented. The data show considerable difference in detection of overdense and underdense trains with changing direction to the train. Changes in echo duration and range distribution as well as in echo amplitude fluctuations lead to the evidence that overdense trains are recorded at any beam direction, irrespective of the condition of its perpendicular orientation to the meteor trail.

104.009 Meteor radar rates and the solar cycle.
B. A. Lindblad.
Nature, Vol. 259, 99 - 101 (1976).

A worldwide increase in meteor echo rates in 1963 was observed in New Zealand, Canada and Sweden and has been widely discussed in the literature. The author reports here further observations, and proposes that the phenomenon can be explained by a solar controlled variation of the atmospheric density gradient at the meteor ablation level, probably caused by a variation in the solar X-ray flux.

104.010 Quadrantid meteors from 41,000 feet.
P. M. Millman.
Sky Telescope, Vol. 51, 225 - 228 (1976).

104.011 Ein Perseide mit ungewöhnlichen spektralen Eigenschaften. W. C. Seitter.
Mitt. Astron. Ges., No. 38, p. 218 (1976). — Abstract.

104.012 Some results of an investigation of ablation of large meteor particles.
P. B. Babadzhanov, V. S. Getman.
Interaction of meteor matter. Conference Dushanbe 1974, (see 012.004), p. 3 - 8 (1975). In Russian. — Abstr. in Referativ. Zhurn. 51. Astron., 3.51.252 (1976).

104.013 Approximate solution of the problem of evaporation of a spherical meteor body. V. G. Kruchinenko.
Interaction of meteor matter. Conference Dushanbe 1974, (see 012.004), p. 21 - 33 (1975). In Russian. — Abstr. in Referativ. Zhurn. 51. Astron., 3.51.253 (1976).

104.014 On the evaporation of a cylindrical meteor body.
A. N. Shajdo.
Interaction of meteor matter. Conference Dushanbe 1974, (see 012.004), p. 34 - 44 (1975). In Russian. — Abstr. in Referativ. Zhurn. 51. Astron., 3.51.254 (1976).

104.015 Luminosity of meteor plasma.
G. G. Novikov, L. N. Rubtsov.

Interaction of meteor matter. Conference Dushanbe 1974, (see 012.004), p. 45 - 50 (1975). In Russian. – Abstr. in Referativ. Zhurn. 51. Astron., 3.51.255 (1976).

104.016 Distribution of meteor matter in the solar system.
O. I. Bel'kovich.
Interaction of meteor matter. Conference Dushanbe 1974, (see 012.004), p. 51 - 66 (1975). In Russian. – Abstr. in Referativ. Zhurn. 51. Astron., 3.51.256; 62. Issled. kosmich. prostranstva, 3.62.166 (1976).

104.017 Astronomical selection of meteors and some methods to take it into account.
V. V. Andreev, O. I. Bel'kovich, V. S. Zabolotnikov.
Interaction of meteor matter. Conference Dushanbe 1974, (see 012.004), p. 67 - 72 (1975). In Russian. – Abstr. in Referativ. Zhurn. 51. Astron., 3.51.257 (1976).

104.018 A model of heliocentric distribution of meteor bodies in the vicinity of the earth's orbit.
V. V. Andreev, O. I. Bel'kovich, V. S. Tokhtas'ev.
Interaction of meteor matter. Conference Dushanbe 1974, (see 012.004), p. 73 - 78 (1975). In Russian. – Abstr. in Referativ. Zhurn. 51. Astron., 3.51.258 (1976).

104.019 Some problems of orbital evolution of meteor bodies under the action of non-gravitational forces.
A. A. Dmitrievskij, K. V. Kostylev.
Interaction of meteor matter. Conference Dushanbe 1974, (see 012.004), p. 79 - 86 (1975). In Russian. – Abstr. in Referativ. Zhurn. 51. Astron., 3.51.259 (1976).

104.020 Effect of the earth's magnetic field on some characteristics of radar meteor echoes.
O. I. Bel'kovich, A. M. Nasyrov, N. I. Sulejmanov, V. S. Tokhtas'ev.
Interaction of meteor matter. Conference Dushanbe 1974, (see 012.004), p. 116 - 127 (1975). In Russian. – Abstr. in Referativ. Zhurn. 51. Astron., 3.51.260 (1976).

104.021 Radar observations of meteor numbers in Frunze.
K. A. Karimov, S. S. Timofeeva.
Interaction of meteor matter. Conference Dushanbe 1974, (see 012.004), p. 157 - 165 (1975). In Russian. – Abstr. in Referativ. Zhurn. 51. Astron., 3.51.261 (1976).

104.022 Some structural peculiarities of Geminid and Quadrantid meteor streams.
O. I. Bel'kovich, V. S. Tokhtas'ev, N. I. Sulejmanov.
Interaction of meteor matter. Conference Dushanbe 1974, (see 012.004), p. 93 - 99 (1975). In Russian. – Abstr. in Referativ. Zhurn. 51. Astron., 3.51.263 (1976).

104.023 On the problem of meteor matter influx on the earth.   N. V. Vasil'ev, A. P. Boyarkina.
Interaction of meteor matter. Conference Dushanbe 1974, (see 012.004), p. 87 - 92 (1975). In Russian. – Abstr. in Referativ. Zhurn. 51. Astron., 3.51.270 (1976).

104.024 Resultaten rekensectie werkgroep meteoren.
H. Betlem, N. de Kort.
Zenit, 3e jaargang, p. 60 - 62, 64 (1976).

104.025 Preliminary orbits from the NASA-NMSU meteor observatory.   E. F. Tedesco, G. A. Harvey.
Bull. American Astron. Soc., Vol. 8, 292 (1976). – Abstr. AAS.

104.026 On photometric referencing of meteor trails to diurnal star trails.   N. N. Izraetskaya, I. S. Shestaka.
Astron. Tsirk., No. 854, p. 6 - 7 (1975). In Russian.

104.027 Recording meteor echoes by FM radio.
K. Suzuki, N. Nagafuji, M. Kinoshita.
Sky Telescope, Vol. 51, 359 - 362 (1976).

104.028 Photographic data on the Leutkirch fireball (EN 300874). (Aug. 30, 1974).
Z. Ceplecha, M. Ježková, J. Boček, T. Kirsten.
Bull. Astron. Inst. Czechoslovakia, Vol. 27, 18 - 23 (1976).
Detailed data on the Leutkirch fireball photographed at 6 stations of the German part of the European All-Sky-Camera Network are given. The fireball flew 113 km of its luminous trajectory in 10.9 seconds. The low initial velocity of 12.6 km/sec, deep penetration of the luminous part of the trajectory down to 25.9 km, where the terminal mass of 14 kg had a directly measured terminal velocity of 4.23 km/sec, and the structural strength and density of the meteoroid, point to a possible meteorite fall. The search was negative.

104.029 On the influence of thermal diffusion on the measured velocity of the drift of meteor traces.
I. A. Delov.
Geomagn. Aeronom., Vol. 16, 191 - 193 (1976). In Russian. Brief information.

104.030 Ablation of meteor bodies in the earth's atmosphere.   V. S. Tokhtas'ev.
Interaction of meteor matter. Conference Dushanbe 1974, (see 012.004), p. 10 - 20 (1975). In Russian. – Abstr. in Referativ. Zhurn. 51. Astron., 4.51.274 (1976).

104.031 Formation and decay of meteor trains.
V. S. Tokhtas'ev.
Interaction of meteor matter. Conference Dushanbe 1974, (see 012.004), p. 100 - 107 (1975). In Russian. – Abstr. in Referativ. Zhurn. 51. Astron., 4.51.275 (1976).

104.032 Effect of the intensity of turbulent atmospheric motions on the distribution of meteor particles in the upper atmosphere.   G. M. Teptin.
Interaction of meteor matter. Conference Dushanbe 1974, (see 012.004), p. 108 - 115 (1975). In Russian. – Abstr. in Referativ. Zhurn. 51. Astron., 4.51.276 (1976).

104.033 Effect of the wind on the behaviour of meteor plasma.   G. G. Novikov, S. F. Tsygankov, L. N. Rubtsov.
Interaction of meteor matter. Conference Dushanbe 1974, (see 012.004), p. 128 - 134 (1975). In Russian. – Abstr. in Referativ. Zhurn. 51. Astron., 4.51.277 (1976).

104.034 Comment on: "The effect of ionic processes on the characteristics of radio-echoes from meteor trains", by L. M. G. Poole and T. F. Nicholson.
W. J. Baggaley, with a reply by L. M. G. Poole.
Planet. Space Sci., Vol. 24, 605 - 606 (1976) – See 14.104. 015.
Attention is drawn to the role of collisional electron-ion recombination in meteoric deionization and to the consequences for radio-echo duration characteristics.

104.035 Observation of faint meteors.
L. N. Rubtsov, B. G. Solovej.
Byull. Inst. Astrofiz., *Dushanbe*, No. 63, p. 33 - 38 (1974). In Russian.
Results of regular annual observations of sporadic meteor numbers down to magnitude +13 are presented. A list of 58 meteor showers is given, the most part of which is more active than the known Geminid shower in the considered magnitude region. Peculiarities of radar observations of small meteors in the wavelength region 15–20 m are discussed.

104.036 Statistical investigation of Leonids according to

observations in 1966–1967 in Dushanbe.
D. Latipov, L. N. Rubtsov.
Byull. Inst. Astrofiz., *Dushanbe*, No. 65, p. 26 - 29 (1975).
In Russian.

104.037 Application of the Halphen-Goryachev method for studying the evolution of the orbits of Quadrantid, δ-Aquarid and α-Capricornid meteor streams.
P. B. Babadzhanov, A. F. Zausaev.
Byull. Inst. Astrofiz., *Dushanbe*, No. 65, p. 40 - 50 (1975).
In Russian.

Calculations of secular perturbations of Quadrantid and δ-Aquarid meteor streams by the Halphen-Goryachev method are presented taking into account perturbations by Jupiter and Saturn. For average Quadrantid, δ-Aquarid and α-Capricornid orbits osculating orbital elements have been calculated down to 6000 years ago.

104.038 On the attachment correction.   P. Pecina.
Bull. Astron. Inst. Czechoslovakia, Vol. 27, 164 - 168 (1976).

An equation for computing the attachment-corrected durations of meteor echoes, derived by Plavcová, is discussed. The influence of new corrected durations is shown on the difference of the mass exponents s for the Geminid shower. It is suggested that the new approach leads to the dependence of the correction on the radiant position, too.

104.039 Longitudinal structure of the Geminid stream.
M. Šimek.
Bull. Astron. Inst. Czechoslovakia, Vol. 27, 168 - 173 (1976).

Long-term observations of the Geminid meteor shower suggest that the meteor density along the Geminid orbit is not homogeneous. Assuming the simplest variation for the longitudinal structure an orbital period near $P = 1.72$ year was established from radio observations for 1958–1974. At least three extremely dense parts in the Geminid stream were found.

104.040 The excitation of the oxygen metastable OI(^{1}S) state in meteors.   W. J. Baggaley.
Bull. Astron. Inst. Czechoslovakia, Vol. 27, 173 - 181 (1976).

The various mechanisms occurring in a meteoric plasma which might yield excited oxygen atoms in the O(^{1}S) state leading to the emission of the forbidden OI$\lambda\lambda$5577 Å are examined.

104.041 Magnitudes of bright Perseids.   I. M. Khaimov.
Dokl. AN TadzhSSR, Vol. 18, No. 6, p. 15 - 19 (1975). In Russian. – Abstr. in Referativ. Zhurn. 51. Astron., 5.51.346 (1976).

104.042 On the determination of sizes of meteor bodies by an optical method.
Yu. A. Medvedev, V. D. Khokhlov.
Probl. fiz. optiki i metrol. Moskva, 1975, p. 64 - 65. In Russian. – Abstr. in Referativ. Zhurn. 51. Astron., 5.51.348 (1976).

104.043 Möglichkeiten praktischer Meteorastronomie.
H. J. Becker.
SuW, 15. Jahrgang, 199 - 202 (1976).

104.044 More precise theoretical equation of meteor light.
I. N. Kovshun.
Astron. Tsirk., No. 895, p. 5 - 6 (1975). In Russian.

104.045 Meteoroid flux from passive seismic experiment data.
F. Duennebier, J. Dorman, D. Lammlein, G. Latham, Y. Nakamura.
Proc. Sixth Lunar Sci. Conference, (see 012.010), Vol. 2,

2417 - 2426 (1975).

104.046 Perseiden 1975.   J. Rendtel.
Sterne, Vol. 52, 115 - 119 (1976).

104.047 Meteor matter from measurements from space vehicles.   T. N. Nazarova, A. K. Rybakov, Z. V. Vasyukova, Yu. D. Vasil'ev.
Kosmich. Issled., Vol. 14, 435 - 444 (1976).   In Russian.

104.048 Flux of hyperbolic meteoroids.   J. S. Dohnanyi.
IAU Colloquium No. 31, (see 012.015), p. 170 - 180 (1976).

104.049 Specific sources of extraterrestrial particles.
J. Rosinski.
IAU Colloquium No. 31, (see 012.015), p. 289 (1976).

104.050 Meteors and interplanetary dust.   P. M. Millman.
IAU Colloquium No. 31, (see 012.015), p. 359 - 372 (1976). – Invited paper.

104.051 Meteoroid densities.   B. A. Lindblad.
IAU Colloquium No. 31, (see 012.015), p. 373 - 378 (1976).

104.052 Possible evidence of meteoroid fragmentation in interplanetary space from grouping of particles in meteor streams.   V. Porubčan.
IAU Colloquium No. 31, (see 012.015), p. 379 - 382 (1976).

104.053 The heliocentric distribution of the meteor bodies at the vicinity of the earth's orbit.
V. V. Andreev, O. I. Belkovich (*Bel'kovich*), V. S. Tokhtas'ev.
IAU Colloquium No. 31, (see 012.015), p. 383 - 384 (1976).

104.054 Fireballs as an atmospheric source of meteoritic dust.   Z. Ceplecha.
IAU Colloquium No. 31, (see 012.015), p. 385 - 388 (1976).

104.055 Meteor radar rates and the solar cycle.
B. A. Lindblad.
IAU Colloquium No. 31, (see 012.015), p. 390 (1976). Abstract of 104.009.

104.056 Radial distribution of meteoric particles in interplanetary space.   J. W. Rhee.
IAU Colloquium No. 31, (see 012.015), p. 448 - 452 (1976).

104.057 Meteors.   Z. Ceplecha.
IAU Colloquium No. 31, (see 012.015), p. 485 - 488 (1976).

104.058 Observations of telescopic meteors with different types of instruments.
J. Kučera, M. Šulc, J. Hollan, J. Žižka.
Contr. Obs. and Planetarium Brno, No. 19, 45 pp. (1975). In Czech and English.

104.059 Photographic fireball networks.
I. T. Zotkin, A. N. Simonenko, V. V. Fedynskij, R. L. Khotinok, E. N. Kramer.
Meteoritika, vyp. (No.) 35, p. 3 - 18 (1976). In Russian.

104.060 SEC Vidicon spectra of Geminid meteors, 1972.
P. M. Millman, K. S. Clifton.
Canadian Journ. Phys., Vol. 53, 1939 - 1947 (1975). – Abstr. in Phys. Abstr., Vol. 79, A027848 (1976).

104.061 Radiowave reflection from meteor trails. Algorithms for calculating the k_1 parameter of amplitude distri-

bution of reflections from meteors.
Yu. I. Voloshchuk, B. L. Kashcheev.
Radiotekhn. Khar'kov, No. 33, p. 3 - 8 (1975). In Russian.
Abstr. in Phys. Abstr., Vol. 79, A027859 (1976).

104.062 Optimum processing of radiometeor information.
Primary data processing during recording of the
quantity of meteors.   Yu. I. Voloshchuk.
Radiotekhn. Khar'kov, No. 33, p. 9 - 13 (1975). In Russian.
Abstr. in Phys. Abstr., Vol. 79, A028059 (1976).

104.063 Study of meteors during December 1974 by for-
ward scatter of VHF radio signals between Dehra
Dun and Waltair.
B. Ramachandra Rao, M. Srirama Rao, S. Raja Ratnam, D. A.
V. Krishna Rao, E. Bhagiratha Rao.
Indian Journ. Radio Space Phys., Vol. 4, 99 - 101 (1975). -
Abstr. in Phys. Abstr., Vol. 79, A037509 (1976).

104.064 Guests from the sky.   H. Korpikiewicz.
Urania Kraków, Vol. 47, 47 - 50, 80 - 83, 106 -
112, 142 - 146 (1976). In Polish.

Observations of the Geminid and Quadrantid
meteors.
Sky Telescope, Vol. 51, 207 - 209 (1976).

Eddy diffusion in the upper atmosphere by global
deposition of meteoroids.   See Abstr. 082.012.

On meteor-generated infrasound.
See Abstr. 082.031.

Results of an experimental investigation of a con-

nection of the E_s-layer with the annual abundance of faint
meteors.   See Abstr. 083.052.

Collision of comets with meteoroids.
See Abstr. 102.004.

The interplanetary and near Jovian dust environment: some experimental results.   See Abstr. 106.018.

The cometary contribution to cosmic dust.
See Abstr. 106.020.

The distribution of dust along the earth's orbit deduced from satellite measurements of zodiacal light.
See Abstr. 106.021.

Evidence for scattering particles in meteor streams.
See Abstr. 106.061.

In-situ records of interplanetary dust particles –
methods and results.   See Abstr. 106.073.

Preliminary results of the micrometeoroid experiment on board Helios A.   See Abstr. 106.074.

Sources of interplanetary dust.
See Abstr. 106.085.

Dynamics of interplanetary dust and related topics.
See Abstr. 106.086.

On the meteor effect in cosmic rays.
See Abstr. 143.039.

105 Meteorites, Meteorite Craters

105.001 On the primordial abundance of argon-40.
F. Begemann, H. W. Weber, H. Hintenberger.
Astrophys. Journ., (*Letters*), Vol. 203, L155 - L157 (1976).

Argon trapped in graphite-diamond-kamacite inclusions from the Haverö ureilite has a $^{40}Ar/^{36}Ar$ ratio of $(1.4 \pm 0.6) \times 10^{-3}$, which is close to the estimated primordial abundance ratio of 2×10^{-4}. Some implications are discussed.

105.002 Indications for a meteoritic impact of early Cambrian age at Conception Bay, Newfoundland: discussion. M. M. Anderson, C. J. Hughes.
Naturwissenschaften, 63. Jahrgang, 87 - 88 (1976). – Short communication.

105.003 Early irradiation of matter in the solar system: magnesium (proton, neutron) scheme.
D. Heymann, M. Dziczkaniec.
Science, Vol. 191, 79 - 81 (1976).

The occurrence of positive and negative ^{26}Mg anomalies in inclusions of the Allende meteorite is explained in terms of proton bombardment of a gas of solar composition. The proton flux required to account for the observed magnesium anomalies is used to investigate possible isotopic anomalies in the elements from oxygen to argon. Detectable isotopic anomalies are predicted only for neon.

105.004 Rb, Sr and strontium isotopic composition, K/Ar age and large ion lithophile trace element abundances in rocks and glasses from the Wanapitei Lake impact structure.
S. R. Winzer, R. K. L. Lum, S. Schuhmann.
Geochim. Cosmochim. Acta, Vol. 40, 51 - 57 (1976).

105.005 Trace elements in primitive meteorites – VI. Abundance patterns of thirteen trace elements and interelement relationships in unequilibrated ordinary chondrites.
C. M. Binz, M. Ikramuddin, P. Rey, M. E. Lipschutz.
Geochim. Cosmochim. Acta, Vol. 40, 59 - 71 (1976).

105.006 Allende inclusions: volatile-element distribution and evidence for incomplete volatilization of presolar solids. C.-L. Chou, P. A. Baedecker, J. T. Wasson.
Geochim. Cosmochim. Acta, Vol. 40, 85 - 94 (1976).

The authors propose that the large size of Allende spheroidal inclusions indicates an origin by incomplete vaporization of presolar solid matter followed by recondensation of refractories on a limited number of condensation nuclei. The low abundance of large refractory inclusions in ordinary and enstatite chondrites reflects complete vaporization of presolar solids at their formation locations; constraints on homogeneous nucleation resulted in the simultaneous condensation of refractories and olivine at these locations.

105.007 Chemical classification of iron meteorites – VIII. Groups IC, IIE, IIIF and 97 other irons.
E. R. D. Scott, J. T. Wasson.
Geochim. Cosmochim. Acta, Vol. 40, 103 - 115 (1976).

Concentrations of Ni, Ga, Ge and Ir in 106 iron meteorites are reported. Three new groups are defined: IC, IIE and IIIF containing 10, 12 and 5 members, respectively, raising the number of independent groups to 12.

105.008 Thermal metamorphism of primitive meteorites – II. Ten trace elements in Abee enstatite chondrite heated at 400–1000°C.
M. Ikramuddin, C. M. Binz, M. E. Lipschutz.
Geochim. Cosmochim. Acta, Vol. 40, 133 - 142 (1976).

105.009 Amoeboid olivine aggregates in the Allende meteorite. L. Grossman, I. M. Steele.
Geochim. Cosmochim. Acta, Vol. 40, 149 - 155 (1976).

105.010 Carbonaceous and non-carbonaceous lithic fragments in the Plainview, Texas, chondrite: origin and history. R. V. Fodor, K. Keil.
Geochim. Cosmochim. Acta, Vol. 40, 177 - 189 (1976).
This paper was first presented as a talk at the 35th annual meeting of the Meteoritical Society, Chicago, Illinois, November, 1972.

105.011 Trace elements in the Allende meteorite – I. Coarse-grained, Ca-rich inclusions.
L. Grossman, R. Ganapathy.
Geochim. Cosmochim. Acta, Vol. 40, 331 - 344 (1976).

105.012 Rare gases and ^{36}Cl in stony-iron meteorites: cosmogenic elemental production rates, exposure ages, diffusion losses and thermal histories.
F. Begemann, H. W. Weber, E. Vilcsek, H. Hintenberger.
Geochim. Cosmochim. Acta, Vol. 40, 353 - 368 (1976).

Metal and silicate portions from 13 mesosiderites, one pallasite, Bencubbin ('unique') and Udei Station ('iron with silicate inclusions') have been analyzed for their content of He, Ne and Ar; in most cases ^{36}Cl could be determined as well.

105.013 First finding of tektites in the USSR.
P. V. Florenskij.
Priroda, 1976, No. 1, p. 85 - 87. In Russian.

105.014 Model ages.
G. J. Wasserburg, D. A. Papanastassiou.
Nature, Vol. 259, 159 - 160, with a reply by R. Hutchison, N. H. Gale and J. W. Arden, p. 160 (1976).

105.015 Formation of iron sulphide in solar nebula.
J. F. Kerridge.
Nature, Vol. 259, 189 - 190 (1976).

Iodine-xenon dating has established that iron sulphide in the Orgueil carbonaceous meteorite is one of the oldest known meteoritic mineral phases, and probably dates from the condensation stage of the early solar system. This sulphide, although generally assumed to be troilite, FeS, is in fact the Fe-deficient monosulphide, pyrrhotite, $(Fe, Ni)_9S_{10}$, containing ~1 weight % Ni. The purpose of this note is to suggest that such mineral chemistry is inconsistent with equilibrium condensation, and that the course of condensation may have been modified by kinetic effects.

105.016 Implications from the absence of a ^{41}K anomaly in an Allende inclusion. F. Begemann, W. Stegmann.
Nature, Vol. 259, 549 - 550 (1976). – Letter.

105.017 Ni and Co content of chondritic metal.
D. W. Sears, H. J. Axon.
Nature, Vol. 260, 34 - 35 (1976). – Letter.

105.018 Spectral reflectance characteristics of the meteorite classes. M. J. Gaffey.
Journ. Geophys. Res., Vol. 81, 905 - 920 (1976).

Spectral reflectance curves ($0.35–2.5\,\mu m$) have been measured for about 150 individual meteorite specimens which represent nearly all the types and subtypes of meteorites present in terrestrial meteorite collections. The spectra of members of each meteorite class are discussed in terms of the composition, abundance, and distribution of the component mineral phases. The important results of this study are as follows: (1) The spectral reflectance curves of a meteorite class, representing a particular mineral assemblage and metamorphic grade, do not differ significantly. (2) The variations between the spectra of different meteorite types are significant. (3) There exist a series of diagnostic features in the spectrum of a meteoritic material (absorption band presence, position, symmetry, and intensity as well as the continuum slope, curvature, and inflection points) which can be utilized in the general interpretation of telescopic spectral reflectance measurements of asteroids. (4) The positions of the 1.0- and 2.0-μm absorption features common in meteoritic materials are sensitive indicators of pyroxene compositions and of the olivine/pyroxene ratio. (5) The physical properties of a surface material (particle size, packing, illumination angle) for a reasonable range of these parameters do not significantly affect the normalized spectral reflectance curve measured for that material.

105.019 **A comparison of the noble gases in three meteorite specimens labeled Springfield.**
E. W. Hennecke, O. K. Manuel.
Zeitschr. Naturforschung, Vol. 31a, 293 - 296 (1976).

The abundance and isotopic composition of the noble gases were measured in three Springfield specimens identified by the Denver Museum of Natural History with numbers 7029, 379.13 and 6040. The latter specimen contains more cosmogenic noble gas isotopes than the other two specimens and the abundance pattern of trapped noble gases in specimen 6040 is distinct from that in the other two specimens. Specimen 7029 contains about seven times as much radiogenic ^{40}Ar and about four times as much radiogenic ^{129}Xe as does specimen 379.13. These results indicate that the three specimens did not come from a single meteoroid.

105.020 **Vorläufige Ergebnisse vom Helios-1-Mikrometeoriten-Experiment.**
E. Grün, J. Kissel, H. Fechtig, P. Gammelin.
Mitt. Astron. Ges., No. 38, p. 148 (1976). − Abstract.

105.021 **Gruppen von Mikrometeoriten im Erde-Mond-System.**   J. S. Dohnanyi.
Mitt. Astron. Ges., No. 38, p. 148 - 149 (1976). − Abstract.

105.022 **Gesamtkatalog der in der Deutschen Demokratischen Republik vorhandenen Meteorite.**   G. Hoppe.
Wiss. Zeitschr. Humboldt-Univ. Berlin, Math.-Nat. Reihe, Vol. 24, 521 - 569 (1975).

105.023 **Can enstatite meteorites form from a nebula of solar composition?**   J. M. Herndon, H. E. Suess.
Geochim. Cosmochim. Acta, Vol. 40, 395 - 399 (1976).

Doubts have been expressed as to whether solar matter is sufficiently reducing to explain the minerals occurring in the enstatite chondrites and in the enstatite achondrites. Thermodynamic calculations on the stabilities of TiN, Si_2N_2O, CaS and silicon-bearing iron metal show that these substances can form under equilibrium conditions from a nebula of solar composition provided that the total pressure exceeds ~1 atm and that thermodynamic equilibria are frozen in at near-formation temperatures.

105.024 **Fission-track ages of four meteorites.**
E. A. Carver, E. Anders.
Geochim. Cosmochim. Acta, Vol. 40, 467 - 477 (1976).

105.025 **Rare earth elements in tektites.**
V. Bouška, Z. Řanda.
Geochim. Cosmochim. Acta, Vol. 40, 486 - 488 (1976).

Four moldavites and one sample each of an australite, billitonite, indochinite, philippinite, thailandite, Ivory Coast tektite, bediasite and a georgianite were analyzed using neutron activation analysis for La, Ce, Sm, Eu, Tb, Yb and Lu. The REE abundances resemble those of sedimentary rocks. Most of the tektites display a depletion of Eu, a characteristic feature of mature Phanerozoic continental sedimentary rocks. However, the Ivory Coast tektite and georgianite are relatively enriched in Eu, possibly due to the presence of plagioclase-rich source rocks.

105.026 **Metallic phase of meteorites and planets.**
A. P. Vinogradov.
Vestn. Mosk. un-ta. Geologiya, 1975, No. 4, p. 8 - 15. In Russian. − Abstr. in Referativ. Zhurn. 51. Astron., 3.51.280 (1976).

105.027 **Waarom heeft de aarde zo weinig inslagkraters?**
J. Van Diggelen.
Zenit, 3e jaargang, p. 110 - 119 (1976).

105.028 **The Tungus event as a small black hole: geophysical considerations.**   J. O. Burns, G. Greenstein, K. L. Verosub.
Monthly Notices Roy. Astron. Soc., Vol. 175, 355 - 357 = Contr. Five College Astron. Dept., *Amherst, Mass.*, No. 203 (1976).

If the Tungus event of 1908 were due to the passage of a small black hole through the earth, extensive thermal alteration of soil and rock would have resulted. In addition enormous seismic activity would have been generated − activity far greater than was observed.

105.029 **The Haverö meteorite.**   G. P. Vdovykin.
Space Sci. Rev., Vol. 18, 749 - 776 (1976).

The meteorite Haverö, which fell on 2 August 1971 in Finland is a representative of a rare, but highly interesting group of stony meteorites − ureilites. Like other ureilites it is enriched in carbon (up to 2%) chiefly represented by diamond and graphite. The meteorite is strongly recrystallized. The peculiarities of the mineral and chemical composition of the meteorite indicate that its material had formed during a disequilibrium process. The content of cosmogenic isotopes witnesses to an unusual orbit of the meteorite.

105.030 **The Bovedy meteorite; mineral chemistry and origin of its Ca-rich glass inclusions.**
A. L. Graham, A. J. Easton, R. Hutchison, D. Y. Jérome.
Geochim. Cosmochim. Acta, Vol. 40, 529 - 535 (1976).

105.031 **Spectrum of carbonaceous-chondrite fission xenon**
D. D. Clayton.
Geochim. Cosmochim. Acta, Vol. 40, 563 - 565 (1976).

Estimations of the fission spectrum in xenon isotopes from the progenitor of the strange carbonaceous-chondrite xenon must take account of p-process nucleosynthesis if the latter is the source of anomalous ^{124,126}Xe. Sample calculations of the p-process yields illustrate the magnitude of the effect, which can greatly increase the estimated ^{132}Xe fission yield.

105.032 **The main crater of the Kaali meteorite.**
V. I. Koval'.
Zemlya i Vselennaya, 1976, No. 2, p. 91 - 93. In Russian.

105.033 **Synthetic phyllosilicates and the matrix material of C1 and C2 chondrites.**   K. L. Day.
Icarus, Vol. 27, 561 - 568 (1976).

Magnesium silicates and gels were synthesized from aqueous solution. A range of products with the Mg/Si ratio ranging from 2 : 1 to 1 : 1 was produced. Their composition and crystallinity were found to vary with the time scale available for formation. Structural similarities between these materials and the matrix silicates of Orgueil and Murchison were revealed through X-ray and infrared analysis.

105.034 On the calculation of cosmic-ray exposure ages of stone meteorites.
P. J. Cressy, Jr., D. D. Bogard.
Geochim. Cosmochim. Acta, Vol. 40, 749 - 762 (1976).

105.035 Amino acids of the Nogoya and Mokoia carbonaceous chondrites. J. R. Cronin, C. B. Moore.
Geochim. Cosmochim. Acta, Vol. 40, 853 - 857 (1976).

Amino acids were found in acid hydrolyzed, hot water extracts of the Nogoya (C2) and Mokoia (C3V) chondrites. About 40 n moles/g of amino acids were found in the Nogoya extract while Mokoia contained less than 1 n mole/g. The amino acid composition of Nogoya differs from that of other C2 chondrites studied earlier. The results from Mokoia are similar to previous data obtained from the C3V chondrite Allende.

105.036 U–Th–Pb and Rb–Sr systematics of Allende and U–Th–Pb systematics of Orgueil.
M. Tatsumoto, D. M. Unruh, G. A. Desborough.
Geochim. Cosmochim. Acta, Vol. 40, 617 - 634 (1976).

105.037 Isotopic lead investigations on the Allende carbonaceous chondrite. J. H. Chen, G. R. Tilton.
Geochim. Cosmochim. Acta, Vol. 40, 635 - 643 (1976).

105.038 Activated release of alkalis during the vesiculation of molten basalts under high vacuum: implications for lunar volcanism. J. L. Gooding, D. W. Muenow.
Geochim. Cosmochim. Acta, Vol. 40, 675 - 686 (1976).

Knudsen cell-quadrupole mass spectrometry was used to study the high-temperature vaporization of Hawaiian basalts, plagioclase, tektites, and samples from the Allende meteorite. The major gases released from the Allende meteorite at 900–1000°C are, in order of decreasing abundance, CO, S_2, CO_2, H_2O, SO_2, and H_2S. It is proposed that nonequilibrium vaporization of alkalis during the vesiculation of lunar lavas was responsible for the production of alkali-rich vapors which subsequently deposited plagioclase crystals in the vugs of lunar rocks. The vesiculative, nonequilibrium vaporization of Na and K during a lunar volcanic eruption should be expected to occur at a high rate upon initial extrusion of the lava into vacuum but then decrease by a factor of approximately three when degassing is nearing completion.

105.039 Olivine microporphyry in the St. Mesmin chondrite.
R. T. Dodd, E. Jarosewich.
Meteoritics, Vol. 11, 1 - 20 (1976).

Olivine microporphyry, of metal- and troilite-poor LL-group composition and composed chiefly of olivine, pyroxenes, and quartz-oligoclase glass, occurs in and locally intrudes clasts in the St. Mesmin chondrite. Its crystallization history resembles that of many Apollo 12 lunar basalts. The microporphyry crystallized from a melt which was emplaced during a period of brecciation or between two episodes of brecciation, but after the thermal metamorphism recorded in other components of the breccia.

105.040 The composition of the Chassigny meteorite.
B. Mason, J. A. Nelen, P. Muir, S. R. Taylor.
Meteoritics, Vol. 11, 21 - 27 (1976).

Spark source mass spectrometric analysis of the Chassigny meteorite has been made. The data, in conjunction with major element composition and mineralogical and textural features, indicate that this meteorite is an olivine-rich cumulate, possibly genetically related to the nakhlites.

105.041 Brazilian meteorites: the Mafra, Santa Catarina State, chondrite. G. R. Levi-Donati, M. Shima,

G. P. Sighinolfi.
Meteoritics, Vol. 11, 29 - 41 (1976).

A chemical analysis and a mineralogical inspection have been performed on about 50 g of the Mafra, Santa Catarina State, Brazil, meteorite. The stone is an H-4 chondrite. It shows some ambiguous characteristics concerning the iron distribution, which are discussed.

105.042 Forms of nonterrestrial dust on the earth.
E. L. Krinov.
Meteoritics, Vol. 11, 43 - 50 (1976).

The author carried out a study of pulverised cosmic matter extracted from the soil at the fall locality of the Sikhote Alin iron meteorite shower. Three forms of dust were distinguishable: meteoritic, sharp-angled, irregular particles from the break-up of the meteorite; meteoric, spherical, magnetic particles from ablation; and micrometeorites. Experiments on the artificial formation of meteoric dust particles (spherules and other spheroidal particles) from meteoritic matter have been done.

105.043 Aïoun el Atrouss: a new hypersthene achondrite with eucritic inclusions. I. S. M. Lomena, F. Touré, E. K. Gibson, Jr., U. S. Clanton, A. M. Reid.
Meteoritics, Vol. 11, 51 - 57 (1976).

Preliminary investigations have been made on two separate pieces from the Aïoun el Atrouss meteorite that fell on April 17, 1974 in southeast Mauritania. The major portion of the meteorite is a brecciated hypersthene achondrite with orthopyroxene as the major phase. Clasts of eucrite occur within the hypersthene achondrite host. No evidence has been found of reaction between the two meteorite types, nor of the presence of any materials intermediate in composition.

105.044 ^{26}Al losses from weathered chondrites.
G. F. Herzog, P. J. Cressy, Jr.
Meteoritics, Vol. 11, 59 - 68 (1976).

105.045 The Meteoritical Bulletin.
Sponsored by the Meteoritical Society,
R. S. Clarke, Jr. (Editor).
Meteoritics, Vol. 11, 69 - 93 (1976).

Place of fall/find, date of fall/find, class and type, number of individual specimens, total weight, and circumstances of fall of 41 meteorites are given.

105.046 Explanation for the very low Ga and Ge concentrations in some iron meteorite groups.
J. T. Wasson, C. M. Wai.
Nature, Vol. 261, 114 - 116 (1976). – Letter.

105.047 Aliphatic amines in the Murchison meteorite.
G. Jungclaus, J. R. Cronin, C. B. Moore, G. U. Yuen.
Nature, Vol. 261, 126 - 128 (1976). – Letter.

105.048 State of the Tunguska meteorite problem at the beginning of 1974. N. V. Vasil'ev.
Problems of meteoritics, Novosibirsk, 1975, (see 003.004), p. 3 - 12. In Russian. – Abstr. in Referativ. Zhurn. 51. Astron., 5.51.365 (1976).

105.049 On shock waves during the flight and explosion of meteorites.
V. P. Korobejnikov, P. I. Chushkin, L. V. Shurshalov.
Problems of meteoritics, Novosibirsk, 1975, (see 003.004), p. 20 - 46. In Russian. – Abstr. in Referativ. Zhurn. 51. Astron., 5.51.363 (1976).

105.050 Calculations of shock waves of the Tunguska meteorite. V. A. Bronshtehn, A. P. Boyarkina.
Problems of meteoritics, Novosibirsk, 1975, (see 003.004),

p. 47 - 63. In Russian. – Abstr. in Referativ. Zhurn. 51. Astron., 5.51.366 (1976).

105.051 On a spatial localization of the source of burn damages in the fall region of the Tunguska meteorite. S. A. Razin, V. G. Fast.
Problems of meteoritics, Novosibirsk, 1975, (see 003.004), p. 64 - 68. In Russian. – Abstr. in Referativ. Zhurn. 51. Astron., 5.51.367 (1976).

105.052 On the origin of forest fire in the fall region of the Tunguska meteorite. N. P. Kurbatskij.
Problems of meteoritics, Novosibirsk, 1975, (see 003.004), p. 69 - 71. In Russian. – Abstr. in Referativ. Zhurn. 51. Astron., 5.51.368 (1976).

105.053 Forest fires in the fall region of the Tunguska meteorite and their effect on forest development.
V. V. Furyaev.
Problems of meteoritics, Novosibirsk, 1975, (see 003.004), p. 72 - 87. In Russian. – Abstr. in Referativ. Zhurn. 51. Astron., 5.51.369 (1976).

105.054 On the verification of the hypothesis of "nuclear explosion" of the Tunguska meteorite from radioactivity of soil of the fall-out track of explosion products.
L. V. Kirichenko.
Problems of meteoritics, Novosibirsk, 1975, (see 003.004), p. 88 - 101. In Russian. – Abstr. in Referativ. Zhurn. 51. Astron., 5.51.370 (1976).

105.055 New verification method of the hypothesis on annihilation and the thermonuclear character of the Tunguska explosion in 1908.
E. M. Kolesnikov, A. K. Lavrukhina, A. V. Fisenko.
Problems of meteoritics, Novosibirsk, 1975, (see 003.004), p. 102 - 110. In Russian. – Abstr. in Referativ. Zhurn. 51. Astron., 5.51.371 (1976).

105.056 On the formation of a local fall-out track after explosion of a cosmic body in 1908.
L. V. Kirichenko.
Problems of meteoritics, Novosibirsk, 1975, (see 003.004), p. 111 - 126. In Russian. – Abstr. in Referativ. Zhurn. 51. Astron., 5.51.372 (1976).

105.057 On the possibility of determination of the explosion nature of the Tunguska cosmic body from neutron activation tracks of the soil in the epicentre of the explosion.
L. V. Kirichenko, I. Ya. Nikolishin.
Problems of meteoritics, Novosibirsk, 1975, (see 003.004), p. 127 - 131. In Russian. – Abstr. in Referativ. Zhurn. 51. Astron., 5.51.373 (1976).

105.058 Once more on the "Siberian darkness" of September 18, 1938.
G. L. Ardakov, N. V. Vasil'ev, P. P. Vaulin, M. I. Dulova, Yu. A. L'vov, T. A. Menyavtseva, N. A. Milyaeva.
Problems of meteoritics, Novosibirsk, 1975, (see 003.004), p. 142 - 143. In Russian. – Abstr. in Referativ. Zhurn. 51. Astron., 5.51.374 (1976).

105.059 Isotopic fractionation in meteoritic cadmium.
K. J. R. Rosman, J. R. De Laeter.
Nature, Vol. 261, 216 - 218 (1976). – Letter.

105.060 Meteoritic material in four terrestrial meteorite craters.
J. W. Morgan, H. Higuchi, R. Ganapathy, E. Anders.
Proc. Sixth Lunar Sci. Conference, (see 012.010), Vol. 2, 1609 - 1623 (1975).

105.061 Volatile elements in Allende inclusions.
L. Grossman, R. Ganapathy.
Proc. Sixth Lunar Sci. Conference, (see 012.010), Vol. 2, 1729 - 1736 (1975).

105.062 Meteor Crater, Arizona, rim drilling with thickness, structural uplift, diameter, depth, volume, and mass-balance calculations.
D. J. Roddy, J. M. Boyce, G. W. Colton, A. L. Dial, Jr.
Proc. Sixth Lunar Sci. Conference, (see 012.010), Vol. 3, 2621 - 2644 (1975).

105.063 A komatiite-like lithic fragment with spinifex texture in the Eva meteorite: origin from a supercooled impact-melt of chondritic parentage.
R. V. Fodor, K. Keil.
Earth Planet. Sci. Letters, Vol. 29, 1 - 6 (1976).

105.064 Major element composition of phyllosilicates in the Orgueil carbonaceous meteorite. J. F. Kerridge.
Earth Planet. Sci. Letters, Vol. 29, 194 - 200 = Publ. Inst. Geophys. Planet. Phys., Univ. Calif., Los Angeles, No. 1491 (1976).

105.065 Thermomagnetic analysis of meteorites, 3. C3 and C4 chondrites.
J. M. Herndon, M. W. Rowe, E. E. Larson, D. E. Watson.
Earth Planet. Sci. Letters, Vol. 29, 283 - 290 (1976).

105.066 The abundances of zirconium and hafnium in the solar system.
R. Ganapathy, G. M. Papia, L. Grossman.
Earth Planet. Sci. Letters, Vol. 29, 302 - 308 (1976).
The concentrations of zirconium and hafnium have been determined in Orgueil, Murchison, Allende, Bruderheim and Alais by RNAA. The mean Zr/Hf weight ratio in the first four of these meteorites is 31.2 ± 2.2 indicating no major fractionation of Zr from Hf. Alais contains anomalously high amounts of many refractory lithophile elements, including Zr and Hf.

105.067 Mafic silicates in the Orgueil carbonaceous meteorite.
J. F. Kerridge, J. D. Macdougall.
Earth Planet. Sci. Letters, Vol. 29, 341 - 348 = Publ. Inst. Geophys. Planet. Phys., Univ. Calif., Los Angeles, No. 1505 (1976).

105.068 A classification of meteorites based on oxygen isotopes. R. N. Clayton, N. Onuma, T. K. Mayeda.
Earth Planet. Sci. Letters, Vol. 30, 10 - 18 (1976).
On the basis of $^{18}O/^{16}O$ and $^{17}O/^{16}O$ ratios, meteorites and planets can be grouped into at least six categories, as follows: (1) the terrestrial group, consisting of the earth, moon, differentiated meteorites and enstatite chondrites; (2) types L and LL ordinary chondrites; (3) type H ordinary chondrites; (4) anhydrous minerals of C2, C3, C4 carbonaceous chondrites; (5) hydrous matrix minerals of C2 carbonaceous chondrites; (6) the ureilites. Objects of one category cannot be derived by fractionation or differentiation from the source materials of any other category.

105.069 La matrice noire et blanche de la chondrite de Tiefschitz (H 3). M. Christophe-Michel-Levy.
Earth Planet. Sci. Letters, Vol. 30, 143 - 150 (1976).

105.070 Depth dependence of spallogenic helium, neon and argon in the St. Severin chondrite.
L. Schultz, P. Signer.
Earth Planet. Sci. Letters, Vol. 30, 191 - 199 (1976).

105.071 Lithification of gas-rich meteorites.
J. R. Ashworth, D. J. Barber.

Earth Planet. Sci. Letters, Vol. 30, 222 - 233 (1976).

The fine structure of four gas-rich meteorites, with particular reference to the cementation and compaction processes that have affected the fine-grained matrix is studied. The observed features are compared with similar effects in lunar breccias. Lithification is attributed to the passage of shock waves through porous aggregates, causing deformation whose intensity varied spatially on a small scale, the most intense deformation and heating effects being concentrated at the edges of large grains and in the matrix between them.

105.072 **Investigation of a new stony meteorite from Mauritania with some additional data on its find site: Aouelloul crater.** R. F. Fudali, P. J. Cressy.
Earth Planet. Sci. Letters, Vol. 30, 262 - 268 (1976).

105.073 **Accretion of the ordinary chondrites.** R. T. Dodd.
Earth Planet. Sci. Letters, Vol. 30, 281 - 291 (1976).

The narrow size distributions of silicate and metal particles in 19 unequilibrated ordinary chondrites and other textural properties of these meteorites strongly suggest that chondritic material was sorted before or during its accumulation in parent bodies. Gravitational sorting during accretion is possible. Aerodynamic sorting can account for the textures of ordinary chondrites.

105.074 **Primordial ^{129}Xe in meteorites.** R. J. Drozd, F. A. Podosek.
Earth Planet. Sci. Letters, Vol. 31, 15 - 30 (1976).

105.075 **^{196}Hg and ^{202}Hg isotopic ratios in chondrites: revisited.** S. Jovanovic, G. W. Reed, Jr.
Earth Planet. Sci. Letters, Vol. 31, 95 - 100 (1976).

105.076 **Trace element content of metals from L-group chondrites.** E. Rambaldi.
Earth Planet. Sci. Letters, Vol. 31, 224 - 238 (1976).

105.077 **Shock-loading meteoritic b.c.c metal above the pressure transition: remnant-magnetisation stability and microstructure.** P. Wasilewski.
Phys. Earth Planet. Interiors, Vol. 11, P5 - P11 (1976).

105.078 **Heavily irradiated grains and neon isotope anomalies in carbonaceous chondrites.** J. Audouze, J. P. Bibring, J. C. Dran, M. Maurette, R. M. Walker.
Astrophys. Journ., (*Letters*), Vol. 206, L185 - L189 (1976).

High-voltage microscope observations of micron-sized grains extracted from a ^{22}Ne-rich fraction of the Orgueil meteorite give strong experimental evidence for their heavy irradiation. It is further suggested that the excess of ^{22}Ne measured in the same mineral separate can be explained by low-energy (< 10 MeV) production by (p, α) reaction on ^{25}Mg. A discussion of the astrophysical sites potentially responsible for the peculiar irradiation history of the meteoritic grains is presented.

105.079 **Mineral composition of meteorite dust from the East Pamirs.** K. P. Yanulov, V. A. Vasil'ev, L. A. Khoroshilova.
Trudy In-t geol. Komi fil. AN SSSR, 1975, vyp. (No.) 21, p. 51 - 58. In Russian. — Abstr. in Referativ. Zhurn. 51. Astron., 6. 51. 348 (1976).

105.080 **Study of the main parameters of tracks observed in olivine crystals from the Ilimaez meteorite.** L. L. Kashkarov, V. S. Pestov.
Struktura i svojstva kristallov. Vyp. (No.) 3. Vladimir, 1975, p. 51 - 56. In Russian. — Abstr. in Referativ. Zhurn. 51. Astron., 6. 51. 354 (1976).

105.081 **Il'inetsk structure — an explosive meteorite crater.** A. A. Val'ter, V. A. Ryabenko.
Geol. zhurn., Vol. 36, No. 1, p. 42 - 53 (1976). In Russian. Abstr. in Referativ. Zhurn. 51. Astron., 6. 51. 359 (1976).

105.082 **Analysis of impact craters from the S-149 Skylab experiment.** D. S. Hallgren, C. L. Hemenway.
IAU Colloquium No. 31, (see 012.015), p. 270 - 274 (1976).

105.083 **Micrometeorite impact craters on Skylab experiment S-149.** K. Nagel, H. Fechtig, E. Schneider, G. Neukum.
IAU Colloquium No. 31, (see 012.015), p. 275 - 278 (1976).

105.084 **Extraterrestrial particles in the stratosphere.** D. E. Brownlee, D. A. Tomandl, P. W. Hodge.
IAU Colloquium No. 31, (see 012.015), p. 279 - 283 (1976).

105.085 **Magellan collections of large cosmic dust particles.** D. S. Hallgren, C. L. Hemenway, R. Wlochowicz.
IAU Colloquium No. 31, (see 012.015), p. 284 - 288 (1976).

105.086 **The isotopic composition and elemental abundance of lutetium in meteorites and terrestrial samples and the ^{176}Lu cosmochronometer.** M. T. McCulloch, J. R. de Laeter, K. J. R. Rosman.
Earth Planet. Sci. Letters, Vol. 28, 308 - 322 (1976).

The isotopic composition of lutetium has been measured in a range of terrestrial and meteoritic materials using solid-source mass spectrometric techniques. The absolute ^{175}Lu/^{176}Lu ratio as determined in this work is 37.36 ± 0.07.

105.087 **Vanadium isotopic composition and contents in gas-rich meteorites.** H. Balsiger, M. D. Mendia, I. Z. Pelly, M. E. Lipschutz.
Earth Planet. Sci. Letters, Vol. 28, 379 - 384 (1976).

105.088 **Iron-silicate fractionation within ordinary chondrite groups.** R. T. Dodd.
Earth Planet. Sci. Letters, Vol. 28, 479 - 484 (1976).

A comparison of recent bulk chemical analyses of fresh, well-classified ordinary chondrites reveals that the unequilibrated H-3 and LL-3 chondrites tend to be iron-poor relative to equilibrated H- and LL-group chondrites (types 4–6). A more complex relationship in the L-group suggests that it consists of two chemical subgroups, in each of which iron is deficient in the lower petrologic types. The available data suggest that the chondrite parent bodies accreted inhomogeneously.

105.089 **Europäisches Netz zur Feuerkugel-Photographie.** Z. Ceplecha.
Sternenbote, 19. Jahrgang, p. 2 - 9 (1976).

105.090 **Discrepancies in the data of the great fireball of August 10, 1972.** A. Carusi, E. Massaro.
Astrophys. Letters, Vol. 17, 113 - 114 (1976).

The published data about the determination of the atmospheric path of the grazing fireball of August, 1972 are not consistent. A new orbit has been computed using only the reported geographical coordinates and times.

105.091 **Paleomagnetism of meteorites.** F. D. Stacey.
Annual Rev. Earth Planet. Sci., Vol. 4, (see 003.012), 147 - 157 (1976).

105.092 **Bestrahlungsalter von Eisenmeteoriten.** W. Hampel, O. A. Schaeffer.
MPI Kernphys., Heidelberg, Jahresbericht 1975, p. 135 - 136.

105.093 **Messung von ^{53}Mn im Mundrabilla-Eisenmeteoriten mit Hilfe der Neutronenaktivierung.**

W. Hampel, T. Kirsten, O. Müller, O. A. Schaeffer.
MPI Kernphys., Heidelberg, Jahresbericht 1975, p. 136 - 137.

**105.094 Neutronen-Tiefenprofile im Mundrabilla-Eisen-
meteoriten.** H. Haag, T. Kirsten.
MPI Kernphys., Heidelberg, Jahresbericht 1975, p. 137 - 138.

**105.095 Bestimmung des räumlichen Gradienten der
kosmischen Strahlung mit Meteoriten.**
G. Heusser, O. A. Schaeffer.
MPI Kernphys., Heidelberg, Jahresbericht 1975, p. 138 - 139.

**105.096 Zur chemischen Zusammensetzung der Metall- und
Sulfidphase des Eisenmeteoriten Mundrabilla
mittels Neutronenaktivierung.** O. Müller.
MPI Kernphys., Heidelberg, Jahresbericht 1975, p. 139.

105.097 Silikateinschlüsse im Mundrabilla-Eisenmeteoriten.
P. Ramdohr, M. Prinz, A. El Goresy.
MPI Kernphys., Heidelberg, Jahresbericht 1975, p. 140.

**105.098 Mikroskopische Untersuchungen am Mundrabilla
Eisenmeteorit und an neuen Steinmeteoriten.**
P. Ramdohr.
MPI Kernphys., Heidelberg, Jahresbericht 1975, p. 140.

**105.099 Morphologie des Rieskraters anhand der Information
aus dem Ries-Tiefbohrkern.** E. C. T. Chao.
MPI Kernphys., Heidelberg, Jahresbericht 1975, p. 159.

105.100 Circumstances of the Gorlovka meteorite fall.
I. T. Zotkin, M. O. Bol'shakov.
Meteoritika, vyp. (No.) 35, p. 19 - 21 (1976). In
Russian.

105.101 The Farmington meteorite. E. Anders,
B. Yu. Levin, A. N. Simonenko.
Meteoritika, vyp. (No.) 35, p. 22 - 36, 145 - 149, 159
(1976). In Russian.

**105.102 The found of the Kifkakhsyagan meteorite on
Chukotka.** R. L. Khotinok, V. I.
Tsvetkov.
Meteoritika, vyp. (No.) 35, p. 37 - 39 (1976). In
Russian.

105.103 The Kifkakhsyagan iron meteorite.
O. A. Kirova, M. I. D'yakonova.
Meteoritika, vyp. (No.), 35, p. 40 - 42 (1976). In
Russian.

105.104 The Chernyj Bor stony meteorite.
O. A. Kirova, M. I. D'yakonova.
Meteoritika, vyp. (No.) 35, p. 43 - 46 (1976). In
Russian.

105.105 The Tobychan iron meteorite.
G. M. Ivanova, I. K. Kuznetsova.
Meteoritika, vyp. (No.) 35, p. 47 - 52, 150 - 155, 159 -
160 (1976). In Russian.

**105.106 The chemical composition of some carbonaceous
chondrites and their comparison with other
C III-type chondrites.** A. A. Yavnel', M. I. D'yako-
nova, V. Ya. Kharitonova.
Meteoritika, vyp. (No.) 35, p. 53 - 58 (1976). In
Russian.

**105.107 Determination of some trace and rare-earth
elements in achondrites and tektites by the
method of instrumental neutron-activation analysis.**

G. M. Kolesov.
Meteoritika, vyp. (No.) 35, p. 59 - 66 (1976). In
Russian.

105.108 Titanium abundance in stony meteorites.
V. Ya. Kharitonova.
Meteoritika, vyp. (No.) 35, p. 67 - 68 (1976). In
Russian.

**105.109 Quantitative analysis of microspherules detected
in Sphagnum peat.** A. P. Boyarkina, N.
V. Vasil'ev, M. K. Nazarenko.
Meteoritika, vyp. (No.) 35, p. 69 - 72 (1976). In
Russian.

**105.110 Composition and structure of magnetic
spherules from the place of the Sikhote-Alin
meteorite fall.** G. M. Kolesov, N. I. Zaslavskaya.
Meteoritika, vyp. (No.) 35, p. 73 - 77 (1976). In Russian.

105.111 Magnetization of carbonaceous chondrites.
E. G. Gus'kova.
Meteoritika, vyp. (No.) 35, p. 78 - 86 (1976). In
Russian.

105.112 On the X-ray structure of tektites.
M. D. Dorfman, A. I. Soklakov, A. V. Podless-
kaya.
Meteoritika, vyp. (No.) 35, p. 87 - 90 (1976). In
Russian.

**105.113 Characteristics of the geological composition
of the craterlike structure of Tabun-Khara-Obo
(South-East Mongolia).** L. M. Shkerin.
Meteoritika, vyp. (No.) 35, p. 97 - 102, 156 - 158,
160 - 161 (1976). In Russian.

105.114 Impactites from the Janisjärvian astrobleme.
V. L. Masajtis, A. S. Sindeev, Yu. G. Staritskij.
Meteoritika, vyp. (No.) 35, p. 103 - 110 (1976). In
Russian.

**105.115 On the Verkhnedneprovsk and Augustinovka
meteorites.** A. A. Yavnel'.
Meteoritika, vyp. (No.) 35, p. 111 - 114 (1976). In
Russian.

**105.116 A short catalogue of the USSR meteorites,
January 1, 1976.** E. L. Krinov.
Meteoritika, vyp. (No.) 35, p. 115 - 135 (1976). In
Russian.

105.117 A list of meteorite-forming minerals.
L. G. Kvasha.
Meteoritika, vyp. (No.) 35, p. 136 - 138 (1976). In
Russian.

**105.118 The archives of the Committee on Meteorites
of the USSR Academy of Sciences.**
T. V. Vodop'yanova.
Meteoritika, vyp. (No.) 35, p. 139 - 144 (1976). In
Russian.

**105.119 Study of the extraterrestrial materials on the ant-
arctica. V. On the Yamato meteorites. II.**
M. Shima, A. Okada, H. Yabuki.
Rep. Inst. Phys. Chem. Res., Vol. 51, No. 4, p. 82 - 92 (1975).
In Japanese. – Abstr. in Phys. Abstr., Vol. 79, A011116
(1976).

105.120 Historical plaque marks the Holleford meteorite

crater. P. B. Robertson.
GEOS, p. 16 - 17 (1975). — Abstr. in Phys. Abstr., Vol. 79, A027375 (1976).

105.121 Visible and near-infrared spectra: X. Stony meteorites.
J. W. Salisbury, G. R. Hunt, C. J. Lenhoff.
Modern Geol., Vol. 5, 115 - 126 (1975). — Abstr. in Phys. Abstr., Vol. 79, A042145 (1976).

105.122 On the study of the Numakai meteorite.
M. Shima, A. Okada, K. Yagi.
Sci. Papers Inst. Phys. Chem. Res., Vol. 69, 136 - 145 (1975). Abstr. in Phys. Abstr., Vol. 79, A042146 (1976).

105.123 Elemental abundances in meteoritic and terrestrial matter. Annual progress report, 1 September 1973 - 31 August 1974. R. A. Schmitt.
Separate print Oregon State Univ., Corvallis, USA, Dept. Chem., 51 pp. (1974).

105.124 Allende meteorite: cosmochemistry's Rosetta Stone.
B. Mason.
Acc. Chem. Res., Vol. 8, 217 - 224 (1975).

105.125 Micrometeoroids and the hyper-velocity impact.
J. C. Mandeville.
Internat. conf. evaluation space environment materials, Toulouse, France, June 17 - 21, 1974, p. 79 - 92 (1974). In French.

105.126 Low-level measurements of the spallation nuclei ^{26}Al and ^{53}Mn and their application to special meteorite problems, as well as on an attempt to determine the nature and global incoming rate of cosmic dust. M. Heimann.
Diss. Köln Univ., F. R. Germany, Math.-Nat. Fak. 114 pp. (1974). In German. (Available from ZAED).

New two-stage accelerator for hypervelocity impact simulation (micrometeoroid environment simulation). See Abstr. 022.100.

Enhanced solar cosmic-ray flux in the early solar system. See Abstr. 078.057.

Geomagnetic disturbances caused by shock waves of large meteoritic bodies. II. See Abstr. 084.255.

The metallic phase of planets and meteorites. See Abstr. 091.024.

Genetic relations between the moon and meteorites. See Abstr. 094.011.

The state of meteoritic material on the moon. See Abstr. 094.138.

Evidence for meteoritic sulfur in the lunar regolith. See Abstr. 094.141.

Monte Carlo simulation of turnover processes in the lunar regolith. See Abstr. 094.144.

The simulated depth history of dust grains in the lunar regolith. See Abstr. 094.145.

Shock effects from a large impact on the moon. See Abstr. 094.159.

Meteoroid storms detected on the moon. See Abstr. 094.180.

Variation in the number of meteoroid impacts on the moon with lunar phase. See Abstr. 094.200.

Clues in the rare gas isotopes to early solar system history. See Abstr. 094.203.

A new type of chondritic meteorite found in lunar soil. See Abstr. 094.584.

Diffusion artifacts in dating by stepwise thermal release of rare gases. See Abstr. 094.589.

Radioactive halos and ion microprobe measurement of Pb isotope ratios. See Abstr. 094.609.

Asteroids as meteorite parent-bodies: the astronomical perspective. See Abstr. 098.038.

Guests from the sky. See Abstr. 104.064.

Cosmic dust influx to the earth. See Abstr. 106.016.

Near-earth cosmic dust fluxes obtained from Skylab experiments. See Abstr. 106.017.

Erste Ergebnisse des Mikrometeoritenexperiments auf Helios A. See Abstr. 106.102.

Simulation of the cosmic dust and micrometeorite particle streams. See Abstr. 106.112.

Graphite grains in the early solar system. See Abstr. 107.006.

Constancy of galactic cosmic rays in time and space. See Abstr. 143.307.

Errata

105.901 Erratum: 'Nickel enrichment of impact melt rocks from Rochechouart' [Meteoritics, Vol. 10, 433 - 436 (1975)]. P. Lambert.
Meteoritics, Vol. 11, 96 (1976).

106 Interplanetary Matter, Interplanetary Magnetic Field, Zodiacal Light

106.001 Interplanetary stream magnetism: kinematic effects.
L. F. Burlaga, E. Barouch.
Astrophys. Journ., Vol. 203, 257 - 267 (1976).

The particle density, and the magnetic field intensity and direction, are calculated for volume elements of the solar wind as a function of the initial magnetic field direction, Φ_0, and the initial speed gradient, $(\partial V/\partial R)_0$. It is assumed that the velocity is constant and radial. Time profiles of n, B, and V are calculated for corotating streams, neglecting effects of pressure gradients. Changes of field direction may be very large, depending on the initial angle; but when the initial angle at 0.1 AU is such that the base of the field line corotates with the sun, the spiral angle is the preferred direction at 1 AU. The theory is also applicable to nonstationary flows.

106.002 The interplanetary acceleration of energetic nucleons. F. B. McDonald, B. J. Teegarden, J. H. Trainor, T. T. von Rosenvinge, W. R. Webber.
Astrophys. Journ., (*Letters*), Vol. 203, L149 - L154 (1976).

Corotating proton and electron streams are the dominant type of low-energy (i.e., 0.1–10 MeV per nucleon) particle event observed at 1 AU. The radial dependence of these events has been studied between 1 and 4 AU. It had been expected that at a given energy the intensity of these streams would decrease rapidly with heliocentric distance due to the effects of interplanetary adiabatic deceleration. Instead it is observed that from event to event the intensity either remains roughly constant or increases significantly. It appears that interplanetary acceleration processes are the most plausible explanation. Several possible acceleration models are discussed.

106.003 Interplanetary dust particles near Jupiter.
S. F. Singer, J. E. Stanley.
Icarus, Vol. 27, 197 - 205 (1976).

The authors calculate the expected counting of a flat micrometeoroid detector of finite sensitivity passing in hyperbolic orbit near a planet. The results of the calculations are then compared with the results returned by Pioneer 10 in its flyby of Jupiter. The observed increase in impact rate near Jupiter can be completely explained in terms of gravitational 'focusing' of particles which are in heliocentric orbits; i.e., they are not in orbit about Jupiter. The absolute concentration of particles near the orbit of Jupiter is of the same order as at 1 AU.

106.004 Polarimetric observations of the zodiacal light in Hawaii from 1969 to 1974.
L. W. Bandermann, R. D. Wolstencroft.
Mem. Roy. Astron. Soc., Vol. 81, 37 - 88 (1976).

Results are presented of polarimeter observations of the zodiacal light at 5080 and 5300 Å made in Hawaii between 1969 April and 1974 November.

106.005 Adiabatic Fermi acceleration of energetic particles between converging interplanetary shock waves.
E. H. Levy, S. P. Duggal, M. A. Pomerantz.
Journ. Geophys. Res., Vol. 81, 51 - 59 (1976).

The process of adiabatic acceleration of relativistic cosmic rays between two converging shocks, through the first-order Fermi mechanism, to which the unusual ground level event of August 4, 1972, was attributed earlier is examined quantitatively, the nature of interplanetary shock waves and their propagation being taken into account.

106.006 A shock surface geometry: the February 15–16,

1967, event. R. P. Lepping, J. K. Chao.
Journ. Geophys. Res., Vol. 81, 60 - 64 (1976).

The flare-associated interplanetary (IP) shock of February 15–16, 1967, observed by Explorer 33 and Pioneer 7 is analyzed to yield an estimation of the ecliptic plane geometry of the shock surface near 1 AU. The average shock speed from the sun to each spacecraft and the local speed at Explorer 33 and their relations to the position of the initiating solar flare are obtained and discussed. In the region of space between the earth and Pioneer 7 the shock surface radius of curvature in the ecliptic plane appears to have been 0.4 AU or less.

106.007 A survey of long-term interplanetary magnetic field variations. J. H. King.
Journ. Geophys. Res., Vol. 81, 653 - 660 (1976).

Interplanetary magnetic field data have been merged into a composite data set spanning 1963–1974. Analysis of the composite data set reveals the following: (1) Although the yearly averaged magnitudes of all field vectors show virtually no solar cycle variation, the yearly averaged magnitudes of positive and negative polarity field vectors show separate solar cycle variations, consistent with variations in the average azimuthal angles of positive and negative polarity field vectors. (2) There is no heliolatitude dependence of long-time average field magnitudes. (3) Field vectors parallel to the earth-sun line are on the average 1 γ less in magnitude than field vectors perpendicular to this line.

106.008 Interplanetary acceleration of relativistic electrons observed with Imp 7.
T. P. Armstrong, S. M. Krimigis.
Journ. Geophys. Res., Vol. 81, 677 - 682 (1976).

The intensities of 0.22- to 0.5-MeV and 0.5- to 0.8-MeV electrons in interplanetary space following the October 29, 1972, solar particle event have been observed with the JHU/APL experiment aboard the Imp 7 satellite. The results are consistent with an energy-dependent acceleration process, which is most effective for electrons of ~ 0.4 MeV. This is the first reported interplanetary shock wave acceleration of relativistic electrons.

106.009 Solar wind electric field modulation in the interplanetary sector structure.
A. Bahnsen, N. D'Angelo.
Journ. Geophys. Res., Vol. 81, 683 - 686 (1976).

The solar wind electric field near the earth has been computed for ~ 4 1/2 solar rotations, utilizing Heos 2 magnetometer and particle velocity data. A clear modulation is apparent as a function of the earth's position within interplanetary sectors. The electric field rises sharply at each sector boundary from a preboundary value of ~ 1 mV/m to peak values between ~ 3 mV/m and ~ 6 mV/m. A correlation is found with several phenomena in the polar cap ionosphere.

106.010 Interpretation of a rocket photometry of the inner zodiacal light.
C. Leinert, H. Link, E. Pitz, R. H. Giese.
Astron. Astrophys., Vol. 47, 221 - 230 (1976).

The results of a previous rocket photometry of the inner zodiacal light are interpreted in terms of spatial distribution and particle properties. It is found that the spatial distribution of the zodiacal cloud is fan-like and within 1 A.U. close to $n(r) \sim r^{-1} \cdot \exp(-2.6 \cdot |\sin^{1.3}\beta_\odot|)$. The plane of symmetry ($\Omega = 66 \pm 11°$, $i = 3.7 \pm 0.6°$) deviates from the ecliptic, apparently more than the invariable plane. Fluxes of interplanetary dust and zodiacal light brightness are found to be compatible.

106.011 Neue Ergebnisse zum optischen Streuverhalten unregelmäßiger Partikel.
K. Weiß, R. Zerull, R. H. Giese.
Mitt. Astron. Ges., No. 38, p. 149 (1976). – Abstract.

106.012 Wirkungsfaktoren für Strahlungsdruck, Absorption und Extinktion für interplanetare Staubteilchen.
G. Schwehm, M. Kröger.
Mitt. Astron. Ges., No. 38, p. 149 (1976). – Abstract.

106.013 A photometry of the zodiacal light in B.
J. Pfleiderer, C. Classen.
Mitt. Astron. Ges., No. 38, p. 152 - 154 (1976). – Short report.

106.014 Über die Existenz von Erde-Mond-Librationswolken.
S. Röser.
Mitt. Astron. Ges., No. 38, p. 190 (1976). – Abstract.

106.015 Characteristics of interplanetary plasma near the earth observed during the solar events of August 1972. F. Cambou, O. L. Vaisberg (*Vajsberg*), H. Espagne, V. V. Temny (*Temnyj*), C. d'Uston, G. N. Zastenker.
Space Research XV, (see 012.003), p. 461 - 469 (1975).

106.016 Cosmic dust influx to the earth. D. W. Hughes.
Space Research XV, (see 012.003), p. 531 - 539 (1975).

The influx of cosmic dust to the earth's surface has been deduced from observations obtained by satellite-borne detectors, from the analysis of backscatter radio echoes off meteor trails, and from visual meteor counts. Observations of zodiacal light have also been used to deduce a value for the density of cosmic dust in space. Recent measurements of the geocentric velocity of particles in the cosmic dust cloud have also been reviewed. These observations are combined to give a value for the total influx of cosmic dust to the earth (in the mass range 10^{-13} g to 10^6 g) and the spatial density of this dust at 1 AU from the sun.

106.017 Near-earth cosmic dust fluxes obtained from Skylab experiments.
C. L. Hemenway, D. S. Hallgren, C. D. Tackett.
Space Research XV, (see 012.003), p. 541 - 547 (1975).

Three exposures (34, 46, and 33 days) of thin films and polished metal plates with a total area of 0.12 m² per exposure were carried out during the Skylab S-149 experiment. Craters and penetration holes have been found ranging from 135 μm diameter to less than 0.5 μm. A cosmic dust flux curve in the mass range $10^{-16}-10^{-7}$ grams is presented. Preliminary composition data concerning the particle residue within the copper crater walls are given.

106.018 The interplanetary and near Jovian dust environment: some experimental results.
J. M. Alvarez, D. H. Humes, W. H. Kinard, R. L. O'Neal.
Space Research XV, (see 012.003), p. 549 - 554 (1975).

The cosmic dust environment in interplanetary and near Jovian space has been measured by Langley Research Center experiments on board Pioneers 10 and 11. The experiments indicate that the asteroid belt does not play a significant role in the production of $10^{-8}-10^{-9}$ g particles. The penetration rates from both experiments decrease with increasing solar distance with no evidence of an increased meteoroid population in the asteroid belt. Analysis of the Pioneer 10 Jupiter encounter data indicates that the two-orders-of-magnitude rise in penetration rate at Jupiter periapsis is apparently due to Jupiter's strong gravitational field.

106.019 Bounds for the interstellar to solar system micro-

particle flux ratio over the mass range $10^{-11}-10^{-13}$ g.
J. A. M. McDonnell, O. E. Berg.
Space Research XV, (see 012.003), p. 555 - 563 (1975).

A statistical analysis of the front film and grid coincidence events recorded over 5 years of observation on deep space probes Pioneer 8 and 9 is presented using computer integrated graphs. The data show a predominance of particles towards the probe apex for mass $\sim 10^{-11}$ g in accordance with a 'sweeping up' of particles in degraded cometary orbits, but for particles $\sim 10^{-13}$ g a predominance of particles incident from an apparent solar direction is observed. The data have been analyzed to investigate the possibility of a component from an interstellar source. A bound on the maximum magnitude of an interstellar component is placed at $<4\%$ for particles of mass $\gtrsim 10^{-13}$ g.

106.020 The cometary contribution to cosmic dust.
D. W. Hughes.
Space Research XV, (see 012.003), p. 565 - 572 (1975).

Recent measurements of the flux of particulate matter into the solar system produced by the break up of comets are reviewed. Processes whereby the resulting meteor stream decays into the solar system dust cloud are investigated.

106.021 The distribution of dust along the earth's orbit deduced from satellite measurements of zodiacal light.
A.-C. Levasseur, J. E. Blamont.
Space Research XV, (see 012.003), p. 573 - 578 (1975).

The zodiacal light intensity has been measured for 26 months from the D2A satellite. The elongation was 90°. Under normal conditions, the intensity remained between 57 S_{10} (vis) at the ecliptic poles and 150 S_{10} (vis) at the equator. It was observed to vary slowly at high ecliptic latitudes with a one-year period. It could also increase rapidly from one day to the next; such increases occurred sometimes over a limited part of the sky. The parallel trends of the evolution with time for selected directions in two consecutive years show that the variations are due to the position of the earth in its orbit. Long-period variations are explained by a zodiacal cloud symmetrical with respect to the solar system invariant plane. Small-period increases may be due to local meteor streams.

106.022 Measurement of the temperature of the interstellar medium around the heliosphere, obtained with the Mars 7 interplanetary probe. J. L. Bertaux, J. E. Blamont, N. Tabarié, N. N. Dementeva (*Dement'eva*), V. G. Kurt, A. S. Smirnov.
Space Research XV, (see 012.003), p. 721 - 726 (1975).

Results of a Lyman-alpha photometer placed on the interplanetary probe Mars 7 are presented. The Lyman alpha emission linewidth of interplanetary hydrogen was analyzed by means of an absorption hydrogen cell, yielding a temperature of 10^4 °K for the local neutral interstellar hydrogen, through which the solar system travels with a velocity of $\simeq 11$ km s^{-1}.

106.023 On the influence of Jupiter on the motion of dust particles of the zodiacal cloud.
N. I. Komarnitskaya.
Interaction of meteor matter. Conference Dushanbe 1974, (see 012.004), p. 135 - 141 (1975). In Russian. – Abstr. in Referativ. Zhurn. 51. Astron., 3.51.265 (1976).

106.024 Temperature of the dust particles of the zodiacal cloud. L. V. Reznova.
Interaction of meteor matter. Conference Dushanbe 1974, (see 012.004), p. 142 - 146 (1975). In Russian. – Abstr. in Referativ. Zhurn. 51. Astron., 3.51.266 (1976).

106.025 Optical properties of interplanetary dust cloud particles based on zodiacal light observations.
S. N. Krylova.

Interaction of meteor matter. Conference Dushanbe 1974, (see 012.004), p. 147 - 156 (1975). In Russian. – Abstr.in Referativ. Zhurn. 51. Astron., 3.51.267 (1976).

106.026 Interplanetary magnetic fields, their fluctuations, and cosmic ray variations.   E. Barouch, J. W. Sari. Journ. Geophys. Res., Vol. 81, 1453 - 1456 (1976).

The cause of Forbush decreases is examined by using neutron monitor data and measurements of the interplanetary magnetic field. It is found that for the period examined (December 15, 1965, to April 23, 1966), large enhancements of the interplanetary magnetic field correlate well with decreases in cosmic ray intensity, while various parameters connected with the fluctuations in the field do not display such good correlation. The inference is drawn that Forbush decreases are not related to the turbulence or random motions in the field but to the large-scale features of the field.

106.027 Discontinuities in the interplanetary medium and scintillations of radio sources.
I. V. Chashej, V. I. Shishov, T. D. Shishova.
Geomagn. Aeronom., Vol. 16, 15 - 20 (1976). In Russian.

106.028 Rotating shock waves in the role of head and back shock waves in streams of interplanetary plasma.
K. G. Ivanov.
Geomagn. Aeronom., Vol. 16, 21 - 24 (1976). In Russian.

106.029 Dynamics of the interplanetary plasma by means of data on time spectra of scintillations.
N. A. Lotova, I. V. Chashej.
Geomagn. Aeronom., Vol. 16, 230 - 238 (1976). In Russian.

106.030 Strong active regions and disturbances of the interplanetary medium.   N. N. Kontor, G. P. Lyubimov, T. G. Khotilovskaya, E. P. Zaborova.
VIIth Leningrad seminar. "Corpuscular streams of the sun and the radiation belts of the earth and Jupiter", 1975, (see 012. 007), p. 159 - 160 (1975). In Russian and English. – Abstr. in Referativ. Zhurn. 51. Astron., 4.51.328; 62. Issled. kosmich. prostranstva, 4.62.184 (1976).

106.031 Electron flux stabilization in interplanetary space.
S. A. Kaplan.
VIIth Leningrad seminar. "Corpuscular streams of the sun and the radiation belts of the earth and Jupiter", 1975, (see 012. 007), p. 179 - 184 (1975). In Russian. – Abstr. in Referativ. Zhurn. 51. Astron., 4.51.330; 62. Issled. kosmich. prostranstva, 4.62.213 (1976).

106.032 The possibility of investigation of the interplanetary magnetic field from polarization data of the zodiacal light.   A. Z. Dolginov, I. G. Mitrofanov.
VIIth Leningrad seminar. "Corpuscular streams of the sun and the radiation belts of the earth and Jupiter", 1975, (see 012. 007), p. 217 - 218 (1975). In Russian and English. – Abstr. in Referativ. Zhurn. 51. Astron., 4.51.335; 62. Issled. kosmich. prostranstva, 4.62.187 (1976).

106.033 Variations of the spectrum of fluctuations of the interplanetary magnetic field and of diffusion coefficients for cosmic radiation in the solar activity cycle.
V. A. Kovalenko, V. N. Malyshkin.
Issled. po geomagnetizmu, aehron. i fiz. Solntsa. Vyp. (No) 37. Moskva, Nauka, 1975, p. 109 - 118. In Russian. – Abstr. in Referativ. Zhurn. 62. Issled. kosmich. prostranstva, 4.62.182 (1976).

106.034 On the continuity of processes in the interplanetary medium conditioned by solar activity during the recent million years.   A. K. Lavrukhina, G. K. Ustinova.

VIIth Leningrad seminar. "Corpuscular streams of the sun and the radiation belts of the earth and Jupiter", 1975, (see 012. 007), p. 376 - 388 (1975). In Russian. – Abstr. in Referativ. Zhurn. 62. Issled. kosmich. prostranstva, 4.62.211 (1976).

106.035 Large-scale features of the interplanetary magnetic field.
V. A. Kovalenko, N. P. Korzhov, V. N. Malyshkin.
Astron. Zhurn. Akad. Nauk SSSR, Vol. 53, 295 - 299 (1976). In Russian. English translation in Soviet Astron., Vol. 20, No. 2.

The effect of temporal solar wind velocity variations on the interplanetary magnetic field configuration is considered. It is shown that these variations may be the cause of long-period (2 days) deviations of the magnetic field vector from a spiral, of configurations of "kinks" type, and, besides, of the decrease of the angle, averaged with regard to the sector, that determines the direction of the interplanetary magnetic field.

106.036 Sun and comets as sources in an external flow.
M. K. Wallis, M. Dryer.
Astrophys. Journ., Vol. 205, 895 - 899 (1976).

The interaction of the largely neutral interstellar gas with the heliosphere, on the one hand, and the interaction of the ionized solar wind with largely neutral cometary comas, on the other, have a number of phenomenological similarities. Both the sun and comets act as sources embedded within an external flow. This communication is an attempt to organize the similarities as well as the differences in a form amenable for further, more detailed, intercomparisons.

106.037 Relations between the solar and interplanetary magnetic field distributions.   V. Bumba.
Bull. Astron. Inst. Czechoslovakia, Vol. 27, 153 - 155 (1976).

The correlation of the negative polarity interplanetary magnetic field·sectors with the negative polarity main body of the solar magnetic field "supergiant" structures is shown.

106.038 Apollo 16 Lyman alpha imagery of the hydrogen geocorona.   G. R. Carruthers, T. Page, R. R. Meier.
Journ. Geophys. Res., Vol. 81, 1664 - 1672 (1976).

Lyman α imagery of the hydrogen geocorona was obtained from the lunar surface during the Apollo 16 mission. The images are of 20° diameter fields, with 2 arc min limiting resolution, centered on the earth and about 12° upsun of the earth. The data confirm that the hydrogen geocorona is detectable above the interplanetary Lyman α background to more than 15 R_E in the upsun direction.

106.039 Comparative analysis of cosmophysical and geophysical phenomena in the 19th and 20th solar activity cycles.
L. I. Dorman, N. S. Kaminer, A. E. Kuz'micheva.
Cosmic rays, No. 15, (see 003.006), p. 24 - 29 (1975). In Russian.

106.040 Density, chemistry, and size distribution of interplanetary dust.
D. E. Brownlee, F. Hörz, J. B. Hartung, D. E. Gault.
Proc. Sixth Lunar Sci. Conference, (see 012.010), Vol. 3, 3409 - 3416 (1975).

106.041 The interplanetary magnetic field structure.
E. H. Levy.
Nature, Vol. 261, 394 - 395 (1976). – Letter.

106.042 Investigation of shock waves and tangential discontinuities of the cosmic plasma by the radio interference technique.   I. P. Stakhanov, J. V. Kovalevsky (*I. V. Kovalevskij*), V. D. Novikov.
Planet. Space Sci., Vol. 24, 617 - 620 (1976).

The possibility of investigation of the cosmic plasma dynamics by the radio interference technique based on a finite time of radio wave propagation between the sounding and responding stations is shown. Interplanetary shock waves and tangential discontinuities were registered and the velocities and plasma density changes on their fronts were determined. By using experimental data one can also obtain information about plasma concentration jump, location and motion of bow shock wave and magnetopause and plasmapause.

106.043 Density and composition of interplanetary dust particles.
K. Nagel, G. Neukum, H. Fechtig, W. Gentner.
Earth Planet. Sci. Letters, Vol. 30, 234 - 240 (1976).

Residual meteoritic material has been detected on the surface of crater interiors on lunar samples 60315,29 and 65315,68. Crater simulation experiments show the dependence of crater diameter to depth ratio on projectile density. Diameter-to-depth measurements yield three groups of micrometeorites in the size range between $1\mu m$ and 1mm: iron-nickel, stony-iron and low-density particles.

106.044 On the behaviour of the pitch angle diffusion coefficients near pitch angles of 90°.
W. Alpers, K. Hasselmann, J. Kunstmann.
14th Intern. Cosmic Ray Conf., (see 012.011), Vol. 3, 888 - 892 (1975).

106.045 A survey of the interplanetary magnetic field.
E. Barouch, J. H. King.
14th Intern. Cosmic Ray Conf., (see 012.011), Vol. 3, 1036 - 1041 (1975).

106.046 Enhanced interplanetary magnetic fields as the cause of Forbush decreases.
L. F. Burlaga, E. Barouch.
14th Intern. Cosmic Ray Conf., (see 012.011), Vol. 3, 1082 - 1085 (1975).

106.047 A model for the acceleration of particles trapped between converging shock waves.
E. H. Levy, S. P. Duggal, M. A. Pomerantz.
14th Intern. Cosmic Ray Conf., (see 012.011), Vol. 5, 1823 - 1828 (1975).

106.048 Analysis of the 31 Oct 1972 interplanetary shock wave and associated unusual phenomena.
F. M. Ipavich, R. P. Lepping.
14th Intern. Cosmic Ray Conf., (see 012.011), Vol. 5, 1829 - 1834 (1975).

106.049 Observations of energetic particles near interplanetary MHD discontinuities.
E. T. Sarris, S. M. Krimigis, T. P. Armstrong.
14th Intern. Cosmic Ray Conf., (see 012.011), Vol. 5, 1835 - 1840 (1975).

106.050 Initial results of the T.U. – Braunschweig fluxgate magnetometer experiment on the macroscopic structure of the interplanetary magnetic field.
F. M. Neubauer, G. Musmann, E. Lammers.
14th Intern. Cosmic Ray Conf., (see 012.011), Vol. 5, 1858 (1975). – Abstract.

106.051 Zum Einfluss des solaren Strahlungsdrucks auf die Bewegung kleiner Teilchen im Sonnensystem.
S. Röser.
Diss. Nat. Gesamtfakultät, Ruprecht-Karl-Univ., Heidelberg, 77 pp. (1975).

106.052 Interaction of interplanetary magnetic fields with cometary plasma.
H. U. Schmidt, R. Wegmann.
Computing in plasma physics and astrophysics, (see 012.014), D4, 2 pp. (1976).

106.053 Investigation of the integral characteristics of cosmic plasma with the method of a dispersion interferometer during the flight of the Mars-4–7 stations.
M. B. Vasil'ev, A. S. Vyshlov, N. A. Savich, V. A. Samovol, L. N. Samoznaev, A. I. Sodorenko.
Kosmich. Issled., Vol. 14, 383 - 391 (1976). In Russian.

106.054 Space observations of the zodiacal light.
J. L. Weinberg.
IAU Colloquium No. 31, (see 012.015), p. 3 - 18 (1976). Invited paper.

A listing and discussion are given of balloon, rocket, satellite, and space probe observations of the zodiacal light.

106.055 Preliminary results of the Helios A zodiacal light experiment.
H. Link, C. Leinert, E. Pitz, N. Salm.
IAU Colloquium No. 31, (see 012.015), p. 24 - 28 (1976).

106.056 Pioneer 10 observations of zodiacal light brightness near the ecliptic: changes with heliocentric distance.
M. S. Hanner, J. G. Sparrow, J. L. Weinberg, D. E. Beeson.
IAU Colloquium No. 31, (see 012.015), p. 29 - 35 (1976).

106.057 The $S_{10}(V)$ unit of surface brightness.
J. G. Sparrow, J. L. Weinberg.
IAU Colloquium No. 31, (see 012.015), p. 41 - 44 (1976).

106.058 Polarization of the zodiacal light: first results from Skylab.
J. G. Sparrow, J. L. Weinberg, R. C. Hahn.
IAU Colloquium No. 31, (see 012.015), p. 45 - 51 (1976).

106.059 Photometry of the zodiacal light with the balloon - borne telescope THISBE.
A. Frey, W. Hofmann, D. Lemke, C. Thum.
IAU Colloquium No. 31, (see 012.015), p. 52 (1976).

106.060 OSO-5 zodiacal light measurements 1969–1975.
G. B. Burnett.
IAU Colloquium No. 31, (see 012.015), p. 53 - 57 (1976).

106.061 Evidence for scattering particles in meteor streams.
A.-C. Levasseur, J. Blamont.
IAU Colloquium No. 31, (see 012.015), p. 58 - 62 (1976).

106.062 The ultraviolet scattering efficiency of interplanetary dust grains. C. F. Lillie.
IAU Colloquium No. 31, (see 012.015), p. 63 (1976). Abstract.

106.063 The color characteristics of the earth–moon libration clouds. J. R. Roach.
IAU Colloquium No. 31, (see 012.015), p. 68 - 72 (1976).

106.064 Visible and UV photometry of the Gegenschein and the Milky Way. A. Llebaria.
IAU Colloquium No. 31, (see 012.015), p. 78 - 82 (1976).

106.065 Ground-based observations of the zodiacal light.
R. Dumont.
IAU Colloquium No. 31, (see 012.015), p. 85 - 100 (1976). Invited paper.

106.066 Polarimetry of the zodiacal light and Milky Way from Hawaii.

R. D. Wolstencroft, L. W. Bandermann.
IAU Colloquium No. 31, (see 012.015), p. 101 - 105 (1976).

106.067 Some formulae to interpret zodiacal light photo-polarimetric data in the ecliptic from ground or space. R. Dumont.
IAU Colloquium No. 31, (see 012.015), p. 115 - 119 (1976).

106.068 Discussion of the rocket photometry of the zodiacal light. C. Leinert, H. Link, E. Pitz.
IAU Colloquium No. 31, (see 012.015), p. 120 (1976).
Abstract.

106.069 Consequences of the inclination of the zodiacal cloud on the ecliptic. R. Robley.
IAU Colloquium No. 31, (see 012.015), p. 121 (1976).
Abstract of 14.106.009.

106.070 Method for the determination of the intensity of scattered sunlight per unit-volume of the interplanetary medium. A. Mujica, F. Sánchez.
IAU Colloquium No. 31, (see 012.015), p. 122 - 123 (1976).

106.071 On the visibility of the libration clouds. S. Röser.
IAU Colloquium No. 31, (see 012.015), p. 124 - 129 (1976).

106.072 The compatibility of recent micrometeoroid flux curves with observations and models of the zodiacal light. R. H. Giese, E. Grün.
IAU Colloquium No. 31, (see 012.015), p. 135 - 139 (1976).

106.073 In-situ records of interplanetary dust particles — methods and results. H. Fechtig.
IAU Colloquium No. 31, (see 012.015), p. 143 - 158 (1976).
Invited paper.

106.074 Preliminary results of the micrometeoroid experiment on board Helios A. E. Grün, J. Kissel, H. Fechtig, P. Gammelin, H.-J. Hoffmann.
IAU Colloquium No. 31, (see 012.015), p. 159 - 163 (1976).

106.075 Composition of impact-plasma measured by a Helios-mircometeoroid-detector.
B.-K. Dalmann, E. Grün, J. Kissel.
IAU Colloquium No. 31, (see 012.015), p. 164 (1976).
Abstract.

106.076 Orbital elements of dust particles intercepted by Pioneers 8 and 9. H. Wolf, J. W. Rhee, O. E. Berg.
IAU Colloquium No. 31, (see 012.015), p. 165 - 169 (1976).

106.077 The cosmic dust environment at earth, Jupiter and interplanetary space: results from Langley experiments on MTS, Pioneer 10, and Pioneer 11.
J. M. Alvarez.
IAU Colloquium No. 31, (see 012.015), p. 181 (1976).
Abstract.

106.078 Dust in the outer solar system—review of early results from Pioneers 10 and 11.
R. K. Soberman, J. M. Alvarez, J. L. Weinberg.
IAU Colloquium No. 31, (see 012.015), p. 182 - 186 (1976).

106.079 Sources of interplanetary dust: asteroids.
J. S. Dohnanyi.
IAU Colloquium No. 31, (see 012.015), p. 187 - 205 (1976).
Invited paper.
Using results from the Pioneer-10 experiments the relative contribution of asteroidal and cometary particles to the zodiacal cloud is estimated using methods developed in earlier studies of meteoroidal collisions (collisional model). It is found that the contribution of asteroidal particles to dust in the asteroidal belt is small compared with the number density of cometary type particles. Similar conclusions apply to the zodiacal cloud between the sun and the asteroid belt. The distribution of asteroidal rotations is analyzed.

106.080 The long term population of interplanetary micro-meteoroids. G. Poupeau, R. M. Walker, E. Zinner, D. Morrison.
IAU Colloquium No. 31, (see 012.015), p. 232 (1976).

106.081 Near-earth fragmentation of cosmic dust. H. Fechtig, C. L. Hemenway.
IAU Colloquium No. 31, (see 012.015), p. 290 - 295 (1976).

106.082 Interplanetary dust in the vicinity of the earth. G. M. Teptin.
IAU Colloquium No. 31, (see 012.015), p. 389 (1976).
Abstract.

106.083 Evolution and detectability of interplanetary dust streams. Ľ. Kresák.
IAU Colloquium No. 31, (see 012.015), p. 391 - 395 (1976).

106.084 On the structure of hyperbolic interplanetary dust streams. L. Kresák, E. M. Pittich.
IAU Colloquium No. 31, (see 012.015), p. 396 - 399 (1976).

106.085 Sources of interplanetary dust. F. L. Whipple.
IAU Colloquium No. 31, (see 012.015), p. 403 - 415 (1976). − Invited paper.

106.086 Dynamics of interplanetary dust and related topics. J. Trulsen.
IAU Colloquium No. 31, (see 012.015), p. 416 - 433 (1976).
Invited paper.

106.087 Modeling of the orbital evolution of vaporizing dust particles near the sun. Z. Sekanina.
IAU Colloquium No. 31, (see 012.015), p. 434 - 436 (1976).

106.088 Orbital evolution of circum-solar dust grains. P. L. Lamy.
IAU Colloquium No. 31, (see 012.015), p. 437 - 442 (1976).

106.089 Temperature distribution and lifetime of interplanetary ice grains. P. L. Lamy, M. F. Jousselme.
IAU Colloquium No. 31, (see 012.015), p. 443 - 447 (1976).

106.090 Rotational bursting of interplanetary dust particles. S. J. Paddack, J. W. Rhee.
IAU Colloquium No. 31, (see 012.015), p. 453 - 457 (1976).

106.091 Lunar ejecta in heliocentric space. W. M. Alexander, M. A. Richards.
IAU Colloquium No. 31, (see 012.015), p. 458 (1976).

106.092 Radiation pressure on interplanetary dust particles. G. Schwehm.
IAU Colloquium No. 31, (see 012.015), p. 459 - 463 (1976).

106.093 Are interplanetary grains crystalline? S. Drapatz, K. W. Michel.
IAU Colloquium No. 31, (see 012.015), p. 464 - 468 (1976).

106.094 A technique for measuring the interstellar component of cosmic dust. D. A. Tomandl.
IAU Colloquium No. 31, (see 012.015), p. 469 - 472 (1976).

106.095 The zodiacal light. H. Elsässer.

IAU Colloquium No. 31, (see 012.015), p. 475 - 477 (1976).

106.096 In situ measurements of dust. O. E. Berg.
IAU Colloquium No. 31, (see 012.015), p. 478 - 480 (1976).

106.097 Interplanetary shocks seen by Ames plasma probe on Pioneer 6 and 7.
B. Abraham-Shrauner, S. H. Yun.
Journ. Geophys. Res., Vol. 81, 2097 - 2102 (1976).
Interplanetary shocks and discontinuities observed by the Ames Research Center plasma probe on Pioneer 6 and 7 are analyzed with Goddard Space Flight Center magnetometer data. Several shock normals are used for the MHD model of a shock where the mixed data shock normals, which use plasma and magnetic field data, give the best agreement with the theoretical requirements.

106.098 Three-dimensional interplanetary stream magnetism and energetic particle motion.
E. Barouch, L. F. Burlaga.
Journ. Geophys. Res., Vol. 81, 2103 - 2110 (1976).
Cosmic rays interact with mesoscale configurations of the interplanetary magnetic field. The authors present a technique for calculating such configurations in the inner solar system, which are due to streams and source conditions near the sun, and construct maps of $|B|$ for some plausible stream and source conditions.

106.099 News about the interplanetary plasma.
Š. Pinter.
Kozmos, Vol. 7, 45 - 47 (1976). In Slovak.

106.100 On the acceleration of energetic particles in the interplanetary medium. L. A. Fisk.
GSFC Document X-660-76-25, Prepr., 18 + A4 + 5pp. (1976).
Fermi-scattering and transit-time damping have been suggested as two possible mechanisms for accelerating low-energy protons (~ 1 MeV) in co-rotating particle streams. In this paper, the requirements and properties of each of these mechanisms are illustrated by means of numerical solutions to the equations which govern particle behavior in such streams.

106.101 The acceleration of energetic particles in the interplanetary medium by transit-time damping.
L. A. Fisk.
GSFC Document X-660-76-26, Prepr., 34 pp. (1976).
It is shown here that particles could be accelerated by transit-time damping propagating fluctuations in the magnitude of the interplanetary magnetic field (e.g. magnetosonic waves). The protons in co-rotating streams may be accelerated by transit-time damping the small-scale variations in the field magnitude that are observed at a low level in the inner solar system. The interstellar ions may be accelerated by transit-time damping large-scale field variations in the outer solar system.

106.102 Erste Ergebnisse des Mikrometeoritenexperiments auf Helios A. G. Dräger, H. Fechtig,
P. Gammelin, E. Grün, J. Kissel.
MPI Kernphys., Heidelberg, Jahresbericht 1975, p. 151- 153.

106.103 Dichte und Zusammensetzung des interplanetaren Staubes aus Untersuchungen an Mikrokratern.
K. Nagel, G. Neukum, H. Fechtig, W. Gentner.
MPI Kernphys., Heidelberg, Jahresbericht 1975, p. 155 - 157.

106.104 Dynamik von geladenen Staubteilchen im interplanetaren Raum.
E. Grün, G. Morfill, H. Völk.
MPI Kernphys., Heidelberg, Jahresbericht 1975, p. 173.

106.105 Solar radiation induced rotational bursting of interplanetary particles. J. G. Sparrow.
Geophys. Res. Letters, Vol. 2, 255 - 257 (1975). — Abstr. in Phys. Abstr., Vol. 79, A006961 (1976).

106.106 On the anisotropies of interplanetary low-energy proton intensities. M. E. Pesses, E. T. Sarris.
Geophys. Res. Letters, Vol. 2, 349 - 352 (1975). — Abstr. in Phys. Abstr., Vol. 79, A023726 (1976).

106.107 Rotational bursting of interplanetary dust particles.
S. J. Paddack, J. W. Rhee.
Geophys. Res. Letters, Vol. 2, 365 - 367 (1975). — Abstr. in Phys. Abstr., Vol. 79, A023727 (1976).

106.108 Fluctuations and the radial variation of the interplanetary magnetic field. J. R. Jokipii.
Geophys. Res. Letters, Vol. 2, 473 - 475 (1975). — Abstr. in Phys. Abstr., Vol. 79, A027862 (1976).

106.109 A correlation method for measurement of variable magnetic fields. A. Anav, S. Cantarano,
P. Cerulli-Irelli, G. V. Pallottino.
IEEE Trans. Geosci. Electronics, Vol. GE-14, 106 - 114 (1976). Abstr. in Phys. Abstr., Vol. 79, A028055 (1976).

106.110 Interplanetary magnetic field data book.
J. H. King.
Separate print National Aeronaut. Space Administr., Greenbelt, Maryland, USA, Goddard Space Flight Center, 382 pp. (1975).

106.111 The structure of plasma density irregularities in the interplanetary medium. II. Observations.
D. G. Singleton.
Separate print Weapons Research Establishment, Salisbury, Australia, 25 pp. (1975).

106.112 Simulation of the cosmic dust and micrometeorite particle streams. A. I. Akishin, V. P. Kiryukhin.
Internat. conf. evaluation space environment materials, Toulouse, France, June 17 - 21, 1974, p. 171 - 180 (1974).

106.113 Shock waves from the sun. L. F. Burlaga.
Symposium significant accomplishments sci. technol., Greenbelt, Maryland, USA, 18 Dec. 1973, p. 100 - 103 (1975).

106.114 Dynamics of interplanetary dust grains.
P. L. Lamy.
Thesis Cornell Univ., Ithaca, New York, USA, 333 pp. (1975). (Available from: Univ. Microfilms, Order No. 75-21,564).

106.115 Comparison of inferred and observed interplanetary magnetic field polarities, 1970 - 1972.
J. M. Wilcox, L. Svalgaard, P. Hedgecock.
Separate print Stanford Univ., California, USA, Inst. Plasma Res., 20 pp. (1974). — See 14.106.027.

The dusty universe. Conference held at Cambridge, Mass., October 1973. See Abstr. 012.027.

Helios zodiacal light experiment.
See Abstr. 032.520.

Presentation of zodiacal light instrument aboard the D2B astronomical satellite. See Abstr. 032.521.

Interplanetary scintillation observations with the Cocoa Cross radio telescope. See Abstr. 033.005.

Magellan: a balloon-borne collection technique for large cosmic dust particles.   See Abstr. 051.025.

Scattering functions of dielectric and absorbing irregular particles.   See Abstr. 063.041.

Observation and analysis of low-energy solar particle propagation from discrete flare events.   See Abstr. 073.009.

Der Einfluß interplanetarer Stoßwellen auf energetische solare Flareteilchen.   See Abstr. 073.082.

A two-region model of the solar wind including azimuthal velocity.   See Abstr. 074.001.

Stochastic scattering and other contributions to the sun's wake in the local hydrogen cloud.   See Abstr. 074.003.

Quasi-exospheric heat flux of solar-wind electrons. See Abstr. 074.007.

October 1972 solar event: the third dimension in solar particle propagation.   See Abstr. 074.010.

Microscale 'Alfvén waves' in the solar wind at 1 AU. See Abstr. 074.011.

Solar wind structure at large heliocentric distances: an interpretation of Pioneer 10 observations. See Abstr. 074.024.

Latitudinal structure of the solar wind and interplanetary magnetic field.   See Abstr. 074.027.

On electron acceleration on the boundary of the solar wind with the interstellar medium. See Abstr. 074.030.

Disturbances of the solar corona and of the interplanetary medium accompanying the activation and disappearance of a prominence.   See Abstr. 074.033.

Coronal holes observed by OSO-7 and interplanetary magnetic sector structure.   See Abstr. 074.041.

Distributions of tangential discontinuity in the corotating solar wind structure.   See Abstr. 074.050.

Summary of observations of the solar corona/inner zodiacal light from Apollo 15, 16, and 17. See Abstr. 074.063.

On the development of the proton region in August 1971.   See Abstr. 077.022.

Propagation of interplanetary shock waves by observations of type II solar radio bursts on IMP-6. See Abstr. 077.024.

Anisotropy of solar protons and inhomogeneities of the interplanetary medium.   See Abstr. 078.013.

Connection of energetic particles during a storm with flare-generated interplanetary shock waves. See Abstr. 078.016.

Some regularities of the connection of solar cosmic rays in auroral regions with sector structure of the interplanetary magnetic field.   See Abstr. 078.022.

The Forbush decrease of November 17, 1966.

See Abstr. 078.036.

Observation using interplanetary scintillations at 34.3 MHz of the effect of a solar wind disturbance on a solar energetic particle event.   See Abstr. 078.066.

On the influence of injection profiles and of interplanetary propagation on the time-intensity- and time-anisotropy-profiles of solar cosmic-rays at 1 AU. See Abstr. 078.072.

On the propagation of low energy solar protons in the interplanetary medium. See Abstr. 078.075.

The effect of interplanetary discontinuities on the angular distributions of energetic particles. See Abstr. 078.085.

Flare-generated interplanetary shock wave deceleration effects in low-energy storm particle events. See Abstr. 078.086.

Effect of interplanetary shock waves on solar cosmic rays.   See Abstr. 078.088.

On the scattering mechanism of nonrelativistic solar electrons in the interplanetary medium. See Abstr. 078.101.

Origin of the twenty year wave in the diurnal variation.   See Abstr. 080.045.

A model of the solar atmosphere and wind. See Abstr. 080.053.

Submicron particles from the sun. See Abstr. 080.054.

Scattering layer of cosmic dust in the upper atmosphere.   See Abstr. 082.077.

Ionospheric radio wave absorption in the auroral zone and the interplanetary magnetic field sector structure. See Abstr. 083.003.

The electric potential of the ionosphere as controlled by the solar magnetic sector structure. See Abstr. 083.032.

On the connection of E_s in the near-polar region with the parameters of the interplanetary magnetic field. See Abstr. 083.039.

Equatorial electrojet and interplanetary plasma parameters.   See Abstr. 083.079.

Flapping motions of the tail plasma sheet induced by the interplanetary magnetic field variations. See Abstr. 084.202.

Geomagnetic field variations. See Abstr. 084.204.

Evidence of magnetic field line merging in the solar wind.   See Abstr. 084.211.

A statistical study of the upstream wave boundary outside the earth's bow shock.   See Abstr. 084.212.

Proton flow measurements in the magnetotail plasma

sheet made with Imp 6.   See Abstr. 084.214.

Dependence of the latitude of the cleft on the interplanetary magnetic field and substorm activity.
See Abstr. 084.220.

Magnetic fields of the magnetosheath.
See Abstr. 084.222.

Magnetospheric convection induced by interplanetary magnetic-field variations.   See Abstr. 084.224.

Multisatellite observations of solar protons penetrating the magnetopause.   See Abstr. 084.229.

On the generation of low-frequency waves in the solar wind in the front of the bow shock.
See Abstr. 084.235.

Magnetic field variations in the polar region during magnetically quiet periods and interplanetary magnetic fields.
See Abstr. 084.241.

Correlation of the intensity of fast electrons in the earth's magnetosphere and in the interplanetary space.
See Abstr. 084.251.

On a possibility of investigation of discontinuities in the magnetosphere and interplanetary space by the method of non-coherent reply.   See Abstr. 084.254.

Day side reconnection between a dipolar geomagnetic field and a uniform interplanetary field.
See Abstr. 084.269.

Magnetospheric convection induced by the positive and negative Z components of the interplanetary magnetic field: quantitative analysis using polar cap magnetic records.
See Abstr. 084.272.

Influence of the interplanetary magnetic field on the occurrence and thickness of the plasma mantle.
See Abstr. 084.279.

Analysis of proton and electron spectrometer data from OGO-5 spacecraft. Final report, 1 July 1973 - 31 July 1974.   See Abstr. 084.296.

Lunar microcraters and interplanetary dust fluxes.
See Abstr. 094.184.

Electrostatic disruption of lunar dust particles.
See Abstr. 094.187.

Dependencies of microcrater formation on impact parameters.   See Abstr. 094.565.

Surface exposure history of individual crystals in the lunar regolith.   See Abstr. 094.566.

Jovian electron bursts: correlation with the interplanetary field direction and hydromagnetic waves.
See Abstr. 099.009.

Orbital evolution of the dust streams released from comets.   See Abstr. 102.003.

Dust in comets and interplanetary matter.
See Abstr. 102.024.

Can short period comets maintain the zodiacal cloud?   See Abstr. 102.026.

Can comets be the only source of interplanetary dust?   See Abstr. 102.030.

Meteors and interplanetary dust.
See Abstr. 104.050.

Radial distribution of meteoric particles in interplanetary space.   See Abstr. 104.056.

Analysis of impact craters from the S-149 Skylab experiment.   See Abstr. 105.082.

Micrometeorite impact craters on Skylab experiment S-149.   See Abstr. 105.083.

Extraterrestrial particles in the stratosphere.
See Abstr. 105.084.

Low-level measurements of the spallation nuclei ^{26}Al and ^{53}Mn and their application to special meteorite problems, as well as on an attempt to determine the nature and global incoming rate of cosmic dust.
See Abstr. 105.126.

Interstellar medium in the vicinity of the sun: a temperature measurement obtained with Mars-7 interplanetary probe.   See Abstr. 131.015.

Interstellares Ionisationsgleichgewicht und interplanetare neutrale Heliumdichten.   See Abstr. 131.067.

Thermal behaviour of the neutral interstellar gas within the solar system.   See Abstr. 131.089.

On the flux and the energy spectrum of interstellar ions in the solar system.   See Abstr. 131.100.

The directional dependence of the primary cosmic rays of energies $10^{11}-10^{12}$ eV.   See Abstr. 143.010.

Anisotropic diffusion of cosmic rays in interplanetary space. I. Density distribution.   See Abstr. 143.014.

Anisotropic diffusion of cosmic rays in the interplanetary space. II. 11-year variation.
See Abstr. 143.016.

Anisotropic diffusion of cosmic rays in the interplanetary space. III. Radial and transversal gradients.
See Abstr. 143.032.

Detection and study of intensity variations of cosmic rays of interplanetary and magnetospheric origin from data of the world network of observatories.
See Abstr. 143.037.

Expected spatial distribution of cosmic rays in interplanetary space including the real heliolatitude distribution of solar activity.   See Abstr. 143.190.

The modulation and energy spectrum of electrons in interstellar space.   See Abstr. 143.209.

Kinetic theory of cosmic ray interaction with powerful interplanetary shock waves.   See Abstr. 143.219.

Anisotropy of cosmic rays associated with high velocity streams of interplanetary plasma.
See Abstr. 143.230.

The cosmic ray diurnal variations and fluctuations of the interplanetary magnetic field.   See Abstr. 143.231.

Distribution of the vector of cosmic ray solar anisotropy in interplanetary space in terms of the asymmetrical model of anisotropic diffusion.   See Abstr. 143.234.

North—south asymmetry of cosmic ray flux and polarity of interplanetary magnetic field. See Abstr. 143.238.

Low energy polar cap electrons during quiet times. See Abstr. 143.253.

Cosmic-ray diurnal variation underground and the interplanetary magnetic field.   See Abstr. 143.269.

Relation of large-scale coronal X-ray structure and cosmic rays: I. Sources of solar wind streams as defined by X-ray emission and Hα absorption features. See Abstr. 143.283.

A quasi-linear kinetic equation for cosmic rays in the interplanetary medium.   See Abstr. 143.341.

Cosmic ray scintillations. 4. The effects of non-field-aligned diffusion.   See Abstr. 143.342.

On the effect of directional medium scale interplanetary variations on the diffusion of galactic cosmic rays and their solar cycle variation.   See Abstr. 143.344.

Errata

106.901   Errata: "Observations of helium in the interplanetary/interstellar wind: the solar-wake effect" [Astrophys. Journ., Vol. 193, 471 - 476 (1974)].
C. S. Weller, R. R. Meier.
Astrophys. Journ., Vol. 203, 769 (1976).

107   Cosmogony of the Planetary System

107.001   **Accretion of the terrestrial planets. II.**
S. J. Weidenschilling.
Icarus, Vol. 27, 161 - 170 (1976).
Some aspects and consequences of the theory of gravitational accretion of the terrestrial planets are examined. The concept of a "closed feeding zone" is somewhat unrealistic, but provides a lower bound on the accretion time. A velocity relation which includes an initial velocity component is suggested. The orbital parameters of the planetesimals and the dimensions of the feeding zone are related to their relative velocities. Mercury, Venus, and the earth have accretion times on the order of 10^8 yr. Mars requires well over 10^9 yr to accrete by the same assumptions. Currently available data do not rule out a late formation of Mars, but the lunar cratering history makes it unlikely. If Mars is as old as the earth, nongravitational forces or a violation of the feeding zone concept is required.

107.002   **An interpretation of Titius-Bode law.**
W.-s. Dai.
Acta Astron. Sinica, Vol. 16, 123 - 130 (1975). In Chinese.
The distance of a planet or a regular satellite from the central body is related to the mass of the planet or regular satellite. The boundary between two adjacent planetary regions (or regular satellite regions) is related to the mass ratio of the two planets (or the regular satellites). When the boundary is properly chosen ($r_{n+1} - r_n$ divided according to the ratio $m_n^{1/3} : m_{n+1}^{1/3}$), it is found that the width Δr of the planetary (regular satellite) region is almost proportional to the size of the gravitational region $2x = 2(m/3M)^{1/3}r$, the ratio $\Delta r/2x$ decreases outwards. This result is a strong support of the view that planets (solid planetary cores in the case of Jupiter and

Saturn) and regular satellites are formed by the accumulation of planetesimals, and that they did not go through the stage of huge proto-planets and proto-satellites.

107.003   **Formation of solar nebula and mass distribution in the planetary system.**   V. Mitra.
Astrophys. Space Sci., Vol. 39, 387 - 396 (1976).
The formation of the solar nebula and the distribution of mass in its planetary system is studied. The underlying idea is that the protosun, fragmented out from an interstellar cloud as a result of cluster formation, gathered the planetary material and, hence, spin angular momentum by gravitational accretion during its orbital motion around the centre of the Galaxy. The study gives the initial angular momentum of the solar nebula nearly equal to the present value of the solar system.

107.004   **Formation of the inner planets.**   J. C. G. Walker.
Monthly Notes Astron. Soc. Southern Africa, Vol. 35, 2 - 8 (1976).

107.005   **The evolution of planetary orbits in the capture theory of the origin of the solar system.**
J. R. Dormand.
Long-time predictions in dynamics, (see 012.005), p. 334 (1976). – Abstract.

107.006   **Graphite grains in the early solar system.**
S. Ramadurai, N. C. Wickramasinghe.
Bull. Astron. Soc. India, Vol. 3, 32 (1975). – Abstract of a paper presented at the A.S.I. meeting 1975.

107.007   **Planet formation: compositional mixing and lunar**

compositional anomalies. W. K. Hartmann.
Icarus, Vol. 27, 553 - 559 (1976).

Significant fractions of each planet's late-accreted mass originated not at its own distance from the sun, but from a neighboring planet's orbit, according to results that follow from calculations by Wetherill (1975). "Late-accreted" refers to a loosely defined period after planets acquired most of their present mass. In an idealized model, Mercury, Venus, earth, and Mars received 47, 45, 37, and 52 % of their late-accreted mass from planetesimals formed closer to other planets. The moon's orbit around earth puts it in a special category: sorting occurs between moon-impacting and earth-impacting material according to approach velocity. In the above model, the moon receives 60 % of its late-accreted mass from planetesimals formed near Venus' orbit.

107.008 The role of gravitation in the formation of celestial bodies. Z. A. Ajtekeeva.
Fizika. Vyp. (No.) 1. Alma-Ata, 1974, p. 14 - 29. In Russian.
Abstr. in Referativ. Zhurn. 51. Astron., 4.51.781 (1976).

107.009 Averaging of eccentricities of orbits of bodies accumulating into a planet.
I. N. Ziglina, V. S. Safronov.
Astron. Zhurn. Akad. Nauk SSSR, Vol. 53, 429 - 435 (1976). In Russian. English translation in Soviet Astron., Vol. 20, No. 2.

The averaging of eccentricities of orbits of preplanetary bodies combining into a planet is investigated. The mathematical expectation of eccentricity of a growing planetary embryo is estimated, the process of its accumulation being treated as a random one. The eccentricities of orbits of preplanetary bodies are assumed to be determined by gravitational perturbations by the growing planet and to increase with time. The eccentricities calculated for three outer planets are comparable with the observed values, while the calculated values for all other planets are considerably less than the observed ones. It seems that the eccentricities of the orbits of the planets resulted essentially not from the accumulation process but rather from mutual gravitational perturbations of the planets during all their life.

107.010 Presolar grains: isotopic clues to solar system origin. A. L. Hammond.
Science, Vol. 192, 772 - 773 (1976). – Research news.

107.011 Palaeontology and the dynamic history of the sun-earth-moon system.
D. H. Weinstein, J. Keeney.
Growth rhythms and the history of the earth's rotation, (see 012.012), p. 377 - 384 (1975).

107.012 Mass distribution of protoplanetary bodies. II. Numerical solution of the generalized coagulation equation. G. V. Pechernikova, V. S. Safronov, E. V. Zvyagina.
Astron. Zhurn. Akad. Nauk SSSR, Vol. 53, 612 - 619 (1976). In Russian. English translation in Soviet Astron., Vol. 20, No. 3.

An analysis is given of numerical solutions of the generalized coagulation equation (with fragmentation) describing the mass distribution of protoplanetary bodies as a function of time.

107.013 On the origin of the planetary system.
M. Hagedušić.
Říše hvězd, Vol. 57, 121 - 124 (1976). In Czech.

107.014 On the evolution of terrestrial planets.
A. S. Monin, V. P. Keondzhyan.
Gerlands Beiträge Geophys., Band 85, 169 - 174 (1976). In Russian.

The process of evolution of the inner planets of the solar system is considered with the aid of a model of the gravitational differentiation of an originally homogeneous and spherically symmetrical planet consisting of heavy "core" and light "mantle" substance. Some structural and energetic parameters of the planets at all evolution stages, as well as dependences of these parameters on time are computed.

107.015 The macroplanetoid? A. Mielnik.
Astron. Rep., Vol. 1, 39 - 40 (1974).

107.016 Origin of the solar system. H. Reeves.
Recherche, Vol. 6, 808 - 817 (1975). In French.
Abstr. in Phys. Abstr., Vol. 79, A015176 (1976).

107.017 Radiometric chronology of the early solar system.
G. W. Wetherill.
Annual Rev. Nuclear Sci., Vol. 25, 282 - 328 (1975). – Abstr. in Phys. Abstr., Vol. 79, A037426 (1976).

107.018 The five forms of Laplace's cosmogony.
S. L. Jaki.
American Journ. Phys., Vol. 44, 4 - 11 (1976). – Abstr. in Phys. Abstr., Vol. 79, A042012 (1976).

107.019 Heterogeneities in the solar nebula.
R. N. Clayton, L. Grossman, T. K. Mayeda, N. Onuma.
Soviet-American conf. cosmochem. moon planets, Moscow, USSR, 4 June 1974, 14 pp. (1974). – See 12.107.020.

107.020 Dynamical studies of satellite origin. A. W. Harris.
Thesis California Univ., Los Angeles, USA, 130 pp. (1975). (Available from: Univ. Microfilms, Order No. 75-16,784).

107.021 Fossil nuclear reactor and plutonium-244 in the early history of the solar system. P. K. Kuroda.
Symposium Oklo phenomenon, Libreville, Gabon, 23 June 1975. Proceedings series, Vienna, IAEA, p. 479 - 487 (1975).

The solar system. Collected articles from the 1975 September issue of The Scientific American.
See Abstr. 003.122.

The principle of least interaction action.
See Abstr. 042.026.

L'évolution des atmosphères des planètes.
See Abstr. 091.002.

La planète d'Olbers et les astéroïdes.
See Abstr. 098.044.

Formation of iron sulphide in solar nebula.
See Abstr. 105.015.

On the origin of galaxy rotation in Ambartsumyan's cosmogony. See Abstr. 158.103.

Stars

111   Stellar Parallaxes

111.001   **An algorithm for the statistically rigorous and non-iterative computation of trigonometric parallaxes.**
H. Eichhorn, J. Russell.
Monthly Notices Roy. Astron. Soc., Vol. 174, 679 - 693 (1976).

A non-iterative algorithm is developed for the statistically rigorous computation of parallaxes from measurements on photographic plates using the initial equations of the central overlap method. For this, it is assumed that no reference star positions and/or proper motions (from a catalogue) are used and that the proper motions of the reference stars are subjected to the condition that they are random except for a systematic component due to solar motion and differential galactic rotation. It is shown that the computation of all unknowns can be achieved so that the largest matrix to be inverted is of the order $2m + 1$ where m is the maximum number of star images measured per plate.

111.002   **Suspected subluminous A stars in the solar neighborhood.**   T. D. Griess.
Astron. Journ., Vol. 81, 53 - 56 (1976).

Parallaxes of 16 A-type stars have been redetermined. The redetermination was made to check the authenticity of a group of A stars in the solar neighborhood which appear subluminous on the basis of the previous parallaxes. The redetermined parallaxes suggest that the existence of the subluminous group is probably spurious and that most of the stars are normal A stars whose former parallaxes constitute the large-parallax tail of a normal distribution.

111.003   **Parallaxes of 31 stars determined from plates taken with the McCormick 26-in. refractor.**
P. A. Ianna, R. G. Probst, G. E. Martin.
Astron. Journ., Vol. 81, 257 - 261 (1976).

Trigonometric parallaxes, proper motions, magnitudes, and colors are presented for 31 stars, 24 of which have no previous published trigonometric parallax. Most of the stars are from the Vyssotsky lists, and include a recently identified subdwarf, an unseen astrometric binary, a new visual binary nearer than 10 pc, and two Hyades members.

111.004   **Les parallaxes dynamiques.**   J. Dommanget.
Ciel et Terre, Vol. 92, 65 - 104 (1976).

A general survey on the various kinds of dynamical parallaxes is given, as well as an intercomparison of the methods used to-day for the computation of the orbital and non-orbital ones. It is recalled that the non-linearity of the mass-luminosity relation does not complicate the present mathematical process used for computing dynamical parallaxes, orbital or non-orbital ones.

111.005   **Photometric parallaxes for faint red dwarf stars.**   D. Weistrop.
Astron. Journ., Vol. 81, 427 - 429 (1976).

Photometric parallaxes have been derived for nine stars selected from the kinematic study by Murray and Sanduleak (1972). The data indicate the stars are more distant than suggested by these authors, and the space density is substantially reduced.

111.006   **Lick parallax program.**   S. Vasilevskis.
Publ. Lick. Obs., Vol. 22, Part 4, 29 pp. (1975).

111.007   **Trigonometric parallaxes measured at Lick Observatory. I.**
S. Vasilevskis, E. A. Harlan, A. R. Klemola, C. A. Wirtanen.
Publ. Lick. Obs., Vol. 22, Part 5, 6 pp. (1975).

The statistically rigorous non-iterative computation of parallaxes.   See Abstr. 031.219.

Absolute proper motions at low galactic latitude. See Abstr. 112.002.

Quelques méthodes de détermination de la magnitude absolue.   See Abstr. 113.052.

The main sequence defined by trigonometrical parallaxes.   See Abstr. 115.003.

112 Proper Motions, Radial Velocities, Space Motions

112.001 **Space velocity of the nearby subdwarf M star AC +54°1646−56.** P. A. Ianna, R. B. Culver.
Astrophys. Journ., (*Letters*), Vol. 203, L137 - L138 (1976).

The space velocity of the recently identified subluminous M dwarf AC +54°1646−56 is calculated from new astrometric and spectroscope data. An additional trigonometric parallax strengthens the determination of the position of the star about 1 mag below the $(M_v, R - I)$ main sequence.

112.002 **Absolute proper motions at low galactic latitude.** K. M. Cudworth.
Astron. Astrophys., Suppl. Ser., Vol. 24, 151 - 158 (1976).

Proper motions referred to "stars of zero parallax" have been derived from field stars in and around four additional clusters with $|b| \leq 8°$. These are combined with results from one field studied earlier to calculate a solar apex close to that of the basic solar motion. The secular parallaxes of these field stars are smaller than those determined by others and are a much stronger function of magnitude. These discrepancies imply that very low galactic latitudes should be treated separately from the usual low latitude zone extending to $|b| \sim 15° - 20°$.

112.003 **On the correlation between the motions and CN anomalies of G and K stars.**
R. B. Shatsova, T. G. Gotska.
Astron. Zhurn. Akad. Nauk SSSR, Vol. 53, 309 - 317 (1976). In Russian. English translation in Soviet Astron., Vol. 20, No. 2.

The correlation between the space velocity and cyanogen anomaly for the G−K stars, discovered earlier by Yoss and Lutz (1971), is partly confirmed. Then it has been examined by means of radial velocities, but is not confirmed for giants, supergiants and dwarfs.

112.004 **Kinematic parameters of the B-supergiants.**
D. K. Karimova, E. D. Pavlovskaya.
Astron. Zhurn. Akad. Nauk SSSR, Vol. 53, 495 - 500 (1976). In Russian. English translation in Soviet Astron., Vol. 20, No. 3.

For an investigation of the kinematics of the B-supergiants, 93 B0 − B9 stars of luminosity classes I and II with the most accurately determined proper motions and well known UBV photometry and V_r were selected. The kinematic parameters were calculated. The parameters of the velocity ellipsoid are given in a table.

112.005 **Proper motion survey with the 48-inch Schmidt telescope. XLVI. On the alleged plethora of nearby M dwarfs with little or no proper motion.**

W. J. Luyten.
Separate print Univ. Minnesota, Minneapolis, Minnesota, p. 1 - 9 (1976).

112.006 **Proper motion survey with the 48-inch Schmidt telescope. XLVII. The South Galactic Pole extension.** W. J. Luyten.
Separate print Univ. Minnesota, Minneapolis, Minnesota, p. 10 - 36 (1976).

112.007 **The North Galactic Pole.** W. J. Luyten.
Separate print Univ. Minnesota, Minneapolis, Minnesota, 182 pp. (1976).

The present catalogue gives data for 10,831 stars in the region of the North Galactic Pole which are believed to have proper motions larger than 0."179 annually. The area covered runs from 10^h to 16^h in right ascencion and from −5° to +60° in declination (equinox of 1950) and this comprises 0.119 of the entire sky.

112.008 **Radial velocities of bright southern stars.**
J. Andersen.
Centre Données Stellaires, Inform. Bull. No. 10, p. 34 (1976). Letter.

112.009 **Systematic motion-phenomena in Selected Areas.**
J. Meurers, E. Fiegweil.
Ann. Univ.-Sternw. Wien, Band 31, (No. 3), 45 - 64 (1974).

112.010 **Standard wavelengths for radial-velocity measurements and the radial velocities of twelve stars.**
A. H. Batten.
Publ. Dominion Astrophys. Obs., Vol. 14, (No. 17), 367 - 377 = National Res. Council Canada, NRC No. 15166 (1976).

Feasibility of a search for planets around solar-type stars with a polarimetric radial velocity meter.
See Abstr. 031.201.

New methods of measuring radial velocities: redshifts of clusters of galaxies and search for planets around stars.
See Abstr. 034.082.

An analysis of the variable radial velocity of Alpha Cygni. See Abstr. 122.065.

Radial-velocity variations of Scorpius X-1 emission lines. See Abstr. 142.095.

113 Stellar Magnitudes, Colors, Photometry

113.001 Rotational-velocity effects in the $uvby$, β systems and on MK spectral types for B stars.
W. H. Warren.
Monthly Notices Roy. Astron. Soc., Vol. 174, 111 - 123 =
Publ. Goethe Link Obs., Indiana Univ., *Bloomington,* No.
178 (1976).

Effects of stellar axial rotation on the intrinsic photometric indices of the $uvby$, β systems have been investigated for early-type members of the Orion OB 1 association. A comparison of the observational results with theoretical predictions for rotating stellar models is made for 78 apparently normal B stars. The deviations in the observed β index are in reasonably good agreement with those predicted. The results show that rapid rotation affects the colour indices of stars near the top of the main sequence and that these effects should be considered when deriving evolutionary (nuclear) ages for stellar aggregates. Axial rotation also appears to affect MK spectral classification for late B stars through a dependence of the ratio He I λ 4471/Mg II λ 4481, as recently suggested by Collins.

113.002 Infrared observations of young stars – VII. Simultaneous optical and infrared monitoring for variability. M. Cohen, R. D. Schwartz.
Monthly Notices Roy. Astron. Soc., Vol. 174, 137 - 155 (1976).

The results of simultaneous photoelectric photometry at U, B, V, R, I and Hα in the optical, and 2, 3, 5, 8, 10 and 11 μ in the infrared are reported for a number of young stars. Of 22 T Tauri and T Tauri-related early-type variables observed, 13 exhibited significant variations on a one-day time scale.

113.003 Galactic centre sequences in B and V.
P. R. Warren, J. E. Penfold, T. G. Hawarden.
Monthly Notices Roy. Astron. Soc., Vol. 174, 213 - 216 (1976).

Photoelectric measures in B and V are presented for three sequences in the galactic centre region.

113.004 Numerical simulation of natural photometric systems. A. W. J. Cousins, D. H. P. Jones.
Mem. Roy. Astron. Soc., Vol. 81, 1 - 23 (1976).

This paper makes use of physical data, measured or assumed, relating to the transmission of the atmosphere, the transmission or reflectivity of telescope optics, the transmission of colour filters and the spectral response of photomultipliers, in conjunction with the flux distributions of a number of stars to compute their magnitudes and colour indices and compare these with observed values. The observed colour transformations for both wide and narrow bands lying between 3800 and 6000 Å can be satisfactorily reproduced by computing but the agreement for the ultraviolet bands is not good. It is believed that the largest source of error is in the adopted flux distributions which lack accurate absolute calibrations at short wavelengths.

113.005 VRI standards in the E regions. A. W. J. Cousins.
Mem. Roy. Astron. Soc., Vol. 81, 25 - 36 (1976).

There is a need for reliable standard stars to define a *VRI* system that can be reproduced with photomultipliers having S-25 and GaAs photocathodes. About 20 stars have now been measured in each of the nine E regions at $-45°$ declination and these regions have been tied together to ensure that the zero points are consistent. The internal se of the individual colour indices is $\pm0\overset{m}{.}002$ and the error of a zero point is unlikely to exceed $0\overset{m}{.}002.$ The results are given in a system that

is similar to Kron's for stars that are hotter than $B-V = 1.0$.

113.006 Emission-free photographic photometry of stars in the Orion nebula cluster.
M. V. Penston, M. F. St J. Mann, M. J. Ward.
Monthly Notices Roy. Astron. Soc., Vol. 174, 449 - 454 (1976).

Photographic photometry of 477 stars in the Orion nebula cluster in wavebands relatively free from nebular emission lines is analysed. A few possible new variables are given and a colour-magnitude diagram is plotted. The distribution of faint stars in this diagram and the presence of stars with $V \sim 14$ near the zero-age main sequence suggest that current models of low-mass contracting stars are wrong. The reddest stars show a concentration centred near the position of the infrared and molecular complex OMC2.

113.007 The bolometric absolute magnitude of S type stars and lithium production.
M. W. Feast, R. M. Catchpole, I. S. Glass.
Monthly Notices Roy. Astron. Soc., Vol. 174, 81P - 85P (1976).

JHKL photometry is given for four S stars with known distances (T Sgr, TT9 and TT12 and π^1 Gruis). Together with published data on R And the authors find a mean bolometric absolute magnitude of -5.4 for these stars. This lies at the predicted threshold for lithium production in a 'hot-bottom' convection zone. The lithium poor star R And may be fainter than the threshold.

113.008 Carbon star photometry: CO and 3.2 micron bands.
T. D. Faÿ, Jr., S. T. Ridgway.
Astrophys. Journ., Vol. 203, 600 - 602 (1976).

This paper reports filter photometry of CO band strengths at 2.36 μ (filter width = 0.18 μ) for 23 carbon stars. Present results are compared with CO depression of 10 stars observed in 1970 at 2.29 μ. The CO index and the scanner index are linearly related. Each of the 23 carbon stars was also observed at the 3.2 μ band (filter width = 0.4 μ). The depression of this unidentified band increases with Na D line strength and C_2 absorption.

113.009 The low excitation spectrum of the peculiar star with infrared excess RX Puppis in 1972 and 1975.
J. P. Swings, M. Klutz.
Astron. Astrophys., Vol. 46, 303 - 311 (1976).

Near infrared photometry of RX Puppis performed in 1972 revealed a prominent excess of radiation probably originating in a circumstellar dust shell surrounding this peculiar object. 1972 and 1975 spectrograms of RX Puppis show essentially emission lines of H (with P Cygni structure variable from night to night), O I, [O I], Ca II, [Ca II], Fe II and [Fe II]. It is suggested that RX Pup may be a nova-like or peculiar emission line-object exhibiting similarities to η Carinae.

113.010 The luminosity dependence of the 1.65 μ flux from K and early M stars. Observations and interpretation.
R. A. Bell, B. Gustafsson, H. L. Nordh, S. G. Olofsson.
Astron. Astrophys., Vol. 46, 391 - 396 (1976).

A number of late-type stars of various luminosity classes have been observed in three infrared colours, to study the luminosity dependence of the "1.65 μ peak" around the minimum of the H$^-$ absorption. An unexpected luminosity dependence (lower peaks) has been found for K and early M supergiants as compared with the giants. Comparisons with model-atmosphere computations, which are in relatively good agreement with the observations, explain the effect as the result of blocking by CO and CN lines in the infrared.

113.011 Space distribution and kinematics of intermediate population II stars. Part I. Photometry and spectroscopy in selected McCormick proper motion fields.
A. Blaauw, C. R. Tolbert, R. M. West, R. A. Bartaya.
Astron. Astrophys., Suppl. Ser., Vol. 23, 393 - 411 (1976).

$uvby$ and Hβ photometry is reported for over 1000 stars brighter than about 13.5 mag. and located generally in the galactic polar regions. Spectral types have also been determined for many of the stars. These data constitute the initial phase of an extensive study of the space distribution and kinematics of the intermediate population II stars.

113.012 Photometric variations of some Ap stars.
P. Renson, J. Manfroid, A. Heck.
Astron. Astrophys., Suppl. Ser., Vol. 23, 413 - 417 (1976).
In French.

Photometric observations allowed to deduce periods of 0.9 d for 72G. Pup, between 2.2 and 2.3 d for 77G. Pup and a little less than 2 d for β Hya. The widths of the lines, known for the former and the latter of these stars, agree with the period − line-width relation. Photometric variations observed for μ Lep, κ^2 Vol, 14 Hya and HD 101065 are also discussed.

113.013 $uvby\beta$ photometry and MK spectral classification.
E. Oblak, S. Considère, M. Chareton.
Astron. Astrophys., Suppl. Ser., Vol. 24, 69 - 88 (1976).

$uvby\beta$ photometry of nearly 2900 stars has been studied as a function of the MK spectral classification. A unique spectral type has been defined for each star. The photometric indices $\beta_r[m_1]$, $[c_1]$, $[u-b]$ which are independent of reddening and their dispersions have been studied for each spectral type and for three groups of luminosity class. For 61 stars, the MK spectral classification differs considerably from what one would expect from the indices of Strömgren's photometry. The authors have also studied the unreddened indices $(b-y)_0$, c_{1_0}, m_{1_0}, as well as their dispersions for the same samples.

113.014 Investigation of light variation of twelve Ap stars in ten spectral regions.
W. Schöneich, G. Hildebrandt, W. Fürtig.
Astron. Nachr., Vol. 297, 39 - 57 (1976). − In German.

Twelve Ap stars with known periods and fairly large amplitudes were photoelectrically investigated in a ten colour system (λ 3400 Å to λ 7600 Å). The amplitude-wavelength relations following from the light curves cannot be explained by the influence of the backwarming effect. These relations point to complicated structures in the atmospheres of the peculiar regions of magnetic stars.

113.015 Hα and Hβ photoelectric photometry of γ Cassiopeiae.
S. L. Baliunas, E. F. Guinan.
Publ. Astron. Soc. Pacific, Vol. 88, 10 - 12 (1976).

Photoelectric observations of γ Cas (B0 IVe) were made using two pairs of wide- and narrow-band interference filters centered near the rest wavelengths of the hydrogen Balmer α and β lines. Variations in the Hα and Hβ indices were found with a range of $0^m 18$ and $0^m 10$, respectively.

113.016 An application of multivariate statistical analysis to a photometric catalogue.
A. Heck.
Astron. Astrophys., Vol. 47, 129 - 135 (1976).

It is shown that the application of methods of statistical multivariate analysis to a photometric catalogue allows to obtain, with a strict minimum of physical considerations, valuable indications on the meaning of photometric indices. In the particular application considered here, one can obtain valuable predictors of spectral type and, through it, of the effective temperature from the $uvby\beta$ photometrical indices, using the catalogue of Lindemann and Hauck (1973).

113.017 Photometry of the Ap-star HD 124224.
W. W. Weiss, R. Albrecht, R. Wieder.
Astron. Astrophys., Vol. 47, 423 - 427 (1976).

HD 124224 has been observed for light variations in the Strömgren system and with a spectrum scanner in the 3500 Å to 4600 Å region. The light curves are asymmetric and show different phases of maximum, depending on the colors. Some evidence could be found for short time fluctuations in Hβ. Θ seems to change from 0.375 to 0.360 and log g from 3.60 to 3.45 (phase of light minimum to phase of maximum).

113.018 The short-term light variations of HD 160529.
C. Sterken.
Astron. Astrophys., Vol. 47, 453 - 455 (1976).

New observations confirm that the A2Ia-O star HD 160529 shows light variations of irregular nature and small amplitude on time scales of a few hours.

113.019 Further evidence concerning the local density of red dwarfs.
D. Weistrop.
Astrophys. Journ., Vol. 204, 113 - 115 (1976).

Recent evidence has suggested the existence of a large number of red dwarfs, highly concentrated to the galactic plane, in the solar neighborhood. These results are reexamined using photoelectrically determined $B-V$ and $(R-I)_W$ colors for a sample of stars near the north galactic pole. Preliminary analysis of the data indicates that the density of the reddest stars is not nearly as great as was previously reported.

113.020 Photometry of two S-type stars near η Carinae.
P. J. Andrews.
Observatory, Vol. 96, 11 - 13 (1976).

113.021 Simultaneous observations of σ Ori E.
D. Groote, U. Haug, K. Hunger.
Mitt. Astron. Ges., No. 38, p. 195 - 197 (1976). − Abstract.

113.022 UBV photometry of the stars in the fields of emission nebulae. II. M17.
K. Ogura, K. Ishida.
Publ. Astron. Soc. Japan, Vol. 28, 35 - 60 (1976).

Three-color photometry is presented for about 700 stars in a 30' × 30' field centered at the radio position of M17. Spectral type is also given for twenty-one of the stars. Analysis of the data reveals six heavily reddened O or possible O stars embedded in the central part of the H II region. A photometric distance of 1.3 kpc was derived for the O- and B-type stars.

113.023 Photoelectric investigations of the Ap star HD 148112.
W. Schöneich, G. Hildebrandt.
Astron. Nachr., Vol. 297, 89 - 90 (1976). In German.

Photoelectric observations of the Ap star HD 148112 in the BV system were obtained from July 1971 to July 1972. The star has a double-wave in the lightcurve with a period of 2.951 days. The amplitudes are 0.04 mag in both spectral ranges.

113.024 Importance of random scatter to variable-extinction analyses.
D. G. Turner.
Astron. Journ., Vol. 81, 97 - 103 (1976).

Becker's (1966) treatment of the effect of random scatter in cluster color–color diagrams on the ZAMS-fitting version of the variable-extinction technique is reanalyzed in detail for stars of spectral types O–F5. The effect of bias on the derived value of R is found to be different from that described by Becker, the presently derived effect being highly dependent on the spectral types of the analyzed stars. An analysis of the importance of this effect using the A0–F5 stars in the Pleiades leads to the conclusion that random scatter in the color–color diagram for this cluster cannot be much larger than that expected solely from normal errors in the photometry.

113.025 O stars and supergiants south of declination −53.0°.
N. Houk, M. R. Hartoog, A. P. Cowley.
Astron. Journ., Vol. 81, 116 - 121 (1976).

Lists are provided of 75 O stars and 281 supergiants from Vol. 1 of the University of Michigan Catalogue of Two-Dimensional Spectral Types for the HD Stars. Twenty-eight percent of the O stars and 44 % of the supergiants are not in present major catalogues. The preponderance of new supergiants are of the later types, with 80 % of the G supergiants being newly identified.

113.026 Standard stars for DDO photometry.
R. D. McClure.
Astron. Journ., Vol. 81, 182 - 208 (1976).

Extensive photometry for 285 stars to be used as standards has been completed on the David Dunlop Observatory (DDO) photometric system. The magnitudes and colors given in a table define the standard system. The matching of filter sets to the DDO system is discussed.

113.027 Photoelectric spectrophotometry of early type stars.
M. K. V. Bappu, K. Nandy, M. Parthasarathy, R. Rajamohan.
Bull. Astron. Soc. India, Vol. 3, 30 (1975). − Abstract of a paper presented at the A.S.I. meeting 1975.

113.028 Photometry of peculiar A stars.
H. Hensberge, C. De Loore, E. J. Zuiderwijk, G. Hammerschlag-Hensberge.
Astron. Astrophys., Vol. 48, 383 - 387 (1976).

The results of $uvby$-photometry of 4 Ap-stars observed during 1975 at ESO, La Silla, are presented. One of the stars, HD 81009, was also observed during 1971 in the UBV system. Periods are proposed for the silicon stars HD 63401 and HD 92664. The three silicon stars in the sample (the two previous ones and HD 73340) are variable in the four channels showing their largest amplitude in u. HD 81009 shows at least variability in the u-channel.

113.029 Three-color photometry of HDE 226868: the optical counterpart of Cygnus X-1.
D. F. Lester, I. G. Nolt, S. A. Stearns, P. Straton, J. V. Radostitz.
Astrophys. Journ., Vol. 205, 855 - 860 (1976).

The results of UBV photoelectric observations of HDE 226868 (Cyg X-1) obtained on fifty nights during 1974 are presented. These observations show a light curve with the following characteristics: (1) two unequal light maxima and minima during the $5\overset{d}{.}6$ orbital period which exhibit a peak-to-peak magnitude change of ~0.07 mag in all colors; (2) a small phase-dependent color variation which results in a reddening in both color indices of ~0.003 mag at times of light minima; (3) erratic light variability on a general time scale of days with an amplitude of ±0.01 to 0.02 mag and no apparent color dependence.

113.030 UBVRI observations of miscellaneous stars.
J. D. Fernie.
Journ. Roy. Astron. Soc. Canada, Vol. 70, 77 - 78 (1976).

Photoelectric photometry on Johnson's $UBVRI$ system is reported for seventeen stars.

113.031 Broad-band photometry of G and K stars: the C, M, T_1, T_2 photometric system.
R. Canterna.
Astron. Journ., Vol. 81, 228 - 244 (1976).

A new broad-band photometric system C, M, T_1, T_2 was developed specifically to obtain accurate temperatures, metal abundances, and a CN index for G and K giants. This new system is more efficient than the $UBVRI$ system. The temperature index $T_1 - T_2$ is a linear function of $R - I$. Metal abundances derived from the index $\Delta(M - T_1)$ are uncertain by a factor of 2. A cyanogen band index $\Delta I(CN)$ can distinguish between strong CN stars (CN +3, +2) and weak CN stars (CN −3, −2) for giants with [Fe/H] ⩾ −1.0. Finally, the applicability of the C, M, T_1, T_2 system for abundance studies of G-type dwarfs is promising.

113.032 The galactic distribution of interstellar absorption as determined from the Celescope catalog of ultraviolet stellar observations and a new catalog of UBV, H-beta photoelectric observations.
W. A. Deutschman, R. J. Davis, R. E. Schild.
Astrophys. Journ., Suppl. Ser., Vol. 30, 97 - 225 (1976).

New UBV data for 2846 stars and Hβ photometry for 2099 stars are presented; all stars had previously been observed in the rocket ultraviolet in project Celescope, and ultraviolet magnitudes for them at U2 or U3 are available. The data are combined in an investigation of the ultraviolet interstellar extinction law, with emphasis placed on finding systematic variations in that law as functions of galactic longitude and distance. The data confirm the previous finding that the ultraviolet extinction law is more variable in U3 (λ_eff = 1590 Å) than in U2 (λ_eff = 2180 Å), and that the U3 extinction is greatest for directions of view along spiral arms.

113.033 On some problems of fundamental stellar photoelectric photometry.
V. B. Nikonov.
Izv. Krymskoj Astrofiz. Obs., Vol. 54, 3 - 23 (1976). In Russian.

A method to obtain fundamental photometric catalogues for photoelectric stellar magnitudes and colours (the "method of control stars") is proposed. It gives the possibility to control currently stability of natural photometric systems. A way to determine relative spectral sensitivity of astronomical photoelectric equipment, including the telescope, is presented. Requirements necessary to secure optimum stability of natural photometric systems in stellar photoelectric photometry are considered.

113.034 Transformation of similar photoelectric systems in variable star investigations.
V. I. Burnashev, V. B. Nikonov.
Izv. Krymskoj Astrofiz. Obs., Vol. 54, 24 - 34 (1976). In Russian.

A procedure to determine the coefficients of formulae relating similar photoelectric systems is proposed. Its main field of application in variable star investigations: reductions of observations to a standard photometric system and account of atmospherical extinction using magnitudes and colours of stars taken from a standard photoelectric catalogue. The accuracy of reduction coefficients and of extinction determinations has been estimated on the basis of photoelectric colour observations. It was found to be quite sufficient for variable stars investigations.

113.035 T Tauri stars and the $(J-H)$, $(H-K)$ diagram.
A. E. Rydgren.
Publ. Astron. Soc. Pacific, Vol. 88, 111 - 115 (1976).

The $(J-H)$, $(H-K)$ infrared two-color diagram has the unusual property of separating the effects of envelope emission and interstellar reddening in T Tauri stars. Applications of this diagram to the study of T Tauri stars include the estimation of $E(B-V)$ for individual stars (provided the reddening law is known), the comparison of T-associations, and the determination of the principal source of brightness variations.

113.036 Further photometry of cepheid-like supergiants.
J. D. Fernie.
Publ. Astron. Soc. Pacific, Vol. 88, 116 - 118 (1976).

Photoelectric photometry on Johnson's *UBVRI* system has been carried out for 19 supergiants having spectral classifications similar to classical cepheids. This is part of a continuing search to find either previously undiscovered low-amplitude cepheids or stable stars within the cepheid instability strip on the H-R diagram. Of the present stars, only one, HD 101947, has been found to show significant variability.

113.037 Photometry of some southern metal-poor stars.
P. Lee, C. L. Perry.
Publ. Astron. Soc. Pacific, Vol. 88, 135 - 136 = Contr. Louisiana State Univ., *Baton Rouge*, No. 111 (1976).
The results of photometry for some selected stars suspected of metal deficiency are presented and discussed in the light of Eggen's calibrations.

113.038 Infrared photometry and polarimetry of cool stars.
III. G. V. Khozov, T. N. Khudyakova, S. N. Nikitin.
Trudy Astron. Obs., *Leningrad*, Vol. 32 (= Uchenye Zapiski Leningr. Un-ta, No. 385 = Seriya matem. nauk, vyp. (No.) 52), p. 61 - 78 (1976). In Russian.
Four-color infrared photometric observations in R, I, H, and K bands are reported for 18 stars. For 6 objects (NML Cyg, VY CMa, NML Tau, IRC+10216, BC Cyg, CIT-13) polarimetric observations are reported. A discussion of the results is also presented.

113.039 Investigation of the photometric system of the
AZT-7 telescope. T. K. Kiseleva.
Byull. Inst. Astrofiz., *Dushanbe*, No. 63, p. 39 - 42 (1974).
In Russian.

113.040 A new Copernican revolution in astronomy. Ultraviolet astronomy with the satellite Copernicus.
T. P. Snow, Jr.
Mercury, (Journ. Astron. Soc. Pacific), Vol. 5, No. 2, p. 26 - 31 (1976).

113.041 Photometric observations at the Special Astrophysical Observatory according to the programmes of
astrophysical researches aboard the orbital station Salyut 4 and experimental flight Soyuz-Apollo. G. N. Alekseev,
V. L. Afanas'ev, G. M. Beskin, S. I. Sinyanskij, V. F. Shvartsman.
Astron. Tsirk., No. 897, p. 3 - 4 (1975). In Russian.

113.042 *uvby* photometry of some Ap stars.
A. Heck, J. Manfroid, P. Renson.
Astron. Astrophys., Suppl. Ser., Vol. 25, 143 - 149 (1976).
In French.
Photometric measurements were carried out at the European Southern Observatory in February 1975 for twelve Ap stars. The results are given with detailed tables for the stars which have been found to be variable during the time of the observations, i.e. one week.

113.043 U Cephei: a mass-transfer event. II. Observations.
E. C. Olson.
Astrophys. Journ., Suppl. Ser., Vol. 31, 1 - 11 (1976).
Four-color *uvby* and β photoelectric observations are given of the 1974 mass-transfer event in U Cep.

113.044 Photographic photometry of six C-type stars.
Z. Alksne, A. Alksnis, V. Ozoliņa.
Investigation of the sun and red stars. 4, (see 003.009), p. 12 - 30 (1976). In Russian.
From photographs taken during 1969—1974 RVB magnitudes are determined for six stars of spectral type C. The light variations of three stars turn out to be cyclic with mean periods of 355, 320 and 376 days respectively. Two other stars show irregular light variations, and one star of spectral

type R seems to be nonvariable. Magnitudes and identification charts are given.

113.045 Ten-color photometry of the four Ap stars HD
27309, HD 119213, HD 170000, and HD 192913.
W. Schöneich, G. Hildebrandt, B. Musielok, E. Želwanowa.
Astron. Nachr., Band 297, 173 - 176 (1976). In German.
The results of a ten-color · photometry of the four Ap stars HD 27309, HD 119213, HD 170000 and HD 192913 are given. The periods were determined to 2.7098, 2.433, 1.716 and 16.3 days. The amplitude-wavelength relation shows a different behaviour for all of them.

113.046 Long-term light variations in magnetic stars.
K. Panov, W. Schöneich.
Astron. Nachr., Band 297, 177 - 180 (1976). In German.
Seven of eleven investigated Ap stars show longtime variations of their magnitude. The Si and Cr-Eu-Sr stars vary in the sense the brighter the star the bluer its colour.

113.047 Photoelectrical investigation of the magnetic field
and the magnitude of the Am stars 15 Vul and
68 Tau. V. M. Kuvshinov, G. Hildebrandt, W. Schöneich.
Astron. Nachr., Band 297, 181 - 188 (1976). In German.
The Am stars 15 Vul and 68 Tau have been observed photoelectrically in the *UBV* system and for determination of the longitudinal magnetic field component. Both the stars show periodical variations of the magnitude and the magnetic field. The periods are determined to 14.0 days for 15 Vul and 57.25 days-for 68 Tau. The variations of these Am stars are the same as those of typical Ap stars. The star 68 Tau was found to be a spectroscopical binary.

113.048 A first comparison between the Celescope magnitudes and ultraviolet spectrometric measurements.
L. Houziaux.
Mem. Soc. Astron. Italiana, Vol. 45, (see 012.017), 717 - 721 (1974).

113.049 Spectroscopic and photometric investigation of the
A21a-O star HD 160529.
B. Wolf, L. Campusano, C. Sterken.
Mem. Soc. Astron. Italiana, Vol. 45, (see 012.017), 733 - 734 (1974). – Abstract.

113.050 How to define a photometric degree of peculiarity
for Ap-stars? H. M. Maitzen.
Mem. Soc. Astron. Italiana, Vol. 45, (see 012.017), 735 - 739 (1974).

113.051 Photometric evidence for the binary nature of Wolf-Rayet stars. A. F. J. Moffat, W. Haupt.
Mem. Soc. Astron. Italiana, Vol. 45, (see 012.017), 811 - 814 (1974).

113.052 Quelques méthodes de détermination de la magnitude absolue. A. Heck.
Separate print Obs. Strasbourg, Centre Données Stellaires, 52 pp. (1976).

113.053 Sensitivity functions of photometric systems.
B. Hauck, M. Mermilliod.
Centre Données Stellaires, Inform. Bull. No. 10, p. 28 - 29 (1976).
In order to have available the maximum of sensitivity functions for the filters used in photometric systems, the authors have collected those published in the literature. All the data they have obtained are now on punched cards and obtainable from the Stellar Data Centre at Strasbourg.

113.054 The development of the UBViyz photometric sys-

tem and its application to the study of metal abundances in galactic G and K giant stars.   P. A. Jennens. Thesis Univ. Rochester, N.Y., USA, 229 pp. (1975). (Available from: Univ. Microfilms, Ann Arbor, Mich., USA. Order No. 75-15207). – Abstr. in Phys. Abstr., Vol. 79, A028064 (1976).

Photometrische Periodensuche bei Ap-Sternen. See Abstr. 031.208.

Grundsätzliche Aspekte der Mehrfarbenphotometrie. See Abstr. 031.209.

Infrared photometry at the Stockholm Observatory. See Abstr. 061.043.

Températures et magnitudes en astrophysique: un aide-mémoire.   See Abstr. 114.038.

Ultraviolet photometry from the Orbiting Astronomical Observatory. XXI. Absolute energy distribution of stars in the ultraviolet.   See Abstr. 114.304.

The photometric and spectrographic histories of HD 245770 $\stackrel{!}{=}$ A0535 + 26, the transient X-ray source. See Abstr. 114.343.

A spectrophotometric atlas of three K-giants. See Abstr. 114.345.

Photoelectrical investigations of seven Ap stars – members of spectroscopic double star systems. See Abstr. 119.018.

Infrared photometry of RT Lacertae. See Abstr. 121.063.

UBV photometry of flare stars in the Pleiades. See Abstr. 122.031.

Multicolour TV photometry of T Tau-stars. I. Methods of observations and preliminary results. See Abstr. 122.044.

Multicolour TV photometry of T Tau-stars. II. On the variability of T Tau-stars.   See Abstr. 122.045.

Five-channel photometry of cepheids and supergiants in the southern Milky Way.   See Abstr. 122.067.

UBV photometry of the RR Lyrae star HX Arae. See Abstr. 122.105.

Five-colour photometry of southern cepheids. A new search for cepheid binaries.   See Abstr. 122.107.

Analysis of broad-band photometry of the long-period variables.   See Abstr. 122.134.

Ultraviolet photometry of Eta Carinae and its interpretation.   See Abstr. 131.036.

Optical observations of WRA 977. See Abstr. 142.016.

Photometric variations of Wray 977 (3U 1223–62?). See Abstr. 142.103.

UBV, Hβ photometry of OB stars in groups: Pup OB2.   See Abstr. 152.002.

Catalogue of UBV photometry and MK spectral types in open clusters.   See Abstr. 153.012.

Three colour photometry of a field in the galactic anticentre section near M37.   See Abstr. 155.005.

Carina arm studies. II. Photometry of faint early-type stars in Carina.   See Abstr. 155.006.

UBV and Hβ photometry of faint early-type stars in Norma.   See Abstr. 155.007.

Relation of NGC 3590, Hogg 10, and Collinder 240 to the structure of the Carina spiral feature. See Abstr. 155.026.

Stellar population samples at the galactic poles. III. $UBVRI$ observations of proper motion stars near the south pole and the luminosity laws for the halo and old disk populations.   See Abstr. 155.027.

Errata

113.901   Errata: " A photoelectric magnitude sequence in the LMC: 0501–6728" [Monthly Notices Roy. Astron. Soc., Vol. 169, 577 - 578 (1974)].
R. P. Olowin, T. G. Hawarden, P. R. Warren.
Monthly Notices Roy. Astron. Soc., Vol. 175, 653 (1976).

114 Stellar Spectra, Temperatures, Spectroscopy, Spectra of Individual Stars

Stellar Spectra, Temperatures, Spectroscopy

114.001 On wavelength shifts of He I lines in rotating B stars. B. E. J. Pagel, J. E. Drew.
Monthly Notices Roy. Astron. Soc., Vol. 174, 13P - 17P (1976).

The wavelength shift between He I triplet and singlet lines in the binary system 68u Her is due to imperfections in the standard wavelength system for B stars rather than to any isotope effect.

114.002 Linear polarization measurements at Hβ of early-type emission line stars.
D. Clarke, I. S. McLean.
Monthly Notices Roy. Astron. Soc., Vol. 174, 335 - 343 (1976).

Linear polarization measurements across the Hβ emission lines of the stars γ Cas, ζ Tau and 48 Per are presented. For the first two stars there is a marked reduction of the polarization at the centre of the line and for γ Cas, this varies from night to night. During the Hβ observations of ζ Tau, a change of polarization over tens of minutes was indicated in a monitor channel tuned to the continuum on the blue side of Hβ. For the fainter star, 48 Per, the uncertainties of the polarimetry were increased in relation to γ Cas and ζ Tau by a factor of about two and at this precision, no differential effects across the line were recorded.

114.003 The violet opacity of carbon stars. A. R. Walker.
Monthly Notices Roy. Astron. Soc., Vol. 174, 609 - 616 (1976).

The equivalent width of the 4866 Å Merrill-Sanford band of SiC$_2$ and the C$_2$(0, 0) Swan band strength have been measured for 40 southern carbon stars. Excellent correlation is found between the violet absorption and the Merrill-Sanford band strength, thus decisively supporting the theory that the excessive violet opacity of carbon stars is due to grains of SiC.

114.004 Empirical effective temperatures and bolometric corrections for early-type stars.
A. D. Code, J. Davis, R. C. Bless, R. Hanbury Brown.
Astrophys. Journ., Vol. 203, 417 - 434 (1976).

An empirical effective temperature for a star can be found by measuring its apparent angular diameter and absolute flux distribution. The angular diameters of 32 bright stars in the spectral range O5f to F8 have recently been measured with the stellar interferometer at Narrabri Observatory, and their absolute flux distributions have been found by combining observations of ultraviolet flux from the Orbiting Astronomical Observatory (OAO-2) with ground-based photometry. In this paper these data have been combined to derive empirical effective temperatures and bolometric corrections for these 32 stars.

114.005 Skylab ultraviolet stellar spectra: cool stars with hot secondaries. S. B. Parsons, J. D. Wray, Y. Kondo, K. G. Henize, G. F. Benedict.
Astrophys. Journ., Vol. 203, 435 - 437 (1976).

A hot companion to the G5 III star HR 3080, a single-line spectroscopic binary, has been discovered from spectra in the vacuum-ultraviolet. The companion must be a subdwarf or pre-white dwarf. A list of previously known systems of the ζ Aur and VV Cep type observed with the ultraviolet spectrograph on Skylab is also given.

114.006 Stellar winds from hot supergiants.
J. B. Hutchings.
Astrophys. Journ., Vol. 203, 438 - 447 = Contr. Dominion Astrophys. Obs., *Victoria*, No. 270 (1976).

Spectrographic data of high dispersion are described from 65 very luminous galactic OB stars. Their absolute magnitudes and masses are estimated. Almost all show evidence for mass outflow, and the flow rate is empirically calibrated from mass-loss models for a few of them. Approximate contours of mass-loss on the H–R diagram are suggested which support the zero point of the individual rates. These are as high as $10^{-5} M_\odot$ yr^{-1} from a 60 $M_\odot$ supergiant. Anomalously high mass-loss rates found in a few stars may be connected with duplicity of the objects.

114.007 C$_3$ as a significant opacity source in Ba II stars.
J. D. Fix.
Astrophys. Journ., Vol. 203, 463 - 465 (1976).

Spectrophotometric scans of Ba II stars have been obtained and compared with those of normal G and K giants. The broad depressions observed near 4000 Å in the spectra of Ba II stars are found to agree in shape and depth with depressions produced by column densities of about 10^{16} cm^{-2} C$_3$ molecules. The temperatures and compositions of Ba II stars are found to be compatible with those necessary to yield the required column densities of C$_3$ molecules.

114.008 A spectral survey of the Southern Milky Way III. O–B9 and M stars $l = 280°$ to $306°$.
L. O. Lodén, K. Lodén, B. Nordström, A. Sundman.
Astron. Astrophys., Suppl. Ser., Vol. 23, 283 - 392 (1976).

A catalogue, part III of the Stockholm Observatory Southern Milky Way Survey, is presented, based on objective prism plates with a dispersion of 240 Å mm^{-1} at Hγ, as well as blue and visual direct plates. The catalogue contains spectral types, blue magnitudes, positions, and finding charts for about 10000 O–B9 and M stars as well as emission-line objects to a limiting magnitude of about 12^{m}5 and 15^m, respectively. The region covered is a 6° wide belt along the galactic equator, $l = 280°$ to 306°.

114.009 Line blocking in the visual and near infrared spectra of G-, K- and M-type giants. J. van Paradijs.
Astron. Astrophys., Suppl. Ser., Vol. 24, 53 - 67 (1976).

Line-blocking coefficients (percentage of light absorbed by spectral lines), partly covering the wavelength range from 5200 to 8700 Å, are given for 50 giant stars with spectral types between G8 and M3. An estimate of the accuracy of these data indicates a possible underestimate by about two per cent. A comparison is made with available theoretical data. It appears that theoretical line blocking based on line statistics from the solar spectrum is deficient throughout the major part of the wavelength interval studied in this paper.

114.010 An explanation of the super-metal-rich phenomenon in field K giants as an effect of a difference in surface temperature. R. Peterson.
Astrophys. Journ, Suppl. Ser., Vol. 30, 61 - 83 (1976).

The author examines the super-metal-rich (SMR) phenomenon in K giants. Three K2 III stars, κ Oph and ι Dra (both relatively normal) and μ Leo (SMR) were observed at high resolution in the yellow and red. The abundances of all elements with weak, unblended atomic lines are the same to a few percent in all three stars. Direct comparison of weak lines also demonstrates that a temperature difference exists in the surface layers but not at continuum depths. The SMR star is cooler near the surface than are the others. A determination of the carbon and nitrogen abundances is made using CN and C_2 lines. Weak CN lines are stronger in the SMR star, but weak C_2 lines are the same strength. Calculations with model atmospheres constructed empirically show that a normal carbon abundance and a factor of 2 increase in the nitrogen abundance of the SMR star match these observations. The implications of this work on estimates of metal abundances in external galaxies is briefly discussed.

114.011 Balmer lines Hα–Hε and the mean electron densities of 94 O–F stars. V. M. Tereshchenko.
Astron. Zhurn. Akad. Nauk SSSR, Vol. 53, 74 - 82 (1976). In Russian. English translation in Soviet Astron., Vol. 20, No. 1.

The equivalent widths W of Balmer lines Hα–Hε and the central depths R of the H_δ line were determined for 94 O9.5–F2-type stars. The obtained W and R were used for determination of the mean electron densities n_e in the atmospheres of the investigated stars.

114.012 Abundance anomalies in Ap stars and the noble gas structure. C. R. Cowley.
Astrophys. Letters, Vol. 17, 3 - 5 = Contr. Dominion Astrophys. Obs., Victoria, No. 271 = NRC No. 14608 (1976).

The author has tested a simple diffusion hypothesis in which the noble gas structure of diffusing ions is the dominant factor. The hypothesis fails to predict large overabundances of yttrium and cerium, but shows promise in predicting the lighter lanthanides in many Ap stars.

114.013 Carbon abundances in G dwarfs. R. A. Bell, D. Branch.
Monthly Notices Roy. Astron. Soc., Vol. 175, 25 - 32 (1976).

Model atmospheres and synthetic spectra are used to calibrate photoelectric measurements of the C_2 Swan bands in spectra of G dwarfs. The assumption that the iron-to-hydrogen ratio can be obtained from the ultraviolet excess and that the abundances of carbon, oxygen and iron vary in unison lead to excellent agreement between prediction and observation for the majority of stars. Super-metal-rich dwarfs are apparent exceptions to the rule, having either small carbon excesses or oxygen deficiencies if the currently-accepted iron abundances are correct.

114.014 The spectra of peculiar Be stars with infrared excesses. D. A. Allen, J. P. Swings.
Astron. Astrophys., Vol. 47, 293 - 302 (1976).

The authors present a spectroscopic survey of Be stars with infrared excesses. The circumstellar dust which produces the infrared excess is found preferentially in the denser envelopes and probably forms in condensations within them. Hot dust, detected at 2 μm, is most often found around stars of spectral types B and early A, but also occurs in a few high density planetary nebulae. Three mechanisms of mass loss may explain the formation of the shells: formation of planetary nebulae as a single event, interaction of an OB star with a late-type companion and direct ejection from a particularly massive Be or Oe star.

114.015 Isotopic abundances of Hg in mercury stars inferred from Hg II λ3984. R. E. White, A. H. Vaughan, Jr., G. W. Preston, J. P. Swings.
Astrophys. Journ., Vol. 204, 131 - 140 (1976).

Wavelengths of Hg II λ3984 in 30 Hg stars are distributed uniformly from the value for the terrestrial mix to a value that corresponds to nearly pure ^{204}Hg. Relative isotopic abundances derived from partially resolved profiles of λ3984 in ι CrB, χ Lup, and HR 4072 suggest that mass-dependent fractionation has occurred in all three stars. The authors suppose that such fractionation occurs in all Hg stars. Theoretical profiles calculated for the derived isotopic composition agree well with high-resolution interferometric profiles obtained for three of the stars.

114.016 Copernicus observations of Betelgeuse and Antares. A. P. Bernat, D. L. Lambert.
Astrophys. Journ., Vol. 204, 830 - 837 (1976).

Copernicus observations of the M supergiants α Ori and α Sco are presented. The Mg II h and k resonance lines are strongly in emission in both stars. The k line is highly asymmetric in both stars but the h line is symmetric. Upper limits for several other resonance lines are given for α Ori. Observations of the Mn I 4030–4033 Å lines are used to show that circumstellar shell absorption is too weak to explain the asymmetry. It is suggested that the absorption occurs in a cool turbulent region between the base of the circumstellar shell and the top of the chromosphere.

114.017 8200 to 11200 Å spectra of peculiar emission-line objects with infrared excess. Y. Andrillat, J. P. Swings.
Astrophys. Journ., (*Letters*), Vol. 204, L123 - L125 (1976).

Spectra of 25 peculiar emission-line stars with infrared excesses were obtained in the 8200–11200 Å region. The strongest emission features shown on the figure are due essentially to lines of the Paschen series, of the Ca II triplet, and of He I λ 10830, O I λ 8446, [S III] λλ 9069 and 9532, and [Fe III] λ 10504; a strong emission present at 9999 Å on most of the spectra remains unidentified.

114.018 Absolute fluxes of K chromospheric emission on the H–R diagram. C. Blanco, S. Catalano, E. Marilli.
Astron. Astrophys., Vol. 48, 19 - 25 (1976).

Absolute fluxes of the K emission line have been evaluated from 10 Å/mm spectrograms of the O.C. Wilson collection for 31 F5–K7 main sequence stars and 172 G2–M5 giants. Previous results on the fluxes of main sequence stars are confirmed. The K line flux of giant stars is found to be about 10^5 erg cm^{-2} s^{-1} for early G type and decreases by two orders of magnitude from spectral type G2 to M5. Differences in the flux are evident for giants earlier than K0, but no clear age dependence could be found. The average flux of giants turns out to be lower by one order of magnitude than the flux of main sequence stars of the same spectral type. The lines of isochromospheric emission on the H–R diagram are roughly vertical, with a slight inclination to the left for earlier spectral types.

114.019 New Hα-emission stars in the region of the nebulae IC 5068-70 and NGC 7000. M. K. Tsvetkov.
Astrofizika, Vol. 11, 579 - 583 (1975). In Russian. English translation in Astrophysics, Vol. 11, No. 4.

Data of 58 new Hα-emission line stars discovered in the region of the nebulae IC 5068-70 and NGC 7000 are presented. The average limiting magnitude of the discovered objects is $18^m.5$ pg. Identification charts on the new Hα-emission line stars are given.

114.020 Fe II emission in late-type stars. A. M. Boesgaard, H. Boesgaard.
Bull. American Astron. Soc., Vol. 8, 291 (1976). – Abstr. AAS.

114.021 On the nature of the broad continuum absorption features of peculiar A stars. S. J. Adelman, S. N. Shore, P. R. Wolken.
Bull. American Astron. Soc., Vol. 8, 291 (1976). – Abstr. AAS.

114.022 Infrared hydrogen emission lines in the spectra of Be stars. L. Luud, M. Ilmas.
Astrophys. Space Sci., Vol. 40, 135 - 139 (1976).

For the interpretation of observed hydrogen emission lines in the spectrum of γ Cas calculations of the relative intensities of spectral lines of the Balmer, Paschen, Brackett and

Pfund series are carried out, by using the theory of moving stellar envelopes. It is shown that in the spectra of Be stars, which have opaque envelopes and therefore a slow Balmer decrement, infrared hydrogen lines must be relatively strong.

114.023 The nature of the objects of Joy: a study of the T Tauri phenomenon.
A. E. Rydgren, S. E. Strom, K. M. Strom.
Astrophys. Journ., Suppl. Ser., Vol. 30, 307 - 336 (1976).

Homogeneous spectroscopic and photometric (optical and infrared) observations of 49 T Tauri stars in the Taurus and ρ Oph clouds are presented; these data form the basis for a study of their envelope properties and evolutionary status. The analysis shows that the T Tauri phenomenon is best understood in terms of a model in which a late-type star is surrounded by a hot ($\sim 20,000$ K) gaseous envelope. Luminosities, together with effective temperatures from the photospheric spectral types, indicate that the T Tauri stars in these two clouds are relatively low-mass stars evolving toward the main sequence along quasi-static equilibrium tracks. Comparison with theoretical evolutionary tracks yields a mass range of $0.5 \lesssim M \lesssim 3 M_{\odot}$ and ages generally less than 10^6 years.

114.024 On the light variability of Ap stars. I. Effects of surface variations of T_{eff} and g_{eff}.
L. I. Snezhko.
Astrofiz. Issled., Izv. Spets. Astrofiz. Obs., Vol. 8, 14 - 19 (1976). In Russian.

Effects of variations of T_{eff} and g_{eff} at a star's surface are considered within the framework of an oblique rotator model. It is shown that the observed values for the amplitudes of the light curves and line variation curves can be reached even at small variation amplitudes of $\Delta T_{eff} \leqslant 700°$K and $\Delta \lg g_{eff} \leqslant 0.5$.

114.025 The OBN and OBC stars.   N. R. Walborn.
Astrophys. Journ., Vol. 205, 419 - 425 (1976).

A review of present knowledge concerning nitrogen and carbon anomalies in absorption-line OB spectra is presented, from the observational point of view. A list of all presently known OBN and OBC stars is given. The problem of more moderate CNO anomalies in OB spectra is examined; an important conclusion is that all of the O9.5–B0.7 supergiants in the Orion Belt and in NGC 6231 (the nucleus of Scorpius OB1) are systematically nitrogen-deficient, relative to the majority of supergiants with the same spectral types. Since the members of these associations have often been used as standards, an extensive list of O9–B1 supergiants with normal CNO spectra is also given.

114.026 Ultraviolet Fe II emission in late-type stars.
A. M. Boesgaard, H. Boesgaard.
Astrophys. Journ., Vol. 205, 448 - 454 (1976).

Spectrograms of 44 K and M giants and supergiants have been taken to determine the presence and intensity of the ultraviolet emission lines of Fe II from multiplets 1, 6, and 7. The emission is nearly universal in the M stars, but not found in any of the K stars; it thus appears to be a natural consequence of stellar surface temperature. The intensity of the emission is well correlated with the strengths of the circumstellar components of the K-line of Ca II and of Hα. There is no apparent relationship between the Fe II emission intensity and the amount of infrared excess at 11 and 20 μ, the indicators of a circumstellar dust shell.

114.027 On Barbier's 'étoiles M, C et à émission nouvelles'.
P. Pesch.
Astrophys. Space Sci., Vol. 40, 351 - 355 (1976).

More than half of the new M, C and emission-line stars discovered on objective-prism plates by Barbier (1975) are shown to be objects with previously published spectral classi-

fications.

114.028 Polarization in the emission lines of Be stars.
G. V. Coyne.
Astron. Astrophys., Vol. 49, 89 - 96 (1976).

Observations of linear polarization in the Hα and Hβ lines of Be stars are presented. A decrease in the polarization across these lines is observed for shell stars and stars with large $V \sin i$. A model is presented in which the continuum polarization is produced by electron scattering in the inner regions (to a distance of about 3 stellar radii) of an extended flattened disk where $N_e \cong 10^{13}$ cm^{-3} and $T_e \cong 10^4$K. The emission flux occurs in an outer region of the disk extending to about 10 stellar radii where $N_e \cong 10^{12}$ cm^{-3}.

114.029 Finding list and spectral classifications for southern luminous stars.
D. J. MacConnell, W. P. Bidelman.
Astron. Journ., Vol. 81, 225 - 227 (1976).

Spectral types of 149 O stars and supergiants determined on objective-prism plates of dispersion 108 Å/mm at Hγ are given. Most of the stars are brighter than m_{pg} 9.5; the majority of the supergiants are in the Henry Draper Catalogue between $-53°$ and the equator and have not previously been recognized as of high luminosity.

114.030 A summary of data on energy distribution in the spectra of early-type stars reduced to the Oke-Schild system.   V. B. Nikonov, G. A. Terez.
Izv. Krymskoj Astrofiz. Obs., Vol. 54, 35 - 41 (1976). In Russian.

The data on the energy distribution in the spectra of 363 O-F0 stars published by many authors have been reduced to the Oke-Schild spectrophotometric system. On the basis of this material, the relative energy distribution normalized at the wavelength 5556 Å has been obtained for a sequence of wavelengths in the spectral range 3300–7100 Å. Absolute values of spectral energy fluxes at 5556 Å are also calculated. Estimates of accuracy obtained for the reduced values of energy distributions show that they could be used for calibration of spectrophotometric observations.

114.031 Luminosity effects in the ultraviolet spectrum of B5–B6 stars.   A. B. Underhill, E. Silversmith.
Astrophys. Journ., Vol. 206, 156 - 162 (1976).

Copernicus U2 spectral tracings of the ultraviolet spectra of η CMa (B5 Ia) and ζ Dra (B6 III) have been compared, and four regions—1330 to 1340 Å, 1340 to 1350 Å, 1390 to 1396 Å, and 1398 to 1405 Å – are found to contain differences due to luminosity effects. The observed profiles are interpreted using synthetic spectra predicted with LTE theory.

114.032 Possible Mg II emission in B stars observed from Copernicus.   Y. Kondo, J. L. Modisette, R. J. Dufour, R. S. Whaley.
Astrophys. Journ., Vol. 206, 163 - 166 (1976).

Four B stars, α Vir, β Cen, α Gru, and β Lib, were observed with the Copernicus Princeton Telescope Spectrometer at a resolution of 0.1 Å in order to investigate the presence of chromospheric emission. Emission was observed in β Cen and α Gru, while the results for α Vir and β Lib were inconclusive.

114.033 Ultraviolet photometry from the Orbiting Astronomical Observatory. XXIII. The resonance lines of triply ionized carbon and silicon in the spectra of hot stars.
R. J. Panek, B. D. Savage.
Astrophys. Journ., Vol. 206, 167 - 181 (1976).

Quantitative measurements of the absorption strength of the Si IV (1400 Å) and C IV (1550 Å) resonance doublets have been made from OAO-2 far-ultraviolet spectra of 118 O- and B-type stars. These lines are generally very strong and a useful measure of their strength can be made. The results are presented and discussed in detail.

114.034 A survey of peculiar and metallic-lined A stars for the actinides.
C. R. Cowley, G. C. L. Aikman, M. R. Hartoog.
Astrophys. Journ., Vol. 206, 196 - 200 = Contr. Dominion Astrophys. Obs., *Victoria*, No. 284 (1976).

In a high-dispersion survey using modern wavelength lists, the authors have failed to find convincing evidence for any transuranic actinide in 51 Ap, Am, and "normal" stars. It is found that stellar wavelengths do occasionally coincide with those of actinide spectra, and that occasionally these coincidences are improbable. It is nevertheless concluded that most of these coincidences have arisen by chance, in this rather extensive search. The authors specifically discuss thorium and uranium, which they believe are present in the rare-earth maximum spectrum of HR 465.

114.035 Lithium and S-type stars.
R. M. Catchpole, M. W. Feast.
Monthly Notices Roy. Astron. Soc., Vol. 175, 501 - 516 (1976).

A survey of 188 S- and 188 C-type stars has revealed two new S-type lithium stars, TT9 and RZ Sgr with lithium equivalent widths of about 1.5 Å. An abundance analysis of RZ Sgr yields $\log (Li/Ca) = -3.59$ and $\log (Zr/Ti) = -1.01$ unusually high values for S-type stars. There is no evidence for stratification effects. Contrary to recent conclusions the authors find that the Zr:Ti ratios for the C stars studied by Kilston are substantially greater than for the S stars of Boesgaard when these investigations are reduced to a common system of oscillator strengths.

114.036 Spectral classification of the bright F stars.
A. P. Cowley.
Publ. Astron. Soc. Pacific, Vol. 88, 95 - 110 (1976).

New uniform two-dimensional spectral types are presented for 575 stars, most of which are F stars. Only a few peculiar spectra, other than composites, were found although care was taken to note stars with unusually strong or weak metal characteristics.

114.037 Spectral energy distributions of O-type stars.
P. Kuan, L. V. Kuhi.
Publ. Astron. Soc. Pacific, Vol. 88, 128 - 134 (1976).

Narrow-band photoelectric spectrum scanner observations of the spectral energy distributions of O and Of stars are presented. After the application of reasonable reddening corrections it is found that both Of and ordinary O stars present evidence for effects of atmospheric extension. These effects are seen primarily in the flatness of the energy distribution which is most extreme in the hottest Of stars.

114.038 Températures et magnitudes en astrophysique: un aide-mémoire. J.-C. Pecker.
L'Astronomie, Vol. 90, 191 - 206 (1976).

114.039 Hα-emission stars in and near NGC 7000. II.
G. Welin.
Astrofizika, Vol. 11, 261 - 268 (1975).

Tentative spectral classifications have been made for most of the hitherto unclassified UHα stars by means of objective-prism plates obtained at the Byurakan Astrophysical Observatory. Together with V and B photometry from Uppsala-Kvistaberg plates, the spectral types make it possible to discern between stars associated with the NGC 7000–IC5070 complex and more distant early-type stars. A new Hα-emission star (UHα 142) has been found.

114.040 Radio observations of eight early-type emission-line stars. K. A. Marsh, C. R. Purton, P. A.
Feldman.
Astron. Astrophys., Vol. 49, 211 - 215 (1976).

Radio continuum measurements at several frequencies have been made of eight early-type emission-line stars which possess strong infrared excesses. Of these, two represent new radio detections. The observations were made at 2.7 and 8.1 GHz. The spectra are in all cases consistent with free-free radiation from a dense circumstellar gas shell.

114.041 A search for helium variability among helium weak stars. H. Pedersen.
Astron. Astrophys., Vol. 49, 217 - 219 (1976).

A small number of helium weak B stars have been measured over several nights to detect variability in the strength of the He I 4026 Å line. Definite variability was found in the case of HR 7185. The observations indicate that the phenomenon is aperiodic with a He-constant phase in between the extremes. The line strength may decrease from maximum to the constant value within 3 h.

114.042 Observations of southern emission-line stars.
K. G. Henize.
Astrophys. Journ., Suppl. Ser., Vol. 30, 491 - 550 (1976).

A catalog is presented of 1929 stars showing emission at Hα. The survey covers the entire southern sky south of declination −25° to a red limiting magnitude of about 11.0. A region of the northern Milky Way approximately 600 square degrees in area is also included in order to allow comparison with northern surveys. Emission-line stars of all spectral classes are listed including 16 M stars, 25 S stars, and 37 carbon stars, as well as 40 confirmed or suspected T Tauri stars. This survey has resulted in the discovery of two new T associations, one in Lupus and one in Chamaeleon. The sky distribution of the emission-line stars shows significant concentrations in the region of the small Sagittarius cloud and in the Carina region.

114.043 Line identifications in the near-ultraviolet for nine bright stars. K. A. van der Hucht, H. J. G. L. M.
Lamers, R. Faraggiana, M. Hack, R. Stalio.
Astron. Astrophys., Suppl. Ser., Vol. 25, 65 - 128 (1976).

Absorption lines in the near ultraviolet spectrum of nine bright stars are identified. The observations are made by the Orbiting Stellar Spectrophotometer S59 in the wavelength ranges λλ2060−2160, 2490−2590 and 2770−2870 Å, with a spectral resolution of 1.8 Å. The stars and their spectral types are: ζ Pup (O4f), ζ Ori (O9.5I), β CMa (B1III−III), o² CMa (B3Ia), ζ Dra (B6III), β Ori (B8Ia), α Cyg (A2Ia), α PsA (A3V) and α CMi (F5IV). With the available spectral resolution, no evidence was found for the presence of elements with $Z > 30$.

114.044 The emission lines in the vicinity of hydrogen-alpha in dMe flare star spectra. S. P. Worden, B. M.
Peterson.
Astrophys. Journ., (*Letters*), Vol. 206, L145 - L147 (1976).

High-resolution spectral data obtained in the vicinity of Hα have been obtained for a number of dMe stars. Centrally reversed Hα emission profiles appear to be a general feature of dMe spectra. Possible mechanisms related to solar phenomena are discussed for forming this type of profile.

114.045 Lists of S, C, MS stars and emission objects revealed in red light observations. M. V. Dolidze.
Byull. Abastumansk. Astrofiz. Obs., No. 47, p. 3 - 144 (1975). In Russian.

A revised list published formerly as separate lists of 90 S, C and MS stars and of one new C star is presented. There is also given a list of 1174 emission objects. Lists of new probable star clusters, associations, groups and clusters related to emission objects, probably variable stars, emission galaxies, Bep-type stars and small diffuse emission and probable planetary nebulae are also given.

114.046 Lists of groupings of emission line stars.
M. V. Dolidze.
Byull. Abastumansk. Astrofiz. Obs., No. 47, p. 145 - 170 (1975). In Russian.
The paper summarizes the revised lists of emission-line star groupings of various types. A general list of possible emission-line star groupings of the Orion population is presented.

114.047 On peculiarities of the apparent distribution of groupings of emission stars. M. V. Dolidze.
Byull. Abastumansk. Astrofiz. Obs., No. 47, p. 205 - 215 (1975). In Russian.
The results of comparison of two sets of possible emission star groupings are discussed. Emission star groupings permitted to reveal the local spiral structure and also two details in the local spiral arm.

114.048 Zvaigžņu spektrum klasifikācija. Z. Alksne.
Zvaigžņotā debess, 1975./76. gada ziema, p. 4 - 8.

114.049 Southern B-type stars with Hα emission.
B. Kucewicz.
Inform. Bull. Southern Hemisph., No. 26, p. 23 - 24, No. 27, p. 24 - 27 (1975).

114.050 Variable polarization in Be stars. G. V. Coyne.
Ric. Astron. Specola Vaticana, *Castel Gandolfo,* Vol. 8, (No. 28), 533 - 541 (1975).

114.051 A survey for Hα emission objects in the Milky Way. II. Cygnus.
G. V. Coyne, W. Wisniewski, C. Corbally.
Vatican Obs. Publ., Specola Vaticana, *Castel Gandolfo,* Vol. 1, (No. 6), 197 - 212 (1975).
This is the second installment of a survey for Hα emission objects in the Milky Way being conducted with a 12-degree objective prism on the Vatican Schmidt telescope. Positions, finding charts and *UBV* photoelectric photometry are given for 117 new emission objects, 45 percent of which are estimated to be Be dwarf stars. Distances are estimated for all of the 120 Be dwarf stars discovered thus far in this survey and their distribution in the Galaxy is plotted.

114.052 Emission stars in a field in Cygnus.
M. F. McCarthy.
Vatican Obs. Publ., Specola Vaticana, *Castel Gandolfo,* Vol. 1, (No. 7), 213 - 223 (1976).
81 new emission Hα objects in a Cygnus region are reported. Most of the newly discovered stars appear to be of early type; some 22 stars show late-type spectral features on infrared sensitive spectral plates; other emission objects in the field include 4 known planetary nebulae plus Nova Cygni 1975 and possibly several T Tauri candidates.

114.053 A survey for Hα emission objects in the Milky Way. III. Cygnus. W. Wisniewski, G. V. Coyne.
Vatican Obs. Publ., Specola Vaticana, *Castel Gandolfo,* Vol. 1, (No. 8), 225 - 236 (1976).
This is the third installment of a survey for Hα emission objects in the Milky Way being conducted with a 12-degree prism on the Vatican Schmidt telescope. Positions, finding charts and *UBVRI* photoelectric photometry are given for 72 new emission objects, 60 percent of which are estimated from the photoelectric measurements to be Be dwarf stars. Distances are estimated for all of the 163 Be dwarf stars discovered thus far in this survey and their distribution in the Galaxy is plotted.

114.054 Results of spectrophotometric studies of the mag-netic and peculiar stars 21 Per, β CrB, γ Boo, γ UMi.
Yu. V. Glagolevskij, K. I. Kozlova, R. N. Kumajgorodskaya, V. S. Lebedev, N. S. Polosukhina, N. M. Chunakova.
Astron. Nachr., Band 297, 189 - 190 (1976). In Russian.
A new method for finding the distribution of chemical elements over the surface of Ap stars is worked out. It is detected that at the surface of 21 Per iron is concentrated in four regions. It is shown that one of the reasons of hydrogen line variability in 21 Per is the inhomogeneous distribution of gravity. Rapid variations of hydrogen lines in some phases and minor changes ($\sim 10\%$) of K Ca II line in γ Boo are detected. Hydrogen lines ($H_9 - H_{12}$) and K Ca II lines are found to vary in γ UMi. Variations of H_α line in β CrB during the period are detected.

114.055 Investigation of the Balmer lines in Ap stars.
I. A. Aslanov, Yu. S. Rustamov, V. M. Khalilov, A. A. Shakir-zade.
Astron. Nachr., Band 297, 191 - 202 (1976). In Russian.
The behaviour of the Balmer lines in the spectra of 13 Ap stars has been investigated. It was found that all the investigated stars show the variation in the hydrogen line contours, equivalent widths and the variation of lg n_e and lg ($N_{02}H$) as well. The curves of the equivalent width variations during the period have several maxima and minima whereas the light curves have no more than two maxima and minima.

114.056 On the formalization of the reversed problem of determination of local line profiles from observed ones in Ap stars spectra.
V. L. Khokhlova.
Astron. Nachr., Band 297, 203 - 206 (1976). In Russian.
An analytical expression for the local spectral line profile is proposed which permits to formalize the problem of determination of local profiles from observed ones in Ap-stars spectrum.

114.057 Observations of luminous early type stars with the ultraviolet sky survey telescope in satellite TD1.
C. M. Humphries, K. Nandy, E. Kontizas.
Mem. Soc. Astron. Italiana, Vol. 45, (see 012.017), 711 - 712 (1974). – Abstract (see also 13.114.003).

114.058 A survey of symbiotic stars in the near infrared.
A. Mammano.
Mem. Soc. Astron. Italiana, Vol. 45, (see 012.017), 727 - 728 (1974). – Abstract.

114.059 On the nature of emission line objects.
A. Cassatella, R. Viotti.
Mem. Soc. Astron. Italiana, Vol. 45, (see 012.017), 741 - 746 (1974).

114.060 Utvärdering av objektivprismespektra av tidiga stjärnor med hjälp av dator. L. Kullberg.
Stockholms Obs. Rep. No. 1, 55 pp. (1973).
A method to reduce objective prism spectrograms of early type stars is presented. All reductions have been carried out with a computer. The line strengths of Hγ and Hδ have been measured with photographic narrow band techniques.

114.061 On the energy distribution of Be stars.
L. Nordh, G. Olofsson.
Stockholms Obs. Rep. No. 7, 71 pp. (1974).
The energy distributions of fifteen Be stars are analysed in the wavelength region $0.35 - 1.0\,\mu$. Generally they are found to exhibit both UV and near IR excesses, a behaviour partly explicable in terms of hydrogen free-bound plus free-free emission from surrounding envelopes superimposed on the stellar continua.

114.062 Observations of the stellar Mg II resonance doublet at 2795 and 2802 Å in late-type giants.
T. H. Morgan.
Proc. Southwest Regional Conf., Vol. 1, (see 012.021), 115 - 140 (1976).

In an ongoing investigation of the Mg II doublet lines (2800 Å) in various type stars, spectrophotometry of 20 stars was performed with the balloon-borne stellar spectrometer on the night of 14−15 March 1975. The author reports observations of the five late-type giants which show pronounced emission in the doublet lines. The variation in the emission as a function of spectral type and absolute magnitude of the stars are discussed. There are noticeable differences in the Mg II emission features among some stars of similar spectral types.

114.063 Spectroscopic measurements of OB supergiants.
J. B. Hutchings.
Publ. Dominion Astrophys. Obs., Vol. 14, (No. 16), 355 - 366 = National Res. Council Canada, NRC No. 15165 (1976).

Detailed measurements are presented of radial velocities and Hγ equivalent widths of 67 very luminous O and B stars, together with other relevant data and derived absolute magnitudes.

114.064 A study of carbon stars whose spectra have shown SiC₂ bands. C. E. Irvine.
Thesis Case Western Reserve Univ., Cleveland, Ohio, USA, 95 pp. (1975). (Available from: Univ. Microfilms, Ann Arbor, Mich., USA. Order No. 75-19215). − Abstr. in Phys. Abstr., Vol. 79, A027972 (1976).

114.065 Orion-2: first scientific results. G. A. Gurzadyan.
Space Sci. Rev., Vol. 18, 95 - 139 (1975).

The ultraviolet spectral images of thousands of faint stars, up to the 13th mag., in the wavelength region of 2000−5000 Å are obtained by means of the space astrophysical observatory Orion-2 aboard the spaceship Soyuz-13. These spectrograms were designed generally for an investigation of the continuous spectra of the stars in ultraviolet. Some of the results obtained are included in the present review.

The observation and analysis of stellar photospheres. See Abstr. 003.049.

On the analysis of blended spectra. See Abstr. 031.216.

The theoretical near ultraviolet spectrum of B type stars. See Abstr. 064.036.

Synthetic spectra of red giants. I. Representative band head profiles of diatomic molecules. See Abstr. 064.037.

Predicted line strengths of the CH infrared bands in stellar spectra. See Abstr. 064.038.

Welcher Stern gleicht der Sonne? See Abstr. 071.013.

On the correlation between the motions and CN anomalies of G and K stars. See Abstr. 112.003.

Rotational-velocity effects in the $uvby$, β systems and on MK spectral types for B stars. See Abstr. 113.001.

Space distribution and kinematics of intermediate population II stars. Part I. Photometry and spectroscopy in selected McCormick proper motion fields. See Abstr. 113.011.

$uvby\beta$ photometry and MK spectral classification. See Abstr. 113.013.

Investigation of light variation of twelve Ap stars in ten spectral regions. See Abstr. 113.014.

How to define a photometric degree of peculiarity for Ap-stars? See Abstr. 113.050.

$V \sin i$ values in the far ultraviolet. See Abstr. 116.004.

MK classification for visual binary components. See Abstr. 118.014.

Limits on the space density of O subdwarfs and hot white dwarfs from a search for extreme ultraviolet sources. See Abstr. 126.009.

Spectral types in the Lacerta OB1 association. See Abstr. 152.010.

A study of Be stars in clusters. See Abstr. 153.010.

Catalogue of UBV photometry and MK spectral types in open clusters. See Abstr. 153.012.

A catalogue of A- and F-type supergiants in the Large Magellanic Cloud. See Abstr. 159.005.

Wolf-Rayet stars in the Large Magellanic Cloud. See Abstr. 159.012.

Spectra of Individual Stars

114.301 Si II equivalent widths in SMC A-type supergiants.
M. W. Feast.
Monthly Notices Roy. Astron. Soc., Vol. 174, 9P - 12P (1976).

Four B9 – A0 supergiants in the SMC (HD 7583, 5030, 5277, R27) have Si II (6347, 6371 Å) equivalent widths which are about a factor 2 smaller than expected from the Rosendhal equivalent width – absolute magnitude relation. Either the Rosendhal relation breaks down or silicon is underabundant in these stars.

114.302 Cobalt in Eta Carinae.   A. D. Thackeray.
Monthly Notices Roy. Astron. Soc., Vol. 174, 59P - 62P (1976).

Support for identifying λ4152.5 in Eta Carinae with [Co II] is found in the presence of seven more possible identifications of weak lines in that region. But computed transition probabilities are required to confirm the suggestion. Several of these weak lines occur in RR Tel with behaviour consistent with identification with [Co II].

114.303 P Cygni profiles in Zeta Ophiuchi and Zeta Puppis.
D. C. Morton.
Astrophys. Journ., Vol. 203, 386 - 398 (1976).

Selected regions of the far UV spectra of ζ Oph (O9.5 V) and ζ Pup (O4 If) obtained with the Copernicus satellite have been plotted to show the P Cygni profiles. In addition, widths and velocity shifts of both emission and absorption features have been tabulated. In both stars, part of most absorption profiles exceeds the escape velocity, showing that mass is being ejected. The short-wavelength edges of the resonance lines average -1590 ± 10 km s^{-1} in ζ Oph and -2660 ± 150 km s^{-1} in ζ Pup, with no significant dependence on ionization potential. The equivalent width of the emission component is always significantly less than the absorption, suggesting that the absorption occurs close to the stellar surface where occultation can hide some of the emission.

114.304 Ultraviolet photometry from the Orbiting Astronomical Observatory. XXI. Absolute energy distribution of stars in the ultraviolet.
R. C. Bless, A. D. Code, E. T. Fairchild.
Astrophys. Journ., Vol. 203, 410 - 416 (1976).

The absolute energy distribution in the ultraviolet is given for the stars α Vir, η UMa, and α Leo. The calibration is based upon absolute heterochromatic photometry between 2920 and 1370 Å carried out with an Aerobee sounding rocket. The fluxes for the three program stars are tabulated in energy per second per centimeter^2 per unit wavelength interval.

114.305 The Ap star HD 224801.
F. Castelli, R. Faraggiana, S. Hvala.
Astron. Astrophys., Vol. 46, 99 - 108 (1976).

The spectral variability of the Si star HD 224801 is studied. The radial velocity curve has an amplitude lower than that found by previous authors. Radial velocities and equivalent widths vary with the photometric period of $3\overset{d}{.}74$. The 2^h04^m period found for small light fluctuations also fits the data. From a model atmosphere analysis, the authors derived $T_e = 11700°$K and $\log g = 3.7$, and these parameters were used for element abundance calculations: these abundances are typical of a Si star. The over-abundance of Si is about 15.8.

114.306 Properties and nature of shell stars. 6. The shell star 4 Herculis as an interacting binary.
P. Harmanec, P. Koubský, J. Krpata, F. Ždárský.
Bull. Astron. Inst. Czechoslovakia, Vol. 27, 47 - 56 (1976).

Rectified intensity profiles of the Hα and Hβ Balmer lines of 4 Her were studied on 86 coudé spectrograms of 1969–

1974. The central intensities of both lines have been found to vary periodically, with the period of velocity changes, 46.194 days, reaching two maxima and minima each cycle. The V/R ratio of the emission peaks of the Hα line also varies in phase with the velocity changes. Definite differences were found in the velocity curves of individual Balmer lines. It is shown that all these data can be reconciled on the assumption that the star is a binary undergoing a variable mass transfer between components.

114.307 Fe XIII line in R Aquarii.   H. Zirin.
Nature, Vol. 259, 466 - 467 (1976).

Gregory and Seaquist have detected radio emission from the interesting object R Aquarii. In the course of reducing plates of this star made in 1970 and 1971 in the course of his He I, $\lambda = 10,830$ Å spectral survey, the author found the emission line of the forbidden coronal transition of Fe XIII at $\lambda = 10,747$ Å. This line ($^3P_1 - ^3P_0$) is often accompanied by the $^3P_2 - ^3P_1$ line at $\lambda = 10,798$ Å at high densities ($N_e > 10^9$). The emission line $\lambda = 10,747$ Å is weak, but clearly seen on two spectra with intensity of ~100 mA.

114.308 The rapidly varying emission spectrum of HD 158503.   A. P. Cowley, N. Houk.
Publ. Astron. Soc. Pacific, Vol. 88, 37 - 40 = Contr. Dominion Astrophys Obs., *Victoria*, No. 257 = NRC No. 14911 (1976).

Spectrograms of the Be star HD 158503 taken during four consecutive nights show that the emission in the hydrogen lines and the Fe II lines varies in intensity and velocity, while the unusual Ca II emission remains constant. It is suggested that this system may be a short-period, interacting binary with the primary below the main sequence.

114.309 The spectrum of FG Sagittae in 1975.
J. Smolinski, J. L. Climenhaga, T. Kipper.
Publ. Astron. Soc. Pacific, Vol. 88, 67 - 68 (1976).

The spectrum of FG Sge in 1975 is compared with spectra of Ia and Ib supergiants of class F5 to G5. The cooling trend described by Langer, Kraft, and Anderson as taking place up to 1972 has continued, but possibly at a faster rate since 1972. The enhancement of rare-earth abundances also is confirmed. Molecular bands are not observed.

114.310 The complex structure of the Ca II (H and K) lines in the spectrum of the A0ep star with infrared excess HD 190073. I. Line profiles and variations during three decades (1943–1974).   J. Surdej, J. P. Swings.
Astron. Astrophys., Vol. 47, 113 - 119 (1976).

Radial velocities and profiles of the components of the H- and K-lines in the spectrum of HD 190073 are analyzed. A 2 to 1 ratio in the radial velocities of some components is shown not to be significant on a spectrum to spectrum basis, contrary to suggestions by Merrill (1951) and Scargle (1973). The details of the H- and K-complex structure are correlated with the profiles of the Balmer lines, especially H$_e$. The same correlation is found in a series of emission-line stars with infrared excess.

114.311 The complex structure of the Ca II (H and K) lines in the spectrum of the A0ep star with infrared excess HD 190073. II. Interpretation: the selective effect of radiative forces.   J. Surdej, J. P. Swings.
Astron. Astrophys., Vol. 47, 121 - 127 (1976).

Radiative forces acting selectively via a resonance scattering mechanism of Ca$^+$ atoms are capable of producing the main features described in paper I (see Abstr. 114.310) concerning the complex profiles of H and K in HD 190073 and in a series of stars exhibiting similarities to HD 190073.

114.312 On the spectrum of FG Sge.
I. Kupo, M. Fishkis.

Astron. Astrophys., Vol. 47, 417 - 422 (1976).

Spectra of FG Sge and of the surrounding nebula Hz I-5 were obtained during 1974–1975. The coincidence in wavelength of most bright features with lines of oxygen, nitrogen, carbon, sulphur, neon ions, forbidden iron transitions found in planetary nebulae, Wolf-Rayet and nova-like stars is indicated. Some arguments to the possible reality of the identified emission lines are given.

114.313 Ultraviolet observations of Gamma-2 Velorum.
A. J. Willis, R. Wilson.
Astron. Astrophys., Vol. 47, 429 - 436 (1976).

New ultraviolet observations of γ^2 Velorum (WC8 + O9I) made with the sky-survey telescope (S2/68) in the TD-1A satellite at two orbital phases are presented. The observed spectra are described and compared to other UV observations of the system. The spectrum of the WC8 component is estimated and is shown to agree well with the UV spectrum observed for a single WC8 star, HD 192103. An estimate of the mass loss from the WC8 component is derived as $2 \times 10^{-6} M_\odot$ yr^{-1}.

114.314 The complex outer shell of Eta Carinae.
N. R. Walborn.
Astrophys. Journ., (Letters), Vol. 204, L17 - L19 (1976).

Short-exposure, interference-filter photographs obtained at the prime focus of the CTIO 4 meter reflector reveal the great complexity and extent of η Carinae's outer shell. Some features have increased their distances from the central object with respect to Thackeray's photographs of a quarter-century ago, and some apparent structural relationships are found between features of the homunculus and the outer shell. A speculative relationship to van den Bergh's quasi-stationary flocculi in Cas A is pointed out.

114.315 GH 7-21: a possible degenerate star with narrow hydrogen lines and strong carbon features.
J. Liebert.
Astrophys. Journ., (Letters), Vol. 204, L93 - L97 (1976).

Image-tube spectrophotometry of the proper-motion star GH 7-21 shows narrow Balmer lines in addition to the strong CH and C_2 bands reported by G. Wegner. If the star lies below the main sequence, the spectrum is consistent with the recent idea that there may be a class of cool, degenerate stars with narrow spectral lines and hydrogen atmospheres. If the star is a Hyades cluster member, it must have a luminosity between the main and degenerate sequences.

114.316 Copernicus ultraviolet spectra of OB supergiants with strong stellar winds. J. B. Hutchings.
Astrophys. Journ., (Letters), Vol. 204, L99 - L102 (1976).

Spectral scans at ~0.2 Å resolution have been obtained in the far-ultraviolet of eight stars which have high mass-loss rates from stellar winds. The P Cygni characteristics of the line profiles appear to vary inversely as the mass flow rate, and in P Cygni itself the C III λ1175 line shows no velocity shift, or emission. It is suggested that higher mass flow rates occur through a denser, slower moving envelope in which collisional interactions are important.

114.317 CD–37°9248, a metal-poor star of high radial velocity. G. Wegner.
Observatory, Vol. 96, 13 - 15 (1976).

114.318 Feinanalysen dreier R CrB-Sterne.
D. Schönberner.
Mitt. Astron. Ges., No. 38, p. 195 (1976). – Abstract.

114.319 Spektrumvariabilität von HD 37776.
I. S. Offick-Clas.
Mitt. Astron. Ges., No. 38, p. 197 - 198 (1976). – Abstract.

114.320 Spektralphotometrie und Feinanalyse des intermediären Heliumsterns HD 133518. M. Gerlach.
Mitt. Astron. Ges., No. 38, p. 198 (1976). – Abstract.

114.321 Analysis of the extreme helium-star BD–9°4395.
J. P. Kaufmann, D. Schönberner.
Mitt. Astron. Ges., No. 38, p. 198 - 200 (1976). – Abstract.

114.322 Ultraviolet spectra of the Wolf-Rayet stars HD 50896 and HD 191765 from the S 2/68 Sky-Survey Telescope in TD-1 A. A. Cucchiaro, C. Jamar, D. Macau-Hercot.
Space Research XV, (see 012.003), p. 657 - 661 (1975).

The ultraviolet spectra of two bright Wolf-Rayet stars HD 50896 and HD 191765 (in the range of 1350–2550 Å) have been reduced from the S 2/S 68 spectrophotometer results. All strong emission lines are satisfactorily explained and identified; a rather strong complex emission near 1990 Å remains unidentified. Effective temperatures of their atmospheric envelopes have been estimated from the emission line intensities. A calculation of the abundances with a fairly plausible ionization temperature has been carried out assuming thermal ionization from excited levels.

114.323 A spectrophotometric study of the Be star HD 164447. S. N. Svolopoulos.
Astron. Nachr., Vol. 297, 87 - 88 (1976).

Spectra of the Be star HD 164447 have been studied. The equivalent widths and the intensities of the measured lines are given. No emission is apparent on the plates. Applying the thin layer theory some parameters of the star's atmosphere were derived.

114.324 Spectrophotometry of two "ultraviolet" stars discovered by Orion 2.
G. A. Gurzadyan, Dzh. B. Oganesyan.
Astrofizika, Vol. 11, 585 - 592 (1975). In Russian. English translation in Astrophysics, Vol. 11, No. 4.

In the course of analysis of the observational material which were obtained during the Orion-2 experiment many "ultraviolet" stars in which the continuous spectrum is strong at least up to 2500 Å were discovered. The results of the spectrophotometric measurements of two such stars – No. 10 and 50 – are derived. Both are fainter than 10^m. It was revealed that the energy distribution in the continuous spectrum of these stars corresponds to a temperature higher than 20000°K, and the estimate of their absolute luminosity does not indicate that they belong to ordinary hot giants.

114.325 The temperature, luminosity, and spectrum of Kapteyn's star.
R. F. Wing, C. A. Dean, D. J. MacConnell.
Astrophys. Journ., Vol. 205, 186 - 193 (1976).

The temperature of Kapteyn's star, a halo-population subdwarf, has been determined from narrow-band photometry in the near infrared. The color temperature is that of a normal dwarf of type M1.0, while the TiO strength indicates type M0.0. The effective temperature is estimated to be 3800 K, and the absolute bolometric magnitude is +9.50. Kapteyn's star lies 1.4 bolometric magnitudes below the mean main sequence for stars in the solar neighborhood.

114.326 Spectroscopic survey of the far-ultraviolet (1160–1700 Å) emissions of Capella.
R. C. Vitz, H. Weiser, H. W. Moos, A. Weinstein, E. S. Warden.
Astrophys. Journ., (Letters), Vol. 205, L35 - L38 (1976).

A far-ultraviolet spectral survey of Capella (α Aur, G5 III + G0 III) has been obtained using a new highly sensitive rocket-borne spectrograph with a microchannel plate detector. The spectral distribution is very similar to that of the sun; however, if the line surface fluxes are due to the primary (G5 III), then, except for Lα, they are about an order of magnitude greater than those of the quiet sun.

114.327   Curve-of-growth analysis of the spectrum of
β Gem.   R. Griffin.
Monthly Notices Roy. Astron. Soc., Vol. 175, 225 - 234 (1976).

A differential curve-of-growth analysis of β Gem (K0 III) relative to α Boo (K2 III) shows that β Gem has solar abundances.

114.328   Changes in the Hα profile of β Lyr.   A. Sanyal.
Bull. American Astron. Soc., Vol. 8, 292 (1976). Abstr. AAS.

114.329   Equivalent widths of absorption lines in the spectra
of 20 stars of luminosity class III.   Z. N. Fenina.
Astrometriya i Astrofizika, *Kiev*, vyp. (No.) 29, (see 003.003), p. 33 - 54 (1976). In Russian.

Equivalent widths of absorption lines in the spectra of 20 stars of luminosity class III were determined from spectrograms with dispersion of 125 Å/mm. Probable errors of one estimate of the equivalent width range from 4 to 25%. The system of equivalent widths corresponds to similar systems of moderate dispersion obtained by other authors.

114.330   Equivalent widths of absorption lines in the spectra
of 16 stars of luminosity class IV in the interval of
spectral classes B9 - F8.   Z. N. Fenina, Yu. S. Romanov.
Astrometriya i Astrofizika, *Kiev*, vyp. (No.) 29, (see 003.003), p. 55 - 65 (1976). In Russian.

Equivalent widths of absorption lines in the spectra of 16 stars of luminosity class IV were determined from spectrograms with dispersion of 125 Å/mm. The results of quantitative spectral classification of these stars are given using separately hydrogen, calcium (K Ca II) and metallic criteria.

114.331   Chemical composition of Procyon. Rough analysis.
I. A. Zenina, O. A. Zenina, V. V. Leushin.
Astrofiz. Issled., Izv. Spets. Astrofiz. Obs., Vol. 8, 25 - 40 (1976). In Russian.

Lines are identified and their equivalent widths are measured for Procyon from spectra with a dispersion of 1.3 Å/mm in the region λλ 3500–7000 Å. The accuracy of W_λ measurement is of the order of ±0.005 Å. The known values for oscillator strengths are analysed and corrections for some lists are presented. The chemical composition of the atmosphere is defined by the curve-of-growth method. The abundance of 70 elements is determined with an accuracy of ±0.2 in the logarithm of the number of atoms. Special consideration has been given to determining the heavy atoms abundance. A comparison with the chemical composition of the sun has been made. Mean values for the physical characteristics of the atmosphere are obtained.

114.332   New evidence for an expanding atmosphere of the
supergiant β Orionis, B8 Ia.   E. L. Chentsov.
Astrofiz. Issled., Izv. Spets. Astrofiz. Obs., Vol. 8, 128 - 131 (1976). In Russian.

In the resonance lines Mg II λλ 2795 and 2802 Å in the spectrum of β Ori there are short-wave components displaced by −95 km/s. The so-far observed resonance line shifts in the ultraviolet spectra of white supergiants do not exceed the maximum line shifts in the visible spectra of these stars, both being an order of magnitude smaller than in OB supergiants. The author could not detect photographically a short-wave component of the resonance line K Ca II in β Ori.

114.333   Organic polymers in carbon stars.
K. S. V. Santhanam, S. M. Chitre, S. Ramadurai, M. S. Vardya, N. C. Wickramasinghe.
Bull. Astron. Soc. India, Vol. 3, 29 (1975). − Abstract of a

paper presented at the A.S.I. meeting 1975.

114.334   Asymmetries in the absorption lines of manganese
stars.   M. A. Smith, S. B. Parsons.
Astrophys. Journ., Vol. 205, 430 - 439 (1976).

In this paper observations are reported on a large number of rather strong unblended lines in a total of three Mn stars (ι CrB, HR 4072, χ Lupi) which appear to have a companion feature at the same location relative to line center as the lines observed by Smith and Parsons, 1975. The authors conclude that the feature can be best understood in terms of material flowing upward in the photosphere at about the sonic velocity.

114.335   MK morphology of a group of Am stars.
H. A. Abt, W. W. Morgan.
Astrophys. Journ., Vol. 205, 446 - 447 (1976).

Among the "classical" metallic-line stars (Roman et al.), some of those showing the greatest difference between K-line type and metallic-line type are found to have markedly brighter luminosity classes in the violet than in the blue spectral region.

114.336   The luminosity of the very red supergiant near the
cluster Tr 27.   C. L. Imhoff, P. C. Keenan.
Astrophys. Journ., Vol. 205, 455 - 461 (1976).

The authors identify the red supergiant to which Albers called attention as CoD −33°12241. Because of its high color index ($B−V$ = 2.83) they used coudé spectrograms in the red region to estimate the luminosity from line widths and relative intensities. The classification of M0 Ia implies $M_v \approx -7.2 \pm 0.5$, which suggests that the star lies somewhat behind the group of OB stars assigned by Thé and Stokes to this rather doubtful cluster. The radial velocity of CoD −33°12241 is −11.4 ± 0.7 km s⁻¹.

114.337   On an investigation of the magnetic variable Ap star
17 Com A.   Yu. S. Rustamov.
Astron. Zhurn. Akad. Nauk SSSR, Vol. 53, 318 - 326 (1976). In Russian. English translation in Soviet Astron., Vol. 20, No. 2.

An investigation of the magnetic variable Ap star 17 Com A has been made. Periodical variations of the equivalent line widths of Ti II, Sr II, Cr II, Ca II, Mg II have been found. It is shown that these variations are due to an atmospheric spot structure. From the comparison of the results obtained with the curve of the magnetic field variation it is concluded that the field has a more complex structure than a dipole.

114.338   On the nature of Sigma Orionis E.
N. R. Walborn, J. E. Hesser.
Astrophys. Journ., *Letters*, Vol. 205, L87 - L91 (1976).

A period of 1ᵈ19 has been found in observations of the Hα emission intensity, He absorption-line strengths, magnitudes, and colors of the helium-rich star σ Ori E. Some characteristics of the variations are reminiscent of the oblique rotator model, while others suggest the presence of a very low-mass companion.

114.339   Mass loss and asymmetries in the lines of Iota
Coronae Borealis.   A. H. Karp.
Astrophys. Journ., *Letters*, Vol. 205, L93 - L96 (1976).

Recent observations by Smith and Parsons of the Hg−Mn star ι CrB have revealed slight asymmetries in the absorption lines. While these asymmetries are near the limits of detection at the resolution used, their appearance in several lines indicates that they are real. It is shown that the asymmetries are consistent with an accelerating radial flow of material in the photosphere of this star. The velocity needed to reproduce the observed asymmetry corresponds to a mass loss rate of roughly $10^{-7} M_\odot$ yr⁻¹.

114.340   A lower limit on the surface ¹²C/¹³C ratio in Alpha

Orionis.
T. N. Gautier III, R. I. Thompson, U. Fink, H. P. Larson.
Astrophys. Journ., Vol. 205, 841 - 847 (1976).

The second overtone CO bands near 1.6μ were analyzed in α Ori using synthetic spectra. No firm identification of ^{13}CO was made which allowed a lower limit of 20 to be set on the $^{12}C/^{13}C$ ratio. A rather low microturbulent velocity of 2 km s^{-1} was found to best match the spectrum of α Ori.

114.341 **The far-ultraviolet (1180–1950 Å) emission spectrum of Arcturus.**
W. R. McKinney, H. W. Moos, J. W. Giles.
Astrophys. Journ., Vol. 205, 848 - 854 (1976).

The far-ultraviolet (1180–1950 Å) emission spectrum of the K2 IIIp star, Arcturus, has been obtained with a rocket-borne multichannel spectrometer. H I λ1216 and O I λ1304 are the only identified emissions, and the observed H I λ1216 flux is low compared with previous observations. A third unidentified feature was observed at 1511 Å. The absence of many lines found in emission from the sun is striking. The absence of certain features implies that the coronal temperature must be either below 50,000 K or above 350,000 K.

114.342 **Spectrophotometry of the flare star BY Draconis.**
J. D. Fix, S. R. Spangler.
Astrophys. Journ.,(*Letters*), Vol. 205, L163 - L164 (1976).

Vogt has proposed that the 1973 $B-V$ variations of BY Draconis were caused by the presence of Ca II emission regions or plages. Spectrophotometric observations of BY Dra do not show Ca II K and H variations capable of accounting for the amplitude of the observed $B-V$ variations. A limit of 0.0016 mag has been set for the variation in B and thus $B-V$ due to H- and K-line changes.

114.343 **The photometric and spectrographic histories of HD 245770 $\stackrel{?}{=}$ A0535 + 26, the transient X-ray source.** M. Stier, W. Liller.
Astrophys. Journ., Vol. 206, 257 - 259 (1976).

The authors have used photographic plates in the Harvard collection dating from 1898 to study the spectrum and photometric properties of the 9th magnitude Bpe star HD 245770. This star, the prime candidate for the transient source A0535 + 26, showed no emission features or other unusual characteristics on early spectrograms covering the region 3800–5000 Å. The extreme B magnitudes of HD 245770 are found to be 9.4 ± 0.1 and 10.1 ± 0.1.

114.344 **Spectroscopic observations of the candidate star coincident with A0620–00.**
T. R. Gull, D. G. York, T. P. Snow, Jr., K. G. Henize.
Astrophys. Journ., Vol. 206, 260 - 264 (1976).

Several spectra of the optical object identified with the X-ray flare source A0620–00 have been obtained at 5 Å mm^{-1}. These spectra are described; a flux distribution is derived, including the effects of reddening; the spectrum is described; and the distance, based on various interstellar indicators, is discussed. Finally, a brief interpretative comment is made.

114.345 **A spectrophotometric atlas of three K-giants.**
L. Gratton, G. Natali, R. Nesci.
Published by Università di Roma, Cattedra di Astrofisica, C.N.R., Laboratorio di Astrofisica Spaziale, Frascati. 9 + 56 pp. (1975).

The authors present here a spectrophotometric atlas of the three K-giants η Cep, γ Tau, δ Tau. Their purpose is to obtain an homogeneous set of data for a sample of stars sufficiently numerous to be statistically significant in order to get some information concerning the following problems: (a) do real differences exist between the abundances of individual elements in "normal" stars? (b) if such differences exist, which

is the variance from the average? (c) is there any correlation between abundances and age as given by kinematical or other age indicators and as required by the general evolution of the Galaxy? The present atlas gives in a graphical form a general information upon the results of measurements of the three stars. The whole observational material was obtained during four successive missions in the years 1972–74. The spectral range covered is λ 3985 - 4810 Å. The 56 photometric tracings of the atlas give the normalized intensity as a function of wavelength. Wavelengths are determined with an accuracy of 0.02 Å.

114.346 **VX Aquilae: the missing link in the evolutionary sequence of S and carbon stars?**
S. Wyckoff, P. A. Wehinger.
Monthly Notices Roy. Astron. Soc., Vol. 175, 587 - 594 (1976).

Spectroscopic observations (4500–9000 Å) of the intermediate CS Mira variable, VX Aql have been obtained which cover nearly two cycles of the star's light variation. The observations appear to support the suggestion of Greene & Wing that the C : O abundance ratio in the atmosphere of VX Aql is closer to 1 than for any known cool SC or CS star. Finally, the possibility that the collective dissociation continua of abundant molecular hydrides with thresholds near 2.8 eV might contribute significantly to the violet depression in SC, CS and S stars is discussed.

114.347 **The structure of the radio emission from the NGC 1579/LkHα 101 region.**
R. L. Brown, J. J. Broderick, G. R. Knapp.
Monthly Notices Roy. Astron. Soc., Vol. 175, 87P - 92P (1976).

Radio-frequency observations at 3.7 and 11 cm of the NGC 1579/LkHα 101 region show that the radio emission arises in a compact, $<1''$ core concentric with a more extended $\sim 1'$ emission region. LkHα 101 appears to be the source of excitation for all of the radio emission; this result, together with the total infrared luminosity, suggests that an appropriate spectral classification for LkHα 101 is B1 IIe.

114.348 **Variable Ca II H and K emission in the spectrum of λ Andromedae.** J. A. Eilek, G. A. H. Walker.
Publ. Astron. Soc. Pacific, Vol. 88, 137 - 140 (1976).

The equivalent width of the H and K emission reversals of λ And showed significant variations relative to the continuum. The variations are not correlated with the 21-day spectroscopic period but do show some correlation with the 56-day photometric period. High-resolution spectra of the K line indicated that there are no short-term variations in excess of 5 % over a period of one hour.

114.349 **Absolute ultraviolet spectrophotometry from the TD-1 satellite. VII. Ultraviolet variations of ζ Tau.**
F. Beeckmans.
Astron. Astrophys., Vol. 49, 263 - 269 (1976).

Magnitude and color variations are observed in the S 2/68 UV spectra of ζ Tau, whereas the depth of the absorption features appears to stay mainly constant. The spectral type deduced from the energy distribution envelope and the depth of the absorption features at 1410, 1555 and 1615 Å is earlier than that deduced from visible observations. On the other hand, ζ Tau shows a large UV defect, probably due to the absorbing shell.

114.350 **Spectral line variations of the early-type star 10 Lacertae in the years 1972–1973.**
D. Chochol, J. Grygar.
Bull. Astron. Inst. Czechoslovakia, Vol. 27, 181 - 190 (1976).

Residual intensities for 81 lines in the spectrum of 10 Lacertae are given. It is shown that some lines, particularly those of O II and O III, may exhibit occasional changes of line depth.

114.351 α CrB en el ultraviolet lejano.
　　　　J. Fernández-Figueroa, R. Fernández.
Urania Barcelona, Año 60, No. 283, p. 3 - 19 (1975).

114.352 **The chemical composition of Gamma Pegasi.**
　　　　G. J. Peters.
Astrophys. Journ., Suppl. Ser., Vol. 30, 551 - 565 (1976).

　　　　Equivalent-width data for 274 spectral lines ($\lambda\lambda 3090$–6700), the profiles of $H\gamma$ and $H\delta$, and the continuous flux distribution ($\lambda\lambda 1400$–8100) of the sharp-lined B2 star γ Peg have been interpreted with the aid of the Princeton ultraviolet line-blanketed model atmospheres and the hydrogen line-broadening theory of Vidal, Cooper, and Smith. New equivalent-width data from 2 Å mm^{-1} coudé plates extend the ground-based observations for γ Peg shortward to $\lambda 3094$. Seventy-one additional lines have been identified, of which 34 could be analyzed. In addition to new ground-based continuum observations ($\lambda\lambda 3300$–8100), flux data in the far-ultraviolet ($\lambda\lambda 1384$–2550) have been obtained with the S2/68 experiment package on board the TD1 satellite. An interpolated model atmosphere of T_{eff} = 21,500 K, log g = 3.7 predicts the observed continuous energy distribution from $\lambda\lambda 2200$–7500, the observed profiles of $H\gamma$ and $H\delta$, and the ionization equilibria of Si III/Si IV and S II/S III. With the exception of neon, chlorine, and argon, the derived abundances are within 0.2 dex of the currently accepted solar values.

114.353 **Identification of novel molecules in the spectrum of 19 Piscium.** R. S. Wojslaw, B. F. Peery, Jr.
Astrophys. Journ., Suppl. Ser., Vol. 31, 75 - 92 = Publ. Goethe Link Obs., Indiana Univ., *Bloomington*, No. 176 (1976).

　　　　In the violet spectrum of 19 Psc (N0; C7, 3) between $\lambda 398$ nm and $\lambda 434$ nm the authors attribute approximately 200 absorption features wholly or in part to lines of numerous bands of CuH, ZnH, GeH, and SnH. The character of one of the bands of CuH is pressure-dependent, and from it they infer log $P_{gas} \approx 2$, appropriate to a cool supergiant of log $g \approx 0$.

114.354 **The ultraviolet spectrum of Beta Lyrae. II.**
　　　　M. Hack, J. B. Hutchings, Y. Kondo, G. E. Mc-Cluskey, M. K. Tulloch.
Astrophys. Journ., Vol. 206, 777 - 784 (1976).

　　　　The far-ultraviolet spectrum of the eclipsing binary system β Lyrae was observed in 1974 with the Copernicus Princeton University spectrometers. The spectrum is completely dominated by low-excitation level emission lines of multi-ionized atoms. The binary is surrounded by a high-temperature plasma giving rise to the emission lines which are believed to be collisionally excited. The mass exchange in this system is nonconservative.

114.355 **Spectral line identification studies of some stars with helium anomalies.** B. Campbell, R. F. Garrison.
Canadian Journ. Phys., Vol. 53, 2170 - 2182 (1975). – Abstr. in Phys. Abstr., Vol. 79, A023786 (1976). – Concerning HD 36256, HD 37140, HD 37321, and HD 37479.

114.356 **The emission-line stars FG and FH Aquilae.**
　　　　N. Sanduleak.
Inform. Bull. Variable Stars, (I.A.U. Commission 27), Konkoly Obs., Budapest, No. 1093 (1976).

114.357 **The photometric activity of o And from October '75 to January '76.**
M. Bossi, G. Guerrero, L. Mantegazza.
Inform. Bull. Variable Stars, (I.A.U. Commission 27), Konkoly Obs., Budapest, No. 1095, 4 pp. (1976).

114.358 CD $-30°5135$. G. Welin.
　　　　Inform. Bull. Variable Stars, (I.A.U. Commission 27), Konkoly Obs., Budapest, No. 1139, 2 pp. (1976).

114.359 o **Andromedae.**   T. Bolton, A. Gulliver.
　　　　IAU Circ., No. 2899 (1976).

114.360 **New carbon stars in several fields of the sky.**
　　　　Z. Alksne, V. Ozoliņa.
Investigation of the sun and red stars. 4, (see 003.009), p. 5 - 11 (1976). In Russian.

　　　　13 new carbon stars have been found in eight fields. Their positions, spectral types, V magnitudes, and identification charts are given.

114.361 **A spectroscopic study of β Monocerotis A.**
　　　　A. Ringuelet, C. Scardamaglia, R. H. Méndez.
Inform. Bull. Southern Hemisph., No. 27, p. 29 - 32 (1975).

114.362 **Nota sobre movimientos en la atmósfera de η Leonis.**
　　　　M. Lopez Arroyo, R. Velasco.
Bol. Astron. Obs. Madrid, Vol. 9, No. 2, p. 37 - 43 (1976).

　　　　Radial velocities from the Balmer lines and from lines of different atoms and ions in η Leonis have been computed to look for relative movements in different depths of the star's atmosphere. These movements have not been observed from the metallic lines, but the Doppler shifts of the Balmer lines seem to indicate an expanding atmosphere of hydrogen with a radial velocity of about -2.4 km/s.

114.363 **Profiles of the Rb I resonance lines in the Arcturus spectrum.**   D. L. Lambert, R. E. Luck.
Observatory, Vol. 96, 100 - 104 (1976).

　　　　High resolution photoelectric line profiles of the Rb I resonance lines at 7800 and 7947 Å in the spectrum of Arcturus are analyzed. Good agreement between the observed and synthesized profiles confirms an earlier identification. An attempt to extract the isotopic abundance ratio ^{85}Rb/^{87}Rb shows that this ratio is terrestrial with a large uncertainty.

114.364 **Ionized manganese in the infrared spectrum of Eta Carinae.**   A. D. Thackeray, R. Velasco.
Observatory, Vol. 96, 104 - 105 (1976).

114.365 **The spectrum of the peculiar A star HD 25354.**
　　　　D. M. Pyper.
Astrophys. Journ., Suppl. Ser., Vol. 31, 249 - 269 (1976).

　　　　Line identifications for the Sr-Cr-Eu Ap star HD 25354 show very strong lines of Si II and Sr II. Many strong lines of Ti II, Cr II, Mn II, and Fe II are also present. Hg and Pt are probably present, making HD 25354 one of four Sr-Cr-Eu stars in which these elements are identified.

114.366 **Spectrophotometry of three stars of the β Cep type.**
　　　　V. S. Popov.
Mem. Soc. Astron. Italiana, Vol. 45, (see 012.017), 791 -792 (1974).

114.367 **Identification of λ 7887 band of C_2H_2 in the spectra of two carbon stars 19 Psc and Y CVn.**
M. Hirai.
Proc. Japan Acad., Vol. 50, 743 - 746 (1974). – Abstr. in Phys. Abstr., Vol. 79, A032957 (1976).

114.368 **Photoelectric spectrophotometry of Wolf-Rayet stars.**   J. D. R. Bahng.
Separate print Northwestern Univ., Evanston, Illinois, USA, Dept. Astron., 15 pp. (1974). – See 13.114.019.

114.369 **Helium emission in the spectrum of κ Canis Majoris.**
　　　　J. D. R. Bahng, E. Hendry.
Separate print Northwestern Univ., Evanston, Illinois, USA, Dept. Astron., 6 pp. (1974). – See 13.114.323.

114.370 **Time variations of UV emission features of Be stars.**

Final report, 1 August 1972 – 31 January 1975.
J. D. R. Bahng.
Separate print Northwestern Univ., Evanston, Illinois, USA,
Dept. Astron., 35 pp. (1975).

114.371 Far ultraviolet spectrum of Arcturus.
W. R. McKinney.
Thesis Johns Hopkins Univ., Baltimore, Maryland, USA, 154
pp. (1974). (Available from: Univ. Microfilms, Order No.
75-12, 991).

The chemical compositions of two subgiant CH stars.
See Abstr. 064.017.

Properties of the chromosphere-corona transition
region in Capella.   See Abstr. 064.034.

Mass-loss from ultraviolet observations of ζ Puppis.
See Abstr. 064.064.

The low excitation spectrum of the peculiar star
with infrared excess RX Puppis in 1972 and 1975.
See Abstr. 113.009.

Photometry of the Ap-star HD 124224.
See Abstr. 113.017.

Simultaneous observations of σ Ori E.
See Abstr. 113.021.

Spectroscopic and photometric investigation of the
A21a-O star HD 160529.   See Abstr. 113.049.

Linear polarization measurements at Hβ of early-type
emission line stars.   See Abstr. 114.002.

Copernicus observations of Betelgeuse and Antares.
See Abstr. 114.016.

HR 7129: A helium variable with a large magnetic
field.   See Abstr. 116.001.

Some spectral peculiarities of the magnetic variable
star β CrB.   See Abstr. 116.007.

HD 196673: a binary system containing a barium
star.   See Abstr. 118.007.

The secondary of U Sagittae.   See Abstr. 121.064.

Spectra and Fabry-Perot interferometry of AG
Carinae and the nebula.   See Abstr. 122.064.

Ultraviolet photometry of Eta Carinae and its interpretation.   See Abstr. 131.036.

Far-ultraviolet extinction in σ Scorpii.
See Abstr. 131.074.

Spectroscopic properties of HZ Herculis in model
calculations.   See Abstr. 142.110.

Errata

114.901 Errata: "Observations of near-infrared C_2 bands in
the spectra of carbon stars" [Astrophys. Journ.,
Vol. 199, 145 - 147 (1975)].   D. P. Gilra.
Astrophys. Journ., Vol. 203, 770 (1976).

114.902 Errata: 'Observations of the Mg II lines near 2800 Å
from Copernicus' [Astrophys. Journ., Vol. 199,
110 - 119 (1975)].   Y. Kondo, J. L. Modisette, G. W. Wolf.
Astrophys. Journ., Vol. 205, 308 (1976).

114.903 Erratum: 'Destruction and production of molecules
in circumstellar regions of T Tauri stars' [Astrophys.
Space Sci., Vol. 32, 297 - 304 (1975)].   G. F. Gahm.
Astrophys. Space Sci., Vol. 41, 253 (1976).

115 Stellar Luminosities, Masses, Diameters, HR-Diagrams and Others

115.001 **Stellar angular diameters and visual surface brightness − I. Late spectral types.**
T. G. Barnes, D. S. Evans.
Monthly Notices Roy. Astron. Soc., Vol. 174, 489 - 502 (1976).

Numerous stellar angular diameters found at occultation permit the relationship between surface brightness and colour index to be defined for stars as late as M8. The best relationship is found for the index $(V-R)$ and is well defined for the entire range of stellar temperatures, without dependence on luminosity class. The relationship is valid for M, S and C stars. The relationship is compared with other calibrations of effective temperature and bolometric correction. A first discussion of an application to a Mira is given with promising results. If correct, the conjecture has extremely wide applicability and intensive work on this is in progress.

115.002 **Stellar angular diameters and visual surface brightness − II. Early and intermediate spectral types.**
T. G. Barnes, D. S. Evans, S. B. Parsons.
Monthly Notices Roy. Astron. Soc., Vol. 174, 503 - 512 (1976).

The relationship between visual surface brightness and the colour index $(V-R)$ found in Paper I (see 115.001) is considered in greater detail in its applications to early type stars. The expectation that the relationship is independent of luminosity class (which is not the case for $(B-V)$) is borne out. The validity of the conjecture that the relationship applies to a variable star during its cycle is strengthened. An investigation of the applicability to δ Cep produces an angular diameter curve in excellent phase concordance with the visual light curve and the radial displacement and a mean angular diameter in excellent agreement with stellar atmosphere models.

115.003 **The main sequence defined by trigonometrical parallaxes.** R. Woolley.
Monthly Notices Roy. Astron. Soc., Vol. 174, 621 - 626 (1976).

Trigonometrical parallaxes with probable errors less than 15 per cent of the quantity sought as given in Roy. Obs. Ann., No. 5 (the Catalogue of stars within 25 pc of the sun) define a main sequence of stars of luminosity class V which stands on its own feet but may be compared with, and lies within about a fifteen hundredths of a magnitude above, the Hyades main sequence as given by Johnson & Knuckles.

115.004 **Digital analysis of speckle photographs: the angular diameter of Arcturus.** S. P. Worden.
Publ. Astron. Soc. Pacific, Vol. 88, 69 - 72 (1976).

Digital reduction of speckle interferometry data for the K2 III giant α Boo (Arcturus) has been done. Angular diameters of $0\rlap{.}''019 \pm 0\rlap{.}''006$ were obtained at 4200 Å for a uniformly illuminated disk, and $0\rlap{.}''027 \pm 0\rlap{.}''010$ for a highly limb-darkened disk. Arcturus is underresolved using the Kitt Peak National Observatory 4-meter telescope. The digital-reduction scheme described here proved ideal for studying such underresolved objects. A discussion of limitations of this technique is also provided.

115.005 **Angular diameters of stars from lunar occultations.**
C. de Vegt.
Astron. Astrophys., Vol. 47, 457 - 459 (1976).

Angular diameters of 6 stars from photoelectric lunar occultation observations are reported, which have been obtained during the period 1969−1975 with the 60 cm refractor

of the Hamburg observatory.

115.006 **Stellar population samples at the galactic poles. IV. Luminosity function for the M-type dwarfs at the south pole.** O. J. Eggen.
Astrophys. Journ., Vol. 204, 101 - 112 (1976).

The $(UBVRI)$ photometry of all M dwarfs which are within 10° of the south galactic pole and brighter than visual magnitude 15, and which have annual proper motions greater than $0\rlap{.}''096$, are discussed. The luminosity function is found to be very similar, in the overlapping sections, to that previously derived spectrophotometrically from the M stars near the sun, and the extension to M_V near +13 mag indicates that this luminosity is near the peak of that function. No support is found in these data for the recently suggested superabundance of low velocity M stars near the sun.

115.007 **Gaps in the blue horizontal branch.**
B. Newell, J. A. Graham.
Astrophys. Journ., Vol. 204, 804 - 809 (1976).

In this paper the authors present additional evidence in support of Newell's (1973) claim that gaps exist in the blue horizontal-branch sequence of the high-latitude faint blue stars. They find that, for 102 stars selected to be predominantly horizontal-branch stars, there are two gaps (or regions of low number density) in the $[c_1, (b-y)]$-diagram; these gaps occur at colors which correspond to $\log T_{eff} \approx 4.11$ (gap 1) and $\log T_{eff} \approx 4.34$ (gap 2). The data lead to the conclusion that several distinct groups of stars exist on the field star horizontal-branch sequence.

115.008 **Remarks on some problems related to O stars.**
G. Goy.
Astron. Astrophys., Vol. 48, 87 - 91 (1976).

The spectral types of O stars are known with greater precision due to the recent works of Walborn and Conti (1970 to 1974). However, this is not sufficient to perceptibly improve the determination of the distances of O stars. The author studies several parameters which determine or influence the distance-modulus, and he shows that the duplicity or multiplicity of stars could play a more important role than has been established until now.

115.009 **Intermediate mass stars in the mass-luminosity diagram.** P. Laques, R. Despiau.
Astron. Astrophys., Vol. 48, 101 - 108 (1976). In French.

The present work consists of plotting in the mass-luminosity diagram a statistical sample of 110 double stars, with $\Delta m \leq 0.3$, from the catalog of Finsen et al. (1970). Assuming that the masses of the components are equal when the binary under consideration has a Δm equal (or nearly) to zero, the authors have determined the individual mass of each component by Kepler's third law. Luminosities, or bolometric magnitudes, are determined from the absolute magnitudes obtained from the color index $(B-V)$, incorporating a corrective term which depends on spectral type. Parallaxes are systematically derived from the color indices obtained in the UBV system (photometric parallaxes). The authors have compared their results with the theoretical M-L relation for main-sequence stars and found a good agreement between observation and theory.

115.010 **Angular diameter of 31 Leonis from a lunar occultation.** I. S. Glass, L. V. Morrison.
Monthly Notices Roy. Astron. Soc., Vol. 175, 57P - 59P (1976).

The angular diameter of the K-type giant 31 Leonis is found to be $0\overset{''}{.}0031 \pm 0\overset{''}{.}0006$ by analysis of the light curve at a lunar occultation. The result suggests that the surface brightness is slightly greater than that expected for a normal star with the same value of $(B-V)_0$.

115.011 A new calibration of Wesselink's correlation for the determination of stellar radii.
M. Marcocci, I. Mazzitelli.
Astrophys. Space Sci., Vol. 40, 141 - 148 (1976).

By means of a new homogeneous set of data from 34 double-lined detached systems, the authors have recalibrated the colour-surface brightness correlation in the interval $-0.3 \leqslant (B-V)_0 \leqslant 0.95$. A comparison is made with Wesselink's calibration. The authors find that Wesselink's method can provide a good estimate of the stellar radii.

115.012 Absolute magnitudes of stars from widths of chromospheric Ca II emission lines. O. C. Wilson.
Astrophys. Journ., Vol. 205, 823 - 840 (1976).

Absolute magnitudes of about 700 late-type subgiants, giants, and supergiants are derived from measures of the widths of the chromospheric emission components of the Ca II K line. They are tabulated along with other pertinent information. The data are essentially complete for all suitable stars down to $m_v = 5.0$ and declination $-25°$.

115.013 On mass determination from T_{eff}, g, and M_{bol}.
M. S. Vardya.
Astrophys. Space Sci., Vol. 41, L1 - L2 (1976).

The mass deduced, using photometrically determined g, gives a lower limit to the mass. Furthermore, mass determination in giant and supergiant stars may be effected by g and T_{eff} being determined for different values of the stellar radius.

115.014 A composite Hertzsprung-Russell diagram for the peculiar red giants. J. M. Scalo.
Astrophys. Journ., Vol. 206, 474 - 489 (1976).

A composite H–R diagram for the peculiar red giants of the disk population is constructed using the available data for stars of types R, N, S, SC, MS, and Ba, along with theoretical evolutionary tracks, The N, S, SC, and MS stars approximately coincide with the sequence of normal M giants in the H–R diagram and can be identified with the helium shell flash phase of evolution. The early R stars and most Ba stars occupy the same region of the H–R diagram as normal K giants; the positions of these stars cannot be reconciled with current stellar evolution calculations.

115.015 Luminosity functions and the evolution of low-mass population I giants. B. M. Tinsley, J. E. Gunn.
Astrophys. Journ., Vol. 206, 525 - 535 (1976).

Luminosity functions in terms of bolometric magnitudes are constructed for M67 and for two samples of old-disk field giants. These are compared with theoretical rates of evolution on the giant branch. M67 has too few stars to give a useful comparison. The field giants show good agreement with theory, and the number of stars at the "clump" suggests that core helium-burning is prolonged by overshoot with semiconvective mixing. The fuel consumption derived from the luminosity functions is consistent with core helium ignition at the theoretically predicted core mass, and with a final core mass in agreement with observed white dwarf masses.

115.016 Star counts in the background sky observed from Pioneer 10. H. Tanabe, K. Mori.
IAU Colloquium No. 31, (see 012.015), p. 36 - 40 (1976).

115.017 A few applications carried out with the files of the Stellar Data Center. A. Heck.
Centre Données Stellaires, Inform. Bull. No. 10, p. 14 - 16 (1976).

Project Daedalus: astronomical data on nearby stellar systems. See Abstr. 051.002.

Results of photoelectric lunar occultation observations obtained at the Hamburg Observatory during 1969–1973. See Abstr. 096.003.

The bolometric absolute magnitude of S type stars and lithium production. See Abstr. 113.007.

Empirical effective temperatures and bolometric corrections for early-type stars. See Abstr. 114.004.

Stellar winds from hot supergiants. See Abstr. 114.006.

Absolute fluxes of K chromospheric emission on the H–R diagram. See Abstr. 114.018.

The temperature, luminosity, and spectrum of Kapteyn's star. See Abstr. 114.325.

The luminosity of the very red supergiant near the cluster Tr 27. See Abstr. 114.336.

Double star masses determined by synchronous stellar area scanning. See Abstr. 117.035.

The distribution of WR stars in galactic clusters and associations and the absolute magnitudes of WR stars. See Abstr. 153.013.

Stars within a radius of 20 light years from the sun. See Abstr. 155.059.

Effects of main-sequence brightening on the luminosity evolution of elliptical galaxies. See Abstr. 158.006.

A catalogue of A- and F-type supergiants in the Large Magellanic Cloud. See Abstr. 159.005.

Errata

115.901 Erratum: 'Apparent radii and other parameters for 116 A5V–F5V stars of the catalogue of Geneva Observatory' [Astrophys. Space Sci., Vol. 35, 313 - 320 (1975)]
M. Fracassini, L. E. Pasinetti, G. Pelagatti.
Astrophys. Space Sci., Vol. 39, 507 (1976).

116   Stellar Magnetic Field, Figure, Rotation

116.001   HR 7129: a helium variable with a large magnetic field.   R. J. Wolff, S. C. Wolff.
Astrophys. Journ., Vol. 203, 171 - 176 (1976).

Zeeman spectroscopic observations show that the helium variable HR 7129 has a longitudinal magnetic field that varies between the limits +7000 gauss and −5000 gauss in a period of 3.670 days. The spectrum and radial velocity variations can be accounted for in terms of the oblique dipole rotator model, as can the conspicuous crossover effect that is also observed in this star. Since the temperature of HR 7129 is approximately 20,000 K, this object represents an extension to much higher temperature of the phenomena that are observed in the classical, magnetic Ap stars.

116.002   The H II region of a rapidly rotating O-star.   R. Kippenhahn, E. Krügel.
Astron. Astrophys., Vol. 46, 179 - 183 (1976).

The effects of rotation on the shape of an H II region around an O star are discussed for the maximum possible (uniform) angular velocity. It is found that for constant density nebulae the polar radius r_p of the H II region may be more than 1.6 times the equatorial radius r_e. If the density peaks towards the centre like r^{-2}, the ratio r_p/r_e may go up to five. The effect is even more pronounced for the He II region. Furthermore, the relation between the total stellar luminosity and the Lyman continuum output is discussed for critically rotating O-stars.

116.003   The period of the magnetic star HD 133 029.   K. Panov, W. Schöneich.
Astron. Nachr., Vol. 297, 33 - 38 (1976). − In German.

The period of 0.741285 days for the light variability of the magnetic star HD 133029 was obtained from UBV observations. The observations of the effective magnetic field by Babcock show variations with a period of 0.7447 days. A small change of the period and a slow change of the magnitude of this magnetic star seems to be present.

116.004 $V \sin i$ values in the far ultraviolet.   J. B. Hutchings.
Publ. Astron. Soc. Pacific, Vol. 88, 5 - 7 = Contr. Dominion Astrophys. Obs., *Victoria*, No. 301 = NRC No. 14988 (1976).

A simple treatment of gravity and limb darkening shows that measured $V \sin i$ values for OB stars are significantly lower for rapid rotators in the far ultraviolet than in the photographic region of the spectrum. This circumstance should help in determining V, i, ω/ω_{crit}, and gravity darkening for these stars.

116.005   The magnetic field and period of the Ap star 45 Leonis.   W. K. Bonsack.
Publ. Astron. Soc. Pacific, Vol. 88, 19 - 21 (1976).

The magnetic field of 45 Leo has been measured in 16 Mauna Kea spectrograms, yielding a range from −230 to + 400 gauss. A period near 7$\overset{d}{.}$9 is found to represent both the magnetic-field data and the photometry of Wolff and Morrison, but not that of Winzer. It is pointed out that the period permits a rigid rotator model for 45 Leo.

116.006   Possible evidence for the occurrence of magnetic fields of order 10 kilogauss in the red dwarf star BY Draconis.   D. J. Mullan, R. A. Bell.
Astrophys. Journ., Vol. 204, 818 - 829 (1976).

Koch and Pfeiffer have reported the presence of linear polarization in broad-band observations of the light from the spotted dwarf BY Dra. The authors compute the polarization expected in a cool ($\theta_{eff} = 1.5$) dwarf observed in $UBVR$ filters.

Comparing the results with observations of BY Dra, the authors find that the surface field strength on this star is at times of order 10 kilogauss. They suggest that the presence of fields of order 10 kilogauss in an old bipolar region is consistent with a recent model of starspots on a cool dwarf.

116.007   Some spectral peculiarities of the magnetic variable star β CrB.   N. S. Polosukhina.
Izv. Krymskoj Astrofiz. Obs., Vol. 54, 120 - 127 (1976). In Russian.

The equivalent widths of Eu II and Ca I lines in the spectra of the magnetic star β CrB were found to be variable. The author tries to reveal a connection between the variations of magnetic field and spectral line intensities using the oblique rotator model. Eu II and Ca I show apparent concentrations in the surface region with positive magnetic polarity. The existence of a second region of Eu II and Ca I concentration connected with negative magnetic polarity is proposed.

116.008   Intrinsic linear polarization of Be stars as a function of $V \sin i$.   R. Poeckert, J. M. Marlborough.
Astrophys. Journ., Vol. 206, 182 - 195 (1976).

The polarization of 48 Be stars has been measured in two bands near Hα with the aim of determining the relation between intrinsic polarization and $V \sin i$. It is found that intrinsic polarization depends strongly on $V \sin i$: stars with low $V \times \sin i$ have little or no polarization. No apparent difference between pole-on and extreme Be stars was obtained. A comparison of the observed relation between the intrinsic polarization and $V \sin i$, and that predicted for an optically thin disk viewed at varying angle i is illustrated, and it is found that an envelope with an electron density of $\leq 5 \times 10^{11}$ cm^{-3} can account for the degree of intrinsic polarization observed in all the program stars.

116.009   The magnetic field of HD 101065.   S. C. Wolff, W. Hagen.
Publ. Astron. Soc. Pacific, Vol. 88, 119 - 121 (1976).

Zeeman spectroscopic observations show that HD 101065 (Przybylski's star) has a magnetic field strength of about −2200 gauss. The limited observations available show no evidence of magnetic or spectrum variability. Observations of the Am star δ Nor were made in order to determine whether systematic errors were present in the determination of H_e, and these data yield an upper limit of 150 gauss for the field of δ Nor.

116.010   Rotating stellar models according to the quasi-dynamic method.
A. Kovetz, G. Shaviv, S. Zisman.
Astrophys. Journ., Vol. 206, 809 - 814 (1976).

There exists a quasi-dynamic method for calculating the axisymmetric equilibrium configuration of a rotating body with prescribed distributions of specific entropy s and axial component of angular momentum. The equation of state is quite general: $P = P(\rho, s)$. The method incorporates a test for dynamical stability that would otherwise have to be performed separately.

Applications of Fourier analysis to broadening of stellar line profiles. II. A saturated line in 32 Aquarii and in Sirius.   See Abstr. 064.004.

Empirical model atmospheres for explanation of the light variations of magnetic stars.   See Abstr. 064.063.

Force-free stellar magnetospheres.

See Abstr. 065.002.

The evolution of magnetic stars.
See Abstr. 065.049.

Spot analysis of rigidly rotating Ap stars.
See Abstr. 065.059.

Angular momentum transport by tidal acoustic wave.
See Abstr. 065.067.

A new theory on the generation of the mangetic field of celestial bodies. III. Aspects of application.
See Abstr. 065.103.

Magnetic properties of neutron-star matter.
See Abstr. 065.109.

Rotational-velocity effects in the *uvby*, β systems and on MK spectral types for B stars. See Abstr. 113.001.

Investigation of light variation of twelve Ap stars in ten spectral regions. See Abstr. 113.014.

Long-term light variations in magnetic stars.
See Abstr. 113.046.

Photoelectrical investigation of the magnetic field and the magnitude of the Am stars 15 Vul and 68 Tau.
See Abstr. 113.047.

On the light variability of Ap stars. I. Effects of surface variations of T_{eff} and g_{eff}.
See Abstr. 114.024.

Results of spectrophotometric studies of the magnetic and peculiar stars 21 Per, β CrB, γ Boo, γ UMi.
See Abstr. 114.054.

On an investigation of the magnetic variable Ap star 17 Com A. See Abstr. 114.337.

Photoelectric observations of β Coronae Borealis in 1970–1971. See Abstr. 122.078.

W Sgr-magnetic field measurements and Strömgren photometry. See Abstr. 122.138.

117 Binary and Multiple Stars, Planetary Companions, Theory

117.001 On the formation of disk around a compact object by two-body tidal encounter. S. Yabushita.
Monthly Notices Roy. Astron. Soc., Vol. 174, 19P - 20P (1976).

Fabian, Pringle & Rees recently argued that close binary systems may be formed by two-body tidal interaction, and that these may account for observed X-ray sources near the centres of globular clusters. In this note it is shown, by making use of Woolfson's results for capture processes, that through tidal interaction, disks can be formed around single compact objects which may emit thermal X-rays for $10^5 \sim 10^6$ yr.

117.002 Effects of tidal distortion on binary-star velocity curves and ellipsoidal variation.
R. E. Wilson, S. Sofia.
Astrophys. Journ., Vol. 203, 182 - 186 (1976).

Radial velocity curves for the more massive components of binaries with extreme mass ratios can show a large distortion due to tides. Binaries in which the effect is large should be rare because nearly all such binaries would be in the rapid phase of mass transfer. However, the optical counterparts of some X-ray binaries may show the effect, which would then serve as a new means of extracting considerable information from the observations. The essential parts of the computational procedure are given. Light curves for ellipsoidal variables with extreme mass ratios were also computed, and were found to be less sinusoidal than those with normal mass ratios.

117.003 Effect of asymmetric explosion on orbital elements of circular binaries. R. Mitalas.

Astron. Astrophys., Vol. 46, 323 - 325 (1976).

A complete, analytic solution for the two body problem of an asymmetric, instantaneous explosion, where both the mass and velocity of each component are altered, is presented. The relations between the periods, semi-major axes, and mass functions of the originally circular and present binaries are given. For PSR 1913 + 16 limits are set on the velocity increments due to the explosion.

117.004 On the astrometric detection of neighboring planetary systems. G. Gatewood.
Icarus, Vol. 27, 1 - 12 (1976).

The range of masses detectable by the astrometric technique in a given hypothetical planetary system is a function of the primary's mass, the system's distance, the orbital periods of its major planets, and the external precision of the positional data. To date, in the opinion of the author, no suggested extrasolar substellar mass (mass $< 0.01 \, M_\odot$) has been confirmed, although some proposed systems still await further study.

117.005 On the possibly overluminous triple star, 13 Ceti. G. Gatewood, S. Sofia.
Publ. Astron. Soc. Pacific, Vol. 88, 50 - 51 (1976).

Values derived for the masses, luminosities, and colors of the principal components of the 13 Cet system are inconsistent with accepted theory. Further observation of the system is suggested.

117.006 An evolutionary model of Beta Lyrae. J. Ziółkowski.

Astrophys. Journ., Vol. 204, 512 - 515 (1976).

The evolution of a binary system with initial parameters $M_1 = 10\,M_\odot$, $M_2 = 3.7\,M_\odot$, $P = 3^d44$ is followed assuming that total mass and orbital angular momentum are conserved. The primary fills its Roche lobe shortly after the exhaustion of hydrogen in its center (case B of binary evolution). After 63,500 years of mass transfer between the components, the parameters of the system match very closely the observational parameters of β Lyrae given by Woolf (1965).

117.007 **Orbital evolution of a singly condensed, close binary, by mass loss from the primary and by accretion drag on the condensed member.**
M. E. Alexander, W. Y. Chau, R. N. Henriksen.
Astrophys. Journ., Vol. 204, 879 - 888 (1976).

The evolution of a close binary system in which one component —the compact one— is embedded in the disturbed and distended atmosphere of the normal companion, is considered. The general equations for the variation of the orbital elements due to the gravitational accretion drag are formulated and then applied to special systems (HD 153919 and Cen X-3). The effects on the tidal stability are also discussed. It is suggested that the kind of system and the evolutionary mechanism considered here can also be related to symbiotic objects.

117.008 **On the masses of cataclysmic binaries.** H. Ritter.
Mitt. Astron. Ges., No. 38, p. 165 - 166 (1976).
Abstract.

117.009 **Synthetic light curves of close binary systems.**
M. Ammann.
Mitt. Astron. Ges., No. 38, p. 225 - 231 (1976). — Short report.

117.010 **Evolution of close binary systems with initial mass of primary 16 $M_\odot$.** Z. T. Krajcheva.
Stellar physics and evolution, (see 003.001), p. 58 - 73 (1974). In Russian.

The evolution of primaries of two close binary systems has been studied. Both double stars are initially composed of a primary of 16 $M_\odot$ and a secondary of 15 $M_\odot$ with separations of 79.44 $R_\odot$ (case I) and 206.8 $R_\odot$ (case II). The primary component fills its Roche lobe when it burns hydrogen in the shell in case I, and helium in the convective core in case II. The Schwarzschild criterion for convective stability was used for the handling of semiconvection. Evolutionary tracks of the primaries are plotted on the H-R diagram. At the end of mass transfer the original primaries have masses equal to 5.37 $M_\odot$ (case I) and 6.65 $M_\odot$ (case II) and separations 241.9 $R_\odot$ and 454.9 $R_\odot$, respectively. The influence of mass loss with a constant rate equal to $\dot{M} = -10^{-6}\,M_\odot$/year and $\dot{M} = -3 \times 10^{-6}$ $M_\odot$/year on the evolution of the remnant of the primary component of case I was investigated.

117.011 **W Ursae Majoris systems with marginal contact.**
L. B. Lucy.
Astrophys. Journ., Vol. 205, 208 - 216 (1976).

Evidence is presented indicating that W UMa binaries of W-type are undergoing thermal relaxation oscillations about a state of marginal contact. This suggestion receives theoretical support from calculations showing that a newly formed contact binary will evolve on a thermal time scale toward a state of marginal contact and that, if contact is then broken, the system will evolve back into contact, again on a thermal time scale. The A-type systems are suggested to be W UMa systems that have been able to achieve thermal equilibrium as a result of nuclear evolution. The relationship of W UMa systems to other short-period binaries is also discussed.

117.012 **A cyclic thermal instability in contact binary stars.**
B. P. Flannery.
Astrophys. Journ., Vol. 205, 217 - 225 (1976).

The author shows analytically that contact binary stars coupled by a common convective envelope, the Lucy model, are almost invariably unstable when subjected to mass transfer: if either component begins to exchange mass, it will continue to do so. The initial instability developes into a cyclic exchange of mass with the mass fraction oscillating between $0.56 \leqslant M_1/(M_1 + M_2) \leqslant 0.62$ with a period of $\sim 10^7$ years. The instability is of a general nature and such oscillating systems can satisfactorily populate the short-period, red region of the period-color relation for W UMa stars.

117.013 **On the masses and the evolution of cataclysmic binaries.** H. Ritter.
Monthly Notices Roy. Astron. Soc., Vol. 175, 279 - 295 (1976).

An evolutionary scheme for cataclysmic binaries (CB's) based on present stellar evolution theories is sketched. According to this scheme, W UMa systems have probably to be excluded as possible progenitors of CB's. However, the typical configuration of a CB could easily be understood as the result of an evolution of type C of a close binary undergoing mass exchange during the primary's red giant phase, connected with substantial losses of mass and angular momentum from the system.

117.014 **Multiplicity among solar-type stars.**
H. A. Abt, S. G. Levy.
Astrophys. Journ., Suppl. Ser., Vol. 30, 273 - 306 (1976).

A search has been made for spectroscopic binaries among 135 F3—G2 IV or V bright field stars. The observed frequencies of singles:doubles:triples:quadruples are 42:46:9:2 percent. It was found to be possible to estimate rather well the number of binaries not revealed by this study. The incompleteness study was based on seven reasonable assumptions and leads to the result that there are really 1.4 companions for each primary star, on the average. This result implies that single stars are rare. The authors conclude that there are two types of binaries: those with periods less than 100 years are fission systems in which a single protostar subdivided because of excessive angular momentum, whereas the longer periods represent pairs of protostars that contracted separately but are gravitationally held to each other. The dividing period of 100 years is such that if two solar masses were distributed over the corresponding volume, the mean density would agree with that assumed for the solar nebula at the time of planetary differentiation.

117.015 **The equilibrium structure of tidally distorted rotating masses in the presence of magnetic fields.**
R. K. Kochhar.
Bull. Astron. Soc. India, Vol. 3, 30 (1975). — Abstract of a paper presented at the A.S.I. meeting 1975.

117.016 **Radial-velocity measurements of the lunar-occultation binary HR 2013.**
R. F. Griffin, H. A. Abt.
Observatory, Vol. 96, 54 - 56 (1976).

117.017 **Precession of the nodes in some triple stellar systems.**
T. Mazeh, J. Shaham.
Astrophys. Journ., (Letters), Vol. 205, L147 - L150 (1976).

Long period variability of apparent binary stellar systems may be caused by the presence of a third companion, via the precession of the nodes phenomenon. The authors show that this phenomenon may indeed have already been observed in the confirmed triplet λ Tauri, as well as in three systems for which a binary structure was confirmed and a possible long period has been observed—Her X-1, HD 217061, and CV Serpentis. Suggestions for some immediate observations of these systems are made.

117.018 Systematic trends in the motions of suspected stellar companions. W. D. Heintz.
Monthly Notices Roy. Astron. Soc., Vol. 175, 533 - 535 (1976).

Orbital elements published as evidence of unseen companions show an unlikely distribution, strongly suggesting systematic errors. No evidence for a real, periodic motion of BD +4°3561 is found.

117.019 The perturbations of G96-45 and G146-72. A. L. Behall, R. S. Harrington.
Publ. Astron. Soc. Pacific, Vol. 88, 204 - 206 (1976).

Each of the star systems, G96-45 and G146-72, has a Keplerian perturbation. G96-45 has a photocentric semimajor axis of $0\overset{''}{.}04$, a period of 7.2 years, and, with an adopted mass of $0.27\odot$ for the primary, the mass of the unseen companion is in the range of $0.10\odot$ to $0.16\odot$. G146-72 has a photocentric semimajor axis of $0\overset{''}{.}03$, a period of 6.7 years, and masses of $0.35\odot$ (adopted) for the primary and possibly $0.16\odot$ for the unseen companion.

117.020 How do binaries arise? Eh. M. Drobyshevskij.
Zemlya i Vselennaya, 1976, No. 3, p. 70 - 76.
In Russian.

117.021 HD 47129: the most massive binary. J. B. Hutchings, A. P. Cowley.
Astrophys. Journ., Vol. 206, 490 - 498 = Dominion Astrophys. Obs., *Victoria*, Contr. No. 283 (1976).

Thirty-two coudé spectrograms of Plaskett's star have been obtained. New orbital elements are presented for lines of different excitation. The mass of each component exceeds 55 $M_\odot$. Neither star fills its Roche limiting surface, but it appears that both lose mass to the system. The secondary is under-luminous for its mass and may be smaller than the primary.

117.022 The accreting component of mass-exchange binaries. R. K. Ulrich, H. L. Burger.
Astrophys. Journ., Vol. 206, 509 - 514 (1976).

The authors have studied the effect of rapid accretion on a moderate-mass stellar model. The surface boundary condition is examined for two modes of accretion — disk and direct impact. It is found that the effective temperature must be increased for disk accretion but not for direct-impact accretion. In the latter mode the excess kinetic energy of the transfer stream is radiated in an optically thick hot spot or in a hot equatorial band. Evolutionary calculations for direct-impact accretion show that the mass-gaining star swells. The present orbital elements of X-ray binaries indicate that these systems went through a contact phase of evolution.

117.023 A universe teeming with planetary systems—perhaps. G. H. Herbig.
Mercury, (Journ. Astron. Soc. Pacific), Vol. 5, No. 2, p. 2 - 7, 31 (1976).

117.024 The nature of optical variability of X-ray binaries Cyg X-2 = V1341 Cyg and Sco X-1 = V818 Sco. V. M. Lyutyj, R. A. Syunyaev.
Astron. Zhurn. Akad. Nauk SSSR, Vol. 53, 511 - 526 (1976). In Russian. English translation in Soviet Astron., Vol. 20, No. 3.

The results of *UBV* observations of Cyg X-2 and Sco X-1 during 1974—1975 are presented. A rapid irregular variability of the optical radiation of Cyg X-2 is detected. The semi-amplitude of the periodical component in the light curve of Cyg X-2 does not exceed $0\overset{m}{.}1$. Such a small amplitude of the periodical component may be a consequence of the existence of an accreting disc around the X-ray source, which contributes sufficiently to the optical brightness of the system.

117.025 Finite size effect in gravitational radiation from very close binary systems. W. Y. Chau.
Astrophys. Letters, Vol. 17, 119 - 121 (1976).

The correction to the gravitational radiation from a very close binary system because of effects of finite size of the components is calculated. It is shown that this would be zero if there is no departure from spherical symmetry, small in "normal" systems but very significant in systems like a white dwarf orbiting about a massive black hole.

117.026 Evolution of carbon-oxygen dwarfs in binary systems. E. V. Ergma, A. V. Tutukov.
Acta Astron., Vol. 26, 69 - 76 (1976).

Evolution of carbon-oxygen dwarf components of binaries is studied. Mass accretion with the rate $\dot{M} < 10^{-7}$ $M_\odot$/year can increase the central density at the start of carbon burning up to $\sim 10^{10}$ g/cm^3. In this way it is possible to form a gravitationally bound body — a neutron star — after the explosion of supernova.

117.027 Remarks concerning the "observed" apsidal constants of semi-detached systems. I. Todoran.
Mem. Soc. Astron. Italiana, Vol. 45, (see 012.017), 817 - 823 (1974).

117.028 Ellipsoid-ellipsoid approximation for close binary stars. V. Ureche.
Mem. Soc. Astron. Italiana, Vol. 45, (see 012.017), 825 - 829 (1974).

117.029 Evolution of massive close binaries and the origin of relativistic objects. A. V. Tutukov, L. R. Yungel'son, Z. T. Kraitcheva.
Mem. Soc. Astron. Italiana, Vol. 45, (see 012.017), 879 - 892 (1974).

117.030 The evolution of a massive close binary up to the X-ray binary stage. C. De Loore, J. P. De Grève, E. P. J. Van den Heuvel, J. P. De Cuyper.
Mem. Soc. Astron. Italiana, Vol. 45, (see 012.017), 893 - 907 (1974).

117.031 Evolutionary aspects of the binary system containing the pulsar 1913+16. T. Daishido, K. Nomoto, D. Sugimoto.
Progr. Theor. Phys., (*Japan*), Vol. 55, 314 - 315 = Tokyo Astron. Obs. Repr. No. 489 (1976).

117.032 Mass accretion onto compact stars and hydromagnetic transfer of angular momentum in close binary systems. H. Inoue.
Publ. Astron. Soc. Japan, Vol. 28, 293 - 305 (1976).

Mass accretion onto a compact star in a close binary system is studied taking account of the magnetohydrodynamic transfer of angular momentum. The material from the companion star accretes onto the compact star with a velocity close to the free fall. The excess angular momentum is transferred inward onto the compact star by the magnetic lines of force. Therefore, the compact star spins up and a certain period of rotation is realized, at which the solution for the steady flow exists marginally. The case with parallel polarities of the magnetic fields is discussed qualitatively. In this case the accretion is shown to take place intermittently.

117.033 Present status of apsidal motion observations. T. J. Herczeg.
Proc. Southwest Regional Conf., Vol. 1, (see 012.021), 79 - 88 (1976).

A review of the available observational evidence for apsidal motion is presented, with Alpha Virginis and the white-dwarf binary system BD+16° 516 discussed in more detail.

117.034   Evolutionary aspects of the binary system containing
          the pulsar 1913+16.
T. Daishido, K. Nomoto, D. Sugimoto.
Progr. Theor. Phys., Vol. 55, 314 - 316 (1976). — Abstr. in
Phys. Abstr., Vol. 79, A042312 (1976).

117.035   Double star masses determined by synchronous
          stellar area scanning.   B. Atwood.
Thesis Wesleyan Univ., Middletown, Connecticut, USA,
176 pp. (1975). (Available from: Univ. Microfilms, Order No.
75-22,996).

117.036   Formation of a contact binary star system.
          E. F. F. Mullen.
Thesis Florida Univ., Gainesville, USA, 154 pp. (1974).
(Available from: Univ. Microfilms, Order No. 75-16,425).

Feasibility of a search for planets around solar-type
stars with a polarimetric radial velocity meter.
See Abstr. 031.201.

Synchronization in binaries and age.
See Abstr. 065.008.

Slow mass transfer in semidetached binaries.
See Abstr. 065.021.

A binary hypothesis for the subdwarf B stars.
See Abstr. 065.024.

Evolution of massive close binaries and formation
of neutron stars and black holes.   See Abstr. 065.058.

On the detectability of gravitational waves from
W Ursae Majoris binary stars.   See Abstr. 066.003.

Interaction of neutron stars with black holes.
See Abstr. 066.004.

A test of post-Newtonian conservation laws in the
binary system PSR 1913+16.   See Abstr. 066.035.

Photometric evidence for the binary nature of Wolf-
Rayet stars.   See Abstr. 113.051.

Properties and nature of shell stars. 6. The shell star
4 Herculis as an interacting binary.   See Abstr. 114.306.

Intermediate mass stars in the mass-luminosity
diagram.   See Abstr. 115.009.

Four-colour light curves of the eclipsing binary TY
Mensae.   See Abstr. 121.016.

An elementary theory of eclipsing depths of the light
curve and its application to Beta Lyrae.   See Abstr. 121.020.

Luminosities of the W UMa-type systems.
See Abstr. 121.113.

Radio emission from the Wolf-Rayet binary γ^2
Velorum.   See Abstr. 141.012.

A search for radio emission from stars.
See Abstr. 141.016.

Determining the stellar masses in the binary system
containing the pulsar PSR 1913+16: is the companion a helium
main-sequence star?   See Abstr. 141.305.

Tidal friction in the binary pulsar system PSR
1913 + 16.   See Abstr. 141.306.

Observational constraints on pulsar binary motion.
See Abstr. 141.323.

The binary pulsar: preexplosion evolution.
See Abstr. 141.333.

Arrival-time analysis for a pulsar in a binary system.
See Abstr. 141.334.

Further observations of the binary pulsar PSR
1913+16.   See Abstr. 141.339.

New relativistic laboratory — a pulsar in a binary
star.   See Abstr. 141.359.

X-ray binaries and asymmetry of supernova explosions.   See Abstr. 142.020.

On X-radiation of double systems containing Wolf-
Rayet-type stars.   See Abstr. 142.023.

The formation of binaries containing black holes by
the exchange of companions and the X-ray sources in globular
clusters.   See Abstr. 142.026.

Spectrophotometric observations of the X-ray binary
HD 153919 = 3U 1700-37.   See Abstr. 142.038.

The long-term intensity behavior of Centaurus X-3.
See Abstr. 142.041.

Black holes in X-ray binaries: marginal existence
and rotation reversals of accretion disks.   See Abstr. 142.044.

On X-ray emission of early massive stars in close
binary systems.   See Abstr. 142.070.

Stellar winds and accretion in massive X-ray binaries.
See Abstr. 142.107.

X-ray binaries.   See Abstr. 142.161.

Sorgenti di raggi X in sistemi binari.
See Abstr. 142.220.

Parameters of X-ray binaries.
See Abstr. 142.236.

Production of triple stars by the dynamical decay
of small stellar systems.   See Abstr. 151.014.

Some results of an investigation of star duplicity in
the T-associations Tau T1, Tau T2, Tau T3 and Ori T2.
See Abstr. 152.005.

118 Visual Double and Multiple Stars

118.001 Two orbits for the visual binary star ADS 3395.
R. H. Wilson.
Monthly Notices Roy. Astron. Soc., Vol. 174, 75P - 79P, with
a correction, Vol. 175, 653 (1976).

Two possible short-period orbits of high inclination, the
first highly eccentric and the second circular, are shown here
for the first time to represent all observations of the visual
binary star ADS 3395 almost equally well. However, the relative radial velocity ephemeris shows that, at the next occultation due to closeness in 1978, Orbit I predicts 53 km s^{-1} while
Orbit II predicts zero, so the orbital ambiguity could be resolved by spectroscopic observations at that time.

**118.002 Measurements of double stars at Meudon (3rd and
last series).**   P. Muller.
Astron. Astrophys., Suppl. Ser., Vol. 23, 205 - 221 (1976).
In French.

451 observations of 249 objects at the 83 cm Meudon
refractor are reported.

118.003 On the nature of the triple-star system 13 Ceti.
G. Gatewood, A. L. Behall.
Astron. Journ., Vol. 80, 1065 - 1070 (1975).

The triple-star system 13 Ceti (ADS 490) appears to be
overluminous when compared with the empirical mass-luminosity relation. One possible explanation is that the masses
may be in error because of a poorly determined parallax. The
present paper redetermines the parallax of the system from 57
Allegheny plates and yields a value of $\pi_{\text{abs}} = 0.''060 \pm 0.''005$.
A reanalysis of the visual orbit has also been carried out and
yields, in part, $P = 6.92 \pm 0.01$ yr and $a = 0.''23 \pm 0.''01$. These
values lead to a total mass, (1.1 ± 0.2) M$_\odot$, which appears to
confirm the overluminosity of the system.

**118.004 UBV photoelectric photometry of the components
of 20 young stellar multiple systems.**
M. L. Burnichon, R. Garnier.
Astron. Astrophys., Suppl. Ser., Vol. 24, 89 - 93 (1976).
In French.

UBV photoelectric photometry of the components of 20
young stellar multiple systems is presented. These multiple
systems are composed of a bright very hot and massive star
(an O star or a blue supergiant) surrounded by less massive
main-sequence stars of type B or sometimes A. They were
selected for a more general study of intrinsic properties of high
luminosity blue stars (Burnichon, 1975).

118.005 The spectroscopic orbit of δ Andromedae.
G. A. Bakos.
Journ. Roy. Astron. Soc. Canada, Vol. 70, 23 - 26 (1976).

From radial velocity determinations made over the last
75 years, the period and the orbital elements of δ And have
been derived.

118.006 Red Sirius.   D. Ya. Martynov.
Zemlya i Vselennaya, 1976, No. 1, p. 36 - 39.
In Russian.

**118.007 HD 196673: a binary system containing a barium
star.**   R. B. Culver, P. A. Ianna.
Publ. Astron. Soc. Pacific, Vol. 88, 41 - 43 (1976).

Photometric and spectroscopic observations are presented
of both members of the binary star system HD 196673 (ADS
14078), whose primary component is a barium star of spectral
class K0-Ba(2). The absolute visual magnitude for the barium
star primary is found to be −0.7 and the distance to the system
is 0.35 kpc. The mass of the barium star is found to lie in the

interval $1.5\, M_\odot < M_{\text{Ba}} < 3\, M_\odot$.

118.008 Measures of southern visual double stars.
F. Holden.
Publ. Astron. Soc. Pacific, Vol. 88, 52 - 57 (1976).

118.009 Otto Struve 21 and some other double stars.
J. Ashbrook.
Sky Telescope, Vol. 51, 15, 25 (1976).

**118.010 Discovery of flare activity in the visual binary
G 208-44/45.**   S. Cristaldi, M. Rodonò.
Astron. Astrophys., Vol. 48, 165 - 166 (1976).

The discovery of flare activity in the nearby visual binary
G 208-44/45 is reported.

**118.011 Dubbelsterren (2) – (4). Meervoudige sterren
eerder regel dan uitzondering.**   E. L. van Dessel.
Zenit, 3e jaargang, p. 20 - 23, 52 - 56, 130 - 134 (1976).

118.012 Radial velocities of the visual binary 70 Ophiuchi.
A. H. Batten, E. L. van Dessel.
Publ. Dominion Astrophys. Obs., Vol. 14, (No. 15), 345 -
353 = National Res. Council Canada, NRC No. 15164 (1976).

118.013 $uvby\,\beta$ study of A- and F-type visual binaries. I.
G. E. Mechler.
Astron. Journ., Vol. 81, 107 - 115 (1976).

$uvby\,\beta$ observations of 65 A- and F-type wide visual
binaries are presented and used to analyze some properties of
the components as well as the photometric system itself. For
the physical pairs, the differences in the absolute magnitudes
derived from the photometry are found to equal the apparent
magnitude differences to within the expected errors. The
components of 35 well-observed physical pairs are plotted in
the H–R diagram using β as the temperature index. For the
F-star pairs whose components are separated in spectral type
by four subclasses or more, the δm_1 indices of the secondaries
seem to be larger.

118.014 MK classification for visual binary components.
T. W. Edwards.
Astron. Journ., Vol. 81, 245 - 249 (1976).

MK classification are presented for both components of
208 visual binaries, most of which appear in Third Catalogue
of Orbits of Visual Binary Stars (Finsen and Worley 1970) and
which have only composite MK spectral types and visual
magnitude differences.

**118.015 Distanzen und Positionswinkel von 291 Doppel-
sternen für die Jahre 1975 bis 2000.**
W. Wepner.
Doppelstern-Ephemeriden 1975 bis 2000. Treugesell-Verlag,
Postfach 140165, 4000 Düsseldorf 14. 96 pp. Price DM 16.80.
Review in SuW, 15. Jahrgang, p. 168 (*H. Vehrenberg*).

118.016 Capella as a close visual binary.   J. Ashbrook.
Sky Telescope, Vol. 51, 322 - 323 (1976).

118.017 New double stars (13th series) discovered at Nice.
P. Couteau.
Astron. Astrophys., Suppl. Ser., Vol. 24, 495 - 505 (1976).
In French.

The author gives a list of 150 double stars discovered at
the 50 and 74 cm refractors.

118.018 Orbita de las estrellas dobles visuales ADS 2578-A

983; ADS 3021-Ho 326; RDS 5484-Rst 769; ADS 13777-A 288 y COU 14-13 Pegasi.
J. M. Costa, C. Morales.
Urania Barcelona, Año 60, No. 283, p. 21 - 37 (1975).

118.019 Orientación de los planos orbitales y distribución de las inclinaciones de las órbitas de las estrellas dobles visuales. J. Pensado, J. F. Lahulla.
Bol. Astron. Obs. Madrid, Vol. 9, No. 2, p. 21 - 35 (1976).
 An examination of the distribution of the inclinations has been made for 619 visual binary stars. Binaries with directly determined inclinations are analyzed.

118.020 Orbites nouvelles. P. Baize, W. D. Heintz.
 Circ. d'Inform. (UAI Commission des Étoiles Doubles), Grasse, France, No. 68 (1976).

118.021 Étoiles doubles découvertes à Nice (Lunette de 50 cm). P. Couteau.
Circ. d'Inform. (UAI Commission des Étoiles Doubles), Grasse, France, No. 68 (1976).

118.022 Orbites nouvelles. W. D. Heintz, P. Muller, W. S. Finsen, D. J. Zulević, R. R. de Freitas Mourão.
Circ. d'Inform. (UAI Commission des Étoiles Doubles), Grasse, France, No. 69 (1976).

118.023 Étoiles doubles nouvelles, Belgrade (Lunette de 65 cm), Nice (Lunette de 50 cm), Sproul Obs. (Lunette de 60 cm). G. M. Popovic, P. Couteau, W. D. Heintz.
Circ. d'Inform. (UAI Commission des Étoiles Doubles), Grasse, France, No. 69 (1976).

Astrophysical parameters of close visual binaries, obtained with an area scanner. See Abstr. 031.266.

Les parallaxes dynamiques. See Abstr. 111.004.

On the possibly overluminous triple star, 13 Ceti. See Abstr. 117.005.

The spectroscopic binary orbit in the triple star ADS 14893. See Abstr. 119.009.

LP 380-5/6: a binary system containing a late-type degenerate star. See Abstr. 126.006.

Polarization of Be stars. See Abstr. 131.178.

Orbits and masses of Hyades visual binaries. See Abstr. 153.005.

119 Spectroscopic Binaries

119.001 Orbital elements of the spectroscopic binary HD 160861 and discussion about the system.
J.-M. Carquillat, A. Pédoussaut, N. Ginestet, R. Nadal.
Astron. Astrophys., Suppl. Ser., Vol. 23, 277 - 281 (1976). In French.
 The authors deduce that the spectroscopic binary HD 160861 has a period of 2.50574d. Orbital elements determined from their observations are: $V_0 = -15,83$ km s^{-1}, $k = 44,99$ km s^{-1}, $\omega = 33°,6$, $e = 0,059$, $T = 2442280,70$ JD, $a \sin i = 1,548 \times 10^6$ km, $f(m) = 0,0236\, m_\odot$. Approximate calculations suggest that the secondary is a dwarf star of a mass between 0.45 and 0,90 $m_\odot$ and that the separation of the two components is about 10 $r_\odot$.

119.002 HD 192276: the spectrographic orbit and possible evidence of a third component. D. P. Hube.
Journ. Roy. Astron. Soc. Canada, Vol. 70, 27 - 30 (1976).
 Seventeen spectrograms obtained between 1962 and 1975 have been employed in deriving orbital elements for the single-lined spectrographic binary HD 192276. Some evidence is present for a third component, but further observations are needed to test its statistical significance.

119.003 HR 5702: a double-lined spectroscopic binary with an Am primary component.
A. H. Batten, M. L. McCall.
Publ. Astron. Soc. Pacific, Vol. 88, 13 - 18 (1976).
 Double lines have been detected in the spectrum of HR 5702, previously known as a single-spectrum binary. The primary component is an Am star and measures of the metal lines in its spectrum lead to the following values for the orbital elements: $P = 3.57725$ days, $T = $ JD2441347.95, $\omega_1 = 39°.3$, $e = 0.093$, $K_1 = 59.4$ km s^{-1}, $V_0 = -22.2$ km s^{-1}. Measures of the hydrogen lines in the secondary spectrum give $K_2 = 74.6$ km s^{-1} and hence $a_1 \sin i = 2.91 \times 10^6$ km, $a_2 \sin i = 3.65 \times 10^6$ km, $M_2/M_1 = 0.80$, $M_1 \sin^3 i = 0.49\, M_\odot$, and $M_2 \sin^3 i = 0.39\, M_\odot$. Both stars lie on or near the main sequence.

119.004 HD 170200: the spectrographic orbit and evidence for mass loss. D. P. Hube.
Publ. Astron. Soc. Pacific, Vol. 88, 58 - 62 (1976).
 Spectrograms obtained between 1924 and 1975 have been used to derive orbital elements for HD 170200. No completely convincing evidence has been found for any change in these elements during the past 51 years, although the most recent velocities appear to lie systematically below the average velocity curve. On several plates sharp, weak absorption lines are found in the violet wings of the primary-star lines. These cannot be identified with the secondary star and are interpreted as evidence for discontinuous mass loss from the primary.

119.005 The orbit of 71 Draconis. D. W. Willmarth.
 Publ. Astron. Soc. Pacific, Vol. 88, 86 - 87 (1976).
 The star 71 Dra is found to be a single-line spectroscopic binary with a period of 5.29807 days and a velocity amplitude

of 53.8 km sec^{-1}.

119.006 On the double-lined spectroscopic binary system
α Equulei. D. J. Stickland.
Monthly Notices Roy. Astron. Soc., Vol. 175, 473 - 479 (1976).
 The orbital elements of the double-lined spectroscopic
binary system α Equulei have been redetermined and the mass
ratio found by Deutsch revised to a more plausible value.

119.007 Spectroscopic binary orbits from photoelectric
radial velocities. Paper 6: HD 183629.
G. A. Radford, R. F. Griffin.
Observatory, Vol. 96, 18 - 21 (1976)

119.008 A tentative envelope model of Phi Persei.
M. Suzuki.
Astrophys. Space Sci., Vol. 39, 495 - 497 (1976). – Research
note.

119.009 The spectroscopic binary orbit in the triple star
ADS 14893. F. R. West.
Astrophys. Journ., Vol. 205, 194 - 207 (1976).
 The orbital elements of the single line spectroscopic
binary ADS 14893 B have been determined. The mass ratio
$q = 2.257$ of the spectroscopic binary to ADS 14893 A
indicates that the unobserved star is possibly the most massive
star in the ADS 14893 system and could be a degenerate star.
The visual binary period is about 6 years. Three preliminary
sets of visual binary orbital elements are determined, from
which three radial velocity ephemerides of ADS 14893 A are
calculated through 1980.

119.010 A new spectroscopic orbit of Delta Librae.
T. Prabhu.
Bull. Astron. Soc. India, Vol. 3, 29 (1975). – Abstract of a
paper presented at the A.S.I. meeting 1975.

119.011 Spectroscopic binary orbits from photoelectric
radial velocities. Paper 7: HD 160952.
G. A. Radford, R. F. Griffin.
Observatory, Vol. 96, 56 - 61 (1976).

119.012 Spectrophotometric investigation of the spectro-
scopic binary system δ Orionis. T. S. Galkina.
Izv. Krymskoj Astrofiz. Obs., Vol. 54, 128 - 158 (1976). In
Russian.
 The composite spectrum of the close binary system δ Ori
is analyzed on the basis of 55 spectrograms obtained with dispersions of 15 Å/mm and 22 spectrograms obtained with dispersion of 37 Å/mm in the regions λλ4900–3650 Å and
λλ6800–5500 Å in the different phases of the orbital period.
The measurements of absorption lines show that the spectral
type of the primary varies with phase from O8.5 to O9.6. The
spectral type of the secondary may be estimated as B1. The
brightness difference between the components is ≈ 2^{m}5. Some
physical parameters of the atmosphere of the primary are determined. A detailed investigation of the contour of the line
λ5696 C III leads to the conclusion on the presence of a shell
surrounding the star. The velocity of the expansion of the
shell is estimated to about 550 km/sec. Comparison with the
data of the investigation of the far ultraviolet spectrum
(Morton et al., 1968) leads to the conclusion that the velocity
of the expansion of the shell increases outward.

119.013 Bright early-type spectroscopic binaries. I. HR 8584.
C. T. Bolton, N. Geffken.
Publ. Astron. Soc. Pacific, Vol. 88, 195 - 197 (1976).
 The authors have redetermined the orbital elements of
HR 8584 using new data. They have been unable to resolve
the discordance between elements obtained by Albitzky
(1933) and by Harper (1933) using virtually simultaneous ob-

servations. The authors find that this system may have apsidal
motion with a period of 260 years.

119.014 Interpretation of a spectrographic observation of
the resolved components of κ Pegasi.
W. R. Beardsley, M. W. King.
Publ. Astron. Soc. Pacific, Vol. 88, 200 - 203 (1976).
 The A and B components of κ Peg were resolved on
1975 May 25 at the coudé feed telescope of the Kitt Peak
National Observatory. An untrailed spectrum of each component was obtained. Star A showed double lines. The radial
velocities, as well as the characteristic appearance of the spectra, suggest that star A is a spectroscopic binary with a period
of 4.77 days; star B is the previously known spectroscopic
binary with a period of 5.97 days.

119.015 Spectroscopic binary orbits from photoelectric
radial velocities. Paper 8: HD 90385.
G. A. Radford, R. F. Griffin.
Observatory, Vol. 96, 98 - 99 (1976).

119.016 Appeal for photoelectric observations of Upsilon
Sagittarii. M. Friedjung.
Inform. Bull. Variable Stars, (I.A.U. Commission 27), Konkoly
Obs., Budapest, No. 1144, 2 pp. (1976).

119.017 The period of AZ Cassiopeiae.
J. T. Bonnell, T. J. Herczeg.
Inform. Bull. Variable Stars, (I.A.U. Commission 27), Konkoly
Obs., Budapest, No. 1146, 4 pp. (1976).

119.018 Photoelectrical investigations of seven Ap stars –
members of spectroscopic double star systems.
B. Musielok, E. Želwanowa, W. Schöneich, G. Hildebrandt.
Astron. Nachr., Band 297, 169 - 172 (1976). In German.
 Preliminary results are given for seven Ap components of
spectroscopic double stars in the B, V system. In all cases the
amplitudes are smaller than 0.025 mag. For no system of these
seven Ap stars there was found a light variation with the period
of the orbit.

119.019 Ultraviolet observation of HD 77581 (= 2U 0900–
40). K. Nandy, W. McD. Napier, G. I. Thompson.
Mem. Soc. Astron. Italiana, Vol. 45, (see 012.017), 713 - 715
(1974). – Abstract (see also 13.119.003).

 The application of a new method for the determina-
tion of relative radial velocities with the Fehrenbach-prism.
See Abstr. 031.205.

 Nomographs for the preliminary solution of spec-
troscopic binary radial-velocity curves.
See Abstr. 031.243.

 Comparison of Lyman alpha and He λ 10830 line
structures and variations in early-type star atmospheres. Final
technical report, January - October 1974.
See Abstr. 064.070.

 On wavelength shifts of He I lines in rotating B
stars. See Abstr. 114.001.

 Skylab ultraviolet stellar spectra: cool stars with hot
secondaries. See Abstr. 114.005.

 CD −30°5135. See Abstr. 114.358.

 Multiplicity among solar-type stars.
See Abstr. 117.014.

 The spectroscopic orbit of δ Andromedae.

See Abstr. 118.005.

BS 1099: a bright variable similar to the radio star UX Arietis. See Abstr. 121.081.

The eclipsing/spectroscopic binary HD 193964 = 71 Dra. See Abstr. 121.090.

Five-colour photometry of southern cepheids. A new search for cepheid binaries. See Abstr. 122.107.

Skylab ultraviolet stellar spectra: a new white dwarf, HD 149499 B. See Abstr. 126.016.

The binary frequency of IC 4665. See Abstr. 153.011.

120   Variable Stars: Catalogues, Ephemerides, Miscellanea

120.001   **Preliminary designations of SVS stars.**
Astron. Tsirk., No. 854, p. 8 (1975). In Russian.

120.002   **Scheinperioden bei veränderlichen Sternen.**
G. Pfeiffer.
BAV Rundbrief, 25. Jahrgang, p. 1 - 15 (1976).

120.003   **Some characteristics of variable stars appearing in infrared sky surveys.**   E. B. Weston.
Journ. American Ass. Variable Star Observers, Vol. 4, 104 (1975/76). — Abstract.

120.004   **Charts for southern variables,** Series 8, 3 pp. + charts 301 - 350 (1975).
F. M. Bateson, M. Morel, R. Winnett.
      Series 8 charts include stars selected because of either their interest, or because the stars concerned are not being sufficiently observed.

120.005   **A propos de la classification internationale des étoiles variables.**   B. V. Kukarkin.
A.F.O.E.V. Bull., Tome 9, 162 - 163 (1975).

121   Eclipsing Variables

121.001   Infrared observations and a model of β Lyr.
R. F. Jameson, A. J. Longmore.
Monthly Notices Roy. Astron. Soc., Vol. 174, 217 - 223 (1976).

The paper presents light curves of the eclipsing binary β Lyr at 1.2, 2.2, 3.6, 4.8 and 8.6 μ. The principal feature of these curves is that at long wavelengths the secondary minimum·becomes deeper than the primary minimum. This can be explained by a plasma cloud around the secondary component of temperature 17500 K and constant electron density of 1.4×10^{12} cm^{-3} out to a distance 1.6×10^7 km. It is suggested that the secondary is a B4 or earlier type star.

121.002   On the photometric variability of V 471 Tau.
B. Cester, M. Pucillo.
Astron. Astrophys., Vol. 46, 197 - 204 (1976).

Photoelectric observations of the unusual binary system V471 Tau made at the end of 1972 and in 1973 show that the light curve is variable. An improvement of the period has been obtained by means of several new epochs of mid-eclipse. From the times of outer and inner contact of eclipse in u-color and from previous spectrographic observations, a new set of elements is proposed. In addition to the explanations already put forward by others, regarding the evolution of the system, another is suggested namely one with mass exchange and subsequent mass loss when both components are in contact.

121.003   Four-colour photometry of eclipsing binaries. IIIb: Zeta Phoenicis, analysis of light curves and determination of absolute dimensions.   J. V. Clausen, K. Gyldenkerne, B. Grønbech.
Astron. Astrophys., Vol. 46, 205 - 212 (1976).

Four-colour $uvby$ observations of the bright early-type double lined eclipsing binary ζ Phe are analysed and discussed. The determination of photometric elements is carried out by simulating models of the system and generating light curves for comparison with the observations. As model type the one used by Wood has been applied taking into account defomation of the components, limb darkening, gravity brightening and reflection. In combination with spectroscopic results obtained by Popper data for radii and masses together with other astrophysical data are given.

121.004   Probable beginning of the chromospheric eclipse of VV Cephei.   R. Faraggiana.
Astron. Astrophys., Vol. 46, 317 - 318 (1976).

Spectrograms of VV Cep (HD 208816), taken since December 1974, show an increasing intensity of the lines of ionized elements at wavelengths shorter than 3600 Å; the number of these lines has been increasing in the last months. The author interprets this as an indication of the beginning of the chromospheric eclipse of this system.

121.005   Four-colour photometry of eclipsing binaries. IIIa. ζ Phe, observations.
J. V. Clausen, K. Gyldenkerne, B. Grønbech.
Astron. Astrophys., Suppl. Ser., Vol. 23, 261 - 275 (1976).

$uvby$ observations of the double lined eclipsing binary ζ Phe are presented. The observations have been obtained by means of the Strömgren spectrograph photometer for simultaneous measurements in the four colours.

121.006   Photometric orbit of the eclipsing binary system X Trianguli.
Ş. Bozkurt, C. İbanoğlu, Ö. Gülmen, N. Güdür.
Astron. Astrophys., Suppl. Ser., Vol. 23, 439 - 446 (1976).

The eclipsing binary X Tri was observed photoelectrically in blue and yellow bands. A total of 282 observations were obtained in each colour. The system is an Algol-type with a deep primary minimum which gives total eclipse. New epochs of minima were obtained and improved light elements were derived. The light curves were analyzed using two different methods. The resulting elements are presented.

121.007   Spectroscopic orbits of the eclipsing binaries DV and DX Aqr.   W. Paffhausen, W. Seggewiss.
Astron. Astrophys., Suppl. Ser., Vol. 24, 29 - 34 (1976).

Orbital elements of the bright eclipsing binaries DV and DX Aqr have been determined from spectrograms taken in 1973 at the European Southern Observatory. Adopting $e = 0$ for the eccentricities of both orbits the authors have derived the velocity amplitudes 95.5 km s^{-1} and 97.9 km s^{-1} and the mean system velocities 10.3 km s^{-1} and 15.0 km s^{-1} for DV and DX Aqr respectively. The radial velocity curve of DX Aqr displays a well-pronounced rotation effect.

121.008   A photometric study of ST Persei.
E. W. Weis, K.-Y. Chen.
Acta Astron., Vol. 26, 15 - 23 = Contr. Rosemary Hill Obs., Univ. Florida, *Gainesville*, No. 61 (1976).

Solutions of 318 observations in R and 309 observations in I are presented. The system appears slightly reddened, and the inferred spectral type of the cool secondary is K1 - 2IV. Difficulties with a previous photoelectric study of this system are pointed out.

121.009   The light-curves of 14 eclipsing variables.
R. Szafraniec.
Acta Astron., Vol. 26, 25 - 54 (1976).

Visual observations were reduced and light-curves determined for 14 eclipsing variables: CU, DF, DK, DO Peg; RT, RV Per; SX Psc; SV, AM, new variable Tau; RR, XZ, AY, BO Vul. These stars do not possess photoelectric light-curves, and some of them have no published light-curve at all. Lists of times of minima and elements are also given.

121.010   Photometric evidence of circumstellar gas and orbital elements for the Algol system.
E. F. Guinan, G. P. McCook, P. J. Bachmann, W. G. Bistline.
Astron. Journ., Vol. 81, 57 - 66 (1976).

Hα and Hβ wide- and narrow-band photometry of Algol was obtained on 23 nights mostly in 1970—1971. A total of 2425 observations was secured. Orbital elements were calculated from the wide-band data and were found to be consistent with previous solutions. Hα and Hβ indices were computed and an analysis of these reveals the presence of circumstellar gas in the vicinity of the hot component. This material is estimated to extend to a distance of 2.6 times the radius of the hot star and probably fills the Roche lobe of this component.

121.011   On the orbital eccentricity of V477 Cygni.
C. D. Scarfe, D. J. Barlow, R. J. Niehaus.
Astrophys. Space Sci., Vol. 39, 129 - 132 (1976).

Recent photoelectric times of minima support the value $e = 0.3$ obtained by O'Connell (1970). The lower value obtained by Budding (1974) is ruled out.

121.012   Orbital solution of the light curve of MY Cyg and absolute dimensions of the system.
J. Tremko, J. Papoušek, M. Veteśnik.
Bull. Astron. Inst. Czechoslovakia, Vol. 27, 125 - 126 (1976).

The orbital elements of the eclipsing binary MY Cyg

derived from the authors' three-colour photoelectric light curves and from the spectroscopic results of Popper are presented in the paper. The system consists of two very similar stars and can be classified as a detached one.

121.013 A radio model for Algol (β Persei).
A. W. Woodsworth, V. A. Hughes.
Monthly Notices Roy. Astron. Soc., Vol. 175, 177 - 190 (1976).

The radio emission from the multiple star system, Algol, varies in intensity from about 30 mJy to more than 1 Jy. Two components can be recognized, a non-thermal flare component with spectral index of unity which is probably associated with the region of mass exchange where the electron density is greater than 10^8 cm^{-3}, and a thermal component from an optically thin H II region which surrounds the optical object to a distance greater than 5.8 AU where the electron density is greater than 6.4×10^6 cm^{-3}.

121.014 On the possibility of apsidal motion in UX Ursae Majoris.   J. Africano, J. Wilson.
Publ. Astron. Soc. Pacific, Vol. 88, 8 - 9 (1976).

Photoelectrically determined times of eclipse minima are given for the eclipsing binary UX UMa. Apsidal motion is discussed as the cause for the (O − C) residuals.

121.015 A three-color photoelectric investigation of the eclipsing binary system RZ Cassiopeiae.
C. R. Chambliss.
Publ. Astron. Soc. Pacific, Vol. 88, 22 - 36 (1976).

The semidetached eclipsing binary system RZ Cas was observed on 25 nights. Approximately 600 observations each were obtained in yellow, blue, and ultraviolet. RZ Cas has long been known to have a variable period, and a new ephemeris based on the author's observations is given. Solutions for its orbital elements were carried out using three different procedures; the Russell-Merrill method, the Kopal-Jurkevich iterative method, and the Wood triaxial ellipsoid model. The Wood model produces orbital elements which differ significantly from those of the spherical model in some cases. Using the available spectroscopic data for RZ Cas, the masses of the components are estimated as 1.75 $M_\odot$ and 0.61 $M_\odot$ and their respective radii as 1.45 $R_\odot$ and 1.83 $R_\odot$.

121.016 Four-colour light curves of the eclipsing binary TY Mensae.   S. I. H. Naqvi, B. Grønbech.
Astron. Astrophys., Vol. 47, 315 - 318 (1976).

The eclipsing binary TY Men of W UMa-type was observed in the Strömgren $uvby$ system and found to have a low metal content, [Fe/H] = −0.9 relative to the Hyades. The light curves are strongly asymmetric, the height of the two maxima differ by up to $0\overset{m}{.}08$ in the v-band. A comparison with VW Cep and AG Vir is made, and a hot spot model is proposed to explain the asymmetries in the light curves.

121.017 Comments on a recent study of VV Orionis.
J. Andersen.
Astron. Astrophys., Vol. 47, 467 - 469 (1976).

The observational basis for the model of VV Orionis recently proposed by Duerbeck (1975), involving bidirectional mass transfer in a detached binary system, is reviewed and certain weak points in the material are discussed. The conclusions do not lend unequivocal support to Duerbeck's results.

121.018 Evidence of circumstellar matter in a detached binary system. A reply to the comments of J. Andersen.   H. W. Duerbeck.
Astron. Astrophys., Vol. 47, 471 - 473 (1976).

The objections of J. Andersen are refuted. The problem of mass transfer and evolution in the eclipsing binary VV Ori

are discussed in some detail.

121.019 U Cephei: a mass-transfer event. I.   E. C. Olson.
Astrophys. Journ., Vol. 204, 141 - 150 (1976).

A major mass-transfer event, with accompanying photometric anomalies, occurred in U Cep in late 1974. Four-color $uvby$ observations during this outburst have been interpreted by a simple quantitative model containing a mass-transferring stream and a hot spot and equatorial disk associated with the primary star. A lower limit to the peak mass-transfer rate of 3×10^{-5} solar masses per year is deduced.

121.020 An elementary theory of eclipsing depths of the light curve and its application to Beta Lyrae.
S.-S. Huang, D. A. Brown.
Astrophys. Journ., Vol. 204, 151 - 159 (1976).

An elementary theory of the ratio of depths of secondary and primary eclipses of a light curve has been proposed for studying the nature of component stars. It has been applied to light curves of Beta Lyrae. The authors have found no trace of the spectrum of primary radiation in the disk and have therefore suggested that LTE is the main radiative process in the disk, which radiates at a temperature of approximately 12,000 K in the portion that undergoes eclipse. A small source corresponding to 14,500 K has also been tentatively detected and may represent a hot spot caused by hydrodynamic flow of matter from the primary component to the disk.

121.021 A note on the Mount Wilson radial velocities of S Cancri.   E. W. Weis.
Observatory, Vol. 96, 9 - 11 (1976).

121.022 Ungewöhnliche Kontaktsysteme frühen Spektraltyps.
M. Ammann, H. Mauder.
Mitt. Astron. Ges., No. 38, p. 232 - 233 (1976).

121.023 Das photometrische Verhalten des W UMa-Systems TZ Bootis.   M. Hoffmann.
Mitt. Astron. Ges., No. 38, p. 234 (1976).

121.024 Four-colour photometry of eclipsing binaries. IV: UX Mensae, light curves, photometric elements and absolute dimensions.   J. V. Clausen, B. Grønbech.
Astron. Astrophys., Vol. 48, 49 - 53 (1976).

The double lined eclipsing binary UX Men was observed in the Strömgren $uvby$ system, and analysed. The photometric elements derived are combined with spectroscopic data by Imbert (1974) to give the absolute dimensions.

121.025 Zum Entwicklungszustand des Doppelsternsystems CW Eri = BV 1000.   H. Mauder, M. Ammann.
Mitt. Astron. Ges., No. 38, p. 231 - 232 (1976). − Abstract.

121.026 Tweede eclipsveranderlijke in Orions trapezium-stelsel.   G. Comello.
Zenit, 3e jaargang, p. 100 - 101 (1976).

121.027 The eclipsing system V444 Cyg (WN 5 + O 6) in the light of emission lines He II 4686, (He II + Hα)
6563, N IV 7112.   Kh. F. Khaliullin, A. M. Cherepashchuk.
Astrofizika, Vol. 11, 593 - 607 (1975). In Russian. English translation in Astrophysics, Vol. 11, No. 4.

The authors obtained about 400 photoelectric intensity measurements for each of several emission lines. This enabled them to find out the role of the proximity effects in the double system and to average the intrinsic fluctuations of the brightness. The contribution of hydrogen to the emission band (He II + Hα) 6563 is probably significant. It seems to be impossible to describe the changes of the intensity of He II 4686 during the secondary minimum with the help of the model of the geometric occultation of the WR star by the O 6 component. The model of the geometric occultation seems to be applicable for the emission feature N IV 7112. The electron temperature in the WR envelope (40000–50000°K) may be described with Beals' model.

121.028 **Photoelectric elements of the binary system AO Monocerotis.** C. D. Kandpal.
Astrophys. Space Sci., Vol. 40, 3 - 14 (1976).

A slightly improved period of $1^d884\,761\,9$, has been given for the system AO Mon. Photoelectric elements have been computed in B and V filters. The orbit is found to be eccentric. Absolute elements have been derived by using the spectroscopic elements given by Struve. The system is found to be a detached one.

121.029 **Photoelectric elements of CD Tau.** J. B. Srivastava.
Astrophys. Space Sci., Vol. 40, 15 - 34 (1976).

Photoelectric and absolute elements of the system CD Tau have been determined in U, B and V filters. The system is a detached one.

121.030 **Fourier analysis of the light curves of eclipsing variables, VI.**
Z. Kopal, V. Markellos, P. Niarchos.
Astrophys. Space Sci., Vol. 40, 183 - 199 (1976).

A numerical study is given of a family of orthonormal polynomials, introduced in a previous communication of this series (Kopal, 1975) for the elimination of the proximity effects from observed light variations of close eclipsing systems: and tables are presented which should facilitate applications of this technique to practical cases.

121.031 **Times of minima of RZ Cassiopeiae.**
M. B. K. Sarma, P. V. Subrahmanyam, C. S. Murthy.
Bull. Astron. Soc. India, Vol. 3, 79 (1975). – Short communication.

121.032 **TX Cancri—which component is hotter?**
R. E. Wilson, P. Biermann.
Astron. Astrophys., Vol. 48, 349 - 357 (1976).

The authors raise the question of whether the W-type light curves of TX Cnc are due to the low mass component being slightly the hotter component, as presumed until now, or whether it might be due to TX Cnc having the relatively strong classical gravity darkening law. To decide this question, they have analysed TX Cnc light curves from four "epochs" showing dissimilar behavior. They find that the polar effective temperature of the low mass star is then about 30°K cooler than that of the high mass star. TX Cnc is the second contact binary, after RZ Com, for which a gravity darkening explanation adequately accounts for a W-type light curve.

121.033 **BM Orionis: the enigmatic eclipsing binary in the Trapezium.** D. M. Popper, M. Plavec.
Astrophys. Journ., Vol. 205, 462 - 471 (1976).

The enigma of BM Ori – namely, the apparent absence of spectral lines of a cooler component in spite of the considerable duration of approximately constant light during primary eclipse – is partially resolved. Weak lines of a secondary star of late A-type are found, and the orbits of both components are obtained. The primary component is a normal B3 star of mass 6 $M_\odot$, lying on the zero-age main sequence of the Orion nebula cluster. The secondary, of mass 1.8 $M_\odot$, conforms in its axial ratio and moderate rotational velocity to models in pre–main-sequence differential rotation (Bodenheimer-Ostriker models), as suggested by D. S. Hall. It lies roughly 2 mag above the main sequence of the cluster.

121.034 **Interpretation of eclipses in frequencies of lines in binary systems with Wolf-Rayet components.**

Kh. F. Khaliullin, A. M. Cherepashchuk.
Astron. Zhurn. Akad. Nauk SSSR, Vol. 53, 327 - 337 (1976). In Russian. English translation in Soviet Astron., Vol. 20, No. 2.

A method of interpretation of eclipses in frequencies of emission lines in double systems with WR components is suggested. From observations of intensities of the emission lines He II 4686 and C II – C IV, N III 4653 in the systems V444 Cyg and CV Ser the distribution of the volumetric coefficient of selective absorption in the WR shell is found. Cases of accelerated, uniform and decelerated expansions are considered.

121.035 **Spectroscopic studies of O-type binaries. I.**
BD + 40°4220: an enigma ripe for resolution.
B. Bohannan, P. S. Conti.
Astrophys. Journ., Vol. 204, 797 - 803 (1976).

A new orbit for BD + 40°4220 (V729 Cygni, Cyg OB2 No 5) is calculated from coudé spectra. While the mass ratio is unchanged from that of Wilson and Abt, the masses ($M_1 \sin^3 i = 47 \pm 7 \ M_\odot$, $M_2 \sin^3 i = 11 \pm 5 \ M_\odot$) and the scale of the system ($a_1 \sin i = 7.6 \pm 0.7 \times 10^6$ km, $a_2 \sin i = 3.3 \pm 0.2 \times 10^7$ km) are larger because the authors are able to resolve the absorption lines of each star. The system is composed of two Of supergiants with equally high luminosity, $M_V \approx -7.1$. The authors propose that BD +40°4220 is on the way to becoming a Wolf-Rayet binary system.

121.036 **Fourier analysis of the light curves of eclipsing variables, VII.** S. A. H. Smith.
Astrophys. Space Sci., Vol. 40, 315 - 324 (1976).

A new method has been developed for the evaluation of the light moments A_{2m}, required for a Fourier analysis of the light curves of eclipsing variables, in terms of the elements of the eclipse – a method simpler and more straightforward than that previously developed in so far as it dispenses with the auxiliary coefficients $a_n^{(l)}$ and $b_n^{(l)}$ used before at the intermediary stage.

121.037 **Fourier analysis of the light curves of eclipsing variables, VIII.** Z. Kopal.
Astrophys. Space Sci., Vol. 40, 461 - 481 (1976).

The main aim of this paper will be to develop explicit form of the moments of the light curves $A_{2m}(r_1, r_2, i)$ required for the solution for the geometrical elements $r_{1,2}$ and i of eclipsing systems exhibiting annular eclipses, as well as partial eclipses. The author demonstrates that—regardless of the type of eclipse and distribution of brightness on the apparent disc of the eclipsed star, or indeed of the shape of the eclipsing as well as eclipsed components—the moments A_{2m} satisfy certain simple functional equations—a fact which relates them to other classes of functions previously studied in applied mathematics.

121.038 **Eclipsing binary system SZ Piscium.**
S. Jakate, G. A. Bakos, J. D. Fernie, J. F. Heard.
Astron. Journ., Vol. 81, 250 - 256 (1976).

SZ Psc is a double-line spectroscopic and eclipsing binary of period about four days, composed of a K1 IV and an F8 V star. The authors present photoelectric photometry and radial velocity data for the system. Comparison with previous photographic photometry reveals a linear rate of decrease in the period. The K1 IV star is found to nearly fill its Roche lobe, and this, together with the existence of strong H and K emission in its spectrum and the period variation, suggests that mass transfer is taking place in the system. The authors derive a rate of $1 \times 10^{-6} M_\odot$ yr^{-1}. SZ Psc appears to be in its earliest phase of interactive evolution as a close binary.

121.039 **Ultraviolet photometry from the Orbiting Astronomical Observatory XXII. Ultraviolet light variation of β Lyrae.**
Y. Kondo, G. E. McCluskey, J. A. Eaton.

Astrophys. Space Sci., Vol. 41, 121 - 137 (1976).

Six-color ultraviolet light curves of the complex eclipsing binary system β Lyr were obtained with the OAO-2 Wisconsin Experiment Package. The filters had a typical width at half maximum of 150 to 200 Å and centered at 1430, 1550, 1910, 2460, 2980 and 3320 Å. The most striking characteristics of the ultraviolet light curves are that the secondary minimum deepens at shorter wavelengths. This indicates that we are not observing the eclipse effect of two stars having roughly a Planckian distribution of energy.

121.040 Observations of KU Cygni, 1926–1975.
M. Taylor.
Journ. American Ass. Variable Star Observers, Vol. 4, 62 - 63 (1975/76).

New photographic estimates on Nantucket patrol plates indicate that the period of this eclipsing binary has been constant from 1926 to 1975 and indicate variability at maximum light.

121.041 Differential UBV photometry of β Lyrae, VI.
H. J. Landis, L. P. Lovell, D. S. Hall.
Journ. American Ass. Variable Star Observers, Vol. 4, 80 - 83 (1975/76).

In a continuing program of photometry of β Lyrae, 219 differential UBV observations were obtained in 1974 at three observatories.

121.042 Minima of eclipsing binary stars, II. M. E. Baldwin.
Journ. American Ass. Variable Star Observers, Vol. 4, 86 - 91 (1975/76).

121.043 La binaria ad eclisse VV Cephei.
L. Baldinelli, S. Ghedini.
Giorn. A.A.B., Anno 11, No. 42, p. 3 - 7 (1976).

121.044 The 3U 0900–40 binary system: orbital elements and masses.
S. Rappaport, P. C. Joss, J. E. McClintock.
Astrophys. Journ., (Letters), Vol. 206, L103 - L106 (1976).

Observations of the 283 s X-ray pulsations from 3U 0900–40 (= Vela X-1) over a 36 day interval have led to a determination of the orbital elements for the X-ray star. These elements yield an X-ray mass function $f(M) = 18.5 \pm 0.8 \, M_\odot$ and, when combined with the results of optical studies of the companion star HD 77581, give a probable lower limit of 1.4 $M_\odot$ to the mass of the X-ray star.

121.045 An unusually strong radio outburst in Algol: VLBI observations.
T. A. Clark, L. K. Hutton, C. Ma, I. I. Shapiro, J. J. Wittels, D. S. Robertson, H. F. Hinteregger, C. A. Knight, A. E. E. Rogers, A. R. Whitney, A. E. Niell, G. M. Resch, W. J. Webster, Jr.
Astrophys. Journ., (Letters), Vol. 206, L107 - L111 (1976).

For 8 hours during a strong radio flare on 1975 January 15 - 16, the close binary star system β Persei (Algol) was observed with a three-station VLBI array operating at 7850 MHz. The observations indicate that, during strong flaring activity, the radio source associated with Algol has a size comparable to those of the individual stars of the close pair and is most likely nonthermal in origin. Any possible expansion of the apparent size of the radio source during the decay of the last flare observed had a velocity no greater than about 100 km s^{-1}.

121.046 Four-colour light curves of the eclipsing binary AE Phe.
B. Grønbech.
Astron. Astrophys., Suppl. Ser., Vol. 24, 399 - 411 (1976).

The eclipsing binary AE Phe of W UMa type was observed in the Strömgren uvby system. An improved ephemeris indicates that the phase of secondary minimum is changing. Asymmetries and variability of the light curves are discussed.

121.047 Narrow-band photoelectric observations of the Wolf-Rayet eclipsing binary V444 Cyg in the emission lines He II 4686, (He II + Hα) 6563, N IV 7112.
Kh. F. Khaliullin, A. M. Cherepashchuk.
Peremennye Zvezdy, Vol. 20, 1 - 12 (1975). In Russian.

Individual observations of the intensity of the emission lines He II 4686, (He II + Hα) 6563, N IV 7112 in the system V444 Cyg are presented in tables. The observations are reduced in mean intensity curves with the orbital elements. The results of investigation of a narrow-band photometer with wedge interference filters are given.

121.048 Gravitational darkening in W UMa-type binary systems. L. N. Ivanov.
Peremennye Zvezdy, Vol. 20, 99 - 101 (1975). In Russian.

An asynchronous rotation of close binary stars influences the efficiency of their gravity darkening. In W UMa systems this value can be as much as 0.3–0.4, while in stars with synchronous rotation it is only 0.08.

121.049 On the period change of TZ Eridani.
N. B. Perova.
Peremennye Zvezdy, Vol. 20, 197 - 198 (1975). In Russian.

Examination of 32 collected minima of the eclipsing variable TZ Eri shows that its period varies. The light elements of GCVS II Min = 2426066.198 + 2^d.606097 × E are valid in the time interval JD 2426000 – 33500. The new light elements Min = 2442414.263 + 2^d.6060653 × E are valid in the time interval JD 2438200–42500. The O–C deviations from these elements are given in a table.

121.050 Preliminary results of a study of the eclipsing variable star EY Orionis. M. M. Zakirov.
Peremennye Zvezdy, Vol. 20, 199 - 206 (1975). In Russian.

121.051 Electrospectrophotometric study of AR Lacertae.
M. B. Babaev.
Peremennye Zvezdy, Vol. 20, 207 - 217 (1975). In Russian.

The spectrum of AR Lac in the region of 3225–7500 Å is studied. The dependences of the energy distribution on the brightness phase are given in tables. The gradient change in the regions λλ 3225–3675, 4025–4525 and 6325–7425 is also studied. The conclusion that all these parameters change with phase is discussed. 6-hour short-period changes of the gradient are observed.

121.052 Photoelectric observations of V502 Ophiuchi.
T. S. Polushina.
Peremennye Zvezdy, Prilozhenie, Vol. 2, 161 - 170 (1975). In Russian.

A total of 193 photoelectric observations in V light, 190 observations in B light and 188 observations in U light of the eclipsing binary V502 Oph of W UMa-type is presented. New elements are determined: Min I J.D.$_\odot$ = 2441174.2288 + 0.45339345 × E.

121.053 Researches of massive close binary systems of early spectral type. 2. Photographic observations and photometric elements of the eclipsing binary GT Cep (BV 374). I. I. Bondarenko, Yu. I. Tokareva.
Peremennye Zvezdy, Prilozhenie, Vol. 2, 171 - 194 (1975). In Russian.

731 photographic observations of the eclipsing binary system GT Cep (BV 374) in blue and 723 ones in yellow were carried out in October 1966 - May 1967. A map of the neighbourhood is given. Photometric elements of the system are calculated by the method of Tsesevich.

121.054 EQ Vulpeculae. G. E. Erleksova.
Peremennye Zvezdy, Prilozhenie, Vol. 2, 258 - 261 (1975). In Russian.

121.055 On the variability of V348 Cygni.
A. Ya. Filin, O. G. Suyarkova.
Peremennye Zvezdy, Prilozhenie, Vol. 2, 263 - 264 (1975).
In Russian.
The variability of V348 Cyg is confirmed. The elements
of the light curve have been determined in the interval J.D.
2441567–595.

121.056 GY Aurigae. E. B. Khotimskaya.
Peremennye Zvezdy, Prilozhenie, Vol. 2, 265 - 267
(1975). In Russian.

121.057 AP Tauri. N. V. Kondrat'ev.
Peremennye Zvezdy, Prilozhenie, Vol. 2, 273 - 276
(1975). In Russian.

121.058 MY Cassiopeiae. G. F. Tokareva.
Peremennye Zvezdy, Prilozhenie, Vol. 2, 287 - 290
(1975). In Russian.

**121.059 Complex of programs for the analysis of the light
curves of eclipsing binary systems by a direct
method.** M. I. Lavrov.
Peremennye Zvezdy, Prilozhenie, Vol. 2, 349 - 373 (1976).
In Russian.
Programs for computer solution of light curves of eclipsing binary systems are described. The problems solved by them
are presented.

**121.060 Bestimmung der Systemkonstanten bei Bedeckungs-
veränderlichen.** W. Bischof.
BAV Rundbrief, 25. Jahrgang, p. 16 - 25 (1976).

**121.061 Choice of the best solution of the light curve of an
eclipsing binary in case of a partial eclipse.**
A. M. Shul'berg.
Astron. Tsirk., No. 892, p. 6 - 8 (1975). In Russian.

121.062 A re-discussion of CW Cep. S. Söderhjelm.
Astron. Astrophys., Suppl. Ser., Vol. 25, 151 - 158
(1976).
Photometric elements for CW Cep are determined from
some 2900 individual observations, including 500 new ones
presented in this paper. Comparison with the spectroscopic
data yields masses and radii in agreement with theoretical
models. The observed apsidal motion constant $k^{(2)}$ indicates
that the presently available opacity calculations may be
further improved.

121.063 Infrared photometry of RT Lacertae.
E. F. Milone.
Astrophys. Journ., Suppl. Ser., Vol. 31, 93 - 109 (1976).
The first of a number of eclipsing binaries with asymmetric light curves has been examined in the near-infrared for
evidence of a distribution of matter in and around the Roche
lobes. The double-lined spectroscopic and peculiar eclipsing
binary RT Lacertae has an apparent excess of radiation at
JHKL over what is to be expected from the component stars,
with due allowance for a reasonable degree of interstellar
extinction. The peak value reached is about 0.7 mag at *K*.

121.064 The secondary of U Sagittae. S. A. Naftilan.
Astrophys. Journ., Vol. 206, 785 - 789 (1976).
Two moderately high-dispersion spectra (15 and 16 Å
mm^{-1}) of the secondary component of U Sagittae were analyzed using a crude curve-of-growth technique and also a more
sophisticated synthetic spectrum analysis. The results show a
mild metal deficiency, [Fe/H] = 6.90. The abundances with
respect to iron resemble those of metallic-line A stars.

121.065 The new photometric elements of BS Draconis.

D. Chis, V. Pop, I. Todoran.
Inform. Bull. Variable Stars, (I.A.U. Commission 27), Konkoly
Obs., Budapest, No. 1079, 2 pp. (1976).

121.066 Observations of the minimum of θ^1 Ori A.
M. F. Walker.
Inform. Bull. Variable Stars, (I.A.U. Commission 27), Konkoly
Obs., Budapest, No. 1080 (1976).

121.067 Photoelectric minima of eclipsing binaries.
T. Z. Dworak.
Inform. Bull. Variable Stars, (I.A.U. Commission 27), Konkoly
Obs., Budapest, No. 1081, 2 pp. (1976).

121.068 Inactive state of HZ Her.
W. Wenzel, R. Hudec.
Inform. Bull. Variable Stars, (I.A.U. Commission 27), Konkoly
Obs., Budapest, No. 1082, 2 pp. (1976).

121.069 Close binaries with H and K emission.
D. M. Popper.
Inform. Bull. Variable Stars, (I.A.U. Commission 27), Konkoly
Obs., Budapest, No. 1083, 3 pp., with a correction, No. 1090
(1976).

121.070 Photoelectric light-curves and minima of TZ Bootis.
N. Güdür, Ö. Gülmen, C. Ibanoglu, S. Bozkurt.
Inform. Bull. Variable Stars, (I.A.U. Commission 27), Konkoly
Obs., Budapest, No. 1086, 3 pp. (1976).

121.071 Photoelectric light curves of CN And.
S. Bozkurt, C. Ibanoglu, Ö. Gülmen, N. Güdür.
Inform. Bull. Variable Stars, (I.A.U. Commission 27), Konkoly
Obs., Budapest, No. 1087, 2 pp. (1976).

**121.072 New observations of the white dwarf eclipsing binary
V 471 Tauri.** C. Ibanoglu.
Inform. Bull. Variable Stars, (I.A.U. Commission 27), Konkoly
Obs., Budapest, No. 1088, 3 pp. (1976).

121.073 Photoelectric light curves of AS Cam.
Ö. Gülmen, C. Ibanoglu, S. Bozkurt, N. Güdür.
Inform. Bull. Variable Stars, (I.A.U. Commission 27), Konkoly
Obs., Budapest, No. 1090, 2 pp. (1976).

**121.074 Period and period change of the eclipsing binary
BV 549 Scorpii.** J. Rahe, E. Schöffel.
Inform. Bull. Variable Stars, (I.A.U. Commission 27), Konkoly
Obs., Budapest, No. 1092, 3 pp. = Veröff. Remeis-Sternw.
Bamberg – Astron. Inst. Univ. Erlangen-Nürnberg, Vol. 10, No.
117 (1976).

**121.075 Photoelectric minima and light curves of BS
Draconis.**
C. Ibanoğlu, S. Bozkurt, N. Güdür, Ö. Gülmen.
Inform. Bull. Variable Stars, (I.A.U. Commission 27), Konkoly
Obs., Budapest, No. 1100, 2 pp. (1976).

121.076 Radio emission from Wolf-Rayet binaries.
D. R. Florkowski, S. T. Gottesman.
Inform. Bull. Variable Stars, (I.A.U. Commission 27), Konkoly
Obs., Budapest, No. 1101 = Rosemary Hill Obs., Univ. Florida,
Gainesville, Contr. No. 65 (1976).

**121.077 Photoelectric V light curve of the eclipsing binary
RT Persei.** S. Mancuso, L. Milano.
Inform. Bull. Variable Stars, (I.A.U. Commission 27), Konkoly
Obs., Budapest, No. 1102, 2 pp. (1976).

**121.078 Further details about the new eclipsing binary
S 10796 = 71 Draconis.**

W. Fürtig, L. Meinunger.
Inform. Bull. Variable Stars, (I.A.U. Commission 27), Konkoly Obs., Budapest, No. 1104, 2 pp. (1976).

121.079 AN Ursae Majoris as an eclipsing binary.
G. S. Mumford.
Inform. Bull. Variable Stars, (I.A.U. Commission 27), Konkoly Obs., Budapest, No. 1109 (1976).

121.080 UBV photoelectric minimum of AO Velorum.
Z. Kvíz.
Inform. Bull. Variable Stars, (I.A.U. Commission 27), Konkoly Obs., Budapest, No. 1111, 4 pp. (1976).

121.081 BS 1099: a bright variable similar to the radio star UX Arietis. H. J. Landis, D. S. Hall.
Inform. Bull. Variable Stars, (I.A.U. Commission 27), Konkoly Obs., Budapest, No. 1113, 2 pp., with a correction, No. 1117 (1976).

121.082 W UMa: a new period variation.
B. Cester, U. Flora, F. Mardirossian, M. Pucillo.
Inform. Bull. Variable Stars, (I.A.U. Commission 27), Konkoly Obs., Budapest, No. 1114 (1976).

121.083 CSV 8871. T. Berthold.
Inform. Bull. Variable Stars, (I.A.U. Commission 27), Konkoly Obs., Budapest, No. 1115 (1976).

121.084 Photometry of eclipsing binary 1073 Cyg.
A. Dumitrescu, R. Dinescu.
Inform. Bull. Variable Stars, (I.A.U. Commission 27), Konkoly Obs., Budapest, No. 1116, 2 pp. (1976).

121.085 BV photometry of BM Cas.
S. Bozkurt, N. Güdür, Ö. Gülmen, C. Ibanoglu.
Inform. Bull. Variable Stars, (I.A.U. Commission 27), Konkoly Obs., Budapest, No. 1117, 3 pp. (1976).

121.086 Photoelectric B, V photometry of AI Hya.
N. Güdür, Ö. Gülmen, C. Ibanoglu, S. Bozkurt.
Inform. Bull. Variable Stars, (I.A.U. Commission 27), Konkoly Obs., Budapest, No. 1118, 2 pp. (1976).

121.087 Photoelectric minima of eclipsing binaries.
H. Brancewicz, J. M. Kreiner.
Inform. Bull. Variable Stars, (I.A.U. Commission 27), Konkoly Obs., Budapest, No. 1119 (1976).

121.088 Minima of RW Persei. H. Busch.
Inform. Bull. Variable Stars, (I.A.U. Commission 27), Konkoly Obs., Budapest, No. 1121 (1976).

121.089 Changes in the period of XZ Andromedae.
J. M. Kreiner.
Inform. Bull. Variable Stars, (I.A.U. Commission 27), Konkoly Obs., Budapest, No. 1122, 2 pp. (1976).

121.090 The eclipsing/spectroscopic binary HD 193964 = 71 Dra. D. P. Hube.
Inform. Bull. Variable Stars, (I.A.U. Commission 27), Konkoly Obs., Budapest, No. 1123 (1976).

121.091 Times of minima of RZ Ophiuchi, B. W. Baldwin.
Inform. Bull. Variable Stars, (I.A.U. Commission 27), Konkoly Obs., Budapest, No. 1127 (1976).

121.092 Recent photoelectric observations of UX Ursae Majoris. G. S. Mumford.
Inform. Bull. Variable Stars, (I.A.U. Commission 27), Konkoly Obs., Budapest, No. 1128, 2 pp. (1976).

121.093 Observe ϑ^1 Ori A! E. Lohsen.
Inform. Bull. Variable Stars, (I.A.U. Commission 27), Konkoly Obs., Budapest, No. 1129, 4 pp. (1976).

121.094 CaII H and K emission of RT Lac.
J. L. Droppo, E. F. Milone.
Inform. Bull. Variable Stars, (I.A.U. Commission 27), Konkoly Obs., Budapest, No. 1130, 2 pp. = Rothney Astrophys. Obs., *Calgary, Alberta, Canada,* Publ. Ser. B, No. 3 (1976).

121.095 Light elements of HD 123732.
R. F. Sisteró, M. E. Castore de Sisteró.
Inform. Bull. Variable Stars, (I.A.U. Commission 27), Konkoly Obs., Budapest, No. 1132 (1976).

121.096 Thirty eclipsing binaries probably 100 pc distant from the sun. T. Z. Dworak.
Inform. Bull. Variable Stars, (I.A.U. Commission 27), Konkoly Obs., Budapest, No. 1136 (1976).

121.097 Photoelectric minima of eclipsing variables.
L. Baldinelli, S. Ghedini.
Inform. Bull. Variable Stars, (I.A.U. Commission 27), Konkoly Obs., Budapest, No. 1143 (1976).

121.098 Hα emission in the eclipsing white dwarf V471 Tau (BD+16°516). H. H. Lanning, P. B. Etzel.
Inform. Bull. Variable Stars, (I.A.U. Commission 27), Konkoly Obs., Budapest, No. 1147, 2 pp. (1976).

121.099 θ^1 Orionis A.
IAU Circ., No. 2905 (1976).

121.100 σ Orionis E. J. E. Hesser, P. Ugarte.
IAU Circ., No. 2911 (1976).

121.101 HR 1099 F. N. Owen.
IAU Circ., No. 2929 (1976).

121.102 Observations of eclipsing binaries in 1974–1975.
Z. Pokorný.
Contr. Nicholas Copernicus Obs. and Planetarium Brno, No. 20, p. 4 - 17 (1976). In Czech and English.

121.103 Troubles with V 1068 Cygni.
K. Brančík, Z. Pokorný, K. Raušal.
Contr. Nicholas Copernicus Obs. and Planetarium Brno, No. 20, p. 18 (1976). In Czech and English.

121.104 Corrected light elements for DG Lac.
L. Kozina.
Contr. Nicholas Copernicus Obs. and Planetarium Brno, No. 20, 19 - 20 (1976). In Czech and English.

121.105 La prossima eclisse della binaria RZ Ophiuchi.
L. Baldinelli, S. Ghedini.
Giorn. A.A.B., Anno 11, No. 42, p. 8 (1976).

121.106 1973 - 75 UBV photometry of the eclipsing binary BM Orionis. C. N. Arnold, D. S. Hall.
Acta Astron., Vol. 26, 91 - 107 (1976).
The authors present new UBV differential observations obtained in 1973 - 75 : 212 in V, 45 in B, 12 in U. The observing technique is discussed in detail. All check star observations, 45 Ori versus θ^2 Ori B, obtained between 1967 and 1975 are presented. A time of minimum derived from the mean light curve in V is JD (hel.) = 2442096.505 ± 0^{d}001.

121.107 A period study for RY Geminorum.
D. S. Hall, T. Stuhlinger.
Acta Astron., Vol. 26, 109 - 116 (1976).

All available times of mid primary eclipse have been collected and analyzed to show that the period decreased around 1950 by $\Delta P/P = -3.8 \times 10^{-5}$. This decrease could have been produced by a sudden outflow of about $10^{-5}\ M_\odot$ from the cooler star, most of which left the system.

121.108 On the effect of time-shift of secondary minimum in eclipsing binaries.
T. Z. Dworak, L. Frasiński.
Astron. Rep., Vol. 1, 3 - 6 (1974).

121.109 Minima of 44 i Bootis.   P. Flin.
Astron. Rep., Vol. 1, 7 - 8 (1974).

121.110 The variable star V 688 Aquilae.
M. Kurpińska.
Astron. Rep., Vol. 1, 25 - 32 (1974).

121.111 Minima of eclipsing variables.
J. Madej, Z. Małas.
Astron. Rep., Vol. 1, 33 - 34 (1974).

121.112 The problem of the period of CO Leonis.
L. Puchalska, K. Rudnicki.
Astron. Rep., Vol. 1, 35 - 37 (1974).

121.113 Luminosities of the W UMa-type systems.
S. M. Rucinski.
Mem. Soc. Astron. Italiana, Vol. 45, (see 012.017), 799 - 804 (1974).

121.114 Possible observational evidence for a black hole in Beta Lyrae system.   M. Hack, J. Hutchings,
Y. Kondo, G. McCluskey, M. Plavec, R. Polidan.
Mem. Soc. Astron. Italiana, Vol. 45, (see 012.017), 805 - 806 (1974). – Abstract.

121.115 The eclipsing binary VV Orionis.
H. W. Duerbeck.
Mem. Soc. Astron. Italiana, Vol. 45, (see 012.017), 807 - 809 (1974).

121.116 Lichtelektrische Messungen und fotometrische Elemente des langperiodischen Bedeckungssternes EE Cephei.   L. Meinunger.
MVS Sonneberg, Vol. 7, 97 - 103 (1976).

121.117 Lichtelektrische Messungen von V640 Aquilae.
W. Wenzel, L. Meinunger.
MVS Sonneberg, Vol. 7, 104 (1976).

121.118 Bemerkungen zum Spektrum des langperiodischen Bedeckungssterns EE Cephei.   V. Brückner.
MVS Sonneberg, Vol. 7, 114 (1976).

121.119 Neue unregelmäßige Periodenänderungen von RW Tauri.   P. Ahnert.
MVS Sonneberg, Vol. 7, 123 - 127 (1976).

121.120 Beitrag zur Apsidendrehung von V477 Cygni.
P. Ahnert.
MVS Sonneberg, Vol. 7, 127 - 130 (1976).

121.121 Periodenänderung des W-Ursae-Maioris-Sterns OO Aquilae.   P. Ahnert.
MVS Sonneberg, Vol. 7, 131 - 135 (1976).

121.122 Lists of minima of eclipsing binaries.
Compiled by R. Diethelm, A. Figer, R. Germann, K. Locher, H. Peter, A. Royer, V. Tuboly, G. Zajàcz, M. Behagle, R. Boninsegna, J. Bourgeois, P. Carnevali, J.-P. Clovin,
P. Doby, A. Fenyvesi, M. Frangeul, J.-F. Le Borgne, R. Leydon, A. Marot, E. Poretti, P. Ralincourt, J. Remis, A. Seretti, M. Wilmet, G. Troispoux, P. Aresi, M. Le Saout.
BBSAG Bull., No. 25, p. 1 - 3; No. 26, p. 1 - 4; No. 27, p. 1 - 4; No. 28, p. 1 - 5 (1976). – 58th - 61st list of Swiss Astronomical Society's Eclipsing Variables Observers.

121.123 BV 1616 Lep: very probably not an eclipsing variable.   K. Locher.
BBSAG Bull,, No. 25, p. 3 - 4 (1976).

121.124 EP Andromedae: definite elements for the reinterpretation.   K. Locher.
BBSAG Bull., No. 25, p. 5 (1976).

121.125 Missed minima of RS Crateris.   K. Locher.
BBSAG Bull., No. 26, p. 5 (1976).

121.126 The true amplitude of AA Ceti.   K. Locher.
BBSAG Bull., No. 26, p. 6 (1976).

121.127 Elements for DM Delphini.   R. Diethelm.
BBSAG Bull., No. 27, p. 5 (1976).

121.128 Period changes in RZ Draconis.   I. D. Howarth.
BBSAG Bull., No. 27, p. 6 (1976).

121.129 AZ Virginis. Unprejudiced surveys support Busch's 'b'-interpretation.   K. Locher.
BBSAG Bull., No. 27, p. 6 - 7 (1976).

121.130 YZ Canum Venaticorum: detection of the period.
K. Locher.
BBSAG Bull., No. 27, p. 7 (1976).

121.131 The rough photometric parameters of V752 Ophiuchi.   K. Locher.
BBSAG Bull., No. 28, p. 5 (1976).

The application of a new method for the determination of relative radial velocities with the Fehrenbach-prism.
See Abstr. 031.205.

Theoretical values of the limb darkening coefficients for dwarf stars in the B, V system.   See Abstr. 064.040.

U Cephei: a mass-transfer event. II. Observations.
See Abstr. 113.043.

On the nature of Sigma Orionis E.
See Abstr. 114.338.

The ultraviolet spectrum of Beta Lyrae. II.
See Abstr. 114.354.

An evolutionary model of Beta Lyrae.
See Abstr. 117.006.

Appeal for photoelectric observations of Upsilon Sagittarii.   See Abstr. 119.016.

The period of AZ Cassiopeiae.
See Abstr. 119.017.

Eclipses of U Geminorum.   See Abstr. 122.109.

An improved spectrosopic orbit for the white dwarf eclipsing binary BD + 16°516.   See Abstr. 126.008.

Mass determination for the X-ray binary system Vela X-1.   See Abstr. 142.034.

Spectrophotometric observations of the X-ray binary HD 153919 = 3U 1700-37. See Abstr. 142.038.

Parameters of Sanduleak 160 (SMC X-1) by differential corrections. See Abstr. 142.043.

Hydrogen and helium lines in theoretical models of Scorpius X-1 and Cygnus X-2. See Abstr. 142.076.

Analysis of a Scorpius X-1 X-ray spectrum obtained with cooled silicon (Li) detectors. See Abstr. 142.077.

Tidal circularization of the binary X-ray sources Hercules X-1 and Centaurus X-3. See Abstr. 142.078.

Radial-velocity variations of Scorpius X-1 emission lines. See Abstr. 142.095.

122 Intrinsic Variables, Flare Stars, Pulsation Theory

122.001 **Red variables in the central bulge of the Galaxy — I. The period distribution of Mira variables.**
T. Lloyd Evans.
Monthly Notices Roy. Astron. Soc., Vol. 174, 169 - 184 (1976).

Baade's three galactic centre windows in the Sagittarius star cloud have been searched for Mira variables using V and I plates. Periods have been determined for 121 red variables of large amplitude. Most of these stars are Mira variables of spectral type M. The dependence of luminosity on period is conspicuous, so that the apparent magnitude limit of the observations imposes a cut-off at long periods. There is a rapid decline in the surface density of Mira variables with increasing angular distance from the galactic centre. The period distribution of the Miras extends to much longer periods than that of those in globular clusters. Several lines of evidence suggest that the presence of a sufficiently large population of stars of age less than 4×10^9 yr in the central bulge of the Galaxy is unlikely, so that the Miras of all periods are presumably older than this.

122.002 **Frequency analysis of the Delta Scuti star, Theta Tucanae.** R. S. Stobie, R. R. Shobbrook.
Monthly Notices Roy. Astron. Soc., Vol. 174, 401 - 409 (1976).

Photometric observations of θ Tucanae were Fourier analysed for their component frequencies. No coherent frequency could be found spanning the complete data set. It is shown that both the frequencies and amplitudes present in θ Tucanae change on a time scale as short as 24 hr.

122.003 **$(R - I)$ colors of cepheids and yellow supergiants in open clusters.** E. G. Schmidt.
Astrophys. Journ., Vol. 203, 466 - 476 (1976).

$(R - I)$ and $(b - y)$ observations of nine cepheids which are members of open clusters, 14 field cepheids, and 34 nonvariable yellow giants in or near open clusters are presented. Color-color relations for the cepheids are found to differ significantly from star to star. Additionally, the nonvariables may differ from the cepheid color-color relation. Pulsational masses are determined for the cepheids using both $(R - I)$ and $(B - V)$ colors as temperature indicators. The cluster

NGC 129 is found to have a nonvariable star within the instability strip in addition to the cepheid DL Cas. Because the absolute magnitude and intrinsic color are well determined from the cluster, this is the best established case of a nonvariable in the cepheid strip.

122.004 **The masses of cataclysmic variables.**
E. L. Robinson.
Astrophys. Journ., Vol. 203, 485 - 489 (1976).

Masses are derived for the individual components of six cataclysmic variables. There is a considerable spread in the masses of the white dwarf components, which range from 0.73 $M_\odot$ in EM Cyg to 1.26 $M_\odot$ in Z Cam. All of the white dwarfs have masses greater than 0.70 $M_\odot$, but there is no evidence that there is any preferred mass. It is found that the morphology of the eruptions (i.e., nova versus dwarf nova) is independent of the mass of both the white dwarf and the late-type star.

122.005 **Simultaneous radio and optical observations of UV Ceti-type flare stars.**
S. R. Spangler, T. J. Moffett.
Astrophys. Journ., Vol. 203, 497 - 508 (1976).

The UV Ceti-type stars, YZ Canis Minoris, AD Leonis, and Wolf 424 AB, were monitored using the 1000 foot radio telescope at Arecibo and the 91 or 76 cm reflector at McDonald Observatory. Radio observations were made at frequencies of 430, 318, and 196 MHz, and optical monitoring was done in the Johnson U or B band, the Strömgren u band, or white light. During the period of simultaneous observations, 62 optical flares were detected. A total of 13 radio frequency flares were independently identified, of which 10 reached maximum flux within 10 min of the peak time of an optical flare. The most probable delay interval between the optical flare maximum and 318 MHz radio peak was found to be 0 to 5 min with the optical flare occurring first. The possible consequences of the radio emission being due to a coherent process are discussed.

122.006 **Radio emission from a source near the flare star AD Leonis.**
R. D. Robinson, O. B. Slee, A. G. Little.
Astrophys. Journ., (Letters), Vol. 203, L91 - L93 (1976).

A weak, probably variable, source has been located $1!85$ southeast of the flare star AD Leo. Characteristics of this source are discussed, and an interpretation is given based on plasma emission from a stellar companion to the flare star.

122.007 Interpretation of the spectral and photometric variations of R Coronae Borealis.
L. Hartmann, J. P. Apruzese.
Astrophys. Journ., Vol. 203, 610 - 615 (1976).

A consistent picture explaining both the spectral and photometric (infrared and visual) variations of R Coronae Borealis is developed. It is found that the extreme infrared maximum and minimum can be fitted by dispersion of the same mass of graphite dust. Random clumps of this dust crossing the line of sight both obscure the visual continuum of the object and reveal an emission spectrum from a circumstellar gas cloud; which satisfactorily accounts for the correlation of the emission with episodes of a faint visual continuum.

122.008 The three radial modes and evolutionary state of AC Andromedae. W. S. Fitch, B. Szeidl.
Astrophys. Journ., Vol. 203, 616 - 624 (1976).

Analysis of 4662 yellow magnitude measures shows that AC And has its fundamental and first and second overtone radial pulsation modes (periods $P_0 = 0\overset{d}{.}71124243$, $P_1 = 0\overset{d}{.}52512677$, and $P_2 = 0\overset{d}{.}421069$) all excited and nonlinearly coupled. The multimode excitation of 25 cepheid-strip stars (comprising eight cepheids, one RR Lyrae star, nine RRs stars, and seven δ Scuti stars) is discussed. AC And is found to have a mass $M \approx 3.1\ M_\odot$ and to be a high-mass analog of the δ Scuti stars, in the hydrogen-shell-burning and helium-core-contraction stage of evolution. Pulsation theory, evolution theory, and observation seem to be in excellent agreement for AC And and δ Sct.

122.009 Photoelectric UBV observations of three RR Lyrae stars in ω Cen. C. R. Sturch.
Astron. Astrophys., Vol. 46, 133 - 134 (1976).

Photoelectric UBV observations of three RR Lyrae stars of type ab in ω Cen are shown to agree with the photographic color indices found by Dickens and Saunders but not with those found by Geyer and Szeidl. The photoelectric data confirm the metal-weakness of the cluster.

122.010 The red irregular variable LV 60 in the direction of the SMC. A. M. van Genderen.
Astron. Astrophys., Vol. 46, 185 - 187 (1976).

Photo-electric observations (Walraven system) of the red irregular variable LV 60 projected onto the SMC are presented. The variation in brightness is nearly 1 mag. The variation in the colours is peculiar. Its relationship with M type variable is discussed. It is unknown whether it is a galactic foreground giant or a member of the SMC, in which case it is a supergiant.

122.011 Photoelectric UBV photometry of six cepheids.
A. A. Wachmann.
Astron. Astrophys., Suppl. Ser., Vol. 23, 249 - 259 (1976).

Photoelectric UBV photometry of six δ Cep stars is discussed. The observations of IR Cep, V924 Cyg, V1154 Cyg, EU Tau, V526 Mon and GQ Ori are presented as well as their mean V light, $(U-B)$ and $(B-V)$ colour curves.

122.012 The study of 4 RR Lyrae type variables.(II) XX and WW Bootis. G. Chiş, D. Chiş, I. Mihoc.
Stud. Univ. Babeş-Bolyai, Math., Anul 20, p. 75 - 80 (1975). In Rumanian.

Results of photographic observations of the RR Lyrae stars XX and WW Bootis are reported. The observations have been carried out at the Astronomical Observatory of the Cluj-Napoca University, during the 1959–66 period. XX Bootis seems to be of RR_{ab} type and WW Bootis seems to be of RR_a type.

122.013 Interpretation of OAO-2 ultraviolet light curves of β Doradus.
J. L. Hutchinson, S. J. Hill, C. F. Lillie.
Astron. Journ., Vol. 80, 1044 - 1049 (1975).

OAO-2 observations of β Doradus are presented. A simple hydrodynamical model containing atmospheric shock waves describes the origin of the bumps of these UV light curves.

122.014 Period changes of RR Lyrae variables in M14.
A. Wehlau, J. Conville, H. Sawyer Hogg.
Astron. Journ., Vol. 80, 1050 - 1058 (1975).

Observations of the globular cluster M14 extending from 1932 to 1974 have been used to study the rate of period change for 32 of the most regular and best observed RR Lyrae variables. While the time span of only 42 years lends some uncertainty to the results there does not appear to be a strong predominence of variables with increasing period such as has been found by others for the Oosterhoff type II clusters ω Centauri and M15. The distribution of period changes appears somewhat similar to that found for the other Oosterhoff type I clusters M3 and M5.

122.015 Survey of pulsation in RR Lyrae star models.
W. H. Spangenberg.
Thesis Colorado Univ., Boulder, USA, 237 pp. (1975).
(Available from Univ. Microfilms, Order No. 75-23,649).

122.016 UV Ceti stars: statistical analysis of observational data.
C. H. Lacy, T. J. Moffett, D. S. Evans.
Astrophys. Journ., Suppl. Ser., Vol. 30, 85 - 96 (1976).

A statistical analysis of 386 flares of eight UV Ceti flare stars observed by Moffett indicates (1) no apparent correlation between times of flare occurrence; (2) no apparent correlation between flare energy and the time interval since the preceding flare; (3) flare energies in the B and V filters and C (no filter) are directly proportional to the flare energy in the U-band; (4) the quiescent luminosity in the U-band is strongly correlated with the mean flare energy, mean rate of energy loss due to flaring, and spectral index. Cumulative distributions of flare energy with frequency of occurrence are constructed, from which the mean rate of energy loss due to flaring is calculated, with the effects of event detection thresholds being taken into account.

122.017 R Coronae Borealis type stars. Part II.
J. Krełowski.
Postępy Astron., Vol. 23, 243 - 255 (1975). In Polish.

The paper includes the main results of spectrophotometric observations and of the multicolour photometry obtained since 1972. It includes also a brief description of polarimetric measurements performed at the time of the R CrB minimum. The model of the R CrB phenomenon is described together with a discussion of its accordance with observational data.

122.018 Period changes of double eruptive stars.
J. M. Kreiner.
Postępy Astron., Vol. 23, 281 - 288 (1975). In Polish.

The $O - C$ diagrams with a short discussion of the observed period changes for 11 double eruptive stars are given.

122.019 Photoelectric observations of cepheids in the galactic centre region. P. R. Warren, G. M. Harvey.
Monthly Notices Roy. Astron. Soc., Vol. 175, 129 - 147 (1976).

Photoelectric B and V observations are presented for 13 cepheids in the galactic centre region. The problem of the population assignments of these variables is discussed.

122.020 **The wavelength dependence of polarization of Mira stars compared with the interstellar polarization in a field near the galactic equator.** J. Materne.
Astron. Astrophys., Vol. 47, 53 - 58 (1976).

The wavelength dependence of the polarization of three Mira stars was observed at different phases. The particle size of the grains in the circumstellar cloud was found to be phase dependent. The existence of an eruption point, as proposed by Fischer (1969) and Shawl (1972) could be confirmed. The amount of interstellar polarization and the wavelength dependence near the galactic equator is discussed as a correction to the stellar polarization.

122.021 **A detailed photometric and spectroscopic study of the short-period classical cepheid SU Cas.**
W. Gieren.
Astron. Astrophys., Vol. 47, 211 - 219 (1976).

Light curves in *U, B* and *V* and a simultaneously obtained radial velocity curve of the short-period classical cepheid SU Cas are presented and discussed. The measurements give strong evidence that SU Cas is pulsating in the first harmonic mode. Consequences for the period-luminosity-colour relation are briefly discussed.

122.022 **The linear polarization of BY Draconis.**
R. H. Koch, R. J. Pfeiffer.
Astrophys. Journ., (*Letters*), Vol. 204, L47 - L49 (1976).

The authors report linear polarization measurements in four bandpasses for the flare star BY Dra. The red polarization is intrinsically variable at a confidence level greater than 99 percent. On a time scale of many months, the variability is not phase-locked to either a rotational or a Keplerian ephemeris. The observations of the three other bandpasses are useful principally to indicate a polarization spectrum rising toward shorter wavelengths.

122.023 **Zeeman observations of the Hα region in BY Draconis.**
C. M. Anderson, L. W. Hartmann, B. W. Bopp.
Astrophys. Journ., (*Letters*), Vol. 204, L51 - L54 (1976).

The results of high-dispersion spectrographic observations of the Hα region in the spotted flare star BY Draconis are presented. A "Zeeman" spectrogram obtained with the starspot near the limb shows obvious differences between the Hα emission profile seen in the different senses of circular polarization. If this is interpreted as longitudinal Zeeman effect, a field of 40 kilogauss is implied. The photospheric absorption-line spectrum shows no gross evidence of large magnetic fields, but the sensitivity of these is such that fields of several kilogauss can not be ruled out.

122.024 **Evolutionary problems of cepheids and other giants investigated with new radiative opacities.**
T. R. Carson, R. Stothers.
Astrophys. Journ., Vol. 204, 461 - 471 (1976).

New evolutionary tracks, pulsation constants, and pulsational stability coefficients have been calculated for stellar models applicable to the problems of classical cepheids, by adopting (*a*) the standard Cox-Stewart opacities and (*b*) the new Carson opacities. The theoretical mass-luminosity relation is slightly fainter for the new opacities. The masses of cepheids inferred from the pulsation constants and from the pulsational period ratios are significantly smaller than the masses inferred from evolutionary theory, for both sets of opacities, but the masses that can be inferred from the secondary bump on the radial-velocity curve are now predicted to agree with the evolutionary masses if the new opacities are used. The present improvements are achieved with a normal helium abundance, no mass loss, and a conventional treatment of convection. The predicted transition between second-overtone instability and fundamental-mode instability agrees excellently with the well-observed transition in the Small Magellanic Cloud.

122.025 **Sympathetic stellar flares and electron precipitation as probes of coronal structure in flare stars.**
D. J. Mullan.
Astrophys. Journ., Vol. 204, 530 - 538, with a correction, Vol. 206, 672 (1976).

The author suggests that short-period (13 s) oscillations observed during a large flare in the Hyades flare star H II 2411 are the whistler analog of sympathetic flares, and are due to electron precipitation triggered by whistlers propagating between starspots at opposite rotation poles of the star. According to this model, if starspot fields are as strong ($\approx 10^4$ gauss) as a recent starspot model suggests, then the coronal density at about $7-8$ stellar radii in H II 2411 is in the range $(0.5-10) \times 10^6$ cm^{-3}. These densities are larger by factors of $20-500$ than the densities in the solar corona at $7-8$ solar radii. The author finds that the coronal temperature in the flare star is lower than in the solar corona by a factor of $3-10$.

122.026 **Spectrophotometry of R Coronae Borealis during the minimum of 1974.**
R. S. Patterson, J. D. Fix, J. S. Neff.
Astrophys. Journ., Vol. 204, 838 - 841 (1976).

Spectrophotometric observations of R Coronae Borealis were obtained as the star returned to normal brightness during the minimum of 1974. Absolute flux distributions and extinction optical thicknesses were determined. A close fit to the rising branch of the visual light curve of R CrB was obtained using a simple model in which the return to maximum brightness of R CrB is caused by the radiation pressure dispersal of a cloud of graphite particles.

122.027 **A search for periodicities in the pulsation of δ Scuti stars. II. The δ Scuti star 44 Tau (HR 1287).**
N. Morguleff, B. Rutily, A. Terzan.
Astron. Astrophys., Suppl. Ser., Vol. 23, 429 - 438 (1976).
In French.

New observations of the δ Scuti variable star 44 Tau were carried out in 1970-1971 simultaneously on the 80 cm photometric telescope and with the coudé-spectrograph of the 152 cm telescope of the Observatoire de Haute-Provence. The analysis of these observations combined with the previously published ones reveals that no periodicities are present in the pulsation of 44 Tau.

122.028 **Scanner observations of M-type Mira variables.**
H. Maehara, Y. Yamashita.
Publ. Astron. Soc. Japan, Vol. 28, 135 - 140 (1976).

The energy distribution in the range $\lambda\lambda 3700-5500$ Å was observed for several Mira variables of M-type. It is shown that the distribution of radiative fluxes measured at several flux peaks is an indicator of the intensities of TiO bands rather than an indicator of temperature. A brief discussion considers the possible correlation between the so-called general line weakening in Mira variables and an enhanced TiO opacity.

122.029 **Preliminary results of observations of IR-spectra of T Tau-type stars.** G. I. Shanin, A. G. Shcherbakov.
Issled. ekhstremal'no molodykh zvezdn. kompleksov.
Tashkent, Fan, 1975, p. 68 - 83. In Russian. − Abstr. in Referativ. Zhurn. 51. Astron., 3.51.432 (1976).

122.030 **Radial pulsations of δ Scuti stars. II. The blue edge of the instability strip: theory in comparison with observations.** A. A. Pamyatnykh.
Stellar physics and evolution, (see 003.001), p. 40 - 57 (1974).
In Russian.

The vibrational stability of δ Sct star models of $1.5-2.25$ $M_\odot$ is studied in linear approximation. The periods of pulsations are computed and the location of the blue edges is determined for the four lowest modes. The theoretical blue edges agree with observations when the helium abundance is between $Y = 0.28$ and 0.38. Assuming that the masses of δ Sct stars are in the range $1.5-2.25\ M_\odot$, the modes of observed pulsations are determined and the distribution of the number of the stars according to the modes is discussed.

122.031 UBV photometry of flare stars in the Pleiades.
　　O. S. Chavushyan, A. T. Garibdzhanyan.
Astrofizika, Vol. 11, 565 - 578 (1975). In Russian. English translation in Astrophysics, Vol. 11, No. 4.

　　The results of UBV photometry of 283 flare stars in the minimum in the Pleiades region are presented. They show that on the diagram (V, B−V) the flare stars are situated on both sides of the main sequence, while on the (U−B, B−V) diagram they lie mainly above the main sequence.

122.032 The light variation of Delta Scuti.
　　A. Muir, W. Wehlau.
Astrophys. Journ., Vol. 205, 155 - 161 (1976).

　　New photoelectric observations of δ Scuti are presented. These and older observations by Fath are analyzed using periodograms constructed with mean light curves. Somewhat different periods are derived from the two sets of data. The best representation of the data is given by a period which is increasing at the rate of 2.2×10^{-7} days yr^{-1}.

122.033 Evidence favoring nonevolutionary cepheid masses.
　　N. R. Simon, E. G. Schmidt.
Astrophys. Journ., Vol. 205, 162 - 164 (1976).

　　It is suggested that the bumps on the velocity curves of nonlinear cepheid models may be understood as the consequence of a resonance between the fundamental and the second overtone modes of the pulsating star. Thus the presence of bumps can be inferred from linear calculations. The region of the observational instability strip containing bumps is compared with the location of the calculated resonances. As was the case for the nonlinear calculations, the authors find that masses considerably less than evolutionary values are required to bring agreement between the observations and the theory.

122.034 Shock-wave interpretation of emission lines in long-period variable stars. I. The velocity of the shock.
L. A. Willson.
Astrophys. Journ., Vol. 205, 172 - 181 (1976).

　　The emission lines seen in long-period variables are classified according to the mechanism by which they are formed. This leads to information on the physical conditions in several regions, and especially to an estimate of the velocity differences between regions. The results are interpreted in terms of a spherical shock wave expanding outward at $v \approx 50$ km s^{-1} and gradually decelerating.

122.035 Periodogram analysis of semi-regular variable stars −
　　(I) S Persei.　　I. D. Howarth.
Journ. British Astron. Ass., Vol. 86, 210 - 213 (1976).

122.036 Some regularities in color variation of long-period variable stars.　　T. K. Kiseleva.
Astron. Tsirk., No. 856, p. 1 - 2 (1975). In Russian.

122.037 Some peculiarities of MS Mira-type star spectra.
　　M. V. Dolidze.
Astron. Tsirk., No. 856, p. 7 - 8 (1975). In Russian.

122.038 Brightness gradients of red variable stars from UBVRI observations.　　A. Eh. Rozenbush.
Astrometriya i Astrofizika, *Kiev,* vyp. (No.) 29, (see 003.003), p. 66 - 70 (1976). In Russian.

　　Brightness gradients are given for 62 variable stars of types M and SR. Relations between any two magnitudes ob-

tained simultaneously proved to be linear-regression ones. Four gradient diagrams are analyzed. The diagram (∇_{UB}, ∇_{VB}) was found to be the most appropriate for studying changes in the spectral distribution of stellar radiation. In this diagram the M and SR variables are more distinct.

122.039 Spectral variations of SS Cyg during rapid light fluctuations in the minimum.
N. F. Vojkhanskaya.
Astrofiz. Issled., Izv. Spets. Astrofiz. Obs., Vol. 8, 3 - 13 (1976). In Russian.

　　The emission lines are identified and their equivalent widths, W, are determined in the spectrum of SS Cyg. The variation of W values with minimum light fluctuations as dependent on the excitation potential of the upper level of a line is detected. The emission line spectrum undergoes rapid changes during the time of the order of minutes. It is shown that brightness fluctuations are mainly due to continuum variations, the hot component of the system being their source. An assumption is made on the existence of an extended envelope of small gaseous nebula type. Hα- and Hβ-line contours are constructed.

122.040 Spectrophotometric study of the unusual variable A stars γ Boo and γ UMi.
R. N. Kumajgorodskaya, N. M. Chunakova.
Astrofiz. Issled., Izv. Spets. Astrofiz. Obs., Vol. 8, 132 - 134 (1976). In Russian.

　　The character of behaviour of different parameters of hydrogen lines and the K line of ionized calcium in the spectra of the unusual variable stars γ Boo, γ UMi is studied, and preliminary conclusions as for their belonging to peculiar stars are drawn.

122.041 A search for slowly varying radio continuum emission from UV Ceti stars.
S. R. Spangler, S. D. Shawhan.
Astrophys. Journ., Vol. 205, 472 - 474 (1976).

　　A search has been made for a 430 MHz radio counterpart to the optical quiescent variations of flare stars for the stars EQ Peg, YZ CMi, and AD Leo using the Arecibo radiotelescope. No discernible emission was detected.

122.042 Spectrophotometric determination of the temperature of the cepheid ζ Geminorum.
B. S. Rautela, S. C. Joshi.
Astrophys. Space Sci., Vol. 40, 455 - 460 (1976).

　　The absolute energy distributions of the cepheid ζ Gem at several phases of the light cycle have been given. By matching these with suitable model atmospheres, the effective temperatures of the star at these phases have been determined. The radius and effective gravity variations as well as the mass have been derived.

122.043 Simultaneous observations of variable stars. II. The Al Velorum star SX Phe.
R. Haefner, K. Metz, R. Schoembs.
Astron. Astrophys., Vol. 49, 107 - 118 = Veröff. Sternw. *München,* Band 7, No. 21 (1976).

　　Simultaneous spectroscopic, polarimetric and photometric observations have been carried out for the Al Velorum star SX Phe. The results partly confirm earlier observations. By applying special observational methods and reduction the authors found some new aspects, but it was not possible to find a model consistent with all observational data.

122.044 Multicolour TV photometry of T Tau-stars. I. Methods of observations and preliminary results.
P. P. Petrov.
Izv. Krymskoj Astrofiz. Obs., Vol. 54, 42 - 68 (1976). In Russian.

Observations of several T Tau-stars had been made 1971 to 1973 with a TV-technique at the Crimean Observatory. A special seven-colour photometric system had been used for measuring the energy distribution within the wavelength range 3650–7250 Å and equivalent widths of Hβ and Hα. The observations show three patterns of variability, which are typical for different groups of stars. It is shown that the groups of T Tau-stars attributed to the different types of variability have different locations in the H–R diagram. Discrepancies between spectral types and observed energy distributions appear to be dependent on the spectral type. This effect can be caused by reddening in circumstellar envelopes.

122.045 Multicolour TV photometry of T Tau-stars. II. On the variability of T Tau-stars. P. P. Petrov.
Izv. Krymskoj Astrofiz. Obs., Vol. 54, 69 - 84 (1976). In Russian.

The analysis of the observational data given in paper I is continued. Absolute fluxes of radiation from photospheres and radiation from gas envelopes in the Balmer continuum (F_{UV}), Hβ and Hα lines ($F\beta$, $F\alpha$) are calculated. Variables attributed to the different groups according to the type of energy distribution variability are found to differ in F_{UV} fluxes. High-luminosity stars have appreciably lower F_{UV} fluxes than stars of low luminosity. The discrepancies between fluxes of radiation from 1 cm² of stellar surface, calculated by different methods, can be explained by non-radiative heating of matter emitting in infrared and/or neutral absorption in circumstellar dust envelopes. Quantitative estimates of the variability of the absolute fluxes F_{UV}, $F\beta$, and $F\alpha$ are derived. It was found that the type of variability of these fluxes is dependent on F_{UV}. A mechanism causing reversed correlation between radiative flux from the photosphere and non-radiative flux exciting gas envelope emission is proposed.

122.046 Magnitudes, colours and photometric period of BY Dra in the years 1973 - 1974.
P. F. Chugajnov.
Izv. Krymskoj Astrofiz. Obs., Vol. 54, 85 - 88 (1976). In Russian.

Photoelectric observations of BY Dra were carried out in the time interval from September 1973 to September 1974. A decrease of the photometric period is found. Variations of V magnitudes and $(B - V)$ colour are represented by the spot model computed by Torres and Mello (1973). Comparison with observations of the years 1965 - 1971 shows that the mean colours $(B - V)$ and $(U - B)$ continue to change.

122.047 On the existence of variable stars similar to BY Dra with masses of the order of solar mass.
P. F. Chugajnov.
Izv. Krymskoj Astrofiz. Obs., Vol. 54, 89 - 98 (1976). In Russian.

The light variability of main-sequence stars of spectral classes G5 Ve – K7 Ve has been studied. Four of ten stars included in the programme of photoelectric observations show variability. The periodicity of light variations and the flare activity of these four stars point out definitely their similarity to the variables of spectral classes K7 Ve – M4 Ve known as BY Dra-type stars. Thus, the explanation that the periodicity of light variations is due to the rotation of a star having a cool spot on its surface is also usable for some stars with masses of the order of solar mass.

122.048 Radial-velocity variations of SZ Lyncis and EH Librae. D. H. McNamara, K. A. Feltz, Jr.
Publ. Astron. Soc. Pacific, Vol. 88, 164 - 167 (1976).

New radial-velocity observations covering complete cycles of the dwarf cepheids (RRs stars) SZ Lyn and EH Lib are reported. The velocity amplitude and mean velocity of

SZ Lyn are $2K$ = 44 km sec⁻¹ and γ = 22 km sec⁻¹; corresponding values for EH Lib are $2K$ = 45.5 km sec⁻¹ and γ = -55 km sec⁻¹. The radial excursions and accelerations of the atmospheres of the two stars are also discussed.

122.049 A photometric and spectrographic study of DE Lacertae. D. H. McNamara, C. D. Laney.
Publ. Astron. Soc. Pacific, Vol. 88, 168 - 173 (1976).

New $uvby\beta$ photometry and radial-velocity measurements of DE Lac are reported. The star appears to be more closely related to the dwarf cepheids (RRs variables) than to the RRc variables.

122.050 Radio sources in the vicinity of flare stars.
S. R. Spangler.
Publ. Astron. Soc. Pacific, Vol. 88, 187 - 191 (1976).

The positions of 22 UV Ceti-type flare stars have been surveyed with the Arecibo radio telescope at a frequency of 430 MHz in a search for close and possibly physically associated radio sources. The active flare stars AD Leo and CN Leo have radio sources within two arc minutes. The probability of this resulting from a random association is estimated to be of the order of 5 %.

122.051 Recent period study of two variables in Libra.
M. Taylor.
Journ. American Ass. Variable Star Observers, Vol. 4, 58 - 61 (1975/76).

New estimates on Harvard patrol plates are used to derive ephemerides for two variables.

122.052 Six long period variables and one peculiar variable in Sagittarius. J. Lukas.
Journ. American Ass. Variable Star Observers, Vol. 4, 64 - 65 (1975/76).

New or updated periods are given for six Mira-type stars, and a new irregular variable is discussed. Finder charts are presented for all.

122.053 Binocular observations of three cepheids.
R. S. Thompson.
Journ. American Ass. Variable Star Observers, Vol. 4, 66 - 67 (1975/76).

The light curves of RV Sco, BF Oph, and AP Sgr are presented with visual comparison sequences. Recent values of O–C and M–m for the stars are included.

122.054 The classical cepheid program JD 2,441,000 - 2,442,000. T. A. Cragg.
Journ. American Ass. Variable Star Observers, Vol. 4, 68 - 79 (1975/76).

122.055 Photoelectric light curves for T Monocerotis and TT Aquilae. H. J. Landis.
Journ. American Ass. Variable Star Observers, Vol. 4, 84 - 85 (1975/76).

This paper presents photoelectric data on two stars that are in the AAVSO cepheid program.

122.056 A revised period of V1828 Sagittarii.
D. Carmichael.
Journ. American Ass. Variable Star Observers, Vol. 4, 95 (1975/76).

New observations indicate a decrease in period but do not distinguish between a smooth and an abrupt change.

122.057 The light curve of V3804 Sagittarii.
M. Brewster.
Journ. American Ass. Variable Star Observers, Vol. 4, 97 - 98 (1975/76).

A light curve spanning the years 1925 to 1975 is presented. The irregularity of the curve is similar to that of some symbiotic stars.

122.058 The changing period of V2526 Sagittarii.
M. Brewster.
Journ. American Ass. Variable Star Observers, Vol. 4, 99 - 100 (1975/76).

Observations for the years 1925 - 1975 of V2526 Sgr indicate that the period has increased: 31.33 days satisfies the observations prior to 1944; 31.45 days satisfies the later observations.

122.059 Two variable stars in Cygnus. P. Guida.
Journ. American Ass. Variable Star Observers, Vol. 4, 101 - 102 (1975/76).

The star BD +40°3673 has been discovered to vary irregularly, and its light curve is presented. Variability of the suspected variable, S5000, is confirmed.

122.060 The secondary period of TU Comae Berenices.
M. McGrath.
Journ. American Ass. Variable Star Observers, Vol. 4, 103 - 104 (1975/76).

Photographic observations between 1928 and 1975 have been analyzed to derive a secular decrease of period. A secondary period of about 40 days in the times of maxima is reported.

122.061 T-Tauri-Sterne – Überblick über Beobachtungen und Deutungsversuche. B. Baschek.
SuW, 15. Jahrgang, p. 151 - 155 (1976).

122.062 Investigation of the light curves of cepheids with periods of 3.0–4.0 days. O. P. Vasil'yanovskaya.
Byull. Inst. Astrofiz., *Dushanbe*, No. 63, p. 21 - 32 (1974). In Russian.

The forms of the light curves of 271 cepheids belonging to different star systems were studied. It is shown that the general form of light changes for classical cepheids with periods $3\overset{d}{.}0$–$4\overset{d}{.}0$ and $0\overset{d}{.}9$–$3\overset{d}{.}0$ is the same. Attention is drawn to an instability of asymmetric light curves of classical cepheids with periods $3\overset{d}{.}0$–$4\overset{d}{.}0$.

122.063 The variability of the object Markarian 388.
K. A. Saakyan.
Astrofizika, Vol. 11, 356 - 358 (1975). In Russian. English translation in Astrophysics, Vol. 11, No. 2.

A variation in brightness of two magnitudes of Markarian 388 in photographic light has been observed. It is suggested that this object is a U Geminorum-type star.

122.064 Spectra and Fabry-Perot interferometry of AG Carinae and the nebula. H. M. Johnson.
Astrophys. Journ., Vol. 206, 469 - 473 (1976).

Observations of the variable P Cygni-type supergiant and the unique ring nebula around it were made in 1974. Line identifications are tabulated, and an estimated $n_e = 1.7 \times 10^3$ cm^{-3} is combined with a radio flux density to derive a nebular mass of $0.18 \pm 0.04\,M_\odot$ on the basis of a distance of 2 kpc. From the data some ideas of the age and kinematics of the nebula are offered, tracing the evolution of P Cygni supergiants from the youthful prototype past the nebular stage of AG Carinae into more typical, relative old age without recognizable nebular envelopes.

122.065 An analysis of the variable radial velocity of Alpha Cygni. L. B. Lucy.
Astrophys. Journ., Vol. 206, 499 - 508 (1976).

On the basis of 447 radial velocities obtained at the Lick Observatory by Paddock in the years 1927–1935, an attempt is made to discover the nature of the semiregular variability of α Cygni (A2 Ia). Harmonic analysis of the 144 velocities obtained in 1931 suggests that this variability is due to the simultaneous excitation of many discrete pulsation modes. The amplitudes and periods of these modes are then determined by least-squares fitting to all the data, and a final solution is obtained that comprises 16 terms with periods from 6.9 to 100.8 days. Reasons are given for believing that most terms represent nonradial oscillations, and this leads to the suggestion that the resulting surface motions are to be identified with macroturbulence. An argument is also given for believing that the pulsational instability persists down to periods at which atmospheric oscillations become progressive, and this leads to the suggestion that such waves are observed as microturbulence and give rise to the observed mass loss.

122.066 Photometry of cepheid variables in the Small Magellanic Cloud. C. J. Butler.
Astron. Astrophys.,Suppl. Ser., Vol. 24, 299 - 356 (1976).

Photographic photometry in B and V is presented for 72 cepheids in the SMC. The following mean observed $P–L$ relations are derived from the data: $<V> = 17.60–2.72 \log P$, $<B> = 17.93–2.45 \log P$. In the period-colour diagram for the SMC cepheids, two features are visible: (1) a much smaller scatter for the short period cepheids than for long period cepheids, (2) an apparent discontinuity in the mean line at $\log P \sim 0.6$.

122.067 Five-channel photometry of cepheids and supergiants in the southern Milky Way. J. W. Pel.
Astron. Astrophys., Suppl. Ser., Vol. 24, 413 - 471 (1976).

Photometry in Walraven's five-channel system has been carried out for about 150 cepheids in the southern Milky Way. In addition, observations have been made of 238 galactic supergiants with spectral types in the range A–K. This paper presents the observational data and details about their reduction.

122.068 Correlation and spectral analysis of the variability of RW Aurigae-type stars.
I. A. Klyus, I. I. Lobusov.
Peremennye Zvezdy, Vol. 20, 13 - 25 (1975). In Russian.

A statistical analysis of the variability of 16 RW Aur-type stars has been carried out. The analysis is based on the formula for calculating the correlation function of series with discontinuities. The consistency of its application is shown both for periodic and nonperiodic variables. Plots of the correlation functions and statistical spectra of the stars analyzed are presented.

122.069 The role of the fragmentation mechanism in the origin of wide pairs of T Tauri-type variable stars.
M. M. Zakirov.
Peremennye Zvezdy, Vol. 20, 27 - 34 (1975). In Russian.

Stability parameters were computed for the model of an evolving binary star after fragmentation. Roche's criterion was extended to the elliptical orbit case.

122.070 Photographic photometry of Orion variables in the region of the T1 Tau and T3 Tau T-associations.
U. A. Nurmanova.
Peremennye Zvezdy, Vol. 20, 35 - 46 (1975). In Russian.

Results of photographic photometry in the UBV system are given for 48 stars in the area of T1 Tau and T3 Tau T-associations. Colour excess ($E_{B-V} = 0\overset{m}{.}4$) and distance modulus ($(m-M)_0 = 7\overset{m}{.}5 \pm 0\overset{m}{.}3$) of these systems are determined.

122.071 Results of observations of T Tauri-type stars according to the joint programme of 1973.
O. Abuladze, R. A. Vardanyan, V. M. Kovalenko, Ya. Kumsishvili, N. D. Melikyan, A. V. Mironov, V. A. Oshchepkov, Dzh. A. Stepanyan, A. Totachava, A. M. Cherepashchuk, G. I.

Shanin, I. V. Shpychka, A. G. Shcherbakov.
Peremennye Zvezdy, Vol. 20, 47 - 61 (1975). In Russian.

Results of UBV and UBVR photometry, two colour polarimetry and spectroscopy in near infrared of the T Tauri-type stars RY Tau, T Tau, AB Aur and NU Ori are presented. A correlation between polarization degree and intensity of emission lines of the infrared triplet Ca II has been found.

122.072 A new list of colour excesses of cepheids.
N. S. Nikolov, G. R. Ivanov.
Peremennye Zvezdy, Vol. 20, 63 - 73 (1975). In Russian.

A new list of color excesses E_{B-V} of 161 cepheids is presented.

122.073 Investigation of the periodicity of light variations of μ Cephei.   T. A. Polyakova.
Peremennye Zvezdy, Vol. 20, 75 - 92 (1975). In Russian.

Light variations of μ Cep which have the same time scale as variations of intrinsic polarization are investigated by the method of spectral analysis of time series. There are oscillations with mean periods of 730^d, 920^d and 1280^d. The amplitudes and periods of these oscillations show slow changes. It is shown that the oscillations are not related to the rotation of the star. The author suggests that these light and polarization variations can be attributed to dust ejection from formations similar to sunspots.

122.074 On the gradient diagram and its interpretation.
E. N. Makarenko.
Peremennye Zvezdy, Vol. 20, 93 - 98 (1975). In Russian.

It is shown that the position of population I cepheids on the gradient diagram (G_U, G_V) can be explained by the hypothesis of an absolutely black body with variable radius and temperature. Three models discussed indicate that the effective radius decreases in all the spectral ranges as the temperature of the stars rises.

122.075 System of criteria of two-dimensional quantitative spectral classification for RR Lyrae-type variable stars.   Z. N. Fenina.
Peremennye Zvezdy, Vol. 20, 103 - 115 (1975). In Russian.

Based on two-dimensional quantitative spectral classification of stable stars of luminosity classes II, III, IV, V in the B8—G0 spectral range a system of criteria for analogous classification of RR Lyrae-type variables has been obtained. 16 criteria have been selected to define the spectral class of a star, 3 more of those to define the absolute magnitude. In the criterion system being obtained the error of determination of the spectral class is of 0.12 spectral subclass for hydrogen lines and 0.28 for metallic lines. The square error of one determination of the absolute magnitude amounts to $0^m.23$ on the average.

122.076 Photometric observations of the emission line star MWC 419 = BD+61°154.   G. A. Ponomareva.
Peremennye Zvezdy, Vol. 20, 117 - 122 (1975). In Russian.

The emission line star MWC 419 = BD+61°154 was investigated photographically in the time interval 1895—1973 and photoelectrically (UBV) in the time interval 1969—1975. The star varies irregularly with amplitude of $0^m.3$ being $11^m.05$ (B) in the middle light. Photographic observations show rare deep weakenings of light to $m_{pg} = 11^m.5$.

122.077 Photographic observations of V1057 Cygni.
O. E. Mandel'.
Peremennye Zvezdy, Vol. 20, 123 - 128 (1975). In Russian.

Light variations of V1057 Cygni have been studied. The observations in pg- and pv-light are given in a table. Comparison with results of other investigators has been made. Rapid rise of light continued from July 1969 to August 1970. After that the star decreased slowly its brightness. On the diagram of brightness gradients the star is located in the region near that populated by nova-type stars at the postmaximum stage.

122.078 Photoelectric observations of β Coronae Borealis in 1970-1971.   Eh. S. Brodskaya.
Peremennye Zvezdy, Vol. 20, 129 - 132 (1975). In Russian.

Photoelectric observations of the Ap star β CrB have been carried out during 1970—71 in a photometric system close to UBV. Resulting photometric curves do not differ significantly from those obtained in 1966 and 1968. Comparison of these photometric curves with the curve of mean stellar surface magnetic field $|Hs|$ determined by Wolff et al. (1970) shows that the phase of the minimum of the brightness does not coincide with the phase of the maximum of $|Hs|$, but the maximum occurs later, during the minimum of He.

122.079 Double mode cepheid V367 Scuti in the open cluster NGC 6649.
Yu. N. Efremov, P. N. Kholopov.
Peremennye Zvezdy, Vol. 20, 133 - 141 (1975). In Russian.

Some years ago the authors have found for V367 Sct the period $6^d.2930$ and suggested that scatter of its light curve is caused by a secondary period. Photographic observations permitted to ascertain that the secondary period is $4^d.3849$.

122.080 FG Sagittae: photometric properties in 1971 - 1975.
V. P. Arkhipova.
Peremennye Zvezdy, Vol. 20, 143 - 151 (1975). In Russian.

The results of photoelectric UBV observations of FG Sge during 64 nights in 1971—75 are given in a table. The decrease of the star brightness was found to be of about 1^m in the U region and of $\sim 0^m.6$ in the B region.

122.081 Variation of the Hα-emission profile in the spectrum of RY Tau observed during night.
E. A. Kolotilov, G. V. Zajtseva.
Peremennye Zvezdy, Vol. 20, 153 - 160 (1975). In Russian.

Observations of the Hα-emission profile of RY Tau with 20Å/mm dispersion have been made during ~ 6.4 and 3 hours on three nights: Jan. 19 - 20, Feb. 17 - 18 and March 8 - 9, 1975. Photoelectric UBV measurements of the star's brightness were made simultaneously with the spectrum observations. The spectrum observations made on Jan. 19 - 20 revealed variations of the Hα-emission profile on a time scale of 10 - 20 min without variations of the star's brightness. There have been differences between profiles of Hα-emission observed during the nights of Feb. 17 - 18 and March 8 - 9, however these profiles have been constant during the observational period on each date.

122.082 Classification of irregular variables in the association T1 Sco.
A. Ya. Filin, V. Satyvoldiev.
Permennye Zvezdy, Vol. 20, 161 - 165 (1975). In Russian.

The classification is based on the form of the light curve and its peculiarities. All variables were divided into 7 types. Description of the types and examples of light curves of each type are given in a table and in figures.

122.083 Variable stars of the association T2 Cep.
O. G. Suyarkova.
Peremennye Zvezdy, Vol. 20, 167 - 171 (1975). In Russian.

The results of 190 photographic observations of the variable stars CSV 8684, CSV 8689, DZ Cep, GL Cep and GM Cep are given. The stars are in the region of nebulosity IC 1396.

122.084 Energy distribution in the spectrum of the red semiregular variable ST Herculis.   R. I. Chuprina.
Peremennye Zvezdy, Vol. 20, 173 - 177 (1975). In Russian.

Estimated absolutized distributions of energy E_λ in the spectrum of the red semiregular variable ST Her in the region 6500 to 10500 Å are presented. The values of E_λ are given in absolute units.

122.085 Photoelectric observations of SU Aurigae according to the cooperative programme. A. F. Pugach.
Peremennye Zvezdy, Prilozhenie, Vol. 2, 195 - 198 (1975). In Russian.

Photoelectric UBV magnitudes of the rapid irregular variable SU Aur are presented. Visual inspection and analysis give evidence for the presence of small amplitude ($\leq 0^m05$) light variations within some dozen minutes.

122.086 Research of the periods of four supershort-period cepheids. L. N. Berdnikov.
Peremennye Zvezdy, Prilozhenie, Vol. 2, 199 - 205 (1975). In Russian.

Results of a period research are given for the four variable stars VZ Cnc, AD CMi, EH Lib and SZ Lyn. The periods of VZ Cnc and AD CMi do not change, new elements are obtained for them. The period of EH Lib varies with a cycle approximately of 1800^d, that of SZ Lyn of 1100^d.

122.087 Investigation of 4 RR Lyrae-type variables. A. A. Batyrev.
Peremennye Zvezdy, Prilozhenie, Vol. 2, 207 - 220 (1975). In Russian. – Concerning UY Cyg, SU Dra, CG Peg, AN Ser.

122.088 On the colours of variable stars in the region of the Scorpius T1 association.
P. P. Petrov, V. Satyvoldiev.
Peremennye Zvezdy, Prilozhenie, Vol. 2, 221 - 224 (1975). In Russian.

Colours and magnitudes of 18 irregular variables and 13 field stars in the region of the Sco T1 association were measured. The position of the variables on the colour-magnitude diagram permits to estimate the distance modulus: $7^m5 \pm 1^m$. This corresponds to an upper limit of the distance of about 300 pc. Probably the variables under consideration belong to the Sco T1 association.

122.089 On the variability type of V564 Ophiuchi and Z Sextantis. G. E. Erleksova.
Peremennye Zvezdy, Prilozhenie, Vol. 2, 250 - 257 (1975). In Russian.

122.090 RR Lyrae-type variable S 10684.
N. V. Kondrat'ev.
Peremennye Zvezdy, Prilozhenie, Vol. 2, 269 - 272 (1975). In Russian.

122.091 Observations of three variable stars.
P. Grigor'ev.
Peremennye Zvezdy, Prilozhenie, Vol. 2, 291 - 300 (1975). In Russian. – Concerning UW And, AV Peg, SY Her.

122.092 Observations of four variable stars.
V. Dem'yanovskij.
Peremennye Zvezdy, Prilozhenie, Vol. 2, 301 - 306 (1975). In Russian. – Concerning W Peg, TW Boo, TV Boo, RR Leo.

122.093 TU Andromedae. V. I. Kirichenko.
Peremennye Zvezdy, Prilozhenie, Vol. 2, 307 - 311 (1975). In Russian.

122.094 XZ Draconis. S. Lebedev.
Peremennye Zvezdy, Prilozhenie, Vol. 2, 313 - 319 (1975). In Russian.

122.095 Determination of the periods of the Blazhko-effect

(amplitude variation) of two **RR Lyrae variables in globular clusters by computers.** V. P. Goranskij.
Peremennye Zvezdy, Prilozhenie, Vol. 2, 323 - 330 (1976). In Russian.

A new algorithm of computer search for secondary periodicity (periodicity cf variations of the light curve form) for variable stars is presented. The computer program was applied to improve the period of the Blazhko-effect of the RR Lyrae variable No. 30 in the globular cluster M53 (NGC 5024) and that for the variable No. 2 in M5 (NGC 5904). These periods are 37^d61 and 132^d38 approximately.

122.096 Results of modelling the light curves of RR Lyrae variables in the globular clusters M3, M5, ω Cen and NGC 3201. T. S. Basharina, E. D. Pavlovskaya, A. A. Filippova.
Peremennye Zvezdy, Prilozhenie, Vol. 2, 331 - 335 (1976). In Russian.

The short-period variables of type a and b with period $P > 0^d4$ in the globular clusters M3, M5, ω Cen, NGC 3201 are investigated by a statistical method. It is shown that the light curves of these stars in clusters M3, M5, NGC 3201 can be considered as realisations of some random process. But in globular cluster ω Cen the light curves of short-period cepheids have individual features.

122.097 Unregelmäßiger Lichtwechsel von R Sct.
J. Bauer.
BAV Rundbrief, 25. Jahrgang, p. 26 - 28 (1976).

122.098 U Geminorum, 1956–69. J. E. Isles.
Journ. British Astron. Ass., Vol. 86, 327 - 332 (1976). – Report of Variable Star Section.

122.099 Absolute magnitudes of SW Andromedae, XZ Cygni and RZ Lyrae. Z. N. Fenina.
Astron. Tsirk., No. 892, p. 1 - 2 (1975). In Russian.

122.100 Flare of SS Cyg. A. V. Mironov.
Astron. Tsirk., No. 896, p. 4 - 5 (1975). In Russian.

122.101 Observations of SS Cyg.
N. N. Kiselev, N. V. Narizhnaya.
Astron. Tsirk., No. 896, p. 5 (1975). In Russian.

122.102 Osculating elements of 11 RR Lyrae-type stars.
G. A. Lange, V. D. Motrich, B. N. Firmanyuk, V. P. Tsesevich.
Astron. Tsirk., No. 900, p. 5 - 6 (1976). In Russian.

122.103 UBV observations of FG Sagittae in 1975.
V. P. Arkhipova, R. I. Noskova.
Astron. Tsirk., No. 901, p. 1 - 2 (1976). In Russian.

122.104 The light curve of the peculiar variable star CH Cyg during 75 years. E. B. Gusev.
Astron. Tsirk., No. 901, p. 2 - 4 (1976). In Russian.

122.105 *UBV* photometry of the RR Lyrae star HX Arae.
H. W. Duerbeck, K. Walter.
Astron. Astrophys., Vol. 49, 471 - 472 (1976).

HX Ara, originally classified as a W UMa star, is found to be an RR Lyrae variable of type c. *UBV* observations of this star are presented.

122.106 Photometric observations of the β CMa star γ Peg.
J.-P. Sareyan, J.-M. Le Contel, J. C. Valtier.
Astron. Astrophys., Suppl. Ser., Vol. 25, 129 - 142 (1976).

The authors present photometric observations of the β CMa star γ Peg secured in 1969, 1970 and 1971. Short period fluctuations superimposed on the variation due to the 0.15

day period occur in the visible continuum, while not present in its UV continuum. A correlation between these light variations and line profile variation exists. A comparison is made, from the available data, with BW Vul, where such a correlation occurs between light and radial velocity variations. This may be a general rule for β CMa variables.

122.107 **Five-colour photometry of southern cepheids. A new search for cepheid binaries.**
E. Janot-Pacheco.
Astron. Astrophys., Suppl. Ser., Vol. 25, 159 - 178 (1976).

UVBGR (Lick system) photoelectric observations are presented for thirteen bright southern cepheids. Mianes' (1963) method for detecting companions of cepheids was applied in the $V-G/G-R$ and $U-G/G-R$ diagrams to sixty-one stars observed in six colours. The zero point for the adopted colour excess was checked using $E(B-V)$ excesses for galactic cluster cepheids. In the two-colour plots twenty stars showed abnormal slopes which can indicate the presence of a companion.

122.108 **High time-resolution observations of UV Ceti stars.**
T. J. Moffett, B. W. Bopp.
Astrophys. Journ., Suppl. Ser., Vol. 31, 61 - 73 (1976).

Simultaneous, high time-resolution spectroscopic and photometric observations of eight flare events on four UV Ceti-type stars are presented. These observations firmly establish the existence of two phases during a flare: the spike phase, which is dominated by continuum emission, and the slow phase, which shows strong emission-line radiation with decreasing continuum radiation. The flare phenomenon is spectroscopically complex, with no apparent simple relation between the various flare parameters.

122.109 **Eclipses of U Geminorum.**
S. Arnold, R. A. Berg, J. G. Duthie.
Astrophys. Journ., Vol. 206, 790 - 794 (1976).

Seven precise eclipse times for the cataclysmic variable star U Geminorum have been obtained during the 1974–1975 observing season. These, together with 65 others, have been used to re-evaluate the eclipse parameters. The authors infer a mass transfer rate of $8.1 \times 10^{-8} M_\odot \mathrm{yr}^{-1}$ away from the primary. This direction of mass transfer is opposite to that required by current theoretical understanding of these systems.

122.110 **Observations of X-rays from flare stars with ANS.**
J. Grindlay, J. Heise.
14th Intern. Cosmic Ray Conf., (see 012.011), Vol. 1, 154 - 158 (1975).

122.111 **A search for X rays from UV Ceti flare stars.**
C. J. Crannell, T. H. Markert, T. J. Moffett, S. R. Spangler.
14th Intern. Cosmic Ray Conf., (see 012.011), Vol. 1, 159 - 161 (1975).

122.112 **Lists of probable long-period variables of M-type revealed in red light by means of spectral features.**
M. V. Dolidze.
Byull. Abastumansk. Astrofiz. Obs., No. 47, p. 171 - 204 (1975). In Russian.

Spectral peculiarities of some long-period variables of spectral type M in red light are described. A list of long-period and probable long-period variables of M type (M, SRa and Lb) revealed in red light by means of spectral features is presented.

122.113 **Simultaneous two channel photoelectric observations of EV Lac.** B. N. Andersen.
Inform. Bull. Variable Stars, (I.A.U. Commission 27), Konkoly Obs., Budapest, No. 1084, 3 pp. (1976).

122.114 **VZ Dor is VW Dor.** J. F. Dean.
Inform. Bull. Variable Stars, (I.A.U. Commission 27), Konkoly Obs., Budapest, No. 1091, 2 pp. (1976).

122.115 **Photoelectric observations of CI Cygni during the outbursts of 1971 and 1973.** P. Tempesti.
Inform. Bull. Variable Stars, (I.A.U. Commission 27), Konkoly Obs., Budapest, No. 1094, 2 pp. (1976).

122.116 **On period-spectrum relation for Delta Scuti stars and dwarf cepheids.** M. S. Frolov.
Inform. Bull. Variable Stars, (I.A.U. Commission 27), Konkoly Obs., Budapest, No. 1096, 2 pp. (1976).

122.117 **"Ultra-violet" Blazhko-effect of X Ari.**
M. S. Frolov.
Inform. Bull. Variable Stars, (I.A.U. Commission 27), Konkoly Obs., Budapest, No. 1097, 2 pp., with a correction, No. 1125 (1976).

122.118 **Photoelectric observations of δ Ceti.** Z. Tunca.
Inform. Bull. Variable Stars, (I.A.U. Commission 27), Konkoly Obs., Budapest, No. 1103. 3 pp. (1976).

122.119 **Flare summary for UV Ceti September 27 - October 14, 1975.** A. H. Jarrett, J. B. Gibson.
Inform. Bull. Variable Stars, (I.A.U. Commission 27), Konkoly Obs., Budapest, No. 1105, 8 pp. (1976).

122.120 **The period of the cepheid variable BD+56°2806.**
L. Szabados.
Inform. Bull. Variable Stars, (I.A.U. Commission 27), Konkoly Obs., Budapest, No. 1107, 2 pp. (1976).

122.121 **V1, the only known cepheid variable in the globular cluster NGC 6752; observations and period.**
A. J. Wesselink.
Inform. Bull. Variable Stars, (I.A.U. Commission 27), Konkoly Obs., Budapest, No. 1110, 3 pp. (1976).

122.122 **Observations of YZ Canis Minoris – November and December 1975.**
A. H. Jarrett, J. B. Gibson.
Inform. Bull. Variable Stars, (I.A.U. Commission 27), Konkoly Obs., Budapest, No. 1112, 4 pp. (1976).

122.123 **Three-colour photographic photometry of V 1057 Cygni.** M. K. Tsvetkov.
Inform. Bull. Variable Stars, (I.A.U. Commission 27), Konkoly Obs., Budapest, No. 1120, 3 pp. (1976).

122.124 **Concerning the relative magnitudes of the SS Cygni components.** K. Krisciunas.
Inform. Bull. Variable Stars, (I.A.U. Commission 27), Konkoly Obs., Budapest, No. 1124 (1976).

122.125 **A study of the period of HR 6684.**
P. Rosenzweig.
Inform. Bull. Variable Stars, (I.A.U. Commission 27), Konkoly Obs., Budapest, No. 1125, 2 pp. (1976).

122.126 **On changes of period of pulsating stars.**
D. Hoffleit.
Inform. Bull. Variable Stars, (I.A.U. Commission 27), Konkoly Obs., Budapest, No. 1131, 2 pp. (1976).

122.127 **Variations d'éclat rapides de BL Lacertae.**
I. A. Dubjago (*Dubyago*), L. A. Ourassine (*Urasin*).
Inform. Bull. Variable Stars, (I.A.U. Commission 27), Konkoly Obs., Budapest, No. 1134, 2 pp. (1976).

122.128 Observations of three U Geminorum stars.
G. Romano, S. Minello.
Inform. Bull. Variable Stars, (I.A.U. Commission 27), Konkoly Obs., Budapest, No. 1140, 3 pp. (1976).

122.129 Photoelectric maxima of VZ Cancri.
I. Todoran.
Inform. Bull. Variable Stars, (I.A.U. Commission 27), Konkoly Obs., Budapest, No. 1141, 2 pp. (1976).

122.130 VY Canis Majoris. G. Wallerstein.
IAU Circ., No. 2902 (1976).

122.131 R CrB variables. P. Williams, J. Mattei.
IAU Circ., No. 2961 (1976).

122.132 Veränderlichenbeobachtung mit einem lichtelektrischen Photometer. E. Heiser.
SuW, 15. Jahrgang, p. 166 - 167 (1976).

122.133 The variations of BW Vulpeculae.
B. A. Goldberg, G. A. H. Walker, G. J. Odgers.
Astron. Journ., Vol. 81, 433 - 444 (1976).

The lines of He I $\lambda4471$, Mg II $\lambda4481$, and Si III $\lambda\lambda4553$, 4568, 4575 in the spectrum of the extreme β Cephei variable BW Vulpeculae (HD 199140) have been observed at high time– and spectral–resolution with a low-light-level digital television system. The line profile variations during certain phases of the 4.8-h pulsation cycle have been clearly resolved for the first time and are described in detail.

122.134 Analysis of broad-band photometry of the long-period variables.
C. Payne-Gaposchkin, C. A. Whitney.
Smithsonian Astrophys. Obs., *Cambridge, Mass.*, Special Rep. 370, 5 + 49 pp. (1976).

Of the 5600 objects in the infrared catalogue of Neugebauer and Leighton, more than 70% have been identified with optically observed stars of late spectrum, and more than 25% are known variable stars (Mira, semiregular, and slow irregular variables in about equal numbers). The variations of color with phase have been studied for 166 Mira stars (spectrum M, S, and C) for which infrared observations and simultaneous visual estimates are available. The variations of color with phase, and of infrared magnitude with phase are related to maximal spectrum and to period.

122.135 Magellanic Cloud cepheids – the Dunsink programme. P. A. Wayman.
Irish Astron. Journ., Vol. 12, 82 - 88 (1975).

122.136 A futile search for two variable stars.
D. Hoffleit.
Irish Astron. Journ., Vol. 12, 103 - 104 (1975).

122.137 On the pulsation mode of s-cepheids.
G. R. Ivanov, N. S. Nikolov.
Astrophys. Letters, Vol. 17, 115 - 117 (1976).

The comparison of the location of galactic cepheids with small amplitudes and almost sinusoidal light curves (s-cepheids) in the H–R diagram with the theoretical blue edges of fundamental and first-harmonic mode pulsators shows that probably s-cepheids are first-harmonic pulsating stars.

122.138 W Sgr-magnetic field measurements and Strömgren photometry. W. W. Weiss, H. J. Wood.
Mem. Soc. Astron. Italiana, Vol. 45, (see 012.017), 729 - 731 (1974). – Abstract (see also 13.122.115).

122.139 Simultaneous spectroscopic and photoelectric observations of the T Tauri star RU Lupi. (II).
L. Nordh, G. Olofsson.
Stockholms Obs. Rep. No. 5, 72 pp. (1974).

122.140 Die Perioden und ihre Veränderungen der veränderlichen Sterne im Kugelhaufen Messier 5 = NGC 5904. H. Wilkens.
MVS Sonneberg, Band 7, 71 - 95 (1976).

122.141 The variability of the Bpe star HDE 245770 (most probable candidate for X-ray source A0535 + 26).
S. Rössiger.
MVS Sonneberg, Vol. 7, 105 - 109 (1976).

122.142 Periodenänderungen des kurzperiodischen Mirasterns SY Herculis. P. Ahnert.
MVS Sonneberg, Vol. 7, 117 - 120 (1976).

122.143 Verbesserte Elemente des Mirasterns RY Ophiuchi.
M. Heß.
MVS Sonneberg, Vol. 7, 122 (1976).

122.144 Photoelectric observations of R Coronae Borealis during its 1972 minimum.
P. Tempesti, R. De Santis.
Mem. Soc. Astron. Italiana, Vol. 46, 443 - 449 (1975).

Photoelectric V and B-V observations of R CrB from March 1972 to August 1973 are reported. The minimum shows the usual features; during the subsequent maximum, fluctuations of semiregular character having $0^{m}15$ of mean amplitude and 45 days of mean period are clearly apparent. The B-V index begins to increase some 20 days after the beginning of the light decline, when the star had faded of more than 5 magnitudes, and reached the greatest value during the light recovery.

122.145 A statistical study of the light curve of R Coronae Borealis. P. Tempesti, R. De Santis.
Mem. Soc. Astron. Italiana, Vol. 46, 451 - 454 (1975).

The photometric history of R CrB, already published until 1960, is here extended up to 1974. A statistical analysis of the light curve from 1853 to 1974 shows that the light minima happen randomly in time; their length however seems to prefer a value comprised between 100 and 300 days.

122.146 Le programme A.F.O.E.V. E. Schweitzer.
A.F.O.E.V. Bull., Tome 9, 166 - 168 (1975).

122.147 La page de l'observateur. M. Duruy.
A.F.O.E.V. Bull., Tome 9, 169 - 170; Tome 10, 23 - 24 (1975/76).

122.148 CI Cygni – CN Cygni – FF Cygni, nouvelles étoiles au programme A.F.O.E.V. E. Schweitzer.
A.F.O.E.V. Bull., Tome 10, 24 - 26 (1976).

122.149 Four Stokes parameter radio frequency polarimetry of a flare from AD Leonis.
S. R. Spangler, J. M. Rankin, S. D. Shawhan.
Separate print Iowa Univ., Iowa City, USA, Dept. Phys. Astron., 19 pp. (1974).

122.150 Ultraviolet observation of classical cepheids by OAO-2. J. L. Hutchinson.
Thesis Wisconsin Univ., Madison, USA, 235 pp. (1974). (Available from Univ. Microfilms, Order No. 74-28,808).

122.151 Combination spectra in long-period variable stars.
C. E. R. Bruce.
Separate print Electrical Res. Assoc. Leatherhead, UK, 11 pp. (1975).

Pulsating stars.   See Abstr. 003.072.

Mass loss from Mira variables by the action of radiation pressure on molecules.   See Abstr. 064.029.

Mass loss from dwarf M stars through stellar flaring. See Abstr. 064.045.

Hydrodynamic models of a cepheid atmosphere. See Abstr. 064.069.

Pulsation of high luminosity helium stars. See Abstr. 065.003.

The cepheid loop as a threshold effect. See Abstr. 065.004.

The evolutionary status of population II cepheids. See Abstr. 065.020.

Excitation of pulsations in the CNO ionization zone of luminous stars.   See Abstr. 065.025.

The mass-luminosity relationship for cepheids in the Small Magellanic Cloud.   See Abstr. 065.064.

Possible properties of pre-outburst FU Orionis stars. See Abstr. 065.066.

Overstability of gravity modes in massive stars with the semiconvective zone.   See Abstr. 065.085.

Infrared observations of young stars – VII. Simultaneous optical and infrared monitoring for variability. See Abstr. 113.002.

Further photometry of cepheid-like supergiants. See Abstr. 113.036.

Photometric observations at the Special Astrophysical Observatory according to the programmes of astrophysical researches aboard the orbital station Salyut 4 and experimental flight Soyuz-Apollo.   See Abstr. 113.041.

Spectrophotometry of the flare star BY Draconis. See Abstr. 114.342.

The emission-line stars FG and FH Aquilae. See Abstr. 114.356.

Spectrophotometry of three stars of the β Cep type. See Abstr. 114.366.

Discovery of flare activity in the visual binary G 208-44/45.   See Abstr. 118.010.

BV 1616 Lep: very probably not an eclipsing variable.   See Abstr. 121.123.

High-speed photometry of luminosity-variable DA dwarfs: R808, GD 99, and G117−B15A. See Abstr. 126.012.

Water emission from infrared stars. See Abstr. 141.621.

NGC 2533 and the cepheid BN Puppis. See Abstr. 153.021.

Search for new variables among UV bright stars in globular clusters.   See Abstr. 154.026.

An analysis of the evolutionary status of stars in the anomalous globular cluster M14.   See Abstr. 154.032.

The ratio between interstellar absorption and reddening A_V/E_{B-V} for the δ Cephei stars in the Magellanic Clouds.   See Abstr. 159.007.

123 Variable Stars: Lists of Observations, Individual Observations

123.001 *UBV* photometric observations of BS Aqr.
E. W. Elst.
Astron. Astrophys., Suppl. Ser., Vol. 23, 419 - 428 (1976).
Photometric observations of BS Aqr in the *UBV* system, obtained during September 1973 at the European Southern Observatory, are presented.

123.002 The Milky Way field around NGC 7635 in Cassiopeia. I. The variable stars.
L. Rosino, A. Bianchini, D. Di Martino.
Astron. Astrophys., Suppl. Ser., Vol. 24, 1 - 27 (1976).
This paper gives the results of a blue and infrared study of new and old variable stars within a field of 30 square degrees centered on the nebula NGC 7635. Twenty-six variables or suspected variables are already known and thirty-two new variable stars have been examined. Types, elements, light curves are given. Most of the new variables belong to Mira type. They are barely visible or invisible in the blue, but fairly strong in the infrared, with $B-I$ colour indices larger than +6. Some general considerations follow.

123.003 Photographic observations of W Comae.
L. T. Markova, S. K. Fomin.
Astron. Tsirk., No. 856, p. 4 - 5 (1975). In Russian.

123.004 Photoelectric observations of R Coronae Borealis.
M. B. Girnyak, V. V. Golovatyj.
Tsirk. Astron. Obs. L'vov, No. 49, p. 6 - 8 (1974). In Russian.

123.005 Electrophotometric observations of R Monocerotis.
I. V. Shpychka.
Tsirk. Astron. Obs. L'vov, No. 49, p. 12 - 17 (1974). In Russian.

123.006 An unusual new variable star. M. Brewster.
Journ. American Ass. Variable Star Observers, Vol. 4, 96 (1975/76).
The light curve of this newly discovered variable shows both short-term and long-term variations during the interval 1895 to the present.

123.007 Stars suspected of variability in the Taurus region.
N. M. Bronnikova.
Peremennye Zvezdy, Prilozhenie, Vol. 2, 225 - 236 (1975). In Russian.

123.008 On three new variable stars. N. B. Perova.
Peremennye Zvezdy, Prilozhenie, Vol. 2, 237 - 240 (1975). In Russian.

123.009 Investigation of the variable stars NP and NR Persei.
P. N. Kholopov.
Peremennye Zvezdy, Prilozhenie, Vol. 2, 241 - 246 (1975). In Russian.

123.010 Photographic observations of two variable stars in the globular cluster M4. G. E. Erleksova.
Peremennye Zvezdy, Prilozhenie, Vol. 2, 247 - 249 (1975). In Russian.

123.011 Investigation of the variable stars GU Cassiopeiae and S 10123. N. K. Mal'shakova.
Peremennye Zvezdy, Prilozhenie, Vol. 2, 277 - 282 (1975). In Russian.

123.012 EM Cassiopeiae. G. A. Morozov.
Peremennye Zvezdy, Prilozhenie, Vol. 2, 283 - 286 (1975). In Russian.

123.013 Three new variable stars. O. G. Suyarkova.
Astron. Tsirk., No. 894, p. 7 - 8 (1975). In Russian.

123.014 New variable SVS 2155 in Ophiuchus.
N. N. Samus'.
Astron. Tsirk., No. 900, p. 6 - 7 (1976). In Russian.

123.015 New variable SVS 2154 in Cygnus.
Kh. M. Rashitov.
Astron. Tsirk., No. 900, p. 7 - 8 (1976). In Russian.

123.016 New variable stars found in galactic clusters.
J. J. Clariá.
Inform. Bull. Variable Stars, (I.A.U. Commission 27), Konkoly Obs., Budapest, No. 1108, 4 pp. (1976).

123.017 Evidence for a bump in the light curve of variable 1 of M13. W. Osborn, P. Rosenzweig.
Inform. Bull. Variable Stars, (I.A.U. Commission 27), Konkoly Obs., Budapest, No. 1126, 2 pp. (1976).

123.018 A revised ephemeris for AN Ursae Majoris.
G. S. Mumford.
Inform. Bull. Variable Stars, (I.A.U. Commission 27), Konkoly Obs., Budapest, No. 1133 (1976).

123.019 Variabilité d'étoiles naines rouges.
M. Petit.
Inform. Bull. Variable Stars, (I.A.U. Commission 27), Konkoly Obs., Budapest, No.1135, 3 pp. (1976).

123.020 The possibility of the search for relativistic objects using the ellipsoidal variability of the run-away stars.
A. M. Cherepashchuk, B. V. Kukarkin.
Inform. Bull. Variable Stars, (I.A.U. Commission 27), Konkoly Obs., Budapest, No. 1137 (1976).

123.021 Two erroneous suspected variables.
W. P. Bidelman.
Inform. Bull. Variable Stars, (I.A.U. Commission 27), Konkoly Obs., Budapest, No. 1138 (1976).

123.022 The period of the cepheid CV Monocerotis.
D. G. Turner.
Inform. Bull. Variable Stars, (I.A.U. Commission 27), Konkoly Obs., Budapest, No. 1142 (1976).

123.023 New variable stars in the region of IC 1396.
F. Gieseking.
Inform. Bull. Variable Stars, (I.A.U. Commission 27), Konkoly Obs., Budapest, No. 1145, 4 pp. (1976).

123.024 R Coronae Borealis.
IAU Circ., No. 2904 (1976).

123.025 CI Cygni.
IAU Circ., No. 2905 (1976).

123.026 VY Canis Majoris.
I. S. Glass, Robertson, Stewart, Williams.
IAU Circ., No. 2911 (1976).

123.027 S Apodis. O. Hull.
IAU Circ., No. 2918 (1976).

123.028 **S Apodis.**   J. Beuning, L. Beuning, B. F. Marino, W. S. G. Walker.
IAU Circ., No. 2935 (1976).

123.029 **γ Cassiopeiae.**   A. D. Mallama, D. R. Skillman.
IAU Circ., No. 2936 (1976).

123.030 **CI Cygni.**   J. Mattei.
IAU Circ., No. 2965 (1976).

123.031 **Maxima and minima of long period variables.**
J. A. Mattei.
AAVSO Bull. 39, 9 pp. (1976). − 1976 annual predictions.

123.032 **Observations of variable stars July−December 1975. Report No. 29.**   L. Plaut, H. Feijth.
Nederlandse Vereniging voor Weer- en Sterrenkunde. Kapteyn Astron. Lab. Groningen−Netherlands. 9 pp. (1976).

123.033 **The observations of Mira Ceti-type variable stars.**
H. Koczot.
Astron. Rep., Vol. 1, 9 - 14 (1974).

123.034 **The visual observations of AE Aurigae.**
H. Koczot.
Astron. Rep., Vol. 1, 15 - 17 (1974).

123.035 **The visual observations of VZ Camelopardalis.**
H. Koczot.
Astron. Rep., Vol. 1, 18 - 20 (1974).

123.036 **The visual observations of Y Canis Venaticorum.**
H. Koczot.
Astron. Rep., Vol. 1, 21 - 23 (1974).

123.037 **Veränderliche Sterne am Südhimmel. Teil V.**
H. Gessner, I. Meinunger.
Veröff. Sternw. Sonneberg, Vol. 8, (No. 5), 247 - 319 (1975).

123.038 **Zwei Maxima des U-Gem-Sterns IR Geminorum.**
L. Meinunger.
MVS Sonneberg, Vol. 7, 113 (1976).

123.039 **Beobachtungen von BF Cygni und AG Pegasi auf Sonneberger Überwachungsplatten.**
E. Splittgerber.
MVS Sonneberg, Vol. 7, 114 - 116 (1976).

123.040 **Maxima von SY Herculis.**   J. Haase.
MVS Sonneberg, Vol. 7, 116 (1976).

123.041 **Bemerkungen zu S7798 = AX Com.**
I. Meinunger.
MVS Sonneberg, Vol. 7, 121 (1976).

123.042 **Bearbeitung von 45 Veränderlichen am Südhimmel. (Feld η Arae, Teil VI).**   H. Geßner.
MVS Sonneberg, Vol. 7, 136 (1976).

123.043 **Maxima von Mirasternen.**   O. Matzek.
MVS Sonneberg, Vol. 7, 137 (1976).

123.044 **18 Maxima von GP Andromedae.**
E. Splittgerber.
MVS Sonneberg, Vol. 7, 137 - 138 (1976).

123.045 **Tableaux des observations faites par les sociétaires de l'AFOEV de septembre 1975 à avril 1976.**
A.F.O.E.V. Bull., Tome 9, 171 - 211; Tome 10, 28 - 57 (1975/76).

123.046 **Variable star notes.**   J. A. Mattei.
Journ. Roy. Astron. Soc. Canada, Vol. 70, 93 - 96 (1976).

124 Novae

124.001 Optical observations of the recurrent nova associated with A0620−00: 1917 - 1975.
L. J. Eachus, E. L. Wright, W. Liller.
Astrophys. Journ., (*Letters*), Vol. 203, L17 - L19 (1976).

A previous outburst of the eruptive star associated with the transient X-ray source A0620−00, first detected in 1975 August, has been discovered on Harvard plates taken in 1917. The authors present the light curve of the earlier event and conclude that this star is a typical recurrent nova with an absolute magnitude $M_V = -5.9 \pm 0.5$ at maximum brightness. The A0620−00 star has no detectable proper motion, and there appears to be little interstellar absorption. The best estimate of its distance is 11 ± 3 kpc. A0620−00 is only $30\overset{.}{.}6$ from the direction of the galactic anticenter and therefore must be located well beyond the detectable limits of our Galaxy.

124.002 Grains of anomalous isotopic composition from novae. D. D. Clayton, F. Hoyle.
Astrophys. Journ., Vol. 203, 490 - 496 (1976).

The authors study the effects of grain formation in nova ejecta with double purpose: (1) to schematically model the optical and infrared luminosities, (2) to identify the anomalies of isotopic composition that should be present in large abundance in these grains. The large carbon concentration makes the rapid and efficient grain formation possible, and accounts for the peculiar luminosities observed in nova Serpentis 1970.

124.003 Magnetic fields and the nova outburst.
W. K. Rose, E. H. Scott.
Astrophys. Journ., Vol. 204, 516 - 518 (1976).

Surface magnetic fields of $\sim 10^6 - 10^7$ gauss have been inferred from polarization observations of the old nova DQ Her. Such strong magnetic fields will probably lead to corotation of the core and envelope of the white dwarf. Assuming a rotation period of 142 s, this corotation will lead to centrifugal forces sufficient to counterbalance gravity as the star's envelope expands during the outburst, and consequently a centrifugal wind that will lead to the loss of mass and angular momentum is likely to be produced. Another probable effect of the high magnetic fields will be to suppress convection in the outer portions of the white dwarf's envelope and thus cause mass loss when the stellar luminosity exceeds the Eddington limit.

124.004 Classical novae − a steady state, constant luminosity, continuous ejection model. G. T. Bath, G. Shaviv.
Monthly Notices Roy. Astron. Soc., Vol. 175, 305 - 322 (1976).

A model of classical novae is considered in which a steady outflow of matter occurs. It is suggested that this outflow is driven by radiation pressure in the continuum. Approximate photospheric relations are derived for such a continuously outflowing region (in which scattering is assumed the dominant opacity source). Using these, six novae are shown to have outflow rates of $\dot m \sim 10^{21} - 10^{22}$ g s^{-1} at maximum light. Physical grounds for this mass-loss rate are given.

124.005 A revised catalogue of pre-telescopic galactic novae and supernovae. F. R. Stephenson.
Quarterly Journ. Roy. Astron. Soc., Vol. 17, 121 - 138 (1976).

124.006 Pre-outbursts of novae and nova Cygni 1975 (V1500 Cyg). E. B. Kostyakova.
Astron. Tsirk., No. 896, p. 1 - 2 (1975). In Russian.

124.007 Kā rodas novas?
E. Grasbergs, N. Cimahoviča.

Zvaigžņotā debess, 1976. gada pavasaris, p. 1 - 4.

124.008 Towards a realistic model for nova envelopes.
M. Friedjung.
Mem. Soc. Astron. Italiana, Vol. 45, (see 012.017), 757 - 761 (1974).

Arguments are given, suggesting that the most probable type of model, is one in which ejection continues after maximum light in the visual region, but with rapid time variations.

124.009 Dwarf novae. G. Bath.
Mem. Soc. Astron. Italiana, Vol. 45, (see 012.017), 793 - 798 (1974).

124.010 Theoretical emission line profiles for novae.
D. R. Bochonko.
Thesis Michigan Univ., Ann Arbor, USA, 139 pp. (1974). (Available from: Univ. Microfilms, Order No. 75-10,135).

The masses of cataclysmic variables.
See Abstr. 122.004.

X-ray nova A0620−00: celestial position and low-energy flux. See Abstr. 142.006.

Optical identification of A0620−00.
See Abstr. 142.007.

Radio emission from the X-ray source A0620−00.
See Abstr. 142.008.

On the X-ray sky − new stars.
See Abstr. 142.089.

124.100 Nova Persei 1974 = V400 Persei

Nova Persei 1974. A. Sh. Khatisov.
Astron. Tsirk., No. 856, p. 2 - 4 (1975). In Russian.

V 400 Persei (nova Persei 1974).
IAU Circ., No. 2906 (1976).

Spectrophotometry of nova Persei 1974.
E. J. Weiler, J. D. R. Bahng.
Monthly Notices Roy. Astron. Soc., Vol. 174, 563 - 569 (1976).

Photoelectric scanner measurements were made on nova Persei 1974 on 13 nights from 1974 November 27 to 1975 March 5. The spectra of the nova show two stages of its evolution: the '4640' and the 'nebular' stages. A relative blue shift was found for several of the lines with respect to the Balmer lines. The emission measures in flux units (erg cm^{-2} s^{-1}) were derived for strong features. From the measurements of [O III] lines, $T_e = 12000$ K was derived for the nova shell, if $N_e \gtrsim 10^8$ is assumed.

124.101 Nova Delphini 1967 = HR Delphini

Spectral variation of nova Delphini before its December 1967 maximum.
H.-r. Hang, R.-l. Liu, F.-x. Hu, Z.-r. Wang, C.-s. Zhu.
Acta Astron. Sinica, Vol. 16, 167 - 179 (1975). In Chinese.

Spectral observations of nova Delphini were made at

Purple Mountain Observatory from July 22 to October 5, 1967. Dispersion of the spectrograms is about 153 Å/mm at Hγ. In this paper its spectral characteristics and changes are described. The expansion velocities of the envelope are derived, and the spectrophotometric measurements are presented.

High-dispersion observations of emission lines in the postnova HR Delphini.   J. S. Gallagher, C. M. Anderson.
Astrophys. Journ., Vol. 203, 625 - 635 (1976).

High-resolution line profiles of Hα; [N II] λ6548, λ6583; and [O III] λ5007 in the postnova HR Del are obtained. The lines show considerable structure with individual subcomponents having velocity widths of 40−70 km s^{-1}, even though the general velocity flow from the nova is ~500 km s^{-1}. It appears that novae can produce well-defined condensations, even though matter is initially ejected at very high velocities. All lines also contain two major velocity groupings which the authors designate as shells A (centered near $V = \pm 100$ km s^{-1}) and B ($V = \pm 400$ km s^{-1}). Physical conditions differ in the two shells. The observations are briefly discussed in terms of predictions of theoretical models for novae, and some possible parallels are noted between novae and other objects containing dynamically moving gas..

124.102   Nova Cygni 1975 = V1500 Cygni

Nova Cygni 1975.
Stellar Divisions of the Peking Observatory and the Purple Mountain Observatory, Academia Sinica.
Acta Astron. Sinica, Vol. 16, 229 - 230 (1975). In Chinese. Research note.

V1500 nova Cygni 1975.   E. Schweitzer.
A.F.O.E.V. Bull., Tome 9, 164 - 165 (1975).

Spectroscopic observations of nova Cygni 1975.
J. Tomkin, J. Woodman, D. L. Lambert.
Astron. Astrophys., Vol. 48, 319 - 326 (1976).

High resolution photoelectric scans of the Ca II H, Ca I, CH$^+$, Na I D$_2$, Li I and K I 7699 Å interstellar lines in the spectrum of nova Cygni 1975 provide a distance estimate that yields an absolute visual magnitude of the nova at maximum of −9.5 to −10.5.

Nova Cygni 1975. I. Beobachtungsmaterial und Entfernungsbestimmung.
E. Bartl, S. Marx, R. Ziener.
Astron. Nachr., Vol. 297, 155 - 157 (1976).

From the strength of the interstellar H- and K-line in the spectrum of nova Cygni 1975 its distance is determined to 1.2 kpc.

Search for radio emission of nova Cygni 1975 (V1500 Cyg).   A. V. Pynzar', V. A. Udal'tsov, S. M. Kutuzov.
Astron. Tsirk., No. 893, p. 2 (1975). In Russian.

Visual light curve of nova Cygni 1975.
L. A. Urasin.
Astron. Tsirk., No. 893, p. 3 (1975). In Russian.

Nova Cygni 1975 = V1500 Cyg.
N. N. Kiselev, N. V. Narizhnaya.
Astron. Tsirk., No. 893, p. 4 - 5 (1975). In Russian.

Brightness variations of N Cyg 1975 (V1500 Cyg).
N. N. Kiselev, N. V. Narizhnaya.
Astron. Tsirk., No. 893, p. 5 - 6 (1975). In Russian.

Polarization of the light of N Cyg 1975 (V1500 Cyg).   N. N. Kiselev, N. V. Narizhnaya.
Astron. Tsirk., No. 893, p. 6 - 8 (1975). In Russian.

Precise positions of nova Cygni (V1500 Cyg).
G. A. Ivanov, S. P. Rybka, A. I. Yatsenko.
Astron. Tsirk., No. 896, p. 2 - 4 (1975). In Russian.

N Cyg 1975 (V1500 Cyg). Determination of the distance and of the interstellar absorption.
V. T. Doroshenko.
Astron. Tsirk., No. 897, p. 1 - 2 (1975). In Russian.

Observations of N Cyg 1975 = V1500 Cyg at the Special Astrophysical Observatory of the USSR Academy of Sciences.   S. M. Morozova, E. A. Kartashova, A. Burenkov.
Astron. Tsirk., No. 898, p. 1 - 6 (1976). In Russian.

Observations of nova Cygni 1975 (V1500 Cyg) in Ussurijsk.   V. A. Golubev.
Astron. Tsirk., No. 899, p. 8 (1976). In Russian.

Spectrophotometric investigations of nova Cyg 1975 (V1500 Cyg) at maximum light.
N. L. Ivanova, S. K. Vinokurov.
Astron. Tsirk., No. 901, p. 5 - 7 (1976). In Russian.

Observations of nova Cygni 1975 (V1500 Cyg).
G. A. Lange.
Astron. Tsirk., No. 902, p. 7 - 8 (1976). In Russian.

Nova Cygni 1975: narrow-band polarimetry and photometry 0.36−1.7 microns.
J. C. Kemp, R. J. Rudy.
Astrophys. Journ., (Letters), Vol. 203, L131 - L135 (1976).

Spectrophotometry and polarization studies are reported for the period 1975 August 30−September 11 over the range 0.36−1.7 μ. The linear polarization is almost invariant and is virtually all interstellar. The circular polarization shows large differences from the usual interstellar curve $q(\lambda)$, indicating magnetic or other intrinsic effects.

The early infrared development of nova Cygni 1975.
J. S. Gallagher, E. P. Ney.
Astrophys. Journ., (Letters), Vol. 204, L35 - L39 (1976).

Broad-band infrared photometry of nova Cygni is presented for 50 days following discovery. During the first three days, the energy distribution is approximately that of a blackbody. A distance of 1.5 ± 0.5 kpc is derived from a measurement of the blackbody-expansion parallax. After the fourth day, the energy spectrum is close to F_ν = constant. A possible mechanism for producing this rapid change is discussed. Although the nova is unusual in several respects, the physical parameters derived from the observations are about normal for a very fast galactic nova.

Fish-eye camera photographs of nova Cygni 1975.
J. Boček, Z. Ceplecha, M. Ježková, M. Novák.
Bull. Astron. Inst. Czechoslovakia, Vol. 27, 190 - 191 (1976).

The V magnitudes of nova Cyg 1975 were measured on thirteen all-sky photographs.

V1500 Cygni.
IAU Circ., Nos. 2902, 2914 (1976).

V1500 Cygni.   A. A. Schoenmaker.
IAU Circ., No. 2926 (1976).

V1500 Cygni.   R. Wood.
IAU Circ., No. 2938 (1976).

V1500 Cygni.   R. Wood, P. J. Andrews.

IAU Circ., No. 2953 (1976).

Photoelectric R, I observations of nova Cygni 1975.
P. S. Thé, M. van der Klis.
Inform. Bull. Variable Stars, (I.A.U. Commission 27), Konkoly Obs., Budapest, No. 1089, 4 pp. (1976).

On the short period light variation of nova Cygni 1975 (V 1500 Cyg). P. Tempesti.
Inform. Bull. Variable Stars, (I.A.U. Commission 27), Konkoly Obs., Budapest, No. 1098 (1976).

V 1500 Cygni (nova Cygni 1975). W. Pfau.
Inform. Bull. Variable Stars, (I.A.U. Commission 27), Konkoly Obs., Budapest, No. 1106, 3 pp. (1976).

Novae through the (convex) looking glass.
L. G. Jacchia.
Journ. American Ass. Variable Star Observers, Vol. 4, 49 - 54 (1975/76). − Concerning nova Cygni 1975.

Properties of the inverse hyperbolic sine are described and are exploited to construct a new form of nova light curve and to display the period-amplitude relation for recurrent novae.

The spectrum and light curve of V 1500 Cygni (nova 1975). J. E. Isles, W. E. Pennell, R. J. Livesey.
Journ. British Astron. Ass., Vol. 86, 245 - 249 (1976).

Die helle Nova Cygni 1975. W. C. Seitter.
Mitt. Astron. Ges., No. 38, p. 218 (1976). − Abstract.

Visuelle Beobachtungen der Nova Cyg 1975 (V 1500 Cygni). D. Böhme.
MVS Sonneberg, Vol. 7, 121 (1976).

UBVRI photometry of nova Cygni 1975.
M. Marcocci, R. Messi, G. Natali, L. Rossi.
Nature, Vol. 259, 186 - 187 (1976).

Multicolour photometric observations have been made of nova Cygni 1975. A periodic variation of ∼3.2 h was detected by Tempesti on 1975 September 14 when the nova was at $m_v = 6$. On 1975 September 18 the authors made a 4-h survey in five colours, UBVRI, to check this variation. The variation was confirmed for all colours, and other significant characteristics of the light curve have been detected.

Coronal lines in near infrared spectrum of nova Cygni 1975. G. L. Grasdalen, R. R. Joyce.
Nature, Vol. 260, 187 - 189 (1976).

The authors have been monitoring the 2- and 3-μm spectral regions of nova Cygni 1975 at irregular intervals. Shortly after maximum, these spectral regions were dominated by the lines of hydrogen and a few of neutral helium superposed on a continuum presumably of free-free radiation. These spectra could be quantitatively interpreted on traditional recombination theory, assuming a temperature of ∼ 10^4 K.

Search for X-ray emission from nova Cygni 1975.
J. A. Hoffman, W. H. G. Lewin, K. Brecher, J. Buff, G. W. Clark, P. C. Joss. T. Matilsky.
Nature, Vol. 261, 208 - 210 (1976).

The authors have used the SAS-3 X-ray observatory to search for X rays from nova Cygni 1975 before, during and after the time of optical maximum (1975 August 30−31). No X rays were detected at any time over the spectral range 0.1 − 50 keV.

The infrared spectrum of nova Cygni 1975.
J. H. Black, J. S. Gallagher.
Nature, Vol. 261, 296 - 298 (1976). − Letter.

Nova Cygni 1975.
A. Woszczyk, S. Krawczyk, A. Strobel.
Postępy Astron., Vol. 24, 51 - 58 (1976). In Polish.

Flare of a nova in Cygnus. V. P. Goranskij.
Priroda, 1976, No. 3, p. 122 - 123. In Russian.

Multi-band photometry of nova Cygni 1975.
K. Kawara, T. Maihara, K. Noguchi, N. Oda, S. Sato, M. Oishi, T. Iijima.
Publ. Astron. Soc. Japan, Vol. 28, 163 - 170 (1976).

Multi-band photometric observations of nova Cygni 1975 were carried out from September 2 to November 9, 1975. Light curves are well expressed by power functions of time. Infrared brightening was not observed, indicating no substantial production of dust particles in the nova explosion.

Interstellar lines in the spectrum of nova Cygni 1975. H. Ando, Y. Yamashita.
Publ. Astron. Soc. Japan, Vol. 28, 171 - 174 (1976).

It is found that in the spectrum of nova Cygni 1975 the interstellar line consists of three components. The radial velocity of the strongest component is −11.0 km s^{-1}, and the equivalent width of interstellar Ca II K line measured for this component amounts to 192 mÅ. This value of equivalent width indicates that the distance modulus of the nova is $m_v - M_v = 10.9 \pm 1.3$ mag.

A photoelectric Hα profile for nova Cygni 1975.
H. J. A. Leparskas.
Publ. Astron. Soc. Pacific, Vol. 88, 154 - 155 (1976).

A coudé photoelectric line scanner was used to obtain an Hα profile measurement for nova Cygni on 1975 September 2/3. The line appears completely in emission with a full half-width corresponding to 3600 km sec^{-1}, and a definite red asymmetry.

Beobachtungen der nova Cygni 1975.
K. Klebert.
SuW, 15. Jahrgang, 208 (1976).

Three-colour photometry of nova Cygni 1975.
K. Ichimura, M. Nakagiri, E. Watanabe, K. Okida, S. Nishimura, Y. Yamashita.
Tokyo Astron. Bull., Second Ser., No. 241, p. 2055 - 2060 (1975).

Observations of nova Cygni 1975 with a spectrum scanner and with a spectrograph of low dispersion.
K. Ichimura, T. Noguchi, Y. Norimoto, K. Nariai.
Tokyo Astron. Bull., Second Ser., No. 242, p. 2061 - 2065 (1976).

New stars and nova Cygni 1975.
A. Woszczyk.
Urania Kraków, Vol. 47, 66 - 74 (1976). In Polish.

Nova Cygni 1975. V. P. Goranskij.
Zemlya i Vselennaya, 1976, No. 3, p. 38 - 41. In Russian.

Novae (4). Waarnemingsresultaten van nova V1500 Cygni. H. Feijth.
Zenit, 3e jaargang, p. 25 - 27 (1976).

Gulbja Nova 1975.
I. Platais, I. Jurģitis.
Zvaigžņotā debess, 1976. gada pavasaris, p. 4 - 8.

Pre-outbursts of novae and nova Cygni 1975 (V 1500 Cyg). See Abstr. 124.006.

124.103　Nova Monocerotis 1975 = V616 Monocerotis = A0620—00

UBV **photometry and an interpretation of nova Monocerotis 1975.**　H. W. Duerbeck, K. Walter.
Astron. Astrophys., Vol. 48, 141 - 144 (1976).
　51 *UBV* observations of nova Mon 1975, the optical counterpart of the transient X-ray source A 0620—00, are presented. The brightness shows a decline of $0\overset{m}{.}5$ during one month, revealing fluctuations with a period of 3.92 days. This period is interpreted as the orbital period of a close X-ray binary.

V616 Monocerotis.
IAU Circ., Nos. 2907, 2918 (1976).

V616 Monocerotis (A 0620—00).
L. J. Kaluzienski, S. S. Holt, E. A. Boldt, P. J. Serlemitsos.
IAU Circ., No. 2935 (1976).

V616 Monocerotis (A0620—00).
IAU Circ., No. 2942 (1976).

V616 Monocerotis (A0620—00).
IAU Circ., No. 2949 (1976).

V616 Monocerotis.　S. Yu. Shugarov.
IAU Circ., No. 2953 (1976).

V616 Monocerotis (A0620—00).
C. Chevalier, S. A. Ilovaisky, H. Mauder.
IAU Circ., No. 2957 (1976).

Optical observations of the recurrent nova associated with A0620—00: 1917 - 1975.
See Abstr. 124.001.

124.104　Nova Scuti 1975 = V373 Scuti

UBV **observations of nova Scuti 1975.**
S. van den Bergh.
Astron. Journ., Vol. 81, 106, 143 (1976).
　UBV observations of nova Scuti 1975 and of nine comparison stars are given. The reddening in the direction of the nova is $E_{B-V} \gtrsim 0.41$ and its distance $D \lesssim 3.4$ kpc.

V373 Scuti.
IAU Circ., No. 2895 (1976).

Characteristics of nova Scuti 1975.
J. A. De Freitas Pacheco.
Rev. Brasil. Fis., Vol. 5, 397 - 403 (1975). — Abstr. in Phys. Abstr., Vol. 79, A042272 (1976).

124.105　Nova Aquilae 1975 = V1301 Aquilae

Light-curve of nova V1301 Aquilae.
I. D. Howarth.
MVS Sonneberg, Vol. 7, 110 - 112 (1976).

125 Supernovae, Supernova Remnants.

125.001 A study of galactic supernova remnants, based on Molonglo-Parkes observational data.
D. H. Clark, J. L. Caswell.
Monthly Notices Roy. Astron. Soc., Vol. 174, 267 - 305 (1976).

Observations with the Molonglo and Parkes radio telescopes have recently produced improved radio frequency data for the southern galactic supernova remnants (SNRs). The authors have now used these observations to investigate the general evolutionary properties of SNRs. Empirical relationships are derived which describe in general terms the expansion of SNRs, at least during the adiabatic phase of their evolution. An improved SNR distance scale is established, based largely on Parkes H I absorption measurements, and the resulting relationship between surface brightness and linear diameter for galactic SNRs is found to be compatible with that determined for the Magellanic Cloud SNRs, contrary to earlier conclusions.

125.002 The population of supernova remnants in the Magellanic Clouds.   J. N. Clarke.
Monthly Notices Roy. Astron. Soc., Vol. 174, 393 - 399 (1976).

Statistical analysis of the Molonglo radio source catalogue suggests that there are few supernova remnants (SNR's) remaining to be discovered in the Magellanic Clouds. If there are no systematic effects, the slope of the number-linear diameter relation in the LMC is 1.0 ± 0.3, significantly less than the slope of ≈ 2.4 for galactic SNR's. The discrepancy may be resolved if there are many undetected SNR's associated with H II regions in the LMC, but the numbers required are implausible.

125.003 An upper limit to microwave pulse emission at the onset of a supernova.
W. P. S. Meikle, R. W. P. Drever, G. A. Baird, T. Delaney, J. V. Jelley, J. H. Fruin, G. G. C. Palumbo, G. Morigi, R. B. Partridge.
Astron. Astrophys., Vol. 46, 477 - 478 (1976).

This paper reports an upper limit at 10 GHz of 4×10^{43} erg in a 40 MHz bandwidth for the microwave pulse emission at the onset of an optically observed supernova.

125.004 Die Supernova vom Jahre 1006.
F. Gondolatsch.
SuW, 15. Jahrgang, p. 48 - 51 (1976).

125.005 Optical observations of supernova remnants; filamentary nebula Simeiz 147.   T. A. Lozinskaya.
Astron. Zhurn. Akad. Nauk SSSR, Vol. 53, 38 - 43 (1976). In Russian. English translation in Soviet Astron., Vol. 20, No. 1.

A large series of Simeiz 147 nebula observations was carried out with a Fabry-Perot interferometer and a contact image converter. A high-velocity gas has been found in the approaching and moving away sides of the shell; the velocity of expansion was evaluated to be 100 km/sec. The kinetic energy of the ejected shell and the age of the supernova were calculated. The agreement of the results obtained with the modern concept of creation and emission of filaments is discussed.

125.006 New observational data on supernovae.
K. Rudnicki.
Postępy Astron., Vol. 23, 257 - 270 (1975). In Polish.

125.007 Optical spectra of supernova remnants.
I. J. Danziger, M. Dennefeld.
Publ. Astron. Soc. Pacific, Vol. 88, 44 - 49 (1976).

Montages of red and blue image-tube spectra are presented for 21 different supernova remnants located in the southern Milky Way and in the Magellanic Clouds. Particular spectral characteristics are described for each remnant and some new emission lines are tentatively identified. Attention is drawn to the apparently radio-quiet objects N70 and N185 in the LMC, and a case is made for their being supernova remnants.

125.008 The origin of radio recombination lines seen toward supernova remnants.   V. Pankonin, D. Downes.
Astron. Astrophys., Vol. 47, 303 - 307 (1976).

New observations have been made of the 166α spectrum in the direction of the supernova remnants G-0.6-0.1, 3C 391 and W 49 B. The variation of the intensity of the hydrogen recombination lines with frequency indicates that the lines arise in extended, low-density H II regions and not in cold clouds.

125.009 High-resolution radio observations of three supernova remnants.   R. H. Becker, M. R. Kundu.
Astrophys. Journ., Vol. 204, 427 - 440 (1976).

The authors present and discuss the radio observations of three supernova remnants (G21.5 − 0.9, G29.7 − 0.3, and 3C 391) made at 3.7 and 11.1 centimeter wavelengths. The supernova remnant G21.5 − 0.9 has an elliptical brightness distribution and is similar to the Crab nebula and 3C 58 in appearance and spectrum. The two remnants G29.7 − 0.3 and 3C 391 both show a broken shell structure typical of most galactic supernova remnants.

125.010 Statistics of extragalactic supernovae.
J. Maza, S. van den Bergh.
Astrophys. Journ., Vol. 204, 519 - 529 (1976).

It is shown that supernovae of type II are concentrated in spiral arms whereas those of type I show no preference for spiral-arm regions. Rediscussion of available supernova statistics suggests that Tammann may have overestimated the dependence of supernova frequency on galaxy inclination. A study of the distribution of supernovae in elliptical galaxies indicates that the supernova rate per unit luminosity may be highest among (metal-poor?) stars in the halos of E galaxies. All galaxies in which supernovae are known to have occurred have been classified on the DDO system.

125.011 Detection of X-ray emission from the remnant of the supernova 1006 A.D.
P. F. Winkler, Jr., F. N. Laird.
Astrophys. Journ., (*Letters*), Vol. 204, L111 - L114 (1976).

Observations from OSO-7 identify SN 1006 as a weak X-ray source, with an intensity of 9×10^{-11} ergs cm^{-2} s^{-1} in the energy range 1 − 10 keV. The spectrum can be represented by a power law with energy index 1.3 or by thermal bremsstrahlung at $kT = 4$ keV. Interpretation of the observations as thermal X-ray emission from an adiabatic blast wave indicates that SN 1006 is about 1.2 kpc distant, and that its initial kinetic energy is of the order of 10^{50} ergs.

125.012 Carbon deflagration supernova, an alternative to carbon detonation.
K. Nomoto, D. Sugimoto, S. Neo.
Astrophys. Space Sci., Vol. 39, L37 - L42 (1976).

As an alternative to the carbon detonation, the authors present a carbon deflagration supernova model by a full hydrodynamic computation. A deflagration wave, which propagates through the core due to convective heat transport, does not grow into detonation. Though it results in a complete disruption of the star, the difficulty of overproduction of iron peak elements can be avoided if the deflagration is relatively slow.

125.013 **Cassiopeia A — an unseen supernova.**
K. Kamper, S. van den Bergh.
Sky Telescope, Vol. 51, 236 - 239 (1976).

125.014 **Zur z-Verteilung von Supernova-Überresten.**
K. Henning, H. J. Wendker.
Mitt. Astron. Ges., No. 38, p. 136 (1976). — Abstract.

125.015 **Supernova-uitbarstingen in melkwegstelsels.**
F. P. Israel.
Zenit, 3e jaargang, p. 39 - 44 (1976).

125.016 **An empirical comparison of X-ray and radio emission from supernova remnants.**
F. Seward, G. Burginyon, R. Grader, R. Hill, T. Palmieri,
P. Stoering, A. Toor.
Astrophys. Journ., Vol. 205, 238 - 246 (1976).

Data from several rocket flights are combined to list upper limits to soft X-ray emission from supernova remnants (SNRs). These limits are compared with observations of SNRs that are known X-ray emitters. There is no strong correlation between measured radio and X-ray flux at the top of the atmosphere. X-ray luminosities and upper limits are calculated for all SNR that are bright radio sources. The ratio between X-ray and radio luminosity is found to be in the range, $20 < L_X/L_R < 2000$. The dependence of SNR luminosity on diameter is discussed.

125.017 **Some results of a recent study of the supernova remnant 3C 400.2 at 49 cm.**
W. M. Goss, S. G. Siddesh, U. J. Schwarz.
Bull. Astron. Soc. India, Vol. 3, 36 (1975). — Abstract of a paper presented at the A.S.I. meeting 1975.

125.018 **Type I supernovae and galactic production of iron.**
R. A. Chevalier.
Nature, Vol. 260, 689 - 690 (1976).

Current models of galactic chemical evolution assume that heavy elements are produced by massive stars, that is, in type II supernovae events. The author shows that Fe may have a different source from other heavy elements—in type I supernovae which result from the deaths of moderately long lived stars.

125.019 **Spectroscopic observations of the supernova remnant candidates 3 C 400.2 and S 91.**
F. Sabbadin, S. D'Odorico.
Astron. Astrophys., Vol. 49, 119 - 123 (1976).

Spectra of 3 C 400.2 and S 91 have been used to measure the emission line ratios Hα/|N II|, Hα/|S II| and 6717/6731 of |S II| in the two nebulae. By comparing the measured ratios with the mean values observed in supernova remnants, in planetary nebulae and in galactic H II regions, it is possible to conclude that the two nebulae are the result of supernova explosions.

125.020 **X-ray spectra of the Puppis A and the Vela supernova remnants.** W. E. Moore, G. P. Garmire.
Astrophys. Journ., Vol. 206, 247 - 253 (1976).

Low-energy X-ray spectra are presented for (1) the entire Vela supernova remnant (SNR), (2) the two most intense sub-regions of the Vela SNR, and (3) the Puppis A SNR. Acceptable χ^2 fits are obtained from the model of a high-temperature, low-density plasma with Allen cosmic abundances when thermal bremsstrahlung, radiative recombination, line emission, and interstellar absorption are included.

125.021 **X-ray spectra of Cassiopeia A and Tycho's supernova observed with Ariel-5.**

P. J. N. Davison, J. L. Culhane, R. J. Mitchell.
Astrophys. Journ., (*Letters*), Vol. 206, L37 - L40 (1976).

X-ray spectra of Cas A and Tycho's supernova from 1.5 to 18 keV and 13 keV, respectively, have been observed. Both spectra are well fitted by two-component thermal models. The higher-temperature component can in each case account for an emission feature, due to Fe XXV, which is observed at about 6.7 keV. Estimates of the iron abundance for each source are presented on the basis of these models.

125.022 **Iron line emission from a high-temperature plasma in Cassiopeia A.**
S. H. Pravdo, R. H. Becker, E. A. Boldt, S. S. Holt, R. E. Rothschild, P. J. Serlemitsos, J. H. Swank.
Astrophys. Journ., (*Letters*), Vol. 206, L41 - L44 (1976).

The X-ray spectrum of Cas A was observed for several days on board OSO-8. The high-energy (> 5 keV) data are well fitted by a thermal spectrum with kT = 3.9 (+0.9, −0.4) keV. A narrow iron line which is predicted by the thermal model is also observed, centered at 6.66 (+0.14, −0.16) keV with an equivalent width of 1270 (+175, −40) eV. Iron abundance in the source relative to normal cosmic abundance is discussed, as in the relation of this observation to shock wave and multicomponent thermal models for supernova remnants.

125.023 **Some implications of the X-ray data from old supernova remnants.** D. H. Clark, J. L. Culhane.
Monthly Notices Roy. Astron. Soc., Vol. 175, 573 - 586 (1976).

Several old supernova remnants emit soft X rays. With certain assumptions, and using a standard adiabatic shock-wave model, values for the initial blast energy of a supernova and the age of its remnant may be estimated. These parameters are evaluated using the most recently available X-ray and radio results for four old supernova remnants. The data imply high ratios of initial blast energy to interstellar density and comparatively young ages for the remnants.

125.024 **Deep Hα photography of the Vela and Puppis supernova remnants.**
K. H. Elliott, C. Goudis, J. Meaburn.
Monthly Notices Roy. Astron. Soc., Vol. 175, 605 - 611 (1976).

A deep Hα photograph with the SRC 48-in. Schmidt telescope of the Vela X and Y and Puppis A radio sources has revealed many new nebulosities. It is presented here with enlargements of particular features. Correlations between these nebulosities, non-thermal radio sources, X-ray sources and the Vela pulsar are discussed.

125.025 **Evidence against a suggested relation between spectral index and z-distribution for supernova remnants.** D. H. Clark.
Monthly Notices Roy. Astron. Soc., Vol. 175, 77P - 80P (1976).

A relation between spectral index and z-distribution for galactic supernova remnants suggested by Becker & Kundu is found not to be substantiated by the most recent radio data for remnants. Although there is no evidence for a population difference between remnants with flat and steep spectra, their suggestion that pulsars are associated with remnants having flat spectra appears worthy of further consideration.

125.026 **Galaxies à fréquentes apparitions de supernovae.**
C. Bertaud.
L'Astronomie, Vol. 90, 175 - 180 (1976).

125.027 **Historical supernovas.**
F. R. Stephenson, D. H. Clark.
Sci. American, Vol. 234, No. 6, p. 100 - 107 (1976).

Early records indicate that seven of these huge stellar explosions were seen over a period of 1,500 years. At the recorded positions of the new stars remnants of the explosions can be observed today.

125.028 Synthesis of the light elements in supernovae.
R. I. Epstein, W. D. Arnett, D. N. Schramm.
Astrophys. Journ., Suppl. Ser., Vol. 31, 111 - 141 (1976).

The results of detailed calculations of possible nucleo-synthesis in supernova shocks are presented. The shock waves are parametrized in such a way as to cover a wide variety of theoretical possibilities. Only if a strong high ion-temperature precursor develops does any appreciable nucleosynthesis occur. For presupernova stars with a population I composition in their hydrogen-rich regions, only ^7Li and ^{11}B can be produced by the shock in interesting amounts. Significant deuterium production requires extreme physical and astronomical con-ditions: (1) the ion precursor would have to develop only when the shock energy exceeds 40 MeV nucleon^{-1}; (2) the total energy of the supernova would have to be greater than $2 \times 10^{52} M_{SN}/M_\odot$ erg, where M_{SN} is the mass of ejected material.

125.029 The Gum nebula: an old supernova remnant ionized by Zeta Puppis and Gamma Velorum?
R. J. Reynolds.
Astrophys. Journ., Vol. 206, 679 - 684 (1976).

A comparison between observations of the Gum nebula and Chevalier's model for an evolving supernova remnant sug-gests that the Gum nebula may be a 1 million year old ex-panding gas shell which was produced by a 5×10^{51} erg ex-plosion in a medium of density 0.25 cm^{-3} and which is now being heated and ionized by the ultraviolet flux from ζ Pup and γ^2 Vel. The stellar wind from ζ Pup may also be an im-portant energy source within the nebula.

125.030 The radio spectrum of Vela supernova and its implication on the propagation of cosmic rays from sources. S. A. Stephens.
14th Intern. Cosmic Ray Conf., (see 012.011), Vol. 2, 481 - 486 (1975).

125.031 Vela gamma rays and the source of cosmic rays revisited. R. E. Lingenfelter, J. C. Higdon.
14th Intern. Cosmic Ray Conf., (see 012.011), Vol. 2, 487 - 488 (1975).

125.032 Supernova remnants. F. D. Kahn.
14th Intern. Cosmic Ray Conf., (see 012.011), Vol. 11, 3566 - 3593 (1975).

125.033 The magnetohydrodynamic rotational model of supernova explosion . G. S. Bisnovatyi-(*Bisnovatyj*)-Kogan, Yu. P. Popov, A. A. Samochin (*Samokhin*)
Astrophys. Space Sci., Vol. 41, 287 - 320, 321 - 356 (1976). In English and Russian.

Calculations of supernova explosion are made, using the one-dimensional nonstationary equations of magnetohydro-dynamics for the case of cylindrical symmetry. The energy source is supposed to be the rotational energy of the system. The magnetic field plays the role of a mechanism of the transfer of rotational momentum. The calculations show that the envelope splits up during the dynamical evolution of the system, the main part of the envelope joins the neutron star and becomes uniformly rotating with it, the outer part of the envelope expands with large velocity, carrying out a con-siderable part of rotational energy and rotational momentum.

125.034 Supernova in anonymous galaxy. M. Lovas.
IAU Circ., No. 2921 (1976).

125.035 Detection of the optical remnant of the supernova of 1006. S. van den Bergh.
IAU Circ., No. 2952 (1976).

125.036 Are supernova explosions driven by magnetic springs? W. Kundt.
Nature, Vol. 261, 673 - 674 (1976).

Supernova explosions are generally believed to be powered by the gravitational energy of a collapsing core of (Chandrasekhar) mass $\sim 1.4 M_\odot$. There is a difficulty, however, with the radial momentum balance: how is the liberated energy converted into the radial motion of an ejected shell? Neither thermal pressure gradients nor neutrino pressures seem to be sufficient to accelerate matter to the observed several per cent of the velocity of light. On the other hand, there is increasing evidence that magnetic neutron stars form at the centre of (at least a large subclass of) supernovae. The author argues that their magnetic pressure is a serious candidate for driving the supernova motion.

125.037 Supernovae in our Galaxy. A. Marks.
Urania Kraków, Vol. 47, 9 - 13 (1976). In Polish.

125.038 Radio observations of the young supernova rem-nants 3C 10 and 3C 358.
R. G. Strom, R. M. Duin.
Mem. Soc. Astron. Italiana, Vol. 45, (see 012.017), 689 - 697 (1974).

125.039 Nucleosynthesis during explosive oxygen and silicon burning. S. E. Woosley.
Explosive nucleosynthesis, (see 012.020), p. 70 - 83 (1973).

125.040 Presupernova evolution. I. Iben, Jr.
Explosive nucleosynthesis, (see 012.020), p. 115 - 138 (1973).

125.041 The carbon detonation supernova model.
J. C. Wheeler.
Explosive nucleosynthesis, (see 012.020), p. 203 - 212 (1973).

125.042 The carbon detonation supernova and associated remnant formation. S. Bruenn.
Explosive nucleosynthesis, (see 012.020), p. 213 - 228 (1973).

125.043 Off-center detonation supernovae.
J.-R. Buchler.
Explosive nucleosynthesis, (see 012.020), p. 229 - 233 (1973).

125.044 Supernova shock waves. S. A. Colgate.
Explosive nucleosynthesis, (see 012.020), p. 248 - 263 (1973).

125.045 Progressi nell'interpretazione e studio delle stelle variabili. L. Rosino.
Coelum, Vol. 44, 97 - 117 (1976).

125.046 High resolution radio observations of supernova remnants. R. H. Becker.
Diss. Graduate School Univ. Maryland, Dept. Phys. & Astron., 156 pp. (1975).

125.047 Supernovae and the origin of cosmic rays.
K. Pimley.
Thesis Univ. Durham, England (1975). – Abstr. in Phys. Abstr, Vol. 79, A006855 (1976).

125.048 Supernova explosion and neutral currents of weak interaction. K. Sato.
Progr. Theor. Phys., Vol. 54, 1325 - 1338 (1975). – Abstr. in Phys. Abstr., Vol. 79, A032950 (1976).

125.049 Ion-ion correlation effect on Freedman's neutrino opacity (supernovae). N. Itoh.
Progr. Theor. Phys., Vol. 54, 1580 - 1581 (1975). – Abstr. in

Phys. Abstr., Vol. 79, A032952 (1976).

125.050 Gravitational radiation from supernova explosions.
R. J. Adler, B. Zeks.
Phys. Rev. D, Vol. 12, 3007 - 3012 (1975). — Abstr. in Phys. Abstr., Vol. 79, A037595 (1976).

125.051 Analytic supernova models.   R. C. Adams.
Thesis Pennsylvania Univ., Philadelphia, USA, 156 pp. (1975). (Available from Univ. Microfilms, Order No. 75-24,037).

125.052 Explosions and light curves of supernovae.
B. Gaffet.
Congress French Physical Soc., Dijon, France, 30 June 1975, 12 pp. (1975). In French.

Importance of isotopic composition of iron in cosmic rays.   See Abstr. 061.006.

Low-energy elastic neutrino-nucleon and nuclear scattering and its relevance for supernovae.
See Abstr. 061.009.

Scattering functions for neutrino transport.
See Abstr. 061.010.

High energy gamma ray astronomy.
See Abstr. 061.013.

Neutrino processes in dense matter.
See Abstr. 061.023.

Explosive carbon burning.   See Abstr. 061.048.

Confirming explosive nucleosynthesis with gamma-ray telescopes.   See Abstr. 061.052.

Radiation from cosmic blast waves.
See Abstr. 062.002.

Coherent scattering of neutrinos from a shock wave at high densities.   See Abstr. 065.009.

Magnetohydrodynamic phenomena in collapsing stellar cores.   See Abstr. 065.026.

Supernova explosions, the new leptons, and right-handed neutrinos.   See Abstr. 065.081.

A neutron star as the main product of supernova explosions.   See Abstr. 065.083.

Early neutron-star matter.   See Abstr. 065.111.

Evolution of carbon-oxygen dwarfs in binary systems.   See Abstr. 117.026.

A revised catalogue of pre-telescopic galactic novae and supernovae.   See Abstr. 124.005.

The formation of interstellar grains in supernovae explosions.   See Abstr. 131.027.

Planetary nebulae, supernova remnants, and the interstellar medium.   See Abstr. 131.143.

Spectrophotometry of planetary nebulae and supernova remnants in the Magellanic Clouds.
See Abstr. 133.025.

W28 — a possible association of supernova remnants and H II regions.   See Abstr. 141.057.

The brightness and polarization structure of the suspected supernova remnant 3 C 58 at centimetre wavelengths.   See Abstr. 141.092.

Electrodynamic coupling between pulsars and surrounding nebulae.   See Abstr. 141.322.

X-ray binaries and asymmetry of supernova explosions.   See Abstr. 142.020.

Soft X-rays from IC 443.   See Abstr. 142.024.

About the possibility of observation of stimulated γ-radiation at the burst of supernova.
See Abstr. 142.154.

A review of some radio and microwave searches for transient phenomena in relation to Vela gamma-ray bursts and supernovae.   See Abstr. 142.197.

Copernicus: spectral studies of Cas-A and Pup-A.
See Abstr. 142.222.

Consistency of cosmic-ray composition, acceleration mechanism, and supernova models.   See Abstr. 143.001.

Consistency of cosmic ray composition, acceleration mechanism and supernova models.   See Abstr. 143.103.

The trapping of cosmic rays around supernovae by plasma instabilities.   See Abstr. 143.108.

Rigidity dependent escape of cosmic rays from supernova remnants.   See Abstr. 143.109.

Origin of cosmic-ray electrons from supernova remnants.   See Abstr. 143.110.

Supernovae as cosmic ray sources.
See Abstr. 143.111.

The character of cosmic ray propagation in the Galaxy.   See Abstr. 143.338.

Origin of cosmic rays.   See Abstr. 143.366.

X-ray astronomy evidence for subrelativistic cosmic rays from supernovae.   See Abstr. 143.367.

The giant spiral galaxy M101: IV. Observations of variable continuum radio emission from supernova 1970g and measurements of the continuum radio structure of the giant H II complex NGC 5455.   See Abstr. 158.060.

Expected rate of transient events from stellar deaths in other galaxies.   See Abstr. 158.082.

Use of supernovae light curves for testing the expansion hypothesis and other cosmological relations.
See Abstr. 162.122.

125.100 Supernova in NGC 4414

The 1974 type I supernova in NGC 4414.
B. Patchett, R. Wood.
Monthly Notices Roy. Astron. Soc., Vol. 175, 595 - 603 (1976).

Spectra of Miss Burgat's supernova in NGC 4414 were taken with the Isaac Newton 2.5-m reflector during 1974 April and May. The spectra cover the period from just before maximum light to 20 days post-maximum, and show many features typical of type I supernovae. In addition secondary features in the spectrum indicate the presence of thin shell or filamentary structure.

125.101 Supernova in NGC 3756

Supernova in NGC 3756. P. Wild.
IAU Circ., No. 2895 (1976).

125.102 Supernova in NGC 4402

Supernova in NGC 4402. M. Lovas.
IAU Circ., No. 2935 (1976).

125.103 Supernova in IC 1231

Supernova in IC 1231. Paparo.
IAU Circ., No. 2959 (1976).

125.104 Supernova in NGC 5253

Supernova 1972e in NGC 5253.
R. P. Kirshner, J. B. Oke.
Separate print Hale Obs. Pasadena, California, USA, 30 pp. (1975). — See 14.125.100.

126 Low-luminosity Stars, Subdwarfs, White Dwarfs

126.001 Spectrophotometry of five magnetic white dwarfs.
G. Wegner.
Monthly Notices Roy. Astron. Soc., Vol. 174, 191 - 202 (1976).

Electronographic spectra of the five magnetic white dwarfs GD 229, Grw + 70° 8247, G 240 − 72, G 195 − 19, G 99 − 47 are described. They show previously unreported substructure in the spectral features. Lists of absorptions are given and possible identifications examined.

126.002 A possible magnetic DA white dwarf.
D. T. Wickramasinghe, M. S. Bessell.
Astrophys. Journ., (*Letters*), Vol. 203, L39 - L41 (1976).

The spectrum of a peculiar southern white dwarf suspect BPM 25114 is described. A possible magnetic interpretation suggests a DA white dwarf with a field of about 10^7 gauss. The star appears to be both a spectrum variable and perhaps light variable.

126.003 On the maximum gravitational redshift of white dwarfs. S. L. Shapiro, S. A. Teukolsky.
Astrophys. Journ., Vol. 203, 697 - 700 (1976).

The stability of uniformly rotating, cold white dwarfs is examined in the framework of the Parametrized Post-Newtonian (PPN) formalism of Will and Nordtvedt. The maximum central density and gravitational redshift of a white dwarf are determined. General relativity predicts that the maximum redshift is 571 km s^{-1} for nonrotating carbon and helium dwarfs, but is lower for stars composed of heavier nuclei. Uniform rotation can increase the maximum redshift to 647 km s^{-1} for carbon stars (the neutronization limit) and to 893 km s^{-1} for helium stars (the uniform rotation limit). The redshift distribu-

tion of a large sample of white dwarfs may help determine the composition of their cores.

126.004 Non explosive collapse of white dwarfs.
R. Canal, E. Schatzman.
Astron. Astrophys., Vol. 46, 229 - 235 (1976).

The neutron stars present in binary X-ray sources are most likely formed by a non explosive process. We show that if a sufficiently cold carbon-oxygen white dwarf, close to the critical mass, accretes matter from a companion in a binary system, the time scale of collapse is long enough to allow neutronization before the onset of pycnonuclear reactions. This can possibly lead to the formation of X-ray sources by a non explosive collapse.

126.005 Evolution of low temperature white dwarfs to thermonuclear runaway.
M. J. Duncan, T. J. Mazurek, R. L. Snell, J. C. Wheeler.
Astrophys. Letters, Vol. 17, 19 - 22 (1976).

High temperature degenerate cores ($T_c \sim 10^8$ K) are susceptible to thermonuclear runaway at the "ignition line" where nuclear energy generation rates are equal to neutrino energy loss rates. In contrast, low temperature white dwarfs ($T_c \sim 10^6$ K) do not necessarily suffer thermal runaway at the ignition line because the nuclear rates are intrinsically low. Subsequent, adiabatic contraction (due to mass accretion) does lead to a thermal runaway before collapse due to electron capture or general relativistic instability can ensue.

126.006 LP 380-5/6: a binary system containing a late-type degenerate star. C. C. Dahn, R. S. Harrington.
Astrophys. Journ., (*Letters*), Vol. 204, L91 - L92 (1976).

Astrometry and photometry of LP 380-5/6 reveal that this binary system contains a very late-type degenerate star and a late-type red dwarf.

126.007 Detection of a He I 4517 Å absorption feature in the DB white dwarf GD 190.
J. Liebert, E. A. Beaver, J. W. Robertson, P. A. Strittmatter.
Astrophys. Journ., (*Letters*), Vol. 204, L119 - L122 (1976).

The spectrum of GD 190 (EG 193), a strong-lined DB white dwarf, is shown to contain an absorption feature at 4517 Å in the redward wing of the He I λ 4471 line. This feature, which is believed to be due to the forbidden $2\,^3P-4\,^3P$ transition in He I, has not been detected previously in DB white dwarfs. It is possible that the λ 4517 feature may provide a valuable gravity discriminant at least for hot DB white dwarfs.

126.008 An improved spectroscopic orbit for the white dwarf eclipsing binary BD + 16°516. A. Young.
Astrophys. Journ., Vol. 205, 182 - 185 (1976).

The original spectrograms plus newly acquired ones have been remeasured and analyzed along with numerous standard stars to provide a more definitive orbit solution and center-of-mass radial velocity for this important binary. Evidence from the Ca II emission lines suggests that a chromospheric "event" may have been observed which is related to mass exchange.

126.009 Limits on the space density of O subdwarfs and hot white dwarfs from a search for extreme ultraviolet sources.
P. Henry, S. Bowyer, M. Lampton, F. Paresce, R. Cruddace.
Astrophys. Journ., Vol. 205, 426 - 429 (1976).

An area of approximately 1350 square degrees toward the galactic anticenter has been searched for sources radiating at extreme ultraviolet wavelengths. Discrete sources within this region were not detected at fluxes above the level set by the instrument sensitivity, 2.9×10^{-8} ergs $cm^{-2}\,s^{-1}$ in the 135–475 Å band. These results, combined with those of a previous search of the north galactic pole, are used to place limits on the space density of O subdwarfs and hot white dwarfs.

126.010 Cooling of white dwarfs with masses higher than Chandrasekhar's limit.
G. S. Adzhyan.
Astron. Zhurn. Akad. Nauk SSSR, Vol. 53, 346 - 348 (1976). In Russian. English translation in Soviet Astron., Vol. 20, No. 2.

The time evolution of hot white dwarfs with masses higher than Chandrasekhar's limit is considered. It is shown that both for static and for uniform rotating white dwarfs the time of existence strongly depends on the mass of the star.

126.011 Gravitational collapse of a cold white dwarf.
A. I. Voropinov, M. A. Podurets.
Astron. Zhurn. Akad. Nauk SSSR, Vol. 53, 349 - 352 (1976). In Russian. English translation in Soviet Astron., Vol. 20, No. 2.

General relativistic equations of motion were computed numerically for the description of the motion of white dwarf matter after its losing of hydrodynamical stability. It is shown that the final result is sensitive to asymptotical properties of the equation of state of matter under high density: for an ultrarelativistic gas with $\epsilon = 3p$ the star undergoes a relativistic collapse, for an extremely rigid equation of state with $\epsilon = p$ a neutron star is formed.

126.012 High-speed photometry of luminosity-variable DA dwarfs: R808, GD 99, and G117 – B15A.
J. T. McGraw, E. L. Robinson.
Astrophys. Journ., (*Letters*), Vol. 205, L155 - L158 (1976).

The authors show that the luminosity-variable white dwarfs make up a homogeneous and definable class of variable stars with the following characteristics: (1) The spectral type is DA. (2) The colors lie in the range $0.16 \le (B-V) \le 0.20$. (3) Power spectra of the light curves indicate that the variations are periodic or pseudo-periodic and have periods in the range of 200–1000 s. The authors suggest that the variations are caused by pulsations, and that the pulsations are driven by the classical cepheid ionization zone mechanism.

126.013 Changing gravitational constant and white dwarfs.
S. C. Vila.
Astrophys. Journ., Vol. 206, 213 - 214 (1976).

If the gravitational constant decreases with time at a constant rate, there is a maximum possible age and a minimum possible luminosity for each white dwarf mass. White dwarfs older and fainter than these limits would, in the past, have exceeded the white dwarf mass limit and have become supernovae.

126.014 On the limiting mass of carbon-oxygen white dwarfs.
J. M. Scalo.
Astrophys. Journ., Vol. 206, 215 - 217 (1976).

The dependence of mass loss rate on stellar parameters suggested by Reimers leads directly to an estimate of the maximum mass of a carbon-oxygen white dwarf, M_L. The quantity M_L is also the minimum mass for degenerate carbon ignition. It is suggested that the present method might best be used to determine the zero-point of the mass loss rate calibration using independent determinations of M_L.

126.015 Sirius B: a thermal soft X-ray source?
H. L. Shipman.
Astrophys. Journ., (*Letters*), Vol. 206, L67 - L69 (1976).

The soft X-rays observed from the Sirius system can be readily explained as thermal emission originating from deep layers of the atmosphere of the white-dwarf star Sirius B, as long as the atmosphere of Sirius B is helium- and metal-poor.

126.016 Skylab ultraviolet stellar spectra: a new white dwarf, HD 149499 B. S. B. Parsons, K. G. Henize,
J. D. Wray, G. F. Benedict, M. Laget.
Astrophys. Journ., (*Letters*), Vol. 206, L71 - L72 (1976).

A strong ultraviolet continuum, seen at the position of HD 149499 (K0 V), is probably due to its 11.8 mag companion. The companion must then be a hot white dwarf.

126.017 The cooling of white dwarfs. G. S. Adzhyan.
Astrofizika, Vol. 11, 347 - 350 (1975). In Russian. English translation in Astrophysics, Vol. 11, No. 2.

A comparatively simple method of calculation of the evolution of white dwarfs is suggested. The results of the calculations for white dwarfs with masses of $1\,M_\odot$, $1.08\,M_\odot$ and $1.2\,M_\odot$ are presented.

126.018 On the white dwarf HZ 43 as an extreme-ultraviolet source. R. H. Durisen, M. P. Savedoff, H. M. Van Horn.
Astrophys. Journ., (*Letters*), Vol. 206, L149 - L152 (1976).

The visual continuum and the recently observed 60–600 Å extreme-ultraviolet spectrum of the white dwarf HZ 43 can be fitted self-consistently with the emergent flux from a hot ($T_{eff} \sim 125{,}000$ K), high-gravity ($\log g \gtrsim 7$) stellar atmosphere. If this interpretation is correct, then comparison with cooling sequences suggests that the mass of the white dwarf must exceed $0.6\,M_\odot$, and the age may be as short as 10^5 years.

126.019 Metal contents in the atmospheres of the subdwarfs.
B. Grabowski.
Acta Astron., Vol. 26, 147 - 182 (1976).

Detailed spectral analysis of six ultra-high-velocity and extremely weak-line subdwarfs, and one slow-velocity late-type dwarf has been performed by the differential weighting-function curve-of-growth method. The main results are: (1) In the investigated sample of stars the overall metal contents range from 1/4 to 1/50 of that of the sun. (2) The metal deficiences derived by the author for the extreme subdwarfs, HD 140283 and HD 19445, are smaller than those obtained in other analyses. (3) The deficiency of the a-, e-, and s-process elements in the limits of the observational and calculation errors in all investigated subdwarfs is the same. (4) The existence of only small turbulent motions in subdwarf atmospheres is confirmed.

126.020 Evolution of crystallizing ^{12}C white dwarfs.
D. Q. Lamb, H. M. Van Horn.
Mem. Soc. Astron. Italiana, Vol. 45, (see 012.017), 769 - 770 (1974). – Abstract.

126.021 X-ray emission from a white dwarf with a strong
magnetic dipole field. H. Inoue.
Progr. Theor. Phys., Vol. 54, 415 - 428 (1975). – Abstr. in Phys. Abstr., Vol. 79, A015453 (1976).

The integral polarization of hydrogen spectral lines in a strong magnetic field. See Abstr. 061.005.

Model atmospheres for cool hydrogen-rich white dwarfs. See Abstr. 064.009.

Konvektive Durchmischung in der Hülle von Weißen Zwergen. See Abstr. 064.021.

Coronas with bremsstrahlung cooling. See Abstr. 064.067.

A binary hypothesis for the subdwarf B stars. See Abstr. 065.024.

Transport properties of dense matter. See Abstr. 065.068.

Cooling times, luminosity functions and progenitor masses of degenerate dwarfs. See Abstr. 065.076.

Non explosive collapse of white dwarfs. See Abstr. 065.084.

Non-adiabatic radial oscillations and pulsational stability of hot degenerate dwarfs. See Abstr. 065.119.

Space velocity of the nearby subdwarf M star AC +54°1646−56. See Abstr. 112.001.

Proper motion survey with the 48-inch Schmidt telescope. XLVI. On the alleged plethora of near-by M dwarfs with little or no proper motion. See Abstr. 112.005.

GH 7-21: a possible degenerate star with narrow hydrogen lines and strong carbon features. See Abstr. 114.315.

The temperature, luminosity, and spectrum of Kapteyn's star. See Abstr. 114.325.

New observations of the white dwarf eclipsing binary V 471 Tauri. See Abstr. 121.072.

Hα emission in the eclipsing white dwarf V471 Tau (BD+16°516). See Abstr. 121.098.

MX 1313 + 29: a compact source of very low energy X-rays in Coma Berenices. See Abstr. 142.009.

An ultrasoft X-ray source in Coma Berenices. See Abstr. 142.010.

Discovery of a nonsolar extreme-ultraviolet source. See Abstr. 142.015.

X-ray emission from accretion on to white dwarfs. See Abstr. 142.025.

On the system 3 U 0352 + 30 − X Persei. See Abstr. 142.102.

Interstellar Matter, Gaseous Nebulae, Planetary Nebulae

131 Interstellar Matter, Polarization of Starlight, H I, H II Regions

Interstellar Matter, Polarization of Starlight

131.001 **Line profiles of the diffuse interstellar lines at 5780 Å, 5797 Å.**
A. C. Danks, D. L. Lambert.
Monthly Notices Roy. Astron. Soc., Vol. 174, 571 - 586 (1976).

Photoelectric coudé scanner observations of the interstellar diffuse lines at 5780 and 5797 Å are reported. With a resolution of about 0.2 Å, line profile differences are seen in the sample of eight stars. Intrinsic profile differences are also suggested. The hypothesis that some of the diffuse lines represent electronic transitions in large molecules is examined. Synthetic spectra are shown which approximate the observed profiles. Excitation of these hypothetical molecules is examined. Their stability against photodissociation by the interstellar ultraviolet radiation field is noted as a key problem for future study.

131.002 **Ground-state OH anomalies in the direction of the galactic centre.** J. B. Whiteoak, F. F. Gardner.
Monthly Notices Roy. Astron. Soc., Vol. 174, 627 - 636 (1976).

The OH clouds along the line of sight towards the galactic centre region produce widespread anomalous emission and absorption in the ground-state satellite-line profiles at 1612 and 1720 MHz. The anomalies are consistent with a transfer of population between the $F = 1$ and the $F = 2$ hyperfine levels of the ground state. For the clouds in the nuclear disk the radiation causing the transfer may be associated with the infrared sources in the galactic nucleus; for the other clouds the radiation must originate farther out from the nucleus.

131.003 **On the existence of symmetrical radial velocity structure in water vapour sources.**
S. H. Knowles, R. A. Batchelor.
Monthly Notices Roy. Astron. Soc., Vol. 174, 69P - 73P (1976).

Several 22 GHz water vapour sources show evidence of a symmetrical structure in their spectra, with one or more pairs of features separated by approximately equal radial velocity displacements of $1-15$ km s^{-1} from a central symmetry axis.

131.004 **Inversion of the OH 1720-MHz line.**
M. Elitzur.
Astrophys. Journ., Vol. 203, 124 - 131 (1976).

It is shown that the OH 1720-MHz line can be strongly inverted by collisions which excite the rotation states, preferentially the $^2\pi_{3/2}$ ladder. It is also argued that radiative pumps (of any wavelength) can strongly invert only the 1612-MHz line. Inversion of the 1720-MHz line with $|\tau| \lesssim 1$ can be achieved by either collisional or radiative pumps at low OH densities ($n_{OH} \lesssim 10^{-3} - 10^{-4}$ cm^{-3}).

131.005 **Interstellar H$_2$: the population of excited rotational states and the infrared response to ultraviolet radiation.** J. H. Black, A. Dalgarno.
Astrophys. Journ., Vol. 203, 132 - 142 (1976).

Molecular hydrogen in interstellar clouds absorbs ultraviolet radiation in lines of the Lyman and Werner systems. The subsequent fluorescence leads to dissociation or to the population of excited rotational-vibrational levels of the ground electronic state. Calculations of the infrared emission spectrum of H$_2$ are given for an illustrative cloud model which includes the processes of fluorescence and hot molecule formation. In favorable circumstances, some of the infrared lines may have detectable intensities. Due to the distribution of lines, it may prove possible to detect interstellar H$_2$ using narrow-band filter photometry at a wavelength of 2.4 μ.

131.006 **Further studies of ionization in interstellar clouds.**
L. M. Hobbs.
Astrophys. Journ., Vol. 203, 143 - 150 (1976).

Interferometric scans of the interstellar $\lambda 7699$ line of K I or of the D_1 line of Na I are reported for 14 stars. Along with results obtained recently by others for the column densities N(H I), N(H$_2$), or N(Na I) toward a larger number of stars, these results are added to a previous compilation of such data, and are used to analyze again the relations among the well-correlated quantities N(K I), N(Na I), and N_H. The estimated depletion factors of both K and Na are about 3 to 4. Photoionization of the heavy elements by starlight can account for the greater part of this ionization, if interstellar carbon is undepleted.

131.007 **The small-scale structure of interstellar hydrogen.**
E. W. Greisen.
Astrophys. Journ., Vol. 203, 371 - 377 (1976).

High-resolution interferometric observations of neutral hydrogen absorption features are presented. The data contain some evidence for small-scale spatial structure ($\sim$ 1 pc) in the absorbing medium, but indicate that scale lengths less than 0.3 pc, if they occur at all, are uncommon.

131.008 **The abundance of deuterium relative to hydrogen in interstellar space.**
D. G. York, J. B. Rogerson, Jr.
Astrophys. Journ., Vol. 203, 378 - 385 (1976).

New observations, made with the Copernicus satellite, of the deuterium and hydrogen Lyman lines in the lines of sight to μ Col, γ^2 Vel, α Cru AB, and α Vir AB are reported. Together with the previously published data for β Cen A, the results yield a value N(D)/N(H) of $1.8 \pm 0.4 \times 10^{-5}$ (m.e.). The results are consistent with a temperature $T_1 \lesssim 6000$ K for the lines of sight studied.

131.009 **CO observations of the bright-rimmed cloud B35.**
C. J. Lada, J. H. Black.
Astrophys. Journ., (Letters), Vol. 203, L75 - L79 (1976).

Detailed observations of millimeter-wave emission from ^{12}CO and ^{13}CO are made throughout the bright-rimmed molecular cloud B35 and compared with optical photographs. It is shown that modification of the cloud conditions by shock waves can explain the observations and account for the observed CO cooling rates in the cloud. Cloud heating by embedded young stars is also possible, and infrared observations are suggested as a means of distinguishing between the two heating models.

131.010 Temperature dependence of mid-infrared silicate absorption.   K. L. Day.

Astrophys. Journ., (*Letters*), Vol. 203, L99 - L101 (1976).

It has been found that two silicates commonly believed to occur in interstellar space and in circumstellar shells show up to a 65 percent enhancement of the 20μ region absorption coefficient upon cooling to 80 K. The apparent sharpness of the absorption spectra is also greatly increased.

131.011 Neutral hydrogen in the W41 region.

C. P. Gordon, K. J. Gordon, M. R. Jacobson.

Astrophys. Journ., Vol. 203, 593 - 599 (1976).

The authors have made observations of neutral-hydrogen emission and absorption in the W41 region, with angular resolution of 10' and velocity resolution of 2 km s^{-1}. The most probable model for the W41 complex is as follows: G22.8−0.3 is at a distance of ∼6 kpc; G23.1−0.3 is composed of nonthermal radiation from a supernova remnant at $4.6 \lesssim d \lesssim 7$ kpc plus possibly a weak extension of the thermal component G22.8−0.3; G23.4−0.2 is not associated with either of these sources and is located at a distance of either ∼8 or ∼11 kpc.

131.012 Small-scale structure in high-velocity clouds.

E. W. Greisen, T. R. Cram.

Astrophys. Journ., (*Letters*), Vol. 203, L119 - L121 (1976).

Low-resolution interferometric observations have revealed small-diameter regions of strong hydrogen emission embedded in high-velocity clouds.

131.013 Detection and significance of the interstellar OH line λ3078.   R. M. Crutcher, W. D. Watson.

Astrophys. Journ., (*Letters*), Vol. 203, L123 - L126 (1976).

The interstellar OH line near 3078 Å has been observed, and the first accurate abundance of OH in diffuse clouds has been obtained. The measured equivalent widths and column densities are $W_\lambda = 3.5 \pm 1.1$ mÅ and log N(OH) = 14.0 for o Per and $W_\lambda = 1.1 \pm 0.7$ mÅ and log N(OH) = 13.5 for ζ Oph.

131.014 OH observations of the dust complex Lynds 1630 and of NGC 2024.

W. M. Goss, A. Winnberg, L. E. B. Johansson, A. Fournier.

Astron. Astrophys., Vol. 46, 1 - 9 (1976).

Approximately 12 square degrees of the dust cloud Lynds 1630 east and north of the H II region NGC 2024 (W12) have been mapped in the 1667 MHz line of OH. The velocity field of the OH associated with L 1630 is essentially constant while the velocity dispersion is quite variable. OH emission associated with the large globule Lynds 1622 has also been detected. Observations of CH at five positions near NGC 2068−71 show emission lines in good agreement with the OH properties.

131.015 Interstellar medium in the vicinity of the sun: a temperature measurement obtained with Mars-7 interplanetary probe.   J. L. Bertaux, J. E. Blamont, N. Tabarié, W. G. (*V. G.*) Kurt, M. C. Bourgin, A. S. Smirnov, N. N. Dementeva.

Astron. Astrophys., Vol. 46, 19 - 29 (1976).

The results of a Lyman-alpha photometer placed on the soviet probe Mars-7, which passed through the interplanetary medium in 1973 and 1974, are presented. As well as the intensity of Lyman-α emission of H-atoms in the solar system (produced by resonance scattering of solar photons), the linewidth and absolute wavelength were analyzed with an absorption hydrogen cell placed in the photometer, yielding the temperature T and bulk velocity $-V_w$ of the atoms. As the atoms are of interstellar origin, T and V_w are parameters which characterize the local interstellar medium. The analysis included a possible extraneous emission I_p. The most likely value of I_p is 0, yielding $T = 12 \pm 1 \times 10^3$ K, $V_w = 19.5 \pm 1.5$ km·s^{-1}. In this case, the interstellar matter, with respect to the local rest

frame, moves at a velocity of $\simeq 16$ km·s^{-1} in the direction $l_{II} = 122°$, $b_{II} = 5.5°$ in galactic coordinates, suggesting that interstellar matter is moving faster than stars in the galactic rotation.

131.016 Dielectronic recombination in the interstellar medium.   P. A. Shaver.

Astron. Astrophys., Vol. 46, 127 - 130 (1976).

Dielectronic recombination onto heavy ions in the hot diffuse interstellar medium could give rise to detectable recombination lines in the range 50−200 MHz. Such lines would appear as absorption features in the spectra of distant galactic radio sources.

131.017 Observations of CO ($J = 2-1$) line emission from sources associated with H II regions and the Rho Ophiuchi dark cloud.

N. J. Cronin, A. R. Gillespie, P. J. Huggins, T. G. Phillips.

Astron. Astrophys., Vol. 46, 135 - 137 (1976).

230 GHz CO ($J = 2-1$) spectra have been obtained for the molecular sources associated with the H II regions M8, M17, W49, W51 and the dark cloud ρ Ophiuchi. It is shown that the ($J = 2-1$) brightness temperatures of all these sources are the same as in the $J = 1-0$ line, implying that the CO is in thermal equilibrium with the molecular hydrogen in both types of source.

131.018 Infrared observations of the H$_2$O maser associated with the H II regions S 255 (IC 2162) and S 257.

J. L. Pipher, B. T. Soifer.

Astron. Astrophys., Vol. 46, 153 - 157 (1976).

Infrared photometric and spectrophotometric observations of the H$_2$O maser 0610 + 18 are presented and compared with other infrared objects that may represent the earliest stages of star formation.

131.019 OH radiation from the interstellar cloud medium.

Nguyen-Q-Rieu, A. Winnberg, J. Guibert, J. R. D. Lépine, L. E. B. Johansson, W. M. Goss.

Astron. Astrophys., Vol. 46, 413 - 428, with a corrigendum in Vol. 49, 157 (1976).

OH radiation has been investigated in H I clouds in front of 22 galactic and extragalactic continuum radio sources. Absorption is detected in the direction of these. OH and H I radial velocities agree to within 1 km s^{-1}. A double absorption feature has been observed in front of 3C 123 in the OH and formaldehyde lines. In the case of H$_2$CO, it is explained by the hyperfine structure. OH spectra often exhibit departures from L.T.E. The excitation temperatures in all four 18 cm OH transitions have been determined for the cloud in front of 3C 123. In the main lines the excitation temperatures T_{ex} are nearly equal and are ∼6 K. A pumping mechanism is proposed. It involves far-infrared radiation from dust and collisions with neutral and charged particles. The four OH excitation temperatures determined from the observations in front of 3C 123 can be explained if the physical conditions inside the cloud are the following: Colour temperature of the far-infrared radiation: ∼20 K; dilution factor: ∼10^{-4}. Kinetic temperature: ∼50 K; density of neutral hydrogen: 10−50 cm^{-3}. Fractional ionization: $\lesssim 3 \times 10^{-5}$.

131.020 A high resolution survey of three high velocity cloud complexes.

R. D. Davies, D. Buhl, J. Jafolla.

Astron. Astrophys., Suppl. Ser., Vol. 23, 181 - 204 (1976).

A survey has been made of the high velocity cloud complexes A IV, M II and C III with high angular resolution and high sensitivity using the NRAO 140 foot and 300 foot radio telescopes. The results are presented in the form of position-velocity diagrams and of maps of the regions at selected velocities. Considerable structure was found in the brighter HVCs having an angular scale of 0.2° to 0.5° and a velocity width of 7 to 20 km s^{-1}. Weaker HVC emission was detected over large areas of the surveys. The structure of the intermediate velocity clouds (IVCs) in these regions was also investigated. The IVCs studied here are significantly larger in angular size and narrower in velocity width than the HVCs.

131.021 **Interstellar circular polarization. II. Northern and southern hemisphere survey results and observational search criteria.**
R. W. Avery, R. A. Stokes, J. J. Michalsky, P. A. Ekstrom.
Astron. Journ., Vol. 80, 1026 - 1030 (1975).

The authors' original survey has been extended to include more northern and southern hemisphere stars; the wavelength coverage was extended and the precision improved for several previously observed stars. Search criteria for interstellar circular polarization based on the linear polarization and reddening have been delineated and refined. Observational data are presented which confirm the theoretical prediction that the wavelength of maximum interstellar linear polarization and the wavelength of the zero crossing of interstellar circular polarization are approximately the same. An intercomparison of the results of various authors is presented and briefly discussed.

131.022 **The absence of systematic kinematics in dust clouds.**
C. Heiles, G. Katz.
Astron. Journ., Vol. 81, 37 - 44 (1976).

Maps of the 6-cm H_2CO line in regions around four dark clouds show no systematic patterns in line intensity, velocity dispersion, or velocity. In general, elongated dust clouds are not rotating.

131.023 **Scintillations of an extended source on inhomogeneities of an interstellar plasma.**
I. V. Chashej, V. I. Shishov.
Astron. Zhurn. Akad. Nauk SSSR, Vol. 53, 26 - 32 (1976). In Russian. English translation in Soviet Astron., Vol. 20, No. 1.

The spatial-frequency correlation function of saturated scintillations of sources with finite angular size is considered in the models of thin phase screen and extended medium. It is shown that the typical scale of diffraction pattern and frequency correlation radius depend weakly on the angular size of the source, while the dependence of these parameters on the size of the source is strong enough in the model of the phase screen. This circumstance opens the possibility for experimental detection of quasars scintillations on irregularities of interstellar plasma by carrying out differential measurements on separated frequencies.

131.024 **Radical formation, chemical processing, and explosion of interstellar grains.** J. M. Greenberg.
Astrophys. Space Sci., Vol. 39, 9 - 18 (1976). — Paper presented at the Symposium on Solid State Astrophysics, held at the University College, Cardiff, Wales, between 9 - 12 July, 1974. — See 131.087.

131.025 **Effects of suprathermal grains.**
S. P. Tarafdar, N. C. Wickramasinghe.
Astrophys. Space Sci., Vol. 39, 19 - 30 (1976). — Paper presented at the Symposium on Solid State Astrophysics, held at the University College, Cardiff, Wales, between 9 - 12 July, 1974. — See 131.088.

131.026 **Diffuse band extinction and polarization in core-mantle grains.** J. M. Greenberg, S.-S. Hong.
Astrophys. Space Sci., Vol. 39, 31 - 40 (1976). — Paper presented at the Symposium on Solid State Astrophysics, held at the University College, Cardiff, Wales, between 9 - 12 July, 1974. — See 131.077.

131.027 **The formation of interstellar grains in supernovae explosions.** S. Simons, I. P. Williams.

Astrophys. Space Sci., Vol. 39, 123 - 127 (1976).

It is shown that the subsequent Brownian coagulation of small molecular clusters produced in the shell following a supernovae explosion leads to grain sizes in reasonable agreement with observations.

131.028 **The optics of spherically stratified graphite grains.**
N. C. Wickramasinghe.
Astrophys. Space Sci., Vol. 39, 151 - 156 (1976).

Extinction and scattering efficiencies are calculated for spherically stratified graphite spheres using formulae which are valid in the small particle limit. The resulting extinction curve for this model is shown to peak at an ultraviolet wavelength $\lambda^{-1} = 4.8 - 5 \, \mu^{-1}$ close to that for the case of the less probable plane stratified model of a graphite sphere. The albedo in the present model is higher than that calculated for the plane stratified case by a factor 1.5 – 2 in the far ultraviolet. Extinction curves are also obtained for the case of a dielectric sphere surrounded by a thin graphite film, and it is shown that there now is an extinction minimum at $\lambda^{-1} \simeq 4.6 \, \mu^{-1}$.

131.029 **Infrared spectra of polyoxymethylene grains.**
A. Cooke.
Astrophys. Space Sci., Vol. 39, L13 - L18 (1976).

The optical constants n and k for polyoxymethylene have been calculated in the spectral range between 7 and 13 μ, using transmission measurements and the Kramers-Kronig dispersion relations. These constants may be of use in making detailed comparisons of observational data such as the interstellar 10 μ band.

131.030 **Comparison of radio and optical studies in a region of the Southern Coalsack.**
J. W. Brooks, M. W. Sinclair, G. A. Manefield.
Monthly Notices Roy. Astron. Soc., Vol. 175, 117 - 127 (1976).

Detailed observations of the 4830 MHz formaldehyde absorption and the 1667 MHz OH emission have been made in the direction of a region in the Southern Coalsack. These are compared with an optical study of the same region, which contains several dark globules for which the interstellar extinction is estimated to be high (>10 mag). Assuming a simple model for the globule complex, the authors have determined transition temperatures and optical depths for each species. From these parameters, estimates of the column density for both the formaldehyde and OH clouds are made.

131.031 **Limits to the broadening of radio recombination lines by electron collisions.**
K. R. Lang, S. D. Lord.
Monthly Notices Roy. Astron. Soc., Vol. 175, 217 - 224 (1976).

Observations of hydrogen 92 α recombination line profiles suggest that the inner regions of Orion A and W^3 A have low electron densities. If a constant electron density, N_e, is assumed, then the observed profiles give 5×10^3 cm$^{-3} \geq N_e \geq 5 \times 10^2$ cm^{-3} for the inner region of Orion A, and 4×10^4 cm$^{-3} \geq N_e \geq 4 \times 10^3$ cm^{-3} for the inner region of W^3 A. For the volume the authors observed in Orion A, forbidden line measurements give an average electron density of $N_e > 10^4$ cm^{-3}.

131.032 **An interstellar H_2 indicator in direction of the Crab nebula.** T. Maccacaro, G. Sironi.
Nature, Vol. 259, 26 - 27 (1976).

Molecular hydrogen is very stable at low temperatures, and is therefore likely to be found in large quantities within interstellar dark clouds which are not penetrated by ultraviolet radiation. The authors have studied the shape of the low energy spectrum of galactic X-ray sources detectable on earth. In the energy range 0.1 – 1 keV the continuous absorption by H_2 is intense. It adds to that produced by other components of the interstellar medium and affects the depth of the absorption edges of elements such as oxygen, neon and others. The oxygen feature at 0.532 keV is very prominent and allows a sensitive probe of the H_2 density between the X-ray source and the observer. The authors therefore calculated how such features in the low energy X-ray spectrum of the Crab nebula would be affected by the interstellar abundance of molecular hydrogen.

131.033 Variations of the infrared polarisation of VY Canis Majoris.
T. Maihara, K. Noguchi, M. Oishi, H. Okuda, S. Sato.
Nature, Vol. 259, 465 - 466 (1976). − Letter.

131.034 On the measurement of the extragalactic background brightness at 4000 Å.   K. Mattila.
Astron. Astrophys., Vol. 47, 77 - 95 (1976).

Photoelectric observations of the night-sky brightness in the area of the high-latitude dark nebula L 134 have been carried out in order to find the extragalactic background light. The dark nebula is used as a "zero point" in which the extragalactic background light is negligible. The difference in surface brightness between the dark nebula and its surroundings is due to two components only: (1) the extragalactic background light, and (2) the diffusely scattered starlight from interstellar dust. The value of the extragalactic background light at 4000 Å is found to be $23 \pm 8 \times 10^{-9}$ erg cm^{-2} s^{-1} sterad^{-1} Å^{-1} (or ~10 stars of 10^m per $\square°$), which is more than ten times higher than the integrated light of galaxies predicted for the conventional models.

131.035 Observations of the 21-cm hydrogen emission line in the direction of 23 southern pulsars.
F. R. Colomb, I. F. Mirabel.
Astron. Astrophys., Vol. 47, 157 - 159 (1976).

The 30 m IAR (Instituto Argentino de Radioastronomía) radiotelescope was used to obtain 21-cm emission line profiles in the direction of 23 southern pulsars. An atlas of the profiles and a table of hydrogen column densities are given.

131.036 Ultraviolet photometry of Eta Carinae and its interpretation.
S. R. Pottasch, P. R. Wesselius, R. J. van Duinen.
Astron. Astrophys., Vol. 47, 443 - 448 (1976).

Measurements of Eta Carinae in the ultraviolet are reported. They are used to determine the extinction. It is concluded that the foreground material will produce the observed extinction, and there is no need to introduce anomalous extra extinction as has been argued by earlier authors. The consequences of this for the infrared radiation is discussed. The intrinsic properties of Eta Carinae are also discussed.

131.037 Interstellar molecular hydrogen toward Zeta Puppis.
D. C. Morton, H. L. Dinerstein.
Astrophys. Journ., Vol. 204, 1 - 11 (1976).

This paper reports the measurement of some 136 interstellar H_2 absorption lines found in a continuous scan of the far-ultraviolet spectrum of ζ Pup at 0.05 Å resolution with the Copernicus telescope. The total H_2 column density is 2.8×10^{14} molecules cm^{-2}; the ratio of H nuclei in H_2 to the number in H I plus molecules is 6×10^{-6} and the population of the J'' levels can be represented by a single excitation temperature of 1120 ± 80 (s.d.).

131.038 Photoelectric heating of the interstellar gas.
M. Jura.
Astrophys. Journ., Vol. 204, 12 - 20 (1976).

The author suggests that the photoelectric effect of very small grains may supply the heating required to maintain diffuse interstellar clouds (i.e., clouds with $E(B-V) \leqslant 0.3$) at temperatures near 80 K. He argues that the photoelectric yield of very small grains may be considerably greater than of bulk materials, and he has calculated several models for interstellar clouds to illustrate the observational consequences of this hypothesis. This heating model can also explain the existence of a hot, neutral intercloud medium.

131.039 OH and H_2O masers in the Monoceros-R2 molecular cloud.   G. R. Knapp, R. L. Brown.
Astrophys. Journ., Vol. 204, 21 - 25 (1976).

Emission from OH and H_2O masers has been detected in the direction of the Monoceros-R2 molecular cloud. Only 1612 and 1665 MHz masers are observed. Two velocity components of the 1665 MHz source are highly linearly polarized (about 100%). This result allows to determine limits on the magnetic field strength in the emission region, $3 \times 10^{-6} \ll B \ll 4 \times 10^{-4}$ gauss. The relation of the maser emission to the embedded H II region and to the infrared sources in this cloud is discussed. The Mon-R2 region is similar to the Orion nebula complex, and appears to represent a somewhat earlier stage of star formation.

131.040 Isotope abundances in interstellar molecular clouds.
P. G. Wannier, A. A. Penzias, R. A. Linke, R. W. Wilson.
Astrophys. Journ., Vol. 204, 26 - 42 (1976).

The authors use the $J = 1 \rightarrow J = 0$ transition of carbon monoxide to study the abundance ratios of carbon and oxygen isotopes in dense molecular clouds. When saturation is properly taken into account, the data from all the regions cluster about a single value of ~ 14 for the $[^{13}C][^{16}O]/[^{12}C][^{18}O]$ abundance ratio. This value is significantly different from the corresponding terrestrial ratio of 5.6, and the authors suggest that the difference results from chemical evolution having occurred in the Galaxy since the birth of the solar system. Results of additional isotope studies which indicate that the value of $[^{12}C]/[^{13}C]$ is ~40 as compared with the terrestrial value of 89 are also presented.

131.041 Observation of the $6_{16}-5_{15}$ transitions of acetaldehyde in Sagittarius B2.   W. Gilmore, M. Morris, D. R. Johnson, F. J. Lovas, B. Zuckerman, B. E. Turner, P. Palmer.
Astrophys. Journ., Vol. 204, 43 - 46 (1976).

The $6_{16}-5_{15}$ transitions of acetaldehyde have been observed in the Sgr B2 molecular cloud. The large line width suggests that both the A and E symmetry states are present with essentially equal intensity. This is the first observation of acetaldehyde at millimeter wavelengths, and the intensity indicates that many other millimeter wave lines of acetaldehyde should be detectable.

131.042 Multiple ionization by low-energy cosmic rays and the abundance of highly ionized interstellar atoms.
W. D. Watson.
Astrophys. Journ., Vol. 204, 47 - 54 (1976).

Calculations are presented and employed in conjunction with available experimental data to estimate cross sections for multiple ionization of atoms by low-energy (MeV nucleon^{-1}) cosmic rays. The efficiency of multiple ionization increases rapidly with the charge of the cosmic ray, so that heavier ($Z \geqslant 6$) nuclei are the dominant contributors for cosmic-ray compositions deduced from higher energy cosmic rays. Multiple ionization dominates in the production of certain highly ionized atoms by cosmic rays under typical interstellar conditions.

131.043 The unusual H_2O maser source near Herbig-Haro object number 11.
K. Y. Lo, M. Morris, J. M. Moran, A. D. Haschick.
Astrophys. Journ., (Letters), Vol. 204, L21 - L24 (1976).

Water emission spectra of an unusual source near Herbig-Haro 11 have been monitored over a 14-month period. Variations in the intensity and the radial velocity of the emission are noticeable on time scales as short as one day. It is suggested that the exciting source is losing mass via a stellar wind, and that the H_2O emission arises in the transition region between the cavity created by the stellar wind and the surrounding molecular medium. The proposed model can be tested by VLBI observations.

131.044   **An almost complete survey of 21 cm line radiation for $|b| \gtrsim 10°$. III. The interdependence of H I, galaxy counts, reddening, and galactic latitude.**   C. Heiles. Astrophys. Journ., Vol. 204, 379 - 402 (1976).

The author compares three extinction indicators: the Shane-Wirtanen galaxy counts, H I column density, and reddening. There is no simple relation between any pair of these three indicators. Each indicator depends on galactic latitude as well as on the other indicator. With reddening considered as the independent variable this latitude dependence occurs for both galactic and extragalactic objects. The H I to extinction ratio varies by a factor of about 2 from one region to another.

131.045   **Thermal-chemical instabilities in CO clouds.** A. E. Glassgold, W. D. Langer. Astrophys. Journ., Vol. 204, 403 - 407 (1976).

The stability of interstellar clouds containing CO is investigated taking into account formation and destruction processes for molecules. Thermal-chemical instabilities are obtained which influence the evolution of clouds.

131.046   **CO observations of the expanding envelope of IRC + 10216.**   T. B. H. Kuiper, G. R. Knapp, S. L. Knapp, R. L. Brown. Astrophys. Journ., Vol. 204, 408 - 414 (1976).

The authors have observed high-sensitivity emission profiles from the $J = 0 \leftarrow 1$ transitions of $^{12}C^{16}O$ and $^{13}C^{16}O$ toward IRC + 10216. It appears that the spherically symmetric uniform mass-outflow model proposed by Morris (1975) is necessary to describe the line profiles. The outlow appears to be slightly accelerated, having a velocity of 15 km s^{-1} at the edges of the CO cloud, compared with 12 km s^{-1} for the more centrally confined molecules.

131.047   **Detection of H_2O maser emission from four infrared sources.**   M. Morris, G. R. Knapp. Astrophys. Journ., Vol. 204, 415 - 419 (1976).

Maser emission from interstellar water vapor has been found toward four infrared sources: OMC-2, OH 231.8 + 4.2 (also known as OH 0739 − 14), S140-IR, and Monoceros R2. Three of these new maser sources appear to be situated in active sites of star formation, while the unusual source OH 231.8 + 4.2 is more likely to be related to a late-type stellar object. The line profiles are relatively simple, showing at most a few velocity components, some of which vary with a time scale as short as a few weeks. The individual sources are discussed in detail.

131.048   **Ultraviolet observations of cool stars. V. The local density of interstellar matter.** W. McClintock, R. C. Henry, H. W. Moos, J. L. Linsky. Astrophys. Journ., (*Letters*), Vol. 204, L103 - L106 (1976).

A high-resolution Copernicus observation of the chromospheric $L\alpha$ emission line of the nearby (3.3 pc) K dwarf ϵ Eri sets limits on the velocity, the velocity dispersion, and the density n_H of atomic hydrogen in the local interstellar medium. Analysis shows that the interstellar $L\alpha$ absorption is on the flat portion of the curve of growth. An upper limit of $n_H = 0.12$ cm^{-3} is derived. The value of n_H is 0.08 ± 0.04 cm^{-3} if the velocity dispersion parameter $b = 9$ km s^{-1}, corresponding to a temperature of 5000 K. Also, the interstellar deuterium $L\alpha$ line may be present in the spectrum.

131.049   **Components in interstellar molecular hydrogen.** L. Spitzer, Jr., W. A. Morton. Astrophys. Journ., Vol. 204, 731 - 749 (1976).

Precise spectrophotometric profiles have been obtained for selected Lyman absorption lines produced by H_2 molecules in various rotational levels, using multiple scans with the Copernicus satellite telescope. For seven stars some of these lines showed complex profiles, and were fitted with two or three separate components at the radial velocities measured from the ground. A least-squares solution for each of 59 lines gave individual values of the column density, $N(J)$, in each component (excluding the highly saturated components). Model calculations by Jura have been used to fit the observed column densities and to determine β, the probability of an upward transition induced by photons in the Lyman and Werner bands of H_2, and n_H, the hydrogen atom particle density. For the components with the most negative velocity, β is some 10 to 30 times greater than the mean interstellar value. The values of n_H, determined for these same approaching components in only three stars, range from 300 to 1000 cm^{-3}.

131.050   **On the existence of molecular hydrogen along lines of sight with low reddening.**   D. G. York. Astrophys. Journ., Vol. 204, 750 - 758 (1976).

The existence of molecular hydrogen along lines of sight with low reddening $[E(B-V) \leqslant 0.03]$ is discussed for the stars HD 28497, μ Col, α Vir, β Cen, and λ Sco. The main observations are of the $R(1)$ features, which for effective rotational temperatures $T > 200$ K are the dominant source of H_2 transitions. If the formation process for molecular hydrogen is similar to that for regions of greater reddening, the observations indicate that the neutral gas outside standard clouds may be partially in small clouds, though a more diffuse component seems to exist in some cases. Extremely high rotational temperatures are inferred for the H_2 toward HD 28497. The possibility of formation of H_2 through the reaction $H^- + H \rightarrow H_2 + e^-$ is briefly discussed. In any case, only small regions seem to contribute to the observed H_2 column density.

131.051   **An analysis of the interstellar material in the line of sight toward Omicron Persei.**   T. P. Snow, Jr. Astrophys. Journ., Vol. 204, 759 - 774 (1976).

Ultraviolet spectrophotometric data obtained with the satellite Copernicus are used to analyze the chemical abundances and physical conditions in the line of sight toward o Per (B1 III, $E(B-V) = 0.32$). The star may be embedded in the near edge of a dense molecular cloud. The far-ultraviolet extinction rise is unusually steep, indicating the probable presence of a high proportion of very small grains. The overall fraction of hydrogen nuclei in H_2 molecules is 0.52, and may be significantly higher in the region containing most of the molecules. The gas-to-dust ratio $N_H/E(B-V) = 4.41 - 5.38 \times 10^{21}$ cm^{-2} mag^{-1} is normal. The density in the region containing the observed C I lines is probably a few hundred cm^{-3}, and may be in excess of 1000 cm^{-3} in the portion of the cloud where the observed molecules exist.

131.052   **A polarization survey of stars near the Orion nebula.**   M. Breger. Astrophys. Journ., Vol. 204, 789 - 796 (1976).

A polarization survey of over 200 young stars near the Orion nebula indicates that 25 percent of the sample shows linear polarization significantly above the average interstellar value for this cluster. Several regions of high polarization and reddening exist in the Orion cluster, although not all the stars in these regions are highly polarized. Position angles of polarization are not randomly distributed, which suggests that an external polarization mechanism operates for most of the polarized stars. Polarization is always accompanied by the presence of infrared excesses. Finally, the amounts and origins of polarization are considered as a function of stellar spectral type and color.

131.053 The detection of interstellar OH absorption in the Zeta Ophiuchi cloud. T. P. Snow, Jr.
Astrophys. Journ., (Letters), Vol. 204, L127 - L130 (1976).

New Copernicus scans have been made of several molecular line wavelengths in the ultraviolet spectrum of ζ Oph. The OH $D\ ^2\Sigma^- - X\ ^2\Pi$ (0,0)Q1(3/2) transition at 1222.071 Å was detected with an equivalent width of 3.5 ± 0.42 Å, implying a column density of $(1.54 - 1.96) \times 10^{14}$ cm^{-2}. Improved upper limits were determined for the remainder of the molecules which were sought. Although the abundance of OH exceeds the values predicted by published calculations using gas-phase ion-molecule reactions, the discrepancy is not large.

131.054 H$_2$CO emission at 2 millimeters in dark clouds. N. J. Evans II, M. L. Kutner.
Astrophys. Journ., (Letters), Vol. 204, L131 - L134 (1976).

The $2_{12} \to 1_{11}$ transition of H$_2$CO at 2 mm wavelength has been detected at three positions in dark clouds. This confirms predictions based on 2 cm H$_2$CO absorption studies. The 2 mm radiation temperatures, together with the 2 cm data and recently computed cross sections for H$_2$—H$_2$CO collisions, are used to deduce cloud conditions. These in turn are used to predict the strengths of emission lines from CS and HCN.

131.055 Isotopic abundances in interstellar carbon monosulfide.
R. W. Wilson, A. A. Penzias, P. G. Wannier, R. A. Linke.
Astrophys. Journ., (Letters), Vol. 204, L135 - L137 (1976).

Measurements of relative abundances of the rare isotopic species $^{13}C^{32}S$ and $^{12}C^{34}S$ have been made by means of their 2 mm rotational line emission in five dense interstellar clouds. The abundance ratio $[^{13}C^{32}S]/[^{12}C^{34}S]$ shows a significant source-to-source variation. Measurements of $^{12}C^{33}S$ in two of the five clouds are consistent with a relative $[^{33}S]/[^{34}S]$ abundance equal to the terrestrial value.

131.056 HCN, X-ogen (HCO$^+$), and U90.66 emission spectra from L134. L. E. Snyder, J. M. Hollis.
Astrophys. Journ., (Letters), Vol. 204, L139 - L142 (1976).

The authors have detected millimeter-wave emission lines from HCN, X-ogen (HCO$^+$), and U90.66 from the direction of the cool dust cloud L134. This is the first reported detection of HCN and U90.66 in a dark nebula. The HCN radial velocity was used to decrease the uncertainties of the previously measured X-ogen (HCO$^+$) and U90.66 rest frequencies. The emission profile of U90.66 was found to be consistent with the expected profile for hydrogen isocyanide (HNC). These observations suggest that L134 and other cool dust clouds should be superb objects for future high-resolution measurements of other interstellar molecules which have never been detected spectroscopically in the laboratory.

131.057 High-density interstellar clouds. R. D. Davies.
Observatory, Vol. 96, 4 - 5 (1976).

131.058 A study of the ρ Ophiuchi molecular cloud. J. Lequeux.
Observatory, Vol. 96, 5 - 6 (1976).

131.059 Revised interstellar neutral helium/hydrogen density ratios and the interstellar UV-radiation field.
P. W. Blum, H. J. Fahr.
Astrophys. Space Sci., Vol. 39, 321 - 334 (1976).

The data deduced from the UV-spectroscope on the Copernicus satellite strongly suggest that the most important ionization source in interstellar space near the solar system is a UV radiation field originating from B-stars. Adopting this hypothesis, we have used the ionization state of several ele-

ments in the interstellar medium observed by Copernicus to determine the required radiation field. From this, the degree of ionization of elements that could not be observed by Copernicus is estimated. It is shown that the ratio of neutral interstellar helium to neutral interstellar hydrogen is likely to be 2 to 3 times as large as the cosmic abundance ratio of these elements. The possibility that this ratio is about 10 times as large, meaning equal interstellar neutral hydrogen and helium densities near the solar system, cannot be ruled out. It would, however, require an interstellar radiation temperature near 9000 K.

131.060 On ionization mechanisms towards γ^2 Vel and ζ Pup. S. P. Tarafdar.
Astrophys. Space Sci., Vol. 39, 419 - 427 (1976).

A perusal of the observed column densities of different ions towards two stars, γ^2 Vel and ζ Pup, has indicated the presence of an ionization mechanism, the rate of which depends exponentially on the ionization potential. An examination of two such mechanisms, collisional ionization and ionization by UV-photon with flux depending exponentially on frequency has shown that the collisional ionization is possibly the dominant one in the intercloud medium towards γ^2 Vel and ζ Pup.

131.061 Models of cosmic maser sources.
S. B. Pikel'ner, V. S. Strel'nitskiy (Strel'nitskij).
Astrophys. Space Sci., Vol. 39, L19 - L24 (1976). — Letter.

131.062 Infrared spectra of small interstellar grains.
W. W. Duley.
Astrophys. Space Sci., Vol. 39, L33 - L36 (1976).

It is shown that surface stresses in small particles will lead to a broadening and weakening of strong resonances in the infrared spectrum of interstellar dust.

131.063 Interstellar matter research with the Copernicus satellite. (Karl-Schwarzschild-Vorlesung 1975).
L. Spitzer, Jr.
Mitt. Astron. Ges., No. 38, p. 27 - 39 (1976). — Karl-Schwarzschild-lecture during the meeting of the Astron. Ges., Berlin 1975.

131.064 Molekulares und thermisches Gleichgewicht in dünnen Neutralgaswolken.
J. Barsuhn, C. M. Walmsley.
Mitt. Astron. Ges., No. 38, p. 79 (1976). — Abstract.

131.065 Energiefluktuationen im interstellaren Staub.
H.-P. Gail, E. Sedlmayr.
Mitt. Astron. Ges., No. 38, p. 79 - 80 (1976). — Abstract.

131.066 Über die Ladungsverteilung von interstellarem Staub.
H.-P. Gail, E. Sedlmayr, G. Traving.
Mitt. Astron. Ges., No. 38, p. 80 (1976). — Abstract.

131.067 Interstellares Ionisationsgleichgewicht und interplanetare neutrale Heliumdichten.
P. W. Blum, H. J. Fahr.
Mitt. Astron. Ges., No. 38, p. 80 - 82 (1976). — Short report.

131.068 Zeitabhängige Rechnungen zur thermischen und Ionisationsbilanz des interstellaren Mediums.
K. P. Brand.
Mitt. Astron. Ges., No. 38, p. 83 - 85 (1976). — Short report.

131.069 Strahlungstransport in kosmischen Masern.
E. Bettwieser.
Mitt. Astron. Ges., No. 38, p. 154 - 155 (1976). — Abstract.

131.070 Beobachtungen von OH-Linien in der Staubwolke Lynds 1630. A. Winnberg.

Mitt. Astron. Ges., No. 38, p. 218-221 (1976). – Short report.

131.071 **Interstellare Absorption bis 1 kpc in Vela.**
U. Haug.
Mitt. Astron. Ges., No. 38, p. 245 (1976). – Abstract.

131.072 **On the ratio of the total to selective absorption.**
W. A. Sherwood.
Astrophys. Space Sci. Library, Vol. 55, (see 012.001), 3 - 10 (1976).

The ratio of total to selective absorption, R, has been found to remain constant as dust is processed in clouds from low to high density, through H II regions and open clusters, and returned to the interstellar medium. R has the same value in dense dust clouds as it has in H II regions of different ages. Variations in R values obtained from stars in H II regions may be due to errors in special type classification. Globular cluster diameters show no tendency to increase with distance from the sun when $R = 3.2$ is used. Large grains evidently do not exist in the interstellar medium. There is no evidence for neutral extinction in the Galaxy at large.

131.073 **Features of the interstellar extinction curve.**
D. H. Morgan.
Astrophys. Space Sci. Library, Vol. 55, (see 012.001), 11 - 17 (1976).

The extinction curves for spherical particles are subject to the errors of the particle material's refractive index. Their sensitivity to these errors has been investigated and is found to be dependent upon wavelength. For graphite, significant errors are produced in the far ultraviolet part of the extinction curve; for silicates, in the near ultraviolet; while for iron the error is relatively small. The wavelength dependence of the 10 μm and 20 μm absorption bands of small silicate spheroids upon their shape and alignment has been studied.

131.074 **Far-ultraviolet extinction in σ Scorpii.**
T. P. Snow, Jr., D. G. York.
Astrophys. Space Sci. Library, Vol. 55, (see 012.001), 19 - 22 (1976).

It was found earlier from OAO-2 data (Bless and Savage, 1972) that considerable variability with direction in space is present in both the shape and level (relative to $B-V$ color excess) of the interstellar extinction curve in the far ultraviolet. The authors have obtained UV data on σ Sco using Copernicus (OAO-3), which has an entrance slit on the order of 10^3 times smaller in projected area than that of OAO-2, so that the contribution to the signal from scattered nebular light would be correspondingly smaller. They find very good agreement with the extinction curve of Bless and Savage.

131.075 **Diffuse interstellar band formation in dense clouds.**
T. P. Snow, Jr., J. G. Cohen.
Astrophys. Space Sci. Library, Vol. 55, (see 012.001), 33 - 38 (1976).

Measurements of the strengths of the diffuse interstellar bands at 4430, 5780 and 5797 Å show that the bands tend to be weak with respect to extinction in dense interstellar clouds. Data on 10 stars in the ρ Ophiuchi cloud complex show further that the diffuse band-producing efficiency of the grains decreases systematically with increasing grain size. It is concluded that the diffuse bands are not formed in the mantles which accrete on the grains in interstellar clouds, but that they could be produced in the cores of grains or in some molecular species.

131.076 **Interstellar extinction and diffuse absorption features.** J. Dorschner.
Astrophys. Space Sci. Library, Vol. 55, (see 012.001), 39 - 47 (1976).

The equivalent width of the λ 2175 Å band, W_{2175}, well

known as the big bump in the interstellar extinction curves, has been found to be closely correlated with the colour excess E_{B-V} as well as with the extinction differences E_{8-6} and E_{9-7} defined to characterize quantitatively the steep slopes of the extinction curves in the far ultraviolet. The results have been qualitatively interpreted in favour of the dust model consisting of a mixture of small silicate grains and larger silicate grains coated by molecular mantles.

131.077 **Diffuse band extinction and polarization in core-mantle grains.**
J. M. Greenberg, S. S. Hong.
Astrophys. Space Sci. Library, Vol. 55, (see 012.001), 49 - 58 (1976).

Diffuse band shapes in both extinction and polarization are calculated for interstellar core-mantle particles for varying size distributions of mantle thickness. It is shown that no matter whether the source of the bands is in the silicate cores or the accreted icy mantles the polarization shapes are highly asymmetric for all mantle thicknesses. The extinction band shapes are significantly less asymmetric although the effect is clearly present. The only apparent possibility for producing symmetric band shapes in the dust grains is in the very small bare particles in interstellar space which, if they are aligned and produce the λ 2200 band, must exhibit a strong polarization effect in this region.

131.078 **Physical adsorption of hydrogen on interstellar graphite grain surfaces.**
R. F. Willis, B. Fitton.
Astrophys. Space Sci. Library, Vol. 55, (see 012.001), 71 - 85 (1976).

The authors review existing single-particle theories concerning parameters of importance which determine the kinetics of hydrogen molecule formation and ejection from cold ($T_g \lesssim 20$ K) graphite grain surfaces. The nature of the single-particle quantum states of low mass gas atoms and molecules in a periodic surface lattice potential is considered. Short-range electron correlation effects at the surface may lead to the formation of a 'quasimolecular state' of adsorbed H_2 with a bond length ~ 3.5 Å and a reduced bond energy ~ 0.075 eV. It is proposed, that one consequence of this dynamical screening of the adsorbed molecules is that they are ejected normal to the grain surface with velocities $\lesssim 20$ km s^{-1} and not necessarily in a high vibrational state.

131.079 **Extinction and polarization models.**
N. C. Wickramasinghe.
Astrophys. Space Sci. Library, Vol. 55, (see 012.001), 87 - 92 (1976).

Currently favoured models for interstellar grains are reviewed in relation to observational criteria which bear on their optical properties.

131.080 **Effects of charged dust grains.**   S. Hayakawa.
Astrophys. Space Sci. Library, Vol. 55, (see 012. 001), 93 - 99 (1976).

Dust grains expelled by radiation pressure of stars are charged to potentials in the range 30–40 V in H I clouds. These grains may be responsible for the following phenomena which are otherwise hardly explicable. (1) A considerable fraction of electrons knocked-out by charged grains of high speeds have energies around 15 eV and produce singly ionized ions but not doubly ionized ones in accord with an ultraviolet observation of interstellar atoms and ions. (2) Transverse momentum transferred to grains by Coulomb scattering of ambient electrons and protons is greater than that by multiple scattering of cosmic ray protons, thus the former being more effective for the grain alignment than the latter. (3) At a shock front charge separation due to a large inertial mass of grains produces an electric field, thus accelerating charged particles and causing a drift of interstellar matter.

131.081 Considerations about the absorption efficiency of dust particles in the infrared.
E. Bussoletti, A. Borghesi, G. Leggieri, A. Blanco.
Astrophys. Space Sci. Library, Vol. 55, (see 012.001), 143 - 149 (1976).

Analytical approximations used often in the literature for calculating energy rates emitted by dust grains in infrared are discussed. Comparisons with correct complete formulations are made for three grain models: (1) pure graphite, (2) ice mantle-graphite core, (3) silicates. λ^{-1} and λ^{-2} dependences for the average effective emissivity of such grains are used. The authors find that for silicate and graphite grains the simplified approximations are valid only when accuracies between 10% and 50% are required and only for grain temperatures higher than 80 K. At lower temperatures the validity of the approximations fails for the graphite particle while it is variable for the silicate dust grain. The ice core mantle particles can instead be treated with approximated formulae without introducing appreciable errors.

131.082 On the presence of phyllosilicate minerals in the interstellar grains.
A. Zaikowski, R. F. Knacke, C. C. Porco.
Astrophys. Space Sci. Library, Vol. 55, (see 012.001), 151 - 169 (1976).

The composition of the interstellar silicate dust is investigated. Condensation or alteration of silicate grains at temperatures of a few hundred degrees, in the presence of H_2O, would result in hydrous or phyllosilicates, the silicate type most abundant in the type I carbonaceous chondrites. The authors propose that the silicates in the interstellar grains are predominantly phyllosilicates and suggest additional spectral tests for this hypothesis.

131.083 The influence of grain mantles on the formation of hydrogen molecules on grain surfaces. T. J. Lee.
Astrophys. Space Sci. Library, Vol. 55, (see 012.001), 171 - 178 (1976).

The physical adsorption energy, E, of hydrogen molecules on various substrates at temperatures between 5 and 30 K and at the lowest practicable gas densities has been measured. Values of E/k are for condensed CO 340 K, CO_2 800 K, H_2O 850 K and for 'dirty' graphite 980 K and 'dirty' copper 800 K. From these measurements temperature ranges in which H atoms might combine on the surface to form H_2 molecules are estimated. Duley has discussed the formation and composition of condensed gas mantles on interstellar grains. The effects of such mantles in promoting and poisoning hydrogen molecule formation are discussed.

131.084 UV radiation fields in dark clouds.
A. P. Whitworth.
Astrophys. Space Sci. Library, Vol. 55, (see 012.001), 207 - 225 (1976).

If interstellar extinction at UV wavelengths is mainly due to scattering with a strongly forward throwing phase-function, the interior of a dark cloud may be much better illuminated at UV wavelengths than its measured extinction would suggest. Computations are made of the radiation fields in (1200, 4500) Å, at the centres of dark clouds with measured visual extinctions. It is found that even in very dark clouds, the radiation energy density in (1200, 1800) Å may be significant, due to the high grain albedo at these short wavelengths.

131.085 The plausibility of silicate-core ice-mantle grains.
M. J. Dempsey, N. C. Wickramasinghe.
Astrophys. Space Sci. Library, Vol. 55, (see 012.001), 227 - 231 (1976).

Extinction curves for silicate-core ice-mantle grains are computed and compared with the infrared spectral data on the BN object in the Orion nebula. A ratio of outer mantle to core radius of 1.3 which best fits this data suggests that silicate-core ice-mantle grains are unlikely to contribute a major part to the total visual extinction coefficient of interstellar material.

131.086 How to make metal-poor stars, redden OB associations and grow mantles on grains.
M. G. Edmunds, N. C. Wickramasinghe.
Astrophys. Space Sci. Library, Vol. 55, (see 012.001), 233 - 238 (1976).

Three consequences of the existence of grains with metal-rich ice mantles are considered: (1) The production of metal-poor stars by expulsion of protostellar grains by radiation pressure during star formation. (2) The effects of these expelled grains in reddening massive stars in an OB association. (3) The production of the icy mantles on grains in OB associations.

131.087 Radical formation, chemical processing, and explosion of interstellar grains. J. M. Greenberg.
Astrophys. Space Sci. Library, Vol. 55, (see 012.001), 239 - 248 (1976).

The ultraviolet radiation in interstellar space is shown to create a sufficient steady state density of free radicals in the grain mantle material consisting of oxygen, carbon, nitrogen, and hydrogen to satisfy the critical condition for initiation of chain reactions. The criterion for minimum critical particle size for maintaining the chain reaction is of the order of the larger grain sizes in a distribution satisfying the average extinction and polarization measures. The triggering of the explosion of interstellar grains leading to the ejection of complex interstellar molecules is shown to be most probable where the grains are largest and where radiation is suddenly introduced; i.e. in regions of new star formation.

131.088 Effects of suprathermal grains.
S. P. Tarafdar, N. C. Wickramasinghe.
Astrophys. Space Sci. Library, Vol. 55, (see 012.001), 249 - 260 (1976).

Grains ejected from stars at velocities of $\sim 10^7$ cm s^{-1} and/or grains accelerated by the pressure of starlight in the inter-cloud medium to velocities in the range $2 \times 10^6 - 10^7$ cm s^{-1} are slowed to velocities of about 2×10^5 cm s^{-1} in a typical interstellar cloud. The interaction of fast grains with gas atoms as they are slowed in clouds could provide (1) the dominant heat source for interstellar clouds; (2) sites for molecule formation; and (3) a mechanism of providing a pressure balance between clouds and the intercloud medium.

131.089 Thermal behaviour of the neutral interstellar gas within the solar system. H. J. Fahr.
Space Research XV, (see 012.003), p. 727 - 731 (1975).

The change of the dynamical and thermal status of interstellar matter that sweeps over the solar system is shown to be adequately described by means of the Boltzmann-Vlasov equation. A solution of this equation is reached by the application of the single particle orbit technique. An explicit form of the velocity distribution function within the solar system is given.

131.090 Two-component structure in the profiles of high velocity clouds. T. R. Cram, R. Giovanelli.
Astron. Astrophys., Vol. 48, 39 - 47 (1976).

Gaussian analysis of 21-cm profiles of high velocity clouds reveal the existence of two well defined components: one with mean velocity widths of about 7 km s^{-1}, and the other with mean widths of about 23 km s^{-1}. Small velocity-width components correlate with small, bright condensations in the high velocity complexes, while larger width components correlate with more extended, faint regions. Analogies with uv interstellar absorption and 21 cm line absorption observations are discussed.

131.091 **Detection of dimethyl ether in SGR B 2.**
G. Winnewisser, F. F. Gardner.
Astron. Astrophys., Vol. 48, 159 - 161 (1976).
Emission has been detected at 9119.7 MHz towards
Sgr B 2 which has been identified as the $2_{02}-1_{11}$ transition of
dimethyl ether, $(CH_3)_2O$. The line, which has a peak brightness
temperature of 0.05 K, agrees with the laboratory spectrum if
the central velocity is 64 ± 4 km s^{-1} and the velocity width
23 km s^{-1}. No line was detected towards the Orion-IR nebula.

131.092 **A bright source of carbon recombination line in the**
Rho Ophiuchi complex.
D. A. Cesarsky, P. J. Encrenaz, E. G. Falgarone, B. Lazareff,
R. Lauqué, R. Lucas, L. Weliachew.
Astron. Astrophys., Vol. 48, 167 - 169 (1976).
The authors present radio observations of a bright carbon
166α recombination line, T_A = 0.20 K, arising in a small region,
θ_s = 4', of the dark cloud near the star ρ-Ophiuchi. A weaker
line, T_A = 0.040 K, which apparently arises in a more extended
region, is identified with the recombination line emission of
sulphur.

131.093 **Identifikation der unbekannten interstellaren Linie**
U 90.7 als Iso-Blausäure, HNC. R. A. Creswell,
E. F. Pearson, M. Winnewisser, G. Winnewisser.
SuW, 15. Jahrgang, p. 118 - 120 (1976).

131.094 **Cyanoacetylene in dense interstellar clouds.**
M. Morris, B. E. Turner, P. Palmer, B. Zuckerman.
Astrophys. Journ., Vol. 205, 82 - 93 (1976).
Cyanoacetylene (H$-$C$\equiv$C$-$C$\equiv$N) has been detected in 17 galactic sources, including six clouds in the vicinity of the galactic
center, nine clouds associated with H II regions and/or far-
infrared sources, one dark dust cloud, and the molecular
envelope of a carbon star. Statistical equilibrium calculations
were performed and matched to the observations of several
sources. Several sources are discussed individually.

131.095 **Radiative association in dense, H$_2$-containing inter-**
stellar clouds. E. Herbst.
Astrophys. Journ., Vol. 205, 94 - 102 (1976).
A method for estimating the order of magnitude of rate
coefficients of radiative association reactions involving polyatomic species is outlined. Calculations are undertaken on reactions of possible importance in dense interstellar clouds.
Several of the calculated rate coefficients are utilized in a
study of the gas phase syntheses of H_2CNH (methanimine) and
H_3CNH_2 (methylamine).

131.096 **High-resolution profiles of the diffuse interstellar**
feature at 5780 Å. B. D. Savage.
Astrophys. Journ., Vol. 205, 122 - 135 (1976).
High-resolution profiles ($\Delta\lambda \approx 0.2$ Å) were obtained of
the diffuse interstellar feature at 5780 Å in 18 heavily reddened
stars. This feature is, in all cases, asymmetrical with its step
side being toward the blue. Good fits to match theoretical
profiles to the observed ones can be obtained for the extinction profiles provided by small ($r \approx 750$ Å), cold grains containing impurities that produce narrow no-phonon absorption
lines. If $\lambda 5780$ is in fact due to this latter process, then the
asymmetry of the feature provides information on the sizes of
interstellar grains, while the width provides information on
the internal temperatures of grains.

131.097 **The distance and mass of the large elephant trunk,**
a CO cloud pointing towards NGC 6231.
W. A. Sherwood, J. Dachs.
Astron. Astrophys., Vol. 48, 187 - 192 (1976).
Photoelectric UBV-Hβ photometry of stars in the region
of the Sco OB1 association indicates that the elephant trunk-
shaped dark cloud noted by Bok et al. (1966) pointing towards

NGC 6231, the central cluster of the association, is indeed
located at about the same distance as the cluster, some 2.0 kpc
away. A strong CO molecular radio source recently detected by
Blair et al. (1975) is located in the trunk. It has a radial velocity similar to that of the H II region and to that of the star cluster. The elephant trunk structure is much larger than elephant
trunks known in other galactic nebulae. Its total mass is likely
to lie between 2×10^3 and 4×10^4 solar masses.

131.098 **New H$_2$O sources associated with late-type stars.**
J. R. D. Lépine, M. H. Paes de Barros, R. H.
Gammon.
Astron. Astrophys., Vol. 48, 269 - 274 (1976).
The authors searched for 1.35 cm water vapor maser
emission in 74 late type stars; five new H_2O sources were detected. The characteristics of the microwave line velocities of
the known H_2O sources associated with visible stars are discussed.

131.099 **Refraction effects and position stability in compo-**
nents of the water source W 49. L. T. Little.
Monthly Notices Roy. Astron. Soc., Vol. 175, 245 - 255 (1976).
Long baseline interferometer observations have shown
that the relative positions of strong components in the water
source W 49 (H_2O) change by angles $\sim 0.''05$ in periods of a
few months (Knowles et al.). Bodily motion of the individual
components cannot explain the apparent changes. It seems
most likely that changes in the location of the pump energy
input to the maser are responsible for the observed behaviour.

131.100 **On the flux and the energy spectrum of interstellar**
ions in the solar system. V. M. Vasyliunas, G. L.
Siscoe.
Journ. Geophys. Res., Vol. 81, 1247 - 1252 (1976).
The flux density of ions created by ionization of interstellar neutral particles in the solar system and picked up by
the solar wind is calculated as a function of the neutral particles. A very broad maximum occurs at an angle of 0 and a
distance that depends on the density and speed of the neutral
particles and on the ionization time but is typically in the
general region of 10 AU. For atomic hydrogen the flux den-
sity is estimated to exceed 10^4 cm^{-2} s^{-1} over the distance range
from a few to nearly 100 AU. If charge exchange is an important contributor to the ionization of hydrogen, the observed
local intensity of interstellar protons should exhibit time variations correlated with the density changes of the solar wind
stream structure.

131.101 **Extinction due to the dark cloud complex in**
Corona Australis. G. S. Rossano.
Bull. American Astron. Soc., Vol. 8, 293 (1976). – Abstr. AAS.

131.102 **Interstellar absorption to the Crab nebula.**
V. V. Golovatyj, O. S. Yatsyk.
Tsirk. Astron. Obs. L'vov, No. 49, p. 9 - 11 (1974). In Russian.

131.103 **B stars and the structure of the interstellar medium.**
J. Lyon.
Astrophys. Letters, Vol. 17, 81 - 86 (1976).
If the low density, intercloud material is cold, as recent
observations suggest, then the H II regions around B stars are
very effective at moving the neutral gas about. A two-dimensional hydrodynamic simulation of such an interstellar medium was performed. On a short time scale, $\simeq 1.5 \times 10^7$ yr, an
initially uniform medium at 1 cm^{-3} separates into low density
($\lesssim 0.1$ cm^{-3}) and high density ($\gtrsim 3$ cm^{-3}) regions. The low
density regions contain both hot, ionized gas and cold, un-
ionized gas. The observed average electron density and the
observed average statistics of clouds are well reproduced by
the model.

131.104 Phase retardation of light due to scattering by interstellar non-spherical grains. G. A. Shah.
Bull. Astron. Soc. India, Vol. 3, 32 (1975). – Abstract of a paper presented at the A.S.I. meeting 1975.

131.105 On the charged states of the low energy (1–30 MeV/n) multiply charged ions in interstellar space. N. Durgaprasad.
Bull. Astron. Soc. India, Vol. 3, 37 (1975). – Abstract of a paper presented at the A.S.I. meeting 1975.

131.106 Radio spectroscopy of the NGC 2023 C II regions. V. Pankonin, C. M. Walmsley.
Astron. Astrophys., Vol. 48, 341 - 348 (1976).

C 157α, C 110α and C 138β lines were observed from the NGC 2023 C II region(s). The carbon profile has two components which are separated by ~ 1 km s^{-1}. The widths of the line components are $\lesssim 1$ km s^{-1}, and the total C II distribution has a diameter of $< 10'$. The physical conditions in the two line sources differ. The ionization and thermal equilibrium of a model C II region which is produced by an early B star are investigated. The ionization structure is strongly influenced by the presence of dust; electron temperatures may vary from ~ 60 K in the core to ~ 10 K at the outer edge of the C II region. A continuum source was observed at 6-cm which coincides with the position of NGC 2023. Formaldehyde and OH lines were observed at several positions toward NGC 2023.

131.107 Time scales for molecule formation by ion-molecule reactions. W. D. Langer, A. E. Glassgold.
Astron. Astrophys., Vol. 48, 395 - 403 (1976).

Analytic solutions are obtained for non-linear differential equations governing the time-dependence of molecular abundances in interstellar clouds. Three gas phase reaction schemes are considered separately for the regions where each dominates. The particular case of CO, and closely related members of the OH and CH families of molecules, is studied for given values of temperature, density, and the radiation field. Non-linear effects and couplings with particular ions are found to be important. The time scales for CO formation range from 10^5 to a few $\times 10^6$ years, depending on the chemistry and regime. The time required for essentially complete conversion of C$^+$ to CO in the region where the H$_3^+$ chemistry dominates is several million years.

131.108 OH–IR stars. II. A model for the 1612 MHz masers. M. Elitzur, P. Goldreich, N. Scoville.
Astrophys. Journ., Vol. 205, 384 - 396 = Contr. California Inst. Techn., Div. Geol. Planet. Sci., *Pasadena, California,* No. 2542 (1976).

The present paper attempts to bring into sharper focus the essential features responsible for the inversion of the 1612 MHz transition in OH–IR stars. A discussion of the principal observational facts is made. Then a brief description of the gas kinematics and the OH abundance in the circumstellar envelopes about OH–IR stars is given. A section is devoted to a detailed description of the population flow in the maser pump cycle and includes the results of numerical solutions of the equations of radiative transfer and statistical equilibrium. A comparison of the theoretical results and the observational facts is made.

131.109 The ionization of cloud and intercloud hydrogen by O and B stars. B. G. Elmergreen.
Astrophys. Journ., Vol. 205, 405 - 418 (1976).

Limitations to the extent of OB-star ionization in a low-density ($\lesssim 0.2$ cm^{-3}) intercloud medium are investigated. The time required for one O star in such a medium to completely ionize or remove by the rocket effect the standard clouds within one-half of the star's Strömgren radius is shown to be comparable with the lifetime of the star. The time required for

the complete ionization of a standard cloud which is exposed to isotropic Lyman continuum radiation in an extensive H II region containing many OB stars is also calculated. A model for the interstellar medium is investigated in which all of the O-star Lyman continuum radiation is converted into Balmer radiation in unit-density Strömgren spheres, and the radiation from B stars and the nuclei of planetary nebulae ionizes a low-density medium which contains standard clouds.

131.110 Polarization of the radiation of stars with non-uniform distribution of luminosity on the star's surface. Yu. N. Gnedin, N. A. Silant'ev.
Astron. Zhurn. Akad. Nauk SSSR, Vol. 53, 338 - 345 (1976). In Russian. English translation in Soviet Astron., Vol. 20, No. 2.

The expressions for the degree of linear polarization of stellar radiation are derived on general assumptions on the distribution of luminosity on the stellar surface. The degree of linear polarization is calculated for accretion onto a compact star with strong magnetic field, for radiation of the optical component of a binary system with X-ray source and for radiation of magnetic and rapidly rotating stars. A qualitative explanation of polarization effects observed from R CrB is given.

131.111 Identification of interstellar X-ogen as HCO$^+$. W. P. Kraemer, G. H. F. Diercksen.
Astrophys. Journ., *Letters*, Vol. 205, L97 - L100 (1976).

Based on the results presented here the interstellar X-ogen line at 89.19 GHz is identified as the lowest rotational transition ($J = 1-0$) of the HCO$^+$ molecular ion. To a comparable accuracy the corresponding data have been obtained for other isotopic species of HCO$^+$, which may serve as a guide for further searches.

131.112 Consequences of a new hot component of the interstellar medium. P. R. Shapiro, G. B. Field.
Astrophys. Journ., Vol. 205, 762 - 765 (1976).

The suggestion that the observed 0.25 keV X-ray background and O VI absorption lines are produced by a new, hot ($T \sim 10^6$ K), diffuse component of the interstellar medium is examined in the context of both a steady-state and a time-dependent model. It is concluded that such a component can explain the two observations only if (1) the pressure of the cooler interstellar medium is ~ 10 times higher than the previously estimated $p/k \approx 2000$ cm^{-3} K; (2) a mechanism such as the proposed convective-radiative "galactic fountain" exists to cool the hot gas; and (3) another component is responsible for the presence of O VI at temperatures below 10^6 K.

131.113 Rotational excitation of CO by collisions with He, H, and H$_2$ under conditions in interstellar clouds. S. Green, P. Thaddeus.
Astrophys. Journ., Vol. 205, 766 - 785 (1976).

Cross sections for rotational excitation of small molecules by low-energy collisions with helium and hydrogen can currently be obtained via accurate numerical solution of the quantum equations that describe both intermolecular forces and collision dynamics. The relevant methods are discussed in some detail and applied to compute excitation rates for carbon monoxide. These calculations also predict collision-induced spectral pressure broadening constants which are in excellent agreement with available experimental data.

131.114 Dynamics of CO molecular clouds in the Galaxy. F. N. Bash, W. L. Peters.
Astrophys. Journ., Vol. 205, 786 - 797 (1976).

The ^{12}C^{16}O spectral line at 115 GHz has been observed at each degree of galactic longitude from $l = 30°$ through $60°$ at $b = 0°$. The radial velocity of the positive-velocity terminus of the profile has been compared with that for the H I, 21 cm line. Explanation of the velocity difference as a function of galactic longitude was attempted on the assumption that the CO has the same kinematics as the H I. The CO was not found to lie in any large, organized patterns. Specifically, the CO is not found to be exclusively in the spiral arms or in any pattern displaced from, but parallel to, the spiral arms.

131.115 **The abundance ratio [^{17}O]/[^{18}O] in dense interstellar clouds.** P. G. Wannier, R. Lucas, R. A. Linke, P. J. Encrenaz, A. A. Penzias, R. W. Wilson.
Astrophys. Journ.,(*Letters*), Vol. 205, L169 - L171 (1976).
The authors have measured the interstellar [^{17}O]/[^{18}O] isotope abundance ratio in eight giant molecule clouds using the $J = 1 \rightarrow J = 0$ transition of carbon monoxide at ~2.7 mm. The average interstellar ratio derived from these data is 0.24, a value which is significantly higher than the corresponding terrestrial abundance ratio of 0.186.

131.116 **Detection of the heavy interstellar molecule cyanodiacetylene.** L. W. Avery, N. W. Broten, J. M. MacLeod, T. Oka, H. W. Kroto.
Astrophys. Journ.,(*Letters*), Vol. 205, L173 - L175 (1976).
The $J = 4 \rightarrow 3$ rotational emission line of cyanodiacetylene $H-C \equiv C-C \equiv C-C \equiv N$ has been detected in Sgr B2. If the molecules are assumed to be in thermal equilibrium at a temperature of 30 K, a column density of 1.5×10^{14} cm^{-2} is obtained. This observation provides further evidence that heavy polyatomic molecules exist in abundance in Sgr B2.

131.117 **Interpretation of low-frequency recombination line observations of the interstellar medium.**
P. A. Shaver.
Astron. Astrophys., Vol. 49, 1 - 16 (1976).
Recombination line observations below 500 MHz are shown to impose severe restrictions on the amount of cold, partially-ionized gas in the interstellar medium (e.g. $EM < 0.5$ pc cm^{-6} for $T_e = 50$ K and a 10 kpc path length). It is also shown that the intensities of galactic ridge recombination lines at centimeter wavelengths can only be explained by high density gas ($n_e \gtrsim 5$ cm^{-3}), in the most cases H II regions.

131.118 **Observations of interstellar magnesium lines in the direction of the stars β, δ, τ Sco and β Cep.**
B. Bates, P. P. D. Carson, P. L. Dufton, C. D. McKeith, A. Boksenberg, B. Kirkham, M. Pettini.
Astron. Astrophys., Vol. 49, 81 - 88 (1976).
Observations of the interstellar lines of Mg^0 and Mg$^+$ in the spectra of the stars β, δ, τ Sco and β Cep have been made. Column densities are derived from a curve-of-growth analysis based on models for the velocity distribution of the line-of-sight gas determined from high resolution scans of the visible Na^0 and Ca$^+$ lines. From the relative Mg/H abundances the authors find that Mg is depleted by approximately a factor 10 for the moderately reddened stars β and δ Sco (E(B$-$V) $\approx$ 0.2) whilst for the stars τ Sco and β Cep (E(B$-$V)$\approx$ 0.05) the Mg/H abundance is closer to the solar value.

131.119 **On the interstellar ionization rate.** P. A. Shaver.
Astron. Astrophys., Vol. 49, 149 - 152 (1976).
It is shown that low frequency (< 500 MHz) recombination line observations can be used together with H I absorption measurements to give direct and reliable estimates of the hydrogen ionization rate for individual interstellar clouds. Values as low as $\zeta_H \lesssim 2 \times 10^{-16}$ s^{-1} have been determined from presently available data.

131.120 **Observations of interstellar silicon monoxide.**
D. F. Dickinson, C. A. Gottlieb, E. W. Gottlieb, M. M. Litvak.
Astrophys. Journ., Vol. 206, 79 - 84 (1976).
The authors report observations of rotational transitions of SiO in the ground vibrational state at 86.8 GHz ($J = 2 \rightarrow 1$)

and 130.3 GHz ($J = 3 \rightarrow 2$). At 86.8 GHz, SiO emission in Orion A peaks in the direction of the Kleinmann-Low nebula. A possible secondary maximum occurs about 80'' southward. The 130.3 GHz line was also detected in Orion A. Mapping of the 86.8 GHz line in Sgr B2 shows the presence of two velocity components and a north—south extent of about 4'. Upper limits to the total HO$_2$ column density in Orion A and Sgr B2 are reported.

131.121 **Abundances of simple oxygen-bearing molecules and ions in interstellar clouds.**
A. E. Glassgold, W. D. Langer.
Astrophys. Journ., Vol. 206, 85 - 99 (1976).
The abundances of simple oxygen-bearing interstellar molecules in warm ($T \gtrsim 40$ K), diffuse, and moderately thick clouds are calculated on the basis of binary gas phase reactions. The most important reactions are ion-molecule, charge exchange, and dissociative recombination reactions. The progenitor of these molecules in diffuse clouds is the cosmic ray produced H$^+$ ion, working through the charge exchange reaction with O. The ionization of H$^+$ and He$^+$ are also discussed. The calculated molecular abundances are consistent with some of the available observational information.

131.122 **Submillimeter observations of NGC 2024, OMC-2, and Mon R-2.** H. S. Hudson, B. T. Soifer.
Astrophys. Journ., Vol. 206, 100 - 108 (1976).
The authors report ground-based submillimeter observations of three molecular clouds; NGC 2024, OMC-2 and Mon R-2. NGC 2024 was extensively mapped at this wavelength, with a 1'.6 beam, while photometry at the peak intensity position is reported for OMC-2 and Mon R-2. The observations strongly suggest that these objects resemble the molecular cloud centered on the KL nebula in Orion.

131.123 **Observations of heavy-element recombination lines in the Rho Ophiuchi dark cloud at 13 centimeters wavelength.** G. R. Knapp, T. B. H. Kuiper, R. L. Brown.
Astrophys. Journ., Vol. 206, 109 - 113 (1976).
The paper discusses observations of the ρ Oph cloud at 13 cm wavelength. Recombination lines due to carbon and sulfur were detected, and again the results suggest depletion of the heavy elements in this dark cloud. The implications of this and Chaisson's result, and the structure of the ionized region, are discussed.

131.124 **Evolution of rotating interstellar clouds. II. The collapse of protostars of 1, 2, and 5 $M_\odot$.**
D. C. Black, P. Bodenheimer.
Astrophys. Journ., Vol. 206, 138 - 149 (1976).
Numerical calculations have been made for the early stages of collapse of axisymmetric, rotating protostars of 1, 2, and 5 $M_\odot$. The principal result of the calculations is that in all cases tried the collapse leads to the formation of a ring structure in the interior of the cloud, with a local density minimum at the center of the cloud. The rings approach equilibrium with a structure consistent with that of previous analytic determinations (Ostriker), after which they undergo further gravitational collapse.

131.125 **Observations of the ^{12}C/^{13}C ratio in four galactic sources of formaldehyde.**
D. N. Matsakis, M. F. Chui, P. F. Goldsmith, C. H. Townes.
Astrophys. Journ., (*Letters*), Vol. 206, L63 - L66 (1976).
Observations of the 6 cm absorption lines of H$_2$^{12}CO and H$_2$^{13}CO have yielded values for the ratio of [^{12}C] to [^{13}C] of 62 in W3, 55 in NGC 2024, and lower limits for this ratio that are close to terrestrial in Cas A and M17. This provides additional evidence for substantial variation of the [^{12}C]/[^{13}C] ratio in interstellar matter. The authors have also observed the H141β recombination line of hydrogen in these sources and

found several new sources of formaldehyde absorption.

131.126 **Excited OH radiation from the $^2\Pi_{3/2}, J = 5/2$ state in southern H II regions.**
S. H. Knowles, J. L. Caswell, W. M. Goss.
Monthly Notices Roy. Astron. Soc., Vol. 175, 537 - 555 (1976).

A survey of more than 50 main-line OH maser sources has been carried out. Eleven new sources of excited OH emission were found. In several of the new sources linear polarization is prominent; this is the first detection of linear polarization in this transition. Three sources have flux densities in the 5 cm line greater than in the main 18 cm lines.

131.127 **The interstellar reddening in the vicinity of 47 Tucanae and the Small Magellanic Cloud.**
J. E. Hesser, A. G. D. Philip.
Publ. Astron. Soc. Pacific, Vol. 88, 89 - 94 (1976).

In order to investigate the growth of reddening at large distances above the galactic plane in the areal vicinity of 47 Tuc, photoelectric photometry has been obtained in the $uvby\beta$ systems for ten distant, early-type stars, including the RR Lyrae variable HV 814. Between 1.2 and 4.6 kpc along the line of sight to 47 Tuc $E(B-V) = 0^m024 \pm 0.021$. For the galactic contribution to the reddening, a value of $E(B-V) = 0^m03$ valid out to $\sim$10 kpc, is recommended for both 47 Tuc and the SMC.

131.128 **Hydrogen molecules in interstellar space.**
L. Spitzer.
Quarterly Journ. Roy. Astron. Soc., Vol. 17, 97 - 120 (1976).
George Darwin Lecture delivered on 1975 December 12.

131.129 **Experimental studies on the 10μm interstellar absorption band.**
J. Dorschner, J. Gürtler, C. Friedemann.
Astron. Nachr., Vol. 297, 159 - 163 (1976).

A curve of growth has been determined for the interstellar 10μm absorption band using laboratory spectra of small olivine grains. The column densities in the direction to several infrared sources are estimated.

131.130 **On the equilibrium and stability of two-sheet hyperboloidal figures of a rotating interstellar medium.**
M. G. Abramyan, S. A. Kaplan.
Astrofizika, Vol. 11, 319 - 334 (1975). In Russian.
English translation in Astrophysics, Vol. 11, No. 2.

The problem of equilibrium and stability of two sheets of a stratum of interstellar medium placed in spheroidal galaxies is investigated, taking into account self-gravitation and the presence of a magnetic field. The study of the stability problem (incompressible model) shows that these figures are unstable. The gravitation of a spheroidal galaxy has a great stabilizing effect on the hyperbolodial equilibrium figures of an interstellar medium layer.

131.131 **Deviation from a Maxwellian velocity distribution in regions of interstellar molecular hydrogen.**
R. J. Gould, M. Levy.
Astrophys. Journ., Vol. 206, 435 - 439 (1976).

In a molecular hydrogen gas the effects of excitation of the $J = 2$ rotational level of para-hydrogen (followed by radiative decay) are investigated to determine the deviation from a Maxwellian distribution. These inelastic collisions deplete the tail of the distribution while elastic collisions try to refill the tail; the resulting steady-state distribution has a small depletion in the tail which slightly reduces the cooling rate for the gas.

131.132 **Accurate positions of OH emission sources.**
N. J. Evans II, R. M. Crutcher, W. J. Wilson.
Astrophys. Journ., Vol. 206, 440 - 442 (1976).

The Owens Valley interferometer has been used to measure accurate positions for nine OH emission sources at 1612 MHz and four sources at 1665 MHz. The baseline was 244 m north-south.

131.133 **CO observations of NGC 1579 (S222) and S239.**
G. R. Knapp, T. B. H. Kuiper, S. L. Knapp, R. L. Brown.
Astrophys. Journ., Vol. 206, 443 - 451 (1976).

The 2.6 mm lines of ^{12}CO and ^{13}CO have been observed toward the small galactic diffuse nebulae NGC 1579 (S222) and S239. The H92α recombination line has also been detected from NGC 1579. Examination of the available data suggests that NGC 1579 is both a reflection nebula (illuminated by LkHα 101), and an obscured H II region excited by a star of spectral type near B1, while S239 is a reflection nebula or a Herbig-Haro object. The authors' studies of these regions show that line widening in CO is at least as important an indicator of star-formation activity as enhanced line emission.

131.134 **Enhanced effects of starlight on the interstellar medium.** H. Gerola, R. A. Schwartz.
Astrophys. Journ., Vol. 206, 452 - 457 (1976).

The photodesorption of molecules and atoms from the surfaces of interstellar grains can be an important source of heating for the interstellar medium and the origin of instabilities which may separate grains and gas. For low densities, the force exerted on the grains is proportional to the gas density and independent of the radiation intensity; for high densities, it is proportional to the radiative flux and independent of the gas density. This force may act differently on grains of different sizes. The photoelectric effect may also be an efficient mechanism for the separation of gas and dust in diffuse clouds.

131.135 **Calculations of the lower electronic states of CH_3^+: a postulated intermediate in interstellar reactions.**
R. J. Blint, R. F. Marshall, W. D. Watson.
Astrophys. Journ., Vol. 206, 627 - 631 (1976).

Molecular-structure calculations are presented to estimate the photodissociation rate for CH_3^+ in the interstellar gas. Restricted Hartree-Fock wave functions for the ground and lowest three singlet states of CH_3^+ are calculated at various molecular geometries. The energies of the states, the force constants for the ground state, and the oscillator strength of the first excitation are determined from these wave functions. In addition, zero-point energies of CH_3^+ and CH_2D^+ are calculated for application in proposed reaction schemes that cause deuterium enhancement in interstellar molecules.

131.136 **A very small interstellar neutral hydrogen cloud observed with VLBI techniques.**
N. H. Dieter, W. J. Welch, J. D. Romney.
Astrophys. Journ., (Letters), Vol. 206, L113 - L115 (1976).

A fluctuation in the density of interstellar neutral hydrogen with an angular scale of $0.''1$ has been observed by a VLBI technique. Its linear diameter is 3×10^{-4} pc and its density 10^5 atoms cm^{-3}, parameters more similar to dense molecular clouds than to "standard" interstellar clouds.

131.137 **Interstellar absorption in intercloud regions. II. Detection of the Na I lines.** L. M. Hobbs.
Astrophys. Journ., (Letters), Vol. 206, L117 - L120 (1976).

The weak interstellar absorption lines of Na I, Ca II, and H I formed in the intercloud medium toward seven nearly unreddened stars are analyzed. Together with the widths of the Na I and Ca II lines, the column-density ratios N(Na I)/N_H determine limits on the temperature of the absorbing gas, while the ratios N(Ca II)/N_H establish limits upon its calcium depletion.

131.138 **Molecular hydrogen formation on interstellar dust grains.** S. A. Cohen.

Nature, Vol. 261, 215 - 216 (1976). — Letter.

131.139 The formation of interstellar molecules.
   E. Herbst, W. Klemperer.
Phys. Today, Vol. 29, No. 6, p. 32 - 39 (1976).

Cosmic rays may provide the energy flux necessary for the continual synthesis of molecules in dense H_2 clouds, according to a detailed model of the carbon, oxygen, hydrogen and nitrogen chemistry of those regions.

131.140 The microwave spectrum of HNC: identification of U90.7. G. L. Blackman, R. D. Brown,
P. D. Godfrey, H. I. Gunn.
Nature, Vol. 261, 395 - 396 (1976). — Letter.

131.141 Neutral hydrogen in a region of Cepheus: nearby galactic structure.
S. C. Simonson III, H. W. van Someren Greve.
Astron. Astrophys., Vol. 49, 343 - 356 (1976).

In the present study several distinct regions are identified that show a significant correlation between neutral hydrogen features and optical features, both stars and dust. These regions are discussed individually. The results are interpreted as indications of the history of the region. Some tentative conclusions are drawn about the effects of star formation on the structure of the interstellar medium during the development of stellar associations. Some consequences for the observational basis of the theory of spiral density waves are pointed out.

131.142 On the polarizing interstellar dust.
   S. Codina-Landaberry, A. M. Magalhães.
Astron. Astrophys., Vol. 49, 407 - 413 (1976).

An interpretation for the parameter k in Serkowski's normalization for the linear polarization is given in relation to a changing alignment model for the interstellar grains. Present observations indicate a correlation between k and dust size changes along the line of sight. Linear and circular polarization measurements seem to be consistent with an absorbing material for the polarizing component of the interstellar grains. Polarization position angle observations are also analyzed.

131.143 Planetary nebulae, supernova remnants, and the interstellar medium. E. E. Salpeter.
Astrophys. Journ., Vol. 206, 673 - 678 (1976). — Text of the Henry Norris Russell Lecture presented in Gainesville, Florida, on 1974 December 11 at the 144th Meeting of the American Astronomical Society.

The formation and destruction of interstellar dust grains is reviewed. Difficulties are pointed out for the "two-phase model" of clouds and intercloud medium in complete pressure equilibrium. Central stars of very old planetary nebulae are advocated for ionizing parts of the intercloud medium. Supernova remnants are advocated for enhancing density variations and for providing bulk kinetic energy to interstellar clouds. A scaled-down version of the "two-phase model" is resurrected, but with a highly heterogeneous intercloud medium.

131.144 Temperature fluctuations in very small interstellar grains. E. M. Purcell.
Astrophys. Journ., Vol. 206, 685 - 690 (1976).

An effect previously discussed by Duley, and by Greenberg and Hong, is analyzed in detail. If an interstellar grain as small as 0.005 μ in diameter absorbs a starlight photon or is struck by a low-energy cosmic ray, its temperature must abruptly rise by some tens of degrees. The temperature-time curves for such grains in a typical unshielded environment are derived.

131.145 Calcium abundance variations in diffuse interstellar clouds. M. Jura.
Astrophys. Journ., Vol. 206, 691 - 698 (1976).

Observations indicate that $N(H_2)/N(Na\,I)$ is relatively constant among clouds, while $N(Na\,I)/N(Ca\,II)$ is variable. The author uses this result to argue that the gas-phase abundance of calcium differs among clouds because varying amounts of calcium are contained within grains. He also discusses the well-known enhancement of $N(Ca\,II)/N(Na\,I)$ with cloud velocity, and he suggests that at least part of this effect can be understood with a model where grains are destroyed in grain-grain collisions.

131.146 The carbon monoxide abundance in interstellar clouds. W. Langer.
Astrophys. Journ., Vol. 206, 699 - 712 (1976).

The author calculates the steady-state abundance of carbon monoxide in interstellar clouds as a function of optical depth, density, and temperature.

131.147 H_2O maser emission associated with T Tauri and other regions of star formation.
G. R. Knapp, M. Morris.
Astrophys. Journ., Vol. 206, 713 - 717 (1976).

The authors report the detection of H_2O maser emission from three new galactic regions: a point 0.9 NE of T Tauri, and toward the H II regions G29.9−0.0 and G0.55−0.85. The T Tauri water source is not coincident with any known optical object; comparison of the H_2O and CO observations of this region suggests the presence of an embedded protostellar region in the molecular cloud. The authors have monitored these sources and those in S140−IR and Mon R2 for time variations and have found that all sources except T Tauri are time-variable on the scale of a month. A model for the varying component of Mon R2 is proposed.

131.148 The irregularity spectrum in interstellar space.
L. C. Lee, J. R. Jokipii.
Astrophys. Journ., Vol. 206, 735 - 743 (1976).

The authors analyze published data on the interstellar scintillations of three pulsars, using a theory based on the Markov approximation for strong scintillations. The data are found to be consistent with either a Gaussian or Kolmogorov-like power-law irregularity spectrum for the interstellar material.

131.149 Strong scintillations in astrophysics. IV. Cross-correlation between different frequencies and finite bandwidth effects. L. C. Lee.
Astrophys. Journ., Vol. 206, 744 - 752 (1976).

The author calculates the cross-correlation of the intensity fluctuations between different frequencies and finite bandwidth effects on the intensity correlations based on the Markov approximation. The results may be applied to quite general turbulence spectra for an extended turbulent medium. Calculations of the cross-correlation function and of finite bandwidth effects are explicitly carried out for both Gaussian and Kolmogorov turbulence spectra.

131.150 A far-infrared map of the Ophiuchus dark cloud region. G. G. Fazio, E. L. Wright, M. Zeilik II,
F. J. Low.
Astrophys. Journ., (*Letters*), Vol. 206, L165 - L169 (1976).

An extensive area of the Ophiuchus dark cloud has been mapped at far-infrared (40–250 μ) wavelengths with high angular resolution (1′). Three sources were detected and resolved. Two of the sources were identified with strong 2 μ sources that were also associated with radio continuum emission. The far-infrared flux is consistent with radiation from dust heated by an early B star.

131.151 Comparison of optical and radio column-density

measurements toward Omicron Persei and Zeta Ophiuchi.   R. M. Crutcher.
Astrophys. Journ., (*Letters*), Vol. 206, L171 - L174 (1976).

Optical and radio measurements of interstellar column densities toward two stars are compared in order to improve the knowledge of diffuse cloud geometry, excitation conditions, and molecular abundances. Mapping of the radio CO line shows that there is a hole in the CO distribution centered on o Per. The CO toward ζ Oph has the same velocity and line width as the strongest neutral atomic constituents. Since the CO and CH$^+$ velocities and line widths are significantly different, the two molecules are not spatially coincident.

131.152   Interstellar reddening and gamma-ray emission as tracers of interstellar matter.
J. L. Puget, C. Ryter, G. Serra, G. Bignami.
14th Intern. Cosmic Ray Conf., (see 012.011), Vol. 1, 52 - 57 (1975).

131.153   Interstellar magnetic fields as tracers of galactic structure.   C. Jäkel, D. Nissen, R. Schlickeiser, K. O. Thielheim.
14th Intern. Cosmic Ray Conf., (see 012.011), Vol. 2, 645 (1975). – Abstract.

131.154   Interstellar gas.   P. G. Mezger.
14th Intern. Cosmic Ray Conf., (see 012.011), Vol. 11, 3643 - 3662 (1975).

131.155   The electric charge distribution on interstellar grains. I. Calculation.   S. Simons.
Astrophys. Space Sci., Vol. 41, 423 - 434 (1976).

It is pointed out that interstellar grains will not all possess the same charge, and that the result given by the Spitzer approach represents only the most probable charge on a grain. A calculation is made of the fluctuations that can occur about this value, and data is given on the proportion of charged and neutral grains in various situations of interest.

131.156   The electric charge distribution on interstellar grains. II. Some consequences.   S. Simons.
Astrophys. Space Sci., Vol. 41, 435 - 445 (1976).

The results of the preceding paper are applied to a consideration of the motion of grains in a magnetic field, the calculation of particle collision cross-sections and charge effects in molecular synthesis.

131.157   Planck mean efficiency factors for six material candidates as interstellar grains.   A. Blanco, A. Borghesi, E. Bussoletti, S. Nicolazzo, A. M. Zambetta.
Astrophys. Space Sci., Vol. 41, 447 - 474 (1976).

Planck mean absorption cross-sections have been computed for spherical grains composed of graphite, iron, ice, olivine, amorphous quartz and a lunar silicate. Experimentally determined infrared optical constants have been used for all these materials. Ice mantle particles and planetesimal particles have also been considered with values of outer and inner radii covering a wide range of astrophysical conditions. The results given both graphically and in a tabular form are discussed and compared with those of other authors. The relationships of mantle and core properties are also critically discussed.

131.158   On a mechanism of destruction of cosmic dust particles.   V. S. Kessel'man.
Astron. Zhurn. Akad. Nauk SSSR, Vol. 53, 544 - 549 (1976). In Russian. English translation in Soviet Astron., Vol. 20, No. 3.

A mechanism of destruction of interstellar solid particles is considered. This mechanism is connected with the influence of a pulse radiation of X-ray sources on such particles.

131.159   Condensation processes in high temperature clouds.   B. Donn.
IAU Colloquium No. 31, (see 012.015), p. 345 (1976).

131.160   Expected distribution of some of the orbital elements of interstellar particles in the solar system.
O. I. Belkovich (*Bel'kovich*), I. N. Potapov.
IAU Colloquium No. 31, (see 012.015), p. 400 (1976).

131.161   New H$_2$O sources associated with late-type stars.   J. R. D. Lépine Paes de Barros, R. H. Gammon.
Inform. Bull. Southern Hemisph., No. 27, p. 28 (1975).

131.162   Theoretical limits on interstellar magnetic poles set by nearby magnetic fields.
S. A. Bludman, M. A. Ruderman.
Phys. Rev. Letters, Vol. 36, 840 - 843 (1976).

131.163   Observations of population inversion between the Λ-doublet states of OH.   J. J. ter Meulen, W. L. Meerts, G. W. M. van Mierlo, A. Dymanus.
Phys. Rev. Letters, Vol. 36, 1031 - 1034 (1976).

131.164   Interstellar extinction: a calibration by planetary nebulae.   J. H. Cahn.
Astron. Journ., Vol. 81, 407 - 418 (1976).

From currently available data, a catalog of temperature-dependent absolute Hβ extinctions and distances of planetary nebulae has been prepared. This data has been used to transform FitzGerald's three-dimensional color excess distribution to one in absolute absorption at the frequency of Hβ (λ 4861). The resulting absorption distribution is adequate to calculate the observed planetary extinctions. It also provides a basis for calculating stellar distances independently of the ratio R total-to-selective absorption. Such calculations lead to a value of R for each object studied.

131.165   A model for a local component of the interstellar hydrogen gas.   K. Grape.
Stockholms Obs. Rep. No. 9, 89 pp. = Diss. Fac. Math. Nat. Sci., Univ. Stockholm (1975).

A shell model and a cloud model for a local component of the interstellar hydrogen gas are discussed.

131.166   Equal-velocity contour diagrams of neutral hydrogen with the 21-cm line radiation in the region of $34.0° < l < 36.3°$, $-1° < b < 1°$.   F. Sato, K. Akabane.
Ann. Tokyo Astron. Obs., Second Ser., Vol. 15, 161 - 208 (1975).

Ninety-one equal-velocity contour diagrams derived from the second edition of the Maryland–Green Bank galactic 21-cm Line Survey are presented for the region of $34.0° < l < 36.3°$, $-1° < b < 1°$, where the well-known supernova remnant W44 lies. Radial velocities from −66 to 114 km s^{-1} are covered, and the contour diagrams are derived with 2 km s^{-1} interval.

131.167   The systematic motion of the interstellar clouds at the galactic shock region.   T. Sawa.
Sci. Rep. Tôhoku Univ., 214 - 218 (1975).

The galactic rotation curve for interstellar clouds is calculated, considering the dynamical interactions between the clouds and the intercloud medium which produced the galactic shock waves.

131.168   Fragmentation of magnetic interstellar clouds by ambipolar diffusion. I.   T. Nakano.
Publ. Astron. Soc. Japan, Vol. 28, 355 - 369 (1976).

It is shown that in a cloud in hydrostatic equilibrium a fluctuation of magnetic field strength, which is accompanied by a density fluctuation, is amplified by ambipolar diffusion and as a result the cloud breaks into fragments. The process of contraction of a fragment is investigated with simplified models.

131.169 Carbon monoxide emission from interstellar dust clouds. A. S. Milman.
Diss. Graduate School Univ. Maryland. 86 pp. (1975).

131.170 Size distribution of grains growing by thermal grain-grain collision. C. Hayashi, Y. Nakagawa.
Progr. Theor. Phys., Vol. 54, 93 - 103 (1975). — Abstr. in Phys. Abstr., Vol. 79, A007078 (1976).

131.171 Interstellar molecules: physics and chemistry of H_2 and CO. J. Lequeux.
Recherche, Vol. 6, 864 - 867 (1975). In French. — Abstr. in Phys. Abstr., Vol. 79, A015522 (1976).

131.172 Theoretical intensities of low frequency recombination lines. P. A. Shaver.
Pramāṇa, Vol. 5, 1 - 28 (1975). — Abstr. in Phys. Abstr., Vol. 79, A015524 (1976).

131.173 Natural masers: maser emission from cosmic objects. W. H. Kegel.
Applied Phys., Vol. 9, 1 - 10 (1976). — Abstr. in Phys. Abstr., Vol. 79, A023888 (1976).

131.174 Comments on growth of solid particles by turbulence in collapsing clouds. G. Horedt.
Progr. Theor. Phys., Vol. 54, 1224 (1975). — Abstr. in Phys. Abstr., Vol. 79, A023904 (1976).

131.175 Cooling of interstellar formaldehyde by collision with helium: an accurate quantum mechanical calculation. B. J. Garrison.
Thesis California Univ., Berkeley, USA, Lawrence Berkeley Lab., 71 pp. (1975).

131.176 Wavelength dependence of interstellar polarization. G. E. Mavko.
Thesis Rensselaer Polytechnic Inst., Troy, New York, USA, 94 pp. (1974). (Available from Univ. Microfilms, Order No. 75-5389).

131.177 Interstellar scattering of pulsar radiation. I. Scintillation. D. C. Backer.
Separate print National Aeronaut. Space Administr., Greenbelt, Maryland, USA, Goddard Space Flight Center, 31 pp. (1974). – See 14.131.111.

131.178 Polarization of Be stars. M. W. Johns.
Thesis Dartmouth Coll., Hanover, New Hampshire, USA, 145 pp. (1975). (Available from: Univ. Microfilms, Order No. 75-23,369).

131.179 Isotope abundances in interstellar molecule clouds. P. G. Wannier.
Thesis Princeton Univ., New Jersey, USA, 79 pp. (1975). (Available from Univ. Microfilms, Order No. 75-23, 249).

131.180 Radiative transfer in circumstellar dust. C. A. Harvel.
Thesis Florida Univ., Gainesville, USA, 98 pp. (1974). (Available from Univ. Microfilms, Order No. 75-16, 390). See 12.131.194.

Atomic and molecular physics and the interstellar matter. (Proceedings of the XXVI Session of the Les Houches Summer School of Theoretical Physics, Grenoble, France, July - August 1974). See Abstr. 012.024.

The dusty universe. Conference held at Cambridge, Mass., October 1973. See Abstr. 012.027.

Glaciations and dense interstellar clouds. See Abstr. 015.013.

Radiative-lifetime measurements for sulfur and silicon transitions observed in interstellar absorption spectra. See Abstr. 022.024.

Results of infrared reflectivity measurements on astronomically interesting silicates. See Abstr. 022.025.

The optical constants of polyoxymethylene. See Abstr. 022.026.

The microwave spectrum of hydrogen isocyanide. See Abstr. 022.031.

Extinction properties of porous spheres. See Abstr. 022.033.

Detection of the millimeter wave spectrum of hydrogen isocyanide, HNC. See Abstr. 022.034.

Optical properties of particulates. See Abstr. 022.038.

Measurement and significance of the reaction $^{13}C^+ + {}^{12}CO \rightleftharpoons {}^{12}C^+ + {}^{13}CO$ for alteration of the $^{13}C/^{12}C$ ratio in interstellar molecules. See Abstr. 022.062.

Molecular processes in interstellar clouds. See Abstr. 022.090.

Rotational excitation of CN molecules by proton impact. See Abstr. 022.096.

R-centroids and Franck-Condon factors of CH^+ molecule. See Abstr. 022.097.

Absolute calibration of millimeter-wavelength spectral lines. See Abstr. 031.221.

Instrumental methods of radio spectroscopy of the interstellar medium. II. Comparative estimate of methods. See Abstr. 031.223.

Laboratory microwave spectrum and rest frequencies of the N_2H^+ ion. See Abstr. 061.019.

Infrared pumping for SiO masers. See Abstr. 061.044.

Nonhomologous contraction and equilibria of self-gravitating, magnetic interstellar clouds embedded in an intercloud medium: star formation. I. Formulation of the problem and method of solution. See Abstr. 062.036.

CO line formation in turbulent interstellar clouds where a small velocity gradient is present. See Abstr. 063.007.

An attempt to determine the circumstellar reddening law. See Abstr. 064.001.

Polarization properties of silicate-like grains in circumstellar envelopes of late-type stars due to temperature variations. See Abstr. 064.028.

Mass loss from dwarf M stars through stellar flaring.
See Abstr. 064.045.

Magnetic braking of collapsing interstellar clouds.
See Abstr. 065.054.

Cloud collapse and star formation.
See Abstr. 065.057.

An observation of the diffuse soft X-ray/extreme-ultraviolet background.   See Abstr. 066.018.

Some aspects of solar wind interaction with the interstellar medium.   See Abstr. 074.047.

On the model of the solar wind — interstellar medium interaction with two shock waves.
See Abstr. 074.060.

Termination of the solar wind in the hot, partially ionized interstellar medium.   See Abstr. 074.090.

A technique for measuring the interstellar component of cosmic dust.   See Abstr. 106.094.

The galactic distribution of interstellar absorption as determined from the Celescope catalog of ultraviolet stellar observations and a new catalog of UBV, H-beta photoelectric observations.   See Abstr. 113.032.

T Tauri stars and the $(J-H)$, $(H-K)$ diagram.
See Abstr. 113.035.

A new Copernican revolution in astronomy. Ultraviolet astronomy with the satellite Copernicus.
See Abstr. 113.040.

Linear polarization measurements at Hβ of early-type emission line stars.   See Abstr. 114.002.

The wavelength dependence of polarization of Mira stars compared with the interstellar polarization in a field near the galactic equator.   See Abstr. 122.020.

Grains of anomalous isotopic composition from novae.   See Abstr. 124.002.

N Cyg 1975 (V1500 Cyg). Determination of the distance and of the interstellar absorption.
See Abstr. 124.102.

Aperture synthesis observations of W3 (OH) (G 133.9 + 1).   See Abstr. 131.507.

Recombination-line observations of W3(OH). I. A model for OH and H$_2$O emission.   See Abstr. 131.517.

A review of recent observations of dust in H II regions.   See Abstr. 131.524.

Grain charging in H II regions.   See Abstr. 131.525.

Giant grains.   See Abstr. 132.005.

Orion nebula: high resolution Pa-12 mapping.
See Abstr. 132.006.

Large-scale ionization fronts and the nature and distribution of light scattering particles in the Orion nebula.
See Abstr. 132.011.

On the number of planetary nebulae in our Galaxy.
See Abstr. 133.014.

Differential deceleration of nebular shells and the displacement of central stars.   See Abstr. 133.015.

The angular broadening of radio sources by scattering in the interstellar medium.   See Abstr. 141.002.

On angular variations of the frequency spectrum of the cosmic radio background along the declination $\delta = 50°30'$.
See Abstr. 141.071.

Interstellar absorption and variable soft X-ray component in Cygnus X-1.   See Abstr. 142.209.

Cross sections for nuclear reactions induced by high energy alpha particles and the influence of interstellar helium in cosmic ray propagation.   See Abstr. 143.115.

Cosmic ray confinement in the Galaxy and the interstellar spectrum of hydromagnetic waves.
See Abstr. 143.327.

Nuclear interactions between cosmic radiation and interstellar gas, and nucleosynthesis of lithium, beryllium, and boron.   See Abstr. 143.371.

On the ratio of total to selective extinction in associations containing reflection nebulae.
See Abstr. 152.004.

Absorbing matter in the region of Tau T1 and Tau T3 associations.   See Abstr. 152.006.

Average properties of molecular clouds in the Galaxy.   See Abstr. 155.046.

Interstellar magnetic fields as tracer of galactic structure.   See Abstr. 155.051.

Optical polarization of stars of galactic latitudes $b < -45°$.   See Abstr. 156.001.

Interstellar absorption and the luminosity function of galaxies.   See Abstr. 158.002.

The spectrum of IR radiation from dust clouds.
See Abstr. 158.112.

The detection of formaldehyde in the Large Magellanic Cloud.   See Abstr. 159.001.

Starlight polarization in the Magellanic Cloud regions.   See Abstr. 159.011.

H I, H II Regions

131.501 Infrared photometry of a heavily reddened association in W 35.
P. R. Jorden, A. D. MacGregor, M. J. Selby, P. A. Whitelock.
Monthly Notices Roy. Astron. Soc., Vol. 174, 1P - 4P (1976).

Seven new 2-μm point-like sources associated with the W 35 H II region have been discovered and are interpreted as an association of late-type giants and supergiants seen through about 10 mag of optical extinction. Their positions and H, K and L magnitudes are reported here.

131.502 Further observations of ionized neon in the galactic centre.
D. K. Aitken, J. Griffiths, B. Jones, J. M. Penman.
Monthly Notices Roy. Astron. Soc., Vol. 174, 41P - 44 P (1976).

The authors report further observations of Sgr A (West) which confirm the presence of the neon fine structure line at 12.8 μ, and firmly establish the source as a thermal H II region of low excitation with a near normal abundance of at least this one heavy element. There is no indication of a northerly extension to the emission, as suggested by radio synthesis maps.

131.503 A survey of H_2O sources for wide-spectrum emission.
W. M. Goss, S. H. Knowles, M. Balister, R. A. Batchelor, K. J. Wellington.
Monthly Notices Roy. Astron. Soc., Vol. 174, 541 - 553 (1976).

Wide-spectrum 1.35 cm H_2O emission has been found in the following seven sources: H_2O 327.3–0.6, H_2O 331.0–0.2, H_2O 331.5–0.1, NGC 6334N, M17, W51 and Orion A. The discovery of these sources indicates that the presence of wide-spectrum emission is a general feature of H_2O sources associated with H II regions. This wide-spectrum emission is indicative of either: (1) a larger range in radial velocities than is usually associated with H II regions or, (2) a scattering phenomenon, with associated frequency shifts, such as the stimulated Raman effect. Maps of the H_2O sources in both Orion and Sgr B2 show evidence of an elongated structure for the masering region.

131.504 Fine radio structure in W3.
S. Harris, C. G. Wynn-Williams.
Monthly Notices Roy. Astron. Soc., Vol. 174, 649 - 659 (1976).

The H II region W3 has been mapped at 5 GHz with 2 arcsec resolution using the Cambridge 5-km telescope. Several new faint sources have been detected. Analysis of the shell source W3(A) shows it to be excited by two highly obscured O-stars and to be strongly ionization-bounded.

131.505 Chemical composition of H II regions in the Small Magellanic Cloud and the pregalactic helium abundance.
M. Peimbert, S. Torres-Peimbert.
Astrophys. Journ., Vol. 203, 581 - 586 (1976).

Photoelectric spectrophotometry of emission lines in the 3700–7400 Å range is presented for NGC 346, NGC 356, and NGC 456 in the Small Magellanic Cloud (SMC). From these observations the chemical abundances of He, N, O, Ne, S, and Ar relative to hydrogen are derived. It is found that N(He)/N(H) = 0.078 ± 0.005. Oxygen and neon are underabundant by a factor of 7, while nitrogen is underabundant by a factor of 15 in the SMC H II regions with respect to the values for the Orion nebula. From the chemical composition of H II regions in our Galaxy and the Magellanic Clouds it is found that the pregalactic value of N(He)/N(H) is 0.074 ± 0.006 (Y = 0.228 ± 0.014). The cosmological implications of this result are briefly discussed.

131.506 Decay of forbidden lines in a recombining H II region.
J. Manfroid.
Astron. Astrophys., Vol. 46, 31 - 39 (1976).

The author has computed the evolution of a high excitation H II region when the ionizing star dies suddenly. The emission of forbidden lines is found to decrease more rapidly than the hydrogen and helium recombination lines. The importance of the charge exchange reactions in the ionization structure of the whole nebula is emphasized. It is shown that the conditions for the detection of such objects are very restrictive.

131.507 Aperture synthesis observations of W3 (OH) (G 133.9 + 1).
R. H. Harten.
Astron. Astrophys., Vol. 46, 109 - 116 (1976).

Observations of the region around the source W3 (OH) made with the Westerbork radio telescope at 6, 21 and 50 cm wavelength show the presence of a complex of compact and extended objects. At least seven sources are found within 30" of the bright compact source identified with the maser W3 (OH). Two other thermal sources have also been detected nearby. The regions in this area are at different stages of their evolution. The exciting stars of all but the brightest source appear to be of early B type or earlier.

131.508 Infrared emission from S 157 A and S 252 A.
M. Zeilik II.
Astron. Astrophys., Vol. 46, 319 - 321 (1976).

The H-α knots and small H II regions S 157 A and S 252 A have been observed from 1.25 to 3.45 μ and mapped at 2.2 μ. S 157 A shows only a slight excess at 2.2 μ, as most of the emission can be accounted for by radiation from the ionized gas and the exciting star. S 252 A has a large excess at 2.2 μ and 3.45 μ that can be attributed to hot dust. Multiaperture observations indicate that the 3.45 – 2.2 μ color increases with radius for S 252 A, implying that the dust has a temperature gradient and so is heated predominantly by the stellar continuum.

131.509 The radio spectrum and physical parameters of the W80 H II complex.
C. Goudis.
Astrophys. Space Sci., Vol. 39, 173 - 180 (1976).

The radio continuum spectrum of the W80 nebular complex (NGC 7000 and IC 5070 nebulae mainly) is established over a wide range of frequencies. The resultant spectrum shows a thermal shape and the spectral index established for the optically thin part of it is −0.04. Certain physical parameters of the complex (rms densities, mass, emission measure, excitation parameter) are derived for adopted models. The possible exciting star is located. Limitations of the models involved and uncertainties of the results are also discussed.

131.510 The radio and infrared spectrum of DR 21.
C. Goudis.
Astrophys. Space Sci., Vol. 39, L1 - L6 (1976).

The radio and infrared spectrum of DR 21 is established over a wide range of frequencies (from 10^2 to 10^8 MHz). Two physical processes, free-free emission from the ionized hydrogen at radio wavelengths and reradiation at infrared wavelengths of the original stellar ultraviolet radiation by dust grains have to be considered in the explanation of the derived spectrum. Physical parameters of the object deduced from its radio emission are also presented.

131.511 Twenty-one centimeter aperture synthesis study of the small-scale structure of the interstellar medium.
D. A. Elliott.
Thesis California Univ., Los Angeles, USA, 417 pp. (1974). (Available from: Univ. Microfilms, Order No. 75-5708).

131.512 New H_2O celestial sources associated with H II regions in the southern hemisphere.

P. Kaufmann, R. H. Gammon, A. L. Ibanez, J. R. D. Lepine, P. Marques dos Santos, M. H. Paes de Barros, E. Scalise, Jr., R. E. Schaal, S. H. Zisk, J. C. Carter, M. L. Meeks, J. M. Sobolewski.
Nature, Vol. 260, 306 - 307 (1976).

The authors report here the first results of a nearly complete, high-sensitivity, water vapour survey of galactic H II regions situated in the southern hemisphere. A typical upper limit of about 40 Jy (1 Jy = 10^{-26} W m^{-2} Hz^{-1}) was attained at the $6_{16} \rightarrow 5_{23}$ rotational transition line of the water vapour molecule.

131.513 Dielectronic recombination in H II regions.
P. A. Shaver.
Astron. Astrophys., Vol. 47, 49 - 52 (1976).

Theoretical b_n-factors and radio recombination line intensities are computed for magnesium and calcium in H II regions. The absence of observable spectral features seems to imply low ionic abundances.

131.514 Helium in the center of the Galaxy.
P. G. Mezger, L. F. Smith.
Astron. Astrophys., Vol. 47, 143 - 151 (1976).

He 109 α lines have been detected for the first time in two giant H II regions, G 0.5−0.1 and G 07−0.0, close to the galactic center. The observed He$^+$-abundances are 0.027 and 0.034, respectively. The He$^+$-abundance in Sgr A West is estimated to be 0.095. Observations of H 137 β, C 109 α and CH$_3$OH ($J_K = 3_1$ K-doublet, type A) lines are also reported. The authors conclude that the He-abundance in the galactic center is at least normal (i.e., 0.1 by number) and an overabundance in Sgr A West, is possible.

131.515 Infrared study of seven possible compact H II regions.
F. Sibille, M. Lunel, J. Bergeat.
Astron. Astrophys., Vol. 47, 161 - 164 (1976).

The authors report observations of seven possible compact H II regions in the infrared between 1.27 μ and 2.37 μ. Measurements are compared with the hydrogen spectrum in order to derive extinction and emission measure. The emission measure is compared with available radio data. For two sources, agreement is found between radio and infrared data. Infrared excess is found in four sources, its origin is discussed. Two sources cannot be interpreted by compact H II regions.

131.516 H 109α aperture synthesis observations of DR 21 and W3.
K. J. Wellington, W. T. Sullivan III, W. M. Goss, H. E. Matthews.
Astron. Astrophys., Vol. 47, 351 - 363 (1976).

Synthesis observations of the compact H II regions DR 21 and W3 have been made in the H 109α recombination line at 5 GHz. The spatial distribution of the H 109α radiation is in general quite similar to that of the radio continuum. The kinematics of both sources have been investigated by assuming that one-component profiles can describe the radiation at each point. For DR 21 an interpretation based on a rotating sphere of gas with its axis substantially inclined towards the observer is considered as a possible model. The velocity contour map for W3 suggests a totally different picture. The lines are narrower (20−30 km s^{-1}) for W 3 A than for W 3 B (35 km s^{-1}). There is a suggestion that the velocity structure of W 3 A is related to the shell-like continuum structure.

131.517 Recombination-line observations of W3(OH). I. A model for OH and H$_2$O emission.
V. A. Hughes, M. R. Viner.
Astrophys. Journ., Vol. 204, 55 - 67 (1976).

This paper reports the discovery of recombination-line emission from the compact H II region W3 (OH) in the H65α and H85α transitions. The differences in velocity of the various molecular lines from the region are summarized and explained

in terms of the shock wave. In particular, this model can account for the production of OH and H$_2$O emissions at different velocities and different angular positions. It is suggested that the compact source may have a magnetic field of the order of 10^{-3} gauss.

131.518 The role of dust in NGC 2024.
C. L. Sarazin.
Astrophys. Journ., Vol. 204, 68 - 72 (1976).

Models for the H II region NGC 2024 containing ultraviolet absorbing dust are presented. It is shown that the presence of ultraviolet absorbing dust in NGC 2024 is not necessarily inconsistent with recent observations of the infrared source 2024 # 2, as has been suggested, and that the effective temperature of 2024 # 2 may have been underestimated. Models for the H II region with dust are compared with models involving a low-temperature exciting star, and some of the radio and infrared observations are shown to favor the former.

131.519 Carbon recombination line observations of the Sharpless 140 region.
G. R. Knapp, R. L. Brown, T. B. H. Kuiper, R. K. Kakar.
Astrophys. Journ., Vol. 204, 781 - 788 (1976).

Carbon recombination line emission has been detected at two frequencies from a dark cloud contiguous with the small H II region Sharpless 140. The observations show the dark cloud to be of unusually low temperature and to have a markedly inhomogeneous density distribution, with localized regions of high density surrounding one or more embedded stars. The carbon is probably ionized by photons from both the exciting star of S140 and the embedded stars. The dark cloud and S140 apparently represent two stages of star formation which have occurred over a period of at least 5 × 10^5 years in adjacent regions of the same dark cloud.

131.520 A classification of the available astrophysical data of particular H II regions. IV: M17: mapping and physical parameters of the object.
C. Goudis.
Astrophys. Space Sci., Vol. 39, 273 - 306 (1976).

Optical, radio, infrared and molecular contour maps of the object are presented. Photographs of the nebula taken through various filters are also shown. The physical parameters of the object are classified and presented either in separate tables or in contour map form.

131.521 Observations of southern nebulae.
A. C. Danks, J. Manfroid, L. Houziaux.
Astrophys. Space Sci., Vol. 39, 307 - 311 (1976).

Spectroscopic observations of 10 H II regions in the southern hemisphere are presented. The observations cover the spectral region 3600−8000 Å.

131.522 Beobachtungen von H II-Regionen hohen Staubgehalts im nahen Infrarot.
M. Beetz, H. Elsässer, C. Poulakos, R. Weinberger.
Mitt. Astron. Ges., No. 38, p. 77 (1976). − Abstract.

131.523 Thermische Eigenschaften von Staub in ausgedehnten Hüllen.
H. J. Staude.
Mitt. Astron. Ges., No. 38, p. 77 (1976). − Abstract.

131.524 A review of recent observations of dust in H II regions.
L. F. Smith.
Astrophys. Space Sci. Library, Vol. 55, (see 012.001), 59 - 65 (1976).

The upper limit for the absorption cross section of dust in H II regions in the wavelength range 912−504 Å derived by Mezger et al. (1974), is compatible with that expected for large dust grains, and a gas-to-dust ratio equal to that in the general interstellar medium. The albedo of the small grains must be high for $\lambda > 504$ Å. This restriction is lifted if the visual extinction cross section of the grains in H II regions is less than that for grains in the general interstellar medium. New observations of the Orion nebula indicate that the visual extinction cross section is within a factor 2 of the value in the general interstellar medium.

131.525   Grain charging in H II regions.
A. F. M. Moorwood, B. Feuerbacher.
Astrophys. Space Sci. Library, Vol. 55, (see 012.001), 179 - 189 (1976).

Equilibrium grain potentials have been calculated as a function of radial position and for a wide range of electron densities in H II regions ionized by stars of spectral type O5 and B0. Results are presented for both graphite (low yield photoemitter) and aluminium oxide (high yield photoemitter) for which laboratory photoemission data has been obtained. The results for aluminium oxide should approximate the behaviour expected of dielectric grains — e.g., silicates which may be present in H II regions. The importance of charging is discussed in relation to the growth and motion of grains in these regions.

131.526   On the electronic temperature of the diffuse galactic nebula IC 434.
J.-L. Vidal.
Astron. Astrophys., Vol. 48, 55 - 61 (1976). In French.

Hα photoelectric measurements for this H II region were obtained during the last few years at the Pic-du-Midi Observatory. The disagreement between the authors' and the former results is such that the comparison of the Hα data with the radio data gives an electron temperature of 8000 K instead of less than 3500 K. It is shown that temperatures as low as 3500 K, compared with a Hα/[N II] ratio equal to 2, result in an unusually high nitrogen abundance (2.1×10^{-3}), or in highly improbable temperature fluctuations, if the abundance is fixed at its normal value.

131.527   Optical and radio observations of the galactic H II region S 206 (NGC 1491).
L. Deharveng, F. P. Israel, M. Maucherat.
Astron. Astrophys., Vol. 48, 63 - 73 (1976).

The galactic emission nebula S 206 has been observed at 1415 MHz and 4995 MHz. Interferograms in Hα, monochromatic photographs in Hβ, [O III], Hα and [N II], and high dispersion spectra near Hα were obtained. S 206 is a high excitation nebula which contains an ionization front in the south and west. The exciting O 5 star is outside the brightest part of the nebula. The brightest structures are an elephant trunk pointing towards the exciting star and globules probably containing neutral dense material. S 206 is an evolved H II region.

131.528   The obscuration of the H II region IC 1318 b, c.
C. Goudis.
Astron. Astrophys., Vol. 48, 145 - 147 (1976).

A contour map of the obscuration of light over the IC 1318 b, c nebular complex is produced by comparing the thermal radio brightness with the Hα brightness of the object. This map clearly shows the presence of the heavily obscuring dust cloud. This cloud might be the one out of which the associated H II region is forming or it might be well in front and quite detached from the H II region. Both possibilities are discussed.

131.529   Additional observations of the unidentified infrared features at 3.28 and 3.4 microns.
G. L. Grasdalen, R. R. Joyce.
Astrophys. Journ., (Letters), Vol. 205, L11 - L14 (1976).

The unidentified emission features at 3.28 and 3.4 μ are reported in two galactic H II regions: the Orion nebula and M17. Observations of NGC 7027 with a spectral resolution of $\lambda/\Delta\lambda \approx 300$ show that the feature at 3.28 μ is not a single spectral line. The feature consists of a continuum or several closely spaced lines. The authors point out that a marginally plausible

identification of the features may be the fundamental bands of CH^+.

131.530   A bright rim in RCW 62.
A. C. Danks, J. Manfroid.
Astron. Astrophys., Vol. 48, 213 - 217 (1976).

Observations of a prominent bright rim in RCW 62 are presented and compared with theoretical calculations. The rim temperature is shown to increase whilst the electron density remains relatively constant. [N II]/Hα ratio indicates a possible abundance anomaly although $N(He)/N(H)$ falls within normal limits. It is suggested that the rim is "old" and shares the same exciting stars as IC 2944.

131.531   A high-resolution map of the Orion nebula at 5 GHz.
A. H. M. Martin, S. F. Gull.
Monthly Notices Roy. Astron. Soc., Vol. 175, 235 - 244 (1976).

The authors present a 5-GHz radio continuum map of the central 4 arcmin region of the Orion nebula made with a resolution of 7 × 20 arcsec. The radio brightness distribution is strikingly similar to that at optical wavelengths and the dark bay to the east and the ionization front to the SE of the Trapezium are clearly seen. The NW boundary of the nebula coincides with a string of H_2O maser sources. Comparison with Hα observations shows that dust in the nebula is well mixed with the ionized gas.

131.532   Observations of five thermal sources at 15 GHz with the 5-km telescope.   S. Harris, P. F. Scott.
Monthly Notices Roy. Astron. Soc., Vol. 175, 371 - 379 (1976).

Five thermal sources have been observed at 15 GHz with an angular resolution of ~ 0".65 arc. For two sources, W3(OH) and NGC 7027, detailed brightness distributions have been obtained; two others NGC 7538 (B) and V 1016 Cyg, are partially resolved. The physical implications of the results are discussed briefly.

131.533   'Cometary' globules and the structure of the Gum nebula.   T. G. Hawarden, P. W. J. L. Brand.
Monthly Notices Roy. Astron. Soc., Vol. 175, 19P - 22P (1976).

Tailed, comet-like objects found in the outskirts of the Gum nebula (and elsewhere) are shown to be $> 10^5$ yr old. This implies that the H II region is much older than the Vela pulsar 0833 – 45. Tailed structure found in NGC 5367 suggests the presence of ionized hydrogen falling into the galactic plane near $l = 320°$.

131.534   Radio continuum measurements of compact H II regions and other sources.   D. N. Matsakis, N. J. Evans II, T. Sato, B. Zuckerman.
Astron. Journ., Vol. 81, 172 - 177 (1976).

The flux densities at wavelengths of 13, 4 and 2 cm have been measured for a variety of sources. The spectra of some previously known sources have been extended to higher frequencies. Physical parameters are derived where possible, assuming a model of thermal emission from ionization-bounded homogeneous, spherical H II regions.

131.535   The radio spectrum, electron temperature and ionizing flux of the H II region IC 5146.
T. B. H. Kuiper, G. R. Knapp, E. N. Rodriguez Kuiper.
Astron. Astrophys., Vol. 48, 475 - 477 (1976).

The authors have made sensitive line and continuum observations of the small ionization-bounded H II region IC 5146 to determine its temperature and excitation parameter. The results show that the stellar Lyman continuum fluxes calculated by Panagia (1973) from non-LTE model atmospheres appear to agree most closely with these observations.

131.536 A classification of the available astrophysical data of particular H II regions. V: M8 and M20: mapping and physical parameters of the objects. C. Goudis.
Astrophys. Space Sci., Vol. 40, 281 - 314 (1976).

Optical and radio contour maps of both the M8 and M20 nebulae are presented. Photographs of both nebulae, taken through different filters, are also shown. Information obtained for M8 from observations made in the infrared range is classified. Physical parameters and their distribution over the objects are also presented.

131.537 A classification of the available astrophysical data of particular H II regions. VI: M16: mapping and physical parameters of the object. C. Goudis.
Astrophys. Space Sci., Vol. 41, 105 - 119 (1976).

Various contour maps of the object are presented. The physical parameters of the nebula are also classified and presented either in separate tables or in contour map form.

131.538 Fine structure in H II regions: III. A search for sub-arc-second structure. B. Balick, R. L. Brown.
Publ. Astron. Soc. Pacific, Vol. 88, 156 - 158 (1976).

The authors have searched for compact structure having an angular size in the range $0''.2 \leq \theta \leq 2''$ at 2.7 and 8.1 GHz in the H II regions SGR B2, M17, M8, W51, DR21, W75N, W3, and M42. No sources smaller than about $2''$ were detected down to a limiting flux of ~10 mfu. These observations support the conclusion that there exists a minimum size scale in the bright radio fine structure in H II regions that is on the order of 1/30 pc.

131.539 A peculiar H I feature at $l = 285°$, $b = -18°$.
E. Bajaja, F. R. Colomb, M. Gil.
Astron. Astrophys., Vol. 49, 259 - 262 (1976).

An extended shell-like H I structure has been observed at $l = 285°$ and $b = -18°$. The measured velocities (LSR) range from -20 to -40 km/s. Several interpretations could be possible for the H I concentration relating it to other large scale features, but none of them is strongly favored.

131.540 High resolution radio observations of a bright rim in IC 1396.
J. W. M. Baars, H. J. Wendker.
Astron. Astrophys., Vol. 49, 473 - 475 (1976).

A bright rim in IC 1396 has been observed at radio wavelengths with high sensitivity and angular resolution. Parts of the rim, particularly the most intense optical feature, are detected at 6 cm. The authors find an H II region coincident with the optically dark region west of the rim, which has recently been shown to contain a molecular cloud.

131.541 H II regions of the northern Milky Way: medium-large-field photographic atlas and catalogue.
R. Dubout-Crillon.
Astron. Astrophys., Suppl. Ser., Vol. 25, 25 - 54 (1976).

Nineteen Hα photographs and a catalogue of H II regions in the northern Milky Way are presented. This atlas reveals 85 new regions of faint emission.

131.542 Interaction of hot stars and of the interstellar medium. VIII. Low-dispersion spectra of galactic nebulae and planetary nebulae.
M. Chopinet, M. C. Lortet-Zuckermann.
Astron. Astrophys., Suppl. Ser., Vol. 25, 179 - 191 (1976).

Low-dispersion spectra in the wavelength range 4300 – 7400 Å were used to study the excitation of 79 galactic nebulae. From these spectra and some additional data on the luminosity of the stars, 24 objects definitely appear to be planetary nebulae, some showing a remarkably high excitation (for instance Sh 2–71); a number of objects are H II regions

or Hα knots inside H II regions.

131.543 The puzzle of the high velocity water vapor features in W49. T. M. Heckman, W. T. Sullivan, III.
Astrophys. Letters, Vol. 17, 105 - 112 (1976).

The primary observational constraints on any acceptable model for the 1.35 cm high velocity water vapor features in W49 are detailed. Several possibilities for models, satisfying various of these constraints, are suggested and analyzed. The most likely interpretation of the low velocity features is that they are associated with a dust shell surrounding a compact H II region.

131.544 Internal motions in H II regions. The radial velocity field of IC 443. P. Pişmiş, M. Rosado.
Mem. Soc. Astron. Italiana, Vol. 45, (see 012.017), 709 (1974).
Abstract (see also 13.131.509).

131.545 Radio mapping observations of M17 and Orion A at 87 GHz.
Y. Chikada, Y. Fukui, J. Inatani, T. Iguchi.
Tokyo Astron. Bull., Second Ser., No. 243, p. 2067 - 2070 (1976).

Radio continuum observations have been made for the two compact H II regions, M17 (W38) and Orion A (W10), at 87 GHz ($\lambda = 3.5$ mm).

131.546 The bright rim IC 2948.
A. C. Danks, J. Manfroid.
Proc. Southwest Regional Conf., Vol. 1, (see 012.021), 157 - 171 (1976).

Observations of the bright rim IC 2948 are presented and compared with theoretical calculations. The rim temperature is shown to increase whilst the electron density remains relatively constant. The Hα/[N II] ratio indicates a possible abundance anomaly although N(He)/N(H) falls with normal limits. It is suggested that the rim is "old" and shares the same exciting stars as IC 2944.

131.547 Radio observations of H II regions in the galactic center. T. A. Pauls.
Thesis New Mexico State Univ., University Park, USA, 93 pp. (1974). (Available from: Univ. Microfilms, Order No. 75-10,825).

131.548 Small diffuse nebulae. C. J. L. Matthews.
Thesis Illinois Univ., Urbana, USA, 82 pp. (1975).
(Available from Univ. Microfilms, Order No. 75-24, 364).

Astrophotographie im Infrarot.
See Abstr. 034.002.

A photometer for the study of H II regions.
See Abstr. 034.006.

The recombination and level populations of ions–I. Hydrogen and hydrogenic ions. See Abstr. 062.001.

On the evolution of ionized gas around hot stars.
See Abstr. 064.049.

UBV photometry of the stars in the fields of emission nebulae. II. M17. See Abstr. 113.022.

The H II region of a rapidly rotating O-star.
See Abstr. 116.002.

The origin of radio recombination lines seen toward supernova remnants. See Abstr. 125.008.

The small-scale structure of interstellar hydrogen.

See Abstr. 131.007.

Observations of CO ($J = 2-1$) line emission from sources associated with H II regions and the Rho Ophiuchi dark cloud.   See Abstr. 131.017.

Infrared observations of the H_2O maser associated with the H II regions S 255 (IC 2162) and S 257.   See Abstr. 131.018.

Radical formation, chemical processing, and explosion of interstellar grains.   See Abstr. 131.024.

On the ratio of the total to selective absorption.   See Abstr. 131.072.

B stars and the structure of the interstellar medium.   See Abstr. 131.103.

The ionization of cloud and intercloud hydrogen by O and B stars.   See Abstr. 131.109.

Interpretation of low-frequency recombination line observations of the interstellar medium.   See Abstr. 131.117.

Excited OH radiation from the $^2\Pi_{3/2}, J = 5/2$ state in southern H II regions.   See Abstr. 131.126.

Interstellar gas.   See Abstr. 131.154.

Observations of the Gum nebula with a Fabry-Perot spectrometer.   See Abstr. 132.002.

Optical and millimeter-wave observations of the M8 region.   See Abstr. 132.003.

Far-ultraviolet brightness of nebulae in Cygnus.   See Abstr. 132.017.

Dynamics of envelopes of planetary nebulae.   See Abstr. 133.012.

Radio structure of the sources near ON2.   See Abstr. 141.001.

Satellite-line anomalies in the 6 GHz transitions of OH associated with G 291.3 − 0.7.   See Abstr. 141.005.

Aperture synthesis observations of galactic H II regions. II. The galactic radio source W 58.   See Abstr. 141.052.

W28 − a possible association of supernova remnants and H II regions.   See Abstr. 141.057.

Infrared observations of M17 S at medium spatial and spectral resolution.   See Abstr. 141.605.

The infrared emission of M17.   See Abstr. 141.615.

Far-infrared spectral observations of M42 and M17.   See Abstr. 141.622.

8 to 13 micron spectrophotometry of compact sources in NGC 7538.   See Abstr. 141.630.

Millimeter observations of galactic sources.   See Abstr. 141.637.

An almost complete survey of 21-cm line radiation for $|b| \geqslant 10°$. V. Photographic presentation and qualitative comparison with other data.   See Abstr. 155.004.

The spiral structure of our Galaxy determined from H II regions.   See Abstr. 155.034.

A first attempt to determine the nitrogen to sulphur abundance gradient across the disk of our Galaxy.   See Abstr. 155.040.

Theoretical aspects of galactic research.   See Abstr. 155.055.

Hot stars and H II regions in our Galaxy.   See Abstr. 155.062.

The giant spiral galaxy M101: IV. Observations of variable continuum radio emission from supernova 1970g and measurements of the continuum radio structure of the giant H II complex NGC 5455.   See Abstr. 158.060.

H II regions in NGC 628. I. Positions and sizes.   See Abstr. 158.085.

The stellar and gaseous content of normal galaxies as derived from their integrated spectra.   See Abstr. 158.130.

The detection of formaldehyde in the Large Magellanic Cloud.   See Abstr. 159.001.

The nebular complexes of the Large and Small Magellanic Clouds.   See Abstr. 159.006.

30 Doradus as the active centre of the Large Magellanic Cloud.   See Abstr. 159.014.

Errata

131.901   Erratum: 'Infrared emission from S 157 A and S 252 A' [Astron. Astrophys., Vol. 46, 319 - 321 (1976)].   M. Zeilik II.
Astron. Astrophys., Vol. 48, 483 (1976).

132 Emission Nebulae, Reflection Nebulae

132.001 The electron temperatures of the [O III] zone of M 42. A. H. Gibbons.
Monthly Notices Roy. Astron. Soc., Vol. 174, 105 - 109 (1976).

Further observations of the line profiles of Hβ and [O III] λ 5007 Å obtained with a single etalon, pressure-scanned Fabry-Perot monochromator have been made and the very accurate temperatures derived in this way are compared with those obtained from the latest [O III] intensity ratio measurements. Agreement between the two methods is shown to exist to within an accuracy limited only by the physical conditions in the nebula rather than the experimental errors.

132.002 Observations of the Gum nebula with a Fabry-Perot spectrometer. R. J. Reynolds.
Astrophys. Journ., Vol. 203, 151 - 158 (1976).

Scans have been made of Hα, [N II] λ6584, [O III] λ5007, and He I λ5876 in selected directions in the Gum nebula. Analyses of the line profiles and line intensities indicate that much of the emitting gas in the Gum nebula is confined to an expanding shell which has a radius of about 125 pc, an expansion velocity of approximately 20 km s^{-1}, an emission measure which ranges from about 15 cm^{-6} pc to about 500 cm^{-6} pc, and a temperature near 11,000 K. The ultraviolet ($h\nu > 13.6$ eV) flux from ζ Pup and γ^2 Vel appears to be capable of producing most of the observed ionization, although the origin of the shell structure and high expansion velocity is not certain.

132.003 Optical and millimeter-wave observations of the M8 region.
C. J. Lada, T. R. Gull, C. A. Gottlieb, E. W. Gottlieb.
Astrophys. Journ., Vol. 203, 159 - 168 (1976).

Millimeter-wave observations made toward the NGC 6530-M8 star-forming complex are compared with high-quality optical interference-filter photographs of that region. Extensive CO observations reveal a large molecular cloud in this direction, within which three bright spots of CO emission are found. The kinematics of the molecular cloud are studied, and we find that the radial velocity of CO emission varies smoothly across the cloud. Analyses of CO, radio continuum, and, especially, optical data indicate that the star Herschel 36 is the most likely ionizing source for most of the M8 H II region. Comparison of optical and radio observations suggests a geometrical model for the region, which places the H II region at the front edge of the molecular cloud.

132.004 Emission-line spectra of individual condensations of Herbig-Haro objects.
K. H. Böhm, W. A. Siegmund, R. D. Schwartz.
Astrophys. Journ., Vol. 203, 399 - 409 (1976).

Emission-line fluxes in the wavelength range 3700 Å < λ < 11000 Å are presented for the northwest part of Herbig-Haro object 1 (HH 1) and for condensations G and H of Herbig-Haro object 2 (HH 2). The condensations G (HH 2G) and H (HH 2H) of HH 2 show quantitatively different spectra. The interpretation of the emission-line spectra of HH 1 and HH 2H leads naturally to a "two-component" model of individual condensations in which one component has a density 30 to 60 times higher than the other one. The high-density component covers only a fraction of a percent of the observed volume of a condensation, while the low-density component fills the volume almost completely. The element abundances in the condensations studied are consistent with typical population I abundances.

132.005 Giant grains. C. D. Andriesse.
Monthly Notices Roy. Astron. Soc., Vol. 175, 13P -

14P (1976).

The arguments for giant grains in the Orion nebula are criticized. The present infrared data do not give a clue for the grain size.

132.006 Orion nebula: high resolution Pa-12 mapping.
C. Barbieri, C. B. Cosmovici, K. W. Michel, T. Nishimura.
Astron. Astrophys., Vol. 47, 255 - 261 (1976).

A tilting-filter Fabry-Perot interferometer has been used to derive a detailed map of Pa-12 (λ=8750 Å) intensity across the surface of the Orion nebula. It is found that over most of the visible nebula the extinction amounts to: $A(H_d) = A(\text{Pa-}12) + 0.60 \pm 0.20$ mag. Beyond the borders of the visible nebula, the column densities of absorbing material can exceed appreciably those of absorbing matter ($n_0 l_0$) in the ionized region. A contour map of "absorbing matter" in the lobes, covering part of the H II region, is derived in terms of $n_0 l_0$.

132.007 One arc-minute resolution maps of the Orion nebula at 20, 50, and 100 microns. M. W. Werner, I. Gatley, D. A. Harper, E. E. Becklin, R. F. Loewenstein, C. M. Telesco, H. A. Thronson.
Astrophys. Journ., Vol. 204, 420 - 426 (1976).

The central 5′ of the Orion nebula has been mapped with 1′ resolution at 50μ and 100μ; a new 20μ map with similar resolution is presented as well. The spatial and temperature resolution of these data allows to reach the following conclusions about the nature of the far-infrared emission from this source: (1) The total luminosity of the infrared cluster is in excess of $1.2 \times 10^5 L_\odot$. (2) Heating by both the infrared cluster and the Trapezium stars is important in producing far-infrared emission associated with the central ridge of the molecular cloud. (3) Far-infrared emission is seen from an optically prominent ionization front to the southeast of the Trapezium.

132.008 The Orion nebula and other regions of star formation. M. V. Penston.
Observatory, Vol. 96, 6 (1976).

132.009 Radio-Beobachtungen des California-Nebels.
H. D. Bohnenstengel, H. J. Wendker.
Mitt. Astron. Ges., No. 38, p. 154 (1976). — Abstract.

132.010 Der Reflexionsnebel um den roten Riesen HR 3126.
J. Dachs, J. Isserstedt.
Mitt. Astron. Ges., No. 38, p. 201 (1976). — Abstract.

132.011 Large-scale ionization fronts and the nature and distribution of light scattering particles in the Orion nebula. M. A. Dopita, S. Isobe, J. Meaburn.
Astrophys. Space Sci. Library, Vol. 55, (see 012.001), 101 - 131 (1976).

Image-tube filter photographs calibrated against photoelectric filter photometry have been used to give maps of M42 in absolute flux units over the central 15 arc min of the nebula in Hα, [N II] (λ 6584 Å), Hβ and continuum at λ 4700 Å. Maps of the ratios Hα/[N II] and (for the first time) of continuum/Hβ have been produced with unprecedented spatial resolution. These show that the gas to dust ratio is high near the exciting stars and falls strongly in the vicinity of large scale ionization fronts marked by minima in the Hα/[N II] ratio.

These results are interpreted in terms of detailed shell models containing either ice or graphite or silicate scattering particles. A schematic model of the Orion nebula is presented to attempt to explain the large scale phenomena observed here. It demonstrates that simple shell models for this nebula are dubious.

132.012   Complex of the large Orion nebula.
V. S. Shevchenko.
Issled. ehkstremal'no molodykh zvezdn. kompleksov.
Tashkent, Fan, 1975, p. 3 - 68. In Russian. – Abstr. in
Referativ. Zhurn. 51. Astron., 3.51.543 (1976).

**132.013   De Carinanevel: een exotische emissienevel op het
zuidelijk halfrond.**   P. S. The, R. Bakker.
Zenit, 3e jaargang, p. 97 - 99, 108 (1976).

132.014   On the physical conditions in NGC 2359.
I. F. Malov, V. S. Artyukh, V. M. Malofeev.
Astrofizika, Vol. 11, 609 - 616 (1975). In Russian. English
translation in Astrophysics, Vol. 11, No. 4.

Radio observations of NGC 2359 are described. The flux
density at 107 MHz is 3×10^{-26} Wm^{-2} Hz^{-1}. It is shown that
optical and radio observations agree with the model of the
radiating envelope. Estimated parameters of the envelope are
given.

132.015   A note on the visible obscuration near ω Centauri.
I. J. Danziger, M. Dennefeld, H. E. Schuster.
Astron. Astrophys., Vol. 48, 479 - 480 (1976).

ESO Schmidt plates reveal the presence of reflection
nebulae and patchy obscuration in the vicinity of ω Centauri.

**132.016   Observations of 10.5 GHz recombination lines to-
ward Orion A.**   I. A. Ahmad.
Astrophys. Journ., Vol. 205, 379 - 383 (1976).

Observations of 85α and 107β lines due to helium and
carbon are reported. Analysis of the carbon lines is consistent
with the Zuckerman-Ball geometry for the source of radiation
at this frequency. Lines apparently due to elements heavier
than carbon have also been detected.

132.017   Far-ultraviolet brightness of nebulae in Cygnus.
G. R. Carruthers, T. Page.
Astrophys. Journ., Vol. 205, 397 - 404 (1976).

A far-ultraviolet electrographic Schmidt camera was
operated on the lunar surface during the Apollo 16 mission
1972. Among the results obtained were imagery, in the 1050–
1600 and 1250–1600 Å wavelength ranges, of a 20° diameter
field in Cygnus containing the Cygnus Loop nebula, the North
America nebula (NGC 7000), and the H II region of the O8
star 68 Cygni.

**132.018   ANS: preliminary X-ray brightness map of the
Cygnus Loop.**   E. H. B. M. Gronenschild, R. Mewe,
J. Heise, A. C. Brinkman, A. J. F. den Boggende, J. Schrijver.
Astron. Astrophys., Vol. 49, 153 - 156 (1976).

With the Utrecht X-ray detectors on ANS the authors
have observed the Cygnus Loop in a sequence of nearly 200
pointings. A preliminary brightness map in the energy band
0.16–0.284 keV is presented. Provisional data of photon
energy fluxes and electron temperatures are given for the bright
regions in the north and northeast filaments and for the
central region.

**132.019   Fabry-Perot observations of peculiar hydrogen-
emission nebulae.**   H. M. Johnson.
Astrophys. Journ., Vol. 206, 243 - 246 (1976).

Kinematical data are illustrated and discussed. Ionized
gas around Sco X-1 is moving with peaks at v_{LSR} = +23 km s^{-1}
and, probably, -22 km s^{-1}. A nebular arc 13$'$ southeast of HD
148937 is kinematically indistinguishable from background.
Velocity distribution north of the O IV-sequence star in the
peculiar nebula G2.4+1.4 is reobserved. The γ Cyg nebular
velocities fall within the range of various radio recombination-
line data, which are themselves in some disagreement.

**132.020   Emission-line variations in the structure of Herbig-
Haro object No. 1.**   R. D. Schwartz.
Publ. Astron. Soc. Pacific, Vol. 88, 159 - 163 = Lick Obs.
Bull., No. 708 (1976).

New spectrophotometric data for the northwestern and
southeastern portions of Herbig-Haro object No. 1 are presented. The [N II] $\lambda 5755/(6548 + 6584)$ intensity ratio implies
$T_e \simeq 13,500$ K $\pm$ 2000 K. These results are discussed in the
context of the shock-wave hypothesis for the origin of Herbig-Haro nebulae.

132.021   A view of the Eta Carinae nebula.
Sky Telescope, Vol. 51, 378 - 379 (1976).

**132.022   Continuous emission from a gaseous nebula beyond
the Lyman limit.**
G. T. Bolgova, G. S. Khromov.
Astrofizika, Vol. 11, 269 - 281 (1975). In Russian.
English translation in Astrophysics, Vol. 11, No. 2.

A set of models of a spherical isothermal hydrogen
nebula with a central exciting star is considered. The spectra
and energies of the diffuse L_c radiation of the nebula as well
as those of the direct ionizing radiation of the exciting star on
the outer boundary of the nebula are computed. It is shown
that the spectrum of the diffuse radiation is practically
invariant with respect to all parameters of the models, except
the electron temperature. The results are applicable to the
problem of the energy balance in a system "exciting star –
nebula" as well as to the computation of the heating of solid
grains and ionization of the surrounding interstellar medium.
They can also be used for the interpretation of observations of
far UV radiation from astronomical objects.

132.023   The excitation of permitted lines in gaseous nebulae.
S. A. Grandi.
Astrophys. Journ., Vol. 206, 658 - 671 (1976).

An analysis of the weak permitted line spectra of the
Orion nebula and of the planetary nebulae NGC 7027 and
NGC 7662 has been undertaken in order to identify the excitation mechanism for each observed line. With a few exceptions,
model calculations show that some combination of recombination, resonance fluorescence by starlight, and resonance fluorescence by other nebular emission lines can successfully
account for the observed line strengths. Among the exceptions
with no satisfactory explanation of their excitation mechanism
are several planetary nebula lines arising from doubly excited
levels.

132.024   Far-infrared spectroscopy of the Orion nebula.
D. B. Ward.
Separate print Cornell Univ., Ithaca, New York, USA, Center
Radiophys. Space Res., 11 pp. (1975).

**132.025   Observation of the Cygnus Loop in X-rays with a
one-dimensional focusing collector.**
H. I. Helava.
Thesis Columbia Univ., New York, USA, 92 pp. (1975).
(Available from: Univ. Microfilms, Order No. 75-25, 688).

**Emission-free photographic photometry of stars in
the Orion nebula cluster.**   See Abstr. 113.006.

UBV **photometry of the stars in the fields of emission nebulae. II. M17.**   See Abstr. 113.022.

**Spectra and Fabry-Perot interferometry of AG
Carinae and the nebula.**   See Abstr. 122.064.

**Optical observations of supernova remnants; fila-
mentary nebula Simeiz 147.**   See Abstr. 125.005.

Limits to the broadening of radio recombination lines by electron collisions.   See Abstr. 131.031.

OH and H_2O masers in the Monoceros-R2 molecular cloud.   See Abstr. 131.039.

The unusual H_2O maser source near Herbig-Haro object number 11.   See Abstr. 131.043.

A polarization survey of stars near the Orion nebula. See Abstr. 131.052.

CO observations of NGC 1579 (S222) and S239. See Abstr. 131.133.

The radio spectrum and physical parameters of the W80 H II complex.   See Abstr. 131.509.

Carbon recombination line observations of the Sharpless 140 region.   See Abstr. 131.519.

A high-resolution map of the Orion nebula at 5 GHz. See Abstr. 131.531.

'Cometary' globules and the structure of the Gum nebula.   See Abstr. 131.533.

A classification of the available astrophysical data of particular H II regions. V: M8 and M20: mapping and physical parameters of the objects.   See Abstr. 131.536.

Interaction of hot stars and of the interstellar medium. VIII. Low-dispersion spectra of galactic nebulae and planetary nebulae.   See Abstr. 131.542.

Electron densities in gaseous nebulae. See Abstr. 133.011.

The infrared emission of M17. See Abstr. 141.615.

On the ratio of total to selective extinction in associations containing reflection nebulae. See Abstr. 152.004.

Sternhaufen und Nebel.   See Abstr. 153.025.

133 Planetary Nebulae

133.001 **On the interpretation of integrated spectra of planetary nebulae.**   B. L. Webster.
Monthly Notices Roy. Astron. Soc., Vol. 174, 157 - 167 (1976).

Model planetary nebulae have been constructed over a range of nebular parameters. The computed emission-line spectra illustrate the dependence of observed quantities such as the excitation class, the [O III] $(N_1 + N_2)/4363$, $H\alpha/[N II]$ 6584 and helium line ratios on the nebular parameters of stellar radiation field, gas density, dilution and abundance. A method for determining the O/H ratio from only the [O III] lines and the excitation class is quantified. As the helium abundance changes the method for deriving stellar temperatures from the helium lines fails before that for calculating He/H.

133.002 **The masses and chemical composition of planetary nebulae in the galactic bulge and Magellanic Clouds.** B. L. Webster.
Monthly Notices Roy. Astron. Soc., Vol. 174, 513 - 529 (1976).

The properties of the brightest planetaries in the Large Magellanic Cloud (seven objects), Small Magellanic Cloud (seven objects) and the galactic bulge (13 objects) have been compared. The resolved galactic objects have been used to investigate the relations between distance, mass, $H\beta$ flux, electron density and angular radius. A mean H II mass/filling factor, M/ϵ, of 0.16 $M_\odot$ is determined for seven of these nebulae and two others may have an M/ϵ of 0.6 $M_\odot$. The LMC planetaries apparently have hotter central stars than those in the SMC. The helium abundance of all the nebulae studied is normal. The oxygen abundance of the galactic bulge planeta-

ries is close to that in the solar neighbourhood. The oxygen abundance of the galactic bulge planetaries is close to that in the solar neighbourhood. The oxygen abundances of the LMC and SMC planetaries are about the same as each other, but a factor of 2 or 3 lower than in those in the Galaxy. The relative abundances of the planetaries and H II regions in the Clouds are considered.

133.003 **Radiative transfer in spherical circumstellar dust envelopes. IV. The infrared emissivity profile and composition of the dust in NGC 7027.**   J. P. Apruzese.
Astrophys. Journ., Vol. 203, 177 - 181 (1976).

Using radiative transfer methods described in the previous papers of this series, dust-emission models of the bright planetary nebula NGC 7027 have been calculated. An analysis of the infrared continuum of NGC 7027 based upon these calculations shows the following: (1) If the spatial dust distribution coincides with the ionized gas distribution as given by Scott in 1973 (within ~3.2″ from the center), the infrared continuum may be produced by an unknown substance whose infrared emissivity declines with increasing wavelength to about 50 μ and then rises. (2) If there exists a cool outer dust shell beyond 5″ − 6″ from the central source, the observed infrared continuum may be produced by iron grains of radius 1 μ or by a mixture of smaller iron and graphite grains.

133.004 **On the abundances of helium, nitrogen, and oxygen in the planetary nebulae of the Magellanic Clouds.**
P. S. Osmer.
Astrophys. Journ., Vol. 203, 352 - 360 (1976).

Photoelectric observations of three planetaries in each of the Magellanic Clouds give evidence for (1) a 40 percent overabundance of helium in five of the six nebulae, (2) a deficiency of a factor of 10 in the abundance of oxygen in the nebulae in the Small Magellanic Cloud (SMC), and (3) ratios of N/O similar to those of galactic planetaries. Observations of five small H II regions in the SMC are consistent with the deficiency of a factor of 4−6 in the oxygen abundance and the low N/O ratio found by other investigators in the more prominent H II regions. Several aspects of the observational situation are discussed.

133.005 **The peculiar object He 2−467.** J. H. Lutz, T. E. Lutz, J. B. Kaler, D. E. Osterbrock, S. A. Gregory.
Astrophys. Journ., Vol. 203, 481 - 484 (1976).

He 2−467 has been classified as a planetary nebula. However, the spectrum of this object shows emission lines superposed on the absorption spectrum of a G-type star. Image tube spectrograph and scanner observations of the emission and absorption features are presented. The emission spectrum is unusual in that no forbidden lines are seen, and the He I singlet to triplet ratios do not agree with recombination theory.

133.006 **A spectrographic survey of 21 planetary nebulae.** J. B. Kaler, L. H. Aller, S. J. Czyzak.
Astrophys. Journ., Vol. 203, 636 - 646 (1976).

The authors present a blue spectral survey and coarse photographic photometry for 21 planetary nebulae. Most of the planetaries studied are of intermediate excitation; five present no detectable He II lines, and two are of high excitation with powerful He II λ 4686. The authors have derived rough electron densities for the nebulae from either or both of the [O II] or [Ar IV] doublet ratios, and electron temperatures from the [O III] lines. Values are generally within the normal range except for Hb 12, for which N_e may be as high as 4×10^5.

133.007 **Two new peculiar southern emission objects.** H.-E. Schuster, R. M. West.
Astron. Astrophys., Vol. 46, 139 - 141 (1976).

Two new peculiar emission objects, both of elliptical shape, have been found on very deep ($22^m - 23^m$) ESO 1m Schmidt sensitization test plates. They are probably planetary nebulae, but spectroscopic verification is needed.

133.008 **Hydrodynamic shaping of planetary nebulae.** R. C. Kirkpatrick.
Astrophys. Letters, Vol. 17, 7 - 9 (1976).

Starting with a slightly oblate spheroidal atmosphere which one would expect for a rotating star, it is shown that a uniform pressure along an equipotential surface transforms the oblate atmosphere into an expanding prolate spheroidal nebular shell, such as those deduced by Weedman for several planetaries.

133.009 **Radio structure and extinction curve for IC 3568.** Gopal Sistla, M. A. Kaftan-Kassim.
Astrophys. Letters, Vol. 17, 49 - 51 (1976).

Radio continuum observations of the planetary nebula IC 3568, with a resolution of 6 and 2 arc sec. are presented. In sharp contrast to the optical appearance of the nebula, the synthesized radio map at the higher resolution exhibits structural details. A value of 0.23 for the total nebular extinction c at H_β is derived and compared with earlier estimates. A systematic deviation from the mean reddening law is found for the higher Balmer lines which may be attributed either to true departures from the theoretical Balmer decrement or to the presence of internal nebular dust.

133.010 **Radio synthesis observations of planetary nebulae. II. A search for sub-arcsecond structure.** B. Balick, Y. Terzian.

Astrophys. Journ., Vol. 204, 441 - 444 (1976).

Observations of 11 planetary nebulae with spatial resolutions from 0.″2 to 2″ at 2695 and 8085 MHz failed to show any very bright structure smaller than about 2″. The observations are shown to be consistent with the present understanding of the temperatures and density distributions thought to typify most planetary nebulae.

133.011 **Electron densities in gaseous nebulae.** L. H. Aller, H. W. Epps.
Astrophys. Journ., Vol. 204, 445 - 451 (1976).

Electron densities derived from new observations of forbidden-line ratios of [S II] and [Cl III] are compared with previously obtained [O II] densities. In general the red [S II] lines indicate higher densities than do either the [O II] or [Cl III] lines by average factors of roughly 3.4 and 1.6 respectively, although there is considerable scatter. It is suggested that the concentrations wherein the singly ionized sulfur radiation is produced are dense, cool regions, while the [O II] and [Cl III] emission may be produced in extended regions as well as in discrete clouds.

133.012 **Dynamics of envelopes of planetary nebulae.** D. G. Wentzel.
Astrophys. Journ., Vol. 204, 452 - 460 (1976).

Faint H II regions surrounding planetary nebulae are receiving increasing attention. Multiple ejection of nebulae can lead to a variety of phenomena. Some of these are evaluated here using the simplest possible analytical hydrodynamics so that the main features appear explicitly. Of special dynamical interest is an older nebula that would have been a planetary except that it was ejected by a cool star. The newly ionized nebula is subject to almost immediate breakup into many small fragments, possibly like globules. Some very thin shells of ionized gas observed around planetaries might be the consequence of double ejections of planetary nebulae. A shell is the 'signature' of the expansion of an older nebula into the ambient gas, with its gas distribution altered by the ejection and expansion of a second nebula.

133.013 **Thermal emission spectra of silicates from planetary nebulae.**
E. Bussoletti, J. P. Baluteau, N. Epchtein.
Astrophys. Space Sci. Library, Vol. 55, (see 012.001), 133 - 141 (1976).

Calculations of the grain equilibrium temperature and of the expected infrared spectra of IC 418, BD + 30°3639, NGC 6572 and NGC 7027 have been performed using dielectric constants of lunar silicates. The results have been compared with previous work on pure graphite and ice-mantle grains. Lα heating of dust followed by thermal re-emission is consistent with the large infrared excesses detected in planetary nebulae. An extra source of heating is, nevertheless, necessary to fit correctly the experimental results. It appears from the calculations that, for each object, it is possible to define theoretically the most probable nature of the emitting dust.

133.014 **On the number of planetary nebulae in our Galaxy.** D. Alloin, C. Cruz-González, M. Peimbert.
Astrophys. Journ., Vol. 205, 74 - 81 (1976).

From several methods it has been estimated that the total number of planetary nebulae in our Galaxy lies in the 4000 to 22,000 range. This result is compared with those derived from other galaxies. It is found that the number of planetary nebulae per unit mass in the nucleus of M31 and in the halo of our Galaxy are smaller than that found in the solar vicinity. It is found that the contribution to the ionization of the interstellar medium due to planetary nebulae is from one to two orders of magnitude smaller than that due to O stars. The mass return to the interstellar medium due to planetary nebulae is investigated, and the birth rate of white dwarfs and planetary nebulae are compared.

133.015 Differential deceleration of nebular shells and the displacement of central stars. H. Smith.
Monthly Notices Roy. Astron. Soc., Vol. 175, 419 - 427 (1976).

The motions of nebulae through the interstellar medium would be expected to cause a shift of the central star towards the leading surface of the nebular shell, essentially a net deceleration of the shell with reference to the central star. On the assumption of a uniform-density interstellar medium, and using simple analytic formulae to describe the deceleration of the nebular shell, the author has examined the development with time of the central star's offset.

133.016 Hydrodynamic shaping of planetary nebulae. R. C. Kirkpatrick.
Bull. American Astron. Soc., Vol. 8, 289 (1976). – Abstr. AAS.

133.017 Recombination line observations of the planetary nebulae NGC 6543, M 1−78, and NGC 7027.
E. Churchwell, Y. Terzian, M. Walmsley.
Astron. Astrophys., Vol. 48, 331 - 339 (1976).

The detection of the H 109α line in the planetary nebulae NGC 6543, M 1−78, and NGC 7027 is reported. In addition, the H 90α and H 113β lines were observed in NGC 7027. The measured line parameters are used together with a simple model to derive the mean physical properties of the nebulae. The analysis of all available radio data for NGC 7027 gives evidence for collision broadening of lines with $n \gtrsim 95$. Condensations of density $\sim 2 \times 10^5$ cm^{-3} and temperature ~ 19000 K are found to be responsible for most of the optically thin radio flux in NGC 7027.

133.018 Absolute spectrophotometry of the planetary nebulae IC 2149, 4593, and NGC 6210 in the near infrared. R. I. Noskova.
Astron. Zhurn. Akad. Nauk SSSR, Vol. 53, 300 - 304 (1976). In Russian. English translation in Soviet Astron., Vol. 20, No. 2.

The absolute monochromatic energy flux was determined for the emission lines of the planetary nebulae IC 2149, 4593 and NGC 6210 in the spectral interval λ 6300 − 11000 Å. The interstellar extinction was estimated by using H I lines of the Paschen and Balmer series. The energy distribution was found in the total continuous spectrum in the interval λ 4000 − 10000 Å. An attempt was made to separate the continuum of the nucleus and the nebula. The theoretical continuous spectra were calculated from λ 3000 Å to the radio range.

133.019 Electron density measurements in NGC 6720. L. H. Aller, H. W. Epps, S. J. Czyzak.
Astrophys. Journ., Vol. 205, 798 - 801 (1976).

Spectral scans of NGC 6720 have been analyzed to obtain electron temperatures from the [N II] $\lambda\lambda5755/6584$ ratio and electron densities from the [S II] $\lambda\lambda6717/6731$ and [Cl III] $\lambda\lambda5517/5537$ ratios. The [S II], [Cl III], and presumably also the [N II] radiation originates in filaments somewhat denser and cooler than the [O II] emitting region, while the [O III] radiation comes primarily from interior zones. It also suggests that the [N II] and [S II] emission observed near the center of the nebula is produced by outer shell material seen in projection.

133.020 Objects common to the Catalogue of Galactic Planetary Nebulae and the General Catalogue of Variable Stars. H. E. Bond.
Publ. Astron. Soc. Pacific, Vol. 88, 192 - 194 = Contr. Louisiana State Univ., *Baton Rouge*, No. 113 (1976).

The Perek-Kohoutek planetary-nebula catalogue was searched for coincidences with the Kukarkin et al. variable-star catalogue. Twenty-six objects appear in both catalogues, but only one, the eclipsing binary UU Sge, appears to be a genuine new variable planetary-nebula central star. The majority of the coincidences are with late-type variable stars erroneously listed as planetary nebulae.

133.021 The spectrum of the planetary nebula IC 4997 in the near infrared. R. I. Noskova.
Astrofizika, Vol. 11, 249 - 259 (1975). In Russian.
English translation in Astrophysics, Vol. 11, No. 2.

The absolute monochromatic energy flux was determined for 22 emission lines in the spectral interval λ 6300− 11000 Å. The interstellar extinction Aβ = 1.^{m}35 was estimated by using spectral lines of S[II] and H I. The energy distribution was found in the continuous spectrum of the nucleus and the nebula. The theoretical continuous spectrum of the nebula was calculated from λ 3000 Å to the radio range. It can be coordinated with radio observations satisfactorily enough under the assumption that the nebula becomes optically thick near $\lambda \sim 2$ cm.

133.022 Time-dependent effects in the nebular shell of FG Sagittae. J. P. Harrington, P. A. Marionni.
Astrophys. Journ., Vol. 206, 458 - 468 (1976).

Theoretical models have been constructed of planetary nebulae illuminated by central stars which are evolving on a time scale comparable with the cooling time of the nebular gas. The emission line spectra of the models have been compared with the spectrum of the planetary nebula surrounding FG Sge. The observed spectrum can be reproduced by that of a nebular shell of density n_H = 160 cm^{-3}, illuminated by a central star which has decreased linearly in temperature at constant luminosity from an initial value of 55,000−75,000 K. The nebula appears to have a N/O ratio which is anomalously high even for planetary nebulae.

133.023 Evolution of planetary nebulae and their nuclei. Analytical review of spectrophotometric observational data. G. S. Khromov.
Astron. Zhurn. Akad. Nauk SSSR, Vol. 53, 534 - 543 (1976). In Russian. English translation in Soviet Astron., Vol. 20, No. 3.

Data are collected on the relative intensities of some 38 lines of atoms and ions of helium and heavier elements in the spectra of 77 planetary nebulae, published during the last 30 years. The average accuracy of the relative intensities of these lines is studied in dependence on the brightness of lines and apparent surface brightness of the nebula. Independently the accuracy of close doublet ratios is analyzed. It is shown that existing spectrophotometric data on planetary nebulae are still unsatisfactory concerning both their completeness and reliability.

133.024 The spectrum of NGC 7027. J. B. Kaler, L. H. Aller, S. J. Czyzak, H. W. Epps.
Astrophys. Journ., Suppl. Ser., Vol. 31, 163 - 186 (1976).

The authors have measured the intensities of the spectrum lines of NGC 7027 between $\lambda\lambda3132$ and 8665 by combining observations made with photographic plates, a Lallemand image tube, the Lick Cassegrain image-tube scanner, and a photoelectric scanner. From the optical data alone, they find that the logarithmic extinction at Hβ, c, = 1.37, based upon the Whitford curve. Analysis of the forbidden lines shows that N_e increases from $\sim 3 \times 10^4$ cm^{-3} at the edge of the nebula to $\gtrsim 3 \times 10^5$ nearer the center. The electron temperature is constant at $\sim 11,500$ K in the outer regions, but higher temperatures near the center cannot be ruled out.

133.025 Spectrophotometry of planetary nebulae and supernova remnants in the Magellanic Clouds.

R. J. Dufour.
Proc. Southwest Regional Conf., Vol. 1, (see 012.021), 31 - 42 (1976).

Relative line strengths are derived for several planetary nebulae and supernova remnants in the Magellanic Clouds from image-tube spectra obtained at CTIO during 1975 October. These are used to derive the physical conditions in the nebulae and estimate the relative concentrations of several astrophysically important elements in each. The results suggest that the planetary nebulae are nitrogen-rich compared with the compositions of the H II regions in each cloud.

133.026 Planetary nebulae — what we see and what we know.   R. C. Kirkpatrick.
Proc. Southwest Regional Conf., Vol. 1, (see 012.021), 43 - 57 (1976).

133.027 Photoelectric scans of the planetary nebula NGC 7027.   R. R. Robbins.
Proc. Southwest Regional Conf., Vol. 1, (see 012.021), 59 - 69 (1976).

First results from an extensive program of photoelectric scans of the brighter planetary nebulae are presented. The observations were obtained with the cassegrain scanner of the 82″ Struve telescope, over a wide range of wavelengths and with sufficient resolution to avoid blending weak emission lines into the nebular continuum. Interpretation of the observations for NGC 7027, the program test object, shows good agreement with previous observations.

133.028 Stellar evolution and planetary nebula ejection. P. R. Wood.
Thesis Australian National Univ., Canberra, (1973).

133.029 Dynamical models of dust-filled planetary nebulae. R. L. Ferch.
Thesis Cornell Univ., Ithaca, New York, USA, 222 pp. (1975). (Available from: Univ. Microfilms, Order No. 75-18,128).

Planetary nebulae, supernova remnants, and the interstellar medium.   See Abstr. 131.143.

Interstellar extinction: a calibration by planetary nebulae.   See Abstr. 131.164.

Additional observations of the unidentified infrared features at 3.28 and 3.4 microns.   See Abstr. 131.529.

Observations of five thermal sources at 15 GHz with the 5-km telescope.   See Abstr. 131.532.

Interaction of hot stars and of the interstellar medium. VIII. Low-dispersion spectra of galactic nebulae and planetary nebulae.   See Abstr. 131.542.

CRL 2688: a post–carbon-star object and probable planetary nebula progenitor.   See Abstr. 141.616.

CRL 2688 and CRL 618: proto–planetary nebulae? See Abstr. 141.617.

Sternhaufen und Nebel.   See Abstr. 153.025.

Pregalactic helium abundance and abundance gradients across our Galaxy from planetary nebulae. See Abstr. 155.012.

Errata

133.901 Erratum: 'Two new peculiar southern emission objects' [Astron. Astrophys., Vol. 46, 139 - 141 (1976)].   H.-E. Schuster, R. M. West.
Astron. Astrophys., Vol. 48, 483 (1976).

134 Crab Nebula

134.001 **Energy dependence of the size of the X-ray source in the Crab nebula.**
W. Ku, H. L. Kestenbaum, R. Novick, R. S. Wolff.
Astrophys. Journ., (*Letters*), Vol. 204, L77 - L81 (1976).

The size of the X-ray emitting region in the Crab nebula observed during the 1974 December 28 (UT) lunar occultation at p.a. = 300° is found to decrease with increasing photon energy. A power law fitted to the source size versus energy with an exponent $\gamma = -0.148 \pm 0.012$ agrees with the optical and X-ray data but does not predict the observed size and energy dependence in the radio region. Energy spectral parameters for different regions of the nebula are also derived from the X-ray data.

134.002 **Lunar occultations of the Crab nebula at 327 MHz.**
T. Velusamy, N. V. G. Sarma.
Bull. Astron. Soc. India, Vol. 3, 35 (1975). – Abstract of paper presented at the A.S.I. meeting 1975.

134.003 **The optical and X-ray surface brightness of the Crab nebula. I. A cosmic-ray diffusion model.**
S. L. Weinberg, J. Silk.
Astrophys. Journ., Vol. 205, 563 - 569 (1976).

Simple cosmic-ray diffusion theory with a finite injection region of approximately constant volume emissivity agrees in detail with the optical isophotal and spectral data on the Crab nebula. An extrapolation yields the predicted X-ray surface brightness.

134.004 **The far side of the Crab nebula: electronographic and spectroscopic observations.** S. Wyckoff,
P. A. Wehinger, R. A. E. Fosbury, D. McMullan.
Astrophys. Journ., Vol. 206, 254 - 256 (1976).

This note describes new electronographic and spectroscopic observations of the Crab nebula. A velocity discontinuity has been detected spectroscopically and isolated with the aid of direct narrow-band electronographs exposed in the light of the redshifted emission component of the [O III] 5007 Å line. In addition, a new measurement is given for the proper motion of the pulsar relative to the center of expansion of the Crab nebula.

134.005 **Observation of the Crab nebula occultation by the moon on September 10, 1974.**
V. I. Altunin, V. P. Ivanov, K. S. Stankevich, V. A. Torkhov.
Astron. Zhurn. Akad. Nauk SSSR, Vol. 53, 453 - 458 (1976). In Russian. English translation in Soviet Astron., Vol. 20, No. 3.

The results of observations of the Crab nebula occultation by the moon at $\nu = 180$ MHz and $\nu = 1646$ MHz are given. The strip brightness distribution, the angular size of the nebula and the position of the center of gravity of radio emission at both frequencies are obtained. The change of brightness distribution and the displacement of the center of gravity of radio emission are discovered when the results of this occultation were compared with those of the occultation in 1964.

The causes of these changes and their probable relationship with the activity of the central region of the nebula in the optical emission are discussed.

Special relativistic mechanics and electrodynamics with applications to synchrotron radiation.
See Abstr. 066.094.

An interstellar H_2 indicator in direction of the Crab nebula.   See Abstr. 131.032.

Interstellar absorption to the Crab nebula.
See Abstr. 131.102.

The brightness and polarization structure of the suspected supernova remnant 3 C 58 at centimetre wavelengths.   See Abstr. 141.092.

Timing of the Crab pulsar. I. Arrival times.
See Abstr. 141.314.

Timing of the Crab pulsar. II. Method of analysis.
See Abstr. 141.315.

Timing of the Crab pulsar. III. The slowing down and the nature of the random process.
See Abstr. 141.316.

Electrodynamic coupling between pulsars and surrounding nebulae.   See Abstr. 141.322.

Emission from pulsars. I. A generalized single-vector polarization model. II. A model for the sub-pulse and integrated pulse behavior.   See Abstr. 141.361.

New X-ray measurements of the Crab spectrum in the range 26 keV–1.2 MeV.   See Abstr. 142.030.

Der Röntgen-Crab: Ergebnisse der Beobachtungen von Mondbedeckungen.   See Abstr. 142.050.

The pulsed fraction of X-rays from the Crab nebula.
See Abstr. 142.146.

The size and position of the high energy X-ray source in the Crab nebula.   See Abstr. 142.147.

A model for the X-ray structure of the Crab nebula.
See Abstr. 142.149.

X-ray astronomy.   See Abstr. 142.157.

Size of the hard X-ray source in the Crab nebula.
See Abstr. 142.215.

SAS-2 high energy gamma-ray observations of galactic structure and the Crab nebula.   See Abstr. 155.066.

Radio Sources, Quasars, Pulsars, Infrared, X-Ray, Gamma-Ray Sources, Cosmic Radiation

141 Radio Sources, Quasars, Pulsars, Infrared Sources

Radio Sources, Quasars

141.001 Radio structure of the sources near ON2.
S. Harris.
Monthly Notices Roy. Astron. Soc., Vol. 174, 1 - 6 (1976).

The extended H II regions in the vicinity of the OH maser source ON2 have been observed with the Cambridge 5-km telescope at a frequency of 5 GHz and a resolution of 2 arcsec. The northern source, G75.84 + 0.4, shows compact structure, but the southern source, G75.77 + 0.34, is extremely diffuse. There is marginal evidence of extended structure in the compact continuum source which coincides with ON2. The sources have also been observed at 408 and 1407 MHz using the One-Mile telescope.

141.002 The angular broadening of radio sources by scattering in the interstellar medium.
P. J. Duffett-Smith, A. C. S. Readhead.
Monthly Notices Roy. Astron. Soc., Vol. 174, 7 - 17, with a correction, Vol. 175, 653 (1976).

The variation with direction of interstellar scattering at 81.5 MHz has been determined by two independent methods based on interplanetary scintillation measurements at 151.5 MHz, presented here, and at 81.5 MHz, already published. The scattering angle, which increases at low galactic latitudes, has a value of $0.''15 \pm 0.''05$ at 81.5 MHz for the lines of sight perpendicular to the galactic plane. The effect of the scattering on the apparent angular diameters of OH and H_2O maser sources is discussed.

141.003 3C268.4 — evidence for the presence of a gravitationally-lensed secondary image. N. Sanitt.
Monthly Notices Roy. Astron. Soc., Vol. 174, 91 - 103 (1976).

Five quasars are chosen which all show evidence for the presence of intervening galaxies. An analysis of the observational data results in one of these — 3C268.4 — being a good candidate for the occurrence of a visible secondary gravitational lens image. The expected apparent magnitude of this companion image is $21.2 ^{+1.6}_{-1.2}$, and if confirmed by observations suggested, would provide an independent measure of the Hubble constant.

141.004 The radio source B2 0055 + 30.   R. Fanti, C. Lari, R. E. Spencer, R. S. Warwick.
Monthly Notices Roy. Astron. Soc., Vol. 174, 5P - 8P (1976).

Observations of the radio source B2 0055 + 30 reveal a compact radio component of angular size less than 0.1 arcsec and a very asymmetric extended structure.

141.005 Satellite-line anomalies in the 6 GHz transitions of OH associated with G 291.3 − 0.7.
J. B. Whiteoak, F. F. Gardner.
Monthly Notices Roy. Astron. Soc., Vol. 174, 21P - 23P (1976).

A study has been made of the four transitions of the $^2\pi_{3/2}, J = 5/2$ excited state of OH in the direction of

G 291.3 − 0.7, the main radio continuum component of the H II region RCW 57. Although the main lines (at 6031 and 6035 MHz) are in absorption with intensities approximately in the LTE ratio of 0.7, the satellite lines are anomalous. For one transition (at 6016 MHz) the absorption is enhanced relative to the LTE value, while for the other (at 6049) the absorption is reduced to below the level of detection.

141.006 Identification of southern radio sources.
A. Savage.
Monthly Notices Roy. Astron. Soc., Vol. 174, 259 - 265 (1976).

Identifications are suggested for 32 radio sources from the southern zones of the Parkes 2700 MHz survey, 18 with galaxies, one with a confirmed and 12 with possible quasi-stellar objects, and one with a supernova remnant in the Large Magellanic Cloud. Accurate optical positions have also been measured for 10 of the objects and for five previously suggested QSOs.

141.007 What docks the tails of radio source components?
C. J. Jenkins, P. A. G. Scheuer.
Monthly Notices Roy. Astron. Soc., Vol. 174, 327 - 333 (1976).

The components of powerful extragalactic radio sources show a characteristic head-tail structure. If synchrotron losses play a major part in limiting the lengths of the tails, the tails should be longer at low frequencies than at high frequencies. New observations of 3C 172 indicate that such differences are not observed, and this conclusion is consistent with published data on other sources.

141.008 Further astrometric observations with the 5-km radio telescope.   B. Elsmore, M. Ryle.
Monthly Notices Roy. Astron. Soc., Vol. 174, 411 - 423 (1976).

Positions of 53 compact extragalactic and two galactic sources have been determined in order to provide a basis for a radio astrometric system for the northern sky. Declinations are measured absolutely with an accuracy of $\approx \pm 0.''03$ arc at $\delta = 45°$ and relative right ascensions are determined to within a few milliseconds. The zero point of right ascension has been established relative to the FK4 position of β Persei with an accuracy of ± 1.5 ms. Earlier 5-km telescope results have been corrected for a mistake in the computation of diurnal aberration and a number of improvements have been incorporated, which include a more refined model for the atmosphere and better corrections for instrumental effects due to temperature variations.

141.009 Confinement of extragalactic radio sources by massive objects.   P. S. Callahan.
Monthly Notices Roy. Astron. Soc., Vol. 174, 587 - 599 (1976).

A model in which the components of extragalactic radio sources are confined by massive objects ($\sim 10^{11} M_\odot$) acting gravitationally on diffuse cold matter throughout each component is investigated. It is found to be consistent with observations only if the material and field are uniform, and if the parameters take somewhat extreme values. A number of difficulties with the model are discussed, and observational tests are proposed.

141.010 Isotropy of radio sources at 1400 MHz.
  G. M. Blake.
Monthly Notices Roy. Astron. Soc., Vol. 174, 63P - 68P (1976).

The claim of Maslowski to have found a statistically significant anisotropy of the radio source counts in a 1400 MHz Greenbank Survey is discussed. It is shown that for flux densities $0.09 < S_{1400} < 0.34$ Jy the survey shows excellent agreement with the assumption of isotropy. For $S_{1400} > 0.34$ Jy the anisotropy found by Maslowski is largely due to the selection of his survey to include the 5C 1 and 5C 2 fields and partly due to an apparent deficit of 12 4C sources in his subregion IV. The latter is shown to be without statistical significance.

141.011 The redshift of 0938+119. E. A. Beaver, R.
  Harms, C. Hazard, H. S. Murdoch, R. F. Carswell,
P. A. Strittmatter.
Astrophys. Journ., (Letters), Vol. 203, L5 - L8 (1976).

Digicon observations of the Molonglo radio source 0938+119, originally classified as a faint radio galaxy, show it to be a QSO with a redshift $z = 3.19$ and a spectrum similar to that of OH 471.

**141.012 Radio emission from the Wolf-Rayet binary γ^2
  Velorum.** E. R. Seaquist.
Astrophys. Journ., (Letters), Vol. 203, L35 - L37 (1976).

The author reports the detection of radio emission from the Wolf-Rayet binary γ^2 Velorum at 5.00, 6.27, and 8.87 GHz. The data are combined with infrared measurements, and the results are interpreted as free-free emission from a circumstellar gas cloud extending to $r \gtrsim 10^{15}$ cm, much larger than the orbital radius of the binary.

**141.013 The distribution of redshifts of quasars and related
  objects.** J. W. Knight, P. A. Sturrock, P. Switzer.
Astrophys. Journ., Vol. 203, 286 - 290 (1976).

The possibility that a short-wavelength periodic modulation is present in the distribution of redshifts of quasars and related objects is evaluated using numerical simulation. This analysis does not support previous claims for such a modulation, particularly when applied to an expanded set of redshifts.

**141.014 Compact radio sources in the directions of rich
  clusters of galaxies.** F. N. Owen, L. Rudnick.
Astrophys. Journ., Vol. 203, 307 - 312 (1976).

Observations with the National Radio Astronomy Observatory (NRAO) interferometer at 2695 and 8085 MHz are reported of 16 radio sources with total angular sizes less than 5″. Flux densities, angular structure down to 0.1″, accurate radio positions, and optical identifications are reported. Only one of these compact sources is possibly identified with a cluster galaxy, while most of the extended sources in the sample are clearly associated with dominant cluster galaxies.

**141.015 3C 303: a source with unusual radio and optical
  properties.** P. P. Kronberg.
Astrophys. Journ., (Letters), Vol. 203, L47 -L48 (1976).

A high-resolution radio map of 3C 303 made with the NRAO interferometer, combined with deep exposure plates of the optical field, show that 3C 303 has an unusual radio structure, and that each of the two main radio components coincides with an optically visible system. The compact, flat spectrum component coincides with the nucleus of a 17th magnitude galaxy, and the extended steep spectrum compo-

nent appears to be associated with a compact trio of faint objects, 2 at the 22 mag level and one of $\sim$ 19.5 mag which has a distinctly nonstellar spectrum.

141.016 A search for radio emission from stars.
  W. J. Altenhoff, L. L. E. Braes, F. M. Olnon, H. J.
Wendker.
Astron. Astrophys., Vol. 46, 11 - 17 (1976).

Radio emission has been searched for in many stars, mainly in close binary systems, emission line stars and nearby stars. Most results were negative. Radio emission was found in a number of stars, mainly emission-line stars. Upper limits are given for the undetected objects. The detected ones are briefly discussed.

**141.017 The radio sources in the nuclei of NGC 3031 and
  NGC 4594.** A. G. de Bruyn, P. C. Crane, R. M.
Price, J. B. Carlson.
Astron. Astrophys., Vol. 46, 243 - 251 (1976).

Radio observations of the nuclei of the nearby spiral galaxies NGC 3031 and NGC 4594 at frequencies ranging from 610 MHz to 8085 MHz reveal the presence of compact radio sources with unusual spectra. Both sources are smaller than 0.1″ in angular size. Interpretation of the sources spectra in terms of internally absorbed synchrotron sources implies extremely small source diameters that are increasing with wavelength. The linear source diameters are in that case comparable to that of the recently discovered sub-arc-second structure in the nucleus of our own galaxy.

**141.018 Ejection speed in the slingshot theory of radio
  sources. I. Newtonian approximation.**
M. J. Valtonen.
Astron. Astrophys., Vol. 46, 429 - 433 (1976).

A study has been made of the maximum possible ejection speed of the recoil binary in the slingshot theory of radio sources, using experiments from the survey of 25000 three-body experiments. Although escape of the binary from a giant elliptical galaxy will occur only if the triple system is relativistic, the Newtonian calculations may be justified for a part of the randomly generated three-body samples. The results for that part show the feasibility of the binary escape in a wide variety of different triple systems especially when the orbits are not restricted to a plane. It is more difficult to achieve symmetric double sources, but a few possible solutions to the problem are proposed.

**141.019 Ejection speed in the slingshot theory of radio
  sources. II. General relativistic approximation.**
M. J. Valtonen.
Astron. Astrophys., Vol. 46, 435 - 440 (1976).

The gravitational three-body problem has been studied using a general relativistic approximation of the equations of motion. It is shown that the maximum possible ejection speed in the slingshot theory of radio sources is much greater than was assumed on the basis of purely Newtonian calculations. Each ejection event produces a large flux of gravitational radiation, which, in principle, should be observable.

141.020 Structure and evolution of compact radio sources.
  G. M. Richter.
Astron. Nachr., Vol. 297, 5 - 21 (1976). − In German.

A model is suggested which accounts for (1) the observed shape and angular variation of compact radio sources (especially the apparent superrelativistic velocities and the absence of contracting sources), (2) the flux variation associated with the angular variation, and (3) all the known cases of apparent occurrence of surface brightness exceeding the theoretical upper limit provided by the inverse Compton effect, preserving the usual premises: cosmological origin of the redshift and incoherent synchrotron radiation of electrons. The second method is applied to BL Lac yielding approximately 6 Mpc. So the underlying galaxy would be a dwarf system of $M \approx -13$. The active nucleus of $M \approx -16$ is rather below the normal quasars. This seems very satisfactory in view of the short time scale of variations in BL Lac compared to the quasars.

141.021 Accurate radio and optical positions for 30 radio sources of small angular size.
T. Edwards, P. P. Kronberg, G. Menard.
Astron. Journ., Vol. 80, 1005 - 1010, 1095 - 1097 (1975).

Accurate radio and optical positions have been obtained for a sample of 30 sources from the 4C and Ohio catalogues, whose radio structure was found to be compact (< 11 arcsec). This permitted optical identifications to be made purely on the criterion of positional coincidence. Eleven sources had no previously published identifications, a further seven are re-identifications which do not confirm previously published results based on inferior positional accuracy, and the remaining 12 confirm previously published identifications. The authors conclude that combined radio-optical errors must be better than ~3 arcsec for reliable identifications to be made down to the limit of the Palomar Sky Survey plates.

141.022 The Parkes 2700 MHz survey (eleventh part): catalogue for declinations −4° to −30°, right ascensions 22^h to 05^h. J. V. Wall, A. E. Wright, J. G. Bolton.
Australian Journ. Phys., Astrophys. Suppl., No. 39, p. 1 - 37 (1976).

A catalogue of 819 radio sources is presented from a 2700 MHz survey of 0.79 sr about the south galactic pole. The catalogue is essentially complete for sources with $S_{2700} \geqslant 0.22$ Jy, corresponding to a source density of 850 sources per steradian. Flux densities were measured for many of the sources at 5009 MHz, including most of the sources with $S \geqslant 0.35$ Jy. The accuracy in both the 2700 and 5009 MHz flux densities is 0.02 Jy or 3%, whichever is the greater. Root-mean-square position errors are $10''$ arc for sources stronger than 0.7 Jy, increasing to $25''$ for the weakest sources in the catalogue. The catalogue includes the results of a search of the Palomar Sky Survey prints for identification data.

141.023 New optical identifications from the eleventh part of the Parkes 2700 MHz survey: declinations −4° to −30° right ascensions 22^h to 05^h. A. Savage, J. V. Wall.
Australian Journ. Phys., Astrophys. Suppl., No. 39, p. 39 - 68 (1976).

Identifications are suggested for 166 radio sources from a survey of 0.8 sr about the south galactic pole, 88 with galaxies, 77 with quasi-stellar objects and one with a planetary nebula. The identifications were made from Palomar Sky Survey prints, supplemented in some cases with plates from the SRC 1.2 m Schmidt telescope and the 3.9 m Anglo-Australian telescope.

141.024 Positions for the optical counterparts of some southern radio sources. J. Vander Haegen.
Australian Journ. Phys., Astrophys. Suppl., No. 39, p. 69 - 71 (1976).

Optical positions have been measured for 36 objects suggested as possible counterparts of radio sources near the south galactic pole. Twenty-nine of these are QSOs or possible QSOs, two are neutral stellar objects and the remainder galaxies. The positions are on the FK4 system and have typical errors of $0.''5$ arc in right ascension and declination.

141.025 Continuum observations of six extragalactic radio sources at 1420 MHz.
C. H. Costain, L. A. Higgs, J. M. MacLeod, R. S. Roger.
Astron. Journ., Vol. 81, 1 - 6 (1976).

Synthesis observations to 2-arcmin resolution of the 1420-MHz continuum emission from six extragalactic sources are described. Two of the sources are unresolved: 3C66A, a

QSO, is found to be 80% more intense than previously reported and may be variable; the other, OM591, is tentatively identified with a stellar-type object. 3C66B is shown to have extended structure very similar to that observed at 408 MHz. OM588, the major component of 4C55.22, itself comprises at least two extended components, one of which is coincident with a 16-mag galaxy. A map of the halo of 3C274 (M87) is given and its relation to the optical and X-ray halos is considered. A suspected halo surrounding 3C103 was not detected at the 4% level.

141.026 Optical behavior of 64 extragalactic radio sources.
R. L. Scott, R. J. Leacock, B. Q. McGimsey, A. G. Smith, P. L. Edwards, K. R. Hackney, R. L. Hackney.
Astron. Journ., Vol. 81, 7 - 19 = Contr. Rosemary Hill Obs., Univ. Florida, *Gainesville*, No. 62 (1976).

A 7-yr photometric study of nearly 200 extragalactic radio sources has shown that their optical variability is a complex and many-faceted phenomenon. In the present paper 64 less-active sources with well-established comparison sequences are discussed. Light curves and photometric data are given for the 23 objects exhibiting variations that are significant at the 95% confidence level. The variability data do not appear to show strong evolutionary effects.

141.027 Neue Erkenntnisse über die rätselhaften Quasare.
K.-F. Hoffmann.
Umschau, 76. Jahrgang, p. 86 - 87 (1976).

Some new results of quasi-stellar object observations are reported. More detailed information about redshifts and the optical picture of these objects could be obtained, and their density in space at great distances was determined. The results confirm the conceptions of cosmological redshifts as well as the "big bang" theory.

141.028 Extragalaktische Doppel-Radioquellen.
C. Möllenhoff.
SuW, 15. Jahrgang, p. 13 - 18 (1976).

141.029 Decay generation of plasma turbulence and nonlinear transfer of electromagnetic radiation in radio sources with high intensity. S. A. Kaplan, R. D. Lomadze.
Astron. Zhurn. Akad. Nauk SSSR, Vol. 53, 20 - 25 (1976). In Russian. English translation in Soviet Astron., Vol. 20, No. 1.

The analogy and difference between Compton scattering and decay processes in nonlinear transfer of radiation are discussed. It is shown that the combined action of both effects contributes to the pumping-over of electromagnetic waves to the low-frequency part of spectra and simultaneous excitation of plasma turbulence with effective temperature by order of magnitude less than the brightness temperature of radiation.

141.030 On the use of lunar occultations of radio sources for investigation of their angular structure. III.
G. L. Abramyan.
Astron. Zhurn. Akad. Nauk SSSR, Vol. 53, 33 - 37 (1976). In Russian. English translation in Soviet Astron., Vol. 20, No. 1.

A dependence of the root mean square error of restoration of the angular structure of radio sources from lunar occultations on the parameters determining the conditions of observations and processing is found.

141.031 High resolution observations of 3C 33 at 327 MHz.
Gopal-Krishna, M. N. Joshi, S. Ananthakrishnan.
Astrophys. Letters, Vol. 17, 11 - 14 (1976).

Lunar occultation and interplanetary scintillation observations of the double radio source 3C 33 at 327 MHz are presented. The occultation observations with a resolution of 2.6 arc sec, which is nearly 100 times smaller than the total extent of the source, have revealed fine structure similar to that found in Cygnus A by Hargrave and Ryle. A faint continuous bridge connecting the two components has also been detected.

141.032 Correlation of scintillation visibility with flux density and angular extent of extragalactic radio sources.
G. Swarup, S. M. Bhandari.
Astrophys. Letters, Vol. 17, 31 - 36 (1976).

The interplanetary scintillation visibility of extragalactic radio sources, defined as the fraction of flux density originating in compact components with angular size less than about 1 arc sec, is statistically shown to increase with decreasing flux density and decreasing overall angular size. These relations can be satisfactorily explained if a majority of extragalactic radio sources have fine structures in between those found in recent high resolution studies of Cygnus A and 3C 33 and if most of the weaker sources are located farther away.

141.033 The radio source Sagittarius A.
Gopal-Krishna, G. Swarup.
Astrophys. Letters, Vol. 17, 45 - 47 (1976).

High resolution observations of the radio source Sgr A made between 160 MHz and 5 GHz are compared. The western peak, known as Sgr A-West, consists of both thermal and non-thermal discrete components. Sgr A-East has a nonthermal origin with the discrete components superimposed on a background component of size $\sim 3 \times 5$ arc min. The discrete non-thermal components in Sgr A have a steeper spectrum than the background component and may be parts of a supernova remnant shell.

141.034 The clustering of radio sources – I. The theory of power-spectrum analysis. A. Webster.
Monthly Notices Roy. Astron. Soc., Vol. 175, 61 - 70 (1976).

The theory of power-spectrum analysis of the clustering of points is described and developed as a sensitive and flexible test for the possible weak clustering of extragalactic sources.

141.035 The clustering of radio sources – II. The 4C, GB and MC1 surveys. A. Webster.
Monthly Notices Roy. Astron. Soc., Vol. 175, 71 - 83 (1976).

The 4C, GB and MC1 catalogues are subjected to power spectrum analysis to investigate possible clustering of the extragalactic radio sources. No evidence of clustering is found, although certain experimental effects are revealed. The results are compared and contrasted with previous work, and a stringent limit is put on the average number of radio sources in any hypothetical clusters.

141.036 Some extended observations of the radio source CL4. A. S. Webster, M. Ryle.
Monthly Notices Roy. Astron. Soc., Vol. 175, 95.- 104 (1976).

CL4 is a radio source having variations of flux density remarkably similar to those of BL Lac, namely quasi-periodic with periods of about two months. The radio spectra are also very similar, increasing to about 3 GHz and flat at higher frequencies. Both sources are associated with stellar objects having non-thermal optical spectra. The absence of detectable λ 21 cm absorption in the spectrum of CL4 suggests that it lies within the Galaxy, and may well be associated with the Cygnus Loop. Recent evidence on BL Lac, however, indicates that this source is extragalactic, so it appears possible that 'BL Lac-type' objects are of two kinds.

141.037 High resolution observations of NGC 1275 with a four-element intercontinental radio interferometer.
I. I. K. Pauliny-Toth, E. Preuss, A. Witzel, K. I. Kellermann, D. B. Shaffer, G. H. Purcell, G. W. Grove, D. L. Jones, M. H. Cohen, A. T. Moffet, J. Romney, R. T. Schilizzi, R. Rinehart.
Nature, Vol. 259, 17 - 20 (1976).

The radio nucleus of NGC 1275 (3C 84) is found to consist of three apparently stationary centres of emission extending over 0.006″ (3 pc) along position angle −9°. One or more of these centres have changed their intensity or size in the last 3 yr. A preferential direction of alignment of both the small scale and large scale radio features as well as the optical features suggests a common cause of alignment operating over linear distances from a few pc to > 100 kpc.

141.038 Radio sources near strong radio galaxies and quasars.
W. van Vliet, R. Harten, G. K. Miley, H. Albers.
Astron. Astrophys., Vol. 47, 345 - 350 (1976).

A catalogue has been compiled of 130 radio sources within 0.7° of 11 strong radio galaxies and 10 quasars observed with the Westerbork Synthesis Radiotelescope. Suggested optical identifications are given for 29 sources. Radio source counts are derived and compared with the published counts of Katgert et al. Previous activity of the radio galaxies and quasars might have resulted in an excess of nearby weak sources. No such effect was found.

141.039 Optical positions and identifications of radio sources.
M. P. Véron, P. Véron, R. L. Adgie, H. Gent.
Astron. Astrophys., Vol. 47, 401 - 405 (1976).

The authors have measured on the Palomar Sky Survey prints the optical positions of 63 objects (24 galaxies and 39 quasars) with an rms error of 0.5−0.6 arc s, using Schlesinger's method. These positions are compared with accurate radio positions, including 38 previously unpublished positions measured with the interferometer at the Royal Radar Establishment. There are 42 new or confirmed identifications and 21 empty fields (including revoked identifications).

141.040 Optical properties of the radio source PKS 0123−01 (3C 40) in Abell 194. S. M. Simkin.
Astrophys. Journ., Vol. 204, 251 - 258 (1976).

An optical bridge between the radio galaxies NGC 547/545 and NGC 541 in A 194 is found to have a $B - V$ color and continuous spectrum which resemble those for stars in the outer regions of elliptical galaxies. Embedded in this bridge is a peculiar galaxy ("Minkowski's object") which is also a radio source. Its redshift is compatible with cluster membership. Its absolute magnitude, UBV colors, and, in some respects, its emission-line spectrum, are similar to those of the metal-poor, isolated extragalactic H II regions sometimes associated with elliptical galaxies.

141.041 The abundance of nitrogen in QSOs. G. A. Shields.
Astrophys. Journ., Vol. 204, 330 - 336 (1976).

Models of photoionized QSO emission-line regions show that measurements of O III]/N IV]/C IV or N III]/C III] can yield the C/N/O ratios to an accuracy of a factor 2 or better. The N III]/C III] intensity ratios observed for the QSO PKS 1756 + 237 ($z = 1.72$) implies a N/C abundance ratio 5 times larger than the solar value. This is comparable with the nitrogen overabundance in the nuclei of nearby galaxies, and it points to advanced chemical evolution in this QSO, with $Z \gtrsim Z_\odot$. Such a large abundance of nitrogen appears to be exceptional; composite spectra indicate that most QSOs have (N/O) approximately one-fourth to one-half the solar value.

141.042 Critique of Bell and Fort's quasar model.
J. M. Barnothy, G. J. Corso.
Astrophys. Journ., Vol. 204, 337 - 340 (1976).

The quantitative local quasar model proposed by Bell and Fort has been subjected to a critical analysis. The authors found that their quasar model requires a large number of sudden increases in the intrinsic luminosity of quasars, occurring in exactly 0.165 magnitude jumps. Such a requirement has no known physical basis. The standard error in the intrinsic redshift component z_x being more than 0.05, no significance can be attached to the feature in the power spectrum of z_x with a 0.1 periodicity. This conclusion has been tested by performing power spectrum analyses on samples with perfect periodicity.

141.043 A spinar model of Cygnus A.
F. M. Flasar, P. Morrison.
Astrophys. Journ., Vol. 204, 352 - 364 (1976).

Compact radio sources lie within each of the giant extended components of Cyg A. The authors suppose that such strong double radio sources begin with a single remarkable event: a compact, gravitationally bound, spinning system within the central optical galaxy breaks cleanly into two similar objects (or systems of objects) of nearly equal mass which are ejected in a straight line. It is these compact ejected objects located within the giant twin clouds which are the current primary energy sources of the radio galaxy.

141.044 Observations of radio sources with variable flux density at 365 and 380 MHz. W. D. Cotton.
Astrophys. Journ., (Letters), Vol. 204, L63 - L66 (1976).

The results of a search for low-frequency variable extragalactic radio sources at 365 and 380 MHz are reported. Six variable and 16 probably variable sources were found. Low-frequency variability is found to be a common phenomenon in a wide variety of sources, including those with extended structure. Some implications are briefly discussed.

141.045 High-frequency structure of Ooty occultation sources. I. Sources with central components.
T. K. Menon.
Astrophys. Journ., Vol. 204, 717 - 730 (1976).

The NRAO three-element interferometer operating simultaneously at 2695 and 8085 MHz has been used to study 39 Ooty occultation sources which have been found to be double at 327 MHz. This paper reports on the comparison of the high- and low-frequency structures of five out of nine of the above sources which show an additional central component at the high frequencies. The spectral indices (below 10 GHz) of the central components vary from being positive, nearly zero to negative values as high as −0.65. The indices of the extended components are all steeper than about −0.9.

141.046 Model calculations for optically thick absorption layers around quasars.
P. R. Preussner, M. Grewing.
Mitt. Astron. Ges., No. 38, p. 96 - 100 (1976). − Short report.

141.047 Radiointerferometrie mit transatlantischen Basislinien: Galaxienkerne und Quasare.
E. Preuss, I. I. K. Pauliny-Toth, A. Witzel.
Mitt. Astron. Ges., No. 38, p. 101 - 102 (1976). − Abstract.

141.048 Explosionsmodell für Doppel-Radioquellen.
C. Möllenhoff.
Mitt. Astron. Ges., No. 38, p. 123 - 124 (1976). − Abstract.

141.049 Radiodurchmusterung von Teilen der nördlichen Hemisphäre bei λ = 6 cm Wellenlänge.
I. I. K. Pauliny-Toth, E. Preuß, A. Witzel, J. E. Baldwin, R. Hills.
Mitt. Astron. Ges., No. 38, p. 241 - 243 (1976). − Abstract.

141.050 Search for the variability of radiation of the QSOs 3C 273 and 3C 279 at 408 MHz.
V. G. Malumyan, V. A. Sanamyan.
Astrofizika, Vol. 11, 699 - 701 (1975). In Russian. English translation in Astrophysics, Vol. 11, No. 4.

3C 273 and 3C 279 have been observed for nearly 2.5 years. The observations have shown that probably the flux density of 3C 273 decreased approximately by 25% from August 1972 to June 1973. The flux density of 3C 279 was practically constant for the period mentioned above.

141.051 Radio sources with wide-angle tails in Abell clusters of galaxies. F. N. Owen, L. Rudnick.
Astrophys. Journ., (Letters), Vol. 205, L1 - L4 (1976).

Observations are presented of six previously unreported radio sources in rich clusters of galaxies. Total intensity maps at 2695 MHz and optical identifications are reported. Physical parameters are also derived for these radio galaxies, and their optical properties are summarized. It is argued that the sources presented here are a simple extension of the head-tail class of radio sources, only with larger angles between their twin tails.

141.052 Aperture synthesis observations of galactic H II regions. II. The galactic radio source W 58.
F. P. Israel.
Astron. Astrophys., Vol. 48, 193 - 212 (1976).

The strong galactic radio source W 58, containing the H II regions S 99, S 100, NGC 6857, the peculiar object K 3−50 and the OH maser source ON−3 has been observed with the Westerbork Synthesis Radio Telescope at wavelengths of 49, 21 and 6 cm. The results, combined with previous work, indicate: (1) W 58 consists of several H II regions with widely varying physical parameters. The present appearance of the complex is best explained in terms of a very rich and very massive OB association, or group of OB associations coming into being. (2) From the presence of large (d = 30 pc) and low-density (n_e = 10−15 cm^{-3}) diffuse emission regions one deduces that the process of star formation started at the near edge of the distant (D = 9 kpc) complex. (3) K 3−50 is a multiple radio source. (4) W 58 as a whole appears to suffer $4^{m}5$ of extinction due to foreground material. If the distance is 9 kpc this corresponds to $0^{m}5$ of extinction per kiloparsec.

141.053 Observations of variable radio sources at λλ3.7 cm and 11.1 cm. D. R. Altschuler, J. F. C. Wardle.
Bull. American Astron. Soc., Vol. 8, 290 (1976). − Abstr. AAS.

141.054 Nonrelativistic Compton scattering and models of quasars. J. I. Katz.
Bull. American Astron. Soc., Vol. 8, 290 (1976). − Abstr. AAS.

141.055 Investigation of optical variability in 3C 66A.
G. H. Folsom, H. R. Miller, D. W. Wingert, R. M. Williamon.
Astron. Journ., Vol. 81, 145 - 146 (1976).

The optical counterpart of the radio source 3C 66A is observed to be variable. The range of the variations is greater than $0^{m}5$. Significant variability is observed on a time scale of days.

141.056 Kinematics of relativistic ejection. C. Behr, E. L. Schucking, C. V. Vishveshwara, W. Wallace.
Astron. Journ., Vol. 81, 147 - 154 (1976).

Radio objects in quasars have been observed with apparent transverse velocities that exceed the speed of light. Ejection of such bodies at relativistic speeds appears to be a possible explanation for the puzzling observational data. A geometrical construction for the observable parameters of relativistic motions in a Friedmann universe is employed to study the optical phenomena associated with relativistic ejection. The results obtained are applied to some specific observations of 3C 279 as an example.

141.057 W28− a possible association of supernova remnants and H II regions. C. Goudis.
Astrophys. Space Sci., Vol. 40, 91 - 110 (1976).

The nature of various components of the W28 complex region is investigated. The radio spectra of W28-A1 (G 6.4−0.2), M20 (G 7.0−0.2), W28-A2 (G 5.9−0.4), W28-A4 (G 5.3−1.1), KE59 (G 6.6−0.3) and G 6.4−0.5 are established over a wide range of frequencies. The W28-A1 (G 6.4−0.2) source is a SNR (sp. index −0.41), the M20, W28-A2 and KE59 seem to be thermal sources (sp. indexes −0.06, −0.15 and −0.04 respectively) whereas the W28-A4 and G 6.4−0.5 are possibly mixed sources containing thermal and non-thermal features. Certain physical parameters of the thermal components are derived by adopting a model. The physical properties of the W28-A1 SNR are investigated. The possibility of a SNR-H II regions association in the W28 region is also discussed.

141.058 Spectra of some Ohio radio sources: list V.
M. R. Gearhart, J. D. Kraus, B. H. Andrew.
Astrophys. Journ., Suppl. Ser., Vol. 30, 337 - 349 (1976).

Flux densities have been measured at six frequencies between 1.4 GHz and 22.2 GHz for 104 radio sources from the Ohio catalog. A position accurate to $20''$ or better is given for each source. Forty-six of the sources have centimeter-excess spectra. The spectrum of OW 316 suggests that it may be a source of X-rays.

141.059 Distance indicators in radio astronomy.
G. Swarup.
Bull. Astron. Soc. India, Vol. 3, 24 - 26 (1975).

141.060 Abundance ratios from the absorption spectrum of the quasar PKS 0237−23. P. K. Mishra.
Bull. Astron. Soc. India, Vol. 3, 29 (1975). − Abstract of a paper presented at the A.S.I. meeting 1975.

141.061 On the distances of quasars. A. K. Sapre.
Bull. Astron. Soc. India, Vol. 3, 32 (1975).
Abstract of a paper presented at the A.S.I. meeting 1975.

141.062 Cosmological implications of the observed angular sizes of extragalactic radio sources. V. K. Kapahi.
Bull. Astron. Soc. India, Vol. 3, 33 (1975). − Abstract of a paper presented at the A.S.I. meeting 1975.

141.063 Relation between scintillation visibility and flux density of extragalactic radio sources.
G. Swarup, S. M. Bhandari.
Bull. Astron. Soc. India, Vol. 3, 35 (1975). − Abstract of a paper presented at the A.S.I. meeting 1975.

141.064 Two dimensional structure of radio sources from interplanetary scintillations.
S. Ananthakrishnan.
Bull. Astron. Soc. India, Vol. 3, 35 (1975). − Abstract of a paper presented at the A.S.I. meeting 1975.

141.065 Lunar occultation studies of nine 3C sources at 327 MHz.
M. N. Joshi, Gopal-Krishna, G. Swarup.
Bull. Astron. Soc. India, Vol. 3, 36 (1975). − Abstract of a paper presented at the A.S.I. meeting 1975.

141.066 High resolution observations of 3C 33 at 327 MHz.
Gopal-Krishna, M. N. Joshi.
Bull. Astron. Soc. India, Vol. 3, 36 (1975). − Abstract of a paper presented at the A.S.I. meeting 1975.

141.067 Interstellar velocities and the 22 GHz spectrum of W49A. V. Radhakrishnan, W. M. Goss.
Bull. Astron. Soc. India, Vol. 3, 36 (1975). − Abstract of a paper presented at the A.S.I. meeting 1975.

141.068 Radiometric studies of sky at centimetre wavelength.
O. P. N. Calla, S. Balasubramanian.
Bull. Astron. Soc. India, Vol. 3, 36 (1975). − Abstract of a paper presented at the A.S.I. meeting 1975.

141.069 The nucleus of Centaurus A. A. C. Fabian, D. Maccagni, M. J. Rees, W. R. Stoeger.

Nature, Vol. 260, 683 - 685 (1976).

The nucleus of Centaurus A contains a very compact radio component, and has recently been discovered to be a variable X-ray source. The authors argue that these phenomena arise from accretion on to a massive black hole. Possible implications for galactic nuclei in general are briefly discussed.

141.070 Fast 8550 MHz survey technique and digital computer processing. M. G. Larionov, M. V. Popov.
Astron. Zhurn. Akad. Nauk SSSR, Vol. 53, 241 - 250 (1976).
In Russian. English translation in Soviet Astron., Vol. 20, No. 2.

A fast survey of the sky region between declinations $0°$ and $+30°$ was carried out at 8550 MHz over an area of 2 sr. The survey technique is described. A magnetic-tape recorder was used for recording the receiver output. The principal receiver parameters are reported. Davies' method of searching for sources was used. Completeness, reliability, flux density and position errors are analyzed. The same parameters are reported for the case of re-observations of the survey sources in selected areas.

141.071 On angular variations of the frequency spectrum of the cosmic radio background along the declination $\delta = 50°30'$. P. P. Belyaev, G. G. Getmantsev, A. F. Tarasov, Yu. V. Tokarev.
Astron. Zhurn. Akad. Nauk SSSR, Vol. 53, 273 - 278 (1976).
In Russian. English translation in Soviet Astron., Vol. 20, No. 2.

The influence of absorption of radio waves and thermal radiation of the interstellar medium on the value of measured angular variations of the total cosmic radio radiation spectrum is considered. It is shown that the mean value of the spectral index of sources of non-thermal galactic radio radiation $\beta_{0\,gal}$ changes along declination $\delta = 50°30'$ within the limits of $2.48 - 2.52$, with the exception of directions to the local arm ($\alpha = 20^h - 21^h$) where an increase of $\beta_{0\,gal}$ is possible up to lower values. An increase of $\beta_{0\,gal}$ with the decrease of α in the region $\alpha = 17^h - 20^h$, where the beam trajectory approaches a spur of Loop III, is marked.

141.072 On the masses of the quasi-stellar objects.
G. Burbidge, J. Perry.
Astrophys. Journ., *Letters*, Vol. 205, L55 - L58 (1976).

If it is assumed that the gas giving rise to the emission and absorption lines in quasi-stellar objects has been driven out of the central object by radiation pressure, arguments based on the dynamics of radiation-driven gas flows enable us to establish limits on the central masses and the rates of mass loss. For QSOs at cosmological distances it is found that the masses of the central objects must lie in the range $5 \times 10^7 M_\odot \lesssim M \lesssim 2 \times 10^9 M_\odot$ and that the mass loss rates should be $\dot{M}/M \approx 10^{-7}$ yr^{-1}. If the QSOs are local objects, the upper limits to the masses are about $2 \times 10^7 M_\odot$.

141.073 The structure and spectrum of nebulosity associated with the QSO 4C 37.43. A. Stockton.
Astrophys. Journ., *(Letters)*, Vol. 205, L113 - L116 (1976).

Spectroscopic observations of nebulosity found surrounding the quasi-stellar object 4C 37.43 show it to have a strong emission-line spectrum and to be similar to the nebulosity around 3C 48. The spectrum is consistent with that expected from a gas photoionized by a synchrotron source. The total mass involved in the emitting regions of the brightest condensation in the nebulosity is of the order of $10^5 M_\odot$. The observations are inconsistent with the nebulosity's being in rotation around the QSO; it is suggested instead that the material has been radiatively expelled from the QSO and that such material may, at a later stage, be responsible for the absorption-line spectra frequently seen in high-redshift QSOs.

141.074 2290-MHz flux densities of 52 high-declination radio sources. A. W. Harris, R. A. Preston,

D. J. Spitzmesser, M. A. Slade, L. J. Skjerve.
Astron. Journ., Vol. 81, 222 - 224 (1976).

The Deep Space Network 64-m antenna at Goldstone was used to obtain total flux-density measurements at 2290 MHz of sources from low-frequency catalogs (principally 4C and 4CP) in a search for potential compact VLBI sources at high declination. Subsequent VLBI observations between Goldstone and Madrid of six candidates identified here have resulted in the discovery of two new high-declination VLBI sources: 4CP 71.07 and OQ 663.

141.075 **Modern models of the source of activity in quasars and in nuclei of galaxies.** L. M. Ozernoj.
Problems of gravitation. Third Soviet Gravitational Conference, Erevan, 1972, (see 012.008), p. 42 - 70 (1975). In Russian.

141.076 **The flux densities of radio source emission at 1.35 cm wavelength.**
V. A. Efanov, I. G. Moiseev, N. S. Nesterov.
Izv. Krymskoj Astrofiz. Obs., Vol. 54, 159 - 164 (1976). In Russian.

The flux densities of 39 radio sources (galaxies, quasars and galactic objects) measured at the wavelength 1.35 cm in June 1973 and March - April 1974 are presented. Radio maps of Sgr A and Sgr B_2 have been obtained with angular resolution of 2.5 min of arc.

141.077 **Rapid increase in the size of 3C 345.**
M. H. Cohen, A. T. Moffet, J. D. Romney, R. T. Schilizzi, G. A. Seielstad, K. I. Kellermann, G. H. Purcell, D. B. Shaffer, I. I. K. Pauliny-Toth, E. Preuss, A. Witzel, R. Rinehart.
Astrophys. Journ., (Letters), Vol. 206, L1 - L3 (1976).

The brightness distribution of the quasar 3C 345 at λ = 2.8 cm is accurately modeled with two Gaussian elliptical components. Observations at four epochs between 1974.15 and 1975.68 show that the separation of the components increased from 1.23 to about 1.61 milli-arcsec, at a rate 0.2 milli-arcsec per year. The apparent transverse velocity is $v \approx 8c$, if one assumes a cosmological interpretation of the redshift.

141.078 **Cygnus A at 8.5 millimeter wavelength.**
O. Hachenberg, E. Fürst, W. Harth, P. Steffen, W. Wilson, W. Hirth.
Astrophys. Journ., (Letters), Vol. 206, L19 - L22 (1976).

The compact central component of Cygnus A was observed at 8.5 mm wavelength with the Effelsberg 100 m telescope. The flux was found to 1.8 ± 0.6 Jy. Including the spectral information already known, a flat or slightly increasing spectrum can be derived in the range from 5 GHz up to 35 GHz. If the measured radiation is interpreted as synchrotron emission, high values of the magnetic field strength and the (relativistic) electron density are needed.

141.079 **The moments of the brightness distribution of a radio source.** B. J. Burn, R. G. Conway.
Monthly Notices Roy. Astron. Soc., Vol. 175, 461 - 471 (1976).

It is often desirable to describe the overall brightness distribution of a radio source by parameters which do not involve assumptions about the form of the distribution. The successive moments of the distribution comprise such a set of parameters, and their use is discussed with reference to (1) a two-dimensional map, and (2) the distribution of polarization. The brightness distribution and polarization of the source 4C 25.33 is discussed by way of an example.

141.080 **Observations of Cygnus A at 15 GHz with the 5-km radio telescope.** P. J. Hargrave, M. Ryle.
Monthly Notices Roy. Astron. Soc., Vol. 175, 481 - 488 (1976).

The compact 'head' regions in the components of Cygnus A have been mapped with the 5-km telescope at 15.4 GHz, the resolution being 0.''65 arc in right ascension and 1.''0 arc in declination.

141.081 **Identification of southern radio sources – II.**
A. Savage, J. G. Bolton, A. E. Wright.
Monthly Notices Roy. Astron. Soc., Vol. 175, 517 - 523 (1976).

Identifications are suggested for 36 radio sources from the southern zones of the Parkes 2700 MHz survey, 28 with galaxies, six with confirmed and two with suggested quasistellar objects. Accurate optical positions have also been measured for nine of the objects and for five previously suggested identifications.

141.082 **On the redshift distribution of quasi-stellar objects.**
D. Wills, R. L. Ricklefs.
Monthly Notices Roy. Astron. Soc., Vol. 175, 81P - 85P (1976).

An updated list of 540 QSO emission-line redshifts has been used to examine earlier suggestions of (1) the existence of periodicities in the redshifts, and (2) anisotropy of the redshift distributions in the two galactic hemispheres. The new data lend no support to either suggestion.

141.083 **Iris photometry of 3C 273.** H. H. Lanning.
Publ. Astron. Soc. Pacific, Vol. 88, 198 - 199 (1976).

Twenty-seven blue plates of 3C 273 originally measured by Zwicky, Karpowicz, and Rudnicki using the Argelander method and measurement of image diameters have been remeasured on an iris photometer to improve their precision. These measures, including two later observations, have an accuracy $\sim \pm 0^m.1$.

141.084 **Possible interpretations of B2 and 5C quasar candidate counts and the "cut off" in quasar density.**
P. Notni.
Astron. Nachr., Vol. 297, 147 - 154 (1976).

Optical quasar candidate counts in the far reaching radio surveys B2 and 5C are consistent either with a luminosity function containing a high percentage of low luminosity objects and a cut off in quasar density or, more probably, with a normal number of quasars at high redshifts and a less steep luminosity function. The absence of high redshifted objects in currently available samples is to be expected if $q_0 \approx o$ and if some of the few quasars observed at $z > 2.2$ are exceptionally bright intrinsically and not typical for the bulk.

141.085 **Photographic photometry of compact extragalactic objects. IV.** M. K. Babadzhanyants, S. K. Vinokurov, V. A. Hagen-Thorn, E. V. Semenova.
Trudy Astron. Obs., Leningrad, Vol. 32 (= Uchenye Zapiski Leningr. Un-ta, No. 385 = Seriya matem. nauk, vyp. (No.) 52), p. 52 - 60 (1976). In Russian.

Results are given of photographic observations made mostly in 1973 of the quasars 3C 351, 3C 454.3, PKS 2135−14, PKS 2344+09, of the starlike object OJ 287 and of the compact galaxy I Zw 0051+12 in the B range as well as UBV observations of the N-galaxy 3C 390.3.

141.086 **Radio and optical observations of the radio source OX 029.** E. R. Craine, J. W. Warner.
Astrophys. Journ., Vol. 206, 359 - 363 (1976).

Radio, optical, and spectroscopic observations of the radio source OX 029 suggest a duplicity of radio sources for which there are two optical counterparts: an elliptical galaxy and a star-like object with a continuous spectrum. The possibility is raised that the latter may be an example of a BL Lacertae-type object which has been ejected from the elliptical galaxy.

141.087 **Redshifts of forty-three radio sources.**

M.-H. Ulrich.
Astrophys. Journ., Vol. 206, 364 - 369 (1976).

New redshifts are given for eight QSOs, 25 radio galaxies including five having head-tail radio structures, and three radio-quiet galaxies in the Coma cluster.

141.088 **Apparent "superrelativistic" expansion of the extragalactic radio source 3C 345.** J. J. Wittels, W. D. Cotton, C. C. Counselman III, I. I. Shapiro, H. F. Hinteregger, C. A. Knight, A. E. E. Rogers, A. R. Whitney, T. A. Clark, L. K. Hutton, B. O. Rönnäng, O. E. H. Rydbeck, A. E. Niell.
Astrophys. Journ., (*Letters*), Vol. 206, L75 - L78 (1976).

The compact extragalactic radio source 3C 345 was observed by very-long-baseline interferometry ($\lambda \approx 3.8$ cm) at 12 epochs distributed over the nearly 4-year period 1971 February to 1974 October. The data were represented adequately by two-component models. The angular separation of the two components increased during this period from about 1.00 ($\pm$ 0.05) to 1.30 ($\pm$ 0.05) milli-arcsec, corresponding to an apparent average speed of expansion of 2.5 ($\pm$ 0.8) c at a fixed position angle of $105° \pm 5°$. These results, coupled with the fact that contraction has never been observed, seem difficult to reconcile with the so-called Christmas-tree model of the "superrelativistic" expansion of extragalactic radio sources.

141.089 **PKS 0116+082 and 3C 330: two distant cluster radio galaxies.**
H. Spinrad, J. Liebert, H. E. Smith, R. Hunstead.
Astrophys. Journ., (*Letters*), Vol. 206, L79 - L82 (1976).

Identification of the radio sources PKS 0116+082 and 3C 330 with faint galaxies has led to the discovery of faint, but rather rich, clusters of galaxies at redshifts $z = 0.5936$ and 0.5490, respectively. Both radio galaxies have relatively strong emission lines; 3C 330 in particular is similar to Cygnus A in both emission-line strength and excitation. The two radio sources show a high intrinsic luminosity in relation to their spectral indices.

141.090 **Observations of high-redshift QSOs from a Molonglo faint source survey.** J. A. Baldwin, H. E. Smith, E. M. Burbidge, C. Hazard, H. S. Murdoch, D. L. Jauncey.
Astrophys. Journ., (*Letters*), Vol. 206, L83 - L86 = Lick Obs. Bull., No. 728 (1976).

The authors present spectrophotometric observations of four objects identified with the Molonglo radio sources 0758+120, 0824+110, 0830+115, and 0938+119. All are QSOs with $z > 2$. The sources 0758+120, 0824+110, and 0830+115 have previously unreported redshifts $z = 2.66, 2.29$, and 2.97, respectively. The optical and radio spectra of these objects are compared with the spectra of other known high-redshift QSOs.

141.091 **Source count, spectral indices and angular sizes of weak radio sources in the 5 C 2 region.**
P. Katgert.
Astron. Astrophys., Vol. 49, 221 - 234 (1976).

The results of the third 1415 MHz Westerbork survey are used to derive the angular size distribution and the source count of a sample, part of which has been observed earlier at 408 MHz. About 25% of the sources stronger than about 10 mJy have an angular size larger than 20 seconds of arc, while about 50% is larger than 10 seconds of arc. The spectral indices between 0.4 and 1.4 GHz of a source sample selected from the original 0.4 GHz observations are used to study possible changes of the spectral index distribution with flux density. No significant changes are detected from about 10 down to about 0.1 Jy. Comparison of the 0.4 and 1.4 GHz counts reveals an inconsistency between counts and spectral index distribution at low flux densities. It is suggested that the effect is due to a population of weak flat-spectrum sources

present in the 1.4 GHz sample but almost absent from the 0.4 GHz sample.

141.092 **The brightness and polarization structure of the suspected supernova remnant 3C 58 at centimetre wavelengths.** A. S. Wilson, K. W. Weiler.
Astron. Astrophys., Vol. 49, 357 - 374 (1976).

Maps of the brightness and polarization distributions over the radio source 3C 58 have been made with the Westerbork Synthesis Radio Telescope at λ's 6, 21 and 50 cm. The high resolution reveals considerable fine structure in both total and polarized emission. The nature of the source is discussed and it is concluded that 3C 58 is probably a galactic supernova remnant, resembling the Crab nebula in its radio properties.

141.093 **Spectroscopy of 206 QSO candidates and radio galaxies.** D. Wills, B. J. Wills.
Astrophys. Journ., Suppl. Ser., Vol. 31, 143 - 162 (1976).

The authors present the results of spectroscopic observations of 206 objects, 113 of which are suggested identifications of radio sources. The remaining 93 objects include 72 from various other lists of radio source identifications, and 21 from lists of optically selected ("radio-quiet") QSO candidates with ultraviolet excess.

141.094 **The optical spectra of 3C 227 and other broad-line radio galaxies.**
D. E. Osterbrock, A. T. Koski, M. M. Phillips.
Astrophys. Journ., Vol. 206, 898 - 909 = Lick Obs. Bull., No. 724 (1976).

Emission line profiles and relative intensities were measured for 3C 227, 3C 382, and 3C 445, and are compared with similar data already published for 3C 390.3. These four radio galaxies all have similar emission line spectra, with broad Balmer lines and much narrower forbidden lines. The Balmer emission lines have weak, narrow components with the same widths and redshifts as the forbidden lines, and the broad components have irregular, nonsymmetric profiles.

141.095 **Nonrelativistic Compton scattering and models of quasars.** J. I. Katz.
Astrophys. Journ., Vol. 206, 910 - 916 (1976).

In order to avoid the problems associated with electron synchrotron models of quasar visible and infrared radiation, the author proposes a nonrelativistic Compton scattering model. Calculations of the emergent flux from such models produce power-law spectra with a wide range of spectral indices. An analogy to reactor theory is pointed out. Fluctuation and polarization behavior resembling that observed is predicted. Possible ranges of parameters and their constraints are discussed.

141.096 **3C 68.1: a very red QSO with an intermediate redshift.** A. Boksenberg, R. F. Carswell, J. B. Oke.
Astrophys. Journ., (*Letters*), Vol. 206, L121 - L123 (1976).

Spectroscopic observations of the red stellar object identified with 3C 68.1 show it to be a QSO with redshift 1.238. The spectral index of the optical continuum is found to be about 6, a value considerably steeper than that previously found for QSOs.

141.097 **A survey for emission-line galaxies and quasars. III. A list of nine new optically selected QSOs with $2.5 < z < 3.1$.** M. G. Smith.
Astrophys. Journ., (*Letters*), Vol. 206, L125 - L127 (1976).

A list is presented of the first nine objects from the Tololo survey to have been confirmed as QSOs with redshift $z > 2.5$. All the objects have continuum magnitudes brighter than 18.5. Extrapolation of the present sample leads one to expect more than 1500 such objects over the whole sky. Some illustrations of potential applications of the sample are given.

141.098   TV spectroscopy of absorption lines in the far-red of PHL 957.   G. R. Gilbert, J. R. P. Angel, S. A. Grandi, G. D. Coleman, P. A. Strittmatter, R. H. Cromwell, E. B. Jensen.
Astrophys. Journ., (Letters), Vol. 206, L129 - L131 (1976).

The authors report the discovery of several absorption lines in the spectrum of PHL 957 in the wavelength range 7300–8050 Å. The absorption lines found confirm the existence of the absorption-line systems at $z = 2.3088$ and 1.7969 which, in conjunction with previous observations of 1331 + 170, provide strong evidence for Lyman line to Lyman continuum locking in QSOs.

141.099   Transient radio source near the galactic centre.   R. D. Davies, D. Walsh, I. W. A. Browne, M. R. Edwards, R. G. Noble.
Nature, Vol. 261, 476 - 478 (1976).

In February 1975 Eyles et al. discovered the transient X-ray source A1742−28 near the galactic nucleus. The authors report here the detection of a radio transient source which they believe is very likely to be associated with this X-ray source. Although they have evidence that this transient radio source lies close to the centre of the Galaxy, it is clearly not to be identified with the nucleus itself.

141.100   The radio spectra of twenty minor radio sources in the Cygnus X region.   C. Goudis.
Astrophys. Space Sci., Vol. 41, 257 - 273 (1976).

The radio spectra of twenty radio sources of the Cygnus X region are established. The sources DR3, DR6, DR7, DR11, DR13 and DR18 show a thermal spectrum whereas DR12 and DR23 show a non-thermal one. Possible thermal sources are also the DR1, DR2, DR9, DR10, DR16, DR17, DR19, DR20, DR22, DR24 and DR26.

141.101   Star on position of PKS 1925–524 variable.   C. J. Smith.
Inform. Bull. Variable Stars, (I.A.U. Commission 27), Konkoly Obs., Budapest, No. 1085, 3 pp. (1976).

141.102   PKS 0422+004.   T. D. Kinman.
IAU Circ., No. 2908 (1976).

141.103   3U 1727–33.   N. E. White, K. O. Mason, P. W. Sanford, C. Chevalier, S. A. Ilovaisky.
IAU Circ., No. 2920 (1976).

141.104   3C 120.   G. Wlérick, R. Garnier, B. Westerlund.
IAU Circ., No. 2930 (1976).

141.105   The identification of radio sources.   R. Minkowski.
Galaxies and the universe, (see 003.010), p. 177 - 197 (1975).

Discovery and early identifications; Surveys of radio sources; Identification of 3C and MSH sources; Precise positions; The identification of the 3CR sources; The identification of Parkes sources; The unidentified sources; Radio sources and clusters of galaxies; The luminosity function.

141.106   Optical identification of 3CR sources (October 1972).   J. Kristian, R. Minkowski.
Galaxies and the universe, (see 003.010), p. 199 - 210 (1975).

141.107   Quasars.   M. Schmidt.
Galaxies and the universe, (see 003.010), p. 283 - 308 (1975).

History; Identifications (Quasi-stellar sources (radio quasars), quasi-stellar objects (quasars), sky distribution);

Emission lines, redshifts; Continuum energy distribution; Nature of quasars (Gravitational hypothesis, local-Doppler hypothesis, cosmological hypothesis, other hypotheses); Absorption lines (Observations, interpretation); Cosmology, evolution; Models, lifetimes, and energy sources.

141.108   An evolutionary model for quasars.   N. V. Zotov, W. Davidson.
Australian Journ. Phys., Vol. 29, 97 - 105 (1976).

A simple, though not necessarily unique, model for the cosmic population dynamics of quasars is presented. This model incorporates a double evolutionary trend: (1) a strong luminosity evolution, and (2) an evolution of total coordinate density. This double trend is the same as that already found to be applicable to the total radio source population, if account is taken only of radio characteristics (Davidson et al. 1971). The predictions of the scheme are compared with five sets of data, covering a wide range of frequencies and flux density limits, and they are found to give good agreement in all cases.

141.109   Astronomía en ondas milimétricas.   J. Pensado.
Bol. Astron. Obs. Madrid, Vol. 9, No. 2, p. 3 - 20 (1976).

141.110   Identification of southern quasi-stellar objects – V.   B. A. Peterson, J. G. Bolton, A. Savage.
Astrophys. Letters, Vol. 17, 137 - 140 (1976).

141.111   Recent Westerbork observations of radio galaxies.   R. G. Strom.
Mem. Soc. Astron. Italiana, Vol. 45, (see 012.017), 523 - 534 (1974). – Invited paper.

141.112   High resolution observations of large and complex radio galaxies.
R. G. Strom, A. G. Willis, A. S. Wilson.
Mem. Soc. Astron. Italiana, Vol. 45, (see 012.017), 535 - 542 (1974).

141.113   Observations of extragalactic radio sources with the Cambridge 5-km telescope.   M. S. Longair.
Mem. Soc. Astron. Italiana, Vol. 45, (see 012.017), 543 - 561 (1974).

141.114   Radio sources in Abell clusters.   J. E. Baldwin.
Mem. Soc. Astron. Italiana, Vol. 45, (see 012.017), 563 - 567 (1974).

141.115   High resolution studies of weak radio sources.   R. E. Spencer, R. S. Warwick.
Mem. Soc. Astron. Italiana, Vol. 45, (see 012.017), 569 - 571 (1974).

141.116   The study of circular polarization in synchrotron continuum radio sources.   K. W. Weiler.
Mem. Soc. Astron. Italiana, Vol. 45, (see 012.017), 573 - 577 (1974).

141.117   Observations of large radio galaxies at 1420 and 2695 MHz.   J. R. Baker.
Mem. Soc. Astron. Italiana, Vol. 45, (see 012.017), 579 - 585 (1974).

The radio galaxies DA 240, 3C227, 3C236 and 3C326 have been mapped in both total power and polarisation using the Bonn 100 m telescope at 1420 and 2695 MHz. DA 240 and 3C236 are known to be radio galaxies of large linear size. 3C326 is probably also such a system, but for 3C227 the evidence remains inconclusive.

141.118   Radio emission of NGC 5128 and some southern

radio sources at 1.35 cm.   P. Kaufmann,
E. Scalise, Jr., P. M. dos Santos, M. A. Bráz, W. G. Fogarty.
Mem. Soc. Astron. Italiana, Vol. 45, (see 012.017), 589 - 591
(1974).

**141.119 Investigation of the structure of extragalactic radio
sources at 3.5 m.**   Yu. V. Volodin, R. D. Dag-
kesamanskii (*Dagkesamanskij*), B. Ya. Losovskii (*Losovskij*),
S. A. Sukhodolskii (*Sukhodolskij*), V. A. Frolov.
Mem. Soc. Astron. Italiana, Vol. 45, (see 012.017), 603 - 606
(1974).

141.120 The structure of NGC 1265.   G. K. Miley.
Mem. Soc. Astron. Italiana, Vol. 45, (see 012.017),
607 - 608 (1974). — Abstract.

**141.121 Catalogue of quasi stellar-objects, (edition of June
28, 1975).**
C. Barbieri, M. Capaccioli, M. Zambon.
Mem. Soc. Astron. Italiana, Vol. 46, 461 - 499 (1975).

**141.122 Possibility of a gravitational effect in the spectra
of quasi-stellar objects. III.**   M. C. Durgapal.
Journ. Phys. A, Vol. 8, 1697 - 1705 (1975). — Abstr. in Phys.
Abstr., Vol. 79, A003364 (1976).

**141.123 Correlation test for clustering of radio sources at
1400 MHz.**
M. P. Kalinkov, K. Y. Stavrev, V. N. Dermendjiev.
Comptes Rendus Acad. Bulg. Sci., Vol. 28, 1009 - 1010
(1975). — Abstr. in Phys. Abstr., Vol. 79, A042281 (1976).

141.124 Low-frequency spectra of compact radio sources.
G. M. Resch.
Thesis Florida State Univ., Tallahassee, USA, 214 pp. (1974).
(Available from Univ. Microfilms, Order No. 75–12, 668).

**141.125 Expanding quasar envelopes. I. Steady radiation-
driven winds.**
R. Kippenhahn, L. Mestel, J. J. Perry.
Separate print Max-Planck-Inst. Phys. Astrophys., München,
F. R. Germany, 50 pp. (1975). (Available from ZAED). In
German. — See 14.141.092.

141.126 Quasars ten years later.   L. Gouguenheim.
Bull. Union Physiciens, Vol. 70, No. 577, p. 5 - 19
(1975). In French.

141.127 Compact radio sources.   D. R. Altschuler.
Thesis Brandeis Univ., Waltham, Massachusetts, USA,
286 pp. (1975). (Available from Univ. Microfilms, Order No.
75-24, 797).

**141.128 Photographic studies of quasi-stellar objects and
other active radio sources.**   R. L. Scott.
Thesis Florida Univ., Gainesville, USA, 376 pp. (1975).
(Available from Univ. Microfilms, Order No. 75-23, 915).

Quasars, pulsars and black holes.
See Abstr. 003.048

**Mathematical cosmology and extragalactic astron-
omy.**   See Abstr. 003.097.

**Radio source tracking by the RATAN-600 radio
telescope with feed on a radial way.**   See Abstr. 033.009.

**Geodetic and astrometric results of very-long-base-
line interferometric measurements of natural radio sources.**
See Abstr. 041.039.

**A new redshift mechanism with possible applica-
tions to astrophysical problems such as quasars.**
See Abstr. 061.059.

On dynamical stability of supermassive objects.
See Abstr. 066.001.

**Radio observations of eight early-type emission-
line stars.**   See Abstr. 114.040.

**The structure of the radio emission from the NGC
1579/LkHα 101 region.**   See Abstr. 114.347.

**An unusually strong radio outburst in Algol: VLBI
observations.**   See Abstr. 121.045.

Close binaries with H and K emission.
See Abstr. 121.069.

Radio emission from Wolf-Rayet binaries.
See Abstr. 121.076.

**Simultaneous radio and optical observations of UV
Ceti-type flare stars.**   See Abstr. 122.005.

**Radio emission from a source near the flare star
AD Leonis.**   See Abstr. 122.006.

**A search for slowly varying radio continuum emis-
sion from UV Ceti stars.**   See Abstr. 122.041.

Radio sources in the vicinity of flare stars.
See Abstr. 122.050.

**The population of supernova remnants in the
Magellanic Clouds.**   See Abstr. 125.002.

**An empirical comparison of X-ray and radio emis-
sion from supernova remnants.**   See Abstr. 125.016.

**Some results of a recent study of the supernova
remnant 3C 400.2 at 49 cm.**   See Abstr. 125.017.

**Deep Hα photography of the Vela and Puppis
supernova remnants.**   See Abstr. 125.024.

Neutral hydrogen in the W41 region.
See Abstr. 131.011.

**Dielectronic recombination in the interstellar medi-
um.**   See Abstr. 131.016.

**Scintillations of an extended source on inhomogenei-
ties of an interstellar plasma.**   See Abstr. 131.023.

**Observation of the $6_{16}-5_{15}$ transitions of acetalde-
hyde in Sagittarius B2.**   See Abstr. 131.041.

Accurate positions of OH emission sources.
See Abstr. 131.132.

Fine radio structure in W3.   See Abstr. 131.504.

**Aperture synthesis observations of W3 (OH)
(G 133.9 + 1).**   See Abstr. 131.507.

**Twenty-one centimeter aperture synthesis study of
the small-scale structure of the interstellar medium.**
See Abstr. 131.511.

Radio continuum measurements of compact H II regions and other sources.   See Abstr. 131.534.

Observations of 10.5 GHz recombination lines toward Orion A.   See Abstr. 132.016.

Further observations of the radio emission from Cygnus X-3 at 5 GHz.   See Abstr. 142.027.

Transient galactic radio and X-ray sources. See Abstr. 142.114.

An aperture synthesis study of H I in the irregular II galaxy NGC 3077.   See Abstr. 158.003.

In situ particle acceleration and physical conditions in radio tail galaxies.   See Abstr. 158.010.

The radio spectra of Markarian galaxies. See Abstr. 158.011.

The variable radio nucleus of M81. See Abstr. 158.018.

The detection of formaldehyde in NGC 5128 (Centaurus A).   See Abstr. 158.028.

Possible collisional enhancement of He I $\lambda5876$ in Seyfert galaxies and QSOs.   See Abstr. 158.038.

CSIRO Division of Radiophysics, Australia: A search for radio emission from compact galaxies. See Abstr. 158.044.

Magnetized accretion disks and the radio outbursts of 3C 120 and Cygnus X-3.   See Abstr. 158.067.

The log N–log S curve for 3CR radio galaxies and the problem of identifying faint radio galaxies. See Abstr. 158.076.

Explosions in galactic nuclei and the formation of double radio sources.   See Abstr. 158.077.

The problem of spiral galaxies and satellite radio sources.   See Abstr. 158.084.

Spectrophotometric observations of N galaxies at large redshift: PKS 0353+027, 3C 99, 3C 467. See Abstr. 158.107.

The distant N galaxy 3C 318. See Abstr. 158.108.

Radio galaxies.   See Abstr. 158.122.

Strong nonthermal radio emission from galaxies. See Abstr. 158.134.

Radio observations of neutral hydrogen in galaxies. See Abstr. 158.135.

M31 at 11 cm.   See Abstr. 158.142.

Millimetre emission from active galaxies. See Abstr. 158.143.

Time dependent emission line profiles in the radially streaming particle model of Seyfert galaxy nuclei and quasistellar objects.   See Abstr. 158.162.

Four-point optical energy distributions for faint BL Lacertae objects.   See Abstr. 158.301.

Photoelectric magnitudes and polarization data for possible BL Lacertae objects.   See Abstr. 158.305.

The peculiar galaxy NGC 404 and its surrounding area.   See Abstr. 158.306.

Variability of AO 0235 + 164 at 2.8 cm. See Abstr. 158.307.

Radio spectrum of the major outburst in the BL Lacertae object AO 0235 + 164.   See Abstr. 158.308.

Photometric and spectroscopic observations of the BL Lacertae object AO 0235 + 164.   See Abstr. 158.309.

OE 110: a new, faint BL Lacertae object. See Abstr. 158.311.

Variation d'éclat de OJ 287. See Abstr. 158.316.

OX-192: A new highly variable BL Lacertae object. See Abstr. 158.317.

Head-tail radio sources in clusters of galaxies. See Abstr. 160.008.

A Westerbork survey of rich clusters of galaxies. II. The luminosity function of bright cluster galaxies at 1415 MHz.   See Abstr. 160.009.

Head-tail radio sources in the cluster of galaxies Abell 1314.   See Abstr. 160.014.

A Westerbork survey of rich clusters of galaxies. III. Observations of the Coma cluster at 610 MHz. See Abstr. 160.032.

Intergalactic extinction and the quasar cut-off. See Abstr. 161.004.

Radio astronomy and cosmology. See Abstr. 162.070.

Pulsars

141.301 On the scattering and absorption of electromagnetic radiation within pulsar magnetospheres.
R. D. Blandford, E. T. Scharlemann.
Monthly Notices Roy. Astron. Soc., Vol. 174, 59 - 85 (1976).

The scattering and absorption of coherent radio emission by ultra-relativistic particles streaming outwards along the open field lines of a pulsar magnetosphere are investigated for incident frequencies in the guiding-centre frame less than or comparable with the particle gyrofrequency. On the basis of these calculations, it is concluded that: (1) induced scattering processes can influence the spectrum and polarization of the radio pulses if they are emitted well within the pulsar magnetosphere. (2) neither spontaneous scattering nor small-pitch-angle synchrotron emission seems to provide a satisfactory explanation for the optical pulses from NP 0532, (3) the avoidance of cyclotron absorption imposes important constraints on models of the radio emission process, and (4) resonant scattering of thermal radiation from the neutron star surface is probably not an important effect.

141.302 Self-consistent equilibria in the pulsar magnetosphere.
V. G. Endean.
Monthly Notices Roy. Astron. Soc., Vol. 174, 125 - 135 (1976).

For a 'collisionless' pulsar magnetosphere the self-consistent equilibrium particle distribution functions are functions of the constants of the motion only. Reasons are given for concluding that to a good approximation they will be functions of the rotating frame Hamiltonian only. This is shown to result in rigid rotation of the plasma, which therefore becomes trapped inside the velocity of light cylinder. The self-consistent field equations are derived, and a method of solving them is illustrated.

141.303 Hydrogen line absorption measurements on four pulsars. R. S. Booth, A. G. Lyne.
Monthly Notices Roy. Astron. Soc., Vol. 174, 53P - 58P (1976).

H I absorption spectra have been obtained for four distant pulsars ($>$ 2 kpc) PSR 0329+54, 0355+54, 1933+16 and 2319+60. The observation on PSR 0329+54 had higher resolution than previous measurements and demonstrated the presence of significant absorption in the Perseus arm, placing the pulsar at a distance of about 2.6 kpc. The mean electron density along the lines of sight to these pulsars is $\leqslant 0.025$ cm^{-3}.

141.304 A possible optical identification for PSR 0833–45.
B. M. Lasker.
Astrophys. Journ., Vol. 203, 193 - 195 (1976).

Deep photographs show a blue star with magnitude $\sim$23.7 about 1.''8 from the Molonglo position for PSR 0833–45. This star does not pulsate like the Crab optical pulsar, but a chance association with a field star is unlikely on the basis of color and star-counts. Most likely the object is physically associated with the pulsar.

141.305 Determining the stellar masses in the binary system containing the pulsar PSR 1913 + 16: is the companion a helium main-sequence star?
D. H. Roberts, A. R. Masters, W. D. Arnett.
Astrophys. Journ., Vol. 203, 196 - 201 (1976).

The authors determine the stellar masses and the angle of inclination for the system containing PSR 1913 + 16 in terms of the observable periastron advance and second-order relativistic effects, for the cases (1) in which classical contributions to the apsidal motion are negligible, and (2) in which the companion is a helium main-sequence star. Black holes, neutron stars, degenerate dwarfs, and helium main-sequence stars remain viable candidates for the companion at present.

141.306 Tidal friction in the binary pulsar system PSR 1913 + 16. S. A. Balbus, K. Brecher.
Astrophys. Journ., Vol. 203, 202 - 205 (1976).

It is shown that in the binary star system containing the pulsar PSR 1913 + 16, orbital-period changes produced by tidal friction may be significant. If the unseen companion is a white dwarf with a turbulent atmosphere, one expects orbital-period changes of order $(P^{-1} dP/dt) \approx 10^{-9}$ yr^{-1}. This is larger than the predicted period change due to gravitational radiation and other higher-order post-Newtonian effects. Detection of an orbital-period increase of this magnitude will almost certainly imply that the unseen companion is a white dwarf.

141.307 Direction of subpulse drifting within pulsar radio emission envelopes. M. Ruderman.
Astrophys. Journ., Vol. 203, 206 - 208 (1976).

It is shown how the direction of drifting subpulses (as well as their subperiods) within pulse envelopes is consistent with observations for certain pulsar models. These have localized electron-positron discharges in polar gaps together with a large angle between the spin axis and the magnetic dipole moment.

141.308 Current flow in pulsar magnetospheres.
A. Cheng, M. Ruderman, P. Sutherland.
Astrophys. Journ., Vol. 203, 209 - 212 (1976).

When particle flow is mainly outward through the light cylinder of a rotating pulsar magnetosphere, the resulting current flow leads to very large potential drops along field lines wherever $\Omega \cdot \mathbf{B}$ vanishes. For the Crab pulsar, the gap potential at breakdown is $\sim 10^{14}$ V, and the resulting radiation is mainly hard X-rays and γ-rays up to 10^{11} eV. The latter are expected to be emitted almost perpendicular to the magnetic axis and probably, therefore, should be observed almost 90$°$ earlier in phase than much lower energy pulsar emission.

141.309 Crust-breaking by neutron superfluids and the Vela pulsar glitches. M. Ruderman.
Astrophys. Journ., Vol. 203, 213 - 222 (1976).

Anderson's suggestion of neutron superfluid vortex pinning and unpinning in the crust lattice of a neutron star as the origin of pulsar timing noise is combined with existing estimates of the lattice breaking strength and creep. For the Vela pulsar such models give large sudden jumps in frequency and in slowing down rates and an interval between glitches all of which are comparable to those observed. The qualitative differences between Vela and Crab glitches can also be understood.

141.310 Stability of a beam-plasma system against the excitation of the longitudinal mode around pulsars.
S. Hinata.
Astrophys. Journ., Vol. 203, 223 - 225 (1976).

The possibility of radio wave emission from the beam-plasma system around pulsars is investigated. The indirect generation of radio waves by first exciting the longitudinal mode is found unlikely, because the beam-plasma system is stable against the excitation of such a mode around the pulsar polar caps. This may be understood if we realize that the rate of excitation of the longitudinal mode is likely to be much less than the coherent curvature-radiation loss rate at any level of excitation.

141.311 Magnetospheric shock discontinuities in pulsars. I. Analysis of the inertial effects at the light cylinder. H. Ardavan.
Astrophys. Journ., Vol. 203, 226 - 232 (1976).

A relativistic version of the generalized Ohm's law is

derived, and on the basis of this equation the conditions for the validity of the magnetohydrodynamic approximation $E + (V/c) \times B = 0$ in the limit where $\gamma = (1 - V^2/c^2)^{-1/2} \gg 1$ are investigated (E and B are the electric and the magnetic fields and V is the streaming velocity of the plasma). It is found that, independently of the relative magnitudes of the inertial and the magnetic forces, this approximation is applicable to the corotating part of the magnetosphere of a pulsar. It is then proposed that a solution of the plasma-electromagnetic field equations consistent with these conditions is one which entails the presence of a steady shock discontinuity at the light cylinder.

141.312 Pulsar geometries. II. Decomposition of the radiation pattern.   L. Oster, W. Sieber.
Astrophys. Journ., Vol. 203, 233 - 244 (1976).

The model proposed in Paper I (see Abstr. 13.141.323) is extended to describe in detail the observed time structure of pulsar emission. It is shown with computer-generated examples that all data can be fitted to small variations of the model parameters. An attempt is made to separate the effect of geometrical parameters and their variations from the effect of differences in the physical emission mechanism.

141.313 New observations of pulsed X-ray emission from NP 0532.   H. L. Kestenbaum, W. Ku, R. Novick, R. S. Wolff.
Astrophys. Journ., (Letters), Vol. 203, L57 - L61 (1976).

Large-area proportional counters sensitive from 0.6 to 23 keV were used to observe about 3 min of pulsed X-ray emission from NP 0532 during two lunar occultation experiments. A detailed pulse profile with 65 μs resolution shows that the X-ray primary pulse shape is essentially identical to its optical counterpart. Spectral data on the pulsed X-ray emission are presented. A pulse-by-pulse examination of the data shows no evidence for temporal variability.

141.314 Timing of the Crab pulsar. I. Arrival times.   E. J. Groth.
Astrophys. Journ., Suppl. Ser., No. 293, Vol. 29, 431 - 442 (1975).

This paper presents the pulse arrival times obtained during five years of observation of the Crab nebula pulsar, PSR 0531 + 21. The reduction from topocentric to barycentric arrival times and errors affecting the results are discussed.

141.315 Timing of the Crab pulsar. II. Method of analysis.   E. J. Groth.
Astrophys. Journ., Suppl. Ser., No. 293, Vol. 29, 443 - 451 (1975).

This paper presents a new method of analysis especially designed for time series such as the Crab pulsar timing data. These data reflect a random process superposed on a smooth behavior described by a low-degree polynomial. In addition, the measurements are not equally spaced and not of uniform quality. With such data, conventional power-spectrum or autocorrelation function techniques cannot be used to study the random process, nor can conventional regression techniques be used to study the polynomial behavior. The method described here allows a relatively clean separation of the polynomial and random components.

141.316 Timing of the Crab pulsar. III. The slowing down and the nature of the random process.   E. J. Groth.
Astrophys. Journ., Suppl. Ser., No. 293, Vol. 29, 453 - 465 (1975).

The Crab pulsar arrival times are analyzed. The data are found to be consistent with a smooth slowing down with a braking index of 2.515 ± 0.005. Superposed on the smooth slowdown is a random process which has the same second

moments as a random walk in the frequency. Neither the braking index nor the strength of the random process show evidence of statistically significant time variations. There is a possibility that the random process contains a small component with the same second moments as a random walk in the phase. If so, a time scale of 3.5 days is indicated.

141.317 Pulsar atmospheric current loops.   E. A. Jackson.
Nature, Vol. 259, 25 - 26 (1976).

The basic difficulty with the theories of pulsar radiation is that there is as yet no self-consistent model of the large scale structure of the plasma atmosphere of the pulsar. The lack of such a model is perhaps most strikingly exemplified by the fact that there is no consistent explanation of how an aligned rotating magnetised neutron star can maintain a net zero current from its surface — that is, how either a finite current loop in its atmosphere, or an 'infinite' loop to the nebula, is established consistent with the space charge and the electromagnetic fields. This letter is limited to a discussion of how current loops are established when the rotational and magnetic axes are aligned.

141.318 Third discontinuity in the Vela pulsar period.   R. N. Manchester, W. M. Goss, P. A. Hamilton.
Nature, Vol. 259, 291 - 292 (1976). – Letter.

141.319 Pulsars and magnetic monopoles.   G. Kalman, D. ter Haar.
Nature, Vol. 259, 467 - 468 (1976). – Letter.

141.320 Direct observation of pulsar microstructure.   D. C. Ferguson, D. A. Graham, B. B. Jones, J. H. Seiradakis, R. Wielebinski.
Nature, Vol. 260, 25 - 27 (1976).

Hankins' observations of pulsars exhibiting structure on time scales as short as a few μs have set stringent requirements on the theories which attempt to explain the pulsar emission mechanism, This letter reports high-frequency observations of PSR 1133 + 16 where the dispersion broadening and the time resolution were short enough to reveal microstructure on a time scale of 14 μs.

141.321 Test of pulsar acceleration mechanisms.   D. Morris, V. Radhakrishnan, C. Shukre.
Nature, Vol. 260, 124 - 126 (1976).

An elegant mechanism to explain the high velocities of pulsars has been proposed by Harrison and Tademaru. Based on the observations that the interpulses seen in some pulsars are located asymmetrically between the main pulses, they assume that pulsar magnetic fields arise in general from off-centred dipoles. The asymmetric radiation reaction produced by such an inclined off-set rotating dipole accelerates the pulsars in a direction parallel to their spin axes and gives the observed velocities. Any verification of the above hypothesis hinges on the possibility of determining, and comparing, the angle projected by the spin axis on the sky S with the observed direction of motion of the pulsar V. According to Harrison and Tademaru's model, these directions would coincide ($V = S$). The authors point out here that a measure of the spin axis projected direction is given by the intrinsic angle of polarisation of the radio emission at the centre of the integrated pulse profile, and that these polarisation data do not support the hypothesis in most cases.

141.322 Electrodynamic coupling between pulsars and surrounding nebulae.   M. Dobrowolny, A. Ferrari.
Astron. Astrophys., Vol. 47, 97 - 104 (1976).

A study is presented of collective plasma processes by which pulsars can energetically support young supernova remnants. The authors show that many of the observed features of the Crab nebula can be adequately interpreted in terms of a parametric interaction between the low-frequency electromagnetic wave emitted by the pulsar in the oblique rotator model and a relativistic wind of charged particle leaking from the pulsar's inner magnetosphere. In particular they show that there is a relativistic parametric resonant coupling of the strong wave with electrostatic and electromagnetic modes, which would be responsible for the formation of wisps.

141.323 Observational constraints on pulsar binary motion.
D. Q. Lamb, F. K. Lamb.
Astrophys. Journ., Vol. 204, 168 - 186 (1976).

The authors have carried out a quantitative investigation of the constraints on pulsar binary motion implied by current timing observations. They show that the high precision of measured periods and period derivatives does not by itself provide evidence against binary motion. The fact that no pulsar for which the period derivative has been measured shows that a secular speeding up excludes, with high probability, binary periods between 10 days and 100 years for companions more massive than 0.1 $M_\odot$. The results are summarized as excluded regions in a plot of companion mass versus binary period. Those future observations which would be most useful for detecting binary motion are indicated.

141.324 Theory of the polarization of pulsar radio radiation.
W. J. Cocke, A. G. Pacholczyk.
Astrophys. Journ., (Letters), Vol. 204, L13 - L15 (1976).

The theory of quasi-transverse propagation of radio radiation in pulsar magnetospheres leads to a model in which the intrinsic pulsar emission is linearly polarized, and circular polarization is produced as an effect of the quasi-transverse propagation. Various other observed features of the polarization are likewise explained.

141.325 The pulsar equation including the inertial term: its first integrals and its Alfvénic singularity.
H. Ardavan.
Astrophys. Journ., Vol. 204, 889 - 895 (1976).

The inertial term has been included in the pulsar equation and the first integrals of two components of the resulting equation which express the conservations of energy and angular momentum are explicitly derived. The Alfvénic critical point of these first integrals is shown to lie within the light cylinder, and the critical conditions which have to be satisfied at the Alfvénic and the light-cylinder singularities are specified. It is concluded that the force-free models of a pulsar magnetosphere are not tenable, and that, at the light cylinder, besides the inertial forces the pressure and the viscous forces must also be included in the pulsar equation.

141.326 Zur räumlichen Verteilung der Pulsare.
P. Biermann, M. Grewing.
Mitt. Astron. Ges., No. 38, p. 132 - 136 (1976). – Short report.

141.327 The generalized single-vector polarization model for pulsars. I. Theory. D. C. Ferguson.
Astrophys. Journ., Vol. 205, 247 - 260 = Louisiana State Univ. Obs. Contr. No. 112 (1976).

Single-vector models for pulsar polarizations are examined from a theoretical standpoint. Limits on the time between linear polarization maxima and minima for some nonrelativistic models are derived. The relativistic models are generalized to allow an arbitrary magnetic field direction in the corotating emission reference frame. Relations between all model parameters are given for the case where the magnetic field vector comes close to the line of sight.

141.328 The luminosity distribution and total space density of pulsars. D. H. Roberts.
Astrophys. Journ., (Letters), Vol. 205, L29 - L33 (1976).

A preliminary analysis has been made of the flux and distance distributions of the Hulse-Taylor sample of pulsars. The existence of a spatial boundary to the sample makes it possible to determine the differential distribution of luminosities $\phi(L)$ from the observed fluxes, essentially independent of the details of the spatial distribution. Knowledge of $\phi(L)$ enables us to estimate the total space density n of active pulsars, $n \approx 30$ kpc^{-3}. For reasonable choices of the lifetimes and scale height of pulsars, the birthrate required to maintain this population is about one per few hundred years in the Galaxy.

141.329 Plasma-wave interaction in non-aligned pulsar models. L. Mestel, G. A. E. Wright, K. C. Westfold.
Monthly Notices Roy. Astron. Soc., Vol. 175, 257 - 278 (1976).

The equations to a cold, non-dissipative plasma in the zero rest-mass limit are applied to rotating systems with simple geometries, and with the magnetic and rotation axes nonaligned. It is conjectured that in realistic models of pulsar magnetospheres, with a Sommerfeld boundary condition at infinity, a quasi-steady state is reached only when the particles have been accelerated to energies high enough for inertial terms to be important: the plasma modifies the vacuum wave, and the wave accelerates the particles to highly relativistic energies. A few comments are made also on current axisymmetric models.

141.330 Ghost remnants around pulsars.
T. Velusamy, M. R. Kundu.
Bull. Astron. Soc. India, Vol. 3, 35 (1975). – Abstract of a paper presented at the A.S.I. meeting 1975.

141.331 Pulsars and high density physics. A. Hewish.
Uspekhi fiz. nauk, Vol. 117, 201 - 210 (1975). In Russian. – Translation of the Nobel lecture on physics in 1974.

141.332 Decametric radio emission from four pulsars.
Yu. M. Bruck, B. Yu. Ustimenko.
Nature, Vol. 260, 766 - 767 (1976). – Letter.

141.333 The binary pulsar: preexplosion evolution.
J. C. Wheeler.
Astrophys. Journ., Vol. 205, 578 - 579 (1976).

Data on PSR 1913 + 16 including the rotation of periastron, $\dot\omega = 4°.24 \pm 0°.04$ yr^{-1}, which is interpreted as a purely relativistic effect, are used to estimate the masses in the system at various stages in its evolution.

141.334 Arrival-time analysis for a pulsar in a binary system.
R. Blandford, S. A. Teukolsky.
Astrophys. Journ., Vol. 205, 580 - 591 (1976).

A method is described for analyzing the arrival times of pulses from the binary pulsar PSR 1913 + 16, in terms of the orbital elements and their possible secular variations. Estimates are given for the times necessary to measure such secular changes and to detect various relativistic effects.

141.335 A modified Rickett diagram.
A. V. Pynzar', V. I. Shishov.
Astron. Zhurn. Akad. Nauk SSSR, Vol. 53, 288 - 290 (1976). In Russian. English translation in Soviet Astron., Vol. 20, No. 2.

The statistical relation between Δf, the characteristic frequency scale of the pulsar scintillation, and the parameter $\lambda^4 (MD)^2$, where λ = wavelength, MD = dispersion measure, is analysed.

141.336 Timing results for seven pulsars. G. E. Gullahorn, R. R. Payne, J. M. Rankin, D. W. Richards.
Astrophys. Journ., (Letters), Vol. 205, L151 - L153 (1976).

New pulse timing results are given for six of the pulsars reported by Davies et al. following one or two years of observations at 430 MHz. Additional observations of pulsar JP

1953 have resulted in a spin-down age measurement ($P/\dot{P}$) of $8.6 \pm 1.3 \times 10^9$ years. A speedup of about one part in 10^9 has occurred in pulsar 1906 + 00, commencing in 1974 April with a relaxation time of about 37 days.

141.337 Pulsar extinction.
P. A. Sturrock, K. Baker, J. S. Turk.
Astrophys. Journ., Vol. 206, 273 - 281 (1976).

Radio emission from pulsars is attributed to an instability associated with the creation of electron-positron pairs from γ-rays. The condition for pair creation therefore leads to an "extinction" condition. The relevant physical processes are analyzed in the context of the "PCFB" model, according to which radiation originates at the polar caps and magnetic field lines change from a closed configuration to an open configuration at the "force-balance" or "corotation" radius.

141.338 Relativistic plasma turbulence and its application to pulsar phenomena.   S. Hinata.
Astrophys. Journ., Vol. 206, 282 - 294 (1976).

A turbulent plasma model of pulsars which has the potential of providing a self-regulatory mechanism for producing an electron-positron plasma over the polar caps, as well as the coherency of the radio wave emission, is analyzed. Turbulent plasma properties including the kinetic and electrostatic energy densities, the wavelength of the most unstable mode, and the effective collision frequency due to the excited electric field, are obtained and applied to the pulsar situation. Since these properties depend on the momentum distribution of the plasma particles, model calculations have been carried out with simple momentum distribution functions. In addition to the above mentioned model, the author examines some wave propagation properties in a relativistic electron-positron plasma immersed in a strong magnetic field.

141.339 Further observations of the binary pulsar PSR 1913+16.   J. H. Taylor, R. A. Hulse, L. A. Fowler, G. E. Gullahorn, J. M. Rankin.
Astrophys. Journ., (Letters), Vol. 206, L53 - L58 (1976).

The authors report the results of more than a year's timing observations of the binary pulsar PSR 1913+16. The pulse arrival times are exactly those expected from a clock of well-defined period, moving in a Keplerian orbit with a constant rate of apsidal advance. The observations have yielded much-improved values for the parameters of the pulsar and its orbit, and are approaching the length of baseline needed to perform certain tests of gravitation theories.

141.340 Gamma-ray emission and nucleosynthesis of lithium by young pulsars.   D. D. Clayton, E. Dwek.
Astrophys. Journ., (Letters), Vol. 206, L59 - L62 (1976).

The authors propose that ^{7}Li is produced in the Galaxy primarily by α-α collisions surrounding newly born pulsars. About 10 percent of the pulsar energy losses are converted to medium-energy α-particles which collide in a dominantly He nebula. The problem of the origin of lithium would be solved by the scenario, and clear-cut tests by nuclear γ-ray astronomy are described.

141.341 Incompatibility of the continuous steady-state models of pulsar magnetospheres with relativistic magnetohydrodynamics.   H. Ardavan.
Monthly Notices Roy. Astron. Soc., Vol. 175, 645 - 651 (1976).

On the basis of the relativistic theory of magnetohydrodynamics, it is shown that, within the framework of the canonical steady-state axisymmetric model of a pulsar's magnetosphere, the conservation laws for angular momentum and energy do not in fact allow of a flux of angular momentum which is everywhere directed away from the neutron star, unless at the Alfvénic critical point of the magnetospheric flow there exists a discontinuity.

141.342 PSR 1055−52 – a pulsar resembling the Crab nebula pulsar.   P. M. McCulloch, P. A. Hamilton, J. G. Ables, M. M. Komesaroff.
Monthly Notices Roy. Astron. Soc., Vol. 175, 71P - 75P (1976).

Mean pulse profiles of the pulsar PSR1055−52 have been measured at frequencies of 308, 400, 635 and 1400 MHz; in addition measurements have been made of the strong linear polarization at 1400 MHz. The pulse shape is found to be complex with a strong interpulse approximately midway between the main components.

141.343 Pulsar signal processing.
T. H. Hankins, B. J. Rickett.
Methods comput. phys., Vol. 14, (see 003.007), 55 - 129 (1975).

Pulsar searches; Dispersion; Sampling, resolution, and average profiles; Polarization; Intensity variations with time; Intensity variations with frequency; Interstellar scattering and scintillation; Timing measurements.

141.344 The existence of an ultrarelativistic plasma beyond the Alfvén cylinder of a pulsar.   H. Ardavan.
Astrophys. Journ., Vol. 206, 822 - 830 (1976).

Previous studies of the steady-state magnetohydrodynamic models of pulsar magnetospheres have indicated that at the Alfvénic critical surface of the flow in such magnetospheres there exists a shock discontinuity. In this paper, the relativistic theory of magnetohydrodynamic shocks is employed to determine the state of the plasma across this discontinuity. It is found that an appreciable fraction of the magnetic energy of the plasma which crosses the shock is thermalized, and to within a first-order approximation, the mean particle energy in the shocked plasma is given.

141.345 A new pulsar atmospheric model. I. Aligned magnetic and rotational axes.   E. A. Jackson.
Astrophys. Journ., Vol. 206, 831 - 841 (1976).

A model is proposed for the plasma atmosphere about a rotating magnetized neutron star with aligned magnetic and rotational axes. The model is generated with the aid of a gedanken experiment in which the rotational velocity of the star is adiabatically increased. By this device it is established that the star will acquire a charge sufficient to stop the charge loss to the nebula, but not the local surface emission. The resulting monopole electric field causes plasma to flow across magnetic surfaces at large distances from the star, where it dominates the dipole magnetic field, thereby forming a current loop from the polar caps to polar annuli.

141.346 Pulsar spin down and cosmologies with varying gravity.   V. N. Mansfield.
Nature, Vol. 261, 560 - 562 (1976).

In this paper the author shows that some cosmologies with weakening gravity are in conflict with the measured spin down of the pulsar JP1953. If gravity is weakening, it must do so even more slowly than the limits set by radar ranging. For group theory cosmologies with strengthening gravity, he also gives a natural explanation for the lack of long period pulsars.

141.347 Particles acceleration by self mode-locked maser in pulsars.   J. C. Jodogne.
14th Intern. Cosmic Ray Conf., (see 012.011), Vol. 2, 435 - 438 (1975).

141.348 Traces for radiative processes in pulsar magnetosphere.   R. Liebermann.
Computing in plasma physics and astrophysics, (see 012.014), P7, 2 pp. (1976).

141.349 VLF electromagnetic radiation from pulsars in the

interstellar medium.
V. P. Dokuchaev, V. V. Tamojkin, Yu. V. Chugunov.
Astron. Zhurn. Akad. Nauk SSSR, Vol. 53, 527 - 533 (1976).
In Russian. English translation in Soviet Astron., Vol. 20,
No. 3.

Whistler mode radiation from pulsars in the interstellar
medium is considered. An oblique magnetic rotator is used as
a possible source of emission. Formulae are derived for the
total power of whistler mode emission from magnetic dipoles
and quadrupoles. Special attention is given to the dependence
of the radiated power upon frequency. This concept is used
for the explanation of the relation established between the
period and its first and second time derivatives for the Crab
pulsar.

141.350 A search for soft X-ray radiation from pulsars with
the Astronomical Netherlands Satellite.
J. Schrijver, J. Heise, A. C. Brinkman, E. H. B. M. Gronenschild, R. Mewe, A. J. F. den Boggende.
Astrophys. Space Sci., Vol. 42, 205 - 210 (1976). – See
012.016.

The authors present the results of a search for X-ray emission in the energy range 0.2–0.28 keV and 1–7 keV from a
number of radio pulsars, including Crab, Vela and the binary
pulsar PSR 1913+16, using the soft X-ray experiment aboard
ANS. Except for the Crab no pulsed flux has been found.
From the Vela pulsar the authors have detected continuous
flux in agreement with earlier observations. Upper limits are
given.

141.351 Comment on "Self-consistent solution for an
axisymmetric pulsar model".
L. G. Kuo-Petravic, M. Petravic.
Phys. Rev. Letters, Vol. 36, 686 - 688 (1976).

Implications of the results described in a former paper
(see 14.141.351) are further discussed and reasons are put
forward for the observed breakdown of the force-free assumption.

141.352 Proper motion of pulsars by radio interferometry.
D. C. Backer, R. A. Sramek.
Astron. Journ., Vol. 81, 430 - 432 (1976).

The authors present the first results of a pulsar proper
motion program using the NRAO 35-km interferometer at 2.7
GHz. PSR 1133 + 16 has a motion of $0\overset{''}{.}40$ yr^{-1} which is consistent with a result derived from pulse timing. PSR 1929 + 10
has a motion of $0\overset{''}{.}15$ yr^{-1}. Limits for three other objects are
given. Space velocities derived from the motion of PSR 1133 +
16 and PSR 1929 + 10 confirm previous indirect evidence
which suggested that pulsars are given a linear momentum
impulse at birth.

141.353 High frequency pulses from oblique magnetic rotators. A. Ferrari, E. Trussoni.
Mem. Soc. Astron. Italiana, Vol. 45, (see 012.017), 783 - 789
(1974).

The authors report about a model of pulsed high-frequency radiation emission by magnetized rotating collapsed
objects, in the following referred to as magnetic rotators. The
study must be considered in the framework of the interpretation of young pulsar and X-star observations; in particular the
authors deal with data on the Crab and Vela pulsars.

141.354 Pulsar slow-down and the temporal change of G.
H. Heintzmann, W. Hillebrandt.
Phys. Letters A, Vol. 54A, 349 - 350 (1975). – Abstr. in Phys.
Abstr., Vol. 79, A011216 (1976).

141.355 Upper limits for the visible counterpart of the
Hulse-Taylor binary pulsar. J. Kristian, K. D.
Clardy, J. A. Westphal.

Astrophys. Journ., (Letters), Vol. 206, L143 - L144 (1976).

The time-averaged intensity of visible pulses from PSR
1913 + 16 is fainter than $V = 23$ mag. The absence of visible
objects at the pulsar position on the Palomar Sky Survey
implies a rough limit of $M > 3$ for the absolute magnitude of
the pulsar's binary companion.

141.356 Photoproduction of gravitational radiation by
pulsars. G. Papini, S.-R. Valluri.
Canadian Journ. Phys., Vol. 53, 2312 - 2314 (1975). – Abstr.
in Phys. Abstr., Vol. 79, A020305 (1976).

141.357 Thermionic emission and pulsar magnetospheres.
N. D. Lubart.
Thesis Univ. Arizona, Tucson, USA, 134 pp. (1975). (Avail-
able from: Univ. Microfilms, Ann Arbor, Mich., USA. Order
No. 75-19585). – Abstr. in Phys. Abstr., Vol. 79, A027984
(1976).

141.358 Spontaneous acceleration of freely rotating helium II
and similar phenomena in pulsars.
J. S. Tsakadze, S. J. Tsakadze.
Proc. 14th internat. conf. on low temperature physics,
Amsterdam, Netherlands, p. 283 - 286 (1975). – Abstr. in
Phys. Abstr., Vol. 79, A030889 (1976).

141.359 New relativistic laboratory – a pulsar in a binary
star. J. Bicak.
Cesk. Casopis Fis., Sek. A, Vol. 25, 628 - 629 (1975). In
Czech. – Abstr. in Phys. Abstr., Vol. 79, A042306 (1976).

141.360 Pulsars. F. C. Auluck.
Journ. Phys. Educ., Vol. 1, No. 4, p. 17 - 20 (1974).

141.361 Emission from pulsars. I. A generalized single-vector
polarization model. II. A model for the sub-pulse
and integrated pulse behavior. D. C. Ferguson.
Thesis Arizona Univ., Tucson, USA, 332 pp. (1974). (Available from Univ. Microfilms, Order No. 75-14,216).

141.362 SAS-2 high-energy gamma-ray observations of the
Vela pulsar. D. J. Thompson, C. E. Fichtel,
D. A. Kniffen, H. B. Ögelman.
Separate print National Aeronaut. Space Administr., Green-
belt, Maryland, USA, Goddard Space Flight Center, 14 pp.
(1975). – See 14.141.318.

141.363 Pulsar magnetospheres: the aligned rotator.
C. D. Morris, Jr.
Thesis Rice Univ., Houston, Texas, USA, 164 pp. (1975).
(Available from: Univ. Microfilms, Order No. 75-22,045).

Quasars, pulsars and black holes.
See Abstr. 003.048.

Teilchenbeschleunigung in starken elektromagnetischen Feldern: theoretische Energiespektren.
See Abstr. 022.036.

Properties of matter in strong and superstrong magnetic fields. See Abstr. 022.037.

On the cascade process in strong magnetic and
electric fields under astrophysical conditions.
See Abstr. 061.027.

Stability of a degenerate electron plasma in a strong
magnetic field. See Abstr. 062.018.

On radiation production in a sheared medium.
See Abstr. 062.041.

Evidence on neutron star structure from pulsars and related objects.   See Abstr. 065.015.

The melting of neutron stars' crystalline cores and gammy-ray bursts.   See Abstr. 065.016.

Solid state physics and cooling of neutron stars. See Abstr. 065.040.

Evidence on neutron star structure from pulsars and related objects.   See Abstr. 065.041.

Relativistic formulation of the neutron starquake theory of pulsar glitches.   See Abstr. 065.099.

Dielectric properties of the magnetic surface of neutron stars.   See Abstr. 065.100.

A test of post-Newtonian conservation laws in the binary system PSR 1913+16.   See Abstr. 066.035.

Is Jupiter's magnetosphere like a pulsar's or earth's? See Abstr. 099.064.

Effect of asymmetric explosion on orbital elements of circular binaries.   See Abstr. 117.003.

Evolutionary aspects of the binary system containing the pulsar 1913+16.   See Abstr. 117.031.

Evolutionary aspects of the binary system containing the pulsar 1913+16.   See Abstr. 117.034.

Observations of the 21-cm hydrogen emission line in the direction of 23 southern pulsars.   See Abstr. 131.035.

The irregularity spectrum in interstellar space. See Abstr. 131.148.

Interstellar scattering of pulsar radiation. I. Scintillation.   See Abstr. 131.177.

The far side of the Crab nebula: electronographic and spectroscopic observations.   See Abstr. 134.004.

Evidence for time-dependent flux of $10^{11}-10^{13}$ eV gamma rays from NP 0532.   See Abstr. 142.135.

NP 0532 X-ray pulse structure. See Abstr. 142.136.

X-ray astronomy.   See Abstr. 142.157.

Gamma ray astronomy.   See Abstr. 142.158.

Spectral variability in the X-ray pulsar GX 1 + 4. See Abstr. 142.233.

Cosmic rays from pulsars.   See Abstr. 143.009.

Generation of cosmic ray electrons by pulsars. See Abstr. 143.102.

Infrared Sources

141.601   **5 GHz radio observations of LkHα 101, M1−82#1 and other infrared sources.**   S. Harris.
Monthly Notices Roy. Astron. Soc., Vol. 174, 601 - 607 (1976).

5 GHz radio continuum observations have been made with the Cambridge 5-km telescope of six small infrared sources and maps are presented for LkHα 101 and M1−82#1, the two detected. All six (except perhaps the one in IC 2087) are thought to be at early evolutionary stages; those not detected at radio wavelengths are probably protostars.

141.602   **Intermediate bandwidth spectrometry in the 10-micron region and its interpretation.**
J. A. Thomas, G. Robinson, A. R. Hyland.
Monthly Notices Roy. Astron. Soc., Vol. 174, 711 - 723 (1976).

Infrared photometry from 1.65 to 20 μ of several interesting sources is presented, including intermediate bandwidth observations at five wavelengths within the 8- to 14-μ window. Three sources have excesses in the 8- to 14-μ region with spectral characteristics quite different from those of 'silicate' emission. It is inferred that these are probably due to carbon-rich material, which is of particular interest in the case of the WC9 star IRC −20417 (Ve 2−45), and the peculiar early M star HR 3126.

141.603   **Deep ice absorption in a peculiar infrared source.**
M. Cohen.
Astrophys. Journ., Vol. 203, 169 - 170 (1976).

2−4 μ spectrophotometry with a resolution $\Delta\lambda/\lambda \approx 0.015$ is presented of an infrared source close to the outer projected edge of the Rosette nebula. A deep absorption feature is seen, centered near 3.0 μ, and this can be approximated by laboratory data on the extinction due to pure, spherical ice grains. With this identification, the central optical depth of the best-fit ice extinction curve is ~2.6, and the column mass of ice is ~2 × 10⁻⁴ gm cm⁻².

141.604   **Polarization studies of the infrared source CRL 2688 at visible wavelengths.**
J. J. Michalsky, R. A. Stokes, P. A. Ekstrom.
Astrophys. Journ., (*Letters*), Vol. 203, L43 - L45, with a correction Vol. 206, L73 (1976).

A study of the polarization of the entire nebulosity surrounding the unusual infrared source CRL 2688 is presented. Wavelength coverage from 3800 to 7800 Å reveals that the linear polarization increases monotonically with wavelength. Circular polarization was discovered and found to be a constant over the same wavelength range. The implications of these findings in terms of the scattering particles producing the polarization are discussed.

141.605   **Infrared observations of M17 S at medium spatial and spectral resolution.**
C. D. Andriesse, J. S. de Vries.
Astron. Astrophys., Vol. 46, 143 - 147 (1976).

Photometric data are presented for M17 S in a beam of 15″ at wavelengths of 8.1, 9.5, 12.2 and 19.6 μm. An unresolved object is found at $\alpha(1950) = 18^h17^m31\overset{s}{.}5$, $\delta(1950) = -16°13'25''$ with a luminosity of at least 400 $L_\odot$. There is circumstantial evidence that this object is a late type cocoon star. The peak brightness temperatures of the extended source are approximately given by $28 + 25\,\nu$ K with the frequency ν in units of 10^{13} Hz. Source profiles are similar for the various wavelengths and the spectrum is similar through the source. This makes it likely that the source is optically thin and that the grains have a constant temperature, probably close to 150 K. Because of its profile the source may be associated with a 10^6 yr old shock front.

141.606 **Classification of 831 Two-Micron Sky Survey sources south of +5°.** O. L. Hansen, V. M. Blanco. Astron. Journ., Vol. 80, 1011 - 1025 (1975).

Eight hundred thirty-one infrared sources located south of +5° have been selected from the Two-Micron Sky Survey (IRC), and studied in an objective-prism survey. The set consists of all 551 sources which were not identified in the IRC, 97 sources whose identification in the IRC is questionable, and 183 sources identified by their variable-star catalog name, but for which no classification is given in the IRC. The great majority (89%) of the sources studied are M stars, but 49 carbon stars and seven S stars as well as 17 peculiar stars with emission near 8600 Å have also been identified. If available, recent classifications or identifications from other investigators are included.

141.607 **The polarization of the infrared source CRL 2688 (the Egg nebula).** S. J. Shawl, M. Tarenghi. Astrophys. Journ., (*Letters*), Vol. 204, L25 - L28 (1976).

Polarization observations of CRL 2688 (the Egg nebula) show the polarization to vary from 40 percent at 3800 Å to 52 percent near 9400 Å. Observations of both components are also given. Polarization and reddening models seem to indicate graphite rather than silicate particles, with a column density of $\sim 10^9$ cm^{-2}.

141.608 **Radio continuum observations of NML Cygni.** P. C. Gregory, E. R. Seaquist. Astrophys. Journ., Vol. 204, 626 - 629 (1976).

An attempt to detect thermal radio emission from a compact circumstellar cloud about the infrared star NML Cyg has been carried out at three frequencies, 2.7, 8.1, and 10.5 GHz. Although positive results were obtained with single-dish observations at 10.5 GHz, the radio emission is not from a circumstellar cloud about NML Cyg. Instead the authors postulate that the emission is from an H II region. The interferometer observations at 2.7 and 8.1 GHz provide an upper limit on the radio emission from any compact circumstellar cloud about NML Cyg of 2.8 mJy.

141.609 **Infrared sources and star formation.** C. G. Wynn-Williams. Observatory, Vol. 96, 6 - 7 (1976).

141.610 **Neue Infrarotquellen in Sgr B2.** C. Thum, D. Lemke. Mitt. Astron. Ges., No. 38, p. 78 - 79 (1976). — Abstract.

141.611 **Relativistic effects in the binary pulsar PSR 1913 + 16.** G. Börner, J. Ehlers, E. Rudolph. Mitt. Astron. Ges., No. 38, p. 117 - 119 (1976). — Short report.

141.612 **Identifizierung von 22 Sternen aus dem IRC.** H. W. Humberg, G. V. Schultz. Mitt. Astron. Ges., No. 38, p. 243 (1976). — Abstract.

141.613 **Identifizierung von "unidentifizierten" OH-Quellen mit IR-Sternen.** G. V. Schultz, E. Kreysa. Mitt. Astron. Ges., No. 38, p. 244 - 245 (1976). — Abstract.

141.614 **Infrared polarization of CRL 2591.** M. Oishi, T. Maihara, K. Noguchi, H. Okuda, S. Sato. Publ. Astron. Soc. Japan, Vol. 28, 175 - 176 (1976).

A large polarization ($9.0 \pm 0.8\%$) at 2.2 μm was found for CRL 2591.

141.615 **The infrared emission of M17.** D. A. Harper, F. J. Low, G. H. Rieke, H. A. Thronson, Jr. Astrophys. Journ., Vol. 205, 136 - 143 (1976).

The infrared emission from M17 is discussed in the light of new middle- and far-infrared observations. The data include a 100 μ, 2.3 resolution map, 12 and 21 μ scans made with a 2.5 × 4.5 beam, and 10.6 μ, 5.7 resolution maps centered on the northern (G15.1−0.7) and southern (G15.0−0.7) source components. Narrow-band photometry of a compact (<5.7 diameter) object found in the 10.6 μ map of G15.0−0.7 shows a deep "silicate" absorption feature.

141.616 **CRL 2688: a post–carbon-star object and probable planetary nebula progenitor.** B. Zuckerman, D. P. Gilra, B. E. Turner, M. Morris, P. Palmer. Astrophys. Journ., (*Letters*), Vol. 205, L15 - L19 (1976).

The authors observed millimeter-wavelength emission toward CRL 2688 from H^{12}CN, H^{13}CN, CS, and HC$_3$N. The similarity of this emission and that from the molecular envelope of the carbon star IRC +10216 establishes that CRL 2688 is a post–carbon-star object. It appears probable that both of these objects will evolve into planetary nebulae. An evolutionary sequence leading from carbon stars to planetary nebulae is outlined.

141.617 **CRL 2688 and CRL 618: proto–planetary nebulae?** K. Y. Lo, K. P. Bechis. Astrophys. Journ., (*Letters*), Vol. 205, L21 - L25 (1976).

Emission from the $J = 1 \rightarrow 0$ transition of ^{12}C^{16}O has been detected at 2.6 mm from the peculiar infrared objects CRL 2688 (the Egg nebula) and CRL 618. The observed parabolic line-shape can be interpreted as optically thick emission from a uniformly expanding molecular envelope with a size smaller than the telescope beam. The presence of mass loss and the rapid evolution of the central stars suggest that CRL 2688 and CRL 618 may be proto–planetary nebulae. It is noted that IRC +10216, CRL 2688, CRL 618, and the planetary nebula NGC 7027 may represent different stages of an evolutionary sequence.

141.618 **A first look at the AFCRL infrared sky survey catalogue.** D. A. Allen, A. R. Hyland, A. J. Longmore. Monthly Notices Roy. Astron. Soc., Vol. 175, 61P - 64P (1976).

Users of the AFCRL infrared sky survey are cautioned that many of the presently unidentified sources at high galactic latitudes do not seem to be detectable at 2.2 μm.

141.619 **Detection of S III fine structure emission at 18.7 μm in galactic sources.** L. T. Greenberg, P. Dyal, T. R. Geballe. Bull. American Astron. Soc., Vol. 8, 289 (1976). — Abstr. AAS.

141.620 **Radio continuum observations of NML Cygni.** P. C. Gregory, E. R. Seaquist. Bull. American Astron. Soc., Vol. 8, 290 - 291 (1976). Abstr. AAS.

141.621 **Water emission from infrared stars.** D. F. Dickinson. Astrophys. Journ., Suppl. Ser., Vol. 30, 259 - 271 (1976).

Twenty-two new infrared stars with microwave water-vapor emission have been found, all but four of which are optically identified long-period variables. They are heavily reddened, late M stars that commonly show time variations. Hydroxyl emission is present in all but a few instances. Excited-state SiO emission is seen in many H$_2$O-infrared stars. Those that are Mira variables always have a visual change of more than 6 mag during their light cycle. Other optical and infrared properties are discussed.

141.622 Far-infrared spectral observations of M42 and M17.
    D. B. Ward, B. Dennison, G. E. Gull, M. Harwit.
Astrophys. Journ., *Letters*, Vol. 205, L75 - L77 (1976).

Coarse-resolution spectra of M17 and M42 between 42 and 115 μ have been obtained at aircraft altitudes. M42 has a very smooth spectrum well fitted by a 100 K gray body, while M17 shows a strong far-infrared excess beyond 70 μ.

141.623 Brackett-α emission in the Becklin-Neugebauer object.   G. L. Grasdalen.
Astrophys. Journ., *Letters*, Vol. 205, L83 - L85 (1976).

The Brackett-α line has been detected in the Becklin-Neugebauer object. A minimum estimate of 17 mag for the visual extinction suffered by the object is derived. Models for the object, assuming the line arises in a proto-H II region or a circumstellar shell, are discussed.

141.624 On the nature of IRC +10420.   E. R. Craine, W. J. Schuster, S. Tapia, F. J. Vrba.
Astrophys. Journ., Vol. 205, 802 - 806 (1976).

The spectral and infrared similarity of IRC +10420 and η Carinae has prompted an investigation into the significance of this resemblance. Long-term light-curve data, photometric and polarimetric observations have been obtained. Analysis of these data strongly suggest that IRC +10420 is not an η Carinae-like object.

141.625 1.3 to 2.5 micron spectra of MWC 349 and LkHα 101.   R. I. Thompson, M. A. Reed.
Astrophys. Journ.,(*Letters*), Vol. 205, L159 - L161 (1976).

Infrared spectra of MWC 349 and LkHα 101 have been obtained which show hydrogen and helium emission lines. The strength of these lines is consistent with H II region models derived from optical observations. No evidence for a cool companion is found in either object. A clumpy model for the dust and gas distribution is postulated.

141.626 A study of the velocity pattern of maser emission from infrared stars.   S. Kwok.
Journ. Roy. Astron. Soc. Canada, Vol. 70, 49 - 66 (1976).

Model spectra for OH and H_2O maser emission associated with infrared stars are constructed under the assumption of total saturation and on the basis of a symmetrically expanding envelope. The OH abundance in the circumstellar envelope is controlled by grain surface reactions in the outer region where chemical exchange reactions have ceased to operate. Observable quantities such as line shapes and angular sizes are related to the pumping parameters.

141.627 A binary model for the infrared source HD 101584.
    R. M. Humphreys.
Astrophys. Journ., Vol. 206, 122 - 127 (1976).

HD 101584 is a peculiar infrared source with the visual spectrum of a luminous F supergiant. Sufficient velocity data are available to yield a preliminary orbit with a period of about 3.5 years and a mass function of 0.25 $M_\odot$. The excess infrared radiation is attributed to an unseen cool companion which has filled its Roche lobe and is losing mass. Limits on the masses and radii of the two stars are derived.

141.628 Search for infrared stars in open star clusters.
    R. A. Vardanyan.
Astrofizika, Vol. 11, 351 - 354 (1975). In Russian.
English translation in Astrophysics, Vol. 11, No. 2.

16 infrared stars with $R-I \geqslant +3.0$ are detected in the open star clusters NGC 225, 654, 6834, 6871, 6913, 7031, 7128, IC 1805 and IC 4996 in a search of 58 open star clusters made with the image tube YM-92. It is shown that infrared stars are found in O-type clusters but are practically not detected in B-clusters.

141.629 Infrared sources in molecular clouds.
    N. Z. Scoville, J. Kwan.
Astrophys. Journ., Vol. 206, 718 - 727 (1976).

As a model for infrared sources in molecular regions, the authors calculate the radiative transfer in a dust cloud surrounding a central star. The grain temperature at each radius r is determined from radiative equilibrium in the radiation of both the star and the other circumstellar dust. In most clouds which are optically thick to the stellar photons, heating by the dust reradiation is dominant.

141.630 8 to 13 micron spectrophotometry of compact sources in NGC 7538.   S. P. Willner.
Astrophys. Journ., Vol. 206, 728 - 734 (1976).

Three infrared and radio sources in NGC 7538 have been observed with 1 percent spectral and 5″ spatial resolution. The very small southern source shows a deep silicate absorption feature at 9.7 μ while the northern extended source shows little or no silicate absorption, but has a 12.8 μ emission line from Ne II. The data imply that the compact source has a gas-to-dust ratio of 75 inside the ionized region.

141.631 An observational study of the AFCRL Infrared Sky Survey. I. Limited ground-based survey and results from preliminary catalog.   F. J. Low, R. F. Kurtz, F. J. Vrba, G. H. Rieke.
Astrophys. Journ., (*Letters*), Vol. 206, L153 - L155 (1976).

Over 1000 square degrees of sky have been surveyed down to mag. -1.2 at 11 μ without finding new sources. Only 27 of 162 sources in the preliminary AFCRL Infrared Sky Survey could be verified from the ground. The 27 objects are concentrated along the galactic plane, and the remainder are widely distributed.

141.632 An observational study of the AFCRL Infrared Sky Survey. II. Present results of a new program to study the final catalog.   M. J. Lebofsky, S. G. Kleinmann, G. H. Rieke, F. J. Low.
Astrophys. Journ., (*Letters*), Vol. 206, L157 - L160 (1976).

An improved ground-based verification program has found only 34 of 281 'new' sources in the final AFCRL catalog. The 34 cold objects are concentrated along the galactic plane, and the remainder are widely distributed.

141.633 A search for anonymous AFCRL infrared sources.
    R. D. Gehrz, J. A. Hackwell.
Astrophys. Journ., (*Letters*), Vol. 206, L161 - L164 (1976).

Ten of 47 anonymous AFCRL infrared sources were confirmed during a recent search. The confirmed sources could represent different stages in the evolution of a single class of stars. It is suggested that these objects might be related to disk population giants or to planetary nebulae.

141.634 Possible infrared counterpart of MXB1730−335.
    W. Wamsteker.
IAU Circ., No. 2954 (1976).

141.635 Possible infrared counterpart of MXB1730−335.
    D. E. Kleinmann.
IAU Circ., No. 2959 (1976).

141.636 Interesants infrasarkanais objekts: oglekla zvaigzne CIT 6 jeb RW LMi.   A. Alksnis.
Zvaigžņotā debess, 1975./76. gada ziema, p. 1 - 3.

141.637 Millimeter observations of galactic sources.
    P. E. Clegg, M. Rowan-Robinson, P. A. R. Ade.
Astron. Journ., Vol. 81, 399 - 406 (1976).

Observations at wavelengths close to 1 mm of 53 galactic source positions, including most of the available strong CO sources and infrared sources associated with compact H II regions, are described. Strip maps across W49 and W51, and a complete map of Orion, are presented, the latter suggesting a strong correlation between far-infrared radiation from dust and molecular line emission. Spectra of some of the sources are shown.

141.638 Infrared photometry of CRL 877 associated with the radio complex in the Monoceros R2 region.
S. Sato, Y. Kobayashi, K. Kawara, T. Maihara, H. Okuda. Publ. Astron. Soc. Japan, Vol. 28, 391 - 396 (1976).
A position of the bright infrared source CRL 877 is determined as R.A. (1950.0) = $6^h 5^m 22^s \pm 0^s.8$, Decl. (1950.0) = $-6°22'30'' \pm 10''$. The source is close to a radio complex in the Mon R2 association, involving the compact radio source G213.7−12.6, OH and H_2O masers, and molecular clouds. The infrared energy distribution of the source is very similar to that of the BN object in the Orion nebula with a deep 10-μm depression.

Infrared: the new astronomy.
See Abstr. 003.013.

Results of infrared reflectivity measurements on astronomically interesting silicates. See Abstr. 022.025.

Algorithm for the restoration of infrared raster scans. See Abstr. 031.220.

Far infra-red background astronomy at ground.
See Abstr. 031.271.

Infrared pumping for SiO masers.
See Abstr. 061.044.

The geometry of VY Canis Majoris derived from SiO maser lines. See Abstr. 064.016.

OH−IR stars. I. Physical properties of circumstellar envelopes. See Abstr. 064.033.

Infrared observations of young stars − VII. Simultaneous optical and infrared monitoring for variability. See Abstr. 113.002.

Infrared photometry and polarimetry of cool stars. III. See Abstr. 113.038.

Ground-state OH anomalies in the direction of the galactic centre. See Abstr. 131.002.

Infrared observations of the H_2O maser associated with the H II regions S 255 (IC 2162) and S 257. See Abstr. 131.018.

Infrared spectra of polyoxymethylene grains.
See Abstr. 131.029.

CO observations of the expanding envelope of IRC + 10216. See Abstr. 131.046.

Detection of H_2O maser emission from four infrared sources. See Abstr. 131.047.

OH−IR stars. II. A model for the 1612 MHz masers.
See Abstr. 131.108.

A far-infrared map of the Ophiuchus dark cloud region. See Abstr. 131.150.

Infrared photometry of a heavily reddened association in W 35. See Abstr. 131.501.

Further observations of ionized neon in the galactic centre. See Abstr. 131.502.

Infrared emission from S 157 A and S 252 A.
See Abstr. 131.508.

The radio and infrared spectrum of DR 21.
See Abstr. 131.510.

Recombination-line observations of W3(OH). I. A model for OH and H_2O emission. See Abstr. 131.517.

The role of dust in NGC 2024.
See Abstr. 131.518.

Thermische Eigenschaften von Staub in ausgedehnten Hüllen. See Abstr. 131.523.

Observations of five thermal sources at 15 GHz with the 5-km telescope. See Abstr. 131.532.

The puzzle of the high velocity water vapor features in W49. See Abstr. 131.543.

One arc-minute resolution maps of the Orion nebula at 20, 50, and 100 microns. See Abstr. 132.007.

High-resolution far-infrared observations of the galactic center. See Abstr. 155.032.

Late-type giants and supergiants in the galactic center. See Abstr. 155.033.

The sizes of the nuclei of galaxies at 10 microns.
See Abstr. 158.096.

The spectrum of IR radiation from dust clouds.
See Abstr. 158.112.

Errata

141.901 Errata: 'Precise positions of radio sources. II. Optical measurements' [Astrophys. Journ., Vol. 162, 391 - 398 (1970)]. J. Kristian, A. Sandage. Astrophys. Journ., Vol. 205, 308 (1976).

141.902 Errata: "Measurements of the linear and circular polarization of some compact radio sources at 5 GHz" [Monthly Notices Roy. Astron. Soc., Vol. 173, 9 - 20 (1975)]. M. Ryle, D. M. Odell, P. C. Waggett. Monthly Notices Roy. Astron. Soc., Vol. 175, 653 (1976).

141.903 Errata: "The radio polarization of quasars" [Monthly Notices Roy. Astron. Soc., Vol. 168, 137 - 162 (1974)]. R. G. Conway, P. Haves, P. P. Kronberg, D. Stannard, J. P. Vallée, J. F. C. Wardle. Monthly Notices Roy. Astron. Soc., Vol. 175, 653 (1976).

141.904 Erratum: 'The new binary pulsar and the observation of gravitational spin precession' [Astrophys. Letters, Vol. 16, 135 - 139 (1975)]. N. D. Hari Dass, V. Radhakrishnan. Astrophys. Letters, Vol. 17, 153 (1976).

142 X-Ray, Gamma-Ray Sources

142.001 Extragalactic X-ray binary systems – I. SMC X 1 = Sk 160. J. A. J. Whelan, D. T. Wickramasinghe. Monthly Notices Roy. Astron. Soc., Vol. 174, 29 - 45, with a correction, Vol. 175, 653 (1976).

The properties of the extragalactic X-ray binary system SMC X1–Sk 160 are discussed using constraints provided by X-ray and optical data, SMC membership and X-ray heating effects. Three different optical light curves are solved and the existence of multiple sets of solutions demonstrated.

142.002 The importance of accretion torques in pulsing X-ray sources. A. C. Fabian, J. E. Pringle. Monthly Notices Roy. Astron. Soc., Vol. 174, 25P - 28P (1976).

The authors compare the rates of change of pulse period, of a binary X-ray source, due to accretion torques and due to orbital motion. They show that these rates of change can be comparable, in particular for the sources with pulse periods of a few minutes. For this reason orbital parameters derived for pulsing transient sources may be subject to gross errors. Conversely, once the orbital motion is known, conclusions can be drawn about the structure of the neutron star itself.

142.003 X-ray observations of NGC 5128. J. P. Stark, P. J. N. Davison, J. L. Culhane. Monthly Notices Roy. Astron. Soc., Vol. 174, 35P - 39P (1976).

Observations of the active galaxy NGC 5128 (Centaurus A) have been made. The observations from Ariel V permit an accurate description of the X-ray spectrum to be made over the band 1.3–28.7 keV. The best-fit spectrum is a power law with photon index of -1.79 ± 0.02 for a line-of-sight hydrogen column density of $1.35 \pm 0.02 \times 10^{23}$ atom cm^{-2}.

142.004 Pulsed X-ray observations of Cen X-3 from Ariel-5. I. R. Tuohy. Monthly Notices Roy. Astron. Soc., Vol. 174, 45P - 50P (1976).

The 4.8-s X-ray pulsations from Centaurus X-3 were monitored by the MSSL collimated proportional counter on board Ariel-5 between 1975 January 18 - 27. Analysis of the source Doppler effect shows that the pulsation period of Cen X-3 decreased by 3.70 ± 0.04 ms during the preceding 2.3 yr. The average heliocentric binary period between 1972 October and 1975 January was 2.087129 ± 0.000007 days. Light curves of the 4.8-s pulsations in the 3–9 keV band are characterized by two pronounced peaks, in contrast with the single peak profiles observed by Uhuru.

142.005 OSO-7 observations of the X-ray nova 3U 1543–47. F. K. Li, G. F. Sprott, G. W. Clark. Astrophys. Journ., Vol. 203, 187 - 192 (1976).

The intensity and spectrum of the transient X-ray source 3U 1543–47 were measured on 13 occasions between 1971 November and 1972 December. The results, together with observations of the same object by the Vela and Uhuru satellites, can be described in terms of four qualitatively distinct phases in the X-ray light curve: (1) outburst (1971 July 26); (2) relatively steady intensity and spectrum (1971 August 1 to 1971 November 20); (3) rapid decrease in intensity and decrease in apparent temperature (1971 November 20 to 1972 January 6); (4) irregular intensity fluctuations with complex spectra (1972 January 6 onward). The term "X-ray nova" is proposed to characterize this and similar transient sources, and a new model for such X-ray novae is proposed.

142.006 X-ray nova A0620–00: celestial position and low- energy flux. R. Doxsey, G. Jernigan, D. Hearn, H. Bradt, J. Buff, G. W. Clark, J. Delvaille, A. Epstein, P. C. Joss, T. Matilsky, W. Mayer, J. McClintock, S. Rappaport, J. Richardson, H. Schnopper. Astrophys. Journ., (Letters), Vol. 203, L9 - L12 (1976).

The X-ray nova A0620–00 (nova Monocerotis 1975) has been observed with the SAS-3 satellite. The 1–10 keV intensity was observed to increase by a factor of 2.3 from August 8 to August 11. It reached and maintained a constant intensity of 1.7×10^{-6} ergs cm^{-2} s^{-1} from August 11 to August 13. A precise position was obtained with the SAS-3 modulation collimators on August 15. This led directly to optical and radio identifications.

142.007 Optical identification of A0620–00. F. Boley, R. Wolfson, H. Bradt, R. Doxsey, G. Jernigan, W. A. Hiltner. Astrophys. Journ., (Letters), Vol. 203, L13 - L14 = Contr. McGraw-Hill Obs., Kitt Peak, Arizona (1976).

An identification of the optical object corresponding to the new, transient X-ray source A0620–00 has been made. The early spectra of the object show no emission or absorption features.

142.008 Radio emission from the X-ray source A0620–00. F. N. Owen, T. J. Balonek, J. Dickey, Y. Terzian, S. T. Gottesman. Astrophys. Journ., (Letters), Vol. 203, L15 - L16 (1976).

Observations are reported of radio emission from the transient X-ray source A0620–00. Data obtained over the period 1975 August 15 to August 22 at frequencies between 1400 and 2695 MHz indicate a decaying radio event. The radio event appears to be quite different from those associated with recent galactic novae and is difficult to reconcile with a standard thermal model. A nonthermal event similar to those associated with Cygnus X-3, however, is consistent with the observations.

142.009 MX 1313 + 29: a compact source of very low energy X-rays in Coma Berenices. D. R. Hearn, J. A. Richardson, H. V. D. Bradt, G. W. Clark, W. H. G. Lewin, W. F. Mayer, J. E. McClintock, F. A. Primini, S. A. Rappaport. Astrophys. Journ., (Letters), Vol. 203, L21 - L24 (1976).

A compact X-ray source near the North Galactic Pole has been detected with the low-energy X-ray telescope aboard the SAS-3 satellite. The position is $\alpha(1950.0) = 13^h 14^m 0 \pm 0^m 5$, $\delta(1950.0) = + 29°15' \pm 18'$. The spectral temperature of the source is less than 10^6 K, and the energy flux at the earth is 3×10^{-10} ergs cm^{-2} s^{-1} above 0.1 keV. The flux appeared to be constant both on short time scales (10–1000 s). The possible identification of this source with the hot white dwarf HZ 43 is discussed.

142.010 An ultrasoft X-ray source in Coma Berenices. B. Margon, R. Malina, S. Bowyer, R. Cruddace, M. Lampton. Astrophys. Journ., (Letters), Vol. 203, L25 - L28 (1976).

The authors have observed an intense soft X-ray source with an extraordinary spectrum in Coma Berenices, 4° northeast of and unassociated with the Coma cluster of galaxies. Two spectra, obtained at different times in a sounding rocket flight, indicate that the source temperature in thermal models is less than 10^6 K; a power-law model requires photon power-law indices steeper than $n = -3$. The lack of bright stars or a supernova remnant in the error box implies that this may be a new class of soft X-ray sources.

142.011 The X-ray behavior of 3U 1700—37.
K. O. Mason, G. Branduardi, P. Sanford.
Astrophys. Journ., (Letters), Vol. 203, L29 - L33 (1976).
An observation with the Copernicus X-ray telescope reveals two kinds of flux variability in 3U1700—37. Fluctuations on a time scale of 1 hour are energy-independent in the 3.5—10.5 keV range, while changes of intensity with binary phase are caused by a variable absorbing column. The authors find that there is relatively more absorbing material in the line of sight following the passage of the X-ray source in front of its companion, but comparison with the light curve of Jones et al. indicates that this must be a variable phenomenon. The length of eclipse during the present observation was significantly shorter than when observed by Jones et al.

**142.012 A search for optical counterparts of nine galactic
X-ray sources.**
A. Davidsen, R. Malina, S. Bowyer.
Astrophys. Journ., Vol. 203, 448 - 454 (1976).
Results of a photographic and spectrophotometric search to $B \sim 18.5$ for the optical counterparts of nine galactic X-ray sources with good positions are reported. The sources included in this survey are 3U 1709—23, 1728—16, 1728—24, 1758—20, 1758—25, 1811—17, 1813—14, 1837+04, and 1908+00. Optical candidates for six of these sources are discussed, including blue objects which may be the optical counterparts of 1709—23 and 1728—16, a distant B star near 1908+00, and a very unusual strong emission line object which is probably associated with 1728—24 (GX 2+5). Coordinates, magnitudes, colors, spectral data, and finding charts are presented.

**142.013 New results from long-term observations of
Cygnus X-1.**   S. S. Holt, E. A. Boldt, P. J.
Serlemitsos, L. J. Kaluzienski.
Astrophys. Journ., (Letters), Vol. 203, L63 - L66 (1976).
Observations of Cyg X-1 between 1974 October and 1975 July reveal a persistent $5\overset{d}{.}6$ modulation of the 3 – 6 keV X-ray intensity, having a minimum in phase with superior conjunction of the HDE 226868 binary system. These data imply that the X-ray emission from Cyg X-1 arises from the compact member of HDE 226868.

**142.014 The 4.8 hour variation of Cygnus X-3 at high
X-ray energies.**   W. Pietsch, E. Kendziorra,
R. Staubert, J. Trümper.
Astrophys. Journ., (Letters), Vol. 203, L67 - L69 (1976).
During a balloon observation of Cygnus X-3 on 1975 February 20, an intensity variation has been found which is in phase with the low-energy X-ray 4.8 hour sinusoidal light curve. The measured relative amplitude in the energy range 32—64 keV is 0.37 (+0.31, −0.29). Compared with the results at lower energies there is no indication for an energy dependence of the relative amplitude up to 64 keV. The encountered low-intensity source spectrum is compared with previous measurements.

142.015 Discovery of a nonsolar extreme-ultraviolet source.
M. Lampton, B. Margon, F. Paresce, R. Stern, S. Bowyer.
Astrophys. Journ., (Letters), Vol. 203, L71 - L74 (1976).
The authors report the first observation of extreme-ultraviolet radiation from an extrasolar object. The data were obtained with a grazing-incidence telescope flown as part of the Apollo-Soyuz mission. The source is located in Coma Berenices, at $\alpha_{1950} = 13^h 13^m$, $\delta_{1950} = +29°$. Positive detections have been made in the 170—620 Å, 114—150 Å, and 55—150 Å wavelength bands. The intensity is 4×10^{-9} ergs cm^{-2} s^{-1} in the 170—620 Å band. Possible identification with the hot white dwarf HZ 43 is discussed.

142.016 Optical observations of WRA 977.
D. J. Bord, D. E. Mook, L. Petro, W. A. Hiltner.
Astrophys. Journ., Vol. 203, 689 - 693 (1976).
UBV photometry of WRA 977 on 36 nights between January and July 1974 reveals activity on a time scale of days at the 0.1 mag level; no activity greater than ~0.01 mag is observed on a time scale of minutes. Correlogram and periodogram analyses of the light curve reveal no persistent periodic variations of significant amplitude.

**142.017 Polarization of X-rays from Cygnus X-1: a test of
the accretion disk model.**
A. P. Lightman, S. L. Shapiro.
Astrophys. Journ., Vol. 203, 701 - 703 (1976).
Measurements of the polarization of X-rays from Cyg X-1 can yield significant information about the structure of the accretion disk which produces the observed spectrum. In particular, both the physical geometry (sensitive to large changes in the viscosity) and also the soft X-ray cutoff of the inner region of the disk have distinguishable effects on the observed polarization.

142.018 A cosmic gamma-ray burst on 1975 May 14.
D. Herzo, B. Dayton, A. D. Zych, R. S. White.
Astrophys. Journ., (Letters), Vol. 203, L115 - L118 (1976).
A cosmic gamma-ray burst is reported that occurred at 29309.11 s UTC, 1975 May 14. The initial rise time of the burst to 90 percent of maximum is 0.015 ± 0.005 s, and the duration is 0.11 s. Time structure down to the 5 ms resolution is seen. An integral energy distribution fitted to a power law $N(>E) = AE^{-\alpha}$ gives $A = 0.24 \pm 0.04$ and $\alpha = 1.3 \pm 0.2$. The total energy in the burst above 0.5 MeV is $2 \pm 0.5 \times 10^{-6}$ ergs cm^{-2}.

**142.019 The log$N$ – log$S$ relationship of extragalactic X-ray
sources and the X-ray background.**
A. Cavaliere, G. Setti.
Astron. Astrophys., Vol. 46, 81 - 85 (1976).
The source counts and the graininess of the background radiation derived from the UHURU survey are found to be incompatible. Either the counts do not extend below 3 ct/s as given, or they are severely contaminated. In either case the bulk of the keV background radiation must be due to sources other than those discovered thus far.

142.020 X-ray binaries and asymmetry of supernova explosions.
P. R. Amnuel (Amnuehl'), O. H. Guseinov (O. Kh. Gusejnov).
Astron. Astrophys., Vol. 46, 163 - 169 (1976).
Only ~ 10^{-3} of all close binaries remain bound following the collapse of one of the components. The bound binaries may show themselves as X-ray sources. About 100 strong X-ray sources and about 1000 weak ones are likely to exist in the Galaxy. The cause of the disruption of binaries is the asymmetry of a supernova explosion. The authors define how the parameters of a system change during an asymmetric explosion of one of the components. The expressions obtained are used to describe the change of the parameters of the Her X-1 and Cen X-3 type systems.

142.021 Binära röntgenkällor.   P. Lindroos.
Astron. Tidssk., Årg. 9, p. 21 - 32 (1976).

142.022 "Filin" investigates X-ray stars.   E. K. Sheffer.
Zemlya i Vselennaya, 1976, No. 1, p. 16 - 22. In Russian.

**142.023 On X-radiation of double systems containing Wolf-
Rayet-type stars.**   O. F. Prilutskij, V. V. Usov.
Astron. Zhurn. Akad. Nauk SSSR, Vol. 53, 6 - 9 (1976). In Russian. English translation in Soviet Astron., Vol. 20, No. 1.
It is shown that close binary systems must be rather in-

tensitive sources of X-radiation one or both components of which are young massive stars with strong outflow of matter from them (Wolf-Rayet stars and OB supergiants). X-radiation of such binary systems is stipulated by gas heating behind the front of shock waves formed as a result of collision of gas outflowing from one component either with the second star surface or with its magnetosphere or with gas outflowing from the second star. The most possible candidates of X-ray sources among double Wolf-Rayet stars are γ_2 Vel and V 444 Cyg.

142.024 Soft X-rays from IC 443.
S. Shulman, S. Naranan, G. Fritz, H. Friedman.
Astrophys. Letters, Vol. 17, 15 - 17 (1976).

A rocket survey, using large area proportional counters, has detected X-ray emission from IC443 in the 0.6–1 keV energy band. The flux is greater than is predicted by the spectra of Winkler and Clark (1974). If the emission is thermal, the low energy data can be made to fit by adding a second, lower temperature component.

142.025 X-ray emission from accretion on to white dwarfs.
A. C. Fabian, J. E. Pringle, M. J. Rees.
Monthly Notices Roy. Astron. Soc., Vol. 175, 43 - 60 (1976).

The authors have extended calculations by Hoshi and by Aizu to produce a self-consistent model for X-radiation from accreting, possibly magnetized, white dwarfs. To generate keV X-rays the flow must be radial on to the stellar surface. The authors expect X-ray luminosities to be in the range 10^{32}–10^{36} erg s^{-1}, and the spectra to be quasi-bremsstrahlung with $kT \sim 30$–100 keV and with a substantial low energy cut-off. The optical (bolometric) luminosity should be comparable to that emitted in the X-rays. If the white dwarf is magnetized a comparable, or greater, amount can be radiated as cyclotron emission in the infrared or optical.

142.026 The formation of binaries containing black holes by the exchange of companions and the X-ray sources in globular clusters.
J. G. Hills.
Monthly Notices Roy. Astron. Soc., Vol. 175, 1P - 4P (1976).

Clark's suggestion that the X-ray sources in globular clusters are binaries containing black holes or neutron stars is viable if one considers their formation by an exchange of companions with primordial binaries rather than by three-star encounters as originally suggested. In this process the black hole or neutron star makes a close encounter with a primordial binary, ejects one of its components, and becomes itself bound to the binary. The model is capable of producing the observed number of X-ray sources in globular clusters if a fraction $f_b = 10^{-3}$–10^{-2} of the ordinary stars in the clusters are binaries with semimajor axes less than 1 AU.

142.027 Further observations of the radio emission from Cygnus X-3 at 5 GHz.
M. McEllin.
Monthly Notices Roy. Astron. Soc., Vol. 175, 5P - 8P (1976).

Observations of Cygnus X-3 at 5 GHz during the periods 1974 September to 1975 January and 1975 June to November have revealed two further outbursts.

142.028 Transient short time periodicities in the optical emission from Cyg X-1.
G. Auriemma, D. Cardini, E. Costa, F. Giovannelli, M. Orciuolo, M. Ranieri.
Nature, Vol. 259, 27 - 29 (1976).

Temporal structure down to the millisecond time scale has been observed in the X-ray emission of Cyg X-1. This has been interpreted as evidence of instabilities in the accretion disk surrounding a black hole[2–4], lasting for approximately the orbital period of the infalling matter. The authors report here observations of modulated optical emission from this source with a period of ~83 ms. The modulated emission was detected in two sets of observations as transient events lasting ~10 min. This is the first evidence of strong emission at optical wavelengths from the X-ray source in the Cyg X-1–HDE 226868 system.

142.029 Transient X-ray source A1118−61.
P. J. N. Davison, P. W. Sanford.
Nature, Vol. 259, 98 (1976).

The transient X-ray source Ariel 1118−61 was discovered late in 1974. It has been suggested that this source may be associated with the long period Mira-type variable RS Cen. The authors have used the collimated proportional counter on board Copernicus to test this suggestion. No signal was detected above background during the whole of the observing session. The 90% confidence upper limit to the source strength in the energy band 2.5–7.5 keV is 3×10^{-11} erg cm^{-2} s^{-1}; this is 1.7% of the peak flux reported by Ariel V in the same energy band. Thus it is likely that the X-ray source A1118−61 is not connected with RS Cen, unless for some reason the X-ray generation was turned off during the time when Copernicus was observing the source.

142.030 New X-ray measurements of the Crab spectrum in the range 26 keV−1.2 MeV.
G. F. Carpenter, M. J. Coe, A. R. Engel.
Nature, Vol. 259, 99 (1976). − Letter.

142.031 B-emission stars and X-ray sources.
L. Maraschi, A. Treves, E. P. J. van den Heuvel.
Nature, Vol. 259, 292 - 293 (1976).

Several transient X-ray sources with decay times of the order of weeks to months have been reported, the discovery of four new sources of this type, in less than 1 yr of operation of the X-ray satellite Ariel V, indicates that these objects are rather common. Important for the understanding of this phenomenon is the optical identification of the sources: it has been proposed that, A1118−61 and A0535+26, coincide with bright B stars with emission lines ('Be' stars), while A0621−00 is identified with a completely different object, which showed a nova-like behaviour optically, simultaneous with the X-ray outburst. Here the authors discuss X-ray emission from systems containing Be stars.

142.032 Anti-correlated hard and soft X-ray intensity variations of the black-hole candidates Cyg X-1 and A0620−00.
M. J. Coe, A. R. Engel, J. J. Quenby.
Nature, Vol. 259, 544 - 545 (1976).

Extended spectral measurements are now available at more than one time epoch for the powerful Monoceros transient, A0620−00, discovered by Elvis et al., and the binary source Cyg X-1. The authors point out a basic similarity in the unusual spectral time variations of these sources. Both are black-hole candidates, Cyg X-1 by virtue of its orbital dynamics and A0620−00 as suggested by arguments based on the Eddington limit and distance estimates of the optical counterpart. Thus the X-ray emission mechanism could be common.

142.033 X-ray transient source at high galactic latitude and suggested extragalactic identification.
M. J. Ricketts, B. A. Cooke, K. A. Pounds.
Nature, Vol. 259, 546 - 547 (1976). − Letter.

142.034 Mass determination for the X-ray binary system Vela X-1.
J. A. van Paradijs, G. Hammerschlag-Hensberge, E. P. J. van den Heuvel, R. J. Takens, E. J. Zuiderwijk, C. De Loore.
Nature, Vol. 259, 547 - 549 (1976). − Letter.

142.035 Origin of the black hole in Cyg X-1.
J. C. Wheeler, G. A. Shields.
Nature, Vol. 259, 642 - 643 (1976).

The evidence that Cygnus X-1 is a black hole of mass $\gtrsim 9 M_\odot$ is not conclusive, but is sufficient to warrant some consideration of the implications of such an object for the late stages of stellar evolution. The authors argue here that such a black hole must form directly by implosion of a star of mass $\gtrsim 30 M_\odot$ rather than by accretion on to a neutron star. The Galaxy may contain $\sim 10^7$ such massive black holes.

142.036 X-ray observations of GX17+2 and GX9+9 by Aryabhata.
K. Kasturirangan, U. R. Rao, D. P. Sharma, M. S. Radha.
Nature, Vol. 260, 226 - 227 (1976).

X-ray observations on GX17+2 and GX9+9 were made using the proportional counter telescope on board the first Indian satellite 'Aryabhata'. The spectrum of GX17+2 can be fitted to an exponential function in the present case, and yields a characteristic kT value of 9.31 ± 0.3 keV. If the X radiation arises from bremsstrahlung in a hot plasma, the equivalent temperature of the emitting body was $(108 \pm 4) \times 10^6$ K at the time of observation. The spectral distribution of GX9+9 seems to be governed by a power law function with an exponent value of ~ 1.2, and is suggestive of a non-thermal mechanism for X-ray emission. In the energy interval of $2.5 - 10$ keV, the integrated energy flux is $(0.57 \pm 0.3) \times 10^8$ erg cm^{-2} s^{-1}. Comparison with other observations leads to the suggestion that X-ray emission from GX9+9 is variable.

142.037 Observations of Cyg X-1 from Aryabhata.
U. R. Rao, K. Kasturirangan, D. P. Sharma, M. S. Radha.
Nature, Vol. 260, 307 - 308 (1976).

Aryabhata, India's first artificial satellite, carried instruments to investigate celestial X rays in the energy range $2.5 - 155$ keV. The authors report here observations of Cyg X-1 made in the range $2.5 - 18.75$ keV. The authors found a hardening in the spectrum.

142.038 Spectrophotometric observations of the X-ray binary HD 153919 = 3U 1700-37. J. Dachs.
Astron. Astrophys., Vol. 47, 19 - 30 (1976).

The binary X-ray source HD 153919 = 3U 1700-37 has been examined for possible periodic variations in the expanding envelope around its O 6.5f-type primary using spectrophotometric scanner measurements of the Hα emission line profile and photographic spectra in the λ 3400 Å to 6600 Å wavelength region. Indications are found for recurrent variations in the wings of the Hα line profile. It is shown that absorption of the low-energy tail of the variable X-radiation from the secondary heating the gas in the envelope, as well as X-ray pressure, are the most probable causes for the weakening of the Hα emission line which frequently occurs when the X-ray source is visible. New measurements of the rotational broadening of the He II λ 4542 Å absorption line strengthen the view that the rotational angular velocity of the primary is only about half the orbital angular velocity of the secondary. This implies that the maxima in the tidal deformation of the star by its companion, observed in the optical light-curve of the system, are propagating at high speed through the photospheric layers along the equator of the primary. The consequences of the observations for models of the binary are discussed.

142.039 A two-temperature accretion disk model for Cygnus X-1: structure and spectrum.
S. L. Shapiro, A. P. Lightman, D. M. Eardley.
Astrophys. Journ., Vol. 204, 187 - 199 (1976).

The authors present a model for Cygnus X-1, involving an accretion disk around a black hole, which can explain the observed X-ray spectrum from 8 to 500 keV. In particular they construct a detailed model of the structure of an accretion disk whose inner region is considerably hotter and geometrically thicker than previous disk models.

142.040 A measurement of fluctuations in the X-ray back-

ground by Uhuru. D. A. Schwartz, S. S. Murray, H. Gursky.
Astrophys. Journ., Vol. 204, 315 - 321 (1976).

The authors have used the data from Uhuru to search for fluctuations in the 2–7 keV X-ray background. Fluctuations, intrinsic to the sky, are observed to be 3.0 percent of the mean X-ray background, over an effective solid angle of 0.004 sr. About one-third of the Uhuru sources with galactic latitude $|b| > 20°$ are identified with extragalactic objects. The authors conclude that a portion of the unidentified sources represents a class of objects at cosmological distances.

142.041 The long-term intensity behavior of Centaurus X-3.
E. J. Schreier, K. Swartz, R. Giacconi, G. Fabbiano, J. Morin.
Astrophys. Journ., Vol. 204, 539 - 547 (1976).

In 3 years of observation from Uhuru (1970 December – 1973 June), the X-ray source Cen X-3 appears to alternate between 'high states', with an intensity of 150 counts s^{-1} (2–6 keV) or greater, and 'low states' where the source is barely detectable. The time scale of this behavior is of the order of months, and no apparent periodicity has been observed. The analysis of two transitions between these states is reported. The data are consistent with a stellar wind accretion model and with different kinds of extended lows caused by increased wind density masking the X-ray emission or by decreased wind density lowering the accretion rate.

142.042 Studies of the average pulse shape of Centaurus X-3 in the 2–20 keV range. M. P. Ulmer.
Astrophys. Journ., Vol. 204, 548 - 550 (1976).

The author reports an analysis of the average pulse shape of Cen X-3 versus energy (2–20 keV) on time scales of hours to days derived from Uhuru observations during 1971 May 5–7 and 1971 December 18–20. The pulsed fraction varied with energy from 0.42 ± 0.02 to 0.80 ± 0.10 over the 2–20 keV range. The 2–6 keV pulsed fraction did not show statistically significant variations ($> 2\sigma$) over the phase of the 2.087 day eclipse cycle. Individual measurements of the 2–6 keV pulsed fraction ranged from 0.45–0.56.

142.043 Parameters of Sanduleak 160 (SMC X-1) by differential corrections. R. E. Wilson, A. T. Wilson.
Astrophys. Journ., Vol. 204, 551 - 554 (1976).

The authors have found the orbital inclination, mass ratio, luminosity ratio, and relative radius (optical component) for Sanduleak 160 (SMC X-1) from two independent sets of optical photometry. They find that the optical star is extremely near to filling its Roche lobe, and they suggest that this circumstance accounts for the high X-ray luminosity by way of an enhancement of the normal stellar wind. The formal computed uncertainty for the mass ratio (for the observations with smaller scatter) shows that the Osmer-Hiltner spectroscopic mass ratio ($M_0/M_x = 6.5$) differs by 2 standard deviations from the photometric value. A mass ratio of 6.5 places the X-ray component about at the borderline of possible neutron star masses. The luminosity ratio found from the X-ray heating of the optical star is in good agreement with the separate X-ray and optical luminosities.

142.044 Black holes in X-ray binaries: marginal existence and rotation reversals of accretion disks.
S. L. Shapiro, A. P. Lightman.
Astrophys. Journ., Vol. 204, 555 - 560 (1976).

The authors adopt the hypothesis that black holes in X-ray binaries are accreting gas from an accretion disk fed by a stellar wind emanating from the primary. The formation of such a disk is found to be marginal in the case of a steady, spherically symmetric stellar wind and may be controlled by random fluctuations in the wind parameters. Such fluctuations may cause reversals in the rotation direction of the disk. Reversals in a disk around a rapidly rotating black hole lead to two distinct intensity states and may be associated with the two states observed for Cyg X-1.

142.045 The transient periodic X-ray source in Taurus, A0535+26. H. Bradt, W. Mayer, J. Buff, G. W. Clark, R. Doxsey, D. Hearn, G. Jernigan, P. C. Joss, B. Laufer, W. Lewin, F. Li, T. Matilsky, J. McClintock, F. Primini, S. Rappaport, H. Schnopper.
Astrophys. Journ., (*Letters*), Vol. 204, L67 - L71 (1976).

Light curves of the 104 s periodicity in the transient X-ray source in Taurus (A0535+26) are presented for six energy intervals in the range 1–35 keV for the period 1975 May 30 - June 2. The pulse structure ranges from an apparently simple modulation at higher energies to a very complex pattern at lower energies. No Doppler shift is observed in the 104 s pulse period during the three days of observations. This places severe constraints upon possible binary orbital motion. Upper limits on the power at other periodicities are $\lesssim 10$ percent for 2 ms – 2 s and $\lesssim 2$ percent for 2 s – 2000 s.

142.046 High-energy X-ray observations of the transient source A0535+26 from a balloon-borne telescope. G. R. Ricker, A. Scheepmaker, J. E. Ballintine, J. P. Doty, G. A. Kriss, S. G. Ryckman, W. H. G. Lewin.
Astrophys. Journ., (*Letters*), Vol. 204, L73 - L76 (1976).

X-ray observations in the energy range 18–150 keV of the transient X-ray source A0535+26 were conducted from a balloon-borne telescope on 1975 June 1. Light curves for the ~103.8 s periodic emission were measured in seven energy bands. The amplitude at minimum decreases with increasing energy, going from ~45 percent of the amplitude at maximum at 20 keV to $\lesssim 10$ percent of the amplitude at maximum at 100 keV. The energy spectrum averaged over the light curve is fitted by an exponential with $kT = 17.6 \pm 0.6$ keV.

142.047 The positions and proper motions of HZ Herculis and 12 neighboring stars. G. Gatewood, S. Sofia.
Astrophys. Journ., (*Letters*), Vol. 204, L89 - L90 (1976).

The positions and proper motions of HZ Her, the optical counterpart of the X-ray source Her X-1, and 12 field stars are determined in the system of the FK4. The proper motion of HZ Her is consistent with that expected from a relatively low peculiar velocity object located at the ~3 kpc distance estimated for Her X-1.

142.048 An opaque shell around Hercules X-1? R. McCray, F. K. Lamb.
Astrophys. Journ., (*Letters*), Vol. 204, L115 - L118 (1976).

The authors suggest that the observations of intense soft X-rays from Her X-1 imply the existence of a centrifugally supported gas shell of radius ~2–7 × 10⁸ cm partially surrounding the neutron star that absorbs a substantial fraction of the hard X-ray luminosity and reradiates it as soft X-rays.

142.049 The spectrum of Cyg X-1 – a theoretical model. J. A. de Freitas Pacheco, J. E. Steiner.
Astrophys. Space Sci., Vol. 39, 487 - 494 (1976).

The authors have calculated the spectrum of Cyg X-1 under the assumption that the radiation originates in a disk around a 11 $M_\odot$ black-hole. Supersonic turbulence prevails in the outer parts of the disk and electron-electron bremsstrahlung appears to be responsible for the maintenance of the temperature at a level less than 10^{10} K near the inner edge of the disk. The theoretical spectrum gives the best fit with the observations if the Reynolds number is about 1200.

142.050 Der Röntgen-Crab: Ergebnisse der Beobachtungen von Mondbedeckungen. E. Kendziorra, R. Staubert, J. Trümper, C. Reppin.

Mitt. Astron. Ges., No. 38, p. 108 - 112 (1976). – Short report.

142.051 The 4.8 hour variation of Cygnus X-3 at high X-ray energies. W. Pietsch, E. Kendziorra, R. Staubert, J. Trümper.
Mitt. Astron. Ges., No. 38, p. 112 (1976). – Abstract.

142.052 Massenbestimmung in Röntgendoppelsternen aus Verformungslichtwechsel und Bedeckungsdauer. H. Mauder.
Mitt. Astron. Ges., No. 38, p. 112 - 117 (1976). – Short report.

142.053 Bremsstrahlung and the spectra of cosmic gamma bursts. U. Anzer, G. Börner.
Mitt. Astron. Ges., No. 38, p. 121 - 123 (1976). – Short report.

142.054 X-ray astronomy with Copernicus. J. H. Parkinson, J. L. Culhane, F. J. Hawkins, P. W. Sanford.
Space Research XV, (see 012.003), p. 663 - 679 (1975).

The Mullard Space Science Laboratory X-ray experiment on the Copernicus satellite has several unique capabilities and this paper reviews some recent results. Observations of Cen X-3, GX 5-1 and X-Per are discussed, and it is shown how new locations have enabled optical and radio identifications to be made. The results of mapping the Cas A and Puppis A supernova remnants, the pulsar PSR 0833-45 and the Perseus cluster are presented; these results are interpreted in terms of physical models for the extended X-ray sources, and the correlations of these models with the radio contours are shown for Cas A and Puppis A. Observations of X-ray variability are described for Cyg X-1, Cyg X-3, Her X-1 and Cen X-3. The Cyg X-1 and Cyg X-3 measurements have uniquely identified these sources with optical and infrared objects.

142.055 Geometry of inverse Compton gamma ray sources. R. Cowsik.
Space Research XV, (see 012.003), p. 715 - 720 (1975).

The author presents the results of a calculation of the energy and space distribution of gamma rays in the bandwidth 0.1–300 MeV arising from various processes in our Galaxy. The processes considered are the Compton scattering of the 2.7°K background photons and of starlight by cosmic ray electrons, and the decay of neutral pions produced in the nuclear interaction of cosmic rays with interstellar matter. Particular emphasis is given to the strong increase in the density of stellar photons close to the galactic centre. This leads to a peaking of the gamma-ray fluxes in the direction from around the centre. The calculated spectral shapes and the angular distributions are in agreement with the available observations in the bandwidth ~10–100 MeV.

142.056 Massa van Vela X-1 pulsar blijkt verrassend hoog. G. Hammerschlag-Hensberge, J. van Paradijs.
Zenit, 3e jaargang, p. 2 - 5 (1976).

142.057 On the ultrasoft X-ray background. A. Levine, S. Rappaport, R. Doxsey, G. Jernigan.
Astrophys. Journ., Vol. 205, 226 - 232 (1976).

The soft X-ray background has been studied down to energies of ~90 eV. The spectrum is found to be extremely soft. The background is both softer and more intense at high northern galactic latitudes than at low latitudes. Measured pulse height spectra are analyzed under the assumption of different source models. The results are consistent with an origin of the background in a hot interstellar plasma with a temperature of ~8 × 10⁵ K and a hot electron density of ~0.01 cm⁻³. Upper limits are set on the soft X-ray emission from several interesting stellar objects.

142.058 The distribution of the galactic compact X-ray

sources: a statistical analysis.
S. Sofia, F. Wesemael.
Astrophys. Journ., Vol. 205, 233 - 237 (1976).

A statistical analysis of the distribution in galactic longitude of the compact galactic X-ray sources listed in the Uhuru catalog shows that the distribution is not significantly inconsistent with a uniform space distribution of the sources throughout the galactic disk. In this scenario, however, there is a significant excess of sources at small distances from the sun. The authors conclude that a significant number of weaker sources should be found by a detector with a better angular resolution than the Uhuru, and that the concentration of sources at longitudes near the galactic center reflects a real concentration of the sources toward the center of the Galaxy.

142.059 **Long-term X-ray studies of Scorpius X-1. I. Search for binary periodicity.**
S. S. Holt, E. A. Boldt, P. J. Serlemitsos, L. J. Kaluzienski.
Astrophys. Journ., (Letters), Vol. 205, L27 - L28 (1976).

No evidence for modulation of the Sco X-1 3–6 keV intensity at the optical period of $0^{d}787313$ is found during one year of quasi-continuous observation. Any persistent X-ray modulation at this period must be less than 1 percent.

142.060 **Features in the brightness distribution and spectra of the soft X-ray background.**
P. A. J. de Korte, J. A. M. Bleeker, A. J. M. Deerenberg, S. Hayakawa, K. Yamashita, Y. Tanaka.
Astron. Astrophys., Vol. 48, 235 - 244 (1976).

Results on the soft X-ray sky background obtained in two rocket experiments have been combined to examine spatial and spectral features with enhanced statistical accuracy. The observed brightness distributions yield no compelling evidence for the presence of an extragalactic or a galactic halo component. Two-colour analysis of the spectral data shows unambiguously that the spectra of the soft background in the range 0.14–1.6 keV cannot be fitted by a thermal bremsstrahlung spectrum of a single temperature, also not when X-ray emission lines are taken into account.

142.061 **Ariel V sky survey observations of X rays from the globular cluster candidates 3U 2131 + 11 and MXO 513 – 40.**   J. P. Pye, B. A. Cooke.
Nature, Vol. 260, 410 - 412 (1976). – Letter.

142.062 **Transient X-ray sources: a discussion of the eccentric binary hypothesis and a model for A0620 – 00.**
Y. Avni, A. C. Fabian, J. E. Pringle.
Monthly Notices Roy. Astron. Soc., Vol. 175, 297 - 304 (1976).

The authors investigate the hypothesis that transient X-ray sources are binary systems with eccentric orbits. They conclude that this is unlikely to be a complete explanation and propose an observational test for the pulsing sources. They suggest an alternative explanation for A0620 – 00.

142.063 **Ariel 5 observations of the X-ray spectrum of the Perseus cluster.**   R. J. Mitchell, J. L. Culhane, P. J. N. Davison, J. C. Ives.
Monthly Notices Roy. Astron. Soc., Vol. 175, 29P - 34P (1976).

An X-ray spectrum of the Perseus cluster in the energy range 1.3–16 keV has been obtained with the collimated proportional counter on Ariel 5. An emission feature has been detected at about 7 keV of strenght 0.0035 ± 0.0004 photon $cm^{-2} s^{-1}$ (equivalent width 360 ± 50 eV). The existence of this feature, which is due to Fe XXV and Fe XXVI transitions, provides strong evidence for the presence of hot plasma in the cluster. The overall spectrum is well described by the bremsstrahlung emitted from an adiabatic hydrostatic atmosphere of hot gas in the gravitational potential well of the cluster.

142.064 **Infrared and X-ray observations of the decline of A 0620 – 00.**   O. Citterio, G. Conti, P. Di Benedetto, E. G. Tanzi, G. C. Perola, N. E. White, P. A. Charles, P. W. Sanford.
Monthly Notices Roy. Astron. Soc., Vol. 175, 35P - 38P (1976).

Measurements of the 1.65- and 2.2-μ flux and the 3 – 9 keV flux of the transient X-ray source A 0620 – 00 show a regular decline. The infrared flux is consistent with an extrapolation of a bremsstrahlung spectrum fitted to the X-rays if the source is self-absorbed in the infrared, a situation similar to Sco X-1.

142.065 **X-ray sources in the Aquila-Serpens-Scutum region.**   F. D. Seward, C. G. Page, M. J. L. Turner, K. A. Pounds.
Monthly Notices Roy. Astron. Soc., Vol. 175, 39P - 46P (1976).

The Ariel-5 sky survey detectors have been used to study X-ray sources within 10° of the galactic plane between longitudes $l^{II} = 20°$ and $l^{II} = 55°$. Seventeen sources have been found and locations and strengths derived. Seven of these sources were already known and this observation yields an improved position for four of them. Ten of the sources are new. One is possibly associated with the globular cluster NGC 6712, and one with the supernova remnant, W 50.

142.066 **Observations of the transient X-ray source at the galactic centre (A 1742 – 28).**   G. Branduardi, J. C. Ives, P. W. Sanford, A. C. Brinkman, L. Maraschi.
Monthly Notices Roy. Astron. Soc., Vol. 175, 47P - 56P (1976).

The transient X-ray source A 1742 – 28, which reached a maximum intensity of 6.5 ph $cm^{-2} s^{-1}$ in the energy range 3–8 keV on 1975 February 17, was repeatedly observed between 1975 February and September by the satellites Ariel 5, ANS and Copernicus. The light curve can be described with successive e-folding times $\tau_1 = 12^d$, for 40 days, $\tau_2 = 90^d$, for 140 days, followed by a more rapid decay. The mean spectrum shows no systematic variations from February to July, being well fitted by a power law spectrum of photon index, $n \simeq 3$, with soft X-ray absorption equivalent to a neutral hydrogen column of $\simeq 10^{23}$ atom cm^{-2}. The similarity with 3U 1543–47 suggests that the source is a binary system containing a compact object with variable mass transfer.

142.067 **X-ray and optical observations of Sco X-1.**   N. E. White, K. O. Mason, P. W. Sanford, C. Chevalier, S. A. Ilovaisky.
Bull. American Astron. Soc., Vol. 8, 293 (1976). – Abstr. AAS.

142.068 **A gamma–ray burst on May 14, 1975 from the Hercules-Cygnus region.**
R. S. White, D. Herzo, B. Dayton, A. D. Zych.
Bull. American Astron. Soc., Vol. 8, 293 (1976). – Abstr. AAS.

142.069 **The optical candidates of the X-ray source LMX-2.**   N. V. Vidal, M. Pakull.
Bull. American Astron. Soc., Vol. 8, 293 (1976). – Abstr. AAS.

142.070 **On X-ray emission of early massive stars in close binary systems.**
O. F. Prilutskij, V. V. Usov.
Astron. Tsirk., No. 854, p. 1 - 3 (1975). In Russian.

142.071 **UBV observations of the X-ray nova in Monoceros.**   S. van den Bergh.
Astron. Journ., Vol. 81, 104 - 105, 141 (1976).

UBV observations are given of the optical counterpart to the X-ray nova during the period 2–6 September 1975. Observations of field stars show that the foreground reddening reaches a value $E_{B-V} \sim 0.3$ at 1 kpc.

142.072 **X-ray emission from γ Cas.**
K. O. Mason, N. E. White, P. W. Sanford.
Nature, Vol. 260, 690 - 691 (1976). – Letter.

142.073 **On gamma-lines of discrete energies of cosmic origin.**
I. V. Ehstulin.
Issled. kosmich. luchej. Moskva, Nauka, 1975, p. 67 - 83. In Russian. – Abstr. in Referativ. Zhurn. 51. Astron., 4.51.630 (1976).

142.074 **Gamma rays from an external galaxy?**
C. J. Cesarsky, M. Cassé, J. Paul.
Astron. Astrophys., Vol. 48, 481 - 482 (1976).
The position of a nearby irregular galaxy discovered by Simonson coincides with that of an unexplained high energy gamma ray source. The authors examine the possibility that the two objects are related.

142.075 **Evidence for a 17-d periodicity from Cyg X-3.**
S. S. Holt, E. A. Boldt, P. J. Serlemitsos, L. J. Kaluzienski, S. H. Pravdo, A. Peacock, M. Elvis, M. G. Watson, K. A. Pounds.
Nature, Vol. 260, 592 - 594 (1976).
Cyg X-3 (3U 2030 + 40) has exhibited phenomena which are observationally unique among identified X-ray sources. A wide variation in X-ray spectra has been observed, including the identification of X-ray emission lines at some times, and consistency with a black body at others. The approximately sinusoidal 4.8-h variation is at a period far in excess of any rotation period which has been ascribed to the compact members of other binary sources, and at least four times shorter than any comparable orbital period. The authors present data indicating that a much longer periodicity of ~ 17 d is also characteristic of Cyg X-3.

142.076 **Hydrogen and helium lines in theoretical models of Scorpius X-1 and Cygnus X-2.**
M. Milgrom, J. I. Katz.
Astrophys. Journ., Vol. 205, 545 - 549 (1976).
The authors calculate the equivalent widths and profiles for the first three Balmer lines and for He II λ 4786 in the spectra of atmospheres heated from above by X-rays. They use parameters appropriate to the systems Sco X-1 and Cyg X-2. The crude features of the observed spectrum are reproduced, notably the fact that certain lines appear in emission and others in absorption. The finer details are shown also to be reproducible with reasonable values of the parameters. These results support the model in which the optical light from the systems is partly due to X-ray heating and reflection.

142.077 **Analysis of a Scorpius X-1 X-ray spectrum obtained with cooled silicon (Li) detectors.**
J. G. Laros, S. Singer.
Astrophys. Journ., Vol. 205, 550 - 555 (1976).
The authors discuss here an X-ray spectrum of Scorpius X-1 over the 0.5–20 keV range obtained with cooled Si(Li) semiconductor detectors during a rocket flight on 1971 May 22. The true source spectrum was essentially unmodified by the detector system above about 1 keV, and the statistical accuracy was also good, with over 5×10^4 source counts. This measurement therefore contains what may be the most reliable determination of spectral shape to date, and thus provides a severe test of Sco X-1 emission models.

142.078 **Tidal circularization of the binary X-ray sources Hercules X-1 and Centaurus X-3.**
M. Lecar, J. C. Wheeler, C. F. McKee.
Astrophys. Journ., Vol. 205, 556 - 562 (1976).
Neutron-star formation in a binary system must result in

an eccentric orbit, whereas the orbits of the X-ray sources Her X-1 and Cen X-3 are circular. The authors show how initially eccentric orbits can be circularized by tidal dissipation in acceptably short times, $\lesssim 10^7$ years for Her X-1 and $\lesssim 10^5$ years for Cen X-3.

142.079 **Long-term X-ray studies of Scorpius X-1. II. Evidence for flare-dominated intensity variations.**
S. S. Holt, E. A. Boldt, P. J. Serlemitsos, L. J. Kaluzienski.
Astrophys. Journ., *Letters*, Vol. 205, L79 - L82 (1976).
Evidence is found for shot-noise character in a large fraction of the 3–6 keV X-ray emission from Sco X-1. Almost all of the emission can be synthesized in terms of ~ 200 flares per day, each with a duration of $\sim 1/3$ day.

142.080 **Observations of linear optical polarization of X-ray sources.** N. M. Shakovskoj, Yu. S. Efimov.
Izv. Krymskoj Astrofiz. Obs., Vol. 54, 99 - 119 (1976). In Russian.
Linear optical polarization of the X-ray sources Sco X-1, Cyg X-1, Cyg X-2 and Her X-1 was measured. Observations have been carried out in 1970–1974 with a single-channel photoelectric polarimeter in U, B, V, O, R colours. A variance analysis was performed to check supposed variations of the observed polarization parameters with time. It is shown that the optical polarization of Sco X-1 and Cyg X-1 seems to be variable within a time scale of few days, superimposed on the interstellar polarization. The amplitude of variations is 0.2–0.4%. The observations available of Cyg X-2 and Her X-1 are insufficient to reveal possible variations of their polarization.

142.081 **Discovery of intense X-ray bursts from the globular cluster NGC 6624.** J. Grindlay, H. Gursky, H. Schnopper, D. R. Parsignault, J. Heise, A. C. Brinkman, J. Schrijver.
Astrophys. Journ., *(Letters)*, Vol. 205, L127 - L130 (1976).
A new type of time variation of cosmic X-ray sources has been found from the Astronomical Netherlands Satellite observations of the source 3U 1820–30 associated with the globular cluster NGC 6624. Two bursts in the ~ 1–30 keV X-ray intensity of this source are reported. Each displayed a rapid rise in flux ($\lesssim 1$ s) by a factor of 20–30 followed by a ~ 8 s exponential decay. Analysis for further source variability, energy spectra, and position is presented. The characteristics of these events may imply the existence of a collapsed core in the globular cluster.

142.082 **Scattering model for X-ray bursts: massive black holes in globular clusters.**
J. Grindlay, H. Gursky.
Astrophys. Journ., *(Letters)*, Vol. 205, L131 - L133 (1976).
A model for the X-ray bursts discovered from 3U 1820–30 (NGC 6624) is presented. The temporal and spectral variations observed in the bursts are interpreted as Compton scattering of a primary pulse in a hot cloud surrounding the X-ray source. The cloud parameters derived imply the presence of a collapsed core in the globular cluster source with a mass greater than several hundred solar masses.

142.083 **Evidence for an 11.2 day periodicity from Cygnus X-2.** S. S. Holt, E. A. Boldt, P. J. Serlemitsos, L. J. Kaluzienski.
Astrophys. Journ., *(Letters)*, Vol. 205, L143 - L145 (1976).
Evidence for a persistent $11^{\mathrm{d}}17 \pm 0^{\mathrm{d}}10$ period from Cyg X-2 is presented from one year of accumulated data from the Ariel-5 All-Sky Monitor. The effect is not a simple sidereal alias of a true source period close to one day.

142.084 **An introduction to the EXOSAT mission.**
B. G. Taylor.
ESRO colloquium X-ray astron. related topics, Noordwijk,

Netherlands, 25 - 26 Feb. 1975, p. 69 - 74 (1975).

142.085 Investigation of discrete sources of hard X-ray radiation aboard the artificial earth satellite Cosmos 428.
L. S. Bratolyubova-Tsulukidze, M. I. Kudryavtsev, A. S. Melioranskij, I. A. Savenko, B. Yu. Yushkov.
Izv. Krymskoj Astrofiz. Obs., Vol. 54, 320 - 323 (1976). In Russian. – Abstract of a paper presented at the seminar on "X-ray and gamma-ray astronomy" at the Crimean Astrophys. Obs., 1974, (see 012.009).

142.086 Galactic X-ray sources (a review).
R. A. Syunyaev.
Izv. Krymskoj Astrofiz. Obs., Vol. 54, 324 - 325 (1976). In Russian. – Abstract of a paper presented at the seminar on "X-ray and gamma-ray astronomy" at the Crimean Astrophys. Obs., 1974, (see 012.009).

142.087 Discrete sources of cosmic gamma radiation.
A. M. Gal'per, V. G. Kirillov-Ugryumov, B. I. Luchkov.
Izv. Krymskoj Astrofiz. Obs., Vol. 54, 328 - 335 (1976). In Russian. – Abstract of a paper presented at the seminar on "X-ray and gamma-ray astronomy" at the Crimean Astrophys. Obs., 1974, (see 012.009).

142.088 Observation of gamma radiation aboard the artificial earth satellite Cosmos 555.
S. A. Volobuev, L. V. Kurnosova, B. I. Luchkov, L. A. Razorenov, V. I. Ryabenkov, M. I. Fradkin.
Izv. Krymskoj Astrofiz. Obs., Vol. 54, 347 - 349 (1976). In Russian. – Abstract of a paper presented at the seminar on "X-ray and gamma-ray astronomy" at the Crimean Astrophys. Obs., 1974, (see 012.009).

142.089 On the X-ray sky – new stars.
G. S. Bisnovatyj-Kogan.
Priroda, 1976, No. 5, p. 49 - 51. In Russian.

142.090 OSO-7 observations of high galactic latitude X-ray sources. T. H. Markert, C. R. Canizares, G. W. Clark, F. K. Li, P. L. Northridge, G. F. Sprott, G. F. Wargo.
Astrophys. Journ., Vol. 206, 265 - 272 (1976).
Six hundred days of observations by the MIT X-ray detectors aboard OSO-7 have been analyzed. All-sky maps of X-ray intensity have been constructed from these data. A sample map is displayed. Seven sources with galactic latitude $|b^{II}| > 10°$, discovered during the mapping process, are reported, and upper limits are set on other high-latitude sources. The OSO-7 results are compared with those of Uhuru and an implication of this comparison, that many of the high-latitude sources may be variable, is discussed.

142.091 Improved position for the X-ray source associated with the globular cluster NGC 6441.
J. E. Grindlay, H. Schnopper, E. Schreier, H. Gursky, D. R. Parsignault.
Astrophys. Journ., (Letters), Vol. 206, L23 - L24 (1976).
The X-ray source 3U 1746−37 was observed by the hard X-ray experiment on the Astronomical Netherlands Satellite in March and September 1975. The one-dimensional position band (±2′, 90% confidence) obtained reduces the Uhuru error box area by a factor of 3 and includes the nucleus of the globular cluster NGC 6441.

142.092 Discovery of X-ray pulsations in SMC X-1.
R. Lucke, D. Yentis, H. Friedman, G. Fritz, S. Shulman.
Astrophys. Journ., (Letters), Vol. 206, L25 - L28 (1976).
Observations of SMC X-1 from an Aerobee rocket and an Apollo spacecraft have detected X-ray pulsations with a period of 0.716 s. The pulsed fraction in the 1.6−10 keV energy range is 25−35 percent. Evidence for significant pulse shape and pulsed fraction changes in the 0.6−1.6 keV range is also presented. The spectrum during both observations is fitted by a photon power law, $0.040 E^{-0.8}$.

142.093 Uhuru observations of the galactic plane in 1970, 1971, and 1972.
W. Forman, C. Jones, H. Tananbaum.
Astrophys. Journ., (Letters), Vol. 206, L29 - L35 (1976).
We have analyzed Uhuru observations of the galactic plane in 1970, 1971, and 1972. The great majority of the galactic X-ray sources are not "transient". Some of the so-called transient sources persist for long periods of time at an intensity of a few percent of their peak values. The data suggest that the transient sources may be quite similar to the other galactic sources with outbursts caused by changes in the accretion rate.

142.094 UCSD OSO-7 observations of the hard X-ray spectrum and variability of Centaurus A.
R. F. Mushotzky, W. A. Baity, W. A. Wheaton, L. E. Peterson.
Astrophys. Journ., (Letters), Vol. 206, L45 - L48 (1976).
Observations from the UCSD OSO-7 experiment show Cen A had a hard $E^{-1.2}$ number spectrum during 1972 July–August and 1973 March. The data indicate a 230 percent increase in the 10−100 keV flux. The data are equally consistent with various Compton-synchrotron models or with a $T > 200$ keV thermal-bremsstrahlung model with cutoff energy less than 4 keV.

142.095 Radial-velocity variations of Scorpius X-1 emission lines. D. J. Bord, R. J. Messina, D. E. Mook, W. A. Hiltner.
Astrophys. Journ., (Letters), Vol. 206, L49 - L52 (1976).
Radial-velocity measurements for the principal emission features in the Sco X-1 spectrograms of Mook, Hiltner, and Lynds are discussed. Plots of the radial velocities of Hα, Hβ, and He II λ4686 as a function of phase for the $0.^{d}787313$ period of Gottlieb, Wright, and Liller are presented. In spite of the large scatter in the data, the time-dependent behavior of the radial velocities for the Balmer lines appears to confirm the binary nature of Sco X-1. In contrast to the report of Cowley and Crampton, no periodic variation in the radial velocity of the He II λ4686 line is observed.

142.096 The Leicester X-ray crystal spectrometer on Ariel V and some early results on Cas A, Tycho and Sco X-1. R. E. Griffiths, B. A. Cooke, A. Peacock, K. A. Pounds, M. J. Ricketts.
Monthly Notices Roy. Astron. Soc., Vol. 175, 449 - 460 (1976).
The Leicester X-ray crystal spectrometer aboard Ariel V is described and details of several early observations given. The supernova remnants Cas A and Tycho, and Sco X-1, have been examined for lines of Si and Fe. The upper limits to narrow-line emission are compared with line strengths inferred from proportional counter data.

142.097 New observations of X-ray clusters of galaxies.
B. A. Cooke, D. Maccagni.
Monthly Notices Roy. Astron. Soc., Vol. 175, 65P - 70P (1976).
The regions of sky containing the 3U sources 1555 + 27 and 1639 + 40 have been scanned several times by the University of Leicester Sky Survey Experiment on board Ariel V. New error boxes are reported confirming the cluster A2142 as a good candidate for 3U 1555 + 27 and identifying A2199 as the counterpart of 3U 1639 + 40. Evidence that A2255, of distance group 3 and richness 2, is also an X-ray source is presented.

142.098 Open transmissietralie voor onderzoek röntgen-straling. J. H. Dijkstra, L. J. Lantwaard.
Zenit, 3e jaargang, p. 167 (1976).

142.099 ANS ontdekt röntgenstoten uit bolvormige sterren-hoop. J. Heise.
Zenit, 3e jaargang, p. 168 - 170 (1976).

142.100 Transient X-ray source A0620—00 observed at 408 MHz. A. G. Little, D. F. Crawford, H. S. Murdoch.
Nature, Vol. 261, 113 - 114 (1976). — Letter.

142.101 Discovery of a 283-second periodic variation in the X-ray source 3U 0900—40. J. E. McClintock, S. Rappaport, P. C. Joss, H. Bradt, J. Buff, G. W. Clark, D. Hearn, W. H. G. Lewin, T. Matilsky, W. Mayer, F. Primini.
Astrophys. Journ., (Letters), Vol. 206, L99 - L102 (1976).

A 283 s periodic pulsation in the X-ray system 3U 0900—40 has been discovered during observations by the SAS-3 X-ray observatory. Pulse profiles of the 283 s periodicity are presented in five energy intervals covering the range 1—30 keV for the period 1975 July 19.4—23.9. The averaged profile is relatively simple at higher energies and is markedly more complex at lower energies. A search for soft X-ray emission ($E < 1$ keV) yielded upper limits in the energy intervals 0.16—0.28 keV and 0.5—0.7 keV, respectively.

142.102 On the system 3 U 0352 + 30 — X Persei. M. Garavoglia, A. Treves.
Astron. Astrophys., Vol. 49, 235 - 237 (1976).

The X-ray source 3 U 0352 + 30 is studied supposing that its optical counterpart is the irregular star X Persei. It is proposed that the companion of X Persei is an accreting white dwarf.

142.103 Photometric variations of Wray 977 (3U 1223—62?). G. Hammerschlag-Hensberge, E. J. Zuiderwijk, E. P. J. van den Heuvel, H. Hensberge.
Astron. Astrophys,, Vol. 49, 321 - 323 (1976).

uvby —photometric observations are presented of candidate stars for the X-ray source 3U 1223—62. The early B supergiant Wray 977 shows light variations with an amplitude of about 0.06 magnitudes in v, b and y. The variations resemble a double-wave with a period of 23 ± 1 days. The A 1 supergiant SAO 251905 inside the error box shows no significant variations.

142.104 X-ray outburst from Circinus X-1. A. M. Wilson, G. F. Carpenter.
Nature, Vol. 261, 295 - 296 (1976).

During observations of the Circinus-Norma region by the rotation modulation collimator experiment on board Ariel V, a sudden strong X-ray outburst was detected from Cir X-1. The source appeared on 1976 February 15 at 0500(UT) at an intensity of 20 counts s^{-1}. It gradually brightened to over 40 counts s^{-1} on February 17 at 2200(UT) and then rapidly declined on February 18. The best position, RA = 229.19°, Dec = —57.00°, with error circle, 0.02° radius (90% confidence), is in excellent agreement with the known position of Cir X-1.

142.105 Accretion-disk scenarios of the precursor peak in X-ray transients. W. R. Stoeger.
Nature, Vol. 261, 211 - 213 (1976). — Letter.

142.106 A return to the pre-1971 intensity level and a 5.6-d modulation for Cyg X-1. S. S. Holt, L. J. Kaluzienski, E. A. Boldt, P. J. Serlemitsos.
Nature, Vol. 261, 213 - 215 (1976).

Cyg X-1 exhibited three pronounced X-ray intensity increases during 1975. The last of these, unlike the previous two, has not rapidly decayed back to its pre-increase value. A 5.6-d modulation synchronised with HDE226868 was observed over the first six months of the Ariel V all-sky monitor data accumulation before the April 1975 increase, but was not detectable immediately afterwards. The authors report here the detection of a similar 5.6-d modulation in the 'high-state' data obtained over the 3-month interval November 1975 — January 1976.

142.107 Stellar winds and accretion in massive X-ray binaries. H. J. G. L. M. Lamers, E. P. J. van den Heuvel, J. A. Petterson.
Astron. Astrophys., Vol. 49, 327 - 335 (1976).

Observational and theoretical data on stellar winds are used to calculate the accretion rates onto compact stars in massive binaries as a function of the spectral type and luminosity of their companions. The velocity profile of a purely radiation-pressure driven wind is assumed.

142.108 Production of galactic X-rays following charge exchange by cosmic-ray nuclei. W. D. Watson.
Astrophys. Journ., Vol. 206, 842 - 846 (1976).

Reconsideration is given to the production processes for X-ray lines (≈ 0.5—10 keV) involving charge exchange between low-energy cosmic rays and the interstellar gas. In addition to recent data on charge exchange, collisional excitation and ionization as well as the accumulation of electrons onto the cosmic ray are included. The production rate for each charge-exchange event is significantly greater than obtained in previous calculations.

142.109 Analysis of periodic optical variability in the compact X-ray source Her X-1/HZ Herculis. J. Deeter, L. Crosa, D. Gerend, P. E. Boynton.
Astrophys. Journ., Vol. 206, 861 - 868 (1976).

Analysis of the large pool of optical photometric data now available on the X-ray binary Her X-1/HZ Her reveals considerable variability on a time scale of hours. These flux variations are found to repeat in a highly regular fashion over the 35^d X-ray on–off cycle. The frequency composition and symmetry properties of this pattern suggest that the time dependence results from the uniformly changing orientation of luminous elements in this system.

142.110 Spectroscopic properties of HZ Herculis in model calculations. M. Milgrom.
Astrophys. Journ., Vol. 206, 869 - 875 (1976).

The emission of HZ Herculis in the first three Balmer lines of hydrogen and in λ4686 of He II was calculated as a function of the orbital phase for different values of various parameters of the system. The results support the picture in which the lines (including the emission lines, in contrast to what is usually assumed) come from the atmosphere of HZ Herculis, heated by the X-ray flux that we observe. The sensitivity of the observed quantities to the assumed mass ratio in the system is studied.

142.111 X-ray heating. II. The reflection effect in Scorpius X-1. S. C. Perrenod.
Astrophys. Journ., Vol. 206, 876 - 882 (1976).

The author discusses theoretically the periodic optical light curve of Sco X-1. A binary model with X-ray heating of the nondegenerate companion fits the observations for an orbital inclination $i \approx 5°$–15°. He rules out orbital eccentricity as the cause of the periodic light variation. If the emission lines are formed in the neighborhood of the degenerate star, then the nondegenerate star does not fill its Roche lobe.

142.112 The discovery of X-ray bursts from a region in the constellation Norma. R. D. Belian, J. P. Conner, W. D. Evans.

Astrophys. Journ., (*Letters*), Vol. 206, L135 - L138 (1976).

A search through the first 15 months of data from X-ray detectors aboard the two Vela-5 satellites has revealed 20 count-rate enhancements at least 15 σ above background. The detectors respond to X-rays in the energy band ~ 3–12 keV. The collimator design permits direction determinations within at 12° X 12° error box. Eleven of the observations above 15 σ can be attributed to 10 X-ray flares (one outburst was observed by both spacecraft) located in a region centered at α ~ 16ʰ, δ ~ −53° in the constellation Norma. Flare durations were observed to be more than 2 s and less than 128 s for all of the events.

142.113 **A fast transient source of hard X-rays at high galactic latitude.** S. Rappaport, J. Buff, G. Clark, W. H. G. Lewin, T. Matilsky, J. McClintock.
Astrophys. Journ., (*Letters*), Vol. 206, L139 - L142 (1976).

An extremely short-lived transient X-ray source has been detected with the SAS-3 satellite at a high galactic latitude ($b^{II} \approx -51°$). The source, designated MX 2346–65, had a duration of between 45 s and 2200 s, and had a very hard spectrum. Possible explanations of this type of event are discussed.

142.114 **Transient galactic radio and X-ray sources.** F. G. Smith.
Nature, Vol. 261, 453 (1976).

142.115 **Steepening of the γ-ray background spectrum from local γ-ray production.** R. Schlickeiser, K. O. Thielheim.
Nature, Vol. 261, 478 - 479 (1976).

Because the solar system lies in a spiral arm of our Galaxy, much of the γ-ray flux arriving at large angles to the galactic plane is of local origin. The authors have tried to calculate this contribution to the flux, and find that the effect is particularly pronounced for γ rays of energy > 100 MeV. This implies a steeper spectrum for extragalactic γ rays than has been hitherto expected.

142.116 **Cosmic X-ray bursts.** K. Brecher.
Nature, Vol. 261, 542 (1976).

142.117 **ANS observations on the X-ray burster MXB1730– 335.** J. Heise, A. C. Brinkman, A. J. F. den Boggende, D. R. Parsignault, J. Grindlay, H. Gursky.
Nature, Vol. 261, 562 - 564 (1976).

A new type of time variability of cosmic X-ray sources ('bursters') was discovered from Astronomical Netherlands Satellite (ANS) observations on the source 3U1820–30, associated with the globular cluster NGC 6624. On March 14, 15 and 16, 1976 the ANS spacecraft was pointed in the direction of the remarkable new X-ray burster MXB1730–335 discovered by SAS-3 in the constellation Scorpius at 5° from the galactic centre in the galactic plane. The source is characterised by rapidly repetitive X-ray bursts of varying intensity. The authors report here an improved position of this source and give the energy spectrum from data obtained in four observations for which quick-look data are available. An upper limit for the steady source level in between the bursts is obtained.

142.118 **Two short lived X-ray transients at high galactic latitude.** B. A. Cooke.
Nature, Vol. 261, 564 - 566 (1976).

Until recently, sources known to exhibit transient behaviour were restricted to a region close to the galactic plane. Now, extended observations at high galactic latitudes by the Ariel V, ANS and SAS-3 X-ray detectors have revealed transients with an impressive variety of temporal behaviour. The author here reports two transients at high galactic

latitudes observed by the Leicester sky survey experiment on Ariel V whose time scales are of the order of hundreds of minutes.

142.119 **Balloon-borne observation of the cosmic hard X-ray/ gamma ray background.** D. D. S. Guo, W. R. Webber, S. V. Damle, M. D. Havey.
14th Intern. Cosmic Ray Conf., (see 012.011), Vol. 1, 1 (1975). – Abstract.

142.120 **Interpretation of various radiation backgrounds observed in the gamma-ray spectrometer experiments carried on the Apollo missions and implications for diffuse gamma-ray measurements.** C. S. Dyer, J. I. Trombka, A. E. Metzger, S. M. Seltzer, M. J. Bielefeld, L. G. Evans.
14th Intern. Cosmic Ray Conf., (see 012.011), Vol. 1, 2 - 7 (1975).

142.121 **New results on the diffuse cosmic gamma radiation with the double Compton telescope.** V. Schönfelder, G. Lichti, J. Daugherty, C. Moyano.
14th Intern. Cosmic Ray Conf., (see 012.011), Vol. 1, 8 - 13 (1975).

142.122 **Diffuse gamma ray measurement above 20 MeV with a balloon borne experiment.** B. Parlier, M. Forichon, T. Montmerle, B. Agrinier, G. Boella, L. Scarsi, M. Niel, R. Palmeira.
14th Intern. Cosmic Ray Conf., (see 012.011), Vol. 1, 14 - 19 (1975).

142.123 **Origin of the diffuse gamma-ray background.** A. W. Strong, J. Wdowczyk.
14th Intern. Cosmic Ray Conf., (see 012.011), Vol. 1, 20 - 22 (1975).

142.124 **The diffuse cosmic gamma rays.** R. R. Daniel, P. J. Lavakare.
14th Intern. Cosmic Ray Conf., (see 012.011), Vol. 1, 23 - 28 (1975).

142.125 **High energy galactic gamma radiation observed by the SAS-2 satellite.** C. E. Fichtel, R. C. Hartman, D. A. Kniffen, D. J. Thompson, G. F. Bignami, H. Ögelman, M. E. Özel, T. Tümer.
14th Intern. Cosmic Ray Conf., (see 012.011), Vol. 1, 29 - 34 (1975).

142.126 **Gamma ray spectrum from the galactic disc.** R. K. Sood, K. Bennett, P. G. Clayton, G. K. Rochester.
14th Intern. Cosmic Ray Conf., (see 012.011), Vol. 1, 35 - 39 (1975).

142.127 **Galactic gamma radiation from cosmic rays concentrated in spiral arms.** D. A. Kniffen, G. F. Bignami, C. E. Fichtel, D. J. Thompson, C. Y. Cheung.
14th Intern. Cosmic Ray Conf., (see 012.011), Vol. 1, 40 - 45 (1975).

142.128 **Detection of nuclear gamma rays from Centaurus A.** R. D. Hall, C. A. Meegan, G. D. Walraven, F. T. Djuth, D. H. Shelton, R. C. Haymes.
14th Intern. Cosmic Ray Conf., (see 012.011), Vol. 1, 84 - 88 (1975).

142.129 **Results of a southern hemisphere search for gamma-ray sources at $E_\gamma \geq 3 \times 10^{11}$ eV.** J. E. Grindlay, H. F. Helmken, R. Hanbury Brown, J. Davis, L. R. Allen.
14th Intern. Cosmic Ray Conf., (see 012.011), Vol. 1, 89 - 94

(1975).

142.130 Research on cosmic gamma-rays with energies above 40 MeV from Cygnus–Cassiopeia region.
A. M. Galper, V. G. Kirillov-Ugryumov, A. V. Kurochkin, N. G. Leikov, B. I. Luchkov, Yu. T. Yurkin.
14th Intern. Cosmic Ray Conf., (see 012.011), Vol. 1, 95 - 99 (1975).

142.131 Distribution of cosmic gamma rays in the galactic anticenter region as observed by SAS-2.
D. A. Kniffen, G. F. Bignami, C. E. Fichtel, R. C. Hartman, H. Ögelman, D. J. Thompson, M. E. Özel, T. Tümer.
14th Intern. Cosmic Ray Conf., (see 012.011), Vol. 1, 100 - 105 (1975).

142.132 High energy gamma rays from Vela and Cygnus.
C. E. Fichtel, D. A. Kniffen, H. B. Ögelman, D. J. Thompson.
14th Intern. Cosmic Ray Conf., (see 012.011), Vol. 1, 106 - 111 (1975).

142.133 γ ray production in nearby cosmogonically active regions as an explanation for localized high energy cosmic γ ray sources. G. Serra, M. Niel.
14th Intern. Cosmic Ray Conf., (see 012.011), Vol. 1, 112 - 117 (1975).

142.134 Emission of gamma-rays with energy $\sim 10^{12}$ eV from Cyg X-3.
B. M. Vladimirsky (*Vladimirskij*), Yu. I. Neshpor, A. A. Stepanian (*Stepanyan*), V. P. Fomin.
14th Intern. Cosmic Ray Conf., (see 012.011), Vol. 1, 118 - 122 (1975).

142.135 Evidence for time-dependent flux of $10^{11} - 10^{13}$ eV gamma rays from NP 0532.
H. F. Helmken, J. E. Grindlay, T. C. Weekes.
14th Intern. Cosmic Ray Conf., (see 012.011), Vol. 1, 123 - 127 (1975).

142.136 NP 0532 X-ray pulse structure. H. Helmken.
14th Intern. Cosmic Ray Conf., (see 012.011), Vol. 1, 128 - 132 (1975).

142.137 A sky survey in gamma rays at 10^{11-13} eV energies.
T. C. Weekes, H. F. Helmken, E. Horine.
14th Intern. Cosmic Ray Conf., (see 012.011), Vol. 1, 133 - 138 (1975).

142.138 Time variation of Cyg X-1.
M. Nakagawa, H. Sakurai, M. Uchida.
14th Intern. Cosmic Ray Conf., (see 012.011), Vol. 1, 139 - 143 (1975).

142.139 A model for Cygnus X-1.
J. A. de Freitas Pacheco, J. E. Steiner.
14th Intern. Cosmic Ray Conf., (see 012.011), Vol. 1, 144 - 147 (1975).

142.140 Low energy spectrum of Sco X-1.
V. S. Iyengar, S. Naranan, B. V. Sreekantan.
14th Intern. Cosmic Ray Conf., (see 012.011), Vol. 1, 148 - 153 (1975).

142.141 Hard X-ray spectra of galactic X-ray sources.
D. D. S. Guo, W. R. Webber.
14th Intern. Cosmic Ray Conf., (see 012.011), Vol. 1, 162 - 167 (1975).

142.142 High energy X-ray observation of Cyg X-3.
E. Kendziorra, W. Pietsch, R. Staubert, J. Trümper.
14th Intern. Cosmic Ray Conf., (see 012.011), Vol. 1, 168 - 173 (1975).

142.143 Ariel-5 hard X-ray measurements of galactic and extra-galactic source spectra.
G. F. Carpenter, M. J. Coe, A. R. Engel, J. J. Quenby.
14th Intern. Cosmic Ray Conf., (see 012.011), Vol. 1, 174 - 179 (1975).

142.144 Some Ariel 5 measurements on time variations of X-ray sources in the range 2 keV – 1.2 MeV.
G. Carpenter, M. Coe, A. R. Engel, J. J. Quenby, J. Ives, P. Sanford.
14th Intern. Cosmic Ray Conf., (see 012.011), Vol. 1, 180 (1975). – Abstract.

142.145 An experiment to study sources of hard, cosmic X-rays.
B. R. Dennis, C. J. Crannell, L. E. Orwig, K. J. Frost, C. S. Dyer.
14th Intern. Cosmic Ray Conf., (see 012.011), Vol. 1, 187 (1975). – Abstract.

142.146 The pulsed fraction of X-rays from the Crab nebula.
R. M. Thomas, K. B. Fenton.
14th Intern. Cosmic Ray Conf., (see 012.011), Vol. 1, 188 - 193 (1975).

142.147 The size and position of the high energy X-ray source in the Crab nebula.
C. Reppin, E. Kendziorra, R. Staubert, J. Trümper, J. A. Hoffman, K. A. Pounds, A. B. Giles.
14th Intern. Cosmic Ray Conf., (see 012.011), Vol. 1, 194 - 199 (1975).

142.148 High energy X-ray observations.
G. R. Ricker, A. Scheepmaker, J. E. Ballintine, G. A. Kriss, J. P. Doty, P. M. Downey, S. G. Ryckman, W. H. G. Lewin.
14th Intern. Cosmic Ray Conf., (see 012.011), Vol. 1, 200 (1975). – Abstract.

142.149 A model for the X-ray structure of the Crab nebula.
B. Aschenbach, W. Brinkmann.
14th Intern. Cosmic Ray Conf., (see 012.011), Vol. 1, 201 - 206 (1975).

142.150 Some aspects concerning accretion and X-ray emission. C. M. Filho.
14th Intern. Cosmic Ray Conf., (see 012.011), Vol. 1, 207 (1975). – Abstract.

142.151 Balloon-borne detector search for small gamma-ray bursts.
A. Bewick, M. J. Coe, J. J. Quenby, R. Wheeler.
14th Intern. Cosmic Ray Conf., (see 012.011), Vol. 1, 208 - 212 (1975).

142.152 On the radiation generation mechanisms of cosmic γ-ray bursts. Yu. E. Charikov, Yu. N. Starbunov
14th Intern. Cosmic Ray Conf., (see 012.011), Vol. 1, 219 - 224 (1975).

142.153 Frequency and spatial distribution of cosmic gamma-ray bursts. M. Yoshimori, M. Kajiwara.
14th Intern. Cosmic Ray Conf., (see 012.011), Vol. 1, 225 - 230 (1975).

142.154 About the possibility of observation of stimulated γ-radiation at the burst of supernova.
E. A. Arutyunyan.

14th Intern. Cosmic Ray Conf., (see 012.011), Vol. 1, 231 - 233 (1975).

142.155 **A model cosmic gamma-ray burst.**
I. B. Strong, R. W. Klebesadel, W. D. Evans.
14th Intern. Cosmic Ray Conf., (see 012.011), Vol. 1, 234 - 236 (1975).

142.156 **Gamma-ray bursts from black hole accretion disks.**
I. B. Strong.
14th Intern. Cosmic Ray Conf., (see 012.011), Vol. 1, 237 - 242 (1975).

142.157 **X-ray astronomy.** J. Trümper.
14th Intern. Cosmic Ray Conf., (see 012.011), Vol. 11, 3663 - 3677 (1975).

142.158 **Gamma ray astronomy.** C. E. Fichtel.
14th Intern. Cosmic Ray Conf., (see 012.011), Vol. 11, 3678 - 3697 (1975).

142.159 **The spectrum of the X and gamma ray diffuse background.** R. Rocchia, R. Ducros.
14th Intern. Cosmic Ray Conf., (see 012.011), Vol. 12, 4072 - 4077 (1975).

142.160 **Some Ariel-5 measurements on time variations of X-ray sources in the range 2 keV−1.2 MeV.**
G. F. Carpenter, M. J. Coe, A. R. Engel, J. J. Quenby, S. J. Bell-Burnell, P. Davison, J. C. Ives, K. O. Mason, P. W. Sanford, P. Murdin.
14th Intern. Cosmic Ray Conf., (see 012.011), Vol. 12, 4085 - 4090 (1975).

142.161 **X-ray binaries.** F. Meyer.
14th Intern. Cosmic Ray Conf., (see 012.011), Vol. 12, 4395 - 4412 (1975).

142.162 **γ Cassiopeiae.**
IAU Circ., No. 2900 (1976).

142.163 **3U 1820−30.**
IAU Circ., No. 2907 (1976).

142.164 **GX 1+4.** J. Doty.
IAU Circ., No. 2910 (1976).

142.165 **X-ray bursts from near the galactic center.**
IAU Circ., No. 2911 (1976).

142.166 **MX0513−40.** C. Jones, W. Forman.
IAU Circ., No. 2913 (1976).

142.167 **Strong X-ray burst.**
IAU Circ., No. 2914 (1976).

142.168 **X-ray burst from Aquila.**
IAU Circ., No. 2915 (1976).

142.169 **X-ray bursts from galactic center region.**
W. H. G. Lewin.
IAU Circ., No. 2918 (1976).

142.170 **Cygnus X-1.**
L. J. Kaluzienski, S. S. Holt, E. A. Boldt, P. J. Serlemitsos.
IAU Circ., No. 2918 (1976).

142.171 **X-ray bursts.** W. H. G. Lewin, G. Clark, J. Doty.
IAU Circ., No. 2922 (1976).

142.172 **X-ray bursts.** D. Hearn, V. A. Hughes, M. R. Viner, E. Argyle, P. Feldman.
IAU Circ., No. 2925 (1976).

142.173 **A1745−36.** P. Davison, J. Burnell, J. Ives, A. Wilson, G. Carpenter.
IAU Circ., No. 2925 (1976).

142.174 **X-ray bursts from galactic center region.**
J. Heise, J. Grindlay, W. Liller.
IAU Circ., No. 2929 (1976).

142.175 **X-ray bursts.** K. O. Mason, S. J. Bell Burnell, N. E. White, J. Grindlay, H. Gursky.
IAU Circ., No. 2932, with a correction, No. 2938 (1976).

142.176 **X-ray sources.** M. Watson.
IAU Circ., No. 2934 (1976).

142.177 **MX0656−07.** L. J. Kaluzienski.
IAU Circ., No. 2935 (1976).

142.178 **X-ray sources.**
J. P. Delvaille, F. K. Li, W. Liller, C. Y. Shao.
IAU Circ., No. 2936 (1976).

142.179 **X-ray bursts.** M. J. Coe, A. R. Engel, J. J. Quenby.
IAU Circ., No. 2938 (1976).

142.180 **Circinus X-1.**
L. J. Kaluzienski, S. S. Holt, E. A. Boldt, P. J. Serlemitsos.
IAU Circ., No. 2939 (1976).

142.181 **X-ray bursts.** E. K. Sheffer, E. I. Moskalenko.
IAU Circ., No. 2946 (1976).

142.182 **WRA 977.** H. Mauder.
IAU Circ., No. 2946 (1976).

142.183 **X-ray bursts.** R. H. Becker, S. H. Pravdo, P. J. Serlemitsos, J. H. Swank, J. Hoffman.
IAU Circ., No. 2953 (1976).

142.184 **MX1803−24.** G. Jernigan.
IAU Circ., No. 2957 (1976).

142.185 **WRA 977 and SAO 251595.**
IAU Circ., No. 2957 (1976).

142.186 **New optical candidate for Circinus X-1.**
S. K. Mayo, J. A. J. Whelan, D. T. Wickramasinghe, T. Hawarden, A. J. Longmore.
IAU Circ., No. 2957 (1976).

142.187 **MX1553−54.**
IAU Circ., No. 2959 (1976).

142.188 **X-ray bursts.** A. S. Melioranskij.
IAU Circ., No. 2959 (1976).

142.189 **X-ray bursts.** J. H. Swank, R. H. Becker, S. H. Pravdo, P. J. Serlemitsos.
IAU Circ., No. 2963 (1976).

142.190 **X-ray sources.** A. M. Wilson, G. F. Carpenter, L. J. Kaluzienski, S. S. Holt, M. Esfandiari, E. A. Boldt, P. J. Serlemitsos.
IAU Circ., No. 2965 (1976).

142.191 Observations of cosmic gamma-ray bursts.
R. W. Klebesadel, I. B. Strong.
Astrophys. Space Sci., Vol. 42, 3 - 15 (1976). – See 012.016.
The present paper is to consider the brief, but intense, transient fluxes of electromagnetic radiation in the high-energy X-ray or gamma-ray region.

142.192 Advances in gamma-ray burst astronomy.
T. L. Cline, U. D. Desai.
Astrophys. Space Sci., Vol. 42, 17 - 27 (1976). – See 012.016.
The gamma-ray instrument on the IMP-7 satellite is presently the most sensitive burst detector still operating in orbit. Its results have shown that all measured event-average energy spectra are consistent with being alike. Using this characteristic spectrum to select IMP-7 candidate events of smaller size than those detected using other spacecraft in coincidence, a size spectrum is constructed which fits the -1.5 index power law down to 2.5×10^{-5} erg cm^{-2} per event, at an occurrence rate of about once per month.

142.193 Discovery of two cosmic X-ray bursts in 1970.
G. H. Share.
Astrophys. Space Sci., Vol. 42, 29 - 33 (1976). – See 012.016.
Two bursts of high-energy photons have been discovered during analysis of 2 1/2 years of data from NRL's solar X-ray detector on OSO-6. The bursts occurred at about 18 087 s UT on 25 January, 1970, and about 56 532 s UT on 1 October, 1970. The October event was also observed by Vela 5A; however, none of the Vela detectors observed the January event which had an intensity of about 2×10^{-5} ergs cm^{-2}. Based on these new data, the number of bursts with intensities above about 10^{-5} ergs cm^{-2} appears to be about 50% higher than the Vela data alone would indicate.

142.194 A search for cosmic gamma-ray bursts with a balloon-borne NaI(T 1) scintillation crystal.
W. N. Johnson, J. D. Kurfess, R. D. Bleach.
Astrophys. Space Sci., Vol. 42, 35 - 42 (1976). – See 012.016.
A search has been made for gamma-ray bursts in 15 hours of data obtained from a balloon-borne gamma-ray detector on 10 October and 21 October, 1970. Searches of the data were made with time resolutions varying from 2 ms to 64 s. Four statistically significant bursts were detected and are considered as possible cosmic gamma-ray burst events. The characteristic duration of all four of the observed events is $\sim$100 ms. The implications of these short duration, low intensity events, if valid gamma-ray bursts, are discussed.

142.195 Analysis of narrow spikes in two cosmic gamma-ray bursts. W. L. Imhof, G. H. Nakano, J. B. Reagan.
Astrophys. Space Sci., Vol. 42, 43 - 47 (1976). – See 012.016.
An analysis has been made of the narrow intensity spikes occurring in two cosmic gamma-ray bursts which were observed with a fast time resolution germanium spectrometer on board the low altitude polar orbiting satellite 1972-076B.

142.196 Fast time spectra of gamma-ray bursts.
G. H. Nakano, W. L. Imhof, J. B. Reagan.
Astrophys. Space Sci., Vol. 42, 49 - 56 (1976). – See 012.016.
Fast time spectral measurements of a gamma-ray burst acquired with a satellite-borne cooled germanium spectrometer during the 18 December, 1972 event indicate significant spectral variations during the course of the event. These data are compared with the results of other experimenters providing additional evidence for spectral variations on short time scales. The fast time spectra are also compared with spectral measurements obtained by others with accumulation periods longer than typical time widths of the structure in the intensity profile and particularly with spectra averaged over the entire duration of the burst event.

142.197 A review of some radio and microwave searches for transient phenomena in relation to Vela gamma-ray bursts and supernovae.   G. A. Baird, W. P. S. Meikle, J. V. Jelley, G. G. C. Palumbo, R. B. Partridge.
Astrophys. Space Sci., Vol. 42, 69 - 72 (1976). – See 012.016.

142.198 Upper limits for high energy γ-rays in association with Vela bursts.   S. O'Brien, N. A. Porter.
Astrophys. Space Sci., Vol. 42, 73 - 76 (1976). – See 012.016.
A sea-level cosmic ray burst detector was in operation at the time of occurrence of 16 Vela bursts, between 1970 and 1973. No events were seen. Upper limits for primary γ-rays of 10^{11} eV or higher were set at 10^{-29} erg cm^{-2} Hz^{-1}.

142.199 A neutron star crustquake origin for γ-ray bursts.
A. C. Fabian, V. Icke, J. E. Pringle.
Astrophys. Space Sci., Vol. 42, 77 - 81 (1976). – See 012.016.
Up to 10^{39} ergs of elastic energy might be released in a neutron star crustquake. The sound waves produced by such a quake will transform into a shock near the surface owing to the dramatic decrease in density. A thin surface layer will then be blown off radiating γ-rays. The authors discuss the temporal structure and spectrum of such an event, and suggest some future observational tests.

142.200 Gamma-ray bursts from degenerate stars.
G. Chanmugam.
Astrophys. Space Sci., Vol. 42, 83 - 87 (1976). – See 012.016.
A number of models have been proposed for the observed cosmic gamma-ray bursts. A class of such models involves the use of magnetic energy as the principal source of energy required for the bursts. In this case, arguments are presented to show that degenerate stars are favored. Mechanisms for magnetohydrodynamic instabilities in white dwarfs and neutron stars are discussed. Preliminary work indicates that magnetic white dwarfs can (but neutron stars probably cannot) account for many of the observed features of the bursts.

142.201 M.I.T. studies of transient X-ray phenomena.
C. R. Canizares.
Astrophys. Space Sci., Vol. 42, 111 - 122 (1976). – See 012.016.
This paper reviews recent studies of transient X-ray phenomena carried out by the X-ray Astronomy Group at the Massachusetts Institute of Technology.

142.202 Temporal X-ray astronomy with a pinhole camera.
S. S. Holt.
Astrophys. Space Sci., Vol. 42, 123 - 141 (1976). – See 012.016.
The first preliminary results from the Ariel-5 All-Sky X-Ray Monitor are presented, along with sufficient experiment details to define the experiment sensitivity. Periodic modulation of the X-ray emission is investigated from three sources with which specific periods have been associated, with the results that the 4.8 h variation from Cyg X-3 is confirmed, a long-term average 5.6 day variation from Cyg X-1 is discovered, and no detectable 0.787 day modulation of Sco X-1 is observed. Consistency of the long-term Sco X-1 emission with a 'shot-noise' model is discussed. A sudden increase in the Cyg X-1 intensity by almost a factor of three on 22 April, 1975 is reported. The light curve of a bright nova-like transient source in Triangulum is presented, and compared with previously observed transient sources.

142.203 Observations of variable and transient X-ray sources with the Ariel V sky survey experiment.
K. A. Pounds, B. A. Cooke, M. J. Ricketts, M. J. Turner, A. Peacock, G. Eadie.
Astrophys. Space Sci., Vol. 42, 143 - 159 (1976). – See 012.016.

Results obtained during the first six months in orbit of Ariel V with the Leicester Sky Survey are reviewed. The light curve of Cen X-3 in a binary cycle shows a dip between phase 0.5 and 0.75, and a secondary maximum at the centre of the dip. The dip and the maximum get progressively weaker in the succeeding cycles. These features are interpreted in terms of the stellar wind accretion model. Four bright transient sources of nova-like light curves have been observed. The light curves and the spectra are given for TrA X-1 (A1524−62) and Tau X-T (A0535+26).

142.204   **Ariel 1118–61 − a very close binary system or a slowly rotating neutron star?**
A. C. Fabian, J. E. Pringle, R. F. Webbink.
Astrophys. Space Sci., Vol. 42, 161 - 164 (1976). − See 012.016.

The transient X-ray source Ariel 1118–61 has a period of 6.75 min. The authors review possible models for the X-ray source and in particular they consider orbital and rotational origins for the periodicity. Finally they discuss the possible identification of Ariel 1118–61 with the Mira-type variable RS Cen.

142.205   **Solar cycle model for oscillating transient X-ray sources.**   K. M. V. Apparao, S. M. Chitre.
Astrophys. Space Sci., Vol. 42, 165 - 167 (1976). − See 012.016.

The oscillations observed recently in the transient X-ray source A1118–61 are attributed to variations in the magnetic activity of a rotating white dwarf similar to the 11-year solar activity cycle. The transient nature, in analogy with the sun, is attributed to the variation in the strength of the cycle.

142.206   **Transient soft X-ray sources.**   S. Hayakawa, T. Murakami, F. Nagase, Y. Tanaka, K. Yamashita.
Astrophys. Space Sci., Vol. 42, 169 - 174 (1976). − See 012.016.
A rocket observation of cosmic soft X-rays suggests the existence of transient, recurrent soft X-ray sources which are found variable during the flight time of the rocket.

142.207   **ANS observations of Cygnus X-1.**
D. R. Parsignault, A. Epstein, J. Grindlay, E. Schreier, H. Schnopper, H. Gursky, Y. Tanaka, A. C. Brinkman, J. Heise, J. Schrijver, R. Mewe, E. Gronenschild, A. den Boggende.
Astrophys. Space Sci., Vol. 42, 175 - 184 (1976). − See 012.016.

The hard and soft X-ray experiment detectors on board the Astronomical Netherlands Satellite (ANS) observed the X-ray source Cygnus X-1 during six consecutive days in November 1974. The authors report a detailed study of the temporal intensity and spectrum variations on a time scale of 64 seconds over the six-day period of observation.

142.208   **On possible long-period pulsations in the hard X-ray flux of Cygnus X-1.**   F. Frontera, F. Fuligni.
Astrophys. Space Sci., Vol. 42, 185 - 188 (1976). − See 012.016.

Fourier analysis performed on the data of three balloon observations (20−200 keV) of Cygnus X-1 has shown a significant peak of the power spectral density at about 17 s. Possible periodic pulsations, sustained over periods of at least one hour, could be responsible for the peak detected. Results of computer simulations in terms of a superposition of randomly occurring pulses give some support to this interpretation.

142.209   **Interstellar absorption and variable soft X-ray component in Cygnus X-1.**   P. Gorenstein.
Astrophys. Space Sci., Vol. 42, 189 - 192 (1976). − See

012.016.

The hydrogen column density along the line of sight to Cyg X-1 is 7×10^{21} cm^{-2} as determined from the extinction of its optical counterpart HD 226868. This value may be used to interpret soft X-ray measurements, including those previously reported, where it is not possible to determine the column density independently from the intrinsic spectral function. The correction for interstellar absorption is larger than previously thought. Application to an old observation suggests that an intense soft X-ray component was present in Cyg X-1. This is consistent with the picture of Cyg X-1 suggested by Price and Thorne, in which transitions in Cyg X-1 are attributed to changes in the high energy cut-off of an intense soft component.

142.210   **Balloon observations of fast intensity fluctuations and flare-like enhancements of X-ray emission from Cygnus X-1.**   U. R. Rao, K. Kasturirangan, D. P. Sharma, A. K. Jain, U. B. Jayanthi.
Astrophys. Space Sci., Vol. 42, 193 - 199 (1976). − See 012.016.

The paper presents the results of the investigation on the short term X-ray emission characteristics of Cyg X-1 in the 20−150 keV range. Fluctuations in the intensity of Cyg X-1 with time scales of the order of minutes have been detected besides short-term flare-like enhancements. The spectral characteristics of the flare emission features are discussed and their relationship to the phase of the binary is examined.

142.211   **Spectral and intensity variations in Cygnus X-3 by the Astronomical Netherlands Satellite.**
A. C. Brinkman, J. Heise, R. Mewe, A. J. F. den Boggende, J. Schrijver, E. Gronenschild, Y. Tanaka, D. R. Parsignault, J. Grindlay, E. Schreier, H. Schnopper, H. Gursky.
Astrophys. Space Sci., Vol. 42, 201 - 204 (1976). − See 012.016.

Measurements obtained with the Utrecht (1−8 keV) and Cambridge (1−28 keV) instruments on board the Astronomical Netherlands Satellite are discussed. Particularly, the 4.8 hr period is investigated.

142.212   **The balloon observations of small cosmic X-ray bursts.**   Y. Ogawara, M. Matsuoka, S. Miyamoto, N. Muranaka, J. Nishimura, M. Oda.
Astrophys. Space Sci., Vol. 42, 211 - 216 (1976). − See 012.016.

In the record of the balloon observation which was performed on 27 September, 1970, a transient burst of X-rays was found. This event is concluded to be a cosmic gamma-ray burst of a smaller size or at a larger distance compared to the Vela bursts observed over the X-ray energy range. The energy spectrum is consistent with that of some of the Vela bursts.

142.213   **A search for soft X-ray emission from stellar sources.**   R. Mewe, J. Heise, E. H. B. M. Gronenschild, A. C. Brinkman, J. Schrijver, A. J. F. den Boggende.
Astrophys. Space Sci., Vol. 42, 217 - 222 (1976). − See 012.016.

With the soft X-ray detector (∼ 0.2−0.284 keV) aboard the Astronomical Netherlands Satellite (ANS) the authors have searched for X-ray emission from hot star coronae and peculiar stars. On Sirius (α CMa) and Capella (α Aur) X-ray emission has been measured at 6σ and 5σ level, respectively, above background. In all other cases the search revealed no evidence for soft X-ray emission. Upper limits to the luminosities of about 25 star coronae (main-sequence stars, (sub)giants, and supergiants) and of 4 peculiar stars (λ Sco, β Lyr, P Cyg, and η Car) have been obtained.

142.214   **Short-term variability of Cyg X-1.**   M. Oda, K. Doi, Y. Ogawara, K. Takagishi, M. Wada.

Astrophys. Space Sci., Vol. 42, 223 - 244 (1976). – See 012.016.

The short-term X-ray variability distinguishes Cyg X-1, which is the most likely candidate for a black hole, from other X-ray sources. The present status of our knowledge on this short-term variation, mainly from the UHURU, the MIT and the GSFC observations, is reviewed. The nature of impulsive variations which compose the time variation exceeding the statistical fluctuation is discussed. There are indications that the energy spectrum of large pulses is harder than the average spectrum, or that the large pulses are the characteristics of the hard component of the spectrum if it is composed of two, soft and hard, components. The substructure of the fluctuations on a time scale of milliseconds suggested by two investigations is also discussed.

142.215 **Size of the hard X-ray source in the Crab nebula.**
Y. Fukada, S. Hayakawa, I. Kasahara, F. Makino, Y. Tanaka, H. Akiyama, M. Matsuoka, J. Nishimura, M. Oda, M. Nakagawa, H. Sakurai, V. S. Iyenger, P. K. Kunte, R. K. Manchanda, B. V. Sreekantan.
Astrophys. Space Sci., Vol. 42, 245 - 248 (1976). – See 012.016.

The size of the hard X-ray source in the Crab nebula was observed with scintillation counters on board two balloons at a lunar occultation on 24 January, 1975. The Gaussian width of the source is $34''$ ($+17''$, $-14''$) and the center thereof is offset from the pulsar by $6'' \pm 4''$ at position angle $102°$.

142.216 **'Copernicus' observations of extragalactic X-ray sources.** A. C. Fabian, R. J. Mitchell, P. J. N. Davison, P. A. Charles, J. L. Culhane.
Astrophys. Space Sci., Vol. 42, 249 - 254 (1976). – See 012.016.

The MSSL X-ray detectors on Copernicus have been used to study a number of extragalactic objects. At least three classes of unresolved sources are found and the authors suggest that accretion may be the dominant mechanism. The mass of the accreting object then determines the X-ray emission properties.

142.217 **Röntgenquellen, Neutronensterne und schwarze Löcher.** R. Breuer.
Umschau, 76. Jahrgang, p. 377 - 382 (1976).

142.218 **Röntgennovor – vad är det?** B. Wennfors.
Astron. Tidssk., Årg. 9, 70 - 75 (1976).

142.219 **Extragalaktiska röntgenkällor.** J. Bystedt.
Astron. Tidssk., Årg. 9, 76 - 81 (1976).

142.220 **Sorgenti di raggi X in sistemi binari.** B. Cester.
Giorn. Astron., Vol. 1, 173 - 188 (1975).

142.221 **A model for the X-ray nova A0620-00.**
A. S. Endal, E. J. Devinney, S. Sofia.
Astrophys. Letters, Vol. 17, 131 - 135 (1976).

The authors propose a model for the transient X-ray source A0620-00 involving a white dwarf accreting mass from a late-type subgiant companion. The transient behavior of the X-ray source is explained by the instability to mass loss of the companion (as in Algol-type binaries). The brightening, spectrum, and decay timescale of the optical counterpart are explained in terms of re-emission of X-radiation intercepted by the subgiant. A0620-00 should provide an excellent test case for numerical models of stellar atmospheres irradiated by an external X-ray flux.

142.222 **Copernicus: spectral studies of Cas-A and Pup-A.**
P. A. Charles, J. L. Culhane, J. C. Zarnecki, A. C. Fabian.

Mem. Soc. Astron. Italiana, Vol. 45, (see 012.017), 699 - 707 (1974).

142.223 **Some recent results in X-ray astronomy.**
J. L. Culhane.
Mem. Soc. Astron. Italiana, Vol. 45, (see 012.017), 831 - 849 (1974). – Review paper.

142.224 **Position measurement and identification of cosmic X-ray sources with the Copernicus satellite.**
F. J. Hawkins, P. W. Sanford.
Mem. Soc. Astron. Italiana, Vol. 45, (see 012.017), 851 - 863 (1974).

142.225 **Copernicus observations of Cygnus X-1.**
P. W. Sanford.
Mem. Soc. Astron. Italiana, Vol. 45, (see 012.017), 865 (1974). Abstract.

142.226 **Observations of Cygnus X-3.** K. O. Mason.
Mem. Soc. Astron. Italiana, Vol. 45, (see 012.017), 867 - 874 (1974).

142.227 **Ultraviolet flickering in HZ Herculis.**
C. Chevalier, S. A. Ilovaisky.
Mem. Soc. Astron. Italiana, Vol. 45, (see 012.017), 875 (1974). Abstract.

142.228 **Rapid spectroscopic changes in HZ Herculis.**
S. A. Ilovaisky, C. Chevalier.
Mem. Soc. Astron. Italiana, Vol. 45, (see 012.017), 876 (1974). Abstract.

142.229 **The identification and UBV photometry of the visible component of the Centaurus X-3 binary system.** W. Krzeminski.
Mem. Soc. Astron. Italiana, Vol. 45, (see 012.017), 877 (1974). Abstract.

142.230 **X-rays from neutron stars heated by accretion.**
S. Tsuruta, M. J. Rees.
Mem. Soc. Astron. Italiana, Vol. 45, (see 012.017), 909 - 916 (1974).

142.231 **Iron line emission from a high temperature plasma in Cas A.** S. H. Pravdo, R. H. Becker, E. A. Boldt, S. S. Holt, R. E. Rothschild, P. J. Serlemitsos, J. H. Swank.
GSFC Document X-661-76-8, Prepr., 15 pp. (1976).

The X-ray spectrum of Cassiopeia A was observed for several days by the GSFC proportional counter experiment on board OSO-8. A narrow iron line which is predicted by the thermal model is also observed. The low energy (2– 5 keV) data show an excess over the high temperature component which is consistent with the presence of an additional low temperature thermal component with $kT \lesssim 0.7$ keV and $N_H \sim 10^{22}$ atoms cm^{-2}. Iron abundance in the source relative to normal cosmic abundance is discussed, as in the relation of this observation to shock wave and multi-component thermal models for supernova remnants.

142.232 **Cyg X-1: a return to the pre-1971 intensity level and a 5.6-day modulation.** S. S. Holt, L. J. Kaluzienski, E. A. Boldt, P. J. Serlemitsos.
GSFC Document X-661-76-37, Prepr., 10 pp. (1976).

142.233 **Spectral variability in the X-ray pulsar GX 1 + 4.**
R. H. Becker, E. A. Boldt, S. S. Holt, S. H. Pravdo, R. E. Rothschild, P. J. Serlemitsos, J. H. Swank.
GSFC Document X-661-76-66, Prepr., 11 pp. (1976).

Observations of the galactic center region, hard X-ray

source GX 1 + 4 by the GSFC X-ray spectroscopy experiment on OSO-8 confirm that GX 1 + 4 is a slow X-ray pulsar. The amount of absorption by cold matter in the spectrum of GX 1 + 4 varies significantly within a 24 hour period, behavior typical of many X-ray binary systems. The light curve for the pulsations from GX 1 + 4 appears to be energy dependent.

142.234 Untersuchungen im Gebiet der Röntgenquelle 3U 1956+11. R. Hudec.
MVS Sonneberg, Vol. 7, 112 - 113 (1976).

142.235 Untersuchungen an HDE 245770, Kandidat für die Röntgenquelle A 0535+26. (Teil 2). R. Hudec.
MVS Sonneberg, Vol. 7, 135 (1976).

142.236 Parameters of X-ray binaries.
H. D. Tananbaum, J. B. Hutchings.
Ann. New York Acad. Sci., Vol. 262, 299 - 311 = Contr. Dominion Astrophys. Obs., Victoria, B.C., No. 267 (1975).
The authors describe the types of X-ray and visible light observations that have been made, give some specific illustrations, and outline the model applied to the observations to determine the system's parameters.

142.237 Evidence for an 11.2d periodicity from Cyg X-2.
S. S. Holt, E. A. Boldt, P. J. Serlemitsos, L. J. Kaluzienski.
GSFC Document X-660-76-1, Prepr., 10 pp. (1976).
Evidence for a persistent 11.17 ±.10d period from Cyg X-2 is presented from one year of accumulated data from the Ariel-5 All-Sky Monitor. The effect is not a simple sidereal alias of a true source period close to one day.

142.238 Rise and fall of an X-ray star. K. Pounds.
New Scient., Vol. 69, 494 - 496 (1976).
Account of Mon X-1.

142.239 The diffuse cosmic gamma rays.
R. R. Daniel, P. J. Lavakare.
Pramāṇa, Vol. 5, 107 - 117 (1975). − Abstr. in Phys. Abstr., Vol. 79, A027687 (1976).

142.240 Simultaneous hard X-ray and optical observations of Sco X-1. M. Matsuoka.
Bull. Inst. Space Aeronaut. Sci., Univ. Tokyo, B, Vol. 11, 477 - 496 (1975). In Japanese. − Abstr. in Phys. Abstr., Vol. 79, A027986 (1976).

142.241 Ionospheric response to gamma-ray bursts of cosmic origin. R. Barletti, G. L. Tagliaferri.
Ann. Géophys., Vol. 31, 297 - 299 (1975). − Abstr. in Phys. Abstr., Vol. 79, A033014 (1976).

142.242 A model for the gamma-ray background: further results. A. W. Strong.
Journ. Phys. A, Vol. 9, 305 - 310 (1976). − Abstr. in Phys. Abstr., Vol. 79, A033015 (1976).

142.243 The X and γ diffuse background.
H. M. Horstman, G. Cavallo, E. Moretti-Horstman.
Nuovo Cimento Rivista, Ser. 2, Vol. 5, 255 - 311 (1975). Abstr. in Phys. Abstr., Vol. 79, A033016 (1976).

142.244 The Richtmyer memorial lecture: progress in X-ray astronomy. R. Giacconi.
American Journ. Phys., Vol. 44, 121 - 134 (1976). − Abstr. in Phys. Abstr., Vol. 79, A042319 (1976).

142.245 Balloon observations of X-rays from celestial sources.
U. R. Rao, K. Kasturirangan, D. P. Sharma, A. K. Jain, U. B. Jayanthi.

Indian Journ. Radio Space Phys., Vol. 4, 199 - 202 (1975). Abstr. in Phys. Abstr., Vol. 79, A042329 (1976).

142.246 Study of optical properties of X-ray sources.
Semiannual report, 1 Jan. - 31 Aug. 1974.
P. A. Vanden Bout.
Separate print Texas Univ., Austin, USA, 5 pp. (1974).

142.247 X-ray and γ-ray astronomy at the turning point.
L. E. Peterson.
7th space conf. Cocoa Beach, Florida, USA, 22. Apr. 1970, p. 8.1 - 8.26 (1970).

142.248 Search for soft stellar X-ray sources.
M. J. Vanderhill.
Thesis Wisconsin Univ., Madison, USA, 124 pp. (1974). (Available from Univ. Microfilms, Order No. 74–27, 768).

142.249 Map of diffuse low energy X-rays from the general direction of the galactic anti-center.
F. O. Williamson.
Thesis Wisconsin Univ., Madison, USA, 104 pp. (1974). (Available from Univ. Microfilms, Order No. 74–28,835).

142.250 Molecular hydrogen in the Galaxy and galactic gamma rays. F. W. Stecker, P. M. Solomon, N. Z. Scoville, C. E. Ryter.
Separate print National Aeronaut. Space Administr., Greenbelt, Maryland, USA, Goddard Space Flight Center, 29 pp. (1975). − See 14.142.096.

142.251 Observations of gamma-ray bursts.
I. B. Strong, R. W. Klebesadel, W. D. Evans.
7th Texas symposium relativistic astrophys., Dallas, Texas, USA, 16 Dec. 1974, 24 pp. (1975).

142.252 Cosmic gamma-ray bursts: NASA observations and their implications. T. L. Cline.
Symposium significant accomplishments sci. technol., Greenbelt, Maryland, USA, 18 Dec. 1973, p. 36 - 39 (1975).

142.253 Uniqueness of the energy spectrum of cosmic gamma-ray bursts, as observed by IMP-6.
U. D. Desai.
Symposium significant accomplishments sci. technol., Greenbelt, Maryland, USA, 18 Dec. 1973, p. 40 - 43 (1975).

142.254 Theoretical interpretation of the SAS-2 high energy gamma-ray observations of the galactic plane in terms of the galactic cosmic-ray and matter distributions.
C. Fichtel.
Symposium significant accomplishments sci. technol., Greenbelt, Maryland, USA, 18 Dec. 1973, p. 56 - 59 (1975).

142.255 Multiple Hercules X-1 periodicities: a simple kinematic explanation. S. Holt.
Symposium significant accomplishments sci. technol., Greenbelt, Maryland, USA, 18 Dec. 1973, p. 67 - 70 (1975).

142.256 Cygnus X-1 temporal microstructure: evidence for a black hole. R. Rothschild.
Symposium significant accomplishments sci. technol., Greenbelt, Maryland, USA, 18 Dec. 1973, p. 63 - 66 (1975).

142.257 High energy diffuse gamma-rays: experimental results and theoretical interpretation.
D. Thompson.
Symposium significant accomplishments sci. technol., Greenbelt, Maryland, USA, 18 Dec. 1973, p. 48 - 51 (1975).

142.258 Low energy spectroscopy with EXOSAT.

J. A. M. Bleeker.
ESRO colloquium X-ray astron.related topics, Noordwijk,
Netherlands, 25 - 26 Feb. 1975, p. 109 - 117 (1975).

142.259   Gamma radiation of galactic origin.   J. Paul.
          Congr. French Physical Soc., Dijon, France, 30 June
1975, 20 pp. (1975). In French.

Five satellites observe short, intense X-ray bursts.
Phys. Today, Vol. 29, No. 4, p. 17, 19 - 20 (1976).

The medium energy experiment for EXOSAT.
See Abstr. 032.540.

The SAS-C X-ray observatory.
See Abstr. 032.541.

Origin of the diffuse γ-ray background.
See Abstr. 061.016.

X-ray astronomy.   See Abstr. 061.036.

X-ray astronomy — a status report.
See Abstr. 061.040.

Radiation from cosmic blast waves.
See Abstr. 062.002.

The hydrodynamics of accretion discs. II: Turbulent
models.   See Abstr. 062.029.

The hydrodynamics of accretion discs. I. Stability.
II. Turbulent models.   See Abstr. 062.095.

Compton reflected spectra of X-ray illuminated
stellar atmospheres.   See Abstr. 063.005.

Transfer of X-rays through a spherically symmetric
gas cloud.   See Abstr. 063.038.

Stellar accretion disks.   See Abstr. 064.043.

Evidence on neutron star structure from pulsars and
related objects.   See Abstr. 065.015.

The melting of neutron stars' crystalline cores and
gamma-ray bursts.   See Abstr. 065.016.

Slow mass transfer in semidetached binaries.
See Abstr. 065.021.

Evidence on neutron star structure from pulsars and
related objects.   See Abstr. 065.041.

The limiting luminosity of accreting neutron stars
with magnetic fields.   See Abstr. 065.055.

Non explosive collapse of white dwarfs.
See Abstr. 065.084.

Non-adiabatic radial oscillations and pulsational
stability of hot degenerate dwarfs.   See Abstr. 065.119.

Obese 'neutron' stars.   See Abstr. 066.012.

The fall of the shell of dust on to a rotating black
hole.   See Abstr. 066.025.

Gamma rays from primordial black holes.
See Abstr. 066.043.

A theory of the instability of disk accretion on to
black holes and the variability of binary X-ray sources, galactic
nuclei and quasars.   See Abstr. 066.045.

Ionospheric effects of transient celestial X-ray and
gamma-ray events.   See Abstr. 083.068.

Ionospheric techniques for the detection of transient X- and γ-ray bursts.   See Abstr. 083.069.

Ionospheric effects of transient celestial X-ray and
gamma ray events.   See Abstr. 083.080.

Lunar occultation of point X-ray sources.
See Abstr. 096.011.

Three-color photometry of HDE 226868: the
optical counterpart of Cygnus X-1.   See Abstr. 113.029.

Photometric observations at the Special Astrophysical Observatory according to the programmes of astrophysical researches aboard the orbital station Salyut 4 and experimental flight Soyuz-Apollo.   See Abstr. 113.041.

The photometric and spectrographic histories of HD
245770 $\stackrel{?}{=}$ A0535 + 26, the transient X-ray source.
See Abstr. 114.343.

Spectroscopic observations of the candidate star
coincident with A0620—00.   See Abstr. 114.344.

On the formation of disk around a compact
object by two-body tidal encounter.   See Abstr. 117.001.

Effects of tidal distortion on binary-star velocity
curves and ellipsoidal variation.   See Abstr. 117.002.

Orbital evolution of a singly condensed, close binary,
by mass loss from the primary and by accretion drag on the
condensed member.   See Abstr. 117.007.

Precession of the nodes in some triple stellar systems.
See Abstr. 117.017.

The nature of optical variability of X-ray binaries
Cyg X-2 = V1341 Cyg and Sco X-1 = V818 Sco.
See Abstr. 117.024.

The evolution of a massive close binary up to the
X-ray binary stage.   See Abstr. 117.030.

Ultraviolet observation of HD 77581 (=2U 0900—
40).   See Abstr. 119.019.

The 3U 0900—40 binary system: orbital elements
and masses.   See Abstr. 121.044.

Inactive state of HZ Her.   See Abstr. 121.068.

Observations of X-rays from flare stars with ANS.
See Abstr. 122.110.

A search for X rays from UV Ceti flare stars.
See Abstr. 122.111.

The variability of the Bpe star HDE 245770 (most
probable candidate for X-ray source A0535 +26).
See Abstr. 122.141.

γ Cassiopeiae.   See Abstr. 123.029.

Optical observations of the recurrent nova associated with A0620−00: 1917 - 1975. See Abstr. 124.001.

V616 Monocerotis (A0620−00). See Abstr. 124.103.

Detection of X-ray emission from the remnant of the supernova 1006 A.D.   See Abstr. 125.011.

An empirical comparison of X-ray and radio emission from supernova remnants.   See Abstr. 125.016.

X-ray spectra of Cassiopeia A and Tycho's supernova observed with Ariel-5.   See Abstr. 125.021.

Iron line emission from a high-temperature plasma in Cassiopeia A.   See Abstr. 125.022.

Some implications of the X-ray data from old supernova remnants.   See Abstr. 125.023.

Deep Hα photography of the Vela and Puppis supernova remnants.   See Abstr. 125.024.

Non explosive collapse of white dwarfs. See Abstr. 126.004.

Sirius B: a thermal soft X-ray source? See Abstr. 126.015.

An interstellar H_2 indicator in direction of the Crab nebula.   See Abstr. 131.032.

Polarization of the radiation of stars with non-uniform distribution of luminosity on the star's surface. See Abstr. 131.110.

Consequences of a new hot component of the interstellar medium.   See Abstr. 131.112.

ANS: preliminary X-ray brightness map of the Cygnus Loop.   See Abstr. 132.018.

Observation of the Cygnus Loop in X-rays with a one-dimensional focusing collector.   See Abstr. 132.025.

Energy dependence of the size of the X-ray source in the Crab nebula.   See Abstr. 134.001.

The optical and X-ray surface brightness of the Crab nebula. I. A cosmic-ray diffusion model. See Abstr. 134.003.

The nucleus of Centaurus A. See Abstr. 141.069.

New observations of pulsed X-ray emission from NP 0532.   See Abstr. 141.313.

A search for soft X-ray radiation from pulsars with the Astronomical Netherlands Satellite. See Abstr. 141.350.

SAS-2 high-energy gamma-ray observations of the Vela pulsar.   See Abstr. 141.362.

Possible infrared counterpart of MXB1730−335. See Abstr. 141.634.

Possible infrared counterpart of MXB1730−335. See Abstr. 141.635.

Cosmic ray sources.   See Abstr. 143.311.

Metal-rich globular clusters in the Galaxy. III. The "X-ray" globular cluster NGC 6441.   See Abstr. 154.001.

Optical structure of the X-ray globular cluster NGC 6624.   See Abstr. 154.009.

The structure of the galactic disk and its implications for gamma-ray astronomy.   See Abstr. 155.039.

Molecular hydrogen in the Galaxy and galactic gamma rays.   See Abstr. 155.041.

Inverse Compton production of gamma rays in interstellar space.   See Abstr. 155.043.

Gamma ray emission from the galactic disc. See Abstr. 155.044.

Average properties of molecular clouds in the Galaxy.   See Abstr. 155.046.

The structure of the galactic disk in the light of gamma ray astronomy.   See Abstr. 155.049.

Balloon-borne hard X-ray study of the southern sky. See Abstr. 155.065.

SAS-2 high energy gamma-ray observations of galactic structure and the Crab nebula.   See Abstr. 155.066.

Orion-arm magnetic monopoles and γ rays. See Abstr. 156.003.

Magnetized accretion disks and the radio outbursts of 3C 120 and Cygnus X-3.   See Abstr. 158.067.

Soft X-rays from the Large Magellanic Cloud: implications on the origin of the diffuse X-ray background. See Abstr. 159.010.

An X-ray red-shift test for clusters of galaxies up to $z \geq 1$.   See Abstr. 160.041.

The dynamics of the intergalactic medium in the vicinity of clusters of galaxies.   See Abstr. 161.001.

X-rays from hot plasma in clusters of galaxies. See Abstr. 161.005.

143 Cosmic Radiation

143.001 Consistency of cosmic-ray composition, acceleration mechanism, and supernova models.
K. L. Hainebach, E. B. Norman, D. N. Schramm.
Astrophys. Journ., Vol. 203, 245 - 256 (1976).

It is shown that a consistent picture of cosmic-ray composition can be obtained by synthesizing the observed cosmic-ray abundances with the acceleration mechanisms of Scott and Chevalier or Jokipii, and the presupernova models of Arnett and Schramm, with explosive processing. If this model is valid, then it is possible to estimate (1) the location of the supernova "mass cut," (2) possible anomalies in the yet to be accurately observed cosmic-ray iron isotopic ratios, (3) the cosmic-ray ratio of r- to s-process elements, and (4) the time between the supernova explosion and the acceleration of the cosmic rays.

143.002 Results on the cosmic ray chemical composition at energies up to 100 GeV/nucl.
W. K. H. Schmidt, K. Atallah, T. F. Cleghorn, W. V. Jones, A. Modlinger, M. Simon.
Astron. Astrophys., Vol. 46, 49 - 59 (1976).

Some aspects of the chemical composition of high energy cosmic rays have been investigated in a balloon flight. The apparatus used was an ionization spectrometer with a geometric factor of 1.01 m^2sr. The charge resolution was sufficient to separate four groups of elements from the CNO group to the iron group. The energy range covered extends up to 100 GeV/nucl. The results indicate that there is no drastic change in abundance ratios at high energies. However, the high energy abundance ratio between the CNO group and the iron group is about a factor of two different from that one that was reported for low energies of about 1 GeV/nucl. by other authors.

143.003 On search for tracks of heavy and superheavy cosmic-ray nuclei in crystals from pallasites.
G. N. Flerov, T. P. Zholud, O. Otgonsuren (*Otgonsurehn*), V. P. Perelygin, H. B. Wiik.
Geochim. Cosmochim. Acta, Vol. 40, 305 - 307 (1976).

The results of a search for far transuranic elements in the primary cosmic radiation are presented. It is shown that olivines from pallasites are very suitable for such investigations. The sensitivity of olivines to charged particles and the fading effect of latent tracks under space conditions have been studied. In the Lipovsky and Marjalahti pallasites, the distributions of tracks of nuclei with $Z > 36$ have been measured, and the abundance of nuclei with $Z \geqslant 70$ in these crystals has been tentatively set at $(2-5) \times 10^{-6}$ with respect to the Fe group nuclei.

143.004 Measurements of hard cosmic radiation in September 1973 aboard the automatic interplanetary stations Mars 4, Mars 5, Mars 7.   E. V. Gorchakov, P. P. Ignat'ev, V. A. Iozenas, T. E. Shvidkovskaya, V. A. Yakovlev, I. V. Getselev, V. I. Tkachenko, G. P. Lyubimov, N. N. Kontor, N. G. Galach'ev.
Kosmich. Issled., Vol. 14, 65 - 72 (1976). In Russian.

143.005 Comments on a closed Galaxy model for cosmic-ray propagation.   D. K. French, J. L. Osborne.
Nature, Vol. 260, 372 (1976).

143.006 Cosmic-ray positron and negatron spectra between 20 and 800 MeV measured in 1974.
R. C. Hartman, C. J. Pellerin.
Astrophys. Journ., Vol. 204, 927 - 933 (1976).

A balloon-borne spark chamber magnetic spectrometer has been used to measure separate spectra of positrons and negatrons. The total electron flux is about 0.03 $m^{-2}\,s^{-1}\,sr^{-1}$ MeV^{-1} between 70 and 800 MeV, and increases toward lower energies. The positron spectrum decreases sharply toward lower energies from a value of about 0.008 $m^{-2}\,s^{-1}\,sr^{-1}\,MeV^{-1}$ at 650 MeV, and only upper limits are obtained for positrons below 200 MeV. The implications of these data are examined with regard to the problem of solar modulation.

143.007 Cosmic rays in the atmosphere and magnetosphere of the earth and in the interplanetary space.
L. I. Dorman.
Geomagnetizm i vysok. sloi atmosfery. Tom (Vol.) 2. Itogi nauki i tekhn. VINITI AN SSSR. Moskva, 1975, p. 7 - 82. In Russian. – Abstr. in Referativ. Zhurn. 62. Issled. kosmich. prostranstva, 3.62.185 (1976).

143.008 The intensity variations of solar and galactic cosmic rays with azimuthal angle in the polar region.
Dj. Heristchi, J.-P. Legrand.
Planet. Space Sci., Vol. 24, 281 - 285 (1976).

A balloon-borne multidirectional detector is used to measure the intensity variation of galactic and solar cosmic rays with the azimuthal angle, the zenith angle being maintained at 60°. In polar regions, the intensity towards the north is found to be 20% larger than that towards the south. It is shown that this anisotropy does not originate in interplanetary space and is not produced by a magnetospheric source. It is suggested that the effect is due to propagation effects within the magnetosphere.

143.009 Cosmic rays from pulsars.
G. S. Saakyan, D. M. Sedrakyan, Eh. V. Chubaryan, R. M. Avakyan, G. P. Alodzhants.
Astrofizika, Vol. 11, 679 - 687 (1975). In Russian. English translation in Astrophysics, Vol. 11, No. 4.

A new mechanism of cosmic ray generation by pulsars is considered. Formulas for the total number of particles in the magnetosphere, the rate of their decrease and the second derivative of the period with respect to time are obtained. The total number of pulsars in the Galaxy are calculated taking into account observational data of cosmic ray flux on the earth.

143.010 The directional dependence of the primary cosmic rays of energies $10^{11}-10^{12}$ eV.
R. G. Marsden, H. Elliot, R. J. Hynds, T. Thambyahpillai.
Nature, Vol. 260, 491 - 495 (1976).

The apparent sidereal variation in the intensity of muons generated by cosmic rays with energies greater than 10^{11} eV showed a marked change in phase at northern hemisphere stations at the time of the recent reversal in sign of the general solar magnetic field. No corresponding change was observed at rather lower energies at Hobart, Tasmania. By adopting a particular model of the large scale interplanetary field, it is possible to account for this behaviour of the sidereal variation in terms of a primary cosmic ray anisotropy which is related to the local direction of the galactic magnetic field.

143.011 Solar influence on galactic cosmic ray anisotropy measurements.   J. Kóta.
Nature, Vol. 260, 507 - 508 (1976).

Cosmic ray particles of energies $\gtrsim 10^{12}$ eV are virtually unaffected by the interplanetary magnetic field, and thus any galactic anisotropy manifests itself in sidereal intensity variation in earth-based measurements. Below 10^{11} eV, on the other hand, solar modulation is of importance. Galactic anisotropy would be smeared out at this energy, and an anisotropy is introduced by the solar modulation effects. In the intermediate $10^{11}-10^{12}$ eV region sidereal variation may originate either in genuine galactic anisotropy or in solar modulation. Here the author considers that the observed variation is a result of the electromagnetic field embedded in the solar wind.

143.012 **A dynamic model for the time evolution of the modulated cosmic ray spectrum.**
J. J. O'Gallagher, G. A. Maslyar III.
Journ. Geophys. Res., Vol. 81, 1319 - 1326 (1976).

A recently developed model predicts an energy dependent phase lag in the modulated cosmic ray density. This model is applied to predict the time evolution of the modulated cosmic ray proton spectrum over a simulated solar cycle. The predicted spectra reproduce most of the features of the so-called hysteresis effect when values of $V = 360$ km/s (solar wind velocity), $R = 60$ AU, and K (the effective average diffusion coefficient) varying between 1.3×10^{22} cm^2/s at solar maximum and 3.5×10^{22} cm^2/s at solar minimum are used. A modulation produced mostly by varying R over the solar cycle is less consistent with the observations.

143.013 **Cosmic ray distribution in the vicinity of the front of a nonstationary shock wave.**
G. F. Krymskij, S. I. Petukhov, I. A. Transkij, V. P. Mamrukova, G. V. Shafer.
Geomagn. Aeronom., Vol. 16, 25 - 29 (1976). In Russian.

143.014 **Anisotropic diffusion of cosmic rays in interplanetary space. I. Density distribution.**
L. I. Dorman, N. P. Milovidova.
Geomagn. Aeronom., Vol. 16, 30 - 36 (1976). In Russian.

143.015 **Theory of the 27-day modulation of galactic cosmic rays.** L. Kh. Shatashvili.
Geomagn. Aeronom., Vol. 16, 37 - 42 (1976). In Russian.

143.016 **Anisotropic diffusion of cosmic rays in the interplanetary space. II. 11-year variation.**
L. I. Dorman, N. P. Milovidova.
Geomagn. Aeronom., Vol. 16, 221 - 224 (1976). In Russian.

143.017 **Frequency spectra of long-period variations of cosmic rays.** T. N. Charakhch'yan, G. A. Bazilevskaya, V. P. Okhlopkov, L. S. Okhlopkova.
Geomagn. Aeronom., Vol. 16, 225 - 229 (1976). In Russian.

143.018 **Observations of low energy cosmic ray ions in the Skylab experiment and their implications.**
S. Biswas, N. Durgaprasad, J. Nevatia, V. S. Venkatavaradan.
Bull. Astron. Soc. India, Vol. 3, 37 (1975). − Abstract of a paper presented at the A.S.I. meeting 1975.

143.019 **Anisotropy of cosmic radiation.** U. R. Rao.
VIIth Leningrad seminar. "Corpuscular streams of the sun and the radiation belts of the earth and Jupiter", 1975, (see 012.007), p. 131 - 152 (1975). − Abstr. in Referativ. Zhurn. 51. Astron., 4.51.325; 62. Issled. kosmich. prostranstva, 4.62.204 (1976).

143.020 **Charge and isotope composition of galactic cosmic rays.** B. M. Kuzhevskij.
Issled. kosmich. luchej. Moskva, Nauka, 1975, p. 84 - 113. In Russian. − Abstr. in Referativ. Zhurn. 51. Astron., 4.51.629 (1976).

143.021 **Isotopic and elemental composition of the anomalous low-energy cosmic-ray fluxes.**
R. A. Mewaldt, E. C. Stone, S. B. Vidor, R. E. Vogt.
Astrophys. Journ., Vol. 205, 931 - 937 (1976).

The authors have measured the quiet-time fluxes of the elements hydrogen through oxygen in the ~4−30 MeV nucleon^{-1} energy interval during the period 1972 October through 1974 October. They find that the low-energy fluxes of Li, Be, B, and C are consistent with those expected from adiabatic deceleration and show no significant evidence for secondary fragmentation products arising from the enhanced nitrogen and oxygen fluxes. In the ~6−12 MeV nucleon^{-1} interval, the observed nitrogen and oxygen nuclei are predominantly ^{14}N and ^{16}O, with upper limits (84% confidence level) of ^{15}N/N $\leqslant 0.26$, ^{17}O/O $\leqslant 0.13$, and ^{18}O/O $\leqslant 0.12$ for the other stable nitrogen and oxygen isotopes.

143.022 **Photonuclear interactions of ultrahigh energy cosmic rays and their astrophysical consequences.**
J. L. Puget, F. W. Stecker, J. H. Bredekamp.
Astrophys. Journ., Vol. 205, 638 - 654 (1976).

The authors present the results of detailed Monte Carlo calculations of the interaction histories of ultrahigh energy cosmic-ray nuclei with intergalactic radiation fields using improved estimates of these fields and empirical determinations of photonuclear cross sections including multinuclear disintegrations for nuclei up to ^{56}Fe. Intergalactic and galactic energy loss rates and nucleon loss rates for nuclei up to ^{56}Fe are also given. The results of the calculations indicate that ultrahigh energy cosmic rays cannot be universal in origin regardless of whether they are protons or nuclei.

143.023 **The isotopic composition of cosmic rays with $5 \leqslant Z \leqslant 26$.**
A. J. Fisher, F. A. Hagen, R. C. Maehl, J. F. Ormes, J. F. Arens.
Astrophys. Journ., Vol. 205, 938 - 946 (1976).

The authors report results obtained from a high-altitude balloon flight in 1973 August. In the energy range 350 MeV per amu $\lesssim T \lesssim 600$ MeV per amu they find ^{16}O/O > 0.9, ^{12}C/C > 0.9, ^{14}N/N = 0.6 ± 0.2. Additionally they find even-Z elements in the $12 \leqslant Z \leqslant 16$ group to be predominantly α-particle nuclei. Neon, on the other hand, seems to have a significant component of neutron-rich isotopes.

143.024 **The scattering of cosmic rays by magnetic bubbles.**
R. F. Flewelling, F. V. Coroniti.
Astrophys. Journ.,(*Letters*), Vol. 205, L135 - L138 (1976).

During large flares, the sun ejects magnetic bubbles which are then convected by the solar wind into the interstellar medium. The bubbles partake of the spherical expansion of the solar wind and grow to spatial scales comparable to the gyroradii of $10^2 - 10^3$ GeV cosmic rays. Assuming that most stars leave behind a wake of magnetic bubbles in the interstellar medium, and that the rate of bubble production is comparable to the solar rate, $10^2 - 10^3$ GeV and possibly higher energy cosmic rays will be well scattered and have mean free paths consistent with the compound diffusion theory.

143.025 **The age distribution of galactic cosmic rays.**
A. J. Owens.
Astrophys. Space Sci., Vol. 40, 357 - 367 (1976).

The general solution for the distribution of ages for primary galactic cosmic rays is given for a class of steady-state, bounded models of cosmic-ray diffusion in the Galaxy. Both one- and three-dimensional models are considered, with point sources and distributed sources. The leaky-box model, with an exponential age distribution, is a good approximation to most diffusive models. It is shown that one-dimensional ('disk') models are consistent with both age and anisotropy data for galactic cosmic rays regardless of whether production takes place near the galactic center or throughout the disk.

143.026 **On the possible annihilation nature of cosmic gamma flares.** P. I. Fomin.

Izv. Krymskoj Astrofiz. Obs., Vol. 54, 340 - 343 (1976). In Russian. – Abstract of a paper presented at the seminar on "X-ray and gamma-ray astronomy" at the Crimean Astrophys. Obs., 1974, (see 012.009).

143.027 Observation of cosmic gamma radiation with energy $E_\gamma \geqslant 100$ MeV aboard the artificial earth satellite Cosmos 561. A. I. Belyaevskij, V. L. Bokov, V. K. Bocharkin, I. F. Bugakov, Yu. G. Derevitskij, B. A. Dmitriev, G. M. Gorodinskij, E. M. Kruglov, E. V. Myakinin, G. A. Pyatigorskij, E. I. Chujkin.
Izv. Krymskoj Astrofiz. Obs., Vol. 54, 343 - 346 (1976). In Russian. – Abstract of a paper presented at the seminar on "X-ray and gamma-ray astronomy" at the Crimean Astrophys. Obs., 1974, (see 012.009).

143.028 Nondiffusive propagation of cosmic rays in the solar system and in extragalactic radio sources.
J. A. Earl.
Astrophys. Journ., Vol. 206, 301 - 311 (1976).

If charged particles are scattered by random magnetic fields while they propagate along the diverging lines of force of a spatially inhomogeneous guiding field, the diffusive mode of transport which occurs when adiabatic focusing is weak compared to scattering, gives way to novel coherent modes when focusing becomes dominant. This paper begins with a nonmathematical discussion of the higher-order transport phenomena that underlie these modes, and goes on to explore some astrophysical implications of their existence. In an interplanetary context, one of the new modes, the supercoherent mode, corresponds exactly to the "scatter-free" propagation of kilovolt solar-flare electrons. Moreover, quasi-diffusive propagation in the presence of moderately strong focusing offers an explanation of several poorly understood aspects of solar cosmic-ray events. On a much larger scale, focused transport provides an interpretation of many observed characteristics of extragalactic radio sources. In particular, their double structure is explained in terms of basic transport phenomena.

143.029 Secondary cosmic-ray $e^{\pm}$ from 1 to 100 GeV in the upper atmosphere and interstellar space, and interpretation of a recent e^{+} flux measurement.
C. D. Orth, A. Buffington.
Astrophys. Journ., Vol. 206, 312 - 332 (1976).

Secondary fluxes of cosmic-ray $e^{\pm}$ from the decay of mesons produced by nuclear interactions are calculated for depths under 10 g cm^{-2} of atmosphere or interstellar space for energies from 1 to 100 GeV. Secondary meson spectra applicable for $e^{\pm}$ energies $\gtrsim 5$ GeV are obtained from the recently measured spectra of Carey et al. using Monte Carlo techniques. An analytic model is presented which identifies all essential parameters and enables easy calculation of $e^{\pm}$ fluxes for various parameter values. This model is used to interpret the e^{+} measurement of Buffington, Orth, and Smoot. The authors find the mean thickness of interstellar and source material to be 4.3 (+1.8, −1.2) g cm^{-2} for cosmic-ray e^{+} above 4 GeV.

143.030 Solar modulation and a galactic origin for the anomalous component observed in low-energy cosmic rays. L. A. Fisk.
Astrophys. Journ., Vol. 206, 333 - 341 (1976).

The current theory of solar modulation can be used to argue that the cosmic-ray component at low energies, which is observed to have an anomalous composition, is not of galactic origin, i.e., it is not a component of the galactic cosmic-ray flux. The current theory predicts, from quite general considerations, that an unreasonably large intensity of cosmic rays, by many orders of magnitude, would be required in the interstellar medium to account for the observed fluxes. Conceivably, the current modulation theory could be modified so that only reasonable interstellar fluxes are predicted. One

such modification involves an unusual scheme for particle diffusion in the interplanetary medium.

143.031 Astronomia con particelle di alta energia: I raggi cosmici. N. Mandolesi, G. G. C. Palumbo.
Coelum, Vol. 44, 1 - 13, 58 - 68 (1976).

143.032 Anisotropic diffusion of cosmic rays in the interplanetary space. III. Radial and transversal gradients.
L. I. Dorman, N. P. Milovidova.
Geomagn. Aeronom., Vol. 16, 401- 406 (1976). In Russian.

143.033 Galactic cosmic-ray intensity 0.99 to 5.26 astronomical units from the sun.
M. F. Thomsen, J. A. Van Allen.
Astrophys. Journ., Vol. 206, 599 - 615 (1976).

The intensity of galactic cosmic radiation in interplanetary space has been observed over the heliocentric radial range $0.99 < r < 5.26$ AU with five Geiger-Müller tubes on the Jupiter-bound spacecraft Pioneer 10 and over the range $1.00 < r < 3.59$ AU with three similar tubes on Pioneer 11, also bound for Jupiter. The observations span the time period 1972 March 3 to 1974 April 29, near the epoch of minimum solar activity. The authors find that the radial gradient of the intensity of particles $E > 80$ MeV nucleon^{-1} is zero to within an experimental uncertainty of less than 2 percent per AU over the range $0.99 < r < 5.26$ AU. Hence, it is less than the theoretically expected value by at least a factor of 5.

143.034 The isotopic composition of hydrogen and helium in low-energy cosmic rays.
R. A. Mewaldt, E. C. Stone, R. E. Vogt.
Astrophys. Journ., Vol. 206, 616 - 621 (1976).

The isotopes ^{2}H and ^{3}He have been identified in low-energy cosmic rays during solar-quiet periods from 1973 January to 1974 October. These observations cover the energy intervals 5–29 MeV per nucleon for ^{2}H and 7–50 MeV per nucleon for ^{3}He. The authors find that the energy spectra of ^{1}H, ^{2}H, and ^{3}He all fall rapidly with decreasing energy, giving ^{2}H/^{1}H and ^{3}He/^{1}H ratios which are essentially independent of energy as expected from current theories of the solar modulation of galactic cosmic rays. The measured ^{4}He spectrum, however, is essentially flat below ~40 MeV per nucleon, suggesting that there may be contributions from a local, nonsolar source of ^{4}He.

143.035 On the origin of cosmic rays (Some problems of high-energy astrophysics).
V. L. Ginzburg, V. S. Ptuskin.
Uspekhi fiz. nauk, Vol. 117, 585 - 636 (1975). In Russian.
Abstr. in Referativ. Zhurn. 51. Astron., 5.51.767 (1976).

143.036 A simple model for the Forbush decrease in the galactic cosmic ray flux. C. J. Bland.
Journ. Geophys. Res., Vol. 81, 1807 - 1811 (1976).
Brief report.

143.037 Detection and study of intensity variations of cosmic rays of interplanetary and magnetospheric origin from data of the world network of observatories.
L. I. Dorman, A. V. Sergeev.
Cosmic rays, No. 15, (see 003.006), p. 5 - 12 (1975). In Russian.

143.038 Solar modulation of galactic cosmic rays of moderate and low energy. Yu. M. Nikolaev.
Cosmic rays, No. 15, (see 003.006), p. 40 - 71 (1975). In Russian.

143.039 On the meteor effect in cosmic rays.
S. A. Bel'skij.

Cosmic rays, No. 15, (see 003.006), p. 75 - 81 (1975). In Russian.

143.040 Modulation of galactic cosmic rays and solar activity distribution over heliocoordinates.
A. K. Efimov, A. G. Zusmanovich, E. V. Kolomeets, Yu. A. Shakhova.
Cosmic Rays, No. 15, (see 003.006), p. 86 - 89 (1975). In Russian.

143.041 North-south asymmetry of cosmic ray intensity.
L. E. Gajnova, G. A. Gonchar, A. G. Zusmanovich, V. I. Ivanov, K. Imazhanova, E. V. Kolomeets, V. T. Pivneva, R. A. Chumbalova, Yu. A. Shakhova;
Cosmic rays, No. 15, (see 003.006), p. 90 - 95 (1975). In Russian.

143.042 Convective transfer effect on relaxation of galactic cosmic rays. V. F. Zakharchenko.
Cosmic rays, No. 15, (see 003.006), p. 96 - 99 (1975). In Russian.

143.043 Cosmic ray intensity variations with a period of 20 - 60 minutes. N. P. Chirkov, V. I. Ipat'ev.
Cosmic rays, No. 15, (see 003.006), p. 100 - 102 (1975). In Russian.

143.044 Diurnal distribution and intensity variations of cosmic rays in the stratosphere above Norilsk.
L. I. Dorman, V. P. Karpov, A. V. Palamarchuk.
Cosmic rays, No. 15, (see 003.006), p. 109 - 117 (1975). In Russian.

143.045 Study of the altitude profile and angular distribution of cosmic ray intensity.
V. N. Aleksandrov, Ya. L. Blokh, I. V. Dorman, L. I. Dorman, V. L. Kashevarov, Yu. N. Kuzin, I. Ya. Libin, Yu. N. Sorokin, Kh. Khamirzov, G. Sh. Shkhalakhov, V. G. Yanke.
Cosmic rays, No. 15, (see 003.006), p. 153 - 163 (1975). In Russian.

143.046 Some problems of cosmic-ray variation study from underground observational data.
A. A. Bishara, L. I. Dorman.
Cosmic rays, No. 15, (see 003.006), p. 198 - 202 (1975). In Russian.

143.047 Synchrotron radiation, high energy gamma rays and the distribution of galactic cosmic rays.
J. Paul, M. Cassé, C. J. Cesarsky.
14th Intern. Cosmic Ray Conf., (see 012.011), Vol. 1, 59 - 64 (1975).

143.048 On the origin of cosmic rays. F. W. Stecker.
14th Intern. Cosmic Ray Conf., (see 012.011), Vol. 1, 70 - 73 (1975).

143.049 Evidence for galactic origin of low energy cosmic rays.
D. Dodds, A. W. Strong, A. W. Wolfendale.
14th Intern. Cosmic Ray Conf., (see 012.011), Vol. 1, 80 - 82 (1975).

143.050 Coincident balloon-borne network for cosmic gamma-ray burst studies.
T. L. Cline, U. D. Desai, B. J. Teegarden, W. K. H. Schmidt.
14th Intern. Cosmic Ray Conf., (see 012.011), Vol. 1, 213 (1975). – Abstract.

143.051 Preliminary results of a search for bursts of 10^{13} to 10^{14} eV gamma rays using spaced cosmic ray sta-

tions in time coincidence.
D. J. Fegan, B. McBreen, C. O'Sullivan, V. Ruddy.
14th Intern. Cosmic Ray Conf., (see 012.011), Vol. 1, 214 - 218 (1975).

143.052 Cosmic ray measurements of light and medium nuclei using a new detector.
A. M. Preszler, J. C. Kish, J. A. Lezniak, G. Simpson, W. R. Webber.
14th Intern. Cosmic Ray Conf., (see 012.011), Vol. 1, 243 (1975). – Abstract.

143.053 Propagation of cosmic rays at medium energies (≈ 100 to 400 MeV/u).
F. W. O'Dell, M. M. Shapiro, R. Silberberg, C. H. Tsao.
14th Intern. Cosmic Ray Conf., (see 012.011), Vol. 1, 244 (1975). – Abstract.

143.054 A balloon measurement of the cosmic ray element abundances. J. F. Ormes, A. Fisher, F. Hagen, R. Maehl, J. F. Arens.
14th Intern. Cosmic Ray Conf., (see 012.011), Vol. 1, 245 - 250 (1975).

143.055 Charge and energy spectra of heavy cosmic rays.
J. C. Benegas, M. H. Israel, J. Klarmann, R. C. Maehl.
14th Intern. Cosmic Ray Conf., (see 012.011), Vol. 1, 251 - 255 (1975).

143.056 Nuclear composition of the cosmic rays at energies between 0.4 and a few GeV/nucleon.
E. Juliusson, P. Meyer.
14th Intern. Cosmic Ray Conf., (see 012.011), Vol. 1, 256 (1975). – Abstract.

143.057 Irregularities in cosmic ray composition.
N. Lund, I. L. Rasmussen, B. Peters, N. J. Westergaard.
14th Intern. Cosmic Ray Conf., (see 012.011), Vol. 1, 257 - 262 (1975).

143.058 Composition changes in the iron group between 0.5 and 10 GeV/n. N. Lund, I. L. Rasmussen, B. Peters, M. Rotenberg, N. J. Westergaard.
14th Intern. Cosmic Ray Conf., (see 012.011), Vol. 1, 263 - 266 (1975).

143.059 High resolution study of nucleonic cosmic rays with $Z \gtrsim 30$. P. H. Fowler, C. Alexandre, V. M. Clapham, D. L. Henshaw, C. O'Ceallaigh, D. O'Sullivan, A. Thompson.
14th Intern. Cosmic Ray Conf., (see 012.011), Vol. 1, 267 (1975). – Abstract.

143.060 The Skylab ultraheavy cosmic ray experiment.
P. B. Price, E. K. Shirk.
14th Intern. Cosmic Ray Conf., (see 012.011), Vol. 1, 268 - 272 (1975).

143.061 Charge composition and energy spectra of primary cosmic ray nuclei between 5 and 100 GeV per nucleon. J. Caldwell, P. Meyer.
14th Intern. Cosmic Ray Conf., (see 012.011), Vol. 1, 273 - 277 (1975).

143.062 An observation of the high charge cosmic ray nuclei.
W. R. Scarlett, P. S. Freier, C. J. Waddington.
14th Intern. Cosmic Ray Conf., (see 012.011), Vol. 1, 278 - 279 (1975).

143.063 Abundance ratios for primary cosmic-ray nuclei

from Be to Fe for 5 to 50 GV/c.
C. D. Orth, A. Buffington, G. F. Smoot.
14th Intern. Cosmic Ray Conf., (see 012.011), Vol. 1, 280 - 284 (1975).

143.064 The energy spectrum of 15–50 BeV/nuc cosmic rays.
J. A. Lezniak, J. Kish, A. Preszler, W. R. Webber.
14th Intern. Cosmic Ray Conf., (see 012.011), Vol. 1, 285 (1975). – Abstract.

143.065 VH particles (20 $\leq$ Z < 30) at high rigidity.
J. M. Kidd, J. P. Wefel.
14th Intern. Cosmic Ray Conf., (see 012.011), Vol. 1, 286 (1975). – Abstract.

143.066 Chemical composition of arriving cosmic rays from Si to Ni in the GeV region from the TD 1 satellite.
C. Julliot, L. Koch, N. Petrou.
14th Intern. Cosmic Ray Conf., (see 012.011), Vol. 1, 287 (1975). – Abstract.

143.067 Rigidity spectrum of iron-group nuclei.
T. Cleghorn, S. A. Stephens, G. G. Badhwar, R. L. Golden.
14th Intern. Cosmic Ray Conf., (see 012.011), Vol. 1, 288 (1975). – Abstract.

143.068 Observation of relativistic cosmic ray particles with Z $\gtrsim$ 50. T. Doke, H. Okamoto, E. Shibamura, T. Hayashi, K. Ito, T. Yanagimachi, S. Kobayashi, M. Miyajima, T. Saito, K. Nagata.
14th Intern. Cosmic Ray Conf., (see 012.011), Vol. 1, 289 (1975). – Abstract.

143.069 Energy spectrum of VH-nuclei observed with emulsion chamber.
T. Matsubayashi, M. Noma, T. Saito, Y. Sato, H. Sugimoto.
14th Intern. Cosmic Ray Conf., (see 012.011), Vol. 1, 290 - 294 (1975).

143.070 Charge and energy spectra of UH nuclei from flights at cutoffs of approximately 2 GV.
W. Z. Osborne, L. S. Pinsky, P. B. Price, E. K. Shirk.
14th Intern. Cosmic Ray Conf., (see 012.011), Vol. 1, 295 (1975). – Abstract.

143.071 Search for antinuclei in the primary cosmic rays. N. S. Ivanova, D. G. Baranov, E. A. Yakubovsky (*Yakubovskij*).
14th Intern. Cosmic Ray Conf., (see 012.011), Vol. 1, 300 - 304 (1975).

143.072 Measurements of the spectrum and isotopic composition of cosmic ray helium and hydrogen nuclei from 100–230 MeV/nucleon.
H. W. Leech, J. J. O'Gallagher.
14th Intern. Cosmic Ray Conf., (see 012.011), Vol. 1, 305 (1975). – Abstract.

143.073 The isotopic composition of hydrogen and helium in low energy cosmic rays.
R. A. Mewaldt, E. C. Stone, R. E. Vogt.
14th Intern. Cosmic Ray Conf., (see 012.011), Vol. 1, 306 - 311 (1975).

143.074 Measurements of ^{3}He and ^{4}He nuclei using a balloon borne telescope. W. R. Webber, N. J. Schofield.
14th Intern. Cosmic Ray Conf., (see 012.011), Vol. 1, 312 - 317 (1975).

143.075 The isotopic composition of galactic cosmic rays – Li-N. F. B. McDonald, N. Lal, B. J. Teegarden, J. H. Trainor, W. R. Webber.
14th Intern. Cosmic Ray Conf., (see 012.011), Vol. 1, 318 (1975). – Abstract.

143.076 The low energy cosmic ray H^2 and He^3 spectra and the anomalous He^4 component.
M. Garcia-Munoz, G. M. Mason, J. A. Simpson.
14th Intern. Cosmic Ray Conf., (see 012.011), Vol. 1, 319 - 324 (1975).

143.077 The isotopic composition of galactic cosmic ray Li, Be, and B.
M. Garcia-Munoz, G. M. Mason, J. A. Simpson.
14th Intern. Cosmic Ray Conf., (see 012.011), Vol. 1, 325 - 330 (1975).

143.078 The relative abundances of galactic cosmic ray Be^7, Be^9, and Be^{10}.
M. Garcia-Munoz, G. M. Mason, J. A. Simpson.
14th Intern. Cosmic Ray Conf., (see 012.011), Vol. 1, 331 - 336 (1975).

143.079 A measurement of the carbon isotopic composition in primary cosmic radiation.
C. Bjarle, N.-Y. Herrström, L. Jacobsson, G. Jönsson, K. Kristiansson.
14th Intern. Cosmic Ray Conf., (see 012.011), Vol. 1, 337 - 342 (1975).

143.080 An experimental study of the isotopic composition of cosmic ray nitrogen and oxygen.
L. Jacobsson, G. Jönsson, K. Kristiansson.
14th Intern. Cosmic Ray Conf., (see 012.011), Vol. 1, 343 - 348 (1975).

143.081 Isotopic composition of the anomalous low energy cosmic ray nitrogen and oxygen.
R. A. Mewaldt, E. C. Stone, S. B. Vidor, R. E. Vogt.
14th Intern. Cosmic Ray Conf., (see 012.011), Vol. 1, 349 - 354 (1975).

143.082 Isotopic measurements of the cosmic ray nuclei at 1.7 GeV/n and 0.5 GeV/n. E. Juliusson.
14th Intern. Cosmic Ray Conf., (see 012.011), Vol. 1, 355 - 360 (1975).

143.083 Results on the isotopic composition of cosmic ray nuclei with 4 $\leq$ Z $\leq$ 10.
F. A. Hagen, A. J. Fisher, J. F. Ormes, J. F. Arens.
14th Intern. Cosmic Ray Conf., (see 012.011), Vol. 1, 361 - 366 (1975).

143.084 Astrophysical implications of the isotopic composition of cosmic rays.
R. Maehl, F. A. Hagen, A. J. Fisher, J. F. Ormes, M. Simon.
14th Intern. Cosmic Ray Conf., (see 012.011), Vol. 1, 367 - 372 (1975).

143.085 A Cerenkov-range analysis of the isotopic composition of cosmic rays with 6 $\leq$ Z $\leq$ 26.
A. J. Fisher, F. A. Hagen, R. Maehl, J. F. Ormes.
14th Intern. Cosmic Ray Conf., (see 012.011), Vol. 1, 373 - 378 (1975).

143.086 Mean isotopic composition of cosmic-ray iron at intermediate energies.
J. C. Benegas, M. H. Israel, J. Klarmann, R. C. Maehl.
14th Intern. Cosmic Ray Conf., (see 012.011), Vol. 1, 379 - 383 (1975).

143.087 **Isotopic composition of cosmic ray iron nuclei.**
K.-P. Bartholomä, G. Siegmon, W. Enge.
14th Intern. Cosmic Ray Conf., (see 012.011), Vol. 1, 384 - 388 (1975).

143.088 **New isotopic measurements of medium and heavy cosmic ray nuclei using a C × E telescope.**
G. Simpson, J. C. Kish, A. M. Preszler, W. R. Webber.
14th Intern. Cosmic Ray Conf., (see 012.011), Vol. 1, 389 (1975). – Abstract.

143.089 **Isotopic composition of cosmic ray nitrogen at 1.5 GeV/amu.** R. Dwyer, P. Meyer.
14th Intern. Cosmic Ray Conf., (see 012.011), Vol. 1, 390 - 394 (1975).

143.090 **Isotope resolution of the iron peak.**
R. P. Henke, E. V. Benton.
14th Intern. Cosmic Ray Conf., (see 012.011), Vol. 1, 395 - 399 (1975).

143.091 **Isotopic composition of cosmic ray nuclei in the iron group.** V. M. Clapham, P. H. Fowler,
C. O'Ceallaigh, D. O'Sullivan, A. Thompson.
14th Intern. Cosmic Ray Conf., (see 012.011), Vol. 1, 400 (1975). – Abstract.

143.092 **The quiet time flux of 0.16−1.6 MeV cosmic ray positrons.** R. A. Mewaldt, E. C. Stone, R. E. Vogt.
14th Intern. Cosmic Ray Conf., (see 012.011), Vol. 1, 401 (1975). – Abstract.

143.093 **Cosmic ray positron and negatron spectra between 20 and 800 MeV measured in 1974.**
R. C. Hartman, C. J. Pellerin.
14th Intern. Cosmic Ray Conf., (see 012.011), Vol. 1, 402 - 407 (1975).

143.094 **Interpretation of recent positron-electron measurements between 20 and 800 MeV.**
C. J. Pellerin, R. C. Hartman.
14th Intern. Cosmic Ray Conf., (see 012.011), Vol. 1, 408 - 410 (1975).

143.095 **Calculations of atmospheric and interstellar secondary $e^{\pm}$ in the 1 to 100 GeV range.**
C. D. Orth, A. Buffington.
14th Intern. Cosmic Ray Conf., (see 012.011), Vol. 1, 411 - 416 (1975).

143.096 **The electron energy spectrum in the 4−100 GeV range using a magnetic spectrometer.**
G. D. Badhwar, R. L. Golden, J. L. Lacy, T. Cleghorn, S. A. Stephens.
14th Intern. Cosmic Ray Conf., (see 012.011), Vol. 1, 417 - 418 (1975).

143.097 **The spectrum of cosmic electrons with energies between 6 and 100 GeV.**
C. A. Meegan, J. A. Earl.
14th Intern. Cosmic Ray Conf., (see 012.011), Vol. 1, 419 - 424 (1975).

143.098 **Intensity of primary electrons above 10 GeV.**
P. Freier, C. Gilman, C. J. Waddington.
14th Intern. Cosmic Ray Conf., (see 012.011), Vol. 1, 425 - 430 (1975).

143.099 **High energy primary electrons observed in the emulsion chamber.** M. Matsuo, J. Nishimura,
T. Kobayashi, K. Niu, E. Aizu, N. Hiraiwa, T. Taira.
14th Intern. Cosmic Ray Conf., (see 012.011), Vol. 1, 431 (1975). – Abstract.

143.100 **Spinning dense objects and the origin of cosmic rays.**
T. Gold.
14th Intern. Cosmic Ray Conf., (see 012.011), Vol. 2, 432 (1975). – Abstract.

143.101 **Cosmic rays from pre-main sequence stars?**
H. Reeves, C. J. Cesarsky.
14th Intern. Cosmic Ray Conf., (see 012.011), Vol. 2, 433 (1975). – Abstract.

143.102 **Generation of cosmic ray electrons by pulsars.**
E. Schrüfer, M. Grewing.
14th Intern. Cosmic Ray Conf., (see 012.011), Vol. 2, 434 (1975). – Abstract.

143.103 **Consistency of cosmic ray composition, acceleration mechanism and supernova models.**
D. N. Schramm, K. L. Hainebach, E. B. Norman.
14th Intern. Cosmic Ray Conf., (see 012.011), Vol. 2, 443 - 447 (1975).

143.104 **Iron isotopic ratios and the cosmic ray source.**
K. L. Hainebach, D. N. Schramm, W. D. Arnett.
14th Intern. Cosmic Ray Conf., (see 012.011), Vol. 2, 448 - 450 (1975).

143.105 **Nucleosynthesis of cosmic-ray nuclei.**
R. Silberberg, M. M. Shapiro, C. H. Tsao.
14th Intern. Cosmic Ray Conf., (see 012.011), Vol. 2, 451 - 454 (1975).

143.106 **Time delay between the nucleosynthesis of cosmic rays and their acceleration to relativistic energies.**
A. Soutoul, M. Cassé, E. Juliusson.
14th Intern. Cosmic Ray Conf., (see 012.011), Vol. 2, 455 - 458 (1975).

143.107 **Neutron bursts and possible anomalous cosmic ray abundances.** J. B. Blake, D. N. Schramm.
14th Intern. Cosmic Ray Conf. (see 012.011), Vol. 2, 459 - 464 (1975).

143.108 **The trapping of cosmic rays around supernovae by plasma instabilities.** R. Kulsrud, E. Zweibel.
14th Intern. Cosmic Ray Conf., (see 012.011), Vol. 2, 465 - 468 (1975).

143.109 **Rigidity dependent escape of cosmic rays from supernova remnants.** J. C. Higdon.
14th Intern. Cosmic Ray Conf., (see 012.011), Vol. 2, 469 - 473 (1975).

143.110 **Origin of cosmic-ray electrons from supernova remnants.** K. P. Beuermann, J. K. Daugherty.
14th Intern. Cosmic Ray Conf., (see 012.011), Vol. 2, 474 (1975). – Abstract.

143.111 **Supernovae as cosmic ray sources.**
R. Cowsik, L. W. Wilson.
14th Intern. Cosmic Ray Conf., (see 012.011), Vol. 2, 475 - 480 (1975).

143.112 **Positron and antiproton abundances as a test for cosmic ray propagation models.**
M. Lachièze-Rey, C. J. Cesarsky.
14th Intern. Cosmic Ray Conf., (see 012.011), Vol. 2, 489 - 494 (1975).

143.113 **Excitation functions for some nuclear reactions relevant to the cosmic ray propagation problem.**
G. M. Raisbeck, F. Yiou.
14th Intern. Cosmic Ray Conf., (see 012.011), Vol. 2, 495 - 498 (1975).

143.114 **Cross sections for the production of Li and Be isotopes in carbon targets irradiated by 300 GeV protons.** G. M. Raisbeck, J. Lestringuez, F. Yiou.
14th Intern. Cosmic Ray Conf., (see 012.011), Vol. 2, 499 - 501 (1975).

143.115 **Cross sections for nuclear reactions induced by high energy alpha particles and the influence of interstellar helium in cosmic ray propagation.**
G. M. Raisbeck, F. Yiou.
14th Intern. Cosmic Ray Conf., (see 012.011), Vol. 2, 502 - 505 (1975).

143.116 **Experimental cross-sections for the production of stable isotopes by high energy spallation of iron.**
C. Perron.
14th Intern. Cosmic Ray Conf., (see 012.011), Vol. 2, 506 - 509 (1975).

143.117 **The fragmentation of 0.5 GeV/u nitrogen-14 ions.**
J. M. Kidd, J. P. Wefel, W. Schimmerling, K. Vosburgh.
14th Intern. Cosmic Ray Conf., (see 012.011), Vol. 2, 510 - 515 (1975).

143.118 **Breakup cross sections of cosmic rays with nuclei in interstellar helium, in air and in detector materials.**
C. H. Tsao, R. Silberberg.
14th Intern. Cosmic Ray Conf., (see 012.011), Vol. 2, 516 - 520 (1975).

143.119 **A chart of cosmic ray isotopes.** C. J. Waddington.
14th Intern. Cosmic Ray Conf., (see 012.011), Vol. 2, 521 - 525 (1975).

143.120 **Effect of new cross-section measurements on the estimate of cosmic-ray "age".**
F. W. O'Dell, M. M. Shapiro, R. Silberberg, C. H. Tsao.
14th Intern. Cosmic Ray Conf., (see 012.011), Vol. 2, 526 - 531 (1975).

143.121 **What new cross sections say about source composition and cosmic-ray propagation.**
M. M. Shapiro, R. Silberberg, C. H. Tsao.
14th Intern. Cosmic Ray Conf., (see 012.011), Vol. 2, 532 - 537 (1975).

143.122 **Cosmic ray chronology from isotopic composition.**
M. M. Shapiro, R. Silberberg.
14th Intern. Cosmic Ray Conf., (see 012.011), Vol. 2, 538 - 543 (1975).

143.123 **Cl^{36} and the age of the cosmic rays.**
M. Cassé, P. Goret, S. Regnier.
14th Intern. Cosmic Ray Conf., (see 012.011), Vol. 2, 544 - 548, with a correction, Vol. 12, 4413 (1975).

143.124 **Cosmic ray origin of the rare odd-odd nuclei and Li, Be, and B.** K. L. Hainebach, D. N. Schramm.
14th Intern. Cosmic Ray Conf., (see 012.011), Vol. 2, 549 - 553 (1975).

143.125 **Galactic light isotopes: significance of the present observations.** J.-P. Meyer.
14th Intern. Cosmic Ray Conf., (see 012.011), Vol. 2, 554 -

559 (1975).

143.126 **Electron capture isotopes as cosmic ray "hydrometers".**
G. M. Raisbeck, G. Comstock, C. Perron, F. Yiou.
14th Intern. Cosmic Ray Conf., (see 012.011), Vol. 2, 560 - 563 (1975).

143.127 **Compositional variation of Fe-group nuclei in cosmic rays due to K-capture effect.**
A. S. Tamahane, V. S. Venkatavaradan, J. N. Goswami, S. K. Gupta.
14th Intern. Cosmic Ray Conf., (see 012.011), Vol. 2, 564 - 569 (1975).

143.128 **Fluxes of gamma rays, antiprotons and deuterons in cosmic rays due to interstellar proton-proton collisions.** S. N. Ganguli, B. V. Sreekantan.
14th Intern. Cosmic Ray Conf., (see 012.011), Vol. 2, 570 - 574 (1975).

143.129 **Investigation of the sidereal anisotropy of muons with energies greater than 2 GeV.**
R. C. Uhr, E. Fähnders, K. Koseck, G. Klemke, H. Jokisch, W. D. Dau.
14th Intern. Cosmic Ray Conf., (see 012.011), Vol. 2, 575 - 577 (1975).

143.130 **Search for anisotropy of muons.**
H. Jokisch, K. Carstensen.
14th Intern. Cosmic Ray Conf., (see 012.011), Vol. 2, 578 - 580 (1975).

143.131 **The Utah 1600 GV anisotropy detector.**
H. E. Bergeson, D. E. Groom, W. J. West.
14th Intern. Cosmic Ray Conf., (see 012.011), Vol. 2, 581 - 584 (1975).

143.132 **Arrival direction of cosmic rays from underground muons in the Mont-Blanc tunnel.**
A. R. Bazer-Bachi, G. Vedrenne, W. R. Sheldon.
14th Intern. Cosmic Ray Conf., (see 012.011), Vol. 2, 585 (1975). – Abstract.

143.133 **Galactic cosmic ray anisotropy at $\approx 6 \times 10^{13}$ eV.**
T. Gombosi, J. Kóta, A. J. Somogyi, A. Varga, B. Betev, L. Katsarski, S. Kavlakov, I. Khirov.
14th Intern. Cosmic Ray Conf., (see 012.011), Vol. 2, 586 - 591 (1975).

143.134 **Fluctuation effects on anisotropy measurements.**
J. Linsley.
14th Intern. Cosmic Ray Conf., (see 012.011), Vol. 2, 592 - 597 (1975).

143.135 **Anisotropy of charged cosmic rays above 10^{17} eV during 1959—1963.** J. Linsley.
14th Intern. Cosmic Ray Conf., (see 012.011), Vol. 2, 598 - 603 (1975).

143.136 **Arrival directions of cosmic rays with energies above 10^{18} eV.**
D. M. Edge, R. J. O. Reid, A. A. Watson, J. G. Wilson.
14th Intern. Cosmic Ray Conf., (see 012.011), Vol. 2, 604 - 608 (1975).

143.137 **The arrival directions of cosmic rays of energy $> 5 \times 10^{18}$ eV.**
A. D. Bray, L. Goorevich, L. Horton, C. B. A. McCusker, L. S. Peak, P. Rapp, J. Ulrichs, M. M. Winn.
14th Intern. Cosmic Ray Conf., (see 012.011), Vol. 2, 609 -

611 (1975).

143.138 **Search for correlations of the arrival directions of ultra-high energy cosmic rays with specific point source candidates.** P. J. Kiraly, J. L. Osborne, M. White, A. W. Wolfendale.
14th Intern. Cosmic Ray Conf., (see 012.011), Vol. 2, 612 - 617 (1975).

143.139 **On the large-scale features of the directional distribution of cosmic rays above 10^{19} eV.**
E. Kiraly, P. Kiraly, J. L. Osborne, M. White.
14th Intern. Cosmic Ray Conf., (see 012.011), Vol. 2, 618 - 622 (1975).

143.140 **Anisotropy of very high energy cosmic rays.** A. M. Hillas, M. Ouldridge.
14th Intern. Cosmic Ray Conf., (see 012.011), Vol. 2, 623 (1975).

143.141 **Possible role of interstellar turbulence in the confinement of galactic cosmic rays.** J. Skilling.
14th Intern. Cosmic Ray Conf., (see 012.011), Vol. 2, 624 - 626 (1975).

143.142 **The range of turbulence in the interstellar medium and the confinement of high energy cosmic rays.**
I. McIvor.
14th Intern. Cosmic Ray Conf., (see 012.011), Vol. 2, 627 - 631 (1975).

143.143 **Energy dependence of galactic cosmic ray propagation and power spectrum of galactic magnetic field irregularities.** A. J. Somogyi.
14th Intern. Cosmic Ray Conf., (see 012.011), Vol. 2, 632 (1975). – Abstract.

143.144 **Cosmic ray confinement in the Galaxy and the interstellar spectrum of hydromagnetic waves.**
C. J. Cesarsky.
14th Intern. Cosmic Ray Conf., (see 012.011), Vol. 2, 633 (1975). – Abstract.

143.145 **Interstellar propagation of low energy cosmic rays.**
C. J. Cesarsky.
14th Intern. Cosmic Ray Conf., (see 012.011), Vol. 2, 634 - 638, with a correction, Vol. 12, 4414 (1975).

143.146 **Hydrostatic equilibrium of the gaseous disk of the Galaxy and the extent of cosmic ray confinement.**
G. D. Badhwar, S. A. Stephens.
14th Intern. Cosmic Ray Conf., (see 012.011), Vol. 2, 639 - 643 (1975).

143.147 **Atomic properties of the elements and cosmic ray composition at the source.**
M. Cassé, P. Goret, C. J. Cesarsky.
14th Intern. Cosmic Ray Conf., (see 012.011), Vol. 2, 646 - 651 (1975).

143.148 **Pathlength distribution and source composition of cosmic ray nuclei.**
M. Meneguzzi, C. J. Cesarsky, J. P. Meyer.
14th Intern. Cosmic Ray Conf., (see 012.011), Vol. 2, 652 (1975). – Abstract.

143.149 **Source composition of cosmic rays at high energy.**
E. Juliusson, C. J. Cesarsky, M. Meneguzzi, M. Cassé.
14th Intern. Cosmic Ray Conf., (see 012.011), Vol. 2, 653 - 658 (1975).

143.150 **The nested leaky-box model for galactic cosmic rays.** R. Cowsik, L. W. Wilson.
14th Intern. Cosmic Ray Conf., (see 012.011), Vol. 2, 659 - 664 (1975).

143.151 **The galactic propagation of cosmic ray nuclei between 10^9 and 10^{13} eV.**
G. J. Dickinson, J. L. Osborne.
14th Intern. Cosmic Ray Conf., (see 012.011), Vol. 2, 665 - 670 (1975).

143.152 **The propagation and confinement of cosmic ray electrons in the Galaxy.**
D. K. French, J. L. Osborne.
14th Intern. Cosmic Ray Conf., (see 012.011), Vol. 2, 671 (1975). – Abstract.

143.153 **A formalism for cosmic ray propagation studies.** R. L. Golden, G. D. Badhwar, S. A. Stephens.
14th Intern. Cosmic Ray Conf., (see 012.011), Vol. 2, 672 - 677 (1975).

143.154 **On the age distribution of galactic cosmic rays.** A. J. Owens.
14th Intern. Cosmic Ray Conf., (see 012.011), Vol. 2, 678 - 683 (1975).

143.155 **A simple model for galactic cosmic rays which is consistent with age and anisotropy data.**
A. J. Owens.
14th Intern. Cosmic Ray Conf., (see 012.011), Vol. 2, 684 - 689 (1975).

143.156 **Propagation of 10^{11}- 10^{14} eV particles in the Galaxy.** T. Gombosi, J. Kóta.
14th Intern. Cosmic Ray Conf., (see 012.011), Vol. 2, 690 - 694 (1975).

143.157 **Some topics connected with the problem of cosmic ray origin.** V. L. Ginzburg, V. S. Ptuskin.
14th Intern. Cosmic Ray Conf., (see 012.011), Vol. 2, 695 - 699 (1975).

143.158 **Nonuniform and energy dependent diffusion of the relativistic electrons in the Galaxy.**
S. V. Bulanov, V. A. Dogiel.
14th Intern. Cosmic Ray Conf., (see 012.011), Vol. 2, 706 - 710 (1975).

143.159 **Metagalactic nuclei of ultra-high energies.** V. S. Berezinsky (*Berezinskij*), S. I. Grigor'eva, G. T. Zatsepin.
14th Intern. Cosmic Ray Conf., (see 012.011), Vol. 2, 711 - 716 (1975).

143.160 **The effect of intergalactic propagation on the energy spectrum of cosmic ray nuclei above 10^{15} eV.**
A. M. Hillas.
14th Intern. Cosmic Ray Conf., (see 012.011), Vol. 2, 717 - 722 (1975).

143.161 **The origin of very high cosmic rays.** S. A. Colgate.
14th Intern. Cosmic Ray Conf., (see 012.011), Vol. 2, 723 - 728 (1975).

143.162 **Mass composition of primary cosmic rays above 10^{18} eV.**
W. Tkaczyk, J. Wdowczyk, A. W. Strong, A. W. Wolfendale.
14th Intern. Cosmic Ray Conf., (see 012.011), Vol. 2, 729 - 733 (1975).

143.163 **The interactions of ultrahigh energy cosmic ray nuclei with intergalactic photon fields: relation between mean atomic mass, age, and origin.**
J. L. Puget, F. W. Stecker.
14th Intern. Cosmic Ray Conf., (see 012.011), Vol. 2, 734 - 738 (1975).

143.164 **Propagation of relativistic dust grains in interstellar space.** K. M. V. Apparao.
14th Intern. Cosmic Ray Conf., (see 012.011), Vol. 2, 739 - 741 (1975).

143.165 **Composition of primaries and atmospheric secondaries for a two-component model of origin.**
K. Sitte.
14th Intern. Cosmic Ray Conf., (see 012.011), Vol. 2, 742 - 747 (1975).

143.166 **On the quiet-time increases of low energy cosmic ray electrons.** J. L'Heureux, P. Meyer.
14th Intern. Cosmic Ray Conf., (see 012.011), Vol. 2, 748 - 751 (1975).

143.167 **Anomalous composition and energy spectra of cosmic rays below 20 MeV/nucleon.**
B. Klecker, D. Hovestadt, G. Gloeckler, C. Y. Fan.
14th Intern. Cosmic Ray Conf., (see 012.011), Vol. 2, 786 - 791 (1975).

143.168 **IMP 6, 7 and 8 observations of the composition and the variations of low energy cosmic rays.**
T. T. von Rosenvinge, F. B. McDonald.
14th Intern. Cosmic Ray Conf., (see 012.011), Vol. 2, 792 - 797 (1975).

143.169 **The elemental composition of 4 - 30 MeV/nuc cosmic ray nuclei with $1 \leqslant Z \leqslant 8$.**
R. A. Mewaldt, E. C. Stone, S. B. Vidor, R. E. Vogt.
14th Intern. Cosmic Ray Conf., (see 012.011), Vol. 2, 798 - 803 (1975).

143.170 **Implications of time variations for the origin of low energy cosmic ray nitrogen and oxygen nuclei.**
R. A. Mewaldt, E. C. Stone, R. E. Vogt.
14th Intern. Cosmic Ray Conf., (see 012.011), Vol. 2, 804 - 809 (1975).

143.171 **On the oxygen and nitrogen enhancements observed in low energy cosmic rays.** L. A. Fisk.
14th Intern. Cosmic Ray Conf., (see 012.011), Vol. 2, 810 - 815 (1975).

143.172 **Observations of low energy C. N, O and Ne ions of unknown origin in the Skylab experiment.**
S. Biswas, N. Durgaprasad, J. Nevatia, V. S. Venkatavaradan.
14th Intern. Cosmic Ray Conf., (see 012.011), Vol. 2, 816 - 821 (1975).

143.173 **Energy spectrum and relative abundances of low energy Ca to Ni ions in the Skylab experiment.**
S. Biswas, N. Durgaprasad, J. Nevatia, V. S. Venkatavaradan.
14th Intern. Cosmic Ray Conf., (see 012.011), Vol. 2, 822 - 826 (1975).

143.174 **Phosphate glass track detectors on Skylab: preliminary results on energy spectra and composition of VH ions between 0.1 - 10 MeV/n.** W. Krätschmer.
14th Intern. Cosmic Ray Conf., (see 012.011), Vol. 2, 827 - 830 (1975).

143.175 **Composition and energy spectra of heavy nuclei of unknown origin detected on Skylab.**
J. H. Chan, P. B. Price.
14th Intern. Cosmic Ray Conf., (see 012.011), Vol. 2, 831 - 836 (1975).

143.176 **Propagation and acceleration of cosmic ray electrons in interplanetary space in presence of inhomogeneous structure of solar wind.**
L. I. Dorman, M. E. Katz (*Kats*), Yu. I. Fedorov, A. K. Yuhimuk (*Yukhimukh*).
14th Intern. Cosmic Ray Conf., (see 012.011), Vol. 3, 837 (1975). — Abstract.

143.177 **Hydromagnetic waves and cosmic ray diffusion theory.** M. A. Lee, H. J. Völk.
14th Intern. Cosmic Ray Conf., (see 012.011), Vol. 3, 838 - 843 (1975).

143.178 **A time-averaged cosmic ray propagation theory.**
A. J. Klimas, G. Sandri.
14th Intern. Cosmic Ray Conf., (see 012.011), Vol. 3, 844 - 849 (1975).

143.179 **An examination of the adiabatic approximation in cosmic ray propagation theory.**
J. Scudder, A. J. Klimas.
14th Intern. Cosmic Ray Conf., (see 012.011), Vol. 3, 850 - 855 (1975).

143.180 **The effect of stochastic averaging of the background magnetic field on the interplanetary cosmic ray diffusion coefficient.**
G. E. Morfill, H. J. Völk, M. A. Lee.
14th Intern. Cosmic Ray Conf., (see 012.011), Vol. 3, 872 - 874 (1975).

143.181 **Field gradient and curvature drifts and cosmic ray transport into the solar system.**
L. R. Barnden, M. Bercovitch.
14th Intern. Cosmic Ray Conf., (see 012.011), Vol. 3, 875 - 880 (1975).

143.182 **A variational principle for cosmic ray transport equations.** I. Lerche.
14th Intern. Cosmic Ray Conf., (see 012.011), Vol. 3, 881 - 886 (1975).

143.183 **Subcosmic rays, their origin, modulation, and role in the space.** L. I. Dorman.
14th Intern. Cosmic Ray Conf., (see 012.011), Vol. 3, 887 (1975). — Abstract.

143.184 **Modulation and spectral redistribution of galactic cosmic rays.** L. J. Gleeson, G. M. Webb.
14th Intern. Cosmic Ray Conf., (see 012.011), Vol. 3, 893 - 898 (1975).

143.185 **Possible theoretical explanations for occasional days of non-field-aligned diffusion at neutron monitor energies.** M. A. Forman.
14th Intern. Cosmic Ray Conf., (see 012.011), Vol. 3, 899 - 903 (1975).

143.186 **Numerical calculations of the spectrum of galactic cosmic rays in interplanetary space.**
L. I. Dorman, Z. Kobyliński.
14th Intern. Cosmic Ray Conf., (see 012.011), Vol. 3, 904 (1975). — Abstract.

143.187 **Possible evidence for latitude-dependent cosmic-ray modulation.** L. A. Fisk.

14th Intern. Cosmic Ray Conf., (see 012.011), Vol. 3, 905 - 909 (1975).

143.188 **Three-dimensional models of the galactic cosmic-ray modulation.** H. Moraal, L. J. Gleeson.
14th Intern. Cosmic Ray Conf., (see 012.011), Vol. 3, 910 (1975). – Abstract.

143.189 **Three-dimensional models of galactic cosmic ray modulation.** S. Cecchini, J. J. Quenby.
14th Intern. Cosmic Ray Conf., (see 012.011), Vol. 3, 911 - 916 (1975).

143.190 **Expected spatial distribution of cosmic rays in interplanetary space including the real heliolatitude distribution of solar activity.** L. I. Dorman, N. P. Milovidova.
14th Intern. Cosmic Ray Conf., (see 012.011), Vol. 3, 917 - 922 (1975).

143.191 **Expected 11-year and annual variations in cosmic rays of various rigidities in terms of the anisotropic diffusion model on the basis of the data on heliolatitude distribution of solar activity.** L. I. Dorman, N. P. Milovidova.
14th Intern. Cosmic Ray Conf., (see 012.011), Vol. 3, 923 - 927 (1975).

143.192 **Cosmic ray transport path as a function of the type and the scale and field-intensity spectra of inhomogeneities.** L. I. Dorman, A. V. Sergeev.
14th Intern. Cosmic Ray Conf., (see 012.011), Vol. 3, 928 (1975). – Abstract.

143.193 **Analytical solution for the problem of the 27-day variations in the cosmic ray intensity and anisotropy.**
A. V. Belov, L. I. Dorman, L. Kh. Shatashvili.
14th Intern. Cosmic Ray Conf., (see 012.011), Vol. 3, 929 (1975). – Abstract.

143.194 **Kinetic theory of fluctuations in cosmic ray distribution function.** L. I. Dorman, M. E. Katz (*Kats*).
14th Intern. Cosmic Ray Conf., (see 012.011), Vol. 3, 930 (1975). – Abstract.

143.195 **On the theory of non-linear cosmic ray modulation by solar wind.** L. I. Dorman, V. Kh. Babayan.
14th Intern. Cosmic Ray Conf., (see 012.011), Vol. 3, 936 (1975). – Abstract.

143.196 **Cosmic ray electron capture isotopes as probes of solar modulation.** G. M. Raisbeck, G. Comstock, C. Perron, F. Yiou.
14th Intern. Cosmic Ray Conf., (see 012.011), Vol. 3, 937 - 940, with a correction Vol. 12, 4415 (1975).

143.197 **Solar effects associated with the cosmic ray modulation spectrum.** R. B. Mendell, S. A. Korff.
14th Intern. Cosmic Ray Conf., (see 012.011), Vol. 3, 941 - 946 (1975).

143.198 **The time and rigidity dependence of the 11 year modulation of cosmic rays.**
H. Moraal, P. H. Stoker.
14th Intern. Cosmic Ray Conf., (see 012.011), Vol. 3, 947 - 951 (1975).

143.199 **The variation in the latitude dependence of cosmic rays at 307 g/cm² during solar cycle No 20.**
P. H. Stoker, C. F. W. Mischke, H. Moraal, P. Oberholzer.
14th Intern. Cosmic Ray Conf., (see 012.011), Vol. 3, 952 - 957 (1975).

143.200 **The solar cycle modulation of the galactic cosmic rays and the solar flare activity.**
N. Iucci, M. Parisi, M. Storini, G. Villoresi.
14th Intern. Cosmic Ray Conf., (see 012.011), Vol. 3, 958 - 963 (1975).

143.201 **Cosmic ray modulation and green coronal line activity.**
N. Iucci, M. Parisi, M. Storini, G. Villoresi, N. L. Zangrilli.
14th Intern. Cosmic Ray Conf., (see 012.011), Vol. 3, 964 - 968 (1975).

143.202 **Generalization of spectrographic method for studying the extraterrestrial and magnetospheric cosmic ray variations including penumbra.**
L. I. Dorman, G. Sh. Shkhalakhov.
14th Intern. Cosmic Ray Conf., (see 012.011), Vol. 3, 969 (1975). – Abstract.

143.203 **Spectrographic method for studying the cosmic ray variations in the medium and high energy ranges in case of underground observations.**
L. I. Dorman, G. Sh. Shkhalakhov.
14th Intern. Cosmic Ray Conf., (see 012.011), Vol. 3, 970 (1975). – Abstract.

143.204 **Expected variations in the effective geomagnetic cutoff rigidity of cosmic rays at the stations of the worldwide network for the 11-year, annual, and 27-day variations; solar anisotropy; Forbush-decreases; increase effects prior to magnetic storms; and solar flares.**
L. I. Dorman, A. A. Shadov.
14th Intern. Cosmic Ray Conf., (see 012.011), Vol. 3, 971 (1975). – Abstract.

143.205 **Observed spectrum of cosmic rays and its origin.** L. I. Dorman.
14th Intern. Cosmic Ray Conf., (see 012.011), Vol. 3, 972 (1975). – Abstract.

143.206 **Dimensions of the modulation region of galactic cosmic rays in 1972 - 1973.**
R. R. Ashirov, A. G. Zusmanovich, E. V. Kolomeets.
14th Intern. Cosmic Ray Conf., (see 012.011), Vol. 3, 973 - 978 (1975).

143.207 **Modulation of low energy electrons and protons near solar maximum.** J. L'Heureux, P. Meyer.
14th Intern. Cosmic Ray Conf., (see 012.011), Vol. 3, 979 - 984 (1975).

143.208 **Anomalous effect in the spectrum of cosmic ray variations in 1971 - 1973.** A. K. Svirzhevskaya, Yu. I. Stozhkov, T. N. Charakhchyan (*Charakhch'yan*).
14th Intern. Cosmic Ray Conf., (see 012.011), Vol. 3, 985 - 989 (1975).

143.209 **The modulation and energy spectrum of electrons in interstellar space.** A. A. Aitmuhambetov (*Ajtmukhambetov*), A. G. Zusmanovich, E. V. Kolomeets.
14th Intern. Cosmic Ray Conf., (see 012.011), Vol. 3, 990 - 994 (1975).

143.210 **Long term modulation of cosmic-ray intensity and solar activity cycle.**
Y. C. Lin, C. Y. Fan, P. E. Damon, E. I. Wallick.
14th Intern. Cosmic Ray Conf., (see 012.011), Vol. 3, 995 - 999 (1975).

143.211 **The cosmic ray electron spectrum in 1973 and**

1974.
J. Caldwell, P. Evenson, S. Jordan, P. Meyer.
14th Intern. Cosmic Ray Conf., (see 012.011), Vol. 3, 1000 - 1004 (1975).

143.212 **Solar modulation and the chemical composition of the cosmic radiation.**
Y. V. Rao, P. S. Young, K. Fukui.
14th Intern. Cosmic Ray Conf., (see 012.011), Vol. 3, 1005 - 1008 (1975).

143.213 **Rate of production of cosmogenic isotopes in the past and solar activity.** G. E. Kocharov, V. A. Dergachev, N. I. Gordeichik (*Gordejchik*),
14 th Intern. Cosmic Ray Conf., (see 012.011), Vol. 3, 1009 - 1014 (1975).

143.214 **The eleven-year cosmic ray cycle according to stratosphere measurements.** S. O. Vernov, A. N. Charakhchyan (*Charakhch'yan*), Yu. I. Stozhkov, T. N. Charakhchyan (*Charakhch'yan*).
14th Intern. Cosmic Ray Conf., (see 012.011), Vol. 3, 1015 - 1019 (1975).

143.215 **Investigation of the long-term variations of cosmic ray latitude effect in the earth atmosphere.**
A. N. Charakhchyan (*Charakhch'yan*), G. A. Bazilevskaya, Yu. I. Stozhkov, T. N. Charakhchyan (*Charakhch'yan*).
14th Intern. Cosmic Ray Conf., (see 012.011), Vol. 3, 1020 - 1024 (1975).

143.216 **Numerical solution of the quasistationary problem about the 27-day modulation of galactic cosmic rays using the perturbation theory.**
L. Kh. Shatashvili.
14th Intern. Cosmic Ray Conf., (see 012.011), Vol. 3, 1046 - 1051, with a correction, Vol. 12, 4415 (1975).

143.217 **The short term modulation of cosmic radiation.** M. A. Pomerantz, S. P. Duggal.
14th Intern. Cosmic Ray Conf., (see 012.011), Vol. 3, 1052 - 1057 (1975).

143.218 **Cosmic ray modulation by corotating solar streams in presence of flare plasma clouds.**
M. M. Bemalkhedkar, H. Razdan, C. L. Kaul.
14th Intern. Cosmic Ray Conf., (see 012.011), Vol. 3, 1058 - 1063 (1975).

143.219 **Kinetic theory of cosmic ray interaction with powerful interplanetary shock waves.**
L. I. Dorman, V. Kh. Shogenov.
14th Intern. Cosmic Ray Conf., (see 012.011), Vol. 3, 1070 (1975). – Abstract.

143.220 **Cosmic ray three-dimensional anisotropies.** R. Anda, N. J. Martinic, N. Penarrieta.
14th Intern. Cosmic Ray Conf., (see 012.011), Vol. 3, 1087 - 1092 (1975).

143.221 **Synthesis of the results of cosmic ray intensity variation during the period of unusual Forbush decrease of August 1972.** S. P. Agrawal.
14th Intern. Cosmic Ray Conf., (see 012.011), Vol. 3, 1093 (1975). – Abstract.

143.222 **The influence of magnetic "corks" upon the galactic cosmic ray distribution.**
S. M. Kamoldinov, V. P. Mamrukova, A. M. Altukhov, P. A. Krivoshapkin, G. F. Krymsky (*Krymskij*).
14th Intern. Cosmic Ray Conf., (see 012.011), Vol. 3, 1102 -

1106 (1975).

143.223 **Two-component model for cosmic-ray fluctuations at neutron monitor energies.**
J. R. Jokipii, A. J. Owens.
14th Intern. Cosmic Ray Conf., (see 012.011), Vol. 3, 1107 - 1112 (1975).

143.224 **The power spectral density of cosmic ray scintillations in a field direction with respect to the earth–sun reference system.** M. R. Attolini, S. Cecchini, M. Galli.
14th Intern. Cosmic Ray Conf., (see 012.011), Vol. 3, 1113 - 1118 (1975).

143.225 **On the amplitude of the transient diurnal variation.** M. Galli, M. R. Attolini, S. Cecchini.
14th Intern. Cosmic Ray Conf., (see 012.011), Vol. 3, 1119 (1975). – Abstract.

143.226 **Short time variations of cosmic ray intensity observed by using balloon.** M. Kodama, T. Sakai, E. Tamai, S. Kogami, M. Kato.
14th Intern. Cosmic Ray Conf., (see 012.011), Vol. 3, 1120 - 1124 (1975).

143.227 **Short-period variations of cosmic rays.**
V. V. Alexeyenko, A. V. Voevodsky (*Voevodskij*), V. A. Dogujaev, V. I. Paramonov, V. G. Sborshikov, A. E. Chudakov.
14th Intern. Cosmic Ray Conf., (see 012.011), Vol. 3, 1125 - 1128 (1975).

143.228 **Power spectra of short-term variations of cosmic ray intensity during the period of shock wave passage.**
V. I. Kozlov, N. P. Chirkov.
14th Intern. Cosmic Ray Conf., (see 012.011), Vol. 3, 1129 - 1132 (1975).

143.229 **The power spectra of cosmic ray intensity and the solar activity indices in the period range of 50 - 1000 days.** T. N. Charakhchyan (*Charakhch'yan*), G. A. Bazilevskaya, V. P. Okhlopkov, L. S. Okhlopkova.
14th Intern. Cosmic Ray Conf., (see 012.011), Vol. 3, 1133 - 1137 (1975).

143.230 **Anisotropy of cosmic rays associated with high velocity streams of interplanetary plasma.**
T. Murayama.
14th Intern. Cosmic Ray Conf., (see 012.011), Vol. 4, 1144 - 1149 (1975).

143.231 **The cosmic ray diurnal variations and fluctuations of the interplanetary magnetic field.**
J. W. Sari, L. J. Lanzerotti, C. G. Maclennan, D. Venkatesan.
14th Intern. Cosmic Ray Conf., (see 012.011), Vol. 4, 1150 - 1155 (1975).

143.232 **The diurnal anisotropy and its relationship to the interplanetary magnetic field.**
A. J. Owens, M. M. Kash.
14th Intern. Cosmic Ray Conf., (see 012.011), Vol. 4, 1156 (1975). – Abstract.

143.233 **Quiet time cosmic ray anomalous diurnal anisotropy with maximum in morning hours.**
M. M. Bemalkhedkar, H. Razdan, C. L. Kaul.
14th Intern. Cosmic Ray Conf., (see 012.011), Vol. 4, 1157 - 1161 (1975).

143.234 **Distribution of the vector of cosmic ray solar**

anisotropy in interplanetary space in terms of the asymmetrical model of anisotropic diffusion.
L. I. Dorman, A. M. Samir-Debish.
14th Intern. Cosmic Ray Conf., (see 012.011), Vol. 4, 1162 - 1166 (1975).

143.235 Space gradients and cosmic ray currents.
G. F. Krymsky (*Krymskij*), A. M. Altukhov, I. S. Samsonov, A. I. Kuzmin, N. P. Chirkov.
14th Intern. Cosmic Ray Conf., (see 012.011), Vol. 4, 1167 - 1171 (1975).

143.236 Enhanced daily waves of muons and nucleons during 5–8 August 1972. R. L. Chasson.
14th Intern. Cosmic Ray Conf., (see 012.011), Vol. 4, 1172 - 1175 (1975).

143.237 Interplanetary shock waves and the north–south anisotropy in cosmic rays.
S. P. Duggal, M. A. Pomerantz.
14th Intern. Cosmic Ray Conf., (see 012.011), Vol. 4, 1176 - 1181 (1975).

143.238 North–south asymmetry of cosmic ray flux and polarity of interplanetary magnetic field.
I. Kondo, Z. Fujii, K. Nagashima.
14th Intern. Cosmic Ray Conf., (see 012.011), Vol. 4, 1182 - 1187 (1975).

143.239 N–S asymmetry and variations of cosmic ray isotropic flux and solar activity.
A. Kh. Bychkovskaya, R. A. Chumbalova, R. B. Zhantuarova, E. V. Kolomeets, Ya. E. Shvartsman.
14th Intern. Cosmic Ray Conf., (see 012.011), Vol. 4, 1188 - 1192 (1975).

143.240 Time variation of the characteristics of the diurnal anisotropy of cosmic radiation.
S. P. Agrawal, R. L. Singh.
14th Intern. Cosmic Ray Conf., (see 012.011), Vol. 4, 1193 - 1198 (1975).

143.241 Cosmic ray anisotropy and its time variations.
S. D. Asylbaeva, G. A. Gonchar, I. D. Leongard, L. A. Mirkin, E. V. Kolomeets, N. B. Slynyaeva.
14th Intern. Cosmic Ray Conf., (see 012.011), Vol. 4, 1199 - 1203 (1975).

143.242 Cosmic-ray diurnal anisotropy and the sun's polar magnetic field. S. E. Forbush, L. Beach.
14th Intern. Cosmic Ray Conf., (see 012.011), Vol. 4, 1204 - 1208 (1975).

143.243 Three-dimensional studies of cosmic ray anisotropies.
P. Chaloupka, W. K. Griffiths.
14th Intern. Cosmic Ray Conf., (see 012.011), Vol. 4, 1225 - 1230 (1975).

143.244 Anisotropy of the semi-diurnal and the third harmonic diurnal variation of cosmic-ray neutron
intensity and their time variation. T. Kanno, Y. Ishida, T. Saito.
14th Intern. Cosmic Ray Conf., (see 012.011), Vol. 4, 1231 - 1235 (1975).

143.245 Spherical zonal harmonic components of cosmic rays in interplanetary space.
H. Takahashi, N. Yahagi, K. Nagashima.
14th Intern. Cosmic Ray Conf., (see 012.011), Vol. 4, 1236 - 1241 (1975).

143.246 Second spherical harmonics of the cosmic ray

angular distribution. J. Kóta.
14th Intern. Cosmic Ray Conf., (see 012.011), Vol. 4, 1242 - 1246 (1975).

143.247 Analysis of three-dimensional cosmic ray anisotropy on hourly basis.
S. Yasue, S. Mori, K. Nagashima.
14th Intern. Cosmic Ray Conf., (see 012.011), Vol. 4, 1247 - 1252 (1975).

143.248 Critical study of the diurnal and semi-diurnal variation of cosmic ray intensity on day-to-day basis.
S. P. Agrawal, R. L. Singh.
14th Intern. Cosmic Ray Conf., (see 012.011), Vol. 4, 1253 - 1257 (1975).

143.249 Some characteristics of the solar tridiurnal variation of cosmic rays. H. S. Ahluwalia.
14th Intern. Cosmic Ray Conf., (see 012.011), Vol. 4, 1258 (1975). – Abstract.

143.250 Non-field-aligned nature of the diffusion vector of cosmic ray diurnal anisotropy. R. P. Kane.
14th Intern. Cosmic Ray Conf., (see 012.011), Vol. 4, 1259 - 1264 (1975).

143.251 Propagation of solar cosmic rays across the earth magnetosheath. R. Gall, S. Bravo.
14th Intern. Cosmic Ray Conf., (see 012.011), Vol. 4, 1265 (1975). – Abstract.

143.252 Calculations of charged particle entry and propagation to the polar regions.
A. J. Masley, W. P. Olson, K. A. Pfitzer.
14th Intern. Cosmic Ray Conf., (see 012.011), Vol. 4, 1266 (1975). – Abstract.

143.253 Low energy polar cap electrons during quiet times.
J. F. Fennell, P. F. Mizera, D. R. Croley.
14th Intern. Cosmic Ray Conf., (see 012.011), Vol. 4, 1267 - 1272 (1975).

143.254 A comparison of vertical cosmic-ray cutoff rigidities as calculated with different geomagnetic field
models. M. A. Shea, D. F. Smart.
14th Intern. Cosmic Ray Conf., (see 012.011), Vol. 4, 1284 - 1288 (1975).

143.255 A preliminary idealized network of neutron monitors for the study of solar modulation.
R. E. Gold, M. A. Shea, D. F. Smart.
14th Intern. Cosmic Ray Conf., (see 012.011), Vol. 4, 1289 - 1293 (1975).

143.256 Variations of the calculated cosmic-ray equator over a 20-year interval. M. A. Shea, D. F. Smart.
14th Intern. Cosmic Ray Conf., (see 012.011), Vol. 4, 1294 - 1297 (1975).

143.257 A five by fifteen degree world grid of calculated cosmic-ray vertical cutoff rigidities for 1965 and
1975. M. A. Shea, D. F. Smart.
14th Intern. Cosmic Ray Conf., (see 012.011), Vol. 4, 1298 - 1303 (1975).

143.258 Cosmic-ray penumbral effects for selected balloon launching locations. D. F. Smart, M. A. Shea.
14th Intern. Cosmic Ray Conf., (see 012.011), Vol. 4, 1304 - 1308 (1975).

143.259 An analysis of trajectory-derived penumbral widths.

D. F. Smart, M. A. Shea.
14th Intern. Cosmic Ray Conf., (see 012.011), Vol. 4, 1309 - 1314 (1975).

143.260 The effect of ionization energy loss on the calculation of rigidity transmittance functions.
J. A. Lezniak, D. F. Smart, M. A. Shea.
14th Intern. Cosmic Ray Conf., (see 012.011), Vol. 4, 1315 - 1319 (1975).

143.261 Geomagnetic cut-offs at 1.7 & 2.6 GV — transmission functions & isotopic analysis of cosmic ray nuclei. W. R. Webber, J. A. Lezniak, M. A. Shea, D. F. Smart.
14th Intern. Cosmic Ray Conf., (see 012.011), Vol. 4, 1320 - 1325 (1975).

143.262 Experimental measurements of charged particle cutoff latitudes. A. J. Masley, P. R. Satterblom.
14th Intern. Cosmic Ray Conf., (see 012.011), Vol. 4, 1326 (1975).

143.263 Cosmic ray variations due to geomagnetic field variations.
Kh. Z. Aldagarova, O. A. Bogdanova, E. V. Kolomeets, V. P. Pivneva.
14th Intern. Cosmic Ray Conf., (see 012.011), Vol. 4, 1327 - 1330 (1975).

143.264 Reverse flux of cosmic ray multicharge nuclei at heights of 250 - 300 km above the earth.
N. A. Dobrotin, L. V. Kurnosova, V. I. Logachev, L. A. Lebedev, M. I. Fradkin.
14th Intern. Cosmic Ray Conf., (see 012.011), Vol. 4, 1346 (1975). – Abstract.

143.265 Cosmic radiation beneath the radiation belts of the earth. Yu. A. Alexandrov, S. N. Kuznetsov, Yu. L. Logachev, V. G. Stolpovsky (*Stolpovskij*).
14th Intern. Cosmic Ray Conf., (see 012.011), Vol. 4, 1347 - 1349 (1975).

143.266 Analysis of hysterisis in cosmic rays relative to variations in the heliolatitude index of solar activity for various epochs of solar activity cycle.
L. I. Dorman, I. A. Pimenov, L. F. Churunova.
14th Intern. Cosmic Ray Conf., (see 012.011), Vol. 4, 1398 (1975). – Abstract.

143.267 The calculation of integral multiplicity of generation of cosmic rays neutron component for the neutron monitor NM-64. A. A. Lusov, N. I. Pakhomov, V. E. Sdobnov, S. E. Chigrinov.
14th Intern. Cosmic Ray Conf., (see 012.011), Vol. 4, 1399 - 1402 (1975).

143.268 Integral multiplicities of various cosmic ray components in the earth's atmosphere as functions of altitude, geomagnetic cutoff rigidity and primary cosmic ray spectrum. O. A. Bogdanova, E. V. Kolomeets, V. L. Shmonin.
14th Intern. Cosmic Ray Conf., (see 012.011), Vol. 4, 1441 - 1445 (1975).

143.269 Cosmic-ray diurnal variation underground and the interplanetary magnetic field. V. H. Regener.
14th Intern. Cosmic Ray Conf., (see 012.011), Vol. 4, 1452 (1975). – Abstract.

143.270 Change in the phase of the apparent sidereal variation. G. Cini-Castagnoli, D. Marocchi,

H. Elliot, R. G. Marsden, T. Thambyahpillai.
14th Intern. Cosmic Ray Conf., (see 012.011), Vol. 4, 1453 - 1457 (1975).

143.271 Seasonal changes in the phase of the diurnal variation measured underground.
G. Cini-Castagnoli, D. Marocchi.
14th Intern. Cosmic Ray Conf., (see 012.011), Vol. 4, 1458 - 1459 (1975).

143.272 Diurnal anisotropies of cosmic ray intensity underground and interplanetary magnetic field directions. G. Cini-Castagnoli, M. A. Dodero.
14th Intern. Cosmic Ray Conf., (see 012.011), Vol. 4, 1460 - 1462 (1975).

143.273 27-day recurrences of enhanced daily variations in the cosmic ray intensity during 1973–1975.
S. Mori, S. Yasue, M. Ichinose, Y. Munakata.
14th Intern. Cosmic Ray Conf., (see 012.011), Vol. 4, 1463 - 1468 (1975).

143.274 Multi-directional measurements of the cosmic ray intensity variation at 30 m.w.e. underground at Misato. S. Mori, S. Yasue, M. Ichinose, S. Akahane.
14th Intern. Cosmic Ray Conf., (see 012.011), Vol. 4, 1469 - 1474 (1975).

143.275 Solar and sidereal anisotropy of cosmic rays observed at 30 m.w.e. underground and its relation to the interplanetary magnetic field.
M. Ichinose, S. Mori, S. Yasue.
14th Intern. Cosmic Ray Conf., (see 012.011), Vol. 4, 1475 - 1480 (1975).

143.276 Sidereal cosmic ray variations at ~365 m.w.e. underground. A. G. Fenton, K. B. Fenton.
14th Intern. Cosmic Ray Conf., (see 012.011), Vol. 4, 1482 - 1484 (1975).

143.277 Observations of cosmic ray density gradients at high rigidities. D. B. Swinson.
14th Intern. Cosmic Ray Conf., (see 012.011), Vol. 4, 1485 (1975). – Abstract.

143.278 The radial gradient of galactic cosmic rays.
F. B. McDonald, N. Lal, B. J. Teegarden, J. H. Trainor, W. R. Webber.
14th Intern. Cosmic Ray Conf., (see 012.011), Vol. 4, 1511 (1975). – Abstract.

143.279 Cosmic ray radial intensity gradients measured by Pioneer 10 and Pioneer 11. R. B. McKibben, K. R. Pyle, J. A. Simpson, A. J. Tuzzolino, J. J. O'Gallagher.
14th Intern. Cosmic Ray Conf., (see 012.011), Vol. 4, 1512 - 1517 (1975).

143.280 Comparisons of ground-based monitor data with Pioneer 10 and 11 observations in 1972–74.
J. A. Lockwood, W. R. Webber.
14th Intern. Cosmic Ray Conf., (see 012.011), Vol. 4, 1518 (1975). – Abstract.

143.281 Measurements of cosmic ray anisotropies from Pioneers 10 & 11. W. I. Axford, W. Fillius, L. J. Gleeson, W.-H. Ip, A. Mogro-Campero.
14th Intern. Cosmic Ray Conf., (see 012.011), Vol. 4, 1519 - 1524 (1975).

143.282 Further studies of the new component of cosmic rays at low energies. W. R. Webber,

F. B. McDonald, J. H. Trainor, B. J. Teegarden.
14th Intern. Cosmic Ray Conf., (see 012.011), Vol. 4, 1525 (1975). – Abstract.

143.283 Relation of large-scale coronal X-ray structure and cosmic rays: I. Sources of solar wind streams as defined by X-ray emission and Hα absorption features.
A. S. Krieger, J. T. Nolte, J. D. Sullivan, A. J. Lazarus, P. S. McIntosh, R. E. Gold, E. C. Roelof.
14th Intern. Cosmic Ray Conf., (see 012.011), Vol. 5, 1698 - 1703 (1975).

143.284 Nucleon radial gradients between 0.45 and 1.0 A.U. from the Mariner-10 mission to Mercury.
S. Christon, S. Daly, J. H. Eraker, J. E. Lamport, G. Lentz, J. A. Simpson.
14th Intern. Cosmic Ray Conf., (see 012.011), Vol. 5, 1848 - 1849 (1975).

143.285 Heliocentric intensity gradient of galactic cosmic ray from Helios A.
M. A. Van Hollebeke, J. A. Trainor, K. G. McCracken.
14th Intern. Cosmic Ray Conf., (see 012.011), Vol. 5, 1850 (1975). – Abstract.

143.286 Cosmic ray anisotropies measured on Helios-1.
G. Green, G. Wibberenz, R. Müller-Mellin, M. Witte, H. Hempe, H. Kunow.
14th Intern. Cosmic Ray Conf., (see 012.011), Vol. 5, 1851 (1975). – Abstract.

143.287 Energy spectra of protons and α-particles measured on Helios-1. M. Witte, H. Hempe, H. Kunow, G. Wibberenz, G. Green, R. Müller-Mellin, B. Iwers.
14th Intern. Cosmic Ray Conf., (see 012.011), Vol. 5, 1852 (1975). – Abstract.

143.288 Interplanetary quiet time differential spectra of protons and alpha particles below 30 MeV/nucleon from Helios A.
J. H. Trainor, M. A. Van Hollebeke, F. B. McDonald, K. G. McCracken.
14th Intern. Cosmic Ray Conf., (see 012.011), Vol. 5, 1853 (1975). – Abstract.

143.289 Variations of cosmic ray intensities during the first part of the Helios-1 mission.
H. Kunow, G. Wibberenz, G. Green, R. Müller-Mellin, M. Witte, H. Hempe, H. G. Hasler.
14th Intern. Cosmic Ray Conf., (see 012.011), Vol. 5, 1854 (1975). – Abstract.

143.290 Relationship of the muon charge ratio to the mass composition of primary cosmic rays.
A. D. Erlykin, L. K. Ng, A. W. Wolfendale.
14th Intern. Cosmic Ray Conf., (see 012.011), Vol. 6, 2003 - 2006 (1975).

143.291 Multiple muons and their relevance to ultra-high energy interactions.
A. Goned, T. R. Stewart, A. W. Wolfendale, J. Wdowczyk.
14th Intern. Cosmic Ray Conf., (see 012.011), Vol. 6, 2007 - 2010 (1975).

143.292 Composition of cosmic rays at 10^{10} to 10^{13} eV/ nucleus. E. Juliusson.
14th Intern. Cosmic Ray Conf., (see 012.011), Vol. 8, 2689 - 2694 (1975).

143.293 Energy spectrum of primary cosmic rays from 5×10^{16} eV to 3×10^{18} eV determined from air

showers observed at Chacaltaya (5,200 m a.s.l.).
T. Kaneko, C. Aguirre, Y. Toyoda, H. Nakatani, S. Jadot, P. K. MacKeown, K. Suga, F. Kakimoto, Y. Mizumoto, K. Murakami, K. Nishi, M. Nagano, K. Kamata.
14th Intern. Cosmic Ray Conf., (see 012.011), Vol. 8, 2695 - 2698 (1975).

143.294 The cosmic ray energy spectrum above 10^{17} eV.
A. R. Clarke, D. M. Edge, A. M. T. Pollock, R. J. O. Reid, A. A. Watson, J. G. Wilson.
14th Intern. Cosmic Ray Conf., (see 012.011), Vol. 8, 2699 - 2703 (1975).

143.295 The energy spectrum of cosmic rays in the region $2 \times 10^{16} - 4 \times 10^{17}$ eV. R. G. Brownlee, A. R. Clarke, D. M. Edge, R. J. O. Reid, A. M. Wray.
14th Intern. Cosmic Ray Conf., (see 012.011), Vol. 8, 2704 - 2707 (1975).

143.296 Primary energy spectrum of cosmic ray particles in the energy range close to 10^{15} eV.
R. A. Antonov, I. P. Ivanenko.
14th Intern. Cosmic Ray Conf., (see 012.011), Vol. 8, 2708 - 2713 (1975).

143.297 The energy spectrum of primary cosmic rays at $10^{14} - 10^{16}$ eV. V. S. Aseikin, V. P. Bobova, A. G. Dubovij, N. V. Kabanova, N. M. Nesterova, N. M. Nikolskaja, S. I. Nikolsky, V. A. Romakhin, I. N. Stamenov, V. D. Ianminchev.
14th Intern. Cosmic Ray Conf., (see 012.011), Vol. 8, 2726 - 2730 (1975).

143.298 Effective mass of the primary cosmic ray particles at about 10^{15} eV. J. P. Hochart, G. Milleret, A. Zawadzki, J. Gawin, J. Wdowczyk.
14th Intern. Cosmic Ray Conf., (see 012.011), Vol. 8, 2736 - 2740 (1975).

143.299 Fluctuation of shower front structure: measurements, $E_p \sim 10^{18}$ eV.
M. L. Barrett, A. A. Watson, P. Wild, J. G. Wilson.
14th Intern. Cosmic Ray Conf., (see 012.011), Vol. 8, 2753 - 2757 (1975).

143.300 Interpretation of fluctuations of shower front structure. $E_p \sim 10^{18}$ eV.
M. L. Barrett, A. A. Watson, P. Wild, J. G. Wilson.
14th Intern. Cosmic Ray Conf., (see 012.011), Vol. 8, 2758 - 2761 (1975).

143.301 Calculation of the depth of the shower maximum development in the energy range of $10^{17} - 10^{19}$ eV.
L. G. Dedenko.
14th Intern. Cosmic Ray Conf., (see 012.011), Vol. 8, 2857 - 2860 (1975).

143.302 Sensitivity of the computed EAS longitudinal development to a variation of the radiation length of high energy electrons.
J. Procureur, M.-F. Bourdeau, J.-N. Capdevielle.
14th Intern. Cosmic Ray Conf., (see 012.011), Vol. 8, 2878 - 2882 (1975).

143.303 Calculated Cerenkov production of the nucleonic component of EAS and dependence on primary mass. J. E. Grindlay.
14th Intern. Cosmic Ray Conf., (see 012.011), Vol. 8, 2915 - 2919 (1975).

143.304 A measurement of the cosmic ray energy spectrum

from 10^{11} to 10^{15} eV.
C. Gerdes, D. Hartman, C. Y. Fan, T. C. Weekes.
14th Intern. Cosmic Ray Conf., (see 012.011), Vol. 8, 3040 - 3045 (1975).

143.305 On the problem of detection of cosmic rays of extremely high energies.
V. D. Volovik, G. B. Khristiansen.
14th Intern. Cosmic Ray Conf., (see 012.011), Vol. 8, 3096 - 3101 (1975).

143.306 Cosmic radiation and fundamental problems in physics. W. Heisenberg.
14th Intern. Cosmic Ray Conf., (see 012.011), Vol. 11, 3461 - 3474 (1975).

143.307 Constancy of galactic cosmic rays in time and space. O. A. Schaeffer.
14th Intern. Cosmic Ray Conf., (see 012.011), Vol. 11, 3508 - 3520 (1975).

143.308 Isotopic composition of cosmic rays. J. P. Meyer.
14th Intern. Cosmic Ray Conf., (see 012.011), Vol. 11, 3698 - 3745 (1975).

143.309 Chemical composition of the cosmic radiation and the electron component. N. Lund.
14th Intern. Cosmic Ray Conf., (see 012.011), Vol. 11, 3746 - 3752 (1975).

143.310 Cosmic ray propagation. R. M. Kulsrud.
14th Intern. Cosmic Ray Conf., (see 012.011), Vol. 11, 3753 - 3757 (1975).

143.311 Cosmic ray sources. R. Cowsik.
14th Intern. Cosmic Ray Conf., (see 012.011), Vol. 11, 3758 - 3783 (1975).

143.312 Modulation and diffusion theory of cosmic rays. M. A. Forman.
14th Intern. Cosmic Ray Conf., (see 012.011), Vol. 11, 3820 - 3832 (1975).

143.313 Short term modulations and anisotropies. M. A. Pomerantz.
14th Intern. Cosmic Ray Conf., (see 012.011), Vol. 11, 3833 - 3866 (1975).

143.314 Geomagnetic effects of cosmic rays. D. F. Smart.
14th Intern. Cosmic Ray Conf., (see 012.011), Vol. 11, 3884 - 3895 (1975).   .

143.315 Modulation observations. H. Moraal.
14th Intern. Cosmic Ray Conf., (see 012.011), Vol. 11, 3896 - 3906 (1975).

143.316 Sidereal variation of cosmic rays and underground measurements. K. B. Fenton.
14th Intern. Cosmic Ray Conf., (see 012.011), Vol. 11, 3907 - 3924 (1975).

143.317 Cosmic ray measurements of light and medium nuclei using a new telescope. A. M. Preszler
J. C. Kish, J. A. Lezniak, G. Simpson, W. R. Webber.
14th Intern. Cosmic Ray Conf., (see 012.011), Vol. 12, 4096 - 4101 (1975).

143.318 The nuclear composition of cosmic rays at their source in a closed galaxy.
I. L. Rasmussen, B. Peters.

14th Intern. Cosmic Ray Conf., (see 012.011), Vol. 12, 4102 - 4106 (1975).

143.319 The energy spectrum of 10–50 BeV/nuc cosmic rays. J. A. Lezniak, W. R. Webber.
14th Intern. Cosmic Ray Conf., (see 012.011), Vol. 12, 4107 - 4111 (1975).

143.320 VH particles ($20 \lesssim Z < 30$) at high rigidity. J. P. Wefel, J. M. Kidd.
14th Intern. Cosmic Ray Conf., (see 012.011), Vol. 12, 4112 - 4117 (1975).

143.321 Satellite measurements of the chemical abundances from Li through Ni in galactic cosmic rays.
C. Julliot, L. Koch, N. Petrou.
14th Intern. Cosmic Ray Conf., (see 012.011), Vol. 12, 4118 - 4122 (1975).

143.322 Observation of relativistic cosmic ray particles with $Z \geqslant 45$. T. Doke, H. Okamoto, E.
Shibamura, T. Hayashi, K. Ito, T. Yanagimachi, S. Kobayashi, M. Miyajima, T. Saito, K. Nagata.
14th Intern. Cosmic Ray Conf., (see 012.011), Vol. 12, 4123 - 4127 (1975).

143.323 Isotopic composition of cosmic ray nuclei in the iron group. V. M. Clapham, P. H. Fowler,
C. O'Ceallaigh, D. O'Sullivan, A. Thompson.
14th Intern. Cosmic Ray Conf., (see 012.011), Vol. 12, 4128 - 4131 (1975).

143.324 High energy primary electrons observed in the emulsion chamber. M. Matsuo, J. Nishimura,
T. Kobayashi, K. Niu, E. Aizu, H. Hiraiwa, T. Taira.
14th Intern. Cosmic Ray Conf., (see 012.011), Vol. 12, 4132 - 4137 (1975).

143.325 Anisotropies in arrival direction of cosmic rays determined from underground muons in the Mont-Blanc tunnel.
A. R. Bazer-Bachi, G. Vedrenne, W. R. Sheldon, J. R. Benbrook.
14th Intern. Cosmic Ray Conf., (see 012.011), Vol. 12, 4151 - 4156 (1975).

143.326 Anisotropy of very high energy cosmic rays. A. M. Hillas, M. Ouldridge.
14th Intern. Cosmic Ray Conf., (see 012.011), Vol. 12, 4160 - 4165 (1975).

143.327 Cosmic ray confinement in the Galaxy and the interstellar spectrum of hydromagnetic waves.
C. J. Cesarsky.
14th Intern. Cosmic Ray Conf., (see 012.011), Vol. 12, 4166 - 4171 (1975).

143.328 Pathlength distribution and source composition of cosmic ray nuclei.
M. Meneguzzi, C. J. Cesarsky, J. P. Meyer.
14th Intern. Cosmic Ray Conf., (see 012.011), Vol. 12, 4183 - 4188 (1975).

143.329 Three-dimensional models of the galactic cosmic-ray modulation. H. Moraal, L. J. Gleeson.
14th Intern. Cosmic Ray Conf., (see 012.011), Vol. 12, 4189 - 4194 (1975).

143.330 Solar effects associated with the cosmic ray modulation spectrum. R. B. Mendell, S. A. Korff.
14th Intern. Cosmic Ray Conf., (see 012.011), Vol. 12,

4195 - 4200 (1975).

143.331 Some characteristics of solar tridiurnal variation of cosmic rays. H. S. Ahluwalia.
14th Intern. Cosmic Ray Conf., (see 012.011), Vol. 12, 4207 - 4212 (1975).

143.332 Sidereal variations observed underground in Tasmania. J. E. Humble, A. G. Fenton.
14th Intern. Cosmic Ray Conf., (see 012.011), Vol. 12, 4226 - 4228 (1975).

143.333 Comparisons of ground-based monitor data with Pioneers 8, 9, 10 and 11 observations in 1968–1974. J. A. Lockwood, W. R. Webber.
14th Intern. Cosmic Ray Conf., (see 012.011), Vol. 12, 4229 - 4232 (1975).

143.334 Further studies of the new component of cosmic rays at low energies. W. R. Webber, F. B. McDonald, J. H. Trainor, B. J. Teegarden, T. T. von Rosenvinge.
14th Intern. Cosmic Ray Conf., (see 012.011), Vol. 12, 4233 - 4238 (1975).

143.335 Cosmic ray intensity in the interplanetary space. Data of observations from August 19 - December 15, 1970.
Mezhduved. geofiz. kom. pri Prezidiume AN SSSR. Materialy Mirovogo tsentra dannykh B. Moskva, 1975. 32 pp. Price 6 Kop. In Russian and English. – Abstr. in Referativ. Zhurn. 62. Issled. kosmich. prostranstva, 6.62.188 (1976).

143.336 Cosmic ray intensity in the interplanetary space. Data of observations from March 27 - July 22, 1972.
Mezhduved. geofiz. kom. pri Prezidiume AN SSSR. Materialy Mirovogo tsentra dannykh B. Moskva, 1975. 36 pp. Price 7 Kop. – Abstr. in Referativ. Zhurn. 62. Issled. kosmich. prostranstva, 6.62.189 (1976).

143.337 Cosmic ray intensity in the interplanetary space. Data of observations from November 10, 1970 - April 20, 1973.
Mezhduved. geofiz. kom. pri Prezidiume AN SSSR. Materialy Mirovogo tsentra dannykh B. Moskva, 1975. 40 pp. Price 6 Kop. – Abstr. in Referativ. Zhurn. 62. Issled. kosmich. prostranstva, 6.62.190 (1976).

143.338 The character of cosmic ray propagation in the Galaxy. K. V. Bychkov.
Astron. Zhurn. Akad. Nauk SSSR, Vol. 53, 501 - 510 (1976). In Russian. English translation in Soviet Astron., Vol. 20, No. 3.
For the case of supernova remnants it is shown that the time needed for transition of cosmic rays (c.r.) from magnetic lines of the source to those of the general galactic field is very long. A "convective" model of c.r. propagation in the Galaxy is suggested according to which the greater part of cosmic rays propagates being frozen in traps created by magnetic lines of the source. Processes of trap formation at different stages of supernova remnant development and the stability of traps are considered. It is shown that the "convective" model does not contradict the observational data available. A program of search for traps around young and old supernova remnants is suggested.

143.339 Field dependent cosmic ray streaming at high rigidities. D. B. Swinson.
Journ. Geophys. Res., Vol. 81, 2075 - 2081 (1976).
Data from underground μ meson telescopes at depths of

25, 40, and 80 mwe covering the period 1965–1973 have been analyzed as a function of interplanetary magnetic field direction. Cosmic ray streaming both in and perpendicular to the ecliptic plane, with directions dependent on the sense of the interplanetary magnetic field, is observed throughout the period at all depths.

143.340 Theory of the solar magnetic cycle wave in the diurnal variation of energetic cosmic rays: physical basis of the anisotropy. E. H. Levy.
Journ. Geophys. Res., Vol. 81, 2082 - 2088 (1976).
It is proposed that the so-called 20-year wave in the diurnal variation of energetic cosmic rays is a consequence of the likely average odd symmetry of the interplanetary magnetic field about the solar equatorial plane. It is assumed that the magnetic field in each hemisphere of the solar magnetic cavity has the same average sense as the polar magnetic field at the corresponding solar pole.

143.341 A quasi-linear kinetic equation for cosmic rays in the interplanetary medium. J. G. Luhmann.
Journ. Geophys. Res., Vol. 81, 2089 - 2093 (1976).
A kinetic equation for interplanetary cosmic rays is set up with the aid of weak plasma turbulence theory for an idealized radially symmetric model of the interplanetary magnetic field. As a starting point this treatment invokes the Vlasov equation instead of the traditional Fokker-Planck equation. Quasi-linear theory is applied to obtain a momentum diffusion equation for the heliocentric frame of reference which describes the interaction of cosmic rays with convecting magnetic irregularities in the solar wind plasma.

143.342 Cosmic ray scintillations. 4. The effects of non-field-aligned diffusion. J. R. Jokipii, A. J. Owens.
Journ. Geophys. Res., Vol. 81, 2094 - 2096 (1976).
The authors generalize their previous discussion of the fluctuations in the cosmic ray intensity near earth by including non-field-aligned diffusion in the model. For frequencies of $\gtrsim 5 \times 10^{-6}$ Hz the fluctuations are interplanetary scintillations caused by the field-aligned anisotropy, as was previously discussed. For lower frequencies the effects of non-field-aligned diffusion dominate and give a power spectral density proportional to f^{-2}. Observations are presented which support the model.

143.343 Hysteresis of primary cosmic rays associated with Forbush decreases. R. S. Rajan.
Australian Journ. Phys., Vol. 29, 89 - 95 (1976).
Regression analysis of primary cosmic ray intensities during Forbush decreases indicates the existence of a differential modulation between high and low rigidity primaries.

143.344 On the effect of directional medium scale interplanetary variations on the diffusion of galactic cosmic rays and their solar cycle variation.
G. E. Morfill, H. J. Völk, M. A. Lee.
Separate print Max-Planck-Inst. Kernphys., Heidelberg, MPI H-1976-V7, 2 + 57 pp. (1976).
It is argued that the interplanetary magnetic field cannot be simply considered as consisting of small short scale resonant fluctuations superposed on the long term average spiral field if one wants to calculate the average transport properties of galactic cosmic rays. The interest is focussed on pitch angle diffusion, where the relevant average field varies strongly on a medium scale exceeding the magnetic correlation length.

143.345 Electrons in a closed galaxy. Model of cosmic rays. R. Ramaty, N. J. Westergaard.
GSFC Document X-660-76-19, Prepr., 34 pp. (1976).
The authors consider the consistency of positrons and

electrons with a propagation model in which the cosmic rays are stopped by nuclear collisions or energy losses before they can escape from the galaxy. This closed-galaxy model predicts steep electron and positron spectra at high energies and predicts that the interstellar electron intensity below a few GeV is larger than that implied by other models. The consequence of this result is that electron bremsstrahlung is responsible for about 50% of the galactic gamma-ray emission at photon energies greater than 100 MeV.

143.346 Solar modulation of galactic cosmic rays. 4. Latitude dependent modulation.   L. A. Fisk.
GSFC Document X-660-76-27, Prepr., 23 pp. (1976).
A numerical method is outlined for solving the equation which describes the solar modulation of cosmic rays in models where interplanetary conditions can vary with heliocentric latitude. As an illustration of the use of this method, it is shown how variations in the modulation with latitude could produce the small radial gradients in the intensity that have been observed from the Pioneers 10 and 11 spacecraft.

143.347 Anätzbare Spuren schwerer Ionen und Elementhäufigkeiten in der kosmischen Strahlung.
W. Krätschmer, I. Schlegel, W. Gentner.
MPI Kernphys., Heidelberg, Jahresbericht 1975, p. 169 - 171.

143.348 Diffusion galaktischer kosmischer Strahlung im Sonnensystem.
M. A. Lee, G. E. Morfill, H. J. Völk.
MPI Kernphys., Heidelberg, Jahresbericht 1975, p. 171 - 172.

143.349 Solar daily variation of cosmic ray intensity.
H. S. Ahluwalia.
Proc. Southwest Regional Conf., Vol. 1, (see 012.021), 141 - 150 (1976).

143.350 The mean energy transfer to black and grey tracks in stars produced by 10 GeV protons. A. Vogel.
Stud. Cerc. Fiz., Vol. 27, 207 - 211 (1975). In Rumanian. Abstr. in Phys. Abstr., Vol. 79, A006862 (1976).

143.351 Annual modulation of solar diurnal variation of the cosmic ray nucleonic component in three-dimensional space. S. Mori.
Journ. Geomagn. Geoelectr., Vol. 27, 1 - 23 (1975). — Abstr. in Phys. Abstr., Vol. 79, A011048 (1976).

143.352 The anisotropy of ultra-high-energy cosmic rays. II. Search for correlations with astronomical objects.
P. Kiraly, J. Kota, J. L. Osborne, M. White, A. W. Wolfendale.
Journ. Phys. A, Vol. 8, 2018 - 2032 (1975). — Abstr. in Phys. Abstr., Vol. 79, A019313 (1976).

143.353 Composition and energy spectrum of a primary cosmic radiation of mixed origin.   K. Sitte.
Nuovo Cimento A, Ser. 11, Vol. 30A, 195 - 211 (1975). Abstr. in Phys. Abstr., Vol. 79, A023576 (1976).

143.354 Observation of heavy primary cosmic ray nuclei with solid state track detectors.   T. Doke,
H. Okamoto, E. Shibamura, T. Hayashi, K. Itoh, T. Yanagimachi, S. Kobayashi, K. Nagata, T. Saito, M. Miyajima.
Bull. Inst. Space Aeronaut. Sci., Univ. Tokyo, B, Vol. 11, 497 - 515 (1975). In Japanese. — Abstr. in Phys. Abstr., Vol. 79, A027685 (1976).

143.355 Cosmic-ray composition measurements with high-energy ionisation spectrometers.
J. F. Arens, J. F. Ormes.
Phys. Rev. D, Vol. 12, 1920 - 1935 (1975). — Abstr. in Phys. Abstr., Vol. 79, A027686 (1976).

143.356 Cosmic-ray electrons and galactic radio noise: some problems.   G. Gavazzi, G. Sironi.
Nuovo Cimento Rivista, Ser. 2, Vol. 5, 155 - 186 (1975). Abstr. in Phys. Abstr., Vol. 79, A032630 (1976).

143.357 Galactic or extragalactic cosmic rays, are we close to the end of the riddle.   J. C. Jodogne.
Publ. Inst. Roy. Meteorol. Belgique, Ser. A, No. 91, p. 153 - 158 (1975). In French. — Abstr. in Phys. Abstr., Vol. 79, A041978 (1976).

143.358 Dynamic spectral analysis of cosmic ray anisotropy by means of high-speed spectral analysis method.
T. Kanno, T. Saito, T. Sakurai, K. Yumoto, Y. Ishida, T. Saito.
Rep. Ionosph. Space Res. Japan, Vol. 29, 118 - 126 (1975). Abstr. in Phys. Abstr., Vol. 79, A041980 (1976).

143.359 Low energy cosmic ray ions observed in the Skylab experiment.
S. Biswas, N. Durgaprasad, J. Nevatia, V. S. Venkatavaradan.
Indian Journ. Radio Space Phys., Vol. 4, 210 - 215 (1975). Abstr. in Phys. Abstr., Vol. 79, A041984 (1976).

143.360 Determination of the low energy cosmic ray gamma photon flux in the atmosphere.
I. M. Martin, S. L. G. Dutra, G. Vedrenne, P. Mandrou.
Rev. Brasil Fis., Vol. 5, No. 1, p. 15 - 23 (1975). In French. Abstr. in Phys. Abstr., Vol. 79, A041986 (1976).

143.361 Primary cosmic radiation.   H. R. Anderson.
2nd conf. natural radiation environment, Houston, Texas, USA, 7 Aug. 1972, p. 1 - 13 (1972).

143.362 Measurement of the primary cosmic electron spectrum from 10 GeV to about 250 GeV.
R. F. Silverberg.
Thesis Maryland Univ., College Park, USA, 188 pp. (1974). (Available from: Univ. Microfilms, Order No. 74-29,766). See 14.143.045.

143.363 Time-dependent diffusion-convection model for the long-term modulation of cosmic rays.
J. J. O'Gallagher.
Separate print Maryland Univ., College Park, USA, Dept. Phys. Astron., 42 pp. (1974). — See 13.143.024.

143.364 Spectrum of cosmic electrons with energies between 6 and 100 GeV.   C. A. Meegan, J. A. Earl.
Separate print Maryland Univ., College Park, USA, Dept. Phys. Astron., 53 pp. (1974).

143.365 Dynamic model for the time evolution of the modulated cosmic-ray spectrum.
J. J. O'Gallagher, G. A. Maslyar III.
Separate print Maryland Univ., College Park, USA, Dept. Phys. Astron., 34 pp. (1975).

143.366 Origin of cosmic rays.   F. W. Stecker.
Separate print National Aeronaut. Space Administr., Greenbelt, Maryland, USA, Goddard Space Flight Center, 14 pp. (1975).

143.367 X-ray astronomy evidence for subrelativistic cosmic rays from supernovae.   E. Boldt.
Symposium significant accomplishments sci. technol., Greenbelt, Maryland, USA, 18 Dec. 1973, p. 71 - 74 (1975).

143.368 What Pioneer-10 means to cosmic-ray modulation theory.   L. Fisk.
Symposium significant accomplishments sci. technol., Greenbelt, Maryland, USA, 18 Dec. 1973, p. 84 - 87 (1975).

143.369 **Balloon observations of cosmic-ray negatrons and positrons.** R. Hartman.
Symposium significant accomplishments sci. technol., Greenbelt, Maryland, USA, 18 Dec. 1973, p. 88 - 91 (1975).

143.370 **Compton-Getting effect for low energy particles.** F. M. Ipavich.
Separate print Maryland Univ., College Park, USA, Dept. Phys. Astron., 16 pp. (1974).

143.371 **Nuclear interactions between cosmic radiation and interstellar gas, and nucleosynthesis of lithium, beryllium, and boron.** M. Meneguzzi.
Thesis Paris-7 Univ., France, (1975). In French.

143.372 **Gamma-ray evidence for galactic Fermi acceleration of cosmic rays.** F. Stecker.
Symposium significant accomplishments sci. technol., Greenbelt, Maryland, USA, 18 Dec. 1973, p. 60 - 62 (1975).

143.373 **Pioneer-10 measurements of the composition of low energy galactic cosmic rays.** B. Teegarden.
Symposium significant accomplishments sci. technol., Greenbelt, Maryland, USA, 18 Dec. 1973, p. 80 - 83 (1975).

143.374 **Origin and propagation of cosmic rays.** C. J. Cesarsky.
ESA/ASE Sci. Tech. Rev., Vol. 1, 23 - 33 = ESRO workshop res. goals cosmic-ray astrophys. in the 1980's, Frascati, Italy, 24 - 25 Oct. 1974, p. 21 - 31 (1975).

143.375 **Primary cosmic ray spectrum from 10^{11} eV to 10^{15} eV.** C. B. Gerdes.
Thesis Arizona Univ., Tucson, USA, 101 pp. (1974). (Available from Univ. Microfilms, Order No. 75-11, 911).

143.376 **Average flux of cosmic-ray nuclei at a distance of 1 AU from the sun.** B. Sitar.
Acta Fac. Rerum Nat. Univ. Comenianae, Phys., Vol. 15, 233 - 248 (1975). In Slovak.

143.377 **Current cosmic-ray programmes in Europe.** P. H. Fowler.
ESRO workshop res. goals cosmic-ray astrophys. in the 1980's, Frascati, Italy, 24 - 25 Oct. 1974, p. 51 - 60 (1975).

143.378 **Review of present knowledge of cosmic rays.** P. H. Fowler.
ESRO workshop res. goals cosmic-ray astrophys. in the 1980's, Frascati, Italy, 24 - 25 Oct. 1974, p. 3 - 10 (1975).

143.379 **A high-energy cosmic-ray composition study as a possible space experiment.**
P. H. Fowler, R. J. Edge, M. R. W. Masheder, M. T. Moses, R. N. F. Walker, A. Morley.
ESRO workshop res. goals cosmic-ray astrophys. in the 1980's, Frascati, Italy, 24 - 25 Oct. 1974, p. 157 - 162 (1975).

143.380 **Electrons in the galactic cosmic radiation.** D. Müller.
ESRO workshop res. goals cosmic-ray astrophys. in the 1980's, Frascati, Italy, 24 - 25 Oct. 1974, p. 13 - 18 (1975).

143.381 **Current cosmic-ray programs in the United States.** A. G. Opp.
ESRO workshop res. goals cosmic-ray astrophys. in the 1980's, Frascati, Italy, 24 - 25 Oct. 1974, p. 47 - 49 (1975).

143.382 **Pulsars and supernovae as sources of cosmic rays.** F. Pacini.
ESRO workshop res. goals cosmic-ray astrophys. in the 1980's, Frascati, Italy, 24 - 25 Oct. 1974, p. 19 (1975). – Abstract.

143.383 **Isotopic analysis of high-energy cosmic-ray nuclei in the geomagnetic field.** B. Peters.
ESRO workshop res. goals cosmic-ray astrophys. in the 1980's, Frascati, Italy, 24 - 25 Oct. 1974, p. 85 - 90 (1975).

143.384 **Cosmic rays and the dynamics of the Galaxy.** K. Pinkau.
ESRO workshop res. goals cosmic-ray astrophys. in the 1980's, Frascati, Italy, 24 - 25 Oct. 1974, p. 33 - 37 (1975).

143.385 **A facility for cosmic-ray astrophysics in connection with Spacelab.** I. L. Rasmussen.
ESRO workshop res. goals cosmic-ray astrophys. in the 1980's, Frascati, Italy, 24 - 25 Oct. 1974, p. 167 - 169 (1975).

143.386 **Interrelations between cosmic rays and other branches of astrophysics.** M. J. Rees.
ESRO workshop res. goals cosmic-ray astrophys. in the 1980's, Frascati, Italy, 24 - 25 Oct. 1974, p. 39 - 44 (1975).

143.387 **Determination of the isotopic composition of high-energy cosmic-ray hydrogen and helium.**
W. K. H. Schmidt, M. Simon.
ESRO workshop res. goals cosmic-ray astrophys. in the 1980's, Frascati, Italy, 24 - 25 Oct. 1974, p. 171 - 173 (1975).

143.388 **Methods for the determination of Z and M using dE/dx, Cerenkov, and total energy measurements.**
E. C. Stone.
ESRO workshop res. goals cosmic-ray astrophys. in the 1980's, Frascati, Italy, 24 - 25 Oct. 1974, p. 63 - 70 (1975).

143.389 **Nitrogen and oxygen isotopes in the low energy cosmic rays.** S. B. Vidor.
Thesis California Inst. Tech., Pasadena, USA, 162 pp. (1975). (Available from Univ. Microfilms, Order No. 75-25, 849).

Ultraschwere kosmische Kerne.
Umschau, 76. Jahrgang, p. 114 - 115 (1976).
  The investigation of ultraheavy cosmic rays (i.e. highly charged nuclei with $Z > 30$) can help us understand the origin of the chemical elements, their production processes in stellar objects (like supermassive stars, supernovae or evolved main sequence stars) and their propagation through the interstellar medium.

Calculation of the cross section for C IV–H charge exchange: significance for interstellar X-rays/cosmic-ray particles. See Abstr. 022.063.

Description of a satellite experiment for isotopic-composition measurement of cosmic nuclei by the slowing-down method. See Abstr. 032.538.

A proposed ultra-heavy cosmic-ray detector for space-shuttle exposure. See Abstr. 032.539.

Cosmic radiation and fundamental problems in physics. See Abstr. 061.002.

Importance of isotopic composition of iron in cosmic rays. See Abstr. 061.006.

Cosmic-ray spallative origin of the rare odd-odd nuclei, consistent with light-element production. See Abstr. 061.021.

Massive stars: a source of cosmic rays. See Abstr. 061.057.

A nonlinear theory of cosmic-ray pitch-angle diffusion in homogeneous magnetostatic turbulence. See Abstr. 062.011.

The effect of adiabatic focusing upon charged-particle propagation in random magnetic fields. See Abstr. 062.026.

On the relation of the development of active regions on the sun to the accompanying phenomena in cosmic ray intensity.   See Abstr. 072.047.

Solar wind structure in the vicinity of the equatorial plane.   See Abstr. 074.051.

The solar cycle variation in the solar wind and the modulation of cosmic rays.   See Abstr. 074.053.

Relation of large-scale coronal X-ray structure and cosmic rays: 4. Amplitude of the diurnal variation in neutron monitors on interplanetary field lines orbiting above coronal holes.   See Abstr. 074.055.

Solar and galactic cosmic ray abundances – a comparison and some comments.   See Abstr. 078.055.

Low energy particle composition. See Abstr. 078.092.

The Forbush decrease of November 17, 1966. See Abstr. 078.094.

Variations of cosmic ray intensities during the first part of the Helios-1 mission.   See Abstr. 078.100.

High energy cosmic ray intensity variations associated with the unusual Forbush decrease of August 1972. See Abstr. 078.104.

Particle track record in lunar silicates: long-term

behavior of solar and galactic VH nuclei and lunar surface dynamics.   See Abstr. 094.202.

The feasibility of ion identification on cosmic-ray tracks in lunar feldspars.   See Abstr. 094.573.

Fossil tracks in lunar samples: information on the abundances of VH nuclei in the ancient cosmic radiation. See Abstr. 094.586.

Bestimmung des räumlichen Gradienten der kosmischen Strahlung mit Meteoriten. See Abstr. 105.095.

Interplanetary magnetic fields, their fluctuations, and cosmic ray variations.   See Abstr. 106.026.

A survey of the interplanetary magnetic field. See Abstr. 106.045.

A model for the acceleration of particles trapped between converging shock waves.   See Abstr. 106.047.

The radio spectrum of Vela supernova and its implication on the propagation of cosmic rays from sources. See Abstr. 125.030.

Vela gamma rays and the source of cosmic rays revisited.   See Abstr. 125.031.

Multiple ionization by low-energy cosmic rays and the abundance of highly ionized interstellar atoms. See Abstr. 131.042.

Production of galactic X-rays following charge exchange by cosmic-ray nuclei.   See Abstr. 142.108.

Galactic gamma radiation from cosmic rays concentrated in spiral arms.   See Abstr. 142.127.

Gamma ray astronomy.   See Abstr. 142.158.

Stellar Systems

151 Kinematics and Dynamics of Stellar Systems

151.001 Dynamical friction in spherical clusters.
S. D. M. White.
Monthly Notices Roy. Astron. Soc., Vol. 174, 19 - 28 (1976).

The effect of dynamical friction on the density profile of the most massive galaxies in a cluster is calculated both for an isothermal model cluster and for Plummer's model. The resulting profiles show a depletion of massive objects near the centre, but this depletion appears unable to produce a secondary maximum. The application of the models to rich clusters of galaxies is discussed and it is found that the evolution should be at least strong enough to produce central objects the size of cD galaxies. Accurate luminosity profiles for clusters of galaxies are shown to be capable of putting constraints on the mass to light ratios of the member galaxies and to give an indication of the form and distribution of the 'missing mass'.

151.002 Elongated equilibrium stellar systems tidally distorted in pairs. G. S. Bisnovatyi (*Bisnovatyj*)-Kogan.
Monthly Notices Roy. Astron. Soc., Vol. 174, 203 - 211 (1976).

The self-consistent solutions for barred stellar configurations under tidal action in pairs are obtained, the gravitational potential being a quadratic function of coordinates. The class of solutions for models under tidal action is wider than the solution for single systems, obtained by Freeman. The solutions for cold, flat elliptical disks and for unbalanced triaxial ellipsoids are obtained, which are absent in the case of a single system.

151.003 A note on the minimum impact parameter for dynamical friction involving spherical clusters.
S. D. M. White.
Monthly Notices Roy. Astron. Soc., Vol. 174, 467 - 469 (1976).

An expression is derived for the minimum impact parameter to be used in calculations of the effect of dynamical friction on massive extended objects moving through a background of lighter particles. The expression is evaluated for a range of cluster models and the result compared with that used by other authors.

151.004 On the mass of the halo population. L. B. Lucy.
Astrophys. Journ., Vol. 203, 75 - 80 (1976).

Implications of the hypothesis that the halo is the major contributor to the Galaxy's interior mass ($R < R_0$) are worked out and compared with observation. The required local density of halo stars is found to be $\sim 0.007\,M_\odot\,pc^{-3}$, and this is shown to be not inconsistent with the observational data. Eddington's model for spherical stellar systems, in which the velocity distribution is everywhere ellipsoidal, is argued to be an acceptable model for the halo and is used to explore the consequences of the halo being an approximately self-gravitating system.

151.005 On density waves in galaxies. II. The turning-point problem at the corotation region.
J. W-K. Mark.
Astrophys. Journ., Vol. 203, 81 - 96 (1976).

A more careful analysis has been made of the galactic density waves near the corotation region where the wave pattern speeds are the same as the local circular speed of the disk. Some ambiguities that occurred in previous solutions due to the assumption of a single wave picture are removed. A reduced form of the wave gravitational potential is now found to satisfy a turning-point differential equation. This equation involves two turning points in a manner analogous to those which occur in the quantum-mechanical problem of wave-tunneling through a parabolic potential barrier.

151.006 The formation of the nuclei of galaxies. II. The local group. S. D. Tremaine.
Astrophys. Journ., Vol. 203, 345 - 351 (1976).

The remnants of clusters which have spiraled to the galactic center through dynamical friction may now be visible as a galactic nucleus. This theory was tested on M31 in Paper I (see Abstr. 13.158.028). Here some of the assumptions made in Paper I are checked using data from the galactic globular cluster system, and the theory is applied to the Galaxy and to nearby elliptical galaxies. The results are consistent with observations, but depend strongly on the assumed initial space distribution of the clusters. The evolution of the cluster population is shown to affect the use of brightest clusters as distance indicators, and an alternative luminosity standard is suggested.

151.007 The dynamic instability of isothermal relativistic star clusters. K. G. Suffern, E. D. Fackerell.
Astrophys. Journ., Vol. 203, 477 - 480 (1976).

Recently Katz, Horwitz, and Klapisch specified two zones of parameter space in which isothermal star clusters may be thermodynamically stable. The authors tested the dynamic stability of a large number of clusters in these zones and found that all such clusters with central redshifts greater than 0.55 are dynamically unstable.

151.008 The escape of stars from isolated clusters. M. Saito.
Astron. Astrophys., Vol. 46, 171 - 178 (1976).

The escape rates both of stars and of energy are computed for several isolated and spherically symmetric clusters with an isotropic velocity distribution in consideration of stellar orbital motions in the smoothed-out gravitational potential and the fluctuation in stellar energy due to the cumulative effect of distant encounters. Theoretical rates are smaller by a factor of 2 to 6 than the experimental ones found in numerical Monte Carlo calculations by Spitzer and Hart (1971). The possible reasons for such results are explained.

151.009 Density maxima formed by trapped orbits.
B. Barbanis.
Astron. Astrophys., Vol. 46, 269 - 274 (1976).

The statistical behaviour of the orbits near the particle resonance of a spiral galaxy is investigated. The numerical integrations were made in a model spiral field which was introduced abruptly and then held at constant amplitude. In this case the density maxima appear near to the maxima of the spiral potential.

151.010 Measuring the dynamical age of N-body systems.
S. von Hoerner.
Astron. Astrophys., Vol. 46, 293 - 302 (1976).

A method is suggested for obtaining the age, and the degree of relaxation, of isolated clusters of galaxies (or stars) from observed properties. The present paper treats equal-mass systems. The relaxed cluster consists of three parts: a small dense center, with an observed projected density ρ_p = constant up to radius R_c; an intermediate core with $\rho_p \sim R^{-1.40}$ between R_c and R_h; and an outer halo with $\rho_p \sim R^{-2.75}$ beyond R_h. A 'modulus of evolution' is defined as $W = \log (R_h/R_c)$. Its time-dependence $W(t)$ is calculated, and a method is given for obtaining W numerically from observations. An error estimate shows that we need $N \gtrsim 26$ for a proper phase distinction on a 2σ level (young versus relaxed cluster); and $N \gtrsim 50$ is needed for finding the age t in years within a factor of two. The projected central density must be at least 10^2 the surrounding background for phase distinction, and at least 10^3 for age determination.

151.011 Some numerical experiments in stellar statistics. G. Lyngå.

Astron. Astrophys., Vol. 46, 369 - 379 (1976).

Numerical methods have been studied which will calculate the expected appearance of star count functions and colour magnitude diagrams for a stellar field assuming a set of parameters that describe the intrinsic distribution of stars, their absolute magnitudes and colours as well as the distribution of interstellar matter in space. A method is also given which will adjust the assumed stellar density function to yield best approximation of observed star count functions. Some examples are given of the use of the methods, and commonly used practices are discussed on the basis of these examples.

151.012 On self gravitating stellar disks I. A new model. J. P. J. Lafon.

Astron. Astrophys., Vol. 46, 461 - 472 (1976).

A new method is given for solving the self-consistent coupled Liouville's and Poisson's equations which govern steady state, infinitesimally thin, axisymmetric self-gravitating systems. A theory of such systems is developed: models are constructed numerically and discussed, including in particular bounded models for which the distribution function has no sharp cut-off at high energies.

151.013 Effect of binary stars on the dynamical evolution of stellar clusters. II. Analytic evolutionary models. J. G. Hills.

Astron. Journ., Vol. 80, 1075 - 1080 (1975).

The author uses analytic models to compute the evolution of the core of a stellar system due simultaneously to stellar evaporation which causes the system (core) to contract and to its binaries which cause it to expand by progressively decreasing its binding energy. The evolution of the system is determined by two parameters: the initial number of stars in the system N_0, and the fraction f_b of its stars which are binaries. Open clusters expand monotonically from the beginning if they have anything approaching average Population I binary frequencies. Globular clusters are highly deficient in binaries in order to have formed and retained the high-density stellar cores observed in most of them. The author estimates that for these systems $f_b \lesssim 0.15$.

151.014 Production of triple stars by the dynamical decay of small stellar systems. R. S. Harrington.

Astron. Journ., Vol. 80, 1081 - 1086 (1975).

Numerical integrations of three-body systems have been carried out to establish the sufficient criteria for the stability of triple stars. In addition, numerical integrations of unstable five-body systems have been performed to examine the statistics of the formation of binaries and stable triples. The number of triples compared to binaries produced by dynamical decay is consistent with the observed frequencies of these systems.

151.015 Suppression of bar instability by a massive halo. F. Hohl.

Astron. Journ., Vol. 81, 30 - 36 (1976).

Numerical experiments are performed to determine the effect of a spherical, uniform-density mass distribution or halo on the evolution of stellar disks. For a halo with radius much smaller than that of the disk, the large-scale bar-forming instability is prevented by a fixed halo (or core) component containing 60% of the total system mass. Similar results are obtained for a halo component corresponding to the Schmidt model of the Galaxy. For a uniform-density spherical halo with radius equal to that of the stellar disk, a halo component containing only 40% of the total mass is sufficient to stabilize the bar-forming mode. These results indicate that a large halo is more effective in stabilizing against bar formation than a more centrally condensed core-halo.

151.016 On a model of star clusters with axial symmetry and discrete mass distribution of stars. V. M. Bagin.

Astron. Zhurn. Akad. Nauk SSSR, Vol. 53, 100 - 105 (1976). In Russian. English translation in Soviet Astron., Vol. 20, No. 1.

A model of a stationary star cluster with axial symmetry and with stellar composition homogeneous with regard to mass is generalized for the case of a model with stellar composition discretely distributed with regard to mass.

151.017 A qualitative analysis of forms of motion of a star in a stationary stellar system with an axisymmetric nucleus. II. V. K. Kajsin.

Astron. Zhurn. Akad. Nauk SSSR, Vol. 53, 106 - 111 (1976). In Russian. English translation in Soviet Astron., Vol. 20, No. 1.

A qualitative analysis of various forms of motion of a star in the corona of a stellar system with prolate nucleus is given. Some particular cases of motions as well as the problem of the influence of the nucleus figure on the form of the trajectory are discussed.

151.018 Kinetics of stars in flat systems. Relaxation. I. L. Genkin, Z. N. Chumak.

Astron. Zhurn. Akad. Nauk SSSR, Vol. 53, 208 - 211 (1976). In Russian. English translation in Soviet Astron., Vol. 20, No. 1.

Relaxation in flat stellar systems is studied. The values of the diffusion tensor in the velocity space and the component of the tensor of third rank describing the asymmetry of the distribution function are found. Expressions of the scattering cross sections not containing Coulomb's logarithm are obtained. The values of the time of relaxation are of the same order of magnitude as the mean orbital time.

151.019 Computer study of galactic spiral arm shapes. N. A. Barricelli, O. Havnes.

Astrophys. Letters, Vol. 17, 37 - 44 (1976).

A galactic explosion model is used in a computer program for a first approach interpretation of spiral arm shapes and velocity distributions in the galaxies M 31 and M 33. A refinement of the model is proposed and its implications for galactic dimensions are presented.

151.020 On the derivation of mass distribution in galaxies from internal rotational velocities. A. Reiz, F. Fabiani.

Astron. Astrophys., Vol. 47, 1 - 3 (1976).

The Volterra integral equation of the first kind relating the observed rotational velocities in a galaxy to the density distribution, is transformed into an integral equation of the second kind; solving this numerically gives the mass distribution. The method has been applied to the Andromeda galaxy, using the rotational velocities measured by Rubin and Ford.

151.021 Dynamical evolution of the triple system of the Galaxy, the Large and Small Magellanic Clouds.

M. Fujimoto, Y. Sofue.
Astron. Astrophys., Vol. 47, 263 - 291 (1976).

Obtained here are some series of orbits of the Large and Small Magellanic Clouds (LMC and SMC) round the Galaxy, along which the two clouds were in a binary state for the last 5 to 10×10^9 years. The peri- and apogalactic distances of the LMC's orbit seem to be about 30 kpc and 60–80 kpc, respectively. The tilt angle of the orbital plane to the galactic disk is 70° to 110° with the ascending and descending nodes respectively at $l \approx 280°$ and 55°. Also discussed in relation to the dynamics of the triple system of the Galaxy, LMC and SMC are the velocity dispersion of common stars, high-velocity H I clouds, high-velocity A-type stars in the Galaxy, distant gaseous spiral arms at 20 to 30 kpc from the galactic center, and north–south asymmetry of our rotation curves.

151.022 **Die Anregung von Spiralarmdichtewellen durch eine Massenasymmetrie im Kerngebiet von Galaxien.**
J. Feitzinger.
Mitt. Astron. Ges., No. 38, p. 104 (1976). – Abstract.

151.023 **Kollisionen sphärischer Galaxien.**   P. Biermann.
Mitt. Astron. Ges., No. 38, p. 104 (1976). – Abstract.

151.024 **Ein Zwei-Komponenten-Modell für die Massenverteilung in Galaxien.**   R. Wielen.
Mitt. Astron. Ges., No. 38, p. 254 - 255 (1976). – Abstract.

151.025 **Massenbestimmungen von Galaxien durch die Dynamik von Systemen von Galaxien.**   J. Materne.
Mitt. Astron. Ges., No. 38, p. 255 - 259 (1976). – Short report.

151.026 **A family of self-gravitating stellar systems with axial symmetry.**   R. Nagai, M. Miyamoto.
Publ. Astron. Soc. Japan, Vol. 28, 1 - 17 (1976).

Toomre's (1963) disk models for flat galaxies have been generalized to yield convenient pairs of three-dimensional potential and density functions. All of the present density functions are non-negative everywhere. For the simplest pair of the present potential and density functions, a velocity distribution of Fricke's (1952) type has been found. Also for that case, a family of self-consistent models for axisymmetric rotating stellar systems with arbitrary oblateness has been constructed.

151.027 **Whence comes the ellipsoidal distribution of star velocities.**   T.-y. Yueh.
Scient. Sinica, Vol. 19, 35 - 44 (1976).

The paper examines the turbulent motion in the protogalaxy cloud, showing that such turbulent motion would lead to the ellipsoidal distribution of velocities. This distribution was first inherited by protostars. Afterwards, the random gravitation of the stellar field would lead merely to the changes in parameters of the ellipsoidal distribution. Thus, an explanation of the original formation of the ellipsoidal distribution of star velocities is provided.

151.028 **Encounters of spherical galaxies. I. Galaxy models with one stellar population.**
P. Biermann, J. Silk.
Astron. Astrophys., Vol. 48, 287 - 293 (1976).

Using models of elliptical galaxies as described by King (1966) for the galaxy NGC 3379, fast encounters of spherical galaxies of equal size are calculated. Results are given for the relative mass loss as well as the relative change of internal energy for the remaining stars. The variation of mass loss and the internal energy change inside a galaxy are also described. The results indicate that the dimensions of galactic halos would have to be $\gtrsim 200$ kpc to cause appreciable mass loss over the history of a galaxy in a rich cluster.

151.029 **Encounters of spherical galaxies. II. Galaxy models with two stellar populations.**   P. Biermann.
Astron. Astrophys., Vol. 48, 295 - 299 (1976).

Refining the models of elliptical galaxies of King (1966) so as to incorporate a separate halo population of stars of high velocity dispersion, fast encounters of spherical galaxies of equal size are calculated. The results indicate that the halo population can be dispersed for sufficiently large galaxies without appreciably effecting the main population.

151.030 **Gravitational attraction of a disk-sphere pair of galaxies.**   G. M. Ballabh.
Bull. Astron. Soc. India, Vol. 3, 30 (1975). – Abstract of a paper presented at the A.S.I. meeting 1975.

151.031 **A classification of galactic collisions.**   K. S. Sastry, S. M. Alladin.
Bull. Astron. Soc. India, Vol. 3, 30 (1975). – Abstract of a paper presented at the A.S.I. meeting 1975.

151.032 **Zero velocity surfaces due to the gravitational field of a pair of disk galaxies.**   P. V. Subramanyam.
Bull. Astron. Soc. India, Vol. 3, 31 (1975). – Abstract of a paper presented at the A.S.I. meeting 1975.

151.033 **Shock wave in the flow against the external gravitational force.**
S. K. Chakrabartty, R. Bondyopadhaya.
Bull. Astron. Soc. India, Vol. 3, 31 (1975). – Abstract of a paper presented at the A.S.I. meeting 1975.

151.034 **Stability of self-gravitating systems with phase space density a function of energy and angular momentum for aspherical modes.**
D. Gillon, J. P. Doremus, G. Baumann.
Astron. Astrophys., Vol. 48, 467 - 474 (1976).

The authors study the stability of spherical stellar systems when the distribution function F depends both on the energy ϵ and on J, the square of the angular momentum. Using the multiple water bag model, they obtain a conserved quantity which can be shown to be always positive for aspherical modes when $\partial F/\partial \epsilon < 0$ and $\partial F/\partial J < 0$. They conclude that under these conditions the systems are always stable.

151.035 **On density waves in galaxies. III. Wave amplification by stimulated emission.**   J. W-K. Mark.
Astrophys. Journ., Vol. 205, 363 - 378 (1976).

Through outward transport of angular momentum, the circular rotation in disk galaxies becomes an abundant "source" for spiral density waves. Increases in wave angular momentum of an order of magnitude are possible. Such wave amplification by stimulated emission could be an integral part of mechanisms for the maintenance of spiral structure. For example, in the suggested "laser analog", even two-fold amplification at corotation can result in temporal growth of spiral modes over the quasi-stationary time-scale of several billion years.

151.036 **On a spherical star system with a collapsed core.**   E. N. Glass, B. Mashhoon.
Astrophys. Journ., Vol. 205, 570 - 577 (1976).

Exact solutions of Einstein's equations for a perfect fluid that describe the gravitational collapse of matter around a collapsed core are presented. The collapsed region grows monotonically until an absolute Schwarzschild horizon is formed. It is found that the minimum redshift of light emitted radially from the boundary of the system and received at infinity is ~4.62. These systems are expected to describe the very last stages in the gravitational collapse of a globular star cluster with a central collapsed core.

151.037 The bar-like objects in the centres of galaxies as a possible generator of spiral density waves. The response in the neighbourhood of the corotation circle.
V. I. Korchagin.
Astron. Zhurn. Akad. Nauk SSSR, Vol. 53, 260 - 267 (1976).
In Russian. English translation in Soviet Astron., Vol. 20, No. 2.

The response of the flat subsystem of a galaxy in the neighbourhood of the corotation circle is considered. The bar generates a ring-like structure in the neighbourhood of the corotation circle.

151.038 The instability of gravitating rotating viscous gaseous systems and the nature of the ring structure of galaxies. Yu. N. Mishurov, V. M. Peftiev, A. A. Suchkov.
Astron. Zhurn. Akad. Nauk SSSR, Vol. 53, 268 - 272 (1976).
In Russian. English translation in Soviet Astron., Vol. 20, No. 2.

The effects of viscosity in a protogalaxy are discussed. The mechanism of viscosity is associated with momentum transport by the gas clouds which have been formed before the violent star formation period. It is shown that in the presence of viscosity a rotating disk (cylinder) is unstable against circular perturbations. It is argued that this instability can lead to the formation of the ring structure of galaxies.

151.039 Dynamics of flat galaxies. II. Biorthonormal surface density-potential pairs for finite disks.
A. J. Kalnajs.
Astrophys. Journ., Vol. 205, 745 - 750 (1976).

Poisson's equation in the flat disk geometry becomes an integral equation, and it can be solved by series expansions in biorthonormal as well as orthonormal functions. A method for generating complete sets of biorthonormal surface density-potential pairs by means of Abel transforms of orthonormal sets is developed. When applied to Jacobi polynomials, it produces surface densities which vanish outside a finite disk, and their corresponding potentials which on the disk are polynomials in the Cartesian coordinates.

151.040 Dynamics of flat galaxies. III. Equilibrium models.
A. J. Kalnajs.
Astrophys. Journ., Vol. 205, 751 - 761 (1976).

The problem of determining equilibrium distributions that give rise to a prescribed axisymmetric surface density can be reduced to inverting a scale-invariant linear integral equation in two variables. The author discusses briefly distribution functions depending only on Jacobi's integral, and more fully distributions of the form $E^{m-1}g[(-2Eh)^{1/2}]$, giving several worked out examples. The techniques developed for the flat distributions can be applied to obtain three-dimensional models.

151.041 Dynamical evolution of two-component star clusters.
M. Saito, M. Yoshizawa.
Astrophys. Space Sci., Vol. 41, 63 - 77 (1976).

The dynamical evolution of two-component star clusters, each of which is enclosed within a perfectly reflecting sphere, is investigated by numerically solving moment equations derived from the Boltzmann equation. One of the two adopted model clusters evolves, starting from a state of no mass segregation, toward an equilibrium state at a quite slow rate. The other one evolves away from an equilibrium state and its central density increases without limit. The different evolutionary behaviors of the two model clusters are explained by the fact that there exists no equilibrium state for such clusters if the total energy is less than a certain critical value.

151.042 The stability of a family of elliptical stellar disks.
S. D. Tremaine.
Monthly Notices Roy. Astron. Soc., Vol. 175, 557 - 571 (1976).

The author describes a method of analysing the stability of the elliptical stellar disks constructed by Freeman as models of the bars in SB galaxies. Most of the systems are found to be stable to first- and second-harmonic coplanar vibrations.

151.043 The stability of a two-dimensional model of a stellar system. V. A. Antonov.
Trudy Astron. Obs., *Leningrad,* Vol. 32 (= Uchenye Zapiski Leningr. Un-ta, No. 385 = Seriya matem. nauk, vyp. (No.) 52), p. 79 - 104 (1976). In Russian.

An exact two-dimensional model of a stellar system is necessarily cylindrical or flat. In the first case, the special model given is stable with respect to two-dimensional perturbations. A special flat Maclaurin disk is also investigated and the corresponding dispersion relation is found. According to Kalnajs, it is unstable. A superposition of a series of disks with different angular velocities can give a stable model.

151.044 A more precise velocity distribution function of stars in clusters.
A. A. V'yuga, V. S. Kaliberda, I. V. Petrovskaya.
Trudy Astron. Obs., *Leningrad,* Vol. 32 (= Uchenye Zapiski Leningr. Un-ta, No. 385 = Seriya matem. nauk, vyp. (No.) 52), p. 105 - 118 (1976). In Russian.

Expressions and tables are given for the probability of a star to change its velocity by a given amount. For the velocity distribution of the field stars the law f(y) is assumed which has been found earlier by the solution of the balance equation. The results are used for the solution of the balance equation in the second approximation. The derived velocity distribution function $f_1(y)$ differs but slightly from f(y).

151.045 On nonlinear spiral density waves in a model with purely circular motions. N. G. Ptitsyna.
Byull. Inst. Astrofiz., *Dushanbe,* No. 63, p. 15 - 20 (1974). In Russian.

It is shown that the existence of nonlinear density waves in a self-gravitating cylinder is incompatible with the assumption of purely circular motions in the system.

151.046 Drift instability of a differentially rotating axisymmetric stellar system. M. N. Maksumov.
Byull. Inst. Astrofiz., *Dushanbe,* No. 64, p. 3 - 15 (1974). In Russian.

Local stability of a disturbance similar to a bar and propagating through a differentially rotating axisymmetric stellar system has been investigated. It is shown that there is specific instability connected with drift motions of stars due to differential rotation. Under the assumption that the spiral structure has wave nature the role of drift instability in its formation is discussed briefly.

151.047 Drift density waves in a differentially rotating axisymmetric disk galaxy.
M. N. Maksumov, Yu. N. Mishurov.
Byull. Inst. Astrofiz., *Dushanbe,* No. 64, p. 16 - 21 (1974). In Russian.

Drift density waves are considered in case of plain geometry. Slow waves are considered only under certain model assumptions. Local instability of a system with respect to slow drift density waves is shown.

151.048 The influence of drift density waves on stellar phase distribution. M. N. Maksumov.
Byull. Inst. Astrofiz., *Dushanbe,* No. 64, p. 22 - 28 (1974). In Russian.

Drift density waves excited in a spatially and kinematically inhomogeneous stellar system are shown to cause stars and motion redistribution within it. The phase distribution function change is described by an equation of diffusion type which coefficients are proportional to the energy density of waves excited. These coefficients may be estimated, at least locally.

151.049 On the evolution of initial disturbances in a stellar system. M. N. Maksumov.
Byull. Inst. Astrofiz., *Dushanbe*, No. 64, p. 29 - 35 (1974). In Russian.

The kinematic evolution of a small initial disturbance in a system with given phase distribution is investigated, and some details connected with density waves excited by this disturbance are discussed. It follows that persistence of not only matter but also the wave spiral pattern depends on the degree of spatial inhomogeneity and differential rotation.

151.050 On the energetic effectiveness of wave processes in galaxies. M. N. Maksumov.
Byull. Inst. Astrofiz., *Dushanbe*, No. 64, p. 36 - 40 (1974). In Russian.

The energetics of wave processes of different dynamical types has been discussed. It is assumed that excitation of waves may be due to Jeans instability, interaction between plane and spherical subsystems, resonant interaction near Lindblad resonances, and drift instability.

151.051 On nonlinear spiral density waves. Eh. Ya. Maldybaeva, N. G. Ptitsyna.
Byull. Inst. Astrofiz., *Dushanbe*, No. 64, p. 41 - 42 (1974). In Russian.

It is shown that nonlinear density waves can exist in a self-gravitating differentially rotating thin disk with a gas distribution corresponding to that in disk galaxies with bulge.

151.052 On star motions in a non-steady gravitational field of a primordial protogalaxy.
V. A. Antonov, L. P. Osipkov, A. D. Chernin.
Astrofizika, Vol. 11, 335 - 345 (1975). In Russian.
English translation in Astrophysics, Vol. 11, No. 2.

Motions of stars of the first generation in the gravitational field of a collapsing gaseous protogalaxy are studied, the collapse is supposed to be a free fall. The protogalaxy is modelled by a uniform sphere with zero total energy. The general solution for the case of purely radial motions of test stars is found and analysed. Conditions of its escape are obtained. It is shown that the energy exchange between escaping stars and the remainder leads to more rapid collapse. The effect of another symmetry and non-uniformity of the system and positivity of its energy is discussed.

151.053 Escape velocities for pairs of galaxies.
S. M. Alladin, A. Potdar, K. S. Sastry.
Bull. Astron. Soc. India, Vol. 3, 31 (1975). — Abstract of a paper presented at the A.S.I. meeting 1975.

151.054 Conditions for galaxy formation from adiabatic fluctuations. S. A. Bonometto, F. Lucchin.
Astrophys. Journ., Vol. 206, 391 - 401 (1976).

An analysis of the conditions that ought to hold in order that primeval adiabatic fluctuations can generate isothermal fluctuations at the decoupling epoch, is carried out. It is found that the minimum mass involved in matter inhomogeneities arising in this way can hardly be smaller than $\sim 3 \times 10^{13} M_{\odot}$, and this value is reached only for very dense universes. There seems to be no possibility of galaxy formation from adiabatic fluctuations if the density parameter Ω is smaller than ~ 0.5.

151.055 On density waves in galaxies. IV. Wave amplification through processes that remove angular momentum from galactic disks. J. W-K. Mark.
Astrophys. Journ., Vol. 206, 418 - 434 (1976).

Spiral density waves are amplified because of a transfer of angular momentum from the rapidly rotating disk to the slowly rotating bulge-halo subsystem of galaxies. This transfer is more efficient, thus amplification is higher, for waves of more open winding. Those waves are amplified whose radial

positions are located within their respective corotation radii.

151.056 The evolution of massive collapsing gas clouds.
S. von Hoerner, W. C. Saslaw.
Astrophys. Journ., Vol. 206, 917 - 933 (1976).

The authors use simple models to explore the basic properties of massive ($10^4 - 10^8 M_{\odot}$) collapsing gas clouds. Turbulent, thermal, radiation, gravitational, nuclear, and magnetic energies all play an important role. Fragmentation can be prevented if the magnetic and thermal energies of the initial cloud are roughly in equipartition, and if there is sufficient ionization. Clouds with masses between about 3×10^5 and $10^6 M_{\odot}$ may explode with double structure or as jets. Exploding clouds with greater or less mass tend to have rough spherical symmetry. These processes may be responsible for some of the activity in exploding galaxies, radio galaxies, and quasars.

151.057 Unstable spiral modes in disk-shaped galaxies.
Y. Y. Lau, C. C. Lin, J. W-K. Mark.
Proc. National Acad. Sci., USA, Vol. 73, 1379 - 1381 (1976).

The mechanisms for the maintenance and the excitation of trailing spiral modes of density waves in disk-shaped galaxies, as proposed by Lin in 1969 and by Mark recently, are substantiated by an analysis of the gas-dynamical model of the galaxy. The self-excitation of the unstable mode is caused by waves propagating outwards from the corotation circle, which carry away angular momentum of a sign opposite to that contained in the wave system inside that circle. Specifically, a simple dispersion relationship is given as a definite integral, which allows the immediate determination of the pattern frequency and the amplification rate, once the basic galactic model is known.

151.058 Velocity field and surface density in the problem of two fixed centres. N. P. Pit'ev.
Vestn. Leningr. un-ta, 1975, No. 19, p. 149 - 155. In Russian. Abstr. in Referativ. Zhurn. 51. Astron., 6. 51. 718 (1976).

151.059 Numerical simulation of a system of colliding bodies in a gravitational field. Astrophysical applications. A. Brahic.
Computing in plasma physics and astrophysics, (see 012.014), B5, 2 pp. (1976).

151.060 Large scale shock waves in barred galaxies.
T. Matsuda, S.-A. Sorensen, M. Fujimoto.
Computing in plasma physics and astrophysics, (see 012.014), P5, 2 pp. (1976).

151.061 On the nature of the ramificated spiral structure of galaxies. Yu. N. Mishurov, A. A. Suchkov.
Astron. Zhurn. Akad. Nauk SSSR, Vol. 53, 488 - 494 (1976). In Russian. English translation in Soviet Astron., Vol. 20, No. 3.

The large-scale ramification of spiral arms in some galaxies is explained in terms of the density-wave theory.

151.062 Solution particulière de l'équation de Vlasov pour les systèmes bidimensionnels.
M. Cantus, J.-P. Doremus, G. Baumann.
Comptes Rendus Acad. Sci. Paris, Sér. B, Vol. 282, 329 - 332 (1976).

Considérant des systèmes stellaires bidimensionnels à symétric cylindrique décrits à l'état stationnaire par une fonction de distribution dépendant uniquement de l'énergie ϵ et du moment angulaire I, les auteurs montrent que le système d'équations Vlasov-Poisson, linéarisées autour de l'équilibre admet toujours une solution particulière indépendant du temps pouvant être comparée à un mode de déplacement.

151.063 Stabilité des systèmes stellaires bidimensionnels en

rotation uniforme. M. Cantus, J.-P. Doremus, G. Baumann.
Comptes Rendus Acad. Sci. Paris, Sér. B, Vol. 282, 369 - 372 (1976).

151.064 The formation and early dynamical history of galaxies. G. B. Field.
Galaxies and the universe, (see 003.010), p. 359 - 407 (1975).

Behavior of density perturbations (Gravitational instability with radiation absent, gravitational instability with radiation present, other types of instability, galaxy formation in other cosmological models, critique of gravitational instability); Contraction of a protogalaxy (Initial conditions, calculation of the collapse, collapse of protogalaxies, instability during collapse); Early evolution (Fragmentation, formation of elliptical systems, formation of disk systems, differentiation between ellipticals and disks); Observational data (Intergalactic matter, observation of young galaxies, history of the Milky Way Galaxy, multiple systems of galaxies).

151.065 Stellar dynamics and the structure of galaxies. K. C. Freeman.
Galaxies and the universe, (see 003.010), p. 409 - 507 (1975).

Foundations of stellar dynamics (The relaxation time, the collisionless Boltzmann equation, the hydrodynamical equation, Jeans's theorem, some simple consequences of Jeans's theorem, adiabatic invariants, the virial theorem); Some basic properties of the Galaxy (The rotation of the Galaxy, stellar motions near the sun, mass models of the Galaxy, epicyclic orbits, populations in the Galaxy); Some general properties of galaxies (Morphological classification of galaxies, absolute luminosities and colors of galaxies, luminosity and color distributions in galaxies, internal motions and masses, the content of galaxies, some problems); The third integral; The self-consistency problem: models for globular clusters and elliptical galaxies; The collapse of the Galaxy; Collective effects and collisionless relaxation in stellar systems (The stability of homogeneous media, phase damping, violent relaxation, time-dependent stellar systems); The disks of spiral and lenticular galaxies (The stability of the disk, Mestel's hypothesis, the exponential disk, the bending of the galactic plane, problems of velocity dispersion in the disk, the problem of spiral structure); Barred spiral galaxies (The formation of the bar, some properties of bar-like stellar systems, formation of spiral structure, formation of ring structure, the Magellanic barred spirals); Formation and evolution of galaxies (The metal-enrichment picture, some kinematic problems, problems of the large-scale mass distribution, morphological-type problems, the origin of angular momentum).

151.066 Non-linear effects on the Jeans criterion for gravitational instability. I. Isothermal case. S. Aoki.
Publ. Astron. Soc. Japan, Vol. 28, 371 - 389 (1976).

The Jeans criterion for the gravitational instability is extended to the case where the density contrast is so large that non-linear effects cannot be neglected. The main results are as follows: (1) The stationary wavelength always increases as the density contrast increases. (2) When the density contrast tends to infinity, then the solution for the non-linearity tends to Ledoux's (1951) solution having only one maximum (or central density concentration). (3) The stability of such non-linear (periodic in space coordinate) solution is given by a statement that it is only unstable for disturbances with longer wavelength than that of (presently obtained) stationary solution, under a physically acceptable initial condition. This concludes that the wavelength of the stationary solution may serve as critical.

151.067 Gamma-dating of the epoch of formation of galaxies. L. M. Ozernoj.
Izv. Krymskoj Astrofiz. Obs., Vol. 54, 326 - 327 (1976). In Russian. — Abstract of a paper presented at the seminar on "X-ray and gamma-ray astronomy" at the Crimean Astrophys. Obs., 1974, (see 012.009).

151.068 Nonlinear density wave (galaxy spiral arms). T. Nakamura, S. Ikeuchi.
Progr. Theor. Phys., Vol. 54, 910 - 911 (1975). — Abstr. in Phys. Abstr., Vol. 79, A019330 (1976).

151.069 On spiral waves in galaxies — A gas dynamic approach. C. C. Lin, Y. Y. Lau.
SIAM Journ. Applied Math., Vol. 29, 352 - 370 (1975). Abstr. in Phys. Abstr., Vol. 79, A023879 (1976).

151.070 Dynamical stability of a supermassive gas-star system. K. Takarada.
Progr. Theor. Phys., Vol. 54, 1318 - 1324 (1975). — Abstr. in Phys. Abstr., Vol. 79, A033065 (1976).

The thermal effects of H_2 molecules in rotating and collapsing spheroidal gas clouds. See Abstr. 065.052.

A dynamical study of NGC 4027. See Abstr. 158.058.

Tidal effects in a cluster of galaxies. See Abstr. 160.020.

152 Stellar Associations

152.001 **Bochum 15, a new young stellar aggregate in Puppis.**
M. P. FitzGerald, R. Hurkens, A. F. J. Moffat.
Astron. Astrophys., Vol. 46, 287 - 291 = Contr. Univ. Water-loo Obs. (1975).

Bochum 15 (centred at $l = 248°.0$, $b = -5°.5$) is found to be a loose aggregate of about 30 OB stars located 4.4 kpc from the sun and 420 pc below the galactic plane. It is very young on the basis of its earliest member (spectral type O7). The aggregate has a mean colour excess $E_{B-V} = 0^m5$ with considerable differential reddening which tends to decrease away from the galactic plane. Its young age makes it a good spiral arm indicator and it is associated with 6 other young spiral tracers located at this distance in Puppis.

152.002 *UBV*, Hβ **photometry of OB stars in groups: Pup OB2.** R. J. Havlen.
Astron. Astrophys., Vol. 47, 193 - 202 (1976).

Evidence is presented in support of the physical reality of the association Pup OB2 at an Hβ derived distance of 4.3 kpc. Pup OB2 is on the outskirts of the local arm of the Galaxy. A possible relationship between Pup OB2 and an apparent velocity bubble in the H I distribution is suggested.

152.003 **An interesting star in the λ Orionis association.**
M. V. Penston, D. A. Allen, C. Lloyd.
Observatory, Vol. 96, 22 - 23 (1976). – Note.

152.004 **On the ratio of total to selective extinction in associations containing reflection nebulae.**
A. F. J. Moffat, T. Schmidt-Kaler.
Astron. Astrophys., Vol. 48, 115 - 119 (1976).

While most stellar associations containing reflection nebulosity (*R*-associations) yield a normal value of the ratio of total to selective extinction $r = 3.3$ (Herbst, 1975), three cases (vdB 130, Ara R1, Car R1) appear to exhibit strongly deviating values. The authors present several arguments which show that, even in the case of these *R*-associations there is no reason not to accept the usual *r* value for the interstellar dust.

152.005 **Some results of an investigation of star duplicity in the T-associations Tau T1, Tau T2, Tau T3 and Ori T2.** M. M. Zakirov.
Issled. ekhstremal'no molodykh zvezdn. kompleksov.
Tashkent, Fan, 1975, p. 95 - 115. In Russian. – Abstr. in Referativ. Zhurn. 51. Astron., 3.51.643 (1976).

152.006 **Absorbing matter in the region of Tau T1 and Tau T3 associations.** V. E. Slutskij.
Issled. ehkstremal'no molodykh. zvezdn. kompleksov. Tashkent, Fan, 1975, p. 115 - 127. In Russian. – Abstr. in Referativ. Zhurn. 51. Astron., 3.51.644 (1976).

152.007 **A possible new OB star aggregate in Puppis and its non-relationship to the cepheid VZ Puppis.**
R. J. Havlen.
Bull. American Astron. Soc., Vol. 8, 289 (1976). – Abstr. AAS.

152.008 **Study of emission stars in Scutum suspected by Pik-Sin The as T-association.** B. Kenzhaev.
Issled. ehkstremal'no molodykh zvezdn. kompleksov. Tashkent, Fan, 1975, p. 138 - 142. In Russian. – Abstr. in Referativ. Zhurn. 51. Astron., 4.51.674 (1976).

152.009 **The stellar ring SR 440 within the open cluster NGC 6709.** T. A. Uranova.
Astron. Zhurn. Akad. Nauk SSSR, Vol. 53, 305 - 308 (1976).

In Russian. English translation in Soviet Astron., Vol. 20, No. 2.

It is concluded that SR 440 is not a real spatial star grouping.

152.010 **Spectral types in the Lacerta OB1 association.**
H. Levato, H. A. Abt.
Publ. Astron. Soc. Pacific, Vol. 88, 141 - 143 (1976).

Spectral types are given for 31 members and 11 probable

152.011 **Stellar rings.** J. Isserstedt.
Vistas Astron., Vol. 19, 123 - 132 (1975).

Importance of random scatter to variable-extinction analyses. See Abstr. 113.024.

T Tauri stars and the $(J-H)$, $(H-K)$ diagram.
See Abstr. 113.035.

The OBN and OBC stars. See Abstr. 114.025.

Observations of southern emission-line stars.
See Abstr. 114.042.

Photographic photometry of Orion variables in the region of the T1 Tau and T3 Tau T-associations.
See Abstr. 122.070.

Classification of irregular variables in the association T1 Sco. See Abstr. 122.082.

Variable stars of the association T2 Cep.
See Abstr. 122.083.

On the colours of variable stars in the region of the Scorpius T1 association. See Abstr. 122.088.

How to make metal-poor stars, redden OB associations and grow mantles on grains. . See Abstr. 131.086.

The distance and mass of the large elephant trunk, a CO cloud pointing towards NGC 6231.
See Abstr. 131.097.

Neutral hydrogen in a region of Cepheus: nearby galactic structure. See Abstr. 131.141.

Infrared photometry of a heavily reddened association in W 35. See Abstr. 131.501.

Aperture synthesis observations of galactic H II regions. II. The galactic radio source W 58.
See Abstr. 141.052.

Infrared photometry of CRL 877 associated with the radio complex in the Monoceros R2 region.
See Abstr. 141.638.

The distribution of WR stars in galactic clusters and associations and the absolute magnitudes of WR stars.
See Abstr. 153.013.

Relation of NGC 3590, Hogg 10, and Collinder 240 to the structure of the Carina spiral feature.
See Abstr. 155.026.

153 Galactic Clusters

153.001 NGC 2204: an old open cluster in the halo.
T. G. Hawarden.
Monthly Notices Roy. Astron. Soc., Vol. 174, 225 - 239 (1976).

Photographic BV and photoelectric UBV photometry of stars in the open cluster NGC 2204 is presented. The cluster has a long giant branch reaching $(B-V)_0 \sim 1^m8$. There is a prominent giant clump at $V = 13^m82$ and the upper main sequence has a gap about 0^m4 below its brightest point. Several blue stragglers are present. The reddening $E(B-V) = 0^m08$ and an ultraviolet excess $(U-B)_{0.6} = 0^m095 \pm 0^m024$ are derived from a two-colour diagram. The cluster appears to be distinctly poor in heavy elements. The turn-off colour indicates an age of about 3×10^9 yr and the adopted distance modulus, $(m-M)_0 = 13^m25 \pm 0^m2$ (estimated uncertainty) implies that the cluster lies about 1250 pc from the galactic plane, an exceptional distance for a Population I system.

153.002 The old open cluster Melotte 66. T. G. Hawarden.
Monthly Notices Roy. Astron. Soc., Vol. 174, 471 - 487 (1976).

Photoelectric and photographic photometry of the open cluster Melotte 66 is presented. Melotte 66 appears to have reddening $E(B-V) = 0^m17$ and ultraviolet excess $\delta(U-B) \sim 0^m1$ corresponding to $[Fe/H]_\odot = -0.3$. The cluster is probably between 6 and 7×10^9 yr old. A distance modulus $(m-M)_0 = 12^m4$ is derived, which implies that the cluster lies about 750 pc from the galactic plane.

153.003 Comparison of photometric and proper motion memberships in open clusters. W. L. Sanders.
Astron. Astrophys., Vol. 46, 131 (1976).

A recent study indicates a roughly 50 % uncertainty in the method of photometric cluster memberships.

153.004 Finding charts for new members of the Hyades.
G. Pels.
Astron. Astrophys., Suppl. Ser., Vol. 23, 223 - 229 (1976).

This is a supplement to the article "New members of the Hyades cluster and a discussion of its structure", by Pels, Oort, and Pels-Kluyver, published in Astron. Astrophys., Vol. 43, 423 - 441 (1975). — See Abstr. 14.153.025.

153.005 Orbits and masses of Hyades visual binaries.
W. C. Wickes.
Astron. Journ., Vol. 80, 1059 - 1064 (1975).

Interferometric measurements of the orbit positions of three Hyades visual binaries have been made as a test of the validity of orbits computed from visual measurements. There is substantial agreement between the observed and predicted positions. However, probable errors computed for the orbital elements yield uncertainties in stellar masses that prevent determination of an accurate mass—luminosity relation for the Hyades despite recent computations of the cluster distance by the convergent-point technique.

153.006 Unresolved binaries as a source of scatter in color-magnitude diagrams. B. M. Schlesinger.
Astron. Journ., Vol. 80, 1071 - 1074 = Publ. Goethe Link Obs., Indiana Univ., *Bloomington*, No. 179 (1975).

Theoretical models have been used to predict the appearance of H—R diagrams of open clusters containing unresolved binary stars. If unresolved binaries are present, use of the luminosity of the brightest main-sequence stars may yield ages that are too low, but no error will be introduced into ages based upon turnoff colors. Binaries will produce scatter on the main sequence only where the slope is gradual — where there is a significant change in color with luminosity. In young clusters, appreciable scatter appears only below the vertical turnup, while in older clusters, scatter appears along the entire main sequence. The presence of unresolved binaries can mask the separation in the H—R diagram of the evolutionary stages of hydrogen exhaustion and thick-shell hydrogen burning.

153.007 Photographic photometry in four clusters.
K. M. Cudworth.
Astron. Astrophys., Suppl. Ser., Vol. 24, 143 - 150 (1976).

Two-colour photographic photometry has been obtained to supplement recent astrometric studies of the open clusters NGC 6530, NGC 6633 and NGC 7062 as well as the globular cluster NGC 6838. The colour-magnitude diagrams of probable members do not show any remarkable features not already discussed by other authors.

153.008 The luminosity function of the cluster M37.
V. M. Archemashvili.
Astron. Zhurn. Akad. Nauk SSSR, Vol. 53, 212 - 214 (1976).
In Russian. English translation in Soviet Astron., Vol. 20, No. 1.

The luminosity functions of stars in M37 to 16^m5 ($M_{pg} = +5.4$) are constructed for regions of the nucleus and corona. It is shown that the luminosity function for the stars in the nucleus is practically constant in the interval of photographic magnitudes from 13^m0 to 16^m0.

153.009 Four-color and Hβ photometry of the galactic cluster NGC 6633. E. G. Schmidt.
Publ. Astron. Soc. Pacific, Vol. 88, 63 - 66 (1976).

Photometric measurements of stars in the intermediate-age galactic cluster NGC 6633 are presented. From these data the color excess is found to vary slightly across the cluster with an average of $E_{b-y} = 0^m124$. The distance modulus is found to be 7^m71 and the turnoff occurs at about $(b-y)_0 = 0^m1$. The cluster is slightly metal poor.

153.010 A study of Be stars in clusters. R. Schild, W. Romanishin.
Astrophys. Journ., Vol. 204, 493 - 501 (1976).

Calibrated spectrograms at Hα of 566 stars in 29 young galactic clusters led to the detection of 41 Be stars in clusters. Using cluster membership, the authors have inferred ages and intrinsic $B - V$ colors of Be stars for a discussion of their evolutionary states. Because of the great strength of Hα emission and the short time duration of the effect, the extreme Be stars would be excellent probes for studies of spiral structure and would also serve as probes for studies of ages and distances of extragalactic systems.

153.011 The binary frequency of IC 4665. D. Crampton, G. Hill, W. A. Fisher.
Astrophys. Journ., Vol. 204, 502 - 511 = Contr. Dominion Astrophys. Obs., *Victoria*, No. 250 (1976).

New radial velocity observations of 13 stars in IC 4665 indicate that the spectroscopic binary frequency in this cluster is $\sim$ 50 percent. It is suggested that the high frequency (95%) previously found by Abt, Bolton, and Levy is a spurious one. The methods of detecting variable velocities and comparing binary frequencies are discussed and applied to IC 4665 and other clusters. On this basis, the authors conclude that the correlation between mean rotational velocity and binary frequency in clusters is marginal.

153.012 Catalogue of UBV photometry and MK spectral types in open clusters. J.-C. Mermilliod.
Astron. Astrophys. Suppl. Ser., Vol. 24, 159 - 297 (1976).

A catalogue of UBV photoelectric photometry in the Johnson and Morgan system and of MK classification of stars in the field of NGC and IC open clusters, as well as in the Hyades, Pleiades, Alpha Persei and Coma Berenices clusters, is presented. All data published until the end of 1974 have been compiled, being 10816 UBV data and 3192 MK types, concerning 200 open clusters. It contains complete cross-identifications for cluster numbering and HD and DM catalogues, and also proper motions studies. The catalogue is available on magnetic tape at the Stellar Data Centre (CDS) at Strasbourg.

153.013 The distribution of WR stars in galactic clusters and associations and the absolute magnitudes of WR stars.
M. D. Popova, V. S. Avedisova.
Stellar physics and evolution, (see 003.001), p. 95 - 112 (1974). In Russian.

A possible relation of WR stars from the list of Smith (1968) with open galactic clusters and associations is studied. Data from the catalogue of star clusters and associations by Alter, Ruprecht and Vanysek (1970) are used. A catalogue of WR stars coincident with OB star aggregates is presented. Estimates of the absolute magnitudes of WR stars — possible members of clusters and associations with known distances — are presented.

153.014 Determination of ages of stars in 20 open clusters,
R. Dinescu, O. B. Dluzhnevskaya, V. V. Muzylev, A. Eh. Piskunov.
Stellar physics and evolution, (see 003.001), p. 113 - 133 (1974). In Russian.

The statistical distribution of stars with regard to age has been studied for 20 open clusters. A special computing procedure is used for determination of masses and ages of every cluster member depending on its location on the H-R diagram.

153.015 Metal abundance in open clusters in the Galaxy and Magellanic Clouds.
G. L. H. Harris, R. G. Deupree.
Bull. American Astron. Soc., Vol. 8, 289 (1976). – Abstr. AAS.

153.016 The frequency of peculiar A and metallic-line stars in open clusters. M. R. Hartoog.
Astrophys. Journ., Vol. 205, 807 - 822 (1976).

MK spectral types were determined for 263 B and A stars in the southern galactic clusters NGC 2287, 2422, 2451, 2516, 2547, 3228, and 3532 in order to investigate the frequency of Ap and Am stars in galactic clusters. In these seven clusters 11 Ap stars and 203 normal B5 to A5 dwarfs were found. There is an indication that the frequency of Ap stars may increase with age, but this result is not statistically significant. The frequency of Am stars in galactic clusters appears to be normal compared with the field. The frequency of Am stars appears to increase with cluster age.

153.017 Photometry of the galactic cluster NGC 2169.
R. Sagar.
Astrophys. Space Sci., Vol. 40, 447 - 454 (1976).

The results of photoelectric U, B, V photometry of the galactic cluster NGC 2169 are presented. The colour excess $E(B-V)$ is $0.^{m}18$. The distance modulus to the cluster and its age are respectively estimated at $9.^{m}6$ and 0.9×10^7 yr.

153.018 Der offene Sternhaufen NGC 5617.
W. Lohmann.
Astrophys. Space Sci., Vol. 41, 27 - 37 = Mitt. Astron. Rechen-Inst., Heidelberg, Serie A (1976).

The open cluster NGC 5617 was investigated by the strip method. The cluster contains about 460 stars with a total mass of 700 $M_{\odot}$. Its radius amounts to 3.7 pc; the star density in the center is 50 stars pc^{-3}; and the mean stellar velocity, 0.89 km s^{-1}. On longer-exposed photographs at a distance of

12.'3 in direction to SSE an unknown open star cluster becomes visible with a radius of $\approx 4.'3$, containing about 150 stars to the limiting magnitude $V \approx 19^m$.

153.019 The Pleiades and the zero-age main sequence.
B. J. McNamara.
Publ. Astron. Soc. Pacific, Vol. 88, 144 - 147 (1976).

New four-color $uvby$ observations of the Pleiades are used to calibrate the ZAMS in the spectral range A2 to F5. Although the Pleiades c_0 versus $(b-y)_0$ calibration is in good agreement with the ZAMS calibration presented by Crawford, its m_0 versus $(b-y)_0$ relation shows a systematic difference from his values. This difference is believed to be caused by the higher mean rotational velocity of the Pleiades stars in comparison to field stars.

153.020 Four-color and Hβ photometry for open clusters. II. NGC 6475. M. S. Snowden.
Publ. Astron. Soc. Pacific, Vol. 88, 174 - 186 (1976).

Extensive photometric observations were conducted in the field of the southern open cluster NGC 6475 with the $uvby$ and Hβ systems. The interstellar reddening for the cluster was found to be nonuniform, and a mean color excess of $E(b-y) = 0.^{m}067$ was found. A distance modulus was obtained. A comparison of the observed data with theoretical models showed this cluster to be 260×10^6 years old.

153.021 NGC 2533 and the cepheid BN Puppis.
R. J. Havlen.
Astron. Astrophys., Vol. 49, 307 - 311 (1976).

New photoelectric UBV, Hβ photometry of the brightest B stars in the galactic cluster NGC 2533 leads to a distance for the group of 1.3 kpc and a fairly uniform mean extinction of $E_{B-V} = 0.20$. Although the 13.7 day period classical cepheid BN Puppis is only 17' away from the cluster center, its distance of 3.8 kpc does not confirm it as a cluster member.

153.022 Collinder 228 and the η Carinae complex.
A. Feinstein, H. G. Marraco, J. C. Forte.
Astron. Astrophys., Suppl. Ser., Vol. 24, 389 - 397 (1976).

Photoelectric UBV observations of 99 stars in the region of the open cluster Cr 228 are reported. It is shown that some of them belong to a group situated at a distance of 2500 pc, but some others seem to be related to the complex Tr 14/Tr 16. The former group is nearer and less reddened. Its age is less than 5×10^6 years. A discussion about the star HD 93206 is also given.

153.023 Evolved stars in open clusters.
G. L. H. Harris.
Astrophys. Journ., Suppl. Ser., Vol. 30, 451 - 490 (1976).

Radial-velocity observations and MK classifications have been used to study evolved stars in 25 open clusters. Published data on stars in 72 additional clusters are rediscussed and combined with the observations made in this investigation to yield positions in the Hertzsprung-Russell diagram for 559 evolved stars in 97 clusters. Ages for the parent clusters were estimated from the main-sequence turnoff points, earliest spectral types, and bluest stars in the clusters themselves. The evolved stars were sorted into six age groups and the composite H-R diagram for each age group was then used to study the evolutionary tracks for stars of various masses. The observational results were found to be in reasonably good agreement with recent theoretical computations. The composite color-magnitude diagrams were found to be strikingly different from those of the rich open clusters in the Magellanic Clouds.

153.024 Kataloge photographischer und photoelektrischer Helligkeiten von 25 galaktischen Sternhaufen im RGU und im U_C BV-System. W. Becker, S. N. Svolo-

poulos, C. Fang.
Separate print Astron. Inst. Univ. Basel, 89 pp. (1976).

153.025 **Sternhaufen und Nebel. Teil I, II.**   W. Meyer.
Veröff. Wilhelm-Foerster-Sternw. No. 40, 41, 328
pp. (1975).

153.026 **Four-color and Hβ photometry for open clusters.**
XI. The Pleiades.   D. L. Crawford, C. L. Perry.
Astron. Journ., Vol. 81, 419 - 426 (1976).
The authors have used photoelectric *uvby* β photometry
for 83 B-, A-, and F-type stars in the Pleiades cluster to derive
an average color excess of $E (b-y)$ = 0^m04 and a distance
modulus of $V_0 - M_v$ = 5^m54. The reddening is variable over
the field. There is a rotational velocity effect on the c_1 index
for the A- and F-type stars.

153.027 **Nomenclature cross-correlation for stars in NGC**
2516 and NGC 6475.   M. S. Snowden.
Centre Données Stellaires, Inform. Bull. No. 10, p. 30 - 33
(1976).

Importance of random scatter to variable-extinction
analyses.   See Abstr. 113.024.

Lists of S, C, MS stars and emission objects revealed
in red light observations.   See Abstr. 114.045.

Utvärdering av objektivprismespektra av tidiga
stjärnor med hjälp av dator.   See Abstr. 114.060.

The luminosity of the very red supergiant near the
cluster Tr 27.   See Abstr. 114.336.

The main sequence defined by trigonometrical
parallaxes.   See Abstr. 115.003.

Luminosity functions and the evolution of low-mass
population I giants.   Seé Abstr. 115.015.

$(R - I)$ **colors of cepheids and yellow supergiants in**
open clusters.   See Abstr. 122.003.

UBV photometry of flare stars in the Pleiades.
See Abstr. 122.031.

New variable stars found in galactic clusters.
See Abstr. 123.016.

New variable stars in the region of IC 1396.
See Abstr. 123.023.

Search for infrared stars in open star clusters.
See Abstr. 141.628.

The dynamic instability of isothermal relativistic
star clusters.   See Abstr. 151.007.

Effect of binary stars on the dynamical evolution of
stellar clusters. II. Analytic evolutionary models.
See Abstr. 151.013.

Relation of NGC 3590, Hogg 10, and Collinder 240
to the structure of the Carina spiral feature.
See Abstr. 155.026.

Errata

153.901 **Erratum: "NGC 3114: another open cluster possibly**
rich in peculiar stars" [Astron. Journ., Vol. 80,
807 - 808 (1975)].   H. Levato, S. Malaroda.
Astron. Journ., Vol. 81, 85 (1976).

154 Globular Clusters

154.001 Metal-rich globular clusters in the Galaxy. III. The "X-ray" globular cluster NGC 6441.
J. E. Hesser, F. D. A. Hartwick.
Astrophys. Journ., Vol. 203, 97 - 112 (1976).

From a photoelectric and photographic investigation in the UBV system of the very compact, low galactic latitude, southern globular cluster NGC 6441, the authors have found characteristics in the C-M diagram that are in accord with observations of the most metal-rich globular clusters in the Galaxy. If $M_{V,HB} = +0.9$ mag, then $(m-M)_0 = 14.7 \pm 0.3$ mag. This makes NGC 6441, one of the intrinsically brightest globular clusters in the Galaxy, consistent with the large mass determined recently by Illingworth. No unusual features were detected in its C-M diagram that might account for its appearing within the error box of the 3U 1746–37 X-ray source. It is also found that NGC 6441 has a normal M/L ratio.

154.002 Metal-rich globular clusters in the Galaxy. IV. A color-magnitude diagram for NGC 6304.
J. E. Hesser, F. D. A. Hartwick.
Astrophys. Journ., Vol. 203, 113 - 123 (1976).

From a study to $V = 19.1$ and $B = 20.8$ mag in the UBV system of ~385 stars near the G2 globular cluster NGC 6304, the authors have found a color-magnitude diagram whose characteristics appear similar to those of 47 Tucanae, supporting the conclusion that NGC 6304 is a moderately metal-rich cluster. The reddening, $E(B-V)$, has been estimated to be 0.58 ± 0.05 mag, and the distance modulus is found, assuming $M_{V,HB} = +0.9 \pm 0.2$ mag, to be $(m-M)_0 = 13.4 \pm 0.2$ mag. The cluster is thus 4.8 kpc from the sun and 450 pc above the galactic plane, and it has an absolute visual magnitude of −6.8.

154.003 Mass loss in globular-cluster red giants.
J. G. Cohen.
Astrophys. Journ., (Letters), Vol. 203, L127 - L129 (1976).

Echelle spectra of the brightest globular-cluster red giants, taken with the 4 m telescope at Kitt Peak National Observatory, reveal the presence of emission features at Hα. These are interpreted as evidence for a circumstellar envelope, which is produced by mass loss. An estimate of the rate of mass loss $(2 \times 10^{-9} M_\odot \, \mathrm{yr}^{-1})$ is obtained.

154.004 UBV photometry of faint globular clusters.
R. Racine.
Astron. Journ., Vol. 80, 1031 - 1036 (1975).

New integrated photometry on the UBV system is reported for 37 faint globular clusters for which previous data were either fragmentary or lacking. The majority of these objects are highly reddened clusters close to the galactic center like NGC 6440, IC 1276, and Terzan 5, or very loose aggregates such as IC 4499, Pal 12, and Pal 13. The following results are established: (1) preliminary analysis of the reddenings is consistent with $E(U-B)/E(B-V) \simeq 1.0$ for late-type clusters; (2) the reddening and distance of the infrared cluster Ter 5 are determined; (3) NGC 6749 may be a highly reddened galactic star cluster; and (4) the peculiar colors of NGC 7492 and Pal 13 are due to a deficiency in red giants.

154.005 New color-magnitude data for twelve globular clusters. W. E. Harris.
Astrophys. Journ. Suppl. Ser., No. 292, Vol. 29, 397 - 429 (1975).

This paper presents results of a major program of BV photoelectric and photographic photometry of southern globular clusters. New color-magnitude diagrams are presented for 11 clusters (NGC 2808, 4590, 5694, 5824, 6325, 6517, 6681,

6715, 6809, 6864, and 7089); and for one additional cluster (NGC 6266) measurements of the variable stars are listed. The features of the diagrams are discussed briefly and used to derive basic information about the clusters, such as their distances, reddenings, and characteristics of the giant and horizontal branches.

154.006 Colour excesses of globular clusters in the bands U, B, V, I. B. V. Kukarkin, N. N. Kireeva.
Astron. Zhurn. Akad. Nauk SSSR, Vol. 53, 83 - 91 (1976). In Russian. English translation in Soviet Astron., Vol. 20, No. 1.

It is shown that practically all the known methods for the determination of colour excesses of globular clusters on the basis of their integrated colour equivalents are not correct. The authors suggest a method for separating the observed colour equivalents into intrinsic colours and colour excesses which is based on the selection of clusters with the smallest excesses, too small to cause any noticeable errors. The values of intrinsic colour equivalents thus obtained strongly correlate with the integrated spectral types and are characterized by a mean error as small as ±0.018. The colour excesses calculated on these grounds made it possible to determine reliable ratios of colour excesses in the bands U, B, V, I. These ratios were used in order to calculate the colour excesses for all the clusters with measured colour equivalents and known integrated spectra. On the basis of the analysis of all the available material finally adopted values of colour excesses, intrinsic colour equivalents, and integrated spectral types for 68 globular clusters have been obtained.

154.007 The axial ratio and position angle of the major axis of several globular clusters.
Z. I. Kadla, N. Richter, A. A. Strugatskaya, W. Högner.
Astron. Zhurn. Akad. Nauk SSSR, Vol. 53, 92 - 99 (1976). In Russian. English translation in Soviet Astron., Vol. 20, No. 1.

The method of integral equidensity curves was applied to the investigation of the large-scale structure of the globular clusters M3, M5, M13 and M15. Previously derived data for M92 are also considered. Altogether 480 equidensity curves were obtained on the basis of U, B, V and R plates taken with the Schmidt camera of the 2-m telescope of the Tautenburg Observatory.

154.008 The masses of globular clusters. II. Velocity dispersions and mass-to-light ratios. G. Illingworth.
Astrophys. Journ., Vol. 204, 73 - 93 (1976).

Dynamical masses have been determined for 10 southern globular clusters. High-dispersion coudé spectra of the integrated light were used to determine the velocity dispersion at the centers of these clusters. The observed velocity dispersions (line-of-sight) range from 7.5 to 18.9 km s^{-1}. These velocity dispersions were incorporated into King's self-consistent globular cluster models to determine the cluster masses. The required length scales were determined from comparisons of King's theoretical surface density distributions with surface brightness and star distributions. The masses of these concentrated clusters range from $1.7 \times 10^5 M_\odot$ to $1.1 \times 10^6 M_\odot$. The $(M/L_v)_\odot$ values range from 0.9 to 2.9 with a mean of 1.6.

154.009 Optical structure of the X-ray globular cluster NGC 6624. N. A. Bahcall.
Astrophys. Journ., (Letters), Vol. 204, L83 - L87 (1976).

The central region of the X-ray globular cluster NGC 6624 is studied optically. A bright unresolved region, about 4″ in size and ~12.5 mag in apparent visual brightness, is observed near the center of the cluster on the shortest-exposure plates. The projected density distribution of the stars is determined from star counts that extend to within about $5''$ (= 0.25 pc × d_{10}, where d_{10} is the distance in units of 10 kpc) of the center of the cluster. The structural parameters of the cluster are derived. The estimated central escape velocity and relaxation time in the cluster fall within the region given by Bahcall and Ostriker as the most likely one for globular clusters having X-ray emission due to the accretion of gas by a massive black hole.

154.010 Thin-prism spectra of Messier 4 in Scorpius.
A. G. D. Philip.
Sky Telescope, Vol. 51, 244 - 246 (1976).

154.011 Evidence for abundance differences among giant branch stars in the globular cluster ω Centauri.
E. A. Mallia.
Astron. Astrophys., Vol. 48, 129 - 132 (1976).
    Abundance differences among stars on the upper giant branch of the globular cluster ω Cen are shown to exist. Their origin appears to lie in the mixing out of the products of nuclear reactions. A weak, slow neutron flux was probably present. Two of the observed stars resemble marginal Ba II stars.

154.012 Bright red giants in the globular cluster M 15.
Z. Kadla, N. Spasova.
Stellar physics and evolution, (see 003.001), p. 74 - 94 (1974). In Russian.
    The present investigation is a continuation of the study of the distribution of bright red giants in globular clusters (1971, 1972) which are selected on the basis of measured proper motions and V magnitudes.

154.013 Stellar collisions in globular clusters.
J. G. Hills, C. A. Day.
Astrophys. Letters, Vol. 17, 87 - 93 (1976).
    The authors have used the recent compilation of data on individual globular clusters by Peterson and King to calculate the frequency of stellar collisions in these systems. If a main sequence star is confined to the dense central cores of each of the globular clusters studied, it has on the average a 3 percent chance of suffering a collision with another main sequence star during the lifetime of the cluster. The most extreme cluster is M80 in which this probability is 41 percent. Considering all the main sequence stars in a cluster and not just those in its core, the authors find that in the 41 globular clusters for which there is sufficient data for an evaluation an average of 0.125 percent of all their main sequence stars have suffered collisions. The median value is 0.036 percent. This corresponds to an average of $\langle N_c \rangle$ = 335 collisions per cluster over a cluster lifetime of 1.2×10^{10} years. The median is 92 collisions per cluster.

154.014 The tidal breakup and capture of stars by black holes in globular clusters. J. G. Hills.
Astrophys. Letters, Vol. 17, 95 - 99 (1976).
    This mechanism is very efficient in globular clusters and may power the X-ray sources found in several of them.

154.015 The masses of globular clusters. I. Surface brightness distributions and star counts.
G. Illingworth, W. Illingworth.
Astrophys. Journ., Suppl. Ser., Vol. 30, 227 - 246 (1976).
    Surface brightness distributions have been determined for 10 globular clusters from photoelectric surface photometry. Star counts have been used to extend the distributions in the outer regions. The observed distributions are compared with the theoretical surface density distributions from King's models of globular clusters. Central surface brightnesses, core radii, tidal radii, and total magnitudes derived from this comparison are tabulated. Reddenings and distance moduli deter-

mined from consideration of all available data are also given.

154.016 The giant populations in globular clusters.
V. Castellani.
Astron. Astrophys., Vol. 48, 461 - 465 (1976).
    A simple approach is suggested for comparing the observed luminosity functions of cluster giants with theoretical predictions. One expects such an approach to give general information about the actual number of stars in the "double shell burning" evolutionary phase. The expected ratio between the number of asymptotic branch stars and the number of red giants is evaluated on the basis of recent evolutionary computations. Observational evidence from galactic globular clusters is discussed in the framework of current theoretical knowledge.

154.017 A very distant globular cluster?
H.-E. Schuster, R. M. West.
Astron. Astrophys., Vol. 49, 129 - 131 (1976).
    A faint object, which is well resolved into individual stars, has been discovered at $\alpha = 1^h 49^m 0$; $\delta = -44°42'$ (1950). It is probably a very distant globular cluster; if so, the distance is estimated at about 100 kpc, the inner diameter is around 100 pc.

154.018 M10 and M12: a couple of interacting twin globular clusters?
R. Buonanno, V. Castellani, F. Smriglio.
Astrophys. Space Sci., Vol. 41, L3 - L7 (1976). – Letter.

154.019 An optical search for ionized hydrogen in globular clusters. M. G. Smith, J. E. Hesser, S. J. Shawl.
Astrophys. Journ., Vol. 206, 66 - 78 (1976).
    An attempt has been made to detect $H\alpha$ emission from ionized gas in 26 globular clusters using a single-etalon, photoelectric Fabry-Perot interferometer. Assuming in each case that any ionized gas present is confined to the core, probable upper limits of the order of a few solar masses of ionized hydrogen have been derived.

154.020 The structure and mass function of the globular cluster M3. G. S. Da Costa, K. C. Freeman.
Astrophys. Journ., Vol. 206, 128 - 137 (1976).
    A detailed dynamical model of the globular cluster M3 is presented. The model is tidally limited and includes a realistic distribution of stellar masses. It is used to derive a total luminosity function for the cluster from the published luminosity function, and to give estimates of its total mass and M/L_V ratio. The model gives an excellent fit to the observed radial distribution of surface brightness and star counts over five decades in surface brightness: a good fit is not possible with the corresponding single-mass models.

154.021 The colour-magnitude diagram of NGC 5053.
M. F. Walker, C. D. Pike, J. D. McGee.
Monthly Notices Roy. Astron. Soc., Vol. 175, 525 - 531 = Contr. Lick Obs., No. 407 (1976).
    The colour-magnitude diagram of NGC 5053 has been derived to V = 21.1 from photographic and electronographic observations. The topology of the horizontal branch is that of clusters with an intermediate metal content and is thus at variance with the mean period of the RR Lyr stars and the unreddened colour of the subgiant branch read at the magnitude level of the horizontal branch, both of which would indicate an extremely low metal content. The reddening of NGC 5053 is E_{B-V} = 0.02 and the apparent distance modulus is $m-M$ = 16.08 ± 0.08.

154.022 Modelling of the horizontal branch of globular clusters. A. M. Ehjgenson, N. N. Samus'.
Astrofizika, Vol. 11, 365 - 368 (1975). In Russian.

English translation in Astrophysics, Vol. 11, No. 2.

Results of numerical calculations of the probability model proposed earlier for the interpretation of the horizontal branch structure of globular clusters are presented. Agreement between model distribution of the parameter α and observations is found.

154.023　The mass of the globular cluster ω Centauri.
　　　　E. V. Naumova.
Astron. Tsirk., No. 896, p. 6 (1975). In Russian.

154.024　Colour-magnitude diagram for the central region of
　　　　M13.　Z. Kadla, M. Antal, F. Zdarsky, N. Spasova.
Astron. Tsirk., No. 897, p. 4 - 6 (1975). In Russian.

154.025　Observation of hard X-rays of the globular cluster
　　　　NGC 6624.　L. B. Bratolyubova-Tsulukidze,
M. I. Kudryavtsev, A. S. Melioranskij, I. A. Savenko,
B. Yu. Yushkov.
Astron. Tsirk., No. 898, p. 6 - 8 (1976). In Russian.

154.026　Search for new variables among UV bright stars in
　　　　globular clusters.　N. N. Samus'.
Astron. Tsirk., No. 899, p. 1 - 2 (1976). In Russian.

154.027　Colour-magnitude diagram of the globular cluster
　　　　M79 (NGC 1904).　V. P. Goranskij.
Astron. Tsirk., No. 902, p. 5 - 7 (1976). In Russian.

154.028　New color-magnitude diagrams for four southern
　　　　globular clusters.
W. E. Harris, R. Racine, J. de Roux.
Astrophys. Journ., Suppl. Ser., Vol. 31, 13 - 31 (1976).

New photoelectric and photographic BV photometry is reported for the globular clusters NGC 6254 (M10), NGC 5286, NGC 5986, and NGC 6273 (M19). At a distance of only 4.4 ± 0.3 kpc from the sun, M10 is one of the nearest globular clusters. The study of the other three clusters leads to distances [assuming M_V(HB) = +0.6] as follows: NGC 5286, 8.7 ± 0.7 kpc; NGC 5986, 11 ± 1.3 kpc; NGC 6273, 10 ± 1.5 kpc. All three appear to be of moderately low metallicity, with well-populated blue horizontal branches.

154.029　On the problem of formation and evolution of
　　　　globular clusters.　S. B. Pikel'ner.
Astron. Zhurn. Akad. Nauk SSSR, Vol. 53, 449 - 452 (1976).
In Russian. English translation in Soviet Astron., Vol. 20,
No. 3.

A qualitative picture of the evolution of globular clusters including their formation as a result of the fragmentation of a protogalaxy is presented. Globular clusters are considered as a very few surviving subsystems which correspond to the most compact fragments of the protogalaxy. Conditions for the formation of less compact systems which can be easier disrupted are suggested.

154.030　Near infrared observations and globular cluster
　　　　stars.　V. Castellani, A. Martini.
Acta Astron., Vol. 26, 83 - 90 (1976).

Eggen's observations (1972) of globular cluster giants are collected in the I, V plane. The location of different evolutionary phases is briefly investigated. The effect of the interstellar reddening as well as intrinsic metal abundance Z for the various clusters is stressed.

154.031　The peculiar space distribution of stars in the
　　　　globular cluster M12.
V. Caloi, V. Castellani, F. D'Antona, R. De Amicis.
Mem. Soc. Astron. Italiana, Vol. 46, 407 - 416 (1975).

The space distribution of the stars in the globular cluster M12 is studied and evidence is given for the occurrence of strong anomalies in the space structure.

154.032　An analysis of the evolutionary status of stars in
　　　　the anomalous globular cluster M14.
F. Caputo, V. Castellani.
Mem. Soc. Astron. Italiana, Vol. 46, 455 - 460 (1975).

The combined evidences from the shape of the HR-diagram and the observed properties of the RR Lyrae pulsators suggest a larger age for the "anomalous" cluster M14, in comparison with the M3-like clusters. The possible occurrence of a larger original helium-content is expected to reduce, but not to avoid, the mentioned difference; in this hypothesis M14 could represent a linkage form between M3 and the intermediate metallicity, RR Lyrae-poor globular clusters like M13. Some observed peculiarities in the period frequency distribution could be connected with this larger helium abundance.

154.033　The horizontal branch stars in the globular cluster
　　　　M14: a peculiar space distribution?
F. Caputo, V. Castellani.
Mem. Soc. Astron. Italiana, Vol. 46, 501 - 504 (1975).
Letter.

154.034　Masses of globular clusters.　G. D. Illingworth.
Thesis Australian National Univ., Canberra (1973).

On the formation of disk around a compact
object by two-body tidal encounter.　See Abstr. 117.001.

Photoelectric UBV observations of three RR Lyrae
stars in ω Cen.　See Abstr. 122.009.

Period changes of RR Lyrae variables in M14.
See Abstr. 122.014.

Determination of the periods of the Blazhko-effect
(amplitude variation) of two RR Lyrae variables in globular
clusters by computers.　See Abstr. 122.095.

Results of modelling the light curves of RR Lyrae
variables in the globular clusters M3, M5, ω Cen and NGC
3201.　See Abstr. 122.096.

V1, the only known cepheid variable in the globular
cluster NGC 6752; observations and period.
See Abstr. 122.121.

Die Perioden und ihre Veränderungen der veränderlichen Sterne im Kugelhaufen Messier 5 = NGC 5904.
See Abstr. 122.140.

Evidence for a bump in the light curve of variable 1
of M13.　See Abstr. 123.017.

On the ratio of the total to selective absorption.
See Abstr. 131.072.

The interstellar reddening in the vicinity of 47
Tucanae and the Small Magellanic Cloud.
See Abstr. 131.127.

A note on the visible obscuration near ω Centauri.
See Abstr. 132.015.

The formation of binaries containing black holes by
the exchange of companions and the X-ray sources in globular
clusters.　See Abstr. 142.026.

Ariel V sky survey observations of X rays from the
globular cluster candidates 3U 2131 + 11 and MXO 513−40.
See Abstr. 142.061.

Discovery of intense X-ray bursts from the globular cluster NGC 6624.   See Abstr. 142.081.

Scattering model for X-ray bursts: massive black holes in globular clusters.   See Abstr. 142.082.

Improved position for the X-ray source associated with the globular cluster NGC 6441.
See Abstr. 142.091.

MX 0513−40.   See Abstr. 142.166.

The formation of the nuclei of galaxies. II. The local group.   See Abstr. 151.006.

Effect of binary stars on the dynamical evolution of stellar clusters. II. Analytic evolutionary models.
See Abstr. 151.013.

On a spherical star system with a collapsed core.
See Abstr. 151.036.

Photographic photometry in four clusters.
See Abstr. 153.007.

Sternhaufen und Nebel.   See Abstr. 153.025.

Clusters of the Magellanic Clouds.
See Abstr. 159.015.

On estimating the unprojected luminosity density within a cluster of galaxies.   See Abstr. 160.004.

Globular clusters in the Hydra I cluster of galaxies.
See Abstr. 160.026.

155 Structure and Evolution of the Galaxy

155.001 Large scale hydrodynamic oscillations of the galactic gas layer. A. H. Nelson.
Monthly Notices Roy. Astron. Soc., Vol. 174, 661 - 669 (1976).

The oscillations of a simple model of the galactic gas layer in the stellar potential well are considered, including motion in the vertical direction. Two types of mode exist, those with a density perturbation which is an even function of the vertical coordinate, and those for which this is an odd function. The simplest mode of the second type is shown to have properties similar to the recently observed corrugations of the 21-cm emission from the inner part of the Galaxy.

155.002 Accretion by the Galaxy: effects of radiative cooling on the flow structure and infall rate.
D. P. Cox, B. W. Smith.
Astrophys. Journ., Vol. 203, 361 - 370 (1976).

The extreme effects of radiative cooling on the structure of an inflow of material into the Galaxy are explored. Significantly, it is found that cooling cannot prevent the formation of a very hot ($T \approx 2 \times 10^6$ K) corona in the flow close to the Galaxy. When it is important, however, cooling in the flow considerably increases the infall rate over the adiabatic maximum value for the same set of boundary conditions. Limits set on the flow rate and coronal gas parameters by observations of the Galaxy are discussed.

155.003 A density-wave map of the galactic spiral structure.
S. C. Simonson III.
Astron. Astrophys., Vol. 46, 261 - 268 (1976).

The large-scale spiral structure is mapped by constructing a model based on density-wave kinematics that reproduces the main features of the 21-cm H I observations. The map shows a basically two-armed spiral pattern with a pitch angle of $6° - 8°$ between the 4-kpc dispersion ring and the solar circle. Near the solar circle, two additional major arms originate, and the pattern outside is multiple-armed, with pitch angles of $\sim 16°$. Comparison of the gas flow in both a linear and a shock solution shows that the shock model gives a better representation of the tangential directions of spiral arms. Comparison with two more open models favors the tightly-wound pattern found here. A synthetic photograph of the Galaxy is produced from the model.

155.004 An almost complete survey of 21-cm line radiation for $|b| \geqslant 10°$. V. Photographic presentation and qualitative comparison with other data.
C. Heiles, E. B. Jenkins.
Astron. Astrophys., Vol. 46, 333 - 360 (1976).

Photographs are presented which show the distribution in galactic coordinates of 21-cm line intensity in 3 velocity ranges—low velocities, and negative and positive intermediate velocities. Most of the HI "clouds" in all velocity ranges appear to be highly elongated and are best described as filaments. The region $l = 250°$ to $330°$, $b > 10°$ contains H I with widely different velocities, which the authors interpret as clouds colliding with relative velocities of 40 km/s or perhaps even 300 km/s. H I is roughly correlated with extinction as revealed by surveys of dark nebulae and by counts of external galaxies. The distributions of H I and radio continuum are extensively compared.

155.005 Three colour photometry of a field in the galactic anticentre section near M37.
W. Becker, S. Svolopoulos.
Astron. Astrophys., Suppl. Ser., Vol. 23, 97 - 107 (1976).

As part of a programme to investigate the stellar content of the galactic anticentre direction, RGU photometry (on plates of the 48" Palomar Schmidt) was carried out for a star field near M37; its size is $0.18 \square°$ and it contains 1234 measurable stars. The interstellar reddening in this direction is caused by two absorbing clouds, one lying between 200 and 400 pc, the other one at 2.5 kpc. Their combined colour excess is $E(G-R) = 0.^m 64$. The luminosity functions for three distance intervals (0.5–1.5; 1.5–2.0; 2.0–3.0 kpc) cover a magnitude range of $0^m < M(G) < 6^m$. The slope of their gradients is similar to the slope of the gradients of the luminosity function for the solar neighbourhood.

155.006 Carina arm studies. II. Photometry of faint early-type stars in Carina. S. Wramdemark.
Astron. Astrophys., Suppl. Ser., Vol. 23, 231 - 247 (1976).

Faint early-type stars in the Carina direction ($l = 290°$) have been detected on three-colour plates. Photoelectric photometry has been made in UBV ($V < 15$) and $H\beta$ ($V < 13$). From combined UBV and $H\beta$ photometry individual distances have been determined. From UBV data alone minimum distances have been determined. The interstellar extinction is low and amounts to about 3^m at $r = 10$ kpc. About ten early-type stars have been found with distances exceeding 10 kpc. The relation between the most distant stars and the galactic structure is examined. Two fields, one at $b = 0°$ and one just below, were investigated. The stellar density seems to be somewhat higher in the plane than below it.

155.007 UBV and $H\beta$ photometry of faint early-type stars in Norma. J. C. Muzzio, J. C. Forte.
Astron. Journ., Vol. 80, 1037 - 1043 (1975).

The authors present UBV and $H\beta$ photoelectric observations of faint early-type stars in the Norma section of the Milky Way. The limiting magnitude is about 15^m for UBV and $13.^m 5$ for $H\beta$. Large color excesses, probably due to dust nearer than 1 or 2 kpc from the sun, are found near $l = 328°$. Many stars near $l = 328°$ belong to the Norma OB1 association and some of them are probably the exciting stars of the H II regions RCW 97 and 98. In the light of these results the radial velocity data obtained by other authors indicate that turbulent motions of the gas may be present in the Norma section of the Sagittarius arm.

155.008 Rediscussion of the local space density of M dwarf stars.
S. M. Faber, D. Burstein, B. M. Tinsley, I. R. King.
Astron. Journ., Vol. 81, 45 - 52 = Lick Obs. Bull., No. 721 (1976).

Further evidence is presented for a systematic error in the photographic colors of Weistrop (1971, 1972) in the sense that the measured colors are too red. Since the colors are used to infer absolute magnitudes, the original uncorrected colors yield a distance scale that is too small. When the counts are corrected for this effect, they are consistent with a low local space density of M dwarfs having a scale height of 300 pc, typical of the old-disk population, and a rather flat luminosity function.

155.009 Motions of near-polar K giants along the z coordinate.
A. N. Balakirev.
Astron. Zhurn. Akad. Nauk SSSR, Vol. 53, 119 - 124 (1976).
In Russian. English translation in Soviet Astron., Vol. 20, No. 1.

The kinematics of near-polar G8–K5 III, IV stars ($|b^{II}| > 60°$) is investigated on the basis of radial velocity analysis. The author found $W_\odot = +6.5 \pm 1.5$ km/sec and $\delta W_\odot / \delta z \approx +0$. In the determination of the dynamic parameter C it is always necessary to take into account the heterogeneity of the material with respect to membership in different stellar population types. $C = 76 \pm 14$ km/sec/kpc was found for the stars of the flat component.

155.010 **The role of population II in spiral structure formation.** Yu. N. Mishurov, A. A. Suchkov.
Astron. Zhurn. Akad. Nauk SSSR, Vol. 53, 206 - 208 (1976). In Russian. English translation in Soviet Astron., Vol. 20, No. 1.

It is shown that because of large velocity dispersion of stars of population II, its contribution to the dynamics of spiral density waves in the Galaxy turns out to be negligible. Therefore the spiral structure of the Galaxy is defined by the parameters of population I.

155.011 **Neutral hydrogen in the galactic centre region — II. Location of the emission features.**
R. J. Cohen, R. D. Davies.
Monthly Notices Roy. Astron. Soc., Vol. 175, 1 - 24 (1976).

An interpretation and discussion is given of Cohen's observations of 21-cm hydrogen line emission from the galactic centre region. Only the high-velocity emission distinct from the main maximum at zero velocity is considered. The distances of many of the individual emission features have been estimated from the variation of velocity with longitude, and these distances used to calculate hydrogen masses and the z-distances of the gas above and below the plane.

155.012 **Pregalactic helium abundance and abundance gradients across our Galaxy from planetary nebulae.**
S. D'Odorico, M. Peimbert.
Astron. Astrophys., Vol. 47, 341 - 344 (1976).

From the observations of planetary nebulae by Peimbert and Torres-Peimbert (1971) the authors have studied the radial gradients across our Galaxy of the helium, oxygen and nitrogen abundances relative to hydrogen. The increase of the oxygen to hydrogen abundance ratio from a radial distance to the galactic center of 14 to 8 kpc is about a factor of 3 while that of the nitrogen to hydrogen ratio is about twice as large. By adopting oxygen as representative of the heavy elements it is found that the helium enrichment is coupled to the heavy metal enrichment by $\Delta Y/\Delta Z \cong 2.9$ in close agreement with the value derived from H II regions. The pregalactic $N(\text{He})/N(\text{H})$ value derived from planetary nebulae is 0.073 ± 0.008 also in agreement with the value derived from H II regions.

155.013 **Rolling motions in an inner spiral arm.**
F. M. Strauss, W. Poeppel.
Astrophys. Journ., Vol. 204, 94 - 100 (1976).

Hydrogen line observations made at low galactic latitudes for $l = 318°$, $326°$, $334°$, and $337°$ show the presence of velocity gradients in latitude in the nearest inner spiral arm, similar to those found by other observers in different regions. Maximum velocity change is about 10 km s^{-1} for $l = 337°$. By generating synthetic line profiles constructed from a model spiral arm, several possible causes of these "rolling motions" were studied, such as a vertical displacement or a tilt of the arm (which failed to account for the observations) and rotation or shearing in the arm. It was further shown that a typical arm can maintain such a motion (≈ 75 km s^{-1} kpc^{-1}) with its own gravitational potential. The results are used to study the origin and tilt of Gould's Belt.

155.014 **Search for neutral hydrogen with high negative velocities ejected from the galactic center.**
I. F. Mirabel.
Astrophys. Space Sci., Vol. 39, 415 - 417 (1976).

A search for neutral hydrogen in the velocity range $-300 > V > -1000$ km s^{-1} has been made in the zone around the galactic nucleus. Observations of 100 points reveal no neutral hydrogen at such high velocities, with brightness temperatures exceeding 0.25 K in the latitude range $|b| < 1°$, and 0.20 K

for $|b| \geqslant 1°$.

155.015 **Sources of ionization in the galactic center.**
K. J. Fricke, R. P. Kudritzki.
Mitt. Astron. Ges., No. 38, p. 100 (1976). – Abstract.

155.016 **Die Dichteabnahme der roten Riesensterne mit dem Abstand vom galaktischen Zentrum.**
B. Loibl, R. Schröder, U. Haug.
Mitt. Astron. Ges., No. 38, p. 245 - 247 (1976). – Short report.

155.017 **The mean eccentricity of the stellar orbits of a sample of stars in the solar neighbourhood chosen on the basis of metallicity.** M. Mayor.
Astron. Astrophys., Vol. 48, 133 - 135 (1976).

The interpretation of the kinematical diagrams for different star groups of given metallicity is ambiguous. A relation as $\bar{e}$ versus [Fe/H] depends not only on the chemical and kinematical history of the Galaxy but is also strongly dependent on the observational errors of [Fe/H] and on criteria used to define the sample.

155.018 **Reobservation of the outer boundary of galactic neutral hydrogen.** P. L. Baker.
Astron. Astrophys., Vol. 48, 163 - 164 (1976).

The discovery by Dieter that the velocity distribution of galactic neutral hydrogen displays a sharp boundary is confirmed for one longitude by high precision measurements. An interaction of the galactic gas disk with the nearby extragalactic environment may be required to explain the result.

155.019 **Propagation of magnetohydrodynamic waves from the active galactic center and their convergence into the 3-kpc arm.** Y. Sofue.
Publ. Astron. Soc. Japan, Vol. 28, 19 - 26 (1976).

Magnetohydrodynamic (MHD) disturbances originating at the nucleus of the Galaxy propagate through the halo, and focus on a ring in the gaseous disk. An efficient ($\gtrsim 90$ percent) convergence of the MHD waves into the ring is suggested as a possible mechanism to drive the 3-kpc arm expansion. The disk gas between the center and the 3-kpc arm is shown to remain undisturbed.

155.020 **Chemical evolution of the galactic disk and the radial metallicity gradient.** M. Mayor.
Astron. Astrophys., Vol. 48, 301 - 315 (1976).

An analysis of the kinematical and photometric properties of about 600 dF stars and 600 gG-gK stars permits the estimation of the radial chemical gradient in the Galaxy. The mean value in the solar neighbourhood obtained for all of these stars is $\partial[\overline{\text{M/H}}]/\partial \varpi = -0.05 \pm 0.01$ kpc^{-1}. For all the samples studied (dF, dG or giants), the order of magnitude for the gradient is the same. However, for the youngest stars in these samples, the metallicity gradient could be larger: $\partial[\overline{\text{M/H}}]/\partial \varpi = -0.10 \pm 0.02$ kpc^{-1}.

155.021 **The fine structure of the spiral arms.**
T. Schmidt-Kaler, F. House.
Astron. Nachr., Vol. 297, 77 - 82 (1976).

An explanation for filaments inclined to the galactic plane observed in the next-inner spiral arm is sought in terms of self-consistent z-oscillations. These filaments or "shingles" are observed to be ~ 1.4 kpc long, 70 pc thick and inclined to the plane by $\sim 12°$. In a collisionless axisymmetric stellar system the authors simplify the Boltzmann and Poisson equations by assuming a constant density normal to the galactic plane up to $|z| \approx 200$ pc and by assuming a constant radial component of the spatial coordinates. The condition for self-consistency in the linear approximation results in shingles 1.3 kpc long, inclined to the plane by 17°. The length of the shingles is independent of the radial distance R from the centre and the inclination depends inversely on R as is observed.

155.022 Energy source for shingle-like structures in spiral arms. F. House, T. Schmidt-Kaler.
Astron. Nachr., Vol. 297, 83 - 86 (1976).

The filaments or shingles observed in neighbouring spiral arms were found in a previous paper (see 155.021) to have a wavelength of ~ 1.3 kpc. The life-time of the shingles was estimated at $\sim 10^8$ years. The energy source to regenerate the shingles is sought in terms of a Jeans-type instability modified by a two stream instability between population I and disk material. From the dispersion relation it is found that an instability of this type occurs at a critical wavelength of 1.9 kpc.

155.023 Spectral and spatial resolution of the 12.8 micron Ne II emission from the galactic center.
E. R. Wollman, T. R. Geballe, J. H. Lacy, C. H. Townes, D. M. Rank.
Astrophys. Journ., (Letters), Vol. 205, L5 - L9 (1976).

High-resolution spectra of the Ne II 12.8μ fine-structure line in emission from the galactic center cloud Sgr A West show a line-center LSR radial velocity of $+75 \pm 20$ km s^{-1} and a velocity dispersion of about 200 km s^{-1}. The radial velocity and dispersion are more or less independent of position and indicate that events as recent as the last 10^4 years have given the ionized gas a systematic motion with respect to the massive stellar component of material at the galactic center. An upper limit of $\sim 4 \times 10^6 M_\odot$ for the mass within 0.8 pc of the galactic center is obtained from the velocity dispersion.

155.024 New galactic population class. F. W. Stecker.
Nature, Vol. 260, 412 - 414 (1976).

Galactic γ rays have been shown to come from regions where H_2 is dense, and where also 'population I' objects are predominantly found. The existence of such regions can be interpreted using the density-wave model of spiral galaxies, and leads the author to propose a change in the Baade classification scheme: 'population I' now to include all the objects associated with the H_2 clouds, and a new class, 'population 0', to refer to the regions, relatively richer in atomic hydrogen, which lie further from the galactic centre.

155.025 Far infrared observations of the galactic center with high spatial resolution.
P. M. Harvey, M. F. Campbell, W. F. Hoffmann.
Bull. American Astron. Soc., Vol. 8, 289 (1976). – Abstr. AAS.

155.026 Relation of NGC 3590, Hogg 10, and Collinder 240 to the structure of the Carina spiral feature.
J. J. Claria.
Astron. Journ., Vol. 81, 155 - 171, 217 - 218 (1976).

Photoelectric measurements in the UBV system are presented for 95 stars, most of them of early type, in the regions of NGC 3590, Hogg 10, and Collinder 240. In addition the intensity of the Hβ line was measured photoelectrically for 40 of the stars. Arguments are presented supporting the idea that Cr 240 is an OB association, perhaps an extension of Carina OB2, with the richest open cluster in the region NGC 3572 being the nucleus. NGC 3590, the small group Hogg 11, and some of the bright stars in the field of Tr 18, as well as the H II regions G 37, An 2, and An 3, would also form part of this complex. Therefore, considerable simplification of the structure of this region in Carina is obtained by accepting this interpretation.

155.027 Stellar population samples at the galactic poles. III.
$UBVRI$ observations of proper motion stars near the south pole and the luminosity laws for the halo and old disk populations. O. J. Eggen.
Astrophys. Journ., Suppl. Ser., Vol. 30, 351 - 396 (1976).

Some 1200 UBV and 650 R, I observations of 1050 stars,

mostly with annual proper motion greater than $0''.096$, brighter than visual magnitude 15, and within $10°$ of the south galactic pole, are presented and discussed. The bluer stars indicate that (1) the slopes of the luminosity laws for old disk and halo stars are fairly similar to M_V near $+6$ mag, (2) the old-disk-population law has an inflection point near $M_V = +7$ mag, (3) the halo-population law may peak near $M_V = +9$ mag on broad plateau that continues to beyond $+10$ mag and drops to zero near $+13$ mag, and (4) the upper limit for the mass density of the halo population near the sun is near $9 \times 10^{-4} M_\odot$ pc^{-3}.

155.028 On the density wave theory of the spiral structure of the Galaxy. A. K. Ray.
Bull. Astron. Soc. India, Vol. 3, 31 (1975). – Abstract of a paper presented at the A.S.I. meeting 1975.

155.029 A restricted three-dimensional study of the spiral structure in the inner region of the Galaxy.
G. Saha.
Bull. Astron. Soc. India, Vol. 3, 31 - 32 (1975). – Abstract of a paper presented at the A.S.I. meeting 1975.

155.030 Mass loss from the nucleus of the Galaxy due to amplification of longitudinal MHD waves.
K. Bandyopadhyay.
Bull. Astron. Soc. India, Vol. 3, 32 (1975). – Abstract of a paper presented at the A.S.I. meeting 1975.

155.031 Steps toward understanding the large-scale structure of the Milky Way: conclusion. H. Weaver.
Mercury (Journ. Astron. Soc. Pacific), Vol. 5, No. 1, p. 19 - 30 (1976).

In this conclusion to his three-part article, the author discusses many of the problems and ideas that are now keeping galactic astronomers busy; the spiral density wave theory, the results on structure of external galaxies as observed with aperture synthesis radio telescopes, and questions relating to the galactic nucleus and halo.

155.032 High-resolution far-infrared observations of the galactic center.
P. M. Harvey, M. F. Campbell, W. F. Hoffmann.
Astrophys. Journ., Letters, Vol. 205, L69 - L73 (1976).

A map at 53 μ with $17''$ resolution and three-color observations at 53μ, 100μ, and 175μ with $\sim 30''$ beams of Sgr A are presented. Sagittarius A is resolved into two main sources, one associated with the cluster of strong 10μ sources and another $\sim 45''$ to the southwest coincident with a weak 10μ source. The dust temperature peaks near the strong 10μ sources, but the 100μ and 175μ fluxes and the far-infrared optical depth are greatest near the southwest source. The amount of dust required to explain the far-infrared emission is comparable to that observed in absorption in the near-infrared.

155.033 Late-type giants and supergiants in the galactic center. G. Neugebauer, E. E. Becklin,
S. Beckwith, K. Matthews, C. G. Wynn-Williams.
Astrophys. Journ.,(Letters), Vol. 205, L139 - L141 (1976).

An absorption feature at 2.3 μ attributed to CO molecules in stellar atmospheres has been found in some, but not all, of the infrared sources within $20''$ of the galactic center. The absorption is characteristic of that found in M-type giant and supergiant stars. The infrared source coincident with the sub-arcsecond nonthermal radio source does not show CO absorption.

155.034 The spiral structure of our Galaxy determined from H II regions. Y. M. Georgelin, Y. P. Georgelin.
Astron. Astrophys., Vol. 49, 57 - 79 (1976).

In order to outline the spiral arms of our Galaxy the

authors have combined their optical observations (distances of exciting stars and Hα radial velocities) with the radio observations of H II regions. The optical detection of very distant H II regions—out to 9 kpc of the sun—has permitted a good overlap between the optical data and the H 109α radio data. In addition, the authors have analyzed the radial velocities and made precise identifications so as to group together all the H 109α sources of a single complex. Eighty per cent of the high-excitation-parameter H II regions fall along two symmetrical pairs of arms (i.e. four altogether) of 12° inclination. The longitudes at which one sees these arms tangentially correspond exactly to the flux maxima in the radio continuum and in the total 21-cm profile integral.

155.035 **Infrared profile of Milky Way at 2.4 μm.**
S. Hayakawa, K. Ito, T. Matsumoto, T. Ono, K. Uyama.
Nature, Vol. 261, 29 - 31(1976).
The diffuse component of galactic radiation has been observed in various wavelength ranges, but few observations are yet available in the near infrared range. Since the interstellar extinction is rather weak in this wavelength range, one can obtain an overview of the galactic structure, such as the distributions of infrared sources and of absorbing matter from this region of the spectrum. The authors have therefore attempted a balloon observation of infrared radiation at the wave length 2.4 μm with a band width 0.1 μm, avoiding intense OH airglow.

155.036 **Model of the galactic centre.**
V. De Sabbata, P. Fortini, C. Gualdi.
Problems of gravitation. Third Soviet Gravitational Conference, Erevan, 1972, (see 012.008), p. 32 - 36 (1975). In Russian.

155.037 **On the nonexpansion of the Galaxy.**
M. W. Ovenden, J. Byl.
Astrophys. Journ., Vol. 206, 57 - 65 (1976).
The radial velocities of more than 1000 O and B stars, cepheids, and open clusters have been analyzed with a view to determining any systematic expansion of the Galaxy. Provided that adequate allowance is made for velocity dispersion, there is an indication of a K term (constant and independent of distance) ≈ -1 km s^{-1}, but no significant expansion term. If a K term is not included in the analysis, a formally significant contraction of the Galaxy is found, ~ 0.6 km s^{-1} kpc^{-1}.

155.038 **Noncircular motions in the Perseus spiral arm.**
R. M. Humphreys.
Astrophys. Journ., Vol. 206, 114 - 121 (1976).
New velocity data are combined with existing velocities to further define the nature of the noncircular motions in the Perseus arm. The results are also compared with the predictions of the density-wave theory and with studies of the kinematics of the neutral hydrogen gas in the same region (Burton and Bania 1974).

155.039 **The structure of the galactic disk and its implications for gamma-ray astronomy.**
B. Fuchs, R. Schlickeiser, K. O. Thielheim.
Astrophys. Journ., Vol. 206, 589 - 598 (1976).
Under the assumption of a static equilibrium existing between the various constituents of our Galaxy the authors derive analytic formulae for the stratification of the stars, the ionized as well as the molecular interstellar gas perpendicular to the galactic plane. Results are compared with observational data and previous theoretical work by Kellman. Predictions are made for the distribution of molecular hydrogen. By using empirical data concerning the radial density distribution of H I and H$_2$ in the Galaxy the intensity of high-energy galactic γ-radiation is calculated.

155.040 **A first attempt to determine the nitrogen to sulphur abundance gradient across the disk of our Galaxy.**   J. P. Sivan.
Astron. Astrophys., Vol. 49, 173 - 177 (1976).
The author presents the first results of an extensive spectrographic survey of the galactic emission regions. The H II regions, selected for the present study, are all large ($d > 40$ pc), low-density ($n_e < 1000$ cm^{-3}) nebulae. He finds a radial increase of the Hα/[N II] λ 6584 line intensity ratio, and a radial decrease of the [N II] λλ 6548, 6584/[S II] λλ 6717, 6731 line intensity ratio, from a radial distance from the galactic center of 8 to 14 kpc.

155.041 **Molecular hydrogen in the Galaxy and galactic gamma rays.**   F. W. Stecker, P. M. Solomon, N. Z. Scoville, C. E. Ryter.
14th Intern. Cosmic Ray Conf., (see 012.011), Vol. 1, 46 - 51 (1975).

155.042 **On the relativistic-particle distribution in the Galaxy.**   K. P. Beuermann, G. Kanbach.
14th Intern. Cosmic Ray Conf., (see 012.011), Vol. 1, 58 (1975). – Abstract.

155.043 **Inverse Compton production of gamma rays in interstellar space.**
P. G. Shukla, M. Cassé, C. J. Cesarsky, J. Paul.
14th Intern. Cosmic Ray Conf., (see 012.011), Vol. 1, 65 - 69 (1975).

155.044 **Gamma ray emission from the galactic disc.**
R. Cowsik, W. Voges.
14th Intern. Cosmic Ray Conf., (see 012.011), Vol. 1, 74 - 79 (1975).

155.045 **The structure of the galactic disk in the light of gamma-ray astronomy.**
B. Fuchs, R. Schlickeiser, K. O. Thielheim.
14th Intern. Cosmic Ray Conf., (see 012.011), Vol. 1, 83 (1975). – Abstract.

155.046 **Average properties of molecular clouds in the Galaxy.**   R. Cowsik, S. Drapatz, K. W. Michel.
14th Intern. Cosmic Ray Conf., (see 012.011), Vol. 1, 181 - 186 (1975).

155.047 **Comments on galactic structure.**
B. Fuchs, K. O. Thielheim.
14th Intern. Cosmic Ray Conf., (see 012.011), Vol. 2, 644 (1975). – Abstract.

155.048 **Dimension of the galactic radio halo.**   V. A. Dogiel, S. V. Bulanov, S. I. Syrovatskii (*Syrovatskij*).
14th Intern. Cosmic Ray Conf., (see 012.011), Vol. 2, 700 - 705 (1975).

155.049 **The structure of the galactic disk in the light of gamma ray astronomy.**
B. Fuchs, R. Schlickeiser, K. O. Thielheim.
14th Intern. Cosmic Ray Conf., (see 012.011), Vol. 12, 4078 - 4084 (1975).

155.050 **Comments on galactic structure.**
B. Fuchs, K. O. Thielheim.
14th Intern. Cosmic Ray Conf., (see 012.011), Vol. 12, 4172 - 4176 (1975).

155.051 **Interstellar magnetic fields as tracer of galactic structure.**
C. E. Jäkel, D. Nissen, R. Schlickeiser, K. O. Thielheim.

14th Intern. Cosmic Ray Conf., (see 012.011), Vol. 12, 4177 - 4182 (1975).

155.052 Generalization of the Fesenkov – Parenago method for determination of the galactic oblateness.
F. A. Tsitsin.
Vestn. Mosk. un-ta. Fiz., astron., Vol. 16, 557 - 562 (1975). In Russian. – Abstr. in Referativ. Zhurn. 51. Astron., 6. 51. 726 (1976).

155.053 Bestimmungen der galaktischen Rotation und der Präzession aus fundamentalen Eigenbewegungen und aus Eigenbewegungen relativ zu Galaxien.   B. du Mont.
Diss. Nat. Gesamtfakultät, Ruprecht-Karl-Univ., Heidelberg, 100 pp. (1975).

155.054 The spiral structure of our Galaxy–a review of current studies.   T. Schmidt-Kaler.
Vistas Astron., Vol. 19, 69 - 89 (1975).

A review of present knowledge of the spiral structure of the Galaxy is given, concentrating on work published since 1970. The principal methods of determining spiral structure in the Galaxy are described and the value of different spiral tracers is assessed. The second part of this review deals with the internal structure of spiral features. In the third part some remarks have been made supplementing this review by considering theoretical possibilities to explain spiral structure apart from the density-wave theory, and summarizing some facts of observation left unexplained by the latter. A density-wave theory using the long-wave mode is sketched.

155.055 Theoretical aspects of galactic research.
W. W. Roberts, Jr.
Vistas Astron., Vol. 19, 91 - 109 (1975).

Our Galaxy is thought not to be greatly different from external spiral galaxies we see; and this review focuses from time to time on one or two external spirals to help us theoretically view our own Galaxy. Because of the great deal of theoretical research in galactic spiral structure accomplished in recent years, it is impossible to cover all the important contributions. There is time to outline only a few of the accomplishments of present theory, but the hope is that the subset of accomplishments selected will shed light on possible research problems for the future and on areas where observational tests would be desirable.

155.056 Radio measurements of galactic spiral structure.
K. Rohlfs.
Vistas Astron., Vol. 19, 111 - 122 (1975).

Problems connected with the determination of the large-scale spiral structure in neutral hydrogen gas from 21-cm line measurements are discussed. The attempts to overcome the difficulty that the distance coordinate cannot be observed directly but must be replaced by radial velocity measurements are classified into three categories: (a) The classical approach. The velocity field is assumed to be that of a given circular galactic rotation model. (b) The model approach. Both a velocity model and a density distribution for the gas are assumed. (c) Determination of the velocity field.

155.057 Popolazioni stellari ed evoluzione galattica.
L. Gratton.
Giorn. Astron., Vol. 1, 13 - 38 (1975).

155.058 La struttura e l'evoluzione della Galassia.
V. Castellani.
Giorn. Astron.,Vol. 1, 189 - 201 (1975).

155.059 Stars within a radius of 20 light years from the sun.
Z. Paprotny.
Urania Kraków, Vol. 47, 172 - 176 (1976). In Polish.

155.060 The surface distributions of O, B and M stars and the space distribution of B stars in the Carina-Crux-Centaurus region $l = 280°$ to $l = 319°$.   A. Sundman.
Stockholms Obs. Rep. No. 2, 39 pp. (1974).

In connection with a spectral survey of the southern Milky Way, the surface and space distributions of more than 13000 stars with spectral types O-B9 and M, N, R, S have been studied.

155.061 Chemical evolution of the Galaxy: coefficients from stellar evolution and alternative solutions to the problem of few metal-poor stars.   R. J. Talbot, Jr.
Explosive nucleosynthesis, (see 012.020), p. 34 - 43 (1973).

155.062 Hot stars and H II regions in our Galaxy.
Y. Georgelin.
Thesis Univ. Provence, Obs. Marseille, 263 pp. (1974). In French.

155.063 Models of galactic nuclei.   V. De Sabbata, P. Fortini, L. Fortini Baroni, C. Gualdi.
Acta Phys. Polonica B, Vol. B7, 99 - 116 (1976). – Abstr. in Phys. Abstr., Vol. 79, A037637 (1976).

155.064 Space density and kinematics of dwarf M stars.
W. G. Smethells.
Thesis Case Western Reserve Univ., Cleveland, Ohio, USA, 109 pp. (1974). (Available from Univ. Microfilms, Order No. 75–5089).

155.065 Balloon-borne hard X-ray study of the southern sky.
D. D. S. Guo.
Thesis New Hampshire Univ., Durham, USA, 206 pp. (1974). (Available from: Univ. Microfilms, Order No. 75-12,008).

155.066 SAS-2 high energy gamma-ray observations of galactic structure and the Crab nebula.
D. Kniffen.
Symposium of significant accomplishments sci. technol., Greenbelt, Maryland, USA, 18 Dec. 1973, p. 52 - 55 (1975).

Stellare Kinematik aus Eigenbewegungen in bezug auf Galaxien.   See Abstr. 041.009.

Enrichment of heavy elements in the Galaxy: a simple formula.   See Abstr. 061.003.

Ionising flux of cosmic background radiation. See Abstr. 061.014.

Nucleosynthesis and star formation of the Galaxy and Magellanic Clouds.   See Abstr. 065.061.

Visible and UV photometry of the Gegenschein and the Milky Way.   See Abstr. 106.064.

Polarimetry of the zodiacal light and Milky Way from Hawaii.   See Abstr. 106.066.

Space distribution and kinematics of intermediate population II stars. Part I. Photometry and spectroscopy in selected McCormick proper motion fields. See Abstr. 113.011.

Further evidence concerning the local density of red dwarfs.   See Abstr. 113.019.

The galactic distribution of interstellar absorption as determined from the Celescope catalog of ultraviolet stellar

observations and a new catalog of *UBV*, H-beta photoelectric observations.　　See Abstr. 113.032.

A spectral survey of the Southern Milky Way III. O−B9 and M stars *l* = 280° to 306°.　　See Abstr. 114.008.

Stellar population samples at the galactic poles. IV. Luminosity function for the M-type dwarfs at the south pole. See Abstr. 115.006.

Red variables in the central bulge of the Galaxy − I. The period distribution of Mira variables.　See Abstr. 122.001.

Neutral hydrogen in the W41 region. See Abstr. 131.011.

An almost complete survey of 21 cm line radiation for $|b| \geq 10°$. III. The interdependence of H I, galaxy counts, reddening, and galactic latitude.　　See Abstr. 131.044.

Dynamics of CO molecular clouds in the Galaxy. See Abstr. 131.114.

Neutral hydrogen in a region of Cepheus: nearby galactic structure.　　See Abstr. 131.141.

Interstellar gas.　　See Abstr. 131.154.

The systematic motion of the interstellar clouds at the galactic shock region.　　See Abstr. 131.167.

Twenty-one centimeter aperture synthesis study of the small-scale structure of the interstellar medium. See Abstr. 131.511.

Radio observations of H II regions in the galactic center.　See Abstr. 131.547.

The masses and chemical composition of planetary nebulae in the galactic bulge and Magellanic Clouds. See Abstr. 133.002.

On the number of planetary nebulae in our Galaxy. See Abstr. 133.014.

Gamma-ray emission and nucleosynthesis of lithium by young pulsars.　　See Abstr. 141.340.

Uhuru observations of the galactic plane in 1970, 1971, and 1972.　　See Abstr. 142.093.

High energy galactic gamma radiation observed by the SAS-2 satellite.　　See Abstr. 142.125.

Gamma ray spectrum from the galactic disc. See Abstr. 142.126.

Galactic gamma radiation from cosmic rays concentrated in spiral arms.　　See Abstr. 142.127.

X-ray bursts from near the galactic center. See Abstr. 142.165.

X-ray bursts from galactic center region. See Abstr. 142.169.

X-ray bursts from galactic center region. See Abstr. 142.174.

X-ray bursts.　　See Abstr. 142.181.

Molecular hydrogen in the Galaxy and galactic gamma rays.　　See Abstr. 142.250.

Theoretical interpretation of the SAS-2 high energy gamma-ray observations of the galactic plane in terms of the galactic cosmic-ray and matter distributions. See Abstr. 142.254.

Gamma radiation of galactic origin. See Abstr. 142.259.

Composition of cosmic rays at 10^{10} to 10^{13} eV/ nucleus.　　See Abstr. 143.292.

The character of cosmic ray propagation in the Galaxy.　　See Abstr. 143.338.

Gamma-ray evidence for galactic Fermi acceleration of cosmic rays.　　See Abstr. 143.372.

Electrons in the galactic cosmic radiation. See Abstr. 143.380.

Cosmic rays and the dynamics of the Galaxy. See Abstr. 143.384.

On the mass of the halo population. See Abstr. 151.004.

On density waves in galaxies. II. The turning-point problem at the corotation region.　　See Abstr. 151.005.

The formation of the nuclei of galaxies. II. The local group.　　See Abstr. 151.006.

Density maxima formed by trapped orbits. See Abstr. 151.009.

Dynamical evolution of the triple system of the Galaxy, the Large and Small Magellanic Clouds. See Abstr. 151.021.

Whence comes the ellipsoidal distribution of star velocities.　　See Abstr. 151.027.

Stellar dynamics and the structure of galaxies. See Abstr. 151.065.

Bochum 15, a new young stellar aggregate in Puppis. See Abstr. 152.001.

UBV, Hβ photometry of OB stars in groups: Pup OB2.　See Abstr. 152.002.

The relation of galactic radio spurs to spiral arms. See Abstr. 157.002.

Inclination of inner radio spurs and horizontal stream of gas in the galactic halo.　　See Abstr. 157.005.

Missing mass in the Local Group of galaxies. See Abstr. 158.154.

The Magellanic Stream and the mass of our hypergalaxy.　See Abstr. 158.155.

The mass of the Local Group of galaxies. See Abstr. 158.156.

Dwarf galaxies and globular clusters in high velocity hydrogen streams.　　See Abstr. 159.002.

The effect of dynamical friction on the orbits of the Magellanic Clouds.   See Abstr. 159.003.

Magellanic stream and the dynamics of our Hyper-galaxy.   See Abstr. 159.013.

Antimatter in the universe.
See Abstr. 162.034.

156 Galactic Magnetic Field

156.001 **Optical polarization of stars of galactic latitudes** $b < -45°$.   R. Schröder.
Astron. Astrophys., Suppl. Ser., Vol. 23, 125 - 137 (1976).
Optical polarization data of about 500 stars of galactic latitudes $b < -45°$ are represented in the following catalogue. The instrumentation is specified and the method of data reduction is described.

156.002 **Structure of the galactic magnetic field.**
A. J. Owens, with a reply by A. J. Somogyi.
Nature, Vol. 259, 344 - 345 (1976).

156.003 **Orion-arm magnetic monopoles and γ rays.**
D. R. Tompkins, Jr., P. F. Rodney.
Phys. Rev. D., Vol. 12, 2610 - 2616 (1975). − Abstr. in Phys. Abstr., Vol. 79, A032629 (1976).

Interstellar magnetic fields as tracers of galactic structure.   See Abstr. 131.153.

The directional dependence of the primary cosmic rays of energies $10^{11} - 10^{12}$ eV.   See Abstr. 143.010.

Energy dependence of galactic cosmic ray propagation and power spectrum of galactic magnetic field irregularities.   See Abstr. 143.143.

Propagation of 10^{11} - 10^{14} eV particles in the Galaxy. See Abstr. 143.156.

Comments on galactic structure.
See Abstr. 155.047.

157   Galactic Radio Radiation

157.001   Recombination line and continuum observations of the galactic center at 10 GHz.
T. Pauls, D. Downes, P. G. Mezger, E. Churchwell.
Astron. Astrophys., Vol. 46, 407 - 412 (1976).

Radio continuum and H85α recombination lines have been observed at 10 GHz within 15′ of the galactic center source Sgr A. The continuum map has a resolution of 77″ and shows new structural details in the radio emission. Sgr A appears to be a core-halo source, as has been shown at lower frequencies. The halo has an angular diameter of 6′ and a flux density of 60×10^{-26} W m^{-2} Hz^{-1}. The radial velocities of the H85α lines are predominantly negative. There is no H85α line toward G0.16−0.15, which may be a flat spectrum, non-thermal source near the galactic center.

157.002   The relation of galactic radio spurs to spiral arms.
Y. Sofue.
Astron. Astrophys., Vol. 48, 1 - 10 (1976).

An isophotal map is constructed for the brightness distribution of the galactic radio continuum background on the assumption that nonthermal radioemission is enhanced in space up to ∼ 1 kpc above the spiral arms. The model map agrees well with observed maps in many characteristics, particularly in the galactic spurs. The existence of Loops II and III is doubtful. Detailed inspection of the observed radio maps reveals a systematic inclination of spur-ridges by 20−30° toward anticenter sides in the inner region of the Galaxy. The inclination suggests some horizontal force on the inner spurs.

157.003   The radio emission of the galactic interstellar medium to the direction l^{II} = 31.°2 and 31.°9.
V. I. Ariskin.
Astron. Zhurn. Akad. Nauk SSSR, Vol. 53, 279 - 285 (1976).
In Russian. English translation in Soviet Astron., Vol. 20, No. 2.

From the results of continuum observation at wavelength 8.2 mm as well as from the data of continuum observations at longer wavelengths and the spectrum of the recombination lines of hydrogen, an investigation of the main physical parameters of the galactic interstellar medium to the longitudes l^{II} = 31.°2 and 31.°9 is carried out.

157.004   A bright spot of the linearly polarized emission at
λ = 1 m.   A. M. Paseka, L. V. Popova, V. A. Razin.
Astron. Zhurn. Akad. Nauk SSSR, Vol. 53, 286 - 287 (1976).
In Russian. English translation in Soviet Astron., Vol. 20, No. 2.

A region of high brightness temperature of the linearly polarized galactic radio emission at 290 MHz (λ = 1.03 m) is discovered. In the center of the region (α = 4^{h}30^m, δ = 59°30′) T_{pol} = (14 ± 0.7) °K. In the vicinity of this region T_{pol} is about 2°K on the average.

157.005   Inclination of inner radio spurs and horizontal stream of gas in the galactic halo.
Y. Sofue, M. Fujimoto, M. Tosa.
Publ. Astron. Soc. Japan, Vol. 28, 317 - 327 (1976).

Inner galactic spurs in the radio continuum located at $l \lesssim 50°$ incline systematically toward the anticenter sides by 20−30°. The spurs are associated with inner spiral arms of H I gas. The systematic inclination suggests the existence of a radial force to push outward the nonthermal emitting region responsible for the spurs.

A peculiar H I feature at l = 285°, b = −18°.
See Abstr. 131.539.

On angular variations of the frequency spectrum of the cosmic radio background along the declination δ = 50°30′.
See Abstr. 141.071.

Transient radio source near the galactic centre.
See Abstr. 141.099.

Cosmic-ray electrons and galactic radio noise: some problems.   See Abstr. 143.356.

An almost complete survey of 21-cm line radiation for $|b| \geqslant 10°$. V. Photographic presentation and qualitative comparison with other data.   See Abstr. 155.004.

Dimension of the galactic radio halo.
See Abstr. 155.048.

Radio measurements of galactic spiral structure.
See Abstr. 155.056.

158 Single und Multiple Galaxies, Peculiar Objects

Single and Multiple Galaxies

158.001 Physical conditions in active nuclei–II. O I λ 8446 fluorescence. H. Netzer, M. V. Penston.
Monthly Notices Roy. Astron. Soc., Vol. 174, 319 - 325 (1976).

The permitted λ 8446 O I line has been found in the Seyfert galaxy NGC 4151 with an intensity 5 per cent that of the broad component Hα; it may be present in other objects. Its strength can be explained as a result of resonant absorption of Lyman β and subsequent cascade if the broad-line emitting region is optically thick in Hα.

158.002 Interstellar absorption and the luminosity function of galaxies. T. Kiang.
Monthly Notices Roy. Astron. Soc., Vol. 174, 425 - 428 (1976).

It is shown that the difference in the zero-point between Kiang's luminosity function of galaxies on the one hand and those by van den Bergh, Shapiro and Huchra & Sargent on the other, is because these authors have overlooked the effect of interstellar absorption in reducing the size of volume surveyed.

158.003 An aperture synthesis study of H I in the irregular II galaxy NGC 3077. G. A. Cottrell.
Monthly Notices Roy. Astron. Soc., Vol. 174, 455 - 466 (1976).

The Cambridge Half-Mile telescope has been used in conjunction with a 160-channel cross-correlation receiver to map the neutral hydrogen emission in NGC 3077. The H I distribution is found to be severely distorted, with the projected centroid lying about 4 kpc SE of the optical centre in a region where star formation may be occurring, and a streamer of H I extending ~10 kpc northwards. The observed H I mass is $4.1 \times 10^8 \, M_\odot$ of which a quarter is in the streamer. A small gradient of radial velocity exists across the galaxy which, if interpreted as rotation, suggests an indicative mass of $(6 \pm 2) \times 10^9 M_\odot$. A computer model is proposed to explain the distorted H I distribution in terms of tidal forces exerted during a close (12 kpc) encounter between NGC 3077 and M81, (2 to 6) $\times 10^8$ yr ago.

158.004 K-corrections for galaxies of different morphological types. W. Pence.
Astrophys. Journ., Vol. 203, 39 - 51 (1976).

Mean energy distributions from 1500 to 8000 Å for different types of galaxies have been derived from satellite ultraviolet filter photometry and from earth-based spectrum scans. K-corrections for the standard U, B, V filters have been calculated from these energy distributions. The K-corrections are used to calculate the mean redshift, mean observed color, and color distribution of galaxies as a function of apparent magnitude and type in a Euclidean universe assuming no galaxy evolution and no intergalactic extinction.

158.005 Evolutionary synthesis of the stellar population in elliptical galaxies. I. Ingredients, broad-band colors, and infrared features. B. M. Tinsley, J. E. Gunn.
Astrophys. Journ., Vol. 203, 52 - 62 (1976).

Broad-band photometric data and infrared line indices have been combined with new results on giant-branch luminosity functions to yield population syntheses for giant elliptical galaxies. If the main-sequence mass function is a power law of slope x, a value $x < 1$ is indicated. This yields rather rapid luminosity evolution and a large correction to the deceleration parameter q_0 as derived from the Hubble diagram for first-ranked cluster ellipticals. The uncertainties are discussed.

158.006 Effects of main-sequence brightening on the luminosity evolution of elliptical galaxies.
B. M. Tinsley.
Astrophys. Journ., Vol. 203, 63 - 65 = Lick Obs. Bull., No. 712 (1976).

The brightening of stars with main-sequence lifetimes longer than the age of the galaxy has significant effects on the evolution of the integrated luminosity of model elliptical galaxies. Analytical and numerical models that treat dwarfs as unevolving may lead to misleading results, possibly even underestimating the rate at which the integrated magnitude becomes fainter.

158.007 Composition gradients across spiral galaxies. II. The stellar mass limit.
G. A. Shields, B. M. Tinsley.
Astrophys. Journ., Vol. 203, 66 - 71 (1976).

The equivalent width of the Hβ emission from H II regions in spiral galaxies increases with distance from the nucleus. The authors interpret this $W(H\beta)$ gradient in terms of a radial gradient in the temperature of the hottest exciting stars (T_u). From Searle's observations of M101, they infer an increase $\Delta \log T_u = 0.02$–0.13 from the intermediate to outermost spiral arms of M101. The authors note also that, even in the absence of changes in the upper mass limit, a T_u gradient is expected because metal-rich stars of a given mass have smaller effective temperatures.

158.008 Cool stellar populations in E/S0 galaxies and a possible outburst in M87. R. W. O'Connell.
Astrophys. Journ., (Letters), Vol. 203, L1 - L4 (1976).

The author discusses a scanner survey of the $0.7-1.1 \, \mu$ energy distributions of the nuclei of E/S0 galaxies ranging over a factor of 100 in intrinsic luminosity. There is a mild (color, absolute magnitude)-correlation. The strength of the λ8542 Ca II + TiO absorption feature appears to reach a maximum near $M_v \approx -20.5$, which may be due to a maximum in the density of low-mass stars in short-lived phases near the tip of the asymptotic giant branch. The color of the nucleus of M87 is anomalously blue for its luminosity, possibly indicating that a nonthermal or supernova outburst has recently occurred.

158.009 An analytic expression for the luminosity function for galaxies. P. Schechter.
Astrophys. Journ., Vol. 203, 297 - 306 (1976).

A new analytic approximation for the luminosity function for galaxies is proposed. The analytic expression is proportional to $L^{-5/4}e^{-L/L^*}$, where L^* is a characteristic luminosity corresponding to a characteristic absolute magnitude $M^*_{B(0)}=-20.6$. For an individual cluster, the characteristic magnitude may be determined with an accuracy of ~0.25 mag. The analytic expression is used to compute an expected richness—absolute magnitude correlation for first ranked cluster galaxies and an expected dispersion, which are compared with the data of Sandage and Hardy.

158.010 In situ particle acceleration and physical conditions in radio tail galaxies.
A. G. Pacholczyk, J. S. Scott.
Astrophys. Journ., Vol. 203, 313 - 322 (1976).

A model for the objects known as radio tail galaxies is presented. Independent plasmons emerging from an active radio galaxy into an intracluster medium become turbulent due to Rayleigh-Taylor and Kelvin-Helmholtz instabilities. The turbulence produces both in situ betatron and second order Fermi acceleration. Predictions of the dependence of spectral index and flux on distance along the tail match observations well. Fitting provides values of physical parameters in the tail. The relevance of this method of particle acceleration for the problem of the origin of X-ray emission in clusters of galaxies is discussed.

158.011 The radio spectra of Markarian galaxies.
G. Kojoian, R. A. Sramek, D. F. Dickinson, H. Tovmassian, C. R. Purton.
Astrophys. Journ., Vol. 203, 323 - 328 (1976).

Radio flux densities between 2.7 and 15.5 GHz are given for 16 Markarian galaxies. Spectra for these objects are obtained from these and previously published data. The Seyfert-type galaxies show power-law spectra with an average index of −0.82, except for MRK 348, which is an active galaxy with a complex spectrum. The Markarian galaxies with featureless optical spectra have rather flat radio spectra, similar to the BL Lacertae-type objects.

158.012 Markarian 376: a Seyfert galaxy with strong Fe II emission.   D. E. Osterbrock.
Astrophys. Journ., Vol. 203, 329 - 334 = Lick Obs. Bull., No. 710 (1976).

In a spectrophotometric survey of Seyfert galaxies, Markarian 376 was found to have strong, broad Fe II emission features of the type previously identified in two quasars and a few other Seyfert galaxies. The relative intensities of the individual emission lines and features in the spectrum of Markarian 376 were measured. A reasonable extrapolation of the observed continuous spectrum has more than enough near-ultraviolet photons in the region $\lambda\lambda 2300-2800$ to produce all the observed Fe II emission by resonance fluorescence. Practically all the Seyfert 1 galaxies observed to date in this survey have Fe II emission in their spectra.

158.013 Profiles of the [O III] nebular lines in two bright Seyfert galaxies.
J. W. Glaspey, J. A. Eilek, G. G. Fahlman, J. R. Auman.
Astrophys. Journ., Vol. 203, 335 - 344 (1976).

Moderate resolution profiles of the [O III] forbidden doublet at $\lambda\lambda 4959, 5007$ have been obtained over a period of about 1 year from the nuclei of the two brightest Seyfert galaxies, NGC 1068 and NGC 4151. No significant changes in the line profiles were apparent during the observational period. There are striking differences in the profiles of these lines in the two galaxies: NGC 1068 has a broad, symmetric, structured profile, while that of NGC 4151 is much narrower and asymmetric.

158.014 Electron temperature in the elliptical galaxy NGC 1052.   A. T. Koski, D. E. Osterbrock.
Astrophys. Journ., (Letters), Vol. 203, L49 - L51 = Lick. Obs. Bull., No. 718 (1976).

Image-dissector scanner observations of the emission-line spectrum of NGC 1052 are discussed. The scan of the elliptical galaxy NGC 584 is used to remove the integrated stellar absorption-line features from the spectrum of NGC 1052. The [O III] lines indicate $T \approx 33,000$ K. The observed line intensities are consistent with shock-wave heating models by Cox.

158.015 Hα emission from the disks of spiral galaxies.
J. G. Cohen.
Astrophys. Journ., Vol. 203, 587 - 592 (1976).

Observations of the amount of Hα emission from the disks of 53 galaxies have been made using large entrance apertures so that the maximum possible fraction of the total area of the galaxy is included. These observations reveal a relationship between $B - V$ color and Hα emission; for a given morphological type the bluer galaxies have more emission at Hα.

From the range in Hα emission, the author obtains the ratio of hot stars to cooler stars corresponding to the observed range in $B - V$ color.

158.016 The semistellar nucleus of M33.
S. van den Bergh.
Astrophys. Journ., Vol. 203, 764 - 765 (1976).

Spectroscopic observations suggest that the light emitted by the semistellar nucleus of M33 is dominated by relatively young stars. This observation indicates that the nuclei of some galaxies may be formed from interstellar gas rather than from an accumulation of globular clusters.

158.017 Have primeval galaxies been detected?
D. L. Meier.
Astrophys. Journ., (Letters), Vol. 203, L103 - L105 (1976).

A theoretical spectrum, synthesized for hydrodynamical models of collapsing galaxies in their brightest phase, bears a strong resemblance to the continuum spectra of OH 471 ($z = 3.40$) and 4C 05.34 ($z = 2.877$). Nonvariability and fuzziness of these objects would be strong tests of the possibility that these are primeval galaxies. Most primeval galaxies may be masquerading as red objects that are quasi-stellar in appearance.

158.018 The variable radio nucleus of M81.
P. C. Crane, T. S. Giuffrida, J. B. Carlson.
Astrophys. Journ., (Letters), Vol. 203, L113 - L114 (1976).

Variations in the radio emission of the nucleus of the normal spiral galaxy M81 are reported. Observations at 2695 MHz between 1967 and 1975 show that the flux density decreased from 120 mJy to 64 mJy.

158.019 Seven new southern galaxies with strong emission lines.   R. M. West.
Astron. Astrophys., Vol. 46, 327 - 331 (1976).

Seven new, southern extragalactic emission-line objects have been found. IC 4687 and 4689 form an interacting pair with IC 4686 on the connecting bridge. IC 4694, ESO 148-IG07, ESO 116-G12 and ESO 116-IG15 are single galaxies. IC 4687 and ESO 116-IG15 appear to be Seyfert galaxies.

158.020 The origin of galaxies: a review of recent theoretical developments and their confrontation with observation.   B. J. T. Jones.
Rev. Modern Phys., Vol. 48, 107 - 149 (1976).

The subject of galaxy formation has advanced considerably during the past decade. On the theoretical side two theories in particular have been developed to the point where confrontation with observation will be possible; these are the "gravitational instability picture" and the "cosmic turbulence theory". These theories are discussed at some length here, with particular attention to the question of the origin of cosmic angular momentum and the nature of the initial conditions. There is now a considerable body of data on galaxies; the problem is in deciding which kind of observation is most relevant to understanding the origin of galaxies. Throughout the review an attempt is made both to put the present research in its historical perspective and to stress the possibilities for future advances towards the goal of understanding the origin of cosmic structure.

158.021 On the mass discrepancy in double galaxies.
H. Oleak.
Astron. Nachr., Vol. 297, 1 - 3 (1976).

Comparing the M/L-distribution function of single spiral and irregular galaxies with that of double galaxies (S,S and E,E) from the list of Karachentsev (1974) it can be shown that the mass discrepancy can be ascribed to the skewedness of this frequency function and thus to the strong dependence of the mean value of M/L on the dispersion. An unavoidable enlargement of the dispersion by errors of measurement and wrong attachment of objects to the sample always raises the mean of M/L. This effect could be then mistaken for a mass discrepancy.

158.022 Photometry of S0 galaxies. III. NGC 524.
P. Hodge, P. Steidl.
Astron. Journ., Vol. 81, 20 - 24 (1976).

Photoelectric and photographic observations of the S0 galaxy NGC 524 are used to establish its luminosity profile and shape. The profile shows luminosity anomalies that are most easily interpreted in terms of luminosity enhancement (rings) at $r \simeq 45$ arcsec and $r \simeq 90$ arcsec. The isophotes indicate that the galaxy is seen not quite face-on, with a position angle determined to be approximately 5° and an eccentricity of approximately 0.03. The color is nearly constant with radius, with a tendency to be redder towards the center.

158.023 The structure of NGC 147.   P. W. Hodge.
Astron. Journ., Vol. 81, 25 - 29, 87 (1976).

NGC 147, a faint distant companion to Andromeda, is found to have a normal elliptical galaxy structure, though of relatively low luminosity gradient. Its total absolute magnitude is $M_V = -14.60$, its ellipticity is 0.44, its core radius is 67 arcsec, its tidal radius is approximately 11 arcmin, and its color averages $B-V = +0.91$. There are four globular clusters, all faint in absolute luminosity.

158.024 The problem of motions in systems of galaxies.
B. I. Fesenko.
Astron. Zhurn. Akad. Nauk SSSR, Vol. 53, 112 - 118 (1976). In Russian. English translation in Soviet Astron., Vol. 20, No. 1.

The mass-luminosity ratio in double galaxies seems to be in accordance with the expected value if one accounts for the errors in radial velocities and excludes optical pairs of galaxies. A considerable admixture of extraneous galaxies exists in the Coma cluster.

158.025 The role of diffuse matter in galactic coronas.
A. Chernin, J. Einasto, E. Saar.
Astrophys. Space Sci., Vol. 39, 41 - 52, 53 - 64 (1976). In Russian and English.

An upper limit to the gas content in the coronas of giant galaxies $M_{gas} < 0.04\ M_{cor}$ has been calculated on the basis of observational data on the emission of clusters of galaxies in different wavelength ranges. An analysis of companion galaxies has revealed a correlation of their morphological types with the distances to central giant galaxies. A gas-dynamical interpretation of this regularity is suggested and a possible picture of the evolution of the galactic coronas is presented.

158.026 Mass-luminosity distribution and physical conditions of ionized gas in NGC 4575.
M. Pastoriza, E. L. Agüero.
Astrophys. Space Sci., Vol. 39, 201 - 211 (1976).

Photographic photometry and spectroscopic observations of NGC 4575 suggest it to be a galaxy of reduced dimensions $D \times d = 14.4 \times 13.5$ kpc and of high luminosity $M = -20.7$. The rotation curve was also determined. Assuming a model of three homogeneous similar spheroids, the authors derived the density and mass distribution, and their total mass was found to be $M_T = 2.33 \times 10^{10}\ M_\odot$. From the emission lines it is found that the electronic density $Ne \sim 100$ cm^{-3} is relatively low in the H II regions. The abundance ratios $N(N)/N(S)$ and $N(N)/N(H)$ for the nucleus and two emission regions were also derived.

158.027 More *JHKL* colours of galaxies.   I. S. Glass.
Monthly Notices Roy. Astron. Soc., Vol. 175, 191 - 195 (1976).

More data on the *JHKL* colours of galaxies is presented

and interpreted in the light of improved knowledge of the infrared colours of late-type stars. It is concluded that most galaxies have similar *JHK* colours but that these yield minimal information as to the giant/dwarf ratio within the stellar population.

158.028 The detection of formaldehyde in NGC 5128 (Centaurus A).   F. F. Gardner, J. B. Whiteoak.
Monthly Notices Roy. Astron. Soc., Vol. 175, 9P - 12P (1976).

Six-centimetre H_2CO absorption has been detected in the direction of the nucleus of NGC 5128. The absorption profile consists of three narrow features (each only 2–3 km s^{-1} wide) with central heliocentric velocities of 541, 546 and 552 km s^{-1}; the absorption may originate in the dust lane that crosses the galaxy. There is a general similarity between the H_2CO absorption and the main H I absorption – the absence of fine structure in the H I profile may be a consequence of thermal broadening.

158.029 New map of the optical polarisation of galaxy M82.
R. G. Bingham, D. McMullan, W. S. Pallister, C. White, D. J. Axon, S. M. Scarrott.
Nature, Vol. 259, 463 - 465 (1976).

The outer parts of the galaxy show linear optical polarisation, discovered by Elvius. This is thought to arise from reflection in an extensive halo of dust particles of the light of the bright nucleus (or nuclear region) and of the galactic disk. The observations reported here were obtained with the 1-m telescope of the Wise Observatory, Israel, during the period February 27–March 15, 1975. The detector was an electronographic camera. The waveband used was B of the UBV system.

158.030 The variability of the compact galaxy Zw 0039.5 + 4003 (IV Zw 29) from 1970 to 1974.
C. Barbieri, G. Romano, L. Rosino.
Astron. Astrophys., Vol. 47, 153 - 155 (1976).

The variability of the compact galaxy Zw 0039.5 + 4003 (IV Zw 29) has been studied during the period 1970–1974 on plates obtained with the Schmidt telescope of Asiago Observatory. An historical light curve is presented for the period from 1935 to 1974 and its significance briefly discussed in relation to other types of extragalactic variable objects.

158.031 Morphological study of Markarian galaxies in pairs. I: Results.   C. Casini, J. Heidmann.
Astron. Astrophys., Vol. 47, 371 - 373 (1976).

The authors present the results of a study of the morphology of 8 pairs of galaxies containing Markarian galaxies, based on large scale electronographs or photographs. They compare their morphology to that of isolated Markarian galaxies obtained by Kalloghlian and Börngen and bring to light a new class of irregular galaxies.

158.032 21-cm line radial velocities of galaxies.
L. Bottinelli, L. Gouguenheim.
Astron. Astrophys., Vol. 47, 381 - 387 (1976).

The systemic radial velocities of 156 galaxies measured from the 21-cm line of neutral hydrogen with the Nançay radiotelescope in the range −341 to 6590 km s^{-1} are investigated. When compared to other 21-cm line determinations, they show good agreement. A comparison with optical determinations leads to a mean value of the difference ΔV between the Nançay and the optical determinations $\langle \Delta V \rangle = (-21 \pm 8$ m.e.) km s^{-1} with a r.m.s. dispersion of 96 km s^{-1}. The sample shows no clear correlation between ΔV and the morphological type.

158.033 The observational effects of explosions in the nuclei of spiral galaxies.   R. H. Sanders, T. M. Bania.
Astrophys. Journ., Vol. 204, 341 - 351 (1976).

Recurring explosive events may occur in the nuclei of

many or most spiral galaxies. Two-dimensional hydrodynamical calculations have been carried out to determine the effects which these events might have on the observable characteristics of spiral galaxies in three respects: (1) the optical appearance of the galaxy; (2) the gas kinematics in the inner regions: and (3) the nonthermal radio morphology of the inner regions. Evidence is presented supporting the suggestion that one particular spiral galaxy, NGC 4736, is a galaxy which may have undergone a recent explosive event in its nucleus.

158.034 **H I in early-type galaxies. II. Mass loss and galactic winds.** S. M. Faber, J. S. Gallagher.
Astrophys. Journ., Vol. 204, 365 - 378 = Lick Obs. Bull., No. 709 (1976).

Implications of observed upper limits to the neutral hydrogen content of E and S0 galaxies are discussed. Several schemes to conceal the mass are proposed, but all seem improbable on one or more grounds. Hence a removal mechanism of some kind seems likely. The most satisfactory candidates for this mechanism appear to be a hot galactic wind (Mathews and Baker 1971) or star formation. The paper concludes with a more speculative discussion of the possible role of hot winds in spiral galaxies. In the course of the discussion the authors summarize several observations which appear to follow logically from the existence of galactic winds: (1) lack of interstellar matter in most early-type galaxies, (2) the existence of young stars and dust in luminous dwarf ellipticals like NGC 205, (3) the fact that radio activity is associated preferentially with massive elliptical galaxies, (4) X-ray sources in clusters of galaxies, and (5) the observed dependence of elliptical metallicity on galaxy mass (Larson 1974).

158.035 **On the formation of spiral and elliptical galaxies.** J. R. Gott III, T. X. Thuan.
Astrophys. Journ., Vol. 204, 649 - 667 (1976).

The authors propose that the key factor distinguishing an elliptical from a spiral galaxy is the amount of gas left over at the point of maximum collapse of the protogalaxy. The gas left depends on the ratio of the star formation time scale τ_s to the collapse time of the protogalaxy τ_c. Observed properties of ellipticals imply that they formed out of relatively larger density perturbations at recombination, giving them a relatively small τ_s/τ_c (if a ρ^2 star formation law is adopted); and thus they are expected to essentially complete their star formation by the time of maximum collapse. Observations of spiral galaxies and our Galaxy imply that they formed out of less dense density perturbations giving them a larger τ_s/τ_c. "Zeroth" order Maclaurin spheroid models for equilibrium spiral and elliptical galaxies are presented, which relate many useful galactic parameters to the original size and angular momentum.

158.036 **Velocity dispersions and mass-to-light ratios for elliptical galaxies.** S. M. Faber, R. E. Jackson.
Astrophys. Journ., Vol. 204, 668 - 683 = Lick Obs. Bull., No. 714 (1976).

Velocity dispersions for 25 galaxies have been measured using conventional and Fourier techniques. Using unpublished data of King, the authors have computed core values of M/L_B. For luminous ellipticals with $M_B < -20$, M/L_B averages 7(H/50 km s^{-1} Mpc^{-1}). This value agrees well with M/L_B for early-type spirals, indicating that there is no large discontinuity in M/L_B between ellipticals and early-type spirals. This result is consistent with the observed small color differences between ellipticals and Sa's. Velocity dispersions increase with luminosity for normal elliptical galaxies of moderate ellipticity. The data also suggest that M/L_B generally increases with luminosity. The close correlation between luminosity and dynamical properties for normal ellipticals is further evidence that the ellipticals are very nearly a one-parameter family with total mass as the most important independent variable.

158.037 **Color and metallicity gradients in E and S0 galaxies.** S. E. Strom, K. M. Strom, J. W. Goad, F. J. Vrba, W. Rice.
Astrophys. Journ., Vol. 204, 684 - 693 (1976).

Photometric maps of the galaxies NGC 3115 (S0), NGC 3377 (E5), NGC 3379 (E0), and NGC 4762 (S0) are presented for four filter bandpasses: U, B, V, and K. In each of the galaxies, the color indices $(U–B)$, $(B–V)$, and $(V–K)$ tend to decrease outward from the nucleus. The observed color gradients are greatest along the minor axes of the flattened systems. Color changes in these and other galaxies of like morphological type appear to be dominated by a single parameter which most logically seems to be the metal-to-hydrogen ratio Z. The $(V–K)$ index places rather severe restrictions on the contribution of M dwarf stars to the integrated colors. If such stars are the major low-luminosity constituent of the galactic mass, it is difficult to justify M/L ratios much in excess of 50.

158.038 **Possible collisional enhancement of He I λ5876 in Seyfert galaxies and QSOs.** G. M. MacAlpine.
Astrophys. Journ., Vol. 204, 694 - 698 (1976).

It is demonstrated that the He I λ5876 line may be enhanced by a combination of scattering and collisional processes in some Seyfert galaxies and QSOs. For an atomic density in excess of a few times 10^9 cm^{-3}, the λ10830 photons can be scattered a sufficient number of times for collisional excitation from $2\,^3P$ to become the principal mode for population of $3\,^3D$. The significance of depopulation of the $2\,^3S$ level through absorption of Lα radiation is discussed briefly.

158.039 **Gas motions in the center of the galaxy NGC 253 from H I line interferometry.**
S. T. Gottesman, R. Lucas, L. Weliachew, M. C. H. Wright.
Astrophys. Journ., Vol. 204, 699 - 702 (1976).

An interferometric study of 21 cm H I absorption against the central radio core in NGC 253 has revealed an excess of blue-shifted velocities. This is best interpreted as absorption in neutral gas moving toward the observer and away from the nucleus, in confirmation of optical studies. While all molecular studies show a feature red-shifted relative to the systematic velocity, the present H I study hardly shows any signal at the same velocities.

158.040 **Aperture synthesis of neutral hydrogen in the galaxy M33.**
D. H. Rogstad, M. C. H. Wright, I. A. Lockhart.
Astrophys. Journ., Vol. 204, 703 - 716 (1976).

Observations of neutral hydrogen in the galaxy M33 with resolutions of 2' and 10 km s^{-1} are described. Detailed modeling has revealed weak evidence for spiral density waves with an arm-interarm contrast of 3 to 2 and with a corotation radius in the range 7' to 12'. It has also disclosed substantial evidence for large-scale warping of the outer plane of this galaxy, similar to that found earlier in M83.

158.041 **Galaxy magnitudes. III: Magnitudes of galaxies in the Zwicky and the Shapley-Ames catalogues.**
G. E. Kron, C. D. Shane.
Astrophys. Space Sci., Vol. 39, 401 - 407 (1976).

Systematic corrections to the magnitudes of galaxies in the Zwicky and the Shapley-Ames catalogues have been determined and are tabulated here. The corrections are based on the photographic measures by Holmberg, published photoelectric measures, and 392 photoelectric measures by Kron and Shane.

158.042 **Southern peculiar galaxies – V.** J. L. Sérsic.
Astrophys. Space Sci., Vol. 39, 477 - 485 (1976).

The author gives astrometric, photometric and spectroscopic information on an apparent triplet of galaxies already discussed by the author. The compact member of the trio has a redshift of one order of magnitude larger than the brighter member and could represent an important case for an investigation of an anomalous redshift.

158.043 **A new look at the Hubble diagram.**
K. R. Lang, G. S. Mumford.
Sky Telescope, Vol. 51, 83 - 87 (1976).

158.044 **CSIRO Division of Radiophysics, Australia: A search for radio emission from compact galaxies.**
W. K. Huchtmeier, O. B. Slee.
Mitt. Astron. Ges., No. 38, p. 87 - 95 (1976). – Short report.

158.045 **Low dispersion spectrophotometry of 6 Markarian galaxies.** K. J. Fricke, J. P. Kaufmann.
Mitt. Astron. Ges., No. 38, p. 102 - 104 (1976). – Short report.

158.046 **A re-examination of the radio continuum "halo" in M 31.** R. Wielebinski.
Astron. Astrophys., Vol. 48, 155 - 158 (1976).

The radio continuum emission from M 31 has originally been interpreted as originating in a disc and in an extended halo (corona). The existence of a halo around the galaxy has been questioned in recent years, but M 31 has not been subject to similar scrutinies. In this paper, a recent 408 MHz survey is re-examined together with additional data at 1420 MHz to show that the halo of M 31 is weaker than that of our Galaxy.

158.047 **Near-infrared profile of M31.**
T. Iijima, K. Ito, T. Matsumoto, K. Uyama.
Publ. Astron. Soc. Japan, Vol. 28, 27 - 33 (1976).

Near-infrared profiles of the central region of M31 were observed at wavelengths $1.0\,\mu$m and $2.2\,\mu$m. The flux with respect to the angular distance from the center is well represented by the $r^{-1/4}$ formula for $r \lesssim 10'$, and the effective radius of the spheroidal component decreases as the wavelength increases. This favors the giant-rich stellar synthesis and the equipartition of stellar kinetic energy; namely, massive stars that have evolved to the giant stage exist more abundantly in the central region.

158.048 **Klassifikation und 'Zustandsgrößen' von Galaxien.**
K.-H. Schmidt.
Astronomie in der Schule, 13. Jahrgang, p. 6 - 10 (1976).

158.049 **Stervorming in het merkwaardige radiostelsel NGC 5128.** S. van den Bergh.
Zenit, 3e jaargang, p. 49 - 51 (1976).

158.050 **A four-colour surface photometry of Markarian galaxies. III. Galaxies No. 11, 12, and 13.**
F. Börngen, A. T. Kalloglyan.
Astrofizika, Vol. 11, 617 - 629 (1975). In Russian. English translation in Astrophysics, Vol. 11, No. 4.

The results of UBVR surface photometry of Markarian galaxies 11, 12 and 13 with ultraviolet continuum are given. The integral brightnesses and U−B, B−V and V−R colours of galaxies have been determined.

158.051 **Spectra of galaxies of high surface brightness.**
V. T. Doroshenko, V. Yu. Terebizh.
Astrofizika, Vol. 11, 631 - 635 (1975). In Russian. English translation in Astrophysics, Vol. 11, No. 4.

Spectra of galaxies of high surface brightness from the list compiled by M. A. Arakelyan have been obtained. Forty of the seventy-three galaxies have emission-line spectra. The redshifts and absolute magnitudes of galaxies with emission lines are determined.

158.052 **Occultation of three Markarian galaxies at 327 MHz by the moon.** V. A. Sanamyan, Gopal-Krishna.
Astrofizika, Vol. 11, 637 - 641 (1975). In Russian. English translation in Astrophysics, Vol. 11, No. 4.

The occultation of three Markarian galaxies by the moon were observed at 327 MHz. Two of these, Markarian 369 and 384 were detected at this frequency. The galaxy Markarian 370 was not detected. The observational and the derived radio characteristics for these galaxies are presented.

158.053 **On a possible mechanism of the optical variability of the nuclei of Seyfert galaxies.**
V. A. Hagen-Thorn.
Astrofizika, Vol. 11, 643 - 649 (1975). In Russian. English translation in Astrophysics, Vol. 11, No. 4.

A family of calculated synchrotron spectra with differing critical frequency ν_c represents reasonably well the results of UBV observations of the variable nuclei of Seyfert galaxies NGC 4151 and NGC 1275. It is shown that variations in ν_c are caused by rather small variations of E_{max}.

158.054 **The distribution of galaxies in the Jagellonian field.**
L. M. Fesenko.
Astrofizika, Vol. 11, 651 - 657 (1975). In Russian. English translation in Astrophysics, Vol. 11, No. 4.

A statistical method is applied for determination of the relative number of galaxies belonging to binary and multiple systems. Counts of galaxies in small regions around 12276 galaxies in the Jagellonian field are made. It is found that about 50% of the galaxies belong to visible multiple systems and about 16% of galaxies belong to binary systems. The characteristic radius of the galaxy systems in the Jagellonian field does not exceed the value of $7'.5$. The typical distance between neighbouring members is lower than $5'-6'$.

158.055 **Far-infrared photometry of NGC 1068.**
C. M. Telesco, D. A. Harper, R. F. Loewenstein.
Astrophys. Journ., (Letters), Vol. 203, L53 - L55 (1976).

The authors have detected the far-infrared flux from NGC 1068 in four passbands spanning the range 28–320 μ. Between 38 and 100 μ the flux density increases with increasing wavelength to a maximum value of (454 ± 94) Jy, but clearly decreases rapidly for $\lambda > 100\,\mu$. The total flux is $\sim 3.0 \times 10^{-11}$ W m^{-2} corresponding to a bolometric luminosity of $3.7 \times 10^{11}\,L_\odot$ at 20 Mpc.

158.056 **Optical and infrared spectrophotometry of 18 Markarian galaxies.**
G. Neugebauer, E. E. Becklin, J. B. Oke, L. Searle.
Astrophys. Journ., Vol. 205, 29 - 43 (1976).

Slit spectra, spectrophotometric scans, and infrared broad-band observations are presented of 18 Markarian galaxies with emission lines. Eight of the program galaxies can be classified as Seyfert galaxies. Arguments are given that thermal, nonthermal, and stellar radiation components are present. The 10 galaxies which are not Seyfert galaxies are shown to be examples of extragalactic H II regions; there is evidence for thermal emission from dust being present at 10 μ in four of these galaxies.

158.057 **The origin of ultraviolet and infrared continuum radiation from Seyfert galaxies.**
W. A. Stein, D. W. Weedman.
Astrophys. Journ., Vol. 205, 44 - 51 (1976).

Observations at 3.5 μ are presented for 39 Seyfert galaxies and are used with existing photometric and spectroscopic data to discuss the continuum radiation from the galaxies. It is concluded that there is an unreddened, nonthermal continuum from class 1 Seyfert galaxies. Limits on the bolometric luminosities of these objects are derived.

158.058 **A dynamical study of NGC 4027.**
J. H. Christiansen, W. H. Jefferys.
Astrophys. Journ., Vol. 205, 52 - 62 (1976).

A model of NGC 4027 incorporating an asymmetric mass

distribution is used in an attempt to explain that galaxy's peculiar rotation curve. A theoretical rotation curve synthesized from stable periodic and quasi-periodic orbits in the potential of the model is found to be in very close agreement with the observed rotation curve. The key to the peculiarities of the observed curve appears to be the presence of a stable equilibrium point in the potential field. The model is also used to simulate other late-type barred spirals by altering the dimensions, mass, and bar rotation rate within the constraints of the dimensionless numbers in the model potential. It appears to represent other asymmetric barred spirals accurately, including NGC 4631.

158.059 **Velocity dispersions in galaxies. V. The nuclei of M31 and M32.** D. C. Morton, B. G. Elmergreen.
Astrophys. Journ., Vol. 205, 63 - 73 (1976).

Stigmatic spectra between 4160 and 4385 Å with 0.7 Å resolution have been obtained of the central regions of M31 and M32. Line-of-sight velocity dispersions of $\sigma = 130 \pm 20$ and 55 $(+10, -15)$ km s^{-1} have been determined for the nuclei of M31 and M32, respectively.

158.060 **The giant spiral galaxy M101: IV. Observations of variable continuum radio emission from supernova 1970g and measurements of the continuum radio structure of the giant H II complex NGC 5455.**
R. J. Allen, W. M. Goss, R. D. Ekers, A. G. de Bruyn.
Astron. Astrophys., Vol. 48, 253 - 261 (1976).

Over the past few years a series of radio observations at $\lambda\lambda$ 2.8, 6.0, 21 and 49.2 cm has been made on the supernova 1970g and on the nearby giant H II complex NGC 5455 in the galaxy M101. The following results have now been obtained: (1) NGC 5455 contributes a non-variable flux density of 7.1 ± 0.6 mJy at λ 21 cm. The radio source is thermal, and coincides both in position and general shape with the optical object. (2) The optical position of the supernova has been redetermined to an accuracy better than 0.″5. (3) In December 1971, variable radio emission was observed near NGC 5455 at a level of 5.6 ± 0.7 mJy at λ 21 cm. The position of this variable source coincides with that of supernova 1970g. (4) A nonthermal radiation mechanism is suggested by the present data.

158.061 **The blue compact galaxy CG 1116 + 51.**
E. M. Jones.
Astron. Astrophys., Vol. 48, 317 (1976).

The gas in CG 1116 + 51 is expected to be compressed to about 10^4 protons cm^{-3} because 94 percent of the volume is comprised of Cox-Smith tunnels.

158.062 **Another possible nearby galaxy.**
R. Weinberger, H. Elsässer, M. Beetz, K. Birkle.
Astron. Astrophys., Vol. 48, 327 - 329 (1976).

Image tube photographs in the I range (0.92 µm) led to the detection of two extended infrared objects in the area of S 119. Object 1 resembles the galaxy Maffei 1. The nature of Object 2 is less clear.

158.063 **Spatial and velocity structure of NGC 7027.**
T. R. Gull, J. Percival.
Bull. American Astron. Soc., Vol. 8, 289 (1976). – Abstr. AAS.

158.064 **Spectrophotometric and photometric observations of two Seyfert galaxies.**
N. J. Woolf, P. A. Strittmatter, G. Rieke, E. A. Beaver.
Bull. American Astron. Soc., Vol. 8, 290 (1976). – Abstr. AAS.

158.065 **A polarimetric study of four Seyfert galaxies.**
J. R. P. Angel, N. J. Woolf, P. G. Martin, E. A. Beaver.
Bull. American Astron. Soc., Vol. 8, 290 (1976). – Abstr. AAS.

158.066 **Tidally interacting galaxies: a model for NGC 5194 with initial spiral structure.** B. S. Crowley.
Bull. American Astron. Soc., Vol. 8, 290 (1976). – Abstr. AAS.

158.067 **Magnetized accretion disks and the radio outbursts of 3C 120 and Cygnus X-3.**
G. A. Shields, J. C. Wheeler.
Astrophys. Letters, Vol. 17, 69 - 76 (1976).

The authors consider models for the radio outbursts of 3C 120 and Cyg X-3 involving accretion disks around black holes. An accretion disk carrying $\sim 0.2\,M_\odot$ yr^{-1} into a $10^{8-9}\,M_\odot$ black hole can store 10^{52} erg of magnetic energy, the amount seen in an outburst in 3C 120, at a radius $\sim 10^{15-16}$ cm. In contrast, an accretion disk around a $\sim 1\,M_\odot$ object in Cyg X-3 cannot store the $\sim 10^{43}$ erg released in the radio outbursts of this object. However, an accretion disk in Cyg X-3 can continuously generate magnetic field rapidly enough to energize the radio outbursts in less than the observed rise time $\sim 10^{5-6}$ s. This requires a 'real time' mechanism for particle acceleration and magnetic field ejection during an active phase at the beginning of the outburst. The authors discuss the particle energies under the assumption of 'real time' acceleration in both 3C 120 and Cyg X-3. Simple considerations of magnetic field reconnection indicate that, for both objects, efficient particle acceleration can occur in the inner region of the disks.

158.068 **UBV photometry of the galaxy NGC 3077.**
B. P. Artamonov, F. Börngen, A. I. Shapovalova.
Astrofiz. Issled., Izv. Spets. Astrofiz. Obs., Vol. 8, 41 - 46 (1976). In Russian.

Results of photographic UBV photometry of the irregular galaxy NGC 3077 from plates obtained by the observers of Tautenburg Observatory with the Schmidt telescope, and those of UBV photoelectric measurements made with the "Zeiss-600" telescope of USSR Academy of Sciences are presented. Distributions of surface brightness and B−V, U−B color indices are obtained along the main axes (photographic measurements) and the radius (photoelectric measurements) of the galaxy. A model is suggested according to which the principal radiation in its central part is due to reddened B stars. The similarity of NGC 3077 and M 82 is pointed out.

158.069 **On a possible classification of the galaxies NGC 4753 and NGC 5363.** A. I. Shapovalova.
Astrofiz. Issled., Izv. Spets. Astrofiz. Obs., Vol. 8, 47 - 52 (1976). In Russian.

An additional analysis of the UBV surface photometry results obtained previously by the author for the galaxies NGC 4753 and 5363 is made. It is noted that these galaxies have been erroneously regarded as belonging to type M 82. Arguments are presented in favour of NGC 4753 to be possibly attributed to type S0−Sa, and NGC 5363 to E type.

158.070 **The mass of M81.** B. Basu.
Bull. Astron. Soc. India, Vol. 3, 31 (1975).
Abstract of a paper presented at the A.S.I. meeting 1975.

158.071 **The peculiar spiral galaxy NGC 3310. I. General properties; far ultraviolet and radio continuum observations.** P. C. van der Kruit, A. G. de Bruyn.
Astron. Astrophys., Vol. 48, 373 - 382 (1976).

This paper is the first in a series on the peculiar spiral galaxy NGC 3310 (Arp 217). The authors give a discussion of the general properties, present far-UV observations made with the Astronomical Netherlands Satellite and radio continuum observations at $\lambda\lambda$ 49, 21 and 6 cm made with the Westerbork Synthesis Radio Telescope. In the inner part of the disk there is a symmetrical, open spiral structure with many bright H II regions. The high luminosity in optical emission lines, early spectral type and the optical colours indicate a relatively high content of early-type stars. This is confirmed by observations in the far UV. NGC 3310 is possibly a member of a loose group of galaxies. The radio emission has a spectral index of -0.69 between 0.6 and 6.6 GHz.

158.072 High resolution studies of spiral and irregular galaxies at 2695 and 8085 MHz I: Maffei 2.
E. R. Seaquist, J. Pfund, R. C. Bignell.
Astron. Astrophys., Vol. 48, 413 - 419 (1976).

The authors present aperture synthesis observations of the late type spiral Maffei 2 at 2695 and 8085 MHz. The disk and nuclear components of the radio emission as well as their relation to the optical features of the galaxy are discussed. The compact non-thermal radio source at the nucleus of Maffei 2 is resolved, and the structure revealed provides evidence for an explosive event at or near the nucleus.

158.073 High resolution studies of spiral and irregular galaxies at 2695 and 8085 MHz II: NGC 1569 and NGC 891.
E. R. Seaquist, R. C. Bignell.
Astron. Astrophys., Vol. 48, 421 - 435 (1976).

The authors present aperture synthesis observations of the irregular galaxy NGC 1569 and the edge-on spiral NGC 891 at 2695 and 8085 MHz. In both cases the radio emission is similar to the optical form of the galaxy, and contains a mixture of thermal and nonthermal components. Furthermore, the brightest source in each galaxy is nonthermal, and is probably synchrotron radiation. The radiation detected from NGC 1569 is predominantly thermal emission from large H II regions some of which are also evident in Hα. The luminosity of the nucleus and the disk emissivity for NGC 891 are larger than that of typical spiral galaxies, but are consistent with van der Kruit's (1973) relation between these two quantities.

158.074 Peculiar motions in Markarian 297.
R. Duflot, J. Lombard, Y. Perrin.
Astron. Astrophys., Vol. 48, 437 - 442 (1976). In French.

A detailed spectroscopic study of Markarian 297 reveals the motions of two distinct objects, and suggests that two recently formed galaxies are involved. The authors determine a total mass $M \gtreqless 2 \times 10^{10} M_{\odot}$ and a kinematic age of 8×10^6 years. The objects observed seem to be superassociations connected with H II regions.

158.075 *UBV* photometry of bright southern galaxies.
M. J. Bucknell, J. V. Peach.
Observatory, Vol. 96, 61 - 64 (1976).

158.076 The log N–log S curve for 3CR radio galaxies and the problem of identifying faint radio galaxies.
G. R. Burbidge, J. V. Narlikar.
Astrophys. Journ., Vol. 205, 329 - 334 (1976).

The authors have shown that the identified galaxies in the 3CR catalog give a log N–log S curve that is Euclidean or sub-Euclidean and is explicable without evolution. If the unidentified sources which steepen the log N–log S curve at large redshifts so that evolution can be invoked, they are so distant that there is little or no chance of proving this observationally in the foreseeable future. Thus, the concept of evolution, at least for radio galaxies, which has been so frequently invoked still remains very difficult to establish observationally.

158.077 Explosions in galactic nuclei and the formation of double radio sources.   R. H. Sanders.
Astrophys. Journ., Vol. 205, 335 - 345 (1976).

Numerical hydrodynamical experiments have been carried out which attempt to define the conditions under which an initially isotropic explosion occurring at the center of a gaseous disk presumed to be in the nucleus of a giant elliptical galaxy will be directed along the rotational axis of the disk by the disk geometry itself. A very general condition for focusing an explosion is that the mass isotropically liberated by the explosion be less than the mass of the ambient disk medium within one vertical scale height of the explosion center. More-over, it is shown that the degree of focusing necessary to account for double radio sources can occur only if the disk is initially in equilibrium with a highly centrally condensed gravitational field.

158.078 Radial velocities and masses of galaxies in groups from 21-centimeter line observations.
H. J. Rood, J. R. Dickel.
Astrophys. Journ., Vol. 205, 346 - 353 (1976).

New 21-cm radial velocities are given for 50 galaxies. Conventional Bottlinger-Lohmann total masses are derived for these galaxies from the width of 21-cm line profiles. The new 21-cm radial velocities are compared with values tabulated in the Reference Catalogue of Bright Galaxies. The 21-cm velocities and galactic masses are used to estimate directly M_{vt} (virial mass) and M (sum of the conventional masses of component galaxies) for six de Vaucouleurs groups.

158.079 Corrected ratios of average mass to average luminosity for double galaxies.   H. J. Rood.
Astrophys. Journ., Vol. 205, 354 - 355 (1976).

Average virial masses of double galaxies are derived with proper allowance for the case where the observed relative velocity of a double galaxy is smaller than its mean error. Revised estimates of the ratio of average mass to average luminosity are given for subsets of Page's sample of double galaxies, supergiant doubles, and compact doubles.

158.080 The H and K lines of Ca II in the nucleus and bulge of M31.   D. C. Morton, C. D. Andereck.
Astrophys. Journ., Vol. 205, 356 - 359 (1976).

A spectrum 34″ wide along the major axis of the nucleus and bulge of M31 has been obtained with about 1 Å resolution in the region of the H and K lines. Four unresolved absorption lines appear superposed on the broad stellar K line, and at least the strongest one is present at the H line. Measurement of the equivalent widths of the stellar H and K lines showed that they increase from the bulge to the nucleus by about 40 percent.

158.081 Color gradients in the nuclear region of M31.
T. X. Thuan, J. B. Oke.
Astrophys. Journ., Vol. 48, 360 - 362 (1976).

Spectrophotometric measurements of the nuclear region of M31 show that the nuclear region is only slightly redder and shows marginally stronger CN blends than the bulge. The data can be interpreted in terms of an increase in heavy-element abundance toward the center of M31 by a factor substantially less than 2.

158.082 Expected rate of transient events from stellar deaths in other galaxies.   R. J. Talbot, Jr.
Astrophys. Journ., Vol. 205, 535 - 540 (1976).

The expected rates for γ-ray, microwave radio, neutrino-antineutrino, or gravitational wave bursts are discussed under the assumption that they are produced with rate proportional to the rate of observed optical supernovae (from Tammann). The γ-ray bursts from the Vela satellites are discussed. The frequency and distribution in the sky are shown to be consistent with their being extragalactic.

158.083 Variability of the emission line spectrum and physical conditions in the gaseous envelope of the nucleus of the Seyfert galaxy NGC 1275.   I. I. Pronik.
Astron. Zhurn. Akad. Nauk SSSR, Vol. 53, 251 - 259 (1976). In Russian. English translation in Soviet Astron., Vol. 20, No. 2.

The physical conditions of the gaseous envelope of the nucleus of the Seyfert galaxy NGC 1275 are discussed on the basis of relative intensities of emission lines. It is shown that the relative intensity variations of H II, [O III], [O II], [S II], [N II] and [O I] emission lines take place with characteristic time of about half a year. These variations may be caused by electron temperature variations within the zones emitting the lines. Data used for discussion show that the structure of the zones under consideration have not been changed essentially during the last 30 years.

158.084 The problem of spiral galaxies and satellite radio sources.
H. Arp, R. Carpenter, S. Gulkis, M. Klein.
Astrophys. Journ., Vol. 205, 721 - 727 (1976).

Regions $2° \times 2°$ in area centered on four spiral galaxies (Shapley-Ames catalog) were scanned with the Goldstone 64 m antenna in search of satellite radio sources at 2295 MHz. A detailed comparison is made between the results of this program and the results of previous investigators. The authors measured 15 sources selected from Tovmasyan's list of 43 satellite sources. Their results confirm his positions and relative flux densities for each of the sources.

158.085 H II regions in NGC 628. I. Positions and sizes.
P. W. Hodge.
Astrophys. Journ., Vol. 205, 728 - 744 (1976).

An identification atlas of 730 H II regions in NGC 628 is presented, and positions in rectangular coordinates are given. Comparison is made with previous Hα studies of this galaxy and objects in common are cross-cataloged. Diameters are given for all H II regions, and the size-frequency relation is compared with those published for NGC 2403 and M33.

158.086 Primeval galaxies: predicted luminosities.
M. Kaufman.
Astrophys. Space Sci., Vol. 40, 369 - 384 (1976).

Absorption by gas and dust in circumstellar H II regions within primeval galaxies could seriously depress the far ultraviolet continuum radiation emitted by primeval galaxies. An appropriate spectral region to search for the redshifted integrated background from primeval galaxies lies between 350μ, where the 2.7 K microwave background radiation becomes important, and 150μ, where other extragalactic discrete sources, such as nearby galactic nuclei, may contribute. The expected IR flux is calculated with Kaufman's (1975) model for the star formation rate in the contracting galaxy.

158.087 The rotation curve of NGC 4096.
R. Barbon, M. Capaccioli.
Astron. Astrophys., Vol. 49, 125 - 127 (1976).

The rotation curve of the Sc galaxy NGC 4096 out to $100''$ from the nucleus has been obtained by measuring the Hα line on three 125 Å/mm spectra. A mass of $8.7 \times 10^9\ M_\odot$ and a lower limit for the mass-to-light ratio of 1.1 have been computed for a distance of 13 Mpc. This latter value has been checked using the light curve of the 1960h type I supernova.

158.088 New redshifts of bright galaxies. II.
A. de Vaucouleurs, R. R. Shobbrook, A. Strobel.
Astron. Journ., Vol. 81, 219 - 221 (1976).

Redshifts of 14 galaxies have been measured. Seven are new determinations for galaxies in nearby groups and clusters. Group membership is discussed. The derived velocities have mean error of ~ 50 km sec^{-1} and no significant systematic difference is found with published values.

158.089 Galactic mass determinations from incomplete rotation curves.
M. M. Schaefer, G. Rybicki, M. Lecar.
Astrophys. Space Sci., Vol. 41, 3 - 14 (1976).

Given an incomplete rotation curve of a spiral galaxy, various assumptions about the Galaxy beyond the last observed point are made: (1) the force falls off as $1/r^2$, (2) the mass density is zero, and (3) the mass density falls off as $1/r^3$. The mass distributions obtained from each of these assumptions are all well behaved, and it is impossible to choose the correct curve from considerations of the resulting mass distributions alone. The correct mass distribution in the disk system of a galaxy cannot be deduced from an incomplete rotation curve.

158.090 On nuclei of galaxies and their activity.
V. A. Ambartsumyan.
Problems of gravitation. Third Soviet Gravitational Conference, Erevan, 1972, (see 012.008), p. 5 - 17 (1975). In Russian.

158.091 On the gaseous structure of the nuclei of Seyfert galaxies.
V. I. Pronik.
Izv. Krymskoj Astrofiz. Obs., Vol. 54, 165 - 170 (1976). In Russian.

It is supposed that a dense gas in Seyfert nuclei, responsible for emitting broad wings of hydrogen lines, is a stable formation not connected with large massive [O III] and [O II] zones. The attempt is made to show that the origin of such zones cannot be due neither to a long outflow of gas nor to numerous explosions and ejections of gas clouds with velocity of about 3000–5000 km/sec.

158.092 Investigation of bright formations in spiral branches of NGC 2903.
N. B. Grigor'eva.
Izv. Krymskoj Astrofiz. Obs., Vol. 54, 171 - 175 (1976). In Russian.

A photometric study of NGC 2903 spiral branches has been carried out to detect H II complexes. Twenty-eight bright details at distances between 1.22 and 5.0 kpc from the galaxy center have been found. The energy distribution in the continuum of 25 spiral arm details has been obtained. Results are given in a table.

158.093 Discrete states of redshift and galaxy dynamics. I. Internal motions in single galaxies.
W. G. Tifft.
Astrophys. Journ., Vol. 206, 38 - 56 (1976).

This is the first of a series of papers developing the concept that the redshift can occur only in specific discrete values. The key to the development is a dual redshift model for individual galaxies. Well-known local galaxies, especially M31 in great detail, are shown to consist of two basic opposed streams of outflowing material which have an intrinsic difference of redshift of 70–75 km s^{-1}. A smooth symmetrical rotation and expansion curve coupled with the multiple redshift model is sufficient to account for all the redshift data. Where definite differences exist in the form of the data as predicted by conventional dynamics and the discrete redshift concept, the data favor the latter.

158.094 The origin of optical polarization in NGC 1068.
J. R. P. Angel, H. S. Stockman, N. J. Woolf, E. A. Beaver, P. G. Martin.
Astrophys. Journ., (Letters), Vol. 206, L5 - L9 (1976).

The polarization of emission lines in the nucleus of NGC 1068 has been measured. It is found that the permitted lines of H and probably also He II are polarized at nearly the same position angle and by the same amount as the neighboring continuum. This argues strongly that a common dust scattering mechanism is responsible for all the polarization. The continuum radiation has been found to be circularly polarized with ellipticity of ~ 5 percent in the red. Such high ellipticity is very unlikely to be of nonthermal origin. The authors take it as evidence that the nuclear dust is in the form of clouds in an asymmetric skew geometry, the polarization then arising from multiple scattering within these clouds.

158.095 Density waves in the disks of two spiral galaxies.

S. E. Strom, E. B. Jensen, K. M. Strom.
Astrophys. Journ., (*Letters*), Vol. 206, L11 - L14 (1976).

The disk galaxies NGC 495 (SBb) and NGC 1268 (SAb-c) both exhibit extremely smooth spiral arms. The ($U-R$) colors in the arm and disk regions are identical to within 0.10 mag. It is suggested that in these galaxies the spiral pattern represents the crests of a density wave in the stellar disk. The observed peak-to-peak amplitude of the density wave for both galaxies is approximately $\pm^1/_3$ of the background disk light. This amplitude is comparable with the largest values observed by Schweizer for late-type spiral systems and is somewhat greater than current theoretical estimates of amplitudes derived from observational data by the use of linear density-wave theory.

158.096 The sizes of the nuclei of galaxies at 10 microns.
    G. H. Rieke.
Astrophys. Journ., (*Letters*), Vol. 206, L15 - L17 (1976).

The nuclei of NGC 2903, 3504, 4536, 5195, and 5236 have diameters of 150–600 pc at 10μ. Although the sizes and shapes of the infrared sources correspond roughly to complexes of hot stars and ionized gas, the absence of strong thermal radio sources indicates that the infrared fluxes are not generated by normal low-density H II regions.

158.097 Radial velocities of southern galaxies.
    W. L. Martin.
Monthly Notices Roy. Astron. Soc., Vol. 175, 633 - 643 (1976).

A combined photographic and spectroscopic study has been made of 112 southern galaxies. Radial velocities of the galaxies have been determined. From the direct photographs a type has been derived which describes the compactness of each galaxy nucleus. The range and distribution (for the star-like and semi-star-like nuclei) of excitation is similar to that of (sharp line emission) Zwicky and Markarian compact galaxies.

158.098 The number-diameter relation of galaxies.
    G. M. Richter.
Astron. Nachr., Vol. 297, 145 - 146 (1976).

Contrary to the number-magnitude relation, the number-angular diameter relation is influenced by the luminosity function. The number-diameter counts by Dodd et al. (1975) are in agreement with the normal luminosity function and a homogeneous space distribution without a local density excess. Almost all the faint small (< 18 arcsec) galaxies may be first-ranked cluster members.

158.099 Some remarks on compact groups of compact galaxies.
V. A. Ambartsumyan, H. C. Arp, A. A. Hoag, L. V. Mirzoyan.
Astrofizika, Vol. 11, 193 - 206 (1975).

Questions connected with the study of groups of compact galaxies discovered at the Byurakan Astrophysical Observatory are discussed. The definition of compactness and sufficient conditions of compactness of faint galaxies are given. It is shown that all galaxies fainter than $17.^m5$ and brighter than $18.^m5$ on the red prints of the Palomar Sky Survey having saturated images are compact. 12 groups of compact galaxies are described from direct photographs. Reproductions of these photographs are presented. New data of observations confirm the real existence of the new-type systems of galaxies named "compact groups of compact galaxies".

158.100 On one form of activity in galaxies.
    K. A. Saakyan, Eh. E. Khachikyan.
Astrofizika, Vol. 11, 207 - 220 (1975). In Russian.
English translation in Astrophysics, Vol. 11, No. 2.

On the basis of surveying of plates obtained with the 21"-Schmidt camera of the Byurakan Observatory, and examination of the Palomar Sky Survey it has been shown that among Markarian galaxies objects occur which represent superassociations physically connected with nearby galaxies. Typical representatives of these types of objects are Markarian 59, 71, 94. A list of Markarian galaxies showing characteristics of superassociations, and brief descriptions of them are presented. About 40 objects of this type are available among the first six lists of Markarian galaxies. Some of the Markarian galaxies contain blue objects like superassociations. Different forms of superassociations are discussed. It has been concluded that one of the forms of activity in galaxies is the formation of superassociations.

158.101 Compact groups of compact galaxies. V.
    F. W. Baier, H. Tiersch.
Astrofizika, Vol. 11, 221 - 227 (1975). In Russian.
English translation in Astrophysics, Vol. 11, No. 2.

The fifth list of compact groups of compact galaxies is presented. The list contains data on 50 new objects of this class. Identification charts for all 50 groups of the list are given.

158.102 Pairs of a Markarian and a compact galaxy.
    J. Heidmann, A. T. Kalloglyan.
Astrofizika, Vol. 11, 229 - 236 (1975).

The existence of close physical pairs consisting of a Markarian and a compact Zwicky galaxy has been statistically shown. A list of 18 such pairs is given. The mean separation of their components (67 kpc) is much smaller than the sizes of groups of galaxies. Four of the examined pairs may be gravitationally bound systems.

158.103 On the origin of galaxy rotation in Ambartsumyan's cosmogony.   R. M. Muradyan.
Astrofizika, Vol. 11, 237 - 248 (1975). In Russian.
English translation in Astrophysics, Vol. 11, No. 2.

The origin of angular momentum of galaxies and of clusters of them is attributed to conservation of the spin of supermassive hadrons in the decay of which galaxies are formed according to the ideas of Ambartsumyan's superdense cosmogony.

158.104 On the problem of fragmentation of galaxies.
    B. A. Vorontsov-Vel'yaminov.
Astrofizika, Vol. 11, 355 - 356 (1975). In Russian.
English translation in Astrophysics, Vol. 11, No. 2.

Examples of photographs of interaction of compact galaxies are given certifying their fragmentation and the decay of groups.

158.105 The Sculptor-type dwarf galaxies in the vicinity of bright galaxies.
F. Börngen, V. E. Karachentseva, I. P. Kostyuk.
Astrofizika, Vol. 11, 358 - 362 (1975). In Russian.
English translation in Astrophysics, Vol. 11, No. 2.

The results of photometry of three dwarf galaxies in the M81 group show that the objects 61 and 64 from Karachentseva's list are probably Sculptor-type dwarf galaxies.

158.106 On a characteristic of spirals with starlike nuclei.
    S. G. Iskudaryan.
Astrofizika, Vol. 11, 362 - 365 (1975). In Russian.
English translation in Astrophysics, Vol. 11, No. 2.

There is an approximate linear correlation between $U-B$ colours of starlike nuclei and integral $U-B$ colours of the surrounding parts of galaxies, which, apparently, indicates long-livedness of the starlike nuclei.

158.107 Spectrophotometric observations of N galaxies at large redshift: PKS 0353+027, 3C 99, 3C 467.
H. E. Smith, H. Spinrad, R. Hunstead.
Astrophys. Journ., Vol. 206, 345 - 354 (1976).

Spectrophotometric observations are presented for three N galaxies identified with the radio sources PKS 0353+027, 3C 99 (PKS 0358+00), and 3C 467 (PKS 2345+18). The systems are spectroscopically similar, showing high excitation and relatively rich emission-line spectra with redshifts of 0.602, 0.426, and 0.631, respectively. The observations are consistent with a composite model for these systems, consisting of a strongly concentrated nonthermal source plus a "normal" stellar component. The authors note that the emission-line strengths for this sample of N galaxies are consistent with models of a gas ionized by the observed nonthermal continuum extrapolated to ionizing energies.

158.108 The distant N galaxy 3C 318.
H. Spinrad, H. E. Smith.
Astrophys. Journ., Vol. 206, 355 - 358 (1976).

The authors have confirmed Wyndham and Véron's suggested optical identification of the compact radio source 3C 318 with a faint red N galaxy. A pair of faint interacting galaxies nearby may be associated with the compact galaxy. The sum of 12 spectroscopic observations of 3C 318 shows two emission features, which the authors identify as Mg II $\lambda 2799$ and [O II] $\lambda 3727$ at $z = 0.752$. A two-component model for 3C 318 is briefly discussed.

158.109 Galaxy spectral synthesis. I. Stellar populations in the nuclei of giant ellipticals. R. W. O'Connell.
Astrophys. Journ., Vol. 206, 370 - 390 (1976).

The author develops an automatic spectral synthesis technique based on linear programming and applies it to narrow-band spectrophotometry covering $\lambda\lambda 3300-10800$ for the nuclei of M31 and three giant elliptical galaxies in Virgo. He describes the data reduction, the general character of the observed energy distributions, the fitting method, and the astrophysical constraints which are imposed on the solutions. Among the results he is particularly interested in examining the history of star formation, the M star population, the chemical composition, the form of the main-sequence luminosity function, and the nature of the interstellar medium in these nuclei and possible differences in these properties between the ellipticals and M31.

158.110 The peculiar spiral galaxy NGC 3310: II. The velocity field of the ionized gas. P. C. van der Kruit.
Astron. Astrophys., Vol. 49, 161 - 171 (1976).

In this paper the velocity field in the inner disk of NGC 3310 (Arp 217), derived from extensive optical spectroscopy, is analysed. Using a purely kinematical approach, it is found that, depending on which side is the nearer, gas is streaming in (if the spiral arms are trailing) or out (leading spiral structure) along the arms, while the regions in between the arms show circular rotation. The velocity field has also been compared to that of a kinematical density wave, which is shown to be also capable of reproducing the observations. It is tentatively concluded that either a very strong density wave, or something approaching it, operates in NGC 3310.

158.111 Abundances and physical conditions in the nucleus of the peculiar galaxy NGC 3310.
Y. Andrillat, S. Collin-Souffrin.
Astron. Astrophys., Vol. 49, 251 - 257 (1976)

The authors present a spectroscopic analysis of the nuclear region ($r \leqslant 7''$) of the peculiar spiral galaxy NGC 3310; the wavelength range covered is 3100 to 8500 Å. The physical conditions and chemical abundances are discussed. A low temperature ($5000 < T < 7000$ K) and a high value of the ratio $A(O)/A(N)$ (~ 30) are obtained. A stellar population analysis using absorption lines gives a large fraction of O and B stars, which explains the intense ultraviolet continuum. It is also possible to identify this contribution (40% at 3500 Å) with a non thermal emission.

158.112 The spectrum of IR radiation from dust clouds.
D. Bollea, A. Cavaliere.
Astron. Astrophys., Vol. 49, 313 - 319 (1976).

An analysis is made of the IR spectrum reradiated by a dust cloud absorbing a primary UV radiation. General features of the spectral distribution are recognized, and an analytical relation is derived for the turn-over frequency, whose validity is confirmed by extensive numerical calculations. The results are discussed in relation to IR emission from galactic nuclei.

158.113 Morphological study of Markarian galaxies in pairs. II. Morphological data. C. Casini, J. Heidmann.
Astron. Astrophys., Suppl. Ser., Vol. 24, 473 - 493 (1976). Paper presented at the third European Astronomical Meeting, Tbilisi, 1975.

The authors present the morphological data for 8 pairs of galaxies containing Markarian galaxies, based on large-scale electronographs or photographs. Isodensity tracings are given for 16 of the galaxies involved.

158.114 Andromedanevel: onze extragalactische buur (1). M. Drummen.
Zenit, 3e jaargang, p. 204 - 209 (1976).

158.115 Explosions of young galaxies and the hidden mass problem. A. A. Suchkov, Yu. A. Shekinov.
Astron. Tsirk., No. 894, p. 1 - 2 (1975). In Russian.

158.116 Optical variability of the nucleus of the Seyfert galaxy MCG 0-14-18 (Arakelian 120).
V. M. Lyutyj.
Astron. Tsirk., No. 902, p. 1 - 4 (1976). In Russian.

158.117 Optical variability of Markarian 509.
K. A. Saakyan.
Astron. Tsirk., No. 902, p. 4 - 5 (1976). In Russian.

158.118 Rotation and mass of NGC 6015. N. Carozzi.
Astron. Astrophys., Vol. 49, 425 - 429 (1976). In French.

Image-tube spectra of the galaxy NGC 6015 have been obtained at a dispersion of 35 Å mm^{-1} in the spectral range 6450 Å –6800 Å. Hα and [N II] 6583 Å emission lines were observed, which gave a rotation curve up to $r = 90''$. The rotation curve has been analyzed to obtain the mass of the galaxy.

158.119 Rotation and mass of NGC 6207. N. Carozzi.
Astron. Astrophys., Vol. 49, 431 - 435 (1976). In French.

The galaxy NGC 6207 has been investigated photographically and spectroscopically. A photograph shows the existence of a nucleus. Two image-tube spectra taken with a nebular spectrograph allow the author to compute the mass of the galaxy. The values found for the mass and the mass to luminosity ratio are low for an Sc galaxy. The author concludes that NGC 6207 is a later type galaxy, Sm rather than Sc.

158.120 A new classification system for galaxies.
S. van den Bergh.
Astrophys. Journ., Vol. 206, 883 - 887 (1976).

(1) A new galaxy classification system is proposed in which normal spirals and lenticulars form parallel sequences within which "early" and "late" systems are distinguished by means of their disk-to-bulge ratios. (2) A sequence of "anemic spirals," which occur most frequently in rich clusters, is found to have characteristics that are intermediate between those of vigorous gas-rich normal spirals and gas-poor systems of type S0. (3) The differences between normal spirals (Sa-Sb-Sc), anemic spirals (Aa-Ab-Ac), and lenticulars (S0a-S0b-S0c) are

tentatively interpreted in terms of the influence of environment on the evolution of flattened galaxies.

158.121 A reinvestigation of the scattering halo of M82 based on polarimetric and isophotal maps.
G. D. Schmidt, J. R. P. Angel, R. H. Cromwell.
Astrophys. Journ., Vol. 206, 888 - 897 (1976).

Maps of the intensity and polarization in the continuum of M82 have been derived from image-tube photographs made with a Polaroid filter. These reveal that the major portion of the halo light can be assigned to a smooth component which is symmetric about the galactic disk, and only a small fraction (~5%) to filamentary structure. The smooth variation in polarization across the halo and the distribution of position angles confirm that the halo radiation is scattered light which originates primarily in the galactic disk and not a bright nucleus.

158.122 Radio galaxies. P. A. G. Scheuer.
14th Intern. Cosmic Ray Conf., (see 012.011), Vol. 11, 3636 - 3642 (1975).

158.123 Surface color photometry of five barred spiral galaxies. S. Okamura, B. Takase.
Astrophys. Space Sci., Vol. 41, 275 - 285 (1976).

Distributions of the surface brightness and the surface color of five barred spiral galaxies expressed in the form of digital maps are presented. This is the first step to determine the composition of the components of barred spiral galaxies — bar, spiral arm, inner ring and outer ring — and to obtain an accurate picture of the dynamical model of a barred spiral galaxy. The authors have found that (a) the bar is redder than the spiral arm and has a color similar to that of the disk and (b) the inner ring of the SB(r) type galaxy is bluer than the bar and rather resembles the spiral arm.

158.124 The maximum and minimum masses of galaxies.
M. Clutton-Brock.
Astrophys. Space Sci., Vol. 41, L9 - L11 (1976).

Galaxies may have formed by fragmentation in a collapsing cloud of very large mass. The most massive galaxies were formed from fragments which were nearly but not quite opaque: the least massive galaxies were formed from fragments about as large as the Jeans mass. If the maximum mass of galaxies is $\sim 10^{13} M_\odot$, then the minimum mass should be $\sim 10^6 M_\odot$.

158.125 Bright phase in the evolution of galaxies and ionization of intergalactic gas.
L. M. Ozernoj, V. V. Chernomordik.
Astron. Zhurn. Akad. Nauk SSSR, Vol. 53, 459 - 474 (1976). In Russian. English translation in Soviet Astron., Vol. 20, No. 3.

The conditions under which the ionization of intergalactic gas with large redshifts is possible by means of the thermal radiation of young galaxies passing through the bright phase of their evolution are investigated.

158.126 On the possibility of the existence of black holes in the centres of galaxies.
G. S. Bisnovatyj-Kogan, S. I. Blinnikov.
Astron. Zhurn. Akad. Nauk SSSR, Vol. 53, 485 - 487 (1976). In Russian. English tranlation in Soviet Astron., Vol. 20, No. 3.

It is argued that the formation of a black hole in the centre of M31 is impossible because of rapid rotation in the uniformly rotating model of a galactic nucleus with an initially smooth density distribution. The upper mass limit of the central black hole in M31 based on the surface brightness observations is $\sim 10^6 M_\odot$ and does not depend on the model.

158.127 Observations de 3C 120.
L. A. Ourassine (Urasin), I. A. Ourassina (Urasina).
Ann. Obs. Astron. Alger, Vol. 4, Fasc. 2, p. 17 - 18 (1975).

158.128 Variations à courte période de 3C 390.3.
L. A. Ourassine (Urasin), I. A. Ourassina (Urasina).
Ann. Obs. Astron. Alger, Vol. 4, Fasc. 2, p. 19 - 22 (1975).

158.129 Classification and stellar content of galaxies obtained from direct photography. A. Sandage.
Galaxies and the universe, (see 003.010), p. 1 - 35 (1975).

Early classification of galaxies; Development of the modern system (Early isolation of the types, the early Hubble system, Hubble's major modification between 1936 and 1950, finer subdivision along the sequence); Revision by de Vaucouleurs (Extension of the sequence beyond Sc, transitions between ordinary and barred spirals, the r and s varieties, graphical representation of the classification, additional features of the de Vaucouleurs revision); Selected illustrations of galaxy types; Van den Bergh's classification; Morgan's classification based on the luminosity concentration of the spheroidal component; System of Vorontsov-Velyaminov; Comparison of the classification systems; Seyfert galaxies, N galaxies, and quasars; Stellar content related to type: formation and evolution.

158.130 The stellar and gaseous content of normal galaxies as derived from their integrated spectra.
H. Spinrad, M. Peimbert.
Galaxies and the universe, (see 003.010), p. 37 - 80 (1975).

The stellar content of normal galaxies (Introductory remarks and a historical review, slit spectra of galaxies — selected observations and interpretations, photoelectric narrow-band measures of galaxies, the Spinrad-Taylor galaxy models, synthesis of other types of galaxies, comments on the stellar content of some spiral galaxies, speculative topics of interest); Emission lines from normal galaxies (Normal H II regions in our Galaxy and in other galaxies, nuclear H II regions, other H II regions).

158.131 The masses of galaxies.
E. M. Burbidge, G. R. Burbidge.
Galaxies and the universe, (see 003.010), p. 81 - 121 (1975).

Theories of mass determination (Rotations of galaxies, masses of spherical galaxies from velocity dispersion of stars, masses from orbital motion of double galaxies, average masses of galaxies in clusters determined by using the virial theorem); Methods of observation (Rotation of galaxies, velocity dispersions and potential energies, orbital motions of double galaxies); Results (Distances, luminosities, uncertainties in results, M31, M33, noncircular motions); Mass-to-light ratios and masses of galaxies in clusters; angular momentum.

158.132 Magnitudes, colors, surface brightness, intensity distributions, absolute luminosities, and diameters of galaxies. E. Holmberg.
Galaxies and the universe, (see 003.010), p. 123 - 157 (1975).

158.133 Integrated energy distribution of galaxies.
A. E. Whitford.
Galaxies and the universe, (see 003.010), p. 159 - 176 (1975).

Color indices; Filter-band spectrophotometry; Absolute energy curves from scanners; Inferences regarding stellar content; The K-correction; Evolutionary effects.

158.134 Strong nonthermal radio emission from galaxies.
A. T. Moffet.
Galaxies and the universe, (see 003.010), p. 211 - 281 (1975).

Synchrotron radiation (Emission from a single particle, emission from an ensemble of particles, modifications of the power-law spectrum, luminosity and energy requirements for a source of synchrotron radiation, Compton losses, expansion losses); Observed radio characteristics (distribution, spectra, angular sizes and brightness distributions, polarization); Intrinsic properties (Optical identification and redshifts, luminosities, linear dimensions, energy requirements, optical properties of radio galaxies); Theories of radio galaxies (Energy sources, symmetrical division, source containment, the end game).

158.135 Radio observations of neutral hydrogen in galaxies.
M. S. Roberts.
Galaxies and the universe, (see 003.010), p. 309 - 357 (1975).

Determination of kinematic properties of galaxies from 21-centimeter observations; Total masses; Systemic radial velocity; The hydrogen content of galaxies; The distribution of hydrogen within a galaxy; Integral properties of galaxies (Total mass and absolute photographic luminosity, total mass and hydrogen mass, hydrogen mass and absolute photographic luminosity, hydrogen mass, absolute photographic luminosity, and color, ordinary and barred spirals, elliptical and radio galaxies); The evolution of galaxies.

158.136 The extragalactic distance scale.
S. van den Bergh.
Galaxies and the universe, (see 003.010), p. 509 - 539 (1975).

Period-luminosity relation of classical cepheids; Novae; RR Lyrae variables; W Virginis stars; Red giants of population II; Globular clusters; Spectral luminosity determinations; Summary of data on the Local Group; Distances beyond the Local Group; Third brightest cluster galaxy; Surface brightness of galaxies; Luminosity classification of galaxies; Diameters of H II regions; Mass-to-light ratios; Brightest nonvariable stars in galaxies; Supernovae; Galaxy diameters; Summary of data on the Hubble constant; Regional variations of the Hubble constant.

158.137 Binary galaxies. T. Page.
Galaxies and the universe, (see 003.010), p. 541 - 556 (1975).

Definition of optical and physical pairs; Observational data; Types and magnitudes of galaxies in pairs; Dimensions of galaxies in pairs; Orientation of axes in pairs of galaxies; The dynamics of binary galaxies; Formation and evolution of binary galaxies.

158.138 Nearby groups of galaxies. G. de Vaucouleurs.
Galaxies and the universe, (see 003.010), p. 557 - 600 (1975).

Definition of a group; Census of nearby groups; Distance moduli; Local Group; The nearer groups within 10 megaparsecs (Sculptor group, M81 group, Canes Venatici I cloud, NGC 5128 group, M101 group, NGC 2841 group, NGC 1023 group, NGC 2997 group, M66 group, Canes Venatici II cloud, M96 group, NGC 3184 group, Coma I cloud, NGC 6300 group); Nearby groups beyond 10 megaparsecs; Statistical properties of nearby groups; Nearby dwarf galaxies; Isolated nearby galaxies; Apparent and space distribution of nearby groups: Local Supercluster.

158.139 Distribution of galaxies. C. D. Shane.
Galaxies and the universe, (see 003.010), p. 647 - 663 (1975).

158.140 Ionized gas in the disk of M33.
R. Dubout, A. Laval, A. Maucherat, G. Monnet, M. Petit, F. Simien.
Astrophys. Letters, Vol. 17, 141 - 145 (1976).

All experimental data show that the disk of M33 is ionized by normal hot stars. The northern disk shows very low excitation, the southern one, moderate excitation, in good qualitative agreement with the corresponding uneven distribution of very hot stars.

158.141 Accurate optical positions of bright galaxies.
L. L. Dressel, J. J. Condon.
Astrophys. Journ., Suppl. Ser., Vol. 31, 187 - 236 (1976).

Optical positions of all galaxies brighter than 14.5 mag and north of declination $-2°30'$ were measured from the Sky Survey prints with computer-generated transparent overlays. Their rms errors are $4''$ in each coordinate. As an aid to locating each galaxy, the serial number of the plate on which it appears and its distances in millimeters from the nearest edges are listed.

158.142 M31 at 11 cm. E. M. Berkhuijsen.
Mem. Soc. Astron. Italiana, Vol. 45, (see 012.017), 587 - 588 (1974).

158.143 Millimetre emission from active galaxies.
M. Rowan-Robinson.
Mem. Soc. Astron. Italiana, Vol. 45, (see 012.017), 593 - 601 (1974).

158.144 Neutral hydrogen study of spiral and irregular dwarf galaxies. C. Balkowski, L. Bottinelli, P. Chamaraux, L. Gouguenheim, J. Heidmann.
Mem. Soc. Astron. Italiana, Vol. 45, (see 012.017), 611 - 617 (1974).

158.145 Sky distribution of Markarian galaxies.
J. Heidmann.
Mem. Soc. Astron. Italiana, Vol. 45, (see 012.017), 619 - 624 (1974).

158.146 Morphology of pairs of Markarian galaxies.
C. Casini, J. Heidmann, G. Lelièvre.
Mem. Soc. Astron. Italiana, Vol. 45, (see 012.017), 625 - 629 (1974).

158.147 A galaxy with a low gas temperature and high oxygen abundance.
Y. Andrillat, S. Collin-Souffrin.
Mem. Soc. Astron. Italiana, Vol. 45, (see 012.017), 631 - 632 (1974).

158.148 Stellar population and M/L ratios in the nuclear region of M31 and M81. M. Joly.
Mem. Soc. Astron. Italiana, Vol. 45, (see 012.017), 633 - 635 (1974).

158.149 Galaxy formation. J. Binney.
Mem. Soc. Astron. Italiana, Vol. 45, (see 012.017), 975 - 985 (1974).

158.150 Surface photometry of M 101, M 51, and NGC 5195.
S. Okamura, T. Kanazawa, K. Kodaira.
Publ. Astron. Soc. Japan, Vol. 28, 329 - 346 (1976).

The surface photometry of M 101 and the M 51 + NGC 5195 system is carried out in a computerized digital method. The results are presented in various sorts of maps and figures, and by the photometric parameters in the de Vaucouleurs system. M 101 is found to be a giant but normal disk galaxy, while M 51 and NGC 5195 show peculiar characteristics, which might be related to the suspected tidal interaction between them.

158.151 Spiral structure in galaxies – analogies.
R. C. Kirkpatrick.
Proc. Southwest Regional Conf., Vol. 1, (see 012.021), 23 - 26 (1976).

158.152 The evolution of disk galaxies. R. J. Talbot, Jr.
Proc. Southwest Regional Conf., Vol. 1, (see 012.021), 151 - 155 (1976).

158.153   The missing mass around galaxies.
J. Einasto, M. Jõeveer, A. Kaasik, J. Vennik.
Astrophys. Obs., *Tartu*, Preprint No. 8, 25 pp. Price 2 Kop.
(1976).

158.154   Missing mass in the Local Group of galaxies.
M. Jõeveer, J. Einasto, A. Kaasik.
Astrophys. Obs., *Tartu,* Preprint No. 9, 6 pp. Price 2 Kop.
(1976).

158.155   The Magellanic Stream and the mass of our hyper-
galaxy.
J. Einasto, U. Haud, M. Jõeveer, A. Kaasik.
Astrophys. Obs., *Tartu,* Preprint No. 10, 31 pp. Price 15 Kop.
(1976).

The kinematics of the Magellanic Stream and of other
high velocity clouds is investigated. Arguments are presented
to show that they are permanent members of the system of
the companions of our Galaxy. The virial theorem mass of
our hypergalaxy (the visible components of the Galaxy, its
companion galaxies and an invisible corona) is estimated:
$M_{tot} = (1.2 \pm 0.5) \times 10^{12} M_\odot$. It is shown that essential observed
properties of the Magellanic Stream are consistent with the
hypothesis that clouds of the stream are moving around the
Galaxy along elliptical orbits.

158.156   The mass of the Local Group of galaxies.
J. Einasto.
Astrophys. Obs., *Tartu,* Preprint No. 11, 6 pp. Price 2 Kop.
(1976). – See 14.158.114.

158.157   The dynamics of aggregates of galaxies as related
to their main galaxies.
J. Einasto, M. Jõeveer, A. Kaasik, J. Vennik.
Astrophys. Obs., *Tartu,* Preprint No. 12, 46 pp. Price 16 Kop.
(1976).

The dynamics of the aggregates of galaxies is compared
with the dynamics of their member galaxies. It is demonstrated that within a factor of 1.5–2 the dispersion of relative line-
of-sight velocities is constant from the nuclei of main galaxies
to the periphery of the aggregates of galaxies. The equality of
the velocity dispersion of stars in galaxies to the velocity
dispersion of galaxies in aggregates concerns only main galaxies. All companion galaxies have a smaller dispersion of stars.
The dynamical evolution of both galaxies and aggregates of
galaxies is very slow. Thus the above data suggest that galaxies
and their aggregates have been formed together.

158.158   The luminosity function for galaxies and the cluster-
ing of galaxies.   P. Schechter.
Thesis California Inst. Techn., Pasadena, USA, 80 pp. (1975).
(Available from: Univ. Microfilms, Ann Arbor, Mich.,USA.
Order No. 75-20012). – Abstr. in Phys. Abstr., Vol. 79,
A028021 (1976).

158.159   Phenomenological analysis of observed relations for
low-redshift galaxies.   J. F. Nicoll, I. E. Segal.
Proc. National Acad. Sci. USA., Vol. 72, 4691 - 4695 (1975).
Abstr. in Phys. Abstr., Vol. 79, A042376 (1976).

158.160   Galaxy angular momentum.   L. A. Thompson.
Thesis Arizona Univ., Tucson, USA, 197 pp. (1974).
(Available from:Univ. Microfilms, Order No. 75–14, 212).

158.161   Collisions of galaxies in dense clusters: morphologi-
cal effects.   D. O. Richstone.
Thesis Princeton Univ., New Jersey, USA, 66 pp. (1974).
(Available from: Univ. Microfilms, Order No. 75-23,235).

158.162   Time dependent emission line profiles in the radially
streaming particle model of Seyfert galaxy nuclei

and quasi-stellar objects.   R. Hubbard.
Separate print Bowling Green State Univ., Ohio, USA, Dept.
Phys., 20 pp. (1974). (Available from NTIS).

158.163   Effect of suprathermal protons on the physical con-
ditions in Seyfert galaxy nuclei.
R. Ptak, R. Stoner.
Separate print Bowling Green State Univ., Ohio, USA, Dept.
Phys., 27 pp. (1974). (Available from NTIS). – See 14.158.053.

Evolution of stars and galaxies.
See Abstr. 003.017.

Mathematical cosmology and extragalactic astron-
omy.   See Abstr. 003.097.

Physical conditions in a hydrogen gas heated by
suprathermal protons.   See Abstr. 022.004.

L'observation des galaxies avec la machine COSMOS.
See Abstr. 031.256.

Origin of the diffuse γ-ray background.
See Abstr. 061.016.

Gravitational imaging by elliptical galaxies.
See Abstr. 066.084.

The variability of the object Markarian 388.
See Abstr. 122.063.

Statistics of extragalactic supernovae.
See Abstr. 125.010.

Galaxies à fréquentes apparitions de supernovae.
See Abstr. 125.026.

On the measurement of the extragalactic background
brightness at 4000 Å.   See Abstr. 131.034.

An almost complete survey of 21 cm line radiation
for $|b| \geq 10°$. III. **The interdependence of H I, galaxy counts,**
reddening, and galactic latitude.   See Abstr. 131.044.

Interstellar gas.   See Abstr. 131.154.

The radio sources in the nuclei of NGC 3031 and
NGC 4594.   See Abstr. 141.017.

Extragalaktische Doppel-Radioquellen.
See Abstr. 141.028.

High resolution observations of NGC 1275 with a
four-element intercontinental radio interferometer.
See Abstr. 141.037.

Optical properties of the radio source PKS 0123–01
(3C 40) in Abell 194.   See Abstr. 141.040.

A spinar model of Cygnus A.
See Abstr. 141.043.

Radiointerferometrie mit transatlantischen Basislini-
en: Galaxienkerne und Quasare.   See Abstr. 141.047.

High resolution observations of 3C 33 at 327 MHz.
See Abstr. 141.066

Modern models of the source of activity in quasars
and in nuclei of galaxies.   See Abstr. 141.075.

Cygnus A at 8.5 millimeter wavelength.
See Abstr. 141.078.

Photographic photometry of compact extragalactic objects. IV.   See Abstr. 141.085.

Radio and optical observations of the radio source OX 029.   See Abstr. 141.086.

Redshifts of forty-three radio sources.
See Abstr. 141.087.

The optical spectra of 3C 227 and other broad-line radio galaxies.   See Abstr. 141.094.

A survey for emission-line galaxies and quasars. III. A list of nine new optically selected QSOs with $2.5 < z < 3.1$. See Abstr. 141.097.

TV spectroscopy of absorption lines in the far-red of PHL 957.   See Abstr. 141.098.

The identification of radio sources.
See Abstr. 141.105.

Optical identification of 3CR sources (October 1972).   See Abstr. 141.106.

Recent Westerbork observations of radio galaxies.
See Abstr. 141.111.

High resolution observations of large and complex radio galaxies.   See Abstr. 141.112.

Observations of large radio galaxies at 1420 and 2695 MHz.   See Abstr. 141.117.

Radio emission of NGC 5128 and some southern radio sources at 1.35 cm.   See Abstr. 141.118.

The structure of NGC 1265.
See Abstr. 141.120.

X-ray observations of NGC 5128.
See Abstr. 142.003.

Gamma rays from an external galaxy?
See Abstr. 142.074.

Elongated equilibrium stellar systems tidally distorted in pairs.   See Abstr. 151.002.

On density waves in galaxies. II. The turning-point problem at the corotation region.   See Abstr. 151.005.

The formation of the nuclei of galaxies. II. The local group.   See Abstr. 151.006.

Density maxima formed by trapped orbits.
See Abstr. 151.009.

Computer study of galactic spiral arm shapes.
See Abstr. 151.019.

On the derivation of mass distribution in galaxies from internal rotational velocities.   See Abstr. 151.020.

A family of self-gravitating stellar systems with axial symmetry.   See Abstr. 151.026.

Encounters of spherical galaxies. I. Galaxy models

with one stellar population.   See Abstr. 151.028.

Encounters of spherical galaxies. II. Galaxy models with two stellar populations.   See Abstr. 151.029.

A classification of galactic collisions.
See Abstr. 151.031.

On density waves in galaxies. III. Wave amplification by stimulated emission.   See Abstr. 151.035.

Escape velocities for pairs of galaxies.
See Abstr. 151.053.

On density waves in galaxies. IV. Wave amplification through processes that remove angular momentum from galactic disks.   See Abstr. 151.055.

The evolution of massive collapsing gas clouds.
See Abstr. 151.056.

Unstable spiral modes in disk-shaped galaxies.
See Abstr. 151.057.

Large scale shock waves in barred galaxies.
See Abstr. 151.060.

The formation and early dynamical history of galaxies.   See Abstr. 151.064.

Stellar dynamics and the structure of galaxies.
See Abstr. 151.065.

Accretion by the Galaxy: effects of radiative cooling on the flow structure and infall rate.   See Abstr. 155.002.

Steps toward understanding the large-scale structure of the Milky Way: conclusion.   See Abstr. 155.031.

Theoretical aspects of galactic research.
See Abstr. 155.055.

La struttura e l'evoluzione della Galassia.
See Abstr. 155.058.

The Hubble diagram for nuclear magnitudes of cluster galaxies.   See Abstr. 160.003.

Head-tail radio sources in clusters of galaxies.
See Abstr. 160.008.

Supergalactic studies. V. The supergalactic anisotropy of the redshift-magnitude relation derived from nearby groups and Sc galaxies.   See Abstr. 160.018.

Counts of galaxies in the region of the 'intergalactic dark cloud' near ι Microscopii.   See Abstr. 161.002.

On apparent associations among astronomical objects.   See Abstr. 162.011.

Variable G: a solution ot the missing mass problem.
See Abstr. 162.056.

The redshift.   See Abstr. 162.071.

Osservazioni in campo ottico richieste per l'interpretazione teorica dei problemi relativi alla Galassia e alla cosmologia.   See Abstr. 162.073.

On a chaotic early universe.

See Abstr. 162.124.

New evidence for an evolutionary universe from

observations of radio galaxies with the Ooty radiotelescope. See Abstr. 162.125.

Peculiar Objects

158.301 Four-point optical energy distributions for faint BL Lacertae objects.
S. Tapia, E. R. Craine, K. Johnson.
Astrophys. Journ., Vol. 203, 291 - 296 (1976).

Broad-band photoelectric observations of 10 radio sources with continuous optical spectra have been reduced to monochromatic flux densities in order to study the shape and variations of their optical energy distributions. The observed spectral indices range between 1 and 3. It is suggested that sources with large optical spectral index may have detectable nebulosity.

158.302 Photoelectric monitoring of BL Lacertae.
P. Véron, M. P. Véron.
Astron. Astrophys., Vol. 47, 319 - 320 (1976).

Photoelectric observations of BL Lac with an integration time of 10 s for about 3 h do not show any significant variations with a time scale in the range 10 to 300 s. The standard deviation for the individual observations was about $\sigma = 0.035$ mag.

158.303 Study of OJ 287 in the optical region.
S. Kikuchi, Y. Mikami, M. Konno, M. Inoue.
Publ. Astron. Soc. Japan, Vol. 28, 117 - 128 (1976).

Results of photometric and polarimetric observations of OJ 287 are presented. It is confirmed that colors remained constant throughout the recent declining phase, although short-term variations of colors were found as in the active phase. The relationship of the polarimetric properties of OJ 287 with those of other BL Lac-type objects, especially with those of BL Lac itself, is also discussed.

158.304 Absence of emission and absorption lines in the spectrum of OJ 287. M. Nishida, J. Jugaku.
Publ. Astron. Soc. Japan, Vol. 28, 129 - 133 (1976).

Spectroscopic observations of OJ 287 were made in February and March, 1974, when this object was supposed to be in the least luminous phase. Neither intrinsic emission lines with equivalent widths wider than ~1 Å nor absorption lines could be found.

158.305 Photoelectric magnitudes and polarization data for possible BL Lacertae objects. T. D. Kinman.
Astrophys. Journ., Vol. 205, 1 - 5 (1976).

Photoelectric UBV magnitudes and polarization data are given for a sample of radio sources with continuous or weak-lined optical spectra which are possible BL Lacertae objects. The majority show optical variability and significant variable

polarization. Objects with and without visible extended components are separated on a plot of spectral index against apparent magnitude. It is shown to be difficult to adopt a useful sharp definition of a BL Lacertae object, and the similarity of this general class to the variable emission-line quasars is stressed. The separation of BL Lacertae objects from field stars on the $(U-V)/(B-R)$ two-color diagram is suggested as a means of identification.

158.306 The peculiar galaxy NGC 404 and its surrounding area. J. W. M. Baars, H. J. Wendker.
Astron. Astrophys., Vol. 48, 405 - 411 (1976).

New high sensitivity radio continuum observations of the galaxy NGC 404 are presented, indicating a very low radio luminosity. It is found that the galaxy exhibits an unusually large neutral hydrogen mass to optical luminosity ratio. Although NGC 404 optically is an S0 galaxy, it possesses several radio peculiarities. It might be a highly peculiar galaxy related to the Local Group. A few radio sources in the immediate vicinity are also discussed.

158.307 Variability of AO 0235 + 164 at 2.8 cm.
J. M. MacLeod, B. H. Andrew, G. A. Harvey.
Nature, Vol. 260, 751 - 752 (1976).

158.308 Radio spectrum of the major outburst in the BL Lacertae object AO 0235 + 164.
J. E. Ledden, H. D. Aller, W. A. Dent.
Nature, Vol. 260, 752 - 754 (1976).

This paper presents an analysis of the evolution of the radio emission of AO 0235 + 164 during the recent large outburst which coincided with the optical and infrared event. The recent event described here offers an unusual opportunity to study an isolated outburst because its amplitude was much larger than the quiescent flux level, good coverage in time at two radio frequencies was obtained, and simultaneous observations over a wide range of radio, infrared, and optical frequencies were made.

158.309 Photometric and spectroscopic observations of the BL Lacertae object AO 0235 + 164.
G. H. Rieke, G. L. Grasdalen, T. D. Kinman, P. Hintzen, B. J. Wills, D. Wills.
Nature, Vol. 260, 754 - 759 (1976).

In this paper, the authors confirm that AO 0235 + 164 is a BL Lacertae object with one absorption-line redshift at $z = 0.524$ and a probable one at $z = 0.852$. Photometry between 0.36 and 21 μm and at 90 and 140 GHz showed a peak luminosity in November 1975 comparable with the most luminous QSOs.

158.310 The absorption-line spectrum of the BL Lacertae object AO 0235+164.   E. M. Burbidge, R. D. Caldwell, H. E. Smith, J. Liebert, H. Spinrad.
Astrophys. Journ.,(*Letters*), Vol. 205, L117 - L120 (1976).

Absorption features identified with Mg II, Mg I, Fe II, and Mn II at a redshift $z = 0.5240 \pm 0.0001$ have been detected in the optical spectrum of the BL Lacertae object AO 0235 + 164 during its recent outburst. An absorption doublet $\lambda\lambda 5176$, 5189 cannot be identified with any likely features at the redshift $z = 0.524$ and is believed to be Mg II $\lambda\lambda 2796$, 2803 at $z = 0.85$. Thus this is the first BL Lacertae object with more than one absorption-line redshift.

158.311 OE 110: a new, faint BL Lacertae object.   R. J. Leacock, A. G. Smith, P. L. Edwards, J. T. Pollock, R. L. Scott, M. R. Gearhart, E. Pacht, J. D. Kraus.
Astrophys. Journ., (*Letters*), Vol. 206, L87 - L89 (1976) = Rosemary Hill Obs., *Univ. Florida, Gainesville*, Contr. No. 66.

The radio source OE 110 has been identified with a BL Lacertae object that is possibly the faintest yet recognized as belonging to this class.

158.312 PG 2337+12.   L. J. Eachus.
IAU Circ., No. 2907 (1976).

158.313 B2 1308+326.   E. W. Gottlieb.
IAU Circ., No. 2939 (1976).

158.314 B2 1308+326.   B. Wills, D. Wills, D. Dickinson, C. R. Purton, P. A. Feldman, R. E. Goodson, A. H. Bridle.
IAU Circ., No. 2954 (1976).

158.315 B2 1308+326.   J. J. Broderick, R. L. Brown.
IAU Circ., No. 2961 (1976).

158.316 Variation d'éclat de OJ 287.   L. A. Ourassine (*Urasin*).
Ann. Obs. Astron. Alger, Vol. 4, Fasc. 2, p. 23 - 24 (1975).

158.317 OX-192: A new highly variable BL Lacertae object.   E. R. Craine, P. A. Strittmatter, S. Tapia, B. H. Andrew, G. A. Harvey, M. R. Gearhart, J. D. Kraus.
Astrophys. Letters, Vol. 17, 123 - 125 (1976).

The radio source OX-192 is highly variable at both radio and optical wavelengths. New observations place this object in the BL Lacertae class with the distinction of the largest known range of optical variability for objects of this type.

Variations d'éclat rapides de BL Lacertae.
See Abstr. 122.127.

Structure and evolution of compact radio sources.
See Abstr. 141.020.

Some extended observations of the radio source CL4.   See Abstr. 141.036.

Investigation of optical variability in 3C 66A.
See Abstr. 141.055.

Photographic photometry of compact extragalactic objects. IV.   See Abstr. 141.085.

PKS 0422+004.   See Abstr. 141.102.

Southern peculiar galaxies – V.
See Abstr. 158.042.

The peculiar spiral galaxy NGC 3310. I. General properties; far ultraviolet and radio continuum observations.
See Abstr. 158.071.

Errata

158.901 Erratum: 'On the origin of S0 galaxies' [Astron. Astrophys., Vol. 41, 441 - 446 (1975)].
P. Biermann, B. M. Tinsley.
Astron. Astrophys., Vol. 46, 151 (1976).

159 Magellanic Clouds

159.001 The detection of formaldehyde in the Large Magellanic Cloud. J. B. Whiteoak, F. F. Gardner.
Monthly Notices Roy. Astron. Soc., Vol. 174, 51P - 52P (1976).

H_2CO absorption has been detected in the direction of N159, an H II region in the LMC, at a velocity (254 km s^{-1}) close to the values for CO and H109α emission.

159.002 Dwarf galaxies and globular clusters in high velocity hydrogen streams. D. Lynden-Bell.
Monthly Notices Roy. Astron. Soc., Vol. 174, 695 - 710 (1976).

The dwarf spheroidal galaxies Draco and Ursa Minor lie in a stream of high velocity clouds, while Sculptor lies within 3° of the Magellanic stream. Of the distant diffuse globular clusters, Palomar 13 lies in the tail of the Magellanic stream, while Palomar I lies in another prominent northern stream. Using the known distances to the optical objects, the parallaxes due to the offset of the sun from galactic centre are calculated. Not only the Magellanic stream, but also Sculptor and the Draco-Ursa Minor stream are then seen to lie in a plane which is presumably the plane of the orbit of the Magellanic Clouds about the galactic centre. With both radial velocities and distances known to so many points on the stream, it seems likely that dynamical modelling will yield a much more accurate total mass for the Galaxy.

159.003 The effect of dynamical friction on the orbits of the Magellanic Clouds. S. D. Tremaine.
Astrophys. Journ., Vol. 203, 72 - 74 (1976).

If our Galaxy has an extended, massive halo, then dynamical friction must have caused substantial decay of the orbits of the Magellanic Clouds over the last 10^{10} yr, during which time the galactic tidal force at perigalacticon must have steadily increased. The present proximity of the Large and Small Clouds can be explained if they were originally a bound system, which was disrupted by tidal forces only at its last, closest perigalacticon. The Large Cloud will be disrupted by the Galaxy in $(2-4) \times 10^9$ yr, increasing the luminosity of the Galaxy by -0.24 mag.

159.004 Star formation and the structure of the Large Magellanic Cloud. A. Ardeberg.
Astron. Astrophys., Vol. 46, 87 - 98 (1976).

Star formation and structure in the Large Magellanic Cloud have been studied by means of data for supergiant stars. It is found that super-luminous stars are formed in super-associations. These super-associations are well defined and contain the great majority of objects belonging to extreme population I. The evolutionary history of the Large Magellanic Cloud has been investigated for the last 2×10^7 years. From data on supergiant stars and clusters evidence is given that star formation has occurred mainly in one burst. This burst of star formation is well defined in time and covers the entire field studied. Neither the structure nor the star-formation processes seem to speak in favour of spiral-structure type generation of stars.

159.005 A catalogue of A- and F-type supergiants in the Large Magellanic Cloud.
J. Stock, W. Osborn, M. Ibañez.
Astron. Astrophys., Suppl. Ser., Vol. 24, 35 - 52 (1976).

A survey for A−F type supergiants in the Large Magellanic Cloud has been carried out using UV objective prism plates. 890 objects were detected and their spectral types, luminosity classes, magnitudes, and precise positions determined. The survey is practically complete to $m_{pg} = 12.5$ and extends for certain types of stars to $m_{pg} = 14$. It is found that the spatial distribution of the A−F supergiants is not correlated with the distribution of the gas and OB stars of the cloud. This is evidence in support of the tentative identification by Stock and Wroblewski of early-type galactic supergiants well off the plane. Several other implications of this result are also discussed.

159.006 The nebular complexes of the Large and Small Magellanic Clouds.
R. D. Davies, K. H. Elliott, J. Meaburn.
Mem. Roy. Astron. Soc., Vol. 81, 89 - 128 (1976).

Long exposures of the complexes of ionized hydrogen in both the LMC and SMC have been taken with the 48-in. SRC Schmidt camera through an Hα+[N II] interference filter of 100 Å bandwidth. These plates and identifying charts are presented in a form in which little information is lost. A catalogue of many individual emission regions in both these galaxies is also compiled. The relationships between the nebulosities and OB associations, 21-cm neutral hydrogen emission and continuum radio emission are discussed, and a number of supernova remnant candidates are listed for further study.

159.007 The ratio between interstellar absorption and reddening A_V/E_{B-V} for the δ Cephei stars in the Magellanic Clouds. J. Isserstedt.
Astron. Astrophys., Vol. 47, 463 - 466 (1976). In German.

The ratio $R = A_V/E_{B-V}$ in the Magellanic Clouds was determined using the photoelectric UBV-photometry by Madore (1975) and the period-luminosity-color (PLC) equations for galactic δ Cephei stars by Tammann (1970). The author obtains $R = 2.95$ (± 0.29) for the cepheids of the LMC. For the SMC he finds $R = 2.15$ (± 0.20). This indicates different grain properties in the SMC to those in our Galaxy. The distance moduli of the LMC and SMC were redetermined using Madore's cepheids. The author finds $18^{m}.6$ and $19^{m}.3$ respectively.

159.008 Sir John Herschel's observations of stars in the Large Magellanic Cloud.
A. P. Fairall and students of the 1974 Astronomy (a) Course. Department of Astronomy, University of Cape Town.
Monthly Notes Astron. Soc. Southern Africa, Vol. 35, 38 - 46 (1976).

C. P. D. identifications are provided for stars catalogued by Herschel. Some sixteen Herschel stars cannot be traced, while twelve modern stars are overlooked. Interesting magnitude discrepancies, which could be attributed to variability, are found for five Cloud members.

159.009 Die Spiralstruktur der Großen Magellanschen Wolke.
T. Schmidt-Kaler, J. Isserstedt.
Astrophys. Space Sci., Vol. 41, 139 - 153 (1976).

The spiral structure of the Large Magellanic Cloud has been investigated using the best spiral indicators. The spiral features emanate from the 30 Doradus H II-complex as centre and are completely unrelated to the LMC bar. The basic structure of the distribution of H I and of the optical spiral tracers is similar: (1) 30 Dor is the centre of density and starting-point of the spiral features, (2) two main complex arms I and II dominate the distribution, (3) the arms are fragmented optically as well as in the H I, start with the same steep pitch angles in directions displaced by about 60° (instead of the usual diametral symmetry of common two-armed spirals) and wind in the same directions. Evidence is presented to support the view that the enormous supergiant H II-complex 30 Doradus is the nucleus of the spiral LMC.

159.010 Soft X-rays from the Large Magellanic Cloud: implications on the origin of the diffuse X-ray background. K. S. Long, P. C. Agrawal, G. P. Garmire. Astrophys. Journ., Vol. 206, 411 - 417 (1976).

A total soft X-ray luminosity of approximately 10^{38} ergs s^{-1} was observed from the Large Magellanic Cloud (LMC) during pointed rocket observations in 1973 November. Simple spectral parameters are derived and discussed. Upper limits at 1 keV to several known LMC point sources are presented. A diffuse bar source reported by Rappaport et al. was not detected. Strong limits are placed on the fraction of halo and extragalactic diffuse X-ray flux observed in the direction of the LMC due to the lack of correlation of the 0.25 keV diffuse flux with interstellar hydrogen in the Galaxy and in the LMC.

159.011 Starlight polarization in the Magellanic Cloud regions. T. Schmidt. Astron. Astrophys., Suppl. Ser., Vol. 24, 357 - 378 (1976).

Optical polarization data of 656 stars in the Magellanic Cloud regions have been analysed (382 being cloud members and 274 galactic foreground stars). The following conclusions can be drawn: (1) Within both clouds no polarimetric evidence of a spiral structure or a symmetry around the rotation axes could be found. (2) A large scale general alignment parallel to the LMC-SMC connection is evident, but in part seriously disturbed by local irregularities. (3) A second large scale alignment parallel to the LMC bar is indicated.

159.012 Wolf-Rayet stars in the Large Magellanic Cloud. C. Fehrenbach, M. Duflot, A. Acker. Astron. Astrophys.,Suppl. Ser., Vol. 24, 379 - 388 (1976). In French.

A catalogue of the 80 known Wolf-Rayet stars in the Large Magellanic Cloud is given. Six of them are new and 30 of them have spectral types unknown up to now.

159.013 Magellanic stream and the dynamics of our Hypergalaxy. J. Einasto, M. Jõeveer, A. Kaasik. U. Haud. Astron. Tsirk., No. 895, p. 1 - 2 (1975). In Russian.

159.014 30 Doradus as the active centre of the Large Magellanic Cloud. T. Schmidt-Kaler, J. V. Feitzinger. Astrophys. Space Sci., Vol. 41, 357 - 370 (1976).

Evidence is presented for the hypothesis that the supergiant H II complex 30 Doradus (NGC 2070) is the mildly active galactic nucleus of the Large Magellanic Cloud. For this purpose the general properties of galactic nuclei and the characteristics of active nuclei are reviewed. Examination of 30 Doradus shows that it plays the same exceptional role among all H II regions of the LMC as Sgr A among those of our Galaxy, and has all the properties of a galactic nucleus.

159.015 Clusters of the Magellanic Clouds. P. W. Hodge. Irish Astron. Journ., Vol. 12, 77 - 81 (1975).

159.016 The old populations in the Magellanic Clouds. J. A. Graham. Irish Astron. Journ., Vol. 12, 138 - 145 (1975).

Color composite photographs of the Magellanic Clouds. See Abstr. 031.269.

Carbon stars in the Large Magellanic Cloud. See Abstr. 065.006.

Nucleosynthesis and star formation of the Galaxy and Magellanic Clouds. See Abstr. 065.061.

The mass-luminosity relationship for cepheids in the Small Magellanic Cloud. See Abstr. 065.064.

Si II equivalent widths in SMC A-type supergiants. See Abstr. 114.301.

The red irregular variable LV 60 in the direction of the SMC. See Abstr. 122.010.

Photometry of cepheid variables in the Small Magellanic Cloud. See Abstr. 122.066.

Magellanic Cloud cepheids – the Dunsink programme. See Abstr. 122.135.

The population of supernova remnants in the Magellanic Clouds. See Abstr. 125.002.

The interstellar reddening in the vicinity of 47 Tucanae and the Small Magellanic Cloud. See Abstr. 131.127.

Chemical composition of H II regions in the Small Magellanic Cloud and the pregalactic helium abundance. See Abstr. 131.505.

The masses and chemical composition of planetary nebulae in the galactic bulge and Magellanic Clouds. See Abstr. 133.002.

On the abundances of helium, nitrogen, and oxygen in the planetary nebulae of the Magellanic Clouds. See Abstr. 133.004.

Spectrophotometry of planetary nebulae and supernova remnants in the Magellanic Clouds. See Abstr. 133.025.

Extragalactic X-ray binary systems – I. SMC X1 = Sk 160. See Abstr. 142.001.

Discovery of X-ray pulsations in SMC X-1. See Abstr. 142.092.

X-ray sources. See Abstr. 142.178.

Dynamical evolution of the triple system of the Galaxy, the Large and Small Magellanic Clouds. See Abstr. 151.021.

Evolved stars in open clusters. See Abstr. 153.023.

The Magellanic Stream and the mass of our hypergalaxy. See Abstr. 158.155.

160 Clusters of Galaxies

160.001 Redshifts of galaxies in the cluster Abell 1367.
R. J. Dickens, C. Moss.
Monthly Notices Roy. Astron. Soc., Vol. 174, 47 - 58 (1976).

Redshifts, 31 of which are new and four previously determined, are presented for 35 galaxies in the region of the cluster of galaxies A 1367. The cluster has a redshift relative to the Local Group of 0.0218 ± 0.0004. A virial mass of 9.74×10^{14} $M_\odot$ and mass-to-light ratio, 295 $M_\odot/L_\odot$ are derived for the cluster. The X-ray luminosity, L_x, for the complete sample of five clusters of richness class 2, within distance class 2, is proportional to $\sigma_v^{3\pm1}$.

160.002 The distribution of matter in the Virgo supercluster.
B. J. T. Jones.
Monthly Notices Roy. Astron. Soc., Vol. 174, 429 - 447 (1976).

A spherically symmetric mass distribution centred on the Virgo cluster is fitted to de Vaucouleurs' survey of the distribution of galaxies within 15 Mpc of the Local Group. Two models are constructed with group distances based respectively on the mean recession velocity of the group and on the bright end of the group luminosity function. Using Shapiro's luminosity function, the mass distribution in consecutive shells centred on the Virgo cluster is determined and fitted with a simple model. Over most of the survey volume, the spatial mass density falls off roughly as r^{-3} from the Virgo cluster. The mean cosmic mass density over the whole survey volume is 2.0×10^{-31} g cm^{-3} and the density of the Virgo supercluster falls to this value at ~5 Mpc from the Virgo cluster centre. On the basis of the model, the mean density of the cluster averaged over a volume of ~2 Mpc radius is 2×10^{-29} g cm^{-3}, and so this is the dynamical radius for the cluster under the present assumptions.

160.003 The Hubble diagram for nuclear magnitudes of cluster galaxies. D. W. Weedman.
Astrophys. Journ., Vol. 203, 6 - 13 (1976).

Luminosities within central diameters of about 5 kpc (defined as the nuclear magnitudes) are given for the 10 brightest galaxies in each of nine rich clusters. Corrections required to use these magnitudes as relative distance indicators are discussed. The resulting Hubble diagram shows no evidence for significant non-Hubble velocities for 1000 km s^{-1} < CZ < 11,000 km s^{-1}. If the mean magnitudes of the five brightest nuclei in each cluster are considered, the $\sigma(\Delta m)$ from the mean Hubble line is only 0.15 mag and the $\sigma(\Delta \log CZ)$ is 0.029.

160.004 On estimating the unprojected luminosity density within a cluster of galaxies. W. H. Press.
Astrophys. Journ., Vol. 203, 14 - 22 (1976).

For an apparently spherical cluster of galaxies, it is desired to estimate the spherical density profile $\rho(r)$ from the positions of individual galaxies projected into the plane of the sky. The classical methods of von Zeipel and Plummer require that the data be smoothed and are sensitive (and unstable) to this smoothing. An alternative method is outlined here: from the observed (projected) positions an integral transform is computed. The transform is then filtered in such a way as to eliminate spurious information due only to the discreteness of the data. Finally, $\rho(r)$ is reconstructed from the filtered transform.

160.005 Is the local supercluster a physical association?
J. N. Bahcall, P. C. Joss.
Astrophys. Journ., Vol. 203, 23 - 32 (1976).

The nearby bright galaxies appear to be concentrated toward a particular great circle on the sky. This distribution has been previously interpreted as implying that our Galaxy is a member of a local supercluster. The authors find that the observed distribution can be explained as the combined result of obscuration by our own Galaxy, purely local clustering on angular scales ≲ 40°, and our proximity to a rather large and populous cluster, the Virgo cluster.

160.006 Is the local supercluster a random clumping accident? G. de Vaucouleurs.
Astrophys. Journ., Vol. 203, 33 - 38 (1976).

The distribution of bright galaxies and nearby groups has been interpreted alternatively as the result of (a) a physical association (the local supercluster hypothesis), or (b) a statistical accident (the random clumping hypothesis). The latter view, as developed by Bahcall and Joss, rests on a presumed demonstration that the "apparent supercluster effect," as measured by a certain statistical index K, can be reduced to a level compatible with hypothesis (b). A critical assessment of the BJ analysis shows that it is faulty in its methodology and that their model is quantitatively untenable in its requirements.

160.007 Method for determining maximum-likelihood distance moduli for groups of galaxies.
P. Schechter, W. H. Press,
Astrophys. Journ., Vol. 203, 557 - 568 (1976).

Two new procedures for estimating distance moduli for groups of galaxies are set forth. Both procedures construct a synthetic, characteristic apparent magnitude m^*, using the measured apparent $m_1, \ldots, m_N$ of the N brightest group members. The distance modulus $m^* - M^*$ then follows from a known universal absolute magnitude M^*. One of the two procedures, useful only for rich clusters, is a generalization of Sandage's "single-brightest" standard candle. The other one is closely related to Abell's "luminosity-function break" method, and is useful for sparse groups as well as for rich clusters. Various tests of the procedures on actual data are shown.

160.008 Head-tail radio sources in clusters of galaxies.
L. Rudnick, F. N. Owen.
Astrophys. Journ., (Letters), Vol. 203, L107 - L111 (1976).

New observations are presented of six head-tail radio galaxies, five of which are previously unreported in the literature. Total intensity maps at 2695 MHz and optical identifications are presented. Some physical parameters of these sources are derived, and are found to be similar to those of the other known head-tail sources. The authors suggest a classification scheme for radio galaxies based on their optical dominance which illustrates the correlation between head-tail sources and galaxies of lower optical luminosity.

160.009 A Westerbork survey of rich clusters of galaxies. II. The luminosity function of bright cluster galaxies at 1415 MHz. W. J. Jaffe, G. C. Perola.
Astron. Astrophys., Vol. 46, 275 - 285 (1976).

From the 1415 MHz WSRT survey in the direction of 5 rich nearby clusters (A 1656, 2147, 2151, 2197 and 2199) the authors have selected those radio sources which are identified with bright cluster galaxies. They summarize the radio and optical data on these galaxies and use these data to derive the bivariate luminosity function for cluster galaxies in the luminosity range $P_{1415} = 10^{20} - 3 \times 10^{22}$ WHz^{-1} sterad^{-1}. The authors discuss briefly the information on the linear extent of the radio sources and on the presence of emission lines in the spectra of these galaxies. Lastly they show that there are few or no radio sources in this power range which are associated with the cluster but not identified with bright cluster members.

160.010 Neutral hydrogen observations of Virgo cluster galaxies.
W. K. Huchtmeier, G. A. Tammann, H. J. Wendker.
Astron. Astrophys., Vol. 46, 381 - 390 (1976).

Neutral hydrogen observations with the Effelsberg radio telescope are reported for 39 Virgo cluster galaxies of various types and luminosities. Hydrogen masses are obtained for 22 galaxies; for the remaining objects upper limits are derived. The 21 cm radial velocities agree well with optical determinations. It appears that the ratio M_H/L increases with decreasing galaxian luminosity. This complicates a comparison of field galaxies with the relatively bright sample of Virgo members. A hydrogen deficiency of Virgo galaxies is suggested but not established beyond doubt.

160.011 A catalogue of southern clusters of galaxies.
J. A. Rose.
Astron. Astrophys., Suppl. Ser., Vol. 23, 109 - 114 (1976).

A catalogue has been compiled of 124 probable clusters of galaxies in selected areas around the south galactic pole. The uniformity and completeness of the catalogue are discussed.

160.012 Études récentes sur le superamas local de galaxies.
G. de Vaucouleurs.
L'Astronomie, Vol. 90, 25 - 32 (1976).

160.013 Possible implications of the Rubin-Ford effect.
H. Karoji, L. Nottale.
Nature, Vol. 259, 31 - 33 (1976).

An anomaly in the distribution of radial velocities was reported for Sc 1 galaxies by Rubin et al. They found different mean velocities in two nearby hemispherical regions, although the mean apparent magnitude was found to be the same for both sets. One interpretation they advanced was that the Hubble constant differs in the ratio $H_{II}/H_I \simeq 1.25$ for the two regions, contradicting the cosmologists' canon. To test this hypothesis the authors divided galaxies into two categories: first, those whose light does not encounter any important cluster of galaxies on its way to the observer (galaxies in region A), and second those situated behind or inside a cluster (region B). The authors tentatively propose two alternative interpretations: (1) Light emitted by distant galaxies is redshifted when passing through clusters of galaxies (an effect which could be connected with 'tired light' theories) or (2) distant sources are more luminous when seen through intermediate clusters of galaxies, which could act, for example, as gravitational lenses.

160.014 Head-tail radio sources in the cluster of galaxies Abell 1314. J. P. Vallée, A. S. Wilson.
Nature, Vol. 259, 451 - 454 (1976).

A high resolution study of the cluster of galaxies Abell 1314 has revealed two 'head-tail' radio sources, associated with the galaxies IC 708 and IC 711. The tail of IC 711 extends fully 820 kpc. The source properties are described and discussed in terms of the 'radio trail' model.

160.015 No anisotropy in angular diameter-redshift relationship. A. Evans, D. Hart.
Nature, Vol. 259, 468 - 469 (1976).

Using data for the brightest cluster galaxies, the authors here consider whether any anisotropy exists in the angular diameter-redshift relationship. Data used in previous examinations of the Hubble anisotropy have been taken from catalogues in which apparent magnitude values have been obtained from a number of sources and it would be difficult to argue that these data are sufficiently homogeneous. Thus any anisotropy in the apparent magnitude-redshift relationship should be regarded as 'not proven' until analysis of a sufficiently homogeneous sample of galaxies shows otherwise.

160.016 Collisions of galaxies in dense clusters. II. Dynamical evolution of cluster galaxies. D. O. Richstone.
Astrophys. Journ., Vol. 204, 642 - 648 (1976).

The application of the results of the numerical and analytic investigation of single collisions to galaxies residing in dense clusters has yielded a number of interesting results. If the galaxies in the core start out with extended halos 500 kpc in extent that encompass all the mass in the cluster, then about 90 percent of the mass and 25 percent of the luminosity are liberated in a Hubble time. The details of the above results depend on the validity of the King model approximation after substantial mass loss has occurred. Finally two-body relaxation in this case galaxy-galaxy encounters – should occur on the same time scale.

160.017 The absolute magnitude of first-ranked cluster galaxies as a function of cluster richness.
A. Sandage.
Astrophys. Journ., Vol. 205, 6 - 12 (1976).

The absolute magnitudes of the brightest several galaxies in clusters show a very shallow dependence on cluster richness. New data for sparse southern groups permit an extension of the richness correlation to aggregates as small as four members. The data seem to require a uniqueness about the first few galaxies in clusters that makes their absolute luminosities nearly independent of cluster population.

160.018 Supergalactic studies. V. The supergalactic anisotropy of the redshift-magnitude relation derived from nearby groups and Sc galaxies. G. de Vaucouleurs.
Astrophys. Journ., Vol. 205, 13 - 28 (1976).

A new analysis of the velocities V_0 and distance moduli μ_0 of nearby groups and Sc galaxies presented by Sandage and Tammann (1975) does not support the view that the local extragalactic velocity field is linear and isotropic. The amplitude of the velocity anisotropy (at constant modulus, $30.5 < \mu_0 < 32.5$) between area E (supergalactic anticenter sector) and area B (center sector) is $\delta \langle \log V_0 \rangle = +0.271 \pm 0.050$, significant at the 5.4 σ level. Similarly, the modulus anisotropy (at constant V_0, $2.8 < \log V_0 < 3.3$) between areas B and E is $\delta \mu_0 = +0.82 \pm 0.22$, significant at the 3.7 σ level. The supergalactic anisotropy of the redshift law is demonstrated with the Sandage-Tammann data for galaxies in the supergalactic equatorial belt $|B| < 30°$ in three intervals of distance moduli ($\langle \mu_0 \rangle = 30, 32, 33$).

160.019 Surface photometry of Virgo cluster galaxies NGC 4321. G. F. Benedict.
Astron. Journ., Vol. 81, 89 - 96, 139 (1976).

A procedure to obtain B and V photographic surface photometry of Virgo cluster galaxies utilizing the well-observed cD galaxy M87 (N 4486) as a calibration source is discussed. Luminosity and color profiles and standard photometric parameters for the galaxy N 4321 [SAB(s)bc] are presented. The colors and surface magnitudes of the two major spiral arm arms are sampled at 14 points from $r = 0.08$ to $r = 4.0$ arcmin. The data suggest that for the spiral component of N 4321 the luminosity generally decreases with r while the color index is low near the center, peaks broadly, and then slowly decreases with r.

160.020 Tidal effects in a cluster of galaxies.
A. Potdar, S. M. Alladin.
Bull. Astron. Soc. India, Vol. 3, 30 - 31 (1975). – Abstract of a paper presented at the A.S.I. meeting 1975.

160.021 Galaxienhaufen. K. H. Schmidt.
Astron. in der Schule, 13. Jahrgang, p. 31 - 33 (1976).

160.022 Clusters of galaxies as gravitational lenses?
C. C. Dyer, R. C. Roeder.

Nature, Vol. 260, 764 - 765 (1976).

Karoji and Nottale have compared galaxies situated behind or inside clusters with those whose light does not encounter any important cluster of galaxies. They suggested that either light emitted by distant galaxies is redshifted when passing through clusters of galaxies, or distant sources appear more luminous when seen through intermediate clusters of galaxies, which could act as gravitational lenses. Here the authors examine the magnitude of the lens effect for clusters of galaxies, and find it unlikely that the observational effect claimed by Karoji and Nottale can be explained in this way.

160.023   The peculiar velocity field in the Local Supercluster.   P. J. E. Peebles.
Astrophys. Journ., Vol. 205, 318 - 328 (1976).

A general method of estimating the peculiar velocity field expected, under the gravitational instability picture, around the outer parts of a mass concentration like a cluster of galaxies is derived and applied to a preliminary analysis of the Sandage-Tammann data on the distances and redshifts of nearby spiral galaxies. It is shown that there is evidence of a small peculiar velocity field of the sort expected. The field, if real, is consistent with a cosmologically flat universe, $\Omega = 1$; equally well, with a low-density cosmological model, $\Omega \approx 0.1$. It is concluded that the data on the local peculiar velocity field do not yet offer a strong constraint on the density parameter Ω.

160.024   The Perseus and Coma clusters of galaxies at energies above 20 keV.
A. Scheepmaker, G. R. Ricker, K. Brecher, S. G. Ryckman, J. E. Ballintine, J. P. Doty, P. M. Downey, W. H. G. Lewin.
Astrophys. Journ., Letters, Vol. 205, L65 - L68 (1976).

Observations of the Perseus and Coma clusters of galaxies were made on 1974 June 21, with a balloon-borne X-ray telescope (energies ~20–150 keV). No positive detection was made. The data favor a thermal bremsstrahlung mechanism for the X-ray production in the Perseus cluster of galaxies over the inverse Compton mechanism. In the case of the Coma cluster of galaxies, the data are inconclusive with respect to determining the origin of the X-rays.

160.025   Direct observations of the large-scale distribution of galaxies.   W. G. Tifft, S. A. Gregory.
Astrophys. Journ., Vol. 205, 696 - 708 (1976).

Complete samples of galaxies in regions $3°$ and $6°$ in radius centered on the Coma cluster are presented. When grouped by redshift and position on the sky, virtually all the galaxies are shown to belong to groups or clusters. Two types of galaxy groupings are found. The first contains a few galaxies in well-localized areas and shows a small redshift dispersion. Galaxies in such groups have a distinct tendency to show some emission lines. The second class of galaxy grouping is the major cluster. Associated with the Coma cluster, there appears to be an extended but highly asymmetrical shred of material which probably shows numerous subconcentrations.

160.026   Globular clusters in the Hydra I cluster of galaxies.   M. G. Smith, D. W. Weedman.
Astrophys. Journ., Vol. 205, 709 - 715 (1976).

Excess faint stellar images with B magnitudes about 24 are found surrounding the elliptical galaxy NGC 3311 in the Hydra I cluster ($V = 3450$ km s^{-1}). The magnitudes and number of these images agree well with those expected if (1) NGC 3311 is surrounded by a system of globular clusters identical to that surrounding M87 in the Virgo cluster and (2) the distances to the Hydra I and Virgo clusters are proportional to their velocities.

160.027   Systematic redshifts in the outer regions of the Coma cluster.   S. A. Gregory, W. G. Tifft.
Astrophys. Journ., Vol. 205, 716 - 720 (1976).

Two statistical tests show that galaxies north and west of the Coma cluster center have higher redshifts than those to the south and east. The significance of the difference in mean redshifts is at the 0.05 level. If rotation is responsible for this systematic effect, the rotation period is at least 2×10^{11} years. Other possible explanations are briefly discussed.

160.028   On the use of correlation functions in finding physical associations of galaxies.   S. M. Fall, M. J. Geller, B. J. T. Jones, S. D. M. White.
Astrophys. Journ.,(Letters), Vol. 205, L121 - L125 (1976).

The authors show that the flatness of the "singles" two-point correlation function found by Turner and Gott is an artifact of the selection criteria used to define a "single" galaxy. Thus their method cannot be used to distinguish a true field from a cluster population of galaxies.

160.029   Covariance function analysis and the clustering of galaxies.   P. S. Wesson.
Astrophys. Space Sci., Vol. 40, 325 - 349 (1976).

Data on a statistic derived from the angular covariance function show that (contrary to the claim of Peebles that galaxies are distributed continuously with no distinct scales), superclusters and the maximum size of clusters are probably defined at scales of 15 and $2.0\,h^{-1}$ Mpc.

160.030   A dynamical condition for a relativistic galaxy cluster model.   D. Trevese, A. Vignato.
Astrophys. Space Sci., Vol. 41, 213 - 219 (1976).

In an attempt to give a coherent interpretation of the secondary maximum in the density distribution of clusters of galaxies the authors use an approximate metric tensor proposed by other authors, with the purpose of building a relativistic generalization of the isothermal models of galaxy clusters. Although such a generalization gives rise to oscillations in the density distribution, the quantitative agreement with the observational data is unsatisfactory.

160.031   The Coma supercluster: analysis of Zwicky-Herzog cluster 16 in field 158.
G. Chincarini, H. J. Rood.
Astrophys. Journ., Vol. 206, 30 - 37 (1976).

Radial velocities for 50 of the 52 galaxies brighter than 15.1 mag within the boundaries of Zwicky-Herzog cluster 16 in field 158 are used to establish that the region is a composite of (1) part of the Coma I cloud of the Local Supercluster, (2) part of the supercluster of which the Coma cluster is a member, (3) the NGC 4169 group, and (4) other galaxies. The Coma supercluster is detected to a radial distance from the center of the Coma cluster of $14°.2$.

160.032   A Westerbork survey of rich clusters of galaxies. III. Observations of the Coma cluster at 610 MHz.
W. J. Jaffe, G. C. Perola, E. A. Valentijn.
Astron. Astrophys., Vol. 49, 179 - 192 (1976).

A WSRT full synthesis observation at 610 MHz of an area $1°.6$ radius about the centre of the Coma cluster of galaxies yielded the detection of twenty cluster members with $m_p \leqslant 17.5$, doubling the number of those detected at 1415 MHz. Eight of these are of elliptical or SO type, twelve are of spiral or irregular type (two are Markarian galaxies).

160.033   Superclusters of galaxies.
J. Einasto, M. Jôeveer, A. Kivila, E. Tago.
Astron. Tsirk., No. 895, p. 2 - 4 (1975). In Russian.

160.034   Gross optical properties of the Coma cluster.
S. A. Gregory, W. G. Tifft.
Astrophys. Journ., Vol. 206, 934 - 938 (1976).

The following properties of the main body of the Coma cluster are found: luminosity function, total apparent and absolute photographic magnitudes, the functional dependence of number and mass density with radius, total mass, average volume mass density, mean redshift and redshift distribution, and the variation of velocity dispersion with radius.

160.035 Bright galaxies in rich clusters: test of a statistical model for magnitude distributions.
M. J. Geller, P. J. E. Peebles.
Astrophys. Journ., Vol. 206, 939 - 957 (1976).

The authors present a test of a statistical model for the magnitude distributions of bright galaxies in rich clusters. The model is based on a luminosity function with an assumed universal shape. To fit the distribution of clusters in Abell richness classes, they allow the normalization parameter of the luminosity function to vary from cluster to cluster according to a power law distribution in cluster "mass." The main test of the model is a comparison of predicted magnitude distributions with the observed distributions of magnitudes of first-, second-, and third-ranked cluster members listed by Sandage and Hardy.

160.036 On the correlation between color and absolute magnitude in first ranked cluster galaxies. P. Crane.
Astrophys. Journ., (Letters), Vol. 206, L133 - L134 (1976).

Evidence for magnitude variations among first ranked cluster galaxies is found in the correlation between magnitude and color of a sample of first ranked galaxies measured by Gunn and Oke.

160.037 Clusters of galaxies. G. O. Abell.
Galaxies and the universe, (see 003.010), p. 601 - 645 (1975).

Numbers and catalogs of clusters; Observed properties of clusters (Types of clusters, galaxian content of clusters, the luminosity function and colors of cluster galaxies, populations of clusters, sizes and structures of clusters, velocity dispersions in clusters); Dynamics of clusters; The distribution of clusters (The evidence for the Local Supercluster, other evidence of second-order clusters, the large-scale distribution of clusters and the mean density of matter in the universe).

160.038 New results of second-order clustering of galaxies. (Optical and radio).
M. Kalinkov, K. Stavrev, V. Dermenjiev.
Mem. Soc. Astron. Italiana, Vol. 45, (see 012.017), 609 (1974). – Abstract.

160.039 High-order clustering of galaxies – new methods.
M. Kalinkov.
Mem. Soc. Astron. Italiana, Vol. 45, (see 012.017), 637 - 662 (1974).

Three new statistical methods for the investigation of the problem for the existence of second-order clusters of galaxies have been developed – generalized χ^2 test, nearest-neighbours test, and correlation methods. Their applicability to Abell and Zwicky clusters has been examined. The most important results of the numerical processing over a region around NGP are presented.

160.040 The distribution of galaxies in the Jagellonian field. P. Flin.
Mem. Soc. Astron. Italiana, Vol. 45, (see 012.017), 663 - 671 (1974).

160.041 An X-ray red-shift test for clusters of galaxies up to $z \geq 1$. E. Boldt.
GSFC Document X-661-76-83, 6 pp. (1976).

Correlated measurements of red-shifted iron line emission and apparent surface brightness are suggested for unambiguously defining intrinsic X-ray characteristics for clusters of galaxies up to $z \geq 1$.

160.042 The definitions and classifications of clusters of galaxies. P. Flin.
Postępy Astron., Vol. 24, 109 - 113 (1976). In Polish.

Neutrino astrophysics.   See Abstr. 065.110.

Compact radio sources in the directions of rich clusters of galaxies.   See Abstr. 141.014.

Radio sources with wide-angle tails in Abell clusters of galaxies.   See Abstr. 141.051.

PKS 0116+082 and 3C 330: two distant cluster radio galaxies.   See Abstr. 141.089.

The identification of radio sources.   See Abstr. 141.105.

Radio sources in Abell clusters.   See Abstr. 141.114.

Ariel 5 observations of the X-ray spectrum of the Perseus cluster.   See Abstr. 142.063.

New observations of X-ray clusters of galaxies.   See Abstr. 142.097.

Dynamical friction in spherical clusters.   See Abstr. 151.001.

Measuring the dynamical age of N-body systems.   See Abstr. 151.010.

Massenbestimmungen von Galaxien durch die Dynamik von Systemen von Galaxien.   See Abstr. 151.025.

Encounters of spherical galaxies. I. Galaxy models with one stellar population.   See Abstr. 151.028.

Encounters of spherical galaxies. II. Galaxy models with two stellar populations.   See Abstr. 151.029.

An analytic expression for the luminosity function for galaxies.   See Abstr. 158.009.

In situ particle acceleration and physical conditions in radio tail galaxies.   See Abstr. 158.010.

The problem of motions in systems of galaxies.   See Abstr. 158.024.

The role of diffuse matter in galactic coronas.   See Abstr. 158.025.

The number-diameter relation of galaxies.   See Abstr. 158.098.

Nearby groups of galaxies.   See Abstr. 158.138.

The luminosity function for galaxies and the clustering of galaxies.   See Abstr. 158.158.

Galaxy angular momentum.   See Abstr. 158.160.

Collisions of galaxies in dense clusters: morphological effects.   See Abstr. 158.161.

The dynamics of the intergalactic medium in the vicinity of clusters of galaxies.   See Abstr. 161.001.

X-rays from hot plasma in clusters of galaxies.
See Abstr. 161.005.

Correlation dynamics in an expanding universe.
See Abstr. 162.001.

The universe as a "non-ideal gas" of galaxies.
See Abstr. 162.014.

The growth of correlations in an expanding universe and the clustering of galaxies.   See Abstr. 162.015.

A "foil" for gravitational clustering investigations.
See Abstr. 162.017.

Modèles cosmologiques non homogènes et Superamas Local de Galaxies.   See Abstr. 162.026.

On the local anisotropy of expansion of the universe.
See Abstr. 162.028.

The extension of the Hubble diagram. I. New redshifts and BVR photometry of remote cluster galaxies, and an improved richness correction.   See Abstr. 162.037.

A cosmic virial theorem.   See Abstr. 162.038.

Variable G: a solution to the missing mass problem.
See Abstr. 162.056.

Neutrino rest mass from cosmology.
See Abstr. 162.059.

Analyse supplémentaire des déplacements vers le rouge liés à la traversée des amas de galaxies par la lumière.
See Abstr. 162.068.

Galaxy clustering: its description and its interpretation.   See Abstr. 162.069.

On the dynamics of clusters of galaxies in a universe with a weak gravitation field.   See Abstr. 162.077.

161 Intergalactic Matter

161.001 The dynamics of the intergalactic medium in the vicinity of clusters of galaxies. S. M. Lea.
Astrophys. Journ., Vol. 203, 569 - 580 (1976).

Numerical solutions to the problem of infall of matter into clusters of galaxies are presented. It is assumed that the universe can be described by a Friedmann cosmological model with $H_0 = 50$ km s^{-1} Mpc^{-1}. It is found that physically reasonable models having $q_0 = 1/2$ lead to excessive X-ray emission from clusters of galaxies. Models including a heat flux due to thermal conduction, heating of the intracluster medium by galaxy motions, and a gas outflow from the galaxies in the cluster are discussed. It is concluded that Ω must be less than 0.2 in order that the predicted X-ray emission not exceed that observed.

161.002 Counts of galaxies in the region of the 'intergalactic dark cloud' near ι Microscopii. I. Meinunger.
Astron. Nachr., Vol. 297, 23 - 24 (1976).

The distribution of the total numbers of galaxies down to about 18th magnitude on 84 squares is largely in agreement with the structure of the hypothetic intergalactic absorbing cloud near ι Microscopii found by C. Hoffmeister. The counts of galaxies were performed on the Whiteoak prints covering that region.

161.003 Existence and amount of intergalactic dust. K.-H. Schmidt.
Astrophys. Space Sci. Library, Vol. 55, (see 012.001), 23 - 31 (1976).

The densities of intergalactic dust are estimated to be between 5×10^{-30} g cm^{-3} (near the centers of clusters of galaxies) and 2×10^{-34} g cm^{-3} (in general intergalactic space). The grains may be formed either in the early phases of the universe ($25 < z < 50$) or may be expelled from galaxies by the radiation pressure. The most effective destruction process seems to be the evaporation by soft cosmic rays.

161.004 Intergalactic extinction and the quasar cut-off. H. Oleak, K.-H. Schmidt.
Astron. Nachr., Vol. 297, 71 - 76 (1976).

The intergalactic extinction in Friedmann (with $\Lambda = 0$) universes homogeneously filled with dust grains is calculated assuming the extinction to be 0.5 mag at $z = 1$ and a λ^{-1} wavelength dependence. With the resulting intergalactic extinction the number of quasars which should be observed at different redshifts are estimated assuming two different luminosity functions and a density evolution of the quasars according to M. Schmidt (1970, 1972). The expected number of quasars decreases rapidly with increasing redshift between $z = 2$ and

$z = 3$. The observed number-magnitude relation by G. A. Richter (1975) is well represented.

161.005 X-rays from hot plasma in clusters of galaxies. A. Cavaliere, R. Fusco-Femiano.
Astron. Astrophys., Vol. 49, 137 - 144 (1976).

To disentangle the X-ray emissions from clusters of galaxies, the authors study the simple limiting model of Bremsstrahlung emission by a static hot plasma in the potential well set up by a cluster. They find a good fit to the existing data on the X-ray luminosity distribution, with no free parameters, for an isothermal gas conforming to the King distribution of galaxies.

161.006 Radio-frequency limits to the abundance of intergalactic neutral hydrogen. K. R. Lang.
Astrophys. Journ., (*Letters*), Vol. 206, L91 - L93 (1976).

Radio-frequency spectra were obtained in order to search for neutral hydrogen in or along the line of sight to the radio galaxies 3C 264, 3C 270, 3C 272.1, and 3C 296. New limits to the volume density of intergalactic neutral hydrogen were obtained. The observations provide additional support for the conclusion that there is insufficient intergalactic hydrogen to stop the expansion of the universe.

161.007 On the Faraday rotation in the intergalactic medium. I. P. Kuznetsova.
Astron. Zhurn. Akad. Nauk SSSR, Vol. 53, 475 - 484 (1976). In Russian. English translation in Soviet Astron., Vol. 20, No. 3.

The Faraday rotation of the polarization plane of radio emission from extragalactic sources is well explained on the assumption that it is due to the interstellar medium in the Galaxy and plasma inside the sources themselves. The data available are insufficient to separate reliably a possible rotation of the polarization plane of radio emission from discrete sources in the intergalactic medium.

Bright phase in the evolution of galaxies and ionization of intergalactic gas. See Abstr. 158.125.

An effect of the pressure gradient on the concentration of matter in an expanding universe. See Abstr. 162.018.

Radio astronomy and cosmology. See Abstr. 162.070.

Die lokale Massendichte und die intergalaktische Materie. See Abstr. 162.074.

162 Structure and Evolution of the Universe, Cosmology

162.001 Correlation dynamics in an expanding universe.
S. M. Fall, G. Severne.
Monthly Notices Roy. Astron. Soc., Vol. 174, 241 - 251 (1976).

The authors present a systematic approach to kinetic theory in homogeneous Newtonian cosmologies, which is based on a correlational picture of the cosmological principle. A kinetic-like equation, including the effects of collisions in the 'weak-coupling' approximation, is derived for the single-particle distribution describing an expanding universe. This equation differs from the standard equation in the adjunction to each binary interaction operator of an expansion factor, so that the effective interaction may be considered time dependent. It predicts the production of kinetic and thus also of correlational energy, in agreement with the fluctuational picture of cosmological clustering, where they can be compared.

162.002 Collisions between galaxies in a symmetric cosmology. N. Bel, P. Martin.
Astron. Astrophys., Vol. 46, 455 - 460 (1976).

The authors calculate the γ-ray flux emitted in a baryon-symmetric universe as a consequence of galaxy-antigalaxy collisions. This flux turns out to be 3 orders of magnitude higher than the observed value: this confirms that the typical extent of an antimatter condensation in such a model cannot be less than the size of a cluster of galaxies.

162.003 Will the universe expand forever?
J. R. Gott III, J. E. Gunn, D. N. Schramm, B. M. Tinsley.
Sci. American, Vol. 234, No. 3, p. 62 - 65, 68 - 72, 77, 79 (1976).

The recession of distant galaxies, the average density of matter, the age of the chemical elements and the abundance of deuterium together suggest that the expansion cannot be halted or reversed.

162.004 Some cosmological models with spin and torsion, I.
B. Kuchowicz.
Astrophys. Space Sci., Vol. 39, 157 - 172 (1976).

The Einstein-Cartan theory, which is a slight modification of the general theory of relativity, is almost indistinguishable in its practical consequences from the latter theory. A characteristic spin-spin repulsive interaction which is of some importance at ultraheavy densities, prevents the singularities occurring in the Einstein-Cartan treatment. It is shown how this mechanism of preventing the singularity applies to cosmological models in which the spins of matter are aligned along some symmetry axis. Some exact solutions without singularities of the relevant set of equations are obtained.

162.005 Evolution of Szekeres's cosmological models.
W. B. Bonnor, N. Tomimura.
Monthly Notices Roy. Astron. Soc., Vol. 175, 85 - 93 (1976).

The models are solutions of Einstein's equations for dust with no Killing vectors. They depend on four arbitrary functions of one variable, and generalize both the Friedmann models and those of Kantowski & Sachs. The possibilities of evolution are diverse. One result is that a Friedmann open model can evolve from a variety of initial states depending on three arbitrary functions of one variable.

162.006 Tachyons and cosmology.
J. V. Narlikar, E. C. G. Sudarshan.
Monthly Notices Roy. Astron. Soc., Vol. 175, 105 - 116 (1976).

The propagation of tachyons in an expanding universe is discussed. It is shown that a primordial tachyon in the big-bang universe cannot survive unless it had very large energy initially. In an indefinitely expanding universe the tachyon trajectory turns back in time. This time barrier is found to exist even in the quantum mechanical discussion of tachyons. This property is used to set limits on the mass of tachyon. The possible astronomical checks on the hypothesis that neutrinos or photons may be tachyonic are also discussed.

162.007 Weak interactions in the big bang.
P. C. W. Davies, with a reply by G. Domokos,
M. M. Janson, S. Kövesi-Domokos.
Nature, Vol. 259, 157 (1976).

162.008 Comments on the "Byurakan direction" in cosmogonical investigation. C. Yu.
Acta Astron. Sinica, Vol. 16, 93 - 100 (1975). In Chinese.

162.009 Die Struktur der.Welt. Ist das Universum positiv oder negativ gekrümmt? H. Dehnen.
Umschau, 76. Jahrgang, p. 209 - 218 (1976).

The methods for determination of the curvature of the space of the universe give different results: The age of the uranium, the magnitude-redshift relation for galaxies and with reservation the distribution of quasars indicate a closed finite space with positive curvature and finite expansion time. On the other hand the mean matter density determined by the masses of the galaxies or the deuterium abundance and the ages of globular clusters point to an infinite space with negative curvature and perpetual expansion.

162.010 Plasma instabilities in the early universe: matter-symmetric cosmologies. G. Benford.
Astron. Astrophys., Vol. 47, 203 - 210 = Technical Report No. 75-30, Univ. Calif., Irvine (1976).

Any anisotropy in the early universe which causes relative drift between leptons and baryons can force unstable growth of current filaments. Charge balance causes electron (positron) filaments to trap protons (antiprotons), thus forcing matter-antimatter separation. In the lepton era ($T < 10^{12}$ °K) growth of these magnetostatic instabilities cannot be suppressed by collisions or thermal effects. The author calculates linear growth rates for fully relativistic, hot, collisional plasma and finds large growth rates comparable to the lepton plasma frequency for a wide range of parameters. It seems unlikely that any other plasma instabilities can cause matter-antimatter separation, because their convective nature mixes charge excesses rapidly.

162.011 On apparent associations among astronomical objects.
P. C. Joss, D. A. Smith, A. B. Solinger.
Astron. Astrophys., Vol. 47, 461 - 462 (1976).

The authors have found an apparent association among bright stars and bright galaxies that has an a posteriori probability for chance occurrence of $\lesssim 8 \times 10^{-3}$. The ability to find such an association is evidence that apparent associations among objects of different redshifts should be viewed with great caution.

162.012 Dynamics of primordial inhomogeneities in model universes. E. P. Liang.
Astrophys. Journ., Vol. 204, 235 - 250 (1976).

The author studies the nonlinear dynamics ($\delta \mu / \mu \gtrsim 1$) of adiabatic density fluctuations in perfect-fluid [$p = (\gamma - 1)\mu$] cylindrical universes in both the regimes $\lambda \ll \lambda_J$ and $\lambda \gg \lambda_J$ ($\lambda_J =$ Jeans length). He clarifies the precise role of the particle horizon and shows that shock waves will form, resulting in the rapid dissipation of the fluctuations whenever $0 < p < \mu$. Various linear perturbation results can be easily retrieved from this analysis in the appropriate limits.

162.013   **A class of Bianchi type VI cosmological models with electromagnetic field.**
K. A. Dunn, B. O. J. Tupper.
Astrophys. Journ., Vol. 204, 322 - 329 (1976).

The "parallel-propagation" solution of the Einstein-Maxwell equations found by Tariq and Tupper is generalized to represent the solution of the field-equations for a perfect-fluid distribution both with and without an electromagnetic field. All the solutions admit a three-parameter group of isometries of Bianchi type VI and form a subclass of the Bianchi type VI models found by Ellis and MacCallum. The perfect-fluid models in this subclass automatically satisfy the Hawking-Penrose strong energy condition, while those models which represent perfect fluid with electromagnetism have current 4-vector which is either spacelike or zero. Properties of these models are discussed and are compared to those of another subclass found by Ellis and MacCallum.

162.014   **The universe as a "non-ideal gas" of galaxies.**
A. Yahil.
Astrophys. Journ., (*Letters*), Vol. 204, L59 - L62 (1976).

The dynamics of correlations among mass-points in an expanding universe is studied using the BBGKY (Bogolyubov, Born, Green, Kirkwood, Yvon) equations. It is shown that the observed spatial correlation function of galaxies, as determined by Peebles and his collaborators, requires that the correlation energy of galaxies be comparable to or greater than their random kinetic energy.

162.015   **The growth of correlations in an expanding universe and the clustering of galaxies.**
S. M. Fall, W. C. Saslaw.
Astrophys. Journ., Vol. 204, 631 - 641 (1976).

From the BBGKY hierarchy of kinetic equations, the authors derive an equation for the growth of pair correlations in an expanding universe of self-gravitating point masses which is correlation-free at some initial time. The authors solve this equation explicitly and, from the full pair correlation function, calculate the spatial correlation function. If t_0 is the time when clustering begins, the solutions are accurate until about $2t_0$. The results suggest that the present degree of galaxy clustering could have arisen from an initially unclustered state at the time of galaxy formation.

162.016   **Kosmologische Beobachtungen und ihre Beziehungen zum Standard-Weltmodell.**   J. Ehlers.
Mitt. Astron. Ges., No. 38, p. 41 - 54 (1976). — Invited paper.

162.017   **A "foil" for gravitational clustering investigations.**
W. H. Press.
Astron. Astrophys., Vol. 48, 149 - 150 (1976).

For the problem of gravitational clustering in an expanding Friedmann cosmology a new equation is derived. It is a first order equation in time which has the same linearized growing modes as the exact second order equation, but it is not equivalent to the exact equation. This "foil" equation may help to answer the question: Is the nature of gravitational clustering determined largely by the linearized growing modes, or are specifically nonlinear processes important?

162.018   **An effect of the pressure gradient on the concentration of matter in an expanding universe.**
K. Sakai.
Publ. Astron. Soc. Japan, Vol. 28, 69 - 75 (1976).

The pressure gradient being taken into account, the fluid-mechanical equations for the flow of matter in an expanding universe are solved by a perturbation method for the three-dimensional case, and by a numerical calculation for the one-dimensional case. The results confirm Kihara's (1967, 1968) suggestion that the density perturbation cannot grow into a concentration of matter unless the wavelength is at least twice as long as the Jeans wavelength. In an application, the existence of expanding intergalactic clouds is suggested.

162.019   **On the density correlations of fluctuations in an expanding universe. I.**   S. Inagaki.
Publ. Astron. Soc. Japan, Vol. 28, 77 - 87 (1976).

The evolution of density correlations of gravitating point masses in an expanding universe is studied by using the Bogolyubov–Born–Green–Kirkwood–Yvon hierarchy equations. The results show that density correlations grow or damp with time in the same manner as the product of density perturbations. That is, the correlations of the density contrast grow with time in proportion to $t^{4/3}$ in the limit of long wavelength, while they damp by phase mixing if their wavelength is sufficiently short.

162.020   **Die extragalaktische Entfernungsskala und der Wert der Hubble-Konstante.**   H.-E. Fröhlich.
Sterne, Vol. 52, 1 - 10 (1976).

162.021   **Die Bestimmung des Beschleunigungsparameters q_0 der Expansion.**   P. Notni.
Sterne, Vol. 52, 11 - 23 (1976).

162.022   **Ist die $N(S)$ - Kurve ein Pfad in die kosmische Vergangenheit?**   G. A. Richter.
Sterne, Vol. 52, 24 - 31 (1976).

162.023   **Frühphasen des Kosmos und die Entstehung der chemischen Elemente.**   K. Fritze.
Sterne, Vol. 52, 32 - 38 (1976).

162.024   **Transformations infinitésimales conformes fermées des espaces cosmologiques de de Sitter et de Robertson-Walker.**   M. Sakoto.
Comptes Rendus Acad. Sci. Paris, Sér. A, Vol. 282, 79 - 82 (1976).

On détermine les transformations infinitésimales conformes fermées des espace-temps cosmologiques de de Sitter et de Robertson-Walker et on donne une propriété des lignes de courant des espaces de Robertson-Walker.

162.025   **Solutions du type de Ricci sans rotation en relativité générale.**   M. Bray.
Comptes Rendus Acad. Sci. Paris, Sér. A, Vol. 282, 83 - 86 (1976).

Solutions de Ricci décrivant un fluide parfait thermo-dynamique sans rotation en cosmologie relativiste.

162.026   **Modèles cosmologiques non homogènes et Superamas Local de Galaxies.**   S. Mavridès.
Comptes Rendus Acad. Sci. Paris, Sér. A, Vol. 282, 451 - 454 (1976).

Le Superamas Local de Galaxies (LSG) est assimilé à une condensation en expansion à l'intérieur d'une vacuole. Celle-ci est considérée comme plongée dans l'univers cosmologique. L'application des résultats déduits de modèles non homogènes en relativité générale, au cas particulier du LSG permet, dans une certaine mesure, de rendre compte des observations.

162.027   **Newtonian cosmology with varying G.**
J. P. Vinti.
Long-time predictions in dynamics, (see 012.005), p. 356 - 357 (1976). — Abstract.

162.028   **On the local anisotropy of expansion of the universe.**
A. G. Doroshkevich, S. F. Shandarin.

Monthly Notices Roy. Astron. Soc., Vol. 175, 15P - 18P (1976).

It is shown that the observed anisotropy of the motion of galaxies is in good agreement with results of the theory of the formation of galaxies based on the non-linear theory of gravitational instability. This result does not contradict the large scale isotropy and homogeneity of the universe because this phenomenon has a local nature and practically vanishes at large distances.

162.029 Some cosmological models with spin and torsion. II: Axially symmetric models with a uniform magnetic field. B. Kuchowicz.
Astrophys. Space Sci., Vol. 40, 167 - 181 (1976).

In axially-symmetric cosmological models of the Einstein-Cartan theory (which may be briefly called 'general relativity plus spin'), the axis of symmetry is at the same time the direction of the magnetic field, and of the aligned spins. The general set of relevant equations is given. Some exact solutions of this set constitute quasi-Euclidean and semiclosed cosmologies with a uniform magnetic field and aligned spinning matter. In contrast to the situation in the framework of general relativity, one may obtain non-singular solutions. Such a behaviour of the solutions of the Einstein-Cartan theory is rendered possible by the specific spin-spin repulsive interaction which is inherent in the theory.

162.030 Cooling of the primordial matter. R. K. Thakur.
Bull. Astron. Soc. India, Vol. 3, 34 (1975).
Abstract of a paper presented at the A.S.I. meeting 1975.

162.031 Observational tests in cosmology. E. R. Harrison.
Nature, Vol. 260, 591 - 592 (1976).

The determination of the deceleration term is one of the main aims of observational cosmology. In the paper this view is questioned, and in a universe of indeterminate density, the third time derivative of the scaling variable is shown to be of greater fundamental importance.

162.032 Exact evolution of photons in an anisotropic cosmology with scattering. W. H. Press.
Astrophys. Journ., Vol. 205, 311 - 317 (1976).

Photons or other zero-rest-mass particles in an anisotropic cosmology are subject to the stochastic processes of scattering, redshifting(in some directions), and blueshifting (in others). The resulting evolution of the stress-energy tensor, which can act back on the metric to help isotropize the expansion, has previously been studied by various approximate methods. Given in this paper is a solution to the problem (in terms of integrals over standard transcendental functions) which is exact when (1) the scattering cross-section is energy-independent and isotropic, (2) the metric is of the Kasner form, and (3) the entropy per scatterer is large. Relaxing most of these model assumptions will probably not change the qualitative nature of the solution.

162.033 Cosmological turbulence reexamined.
A. M. Anile, L. Danese, G. De Zotti, S. Motta.
Astrophys. Journ., *Letters*, Vol. 205, L59 - L63 (1976).

The authors have carried out a detailed analysis of the constraints that must be imposed on the parameters of the theory of cosmological turbulence. They show that a rather stringent upper limit to the vortex velocity on the maximum scale arises from the necessity of avoiding small-scale fluctuations of the relict radiation exceeding the observational limits. They also show that if the exact time evolution of the degree of ionization of the matter is taken into account, the damping of the eddy motions in the recombination process is more effective than previously considered. The main conclusions are: (i) the primeval turbulence theory is not compatible with observations in a low-density universe; (ii) it cannot account

for the large angular velocity of the Local Supercluster claimed by de Vaucouleurs.

162.034 Antimatter in the universe.
O. F. Prilutskij, I. L. Rozental'.
Priroda, 1976, No. 4, p. 100 - 106. In Russian.

162.035 Inhomogeneities in the early universe.
V. Canuto.
Astrophys. Journ., Vol. 205, 659 - 673 (1976).

In big-bang cosmology it is traditional to assume that before decoupling, matter was so tightly bound to radiation as to be prevented from forming any permanent structure. In particular, galaxies were difficult to form before t_D. It is the goal of this paper to point out that the previous arguments are based on linearized theories for both radiation and matter, as well as for their interaction, and that if nonlinear effects are taken into account, the situation can change considerably.

162.036 Multiple image probabilities for a spheroidal gravitational lens. R. R. Bourassa, R. Kantowski.
Astrophys. Journ., Vol. 205, 674 - 687 (1976).

The authors present the probabilities of a point source at large redshift appearing as multiple images due to single scattering from transparent spheroidal masses uniformly distributed in a Friedmann universe. The spheroids used have volume mass densities $\rho \propto 1/a$. In addition the authors present the expected minimal separations of the multiple images. The double image probabilities and separations for point mass deflectors are derived analytically and compared with the corresponding spheroidal quantities in the appropriate limit.

162.037 The extension of the Hubble diagram. I. New redshifts and BVR photometry of remote cluster galaxies, and an improved richness correction.
A. Sandage, J. Kristian, J. A. Westphal.
Astrophys. Journ., Vol. 205, 688 - 695 (1976).

Absorption-line redshifts for 37 galaxies in 31 remote clusters have been measured. Twenty-five galaxies in the sample have redshifts larger than $z = 0.20$. New photoelectric BVR photometry has been obtained for 16 of the clusters. The data permit an improvement in the statistics of the Hubble diagram to $z = 0.28$ and a new determination of the dependence of the absolute magnitude of the brightest cluster galaxy on cluster richness. The new points in the Hubble diagram give no indication that the (m, z) relation deviates from the q_0(formal) = + 1 line.

162.038 A cosmic virial theorem. P. J. E. Peebles.
Astrophys. Journ.,(*Letters*), Vol. 205, L109 - L111 (1976).

A "cosmic virial theorem" involving the two-point and three-point galaxy correlation functions is described. It is based on a stability condition, that the time scale for evolution of the correlation functions is long compared to a crossing time. This yields a condition on the mean mass density in galaxies and the mean square relative velocity of pairs of galaxies as a function of separation. Tentative results from available data and the prospects for a more detailed test are discussed.

162.039 Baryon number instability and the matter content of a charge-symmetric universe.
E. Etim, A. F. Grillo, M. Grilli, L. Donazzolo, F. Occhionero.
Astron. Astrophys., Vol. 49, 97 - 102 (1976).

It is proposed that, as a charge-symmetric universe cools through the hadron era, it undergoes a phase transition which breaks the underlying charge conjugation symmetry giving rise thereby to separate regions of matter and antimatter. The matter content of the universe at the critical temperature is calculated using the statistical bootstrap-model. The observed value of the entropy per baryon is deduced therefrom and shown to be related to the ratio of electromagnetic to strong interaction couplings.

162.040 **Thermal radiation produced by the expansion of the universe.** L. Parker.
Nature, Vol. 261, 20 - 23 (1976).

The author shows that the expansion of the universe can create particles with a black-body distribution. He gives a simple model in which the creation of particles in an initially empty spacetime leads at very early times to a Friedmann expansion dominated by hot relativistic particles (probably including gravitons) having a black-body spectrum, even before classical thermalising processes have had a chance to act. If the particle creation occurs near the Planck time, the entropy produced is consistent, in order of magnitude, with that required by the Einstein equations. The conditions in which one achieves consistency with the observed 3K black-body radiation are also discussed.

162.041 **The rate of growth of density perturbations in simple big-bang model universes.**
D. Edwards, D. Heath.
Astrophys. Space Sci., Vol. 41, 183 - 193 (1976).

The authors present exact solutions to the density perturbation equation derived by Bonnor for the cases where $\Lambda = \Lambda_c$, $k = 1$ and $\Lambda = -\Lambda_c$, $k = -1$. The solutions when $\Lambda = 0$, $k = 1$ and $\Lambda = 0$, $k = -1$ have been previously published. Using these solutions a quantitative analysis has been carried out that has enabled to estimate the size of the fluctuations that must be postulated at decoupling in order to explain the formation of the galaxies in these model universes.

162.042 **General cosmological solution of Einstein equations with singularity in time.**
V. A. Belinskij, E. M. Livshits, I. M. Khalatnikov.
Problems of gravitation. Third Soviet Gravitational Conference, Erevan, 1972, (see 012.008), p. 136 - 164 (1975). In Russian.

162.043 **Some cosmological consequences of primordial black-hole evaporations.** B. J. Carr.
Astrophys. Journ., Vol. 206, 8 - 25 (1976).

According to Hawking, primordial black holes of less than 10^{15} g would have evaporated by now. This paper examines the way in which small primordial black holes could thereby have contributed to the background density of photons, nucleons, neutrinos, electrons, and gravitons in the universe. The author predicts the spectrum of neutrinos, electrons, and gravitons which should result from primordial black-hole evaporations and shows that the observational limits on the background electron flux might place a stronger limitation on the number of 10^{15} g primordial black holes than the γ-ray observations. Finally, he examines the limits that various observations place on the strength of any long-range baryonic field whose existence might be hypothesized as a means of preserving baryon number in black-hole evaporations.

162.044 **Big-bang nucleosynthesis with nonzero lepton numbers.** A. Yahil, G. Beaudet.
Astrophys. Journ., Vol. 206, 26 - 29 (1976).

Big-bang nucleosynthesis has been calculated for a universe with nonzero and unequal electron and muon lepton numbers. A comparison of the observed abundances of ^{4}He and D with the values obtained in the calculations shows that the introduction of nonzero lepton numbers allows a wide margin of mean mass-energy densities at the present epoch, 10^{-31} g cm$^{-3} \lesssim \rho_0 \lesssim 10^{-28}$ g cm^{-3}. Hence the deuterium abundance is not as sensitive a gauge of ρ_0 as has been thought. The permissible ranges of all the quantum numbers of the universe are discussed.

162.045 **Boltzmanns Kosmogonie und die hierarchische Struktur des Kosmos.** H.-J. Treder.
Astron. Nachr., Vol. 297, 117 - 126 (1976).

Boltzmann's cosmogony of quasi-isolated space domains means that the manifold of the possible pre-histories of an isolated system is a measure of the statistic weight for the momentary state of this system itself. This statement clears Poincaré's and Zermelo's „Wiederkehreinwand" against the physical meaning of Boltzmann's H-theorem and proves the epistemological and methodological meanings of "documents" for the historical views in sciences. By this, the universe is a hierarchy of systems without limits in space or in time. – The metagalaxy of the relativistic cosmology is one system in this infinite manifold of quasi-isolated space domains, only.

162.046 **The inversion of the Einstein-Straus theorem.** U. Kasper.
Astron. Nachr., Vol. 297, 127 - 129 (1976).

We consider the inversion of a problem put by A. Einstein and E. G. Straus, that is, we ask for restrictions on the scaling factor $R(t)$ of the Robertson-Walker metric and the functions $H^2(r')$ and $A^2(r')$ of a spherically symmetric and static vacuum metric, which are consequences of the requirement that the vacuum metric shall pass continuously differentiable into the Robertson-Walker metric at a certain value r_b of the comoving radial coordinate r.

162.047 **On the possibility of a finite model describing the universe.** G. Järnefelt.
Astron. Nachr., Vol. 297, 131 - 139 (1976). – Summary of a letter from G. Järnefelt to H.-J. Treder (August 1975).

In his book "The expanding universe" Eddington (1933) suggested the idea of an astronomical world model containing a finite number of material particles. Starting from this idea the author has speculated about the possibility to create a cosmological model where in addition also the number of geometrical points and time-instants is finite. The ultimate goal would be a world model based essentially upon Galois fields. In order to explore such a possibility the most simple actually existing physical system, the structure describing a nonrelativistic free particle in a one-dimensional Euclidean space, is treated here.

162.048 **Hydrodynamics of the universe.** A. D. Chernin.
Priroda, 1976, No. 6, p. 108 - 118. In Russian.

162.049 **A matter and radiation filled universe: consequences of the astronomical observations.** E. Nowotny.
Astrophys. Journ., Vol. 206, 402 - 410 (1976).

The Friedmann equations are solved for a universe containing matter and radiation in an arbitrary ratio without interaction between these two constituents. The consequences on the age of the universe, the Hubble relation, the counting method, and the diameter of galaxies are derived with dependence on a cosmological parameter and a matter parameter.

162.050 **The arrow of time.** D. Layzer.
Astrophys. Journ., Vol. 206, 559 - 569 (1976).

Four distinct classes of physical processes define a preferred direction in time: entropy-generating processes, which define the thermodynamic arrow; information-generating processes, which define the historical arrow; the cosmic expansion, which defines the cosmological arrow; and the decay of neutral kaons, which defines the microscopic arrow. The theory presented here shows that the thermodynamic, historical, and cosmological arrows may be derived from a pair of closely related cosmological postulates.

162.051 **A new cosmological test for q_0.** D. A. Schwartz.
Astrophys. Journ., (Letters), Vol. 206, L95 - L97

(1976).

Many difficulties encountered with the classical global measurements of q_0 may be relieved by using clusters of galaxies instead of individual galaxies as the observational object, and by using the theoretical relation for the differential number of objects at a given redshift instead of the relations for apparent magnitude or apparent size. The author shows that such a test is feasible, and considers briefly the effects of time evolution on the practicality of the test.

162.052 Vorticity perturbations and isotropy of the cosmic microwave background.
A. M. Anile, S. Motta.
Astron. Astrophys., Vol. 49, 205 - 209 (1976).

The authors investigate the effect of vorticity perturbations of an arbitrary Robertson-Walker universe on the isotropy of the cosmic microwave background. They obtain an upper limit on the present vorticity on scales $L \sim 10$ Mpc which is only marginally consistent with the value suggested by de Vaucouleurs (1971), de Vaucouleurs and Peters (1968).

162.053 Particle generation in the vortex cosmological model.
V. N. Lukash, I. D. Novikov, A. A. Starobinskij.
Zhurn. ehksperim. i teor. fiz., Vol. 69, 1484 - 1500 (1975).
In Russian. – Abstr. in Referativ. Zhurn. 51. Astron., 5.51.951 (1976).

162.054 Hydrodynamics of the universe.   Ya. B. Zel'dovich.
Uspekhi mat. nauk, Vol. 30, 204 (1975). In
Russian. – Abstr. in Referativ. Zhurn. 51. Astron., 5.51.952 (1976).

162.055 Dirac's continuous creation cosmology and the temperature of the earth.   I. W. Roxburgh.
Nature, Vol. 261, 301 - 302 (1976). – Letter.

162.056 Variable G: a solution to the missing mass problem.
B. M. Lewis.
Nature, Vol. 261, 302 - 304 (1976).

Dirac suggested in 1937 that the gravitational 'constant' G might vary with time. The hypothesis of a variable gravitational constant can completely solve the 'missing mass' problem, and promises to assist the study of the formation, structure and evolution of galaxies.

162.057 Why do we know so little about the universe?
S. von Hoerner.
Naturwissenschaften, 63. Jahrgang, p. 212 - 217 (1976).

162.058 Leptonic numbers and the neutron to proton ratio in the hot Big Bang model.
G. Beaudet, P. Goret.
Astron. Astrophys., Vol. 49, 415 - 419 (1976).

The neutron to proton (n/p) ratio prior to the nucleosynthesis era ($T > 10^9 \,^\circ K$) is calculated as a function of the temperature in a hot Big Bang model with non zero electron and muon leptonic numbers.

162.059 Neutrino rest mass from cosmology.
A. S. Szalay, G. Marx.
Astron. Astrophys., Vol. 49, 437 - 441 (1976).

In standard cosmological models, the overall mass density of the universe can be calculated from the observed value of the Hubble constant H_0 and the deceleration parameter q_0. Their most recent values suggest a density considerably higher than the estimated density of the known matter in the universe. The missing mass may be explained by the relict cosmological neutrinos, produced in the hot era following the Big Bang, if nonvanishing neutrino and neutretto rest masses are assumed. The cosmological evolution of the universe has

been calculated in this model. The observed values of H_0, q_0 and t_0 (age of the universe) agree with the cosmological model if one chooses an appropriate value for the neutrino mass m. The upper limit on the neutrino and neutretto rest mass obtained in this way is $m = 13.5$ eV. The relict neutrinos with a rest mass could form a halo around clusters of galaxies: this halo would influence the density profile of the cluster in the outer region. The final conclusion is that a neutrino or neutretto rest mass larger than 15 eV would contradict the astrophysical evidence.

162.060 Zur Grundaufgabe der Kosmologie.
D.-E. Liebscher.
Sterne, Vol. 52, 65 - 76 (1976).

162.061 Beobachtungsparameter und Weltmodelle.
H. Oleak.
Sterne, Vol. 52, 77 - 82 (1976).

162.062 Particle creation and Dirac's large numbers hypothesis.   G. Steigman.
Nature, Vol. 261, 479 - 480 (1976).

In connection with cosmologies based on the large numbers hypothesis (LNH), Dirac has suggested that continuous creation of matter is required. The author demonstrates here that, within the context of the LNH, the number of particles in the universe varies as expected ($N \propto t^2$) so that particle creation is unnecessary. Some unpleasant features of cosmologies based on the LNH are also discussed.

162.063 Quarks in the early universe.   G. F. Chapline.
Nature, Vol. 261, 550 - 551 (1976).

The quark theory of hadrons provides a valuable tool for interpreting conditions during early epochs of the universe. In particular, an asymptotically free theory of quarks seems to rule out the possibility that the present-day entropy of the universe could be due to dissipation of low level fluctuations. Apparently either the early universe was inhomogeneous on small mass scales or anisotropic.

162.064 Long gravitational waves in a closed universe.
L. P. Grishchuk, A. G. Doroshkevich, V. M. Yudin.
Zhurn. ehksperim. i teor. fiz., Vol. 69, 1857 - 1871 (1975). In Russian. – Abstr. in Referativ. Zhurn. 51. Astron., 6. 51. 807 (1976).

162.065 On some properties of the type IX cosmological model with moving matter.   O. I. Bogoyavlenskij.
Zhurn. ehksperim. i teor. fiz., Vol. 70, 361 - 373 (1976). In Russian. – Abstr. in Referativ. Zhurn. 51. Astron., 6. 51. 804 (1976).

162.066 Creation of gravitons in a Friedmann universe.
B. V. Vajner, P. D. Nasel'skij.
Pis'ma v ZhurnEhTF, Vol. 23, 141 - 145 (1976). In Russian. Abstr. in Referativ. Zhurn. 51. Astron., 6. 51. 808 (1976).

162.067 Forces cosmologiques répulsives: un mécanisme de fragmentation de l'univers?   J. Eisenstaedt.
Comptes Rendus Acad. Sci. Paris, Sér. A, Vol. 282, 1063 - 1065 (1976).

Les équations des géodésiques des solutions cosmologiques inhomogènes permettent de mettre en évidence des forces répulsives. A partir des fluctuations statistiques de la densité cosmologique, un tel mécanisme pourrait contribuer à expliquer la formation des amas de matière actuellement observés.

162.068 Analyse supplémentaire des déplacements vers le rouge liés à la traversée des amas de galaxies par la lumière.   L. Nottale.

Comptes Rendus Acad. Sci. Paris, Sér. B, Vol. 282, 519 - 522 (1976).

L'interprétation de l'excès de module de Hubble (h_m = log V − 0.2m) présenté par les galaxies situées derrière un amas, en terme d'un accroissement du déplacement vers le rouge de leur spectre est confirmée.

162.069 Galaxy clustering: its description and its interpretation. D. Layzer.
Galaxies and the universe, (see 003.010), p. 665 - 723 (1975).
The analysis of galaxy counts; Dynamics of clustering; Origin of clustering.

162.070 Radio astronomy and cosmology.
P. A. G. Scheuer.
Galaxies and the universe, (see 003.010), p. 725 - 760 (1975).
Summary of basic formulae; Attempts to detect intergalactic atomic hydrogen; Attempts to detect intergalactic ionized hydrogen; The microwave background; Counts of radio sources; The evolution of the luminosity function of quasi-stellar objects.

162.071 The redshift. A. Sandage.
Galaxies and the universe, (see 003.010), p. 761 - 785 (1975).
Early results on the spectra and redshifts of galaxies (The first observations, De Sitter's static solution, Hubble and Humason's observational extension of the redshift law); Aids in the practical measurement of redshifts by photographic methods; Redshifts of bright galaxies; constancy of $\Delta\lambda/\lambda_0$; systematic errors; Corrections to measured redshifts; Absolute luminosity and intrinsic diameter as functions of observables in specific models; The observations and world models.

162.072 Heaven and its properties. E. T. Newman.
General Relativ. Gravitation, Vol. 7, 107 - 111 (1976).

162.073 Osservazioni in campo ottico richieste per l'interpretazione teorica dei problemi relativi alla Galassia e alla cosmologia. N. Dallaporta.
Giorn. Astron., Vol. 1, 109 - 117 (1975).

162.074 Die lokale Massendichte und die intergalaktische Materie. S. Marx, K.-H. Schmidt.
Sterne, Vol. 52, 155 - 161 (1976).

162.075 Why is the universe like what it is? W. H. McCrea.
Irish Astron. Journ., Vol. 12, 95 - 102 (1975).

162.076 Big bang cosmology and the cosmic black-body radiation. R. A. Alpher, R. Herman.
Proc. American Phil. Soc., Vol. 119, 325 - 348 (1975).

162.077 On the dynamics of clusters of galaxies in a universe with a weak gravitation field. T. B. Omarov.
Mem. Soc. Astron. Italiana, Vol. 45, (see 012.017), 687 - 688 (1974).

162.078 Kosmologiske modeller − hvilket univers lever vi i?
Y. Hartvigsen.
Naturen, Årg. 100, 17 - 24 = Inst. Teor. Astrofys., Blindern−Oslo, Småtr. No. 86 (1976).

162.079 Cosmological models with the cosmical constant containing matter and radiation. II. T. Fukui.
Sci. Rep. Tôhoku Univ., 200 - 213 (1975).
Various cosmological quantities and their relations are computed in order to find out suitable ranges of the parameters appeared in our assumption which explain the observational data as a whole. The range of possible values of the

cosmical constant Λ is also discussed.

162.080 Struktur des Kosmos. H. Oleak.
Astron. in der Schule, 13. Jahrgang, p. 51 - 53 (1976).

162.081 Solutions non statiques en théorie de Jordan-Thiry.
P. Pigeaud.
Comptes Rendus Acad. Sci. Paris, Sér. A, Vol. 282, 1387 - 1390 (1976).

162.082 Les modèles d'univers cosmologiques.
D. Proust.
A.F.O.E.V. Bull., Tome 10, 17 - 22 (1976).

162.083 Il modello di Klein e la cosmologia.
G. Arcidiacono.
Collect. math., Vol. 25, 159 - 184 (1974). − Abstr. in Zentralbl. Math. Grenzgebiete − Math. Abstr.,Band 308, No. 76073 (1976).

162.084 On the initial singularity in the scalar-tensor anisotropic cosmology.
V. A. Ruban, A. M. Finkelstein (*Finkel'shtejn*).
General Relativ. Gravitation, Vol. 6, 601 - 638 (1975).
In connection with the problem of the initial singularity in the scalar-tensor anisotropic cosmology of Jordan-Brans-Dicke, the dynamics of homogeneous models of Bianchi type I is examined on the basis of the general analytic solutions in vacuo and in the presence of gravitating matter with state equations $P = n\epsilon$ $(0 \leqslant n \leqslant 1)$.

162.085 Can the effect of distant matter on physical observables be observed.
T. Ohta, T. Kimura, K. Hiida.
Nuovo Cimento B, Ser. 11, Vol. 27B, 103 - 120 (1975).
Abstr. in Phys. Abstr., Vol. 79, A003239 (1976).

162.086 Relativistic invariance and the expansion of the universe. A. Bergstrom.
Nuovo Cimento B, Ser. 11, Vol. 27B, 145 - 160 (1975).
Abstr. in Phys. Abstr., Vol. 79, A003243 (1976).

162.087 On empty horizons in cosmological models.
A. R. King.
Phys. Letters A, Vol. 54A, 115 - 116 (1972). − Abstr. in Phys. Abstr., Vol. 79, A003244 (1976).

162.088 Mach's principle in Newtonian cosmology.
A. M. Anile.
Nuovo Cimento B, Ser. 11, Vol. 29B, 31 - 38 (1975). − Abstr. in Phys. Abstr., Vol. 79, A006880 (1976).

162.089 Entropy in an oscillating universe.
P. T. Landsberg, D. Park.
Proc. Roy. Soc. London, Ser. A, Vol. 346, 485 - 495 (1975).
Abstr. in Phys. Abstr., Vol. 79, A006881 (1976).

162.090 Observational validation of the chronometric cosmology. I. Preliminaries and the redshift-magnitude relation. I. E. Segal.
Proc. National Acad. Sci. USA, Vol. 72, 2473 - 2477 (1975).
Abstr. in Phys. Abstr., Vol. 79, A011081 (1976).

162.091 Cosmological implications of anomalous redshifts − a possible working hypothesis. T. Jaakkola,
M. Moles, J. P. Vigier, J.-C. Pecker, W. Yourgrau.
Found. Phys., Vol. 5, 257 - 269 (1975). − Abstr. in Phys. Abstr., Vol. 79, A011262 (1976).

162.092 Mach's principle: micro- or macrophysical.

J. F. Woodward, W. Yourgrau.
British Journ. Philosophy Sci., Vol. 26, 137 - 141 (1975).
Abstr. in Phys. Abstr., Vol. 79, A012106 (1976).

162.093 Evolution of irregularities in a chaotic early universe.
K. Tomita.
Progr. Theor. Phys., Vol. 54, 730 - 739 (1975). — Abstr. in
Phys. Abstr., Vol. 79, A015173 (1976).

162.094 Scalar particle creation in an anisotropic universe.
B. K. Berger.
Phys. Rev. D, Vol. 12, 368 - 370 (1975). — Abstr. in Phys.
Abstr., Vol. 79, A015175 (1976).

162.095 On the acoustic decay of primordial cosmic turbulence.
K. Tanabe, H. Nariai, T. Matsuda, H. Takeda.
Progr. Theor. Phys., Vol. 54, 719 - 729 (1975). — Abstr. in
Phys. Abstr., Vol. 79, A019337 (1976).

162.096 Whimper singularities in comoving coordinates.
A. B. Evans.
Phys. Letters A, Vol. 55A, 271 - 272 (1975). — Abstr. in Phys.
Abstr., Vol. 79, A020298 (1976).

162.097 A static universe filled with spinning matter and magnetic field.   B. Kuchowicz.
Current Sci., Vol. 44, 537 - 538 (1975). — Abstr. in Phys.
Abstr., Vol. 79, A023613 (1976).

162.098 Curvature collineation for the plane symmetric cosmological model.   K. P. Singh, Shri Ram.
Indian Journ. Pure Appl. Math., Vol. 5, 241 - 245 (1974).
Abstr. in Phys. Abstr., Vol. 79, A023614 (1976).

162.099 Cosmological model with expansion, shear, and vorticity.   N. Batakis, J. M. Cohen.
Phys. Rev. D, Vol. 12, 1544 - 1550 (1975). — Abstr. in Phys.
Abstr., Vol. 79, A023615 (1976).

162.100 Numerical examples from perturbation analysis of the mixmaster universe.   B. L. Hu.
Phys. Rev. D, Vol. 12, 1551 - 1562 (1975). — Abstr. in Phys.
Abstr., Vol. 79, A023616 (1976).

162.101 Spherical mass immersed in a cosmological universe: a class of solutions. II.   J. Eisenstaedt.
Phys. Rev. D, Vol. 12, 1573 - 1575 (1975). — Abstr. in Phys.
Abstr., Vol. 79, A023617 (1976).

162.102 Gravitating 't Hooft monopoles.
Y. M. Cho, P. G. O. Freund.
Phys. Rev. D, Vol. 12, 1588 - 1589 (1975). — Abstr. in Phys.
Abstr., Vol. 79, A023618 (1976).

162.103 Sound velocity in a curved space-time.
V. V. Petrov.
Izv. vyssh. ucheb. zavedenij fiz., No. 11, p. 73 - 75 (1975).
In Russian. — Abstr. in Phys. Abstr., Vol. 79, A027704 (1976).

162.104 Einstein-Cartan cosmologies with a magnetic field.
A. K. Raychaudhuri.
Phys. Rev. D, Vol. 12, 952 - 955 (1975). — Abstr. in Phys.
Abstr., Vol. 79, A027705 (1976).

162.105 The rotation of the universe.   A. J. Fennelly.
Thesis Yeshiva Univ., New York, USA, 143 pp.
(1975). (Available from: Univ. Microfilms, Ann Arbor, Mich.,
USA. Order No. 75-20584). — Abstr. in Phys. Abstr., Vol. 79,
A027709 (1976).

162.106 Solutions for the general cylindrically symmetric stationary dust model.   J. C. Zimmerman.
Journ. Math. Phys., Vol. 16, 2458 - 2460 (1975). — Abstr. in
Phys. Abstr., Vol. 79, A028746 (1976).

162.107 Global time problem in relativistic cosmology.
M. Heller.
Ann. Soc. Sci. Bruxelles, Ser. I, Vol. 89, 522 - 532 (1975).
Abstr. in Phys. Abstr., Vol. 79, A032679 (1976).

162.108 Hamiltonian cosmology: a further investigation.
G. E. Sneddon.
Journ. Phys. A, Vol. 9, 229 - 238 (1976). — Abstr. in Phys.
Abstr., Vol. 79, A032680 (1976).

162.109 A cylindrically symmetric expanding universe.
S. R. Roy, P. N. Singh.
Journ. Phys. A, Vol. 9, 255 - 259 (1976). — Abstr. in Phys.
Abstr., Vol. 79, A032681 (1976).

162.110 Some viscous fluid cosmological models of plane symmetry.   S. R. Roy. S. Prakash.
Journ. Phys. A, Vol. 9, 261 - 267 (1976). — Abstr. in Phys.
Abstr., Vol. 79, A032682 (1976).

162.111 On the difference in creation of particles with spin 0 and 1/2 in isotropic cosmologies.
V. M. Frolov, S. G. Mamayev, V. M. Mostepanenko.
Phys. Letters A, Vol. 55A, 389 - 390 (1976). — Abstr. in Phys.
Abstr., Vol. 79, A032683 (1976).

162.112 On the average effect of a highly turbulent gravito-hydrodynamic field in the hadron era of the universe. III. Complete set of dynamical equations.   H. Nariai.
Progr. Theor. Phys., Vol. 54, 1356 - 1367 (1975). — Abstr. in
Phys. Abstr., Vol. 79, A032684 (1976).

162.113 f-gravity and Dirac's large numbers hypothesis.
C. Sivaram, K. P. Sinha.
Phys. Letters B, Vol. 60B, 181 - 182 (1976). — Abstr. in Phys.
Abstr., Vol. 79, A032685 (1976).

162.114 Cosmology and particle pair production via gravitational spin-spin interaction in the Einstein-Cartan-Sciama-Kibble theory of gravity.   G. D. Kerlick.
Phys. Rev. D, Vol. 12, 3004 - 3006 (1975). — Abstr. in Phys.
Abstr., Vol. 79, A037414 (1976).

162.115 The Einstein-Cartan equations in astrophysically interesting situations. II. Homogeneous cosmological models of axial symmetry.   B. Kuchowicz.
Acta Phys. Polonica B, Vol. B7, 81 - 97 (1976). — Abstr. in
Phys. Abstr., Vol. 79, A037420 (1976).

162.116 On the b-boundary of the closed Friedmann model.
B. Bosshard.
Commun. Math. Phys., Vol. 46, 263 - 268 (1976). — Abstr. in
Phys. Abstr., Vol. 79, A037423 (1976).

162.117 A cosmological characteristic initial value problem.   D. S. Chellone.
Journ. Phys. A, Vol. 9, 337 - 342 (1976). — Abstr. in
Phys. Abstr., Vol. 79, A038482 (1976).

162.118 The relation of cosmology to the microcosm.
D. D. Ivanenko.
Izv. vyssh. ucheb. zavedenij fiz., No. 1, 49 - 51 (1976). In Russian. — Abstr. in Phys. Abstr., Vol. 79, A042017 (1976).

162.119 Strong (f) gravity, Dirac's large numbers hypothesis and the early hadron era of the big-bang universe.

C. Sivaram, K. P. Sinha.
Journ. Indian Inst. Sci., Vol. 57, 257 - 269 (1975). – Abstr. in Phys. Abstr., Vol. 79, A042018 (1976).

162.120 On the Hamiltonian treatment of a quantum scalar field in a Bianchi I universe. I. Time-dependent operator theory.   E. Pessa.
Nuovo Cimento Lettere, Ser. 2, Vol. 15, 291 - 294 (1976). Abstr. in Phys. Abstr., Vol. 79, A042019 (1976).

162.121 Exact solution of the Lifshitz equations governing the growth of fluctuations in cosmology.
P. J. Adams, V. Canuto.
Phys. Rev. D, Vol. 12, 3793 - 3799 (1975). – Abstr. in Phys. Abstr., Vol. 79, A042021 (1976).

162.122 Use of supernovae light curves for testing the expansion hypothesis and other cosmological relations.
B. W. Rust.
Thesis Illinois Univ., Urbana, USA, 382 pp. (1974). (Available from Univ. Microfilms, Order No. 75-11,539).

162.123 On characteristic turbulent quantities at the bounce epoch in our hadron-dominated model of the universe.   H. Nariai.
Separate print Hiroshima Univ., Takehara, Japan, Res. Inst. Theoret. Phys., 4 pp. (1974). – See 13.162.099).

162.124 On a chaotic early universe.   K. Tomita.
Separate print Hiroshima Univ., Takehara, Japan, Res. Inst. Theoret. Phys., 15 pp. (1974).

162.125 New evidence for an evolutionary universe from observations of radio galaxies with the Ooty radiotelescope.   V. K. Kapahi.
Phys. News, Vol. 6, No. 4, p. 163 (1975). – Summary.

The universe and its structure.
See Abstr. 003.032.

The universe: its beginning and end.
See Abstr. 003.084.

Mathematical cosmology and extragalactic astronomy.   See Abstr. 003.097.

Limit on the variation of the proper mass of an electron with time.   See Abstr. 022.057.

Light elements and the isotropy of the universe.
See Abstr. 061.015.

Possible role of strongly interacting fields in astrophysics and relativistic cosmology.   See Abstr. 061.029.

On the average effect of a highly turbulent gravitohydrodynamic field in the hadron era of the universe. 2. Consideration of dissipative processes.   See Abstr. 062.096.

The electromagnetic background: limitations on models of unseen matter.   See Abstr. 066.002.

Is gravity getting weaker?
See Abstr. 066.007.

Entropy production by black holes.
See Abstr. 066.010.

Annihilation of matter and antimatter and the cosmic X-ray background.   See Abstr. 066.013.

The gravitational perturbation of the cosmic background radiation by density concentrations.
See Abstr. 066.026.

Theoretical frameworks for testing relativistic gravity. V. Post-Newtonian limit of Rosen's theory.
See Abstr. 066.049.

The gravitational collapse of a matter-antimatter symmetric gas sphere.   See Abstr. 066.059.

New test of the equivalence principle from lunar laser ranging.   See Abstr. 066.071.

Radiation from a moving mirror in two dimensional space-time: conformal anomaly.   See Abstr. 066.129.

On the Hamiltonian treatment of a quantum scalar field in a Bianchi I universe. II. The explicit construction of the Hamiltonian.   See Abstr. 066.134.

Black holes – a way out of the universe.
See Abstr. 066.156.

Annihilation of matter and antimatter and the cosmic X-ray background.   See Abstr. 066.158.

Gravity and the earth's rotation.
See Abstr. 081.023.

New observational data on supernovae.
See Abstr. 125.006.

On the measurement of the extragalactic background brightness at 4000 Å.   See Abstr. 131.034.

Structure and evolution of compact radio sources.
See Abstr. 141.020.

Neue Erkenntnisse über die rätselhaften Quasare.
See Abstr. 141.027.

Kinematics of relativistic ejection.
See Abstr. 141.056.

A survey for emission-line galaxies and quasars. III. A list of nine new optically selected QSOs with $2.5 < z < 3.1$.
See Abstr. 141.097.

Pulsar spin down and cosmologies with varying gravity.   See Abstr. 141.346.

Gamma-dating of the epoch of formation of galaxies.   See Abstr. 151.067.

Evolutionary synthesis of the stellar population in elliptical galaxies. I. Ingredients, broad-band colors, and infrared features.   See Abstr. 158.005.

The origin of galaxies: a review of recent theoretical developments and their confrontation with observation.
See Abstr. 158.020.

The Hubble diagram for nuclear magnitudes of cluster galaxies.   See Abstr. 160.003.

Is the local supercluster a physical association?
See Abstr. 160.005.

Possible implications of the Rubin-Ford effect.
See Abstr. 160.013.

Supergalactic studies. V. The supergalactic anisotropy of the redshift-magnitude relation derived from nearby groups and Sc galaxies.   See Abstr. 160.018.

The peculiar velocity field in the Local Supercluster. See Abstr. 160.023.

Globular clusters in the Hydra I cluster of galaxies. See Abstr. 160.026.

Covariance function analysis and the clustering of galaxies.   See Abstr. 160.029.

On the correlation between color and absolute magnitude in first ranked cluster galaxies.   See Abstr. 160.036.

High-order clustering of galaxies — new methods. See Abstr. 160.039.

The dynamics of the intergalactic medium in the vicinity of clusters of galaxies.   See Abstr. 161.001.

Intergalactic extinction and the quasar cut-off. See Abstr. 161.004.

Author Index

The authors are listed in alphabetical order
according to the initial letter following the first names.

Branch, D.
022.018
114.013
Brancik, K.
121.103
Brand, K. P.
131.068
Brand, P. W. J. L.
131.533
Brandt, L.
004.001
Branduardi, G.
142.011 .066
Brannon, J. C.
094.442 .476 .526 .528
Bratijchuk, M. V.
054.005 .006
Bratolyubova-Tsulukidze,
L. B.
154.025
Bratolyubova-Tsulukidze,
L. S.
142.085
Bravo, S.
143.251
Bray, A. D.
143.137
Bray, M.
162.025
Braz, M. A.
141.118
Breakwell, J. V.
052.043 .058 .061
Breazeal, J.
062.077
Brecher, A.
094.442 .543 .579
Brecher, K.
066.012
124.102
141.306
142.116
160.024
Breckinridge, J. B.
031.234
082.048
Bredekamp, J. H.
143.022
Breger, M.
131.052
Bregman, J. D.
033.001
Breitenecker, M.
042.090
Breuer, R.
066.021
142.217
Breuer, R. A.
003.026
Breus, T. K.
097.052
Brewer, S. H.
031.014
Brewster, M.
122.057 .058
123.006
Bricard, J.
082.079
Bridge, H. S.
093.010

Bridle, A. H.
158.314
Brinkman, A. C.
132.018
141.350
142.066 .081 .117 .207
.211 .213
Brinkmann, R. T.
099.201
Brinkmann, W.
142.149
Bristeau, P.
094.139 .207
Britkov, V. B.
052.030
Broadfoot, A. L.
082.033
093.029
Broderick, J. J.
114.347
158.315
Brodskaya, Eh. S.
122.078
Brodzinski, R. L.
094.175
Bronnikova, N.
098.041
Bronnikova, N. M.
103.103 .106
123.007
Bronshtehn, V. A.
003.027
098.037
105.050
Bronshtejn, Yu. L.
031.002
Bronshten
See Bronshtehn
Brookes, J. R.
080.009
Brooks, J. W.
131.030
Broqua
098.001
Broten, N. W.
131.116
Broucke, R.
042.021
Brouw, W. N.
031.253
Brown, D. A.
121.020
Brown, G. M.
072.001
094.415
Brown, J. C.
061.028
Brown, N.
104.004
Brown, P. L.
003.028
Brown, R. A.
099.047
Brown, R. D.
081.037
131.140
Brown, R. Hanbury
See Hanbury Brown, R.
Brown, R. L.
114.347
131.039 .046 .123 .133

Brown, R. L.
131.519 .538
158.315
Brown, R. W.
094.528 .607
Brown, S. E.
066.143
Brown III, W. E.
083.018
Brown Jr., W. E.
094.106 .108
Brownd, J. E.
081.038
Browne, G. C.
099.012
Browne, I. W. A.
141.099
Brownlee, D. E.
094.179
105.084
106.040
Brownlee, R. G.
143.295
Bruce, C. E. R.
122.151
Bruck, Yu. M.
141.332
Brucker, A.
003.057
Brueck, H. A.
005.025
Brueckner, G. E.
073.036
074.018
Brueckner, V.
121.118
Brueckner, W.
072.018
Bruenn, S.
125.042
Bruevich, A. N.
033.033
Bruk, Ju. M.
See Bruk, Yu. M.
Bruk, Yu. M.
065.016
Brun, R.
061.067
Brune Jr., R. A.
066.078
Bruwer, J. A.
103.106 .131
Bryan, W. B.
094.538
Bryant Jr., W. C.
052.057
Brzostkiewicz, S. R.
032.029
097.063
Bucci, P.
065.091 .093
Buchler, J.-R.
061.010 .023
125.043
Buck, R. M.
078.002
084.410
Bucknell, M. J.
158.075
Budden, K. G.
084.403

Dixon, T. A.
061.019
Dixon, W. J.
053.028
Djurovic, D.
044.002 .013 .028 .029
Djuth, F. T.
142.128
Dluzhnevskaya, O. B.
153.014
Dmaiao Soares, I.
061.058
Dmitriev, B. A.
143.027
Dmitriev, V. I.
094.183 .411
Dmitrievskij, A. A.
104.019
Dobaczewska, W.
011.002
052.019
Dobrotin, N. A.
078.056 .083
143.264
Dobrovol'skij, A. V.
034.018
Dobrowclny, M.
074.027
084.211
141.322
Doby, P.
121.122
Dodd, R. T.
105.039 .073 .088
Dodds, D.
143.049
Dodero, M. A.
143.272
Dodgen, D.
032.041
Dodon, G.
084.257
D'Odorico, S.
125.019
155.012
Dodson, W. H.
083.001
Doebele, H. F.
031.227
Dogiel, V. A.
143.158
155.048
Dogujaev, V. A.
143.227
Dohnanyi, J. S.
104.048
105.021
106.079
Doi, K.
142.214
Doke, T.
143.068 .322 .354
Dokuchaev, V. P.
141.349
Dolginov, A. Z.
106.032
Dolginov, Sh. Sh.
094.401
Dolgoarshinnykh, B. G.
072.030

Dolidze, M. V.
114.045 .046 .047
122.037 .112
Dolique, J. M.
062.074
Dollfus, A.
094.540
Domingo, V.
074.010
078.075
084.229
Dominski, I.
044.017
Dommanget, J.
111.004
Domokos, G.
162.007
Donahue, T. M.
091.016
097.038
Donaldson, C. H.
094.418 .460 .469 .528
Donath, F. A.
003.012
Donazzolo, L.
162.039
Dongen, L. V.
033.028
Donn, B.
131.159
Donn, B. D.
103.100 .101
Donnison, J. R.
064.062
Dopita, M. A.
132.011
Doremus, J. P.
151.034 .062 .063
Dorfman, M. D.
105.112
Dorman, H. J.
094.180
Dorman, I. V.
078.044
143.045
Dorman, J.
104.045
Dorman, L. I.
003.006
078.015 .031 .033 .034
    .035 .044 .045
106.039
143.007 .014 .016 .032
    .037 .044 .045 .046
    .176 .183 .186 .190
    .191 .192 .193 .194
    .195 .202 .203 .204
    .205 .219 .234 .266
Dormand, J. R.
107.005
Doroshenko, V. T.
124.102
158.051
Doroshkevich, A. G.
065.013
162.028 .064
Dorschner, J.
003.038
061.003
131.076 .129

Dory, R. A.
062.047
Dos Santos, P. M.
141.118
Doschek, G. A.
071.001 .028
Dott Jr., R. H.
003.039
Doty, J.
142.164 .171
Doty, J. P.
142.046 .148
160.024
Dougherty, L. M.
035.003
Douglas, A. V.
004.020
008.052
Douglas, B. C.
046.012 .049
081.036
Douglas, B. E.
052.009
Doulade, C.
051.022
Downes, D.
012.038
125.008
157.001
Downey, P. M.
142.148
160.024
Dowty, E.
094.601
Doxsey, R.
142.006 .007 .045 .057
Draeger, G.
106.102
Dragesco, J.
032.017
Dragon, J. C.
094.514
Dragon, R.
098.033
Dragt, A. J.
084.411
Drake, M. J.
094.117 .430
Drake, N. A.
071.033
Dran, J. C.
105.078
Drapatz, S.
106.093
155.046
Drawin, H. W.
022.069
Drayson, S. R.
063.031
Dreibus, G.
094.479
Drescher, A.
082.022
Dressel, L. L.
158.141
Drever, R. W. P.
125.003
Drew, J. E.
114.001
Driatskij, V. M.
078.039

Dymek, R. F.
094.417 .431
Dymnikova, I. G.
066.020
Dyuston, K.
074.062
Dzhordzhio, N. V.
084.016
Dziczkaniec, M.
105.003

Eachus, L. J.
124.001
158.312
Eades Jr., J. B.
052.040
Eadie, G.
142.203
Eardley, D. M.
066.133
142.039
Earl, J. A.
062.026 .038
143.028 .097 .364
Easton, A. J.
105.030
Eaton, D.
103.103
Eaton, J. A.
121.039
Ebel, A.
083.012
Eberhardt, P.
094.512
Eberst, R. D.
054.012
Eckhardt, D. H.
066.071
Eddy, J. A.
072.049
Edelman, C.
098.012
Edgar, R.
051.001
053.001
Edge, D. M.
143.136 .294 .295
Edge, R. J.
143.379
Edmondson, F. K.
008.018
Edmunds, M. G.
064.057
131.086
Edrich, J.
033.023
Edwards, D.
162.041
Edwards, M. R.
141.099
Edwards, P. L.
141.026
158.311
Edwards, T.
141.021
Edwards, T. W.
118.014
Efanov, V. A.
141.076

Efimenko, V. I.
073.041
Efimenko, V. M.
072.024
073.041
Efimov, A. B.
042.069
Efimov, A. K.
143.040
Efimov, Yu. S.
142.080
Efremov, Yu. N.
122.079
Egan, W. G.
097.003
Egawa, Y.
065.102
Egeland, A.
084.025
Eggen, O. J.
115.006
155.027
Eggleton, P. P.
065.005
Eglinton, G.
094.142
Egorov, A. D.
031.248
Egorov, S. T.
097.011
Egorov, Yu. A.
009.014
Egortsev, E. Ya.
052.006
Ehfendiev, Ch. A.
031.247
Ehfendieva, S. A.
072.026 .027
Ehjdman, V. Ya.
061.027
Ehjgenson, A. M.
154.022
Ehlers, J.
141.611
162.016
Ehl'tekov, V. A.
084.261
Ehl'yasberg, P. E.
052.004
082.025
Ehmann, W. D.
094.481
Ehramzhyan, R. A.
061.018
Ehrgma, Eh.
031.007
Ehstulin, I. V.
032.523
142.073
Eichhorn, G.
022.083
094.565
Eichhorn, H.
031.219
111.001
Eichler, D.
066.002
Eilek, J. A.
114.348
158.013

Einasto, J.
158.025 .153 .154 .155
.156 .157
159.013
160.033
Einfeld, D.
022.035
Eisenstaedt, J.
162.067 .101
Eisner, S.
004.018
Ekberg, J. O.
022.007
Ekers, R. D.
158.060
Ekstrom, P. A.
131.021
141.604
El Goresy, A.
094.452 .595
105.097
El-Baz, F.
094.154
El-Gowhari, A.
062.086
El-Khishen, M.
062.086
El-Lakani, A.
062.086
El-Raey, M.
080.021
Elachi, C.
094.106 .108
Eldridge, J. S.
094.486
Elgaroy, Oe.
003.040
Elias, D. P.
079.102
Eliseev, G. F.
077.029 .032 .033
Elitzur, M.
131.004 .108
Elliot, H.
143.010 .270
Elliot, J. L.
097.056
099.017 .202
Elliott, D. A.
131.511
Elliott, K. H.
034.007
125.024
159.006
Ellis, G. F. R.
008.029
Elmergreen, B. G.
131.109
158.059
Elmore, D.
103.103
Elokhov, A. S.
084.017
Elsaesser, H.
012.015
034.002
082.001
106.095
131.522
158.062

Findlay, J. W.
033.030 .077
Fink, U.
098.029
099.210
114.340
Finkelman, R. B.
094.482 .484
Finkel'shtejn, A. M.
094.008
162.084
Finkelstein
See Finkel'shtejn
Finkenzeller, H. U.
104.005
Finn, G. D.
063.032
Finn, J. M.
084.411
Finsen, W. S.
118.022
Fiorito, G.
003.037
Fireman, E. L.
074.022 .049
Firmanyuk, B. N.
122.102
Fiscella, B.
034.065
Fischel, D.
031.405
Fischer, G.
008.088
Fischer, H.
046.033
Fischer, P. L.
103.106
Fischer, W.
076.012
Fisenko, A. V.
105.055
Fisenko, M. I.
073.077
Fisher, A.
143.054
Fisher, A. J.
143.023 .083 .084 .085
Fisher, M. P.
084.283
Fisher, R. M.
094.544
Fisher, W. A.
153.011
Fishkis, M.
114.312
Fishkova, L. M.
082.060
Fisk, L.
143.368
Fisk, L. A.
106.100 .101
143.030 .171 .187 .346
Fiskina, M. V.
084.243
Fitch, W. S.
122.008
Fitton, B.
131.078
FitzGerald, M. P.
152.001

Fix, J. D.
101.001
114.007 .342
122.026
Flannery, B. P.
117.012
Flasar, F. M.
097.005
141.043
Flavin, R. K.
033.035 .047
Fleck, R. C.
065.054
Fleming, B. K.
094.168
Fleming, J. W.
022.013
Flerov, G. N.
143.003
Flesch, T. R.
031.243
Flewelling, R. F.
143.024
Fliegel, H. F.
031.203
Fligel', M. D.
083.045
Flin, P.
121.109
160.040 .042
Flora, U.
121.082
Florenskij, K. P.
094.480
Florenskij, P. V.
105,013
Flores, J.
042.087
Florkowski, D. R.
121.076
Florsch, G.
061.054
095.001
Flowers, E.
065.062 .068
Flowers, E. G.
065.095
Floyd, J. E.
066.078
Flueckiger, E.
078.041
Fodor, R. V.
105.010 .063
Fogarty, W. G.
141.118
Fogliani, R.
052.021
Folomeshkin, V. N.
061.018
Folsom, G. H.
141.055
Fomalont, E. B.
066.077
Fomenko, A. F.
032.009
Fomichev, V. V.
077.024
Fomin, P. I.
143.026
Fomin, S. K.
123.003

Fomin, V. P.
142.134
Fomin, V. S.
094.115
Fominov, A. M.
052.018
Fonti, S.
031.271
Forbes, E. G.
003.043
004.038
Forbush, S. E.
143.242
Ford, C. B.
010.001
Ford, L. H.
066.138
Forichon, M.
142.122
Forman, M. A.
143.185 .312
Forman, W.
142.093 .166
Formisano, V.
084.212 .225
Forrest, D. J.
061.033 .034
076.021
Forslund, D. W.
074.023 .074
Forte, J. C.
153.022
155.007
Fortini, P.
155.036 .063
Fortini Baroni, L.
155.063
Fosbury, R. A. E.
134.004
Foster, J. C.
084.270
Foukal, P.
071.004
076.016
Foukal, P. V.
073.001
076.029
Fournier, A.
131.014
Fowler, L. A.
141.339
Fowler, P. H.
143.059 .091 .323 .377
.378 .379
Fowler, W. A.
022.086 .094
061.035
Fowler, W. T.
052.051
Fox, W. E.
010.012
Fracassini, M.
115.901
Fracastoro, M. G.
047.010
066.047
Fradkin, M. I.
078.005 .006 .012 .056
.083
142.088
143.264

Graham, A. L.
105.030
Graham, D. A.
141.320
Graham, J. A.
115.007
159.016
Graham, W. R. M.
022.029
Grallath, E.
094.594
Granat, L.
082.054
Grandi, S. A.
132.023
141.098
Granitskij, L. V.
034.025 .030 .031
Grant, R. W.
094.554
Grape, K.
131.165
Grard, R. J. L.
084.230
Grasbergs, E.
124.007
Grasdalen, G. L.
124.102
131.529
141.623
158.309
Gratton, L.
114.345
155.057
Gray, D. F.
003.049
Grebenikov, E. A.
003.050
Grebowsky, J. M.
084.262
Grec, G.
034.006
Grechnev, K. V.
097.031
Greeley, R.
097.008
Green, D. H.
094.461
Green, G.
078.098 .099 .100
143.286 .287 .289
Green, R.
084.203
Green, S.
131.113
Greenberg, J. M.
063.037
131.024 .026 .077 .087
Greenberg, L. T.
141.619
Greenberg, R.
042.093
101.002
Greene, G. M.
094.440
Greenspan, D.
021.013
Greenstadt, E. W.
084.212 .260
Greenstein, G.
105.028

Greenstein, J. L.
015.020
Gregory, P. C.
141.608 .620
Gregory, S. A.
133.005
160.025 .027 .034
Greisen, E. W.
131.007 .012
Gretskij, A. M.
100.010
Greve, A.
031.032
Grevesse, N.
022.024
Grew, S.
005.026
Grewing, M.
011.020
022.036
141.046 .326
143.102
Grib, S. A.
074.048
Gribbin, J.
003.051
015.016
085.017
099.039
Gribov, Yu. A.
034.015
Griess, T. D.
111.002
Griffin, R.
114.327
Griffin, R. F.
117.016
119.007 .011 .015
Griffiths, G. M.
061.064
Griffiths, J.
131.502
Griffiths, R. E.
142.096
Griffiths, W. K.
143.243
Grigor'ev, P.
122.091
Grigor'ev, V. M.
034.026 .027
Grigor'eva, N. B.
158.092
Grigor'eva, S. I.
143.159
Grigor'eva, V. P.
073.046
Grigorevskij, V. M.
034.018
Grigorov, N. L.
032.512
034.012 .013 .014
Grigor'yan, A. G.
083.035
Grilli, M.
162.039
Grillo, A. F.
162.039
Grindlay, J.
142.081 .082 .117 .174
.175 .207 .211

Grindlay, J. E.
122.110
142.091 .129 .135
143.303
Grineva, Yu. I.
073.023
076.014
Grinevitskaya, L. K.
052.029
Grinin, V. P.
064.050
Griscom, D. L.
094.402 .542
Grishchuk, L. P.
162.064
Grishin, N. I.
003.027
Groegler, N.
094.512
Groenbech, B.
121.003 .005 .016 .024
.046
Grognard, R. J.-M.
062.006
Grokhovskij, V. I.
094.597 .605
Gromov, S. V.
081.027
Gromov, V. V.
094.408
Gronenschild, E.
142.207 .211
Gronenschild,
E. H. B. M.
132.018
141.350
142.213
Groom, D. E.
143.131
Groote, D.
113.021
Gross, P. G.
065.024 .079
Grossman, L.
105.009 .011 .061 .066
107.019
Groten, E.
046.042
Groth, E. J.
141.314 .315 .316
Grove, G. W.
141.037
Gruber, R.
021.009
062.049
Grudler, P.
041.001
Grudzinska, S.
103.144
Gruen, E.
103.102
105.020
106.072 .074 .075 .102
.104
Gruen, M.
051.016 .018
Gruenwaldt, H.
074.067
Gruschel, W.
036.005

Ma, C.
121.045

Ma, M.-S.
094.481

Ma Sung, L. S.
078.076 .105

Maas, L.
033.028

Macak, P.
011.035

MacAlpine, G. M.
158.038

Macau-Hercot, D.
114.322

Maccacaro, T.
131.032

Maccagni, D.
141.069
142.097

Macchetto, F.
064.064

MacConnell, D. J.
114.029 .325

MacDoran, P. F.
031.203

Macdougall, D.
094.567

Macdougall, J. D.
105.067

MacGregor, A. D.
034.060 .066
131.501

Mach, J.
103.106

Machackova, K.
046.025

Mache, C.
066.104

Machlan, L. A.
094.504

Machol, R. E.
015.003

Maciel, W. J.
064.029

MacKeown, P. K.
143.293

Maclennan, C. G.
143.231

MacLeod, J. M.
131.116
141.025
158.307

MacRae, D. A.
008.108 .135

Macris, C.
008.009

Macris, C. J.
073.079 .080

Macris, G.
084.237

Madej, J.
121.111

Madelaine, G.
082.079

Maeda, H.
084.209

Maeda, K.
084.249

Maeder, A.
065.018

Maegley, W. J.
097.017 .062

Maehara, H.
122.028

Maehl, R. C.
143.023 .054 .055 .084
.085 .086

Maezawa, K.
084.224 .272

Maffei, P.
004.043

Magalhaes, A. M.
131.142

Magerramov, V. A.
080.005

Maggs, J. E.
101.004

Magill, J.
062.065

Magnan, C.
063.014

Mahafey, J. H.
065.112

Mahanthappa, K. T.
061.026

Maier, E. J.
083.001 .029

Maihara, T.
124.102
131.033
141.614 .638

Maitan, A.
008.019

Maitzen, H. M.
031.208
098.003
113.050

Makarenko, E. N.
122.074

Makarov, V. I.
073.058

Makarova, E. A.
074.901

Makino, F.
142.215

Makovetskij, P. V.
015.001

Maksimov, V. P.
073.007

Maksimov, Yu. M.
033.008

Maksumov, M. N.
151.046 .047 .048 .049
.050

Maksyukov, N. I.
066.055

Malafeev, L. I.
097.011

Malaise, D.
031.255

Malaroda, S.
153.901

Malas, Z.
121.111

Maldybaeva, Eh. Ya.
151.051

Maley, P.
103.103

Malin, S. R. C.
084.208

Malina, R.
066.018
142.010 .012

Mallama, A. D.
123.029

Mallia, E. A.
154.011

Mallow, J. V.
022.054

Malofeev, V. M.
132.014

Malov, I. F.
132.014

Mal'shakova, N. K.
123.011

Malumyan, V. G.
141.050

Malyshkin, V. N.
106.033 .035

Mamayev, S. G.
162.111

Mammano, A.
114.058

Mamrukova, V. P.
074.051

Manchanda, R. K.
142.215

Manchester, R. N.
141.318

Mancuso, S.
121.077

Mandel', O. E.
122.077

Mandel'shtam, S. L.
073.023
076.014

Mandel'stam
See Mandel'shtam

Mandeville, J. C.
094.564
105.125

Mandolesi, N.
143.031

Mandrou, P.
143.360

Mandrykina, T. L.
072.021

Manefield, G. A.
131.030

Manfroid, J.
064.049
113.012 .042
131.506 .521 .530 .546

Mangeney, A.
077.008 .009

Mann, F. I.
051.023

Mann, F. M.
061.068

Mann, M. F. St J.
113.006

Mansfield, V. N.
015.013
141.346

Mansinha, L.
081.020

Manson, J. E.
076.020

Mansurov, S. M.
078.022

Roger, R. S.
141.025
Rogers, A. E. E.
121.045
141.088
Rogers, A. J.
054.008
Rogerson Jr., J. B.
131.007
Rogovin, D.
022.051
Rogozhkina, S. P.
083.050
Rogstad, D. H.
158.040
Rohat-Jullien, L.
044.023
Rohlfs, K.
155.056
Rohr, H.
003.092
Rokoske, T. L.
103.103
Rolfs, C.
061.056
Romakhin, V. A.
143.297
Romanchuk, P. R.
072.022 .023 .024 .029
Romanishin, W.
153.010
Romankevich, G. N.
094.120
Romano, G.
122.128
158.030
Romanov, A. M.
032.019
Romanov, Yu. S.
114.330
Romanovskij, Yu. A.
083.023 .024
Romanowicz, B. A.
052.031
Romick, G. J.
093.017
Romig, J. H.
066.094
Romney, J.
141.037
Romney, J. D.
131.136
141.077
Rood, H. J.
158.078 .079
160.031
Rood, R. T.
065.901
Rosado, M.
131.544
Rose, C. A.
031.041
Rose, J. A.
160.011
Rose, M. F.
094.545
Rose, W. K.
124.003
Rose Jr., H. J.
094.482 .484

Rosen, J.
012.034
033.043
Rosenbauer, H.
062.040
074.067
084.279
Rosenberg, G. D.
012.012
Rosenberg, H.
011.050
Rosenberg, T. J.
084.270
Rosenblum, M.
062.066
Rosenkrantz, M.
103.103
Rosenvinge, T. T. Von
See Von Rosenvinge, T. T.
Rosenzweig, P.
122.125
123.017
Rosino, L.
008.094
123.002
125.045
158.030
Rosinski, J.
104.049
Rosman, K. J. R.
105.059 .086
Rosolen, C.
077.044
Ross, C. L.
074.063
094.014
Ross, J. E.
064.037
080.022
Ross, M.
094.441
Rossano, G. S.
131.101
Rossi, L.
124.102
Rostoker, G.
084.026
Rotenberg, M.
003.007
143.058
Roth, E. A.
066.100
Roth, G. D.
007.000
Roth, L.
094.106 .108
Rothschild, R.
142.256
Rothschild, R. E.
125.022
142.231 .233
Rothwell, P. L.
073.010
Roueff, E.
071.901
Roughton, N. A.
022.048
Rountree, S. P.
022.056
Roux, F.
022.030

Rovera, G.
031.267
044.027
Rowan-Robinson, M.
012.033
141.637
158.143
Rowe, M. W.
105.065
Roxburgh, I. W.
162.055
Roy, J.-R.
076.025 .026
Roy, S. R.
162.109 .110
Royer, A.
121.122
Rozantsev, I. N.
073.004
Roze, L.
011.028
Rozenbush, A. Eh.
122.038
Rozenfel'd, B. A.
004.049
Rozental', I. L.
162.034
Ruban, V. A.
162.084
Rubashevskij, A. A.
064.040 .041
Ruben, G. V.
065.049
Rubenstein, D. M.
074.078
Rubtsov, L. N.
083.047 .048 .050 .052
104.015 .033 .035 .036
Rucinski, S. M.
121.113
Ruddy, V.
031.272
143.051
Rudenko, V. N.
066.036
Ruderfer, M.
015.024
Ruderman, M.
065.062
141.307 .308 .309
Ruderman, M. A.
131.162
Ruderman, M. S.
074.060
Rudnick, L.
141.014 .051
160.008
Rudnicki, K.
011.048
121.112
125.006
Rudolph, D.
034.003
Rudolph, E.
066.116
141.611
Rudy, R. J.
124.102
Ruediger, G.
062.013 .027

Ruehle, G.
082.083
Ruffini, R.
066.119 .135 .137 .146
Rugge, H. R.
074.002
Rumball-Petre, H.
034.053
Rumsey, N. J.
010.024
Rumyantsev, A. A.
062.078
Rumyantsev, S. A.
062.021
Runcorn, S. K.
012.012
044.008
094.167 .177 .198
Ruppel, H. M.
062.094
Rusch, D. W.
082.901
Rusch, W. V. T.
033.018 .067
Russell, C. T.
074.013
084.216 .223
094.116 .168 .172
Russell, J.
031.219
111.001
Russo, A.
046.008
Rust, B. W.
162.122
Rustamov, Yu. S.
114.055 .337
Rutherford, M. J.
094.462
Rutherford, P. H.
062.061
Rutily, B.
122.027
Ruze, J.
033.019
Ruzickova-Topolova, B.
076.003
Ruzmajkin, A. A.
061.038
072.034
Ruzskij, E. G.
094.115 .588
Ryabenko, V. A.
105.081
Ryabenkov, V. I.
142.088
Ryabov, Yu. A.
003.050
Ryan, M. P.
066.078
Rybakov, A. K.
104.047
Rybicki, G.
158.089
Rybicki, G. B.
064.901
Rybka, S. P.
124.102
Ryckman, S. G.
142.046 .148
160.024

Rycroft, M. J.
012.003
Rydbeck, O. E. H.
141.088
Ryder, G.
094.436 .580
Rydgren, A. E.
113.035
114.023
Ryerson, F. J.
094.462
Rykhlova, L. V.
012.006
045.003
Rylaarsdam, J. C.
094.170
Ryle, M.
033.007
141.008 .036 .080 .902
Rylov, V. S.
034.019 .020
036.002
Rylov, Yu. A.
065.014
Ryter, C. E.
131.152
142.250
155.041
Rytova, Z. A.
031.248
Ryvkin, B. A.
085.013
Ryzhkov, N. F.
031.223
Rzhevskij, V. V.
094.410

Saakyan, G. S.
012.008
061.024
143.009
Saakyan, K. A.
122.063
158.100 .117
Saar, E.
158.025
Sabbadin, F.
125.019
Sackmann, I.-J.
065.086
Sadler, D. H.
007.000
046.008
Safronov, V. S.
107.009 .012
Safronov, Yu. I.
041.013
Safronova, N. M.
034.014
Safronova, U. I.
073.023
Sagan, C.
015.010
093.016
097.008 .042
Sagar, R.
153.017
Sagdeev, R. Z.
003.093

Sagot, R.
010.028
Saha, G.
155.029
Sahai, Y.
082.022
Saint-Marc, A.
062.024
Saisse, M.
032.527
Saito, M.
151.008 .041
Saito, T.
084.015
143.068 .069 .244 .322
.354 .358
Sajdov, P. I.
042.049
Sajdov, Yu. P.
042.049
Saka, O.
074.050 .083
Sakai, K.
162.018
Sakai, T.
143.226
Sakoto, M.
162.024
Sakurai, H.
142.138 .215
Sakurai, K.
073.069
099.045 .046
Sakurai, T.
065.067
073.083
084.015
143.358
Salajczyk, H.
022.095
Sale, R. D.
003.039
Saliba, G. A.
004.006
Salimov, G. R.
052.027
Salimzibarov, R. B.
074.047
Salio, G.
082.068
Salisbury, J. W.
002.006
105.121
Salm, N.
032.520
106.055
Salmistraro, F.
091.023
Salomone, M.
097.025
Salpeter, E. E.
064.026 .027
131.143
Salukvadze, G. N.
011.022
Salzwedel, H.
052.058
Samir-Debish, A. M.
143.234
Samochin
See Samokhin

Scalo, J. M.
  064.037
  065.007
  115.014
  126.014
Scaltriti, F.
  098.002 .020
Scardamaglia, C.
  114.361
Scarf, F. L.
  099.043
Scarfe, C. D.
  121.011
Scarlett, W. R.
  143.062
Scarrott, S. M.
  158.029
Scarsi, L.
  142.122
Schaal, R. E.
  131.512
Schaber, G. G.
  094.419
Schaechter, D.
  052.043
Schaechter, D. B.
  052.061
Schaefer, M. M.
  158.089
Schaeffer, O. A.
  084.412
  105.092 .093 .095
  143.307
Schamel, H.
  062.060
Schamoni, P.
  003.041
Scharlemann, E. T.
  141.301
Schatzman, E.
  065.084
  126.004
Schechter, P.
  158.009 .158
  160.007
Scheepmaker, A.
  142.046 .148
  160.024
Scheifele, G.
  003.104
Scheinin, N. B.
  094.487
Scherb, F.
  084.009
Scherer, F. M.
  046.030
Scherk, J.
  066.093
Scherrer, V. E.
  073.036
  074.018
Scheuer, P. A. G.
  141.007
  158.122
  162.070
Schielicke, R.
  034.016
Schild, R.
  153.010
Schild, R. E.
  113.032

Schilizzi, R. T.
  141.037 .077
Schimmerling, W.
  143.117
Schlegel, I.
  143.347
Schleicher, H.
  072.013
  073.013
Schlesinger, B. M.
  065.004
  153.006
Schlickeiser, R.
  131.153
  142.115
  155.039 .045 .049 .051
Schloerb, F. P.
  093.002
Schlueter, A.
  062.048
Schmadel, L. D.
  002.003
Schmahl, E. J.
  073.001
  076.029
Schmahl, G.
  034.003
Schmid, H. H.
  046.032
Schmid-Burgk, J.
  064.020
  065.048 .074
Schmidt, D. S.
  042.033
Schmidt, E. G.
  122.003 .033
  153.009
Schmidt, G.
  083.012
Schmidt, G. D.
  103.102
  158.121
Schmidt, H.
  004.021
Schmidt, H. U.
  106.052
Schmidt, K.-H.
  158.048
  160.021
  161.003 .004
  162.074
Schmidt, M.
  141.107
Schmidt, T.
  159.011
Schmidt, W. K. H.
  143.002 .050 .387
Schmidt-Kaler, T.
  061.039
  152.004
  155.021 .022 .054
  159.009 .014
Schmidtke, G.
  076.012
Schmitt, H. H.
  094.017
Schmitt, R. A.
  094.425 .475
  105.123
Schmitz, F.
  064.019

Schmutzer, E.
  003.094
  065.103
  066.038
Schneider, E.
  022.082
  094.188 .565
  105.083
Schneider, H.
  082.083
Schneider, M. V.
  033.046
Schneider, W.
  021.006
  062.051
Schneiderman, A. M.
  031.240
Schnepfe, M. M.
  094.482 .484
Schnopper, H.
  142.006 .045 .081 .091
    .207 .211
Schnur, G. F. O.
  082.001
Schober, H. J.
  098.011
Schoeffel, E.
  121.074
Schoembs, R.
  122.043
Schoenberner, D.
  114.318 .321
Schoeneich, W.
  032.030
  064.063
  113.014 .023 .045 .046
    .047
  116.003
  119.018
Schoenfelder, V.
  142.121
Schoenmaker, A. A.
  124.102
Schofield, N. J.
  143.074
Scholer, M.
  073.082
  078.088
  084.259
Scholl, H.
  002.003
  098.016
Scholz, D.
  075.006
Scholz, M.
  064.020
Schonfeld, E.
  094.483
Schove, D. J.
  004.029
Schrage, D.
  032.011
Schramm, D.
  064.056
Schramm, D. N.
  012.020
  061.021 .049 .901
  065.005 .023 .026
  125.028
  143.001 .103 .104 .107
    .124

Subject Index

ASTRONOMY
AND ASTROPHYSICS
ABSTRACTS

A Publication of the
Astronomisches Rechen-Institut Heidelberg
Member of the Abstracting Board
of the International Council of Scientific Unions

Editors:
S. Böhme, U. Esser, W. Fricke, U. Güntzel-Lingner, I. Heinrich,
F. Henn, D. Krahn, L. D. Schmadel, H. Scholl, G. Zech

K. R. Lang

Astrophysical Formulae

A Compendium for the Physicist and Astrophysicist

1974. 46 figures. XXV, 735 pages

Contents: Continuum Radiation. Monochromatic (Line) Radiation. Gas Processes. Nuclear Astrophysics and High Energy Particles. Astrometry and Cosmology.

Problems in Stellar Atmospheres and Envelopes

Editors: *B. Baschek, W. H. Kegel, G. Traving*

1975. 75 Figures. XVII, 375 pages

Contents: The Energy Flux of the Sun. Model Stellar Atmospheres and Heavy Element Abundances. Properties and Problems of Helium Stars. Abundance Annomalies in Early-Type Stars. A-Type Horizontal-Branch Stars. White Dwarfs: Composition, Mass Budget and Galactic Evolution. Herbig-Haro Objects and T Tauri Nebulae. Circumstellar Envelopes and Mass Loss of Red Giant Stars. Cosmic Masers. Radio Emission from Stellar and Circumstellar Atmospheres. Line Formation in Turbulent Media.

Galactic and Extra-Galactic Radio Astronomy

By the Staff of the National Radio Astronomy Observatory

Editors: *G. L. Verschuur, K. I. Kellermann*
With the Assistance of *V. Van Brunt*

1974. 127 figures. X, 402 pages

Contents: Galactic Nonthermal Continuum Emission. Interstellar Neutral Hydrogen and Its Small-Scale Structure. The Radio Characteristics of HII Regions and the Diffuse Thermal Background. The Large-Scale Distribution of Neutral Hydrogen in the Galaxy. Supernova Remnants. Pulsars. Radio Stars. The Galactic Magnetic Field. Interstellar Molecules. Interferometry and Aperture Synthesis. Mapping Neutral Hydrogen in External Galaxies. Radio Galaxies and Quasars. Cosmology.

Springer-Verlag Berlin Heidelberg New York